ELECTRONIC I
Circuits and Syst

ELECTRONIC DESIGN

Circuits and Systems

SECOND EDITION

C. J. SAVANT, Jr.
California State University, Los Angeles

MARTIN S. RODEN
California State University, Los Angeles

GORDON L. CARPENTER
California State University, Long Beach

ADDISON-WESLEY WORLD STUDENT SERIES

The Benjamin/Cummings Publishing Company, Inc.

Redwood City, California • Fort Collins, Colorado • Menlo Park, California
Reading, Massachusetts • New York • Don Mills, Ontario
Wokingham, U.K. • Amsterdam • Bonn • Sydney • Singapore • Tokyo
Madrid • San Juan

WORLD STUDENT SERIES EDITION

To our loving wives,
Elnora Carpenter and Barbara Savant.

Sponsoring Team: Alan Apt, Mary Ann Telatnik
Production Supervisor: John Walker
Copy Editor: Mary Prescott
Illustrations: Art by AYXA, Carl Brown, Becky and Al McCahon, Rolin Graphics
Compositor: Beacon Graphics
Text Design: Bruce Kortebein, Mark Ong
Cover Design: Mark Ong
Cover Photo: Photo courtesy of Hewlett-Packard Company.

The basic text of this book was designed using the Modular Design System, as developed by Wendy Earl and Design Office Bruce Kortebein. The text design for the second edition was modified by Mark Ong.

The Benjamin/Cummings Publishing Company, Inc.
390 Bridge Parkway
Redwood City, California 94065

ISBN 0-8053-0292-1
1 2 3 4 5 6 7 8 9 10-DO-96 95 94 93 92 91

ABOUT THE AUTHORS

C. J. SAVANT, JR. is a dedicated engineering educator. He received his Ph.D. *cum laude* from California Institute of Technology and has taught in the California State University system at both the Long Beach and the Los Angeles campuses. Dr. Savant is the recipient of the California State University "Outstanding Professor Award." He has also received the IEEE Outstanding Contribution to Engineering Award, and the AT&T Foundation award for excellence in teaching. In addition to his teaching, Professor Savant has completed considerable research and has owned an electronics engineering firm.

MARTIN S. RODEN is the Chairman of the Department of Electrical and Computer Engineering at California State University, Los Angeles. Dr. Roden received his BSEE *summa cum laude* from Polytechnic Institute of Brooklyn, and then went on to spend five years doing research at Bell Labs, the birthplace of the transistor. His interest in teaching led him back to academia, where he has gone on to hold positions as Department Chair, Associate Dean, Dean and Associate Vice President at various times. However, Professor Roden's first love remains teaching, for which he was awarded the University's Outstanding Professor Award and the AT&T Foundation award for excellence in teaching. He is very active in IEEE, has earned the IEEE's Outstanding Adviser Award, and is a Fellow of the Institute for the Advancement of Engineering.

GORDON CARPENTER retired with the rank of Lt. Colonel from the U.S. Air Force, where he had over twenty years of experience in the design and development of high technology USAF equipment. This experience has made Professor Carpenter very realistic in his approach to the education of future engineers. In his Air Force career as an R&D manager, he trained new engineers to develop hardware specifications from system requirements, and to insure that hardware could be built to meet those specifications. Colonel Carpenter is a strong proponent of design-oriented education, and his practical experience is essential to this text. Professor Carpenter has earned the IEEE's Outstanding Advisor Award, the Orange County Outstanding Engineering Educator Award, and is a Fellow of the Institute for the Advancement of Engineering.

PREFACE

Electronic Design: Circuits and Systems, Second Edition is written for use in the core electronics courses in undergraduate programs in electrical engineering. The book provides coverage of three areas: discrete devices, linear integrated circuits, and digital integrated circuits. A practicing engineer looking for a current reference for self-study will also find this book valuable. The only prerequisite for understanding the material in this text is a basic knowledge of circuit analysis.

Why This Book?

With literally dozens of books in the field of analog and digital electronics to chose from, you may wonder why we have written yet another book on the subject. Our principal goal in writing this text is to relieve our frustrations. We had attempted to teach electronics to undergraduates using existing texts. These older texts look at the field from a theoretical point of view, emphasizing analysis and the physics of semiconductors, but ignoring the important and exciting design applications. Dealing only with fundamentals detracts from the excitement of the subject, and, indeed, the student may never develop the design skills required for a career in electronics. While this book covers the fundamentals in a thorough and direct fashion, it goes one step further toward a balanced approach to designing electronic systems. With both the Accreditation Board for Engineering and Technology (ABET) and the industry demanding more design in engineering programs, we feel that it is time to balance the fundamentals with a strong taste of design. It is our hope that this book will inspire the imaginations of tomorrow's engineers, who will be called upon to *design,* not just to analyze, electronic systems.

Retained from the First Edition

We have retained many of the outstanding qualities of the first edition that helped make it a success, such as:

- Heavy emphasis on design.
- Numerous design examples that provide a real-world flavor. These are taken from the years of engineering experience of the authors.
- A readable and comprehensible writing style that results from class testing and student evaluation of the manuscript.
- A proper balance among the three areas of the text: discrete devices, linear integrated circuits, and digital integrated circuits.
- Extensive use of Drill Problems. The first edition contained a large number of drill problems that are meant to offer immediate reinforcement for students. The answers are found along with the problems.

New to the Second Edition

Suggestions from many colleagues and our own classroom use of the first edition has lead us to make the following improvements:

- The emphasis upon a systems design approach has been increased.
- The system design approach has been enhanced with the use of the computer. Throughout the second edition, computer modeling programs are used to aid in the design process and to verify the results.
- Organization and pedagogy have been improved. The authors and our reviewers have taught from the first edition for several years. As a result, the second edition is reorganized to improve the order of presentation of the topics, and to make the book even easier to teach from.
- A Bode plot appendix has been added. The placement of this material in an appendix makes it easier to introduce the Bode plot concepts at various points within the study of electronic systems.
- Semiconductor physics, which previously appeared as an appendix, has been integrated throughout the text, thereby giving students reinforcement and review.
- New components have been added. Among these are switched-capacitor Butterworth filters and power operational amplifiers.
- New design examples have been added.

Uniqueness of the System Design Approach

Contact with practicing engineers and engineering recruiters has led us to place significant emphasis upon *design* of electronic *systems.* The new engineer will be asked to design systems using an ever-increasing inventory of new linear and digital ICs, discrete components, and electromechanical devices. Thus, we attempt to teach engineering students to *think* as system designers, rather than to mimic just a few design approaches. Our goal is to "educate" rather than to "train."

Elementary design procedures are introduced early in the text to motivate the reader. It is our experience that electronics design is best comprehended through a "learn by doing" approach. Thus, topics such as small signal analysis have been presented immediately following dc analysis to allow for presentation of some early meaningful and realistic design problems.

Wherever possible, a step-by-step procedure is developed, which should give the student confidence. Rather than replacing the theory, such procedures reinforce and clarify it. We emphasize the techniques of developing procedures from the theory so the student can extend the concepts. This approach culminates in Chapter 17, which presents system design procedures that can be applied to both analog and digital systems.

SPICE Appendix

A comprehensive appendix on the SPICE circuit modeling program is included. This appendix also explores the alternate use of MICRO CAP II. Instructors have the flexibility of introducing SPICE at any point in the course. Examples of SPICE program printouts are included as models for the student.

Other Appendices

The book includes appendices covering:

- Bode Plots
- Laplace Transforms
- Manufacturers' Data Sheets (for selected devices)
- Noise in Electronic Systems
- Answers to Selected Problems

Accuracy of the Book

Every effort has been made to write and publish an *accurate* book. Many reviewers were used for both the first and second editions, and a number of users of the first edition helped us to assure a high level of accuracy in the second edition. This second edition evolved out of extensive use in the classroom environment, and has been extensively checked and class-tested with students. Many of the primary reviewers and "error checkers" are acknowledged later in this preface.

Instructional Adjuncts

Overhead projector **Transparency Masters** of the important figures in the text are available for instructors. Also available for instructors is an **Instructor's Manual,** containing complete solutions to all the drill and end-of-chapter problems in the book. Many of the design problems, particularly those in Chapter 17, have been drawn from actual industrial applications, where the resulting systems have been implemented.

The authors recommend that the latest manufacturers' data books be used in conjunction with this text. For example, the National Semiconductor *Linear Data Book* is a good supplement for the linear IC portion of the text. The Texas Instruments *TTL Data Book, Volume II* and the National Semiconductor *CMOS Data Book* would be suitable companions to the third portion of our book on digital ICs.

Another new concept with the second edition is the **Exam Data Bank.** Instructors teaching from the text are invited to contribute to an Exam Data Bank, which will be compiled and distributed to all participants. Details appear in the Instructor's Manual.

Guide for Classroom Use

The material in this book can be presented in a series of two or three one-semester courses or three one-quarter courses in the junior and/or senior years.

Chapters 1 through 7 cover the analysis and design of discrete devices such as diode circuits, bipolar junction and field effect transistors, and audio frequency power amplifiers. Some frequency response concepts are needed for an understanding of Chapter 7. A new appendix on Bode plots is included.

The second portion of the book (Chapters 8 through 13) is devoted to linear integrated circuits. This material includes thorough coverage of ideal and actual operational amplifiers, feedback and stability, and transfer characteristic design. Chapter 12 covers the subject of active filters, including a design procedure for Butterworth and Chebyshev filters. Some background in the Laplace transform method is helpful for covering this part of the text. With this in mind, an appendix is included for the review of Laplace transforms. Chapter 13 covers various quasi-linear systems, such as rectifiers, feedback limiters, comparators, and Schmitt triggers.

The final four chapters in the book (Chapters 14 through 17) are devoted to a study of digital ICs and electronic systems. Steady state analysis of pulse driven circuits and relaxation oscillators is presented. Three logic families—TTL, CMOS, and ECL—are compared and analyzed. Numerous unclocked and clocked ICs are studied with many practical applications included.

Chapter 17 is unique, presenting an electronic system design methodology that can be used with digital and analog ICs. Many of the examples of design in this chapter have been drawn from industry.

Acknowledgments

Some material on operational amplifiers (including active filter design) was derived from class notes originally developed by (the late) Professor Gene H. Hostetter while he was at California State University, Long Beach. Some material on active filter design also appears in Dr. Hostetter's text, *Engineering Network Analysis* (New York: Harper & Row, 1984). We are grateful to Dr. Hostetter and to Harper & Row for permitting this joint usage.

We would like to express our appreciation to the students in the various electronic design classes the authors have taught while using the early versions of this text. Sincere thanks are extended to our colleagues, Professor Hassan Babaie, Lou Balin, Roy Barnett, Ed Evans, Mike Hassul, Ken James, and George Killinger for their comments and assistance with various portions of the manuscript.

We would like to thank Paul Van Halen of Portland State University, who supplied the first draft of the material on Semiconductor Physics (an appendix in the first edition), and Mahmoud El Nokali, who made many valuable suggestions for the first edition. Special thanks also go to Bernhard Schmidt of the University of Dayton, who carefully checked the illustrations and problem sets for the first edition.

Every book is the result of a number of iterations and revisions based upon classroom experience and the expert advice of reviewers. We were fortunate to have thirty-three readers review all or part of the manuscripts for the first and second editions. We would like to thank the following reviewers, and the many others who are not mentioned by name, for their efforts:

H. Jack Allison, Oklahoma State University

Kay D. Baker, Utah State University

W. L. Beasley, Texas A&M University

Robert L. Bernick, California Polytechnic State University, Pomona

Raymond Black, New Mexico State University

T. V. Blalock, University of Tennessee

Frank Brands, Washington State University

John Churchill, University of California, Davis

R. G. Deshmukh, Florida Institute of Technology

Mahmoud El Nokali, University of Pittsburgh

E. L. Gerber, Drexel University

Ward Helms, University of Washington

George W. Hoyle, Northern Arizona University

Alfred T. Johnson, Jr., Widener University

B. Lalevic, Rutgers University

Hung Chang Lin, University of Maryland

John Lowell, Texas Tech University

Edward W. Maby, Rensselaer Polytechnic

Eugene Manus, Virginia Polytechnic Institute and State University

Donald C. Moore, South Dakota State University

Richard Morris, University of Portland

David A. Navon, University of Massachusetts, Amherst

Harry Neinhaus, University of South Florida

Charles Nelson, California State University, Sacramento

David Perlman, Rochester Institute of Technology

William Sayle, Georgia Institute of Technology

Bernhard Schmidt, University of Dayton

Paul Van Halen, Portland State University

Darrell L. Vines, Texas Tech University

J. L. Yeh, Rutgers University

Carl R. Zimmer, Arizona State University

Reza Zoughi, Colorado State University

We also thank the following for responding to a survey questionnaire regarding the second edition:

R. G. Deshmukh, Florida Institute of Technology

Aswartha Narayana, St. Cloud State College

V. Ramachandran, Concordia University

Joseph L. Kozikowski, Villanova University

In addition to our colleagues and reviewers, many students helped us out along the way. The following students deserve our appreciation for their special assistance: Gabriel Cocco, Ted Curmi, Jim Eckman, Kevin Kean, Lyle Mattes, Bob McBride, Mark Pendleton, Steve Phillips, Gloria Quinn, Bob Topper, Bob Tran, Phil Vrbancic, and Ann Weichbrod.

We truly hope that each of the people who contributed to this book and had a hand in its development are as satisfied with the finished product as we are.

Gordon L. Carpenter
Martin S. Roden
C. J. Savant, Jr.

INTRODUCTION TO THE STUDENT

Electronics is the backbone of electrical engineering. Whether you specialize in solid state electronic design, or in diverse areas such as power, computers, controls, or communications, you must first become familiar with the basics of the design and analysis of electronic circuits and systems.

This is not an easy task, since the field is changing at a rapid pace. You must be careful to concentrate upon *education,* rather than *training,* in the area of electronics. Those who were *trained* in vacuum tube electronic design during the 1950s found their training to be useless a decade later when transistors replaced vacuum tubes in all but a few high-power or high-frequency applications. Likewise, those who were *trained* in transistor design during the 1960s and early 1970s found that training to be obsolete with the advent of integrated circuits and op-amp systems. It is therefore important that you prepare yourself for the next revolution by both learning the fundamentals and "learning how to learn."

Many texts approach this challenge by overemphasizing the theory and completely avoiding applications. You will find this not to be the case in this text. A sterile theoretical presentation could leave you with some basic knowledge that you could someday apply. However, you would probably not experience the excitement of applying this knowledge to practical situations as you learn. Indeed, you would not even know if you are capable of it.

For that reason, this text is heavily *design oriented.* You will be led through many practical applications of the theory—and we do mean *practical.* We emphasize that the practical design procedures presented in this text represent one way of reducing the theory to practice. True design requires many tradeoff decisions, and there is more than one correct answer given the system constraints. For this reason, the procedures we illustrate are **not** the only ways in which the theory can be applied to practical situations. You are encouraged to seek other procedures, either independently or with the help of your professor. We hope you will be motivated to construct some of the systems you will design on paper, for that will truly "close the loop" and make your education more meaningful.

Some of the problems at the back of the chapters might seem overwhelming at first glance. Learning design is gradual, so don't become discouraged. You will find you are capable of making progress on even the most complex design problems. Your professor should be able to provide guidance.

Most of all, enjoy the subject material. You have chosen an exciting career, but the same factors that make it exciting also make it challenging. You must sometimes strain to the limits of your mental abilities if you are to succeed, but the rewards of success will be fine compensation.

If you have any comments or suggestions about the text, please feel free to communicate them to any of the three authors. Professors Roden and Savant are at California State University, Los Angeles, and Professor Carpenter is at California State University, Long Beach. Because we take a genuine interest in engineering education, we welcome all your comments and suggestions.

BRIEF CONTENTS

CHAPTER 1 SEMICONDUCTOR DIODE CIRCUIT ANALYSIS 1

CHAPTER 2 BIPOLAR JUNCTION TRANSISTOR AMPLIFIERS 69

CHAPTER 3 DESIGN OF BIPOLAR JUNCTION TRANSISTOR AMPLIFIERS 118

CHAPTER 4 FIELD-EFFECT TRANSISTOR AMPLIFIERS 171

CHAPTER 5 BIAS STABILITY OF TRANSISTOR AMPLIFIERS 213

CHAPTER 6 POWER AMPLIFIERS AND POWER SUPPLIES 238

CHAPTER 7 FREQUENCY RESPONSE CHARACTERISTICS 283

CHAPTER 8 IDEAL OPERATIONAL AMPLIFIERS 325

CHAPTER 9 OPERATIONAL AMPLIFIERS AS INTEGRATED CIRCUITS 366

CHAPTER 10 PRACTICAL OPERATIONAL AMPLIFIERS 405

CHAPTER 11 FEEDBACK AND STABILITY 468

CHAPTER 12 ACTIVE FILTERS 519

CHAPTER 13 QUASI-LINEAR CIRCUITS 586

CHAPTER 14 PULSED WAVEFORMS AND TIMING CIRCUITS 635

CHAPTER 15 DIGITAL LOGIC FAMILIES 679

CHAPTER 16 DIGITAL ICs 724

CHAPTER 17 ELECTRONIC SYSTEM DESIGN 784

APPENDIX A SPICE A-1

APPENDIX B BODE PLOTS A-36

APPENDIX C THE LAPLACE TRANSFORM A-60

APPENDIX D MANUFACTURERS' DATA SHEETS A-74
APPENDIX E STANDARD COMPONENT VALUES A-126
APPENDIX F NOISE IN ELECTRONIC SYSTEMS A-128
APPENDIX G ANSWERS TO SELECTED PROBLEMS A-136
REFERENCES AND SOURCES FOR FURTHER STUDY A-143
INDEX I-1

DETAILED CONTENTS

CHAPTER 1 SEMICONDUCTOR DIODE CIRCUIT ANALYSIS 1

1.0 Introduction 1

1.1 Theory of Semiconductors 2

1.1.1 Conduction in Materials 4
1.1.2 Conduction in Semiconductor Materials 5
1.1.3 Intrinsic Semiconductors 7
1.1.4 Doped Semiconductors 9
1.1.5 n-type Semiconductors 10
1.1.6 p-type Semiconductors 11
1.1.7 Carrier Concentrations 11
1.1.8 Excess Carriers 12
1.1.9 Recombination and Generation of Excess Carriers 13
1.1.10 Transport of Electric Current 13
1.1.11 Drift in an Electric Field 14

1.2 Semiconductor Diodes 17

1.2.1 Physics of Solid-State Diodes 18
1.2.2 Diode Construction 19
1.2.3 Relationship Between Diode Current and Diode Voltage 20
1.2.4 Diode Operation 22
1.2.5 Temperature Effects 24
1.2.6 Diode Equivalent Circuit Models 25
1.2.7 Diode Circuit Analysis 26
Computer Simulations 27
1.2.8 Power-Handling Capability 35
1.2.9 Diode Capacitance 35

1.3 Rectification 36
1.3.1 Half-Wave Rectification 37
1.3.2 Full-Wave Rectification 38
1.3.3 Filtering 39
1.3.4 Voltage-Doubling Circuit 43
1.4 Zener Diodes 44
1.4.1 Zener Regulator 46
1.4.2 Practical Zener Diodes and Percent Regulation 49
1.5 Clippers and Clampers 51
1.5.1 Clippers 51
1.5.2 Clampers 56
1.6 Alternate Types of Diodes 58
1.6.1 Schottky Diodes 58
1.6.2 Light-Emitting Diodes 59
1.6.3 Photodiodes 60
1.7 Manufacturers' Specifications 60
Problems 62

CHAPTER 2 BIPOLAR JUNCTION TRANSISTOR AMPLIFIERS 69

2.0 Introduction 69
2.1 Dependent Sources 70
2.2 Bipolar Transistors 71
2.3 Transistor Models 75
2.3.1 Ebers-Moll Model 76
2.3.2 Hybrid-π and h-parameter Model 78
2.4 Transistor Circuits 79
2.4.1 Common Circuit Configurations 79
2.4.2 Characteristic Curves 80
2.5 The Common-Emitter Amplifier 84
2.5.1 Common-Emitter Amplifier with Emitter Resistor 86
2.6 Power Considerations 90
2.6.1 Derivation of Power Equations 90
2.7 Bypass and Coupling Capacitors 92
2.7.1 Bypass Capacitors 92
2.7.2 Coupling Capacitors 93
2.8 ac Load Line for CE Configuration 93
2.8.1 ac Load Line Through Any Q-Point 94
2.8.2 Choice if ac Load Line for Maximum Output Swing 95
2.9 ac Analysis and Design 97
2.9.1 Analysis Procedure 97
2.9.2 Design Procedure 100
2.9.3 Amplifier Power Sources 105
2.10 The Emitter-Follower (Common Collector) Amplifier 106
Problems 110

CHAPTER 3 DESIGN OF BIPOLAR JUNCTION TRANSISTOR AMPLIFIERS 118

3.0 **Introduction 118**
3.1 **Analysis of Two-Port Networks—Hybrid Parameters 118**
3.2 **Short-Circuit Input Resistance 122**
3.3 **CE Parameters 123**
3.3.1 Input Resistance, R_{in} 123
3.3.2 Current Gain, A_i 126
3.3.3 Voltage Gain, A_v 127
3.3.4 Gain Impedance Formula 128
3.3.5 Output Resistance, R_o 129
3.4 **Nonlinearities of Bipolar Junction Transistors 135**
3.5 **Parameters for CC (EF) Amplifier 138**
3.5.1 Input Resistance, R_{in} 138
3.5.2 Current Gain, A_i 139
3.5.3 Voltage Gain, A_v 139
3.5.4 Output Resistance, R_o 140
3.6 **Parameters for CB Amplifier 143**
3.6.1 Input Resistance, R_{in} 143
3.6.2 Current Gain, A_i 144
3.6.3 Voltage Gain, A_v 145
3.6.4 Output Resistance, R_o 145
3.7 **Transistor Amplifier Applications 149**
3.8 **Amplifier Coupling 149**
3.8.1 Capacitive Coupling 150
3.8.2 Direct Coupling 150
3.8.3 Transformer Coupling 152
3.8.4 Optical Coupling 154
3.9 **Phase Splitter 157**
3.10 **Multistage Amplifier Analysis 158**
3.11 **Cascode Configuration 161**
Problems 163

CHAPTER 4 FIELD-EFFECT TRANSISTOR AMPLIFIERS 171

4.0 **Introduction 171**
4.1 **Advantages and Disadvantages of the FET 172**
4.2 **Types of FETs 172**
4.3 **JFET Operation and Construction 172**
4.3.1 JFET Gate-to-Source Voltage Variation 174
4.3.2 JFET Transfer Characteristics 175
4.3.3 Equivalent Circuit, g_m and r_{DS} 177

4.4 MOSFET Operation and Construction 181
4.4.1 Depletion MOSFET 182
4.4.2 Enhancement MOSFET 184
4.5 Biasing of FETs 186
4.6 Analysis of CS Amplifiers 188
4.7 Design of CS Amplifiers 191
4.8 Selection of Components 194
4.9 Analysis of CD (SF) Amplifiers 199
4.10 Design of CD Amplifiers 200
4.11 SF Bootstrap Amplifiers 203
4.12 Metal Semiconductor Barrier Junction Transistor 206
4.13 Other Devices 207
4.13.1 VMOSFET (VMOS) 207
4.13.2 Other MOS Devices 208
Problems 208

CHAPTER 5 BIAS STABILITY OF TRANSISTOR AMPLIFIERS 213

5.0 Introduction 213
5.1 Types of Biasing 213
5.1.1 Current Biasing 214
5.1.2 Voltage and Current Biasing 215
5.2 Effects of Parameter Changes—Bias Stability 217
5.2.1 CE Configuration 219
5.2.2 EF Configuration 224
5.3 Diode Compensation 227
5.4 Designing for BJT Amplifier Bias Stability 229
5.5 FET Temperature Effects 229
5.6 Reducing Temperature Variations 232
Problems 234

CHAPTER 6 POWER AMPLIFIERS AND POWER SUPPLIES 238

6.0 Introduction 238
6.1 Classes of Amplifiers 238
6.1.1 Class-A Operation 239
6.1.2 Class-B Operation 239
6.1.3 Class-AB Operation 241
6.1.4 Class-C Operation 242
6.2 Power Amplifier Circuits—Class-A Operation 242
6.2.1 Inductively-Coupled Amplifier 243
6.2.2 Transformer-Coupled Power Amplifier 245
6.3 Power Amplifier Circuits—Class-B Operation 248
6.3.1 Complementary-Symmetry Class-B Power Amplifier 248

6.3.2 Design of Diode-Compensated Complimentary-Symmetry Class-B Power Amplifier 252
6.3.3 Power Calculations for Class-B Push-Pull Amplifier 255
6.4 Darlington Circuit 262
6.5 Power Supply Using Power Transistors 267
6.5.1 Power Supply Using Discrete Components 267
6.5.2 Power Supply Using IC Regulator (Three-Terminal Regulator) 270
6.5.3 Power Supply Using Three-Terminal Adjustable Regulator 274
6.5.4 Higher-Current Regulator 275
6.6 Switching Regulators 276
6.6.1 Efficiency of Switching Regulators 278
Problems 278

CHAPTER 7 FREQUENCY RESPONSE CHARACTERISTICS 283

7.0 Introduction 283
7.1 Low-Frequency Response—BJT Amplifier 285
7.1.1 Low-Frequency Response—CE Amplifier 285
7.1.2 Design for a Given Frequency Characteristic 290
7.1.3 Emitter-Resistor Bypass Capacitor 293
7.1.4 Low-Frequency Response—EF Amplifier 295
7.1.5 Low-Frequency Response—CB Amplifier 296
7.2 Low-Frequency Response—FET Amplifier 297
7.2.1 Low-Frequency Response—CS Amplifier 297
7.2.2 Low-Frequency Response—CD Amplifier 302
7.3 High-Frequency Response—BJT Amplifier 303
7.3.1 High-Frequency Response—CE Amplifier 304
7.3.2 High-Frequency Response—EF Amplifier 312
7.3.3 High-Frequency Response—CB Amplifier 313
7.4 High-Frequency Response—FET Amplifier 315
7.4.1 High-Frequency Response—CS Amplifier 315
7.4.2 High-Frequency Response—CD Amplifier 318
7.5 High-Frequency Amplifier Design 320
7.6 Concluding Remarks 320
Problems 321

CHAPTER 8 IDEAL OPERATIONAL AMPLIFIERS 325

8.0 Introduction 325
8.1 Ideal Op-Amps 326
8.1.1 Analysis Method 328
8.2 The Inverting Amplifier 328
8.3 The Noninverting Amplifier 331

8.4 **Input Resistance of Op-Amp Circuits 338**
8.5 **Combined Inverting and Noninverting Inputs 340**
8.6 **Design of Op-Amp Circuits 342**
8.7 **Other Op-Amp Applications 349**
8.7.1 Negative-Impedance Circuit 349
8.7.2 Dependent-Current Generator 350
8.7.3 Current-to-Voltage Converter 351
8.7.4 Voltage-to-Current Converter 352
8.7.5 Inverting Amplifier with Impedances 353
8.7.6 Analog Computer Applications 354
8.7.7 Noninverting Miller Integrator 356
8.7.8 Op-Amps and Diodes 357
Problems 360

CHAPTER 9 OPERATIONAL AMPLIFIERS AS INTEGRATED CIRCUITS **366**

9.0 **Introduction 366**
9.1 **Integrated Circuit Fabrication 367**
9.1.1 Transistors and Diodes 367
9.1.2 Resistors 368
9.1.3 Capacitors 368
9.1.4 Lateral Transistors 368
9.2 **Differential Amplifiers 369**
9.2.1 dc Transfer Characteristics 370
9.2.2 Common-Mode and Differential-Mode Gains 371
9.2.3 Differential Amplifier with Constant-Current Source 374
9.2.4 Differential Amplifier with Single-Ended Input and Output 377
9.3 **Current Sources, Active Loads, and Level Shifters 380**
9.3.1 A Simple Current Source 380
9.3.2 Widlar Current Source 381
9.3.3 Wilson Current Source 384
9.3.4 Current Mirrors 387
9.3.5 Current Sources as Active Loads 387
9.3.6 Level Shifters 390
9.4 **Typical Op-Amps 394**
9.5 **Op-Amp Packaging 394**
9.5.1 Power Requirements 395
9.6 **The 741 Op-Amp 396**
9.6.1 Bias Circuits 396
9.6.2 Short-Circuit Protection 397
9.6.3 Input Stage 397
9.6.4 Intermediate Stage 398
9.6.5 Output Stage 398

9.7 Manufacturers' Specifications 399
Problems 399

CHAPTER 10 PRACTICAL OPERATIONAL AMPLIFIERS 405

10.0 Introduction 405
10.1 Op-Amp Frequency Response 405
10.1.1 Open-Loop and Closed-Loop Op-Amp Response 409
10.2 Practical Op-Amps 412
10.2.1 Open-Loop Voltage Gain (G) 413
10.2.2 Modified Op-Amp Model 413
10.2.3 Input Offset Voltage (V_{io}) 414
10.2.4 Input Bias Current (I_{Bias}) 415
10.2.5 Common-Mode Rejection 418
10.2.6 Power Supply Rejection Ratio 419
10.2.7 Phase Shift 420
10.2.8 Slew Rate (SR) 420
10.2.9 Output Resistance 423
10.3 Noninverting Amplifier 426
10.3.1 Input and Output Resistances 426
10.3.2 Voltage Gain 429
10.3.3 Bandwidth 431
10.3.4 Multiple-Input Amplifiers 434
10.4 Inverting Amplifier 438
10.4.1 Input and Output Resistance 438
10.4.2 Voltage Gain 439
10.4.3 Bandwidth 440
10.4.4 Multiple-Input Amplifiers 442
10.5 Differential Summing 445
10.6 Designing Amplifiers Using Multiple Op-Amps 448
10.7 101 Amplifier 452
10.7.1 Noninverting 101 Amplifiers 454
10.7.2 Inverting 101 Amplifiers 455
10.8 Amplifiers with Balanced Inputs or Outputs 456
10.9 Coupling Between Multiple Inputs 459
10.10 Power Audio Op-Amps 460
10.10.1 Bridge Power Op-Amp 462
10.10.2 Intercom 463
Problems 463

CHAPTER 11 FEEDBACK AND STABILITY 468

11.0 Introduction 468
11.1 Feedback Amplifier Considerations 469
11.2 Types of Feedback 470

11.3 Feedback Amplifiers 472
11.3.1 Current Feedback—Voltage Subtraction for Discrete Amplifiers 472
11.3.2 Voltage Feedback—Current Subtraction for a Discrete Amplifier 475
11.4 Multistage Feedback Amplifiers 478
11.5 Feedback in Operational Amplifiers 479
11.6 Stability of Feedback Amplifiers 483
11.6.1 System Stability and Frequency Response 485
11.6.2 Bode Plots and System Stability 488
11.7 Frequency Response—Feedback Amplifier 493
11.7.1 Single-Pole Amplifier 493
11.7.2 Two-Pole Amplifier 494
11.8 Design of a Three-Pole Amplifier With a Lead Equalizer 499
11.9 Phase-Lag Equalizer 504
11.10 Effects of Capacitive Loading 506
11.11 Oscillators 506
11.11.1 The Colpitts and Hartley Oscillator 507
11.11.2 Wien Bridge Oscillator 508
11.11.3 The Phase Shift Oscillator 509
11.11.4 The Crystal Oscillator 511
11.11.5 Touch-Tone Generator 512
Problems 514

CHAPTER 12 ACTIVE FILTERS 519

12.0 Introduction 519
12.1 Integrators and Differentiators 520
12.2 Active-Network Design 524
12.3 Active Filters 527
12.3.1 Filter Properties and Classification 528
12.3.2 First-Order Active Filters 534
12.4 Single Amplifier—General Type 547
12.5 Classical Analog Filters 549
12.5.1 Butterworth Filters 550
12.5.2 Chebyshev Filters 553
12.6 Transformations 555
12.6.1 Low-Pass to High-Pass Transformation 555
12.6.2 Low-Pass to Band-Pass Transformation 557
12.7 Design Procedure for Butterworth and Chebyshev Filters 558
12.7.1 Low-Pass Filter Design 558
12.7.2 Filter Order 559
12.7.3 Parameter Scale Factor 561
12.7.4 High-Pass Filter 569
12.7.5 Band-Pass and Band-Stop Filter Design 571

12.8 Integrated Circuit Filters 575
12.8.1 Switched-Capacitor Filters 575
12.8.2 A Sixth-Order Switched-Capacitor Butterworth Low-Pass Filter 578
12.9 Concluding Remarks 580
Problems 581

CHAPTER 13 QUASI-LINEAR CIRCUITS 586

13.0 Introduction 586
13.1 Rectifiers 586
13.2 Feedback Limiters 597
13.3 Comparators 612
13.4 Schmitt Triggers 616
13.4.1 Schmitt Triggers with Limiters 620
13.5 Integrated Circuit Schmitt Trigger 627
Problems 628

CHAPTER 14 PULSED WAVEFORMS AND TIMING CIRCUITS 635

14.0 Introduction 635
14.1 High-Pass RC Network 637
14.1.1 Steady-State Response—High-Pass Network 641
14.2 Low-Pass RC Network 647
14.3 Diodes 652
14.3.1 Steady-State Response of Diode Circuits to Pulse Train 652
14.4 Trigger Circuits 657
14.4.1 Pulse Train Response 657
14.5 The 555 Pulse Generator 660
14.5.1 The Relaxation Oscillator 660
14.5.2 The 555 as an Oscillator 664
14.5.3 The 555 as a Monostable Circuit 671
Problems 672

CHAPTER 15 DIGITAL LOGIC FAMILIES 679

15.0 Introduction 679
15.1 Basic Concepts of Digital Logic 680
15.1.1 State Definitions—Positive and Negative Logic 681
15.1.2 Time-Independent or Unclocked Logic 682
15.1.3 Time-Dependent or Clocked Logic 682
15.1.4 Elementary Logic Functions 682
15.1.5 Boolean Algebra 686
15.2 IC Construction and Packaging 688
15.3 Practical Considerations in Digital Design 690
15.4 Digital Circuits Characteristics of BJTs 693
15.5 Bipolar Logic Families 694

15.6 Transistor-Transistor Logic (TTL) 695
15.6.1 Open Collector 697
15.6.2 Active Pull-Up 699
15.6.3 H-TTL and LP-TTL Gates 700
15.6.4 Schottky TTL Gates 700
15.6.5 Tri-State Gates 701
15.6.6 Device Listings 703
15.7 Emitter-Coupled Logic (ECL) 704
15.7.1 Device Listings 706
15.8 Digital Circuit Characteristics of FETs 706
15.8.1 n-Channel Enhancement MOSFET 706
15.8.2 p-Channel Enhancement MOSFET 707
15.9 FET Transistor Families 707
15.9.1 n-Channel MOS 708
15.9.2 p-Channel MOS 708
15.10 Complementary MOS (CMOS) 709
15.10.1 CMOS Analog Switch 711
15.10.2 CMOS Device Listings and Usage Rules 713
15.11 Comparison of Logic Families 713
15.12 Concluding Remarks 714
Problems 718

CHAPTER 16 DIGITAL ICs 724

16.0 Introduction 724
16.1 Decoders and Encoders 725
16.1.1 Keyboard Encoders/Decoders 730
16.1.2 Parity Generators/Checkers 732
16.2 Drivers and Associated Systems 733
16.3 Flip-Flops, Latches, and Shift Registers 735
16.3.1 Flip-Flops 737
16.3.2 Latches and Memories 742
16.3.3 Shift Registers 745
16.4 Counters 747
16.4.1 Frequency Measurement 753
16.5 Clocks 756
16.5.1 Voltage-Controlled Oscillator (VCO) 757
16.6 Conversion Between Analog and Digital 762
16.6.1 Digital-to-Analog (D/A) Converter 762
16.6.2 Analog-to-Digital (A/D) Converter 763
16.6.3 The 3 1/2-Digit A/D Converter 765
16.6.4 Liquid Crystal Display (LCD) 767
16.7 Memories 768
16.7.1 Serial Memories 768

16.7.2 Random-Access Memories (RAM) 771
16.7.3 ROMs and PROMs 772
16.7.4 EPROMs 772
16.8 **More Complex Circuits 774**
16.8.1 Arithmetic Logic Unit (ALU) 774
16.8.2 Full Adders 776
16.8.3 Look-Ahead Carry Generators 776
16.8.4 Magnitude Comparators 778
16.9 **Programmable Array Logic (PAL) 779**
Problems 780

CHAPTER 17 ELECTRONIC SYSTEM DESIGN 784

17.0 **Introduction 784**
17.1 **Principles of Design 784**
17.2 **Problem Definition 786**
17.3 **Subdividing the Problem 786**
17.4 **Documentation 787**
17.4.1 The Schematic Diagram 787
17.4.2 The Parts List 788
17.4.3 Running Lists and Other Documentation 788
17.4.4 Using Documents 789
17.5 **Design Checklist 790**
17.6 **Prototyping Circuits 791**
17.7 **Design Examples 793**
17.8 **Introduction to Problems 812**
17.8.1 Generating Random Numbers 812
17.8.2 Measurement of Mechanical Angle or Velocity 813
17.8.3 The Hall-Effect Switch 813
17.8.4 Use of Timing Windows 814
17.9 **Concluding Remarks 815**
Problems 816

APPENDIX A SPICE

A.0 **Introduction**
A.1 **Programming Information A-2**
A.1.1 Format A-3
A.1.2 Circuit Description A-3
A.2 **Input Data A-4**
A.2.1 Element Description A-4
A.2.1.1 Passive Elements A-4
A.2.1.2 Active Components A-5
A.2.1.3 Semiconductor Devices A-6

A.2.2 Source Description A-8
A.2.2.1 Dependent Linear Sources A-9
A.2.2.2 Independent Linear Sources A-10
A.2.3 Subcircuits A-11
A.2.4 Analysis Request A-12
A.2.4.1 dc Analysis A-12
A.2.4.2 ac Analysis A-12
A.2.4.3 Transient Analysis A-13
A.2.5 Output Request A-13
A.2.5.1 Print A-13
A.2.5.2 Plot A-14
A.2.5.3 Options Statement A-15

A.3 Examples of Programs A-15

A.3.1 CE Amplifier A-15
A.3.2 CS Amplifier A-20
A.3.3 Full-Wave Rectifier A-23
A.3.4 Fourth-Order Chebyshev Low-Pass Filter A-28
A.3.4.1 MICRO-CAP II A-30
A.3.5 Two-Input NAND Gate A-32

APPENDIX B BODE PLOTS A-36

B.1 Bode Plots A-36

B.1.1 G(s)H(s) Function Terms A-41
B.1.2 The Asymptotic Approximation A-41
B.1.2.1 Frequency-Invariant Factors A-42
B.1.2.2 Zeros and Poles at the Origin A-42
B.1.2.3 Zeros A-44
B.1.2.4 Poles A-47
B.1.2.5 Quadratic Poles A-48
B.1.3 Examples of Bode Plots A-51

Problems A-56

APPENDIX C THE LAPLACE TRANSFORM A-60

C.0 Introduction A-60

C.1 The Laplace Transform of Functions A-61

C.2 The Laplace Transform of Operations A-61

C.3 Solution of Ordinary Linear Differential Equations A-64

C.4 Partial-Fraction Expansion A-67

C.5 Additional Properties of the Laplace Transform A-70

C.5.1 Real Translation A-71
C.5.2 Second Independent Variable A-71
C.5.3 Final Value and Initial Value Theorems A-72
C.5.4 The Convolution Theorem A-73

APPENDIX D MANUFACTURERS' DATA SHEETS **A-74**

D.1 **Diodes A-76**

D.1.1 1N4001 through 1N4007 Diode A-76

D.1.2 1N746 through 1N759 Zener Diode A-80

D.2 **Transistors A-86**

D.2.1 2N3903 and 2N3904 npn Silicon Transistor A-86

D.2.2 2N3905 and 2N3906 pnp Silicon Transistor A-92

D.2.3 2N2217 through 2N2222 npn General Purpose Transistor A-94

D.2.4 2N3821, 2N3822 and 2N3824 n-Channel JFET A-98

D.2.5 3N128 n-Channel MOSFET A-100

D.3 **Miscellaneous Devices A-102**

D.3.1 MC7800 IC Voltage Regulator A-102

D.3.2 LM 139 Voltage Comparator A-109

D.4 **Op-Amps A-112**

D.4.1 μA741 Op-Amp A-112

D.4.2 LM101A Op-Amp A-121

APPENDIX E STANDARD COMPONENT VALUES **A-126**

E.0 **Introduction A-126**

E.1 **Resistors A-126**

E.2 **Capacitors A-127**

APPENDIX F NOISE IN ELECTRONIC SYSTEMS **A-128**

F.0 **Introduction A-128**

F.1 **Sources of Noise A-128**

F.1.1 Resistor Noise A-129

F.1.2 Other Types of Noise A-130

F.1.3 Diode Noise A-131

F.1.4 BJT Noise A-131

F.1.5 FET Noise A-132

F.2 **Noise in Op-Amps A-132**

F.2.1 Signal-to-Noise Ratio A-133

F.2.2 Noise Figure A-134

F.2.3 Noise-Reduction Considerations A-134

APPENDIX G ANSWERS TO SELECTED PROBLEMS **A-136**

REFERENCES AND SOURCES FOR FURTHER STUDY **A-143**

INDEX **I-1**

ELECTRONIC DESIGN
Circuits and Systems

1

SEMICONDUCTOR DIODE CIRCUIT ANALYSIS

1.0 INTRODUCTION

The simplest nonlinear electronic device is known as a *diode.* A solid-state diode is composed of two different materials placed together in a manner such that charge flows easily in one direction but is impeded in the other direction. This device was developed by Henry Dunwoody in 1906 when he placed a piece of electric furnace carborundum between two brass holders. Later that year, Greenleaf Pickard developed a crystal radio detector in the form of a cat whisker in contact with a crystal. Various studies conducted during the period from 1906 to 1940 indicated that silicon and germanium were excellent materials to use in the construction of these devices.

Many problems had to be overcome in the construction and fabrication of diodes. It took until the mid-1950s for engineers to solve the most critical of these problems. During the technological explosion of the late 1950s and early 1960s, solid-state technology received a great deal of attention. This was due to the need for lightweight, small, and low-power-consuming electronic components for use in the development of intercontinental missiles and space vehicles. Emphasis was placed on fabricating solid-state devices that could achieve high reliability in applications where maintenance would be impossible. The result was the development of solid-state components that are more economical and more reliable than vacuum tubes.

This chapter provides an introduction to the operation and applications of the solid-state diode. This two-terminal device, which is often smaller than a grain of rice, is *nonlinear.* This means that applying the sum of two voltages produces a current that is *not* the sum of the two individual resulting currents. The diode behavior depends on the polarity of the applied voltage. The diode's nonlinear characteristic is the reason it finds so many applications in electronics.

We first consider the basic physical concepts of semiconductors. The silicon junction diode is analyzed and an equivalent circuit developed. Some important diode applications are then discussed.

The *zener* diode is presented and its use for voltage regulation is investigated. A design technique is then developed.

A number of special-purpose diodes such as the *Schottky diode, varactor diode, light-emitting diode (LED),* and *photodiode* are discussed.

1.1 THEORY OF SEMICONDUCTORS

It is helpful to have an exposure to the physics of diodes in order to appreciate the origins of the equivalent circuits and to understand their limitations.

An *atom* consists of a *nucleus,* which has a positive charge, and *electrons,* which have negative charges and move around the nucleus in elliptic paths. These electrons distribute themselves in *shells.* Electrons in the outermost shell are known as *valence electrons.*

When extremely pure elements, such as silicon and germanium, are cooled from the liquid state, their atoms arrange themselves in orderly patterns called *crystals,* as illustrated in Figure 1.1. The valence electrons determine the exact shape or lattice structure of the resulting crystal.

Silicon and germanium atoms each have four valence electrons; however, the germanium atom has one more filled outer ring than does the silicon atom, and this germanium outer ring is at a greater distance from the atomic nucleus than is the silicon outer ring. This is illustrated in Figure 1.2.

The atoms are bound in a lattice structure such that each atom "shares" its four valence electrons with neighboring atoms in the form of *covalent bonds.* The covalent bonds hold the lattice together.

Although the valence electrons are bound tightly in the crystalline structure, it is possible for these electrons to break their bonds and thus be capable of moving about as nonlocalized *conduction electrons.* This will happen if sufficient external energy is supplied (e.g., from light or heat).

Because of interaction among the atoms in a crystal, it is possible for the valence electrons to possess energy levels within a range of values. The farther an electron is from atomic nuclei, the higher the energy level. Thus, as the crystal becomes "tighter," the levels decrease.

Just as there is a range, or band, of energies for the valence electrons, there is another range of energy values for the *conduction electrons*—those that have broken loose to participate in the conduction process. The two energy bands may or may not overlap.

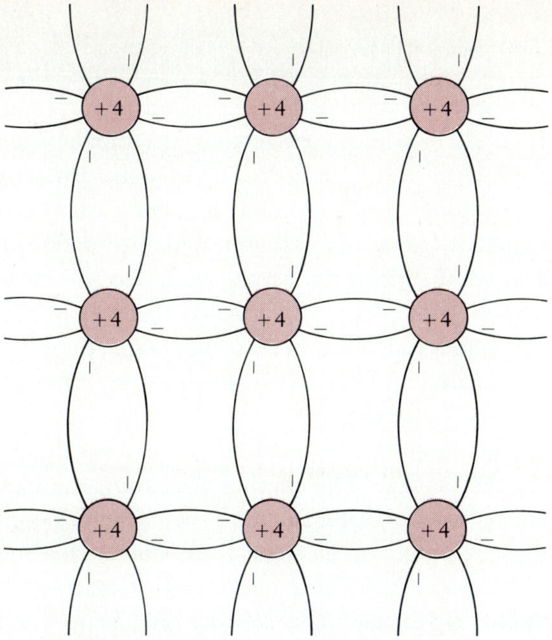

Figure 1.1
Crystal structure.

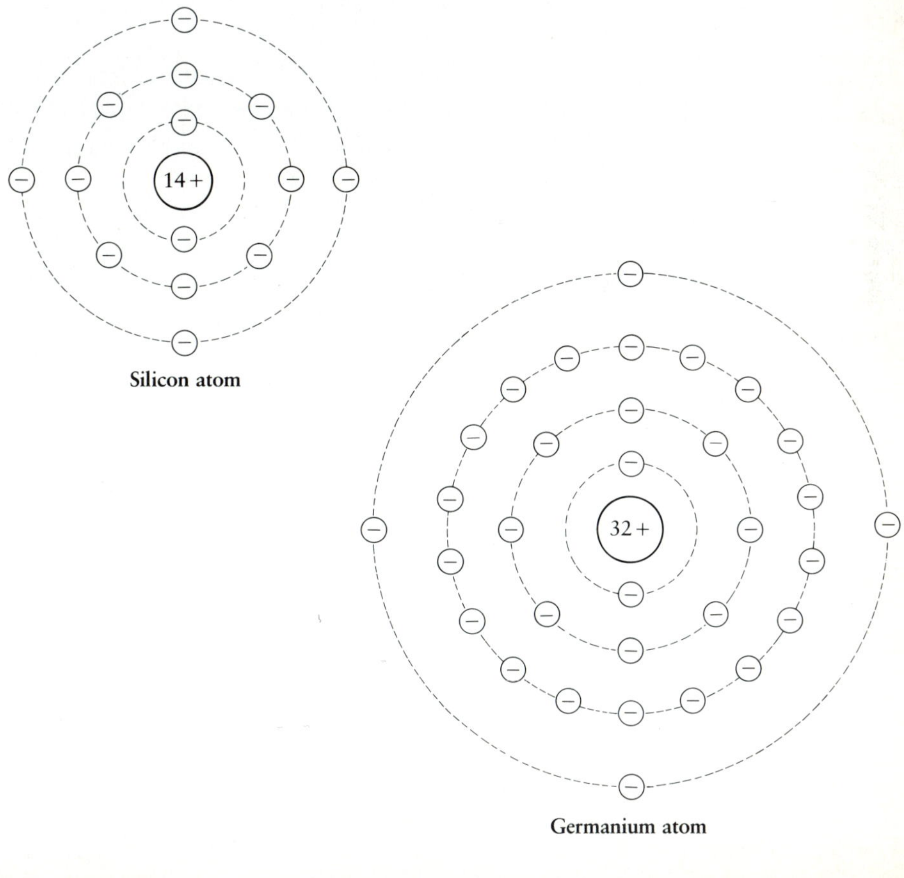

Figure 1.2
Atomic structure of silicon and germanium.

1.1.1 Conduction in Materials

Figure 1.3 presents three *energy level* diagrams. In Figure 1.3(a) the energy bands are widely separated. The unshaded region represents a *forbidden band* of energy levels in which no electrons are found. When this band is relatively large, as shown in the figure, the result is an *insulator.*

If the forbidden band is relatively small, on the order of 1 electron volt (eV),* the result is a *semiconductor.* This is illustrated in Figure 1.3(b).

A metal results when the bands overlap as shown in Figure 1.3(c). There is no gap between the energy of a valence electron and that of a conduction electron in a conductor. This means that a particular valence electron is not strongly associated with its own nucleus. It is therefore free to move about throughout the structure.

The energy required to break a covalent bond is a function of the atomic spacing in the crystal. The smaller the atom, the closer the spacing and the greater the energy required to break the covalent bonds. It is more difficult to break loose a conductive electron from silicon than one from germanium because the silicon crystals have closer lattice spacing.

This point is further illustrated by comparing the *energy gaps* of the two materials, as shown in Figure 1.4. Germanium has a smaller energy gap separating its valence and conduction bands, so less energy is required to cross the gap between bands.

Electrons in materials can be raised to higher energy bands by the application of heat, which causes vibration of the lattice. Materials that are insulators at room temperature can become conductors when the temperature is raised to a high enough level. This causes some electrons to move to a higher energy band, where they become available for conduction. The energy band diagram of Figure 1.5 is used to illustrate the amount of energy required for electrons to reach the conduction band. The abscissa of this graph is the atomic spacing of the crystal. As the spacing increases, the nuclei exert a smaller force on the valence electrons. The axis is marked with the atomic spacing for four materi-

Figure 1.3
Energy levels.

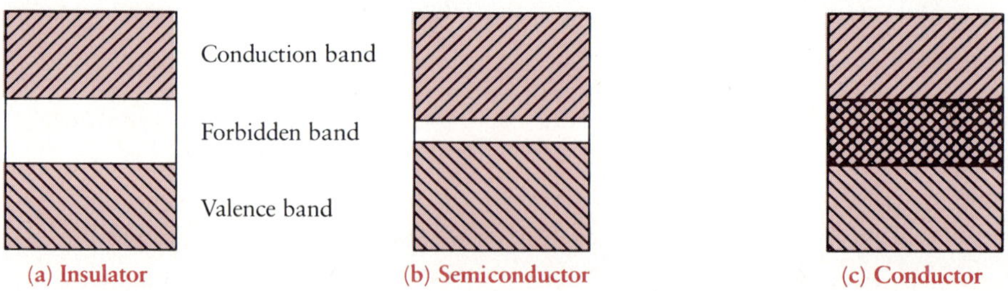

* An electron volt is the amount of kinetic energy by which an electron increases when it falls through a potential of 1 V. It is equal to 1.6×10^{-19} J.

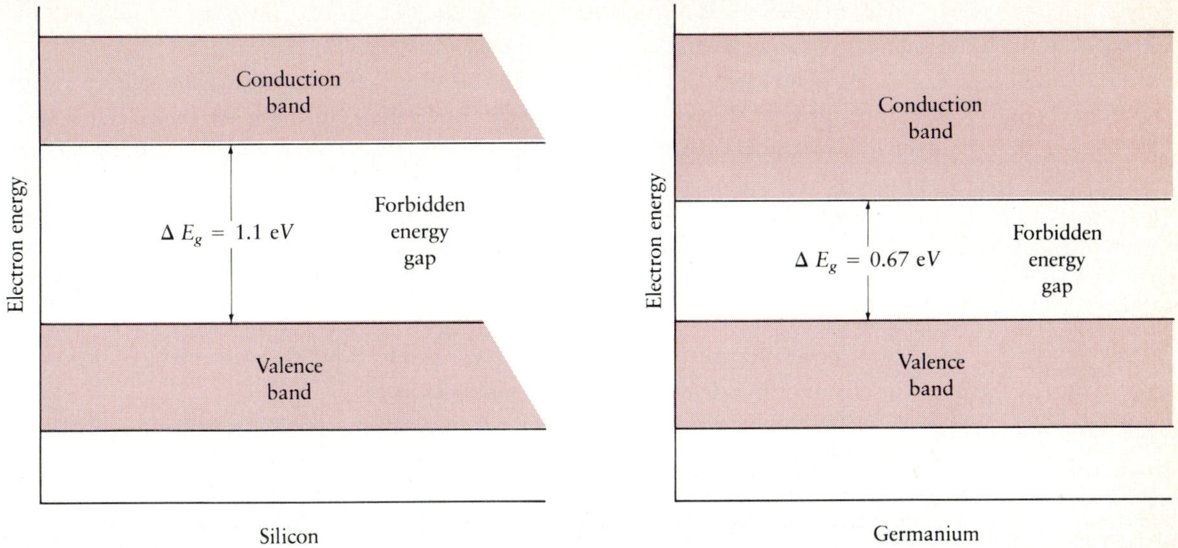

Figure 1.4 Energy gaps for germanium and silicon.

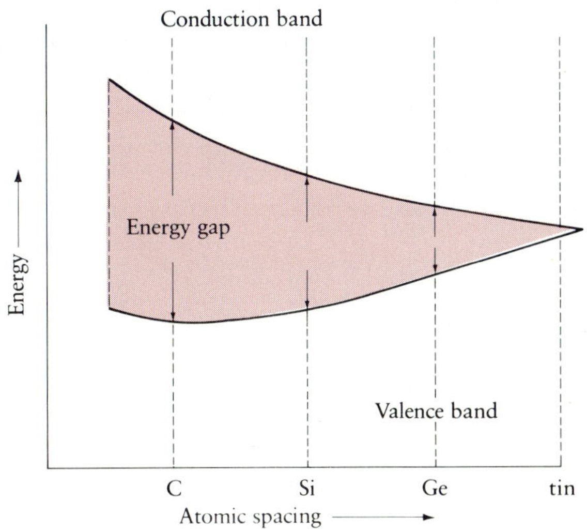

Figure 1.5 Energy band diagram.

als. Carbon (C) is an insulator in crystalline form (diamond). Silicon (Si) and germanium (Ge) are semiconductors, and tin is a conductor. The energy gap shown on the figure represents the amount of external energy required to move the valence electrons into the conduction band.

1.1.2 Conduction in Semiconductor Materials

Electrons are tightly held together in silicon and germanium atoms. The inner electrons are buried deep within the atom, whereas the valence electrons are part of the covalent bonding. That is, they cannot break away without receiving a considerable amount of energy.

At a temperature of absolute zero, there is no thermally induced vibration in the crystal. No covalent bonds can be broken, so there are no available electrons in the conduction band. Hence, current is zero and the semiconductor acts as an insulator.

Heat and other sources of energy cause valence band electrons to break their covalent bonds and become free electrons in the conduction band. As an electron leaves the valence band, it leaves a *hole*. A nearby valence band electron can move in and fill the hole, thus creating a relocated hole with practically no exchange of energy. Figure 1.6 shows how the movement of electrons in covalent bonds contributes to conduction.

The conduction caused by the electrons in the conduction band is different from the conduction due to the holes left in the valence band. In pure semiconductors there are as many holes as there are free electrons.

Internal thermal energy increases the activity of electrons, thus moving valence electrons out of the influence of the covalent bond into the conduction band. In this way, a limited number of conduction band electrons are under the influence of any applied external force (e.g., an electric field); these electrons move in one direction and establish a current, as shown in Figure 1.7.

Each time an electron is raised to a higher band, a hole is created in the valence band. The motion of holes is opposite in direction from that of electrons

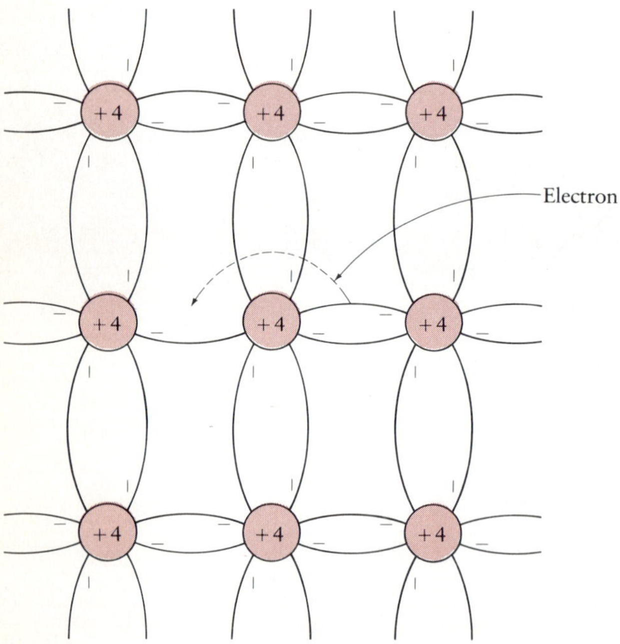

Figure 1.6 Conduction from broken covalent bond.

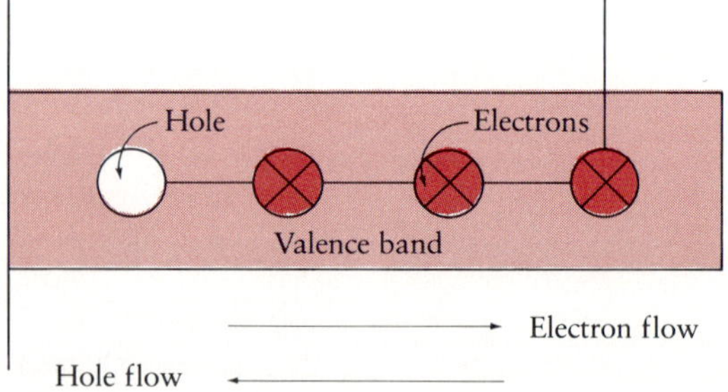

Figure 1.7 Current flow.

and is known as *hole current*. The holes act as if they are positive particles and contribute to the overall current flow.

The two methods by which holes and electrons move through a silicon crystal are *diffusion* and *drift*. Thermal agitation causes random electron movement in a semiconductor. Although this phenomenon can be related to diffusion, it does not result in any net flow of charge. However, if some other mechanism causes an accumulation of electrons at one end of the semiconductor, thereby creating a gradient, the electrons will diffuse to the other end. This gives rise to a net charge flow, known as *diffusion current*. The other method of movement, *drift*, results when an electric field is applied to the semiconductor and the free holes and electrons are accelerated in the electric field. The average velocity of this movement is called *drift velocity*, and the movement results in *drift current*. The drift current is proportional to the applied electric field (Ohm's law).

1.1.3 Intrinsic Semiconductors

Semiconductors are materials in which there is a forbidden energy gap—a gap in the energy levels where no electrons can exist. The valence electrons of the atoms of a semiconductor are not free to move throughout the volume of the material. Instead, they participate in the covalent bonds that hold the assembly of atoms together in a periodic crystalline structure.

Figure 1.8 shows the crystal structure of silicon. Other semiconductors, such as germanium, have the same crystal structure. Silicon is by far the most extensively used semiconductor. Germanium is used in specialized applications

Figure 1.8
Crystal structure of Si.

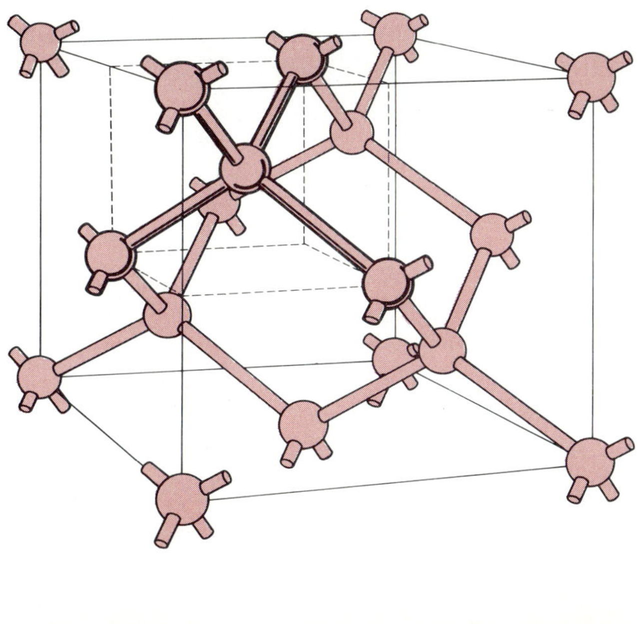

(e.g., high-current implementations). Gallium arsenide (GaAs) operates much faster than silicon, and although it is difficult to process, it is increasingly being used in integrated circuits.

In the lattice structure of Figure 1.8, every atom has four nearest neighbors and shares its valence electrons with those neighbors. The atoms are in the center of a tetrahedron. The lattice spacing of silicon is constant at 5.4 Å. The connecting rods between the atoms in Figure 1.8 can be thought of as indicating the spatial location of the valence electrons making up the covalent bonds.

At 0°K all electrons are constrained to their covalent bonds. No free electrons are available and no conduction is possible. The material behaves as an insulator. At room temperature, some covalent bonds will be broken. These broken bonds result from the random thermal vibration of the atoms and of the valence electrons. A few electrons acquire enough energy to "shake loose" from the bonds and become free. The energy required to break a covalent bond in silicon at room temperature is about 1.1 eV. The number of broken bonds is relatively low—about $10^{11}/\text{cm}^3$ at room temperature. Since there are about 10^{23} atoms/cm^3, only one out of every trillion atoms has a broken bond.

By comparison, diamond, which has the same crystalline structure, requires much higher energy to break a bond. Only 10^8 atoms/cm^3 have broken bonds at room temperature. Diamond is thus an insulator.

There are two independent groups of charge carriers: conduction electrons and holes. When a bond is broken, an electron is free to move through the crystal. As the negatively charged electron moves away, a positive charge remains, since the lattice is electrically neutral. The broken bond remains behind and is positively charged. This bond attempts to restore itself to its normal completed state by capturing an electron. Even though the bound valence electrons are not free to move around through the crystal, it is possible for them to move from one bond to another, provided the destination bond is incomplete. The net result is that the broken bond still exists but has now moved in the opposite direction from the valence electron. A mobile positive charge, or *hole,* results. A hole has properties similar to those of a free electron (e.g., mass, mobility, lifetime) but has exactly the opposite charge.

In a pure (intrinsic) semiconductor, the number of holes is equal to the number of free electrons. This is true because the generating processes create both a hole and an electron.

Hole-electron pairs can also disappear. This process is called *recombination.* Recombination takes place when a free electron moves close to a hole and is recaptured in the broken bond. At that point the electron ceases to be free and again becomes part of a covalent bond. Both a conduction electron and a hole disappear at that point.

Since generation and recombination affect holes and electrons as pairs, the hole concentration (p) must equal the electron concentration (n).

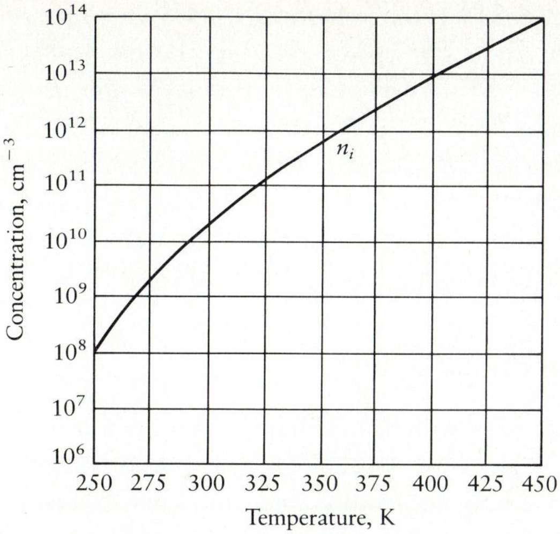

Figure 1.9 Temperature dependence of n_i.

$$n = p = n_i \tag{1.1}$$

n_i is called the *intrinsic concentration.* The product, *np*, is then given by

$$np = n_i^2 \tag{1.2}$$

The intrinsic concentration, n_i, is a function of temperature. For example, there are no broken bonds at 0°K and $10^{11}/cm^3$ at room temperature. This temperature relationship is shown in Figure 1.9. The formula for the temperature dependence is

$$n_i^2 = AT^3 e^{-E_G/kT} \tag{1.3}$$

A is a proportionality constant, T is the absolute temperature, k is Boltzmann's constant (1.38066×10^{-23} J/°K), and E_G is the band gap energy—the minimum energy required to break a covalent bond (1.12 eV for silicon at room temperature). For silicon the constant A equals 5.06×10^{43} m^{-6} $°K^{-3}$.

1.1.4 Doped Semiconductors

Currents induced in pure semiconductors are relatively small (typically less than 10^{-9} A). The conductivity of a semiconductor can be greatly increased when small amounts of specific impurities are introduced into the crystal. This procedure is called *doping.* If the doping substance has excess free electrons (more electrons than holes), it is known as a *donor,* and the doped semiconductor will be *n-type.* Since there are more electrons than holes, the *majority carriers* are electrons and the *minority carriers* are holes. We discuss *n*-type semiconductors in Section 1.1.5.

If the doping substance has extra holes, it is known as an *acceptor,* and the doped semiconductor will be *p-type.* The majority carriers then become the holes and the minority carriers electrons. We discuss *p*-type semiconductors in Section 1.1.6. Figure 1.10 illustrates the crystal structure of *n*-type and *p*-type semiconductors. The doped materials are known as *extrinsic semiconductors,* while the pure substances are *intrinsic semiconductors.*

The resistance of a semiconductor is known as *bulk resistance.* A lightly doped semiconductor has a high bulk resistance.

1.1.5 *n*-type Semiconductors

Phosphorus, arsenic, and antimony are commonly used as donor impurities in silicon. They have five valence electrons instead of the four that silicon has. Their atoms fit readily into the silicon crystal structure without disturbing the lattice.

After covalent bonding of a pentavalent atom into the silicon lattice, the fifth valence electron is loosely bound and needs only 0.05 eV to be detached. In other words, at room temperature we create a free electron. The donor ion is bound in the lattice structure and thus donates free electrons without contributing free holes. Donor-doped silicon is referred to as *n-type silicon* or *n-silicon.*

Figure 1.10 Crystal structure of doped semiconductors.

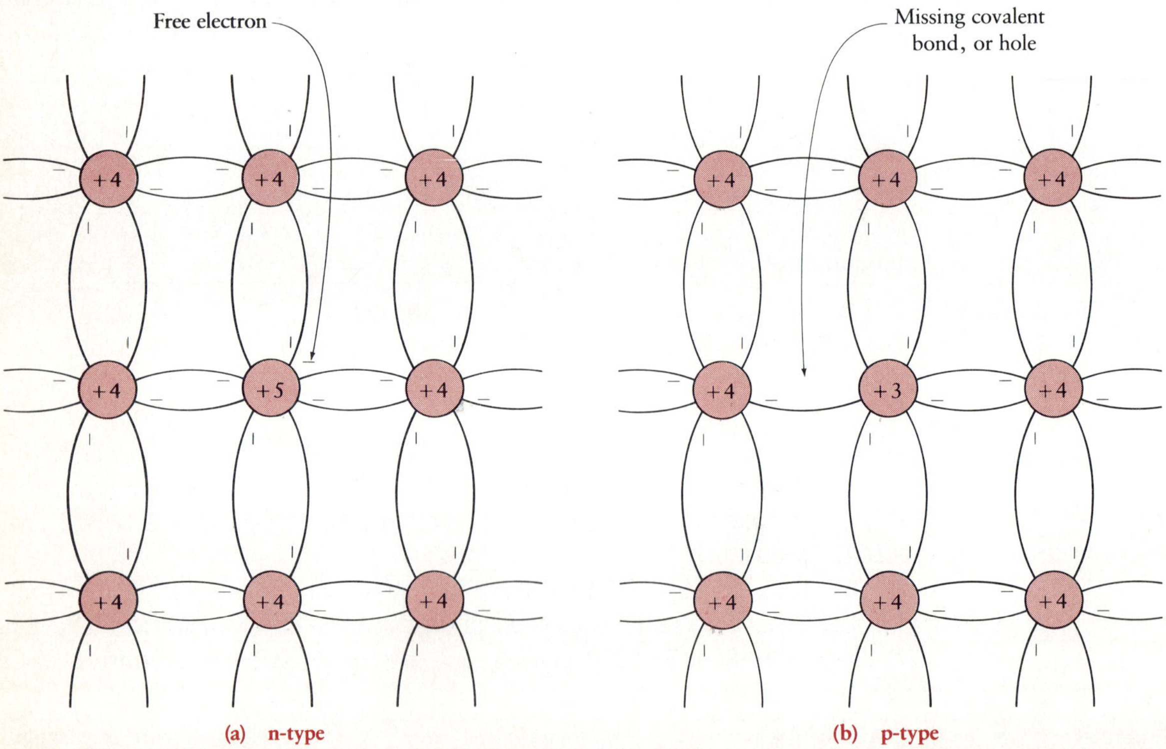

1.1.6 *p*-type Semiconductors

Acceptor impurities have three valence electrons. Boron is the most commonly used acceptor in silicon. Since it lacks one valence electron, it leaves a vacancy in the bond structure—a hole. This vacancy can accept electrons, but in the process a new vacancy is created. In other words, the hole is free to move around—an acceptor creates a mobile positive charge. The negatively charged boron atom is tied into the lattice and cannot move. An acceptor impurity thus contributes only a free hole to the conduction process.

Acceptor-doped silicon is referred to as *p-type silicon* or *p-silicon.*

1.1.7 Carrier Concentrations

We now consider the manner in which hole and electron concentrations depend on donor-acceptor concentrations and on other semiconductor parameters.

We first examine the equilibrium situation in which a semiconductor is undisturbed in its thermal environment. Under equilibrium conditions, there must be no net flow of holes or electrons. Four classes of charged particles exist in a semiconductor:

Particles with a positive charge: mobile holes (p) and immobile donors (N_D)

Particles with a negative charge: mobile electrons (n) and immobile acceptors (N_A)

In each case the symbol in parentheses represents the volume concentration of the particular particle. Each of these particles carries a charge of magnitude q, the electron charge. The local charge density, ρ, can be written as

$$\rho = q(p + N_D - n - N_A) \tag{1.4}$$

Under equilibrium conditions ρ is equal to zero. Local neutrality must be preserved as a consequence of Gauss's law.

With the constraint that ρ equals zero, we can rewrite equation (1.4) as follows:

$$n - p = N_D - N_A \tag{1.5}$$

Impurity concentrations N_D and N_A are determined by processing parameters of the semiconductor, so equation (1.5) provides one constraint on the concentration of holes and electrons. Note that with no doping $n = p$, as we discussed earlier.

A second constraint follows from the fact that the product of the equilibrium carrier concentrations is independent of the donor and acceptor impurity concentrations and is dependent only on the absolute temperature.

$$n_o p_o = n_i^2 \tag{1.6}$$

In equation (1.6), n_o and p_o are the equilibrium carrier concentrations. The notation n_i^2 evolves from the intrinsic situation, where we observed that $n = p = n_i$. Since we postulated that the electron and hole concentrations are independent of doping in extrinsic semiconductors as well, we write the product as n_i^2.

As long as the doping levels are not excessive, all impurity atoms are ionized. Every donor will contribute one free electron; every acceptor will contribute one free hole. This would not be true at very low temperatures (temperatures below that of liquid nitrogen). At such temperatures, not all dopants would be ionized.

We can now calculate the electron and hole concentrations as a function of the doping concentrations. The explicit solution can be simplified in most circumstances by noting that there will be orders of magnitude difference between N_D, N_A, and n_i. In n-type materials, N_D will be large compared with N_A and n_i, yielding

$$n \approx N_D \tag{1.7}$$

$$p \approx \frac{n_i^2}{N_D} \tag{1.8}$$

In p-type materials, on the other hand, N_A will be large compared with N_D and n_i, giving

$$p \approx N_A \tag{1.9}$$

$$n \approx \frac{n_i^2}{N_A} \tag{1.10}$$

In silicon, the hole concentration is $5 \times 10^{21}/\text{m}^3$, and the electron concentration is $4.205 \times 10^{10}/\text{m}^3$. The intrinsic concentration at room temperature is $n = p = n_i = 1.5 \times 10^{16}/\text{m}^3$.

It is important to note that this minute amount of doping does not affect the chemical or mechanical properties of the semiconductor. It does, however, have an enormous effect on the electrical properties.

1.1.8 Excess Carriers

We showed the relationship between donor-acceptor concentrations and hole-electron concentrations in Section 1.1.7. An important difference between metals and semiconductors is that the local carrier concentration can be disturbed significantly without producing significant deviations from electrical neutrality.

Because of the complementary charge of electrons and holes, equal changes in n and p will *not* produce a local charge density.

$$n = n_o + n' \tag{1.11}$$

$$p = p_o + p' \tag{1.12}$$

n' and p' are the excess concentrations. The charge density is then given by

$$\rho = q(p_o + N_D - n_o - N_A) + q(p' - n') \tag{1.13}$$

If $n' \approx p'$ we have the quasi-neutral condition. Note that ρ is still zero, but the np product in this case is no longer equal to n_i^2.

1.1.9 Recombination and Generation of Excess Carriers

A nonzero excess-carrier concentration is a deviation from equilibrium. Physical mechanisms will endeavor to restore the equilibrium.

If n' and p' are positive, meaning charge concentrations greater than equilibrium, recombination will take place. If n' and p' are negative, meaning concentrations lower than equilibrium, generation will take place.

The rate of recombination can be written as

$$R = an' + bp' \tag{1.14}$$

where a and b are proportionality constants. Under quasi-neutral conditions (i.e., $n' = p'$) we can write

$$R = (a + b)n' = (a + b)p' \tag{1.15}$$

The rate of the recombination process is therefore proportional to the excess concentration.

$$R = \frac{n'}{\tau} = \frac{p'}{\tau} \tag{1.16}$$

The parameter τ, which has dimensions of time, is called the *carrier lifetime* of the excess carriers. The lifetime will typically vary between nanoseconds and milliseconds.

Carrier lifetime is not constant for a given semiconductor. It depends strongly on the preparation of the semiconductor and the processing history of the material as well as the doping levels in the material.

1.1.10 Transport of Electric Current

We now focus on the two mechanisms through which current can flow in a material: drift and diffusion.

At all nonzero temperatures, atoms in the crystal lattice possess kinetic energy in the form of vibrations about neutral positions in the lattice. There is a continuous interchange of energy between the vibrating ions and the free electrons in the form of elastic and inelastic collisions. The resulting electron motion is random; there is a zero net motion, so the net current is zero.

The thermal speed of carriers is given by

$$v_{\text{th}} = \sqrt{\frac{3kT}{m^*}} \text{ m/s} \tag{1.17}$$

In equation (1.17), m^* is the effective mass of the carrier, k is Boltzmann's constant, and T is the absolute temperature. Because of the influence of neighboring atoms, a charge carrier does not behave as if it had the free-electron mass (9.1×10^{-31} kg). A proportionality factor is used to compensate for this behavior.

At room temperature, the mean thermal speed of electrons and holes is 10^5 m/s, and they have collision frequencies of about 10^{11} s^{-1}. (Note that equation (1.17) yields a velocity of 1.17×10^5 m/s at room temperature if the free-electron mass is substituted for m^*.)

1.1.11 Drift in an Electric Field

Equilibrium can be disturbed by the application of electric fields. An electric field will cause a net movement of carriers, which is superimposed on the random thermal movement, as shown in Figure 1.11. This net effect is called *drift*. The *drift velocity* is defined as follows.

$$v_h = \mu_h \xi, \qquad v_e = -\mu_e \xi \tag{1.18}$$

where μ_h and μ_e are the hole and electron mobilities in m^2/V · s and ξ is the *electric field intensity* (the force exerted on a unit positive charge). The elec-

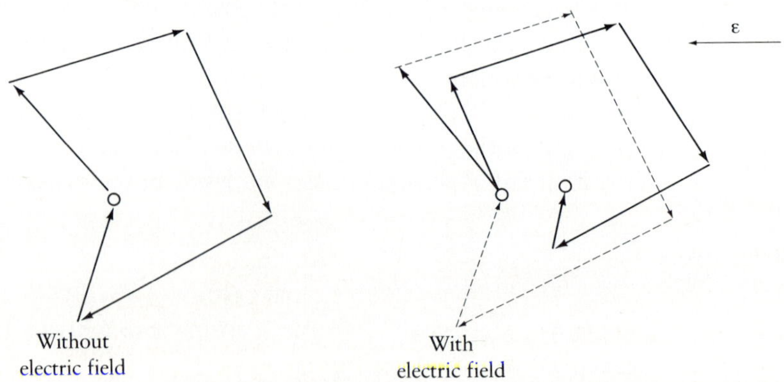

Figure 1.11 Random thermal motion of charged particles.

tron mobility for silicon is about 0.135 $m^2/V \cdot s$ and the hole mobility is approximately 0.045 $m^2/V \cdot s$.

These mobilities are not constant. They have complex dependences on both temperature and doping densities. The movement of the carriers is determined by scattering mechanisms. An electron that collides with another electron, an Si atom, or a doping atom will be scattered and change direction. This scattering process depends on the temperature and the amount of ions.

The electric current associated with drift is calculated in the following way. Assume there are N electrons contained in a length L of conductor, as shown in Figure 1.12. Further assume that it takes an electron t_o seconds to travel a distance of L meters in the conductor. The total number of electrons passing through any cross section of wire in a unit of time is N/t_o. The current, in amperes, is then

$$I = \frac{N}{t_o} q = \frac{Nq}{t_o} = \frac{Nq(L/T)}{L} = \frac{Nqv}{L}$$

where q is the electron charge.

Also by definition, the current density, J (A/m^2), is

$$J = \frac{I}{A} = \frac{Nqv}{LA} \tag{1.19}$$

With N/LA being the electron concentration, n, in units of electrons/m^3, we obtain

$$J = nqv \tag{1.20}$$

It should be noted that this derivation is independent of the form of the conducting medium.

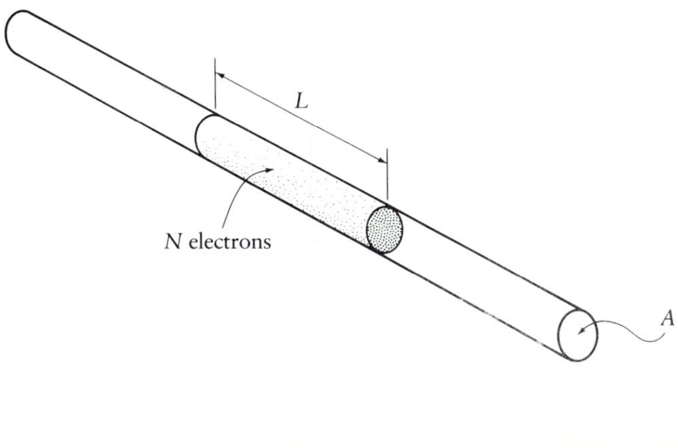

Figure 1.12
Carrier flow in a conductor.

From equations (1.18) and (1.20), we find

$$J = nqv = nq\mu\xi = \sigma\xi$$

where $\sigma = nq\mu$ is the conductivity in $(\Omega \cdot \text{m})^{-1}$. Combining these equations yields

$$I = JA = \frac{\sigma AL\xi}{L} = \frac{\sigma AV}{L} = \frac{V}{R} \tag{1.21}$$

which is recognized as Ohm's law.

For holes and electrons in silicon we can now write

$$J_h = qp\mu_h\xi, \qquad J_e = qn\mu_h\xi \tag{1.22}$$

so the total drift current is

$$J = q(p\mu_h + n\mu_e)\xi \tag{1.23}$$

Example 1.1

Find the average drift velocity in a silicon semiconductor doped with phosphorus. Assume a doping density of 1 per million, a cross-sectional area of 10^{-4} m^2, a current of 4 A, and the concentration of silicon atoms in a silicon crystal is

$$n_{\text{Si}} = 4.97 \times 10^{28} \text{ atoms/m}^3$$

SOLUTION With 1 per million phosphorus atoms and $n = N_d$ we have $n = 4.97 \times 10^{22}$ electrons/m^3 and a negligible density of holes.

$$v_e = \frac{J_e}{nq} = \frac{I/A}{nq} = \frac{4 \times 10^4}{(4.97 \times 10^{22})(1.602 \times 10^{-19})} = 5 \text{ m/s}$$

The average drift velocity is thus orders of magnitude lower than the random thermal velocity.

Example 1.2

What is the resistivity of intrinsic silicon at room temperature? What is the resistivity of silicon doped with 1/million boron or phosphorus at room temperature?

SOLUTION For intrinsic material:
From Section 1.1.7, we find

$$n = p = n_i = 1.5 \times 10^{16}/\mathrm{m}^3$$

From equation (1.23),

$$\sigma_i = q(\mu_e n + \mu_h p) = (0.135 + 0.045)(1.5 \times 10^{16})(1.609 \times 10^{-19})$$
$$= 4.34 \times 10^{-4}\ (\Omega \cdot \mathrm{m})^{-1}$$

$$\rho_i = \frac{1}{\sigma_i} = 2.3 \times 10^3\ \Omega \cdot \mathrm{m}$$

For the phosphorus-doped n-Si

$$\rho = 9.3 \times 10^{-4}\ \Omega \cdot \mathrm{m}$$

and for the boron-doped p-Si

$$\rho = 2.78 \times 10^{-3}\ \Omega \cdot \mathrm{m}$$

These values show the tremendous effect of doping on the electrical properties of silicon. The resistivity values are only approximations: The mobilities are doping dependent.

1.2 SEMICONDUCTOR DIODES

The simplest *linear* circuit element is the resistor. The voltage across this element is related to the current through it by Ohm's law. This relationship is graphically depicted by a straight line as shown in Figure 1.13. The slope of the line is the *conductance* of the resistor, that is, the ratio of current to voltage. The reciprocal of this slope is the *resistance,* in ohms. If the resistor is connected in any circuit, the operating point must fall somewhere on this curve. The actual operating point location is determined by constraints provided by the external circuitry.

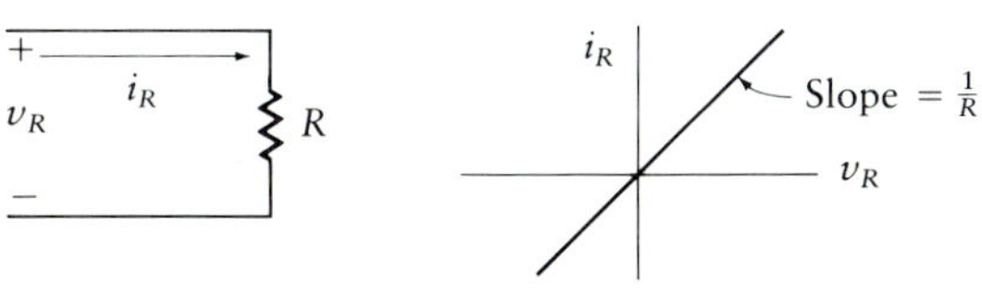

Figure 1.13 Operating curve for resistor.

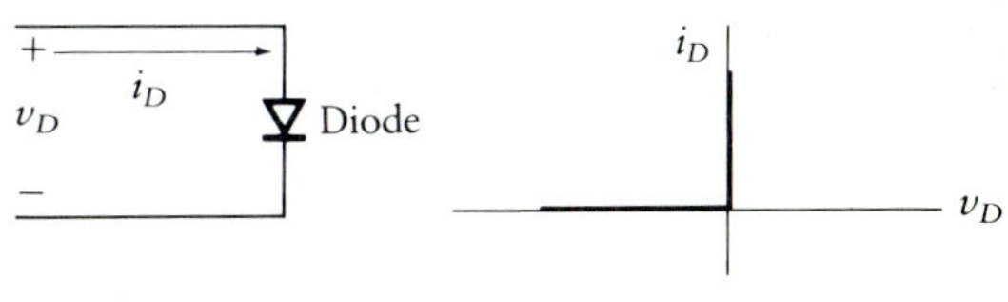

Figure 1.14 Operating curve for ideal diode.

The ideal diode is a *nonlinear* device with a current versus voltage characteristic as shown in Figure 1.14. This characteristic is referred to as *piecewise linear,* since the curve is constructed from segments of straight lines. Note that as we attempt to impose a positive or forward voltage on the diode, we are not successful and the voltage is limited to zero. The slope of the curve approaches infinity. Therefore, under this condition the resistance is zero and the diode behaves as a short circuit. If we place a negative voltage across the diode, the current is zero and the slope of the curve is zero. Thus, the diode is now behaving as an infinite resistance, or open circuit.

1.2.1 Physics of Solid-State Diodes

Figure 1.15 shows a *p*-type material and an *n*-type material placed together to form a junction. This represents a simplified model of diode construction. The model ignores gradual changes in concentration of the impurities in the material. Practical diodes are constructed as a single piece of semiconductor material with one side doped with *p*-type material and the other side with *n*-type material.

Also shown in Figure 1.15 is the schematic symbol of the diode. Note that the "arrow" in this symbol points from the *p*- to the *n*-type material.

When the *p*- and *n*-type materials exist together in a crystal, a charge redistribution occurs. Some of the free electrons from the *n*-type material migrate across the junction and recombine with the free holes in the *p*-type material. Similarly, some of the free holes from the *p* material migrate across the junction and recombine with free electrons in the *n* material. As a result of this charge redistribution, the *p* material acquires a net negative charge and the *n* material acquires a net positive charge. These charges create an electric field and a potential difference between the two types of material that will inhibit any further charge movement. The result is a reduction in the number of current carriers near the junction. This happens in an area known as the *depletion region.* The resulting electric field provides a *potential barrier* or *hill* in a direction that inhibits the migration of carriers across the junction. This is shown in Figure 1.16. In order to produce a current across the junction, we must reduce the potential barrier or hill by applying a voltage of the proper polarity across the diode.

Figure 1.15 Simplified diode model.

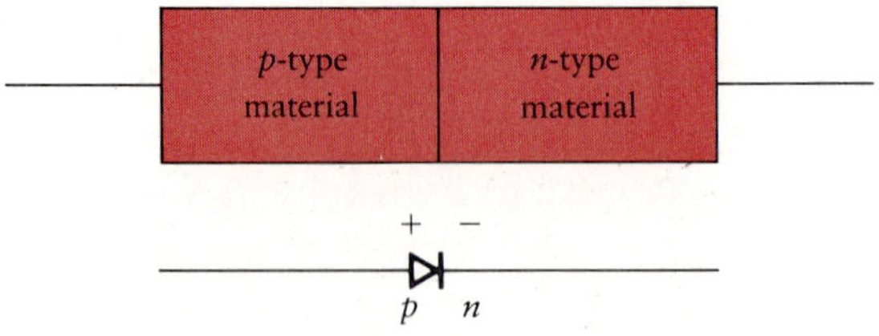

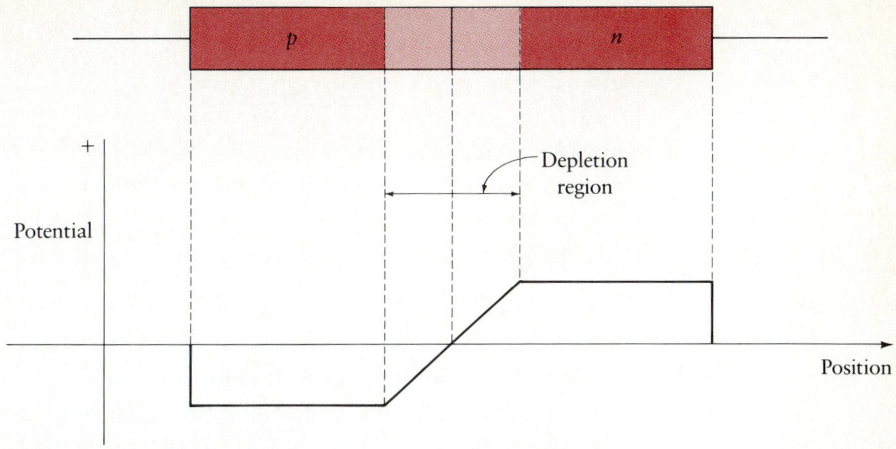

Figure 1.16
Barrier potentials.

1.2.2 Diode Construction

Three different materials are commonly used in the construction of diodes: germanium, silicon, and gallium arsenide. Silicon has generally replaced germanium for diodes. Gallium arsenide is particularly useful in microwave and high-frequency applications.

The precise distance over which the change from *p*- to *n*-type material occurs within the crystal varies with the fabrication technique. One feature of the *pn* junction is that the change in impurity concentration must occur in a relatively short distance. There are cases where the *pn* junction cannot be treated as an abrupt change in material type, notably when the diode is formed by diffusion. This causes the doping near the junction to be *graded*—that is, the donor and acceptor concentrations are a function of distance across the junction [2, 44].

A *depletion region* will exist in the vicinity of the junction, as shown in Figure 1.17(a). This phenomenon is due to a combination of electrons and holes where the materials join. This depletion region will have few carriers. The minority carriers on each side of the depletion region (electrons in the *p* region and holes in the *n* region) will migrate to the other side and combine with ions in that material. Likewise, the majority carriers (electrons in the *n* region and holes in the *p* region) will migrate across the junction.

The two components of the current formed by the hole and electron movements across the junction add together to form the diffusion current, I_D. The direction of this current is from the *p* side to the *n* side. An additional current exists due to the minority-carrier drift across the junction, and this is referred to as I_S. Thermally generated minority carriers (holes in the *n* material and electrons in the *p* material) diffuse to the edge of the depletion region. There they experience the electric field and are swept across the depletion region.

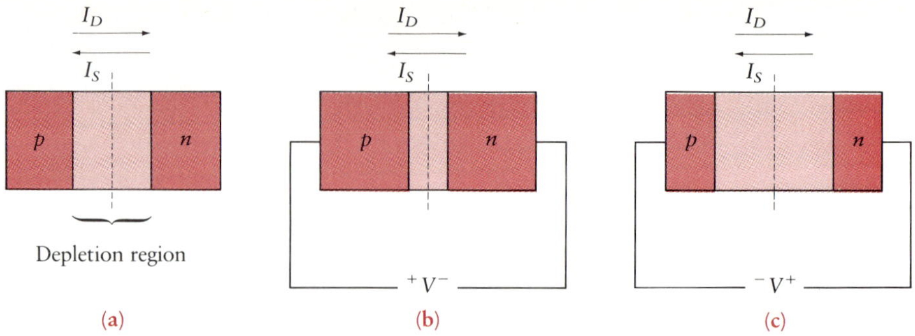

Figure 1.17 Depletion regions.

The components of these actions combine to form the drift current, I_S. During open-circuit equilibrium conditions, the diffusion current is equal in magnitude (and opposite in direction) to the drift current.

If we now apply a positive potential to the p material relative to the n material, as shown in Figure 1.17(b), the diode is said to be *forward-biased.* The depletion region shrinks in size because of the attraction of majority carriers to the opposite side. That is, the negative potential at the right attracts holes in the p region and vice versa. With a lower barrier, current can flow more readily. When forward-biased, $I_D - I_S = I$ after equilibrium is achieved, where I is the current through the junction.

Alternatively, if the applied voltage is as shown in Figure 1.17(c), the diode is *reverse-biased.* Free electrons are drawn from the n material toward the right, and holes are drawn to the left. The depletion region gets wider and the diode acts as an insulator.

1.2.3 Relationship Between Diode Current and Diode Voltage

An exponential relationship exists between the carrier density and the applied potential. It is possible to write a single expression for the density distribution and account for both the forward- and reverse-bias conditions. The expression applies as long as the voltage does not exceed the breakdown voltage. The relationship is described by equation (1.24).

$$i_D = I_o\left[\exp\left(\frac{qv_D}{nkT}\right) - 1\right] \tag{1.24}$$

The terms in equation (1.24) are defined as follows:

i_D = current in the diode (amperes)

v_D = potential difference across the diode (volts)

I_o = reverse saturation current (amperes)

q = electron charge, 1.6×10^{-19} J/V

k = Boltzmann's constant, 1.38×10^{-23} J/°K

T = absolute temperature (degrees Kelvin)

n = empirical constant between 1 and 2, sometimes referred to as the exponential ideality factor

Equation (1.24) can be simplified by defining

$$V_T = \frac{kT}{q}$$

This yields

$$i_D = I_o\left[\exp\left(\frac{v_D}{nV_T}\right) - 1\right] \tag{1.25}$$

If we operate at room temperature (25°C) and only in the forward-bias region, the first term in the brackets predominates and the current is given approximately by

$$i_D \approx I_o \exp\left(\frac{v_D}{nV_T}\right) \tag{1.26}$$

These equations are illustrated in Figure 1.18.

The reverse saturation current, I_o, is a function of the doping, the geometry of the diode, and the temperature.

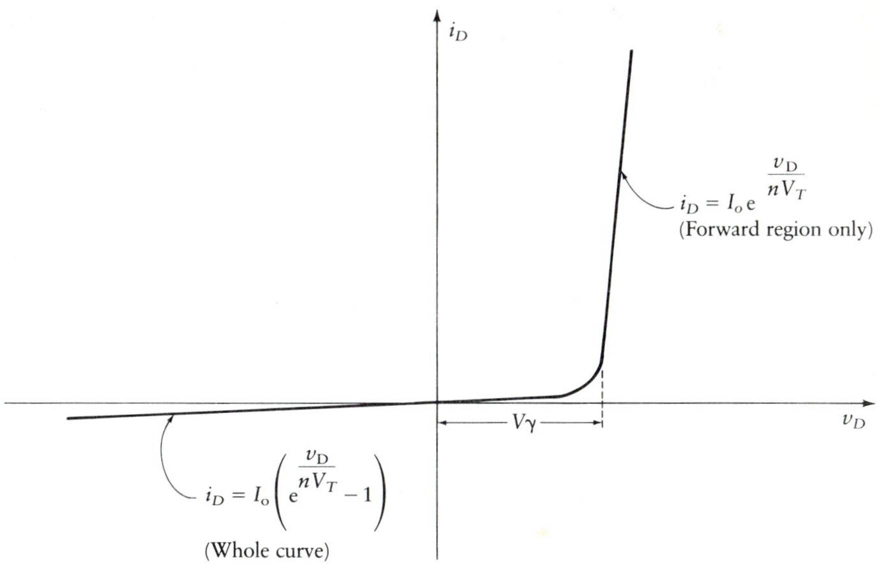

Figure 1.18
Diode voltage-current relationship.

The empirical constant, n, is a number that can vary in accordance with the voltage and current levels and depends on electron drift, diffusion, and carrier recombination in the depletion region. It approaches 2 as the number of electron-hole recombinations in the depletion region increases. If $n = 1$, the value of nV_T is 26 mV at 25°C. When $n = 2$, nV_T becomes 52 mV.

We now have the necessary information to evaluate the relationship between current and voltage at an operating point, Q. The slope of the curves of Figure 1.18 changes as current changes; it follows an exponential relationship. We can differentiate the expression of equation (1.25) to find the slope at any fixed i_D. This slope is the equivalent conductance of the device.

$$\frac{di_D}{dv_D} = \frac{I_o[\exp(v_D/nV_T)]}{nV_T} \tag{1.27}$$

To eliminate the exponential function, we solve the basic diode equation, equation (1.25), to obtain

$$\exp\left(\frac{v_D}{nV_T}\right) = \frac{i_D}{I_o} + 1$$

Substituting this expression into the slope equation yields

$$\frac{di_D}{dv_D} = \frac{i_D + I_o}{nV_T} \tag{1.28}$$

The dynamic resistance is the reciprocal of this expression, or

$$r_d = \frac{nV_T}{i_D + I_o} \approx \frac{nV_T}{i_D} \tag{1.29}$$

since $I_o << i_D$. Although we know r_d changes when i_D changes, we often assume it is fixed for a specific operating range. That is, we choose a value within the range of varying resistances. We use the term R_f to denote diode forward resistance, which is composed of r_d and the contact resistance.

1.2.4 Diode Operation

Figure 1.19 illustrates the operating characteristics of a *practical* diode. The curve differs from the ideal characteristic of Figure 1.14 in the following ways. As the forward voltage increases beyond zero, current does not immediately start to flow. It takes a minimum voltage, denoted V_γ (0.2 or 0.7 V in the fig-

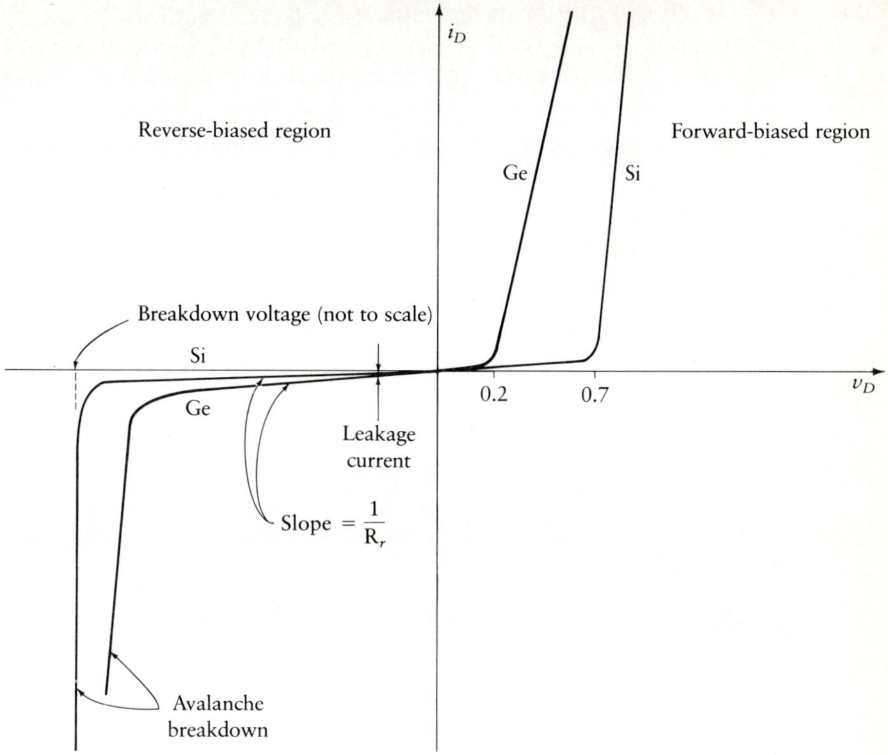

Figure 1.19
Diode operating characteristics.

ure), to obtain any noticeable current. As the voltage tries to exceed V_γ, the current increases rapidly. The slope of the characteristic curve is large, but not infinite as is the case with the ideal diode.

The minimum voltage required to obtain noticeable current, V_γ, is approximately 0.7 V for silicon semiconductors (at room temperature), 0.2 V for germanium semiconductors.

When the diode is reverse-biased, there exists a small leakage current. This current occurs provided that the reverse voltage is less than the voltage required to break down the junction. The leakage current is much greater for germanium diodes than for silicon or gallium arsenide diodes. If the negative voltage becomes large enough to be in the breakdown region, a normal diode may be destroyed. This breakdown voltage is defined as the *peak inverse voltage* (PIV) in manufacturers' specifications. (Appendix D shows representative specification sheets. We refer to these often in this text, so you should take a minute to locate them at this time.) The damage to the normal diode at breakdown is due to the avalanche of electrons that flow across the junction, with the result that the diode overheats. The large current can cause destruction of the diode if excessive heat builds up. This breakdown is sometimes referred to as the diode breakdown voltage (V_{BR}).

1.2.5 Temperature Effects

Temperature plays an important role in determining operational characteristics of diodes. Changes in the diode characteristics caused by changing temperature may require adjustments in the design and packaging of circuits.

As temperature increases, the turn-on voltage, V_γ, decreases. Alternatively, a decrease in temperature results in an increase in V_γ. This is illustrated in Figure 1.20, where V_γ varies linearly with temperature according to the following equation (we assume that the diode current is held constant).

$$V_\gamma(T_1) - V_\gamma(T_0) = k_T(T_1 - T_0) \tag{1.30}$$

where

T_0 = room temperature, or 25°C

T_1 = new temperature of diode (°C)

$V_\gamma(T_0)$ = diode voltage at room temperature (volts)

$V_\gamma(T_1)$ = diode voltage at new temperature (volts)

k_T = temperature coefficient (V/°C)

Values of k_T for the various types of diodes are as follows ([50], Section 1.11):

$k_T = -2.5$ mV/°C for germanium diodes

$k_T = -2.0$ mV/°C for silicon diodes

$V_\gamma(T_0)$ is equal to the value given below:

Silicon diodes 0.7 V

Germanium diodes 0.2 V

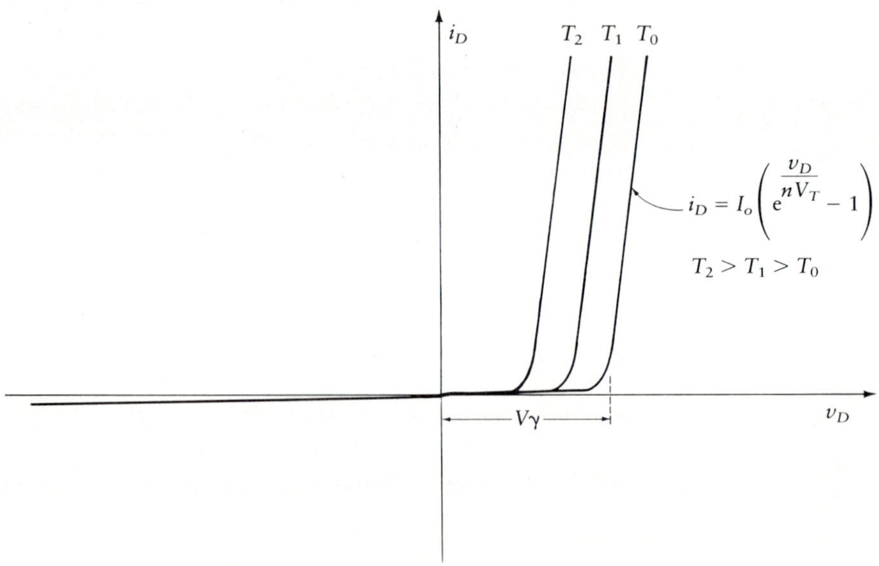

Figure 1.20 Dependence on temperature.

The reverse saturation current, I_o, is another diode parameter that depends on temperature. It increases approximately 7.2%/°C for both silicon and germanium diodes. In other words, I_o approximately doubles for every 10°C increase in temperature. The expression for the reverse saturation current as a function of temperature is

$$I_o(T_2) = I_o(T_1)\exp[k_i(T_2 - T_1)] \tag{1.31}$$

where $k_i = 0.07/°\text{C}$ and T_1 and T_2 are two different temperatures. This expression can be approximated and simplified as follows ([25], Section 2.4):

$$I_o(T_2) = I_o(T_1)2^{(T_2-T_1)/10} \tag{1.32}$$

This approximation is possible because

$$e^{0.7} \approx 2$$

Drill Problems

D1.1 When a silicon diode is conducting at a temperature of 25°C, a 0.7-V drop exists across its terminals. What is the voltage, V_γ, across the diode at 100°C?

Ans: $V_\gamma = 0.55$ V

D1.2 The diode described in Drill Problem D1.1 is cooled to −100°C. What is the voltage required across the diode to establish a noticeable current at the new temperature?

Ans: $V_\gamma = 0.95$ V

1.2.6 Diode Equivalent Circuit Models

The circuit shown in Figure 1.21(a) represents a simplified model of the silicon diode under both forward and reverse dc operating conditions. The relationships for this model approximate the diode operating curve of Figure 1.19. The resistor R_r represents the reverse-bias resistance of the diode and is usually several megohms. The resistor R_f represents the contact and bulk resistance of the diode and is usually less than 50 Ω. When forward-biased, the ideal diode is a short circuit, or zero resistance. The circuit resistance of the forward-biased practical diode modeled in Figure 1.21(a) is the terminal resistance with the ideal diode short-circuited, or

$$R_r \parallel R_f \approx R_f$$

Figure 1.21
Diode models.

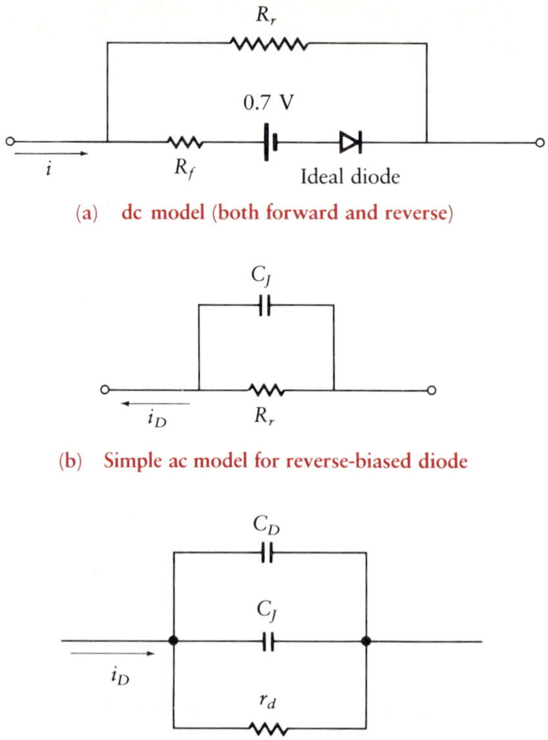

(a) dc model (both forward and reverse)

(b) Simple ac model for reverse-biased diode

(c) ac model for forward-biased diode

Under reverse-bias conditions, the ideal diode has infinite resistance (open circuit) and the circuit resistance of the practical model is R_r. The ideal diode that is part of the model of Figure 1.21(a) is forward-biased when the terminal voltage exceeds 0.7 V.

The *ac* circuit models are more complex because the diode operation depends on frequency. A simple ac model for a reverse-biased diode is shown in Figure 1.21(b). The capacitor C_J represents the *junction capacitance.* This capacitance arises since the depletion region acts as a capacitor. Figure 1.21(c) shows the ac equivalent circuit for a forward-biased diode. The model includes two capacitors, the *diffusion capacitor* C_D and the *junction capacitor* C_J. Diffusion involves movement of carriers and leads to a condition comparable to charge storage. Therefore, the consequences of diffusion include capacitive effects. The diffusion capacitance, C_D, approaches zero for reverse-biased diodes. The *dynamic resistance* is r_d. At low frequencies the capacitive effects are small and R_f is the only significant element.

1.2.7 Diode Circuit Analysis

You now have the necessary background to analyze circuits that include diodes. Suppose you are given a circuit containing passive linear devices, sources and diodes, and you are asked to solve for a particular current or voltage. How would you approach such a problem?

You might decide to go to the electronics laboratory, select the proper components, and wire up the circuit. You could then measure voltages or currents using meters and/or oscilloscopes. You would, of course, have to be certain that the laboratory conditions properly mirror the conditions of the given problem. In particular, you would have to be certain that the very process of measuring quantities is not disturbing the circuit. This technique does have advantages over a theoretical solution, since there is no uncertainty regarding the correct modeling of devices. That is, suppose the models of the previous section do not adequately describe the physical devices. In such cases, no amount of "paper solution" is going to yield useful results. Indeed, our purpose throughout this study is to be able to predict and explain real physical behavior.

If you did not wish to go through the time, expense, and uncertainty of a hardware solution, you might conduct a theoretical analysis using the equations for each element (e.g., Ohm's law and the diode equation). Alternatively, if you trust the models of the previous section, you might use these in place of the diodes and then perform a standard circuit analysis. Such an analysis necessarily contains approximations, since the models themselves are approximations. You might also not take into account many physical conditions, such as temperature variation and tolerance of components.

Computer Simulations An attractive alternative to the above methods is computer simulation. A computer is capable of using detailed models of various components in the performance of circuit analysis. It can factor in such physical conditions as temperature dependence and device tolerances. With the proliferation of personal computers and associated software, computer simulation has become the technique of choice for many areas of circuit analysis and design.

One of the earliest and most widely used circuit analysis programs is *SPICE* (simulated program with integrated circuit emphasis). This program was developed at the University of California, Berkeley, and was first released about 1973. It has been continuously refined since that time and is still accepted as the de facto standard in analog network simulation programs. We include a thorough discussion of SPICE in Appendix A. One derivative of SPICE, PSPICE, from MicroSim Corporation, is available for use in the IBM personal computer. Another popular derivative is MICRO-CAP II.

MICRO-CAP II is a SPICE-based computer simulation program that is available in a low-cost student version (Addison-Wesley Publishing Company). We will refer to this program in several of the examples throughout this text. A major difference between the various simulation programs is in the method of drawing circuits. SPICE typically requires that the user specify components in terms of the nodes to which they are connected. Alternatively, MICRO-CAP II allows the user to draw the circuit directly on the computer display screen.

Both programs use the same transistor models. PSPICE uses an input file and is written in C language; MICRO-CAP II uses a schematic input and is written in Microsoft Basic. Although this gives MICRO-CAP II some advantages regarding circuit alterations (a new run with modified file is not required each time something is changed), it does give PSPICE an overall speed advantage. Such a speed advantage could become significant when simulations are used for design rather than analysis.

The computer results will vary from our calculated results because of the use of more detailed models. For example, in many of our calculations, we will assume that parameters of electronic devices remain constant at their nominal values. The computer simulation varies these parameters according to operating point. If the computer simulation yields a result far different from the "paper solution," we will have to reconcile these discrepancies.

Computer simulation is becoming popular on PCs. The capacity and speed of the PC are such that we usually use the simulation for *analysis* rather than *design*. That is, we usually *verify the performance* of a circuit in which the various components have *already been selected*.

Simulation programs can be used for design by employing an *iteration* technique. For example, if we wish to choose a resistor value, we can analyze the circuit for various values and choose the one that achieves the desired specifications.

The computer simulation is also useful in performing worst-case and temperature analysis. The programs assign a tolerance to each parameter (e.g., component values, transistor gain). In a worst-case analysis, the computer runs the simulation many times and, between runs, randomly varies parameters within their specified ranges. This produces a family of performance curves. The program can also step temperature through a specified range, thereby developing another family of curves. These results are critical to the circuit designer. Even though a particular design may nominally meet specifications, it may not do so in actual application because of temperature variations and tolerances. It is important for the designer to know this and to make appropriate adjustments.

Example 1.3

Find the output current for the circuit shown in Figure 1.22(a). Check your result using a computer simulation program.

SOLUTION Theoretical solutions of problems of this type can be obtained graphically or by writing circuit equations. When time-varying sources exist in a circuit, the diode often changes state between forward- and reverse-biased.

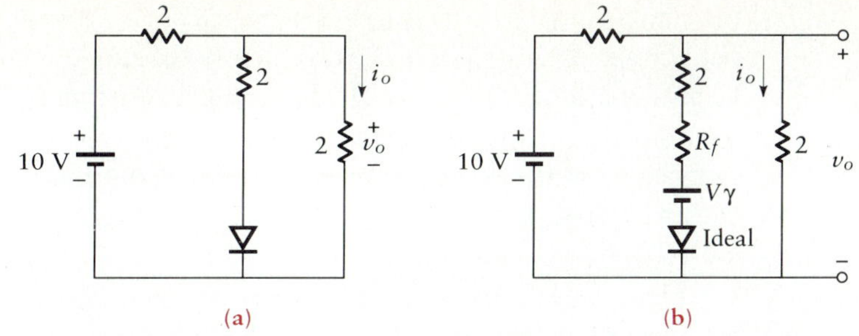

Figure 1.22 Circuit for Example 1.3.

In such cases, the graphical technique usually proves simpler. We discuss this technique in the next section. Since the problem contains only a dc source, we use the diode equivalent circuit as shown in Figure 1.22(b). Once we determine the state of the ideal diode in this model (i.e., either open circuit or short circuit), the problem becomes one of simple dc circuit analysis.

The state of this ideal diode can either be reasoned based on the direction of sources or, if this is too complex, can be guessed. For example, if we guess that the diode is forward-biased, we then solve the circuit substituting a short circuit for the diode. If the resulting current through the shorted diode is in the reverse direction, we know the guess was incorrect. Alternatively, if we guess that the diode is reverse-biased, an open circuit is substituted. If the solution shows a forward voltage across the open circuit, we know the guess was incorrect.

For this particular problem, it is reasonable to assume that the diode is forward-biased and can be replaced by a short circuit. This is true because the only external source is 10 V, which clearly exceeds the turn-on voltage of the diode, even taking the voltage division into account. The equivalent circuit then becomes that of Figure 1.22(b) with the diode replaced by a short circuit. We use simple circuit analysis (resistor combinations, superposition, loop or nodal analysis) to find that

$$v_o = \frac{10 + V_\gamma + 5R_f}{3 + R_f} \tag{1.33}$$

We now repeat the analysis using the MICRO-CAP II simulation program (other simulation programs, such as PSPICE, may be used). We enter the circuit of Figure 1.22(a) using a battery for the dc source. Since the problem did not specify a particular diode, we ran the simulation using a 1N3492 diode (a common silicon diode). The parameters for this diode can be read from the library that forms a part of the program. The battery in the simulation contains a 0.001-Ω source resistance. The MICRO-CAP program provides for several types of analysis of this circuit. We can perform a dc analysis, which plots output versus input voltage. We can then read the result from the graph for an input of 10 V. An alternative method is to run a transient analysis and to display the final values of the voltages. The result of this transient simulation is an output voltage of 3.64 V. The voltage across the diode is 0.91 V, which represents V_γ plus the drop across the diode forward resistance. To compare this to the theoretical solution, we need to assume values for V_γ and R_f. As an example, suppose $V_\gamma = 0.7$ V and $R_f = 0.2\ \Omega$. Then the theoretical solution yields an output of 3.66 V.

Diode Load Lines Because the diode is a nonlinear device, standard circuit analysis techniques must be modified. We cannot simply write equations and solve for the variables, since the equations hold only within a particular operating region.

A circuit often contains both dc supply voltages and time-varying sources. If we set the time-varying sources equal to zero, the only energy supplied to the circuit comes from the dc supply voltages. With the time-varying sources out of the circuit, the diode voltage and current define what is known as the *quiescent operating point (Q-point).*

Before looking at the diode analysis problem, let us illustrate the technique of graphical circuit analysis. Suppose we wish to solve for I_o and V_o in the circuit of Figure 1.23(a). Of course, solution of this circuit is trivial, but suppose we insisted on forming two simultaneous equations and then using graphical analysis. The constraint provided by the load resistor is that

$$V_o = I_o(R_L) = I_o(1\ \Omega)$$

The source battery and source resistance provide the constraint

$$V_o = 1\text{ V} - I_o(1\ \Omega)$$

These two equations must be solved simultaneously. This can be done graphically, as shown in Figure 1.23(b). Each of the equations is plotted as a straight line, and the simultaneous solution is the intersection of the two lines. The result is that

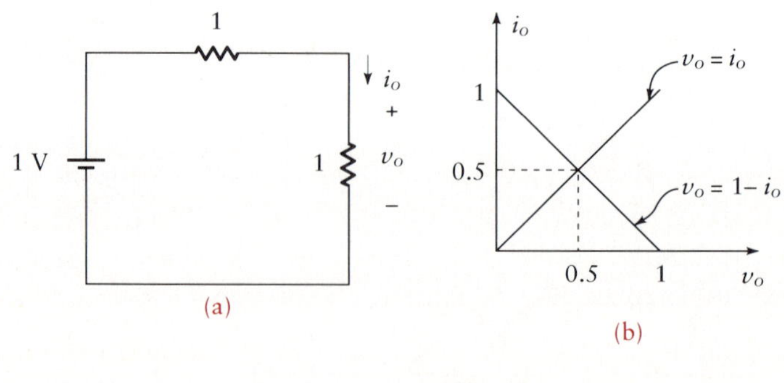

Figure 1.23 Graphical circuit analysis.

$$I_o = 0.5 \text{ A}$$

$$V_o = 0.5 \text{ V}$$

The graphical solution of such a simple problem seems to be creating extra work. However, if one or both of the equations becomes nonlinear, we can often perform the analysis graphically more easily than by using other techniques.

Figure 1.24(a) illustrates a circuit with a diode, capacitor, source, and two resistors. If we designate the diode current and diode voltage as the two circuit unknowns, we need two independent equations involving these unknowns to solve uniquely for the operating point. One of these equations is the constraint provided by the circuitry connected to the diode. The second is the actual diode voltage-current relationship. These two equations must be solved simultaneously to yield the diode voltage and current. This simultaneous solution can be done graphically.

Figure 1.24
Circuit with diode.

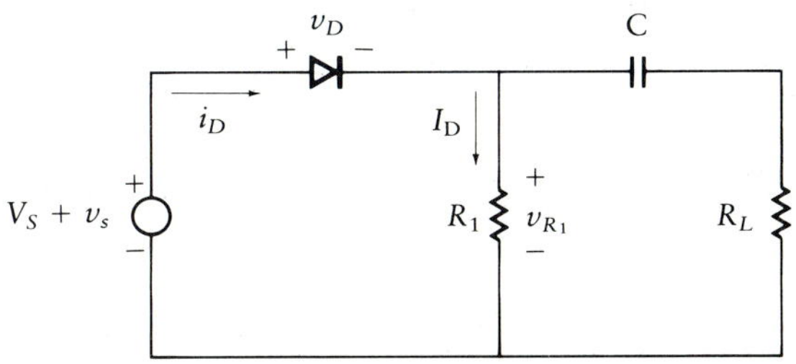

(a) Diode circuit

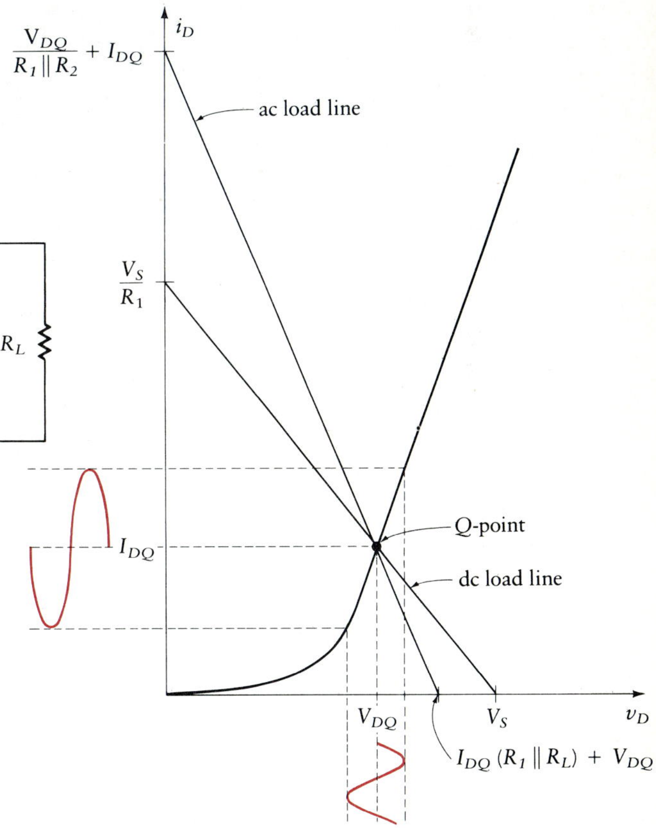

(b) Diode load lines

If we first look at the dc condition, the voltage source becomes simply V_S, and the capacitor is an open circuit (i.e., the impedance of the capacitor is infinite at a frequency of zero). Thus, the loop equation for dc can be written as

$$V_S = V_D + V_{R1} = V_D + I_D R_1 \tag{1.34}$$

or

$$V_D = -R_1 I_D + V_S \tag{1.35}$$

This is the first of two simultaneous equations involving the diode voltage and current. We need to combine this with the diode characteristic in order to solve for the operating point. The graph of this equation is shown in Figure 1.24(b) and is labeled "dc load line." The graph of the diode characteristic is also shown on the same set of axes. The intersection of the two plots yields the simultaneous solution of the two equations and is labeled "Q-point" on the figure. This is the point at which the circuit will operate with the time-varying inputs set to zero. The "Q" denotes the "quiescent" or rest condition.

If a time-varying signal is now applied in addition to the dc input, one of the two simultaneous equations changes. If we assume that the time-varying input is of a high enough frequency to allow approximation of the capacitor as a short circuit, the new equation is given by

$$v_s = v_d + i_d(R_1 \parallel R_L) \tag{1.36}$$

or

$$v_d = -(R_1 \parallel R_L) i_d + v_s \tag{1.37}$$

We are considering only the time-varying components of the various parameters. (Note the use of lowercase letters for the variables. Refer to the labeling convention presented at the beginning of this text.) Thus, the total parameter values are given by

$$v_D = v_d + V_{DQ}$$

$$i_D = i_d + I_{DQ}$$

and equation (1.37) becomes

$$v_D - V_{DQ} = -(R_1 \parallel R_L)(i_D - I_{DQ}) + v_s$$

This last equation is labeled "ac load line" in Figure 1.24(b). Since the equation deals only with time-varying quantities, the axis intercept is not known. However, this ac load line must pass through the Q-point, since at the times when the time-varying part of the input goes to zero, the two operating conditions (dc and ac) must coincide. Thus, the ac load line is uniquely determined.

Example 1.4

Given the circuit of Figure 1.25(a) and a source voltage of

$$v_s = 1.1 + 0.1 \sin 1000t$$

Figure 1.25
Circuit and simulation for Example 1.4.

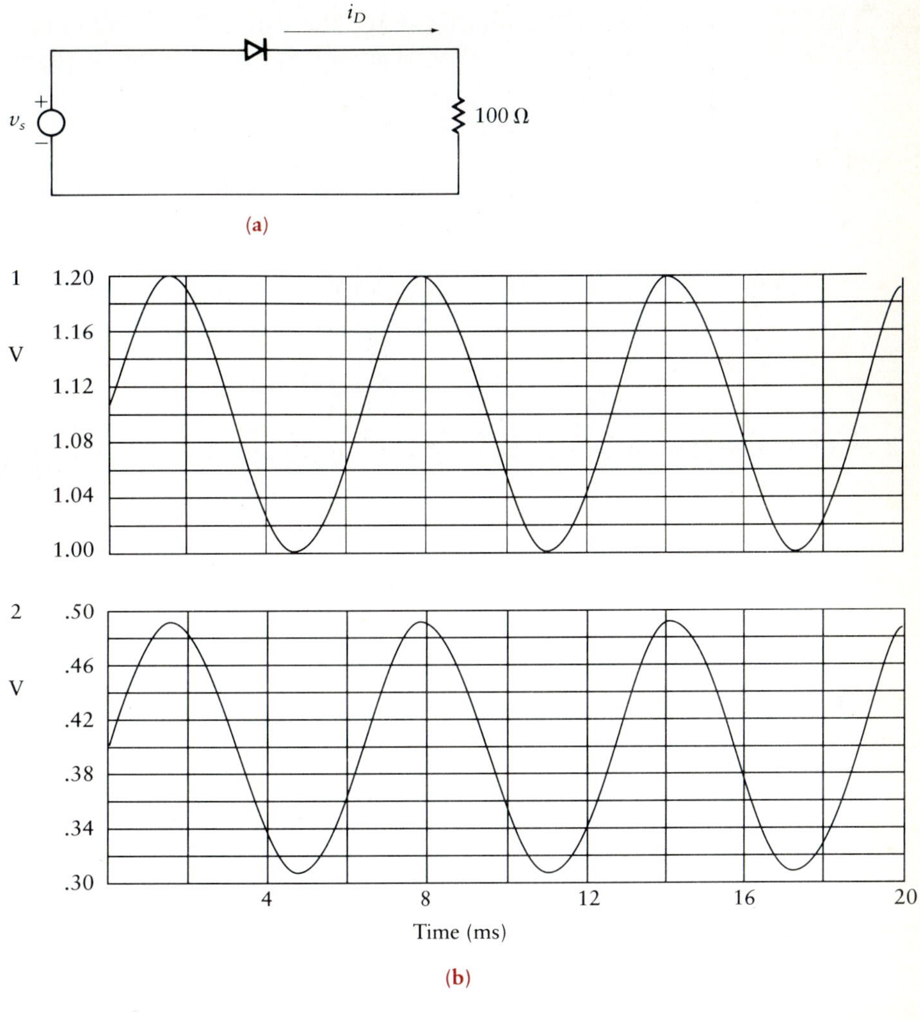

find the current, i_D. Assume that

$$nV_T = 40 \text{ mV}$$
$$V_\gamma = 0.7 \text{ V}$$

Repeat your analysis using a computer simulation program.

SOLUTION We use Kirchhoff's voltage law (KVL) for the dc equation to yield

$$V_S = V_\gamma + I_D R_L$$
$$I_D = \frac{V_S - V_\gamma}{R_L} = 4 \text{ mA}$$

This sets the dc operating point of the diode. We need to determine the dynamic resistance (we use the symbol R_f instead of r_d since R_f includes the contact resistance) so that we can establish the resistance of the forward-biased junction for the ac signal.

Assuming that the contact resistance is negligible, we have

$$R_f = \frac{nV_T}{I_D} = 10 \ \Omega$$

Now we can replace the diode with a 10-Ω resistor, provided it remains forward-biased during the entire period of the ac signal. Again using KVL, we have

$$v_s = R_f i_d + R_L i_d$$
$$i_d = \frac{v_s}{R_f + R_L} = 0.91 \sin 1000t \text{ mA}$$

The diode current is then given by

$$i_D = 4 + 0.91 \sin 1000t \text{ mA}$$

Since i_D is always positive, the diode is always forward-biased, and the solution is complete.

If the ac current amplitude becomes greater than the dc current value, i_D will not always be positive, and the assumption that the diode is forward-biased is incorrect. Therefore the solution must be modified. In that case, when the ac current amplitude in the negative direction becomes larger than the dc value, the diode becomes reverse-biased and the current is cut off. This case is covered in Section 1.5.

We now repeat the analysis using a computer simulation program (we ran this on MICRO-CAP II, Student Version). We can use the sinusoidal source component to model the offset source. We do this by modifying a library entry with the appropriate parameters. An alternative would be to place a sinusoidal source in series with a battery. The program does not permit zero source resistance, so we set this equal to 0.01 Ω. Since the problem did not specify a specific diode, we chose the 1N3492, which has parameters close to those given in the problem. Since the period of the input sinusoid is $2\pi/1000$, we set the simulation time at 0.02 second in order to view slightly more than three complete periods of the input and output. Figure 1.25(b) shows the results of the simulation. The top curve is the input voltage and the bottom curve is the output voltage (across the 100-Ω resistor). We note that the output has a dc value of 0.4 V and an amplitude of 0.095 (actual values are printed out during the simulation, so it is not necessary to read the graph accurately). To find the diode current, we simply use Ohm's law and divide this result by 100. We note that the simulation yielded a result of

$$i_D = 4 + 0.95 \sin 1000t \text{ mA}$$

1.2.8 Power-Handling Capability

Diodes are rated according to their power-handling capability. The ratings are determined by the physical construction of the diode (e.g., size of junction, type of packaging, and size of diode). The manufacturer's specifications are used to determine the power capability of a diode for certain temperature ranges. Some diodes, such as power diodes, are rated by their current-carrying capacity.

The instantaneous power dissipated by a diode is defined by the expression in equation (1.38).

$$p_D = v_D i_D \tag{1.38}$$

Since diodes sometimes conduct relatively large currents, they must be mounted so that the heat generated in the diode can be dissipated away from the diode. In order to dissipate the internal heat generated, the diodes are mounted with heat sinks, as discussed in Chapter 5.

1.2.9 Diode Capacitance

The equivalent circuit of a diode includes a small capacitor. The size of this capacitor depends on the magnitude and polarity of the voltage applied to the diode as well as the characteristics of the junction formed during manufacture.

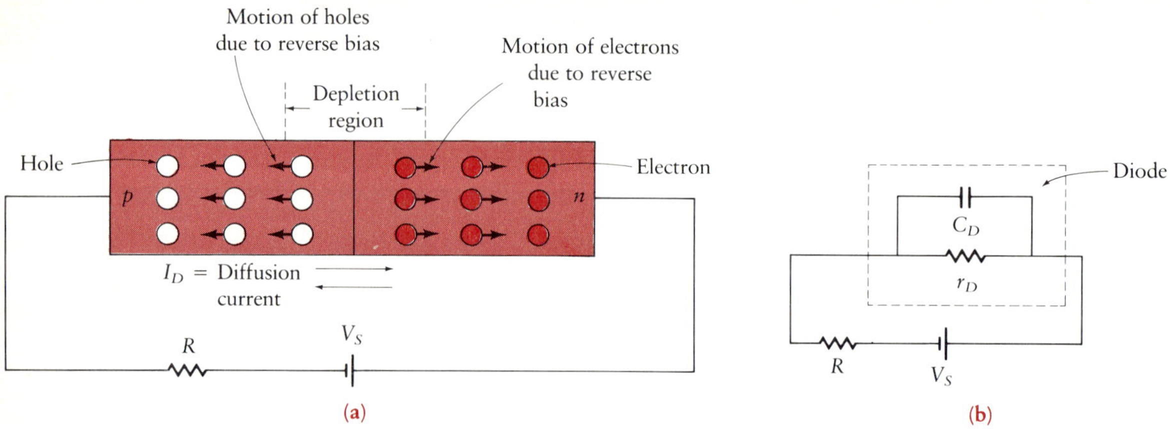

Figure 1.26
Diode model and equivalent circuit.

In the simple model of a diode junction shown in Figure 1.26, the region at the junction is depleted of both electrons and holes. On the p side of the junction there is a high concentration of holes, and on the n side there is a high concentration of electrons. Diffusion of these electrons and holes occurs close to the junction, causing an initial *diffusion current.* When the holes diffuse across the junction into the n region, they quickly combine with the majority electrons present in that area and disappear. Likewise, electrons diffuse across the junction, recombine, and disappear. This causes a *depletion region* (sometimes called the *space charge region*) near the junction due to the depletion of electrons and holes. As a reverse voltage is applied across the junction, this region widens, causing the depletion region to increase in size.

The depletion region acts like an insulator. Thus, a reverse-biased diode acts like a capacitor with a capacitance that varies inversely with the square root of the voltage drop across the semiconductor material.

The equivalent capacitance for high-frequency diodes is less than 5 pF. This capacitance can become as large as 500 pF in high-current (low-frequency) diodes. The manufacturer's specifications should be consulted to determine the anticipated amount of capacitance for a given operating condition.

1.3 RECTIFICATION

We are now ready to see how the diode is configured to perform a useful function. The first major application we consider is that of *rectification.*

Rectification is the process of turning an alternating signal (ac) into one that is restricted to only one direction (dc). Rectification is classified as either *half-wave* or *full-wave.*

1.3.1 Half-Wave Rectification

Since an ideal diode can sustain current flow in only one direction, it can be used to change an ac signal into a dc signal.

Figure 1.27 illustrates a simple *half-wave rectifier* circuit. When the input voltage is positive, the diode is forward-biased and can be replaced (assume it is ideal) by a short circuit. When the input voltage is negative, the diode is reverse-biased and can be replaced by an open circuit. Thus, when the diode is forward-biased, the output voltage across the load resistor can be found from the voltage-divider relationship. Alternatively, in the reverse-biased condition, the current is zero so the output voltage is also zero.

Figure 1.27 shows an example of the output waveform assuming a sinusoidal input of 100-V amplitude, $R_s = 10\ \Omega$, and $R_L = 90\ \Omega$.

The *average* of a periodic function is defined as the integral of the function over one period divided by the period. It is equal to the first term in a Fourier series expansion of the function. Note that while the input sinusoid has an average value of zero, the output waveform has an average of

$$V_{o\,\text{avg}} = \frac{1}{T}\int_0^{T/2} 90 \sin\frac{2\pi t}{T}\,dt = \frac{90}{\pi}$$

The half-wave rectifier can be used to create an almost constant dc output if the resulting waveform of Figure 1.27 is filtered. The filtering operation is discussed in Section 1.3.3. We note at this time that the half-wave rectifier is not very efficient. During one half of each cycle, the input is completely blocked from the output. If we could transfer input energy to the output during this half-cycle, we would increase output power.

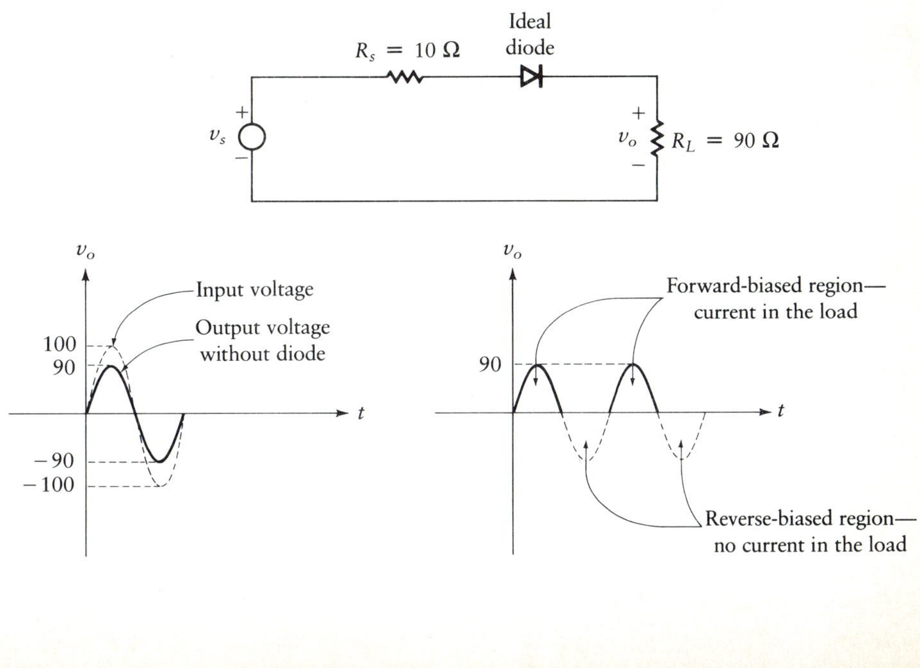

Figure 1.27
Half-wave rectifier.

1.3.2 Full-Wave Rectification

A *full-wave rectifier* transfers input energy to the output during both halves of the input cycle and provides increased average current per cycle over that obtained using the half-wave rectifier. A transformer can be used in constructing a full-wave rectifier in order to obtain the positive and negative polarities. A representative circuit and the output voltage curve are shown in Figure 1.28. The full-wave rectifier produces *twice* the average current of the half-wave rectifier. (You should verify this statement!)

Full-wave rectification is possible without the use of a transformer. For example, the *bridge rectifier* of Figure 1.29 accomplishes full-wave rectification. When the source voltage is positive, diodes 1 and 4 conduct and diodes 2 and 3 are open circuit. When the source voltage goes negative, the reverse situation occurs and diodes 2 and 3 conduct. This is indicated in Figure 1.30. Study of Figure 1.29 indicates a possible practical shortcoming of the bridge rectifier circuit. If one terminal of the source is grounded, neither terminal of the load resistor can be grounded. To do so would cause a *ground loop,* which would effectively short out one of the diodes. Therefore, it may be necessary to add a transformer to this circuit in order to isolate the two grounds from each other. In this case the transformer does not need a center tap, as does the full-wave rectifier of Figure 1.28. Notice also that since there are two diodes conducting in series, the diode voltage drop is $2V_\gamma$.

Figure 1.28
Full-wave rectifier.

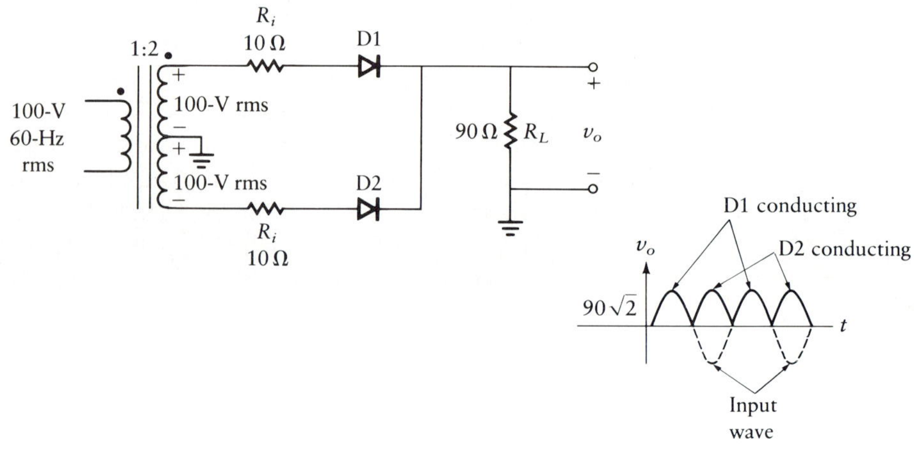

Figure 1.29
Full-wave bridge rectifier.

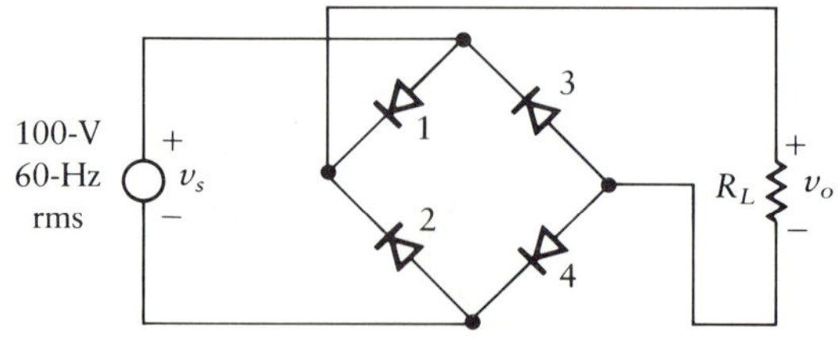

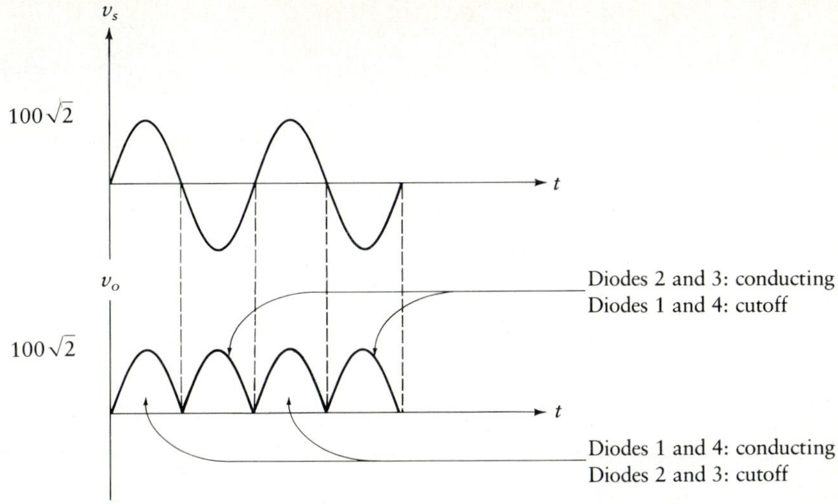

Figure 1.30
Bridge rectifier diode conduction times.

1.3.3 Filtering

The rectifier circuits of the previous section provide a *pulsating* dc voltage at the output. These pulsations, known as output *ripple,* can be reduced considerably by filtering the rectifier output.

The most common type of filter employs a single capacitor. Figure 1.31(a) shows the full-wave rectifier with a capacitor added in parallel with the load resistor. The modified output voltage is shown in Figure 1.32.

In a practical application, the diodes should be reversed and placed close to ground potential as shown in Figure 1.31(b). This allows the cathode to be at ground potential, so the diodes can be attached to a ground plate, thereby allowing dissipation of heat for high-power rectifier circuits.

The capacitor will charge to the highest voltage ($V_{\max}$) when the input peaks at its most positive or negative value. When the input voltage falls below that value, the capacitor cannot discharge through either diode. Therefore, discharge takes place through R_L. This leads to an exponential decay given by the equation

$$v(t) = V_{\max}e^{-t/\tau} = V_{\max}e^{-t/R_LC} \tag{1.39}$$

Design of this filter consists of choosing the capacitor value, C. For example, let us assume that the input is a sinusoid with amplitude 100 V and that the lowest output voltage we can accept in a given application is 95 V. Then

$$95 = 100e^{-T'/R_LC}$$

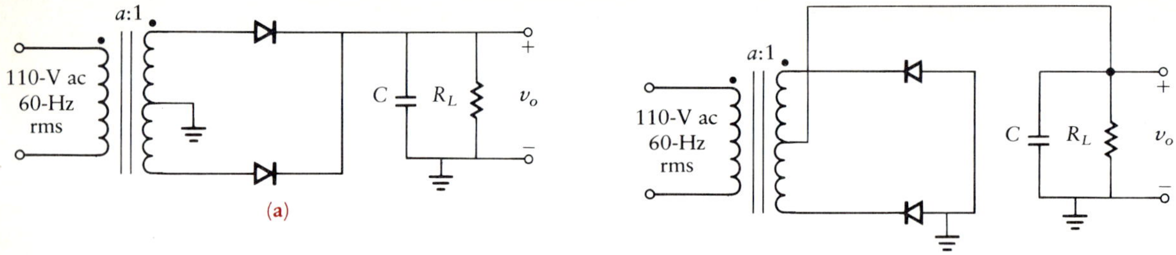

Figure 1.31 Full-wave rectifier with filter.

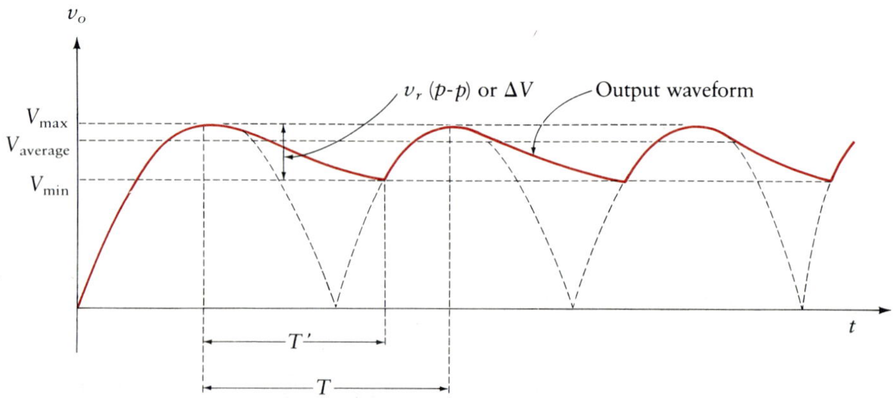

Figure 1.32 Filtered output waveform.

where T' is the discharge time available as indicated in Figure 1.32. We can solve for C in terms of T' and R_L as follows:

$$95 = 100e^{-T'/R_L C}$$

$$\ln 1.05 = \frac{T'}{R_L C}$$

and finally

$$C = 19.4\frac{T'}{R_L}$$

This formula is difficult to use in filter design, since T' is dependent on the $R_L C$ time constant and therefore on the unknown, C. We shall make some approximations. Certainly it is known that

$$T' < T$$

For a 60-Hz input, the output fundamental frequency is 120 Hz. Therefore,

$$T = \frac{1}{f} = \frac{1}{120} = 8.33\,\text{ms}$$

We can estimate the value of the filter capacitor needed for a particular load by using a straight-line approximation, as shown in Figure 1.33. We solve for C with the straight-line approximations.

The initial slope of the exponential of equation (1.39) is

$$m_1 = \frac{-V_{\max}}{R_L C}$$

which is the slope of line A in the figure. The slope of line B of Figure 1.33 is

$$m_2 = \frac{V_{\max}}{T/2}$$

Then

$$t_1 = \frac{-\Delta V}{m_1} = \frac{R_L C\,\Delta V}{V_{\max}}$$

Using corresponding triangles, we find

$$t_1 = \frac{T}{2} + t_2 = \frac{T}{2} + \frac{T V_{\min}}{2V_{\max}}$$

and

$$t_1 = \frac{R_L C\,\Delta V}{V_{\max}} = \frac{T(2 - \Delta V/V_{\max})}{2}$$

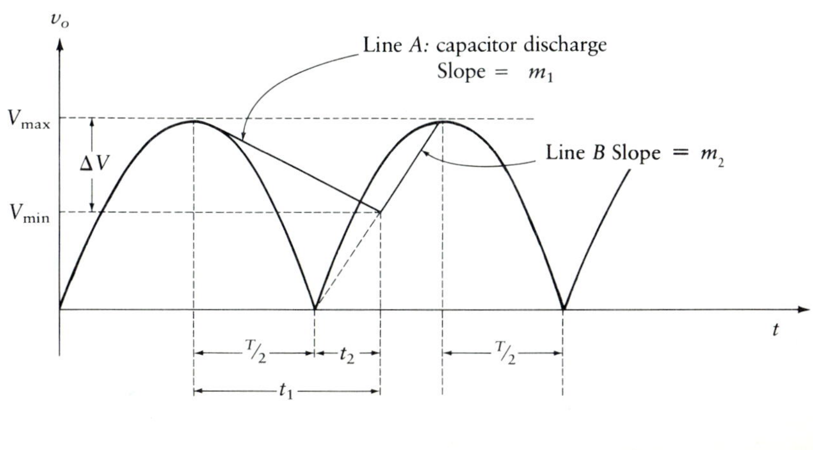

Figure 1.33
Discharge time approximation.

Substituting $T = 1/f_p$, where f_p is the number of pulses per second (twice the original frequency), we have

$$R_L C \frac{\Delta V}{V_{max}} = \frac{1}{2f_p}\left(2 - \frac{\Delta V}{V_{max}}\right) = \frac{1}{f_p}\left(1 - \frac{\Delta V}{2V_{max}}\right) \tag{1.40}$$

In most filter designs, we wish the ripple to be much less than the dc amplitude. Therefore,

$$\frac{\Delta V}{2V_{max}} << 1$$

and equation (1.40) becomes

$$C = \frac{V_{max}}{\Delta V f_p R_L} \tag{1.41}$$

This formula represents a conservative solution to the design problem. This is true because, if the straight line never goes below V_{min}, the exponential curve certainly will stay above this value.

We now use equation (1.41) to solve for the capacitance in the example presented earlier. We assume that the input is a 60-Hz sinusoid of 100-V amplitude and that the lowest acceptable output voltage is 95 V. Thus, for this example, $V_{max} = 100$ V, $\Delta V = 5$ V, and the frequency after full-wave rectification is

$$f_p = 120$$

For half-wave rectification,

$$f_p = 60$$

Thus, from equation (1.41),

$$C = \frac{100}{5 \times 120 R_L} = \frac{0.167}{R_L}$$

The amount of ripple does not follow any standard shape (e.g., sinusoidal or sawtooth), and we need some way to characterize its size. The ripple voltage, V_r(rms), is given by

$$V_r(\text{rms}) = \frac{V_{max} - V_{min}}{2\sqrt{3}} \tag{1.42}$$

Note that we use $\sqrt{3}$ rather than $\sqrt{2}$ in the denominator. The latter figure would be used to find the rms value of a sinusoid, which is the amplitude divided by $\sqrt{2}$. For a sawtooth wave, the rms is the amplitude divided by $\sqrt{3}$. These figures are verified by taking the square root of the average of the square of the waveform over one period. The shape of the ripple is closer to that of a sawtooth waveform than it is to a sinusoid. The average value of the ripple voltage is assumed to be the midpoint of the waveform (this is an approximation). The *ripple factor* is defined as

$$\text{Ripple factor} = \frac{V_r(\text{rms})}{V_{dc}}$$

1.3.4 Voltage-Doubling Circuit

Figure 1.34 shows a circuit that produces a voltage equal to approximately twice the maximum peak output of the transformer (at no load). This is known as a *voltage doubler circuit.* Notice that it is the same as the full-wave bridge rectifier of Figure 1.29 except that two diodes have been replaced by capacitors. When the input voltage has the polarity shown in the figure, there are two paths for the current leaving diode D1. One path is through C_2, and this capacitor charges toward V_{max}. The other path is through the load resistor and C_1. If C_1 had already been charged to V_{max} during the previous cycle, it would effectively place another voltage source of V_{max} in series with the transformer output voltage. This results in doubling the maximum voltage to the load. The capacitors also act to reduce the amount of the voltage ripple at the output.

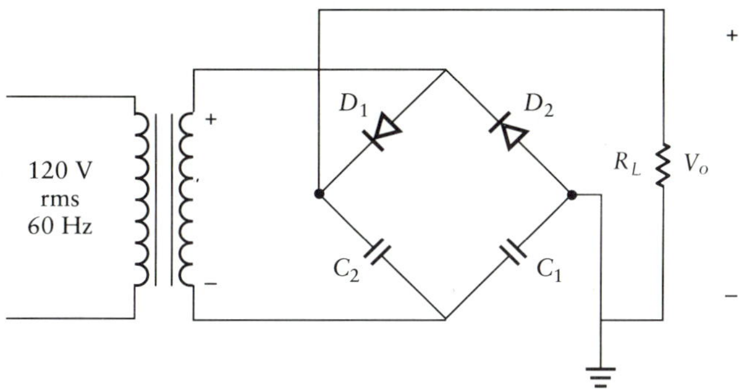

Figure 1.34 Voltage doubler circuit.

Drill Problems

D1.3 The circuit of Figure 1.31(a) is used to rectify a sinusoid of 100 V rms and 60 Hz. The minimum voltage output cannot drop below 70 V and the transformer has a turns ratio of 1:2. The load resistance is 2 kΩ. What size capacitor is needed across R_L?

Ans: 8.25 μF

D1.4 A half-wave rectifier output has 50-V amplitude at 60 Hz. Assuming no forward resistance in the diode, what minimum load could be added to the circuit when using a 50-μF capacitor to maintain the minimum voltage above 40 V?

Ans: 1.67 kΩ

D1.5 A full-wave rectifier similar to the one shown in Figure 1.31(a) has a transformer with a 5:1 turns ratio. What capacitance would be required to maintain a 10-V minimum voltage across a 100-Ω load?

Ans: 233 μF

D1.6 If the voltage input in Drill Problem D1.5 varies between 110 and 120 V rms at 60 Hz, what capacitance is needed?

Ans: 233 μF

1.4 ZENER DIODES

A *zener diode* is a device in which the doping is performed so as to make the *avalanche* or *breakdown* voltage characteristic very steep. If the reverse voltage exceeds the breakdown voltage, the diode will normally not be destroyed. This is true as long as the current does not exceed a predetermined maximum value and the device does not overheat.

When a thermally generated carrier (part of the reverse saturation current) falls down the junction barrier (see Figure 1.16) and acquires energy of the applied potential, the carrier collides with crystal ions and imparts sufficient energy to disrupt a covalent bond. In addition to the original carrier, a new electron-hole pair is generated. The new pair may pick up sufficient energy from the applied field to collide with another crystal ion and create still another electron-hole pair. This action continues and thereby disrupts the covalent bonds; the process is referred to as *avalanche multiplication* or *avalanche breakdown.*

There is a second mechanism that disrupts the covalent bonds. Use of a sufficiently strong electric field at the junction can cause direct rupture of the bond. If the electric field exerts a strong force on a bound electron, the electron can be torn from the covalent bond, thus causing the number of electron-

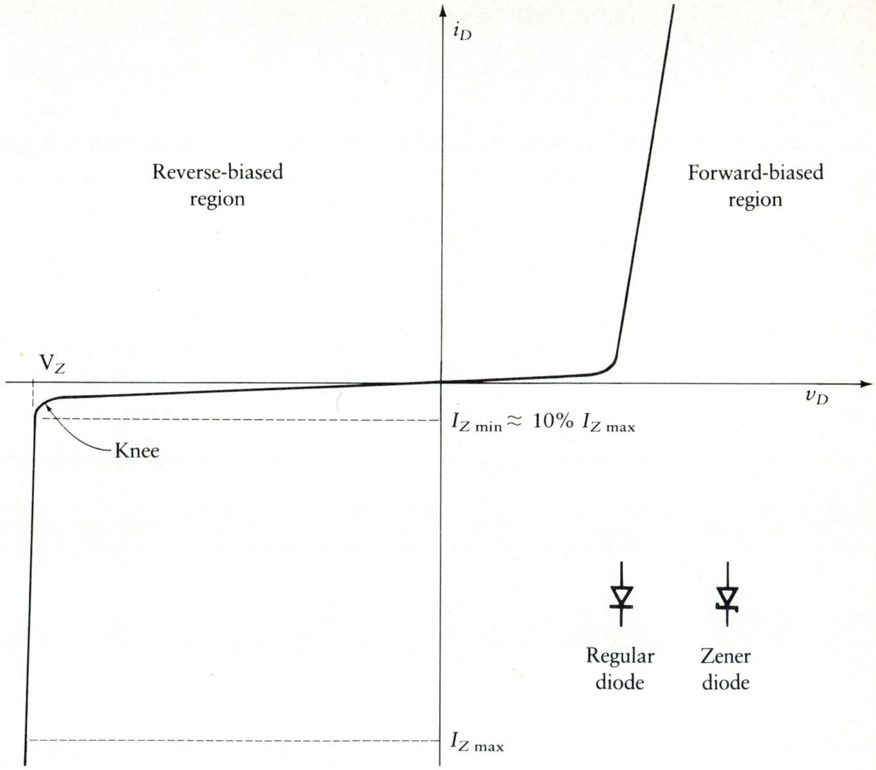

Figure 1.35 Zener diode characteristic.

hole pair combinations to multiply. This mechanism is called *zener breakdown.* The value of reverse voltage at which this occurs is controlled by the amount of doping of the diode. A heavily doped diode has a low zener breakdown voltage, whereas a lightly doped diode has a high zener breakdown voltage.

Although we describe two distinctly different mechanisms that effect breakdown, they are commonly interchanged. At voltages above approximately 10 V, the predominant mechanism is the avalanche breakdown ([25], Section 2.9). Since the zener effect (avalanche) occurs at a predictable point, the diode can be used as a voltage reference. The reverse voltage at which the avalanche occurs is called the *zener voltage.*

A typical zener diode characteristic is shown in Figure 1.35. The circuit symbol for the zener diode is different from that of a regular diode and is illustrated in the figure.

The maximum reverse current, $I_{Z\max}$, which the zener diode can withstand is dependent on the design and construction of the diode. We will assume that the minimum zener current at which the characteristic curve remains at V_Z (near the knee of the curve) is $0.1I_{Z\max}$. The amount of power the zener diode can withstand ($V_Z I_{Z\max}$) is a limiting factor in power supply design.

1.4.1 Zener Regulator

A zener diode can be used as a voltage regulator in the configuration shown in Figure 1.36. The figure illustrates a varying load current that is represented by a variable load resistance. The circuit is designed so that the diode operates in the breakdown region, thereby approximating an ideal voltage source. In practical application the source voltage, v_S, varies and the load current also varies. The design challenge is to choose a value of R_i that permits the diode to maintain a relatively constant output voltage, even when the input source voltage varies and the load current also varies.

We shall now analyze the circuit of Figure 1.36 to determine the proper choice of R_i. The node equation for the circuit is

$$R_i = \frac{v_S - V_Z}{i_R} = \frac{v_S - V_Z}{i_Z + i_L} \tag{1.43}$$

This equation can be solved for the zener current, i_Z, to yield

$$i_Z = \frac{v_S - V_Z}{R_i} - i_L \tag{1.44}$$

The variable quantities in equation (1.44) are v_S and i_L. To ensure that the diode remains in the constant-voltage (breakdown) region, we examine the two extremes of input/output conditions, as follows:

1. The current through the diode, i_Z, is a minimum ($I_{Z\min}$) when the load current, i_L, is maximum ($I_{L\max}$) and the source voltage, v_S, is minimum ($V_{S\min}$).
2. The current through the diode, i_Z, is a maximum ($I_{Z\max}$) when the load current, i_L, is minimum ($I_{L\min}$) and the source voltage, v_S, is maximum ($V_{S\max}$).

When these characteristics of the two extremes are inserted into equation (1.43), we find

$$\text{Condition 1:} \quad R_i = \frac{V_{S\min} - V_Z}{I_{L\max} + I_{Z\min}} \tag{1.45}$$

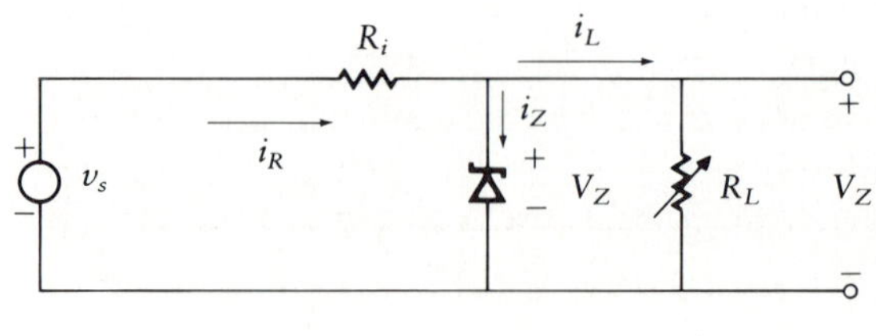

Figure 1.36
Zener regulator.

$$\text{Condition 2:} \quad R_i = \frac{V_{S\max} - V_Z}{I_{L\min} + I_{Z\max}} \tag{1.46}$$

Since the value of R_i is the same in both equation (1.45) and equation (1.46), we can equate these two expressions to obtain

$$(V_{S\min} - V_Z)(I_{L\min} + I_{Z\max}) = (V_{S\max} - V_Z)(I_{L\max} + I_{Z\min}) \tag{1.47}$$

In a practical problem, it is reasonable to assume that we know the range of input voltages, the range of output load currents, and the desired zener voltage. Equation (1.47) thus represents one equation in two unknowns, the maximum and minimum zener currents. A second equation is found by examining Figure 1.35. To avoid the nonconstant portion of the characteristic curve, we use a rule of thumb that the minimum zener current should be 0.1 times the maximum; that is,

$$I_{Z\min} = 0.1 I_{Z\max}$$

We now solve equation (1.47) for $I_{Z\max}$, where we use the design criterion presented above.

$$I_{Z\max} = \frac{I_{L\min}(V_Z - V_{S\min}) + I_{L\max}(V_{S\max} - V_Z)}{V_{S\min} - 0.9V_Z - 0.1V_{S\max}} \tag{1.48}$$

Now that we can solve for the maximum zener current, the value of R_i is calculated from either equation (1.45) or equation (1.46).

Example 1.5 Zener Regulator Design

Design a 10-V zener regulator (Figure 1.37) for the following conditions:
a. The load current ranges from 100 to 200 mA and the source voltage ranges from 14 to 20 V.

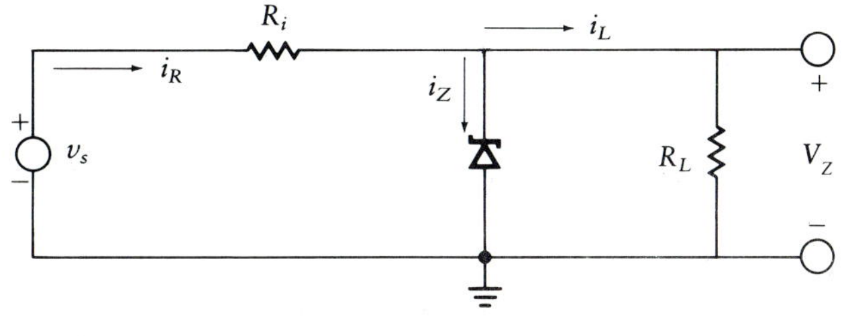

Figure 1.37
Zener diode regulator.

b. The load current ranges from 20 to 200 mA and the source voltage ranges from 10.2 to 14 V.
Use a 10-V zener diode in both cases.

SOLUTION a. The design consists of choosing the proper value of resistance, R_i, and power rating for the zener. We use the equations from the previous section to calculate the maximum current in the zener diode and then to find the input resistor value. From equation (1.48), we have

$$I_{Z\max} = \frac{0.1(10 - 14) + 0.2(20 - 10)}{14 - 0.9(10) - 0.1(20)} = 0.533 \text{ A}$$

Then, from equation (1.46), we find R_i as follows:

$$R_i = \frac{V_{S\max} - V_Z}{I_{Z\max} + I_{L\min}} = \frac{20 - 10}{0.533 + 0.1} = 15.8\ \Omega$$

It is not sufficient to specify only the resistance R_i. We must also select the proper resistor power rating. The maximum power is given by the product of voltage with current, where we use the maximum for each value.

$$P_R = (I_{Z\max} - I_{L\min})(V_{S\max} - V_Z) = 6.3 \text{ W}$$

Finally, we must determine the power rating of the zener diode. The maximum power dissipated in the zener diode is given by the product of voltage and current.

$$P_Z = V_Z I_{Z\max} = 0.53 \times 10 = 5.3 \text{ W}$$

b. Repeating these steps for the parameters of part b yields

$$I_{Z\max} = \frac{0.02(10 - 10.2) + 0.2(14 - 10)}{10.2 - 0.9(10) - 0.1(14)} = -4 \text{ A}$$

The negative value of $I_{Z\max}$ indicates that the margin between $V_{S\min}$ and V_Z is not large enough to allow for the variation in load current. That is, under the worst-case condition of a 10.2-V input and 200-mA load current, the zener cannot possibly sustain 10 V across its terminals. Therefore, the regulator will not operate correctly for any choice of resistance. We would have to raise the source voltage or reduce the output current requirements.

The zener regulator of Figure 1.37 can be combined with the full-wave rectifier of Figure 1.28 to yield the full-wave zener regulator of Figure 1.38.

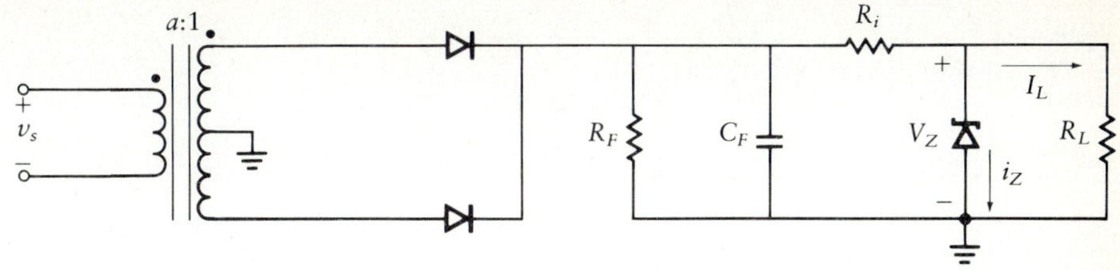

Figure 1.38
Full-wave zener regulator.

The component R_F is called a *bleeder resistor* and is used to provide a discharge path for the capacitor when the load is removed. Bleeder resistors normally have high resistances so that they do not absorb significant power when the circuit is operating. The value of C_F is found by adapting equation (1.41) to this situation. The resistance in the equation is the equivalent resistance across C_F. The zener diode is replaced by a voltage source, V_Z. The equivalent resistance is then the parallel combination of R_F with R_i. The zener diode effectively shorts R_L. Since R_F is much larger than R_i, the parallel resistance is approximately equal to R_i. Since the voltage across R_i does not go to zero as it does for the full-wave rectifier, the V_{max} in equation (1.41) must be replaced by the total voltage swing. Thus, the capacitor is approximated by equation (1.49), where we are assuming a, the transformer ratio, is $\frac{1}{2}$.

$$C_F = \frac{V_{S\max} - V_Z}{\Delta V f_p R_i} \tag{1.49}$$

The largest voltage imposed on the regulator is $V_{S\max}$. As before, ΔV is the peak-to-peak ripple and f_p is the fundamental frequency of the rectified waveform (i.e., twice the original frequency for full-wave rectification).

1.4.2 Practical Zener Diodes and Percent Regulation

In the previous section we assumed that the zener diode is ideal. That is, in the avalanche breakdown region, the diode behaves as a constant-voltage source. This assumption means that the curve of Figure 1.35 is a vertical line in the breakdown region. In practice, this curve is not precisely vertical, and the noninfinite slope results in a nonzero series resistance. The breakdown voltage is then a function of current instead of a constant. We model the practical zener diode as shown in Figure 1.39. This model replaces the practical zener diode with an ideal diode in series with a resistance, R_Z.

To show the effects of this series resistor, we assume that a practical zener diode is incorporated into Example 1.5(a) with the diode resistance $R_Z = 2\ \Omega$. We assume that $I_{Z\min}$ is 10% of $I_{Z\max}$, or 0.053 A. The output voltage (across the load) is no longer a constant 10 V because of R_Z. We find the minimum and maximum values of this voltage from Figure 1.38 using the minimum and

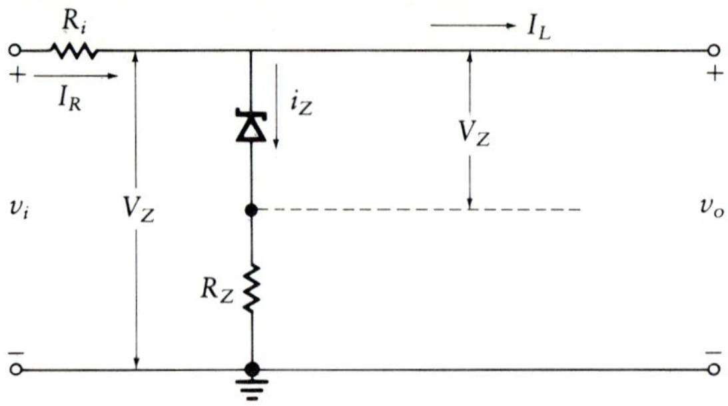

Figure 1.39 Zener equivalent circuit.

maximum current values. The voltage across the ideal diode of Figure 1.39 is 10 V, so we can write

$$V_{o\,\min} = 10 + (0.053 \times 2) = 10.1 \text{ V}$$

$$V_{o\,\max} = 10 + (0.53 \times 2) = 11.1 \text{ V}$$

The *percent regulation* is defined as the total voltage swing divided by the nominal (or ideal regulated) voltage. The smaller the percent regulation, the better the regulator. Therefore, for this example,

$$\%\text{ Reg} = \frac{V_{o\,\max} - V_{o\,\min}}{V_{o\,\text{nominal}}} = \frac{11.1 - 10.1}{10}$$

$$= 0.1 \text{ or } 10\% \tag{1.50}$$

This regulation is considered poor for most applications. The regulation could be improved by limiting the zener current to a smaller value. This is accomplished by using an amplifier in series with the load. The effect of this amplifier is to limit the variations of current through the zener diode. We study such amplifiers in Chapter 6.

Drill Problems

D1.7 A zener diode regulator circuit (see Figure 1.37) has an input whose voltage varies between 10 and 15 V and a load whose current varies between 100 and 500 mA. Find the values of R_i and $I_{Z\max}$ assuming that a 6-V zener is used.

Ans: 6.33 Ω; 1.32 A

D1.8 In Drill Problem D1.7, find the power ratings for the zener diode and for the input resistor.

Ans: 7.92 W; 12.8 W

D1.9 In Drill Problem D1.7, find the value of capacitor required if the source is a half-wave rectifier output with a 60-Hz input.

Ans: 4740 μF

D1.10 If no resistor, R_F, was used in the circuit of Figure 1.37 and the transformer was a 4:1 center-tapped transformer with an input of 120 V rms, 60 Hz, what value of R_i would be needed to maintain 10 V across a load whose current varies from 50 to 200 mA? Assume that the minimum voltage allowed at the regulator input is 14 V.

Ans: 14.8 Ω

D1.11 What value of capacitor is needed in the regulator of Drill Problem D1.10 to maintain a minimum voltage of 14 V?

Ans: 875 μF

D1.12 In the circuit of Drill Problem D1.10, assume that the input voltage varies from 110 to 120 V rms at 60 Hz. Select a value of capacitance that will accommodate both a load current variation of 50 to 200 mA and the specified input voltage variation.

Ans: $C = 1160\ \mu$F

1.5 CLIPPERS AND CLAMPERS

Diodes can be used to clip an input signal or to limit parts of the signal. Diodes are also used in restoring a dc level to an input signal.

1.5.1 Clippers

Clipping circuits are used to eliminate a part of a waveform that lies above or below some reference level. Clipping circuits are sometimes referred to as *limiters, amplitude selectors,* or *slicers.* The half-wave rectification circuit of the previous section uses clipping action at the zero level. If a battery is added in series with the diode, a rectification circuit will clip everything above or below the battery voltage, depending on the orientation of the diode. This is illustrated in Figure 1.40.

The output waveforms indicated in Figure 1.40 assume that the diodes are ideal. We can relax this assumption by including two additional parameters in the diode model. First, we assume that a voltage of V_γ must be overcome before the diode will conduct. Second, when the diode is conducting, we include a forward resistance, R_f. Figure 1.41(a) shows the modified circuit. The effect of V_γ

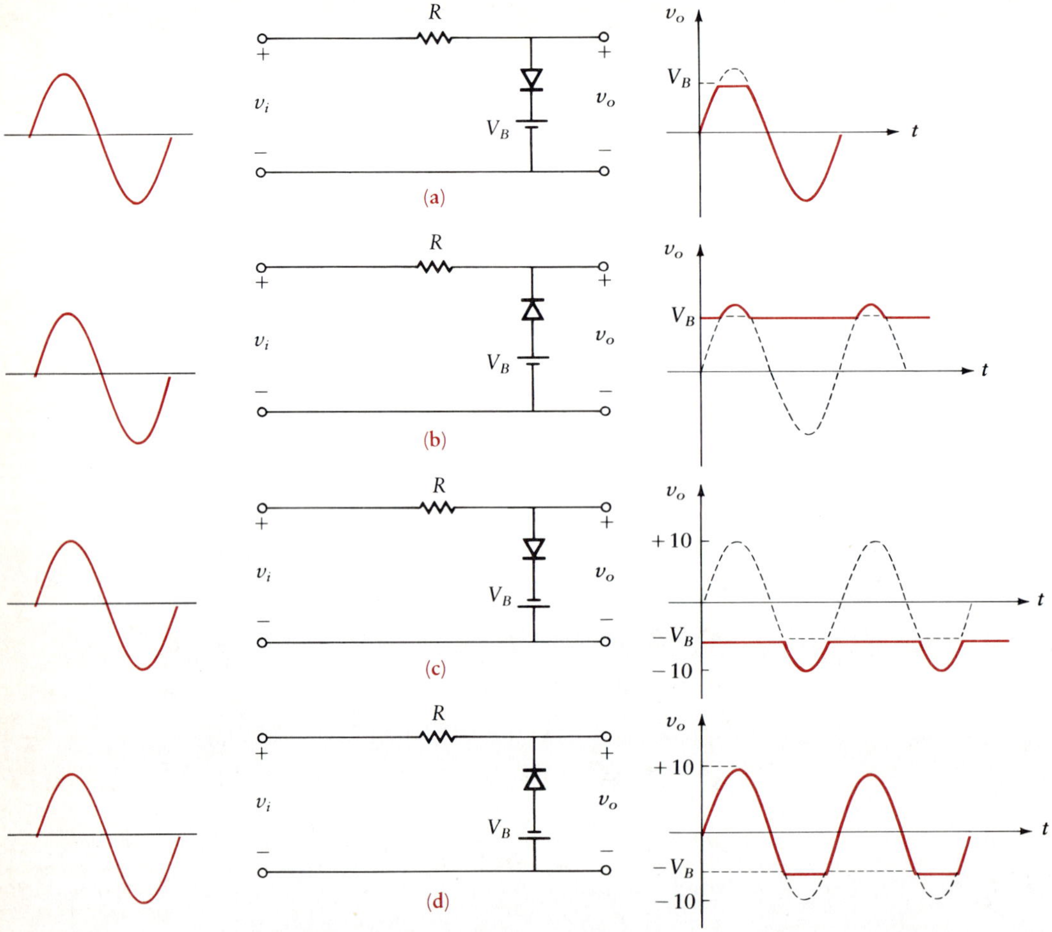

Figure 1.40
Ideal clipping circuits.

is to make the clipping level $V_\gamma + V_B$ instead of V_B. The effect of the resistance is to change the flat clipping action to one that proportionately follows the input voltage (i.e., a voltage-divider effect). The resulting output is calculated as follows and is illustrated in Figure 1.41(b). For

$$v_i < V_B + V_\gamma, \qquad v_o = v_i$$

For

$$v_i > V_B + V_\gamma, \qquad v_o = v_i\frac{R_f}{R + R_f} + (V_B + V_\gamma)\frac{R}{R + R_f}$$

Positive clipping and negative clipping can be performed simultaneously. The result is a *parallel-biased clipper,* which is designed by using two diodes and two voltage sources oriented in opposite directions. The circuit produces the output wave shape as shown in Figure 1.42 when ideal diodes are assumed.

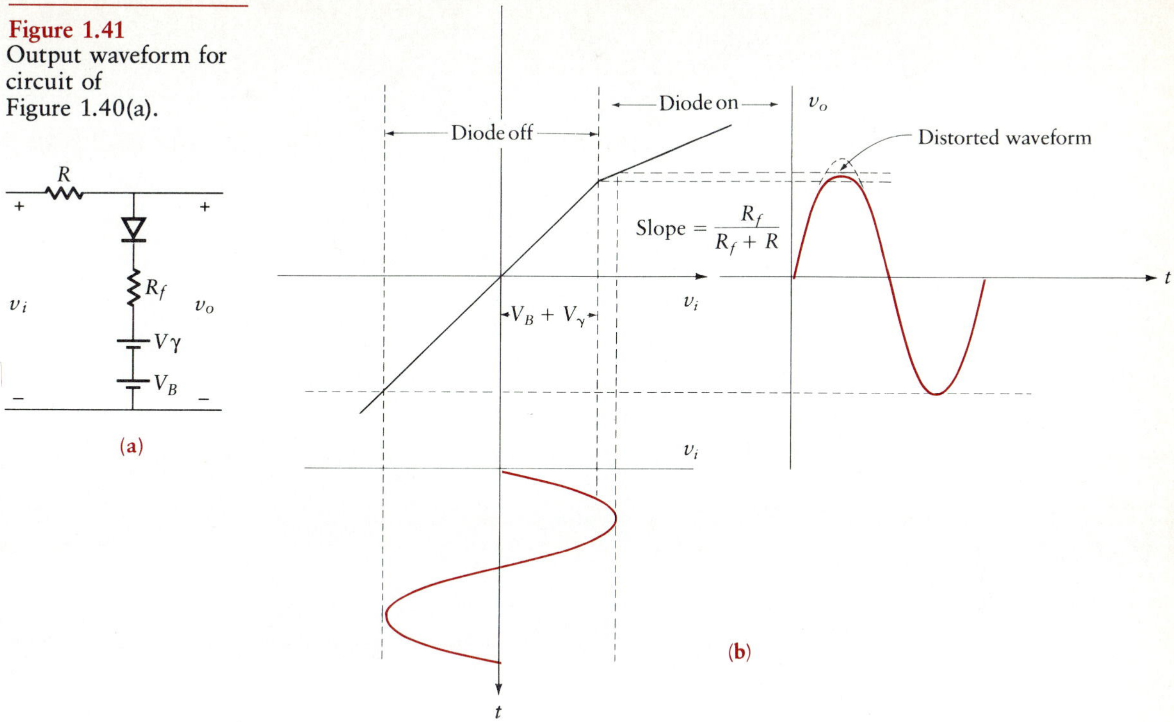

Figure 1.41 Output waveform for circuit of Figure 1.40(a).

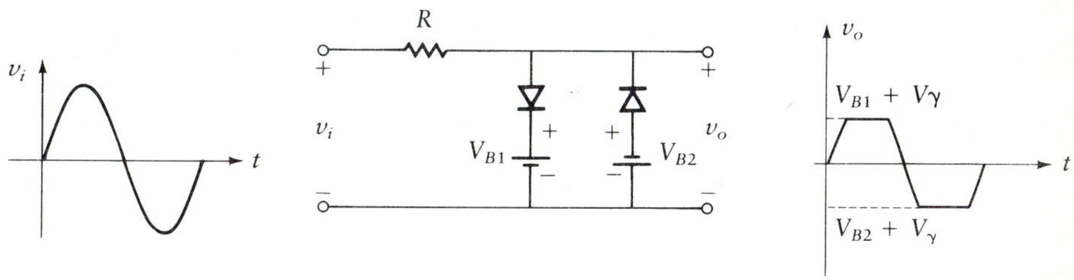

Figure 1.42 Parallel-biased clipper.

The extension to practical diodes parallels the analysis leading to the results in Figure 1.41.

Another type of clipper is the *series-biased* clipper, which is shown in Figure 1.43. The 1-V battery in series with the input causes the input signal to be superimposed on a dc voltage of −1 V rather than being symmetrical about the zero axis. Assuming that this system uses an ideal diode, we find that the diode of Figure 1.43(a) will conduct only during the negative-going shifted input signal. When the diode is conducting, the output is zero. We have a nonzero output when the diode is not conducting. In Figure 1.43(b) the reverse is true. When the conditioned signal is positive, the diode conducts and an output signal exists, but when the diode is off, no output occurs. Although the operation of the two circuits is different, the two outputs are identical. In Figure 1.43(c) and (d) we reverse the polarity of the battery and obtain output waveforms as shown.

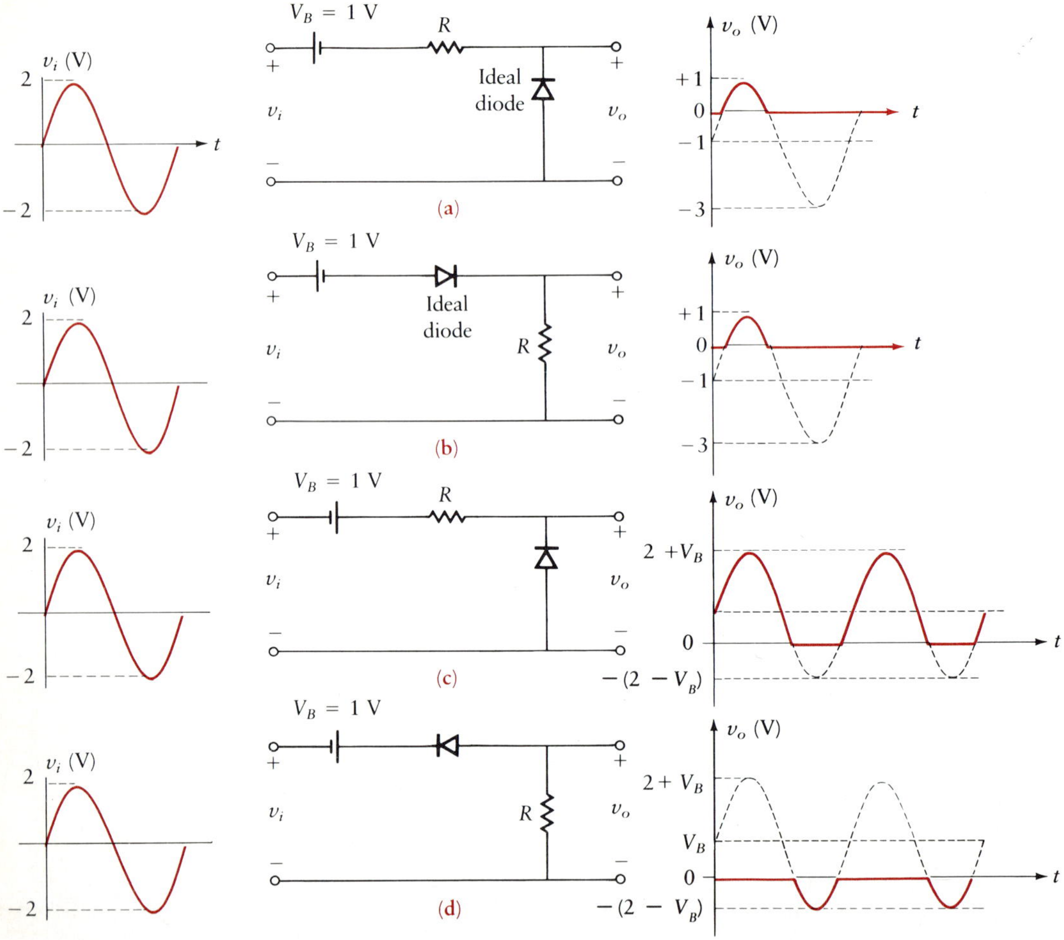

Figure 1.43
Series-biased clipper.

Example 1.6

Find the output of the clipper circuit of Figure 1.44(a), assuming that a. $V_\gamma = 0$ and b. $V_\gamma = 0.7$ V. $R_f = 0$ for both cases.

SOLUTION a. When $V_\gamma = 0$ with v_i positive and $v_i < 3$ V, then

$$v_i = v_o$$

When v_i is positive and $v_i > 3$ V, then

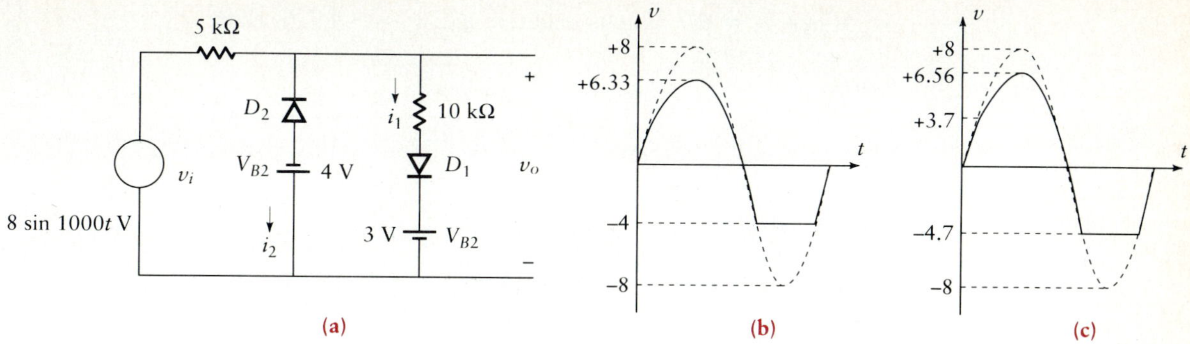

Figure 1.44 Clipper for Example 1.6.

$$i_1 = \frac{v_i - 3}{1.5 \times 10^4}$$

$$v_o = 10^4 i_1 + 3$$

and

$$v_o = \tfrac{2}{3} v_i + 1$$

For $v_i = 8$ V, $v_o = 6.33$ V.

When v_i is negative and $v_i > -4$ V, then

$$v_i = v_o$$

When v_i is negative and $v_i < -4$ V, then

$$v_o = -4 \text{ V}$$

The resulting output wave shape is shown in Figure 1.44(b).

b. When $V_\gamma = 0.7$ V, v_i is positive, and $v_i < (3.7$ V), then

$$v_i = v_o$$

When $v_i > (3.7$ V), then $i_1 = (v_i - 3.7 \text{ V})/1.5 \times 10^4$,

$$v_o = 10^4 i_1 + 3.7$$

and

$$v_o = \tfrac{2}{3} v_i + 1.23$$

For $v_i = 8$ V, $v_o = 6.56$ V.

When $V_\gamma = 0.7$ V, v_i is negative and $v_i > -4.7$ V, then

$$v_i = v_o$$

When $v_i < -4.7$ V, then

$$v_o = -4.7 \text{ V}$$

The resulting output wave shape is shown in Figure 1.44(c).

1.5.2 Clampers

A voltage waveform can be shifted by adding an independent voltage source, either a constant or a time function, in series with the waveform. *Clamping* is a shifting operation, but the additive source is no longer independent of the waveform. The amount of shift depends on the actual waveform. Figure 1.45 shows an example of clamping.

The input waveform in Figure 1.45 is shifted by an amount that makes the shifted waveform peak at a value of V_B. Thus, the amount of shift is the exact amount necessary to change the original maximum, V_m, to a new maximum, V_B. The waveform is "clamped" to a value of V_B. If we knew the exact value of the original maximum, V_m, we could accomplish this shift with an independent dc source in series with the waveform. The distinguishing feature of a clamper is that it adjusts the waveform without needing the initial knowledge of the exact shape. The amount of shift is determined by the actual wave shape. If the input waveform changes, the amount of shift will change such that the output is always clamped to V_B. The clamping circuit thus provides a dc component in an amount necessary to achieve the desired clamping level. For this particular example, the capacitor of Figure 1.45 charges to a value equal to the difference between the original waveform peak and V_B. The capacitor then acts like a series battery of this voltage magnitude, thereby shifting the waveform to the value shown in Figure 1.45(c).

Figure 1.45
Clamping circuit.

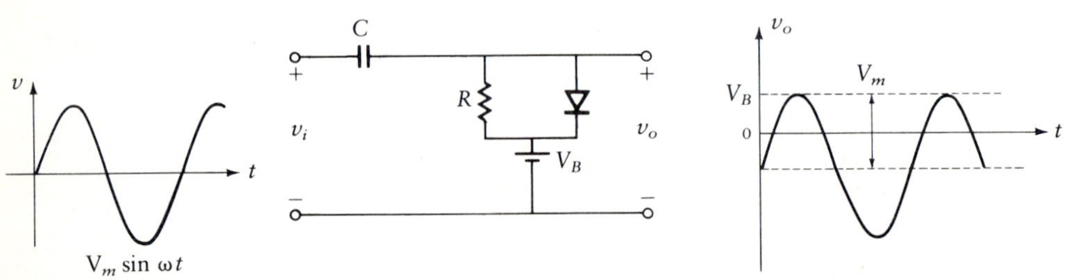

(a) Input (b) Clamping circuit (c) Output when $V_B < V_m$

A clamping circuit is composed of a battery (or dc supply), diode, capacitor, and resistor. The resistor and capacitor are chosen so that the time constant is large. We want the capacitor to charge to a constant value and remain at that value throughout the period of the input waveform. If the capacitor voltage did not remain approximately constant, the result would be a distortion of the waveform rather than a simple shifting. If this condition is met and the forward resistance of the diode is assumed to be zero, the output is a reproduction of the input with the appropriate shift. Whenever the output tries to exceed V_B, the diode forward-biases and the output is limited to V_B. During these times, the capacitor charges. When steady state is reached, the capacitor will be charged to a value of

$$V_C = V_m - V_B$$

Figure 1.46 illustrates a clamping circuit for which the output is clamped to zero (i.e., there is no battery so $V_B = 0$). Because the diode is in the reverse direction from that of the previous circuit, the *minimum* rather than *maximum* of the output is clamped. That is, the capacitor can charge only in a direction that will add to the input voltage. The circuit is shown with a square wave as input. It is important that the voltage across the capacitor remain approximately constant during the half-period of the input waveform. A design rule of thumb is to make the RC time constant at least five times the duration of the half-period (i.e., five times either $t_1 - t_0$ or $t_2 - t_1$). If this design rule is followed, the RC circuit has less than 20% of a time constant to charge or discharge during the half-period. This places the final value within 18% of the starting value (i.e., $\exp(-0.2) = 0.82$). If the time constant is too small, the waveform will be distorted as illustrated in Figure 1.46(c). To reduce the error to less than 18%, the time constant can be increased (e.g., to 10 times the half-period duration). The square-wave input represents a *worst-case* situation since it places the greatest demands on the clamping circuit.

Figure 1.46
Clamping at zero.

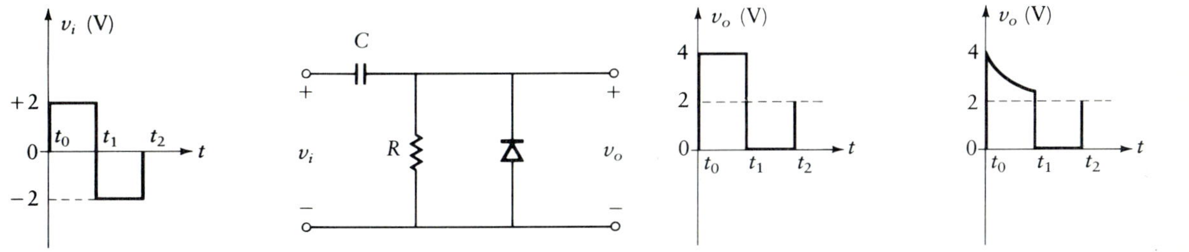

(a) **Input** (b) **Clamper with long time constant** (c) **Output with short time constant**

1.6 ALTERNATIVE TYPES OF DIODES

This section briefly presents the following types of diodes:

- Schottky diode
- Light-emitting diode
- Photodiode

1.6.1 Schottky Diodes

A *Schottky diode* (or *Schottky barrier diode*) is formed by bonding a metal, such as aluminum or platinum, to *n*- or *p*-type silicon. It is often used in integrated circuits for high-speed switching applications. Its symbol and construction are shown in Figure 1.47. The Schottky diode has a voltage-current characteristic similar to that of the silicon *pn* junction diode, except that the forward break voltage, V_γ, is in the range from 0.3 V to 0.6 V. The capacitance associated with this diode is extremely small. When the Schottky diode is operated in the forward mode, current is induced by the movement of electrons from the *n*-type silicon across the barrier and through the metal. Since electrons recombine relatively quickly through metals, the recombination time is short, on the order of 10 ps. This is much faster than an ordinary *pn* junction diode. Therefore, the Schottky diode is of great value in high-speed switching applications.

The metallic material in contact 1 and the lightly doped *n* region form a rectifying junction, while the heavily doped *n* region and contact 2 form an ohmic contact. The forward-direction electrons from the *n*-type silicon cross the barrier into the metal, where there is a high concentration of electrons. This results in a *majority* carrier device.

Schottky diodes are useful in integrated circuit technology because they are easy to fabricate and can be manufactured at the same time as the other components on the chip. Fabrication of a *pn* junction diode requires one more *p*-type diffusion than does a Schottky diode. However, fabrication of a Schottky diode may require an extra metallization step. The low-noise characteristics of the Schottky diode make it ideal for application in power monitoring of low-level radio frequencies, detectors for high frequency, and Doppler radar mixers.

Figure 1.47
Schottky diode.

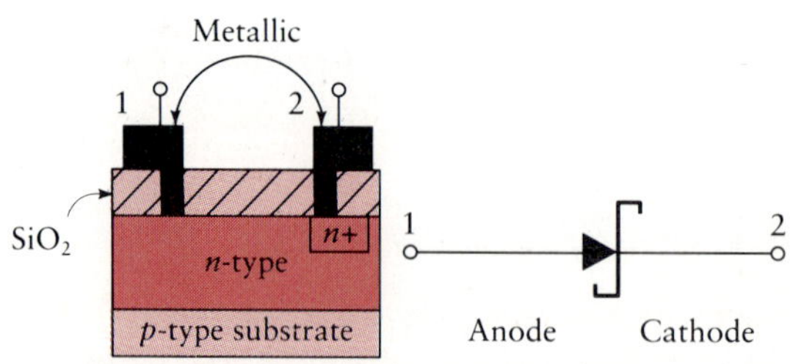

(a) Conduction (b) Symbol

1.6.2 Light-Emitting Diodes

Certain types of diodes are capable of changing electric energy into light energy. The *light-emitting diode (LED)* transforms electric current into light. It is useful in various types of displays and can sometimes be used as a light source for optical fiber communication applications.

An electron can fall from the conduction band into a hole and give up energy in the form of a photon of light. The momentum and energy relationships in silicon and germanium are such that the electron gives up its energy as heat when it returns from the conduction band to the valence band. However, the electron in a gallium arsenide crystal produces a photon when it returns from the conduction band to the valence band. Although there are not enough electrons in a crystal to produce visible light, when a forward bias is applied, large numbers of electrons are injected from the n material into the p material. These electrons combine with holes in the p material at the valence band energy level, and photons are released. The light intensity is proportional to the rate of recombination of electrons and, therefore, proportional to the diode current. The gallium arsenide diode emits light waves at a wavelength near the infrared band. To produce light in the visible range, gallium phosphide must be mixed with the gallium arsenide.

When an LED is conducting, the forward voltage drop is approximately 1.7 V. The amount of light emitted by the LED depends on the current through the diode; the greater the current, the more light is emitted. A current-limiting resistor must be placed in series with the LED to prevent destruction of the diode. The magnitude of this current-limiting resistor is easily calculated by limiting the LED on-current to approximately 10 mA with a diode on-voltage of approximately 1.7 V. Example 1.7 illustrates the calculation.

Example 1.7

Select a current-limiting resistor to properly light an LED when 5 V dc is applied to the LED as shown in Figure 1.48(a).

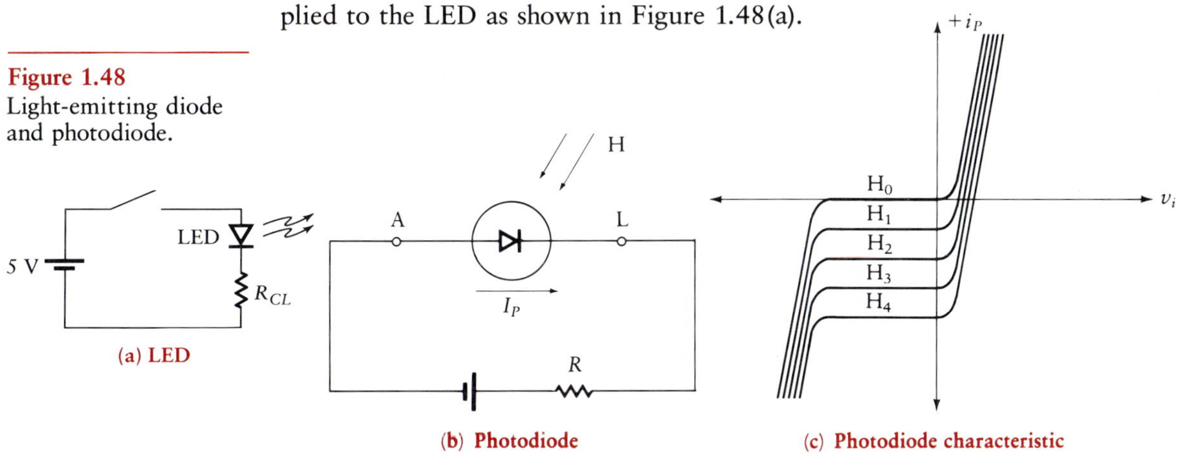

Figure 1.48 Light-emitting diode and photodiode.

SOLUTION We wish to limit the current through the LED to approximately 10 mA when the LED is conducting and when the on-voltage is 1.7 V. The voltage to be dropped across R_{CL} is $5 - 1.7 = 3.3$ V. With 10-mA maximum current, the value of $R_{CL} = 330\ \Omega$. This is a typical resistor value for use with 5-V LED displays.

1.6.3 Photodiodes

A *photodiode* performs the inverse of an LED. That is, it transforms light energy into an electric current. Reverse bias is applied to the photodiode and the reverse saturation current is controlled by the light intensity that shines on the diode. The light generates electron-hole pairs, which induce current. The result is a *photocurrent* in the external circuit, which is proportional to the effective light intensity on the device. The diode behaves as a constant-current generator as long as the voltage does not exceed the avalanche voltage. The response times are less than 1 μs. The sensitivity of the diode can be increased if the junction area is made larger since more photons are collected, but this will also increase the response time since the junction capacitance (and therefore the RC time constant) increases.

Figure 1.48(b) shows a photodiode circuit, and Figure 1.48(c) shows its characteristic curve for various light intensities, H. Photodiode current, I_P, can be estimated from the following equation:

$$I_P = \eta q H$$

where

η = quantum efficiency

q = charge on an electron, 1.6×10^{-19} C

$H = \phi \times A$ = light intensity (photons per second)

ϕ = photon flux density (photons per second per square centimeter)

A = junction area (square centimeters)

Most silicon light detectors consist of a photodiode junction and an amplifier, frequently on a single chip. We defer discussion of this type of device until Chapter 3.

1.7 MANUFACTURERS' SPECIFICATIONS

The construction of a diode determines the amount of current it is capable of handling, the amount of power it can dissipate, and the peak inverse voltage it will withstand without damage. Each manufacturer develops these criteria in specification sheets for the device. Examples of manufacturer specifications are given in Appendix D.

Listed next are the principal parameters that are found on a manufacturer's specification sheet for a rectifier diode:

1. Type of device with generic number or manufacturer's numbers.
2. Peak inverse voltage (PIV).
3. Maximum reverse current at PIV.
4. Maximum dc forward voltage.
5. Average half-wave rectified forward current.
6. Maximum junction temperature.
7. Current derating curves.
8. Characteristic curves for changes in temperature so that the device can be *derated* for higher temperatures.

In the case of zener diodes, the following parameters usually appear on the specification sheets:

1. Type of device with generic number or manufacturer's number.
2. Nominal zener voltage (avalanche breakdown voltage).
3. Voltage tolerance.
4. Maximum power dissipation (at 25°C).
5. Test current, I_{ZT}.
6. Dynamic impedance at I_{ZT}.
7. Knee current.
8. Maximum junction temperature.
9. Temperature coefficient.
10. Derating curves for higher temperatures.

Let us select an example specification and view the information given in the specification sheet. Using the 1N4001 rectifier diode shown in Appendix D, we find listings as follows:

1. PIV = 50 V.
2. Maximum reverse current (at rated dc voltage) at 25°C is 10 μA. At 100°C the maximum current is 50 μA.
3. Maximum instantaneous forward voltage drop at 25°C is 1.1 V.
4. Average rectified forward current at 25°C is 1 A.
5. Operating and storage junction temperature range (T_j) is $-65°$ to $+175$°C.

Figure 1.49 shows a typical *current derating curve*. This curve indicates the required adjustment in rated current as the temperature increases beyond ambient. A similar curve is often given for *power derating*.

Figure 1.49
Current derating curve.

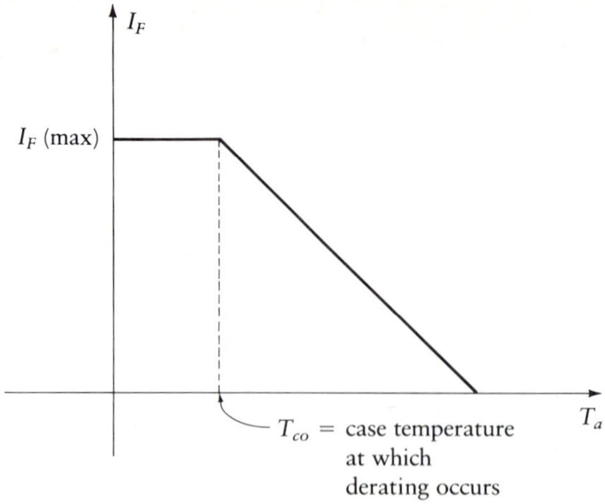

PROBLEMS

1.1 Sketch the output of the circuit shown in Figure P1.1 when the input, v_s, is a 100 V peak-to-peak symmetrical square wave having a period of 2 s. Assume that the diode is ideal.

1.2 Sketch the output of the circuit shown in Figure P1.2 (the diode is ideal) when v_s is a

a. Symmetrical square wave of 100 V peak to peak with a period of 2 s.
b. Sine wave of 100 V peak to peak with a period of 2 s.
c. Symmetrical triangular wave of 40 V peak-to-peak with a period of 2 s.

1.3 Sketch the output of the circuit shown in Figure P1.3 when v_s is a 100-V peak-to-peak sine wave with a period of 2 s. Assume that the diode is ideal, and C = 0.

1.4 Plot I_D versus V_D for a silicon diode if the reverse saturation current $I_o = 0.1\ \mu A$, using $n = 1.5$ for silicon. Also determine the turn-on voltage for the diode.

1.5 Plot I_D versus V_D for a germanium diode if the reverse saturation current $I_o = 0.01$ mA. Also determine the turn-on voltage for the diode (this curve can be plotted on the same graph as that of Problem 1.4).

1.6 A particular diode has a reverse saturation current of 0.2 μA, $n = 1.6$, and $V_T = 26$ mV. Determine the diode current when the voltage across the diode is 0.4 V. Also determine the forward resistance of the diode at this operating point.

1.7 For the circuit shown in Figure P1.4, determine the current through the diode when the dc voltage across the diode is 0.6 V for this range of current and $nV_T = 40$ mV.

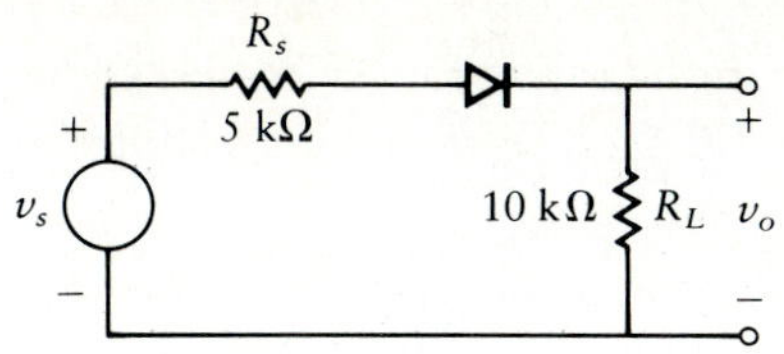

Figure P1.1

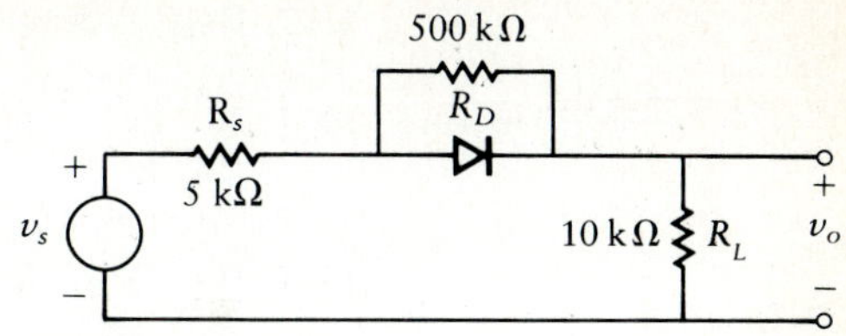

Figure P1.2

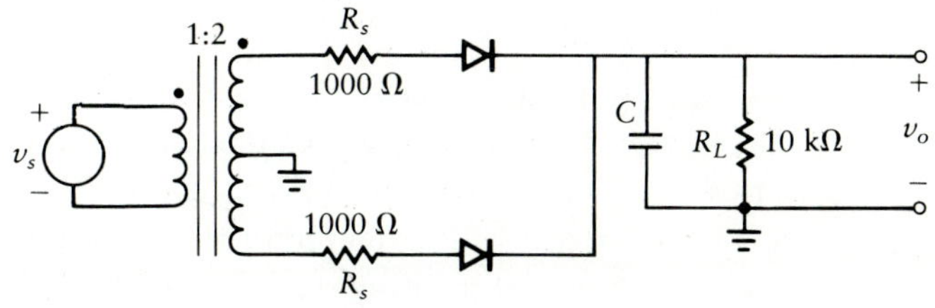

Figure P1.3

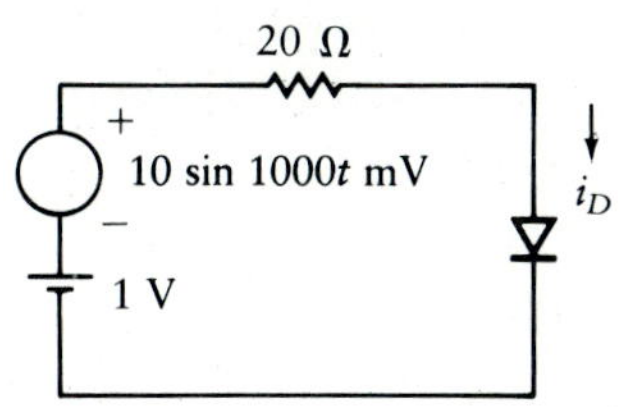

Figure P1.4

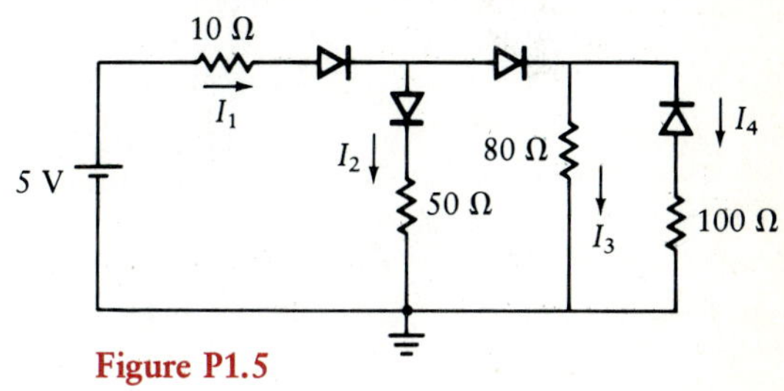

Figure P1.5

1.8 For the circuit shown in Figure P1.5, determine I_3

a. When the diodes are considered ideal.

b. When the diodes are considered nonideal with $R_f = 10\ \Omega$ and $V_\gamma = 0.7$ V. Ignore the reverse saturation current.

1.9 If the output load of a half-wave rectifier is 10 kΩ, what value of capacitor is required to obtain an output voltage that will not vary more than 5%? The input voltage is 100 V rms at 60 Hz. Refer to Figure P1.1. Draw the output waveform.

1.10 Design a half-wave rectifier power supply that has an input of 120 V rms at 60 Hz and requires a maximum voltage output of 17 V and a minimum output of 12 V. The power supply will provide power to an electronic circuit that requires a constant current of 1 A. Determine the circuit configuration, transformer winding ratio, and capacitor size. Assume the diodes and transformer are ideal.

1.11 If the output load of a full-wave rectifier is 10 kΩ, what value of capacitor is required to maintain an output voltage that will not vary more than 10%? The input is 110 V rms at 60 Hz. Refer to Figure P1.3. Draw the output waveform.

1.12 Repeat Problem 1.10 for a full-wave rectifier power supply.

1.13 Determine the capacitor size for the circuit shown in Figure 1.31(a) when $a = 6$ and $R_L = 50\ \Omega$. The minimum voltage to the load must not drop by more than 20%.

1.14 If a zener diode is connected in the circuit shown in Figure P1.6, what is the value of resistor R_i that will maintain the load voltage at 10 V (V_z) when the load varies from 50 to 500 mA and the input voltage varies from 15 to 20 V? Determine the power rating required for the resistor and zener diode.

1.15 The zener regulator shown in Figure P1.6 uses a 20-V zener diode to maintain a constant 20 V across the load resistor, R_L. If the input voltage varies from 32 to 43 V and the load current varies from 200 to 400 mA, select the value of R_i to maintain the constant voltage across the load. Determine the power rating required for the resistor and the zener diode.

1.16 A zener regulator as shown in Figure P1.7 uses a 9-V zener diode to maintain a constant 9 V across the load with the input varying from 18 to 25 V and the output varying from 400 to 800 mA. Assume $R_Z = 0$.

a. Select the value for R_i needed and determine its minimum power requirement.

b. Determine the power rating of the zener diode.

c. Calculate the peak-to-peak output variation if $R_Z = 1\ \Omega$.

1.17 Assuming no loss in the rectifier diodes of the full-wave rectifier (Figure P1.8) with $n = 2$, what is the value of R_i needed to maintain V_L at 16 V with a load current of 500 mA using a 16-V zener? V_S varies between 110 and 120 V rms, 60 Hz. Assume $R_Z = 0$. The voltage to the regulator should not drop to more than 8 V above V_Z.

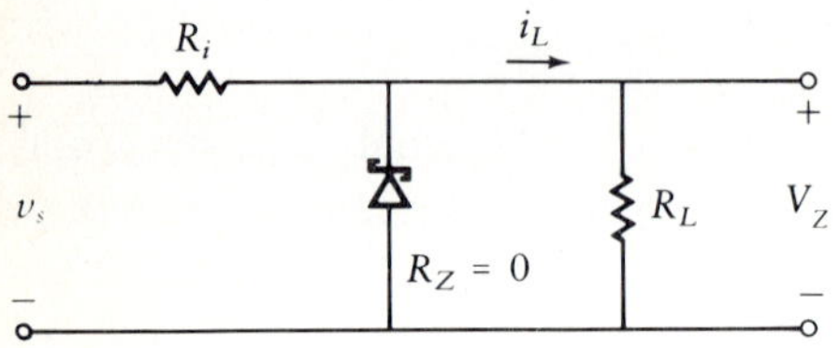

Figure P1.6

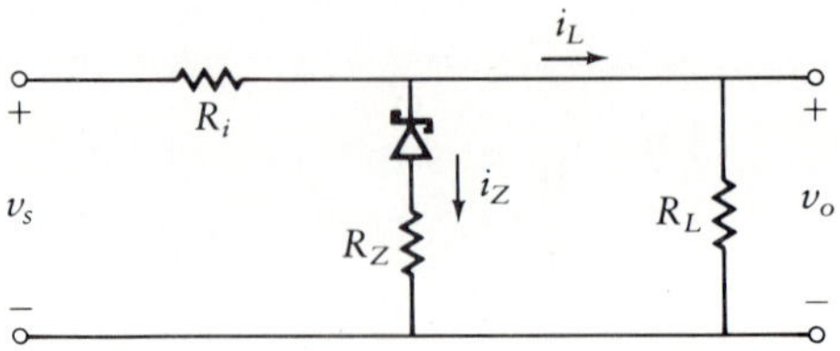

Figure P1.7

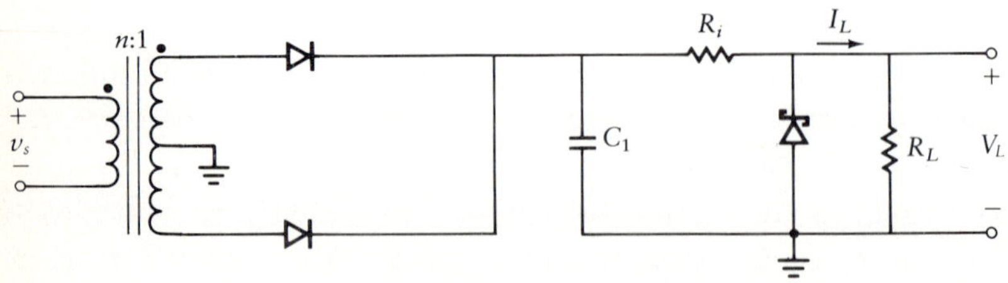

Figure P1.8

1.18 Assuming no voltage drop in the rectifier diodes of Figure P1.8 and $n = 2$, what is the value of R_i necessary to maintain V_L at 16 V with a 500-mA load? The input voltage to the transformer is 110 to 120 V rms, 60 Hz. The filtered output from the rectifier may not vary more than ±5 V. Determine the power rating needed for the resistor and zener diode.

1.19 Design a full-wave regulated power supply when using a 4:1 center-tapped transformer and an 8-V, 1-W zener diode that will provide a constant 8 V to a load varying from 200 to 500 Ω. The input voltage to the transformer is 120 V rms, 60 Hz. Ignore the losses in the transformer and diodes. Determine:

a. $I_{Z\max}$ and $I_{Z\min}$
b. R_i and $V_{s\min}$
c. Size of capacitor needed
d. Percent regulation when $R_Z = 2\ \Omega$

1.20 Design a full-wave regulated power supply using a 5:1 center-tapped transformer and an 8-V, 2-W zener diode that will provide a constant 8 V to a load varying from 100 to 500 Ω. The input voltage to the transformer is 120 V rms, 60 Hz. Ignore the losses in the transformer and diodes. Determine:

a. $I_{Z\max}$ and $I_{Z\min}$
b. R_i and $V_{s\min}$
c. Size of capacitor needed
d. Percent regulation when $R_Z = 2\ \Omega$
e. Power rating of R_i

1.21 Using the values for the input voltage to R_i of Problem 1.18 but using a 12-V zener, what would the value of R_i be to maintain 12 V at the output if the load varied from 20 to 600 mA? What capacitor size is needed?

1.22 Using the circuit of Figure P1.8 and assuming no loss in the rectifier diodes, what is the value of R_i to maintain 12 V across the load using a 12-V zener when V_s is 105 to 125 V rms at 60 Hz? The output of the rectifier drops 20% due to the size of the capacitor, C_1, and the load varies from 50 to 500 mA. What is the size of the capacitor? Set $n = 2$.

1.23 With an input waveform of 10 sin ωt, what is the output waveform for the clipping circuits shown in Figure P1.9? Assume that all diodes are ideal with $V_\gamma = 0$ and $R_f = 0$.

1.24 In the diode clipping circuits of Figure P1.10, $v_i = 20 \sin \omega t$, $R = 2\ \text{k}\Omega$, and $V_R = 10$ V. The reference voltage is obtained from a tap on a 10-kΩ divider connected to a 100-V source. Neglect all capacitances. The diode forward resistance is 100 Ω, $R_r \to \infty$, and $V_\gamma = 0$. Draw the input and output waveforms. Apply Thevenin's theorem to the reference voltage-divider network.

Figure P1.9

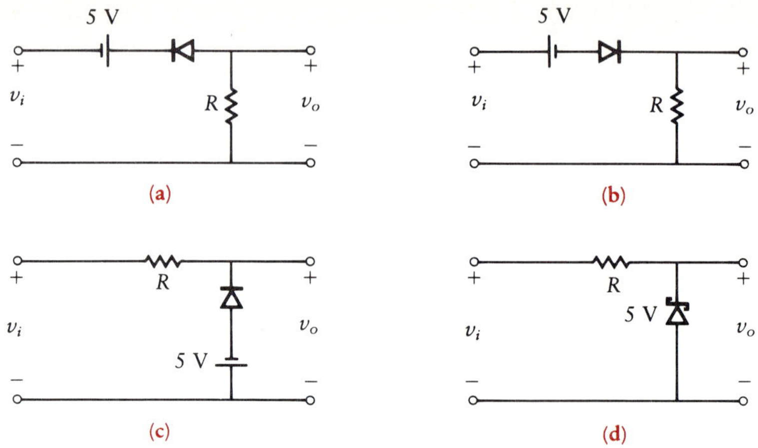

Figure P1.10

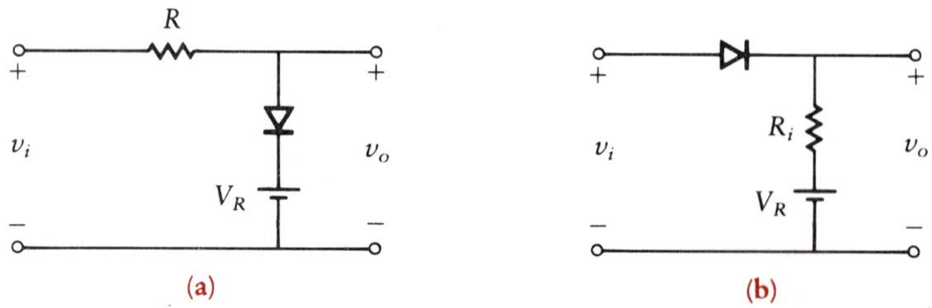

Figure P1.11

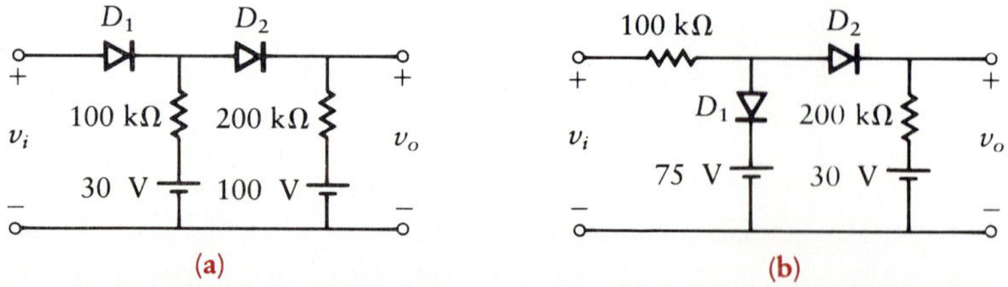

1.25 a. The input voltage, v_i, to the clipper shown in Figure P1.11(a) varies linearly from 0 to 150 V. Sketch the output voltage, v_o, on the same time plot as the input. Assume ideal diodes.

b. Repeat part a for the circuit of Figure P1.11(b).

1.26 a. Sketch the output waveform of the circuit shown in Figure P1.12(a) when $v_i = 9 \sin 1000t$ V. Show the maximum and minimum values on the sketch and the equation for the curves at different time intervals. Assume the diodes are ideal.

b. Repeat part a for Figure P1.12(b).

1.27 Design a clipper circuit to obtain the output shown in Figure P1.13 from an input symmetrical square wave of ±10 V. Assume $V_\gamma = 0.7$ V.

1.28 What type of clipper is needed to obtain the waveforms illustrated in Figure P1.14? Assume the input is 10 sin t V. Draw the circuits and label them.

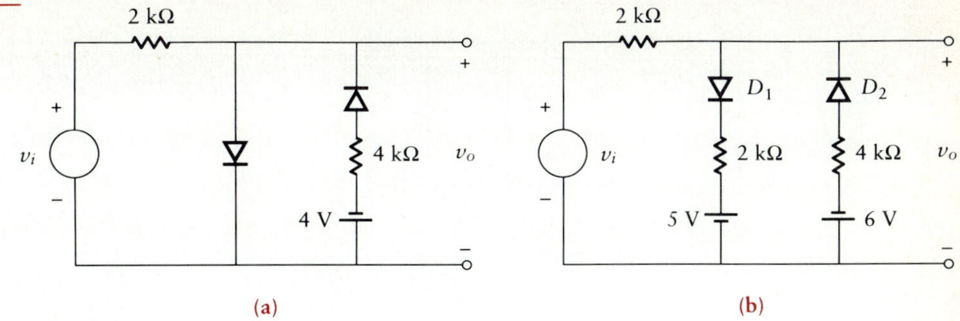

Figure P1.12

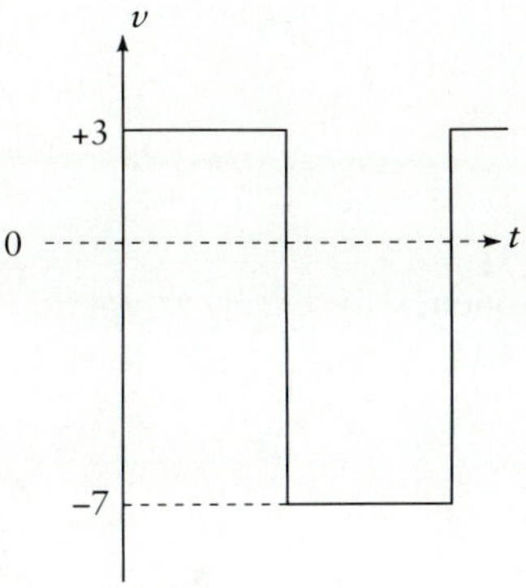

Figure P1.13

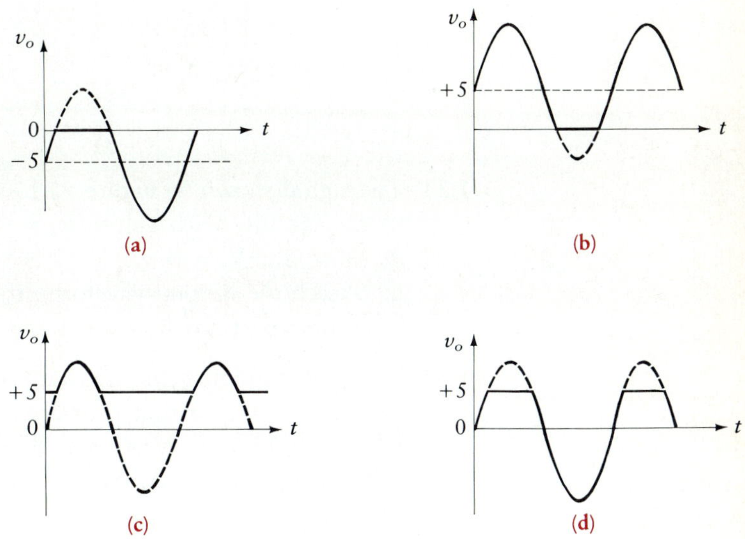

Figure P1.14

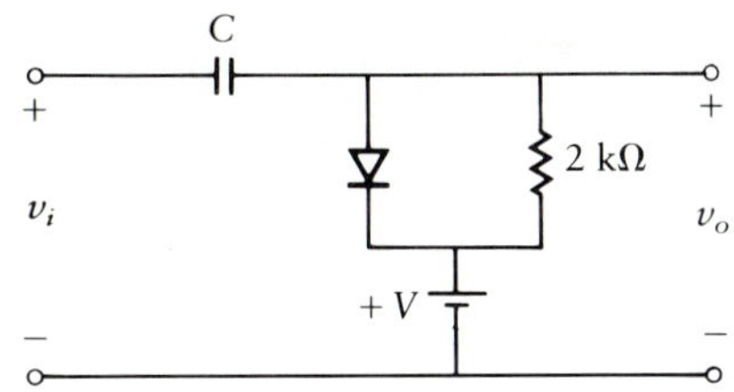

Figure P1.15

1.29 Design a clamper that will provide a +2-V clamped level to a square-wave input for the circuit shown in Figure P1.15. The symmetrical square-wave input peak-to-peak amplitude is 4 V with a period of 100 μs.

1.30 An ideal 10-kHz sinusoidal voltage source whose peak excursions are 10 V with respect to ground is applied to the diode clamping circuit of Figure P1.16. Assume $R \to \infty$, $R_s = 0$, $C = 1\ \mu$F, the diode has $R_r \to \infty$, $R_f = 0$, and $V_\gamma = 0$. Sketch the output waveform.

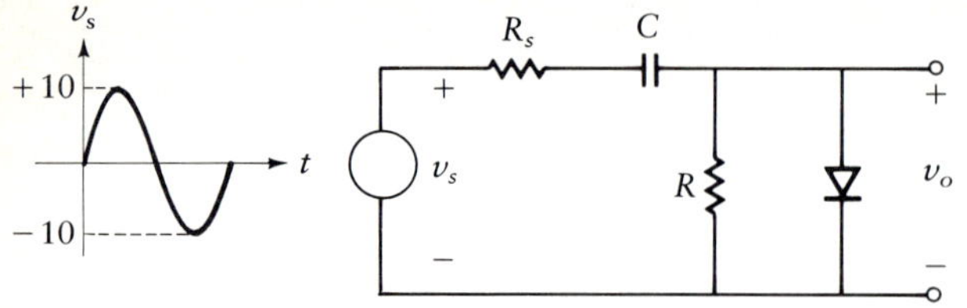

Figure P1.16

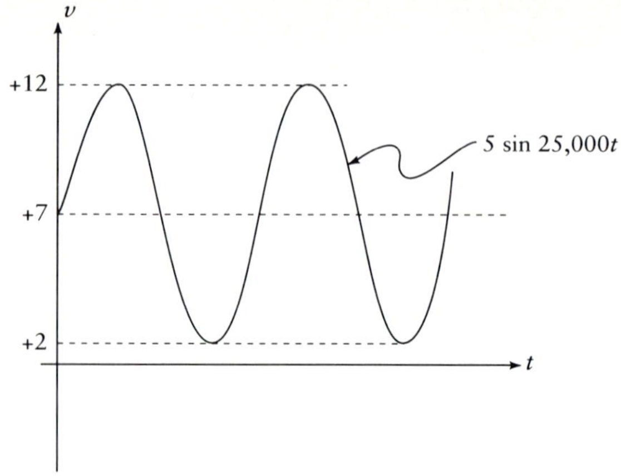

Figure P1.17

1.31 The signal shown in Figure P1.16 with a frequency of 1 kHz is applied to the circuit, with values $R_s = 0$, $R = 10\ \text{k}\Omega$, $C = 0.01\ \mu\text{F}$, $R_f = 0$, $R_r \to \infty$, and $V_\gamma = 0$.

a. Sketch the output waveform, v_o.

b. Repeat part a if $R = 1\ \text{k}\Omega$ and $C = 0.001\ \mu\text{F}$.

1.32 Design a clamping circuit that will provide the output shown in Figure P1.17. Assume that the capacitor available is 0.1 μF and the input $v_i = 5 \sin 25{,}000\ t$ V. Assume $V_\gamma = 0.7$ V.

1.33 Select a current-limiting resistor to properly light an LED when 8 V is applied to the LED.

2

BIPOLAR JUNCTION TRANSISTOR AMPLIFIERS

2.0 INTRODUCTION

Basic circuit analysis is the study of interconnections of passive devices and sources. The *passive linear devices* include resistors, capacitors, and inductors. These devices perform the linear operations of proportional multiplication, integration, and differentiation. *Independent sources* are either voltage or current, and their output values do not depend on any other quantity in the circuit. *Dependent sources* are characterized by an output voltage or current that is a function of a parameter in some branch of the circuit separate from that of the source. Dependent sources usually arise from modeling of active devices such as bipolar junction transistors, field-effect transistors, and operational amplifiers.

We begin our study of active devices with the *bipolar junction transistor.* In 1948, Drs. Bardeen, Brattain, and Shockley of Bell Telephone Laboratories built and tested the first transistor. It was a crude device with low amplification—really not of much use for purposes other than laboratory experiments. Meanwhile, in industry, the vacuum tube reigned supreme in applications ranging from consumer goods to military hardware. However, there were some roles which the tube could not fulfill without a great deal of expense. Even worse, certain applications were impossible to achieve using vacuum tubes. Soon after the invention of the transistor, engineers saw its advantages for small portable devices and set out to improve performance. A continuing evolution resulted, which has led to the transistor of today. During the 1960s, manufacturing processes and methods improved so that the transistor could be built reliably. This brought about a boom in the electronics industry as many consumer products could be inexpensively constructed. Power-handling ability and maximum operating frequency improved steadily throughout this period. The transistor has now almost completely replaced the vacuum tube except in some high-power and high-frequency applications.

In order to understand transistor operation, we must first gain some familiarity with dependent sources.

2.1 DEPENDENT SOURCES

Dependent sources produce a voltage or current whose value is determined by a voltage or current existing in some other location in the circuit. Note that passive devices produce a voltage or current whose value is determined by a voltage or current existing at the *same* location in the circuit. Both independent and dependent voltage and current sources are *active* elements. That is, they are capable of delivering power to some external device. *Passive* elements are not capable of generating power, although they can store finite amounts of energy for delivery at a later time, as is the case with capacitors and inductors.

Figure 2.1 illustrates a configuration that occurs in the process of solid-state circuit analysis. We will find the voltage and current gain of this system. *Voltage gain* is defined as the ratio of output to input voltage. Likewise, *current gain* is the ratio of output to input current. The input current is

$$i_{in} = \frac{v_i}{R_{in}} = \frac{20 \sin \omega t \text{ mV}}{1 \text{ k}\Omega} = 20 \sin \omega t \ \mu\text{A}$$

A current-divider relationship can be used to find i_1.

$$i_1 = \frac{2000(20 \sin \omega t \ \mu\text{A})}{2000 + 2000} = 10 \sin \omega t \ \mu\text{A}$$

The output voltage is then given by

$$v_o = -100 i_1 \times (10 \text{ k}\Omega \parallel 10 \text{ k}\Omega)$$

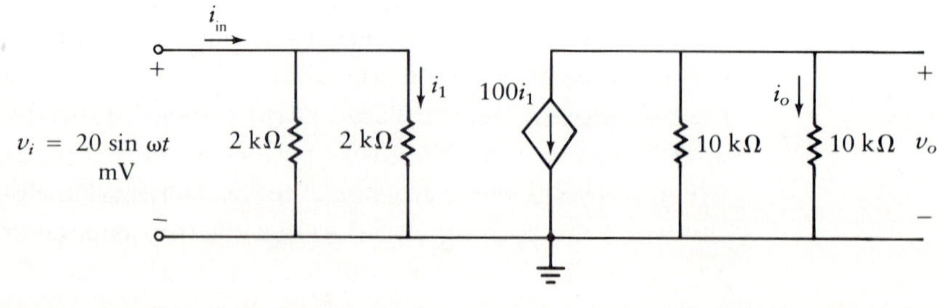

Figure 2.1
Solid-state equivalent circuit.

where ∥ indicates a parallel combination of resistors.

$$v_o = -(5 \times 10^5)i_1 = -5 \times 10^5(10 \times 10^{-6} \sin \omega t) = -5 \sin \omega t \text{ V}$$

Then, using a current-divider relationship, the output current is found to be

$$i_o = \frac{10^4(-100i_1)}{10^4 + 10^4} = -50i_1$$

The voltage gain is

$$\frac{v_o}{v_i} = \frac{-5 \sin \omega t}{0.02 \sin \omega t} = -250$$

The current gain is

$$\frac{i_o}{i_{in}} = \frac{-50(10\ \mu\text{A})}{20\ \mu\text{A}} = -25$$

2.2 BIPOLAR TRANSISTORS

The transistor is a three-terminal device, in contrast to the diode, which is a two-terminal device. The diode consists of a *p*-type material and an *n*-type material; the transistor consists of two *n*-type materials separated by a *p*-type material (*npn* transistor) or two *p*-type materials separated by an *n*-type material (*pnp* transistor). Figure 2.2(a) is a schematic representation of a transistor.

The three different layers or sections are identified as emitter, base, and collector. The *emitter* is a heavily doped, medium-sized layer designed to emit or inject electrons. The *base* is a medium doped, small layer designed to pass electrons. The *collector* is a lightly doped, large layer designed to collect electrons.

The transistor can be idealized as two *pn* junctions placed back to back; these are termed *bipolar junction transistors (BJTs)*.

A simple but effective explanation of the operation of the *npn* transistor is developed using the potential-hill diagram illustrated in Figure 2.2(b). This approach illustrates a simplified visual picture of the basic operation of a bipolar transistor so that simple circuit applications can be understood. When the emitter-base junction is biased in the forward direction and the collector-base junction is biased in the reverse direction, electrons leaving the *n* material of the emitter will see only a small potential hill at the *np* junction. Since the potential hill is small, most of the electrons have enough energy to progress to the top of the hill. Once on top of the potential hill, the electrons move easily

Figure 2.2
The bipolar transistor.

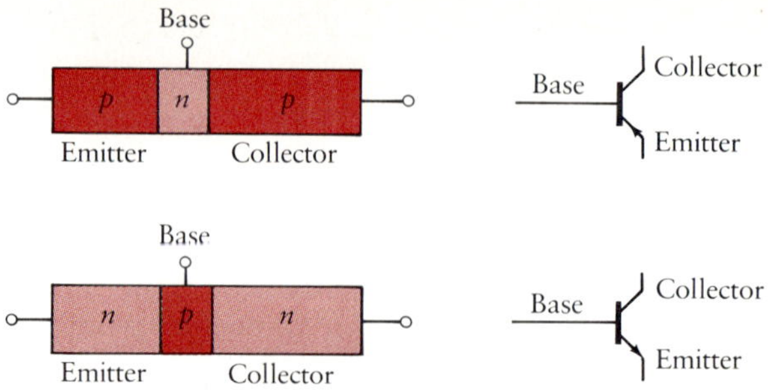

(a) Circuit symbols

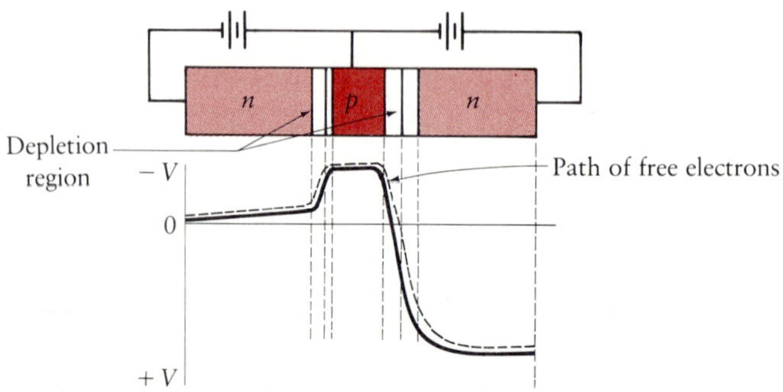

(b) Potential-hill diagram

through the *p* material (base) to the *pn* (base-collector) junction. When they approach that junction, the electrons are under the influence of the positive supply voltage and move forward rapidly as they move down the potential hill. If the forward bias on the base-emitter junction is reduced, the height of the potential hill is raised. Electrons leaving the emitter will have more difficulty in reaching the top of the hill. The electrons reaching the top are the ones with the highest amount of energy, and these will progress to the collector. The reduction of forward bias thus causes the current through the transistor to be considerably reduced. On the other hand, increasing the forward bias on the base-emitter junction will reduce the potential hill and allow more emitter electrons to flow through the transistor.

The current in a junction transistor can also be understood by examining charge carrier behavior and the depletion regions. The depletion regions have been indicated on Figure 2.2(b). Note that since the base-emitter junction is forward-biased, the depletion region is relatively narrow. The reverse is true for the base-collector junction. A large number of majority carriers (electrons) will diffuse across the base-emitter junction, since this is forward-biased. These electrons then enter the base region and have two choices. They may exit this region through the connection to the voltage sources, or they may continue flowing to the collector region across the wide depletion region of the reverse-biased junction. We would normally expect the major portion of this current to return to the source, except for the following observations. Since the base region is so thin, these electrons need to travel less distance to be attracted to the positive potential of the collector connection. In reality, a small fraction of the electrons leave the base through the source connection. The major portion of current does flow into the collector.

Let us now examine current in greater detail. Refer to Figure 2.3(a) during the following discussion.

The base-emitter junction acts as a forward-biased diode with a forward current of $i_B + i_C$. The collector-base junction is reverse-biased and exhibits a small leakage current, I_{CBO}, in addition to a larger current caused by the interaction of currents in the base.

Looking at the currents in the three external connections, we see that

$$i_E = i_C + i_B \tag{2.1}$$

Note that the positive direction for i_C and i_B is defined to be *into* the transistor, while the reverse is true for i_E.

We shall be developing equivalent circuits for the transistor, and these equivalent circuits will contain dependent sources as introduced in Section 2.1. We therefore must establish the dependence of one parameter on another. To do so, we define several gain constants that characterize the manner in which one parameter affects another parameter.

Figure 2.3
Internal currents in a transistor.

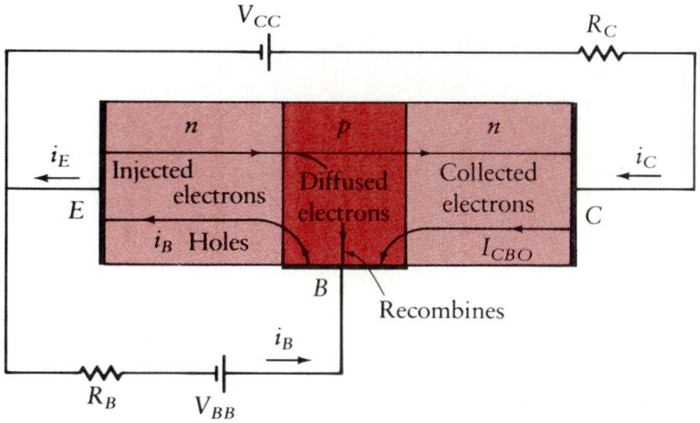

(a) Detailed inside view

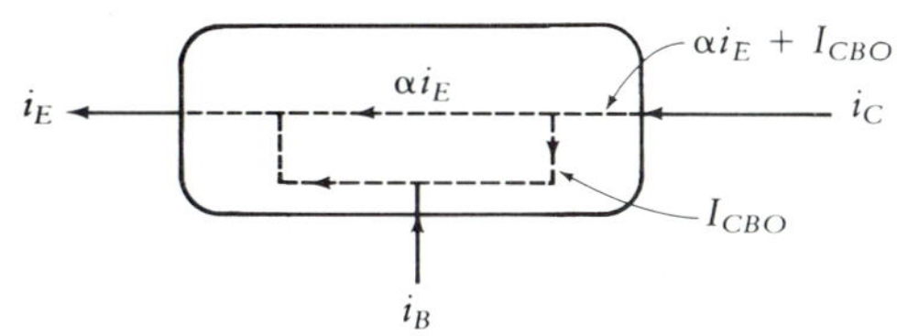

(b) Simplified view

The *common-base current gain,* α, is defined as the ratio of the change in collector current to the change in emitter current, assuming that the voltage between collector and base is a constant. Thus,

$$\alpha = \left.\frac{\Delta i_C}{\Delta i_E}\right|_{v_{CB}=\text{constant}}$$

This is shown in Figure 2.3(b), where I_{CBO} is the leakage current between base and collector.

We wish to find a relationship between the collector and base currents. The collector current is found by viewing Figure 2.3(b).

$$i_C = \alpha i_E + I_{CBO} \tag{2.2}$$

Combining equations (2.1) and (2.2) yields the emitter current,

$$i_E = \alpha i_E + I_{CBO} + i_B$$

and solving for the base current,

$$i_B = i_E(1 - \alpha) - I_{CBO} \tag{2.3}$$

We can eliminate i_E from equation (2.3) by rewriting equation (2.2) as

$$i_E = \frac{i_C - I_{CBO}}{\alpha}$$

Finally, this is substituted in equation (2.3) to yield a relationship between i_B, i_C, and I_{CBO}.

$$i_B = \frac{(i_C - I_{CBO})(1 - \alpha)}{\alpha} - I_{CBO}$$

$$= \frac{(1 - \alpha)i_C}{\alpha} - \frac{I_{CBO}}{\alpha} \tag{2.4}$$

The common-base current gain, α, usually lies in the range 0.9 to 0.999. The reciprocal can therefore often be approximated as unity, thus yielding the approximation

$$i_B = \frac{(1 - \alpha)i_C}{\alpha} - I_{CBO} \tag{2.5}$$

We define a second gain factor, β, as the ratio of changes in collector current to changes in base current. That is,

$$\beta = \frac{\Delta i_C}{\Delta i_B}$$

Therefore,

$$\beta = \frac{\alpha}{1 - \alpha}$$

Typical values of β range from 20 to 800. Making the substitution for β in equation (2.5) yields the very important result

$$i_B = \frac{i_C}{\beta} - I_{CBO}$$

We can usually neglect I_{CBO} since it is small in magnitude compared to i_C/β. Thus,

$$i_C \approx \beta i_B \tag{2.6}$$

The result of equation (2.6) indicates the reason why β is referred to as the *large-signal amplification factor* or the *direct-current amplification factor.* In practice, the value of β varies with base current.

Design challenges exist because β varies with changes in the transistor current. In addition, during the fabrication of the transistor, variation of the value of β occurs within a single production run. Thus two transistors fabricated at the same time will have different values of β, even at the same current levels. This leads us to develop a design procedure that makes the value of collector current relatively independent of changes in β. These methods are discussed in Section 2.5.

Another simplifying assumption that is often made is that the collector current is approximately equal to the emitter current. That is, since I_{CBO} is small compared to i_C and since α ranges from 0.9 to 0.999, we have

$$i_C \approx i_E$$

2.3 TRANSISTOR MODELS

We will be looking at transistor operating curves in Section 2.4.2. As with the diode of the previous chapter, these curves can be used to establish operating points and to perform graphical analysis of transistor circuit operation. In the current section, we will develop a set of transistor models, much as we did for the diode of the previous chapter. We will confine our analysis to low frequencies. In Chapter 7, we expand the model to include additional components so that we can analyze transistor operation at high frequency.

2.3.1 Ebers-Moll Model

We begin with the *Ebers-Moll model* of the transistor. Models of this type are used in SPICE computer simulation programs. The model is presented in Figure 2.4.

This model takes into account the fact that the current between emitter and base consists of two components. The first is associated with the emitter-base diode, and the second is a fraction of the collector current that is coupled through the base to the emitter. Similarly, the current between collector and base consists of a diode component and a component of the emitter current coupled through the base. Since the emitter-base and collector-base voltages are both measured relative to the base, this is known as a *common-base configuration.* We explore other configurations in the next section.

Figure 2.4 presents the *transport* model version of the Ebers-Moll model. The controlled sources depend on *terminal currents* that physically exist in wires. There are other forms of the model. The *injection* model results directly from the *Ebers-Moll equations.* This model contains controlled sources that depend on *internal* diode currents. As such, it is more difficult to use in the analysis of circuits since these internal diode currents cannot be directly measured.

Notice that the model contains two current gain factors, α_R and α_F. The subscripts refer to *reverse* and *forward.* α_F is the *common-base forward short-circuit current gain* and is the rate of change of collector current as a function of emitter current. α_R is the *common-base reverse short-circuit current gain* and is the rate of change of emitter current as a function of collector current.

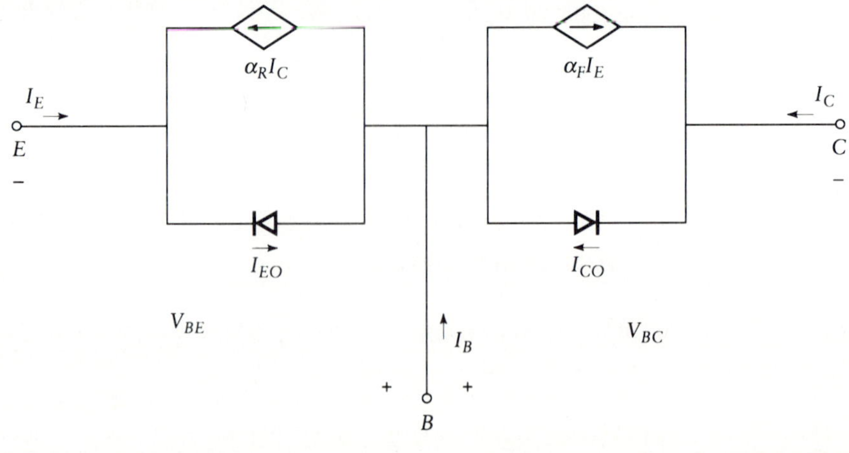

Figure 2.4 Ebers-Moll *npn* transistor model.

The value of α_F is close to unity, while typical values of α_R are less than 0.5. I_{EO} is the reverse emitter current that flows when $I_C = 0$. Similarly, I_{CO} is the reverse collector current that flows when $I_E = 0$. These two saturation currents are known as *open-circuit saturation currents.*

The model of Figure 2.4 accounts for the various operating regions of the transistor (i.e., the states of the equivalent diodes). In normal operation, we will find that V_{BC} is negative and V_{BE} is positive. That is, the base-emitter junction is forward-biased, while the base-collector junction is back-biased. In this case, the model can be simplified to that shown in Figure 2.5(a). Since the saturation currents are small relative to other currents in the circuit, the model of Figure 2.5(a) can be further simplified to that of Figure 2.5(b). Note that the value of α_F is normally near unity.

Viewing Figure 2.5(b), we see that

$$I_B + I_E = \alpha_F I_E = -I_C$$

and, solving for I_C in terms of I_B, we find

$$I_C = \frac{\alpha_F}{1 - \alpha_F} I_B = \beta I_B$$

Figure 2.5
Model of *npn* transistor for active region.

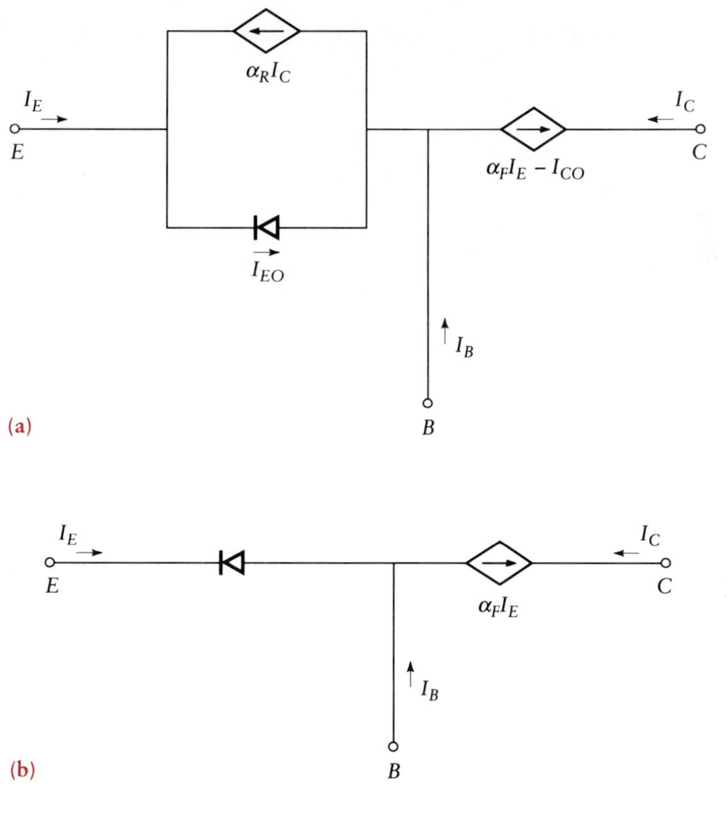

We have used the same definition of β, the large-signal amplification factor, as in Section 2.2 [see equation (2.6)].

Although we presented the Ebers-Moll model for an *npn* transistor, the extension to *pnp* transistors is simple. We need only realize that the two diodes are reversed. We will cover *pnp* transistors as needed in the following work.

2.3.2 Hybrid-π and h-parameter Model

The hybrid-π model of the transistor will prove useful to us as we examine transistors in greater detail, particularly with respect to high-frequency operation. This type of model can be used for a wide variety of *two-port networks.* The model represents a variation of the hybrid, or h-parameter, model, which we use later in this text. The model accounts for the fact that each measurable parameter in the network consists of two components: one due to the driving forces at the first port and a second due to the driving forces at the second port. This is illustrated in Figure 2.6, where we use parameters to indicate a *common-emitter* transistor. The reason for this terminology is that the emitter (E) is common to both input and output. We use lowercase letters since the model is used for *small-signal analysis.* That is, we will deal with the variations in input and output rather than total (dc plus ac) values. Assuming linearity, we can express the input voltage and output current by the following two equations.

$$v_{be} = h_{11}i_b + h_{12}v_{ce} \tag{2.7a}$$

$$i_c = h_{21}i_b + h_{22}v_{ce} \tag{2.7b}$$

Figure 2.6 Two-port common-emitter network configuration.

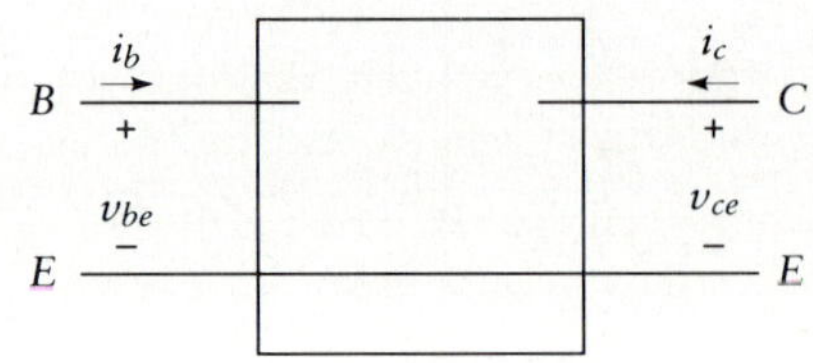

(a)

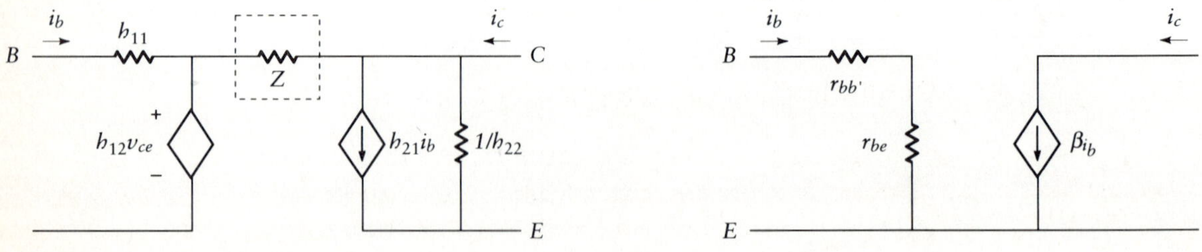

(b) (c)

These equations express the observation made earlier that each parameter consists of two components. The four constants, h_{ij}, characterize the circuit inside the box of Figure 2.6(a). The equivalent circuit modeling equation (2.7) is shown in Figure 2.6(b). This is the h-parameter model of the transistor. If the output and input are coupled through an impedance, as shown in the dashed box in Figure 2.6(b), the circuit becomes the hybrid-π model. The impedance consists of resistance and capacitance. We will be concerned with the capacitance when we deal with the high-frequency model, since capacitors approach short circuits as the frequency approaches infinity. At low frequency, the impedance is relatively high and is often ignored. By ignoring the impedance (i.e., not drawing Z as in Figure 2.6(b)), the model is really an h-parameter rather than a hybrid-π model. However, we will follow the tradition of continuing to label this as the hybrid-π transistor model. In the hybrid-π transistor model, these parameters are given more descriptive names in the following manner. The second subscript letter, e, is related to the common-emitter configuration.

$h_{11} = h_{ie}$ = input resistance—units = ohms

$h_{12} = h_{re}$ = reciprocal voltage gain—dimensionless

$h_{21} = h_{fe}$ = forward current gain—dimensionless

$h_{22} = h_{oe}$ = output conductance—units = siemens (1/ohms)

The h-parameters are given by partial derivatives of equations (2.7a) and (2.7b). For example, we see that h_{fe} is what we have called β. h_{re} is typically near zero. h_{ie} is the input resistance to the base of the transistor. It is usually split into two portions—$r_{bb'}$, the *base-spreading resistance,* and $r_{b'e}$ (sometimes denoted r_π), a resistance related to the base-emitter diode. $r_{bb'}$ represents the resistance between the base contact and the base region and is typically around 100 Ω. $r_{b'e}$ can be as high as several thousand ohms. The output resistance, $1/h_{oe}$, is typically large. In fact, when a load resistor is placed across the output, the parallel combination is usually approximated as simply the load resistance. Therefore, we often ignore h_{oe}. Putting all of these observations together leads to the simplified approximate model of Figure 2.6(c).

2.4 TRANSISTOR CIRCUITS

2.4.1 Common Circuit Configurations

Three general configurations are utilized in transistor circuits. The most often used is the *common-emitter (CE) amplifier,* so called because the emitter is in both the input and output loops. The next most widely used circuit is the *common-collector* (CC) configuration, also known as the *emitter follower.*

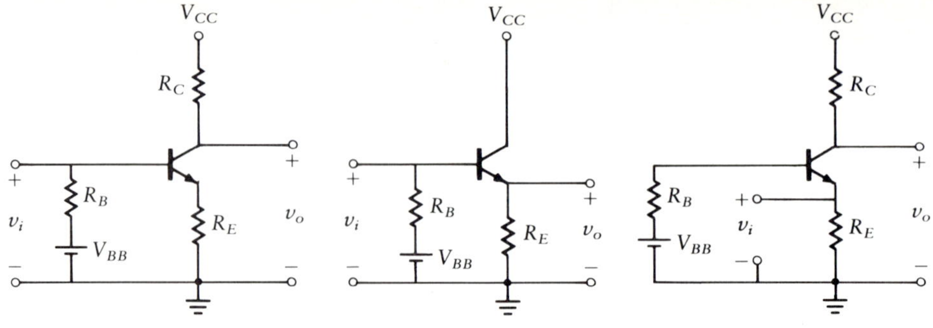

(a) Common emitter (b) Common collector (emitter follower) (c) Common base

Figure 2.7
Amplifier circuits.

The third configuration is the *common-base* (CB) circuit, as we saw in the Ebers-Moll model of the previous section. Examples of these amplifier configurations are shown in Figure 2.7, where we have illustrated the circuits using *npn* transistors.

Note that each of the configurations contains dc sources labeled V_{BB} and V_{CC}. If we attempted to use a transistor for amplification of signals by simply injecting the original signal into one of the terminals, we would be confronted with two major problems. Because most real-life signals have both positive and negative values (i.e., they have a zero average value), we would be operating the transistor over multiple states of the equivalent diodes. They would sometimes be forward-biased and sometimes reverse-biased. This leads to severe distortion of the input signal. Indeed, we would see a type of clipping experienced in Chapter 1 with clipping circuits. Even if we remained in one particular operating region, the low input signal values would experience the nonlinear portions of the diode characteristic, also leading to distortion. For these reasons, we use dc sources to move the zero input signal operation to a relatively linear portion of the transistor curves. Then, as the time-varying portion of the input moves operation away from this point, we can still operate within the linear portion, thereby reducing output distortion. The process of moving the operating point is known as *biasing,* and the offset operating point is known as the *quiescent point,* or simply *Q-point.*

In this chapter we consider the design of the bias, or dc, circuit. The bias technique for the CE amplifier is the same as that for the CB configuration, so these are considered together. The CC configuration is considered separately.

When we use *pnp* transistors, the voltage polarities of V_{BB} and V_{CC} are reversed, but the ac equivalent circuits we have developed remain the same.

2.4.2 Characteristic Curves

Since the transistor is a nonlinear device, one way to define its operation is with a series of characteristic curves in a manner similar to that used for

Figure 2.8 Transistor characteristic curves.

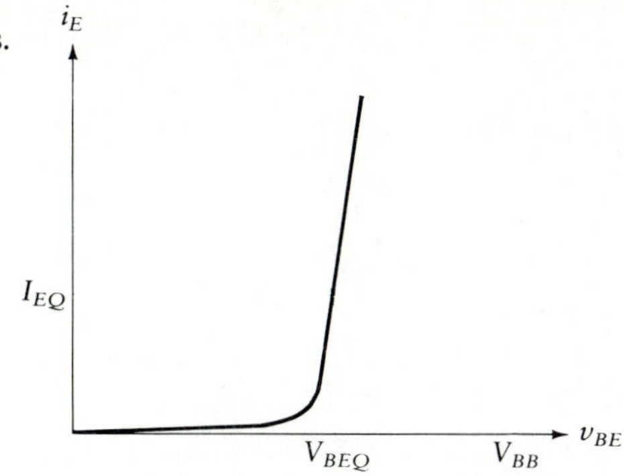

(a) Emitter-base characteristics

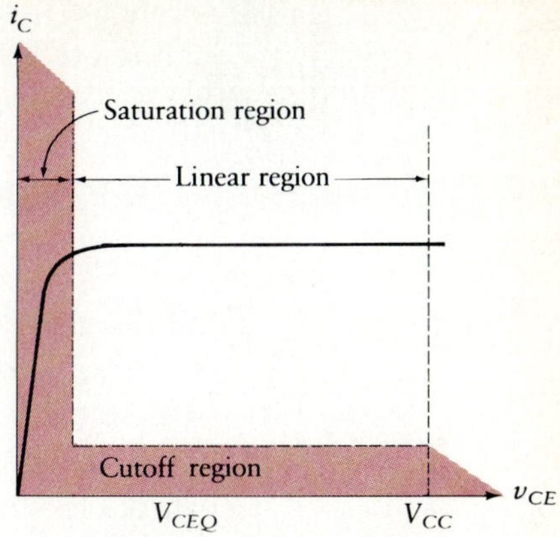

(b) Collector-emitter characteristics

diodes in the previous chapter. There is a set of curves for each type of transistor. Since we are no longer dealing with two-terminal devices, equations involve at least three variables. Therefore, *parametric curves* are usually used to describe transistor behavior. Figure 2.8 shows two typical plots.

Figure 2.8(a) shows the emitter current as a function of the voltage between base and emitter when v_{CE} is held constant. Note that, as we might have expected, this curve is similar to the curve for a diode, since it is the characteristic of the current in the single junction. The slope of the characteristic curve is $1/r_d$, where r_d is the *dynamic resistance* of the transistor emitter-base junction. This slope can be calculated from equation (1.26) and the simplifications that follow that equation. Since this is a *pn* junction, $nV_T \approx 26$ mV (assuming a junction where $n \approx 1$). Taking the derivative of equation (1.26) at the Q-point, we find the dynamic resistance to equal

$$r_d = \frac{dv_D}{di_D} \approx \frac{nV_T}{I_o \exp(v_D/nV_T)}$$

But

$$I_o \exp\left(\frac{v_D}{nV_T}\right) = I_{EQ}$$

so

$$r_d \approx \frac{0.026 \text{ V}}{I_{EQ}} \approx \frac{26 \text{ mV}}{I_{CQ}}$$

where I_{EQ} is the emitter current at the Q-point (the rest condition with zero input signal). Remember that $I_{EQ} \approx I_{CQ}$.

Since $i_B = i_C/\beta$, the base-emitter junction is similar to that of a diode. Therefore, for the forward-biased junction,

$$i_B = \frac{I_o}{\beta}\exp\left(\frac{v_{BE}}{nV_T}\right)$$

In this text, we will use $n = 1$ and $nV_T = 26$ mV for silicon transistors.

A straight-line extension of the characteristic curve would intersect the v_{BE} axis at 0.7 V for silicon transistors, 0.2 V for germanium.

If we now hold i_B constant, the collector-emitter junction is defined by the curve of i_C versus v_{CE} shown in Figure 2.8(b). As can be seen from this typical curve, the collector current is almost independent of the voltage between the collector and the emitter, v_{CE}, throughout the "linear range" of operation. When i_B is near zero, i_C approaches zero in a nonlinear manner. This is known as the *cutoff region* of operation. For the section of the characteristic curves where v_{CE} is near zero, i_C is maximum. This region, known as the *saturation region,* is also not usable for amplification because of nonlinear operation.

Collector-emitter characteristic curves are parametric curves of i_C versus v_{CE} where i_B is a parameter. Figure 2.9 shows an example of such curves. Each transistor type has its own unique set of characteristic curves.

As an example of the use of the characteristic curves, we shall analyze the circuit of Figure 2.10. This circuit contains resistors, dc sources, and the transistor. We shall conduct a graphical analysis using techniques developed in Chapter 1. That is, we will be solving a set of two simultaneous equations in i_C and v_{CE}. The first of these simultaneous equations is given by the set of curves in Figure 2.9. The second equation is found from standard circuit analysis, as follows.

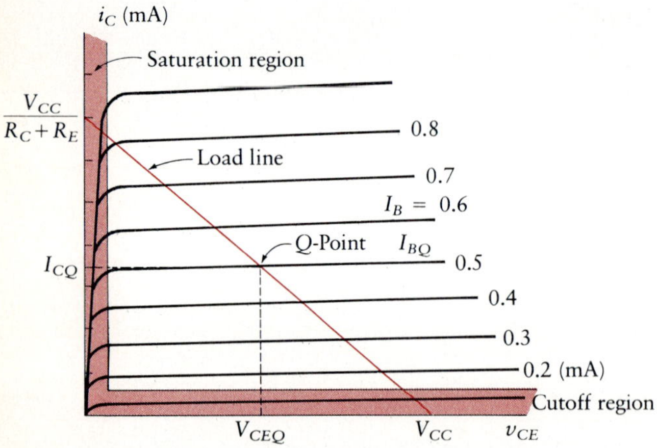

Figure 2.9 Transistor characteristic curves.

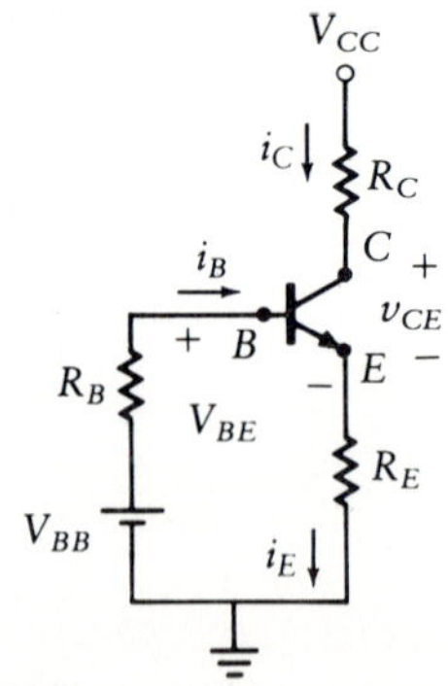

Figure 2.10 Simple transistor circuit.

Applying Kirchhoff's voltage law around the collector-to-emitter loop, we obtain

$$V_{CC} = i_C R_C + v_{CE} + i_E R_E \tag{2.8}$$

Since i_E is approximately equal to i_C, equation (2.8) can be simplified as in equation (2.9).

$$V_{CC} = i_C(R_C + R_E) + v_{CE} \tag{2.9}$$

Equation (2.9) defines a straight-line relationship between i_C and v_{CE}. That is,

$$i_C = \frac{V_{CC} - v_{CE}}{R_C + R_E} = -\frac{1}{R_C + R_E} v_{CE} + \frac{V_{CC}}{R_C + R_E} \tag{2.10}$$

This straight line is known as the *load line,* and we wish to plot it on the same set of axes with the transistor operating curves. One way to plot this straight line is to solve for the two axis intercepts. When $i_C = 0$, $v_{CE} = V_{CC}$, and when $v_{CE} = 0$ then

$$i_C = \frac{V_{CC}}{R_C + R_E}$$

The dc load line is plotted on the characteristic curves of Figure 2.9. When we discuss design, we will see how properly to select the circuit parameters. For now, we assume that the operating point (the Q-point or quiescent point) can be selected anywhere on this load line. The actual location of the Q-point depends on the quiescent value of base current, I_{BQ}, which is determined by V_{BB} and R_B. The Q-point will have coordinates of V_{CEQ} and I_{CQ}, the quiescent values of v_{CE} and i_C, respectively. The quiescent point is the zero-signal value of v_{CE} and i_C.

If the flat parts of the characteristic curves are extended to the left in Figure 2.9, they will all cross the v_{CE} axis at a point as shown in Figure 2.11. The value of v_{CE} at that point is defined as V_A or is known as the *early voltage.* The magnitude of this voltage can exceed 100 V for a discrete transistor (the word

Figure 2.11
Early voltage.

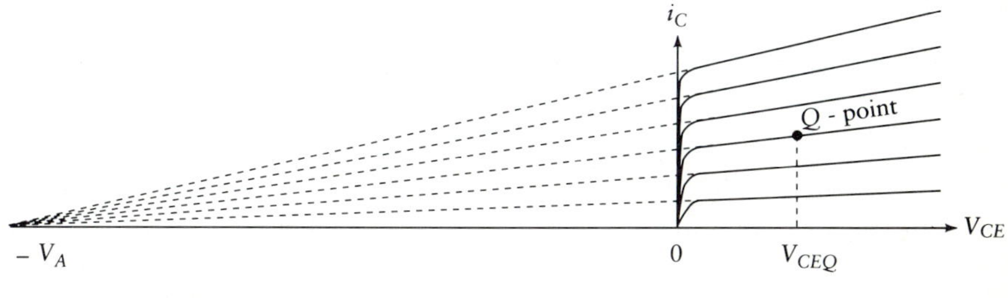

discrete distinguishes this from a transistor fabricated as part of an integrated circuit, to be discussed later in the book). Using this value of V_A, we can determine the output impedance of the transistor when operating at a specific collector current. We define r_o as the internal impedance of a transistor at a specific operating point.

$$r_o = \frac{V_A + V_{CEQ}}{I_{CQ}} \approx \frac{V_A}{I_{CQ}}$$

2.5 THE COMMON-EMITTER AMPLIFIER

The CE, or *common-emitter,* transistor amplifier is so called because the base and collector have a common connection separated only by resistors. Figure 2.12 shows the configuration of the transistor, where an *npn* transistor has been selected for illustration.

We first analyze the circuit of Figure 2.12 under dc conditions. The variable source, v_s, is set equal to zero. Kirchhoff's voltage law (KVL) around the base loop is written as follows:

$$I_{BQ}R_B + V_{BE} - V_{BB} = 0$$

The quiescent point is then given by

$$I_{BQ} = \frac{V_{BB} - V_{BE}}{R_B}$$

We now write KVL around the collector-emitter loop in order to determine the load line:

$$V_{CC} = R_C I_C + V_{CE}$$

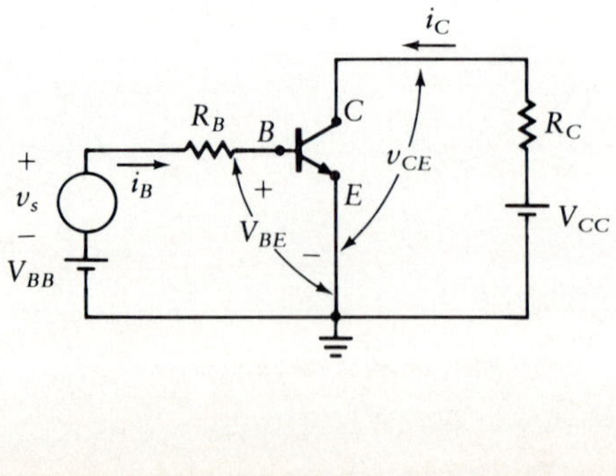

Figure 2.12
Common-emitter *npn* transistor amplifier.

Then

$$I_C = \frac{V_{CC} - V_{CE}}{R_C} \tag{2.11}$$

Equation (2.11) defines the load line, which is drawn on the characteristic curves in Figure 2.13(a). A *Q*-point, or operating point, that lies on the load line and is defined as the zero-signal point can now be selected. Although we do not wish to assign specific values to V_{BB} and R_B at this time, we have selected $I_{BQ} = 200\ \mu A$ for purposes of illustration in Figure 2.13(a). Now if we assume an ac input of

$$v_s = V \sin \omega t$$

Figure 2.13
Characteristic curves for CE amplifier.

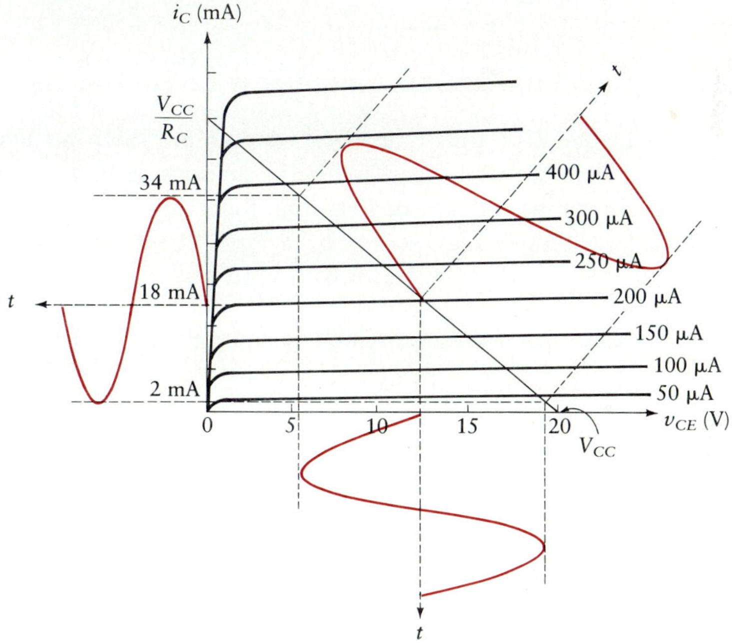

(a) Transistor characteristics curve

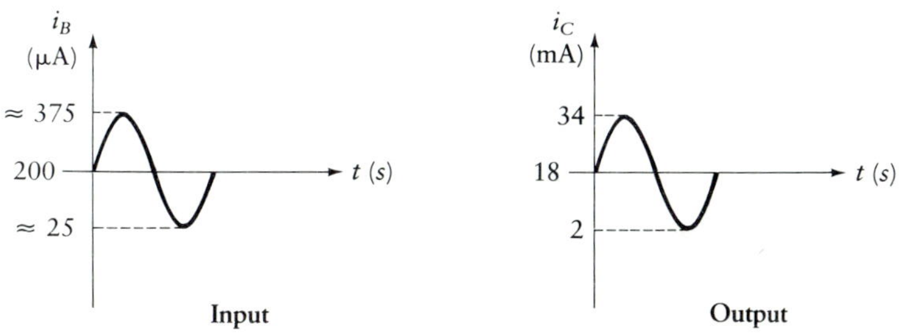

(b) Input-output current curves

the output wave can be found graphically. By moving the operating point up and down along the load line as i_B varies, we can plot i_C, i_B, and v_{CE} as shown in Figure 2.13(a). Again, for purposes of illustration, we are assuming that the base current varies sinusoidally between a maximum of about 400 μA and a minimum of zero. The base current and resulting collector current are shown in Figure 2.13(b).

Let us determine the change in collector current for a given change in base current. This ratio is the *current gain,* which is defined as

$$|A_i| = \frac{\Delta i_C}{\Delta i_B} = \frac{32 \text{ mA}}{350\ \mu\text{A}} = 91.4$$

Δi_C and Δi_B are read from Figure 2.13(b) as the swings in these parameter values. It is this current gain that makes the device important for many engineering applications.

2.5.1 Common-Emitter Amplifier with Emitter Resistor

Figure 2.14 illustrates a CE circuit to which an emitter resistor has been added. We write the Kirchhoff equations around the emitter-collector loop to determine the new dc load line. Referring to Figure 2.14(a), we find

$$V_{CC} = R_C I_C + V_{CE} + R_E I_E$$

Now since I_C is approximately equal to I_E, we have

$$I_C = \frac{V_{CC} - V_{CE}}{R_E + R_C} \tag{2.12}$$

Figure 2.14
CE amplifier with emitter resistor.

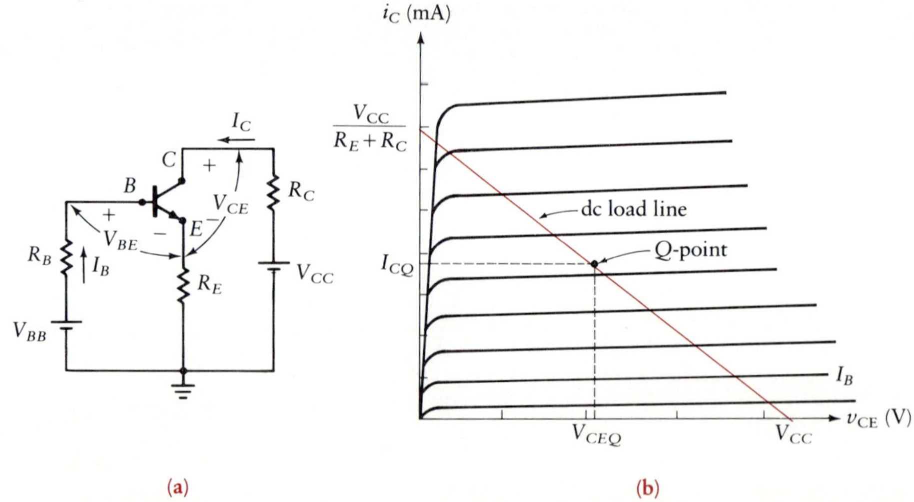

To plot the load line, we find the two intercepts as follows. With $I_C = 0$,

$$V_{CE} = V_{CC}$$

and when $V_{CE} = 0$,

$$I_C = \frac{V_{CC}}{R_E + R_C}$$

The resulting load line is as drawn in Figure 2.14(b).

When using a transistor in the common-emitter mode, we avoid the nonlinear region of the characteristic curves occurring at low values of i_C (cutoff) and at low values of v_{CE} (saturation). In designing a transistor amplifier, we usually desire maximum undistorted output swing. If the ac input signal is symmetrical about zero, we can achieve maximum swing by placing the Q-point in the center of the load line. Thus,

$$V_{CEQ} = V_{CC}/2$$

This equation establishes V_{CEQ} and I_{CQ}. In addition, since the base-emitter junction acts as a diode,

$$V_{BE} = V_\gamma$$

Writing KVL equations around the base loop, we obtain

$$V_{BB} = R_B i_B + v_{BE} + i_C R_E$$

Note that we are using lowercase letters and uppercase subscripts for the variables. This indicates total (dc + ac) values. (This would be an appropriate time to review the notation conventions presented on the inside of the front cover.) Now since

$$i_C = \beta i_B$$

we have

$$V_{BB} = \frac{R_B i_C}{\beta} + V_{BE} + i_C R_E$$

and at the quiescent point,

$$I_{CQ} = \frac{V_{BB} - V_{BE}}{R_B/\beta + R_E} \tag{2.13}$$

The voltage V_{BE} is considered to be a constant at room temperature (25°C) and has a value of about 0.7 V for silicon transistors.

To avoid using two separate dc sources (V_{BB} and V_{CC}), a voltage-divider network can be used to provide the dc source for the base circuit, as shown in Figure 2.15(a). The values for R_1 and R_2 determine the location of the Q-point. If the resistor and source voltage combination connected to the base in Figure 2.15(a) is replaced by the Thevenin equivalent of Figure 2.15(b), the new circuit is identical to that of Figure 2.14(a). Therefore it is necessary only to choose R_1 and R_2 properly.

The Thevenin equivalent voltage and resistance from base to ground are

$$V_{TH} = V_{BB} = \frac{R_1 V_{CC}}{R_1 + R_2} \tag{2.14}$$

$$R_{TH} = R_1 \parallel R_2 = R_B = \frac{R_1 R_2}{R_1 + R_2} \tag{2.15}$$

If we have already determined the required values of V_{BB} and R_B in order to place the Q-point properly, we can solve for R_1 and R_2 by substituting equation (2.14) into equation (2.15).

$$R_1 = \frac{R_B V_{CC}}{V_{CC} - V_{BB}} = \frac{R_B}{1 - V_{BB}/V_{CC}} \tag{2.16}$$

$$R_2 = \frac{V_{CC} R_B}{V_{BB}} \tag{2.17}$$

The values of R_1 and R_2 must be determined to establish the required bias point.

Figure 2.15 Transistor circuit using one source.

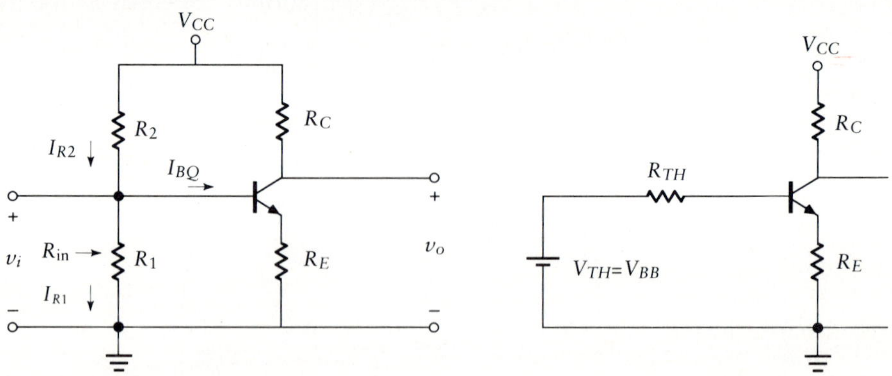

(a) (b)

As we get closer to designing amplifiers, we need to think about conditions for choosing circuit parameters. The purpose of R_B and V_{BB} is to set the operating point, I_{BQ}, to the proper location. That is, we specify I_{BQ} so that the input variations provide the desired output. This represents one constraint on two variables. In order to select R_B and V_{BB}, we will need an additional constraint. This is provided by what is known as *bias stability*. Stability is related to the sensitivity of certain operating results to changes in input parameters. We defer detailed discussion of bias stability to Chapter 5. For now, we ask you to accept certain results.

For the circuit under consideration, we wish to have approximately 10% of the input current going into the base and about 90% shunted through the equivalent external resistor, R_B. This provides bias stability. Hence the current in R_B should be about 10 times the base current. We express this constraint in terms of β, the large-signal amplification factor. From equation (2.6), we see that

$$i_C \approx i_E \approx \beta i_B$$

The time-varying current in R_B is given by v_b/R_B, while the current into the base is approximately equal to

$$i_b \approx \frac{i_E}{\beta} = \frac{v_b}{R_E \beta}$$

We neglect V_{BE} since we are dealing only with time-varying components. The current restriction then becomes

$$R_B < 0.1\beta R_E \tag{2.18}$$

or

$$\frac{R_B}{\beta} < 0.1R_E$$

This prevents variation in β from significantly affecting the dc operating point of the stage. We will have more to say about this in Chapter 5.

We can now use equation (2.13) to solve for the quiescent collector current. Letting R_B equal $0.1\beta R_E$, we have

$$I_{CQ} = \frac{V_{BB} - V_{BE}}{0.1\beta R_E/\beta + R_E} = \frac{V_{BB} - V_{BE}}{1.1R_E} \tag{2.19}$$

Equation (2.19) is used in the design process.

2.6 POWER CONSIDERATIONS

Power rating is an important consideration in selecting resistors. The resistors must be capable of withstanding the maximum anticipated power without overheating. Power considerations also affect transistor selection. Designers normally select components that have the lowest power-handling capability suitable for the design. Frequently, *derating* is used to improve the reliability of a device. This is similar to using safety factors in the design of mechanical systems where the system is designed to withstand values which exceed the maximum.

2.6.1 Derivation of Power Equations

Average power is calculated as follows:

$$\text{For dc:}\quad P = VI = I^2R = \frac{V^2}{R}\ \text{W}$$

$$\text{For ac:}\quad P = \frac{1}{T}\int_0^T v(t)i(t)\,dt\ \text{W}$$

In the ac equation, T is one period of the waveform. If the signal is not periodic, we let T approach infinity. The power supplied by the power source to the CE amplifier is dissipated either in the bias circuitry or in the transistor. The bias circuitry consists of R_1 and R_2, and the power in these resistors is found as follows (refer to Figure 2.15):

$$P_{(\text{bias})} = I_{R2}^2 R_2 + I_{R1}^2 R_1$$

where I_{R1} and I_{R2} are the (downward) currents in the two resistors. Kirchhoff's current law (KCL) yields a relationship between these two currents and the base quiescent current.

$$I_{R1} = I_{R2} - I_{BQ}$$

KVL yields the base loop equation,

$$I_{R2}R_2 + I_{R1}R_1 = V_{CC}$$

These two equations can be solved for the currents to yield

$$I_{R1} = \frac{V_{CC} - R_2 I_{BQ}}{R_2 + R_1}$$

$$I_{R2} = \frac{V_{CC} + R_1 I_{BQ}}{R_2 + R_1}$$

We can then combine these equations to solve for the power dissipated in the bias circuitry. In most practical circuits, the power due to I_{BQ} is negligible relative to the power dissipated in the transistor and in R_1 and R_2. We will therefore assume that the power supplied by the source is approximately equal to the power dissipated in the transistor and in R_1 and R_2. This quantity is given by

$$P = \frac{1}{T}\int_0^T V_{CC}[I_{CQ} + i_c(t)]\,dt + P_{\text{(bias circuit)}} = V_{CC}I_{CQ} + \frac{V_{CC}^2}{R_1 + R_2} \tag{2.20}$$

We have assumed that the average value of $i_c(t)$ is zero. For example, if the input ac signal is a sinusoid,

$$i_c(t) = A \sin \omega t$$

Then

$$\int_0^T A \sin \omega t\, dt = 0$$

where $T = 2\pi/\omega$.

The average power dissipated by the transistor is

$$P_{\text{(transistor)}} = \frac{1}{T}\int_0^T v_{CE}(t) i_C(t)\, dt$$

For zero signal input, this becomes

$$P_{\text{(transistor)}} = V_{CEQ} I_{CQ}$$

For an input signal with maximum possible swing,

$$\begin{aligned}
v_{CE}(t) &= V_{CEQ} - V_{CEQ} \sin \omega t = V_{CEQ}(1 - \sin \omega t) \\
i_C(t) &= I_{CQ} + I_{CQ} \sin \omega t = I_{CQ}(1 + \sin \omega t) \\
P_{\text{(transistor)}} &= \frac{1}{T}\int_0^T V_{CEQ} I_{CQ}(1 - \sin \omega t)(1 + \sin \omega t)\, dt \\
&= \frac{V_{CEQ} I_{CQ}}{T}\int_0^T (1 - \sin^2 \omega t)\, dt \\
&= \frac{V_{CEQ} I_{CQ}}{T}\int_0^T \cos^2 \omega t\, dt \\
&= \frac{V_{CEQ} I_{CQ}}{2T}\int_0^T (1 + \cos 2\omega t)\, dt \\
&= \frac{V_{CEQ} I_{CQ}}{2}
\end{aligned} \tag{2.21}$$

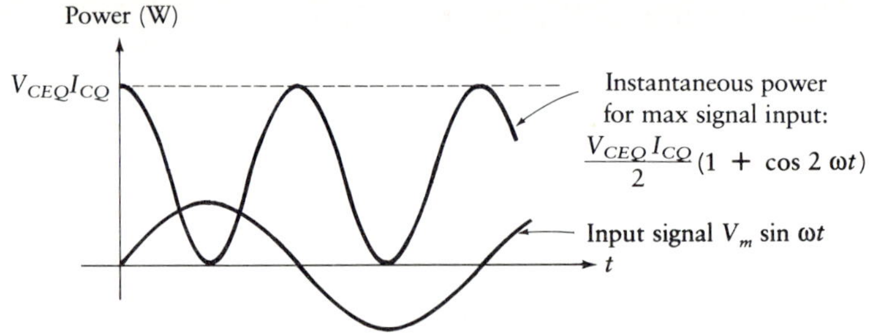

Figure 2.16 Instantaneous transistor power.

From the above derivation, we see that the transistor dissipates its maximum power when no ac signal input is applied. This is shown in Figure 2.16, where we note that the frequency of the instantaneous power sinusoid is 2ω. Depending on the amplitude of the input signal, the transistor will dissipate an average power between $V_{CEQ}I_{CQ}$ and one-half of this value. Therefore, the transistor is selected for zero input signal so that it will handle the maximum power as follows:

$$P_{\text{(transistor-max average)}} = V_{CEQ}I_{CQ} \tag{2.22}$$

We will need a measure of efficiency to determine how much of the power delivered by the source appears as signal power at the output. We define *conversion efficiency* as

$$\eta = \frac{P_o(\text{ac})}{P_{\text{in}}(\text{dc})} \times 100$$

2.7 BYPASS AND COUPLING CAPACITORS

Capacitors are approximated as short circuits for ac signals and open circuits for dc signals. *Bypass capacitors* are therefore used to effectively eliminate (short out) resistors during ac operation. *Coupling capacitors* are used to block the direct current yet allow the ac signal to pass.

2.7.1 Bypass Capacitors

Capacitors can be used to bypass the emitter resistor, thus increasing the voltage gain of an amplifier. To accomplish this, a capacitor is selected so that its impedance at operating frequencies is much less than the resistance of the emitter resistor. Since impedance increases with decreasing frequency, the capacitor impedance should be much less than the value of the equivalent resistance across the capacitance at the lowest operating frequency of the amplifier.

2.7.2 Coupling Capacitors

Each pair of stages of a multistage amplifier can be coupled together with a capacitor. The input impedance of the following stage is the load of the previous stage. A coupling capacitor is necessary to prevent interactions of dc currents between adjacent stages. A single-stage transistor amplifier has the form shown in Figure 2.17, where R_L is the equivalent input resistance of the next stage. The capacitors are open circuits at dc, and they are short circuits over amplifier operating frequencies.

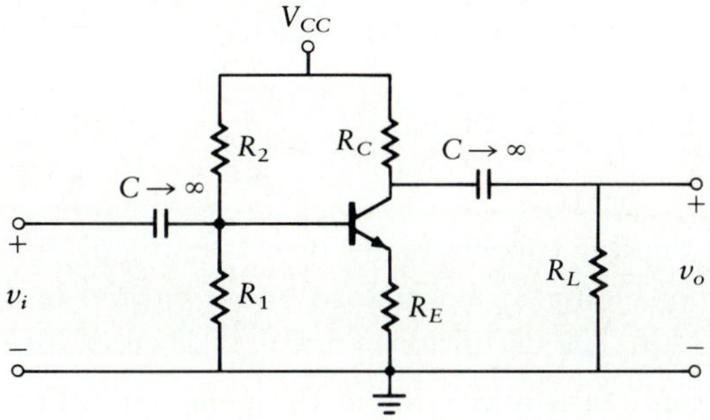

Figure 2.17 Common-emitter ac amplifier stage.

2.8 ac LOAD LINE FOR CE CONFIGURATION

Before beginning discussion of load lines for the CE amplifier, we note that the bias methods for CE and CB configurations are identical. Thus, although we are presenting the theory for the CE, we use the same concepts for both CE and CB.

The resistance in the emitter-collector circuit for dc operation is $R_C + R_E$. Note that we define this quantity as R_{dc}. When a load is coupled to the transistor through a capacitor, the ac resistance is different. Under ac conditions, the resistance in the emitter-collector circuit is now

$$R_{ac} = (R_L \parallel R_C) + R_E$$

If the emitter resistor is bypassed with a capacitor, the ac resistance is reduced to

$$R_{ac} = R_L \parallel R_C$$

The ac load line has a slope of $-1/R_{ac}$. Since a zero value for the ac input places the operation at the Q-point, the ac load line intersects the dc load line at the Q-point. If the input signal is small, the Q-point should normally be located to minimize the quiescent collector current (i.e., to lower total power dissipation).

In designing such circuits, we raise I_{CQ} above the zero point just enough to allow linear reproduction of the input signal (i.e., no distortion by entering the cutoff region). Under this condition, the transistor dissipates less power than if the Q-point is placed in the middle of the ac load line. We investigate this design procedure in Section 2.9.2.

2.8.1 ac Load Line Through Any Q-Point

We determined the dc load line from equation (2.12). This is then given by the equation

$$i_C = \frac{-v_{CE}}{R_E + R_C} + \frac{V_{CC}}{R_E + R_C}$$

Since the coupling capacitors are open circuits to dc, this load line applies to the circuit of Figure 2.17. The load line is plotted on the characteristic curves of Figure 2.18. The definitions of ac and dc resistance are repeated below:

R_{dc} = total resistance around the collector-emitter loop under dc conditions (capacitors considered open circuits)

R_{ac} = total resistance around the collector-emitter loop under ac conditions (capacitors considered short circuits)

For the circuit in Figure 2.17, we have

$$R_{dc} = R_E + R_C \tag{2.23}$$

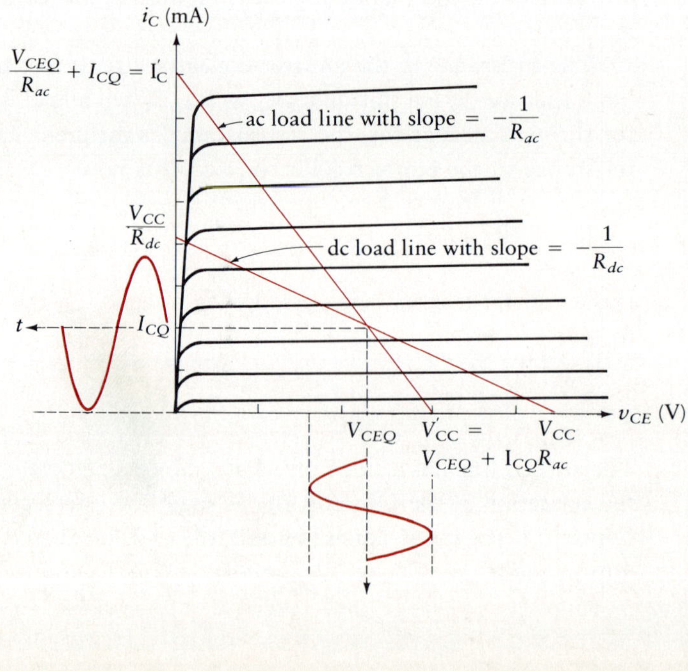

Figure 2.18
Characteristic curves.

It is important to realize that the dc source voltage is provided by an ideal voltage source of magnitude V_{CC}. Since an ideal voltage source has zero internal impedance, for ac signals the point labeled V_{CC} is ground. As a result, R_L and R_C are in parallel for ac, and

$$R_{ac} = R_L \parallel R_C + R_E \tag{2.24}$$

The equation for the dc load line is then

$$i_C = \frac{V_{CC}}{R_{dc}} - \frac{v_{CE}}{R_{dc}} = \frac{1}{R_{dc}}(V_{CC} - v_{CE})$$

The Q-point, which is specified for zero signal value, is on both the ac and dc load lines. The ac load line goes through the Q-point and has a slope of $-1/R_{ac}$. This slope is greater in magnitude than that of the dc load line, since

$$(R_L \parallel R_C + R_E) < (R_E + R_C)$$

The ac load line is plotted in Figure 2.18. The intersections with the i_C axis and the v_{CE} axis can be obtained from the equation for a straight line through a given point (x_1, y_1) with known slope (m) as follows:

$$(y - y_1) = m(x - x_1)$$

$$(i_C - I_{CQ}) = \frac{-(v_{CE} - V_{CEQ})}{R_{ac}}$$

$$i_C = \frac{-v_{CE}}{R_{ac}} + \frac{V_{CEQ}}{R_{ac}} + I_{CQ}$$

The intersection of the ac load line with the i_C axis is then

$$I_C' = \frac{V_{CEQ}}{R_{ac}} + I_{CQ}$$

The intersection of the ac load line with the v_{CE} axis is

$$V_{CC}' = V_{CEQ} + I_{CQ}R_{ac} \tag{2.25}$$

2.8.2 Choice of ac Load Line for Maximum Output Swing

We sometimes wish to design for the maximum output voltage swing from the amplifier. In that case, the Q-point must be placed in the center of the ac load line. Figure 2.19 shows the load lines for the circuit in Figure 2.17. The

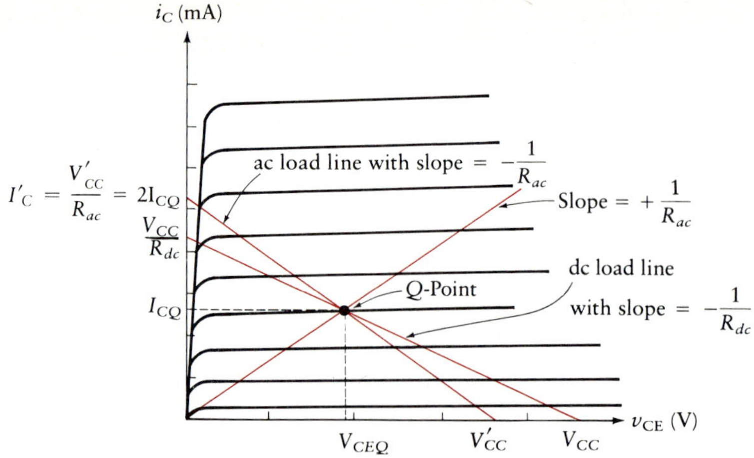

Figure 2.19
Load lines for maximum ac swing.

maximum collector current, I'_C, must be twice the quiescent collector current, yielding

$$I'_C = \frac{V_{CEQ}}{R_{ac}} + I_{CQ} = 2I_{CQ}$$

which reduces to

$$I_{CQ} = \frac{V_{CEQ}}{R_{ac}} \tag{2.26}$$

Equation (2.26) represents one equation in two unknowns for specifying the Q-point location for maximum output swing. The second equation is derived by using the dc load line equation. The dc and ac load line equations are repeated as equations (2.27) and (2.28):

$$i_C = \frac{1}{R_{dc}}(V_{CC} + v_{CE}) \tag{2.27}$$

$$i_C = -\frac{v_{CE}}{R_{ac}} + \frac{V_{CEQ}}{R_{ac}} + I_{CQ} \tag{2.28}$$

Combining equations (2.26), (2.27), and (2.28) yields

$$V_{CEQ} = \frac{V_{CC}}{1 + R_{dc}/R_{ac}} \tag{2.29}$$

Finally, the Q-point is specified by

$$I_{CQ} = \frac{V_{CC}}{R_{ac} + R_{dc}} \tag{2.30}$$

V'_{CC} is the intercept of the ac load line with the v_{CE} axis, as shown in Figure 2.19. This is given by

$$V'_{CC} = V_{CEQ} + I_{CQ}R_{ac}$$

but from equations (2.29) and (2.30) this becomes

$$V'_{CC} = \frac{2V_{CC}}{1 + R_{dc}/R_{ac}} = 2V_{CEQ}$$

as expected.

2.9 ANALYSIS AND DESIGN

We now have the necessary tools to permit analysis and design of amplifier circuits. It is necessary only to pull together the results derived in previous sections.

In *analyzing* an ac amplifier, the circuit components are specified. We begin the solution by examining the dc bias. The Thevenin equivalent circuit for the base-emitter loop is first derived. This provides the values needed to solve the bias equation for I_{CQ}. The dc and ac load lines are constructed next. If I_{CQ} is in the transistor operating region (i.e., not in the cutoff or saturation region), the maximum undistorted output ac voltage swing of the amplifier is determined by examining the ac load line.

In *designing* an amplifier, the situation is reversed, since the designer must select the circuit components and has the option of selecting I_{CQ}. If a maximum output voltage swing is desired, I_{CQ} is placed in the center of the ac load line. On the other hand, if the input signal is small, I_{CQ} can be made just large enough so that the ac signal output will not be clipped during the input signal maximum. In designing, the engineer starts calculations at the collector-emitter side of the amplifier rather than at the base-emitter side. After I_{CQ} has been determined, the bias equation is used to determine the values of R_1 and R_2 required for the transistor to operate at the selected I_{CQ}.

2.9.1 Analysis Procedure

In analysis, the values of R_1 and R_2 are given. The analysis is therefore conducted starting from the base side of the amplifier. V_{CC}, V_{BE}, R_1, R_2, R_E, R_C, R_L, and β are all given.

We now present an organized procedure for analysis. The equations used have been derived earlier in this chapter, and we cite references so that the derivations can be consulted. We strongly recommend that you consult these derivations, since it is important to be aware of the various assumptions. Our purposes in presenting this analysis procedure are not confined to teaching you the art of amplifier analysis. It is more important for you to appreciate the methodology of reducing theory to a step-by-step procedure. In this manner, you will be able to deal with new situations as they arise. *This last point is critical!* The step-by-step procedure is far less important than learning to develop such a procedure from the theory!

Step 1 Use R_1 and R_2 to determine V_{BB} and R_B from the following equations:

$$V_{BB} = \frac{R_1 V_{CC}}{R_1 + R_2}$$

$$R_B = R_1 \| R_2 = \frac{R_1 R_2}{R_1 + R_2}$$

(*Reference* equations (2.14) and (2.15))

Step 2 Use the bias equation to calculate I_{CQ}.

$$I_{CQ} = \frac{V_{BB} - V_{BE}}{R_B/\beta + R_E}$$

(*Reference* equation (2.13))

Step 3 Use the dc load line equation to determine V_{CEQ}.

$$V_{CEQ} = V_{CC} - (R_E + R_C)I_{CQ} = V_{CC} - R_{dc}I_{CQ}$$

(*Reference* equation (2.27))

Step 4 Construct the dc load line on the characteristic curves. Since we know that the ac load line intersects the dc load line at the Q-point, the ac load line is constructed from the equation

$$V'_{CC} = V_{CEQ} + I_{CQ}R_{ac}$$

where R_{ac} is the ac equivalent resistance in the collector-emitter loop.

(*Reference* equation (2.25))

Step 5 Determine the maximum possible symmetrical output voltage swing using the load line construction on the characteristic curves. If the Q-point is on the upper half of the ac load line, I_{CQ} is subtracted from the maximum value of i_C (the point where the ac load line intersects the i_C axis). This will provide the maximum-amplitude ac output current of the transistor. Alternatively, if the Q-point is on the lower half of the ac load line, I_{CQ} is the maximum-amplitude ac output current of the transistor. Then the maximum possible symmetrical output voltage swing is given by

$$2I_C(\text{maximum amplitude}) \times (R_C \parallel R_L)$$

Example 2.1 (Analysis)

Determine the Q-point for the circuit given in Figure 2.20 if $R_1 = 1.5\ \text{k}\Omega$ and $R_2 = 6\ \text{k}\Omega$. A 2N3903 transistor (see Appendix D) is used with $\beta = 180$, $R_E = 100\ \Omega$, and $R_C = R_L = 1\ \text{k}\Omega$. Also determine the $P_o(\text{ac})$ and the dc power delivered to the circuit.

SOLUTION Using the step-by-step procedure of Section 2.9.1, we obtain

$$V_{BB} = \frac{R_1 V_{CC}}{R_1 + R_2} = \frac{1500 \times 5}{1500 + 6000} = 1\ \text{V}$$

$$R_B = \frac{R_1 R_2}{R_1 + R_2} = 1.2\ \text{k}\Omega$$

We determine whether the amplifier maintains bias stability with changes in β by checking $R_B < 0.1\beta R_E = 0.1(180)(100) = 1800\ \Omega$. Since the inequality holds, bias stability is maintained. We find the Q-point as follows:

$$I_{CQ} = \frac{V_{BB} - V_{BE}}{R_B/\beta + R_E} = \frac{1 - 0.7}{1200/180 + 100} = 2.81\ \text{mA}$$

We find $R_{ac} = R_C \parallel R_L = 500\ \Omega$ and $R_{dc} = R_C + R_E = 1.1\ \text{k}\Omega$. V_{CEQ} is found as in Step 3.

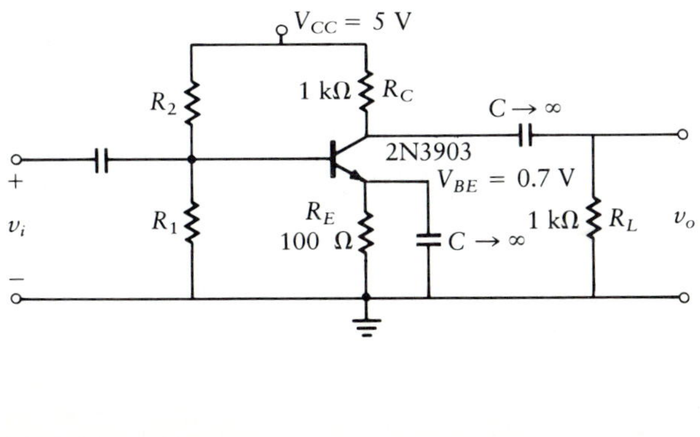

Figure 2.20 Circuit for Example 2.1.

$$V_{CEQ} = V_{CC} - I_{CQ}R_{dc} = 5 - (2.81 \times 10^{-3})(1.1 \times 10^3) = 1.91\text{ V}$$

Then

$$V'_{CC} = V_{CEQ} + I_{CQ}R_{ac} = 1.91 + (2.81 \times 10^{-3})(500) = 3.32\text{ V}$$

Since the Q-point is on the lower half of the ac load line, the maximum possible symmetrical output voltage swing is then

$$2I_{CQ}(R_C \parallel R_L) = 2(2.81 \times 10^{-3})(500) = 2.81\text{ V}$$

We now calculate the power as follows:

$$\begin{aligned}
P_o(\text{ac}) &= \frac{1}{2} i_L^2 R_L \\
&= \frac{1}{2}\left(2.81 \times 10^{-3} \times \frac{1000}{2000}\right)^2 \times 1000 \\
&= 0.987\text{ mW} \\
P_{V_{CC}}(\text{dc}) &= I_{CQ}V_{CC} + V_{CC}^2(R_1 + R_2) \\
&= 2.81 \times 10^{-3} \times 5 + \frac{25}{1500 + 6000} \\
&= 17.4\text{ mW}
\end{aligned}$$

The Q-point in this example is not in the middle of the load line, so output swing is not a maximum. However, if the input signal is small and maximum output is not required, a small I_{CQ} can be used to reduce the power dissipated in the circuit.

2.9.2 Design Procedure

In design problems, we work first with the collector-emitter side of the transistor rather than with the base side. We wish to place the Q-point at a specified location on the ac load line. For this approach, R_C and R_E are assumed to be given, and we need to determine the values of R_1 and R_2. The more general design challenge is considered in Chapter 3.

As discussed earlier, it is not always desirable to design an amplifier for maximum possible output swing. If the input signal is small, the operating point may move only a relatively small distance on either side of the Q-point and never get near saturation or cutoff. In that case, designing an amplifier with the Q-point in the middle of the load line wastes power. The power dissipated during the rest condition is greater than necessary for undistorted operation. In this section, we present a design procedure for placement of the Q-point anywhere along the load line. Following this, we modify the procedure for placement of the Q-point in the center of the load line, thereby yielding maximum undistorted output.

Step 1 Find I_{CQ} in terms of ac and dc resistance and V_{CC}. Suppose we wish to design for a quiescent current,

$$I_{CQ} = \delta I'_C \tag{2.31}$$

where I'_C is the intersection of the ac load line with the i_C axis and δ is a number between 0 and 1. It is equal to 0.5 for the maximum symmetrical swing case. We found earlier that I'_C is given by (refer to Figure 2.19)

$$I'_C = \frac{V'_{CC}}{R_{ac}} = I_{CQ} + \frac{V_{CEQ}}{R_{ac}} \tag{2.32}$$

We can combine equations (2.31) and (2.32) to solve for V_{CEQ}.

$$V_{CEQ} = \frac{(1 - \delta)I_{CQ}R_{ac}}{\delta} \tag{2.33}$$

Since the Q-point must also lie on the dc load line, we have

$$V_{CEQ} = V_{CC} - I_{CQ}R_{dc} \tag{2.34}$$

Setting equation (2.33) equal to equation (2.34) yields

$$V_{CEQ} = \frac{(1 - \delta)I_{CQ}R_{ac}}{\delta} = V_{CC} - I_{CQ}R_{dc}$$

and solving for I_{CQ}, we obtain

$$I_{CQ} = \frac{V_{CC}}{(1 - \delta)R_{ac}/\delta + R_{dc}} \tag{2.35}$$

Step 2 Use the ac load line equation to determine V_{CEQ}.

$$V_{CEQ} = \frac{V'_{CC}}{1 - \delta}$$

where

$$V'_{CC} = 2I_{CQ}R_{ac}$$

(*Reference* equation (2.25))

Step 3 If no other restrictions exist, select R_B for bias stability.

$$R_B = 0.1\beta R_E$$

(*Reference* equation (2.18))

Step 4 Use the bias equation to determine V_{BB}.

$$V_{BB} = V_{BE} + I_{CQ}(1.1R_E)$$

(*Reference* equation (2.19))

Step 5 Find R_1 and R_2 from R_B and V_{BB}.

$$R_1 = \frac{R_B}{1 - V_{BB}/V_{CC}}$$

$$R_2 = \frac{V_{CC}R_B}{V_{BB}}$$

(*Reference* equations (2.16) and (2.17))

Step 6 Determine V_o (maximum symmetrical output) as in Step 5 of the analysis procedure. Assuming that the Q-point is on the lower half of the ac load line (i.e., $\delta < 0.5$),

$$V_o = 2I_C(\text{maximum amplitude}) \times (R_C \parallel R_L)$$

If we wish to design for maximum output voltage swing, we must place the Q-point in the center of the ac load line. In that case, δ is 0.5 and equation (2.35) reduces to

$$I_{CQ} = \frac{V_{CC}}{R_{ac} + R_{dc}}$$

This expression replaces equation (2.35) in Step 1 of the design procedure. The remaining steps stay unchanged.

Example 2.2 **(Design)**

Select R_1 and R_2 for maximum output voltage swing in the circuit shown in Fig. 2.20.

SOLUTION Following the design steps of Section 2.9.2, we first determine I_{CQ} for the circuit.

$$I_{CQ} = \frac{V_{CC}}{R_{ac} + R_{dc}} = \frac{5}{500 + 1100} = 3.13 \text{ mA}$$

since

$$R_{ac} = R_C \parallel R_L = 500\ \Omega$$

and

$$R_{dc} = R_E + R_C = 1100\ \Omega$$

For maximum swing,

$$V'_{CC} = 2V_{CEQ}$$

V_{CEQ} is then given by

$$V_{CEQ} = (3.13 \text{ mA})(500\ \Omega) = 1.56 \text{ V}$$

The intersection of the ac load line on the v_{CE} axis is V'_{CC}. Since

$$V_{CEQ} = \frac{V'_{CC}}{2}$$

then

$$V'_{CC} = 3.13 \text{ V}$$

From the manufacturer's specification in Appendix D, the maximum β for the 2N3903 is 200. R_B is set equal to $0.1\beta R_E$, so

$$R_B = 0.1(200)(100) = 2 \text{ k}\Omega$$

$$V_{BB} = (3.13 \times 10^{-3})\left(\frac{2000}{200} + 100\right) + 0.7 = 1.044 \text{ V}$$

Figure 2.21
Load lines for Example 2.2.

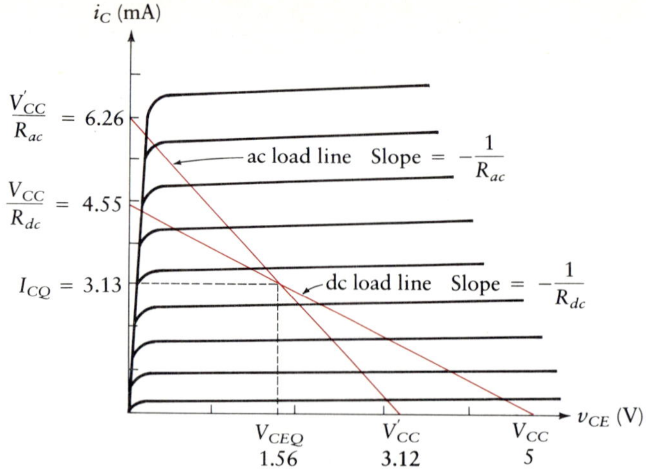

Since we know V_{BB} and R_B, we find R_1 and R_2:

$$R_1 = \frac{R_B}{1 - V_{BB}/V_{CC}} = \frac{2000}{1 - 1.044/5} = 2.53\ \text{k}\Omega$$

$$R_2 = \frac{R_B V_{CC}}{V_{BB}} = \frac{2000 \times 5}{1.044} = 9.6\ \text{k}\Omega$$

The maximum output voltage swing, ignoring the nonlinearities at saturation and cutoff, would then be

$$\begin{aligned}\text{Maximum output swing} &= 2I_{CQ}(R_C \parallel R_L)\\ &= 2(3.13\ \text{mA})\,(500\ \Omega) = 3.13\ \text{V}\end{aligned}$$

The load lines are shown on the characteristic curves of Figure 2.21.

We check the maximum power dissipated by the transistor to assure that it will not exceed the specifications. From equation (2.22), we have

$$\begin{aligned}P_{(\text{transistor})} &= V_{CEQ} I_{CQ}\\ &= (1.56\ \text{V})\,(3.13\ \text{mA}) = 4.87\ \text{mW}\end{aligned}$$

This is well within the 350-mW maximum given on the specification sheet. The maximum conversion efficiency is

$$\eta = \frac{P_o(\text{ac})}{P_{VCC}(\text{dc})} = \frac{(3.13 \times 10^{-3}/2)^2 \times 1000/2 \times 100}{5 \times 3.13 \times 10^{-3} + 5^2/12.13 \times 10^3} = 6.87\%$$

Drill Problems

D2.1 Find the peak-to-peak output voltage swing of the circuit of Figure 2.17 when $R_1 = 2\ \text{k}\Omega$, $R_2 = 15\ \text{k}\Omega$, $R_E = 200\ \Omega$, $R_C = 2\ \text{k}\Omega$, $R_L = 2\ \text{k}\Omega$, $\beta = 200$, $V_{BE} = 0.7$ V, and $V_{CC} = 15$ V.

Ans: 6.3 V peak-to-peak

D2.2 In Drill Problem D2.1, design the amplifier for maximum symmetrical swing. Find the values of R_1 and R_2.

Ans: $R_1 = 4.5\ \text{k}\Omega$; $R_2 = 36\ \text{k}\Omega$

D2.3 What is the maximum symmetrical voltage swing for the configuration of Drill Problem D2.2?

Ans: 8.8 V peak-to-peak

D2.4 What is the output power of the amplifier of Drill Problem D2.2? What is the power supplied to the amplifier?

Ans: 4.9 mW; 71.7 mW

2.9.3 Amplifier Power Sources

The power supplied to an amplifier can be placed above the reference point (commonly referred to as positive with respect to ground) or below the reference point. The only criterion is that the polarity of the voltage must be proper for the operation of the transistor. In other words, the voltage must be positive to negative from the collector to the emitter for *npn* transistors or vice versa for *pnp* transistors. This allows us to use both positive and negative power

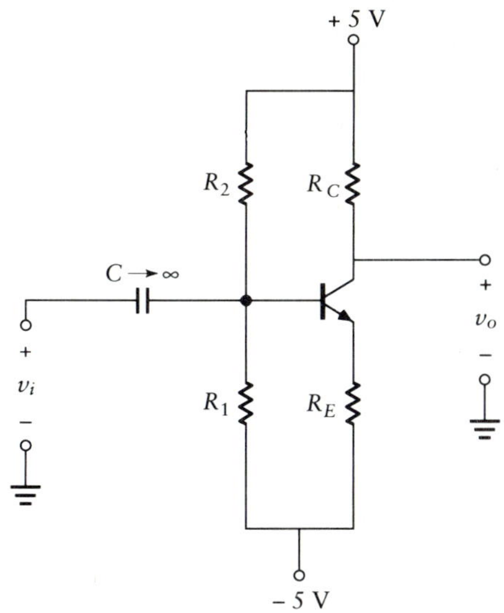

Figure 2.22 Amplifier with two power supplies.

sources. By doing so, we can set the dc value of the output to any desired value, including zero. We illustrate the amplifier in Figure 2.22. By setting

$$I_{CQ}R_C = 5 \text{ V}$$

the dc output taken from the collector is zero. The value of I_{CQ} is easily adjusted to whatever value is needed by changing the values of R_1 and R_2. We will be using this concept when working with direct coupled amplifiers in Chapter 3.

Drill Problems

D2.5 Design a dc bias-stable common-emitter amplifier using the circuit shown in Figure 2.22 to obtain a quiescent output voltage of zero. Let $\beta = 150$, $V_{BE} = 0.7$, $R_E = 100\ \Omega$, and $R_C = 1\ \text{k}\Omega$.

Ans: $R_1 = 1.71\ \text{k}\Omega$; $R_2 = 12\ \text{k}\Omega$

D2.6 Using the amplifier specified in Problem D2.1, design the amplifier to provide an output for minimum current drain when $\delta = 0.2$.

Ans: $R_1 = 4.34\ \text{k}\Omega$; $R_2 = 51.2\ \text{k}\Omega$

2.10 THE EMITTER-FOLLOWER (COMMON-COLLECTOR) AMPLIFIER

The *emitter-follower (EF)* or *common-collector (CC)* amplifier is illustrated in Figure 2.23. Its output is developed from the emitter to ground rather than from the collector to ground, as in the case of the common emitter. This type of amplifier configuration is used to obtain *current gain* and *power gain.*

The common emitter has a 180° phase shift between the base and collector voltages. That is, as the input signal increases in value, the output signal decreases. Alternatively, for an emitter follower, the output signal is in phase with the input signal. The amplifier has a voltage gain of slightly less than

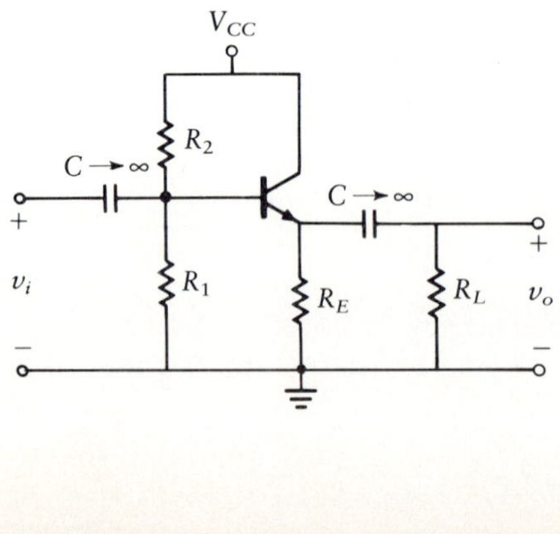

Figure 2.23
Emitter follower.

unity. On the other hand, the current gain is significantly greater than one. Note that the collector needs no resistor ($R_C = 0$), and no emitter bypass capacitor is required.

We analyze this circuit in the same manner as we did the common emitter. The only differences are the values we use for R_{ac} and R_{dc}. For the emitter follower of Figure 2.23,

$$R_{ac} = R_E \parallel R_L$$

and

$$R_{dc} = R_E$$

and the dc load line (refer ahead to Figure 2.24(b) for a typical sketch of the load line) is given by the equation

$$i_C = \frac{V_{CC} - v_{CE}}{R_{dc}}$$

For maximum swing, the *Q*-point is located at

$$I_{CQ} = \frac{V_{CC}}{R_{ac} + R_{dc}} = \frac{V_{CC}}{R_E \parallel R_L + R_E}$$

and

$$V_{CEQ} = I_{CQ} R_{ac} = I_{CQ}(R_E \parallel R_L)$$

Example 2.3 **(Design)**

In the circuit of Figure 2.24(a), find the values of R_1 and R_2 that yield maximum symmetrical output swing as shown in Figure 2.24(b). Assume that a 2N2222 transistor is used (see Appendix D for data sheets) with an average β of 100.

SOLUTION

$$R_{dc} = R_E = 600\ \Omega$$

$$R_{ac} = R_E \parallel R_L = 300\ \Omega$$

$$I_{CQ} = \frac{V_{CC}}{R_{ac} + R_{dc}} = \frac{12}{600 + 300} = 13.3\ \text{mA}$$

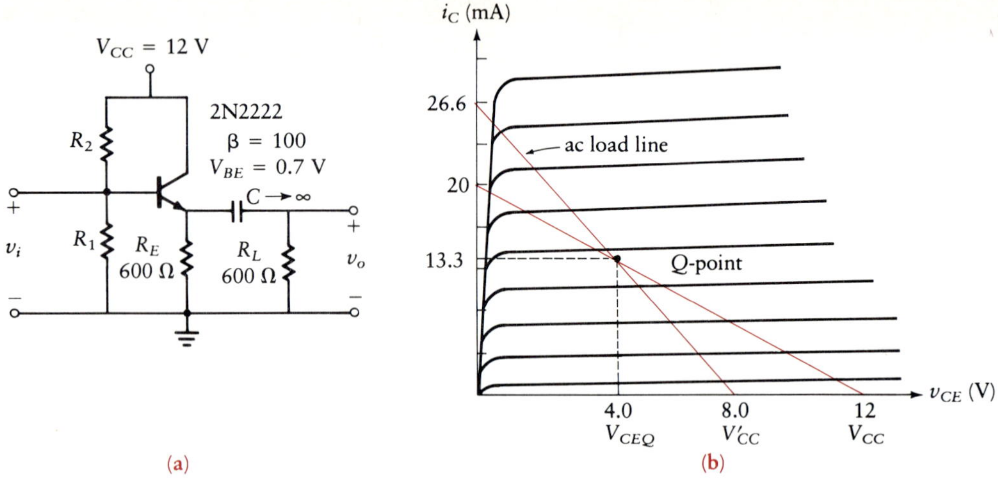

Figure 2.24 EF amplifier for Example 2.3.

Then

$$V_{CEQ} = I_{CQ} R_{ac} = (13.3 \times 10^{-3})(300) = 4 \text{ V}$$

In order to reduce the effects of variations in β, we choose

$$R_B = 0.1\beta R_E = 0.1(100 \times 600) = 6 \text{ k}\Omega$$

$$V_{BB} = V_{BE} + I_{CE}\left(\frac{R_B}{\beta} + R_E\right) = 9.48 \text{ V}$$

From equations (2.16) and (2.17), we obtain

$$R_1 = \frac{R_B}{1 - V_{BB}/V_{CC}} = \frac{6000}{1 - 9.48/12} = 28.5 \text{ k}\Omega$$

$$R_2 = \frac{R_B V_{CC}}{V_{BB}} = \frac{6000 \times 12}{9.48} = 7.6 \text{ k}\Omega$$

From equation (2.36), we find

$$\text{Maximum output swing} = 2I_{CQ}(R_E \parallel R_L)$$

$$= 2(0.0133)(300) = 8.0 \text{ V}$$

Example 2.4 (Analysis)

Find the Q-point and output voltage swing of the circuit of Figure 2.24(a) with R_1 = 10 kΩ and R_2 = 20 kΩ.

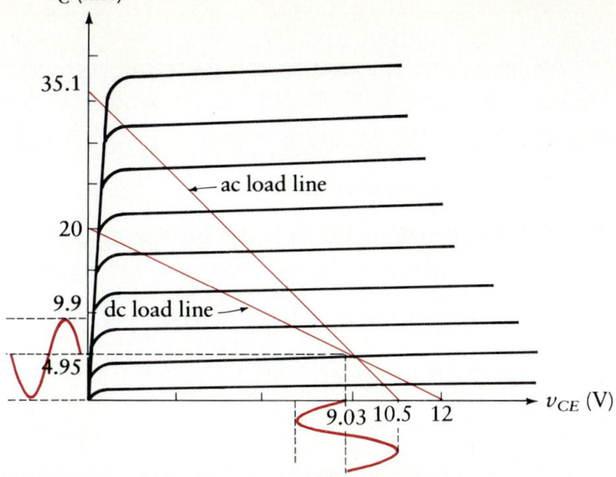

Figure 2.25 Load lines for Example 2.4.

SOLUTION Using equations (2.17) and (2.18), we have

$$R_B = R_1 \parallel R_2 = 6.67 \text{ k}\Omega$$

$$V_{BB} = \frac{R_1 V_{CC}}{R_1 + R_2} = \frac{12(10 \times 10^3)}{30 \times 10^3} = 4 \text{ V}$$

From equation (2.13), we have

$$I_{CQ} = \frac{V_{BB} - V_{BE}}{R_B/\beta + R_E} = \frac{4 - 0.7}{6670/100 + 600} = 4.95 \text{ mA}$$

The output swing is then given by

$$\begin{aligned}\text{Output swing} &= 2I_{CQ}(R_E \parallel R_L) \\ &= 2(4.95 \times 10^{-3})(300) = 2.97 \text{ V}\end{aligned}$$

This is less than the maximum possible output swing. Continuing the analysis,

$$V_{CEQ} = V_{CC} - I_{CQ}R_E = 9.03 \text{ V}$$

$$V'_{CC} = V_{CEQ} + I_{CQ}(R_E \parallel R_L) = 10.5 \text{ V}$$

$$i_{C\max} = \frac{10.5}{300} = 35.1 \text{ mA}$$

The load lines for this problem are shown in Figure 2.25.

Drill Problems

D2.7 What is the maximum symmetrical voltage swing for the amplifier of Figure 2.23 where $V_{CC} = 15$ V, $R_1 = 8$ kΩ, $R_2 = 2$ kΩ, $R_E = 1$ kΩ, $R_L = 1$ kΩ, $V_{BE} = 0.7$ V, and $\beta = 80$?

Ans: 7.8 V peak-to-peak

D2.8 In Drill Problem D2.7, redesign the amplifier for the maximum symmetrical voltage swing. What are the new values of R_1, R_2, and V_o?

Ans: 36.4 kΩ; 10.3 kΩ; 10 V peak-to-peak

D2.9 What is the conversion efficiency of the amplifier design in Drill Problem D2.8?

Ans: 8.4%

PROBLEMS

2.1 Find the values of R_1 and R_2 necessary to place the Q-point of the circuit of Figure P2.1(a) in the center of the dc load line. Assume that $V_{CC} = -25$ V, $R_C = 2$ kΩ, $R_E = 1$ kΩ, and β has the following values.
a. $\beta = 150$
b. $\beta = 100$
c. $\beta = 50$

2.2 Find the maximum peak-to-peak amplitude swing of i_C in the circuit of Figure P2.1(b). Assume that $V_{CC} = 24$ V, $R_C = 2$ kΩ, $R_E = 400$ Ω, and $\beta = 100$. Draw the dc load line when
a. $R_1 = 1$ kΩ; $R_2 = 7$ kΩ
b. $R_1 = 1$ kΩ; $R_2 = 35$ kΩ
c. $R_1 = 1$ kΩ; $R_2 = 3$ kΩ

Figure P2.1

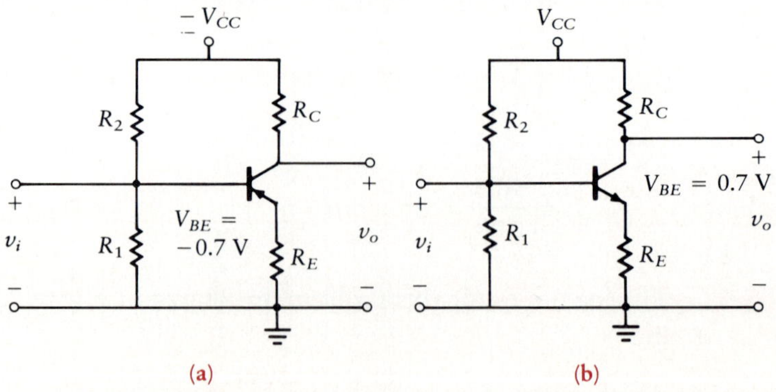

2.3 Find the following for the amplifier of Figure P2.2:
a. The values of R_1 and R_2 to achieve $I_{CQ} = 10$ mA.
b. The output symmetrical swing for the resistors of part a.
c. Draw the ac and dc load lines.
d. Sketch the waveforms for i_C and v_{CE}.

2.4 For the amplifier of Figure P2.2,
a. Find the values of R_1 and R_2 to achieve maximum symmetrical swing.
b. Determine the value of maximum symmetrical swing achieved in part a.
c. Draw the ac and dc load lines.
d. Sketch waveforms for i_C and v_{CE}.

2.5 Find the output peak-to-peak symmetrical swing of i_C in the circuit of Figure P2.3 when $R_1 = 5$ kΩ, $R_2 = 50$ kΩ, $V_{CC} = 12$ V, $V_{BE} = 0.7$ V, $R_E = 300$ Ω, $\beta = 200$, and $R_C = R_L = 5$ kΩ.

2.6 With the circuit of Problem 2.5, find the values of R_1 and R_2 that yield the maximum possible symmetrical peak-to-peak swing of i_C. Draw the load lines.

2.7 For the amplifier of Problem 2.5, calculate the following:
a. Power supplied by the battery.
b. Power dissipated by R_1, R_2, R_E, and R_C.
c. Power dissipated by the collector junction.

2.8 For the amplifier of Problem 2.6, calculate the following:
a. Power supplied by the battery.
b. Power dissipated by R_1, R_2, R_E, and R_C.
c. Power dissipated by the collector junction.

Compare your answers with those of Problem 2.7.

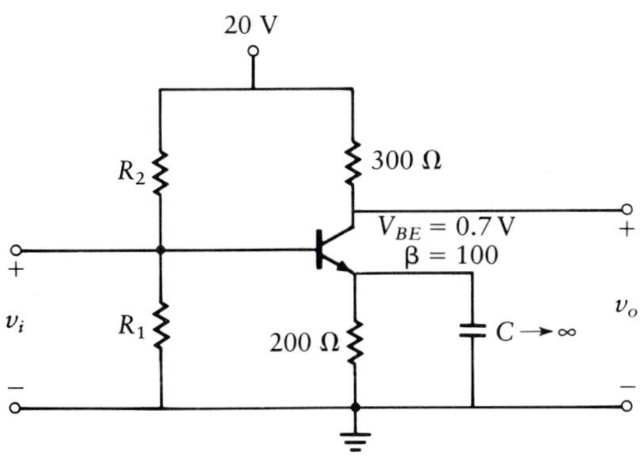

Figure P2.2

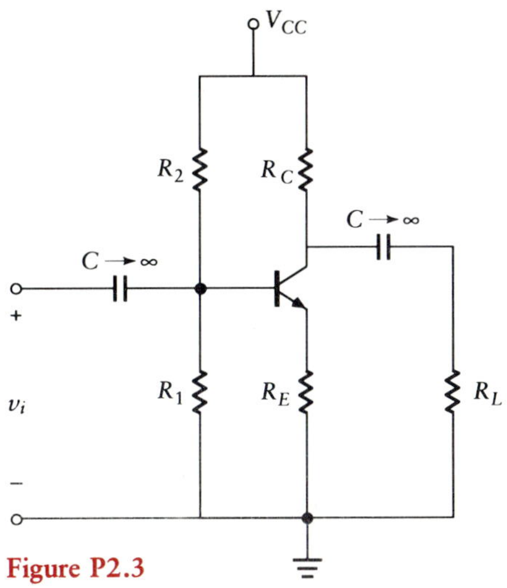

Figure P2.3

2.9 For the amplifier of Figure P2.3 where $R_1 = 3$ kΩ, $R_2 = 20$ kΩ, $R_C = R_L = 1$ kΩ, $R_E = 200$ Ω, $\beta = 100$, and $V_{CC} = 20$ V, find the location of the Q-point. The transistor is replaced with another of different β. Find the minimum required value of β so that I_{CQ} does not change by more than 10%.

2.10 For the amplifier of Figure P2.4.

a. Find the values of R_1 and R_2 for $I_{CQ} = 8$ mA.
b. Determine the symmetrical output voltage swing for the values of part a.
c. Draw the ac and dc load lines.
d. Determine the power dissipated by the transistor and the power dissipated by R_L.

2.11 For the amplifier shown in Figure P2.4,

a. Find the values of R_1 and R_2 for $I_{CQ} = 4$ mA.
b. Determine the symmetrical output voltage swing for the values of part a.
c. Draw the ac and dc load lines.
d. Determine the power dissipated by the transistor and the power dissipated by R_L.

2.12 For the amplifier of Figure P2.4,

a. Find the values of R_1 and R_2 needed to achieve maximum symmetrical swing.
b. Determine the symmetrical output voltage swing for the values of part a.
c. Draw the ac and dc load lines.
d. Determine the power dissipated by the transistor and the power dissipated by R_L.

2.13 Determine the value of R_C for maximum symmetrical output swing for the circuit of Figure P2.5. Assume that a *pnp* transistor is used. Draw the

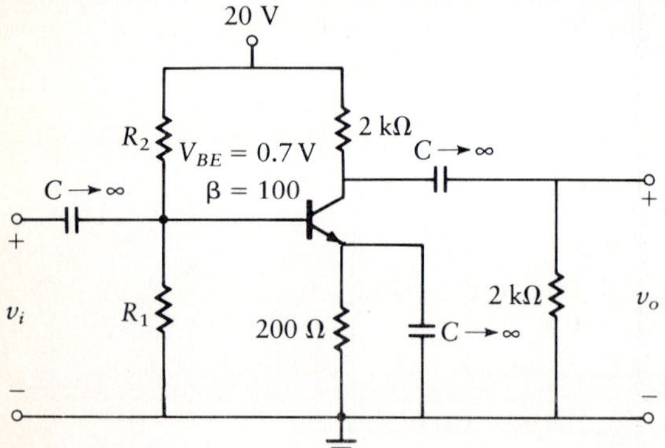

Figure P2.4

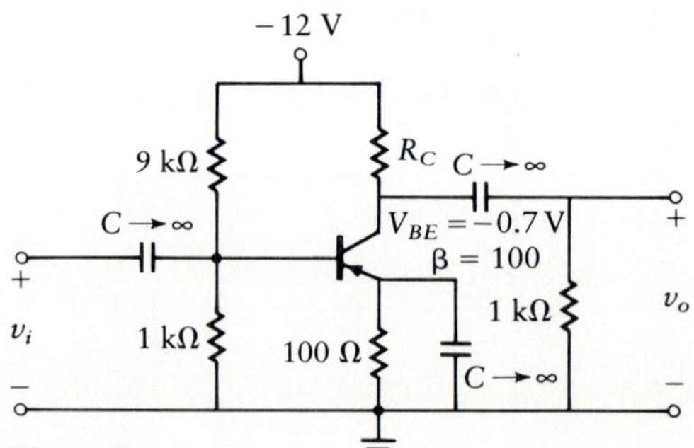

Figure P2.5

dc and ac load lines. What is the peak-to-peak value of maximum symmetrical output voltage?

2.14 Select I_{CQ} and V_{CEQ} for maximum symmetrical output voltage swing for the circuit of Figure P2.6.

a. Determine the values of R_1 and R_2 in order to achieve this operating point.
b. Find the maximum symmetrical output voltage swing.
c. Determine the power dissipated by the transistor and the power delivered to the load.

2.15 For the circuit of Figure P2.7,

a. Find I_{CQ} and V_{CEQ}.
b. Determine whether the amplifier is stable for large changes in β. You may assume that β is in the range $150 < \beta < 250$.
c. Draw the load lines.
d. Determine the symmetrical output voltage swing.

2.16 An ac voltage source is applied directly to the base of an *npn* transistor as shown in Figure P2.8. The internal resistance of the ac voltage source is shown as R_i. Determine the I_{CQ}, V_{CEQ}, and V_o when the ac input is zero. (Assume $\beta = 100$ and $V_{BE} = 0.7$ V.)

Figure P2.6

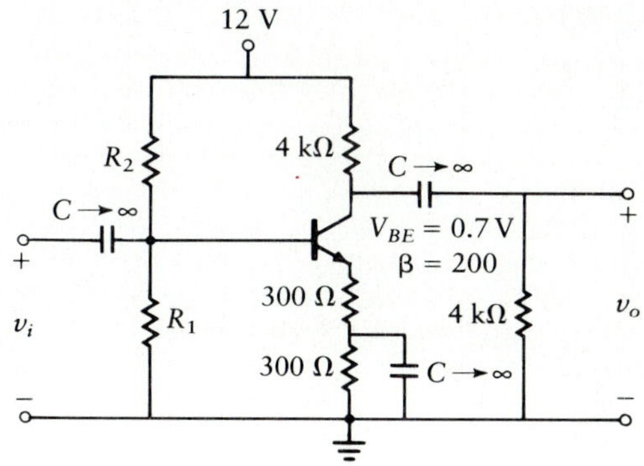

Figure P2.7

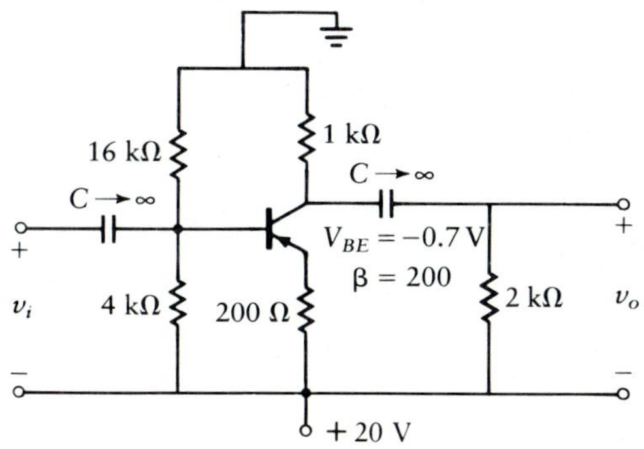

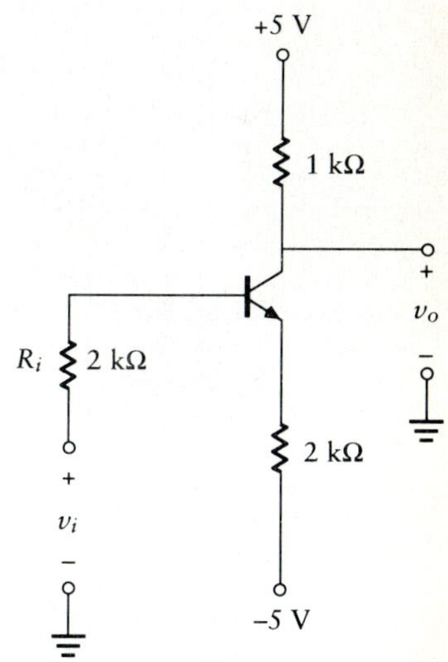

Figure P2.8

2.17 In Problem 2.16, what value of resistor will be needed to add to the internal resistance of the source to make the output of the circuit go to 3 V when $v_i = 0$?

2.18 Analyze the circuit shown in Figure P2.9 and determine the following (use $V_{BE} = 0.7$ V and $\beta = 100$):

a. I_{CQ} and V_{CEQ}.
b. Symmetrical output voltage swing.
c. Power supplied from the battery.
d. ac power output.
e. Load lines for the amplifier.

2.19 Design a common-emitter amplifier using the circuit shown in Figure P2.10 to obtain the maximum symmetrical output voltage swing.

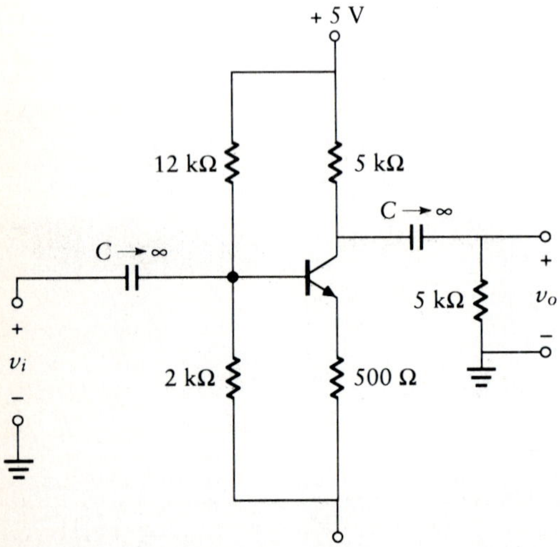

Figure P2.9

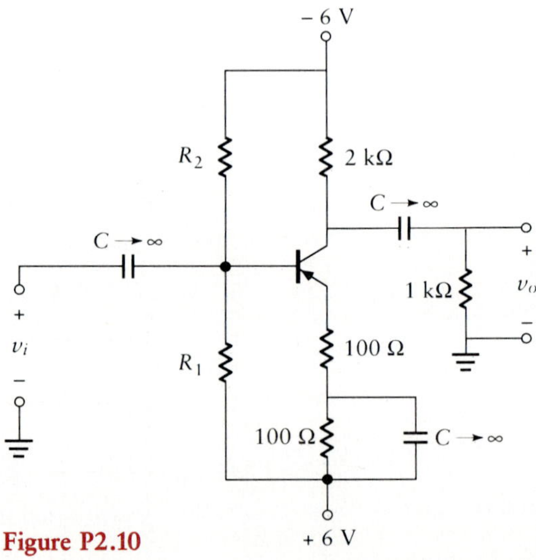

Figure P2.10

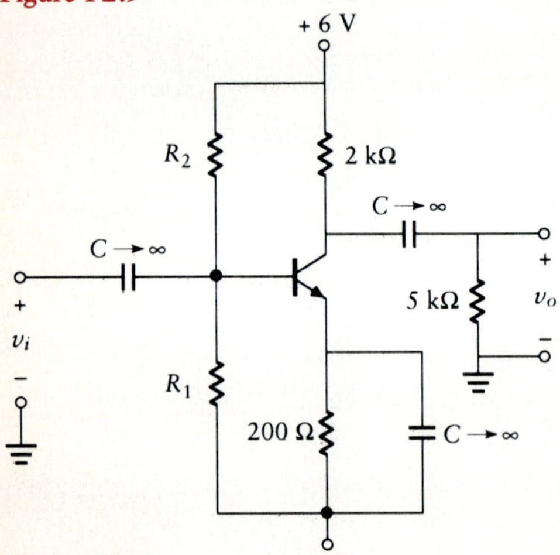

Figure P2.11

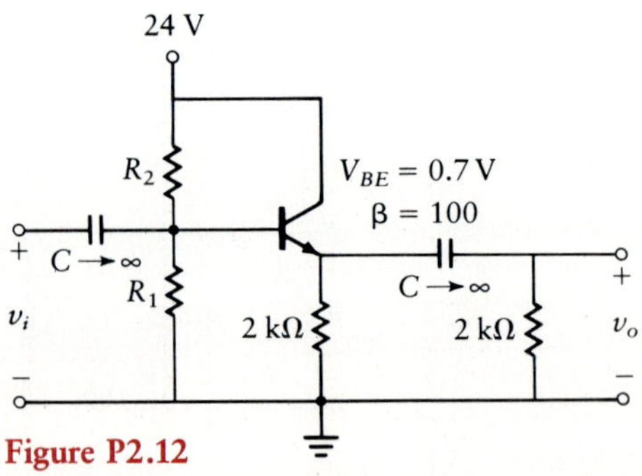

Figure P2.12

The design will be dc bias stable. (Use $V_{BE} = -0.6$ V and $\beta = 200$.) Determine:

a. I_{CQ} and V_{CEQ}.
b. R_1 and R_2.
c. Maximum symmetrical output voltage swing.
d. Power rating of the transistor needed.
e. ac power output of the amplifier.

2.20 Design a dc bias-stable common-emitter amplifier using the circuit shown in Figure P2.11 to obtain an output voltage of 1 V zero-peak. (Use $V_{BE} = 0.7$ V and $\beta = 200$.) This amplifier will use minimum power from the battery. Determine:

a. I_{CQ} and V_{CEQ}
b. R_1 and R_2

2.21 For the circuit of Figure P2.12:

a. Find the values of R_1 and R_2 if $I_{CQ} = 6$ mA.
b. Draw the ac and dc load lines.
c. Determine the symmetrical output voltage swing.
d. Find the power dissipated by the transistor and the power delivered to the load.

2.22 For the circuit shown in Figure P2.12:

a. Find the values of R_1 and R_2 if $I_{CQ} = 10$ mA.
b. Draw the ac and dc load lines.
c. Determine the symmetrical output voltage swing.
d. Find the power dissipated by the transistor and the power delivered to the load.

2.23 For the circuit shown in Figure P2.12:

a. Find the values of R_1 and R_2 needed to achieve the maximum possible symmetrical swing.
b. Draw the ac and dc load lines.
c. Determine the maximum symmetrical output voltage swing.
d. Find the power dissipated by the transistor and the power delivered to the load.

Figure P2.13

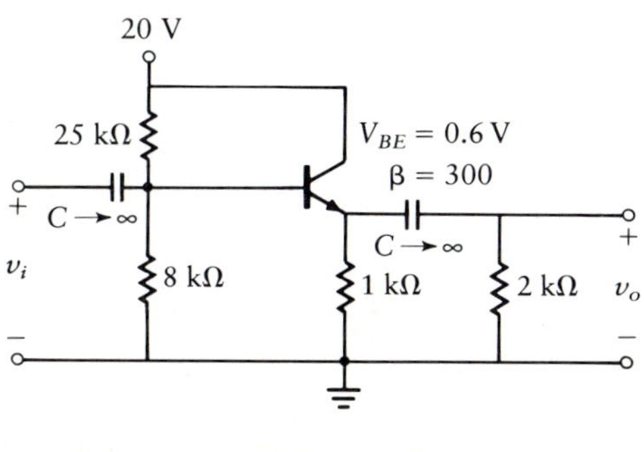

Figure P2.14

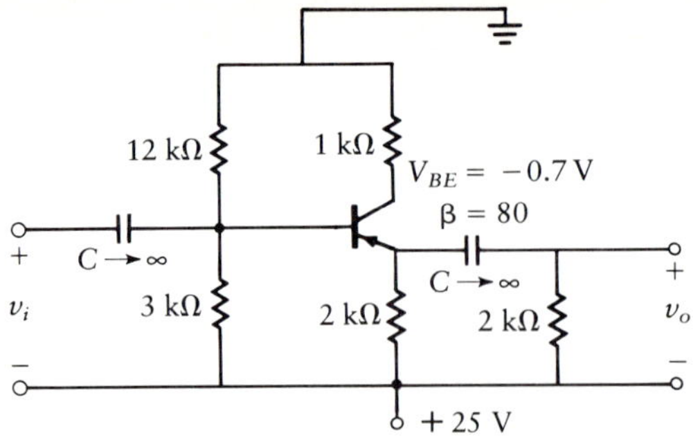

2.24 For the emitter-follower amplifier of Figure P2.13:

a. Determine the values of V_{CEQ} and I_{CQ}.

b. Find the output voltage swing.

c. Find the power delivered to the load and the required power rating of the transistor.

2.25 For the emitter-follower amplifier shown in Figure P2.14:

a. Determine the values of V_{CEQ} and I_{CQ}.

b. Draw the dc and ac load lines.

c. Determine the value of symmetrical output voltage swing.

d. The 1-kΩ resistor is now bypassed with a capacitor. Describe the changes that occur in the operation of the circuit.

2.26 Determine the maximum symmetrical output voltage swing for the circuit of Figure P2.14 by selecting new values of R_1 (shown as 3 kΩ) and R_2 (shown as 12 kΩ). What are the values of these resistors? Draw the new load lines.

2.27 The collector resistor, R_C (shown as 1 kΩ), is bypassed by a capacitor in the circuit of Figure P2.14. Determine the maximum symmetrical output voltage swing by selecting new values of R_1 (shown as 3 kΩ) and R_2 (shown as 12 kΩ). What are the values of these resistors? Find the power delivered to the load and the necessary power rating of the transistor.

2.28 By selecting new values of R_1 and R_2, determine the maximum symmetrical output voltage swing for the circuit of Figure P2.12 if the load resistor is reduced to 500 Ω. Find the power delivered to the load and the required power rating of the transistor.

2.29 Design an emitter-follower amplifier using an *npn* transistor for maximum output symmetrical swing with the following specifications: $R_B = 250\ \Omega$, $V_{CC} = 12$ V, $R_E = R_L = 8\ \Omega$, $V_{BE} = 0.7$ V, and $\beta = 200$. Also, determine $P_o(\text{ac})$, the power supplied from the battery, and the power required to be dissipated by the transistor.

2.30 Analyze the circuit shown in Figure P2.15 and determine the following when $\beta = 200$ and $V_{BE} = 0.7$ V:

Figure P2.15

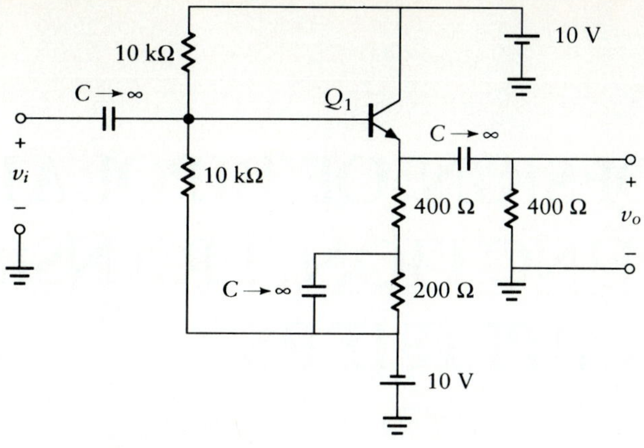

a. I_{CQ} and V_{CEQ}.
b. Symmetrical voltage output swing.
c. Power supplied from the battery.
d. ac power output.
e. Power rating of the transistor needed.

3

DESIGN OF BIPOLAR JUNCTION TRANSISTOR AMPLIFIERS

3.0 INTRODUCTION

In Chapter 2 we discuss the biasing, or dc operation, of bipolar junction transistor circuits. The transistor is biased to obtain the required peak-to-peak output voltage swing. In the current chapter we concentrate on *small-signal analysis* through the use of equivalent-circuit techniques. The transistor parameters required to accomplish this analysis can be obtained from manufacturers' data sheets. The data are provided by the manufacturers in a format as shown in the examples of Appendix D. The design method presented here reduces the dependence of the circuit on the variations in the transistor parameters.

The chapter begins with an introduction to hybrid parameters, which are used to develop a transistor mathematical model. We derive the equations for input resistance, voltage gain, current gain, and output resistance for the various amplifier configurations (i.e., CE, CC, and CB). Design examples are presented for each case. Multistage amplifier analysis is also discussed and example problems are given.

3.1 ANALYSIS OF TWO-PORT NETWORKS—HYBRID PARAMETERS

There are a number of ways to characterize four-terminal networks. In a four-terminal system, there are four circuit variables: the input voltage and current and the output voltage and current. These four variables can be related by various equations, depending on which variables are considered to be independent and which are dependent.

When h-parameters are used to describe a transistor network, the equation pair is written as follows:

$$v_1 = h_i i_1 + h_r v_2 \tag{3.1}$$

$$i_2 = h_f i_1 + h_o v_2 \tag{3.2}$$

where the h-parameters are defined as follows:

$h_i = h_{11} =$ input resistance of transistor

$h_r = h_{12} =$ reciprocal voltage gain of transistor

$h_f = h_{21} =$ forward current gain of transistor

$h_o = h_{22} =$ output conductance of transistor

When h-parameters are applied to transistor networks, the parameters take on a practical significance related to the transistor performance. The circuit developed using the h-parameters is shown in Figure 3.1. A simple application of Kirchhoff's laws to the circuit of Figure 3.1 shows that it satisfies equations (3.1) and (3.2).

When the input and output parameters are individually set equal to zero, each hybrid parameter represents either a resistance, a conductance, a ratio of two voltages, or a ratio of two currents. The following equations are derived from equations (3.1) and (3.2). Following each equation are the units associated with the parameter and the name given to it.

$R_{in} = h_i = \left.\dfrac{v_i}{i_1}\right|_{v_2=0}$ *ohms* input resistance with v_2 shorted

$h_f = \left.\dfrac{i_2}{i_1}\right|_{v_2=0}$ *dimensionless* forward current gain with v_2 shorted

$h_r = \left.\dfrac{v_1}{v_2}\right|_{i_1=0}$ *dimensionless* reverse voltage gain with i_1 open-circuited

$Y_{out} = h_o = \left.\dfrac{i_2}{v_2}\right|_{i_1=0}$ *siemens* (or mhos) output conductance with i_1 open-circuited

Although these parameters are ideally constant for a specific operating point, their numerical values depend on the transistor configuration. For ex-

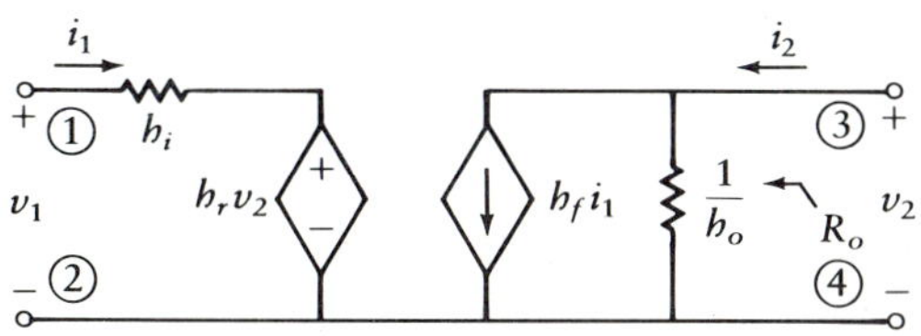

Figure 3.1
Equivalent circuit for h-parameters.

ample, if terminal 1 in Figure 3.1 is the base, 2 is the emitter, and 3 is the collector, the circuit represents a CE configuration. Similarly, the transistor can be modeled as a CB configuration if terminals 1, 2, and 3 are the emitter, base, and collector, respectively. Note that terminal 2 is common to terminal 4.

It is helpful to have a way to distinguish between the three wiring configurations, i.e., CE, CC, and CB. A second subscript is added to each hybrid parameter to provide this bookkeeping distinction. For example, a CE circuit would normally have h_i in the base circuit, and it would be renamed h_{ie}. Similarly, for the CB, h_i is renamed h_{ib}, and for the CC, it is named h_{ic}. The three values of this short-circuit input impedance are related to each other as follows:

$$h_{ib} = \left.\frac{\Delta v_{eb}}{\Delta i_e}\right|_{v_{cb}=\text{constant}} = \left.\frac{\Delta v_{eb}}{-(1+\beta)\Delta i_b}\right|_{v_{cb}=\text{constant}} = \frac{h_{ie}}{1+\beta}$$

Similarly, we can show that $h_{ic} = h_{ie}$. Therefore, in summary,

$$h_{ie} = (1+\beta)h_{ib} \approx \beta h_{ib} = h_{ic} \tag{3.3}$$

Figure 3.2 shows a CE amplifier circuit with two different equivalent circuits. Although the h-parameter model defines the second subscript as associated with the type of amplifier configuration, h_{ib} and h_{ie} are values of resistance that are based on the operating point of the amplifier and the location of these resistances within the equivalent circuit. For this reason, we will sometimes see the hybrid notation for one type of circuit used in a different type. For example, h_{ie} might appear in a CC circuit. The same concept is also applied to h_{fe}, which refers to β regardless of how the transistor is placed within the amplifier configuration.

In each equivalent circuit, we have made the (usually reasonable) simplification that $h_r = h_o = 0$. Figure 3.2(b) uses the CE model, where the transistor

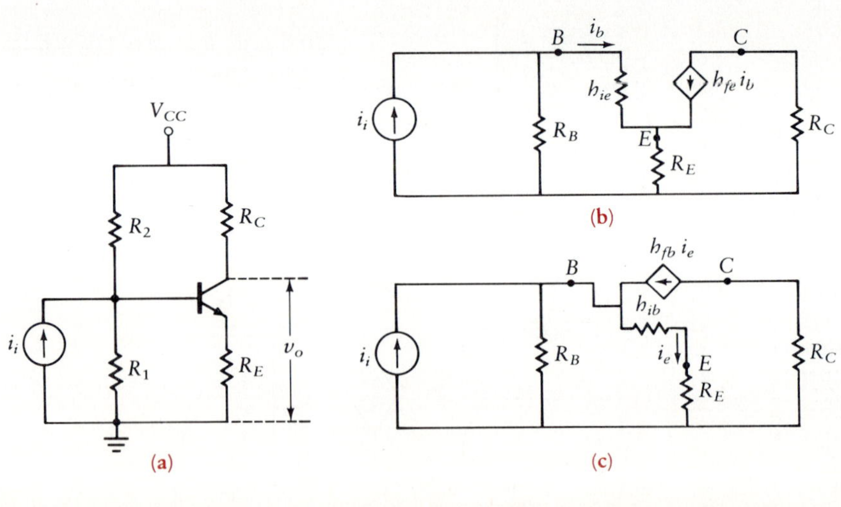

Figure 3.2
CE and equivalent circuits.

has been replaced by the circuit of Figure 3.1 with terminal 1 as the base, 2 as the emitter, and 3 as the collector. In Figure 3.2(c) the transistor is replaced by the CB model. That is, using Figure 3.1, terminal 1 is the emitter, 2 the base, and 3 the collector.

For small-signal current, we see that h_{fe} is the ratio of the change in output current (Δi_c) to the change in input current (Δi_b). Recall that this relationship is also the defining expression for β. As a result,

$$h_{fe} = \beta = \left.\frac{\Delta i_c}{\Delta i_b}\right|_{v_{CE}=\text{constant}}$$

The actual value of β is a function of the operating point (I_{CQ}) of the transistor. In the flat portion of the i_C versus v_{CE} curve with i_B constant, there is little change in β. As the transistor approaches saturation, β starts dropping. As the transistor approaches cutoff, β also approaches zero. Manufacturers' specifications often present a graph of h_{fe} as a function of i_C.

The output admittance, h_{oe}, of the transistor is usually small. Therefore the output resistance, r_o, is usually large. As an example, let us examine the manufacturer's specification sheet for the 2N3903 (see Appendix D). From the dynamic characteristics of the 2N3903 (Figure 18 of Appendix D.2), we see that the output admittance varies from 3 to 40 μmhos as i_C varies from 0.2 to 15 mA. This means that the output resistance varies from 25 to 333 kΩ. As an example of a typical value, when I_{CQ} is equal to 1 mA, the transistor output resistance is approximately 167 kΩ. With a typical load (R_L) of 4 kΩ, the parallel combination of r_o and R_L is approximately 4 kΩ (since 167 kΩ is much larger than 4 kΩ) and r_o can be assumed to be infinite. The reverse voltage gain of the network, h_r, is also small and is usually ignored in the transistor equivalent circuit.

Another two-port model that is used in the study of transistor circuits is the *hybrid-π* model, which is important when the transistor is used at high frequency. This model was introduced in Chapter 2. It includes the effects of parameters that become significant at high frequency (this is discussed in Chapter 7).

A low-frequency small-signal hybrid-π transistor model is shown in Figure 3.3. We discussed the approximations leading to this simplified equivalent in Chapter 2. In particular, it should be noted that the impedance coupling the collector to the base has been omitted. The model for low frequency is similar

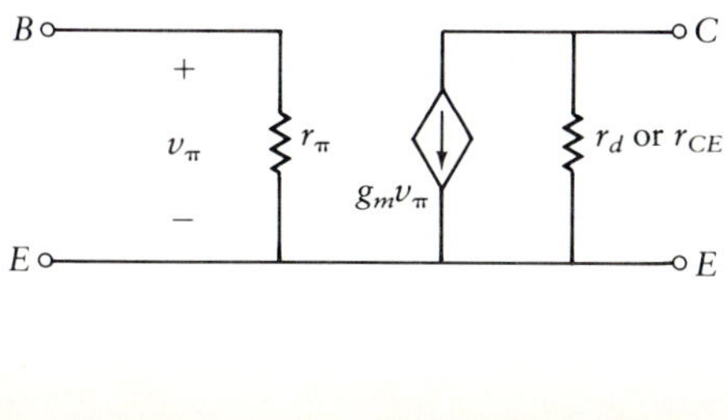

Figure 3.3 Hybrid-π model equivalent circuit.

to that of the h-parameter model for the CE. In fact, a comparison of the parameters is easily accomplished as follows:

$$g_m = \frac{1}{h_{ib}}$$

$$r_\pi = h_{ie}$$

$$r_{CE} = \frac{1}{h_o} = r_o$$

g_m is known as the *transconductance* of the transistor.

The choice of model is often dictated by the form in which parameters are specified on the manufacturer's data sheets.

3.2 SHORT-CIRCUIT INPUT RESISTANCE

We explore the parameter values before discussing the actual use of equivalent circuits for design and analysis. We first develop equations for h_{ie} and h_{ib} that display the dependence of these parameters on the location of the operating point.

We begin with the equation for the operating characteristics of the base-emitter junction as presented in Chapter 1, with $n = 1$.

$$i_B = I_o\left[\exp\left(\frac{v_{BE}}{V_T}\right) - 1\right]$$

This equation is now differentiated with respect to v_{BE} to obtain

$$\frac{di_B}{dv_{BE}} = \frac{I_o}{V_T}\exp\left(\frac{v_{BE}}{V_T}\right)$$

In the forward-bias region, i_B is approximately given by

$$i_B = I_o \exp\left(\frac{v_{BE}}{V_T}\right)$$

Then

$$\frac{di_B}{dv_{BE}} = \frac{i_B}{V_T}$$

But from the definition of h_{ie}, we obtain (see Figure 3.1)

$$h_{ie} = \left.\frac{\Delta v_1}{\Delta i_1}\right|_{v_2=0} = \frac{dv_{BE}}{di_B} = \frac{V_T}{I_{BQ}}$$

Now recall that

$$h_{ib} = \frac{h_{ie}}{\beta}$$

Finally,

$$h_{ib} = \frac{V_T}{|\beta I_{BQ}|} = \frac{V_T}{|I_{CQ}|} \tag{3.4}$$

Equation (3.4) is known as the *Shockley equation.* Using the approximation $V_T = 26$ mV, which applies to the bipolar junction transistor, equation (3.4) becomes

$$h_{ib} = \frac{0.026}{|I_{CQ}|} \tag{3.5}$$

Equation (3.5) is useful in estimating the value of h_{ib} to be used in the equivalent circuit of Figure 3.2(c) (or g_m in the equivalent circuit of Figure 3.3 since $g_m = 1/h_{ib}$).

3.3 CE PARAMETERS

The equations that define ac amplifier parameters are summarized in Table 3.1 and are derived in the following sections. Table 3.2 summarizes the equivalent circuits used in the derivations.

3.3.1 Input Resistance, R_{in}

The hybrid parameter circuit is used to derive the input resistance equation for each type of amplifier configuration. Figure 3.4 modifies the common-emitter amplifier of Figure 3.2 by adding a capacitor-coupled load resistance. The basic circuit is shown in Figure 3.4(a) and two forms of equivalent circuit are shown in Figure 3.4(b) and (c). Note that we have omitted the reverse voltage gain, h_r, and the output admittance, h_o, from the model.

Table 3.1 Formula for Different Amplifier Configurations

Type		Voltage Gain (A_v)	Current Gain (A_i)	Input Resistance (R_{in})
Common emitter	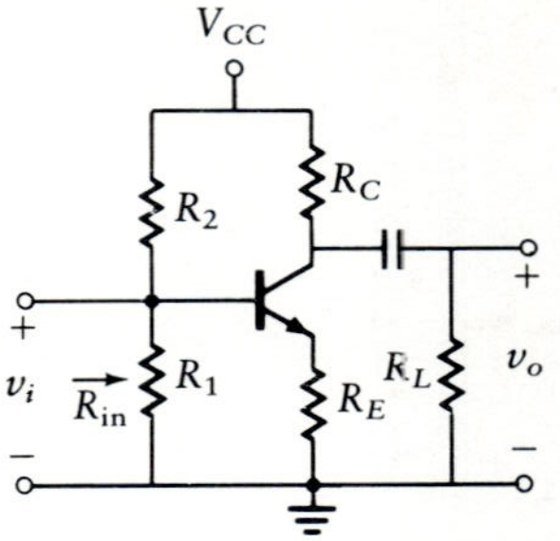	$\dfrac{-(R_L \parallel R_C)}{h_{ib} + R_E}$	$\dfrac{-R_B}{\dfrac{R_B}{\beta} + h_{ib} + R_E} \dfrac{R_C}{R_L + R_C}$	$\dfrac{R_B(h_{ib} + R_E)}{\dfrac{R_B}{\beta} + h_{ib} + R_E}$
Common collector (Emitter follower)	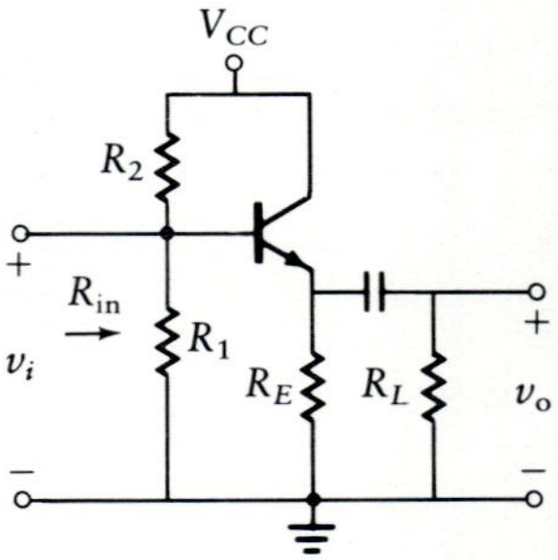	$\dfrac{R_E \parallel R_L}{h_{ib} + (R_E \parallel R_L)}$	$\dfrac{R_B}{\dfrac{R_B}{\beta} + h_{ib} + (R_E \parallel R_L)} \dfrac{R_E}{R_E + R_L}$	$\dfrac{R_B[h_{ib} + (R_E \parallel R_L)]}{\dfrac{R_B}{\beta} + h_{ib} + (R_E \parallel R_L)}$
Common base	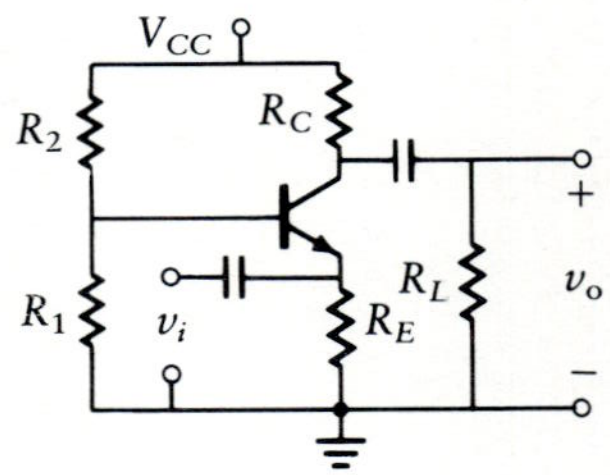	$\dfrac{R_C \parallel R_L}{h_{ib} + \dfrac{R_B}{\beta}}$	$\dfrac{+R_C}{R_C + R_L} \dfrac{R_E}{R_E + h_{ib} + \dfrac{R_B}{\beta}}$	$R_E \Bigg\Vert \left(h_{ib} + \dfrac{R_B}{\beta}\right)$

Table 3.2 Equivalent Circuits for Different Amplifier Configurations

Type	Circuit	Equivalent Circuit	Equivalent Circuit with R_E Bypassed
Common emitter	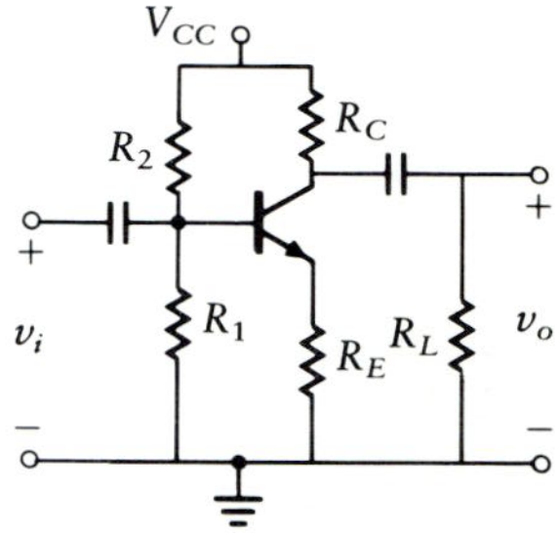	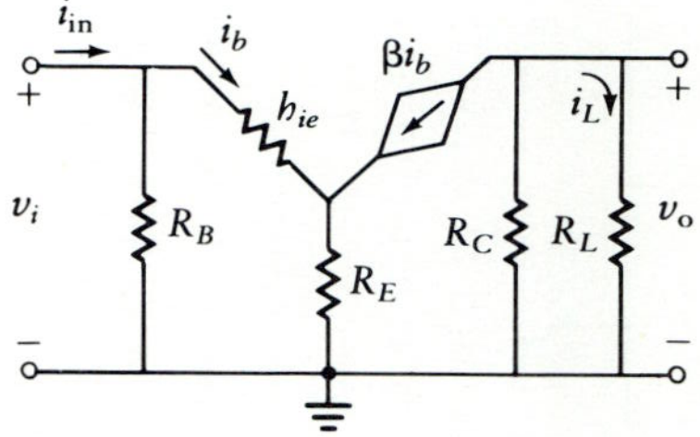	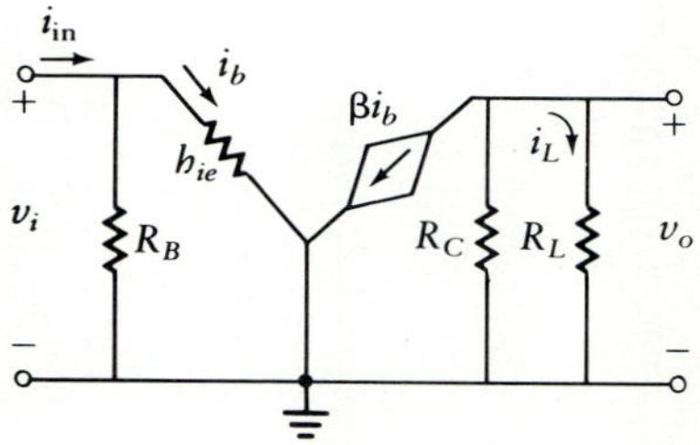
Common collector	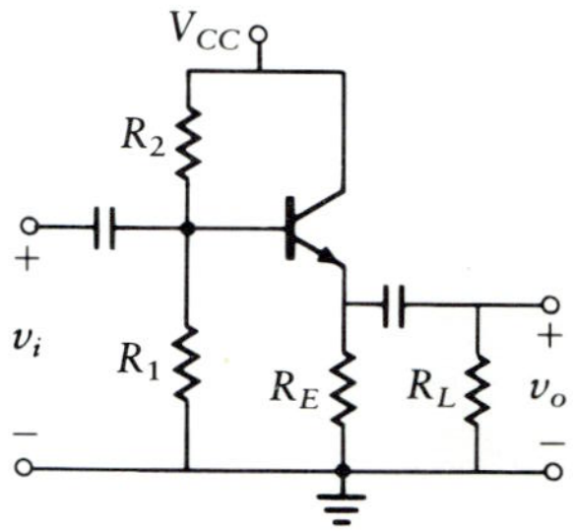	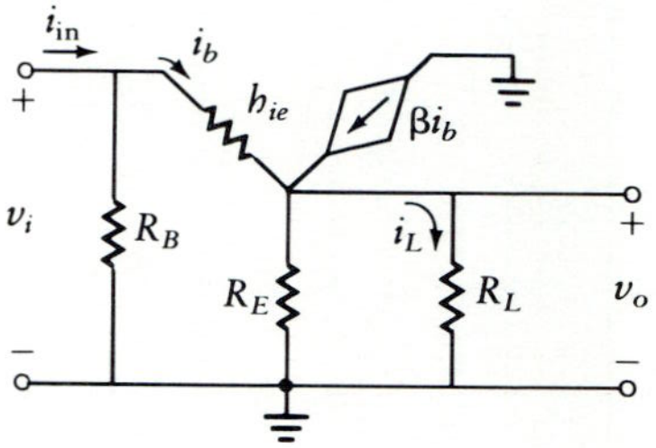	
Common base	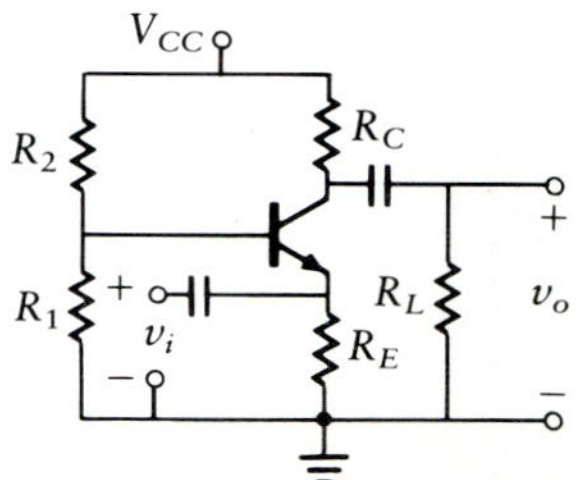	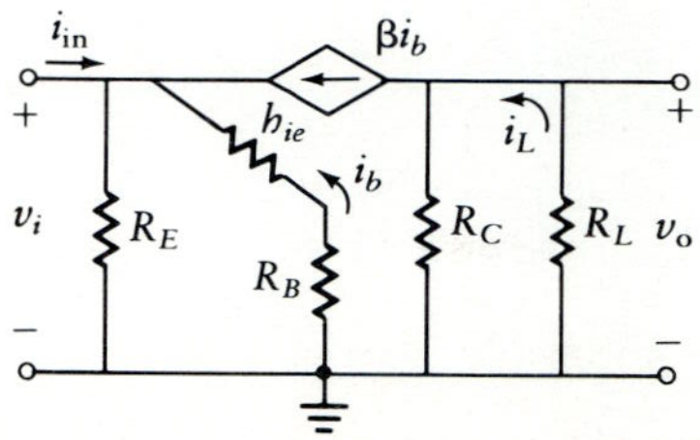	With R_B bypassed 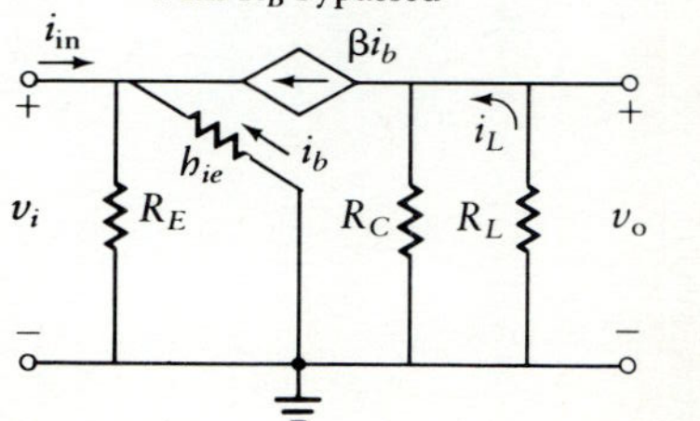

Figure 3.4
CE configuration.

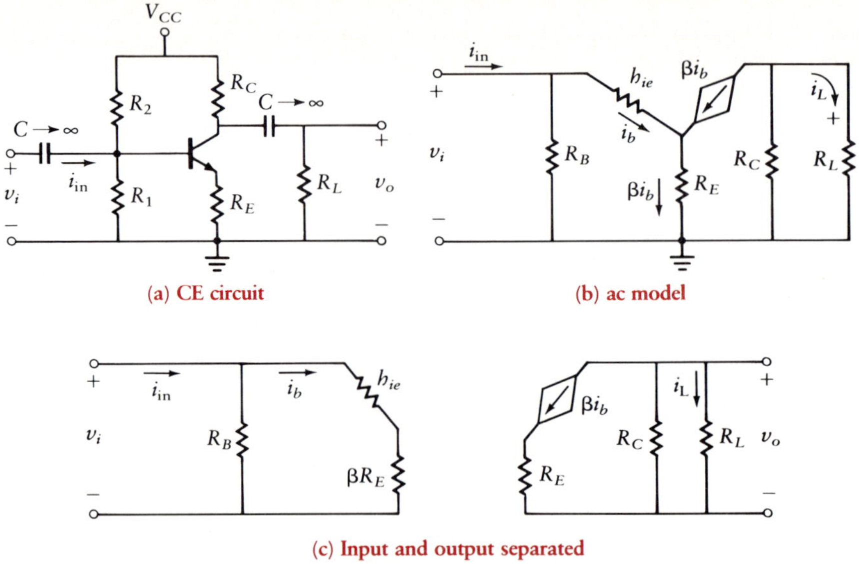

The equivalent circuit of Figure 3.4(b) is used to derive the input resistance, R_{in}. Usually β is large enough so that we can approximate $1 + \beta$ as β. The current in R_E is therefore approximately equal to βi_b. If the circuit is now split as in Figure 3.4(c), the current through the resistor in series with h_{ie} in the input loop is simply i_b. Thus, to keep the voltage at the same value as in the original circuit, we must change the resistor value to βR_E. The input resistance is then found by writing Kirchhoff's voltage and current law equations for the input loop.

$$R_{in} = \frac{v_i}{i_{in}} = R_B \parallel (h_{ie} + \beta R_E) = \frac{R_B(h_{ie} + \beta R_E)}{R_B + h_{ie} + \beta R_E}$$

We substitute $h_{ie} = \beta h_{ib}$ to obtain

$$R_{in} = \frac{R_B(h_{ib} + R_E)}{R_B/\beta + h_{ib} + R_E} \tag{3.6}$$

Equation (3.6) is the input resistance equation. It requires only the approximation that $\beta >> 1$.

3.3.2 Current Gain, A_i

The current gain of the CE amplifier is defined as the ratio of load current to input current. Thus,

$$A_i = \frac{i_L}{i_{\text{in}}}$$

where the currents are shown in Figure 3.4(b). To obtain an expression for current gain in terms of the circuit parameters, we must derive expressions for these two currents. We can use current division in the base circuit to relate the input current to base current.

$$i_b = \frac{R_B i_{\text{in}}}{R_B + h_{ie} + \beta R_E} \tag{3.7a}$$

Current division can be applied to the output circuit to obtain an expression for load current.

$$i_L = \frac{-R_C \beta i_b}{R_L + R_C} \tag{3.7b}$$

The negative sign results from the direction of βi_b being opposite to that of i_L. We combine equations (3.7a) and (3.7b) to find the current gain, as follows.

$$A_i = \frac{i_L}{i_{\text{in}}} = -\frac{R_B R_C}{(R_B/\beta + h_{ib} + R_E)(R_C + R_L)} \tag{3.8}$$

3.3.3 Voltage Gain, A_v

The voltage gain is defined as the ratio of output voltage to input voltage. Thus,

$$A_v = \frac{v_o}{v_i}$$

where the voltages are shown in Figure 3.4(b). We can save time in deriving an expression for voltage gain if we borrow the results of the previous section. We relate current to voltage using Ohm's law:

$$v_o = i_L R_L \qquad \text{and} \qquad v_i = i_{\text{in}} R_{\text{in}}$$

so

$$A_v = \frac{i_L}{i_{\text{in}}} \frac{R_L}{R_{\text{in}}} = A_i \frac{R_L}{R_{\text{in}}} \tag{3.9}$$

since $A_i = i_L/i_{\text{in}}$. We substitute equations (3.6) and (3.8) into equation (3.9) with the result

$$A_v = \frac{-R_B R_C R_L}{(R_B/\beta + h_{ib} + R_E)(R_C + R_L)} \frac{R_B/\beta + h_{ib} + R_E}{R_B(h_{ie} + R_E)} \tag{3.10}$$

When like terms are canceled from the numerator and denominator and we recognize that $h_{ie} = \beta h_{ib}$, equation (3.10) simplifies to

$$A_v = \frac{-\beta(R_L \parallel R_C)}{h_{ie} + \beta R_E} = \frac{-R_L \parallel R_C}{h_{ib} + R_E} \tag{3.11}$$

This is the general expression for voltage gain. We now examine two special cases.

If R_E is bypassed with a large capacitor, the voltage gain equation becomes

$$A_v = \frac{-(R_L \parallel R_C)}{h_{ib}}$$

If $h_{ib} << R_E$ and R_E is not bypassed, the voltage gain can be written as

$$A_v = -\frac{R_L \parallel R_C}{R_E}$$

These equations for the common-emitter amplifier are summarized in Table 3.1.

3.3.4 Gain Impedance Formula

We derive an important relationship between the ac quantities of voltage gain, A_v, and current gain, A_i. Figure 3.5 shows a block diagram of a four-terminal (two-port) network with input resistance R_{in} and load resistance R_L. Although we are assuming these to be resistors, they can be complex impedances.

The relationships between the input variables, v_i and i_{in}, and the output variables, v_o and i_o, are derived directly from Ohm's law. That is,

$$v_o = i_o R_L$$

$$v_i = i_{in} R_{in}$$

Forming the ratio of these two equations yields

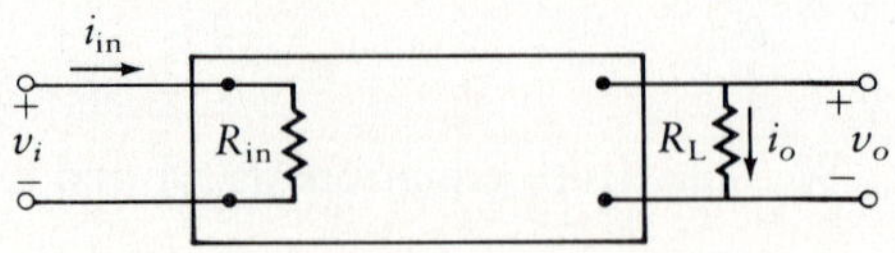

Figure 3.5
Two-port network.

$$\frac{v_o}{v_i} = \frac{i_o}{i_{\text{in}}} \frac{R_L}{R_{\text{in}}}$$

Voltage gain is defined as

$$A_v = \frac{v_o}{v_i}$$

and current gain is defined as

$$A_i = \frac{i_o}{i_{\text{in}}}$$

Combining the various equations, we find

$$A_v = \frac{A_i R_L}{R_{\text{in}}} \tag{3.12}$$

Equation (3.12) is called the *gain impedance formula* and is used throughout this text.

3.3.5 Output Resistance, R_o

In the equivalent circuit for the transistor as shown in Figure 3.1, the output circuit contains an ideal current generator in parallel with a resistance of value $r_o = 1/h_o$. The ideal current source exhibits an infinite impedance since we measure output resistance with the input open-circuited (i.e., $i_b = 0$). The output resistance for the CE transistor is then

$$r_o = \frac{v_2}{i_2} = \frac{1}{h_{oe}}$$

The parameter h_{oe} is usually small enough to be neglected in calculations, so the output resistance of the transistor approaches infinity. The value of h_{oe} can be determined by consulting the transistor specifications. The output resistance of a CE amplifier is the value of R_C when r_o is large.

Example 3.1 **CE Amplifier Design**

Design the CE amplifier of Figure 3.6 to drive a 2-kΩ load using a *pnp* silicon transistor, $V_{CC} = -24$ V, $\beta = 200$, $A_v = -10$, and $V_{BE} = -0.7$ V. Determine

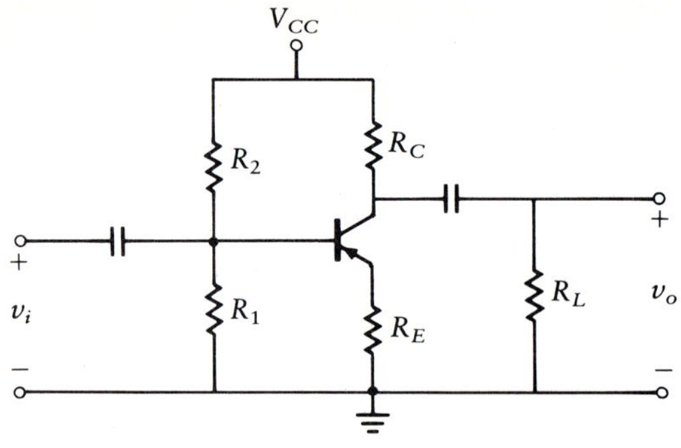

Figure 3.6 Amplifier for Example 3.1.

all element values and calculate A_i, R_{in}, I_{CQ}, and the maximum undistorted symmetrical output voltage swing for three values of R_C as given below:

a. $R_C = R_L$

b. $R_C = 0.1R_L$

c. $R_C = 10R_L$

SOLUTION a. ($R_C = R_L$) We use the various equations of Section 3.3 to derive the parameters of the circuit of Figure 3.6. Beginning with equation (3.8) for the voltage gain, we can solve for R_E'.

$$A_v = -10 = \frac{-R_L \parallel R_C}{h_{ib} + R_E} = \frac{-2\ \text{k}\Omega \parallel 2\ \text{k}\Omega}{h_{ib} + R_E}$$

so

$$R_E' = h_{ib} + R_E = 100\ \Omega$$

We can find I_{CQ} from the collector-emitter loop using equation (2.30) for the condition of maximum output swing.

$$I_{CQ} = \frac{V_{CC}}{R_{dc} + R_{ac}} = -7.5\ \text{mA}$$

Now, using equation (3.5), we find that h_{ib} is 3.5 Ω. This is small enough that we shall ignore it to find that $R_E = 100\ \Omega$. Since we now know β and R_E, we can use the design guideline,

$$R_B = 0.1\beta R_E$$

to find that $R_B = 2\ \text{k}\Omega$.

The design discussion of the previous chapter can now be used to specify the biasing circuitry. You may wish to refer back to Example 2.3 for details. We find

$$V_{BB} = -1.52 \text{ V}$$

$$R_1 = 2.14 \text{ k}\Omega$$

$$R_2 = 31.6 \text{ k}\Omega$$

The maximum symmetrical peak-to-peak output swing is then

$$v_o(p\text{-}p) = 2I_{CQ}(R_L \parallel R_C) = 15 \text{ V}$$

We now have all the necessary parameters to use equation (3.12) for current gain and equation (3.6) for input resistance. This yields

$$A_i = -9.1$$

$$R_{\text{in}} = 1.82 \text{ k}\Omega$$

b. ($R_C = 0.1R_L$) We repeat the steps of part a to find

$$R_C = 200\ \Omega \qquad R_1 = 390\ \Omega$$

$$I_{CQ} = -57.4 \text{ mA} \qquad R_2 = 4.7 \text{ k}\Omega$$

$$h_{ib} = 0.45\ \Omega \qquad v_o(p\text{-}p) = 20.8 \text{ V}$$

$$R_B = 360\ \Omega \qquad A_i = -1.64$$

$$V_{BB} = -1.84 \text{ V} \qquad R_{\text{in}} = 327\ \Omega$$

c. ($R_C = 10R_L$) Once again, we follow the steps of part a to find

$$R_C = 20 \text{ k}\Omega \qquad R_1 = 3.28 \text{ k}\Omega$$

$$I_{CQ} = -1.07 \text{ mA} \qquad R_2 = 85.6 \text{ k}\Omega$$

$$h_{ib} = 24.2\ \Omega \qquad v_o(p\text{-}p) = 3.9 \text{ V}$$

$$R_B = 3.16 \text{ k}\Omega \qquad A_i = -14.5$$

$$V_{BB} = -0.886 \text{ V} \qquad R_{\text{in}} = 2.91 \text{ k}\Omega$$

We compare the results of this example in Table 3.3 for the purpose of making the optimum choice for R_C. Note that when $R_C = 0.1R_L$, the value of A_i is too small. Also, the value of R_{in} is so low that the stage would unnecessarily load the previous stage. When $R_C = 10R_L$, the values of A_i and R_{in} are satisfactory, but the amplifier stage has an output peak-to-peak voltage of only 3.9 V. By selecting $R_C = R_L$, the values of A_i and R_{in} are 5.5 times the equivalent values for the case of $R_C = 0.1R_L$. Also, the value of peak-to-peak voltage is 3.8 times the equivalent value for the case of $R_C = 10R_L$.

Table 3.3 Comparison for Three Selections of R_C

	I_{CQ}	A_i	R_{in}	$v_o(p\text{-}p)$
$R_C = R_L$	−7.5 mA	−9.1	1.82 kΩ	15 V
$R_C = 0.1R_L$	−57.4 mA	−1.64	327 Ω	20.8 V
$R_C = 10R_L$	−1.07 mA	−14.5	2.91 kΩ	3.9 V

Example 3.1 indicates that, of the three given ratios of R_C to R_L, $R_C = R_L$ has the most desirable performance in the CE amplifier stage. We shall use this as a guide to develop a reasonable starting point in our designs. In most cases, this choice will provide performance that meets specifications. In some applications, it may be necessary to do additional analysis to find the "optimum" ratio of R_C to R_L.

Example 3.2 Capacitor-Coupled CE Amplifier (Design)

Design a CE amplifier (see Figure 3.7) with $A_v = -10$, $\beta = 200$, and $R_L = 1$ kΩ. A *pnp* transistor is used and maximum symmetrical output swing is required. Check the value of A_v using a computer simulation.

SOLUTION Refer to Figure 3.8 during this derivation. As a result of Example 3.1, we shall choose $R_C = R_L = 1$ kΩ. Using the equation for A_v to solve for R_E,

$$A_v = \frac{-(R_L \parallel R_C)}{R'_E} \qquad \text{where } R'_E = R_E + h_{ib}$$

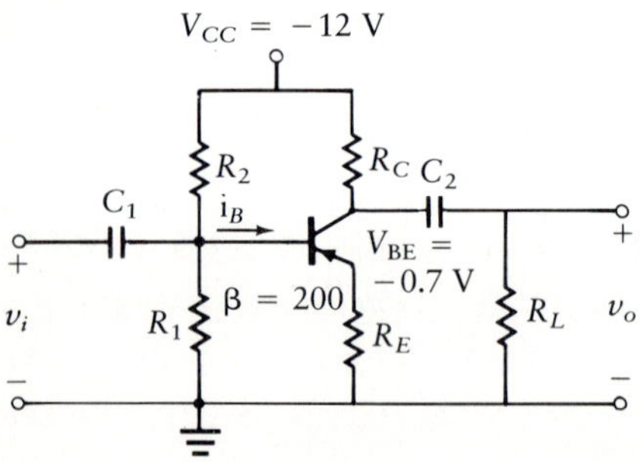

Figure 3.7 CE amplifier.

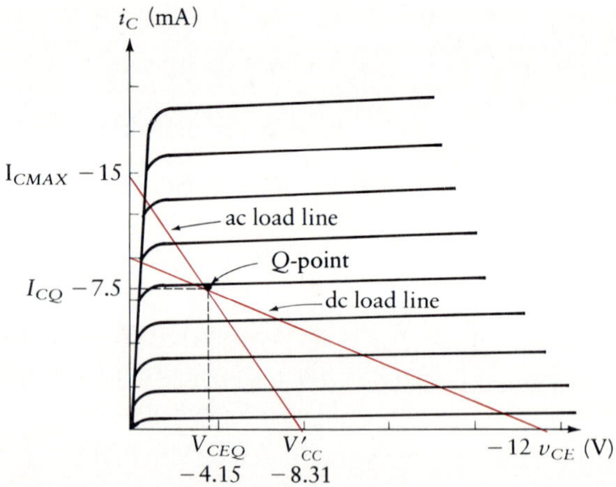

Figure 3.8 Load lines for Example 3.2.

When known values are substituted into this equation, we find $R'_E = 50\ \Omega$. We need to know the value of h_{ib} in order to find R_E. We first find R_{ac} and R_{dc} and then calculate the Q-point as follows:

$$R_{ac} = R_E + R_C \parallel R_L = 550\ \Omega$$

$$R_{dc} = R_E + R_C = 1050\ \Omega$$

With R_{ac} and R_{dc} determined, the design of this circuit now parallels the step-by-step procedure for the design given in Section 2.9.2. The first step is to calculate the quiescent collector current needed to place the Q-point in the center of the ac load line (i.e., maximum swing). The equation is

$$I_{CQ} = \frac{-V_{CC}}{R_{ac} + R_{dc}} = -7.5\text{ mA}$$

The quantity h_{ib} is found from equation (3.5) as follows.

$$h_{ib} = \frac{26\text{ mV}}{|I_{CQ}|} = \frac{26\text{ mV}}{7.5\text{ mA}} = 3.47\ \Omega$$

Then

$$R_E = 50 - h_{ib} = 46.5\ \Omega$$

If there were a current gain or input resistance specification for this design, we would use it to solve for the value of R_B. Since there is no such specification, we use the expression

$$R_B = 0.1\beta R_E = 0.1(200)(46.5) = 930\ \Omega$$

Then continuing with the design steps,

$$A_i = \frac{-R_B}{R_B/\beta + h_{ib} + R_E}\,\frac{R_C}{R_C + R_L} = -8.5$$

$$V_{CEQ} = V_{CC} - (R_C + R_E)I_{CQ}$$

$$= -12 - (1046)(-0.0075) = -4.15\text{ V}$$

$$V_{BB} = I_{CQ}\left(R_E + \frac{R_B}{\beta}\right) + V_{BE}$$

$$= (-0.0075)\left(46.5 + \frac{930}{200}\right) + (-0.7) = -1.08\text{ V}$$

$$R_1 = \frac{R_B}{1 - V_{BB}/V_{CC}} = \frac{930}{1 - 1.08/12} = 1.02\ \text{k}\Omega$$

$$R_2 = \frac{R_B V_{CC}}{V_{BB}} = \frac{(930)(-12)}{-1.08} = 10.3\ \text{k}\Omega$$

$$R_{\text{in}} = \frac{R_B(h_{ib} + R_E)}{R_B/\beta + h_{ib} + R_E} = \frac{930(50)}{(930/200) + 50} = 851\ \Omega$$

$$R_o = R_C = 1\ \text{k}\Omega \qquad \text{(assuming } r_o \text{ is large compared to } R_C\text{)}$$

The maximum peak-to-peak output swing is given by

$$2|I_{CQ}|(R_C \parallel R_L) = 2(0.0075)(500) = 7.5\ \text{V}$$

The power delivered into the load and the maximum power dissipated by the transistor are found using equations from Sections 2.6 and 2.9.

$$P = \frac{1}{2}\left(I_{CQ}\frac{R_C}{R_C + R_L}\right)^2 R_L = \frac{(I_{CQ})^2 R_L}{8} = 7\ \text{mW}$$

The load lines for this circuit are shown in Figure 3.8. If R_{in} or A_i had been specified (instead of A_v), the equation for R_{in} or A_i must be used to determine R_B.

We use MICRO-CAP II to check the results of this design. The first task is to input the circuit. MICRO-CAP provides a drawing board for this purpose. The only decisions to be made involve the capacitors and the transistor. The capacitors must act as short circuits for the signals of interest and open circuits for the dc bias voltages. Since we are not concerned with frequency response at this time, we can set the capacitors at unrealistically large values, such as 1000 μF. The only transistor specifications are that β is 200 and the base-emitter voltage is -0.7 V. The transistor models used by the simulation program arc considerably more sophisticated and include many parameters. We can either custom design a library entry with the given parameters or choose a transistor that comes close to having these values. We chose to use the 2N3906 transistor, which has a nominal β of 200 and a V_{BE} of -0.75.

Most simulation programs offer at least three forms of analysis for electronic circuits in addition to other features. The three common forms are dc, transient, and ac analysis. The *dc* analysis provides a plot of output versus input voltage under dc conditions. With proper designation of input and output, it can be used to check quiescent conditions. The *transient* analysis provides a time domain plot of any designated parameter(s). Following the execution of a transient analysis, the program can display the "final" values of all node voltages. This mode can easily be used to check the quiescent operating point. For the given circuit, the results yield a base voltage of -1.05 V, collector voltage of −5.3 V, and emitter voltage of −0.31. The quiescent voltages are therefore

$$V_{BEQ} = -0.74$$

$$V_{CEQ} = -4.99$$

compared with calculated values of −0.7 and −4.155, respectively. The variation is due to the transistor model (variation in β and V_{BE} with operating point) and to the tolerance of resistor values.

There are two ways in which we can find the gain. We can use an input source to insert a sinusoidal waveform and plot the output waveform. The gain is found by comparing amplitudes. The second, and easier, method is to use the *ac* analysis feature of the simulation program. This module plots output gain as a function of frequency (it also plots other parameters, including phase, but we will deal with these in later chapters). We need only run the ac analysis over a reasonable frequency range (not so low as to make the coupling capacitors deviate from short circuits and not high enough to force consideration of coupling capacitance). The results of the simulation for this example yield a gain of 19.899 dB. The required gain of 10 is equivalent to 20 dB, so the simulation has verified the design.

3.4 NONLINEARITIES OF BIPOLAR JUNCTION TRANSISTORS

In Section 2.4.2 we learned that a transistor operates in a linear manner except in the saturation and cutoff regions. Operating in or near these regions causes distorted reproduction of an input signal. Therefore, the shaded regions shown in Figure 3.9 should be avoided. Designers frequently discard 5%

Figure 3.9
Nonlinear portions of characteristic curve.

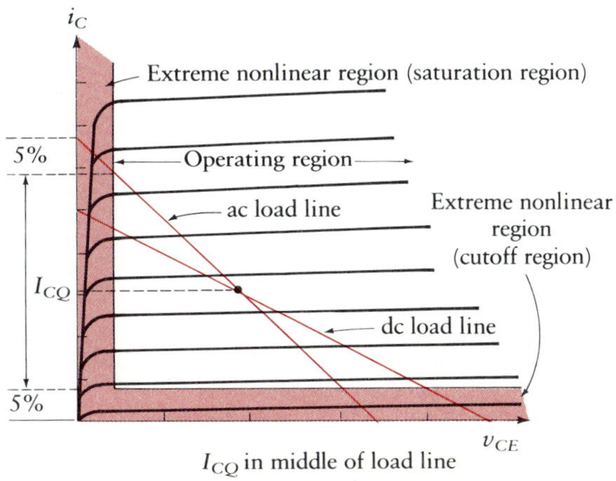

of the characteristic curve in the vicinity of the saturation region and 5% of the curve near the cutoff region. The actual size of the nonlinear region is a function of current, so the 5% simply serves as a rule of thumb and a starting point in a design. If increased output is needed, or if there are very "tight" constraints on distortion due to the nonlinear operation, this guideline may have to be modified.

Using these guidelines and assuming the I_{CQ} has been placed in the center of the ac load line, the undistorted peak-to-peak output voltage is given by equation (3.13).

$$v_o(p\text{-}p) = 0.9 \times 2|I_{CQ}|\,(R_L \parallel R_C)$$
$$= 1.8|I_{CQ}|\,(R_L \parallel R_C) \qquad (3.13)$$

Suppose now that I_{CQ} is not in the middle of the load line. The circuit will have a reduced output swing for a symmetrical input signal. Figure 3.10(a) and (b) graphically illustrates this reduced swing. The maximum symmetrical output swing can be determined as follows. Let us assume that the input is a sinusoid with peak, I_{Cm}.

$$i_C(t) = I_{Cm} \sin \omega t$$

Then

$$v_o(\text{max undistorted swing}) = 1.8 I_{Cm}(R_L \parallel R_C)$$

For the case where I_{CQ} is below the center of the ac load line as in Figure 3.10(a),

$$I_{Cm} = I_{CQ} - 0.05 \times I_{C\max}$$

Figure 3.10 Reduced swing on load line.

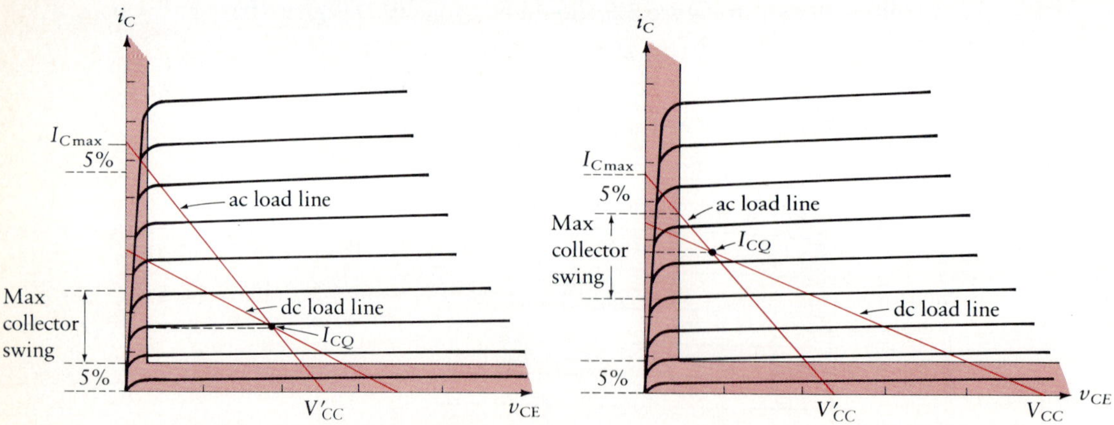

(a) I_{CQ} below middle of ac load line

(b) I_{CQ} above middle of ac load line

where from Figure 3.10(a), $I_{C\max}$ is the intercept of the ac load line with the i_C axis.

$$I_{C\max} = \frac{V'_{CC}}{R_{ac}}$$

Then

$$v_o(\text{max undistorted swing}) = 2(I_{CQ} - 0.05I_{C\max})\,(R_L \parallel R_C) \tag{3.14}$$

For the case where I_{CQ} is above the center of the ac load line as in Figure 3.10(b),

$$I_{CM} = 0.95I_{C\max} - I_{CQ}$$

Then

$$v_o(\text{max undistorted swing}) = 2(0.95I_{C\max} - I_{CQ})\,(R_L \parallel R_C) \tag{3.15}$$

Drill Problems

D3.1 The CE amplifier of Figure 3.7 has $V_{CC} = -15$ V, $R_L \to \infty$, $R_C = 5$ kΩ, $R_E = 500$ Ω, and $\beta = 200$. Determine R_1, R_2, A_v, A_i, and the maximum undistorted symmetrical output voltage.

Ans: $R_1 = 11.1$ kΩ; $R_2 = 103$ kΩ; $A_v = -9.63$;
$A_i = -17.6$; $v_o(p\text{-}p) = 12.3$ V

D3.2 In Drill Problem D3.1, R_E is bypassed with a capacitor. What are the new values of R_1, R_2, A_v, A_i, and R_o?

Ans: $R_1 = 11.1$ kΩ; $R_2 = 101$ kΩ; $A_v = -275$;
$A_i = -147$; $R_{\text{in}} = 2.67$ kΩ; $R_o = 5$ kΩ

D3.3 The *pnp* transistor amplifier shown in Figure 3.7 requires a voltage gain $A_v = v_o/v_i = -5$ and an input resistance $R_{\text{in}} = 1$ kΩ. $R_L = 5$ kΩ, $V_{CC} = -12$, $V_{BE} = -0.7$ V, and $\beta = 200$. Determine the current gain, maximum output swing, and other resistor values.

Ans: $A_i = -1$; $v_{o\max} = 6.35$ V;
$R_1 = 1.1$ kΩ; $R_2 = 8.6$ kΩ

D3.4 Design a CE amplifier (Figure 3.7) with a voltage gain of $A_v = -60$, $R_L = 5$ kΩ, and $R_{\text{in}} = 5$ kΩ. Design for a maximum voltage output swing.

Ans: $R_E = 25.2$ Ω; $R_1 = 13.4$ kΩ; $R_2 = 179$ kΩ;
$A_i = -60$; $v_o(p\text{-}p) = 7.12$ V
(Note that the amplifier is not bias stable.)

D3.5 Design a CE amplifier (Figure 3.7) for maximum output swing with $R_L = 6\ k\Omega$, $A_v = -60$, and $A_i = -20$.

Ans: $R_E = 30\ \Omega$; $R_1 = 2.67\ k\Omega$; $R_2 = 40\ k\Omega$;
$R_{in} = 2\ k\Omega$; $v_o(p\text{-}p) = 7.12\ V$

3.5 PARAMETERS FOR CC (EF) AMPLIFIER

3.5.1 Input Resistance, R_{in}

The EF circuit is shown in Figure 3.11. As before, the capacitors, C, are considered to be short circuits for midrange frequencies. We first wish to solve for the input resistance. Let us begin by finding the resistance to the right of the dashed line in Figure 3.11(b), R'.

$$R' = \frac{v_i}{i_b}$$

$$v_i = i_b h_{ie} + (1 + \beta) i_b (R_E \parallel R_L)$$

Now assuming that β is large (i.e., $1 + \beta \approx \beta$), we have

$$R' = h_{ie} + \beta(R_E \parallel R_L)$$

Now solving for R_{in}, we have

$$R_{in} = R_B \parallel [\beta h_{ib} + \beta(R_E \parallel R_L)]$$

$$= \frac{R_B[h_{ib} + (R_E \parallel R_L)]}{R_B/\beta + h_{ib} + (R_E \parallel R_L)} \tag{3.16}$$

In EF circuits where R_C is usually zero, the collector current is large. Then h_{ib} is small compared with $R_E \parallel R_L$, so h_{ib} can be ignored, and

Figure 3.11
EF amplifier.

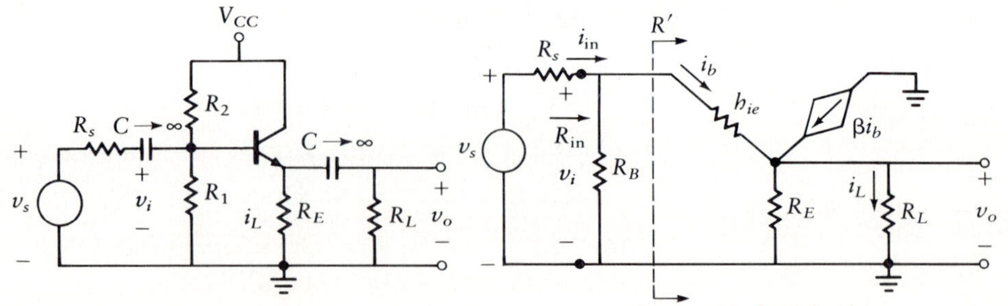

(a) CC circuit (b) ac equivalent circuit

$$R_{\text{in}} = R_B \parallel [\beta(R_E \parallel R_L)] \tag{3.17}$$

3.5.2 Current Gain, A_i

To obtain A_i, we find i_{in} and i_o and take the ratio. Current division in the output of Figure 3.11(b) yields

$$i_o = i_L = \frac{i_b \beta R_E}{R_E + R_L} \tag{3.18}$$

and current division in the input circuit yields

$$i_b = i_{\text{in}} \frac{R_B}{R_B + h_{ie} + \beta(R_E \parallel R_L)} \tag{3.19}$$

The current gain is found by combining equations (3.18) and (3.19) to obtain

$$A_i = \frac{i_o}{i_{\text{in}}} = \frac{R_B}{R_B/\beta + h_{ib} + (R_E \parallel R_L)} \frac{R_E}{R_E + R_L} \tag{3.20}$$

Equation (3.20) is the general expression for current gain. If h_{ib} is much smaller than the parallel combination of R_E and R_L, the following approximation results (if $R_B << \beta(R_E \parallel R_L)$):

$$A_i = \frac{R_B}{R_B/\beta + (R_E \parallel R_L)} \frac{R_E}{R_E + R_L} \approx \frac{R_B}{R_L} \tag{3.21}$$

Note that current gain is positive for the EF amplifier, since i_{in} is in phase with i_o.

3.5.3 Voltage Gain, A_v

We find the voltage gain from the current gain and the gain impedance formula (equation (3.11)). We substitute for R_{in} and A_i from equations (3.16) and (3.20) as follows:

$$A_v = A_i \frac{R_L}{R_{\text{in}}}$$

$$= \frac{R_B}{R_B/\beta + h_{ib} + (R_E \parallel R_L)} \frac{R_E}{R_E + R_L} \frac{R_L[R_B + h_{ie} + \beta(R_E \parallel R_L)]}{R_B[h_{ie} + \beta(R_E \parallel R_L)]} \tag{3.22}$$

When like terms are canceled from the numerator and denominator and we note that $h_{ie} = \beta h_{ib}$, equation (3.22) simplifies to

$$A_v = \frac{\beta(R_E \parallel R_L)}{h_{ie} + \beta(R_E \parallel R_L)} = \frac{R_E \parallel R_L}{h_{ib} + (R_E \parallel R_L)} \tag{3.23}$$

Since h_{ib} is usually small compared to $R_E \parallel R_L$, we can approximate the gain as

$$A_v = 1 \tag{3.24}$$

Notice that the gain is positive since v_i is in phase with v_o.

3.5.4 Output Resistance, R_o

An alternative equivalent circuit for an EF amplifier is shown in Figure 3.12(a). Here we have used the CB model of the transistor instead of the CE as employed in Figure 3.11. The resistance of the input voltage source is shown as R_s.

One technique for finding the output resistance is to assume an output voltage, solve for the output current, and take the ratio. In Figure 3.12(b) we take this approach, labeling the output current and voltage as i_T and v_T, respectively. We write the equations for the circuit as follows:

$$i_b = \frac{-V_1}{R_s \parallel R_B} \tag{3.25}$$

$$V_1 - V_T = (1 + \beta)i_b h_{ib}$$

$$\frac{V_T - V_1}{h_{ib}} + \frac{V_T}{R_E} = i_T$$

We combine these equations to find

$$R_o = \frac{v_T}{i_T}$$

$$= \left(h_{ib} + \frac{R_s \parallel R_B}{\beta}\right) \Bigg\| R_E \tag{3.26}$$

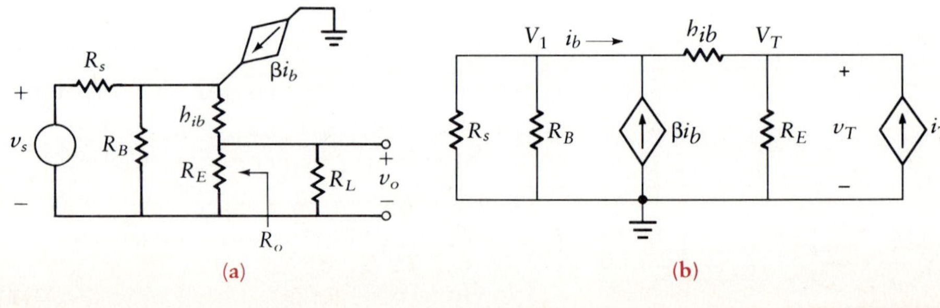

Figure 3.12
Output resistance of EF configuration.

The output resistance is dependent on the input parameters R_s and R_B. This is in contrast to the result for the CE amplifier, where R_o depends only on R_C (see Section 3.3.5).

Example 3.3 Capacitor-Coupled Common Collector (Design)

Design a single-stage *npn* EF amplifier (Figure 3.11a) with $\beta = 60$, $V_{BE} = 0.7$ V, $R_i = 1$ kΩ, and $V_{CC} = 12$ V. Determine the circuit element values for the stage to achieve $A_i = 10$ with a 100-Ω load.

SOLUTION We must select R_1, R_2, and R_E but, again, we have only two equations. These two equations are specified by the current gain and the placement of the Q-point. Example 3.1 showed that the best choice for a CE amplifier is to make $R_C = R_L$. We could derive a similar result for R_E and R_L in the CC amplifier (Problem 3.18 develops this result). We shall therefore begin by constraining R_E to be equal to R_L. This yields a third equation,

$$R_E = R_L = 100\ \Omega$$

Now finding the load line slopes,

$$R_{ac} = R_L \parallel R_E = 50\ \Omega$$

$$R_{dc} = R_E = 100\ \Omega$$

We once again use the design procedure introduced in Section 2.10. Since the amplitude of the input is not specified, we choose the quiescent current to place the Q-point in the center of the ac load line for maximum swing.

$$I_{CQ} = \frac{V_{CC}}{R_{ac} + R_{dc}} = 80\ \text{mA}$$

$$V_{CEQ} = I_{CQ} R_{ac} = 4\ \text{V}$$

We now find the value of h_{ib}.

$$h_{ib} = \frac{26\ \text{mV}}{|I_{CQ}|} = \frac{26\ \text{mV}}{80\ \text{mA}} = 0.33\ \Omega$$

Since h_{ib} is insignificant compared to $R_E \parallel R_L$, it can be ignored, which is usually the case for EF circuits.

Using the equation for current gain (equation (3.21)), we find

$$A_i = \frac{\beta R_E R_B}{(R_E + R_L)\,[R_B + (R_E \parallel R_L)\beta]}$$

Everything in this equation is known except R_B. We solve for R_B with the result

$$R_B = 1500\ \Omega$$

V_{BB} is found from the base loop.

$$V_{BB} = V_{BE} + I_{CQ}\left(\frac{R_B}{\beta} + R_E\right) = 10.7\ \text{V}$$

Continuing with the design as presented earlier, we find

$$R_1 = 13.8\ \text{k}\Omega$$

and

$$R_2 = 1.68\ \text{k}\Omega$$

The voltage gain of the CC amplifier is approximately unity. The input resistance is found from equations derived in Section 3.5.1.

$$R_{\text{in}} = R_B \parallel [\beta(R_E \parallel R_L)] = 1\ \text{k}\Omega$$

The output resistance is found from equations derived in Section 3.5.4.

$$R_o = \left(h_{ib} + \frac{R_i \parallel R_b}{\beta}\right) \Bigg\| R_E = 9.36\ \Omega$$

The maximum peak-to-peak symmetrical output swing is given by equations derived in Section 3.4.

$$v_o(p\text{-}p) = 1.8|I_{CQ}|\,(R_E \parallel R_L) = 7.2\ \text{V}$$

The power dissipated in the load, P_L, and the maximum power required of the transistor, P_T, are

$$P_L = \frac{(0.9I_{CQ}/2)^2 R_L}{2} = 64.8\ \text{mW}$$

$$P_T = I_{CQ}V_{CEQ} = 320\ \text{mW}$$

Drill Problems

D3.6 Design an EF amplifier (Figure 3.11) with $A_i = 15$, $V_{CC} = 18$ V, $\beta = 100$, $V_{BE} = 0.7$ V, and $R_L = 200\ \Omega$. Find the peak-to-peak undistorted output voltage.

Ans: $R_1 = 28.3\ \text{k}\Omega$; $R_2 = 5.1\ \text{k}\Omega$; $R_E = 200\ \Omega$;
$v_o(p\text{-}p) = 10.8\ \text{V}$

D3.7 Design an EF amplifier that has $R_{\text{in}} = 2\ \text{k}\Omega$; $R_C = 100\ \Omega$, $V_{CC} = 18$ V, $\beta = 100$, $V_{BE} = 0.7$ V, and $R_L = 200\ \Omega$.

Ans: $R_E = 200\ \Omega$; $R_1 = 4.89\ \text{k}\Omega$; $R_2 = 5.1\ \text{k}\Omega$;
$A_i = 10$; $v_o(p\text{-}p) = 6.48\ \text{V}$

3.6 PARAMETERS FOR CB AMPLIFIER

The CB amplifier is shown in Figure 3.13(a). The circuit is redrawn in Figure 3.13(b), where it has been reoriented to simplify the analysis.

3.6.1 Input Resistance, R_{in}

The equivalent hybrid circuit for the CB is shown in Figure 3.14, where the CE hybrid model is used. R_B is the parallel combination of R_1 and R_2. We derive R_{in} by taking the ratio of input voltage to current. The current in R_E is $i_{\text{in}} + (1 + \beta)i_b$. In the following analysis, as throughout this text, we assume that β is large enough to permit substituting β for $1 + \beta$. Thus,

$$v_i = R_E(i_{\text{in}} + \beta i_b)$$

Figure 3.13 CB amplifier.

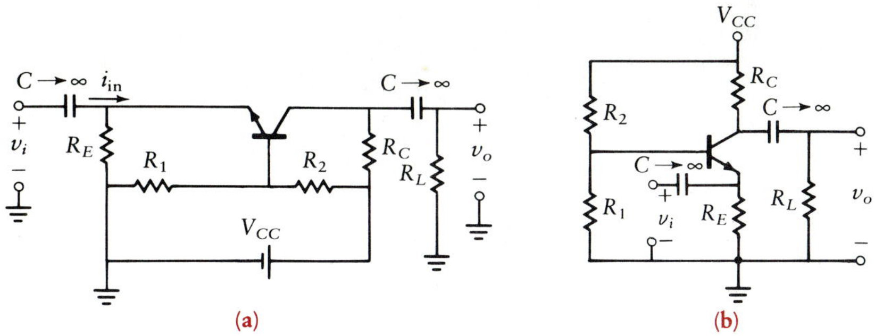

Figure 3.14 CB amplifier equivalent circuit.

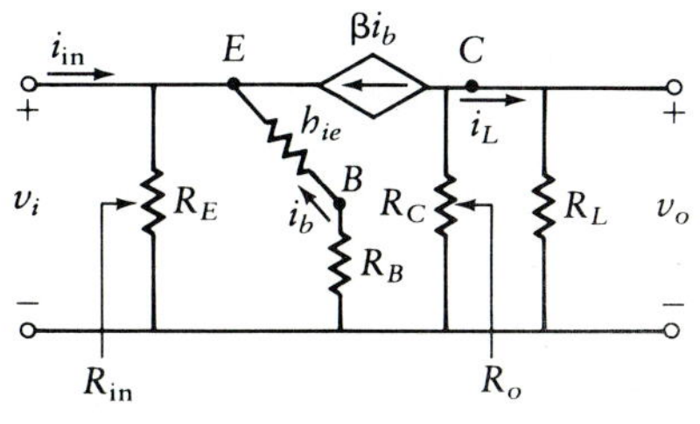

To find v_i/i_{in}, we must eliminate i_b from this equation. We see from Figure 3.14 that

$$i_b = -\frac{v_i}{h_{ie} + R_B}$$

Substituting this in the equation for v_i, we have

$$v_i = R_E\left(i_{in} - \frac{\beta v_i}{h_{ie} + R_B}\right)$$

Finally, forming the ratio of input voltage to current, we have

$$R_{in} = \frac{v_i}{i_{in}} = \frac{R_E(h_{ib} + R_B/\beta)}{h_{ib} + R_B/\beta + R_E} = R_E \,\Big\|\, \left(h_{ib} + \frac{R_B}{\beta}\right) \tag{3.27}$$

Both h_{ib} and R_B/β are usually only a few ohms, so R_{in} is quite small. This low input resistance represents a serious limitation of the CB configuration. If R_1 is bypassed by a capacitor, R_B (the parallel combination of R_1 and R_2) is zero and the input resistance is

$$R_{in} = R_E \,\|\, h_{ib}$$

3.6.2 Current Gain, A_i

The current gain for the circuit of Figure 3.14 is found as follows:

$$A_i = \frac{i_L}{i_{in}} = -\frac{\beta i_b}{i_{in}} \frac{R_C}{R_C + R_L}$$

From the previous derivation for R_{in}, we have

$$i_b = \frac{-v_i}{h_{ie} + R_B} = \frac{-i_{in}R_{in}}{h_{ie} + R_B}$$

$$= \frac{-R_E i_{in}}{\beta R_E + h_{ie} + R_B}$$

We combine these equations to find the current gain.

$$A_i = \frac{R_E}{R_E + h_{ib} + R_B/\beta} \frac{R_C}{R_C + R_L} \tag{3.28}$$

3.6.3 Voltage Gain, A_v

The *gain impedance formula* is used to find A_v. We use A_i from equation (3.28) and R_{in} from equation (3.27) to obtain the expression for voltage gain.

$$A_v = A_i \frac{R_L}{R_{in}} = \frac{R_C \parallel R_L}{h_{ib} + R_B/\beta} \tag{3.29}$$

If a bypass capacitor is added between base and ground, R_B/β is deleted from equation (3.29) and the expression simplifies to

$$A_v = \frac{R_L \parallel R_C}{h_{ib}}$$

Note that the voltage gain increases significantly with the addition of this bypass capacitor.

3.6.4 Output Resistance, R_o

As in the case of the CE amplifier, the dependent current generator, βi_b, exhibits a high resistance. Therefore,

$$R_o = R_C.$$

Example 3.4 **Capacitor-Coupled CB (Design)**

Design a CB amplifier using an *npn* transistor (Figure 3.13(a)) with $\beta = 100$, $V_{CC} = 24$ V, $R_L = 2$ kΩ, $R_E = 400$ Ω, and $V_{BE} = 0.7$ V. Design this amplifier for a voltage gain of 20. Verify your design using a computer simulation.

SOLUTION We need an additional constraint, so we set

$$R_C = R_L = 2 \text{ k}\Omega$$

in accordance with the results of Example 3.1. Then we have

$$h_{ib} + \frac{R_B}{\beta} = \frac{R_C \parallel R_L}{A_v} = 50 \ \Omega$$

$$R_{ac} = 1.4 \text{ k}\Omega$$

$$R_{dc} = 2.4 \text{ k}\Omega$$

For maximum swing, we set I_{CQ} to

$$I_{CQ} = \frac{V_{CC}}{R_{ac} + R_{dc}} = 6.3 \text{ mA}$$

$$h_{ib} = \frac{0.026}{I_{CQ}} = 4.13$$

h_{ib} is much less than R_E, so

$$R_B = \beta(50 - 4.13) = 4.59 \text{ k}\Omega$$

Since we have already found $h_{ib} + R_B/\beta = 50$, we can substitute this into the current gain equation to yield

$$A_i = \frac{400}{400 + 50} \frac{2000}{2000 + 2000} = 0.44$$

$$R_{\text{in}} = R_E \left\| \left(h_{ib} + \frac{R_B}{\beta} \right) \right. = 44 \ \Omega$$

We use the bias equation (equation 2.13)) to find the parameters of the input bias circuitry.

$$V_{BB} = V_{BE} + I_{CQ}\left(\frac{R_B}{\beta} + R_E\right) = 3.52 \text{ V}$$

The bias resistors are then given by

$$R_1 = \frac{R_B}{1 - V_{BB}/V_{CC}} = 5.38 \text{ k}\Omega$$

$$R_2 = \frac{R_B V_{CC}}{V_{BB}} = 31.3 \text{ k}\Omega$$

The maximum peak-to-peak undistorted output voltage is

$$v_o(p\text{-}p) = 1.8|I_{CQ}|\,(R_C \parallel R_L) = 11.4 \text{ V}$$

We now use a computer simulation to verify this design. As in Example 3.2, our first job is to input the circuit. Once again, we can either add an entry to the transistor library to match the given parameters or choose a transistor that comes close to these parameters. The 2N2369 has a nominal gain of 100 and a nominal V_{BE} of 0.75 V. We choose this transistor and set the coupling capacitors to a very high value of 1000 μF. The ac analysis module of the simulation program yields a gain of 26.7 dB. The required gain is 20, which is equivalent to 26 dB, so we have met the specifications.

Example 3.5 Capacitor-Coupled CB (Analysis)

An *npn* transistor is connected in a CB configuration as shown in Figure 3.13(b). The voltage source is $V_{CC} = 20$ V, $\beta = 200$, $V_{BE} = 0.7$ V, $R_E = 200\ \Omega$, $R_1 = 5$ kΩ, $R_2 = 80$ kΩ, and $R_C = R_L = 5$ kΩ. A large capacitor is placed between the base of the transistor and ground. Determine the voltage gain, current gain, input impedance, and maximum undistorted symmetrical output voltage swing.

SOLUTION We find the Thevenin equivalent of the base circuitry.

$$R_B = R_1 \parallel R_2 = 4.7\ \text{k}\Omega$$

$$V_{BB} = 20 \times \frac{5 \times 10^3}{8.5 \times 10^4} = 1.18\ \text{V}$$

The *Q*-point location is found by writing a KVL equation around the base-emitter loop.

$$V_{BE} + I_{CQ}\left(R_E + \frac{R_B}{\beta}\right) = V_{BB}$$

where we assume that $I_C = I_E$. Substituting values and solving for I_{CQ} yields

$$I_{CQ} = 2.13\ \text{mA}$$

Then

$$h_{ib} = \frac{26\ \text{mV}}{|I_{CQ}|} = 12.2\ \Omega$$

A_i and A_v are found from equations derived in Sections 3.6.2 and 3.6.3.

$$A_i = \frac{R_E}{R_E + h_{ib}} \frac{R_C}{R_C + R_L} = 0.47$$

$$A_v = \frac{R_L \parallel R_C}{h_{ib}} = 205$$

Note that we have not included R_B in the formula for A_i since it is bypassed by a large capacitor.

Solving for R_{in} using equations derived in Section 3.6.1, we obtain

$$R_{in} = R_E \parallel h_{ib} = 11.5\ \Omega$$

To determine the maximum output voltage swing, we evaluate the ac and dc load line equations to determine whether I_{CQ} is above or below the center of the ac load line.

$$V_{CEQ} = V_{CC} = (R_C + R_E)I_{CQ}$$
$$= 20 - 5.2\ \text{k}\Omega(2.13\ \text{mA}) = 8.9\ \text{V}$$

The ac load line intersects the axis at

$$V'_{CC} = V_{CEQ} + I_{CQ}R_{ac}$$

where

$$R_{ac} = R_C \parallel R_L + R_E = 2.7\ \text{k}\Omega$$

Then

$$V'_{CC} = 8.9 + 2.13\ \text{mA}(2.7\ \text{k}\Omega) = 14.7\ \text{V}$$

$$I'_C = \frac{V'_{CC}}{R_{ac}} = 5.4\ \text{mA}$$

Note that the Q-point is below the center of the load line ($I_{CQ} = 2.13$ mA and the center is at $5.4/2 = 2.7$ mA). Therefore, using the appropriate equation from Section 3.4, we obtain the peak-to-peak output voltage swing as follows:

$$v_o(p\text{-}p) = 2(I_{CQ} - 0.05I'_{C\max})(R_L \parallel R_C) = 9.3\ \text{V}$$

Drill Problems

D3.8 Determine the voltage gain of a CB amplifier (Figure 3.13(a)) with $R_L = 3\ \text{k}\Omega$, $R_E = 500\ \Omega$, $V_{CC} = 15$ V, $V_{BE} = 0.7$ V, $R_B = 6\ \text{k}\Omega$, and $\beta = 200$. The circuit is designed for maximum voltage output swing.

Ans: $A_v = 37.9$

D3.9 Repeat Drill Problem D3.8 assuming that a large capacitor is added from the base to ground.

Ans: $A_v = 157$

D3.10 Design a CB amplifier (Figure 3.13(a)) that has a voltage gain of 40. Determine the component values when $V_{CC} = 20$ V, $R_L = 4$ kΩ, $R_E = 500$ Ω, $V_{BE} = 0.7$ V, and $\beta = 100$.

Ans: $R_1 = 4.6$ kΩ; $R_2 = 36.4$ kΩ;
$R_C = 4$ kΩ; $R_{in} = 45$ Ω

3.7 TRANSISTOR AMPLIFIER APPLICATIONS

In this section, we summarize the results of the previous sections and suggest applications for the three types of amplifier configuration based on their properties.

The CE amplifier has significant current and voltage gain with moderate input and output impedance. The high input impedance is desirable, but the high output impedance poses some problems. The higher the output impedance, the less current can be drawn from an amplifier without a significant drop in output voltage. The CE is used most often for voltage amplification. It can provide a large output voltage swing. In multistage systems, this output becomes the input of the next stage of the system.

The EF (CC) amplifier provides a high current gain with a low output impedance. It has a high input impedance and a voltage gain near unity. Clearly, it is not used for voltage amplification. The low output impedance makes this circuit useful for driving high-current devices. It can be used as a *buffer* between a CE amplifier and a current-drawing load. The CC amplifier is frequently used as a power amplifier and also in impedance matching applications. This amplifier is normally found in the final output stage of a signal amplifier, as it not only drops the impedance to a low value but also provides the necessary power to drive the load.

The CB amplifier has a low input impedance and a relatively high output impedance. These properties are less desirable for signal amplification. If the base is bypassed to ground with a capacitor, the amplifier has high voltage gain but the current gain is less than unity. Even without the bypass capacitor, the voltage gain is higher than that of the common emitter. Thus, if the source driving the amplifier has a low impedance and the load is drawing little current, the CB can be used as a voltage amplifier.

3.8 AMPLIFIER COUPLING

When a system is composed of more than one transistor stage, it is necessary to connect, or couple, the transistor amplifiers to each other. There are several common ways of accomplishing this interconnection between amplifiers. In the following sections, we discuss capacitive, direct, transformer, and optical coupling.

3.8.1 Capacitive Coupling

Capacitive coupling is the type that is illustrated in the designs of this chapter. It is the simplest and most effective way of decoupling the effects of the dc level of the first amplifier stage from those of the second stage. The capacitor removes the dc component from the ac signal. Thus, the biasing of the next stage is not affected by the previous stage. To ensure that the signal is not significantly changed by the addition of a capacitor, it is necessary for the capacitor to look like a short circuit to all frequencies being amplified. The specific criteria for choosing capacitor size are discussed in Chapter 7. For the present analysis, we assume that the capacitor is large.

3.8.2 Direct Coupling

Two amplifiers are *direct-coupled* if the output of the first amplifier is connected to the input of the second without the use of capacitors. An example is shown in Figure 3.15(a). Note that there are three different dc power supply voltages, labeled V_{CC}, V'_{CC}, and V_{EE}. The ac output of the first stage is superimposed on the dc quiescent level of the second stage. The dc voltage level of the output from the previous stage interacts with the dc voltages of the second stage, thereby affecting the bias conditions. To compensate for the changing of bias levels, the amplifier uses negative and positive dc voltage sources instead of a single V_{CC} source. A level shifter, as discussed in Chapter 7, can provide the bias voltage change.

Direct coupling can be used effectively when coupling an EF amplifier to a CE amplifier, as shown in Figure 3.15(b). Direct coupling eliminates the need for the coupling capacitor and the resistors R_1 and R_2 of the second stage. The

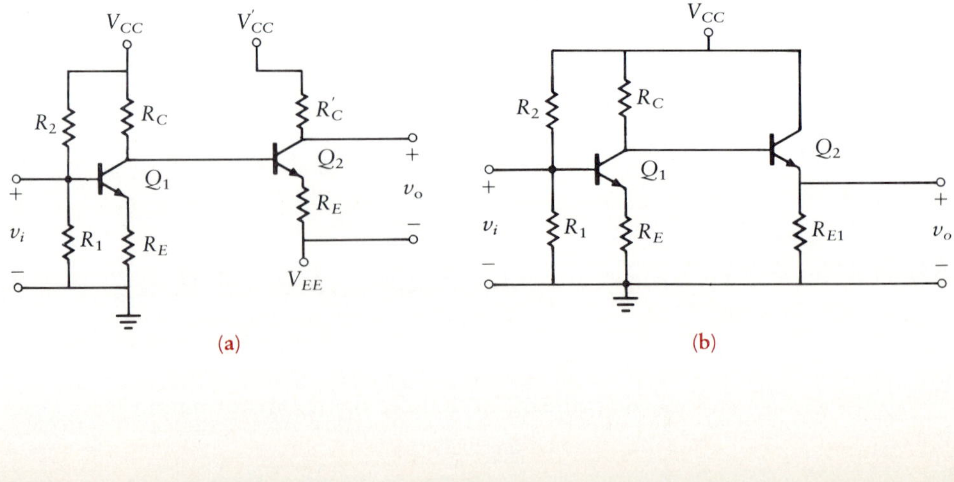

Figure 3.15 Direct-coupled amplifier.

directly coupled amplifier has good frequency response, since there are no series capacitors (i.e., frequency-sensitive elements) to affect the output signal at low frequency.

Direct coupling is commonly used in the design of integrated circuits. There are several reasons for its use. The resulting amplifier has excellent low-frequency response and can amplify dc signals. It is also simpler to fabricate on a chip, since there is no need for capacitors. However, direct coupling has the disadvantages of requiring additional power supplies or level shifters and of being sensitive to drift. Drift is a slowly varying dc level change that is normally rejected by capacitive coupling.

We note from Figure 3.15(b) that V_C of the first amplifier is V_{BB} of the second amplifier. This causes problems if the CE amplifier follows the CE amplifier, since CE amplifiers have a low V_{BB}, just above the 0.7 V needed to cause the transistor to conduct. We must therefore reduce the quiescent voltage, V_C, to a low level, thus significantly reducing the output voltage swing. Later in this text (Chapter 7), we will explore techniques for maintaining large output swings. Alternatively, we can easily couple a CE amplifier to an EF amplifier (as shown in the figure), since V_{BB} for EF amplifiers can be reasonably large. This is illustrated in the following example.

Example 3.6

Design a direct-coupled amplifier (Figure 3.15(b)) to have an 8-V peak-to-peak output voltage swing from Q_2. Assume $\beta = 200$ and $V_{BE} = 0.7$ V for both transistors, $R_{E1} = 2$ kΩ, $R_C = 4$ kΩ, $V_{CC} = 10$ V, $R_{in} = 5$ kΩ, and $A_v = -20$.

SOLUTION At Q_2, the voltage drop across R_{E1} must be 4 V 0-to-peak for the negative-going signal. Therefore,

$$V_{CEQ}(Q_2) = 10 - 4 = 6 \text{ V}$$

$$I_{CQ}(Q_2) = \frac{4}{2000} = 2 \text{ mA}$$

$$V_C(Q_1) = V_B(Q_2) = 4.0 + 0.7 = 4.7 \text{ V}$$

$$I_{BQ}(Q_2) = \frac{0.002}{200} = 10\ \mu\text{A}$$

$$V_C(Q_1) = 4.7 = 10 - [I_{CQ}(Q_1) + I_{BQ}(Q_2)]\,[4\text{ k}\Omega]$$

$$I_{CQ}(Q_1) + I_{BQ}(Q_2) = 1.33 \text{ mA}$$

Then

$$I_{CQ}(Q_1) = 1.32 \text{ mA}$$

$$R_L(Q_1) = h_{ie}(Q_2) + \beta R_{E1} = \beta\left[\frac{26}{2} + 2000\right] = 403 \text{ k}\Omega$$

Since the EF amplifier gain is unity, we use the A_v equation for the CE amplifier to find

$$-20 = -\frac{(4 \text{ k}\Omega \parallel 403 \text{ k}\Omega)}{R'_E}$$

$$R'_E = 198 \ \Omega$$

Since $I_{CQ} = 1.32$ mA and $h_{ib} = 19.7\ \Omega$

$$R_E = 178\ \Omega$$

$$R_{\text{in}} = 5 \text{ k}\Omega = \frac{R_B \times 198}{R_B/200 + 198}$$

$$R_B = 5.72 \text{ k}\Omega$$

$$V_{BB} = 0.7 + 0.00132 \times \left(\frac{5720}{200} + 178\right) = 0.973 \text{ V}$$

$$R_1 = \frac{5720}{1 - 0.973/10} = 6.34 \text{ k}\Omega$$

$$R_2 = 5720 \times \frac{10}{0.973} = 58.8 \text{ k}\Omega$$

This result illustrates that CE and EF amplifiers can be cascaded without creating problems, except that maximum output swing cannot be attained. In this particular example, only 8-V peak-to-peak signal output is required. The dc coupling does not affect the output as long as that output does not exceed the limit of ±4 V. The coupling provides the ability to pass a dc signal.

3.8.3 Transformer Coupling

A transformer can be used to couple two amplifier stages. This type of coupling is often used when high frequencies are being amplified. Although transformers are more costly than capacitors, their advantages can justify the additional cost. Through appropriate choice of turns ratio, a transformer can be used to increase either the voltage or current gain. For example, in the out-

Figure 3.16
Transformer coupling into a speaker.

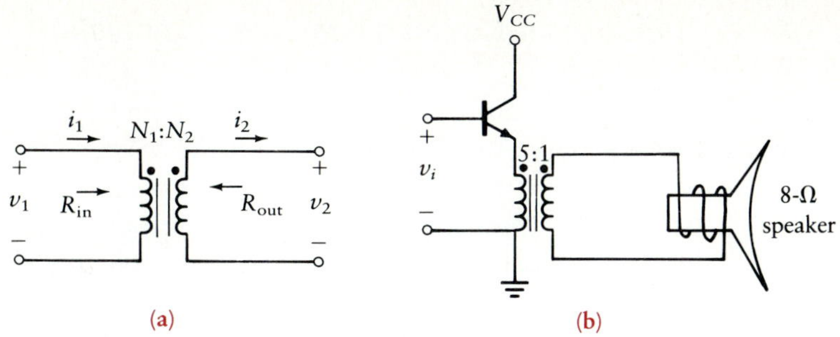

put stage of a power amplifier, the transformer is used to increase the current gain. There are other benefits associated with the use of a transformer. For example, the transformer can be tuned to resonance so that it becomes a band-pass filter (a filter that passes desired frequencies while attenuating frequencies outside the desired band).

Coupling the output stage to the load in an emitter follower can be accomplished by using a transformer. Figure 3.16 illustrates this technique. We refer to Figure 3.16 in reviewing the operation of a transformer. The input and output voltages are proportional to the transformer turns ratio as follows:

$$v_2 = v_1\left(\frac{N_2}{N_1}\right)$$

where N_1 is the number of turns in the primary coil and N_2 is the number of turns in the secondary coil. The input and output currents are related inversely to the voltage, since power must be conserved. Thus,

$$i_2 = i_1\left(\frac{N_1}{N_2}\right)$$

Taking the ratio of voltage to current yields the impedance relationship,

$$Z_1 = Z_2\left(\frac{N_1}{N_2}\right)^2$$

Figure 3.16(b) illustrates an application of these results where the transformer is used to drive an 8-Ω speaker. If the transformer turns ratio is 5:1, the equivalent resistance seen by the transistor emitter is 200 Ω. If v_i is a sinusoid of amplitude 10 V, the emitter voltage is approximately the same value since the gain of the EF amplifier is unity. The voltage at the speaker is one-fifth of this, or a 2-V amplitude sinusoid. The current at the speaker is a sinusoid of 250 mA amplitude (i.e., use Ohm's law at the speaker terminals), and the current in the transistor emitter is a sinusoid of 50 mA amplitude. The biasing of these circuits will be deferred to Chapter 6, where we discuss the details of power amplifiers.

Example 3.7 Transformer-Coupled Amplifier (Analysis)

Calculate the current gain, voltage gain, and input resistance for the transformer-coupled amplifier of Figure 3.17.

SOLUTION Note that the total amplification for the stage is obtained by taking the products of the gains of each section (the sections are separated by dashed lines in the figure).

Since the dc operating parameters are not specified, we will make approximations of A_v, A_i, and R_{in} as shown in Figure 3.17. We will be studying these types of amplifiers in more detail in Chapter 6.

The voltage gain of the transistor is found from the equation for the CE amplifier, where the emitter resistance is found by reflecting the 500-Ω load resistance back through the transformer. The overall voltage gain, A_v, includes the voltage-scaling effects of the two transformers. The results are shown directly on the figure.

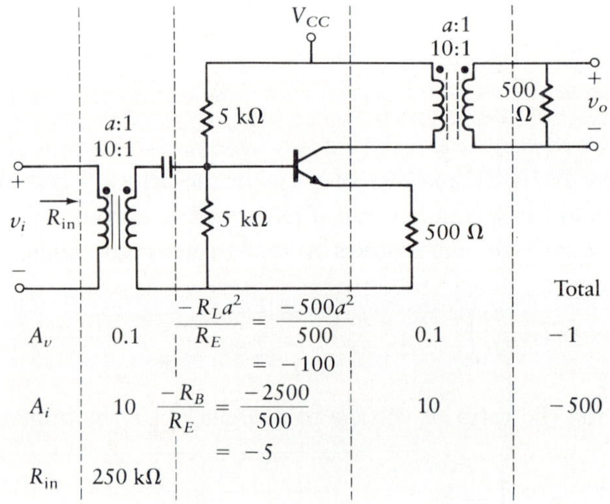

Figure 3.17 Transformer-coupled amplifier of Example 3.7.

The current gain, A_i, is found similarly. Note that there is only one resistor in the collector circuit, that being the load reflected through the transformer.

The input resistance to the transistor is R_B, which is the parallel combination of R_1 and R_2, or 2500 Ω. This is reflected through the transformer to obtain R_{in}.

3.8.4 Optical Coupling

Numerous applications require optical coupling of electronic circuits. These applications can be categorized as follows:

- Light-sensitive and light-emitting devices.
- Discrete detectors and emitters for fiber-optic systems.
- Interrupter/reflector modules that detect objects by modifying the light path.
- Isolators/couplers that transmit electrical signals without wire connections.

As an example of this last application, suppose we wish to use the 60-Hz power line as a driver for a clock. Because of the 15 to 25 A current available from the power line, we do not wish to make a wire connection for our timing needs but choose instead an optical connection. In the event of a component failure (e.g., a capacitor or transformer shorting), an optical coupler would prevent a dangerous, perhaps fatal, connection of the operator to the 110-V, 60-Hz power line. We discuss some of the major optical devices in the following paragraphs.

Optoelectronic Detectors and Emitters The light-sensitive diode operates such that as the light intensity, H, increases, the current in the external circuit also increases. This is the same as the phenomenon that occurs as we increase the base current in a transistor. Such a device is a *phototransistor* and is illustrated in Figure 3.18.

Optoelectronic components require packaging that allows light to pass through the package to the chip and also protects the chip. The semiconductor package "window" can be modified to provide lens action, which gives improved response along the optical axis of the lens and greater directional sensitivity. A typical package configuration is shown in Figure 3.19. Communication systems (e.g., telephone lines) using optical fibers have replaced copper wire systems. The light is emitted into and out of the optical fiber with devices such as that shown in Figure 3.19.

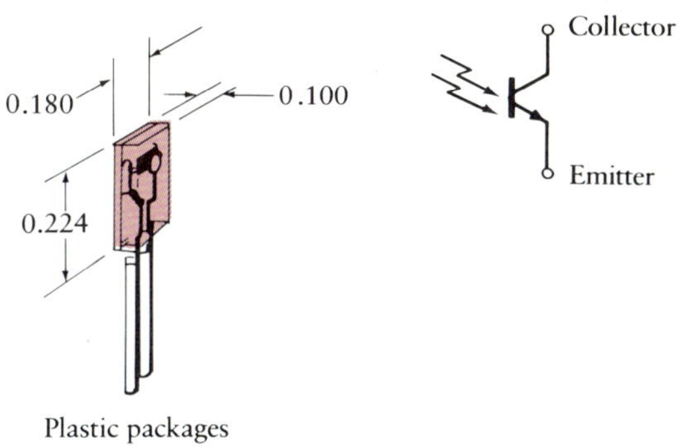

Figure 3.18 Phototransistor.

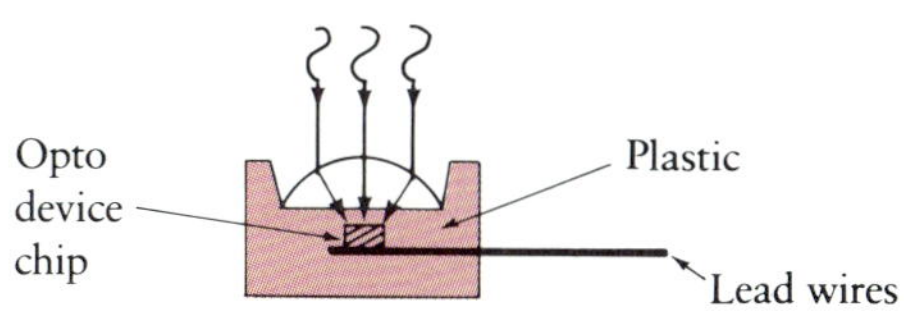

Figure 3.19 Optoelectronic package configuration. Courtesy of Power Electronics Semiconductor Department, General Electric Co.

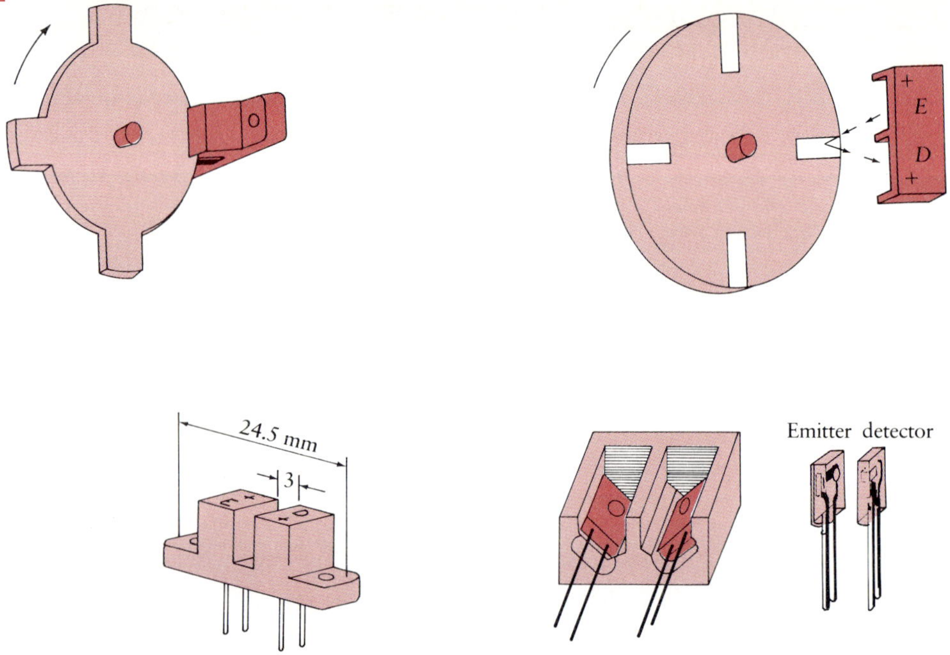

(a) (b)

Figure 3.20 Interrupter/reflector modules. Courtesy of Power Electronics Semiconductor Department, General Electric Co.

Figure 3.21 Optoisolator data sheet. Courtesy of Power Electronics Semiconductor Department, General Electric Co.

Photon Coupled Isolator H11A1, H11A2

The H11A1 and H11A2 are gallium arsenide infrared emitting diodes coupled with a silicon phototransistor in a dual in-line package, with 6 terminals.

Absolute maximum ratings: (25°C)

Infrared Emitting Diode

Power Dissipation	*100	milliwatts
Forward Current (Continuous)	60	milliamps
Forward Current (Peak) (Pulse width 1 μsec 300 P Ps)	3	ampere
Reverse Voltage	3	volts

*Derate 1.33 mW/°C above 25°C ambient

Phototransistor

Power Dissipation	150	milliwatts
V_{CEO}	30	volts
V_{CBO}	70	volts
V_{ECO}	7	volts
Collector Current (Continuous)	100	milliamps

Interrupter/Reflector Modules In many applications, it is necessary to determine the mechanical position or velocity of a shaft. Use of a light emitter and detector in either an interrupter mode, as shown in Figure 3.20(a), or a reflector mode, as shown in Figure 3.20(b), allows the engineer to measure mechanical shaft motion.

Optocouplers When we wish to couple two electrical circuits without making any direct wiring connections, we can use optocouplers (also termed *optoisolators*), which have no wire connection between input and output. The light path, emitter to detector, is totally enclosed in the component and cannot be modified externally. The degree of electrical isolation between the two devices is controlled by the materials in the light path and by the physical distance between the emitter and detector. The greater that distance, the better the isolation.

A portion of a data sheet for an optoisolator is shown in Figure 3.21. Notice that this device will isolate 1500 V peak (1060 V rms) between input and output.

3.9 PHASE SPLITTER

When two signals of opposite polarity are required, we use a *phase splitter,* shown in Figure 3.22. This amplifier is simultaneously a common emitter and a common collector. We choose $R_C = R_E = R_L$ so that the output voltage at the collector is equal in magnitude to the output voltage at the emitter, but these voltages are 180° out of phase. The two signal outputs from this circuit are approximately equal to the input signal in amplitude; that is, the voltage gain ratios, v_1/v_i and v_2/v_i, are approximately equal to unity in magnitude. The two outputs resulting from a sinusoidal input are sketched on the figure. The output at the emitter is in phase with the input signal, while the output of the collector is 180° out of phase with the input signal.

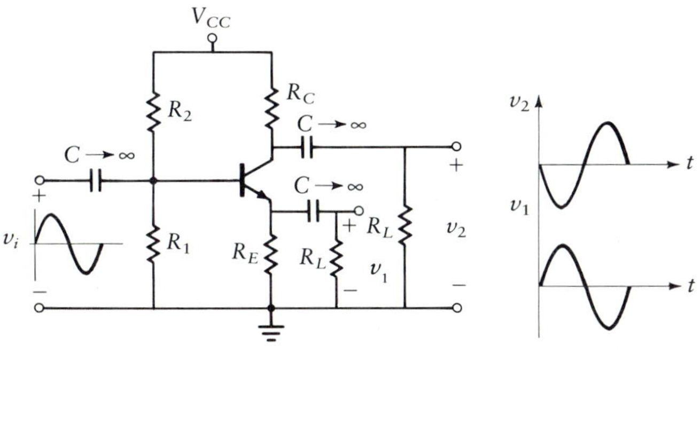

Figure 3.22
Phase splitter.

Drill Problem

D3.11 The outputs of Figure 3.22 are each connected to a 2-kΩ load. What is the output voltage swing of the phase splitter when $V_{CC} = 20$ V? Also determine R_C, R_E, R_1, and R_2 for maximum output swing when $\beta = 200$ and $V_{BE} = 0.7$ V.

Ans: $R_C = R_E = 2$ kΩ; $R_1 = 66.9$ kΩ; $R_2 = 99.6$ kΩ;
$v_o(p\text{-}p) = 6$ V (each output)

3.10 MULTISTAGE AMPLIFIER ANALYSIS

Amplifiers are often connected in series (cascaded) as shown in Figure 3.23. The load on the first amplifier is the input resistance of the second amplifier. The various stages need not have the same voltage and current gain. In practice, the earlier stages are often voltage amplifiers and the last one or two stages are current amplifiers. The voltage-amplifier stages ensure that the current stages have the proper input swings. The amount of gain in a stage is determined by the load on the amplifier stage, which is governed by the input resistance to the next stage. Therefore, in designing or analyzing multistage amplifiers, we start at the output and proceed toward the input.

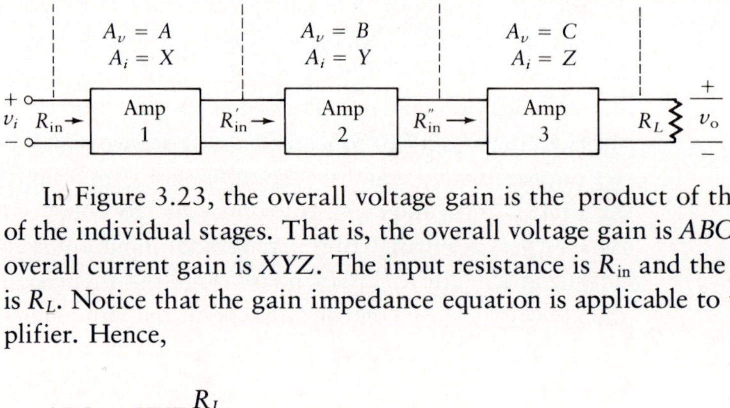

Figure 3.23
Multistage amplifier.

In Figure 3.23, the overall voltage gain is the product of the voltage gains of the individual stages. That is, the overall voltage gain is ABC. Similarly, the overall current gain is XYZ. The input resistance is R_{in} and the load resistance is R_L. Notice that the gain impedance equation is applicable to the overall amplifier. Hence,

$$ABC = XYZ\frac{R_L}{R_{in}} \tag{3.30}$$

We present the analysis of a multistage amplifier in the following example.

Example 3.8 Multistage Amplifier (Analysis)

Determine the current and voltage gains for the two-stage capacitor-coupled amplifier shown in Figure 3.24. Each capacitor is so large that it can be considered to be a short circuit to the ac signal.

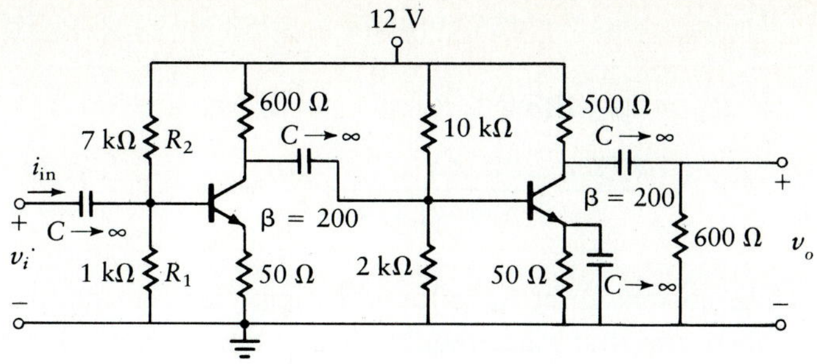

Figure 3.24
Multistage amplifier of Example 3.8.

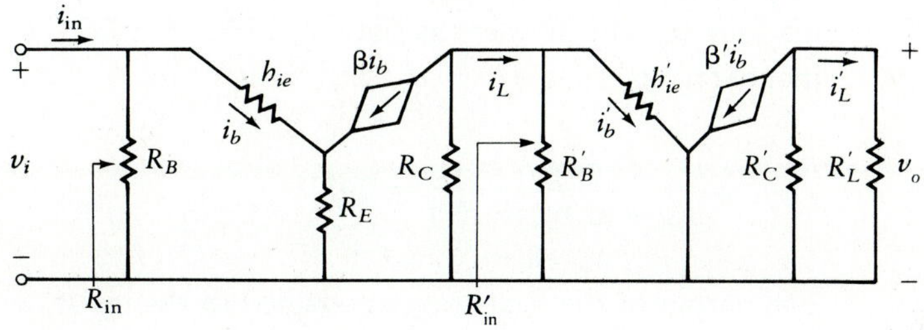

Figure 3.25
Equivalent circuit for Example 3.8.

SOLUTION We develop the hybrid equivalent circuit for the multistage amplifier. This equivalent is shown in Figure 3.25. Primed variables denote output stage quantities and unprimed variables denote input stage quantities. Calculations for the output stages are

$$R'_B = \frac{10^4 \times 2 \times 10^3}{10^4 + 2 \times 10^3} = 1.67 \text{ k}\Omega$$

$$V'_{BB} = \frac{12 \times 2 \times 10^3}{10^4 + 2 \times 10^3} = 2 \text{ V}$$

$$I'_{CQ} = \frac{V'_{BB} - V_{BE}}{R'_B/\beta + R'_E} = 22 \text{ mA}$$

$$h'_{ib} = \frac{26 \text{ mV}}{I'_{CQ}} = 1.17 \ \Omega$$

For the input stage,

$$R_B = \frac{7000 \times 1000}{7000 + 1000} = 875 \ \Omega$$

$$V_{BB} = \frac{12 \times 1000}{7000 + 1000} = 1.5 \text{ V}$$

$$I_{CQ} = \frac{1.5 - 0.7}{875/200 + 50} = 14.7 \text{ mA}$$

$$h_{ib} = \frac{26 \text{ mV}}{14.7 \text{ mA}} = 1.77 \ \Omega$$

We determine the input resistance using the equation from Table 3.1 as follows:

$$R_{\text{in}} = R_B \parallel \beta(h_{ib} + R_E)$$

$$= \frac{875 \times 200 \times (1.77 + 50)}{875 + 10{,}354} = 807\ \Omega$$

The current gain, A_i, is found by applying the equation from Table 3.1 twice, where the first stage requires using the correct value for R_L derived from the value of R_{in} to the next stage. Alternatively, we analyze Figure 3.25 by extracting four current dividers as shown in Figure 3.26. The current division of the input stage is

$$i_b = \frac{R_B i_{\text{in}}}{R_B + \beta(h_{ib} + R_E)} = 0.078 i_{\text{in}}$$

The output of the first stage is coupled to the input of the second stage in Figure 3.26(b). The input resistance of the second stage is

$$R'_{\text{in}} = R'_B \parallel h'_{ie} = 205\ \Omega$$

The current in R'_{in} is i_L and is given by

$$i_L = \frac{15.6 i_{\text{in}} \times 600}{805} = 11.6 i_{\text{in}}$$

Again, i_L is current-divided at the input to the second stage. Thus,

$$i'_b = \frac{-R'_B i_L}{R'_B + h'_{ie}} = -10.2 i_{\text{in}}$$

Figure 3.26
Current dividers for circuit of Figure 3.25.

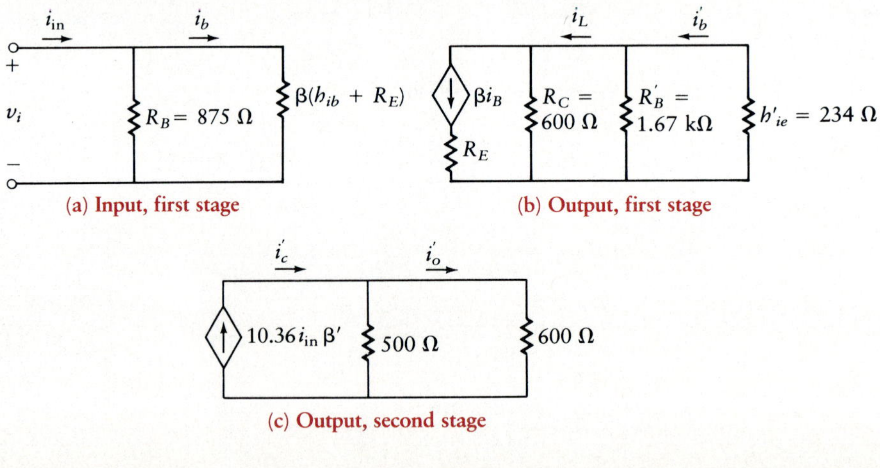

The output current is found from Figure 3.26(c):

$$i_o' = \frac{10.2i_{\text{in}} \times 200 \times 500}{500 + 600} = 927i_{\text{in}}$$

The current gain is then

$$A_i = 927$$

Now, using the gain impedance formula, we find the voltage gain as follows:

$$A_v = \frac{927 \times 600}{807} = 689$$

3.11 CASCODE CONFIGURATION

An interesting and useful multistage amplifier is the *cascode* configuration, which is shown in Figure 3.27(a). This two-stage amplifier comprises a CE amplifier driving a CB amplifier. Transistor Q_1 forms the CE amplifier, and the CB amplifier utilizes Q_2. This configuration has the advantages of increased output resistance and wider frequency response while maintaining high voltage gain.

The input resistance of the CE stage is relatively high. The low input impedance of the CB circuit forms the load resistance for the CE stage. The collector current of Q_2 is almost equal to the collector current of Q_1, which in turn drives the load ($R_L \parallel R_C$). This can be seen from the equivalent circuit of Figure 3.27(b). In this circuit, the capacitors are considered short circuits. The dc source, V_{CC}, is set to ground potential, since we are interested in the ac component of the signals. Since R_E is bypassed by C_E, it is also eliminated from the equivalent circuit. Figure 3.27(b) can be verified by referring to the CE and CB equivalent circuits of Table 3.2.

In the analysis of Figure 3.27(b), we let h_{ie} be much smaller than $R_1 \parallel R_2$, so

$$i_b = i_i \frac{R_1 \parallel R_2}{h_{ie} + R_1 \parallel R_2} \approx i_i$$

Since the collector currents in the transistors are equal, $\beta i_b = \beta' i_b'$. The load current, i_L, is given by equation (3.31), where we have used current division.

$$i_L = \frac{-\beta' i_b' R_C}{R_C + R_L} = \frac{-\beta i_b R_C}{R_C + R_L} = \frac{-\beta i_i R_C}{R_C + R_L} \tag{3.31}$$

Figure 3.27
Cascode configuration.

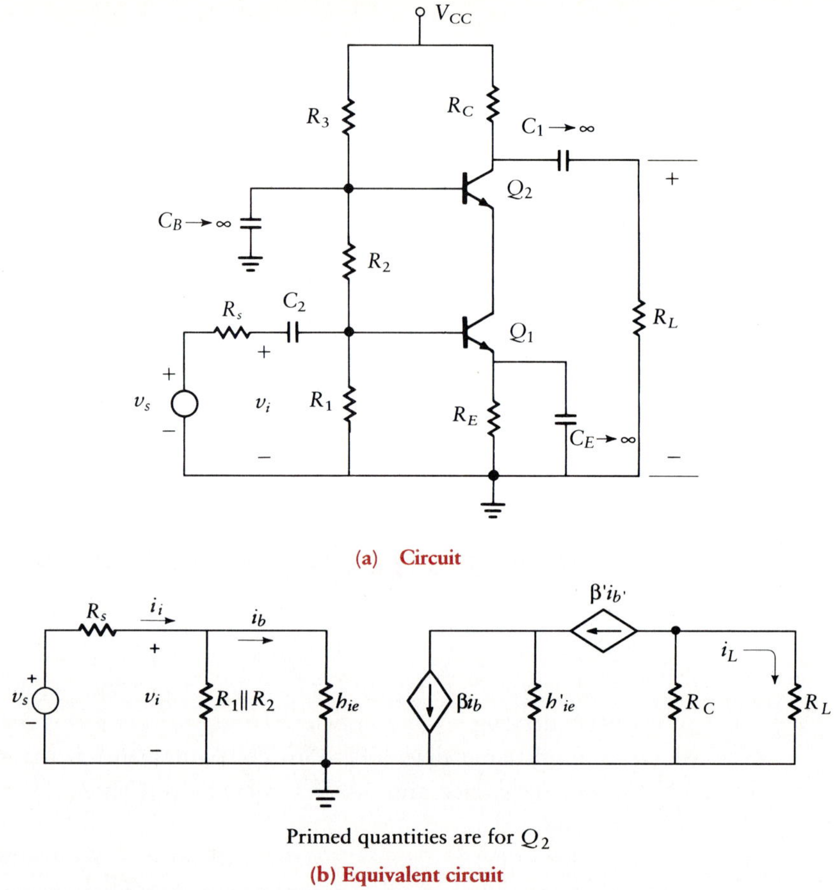

The current gain, A_i, for the two-stage amplifier is

$$A_i = \frac{i_L}{i_i} = \frac{-\beta R_C}{R_C + R_L}$$

The input resistance is

$$R_{\text{in}} = (R_1 \| R_2) \| h_{ie}$$

which is approximately h_{ie} since $(R_1 \| R_2) << h_{ie}$. The voltage gain is found from the gain impedance formula.

$$A_v = A_i \frac{R_L}{R_{\text{in}}} = -\frac{\beta R_C}{R_C + R_L} \frac{R_L}{h_{ie}}$$

$$= -\frac{R_C \| R_L}{h_{ie}/\beta} = \frac{-R_C \| R_L}{h_{ib}} \tag{3.32}$$

Equation (3.32) shows that the voltage gain of the cascode configuration is the same as that of a CE amplifier with a bypassed emitter resistor (R_E).

PROBLEMS

3.1 Derive equations for A_v, A_i, and R_{in} for the CE amplifier shown in Figure P3.1.

3.2 Calculate R_{in}, A_v, and A_i when $R_B = R_L = 5$ kΩ, $R_E = 1$ kΩ, and $h_{ie} = 0$ for the CE amplifier shown in Figure P3.2. Let β be given by:
a. $\beta = 200$
b. $\beta = 100$
c. $\beta = 10$

3.3 Determine A_v, A_i, and R_{in} for the amplifier shown in Figure P3.2 when $R_L = R_B = 5$ kΩ, $h_{ib} = 40$ Ω, $\beta = 300$, and R_E is given by:
a. $R_E = 1000$ Ω
b. $R_E = 500$ Ω
c. $R_E = 100$ Ω
d. $R_E = 0$

Discuss the effects of changing R_E.

3.4 For the common-emitter amplifier shown in Figure P3.1, $V_{BE} = 0.6$ V, $V_{CC} = 12$ V, $\beta = 300$, P_L(max average) = 100 mW, and $A_v = -10$. Determine R_1, R_2, R_{in}, and A_i. How much power is dissipated in the transistor?

3.5 Find A_v for the amplifier shown in Figure P3.3, where $h_{ie} = 2$ kΩ, $h_{re} = 0$, $h_{fe} = 200$, and $1/h_{oc} = 8$ kΩ.

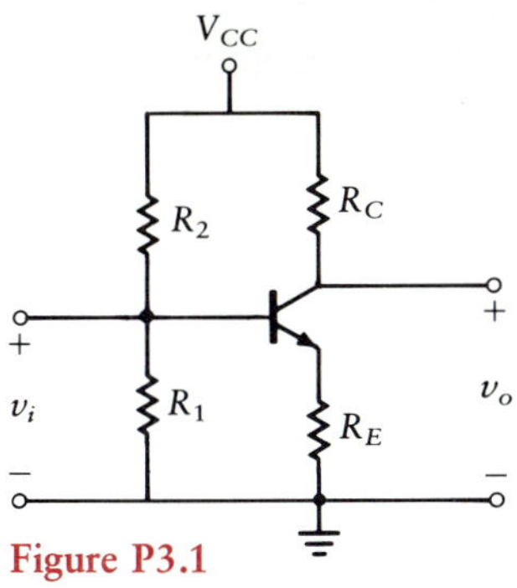

Figure P3.1

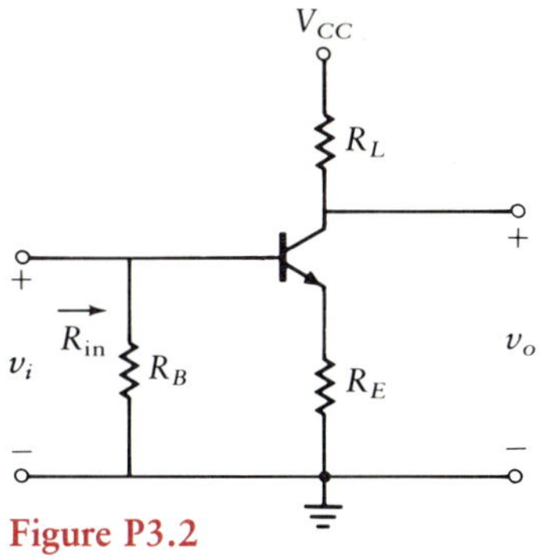

Figure P3.2

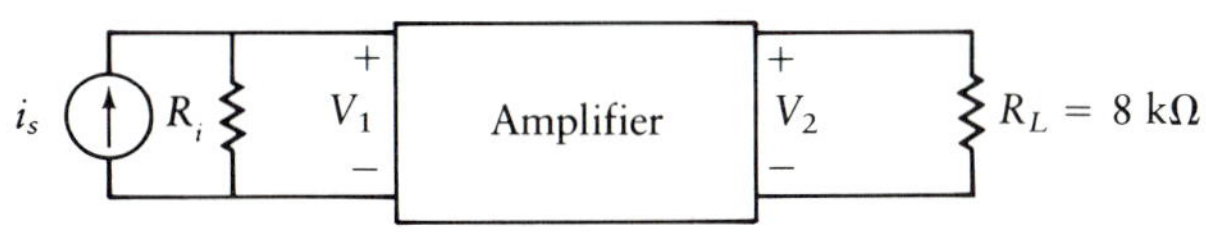

Figure P3.3

3.6 For the CE amplifier shown in Figure P3.4, where $h_{ie} = 1\ \text{k}\Omega$, $h_{oe} = 10\ \mu\text{S}$, and $h_{fe} = 50$, plot each of the following.

a. $A_i = i_L/i_{\text{in}}$, assuming $R_B << h_{ie}$, as a function of the value of R_L. Let R_L vary from 0 to 500 kΩ.

b. A_i as a function of R_L but assume $h_{re} = 0 = h_{oe}$.

3.7 For the CE amplifier shown in Figure P3.5, determine the variation of A_i and R_{in} if h_{fe} varies from 50 to 150 for the silicon transistor.

3.8 Determine h_{ie}, A_i, R_{in}, v_o/v_i, and R_o for the CE amplifier shown in Figure P3.6 if $h_{fe} = 100$ and $h_{re} = h_{oe} = 0$.

3.9 Compare input resistances and voltage gains for the ac equivalent amplifier circuits shown in Figure P3.7.

3.10 Design a CE amplifier as shown in Figure P3.8 using a *pnp* transistor when $R_L = 3\ \text{k}\Omega$, $A_v = -10$, $V_{BE} = -0.7$ V, $\beta = 200$, $A_i = -10$, and $V_{CC} = -12$ V. Determine all element values, R_{in}, and the maximum voltage swing across R_L.

3.11 Design a CE amplifier as shown in Figure P3.8 using a *pnp* transistor when $R_L = 4\ \text{k}\Omega$, $A_v = -15$, $R_{\text{in}} = 20\ \text{k}\Omega$, $V_{CC} = -20$ V, $\beta = 300$, and $V_{BE} = -0.6$ V. Determine all element values and the maximum peak-to-peak output voltage swing.

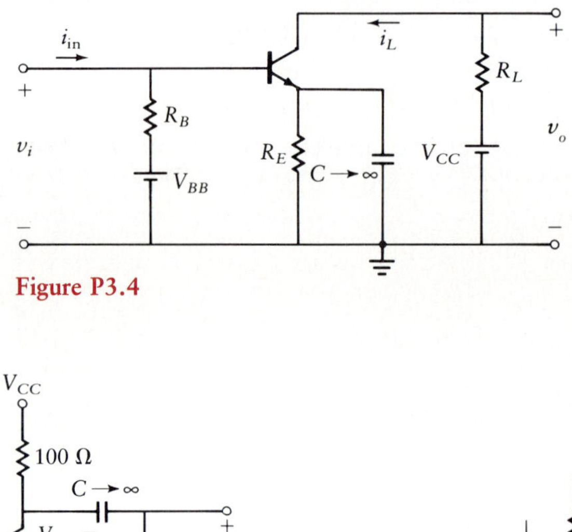

Figure P3.4

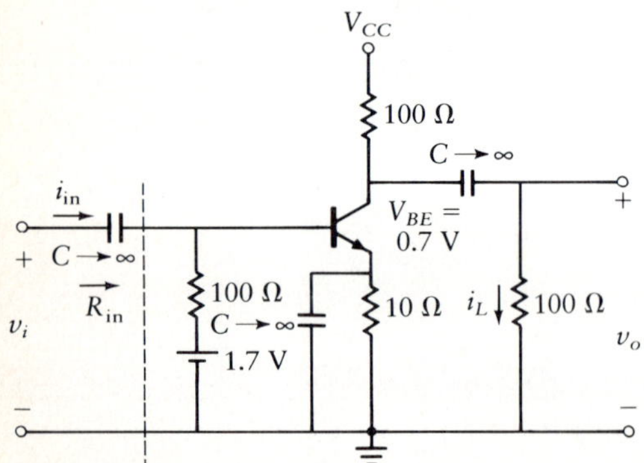

Figure P3.5

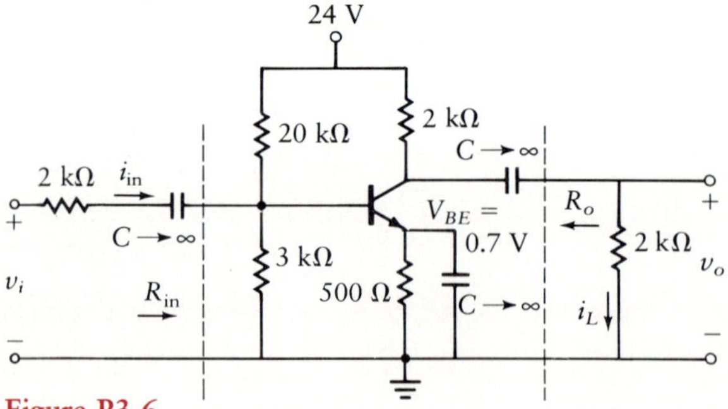

Figure P3.6

Figure P3.7

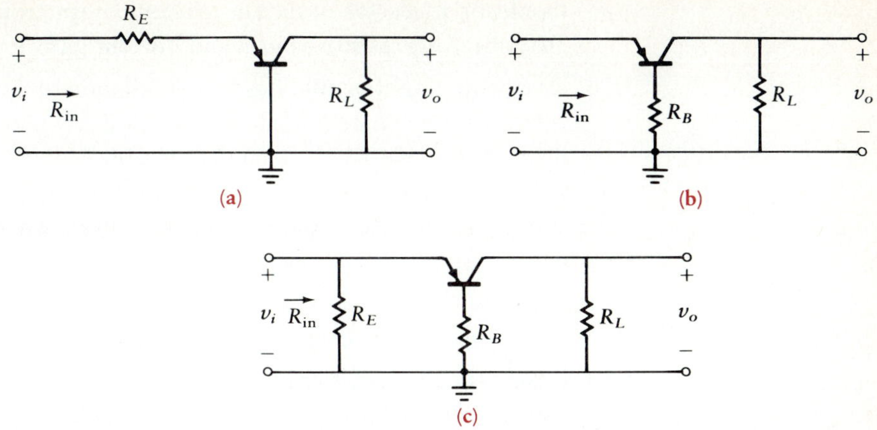

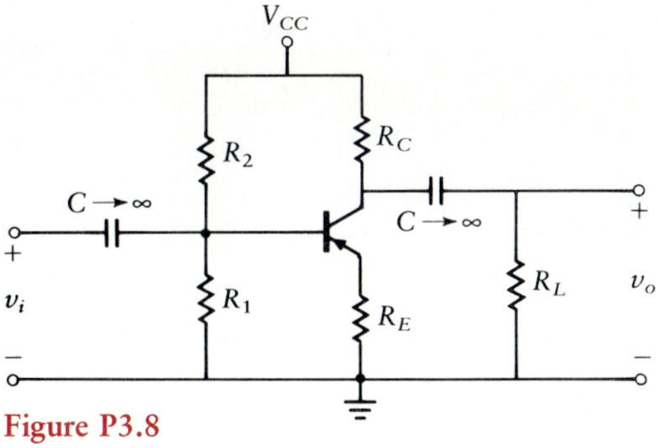

Figure P3.8

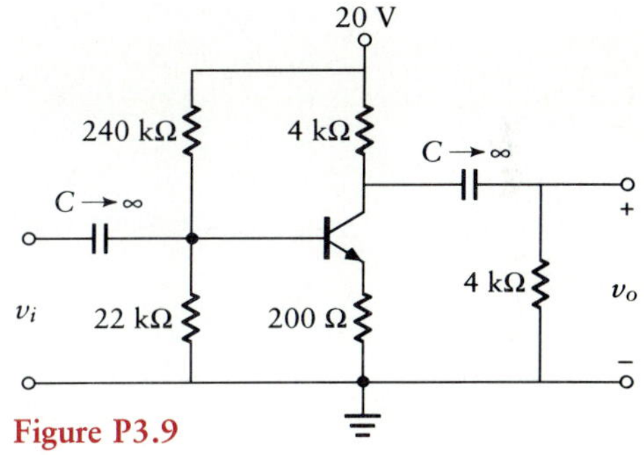

Figure P3.9

3.12 Design a CE amplifier as shown in Figure 3.4 using an *npn* transistor when $R_L = 9\ \text{k}\Omega$, $A_v = -10$, $A_i = -10$, $V_{BE} = 0.7$ V, $\beta = 200$, and $V_{CC} = 15$ V. Determine all element values, R_{in}, and the maximum peak-to-peak output voltage swing.

3.13 Design a CE amplifier to obtain a voltage gain of -25 when $R_{in} = 5\ \text{k}\Omega$, $R_L = 5\ \text{k}\Omega$, $V_{CC} = 12$ V, $\beta = 200$, and $V_{BE} = 0.7$ V. Determine all resistor values, current gain, and maximum output voltage swing. Use the circuit of Figure P3.8 except with an *npn* transistor.

3.14 Analyze the circuit shown in Figure P3.9 and determine the following when $\beta = 300$ and $V_{BE} = 0.6$ V.

a. I_{CEQ} and V_{CEQ}
b. Undistorted output voltage swing
c. Power supplied from power supply
d. Voltage gain
e. Load lines

3.15 Design a CE amplifier to obtain a voltage gain of -10 when $R_{in} = 2\ \text{k}\Omega$, $R_L = 4\ \text{k}\Omega$, $V_{CC} = 15$ V, $V_{BE} = 0.6$ V, and $\beta = 300$. This amplifier requires an output swing of 2 V undistorted peak-to-peak, so the design should be made for minimum current drain from the dc power source. Determine all resistor values and current gain.

3.16 Design an amplifier that has an overall gain of $A_v = -15$ when the input voltage has a source impedance (R_i) of 2 kΩ and the amplifier itself has $R_{\text{in}} = 4$ kΩ, $V_{BE} = 0.7$ V, and $\beta = 200$ (see Figure P3.10). The amplifier requires a maximum output voltage swing. Determine all resistor values, A_i, and the maximum output voltage swing.

3.17 Design an amplifier as shown in Figure P3.11 to obtain a voltage gain of -200 with an input resistance of 1 kΩ. Determine all resistor values and the maximum output voltage swing when $\beta = 400$ and $V_{BE} = 0.7$ V.

3.18 Design the EF amplifier in Figure P3.12 to drive a 200-Ω load using a *pnp* silicon transistor. $V_{CC} = -24$ V, $\beta = 200$, $A_i = 10$, and $V_{BE} = -0.7$ V. Determine all element values and calculate R_{in}, I_{CQ}, and the undistorted symmetrical output voltage swing for each R_E given below:

a. $R_E = R_L$
b. $R_E = 0.2R_L$
c. $R_E = 5R_L$

Compare your results by using a chart.

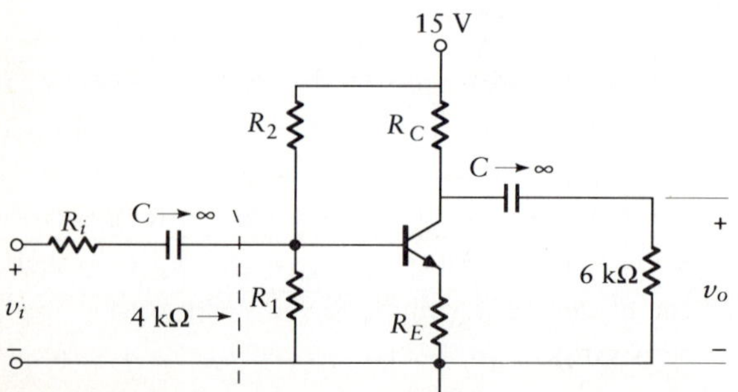

Figure P3.10

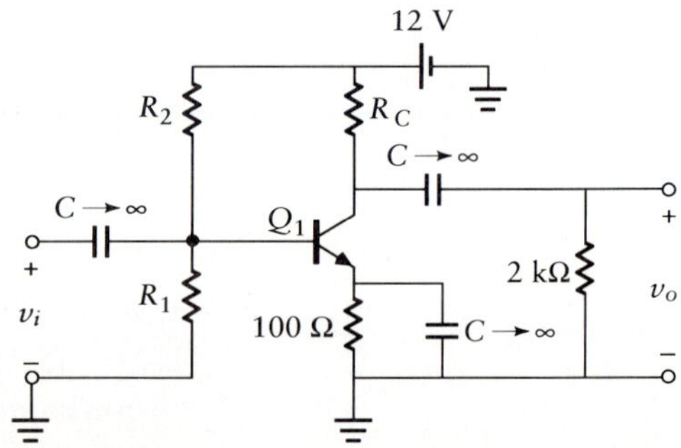

Figure P3.11

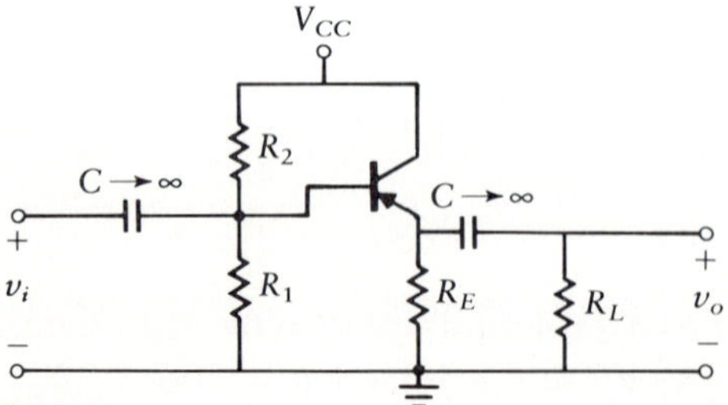

Figure P3.12

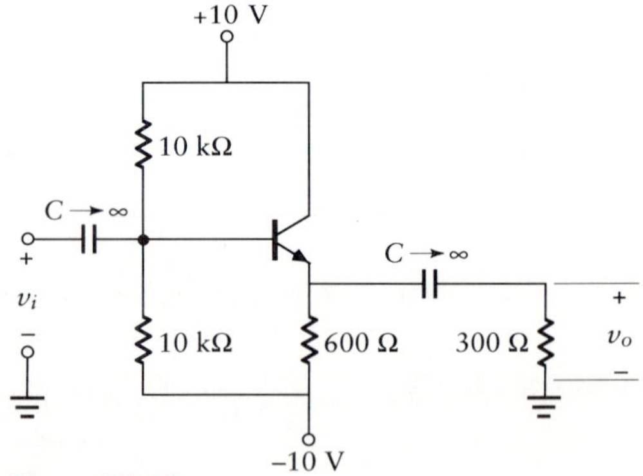

Figure P3.13

3.19 Design an EF amplifier as shown in Figure 3.11(a) using an *npn* transistor with $R_L = 500\ \Omega$, $V_{BE} = 0.7$ V, $A_i = 25$, $\beta = 200$, and $V_{CC} = 15$ V. Determine all element values, R_{in}, A_v, and the maximum output voltage swing.

3.20 Design an EF amplifier to drive an 8-Ω load when $\beta = 60$, $V_{CC} = 24$ V, $V_{BE} = 0.7$ V, $A_v = 1$, and $A_i = 10$. Use the circuit of Figure 3.11(a). Determine all element values, output voltage swing, and R_{in}.

3.21 Design an EF amplifier as shown in Figure 3.11(a) using an *npn* transistor when $R_L = 1500\ \Omega$, $V_{BE} = 0.7$ V, $A_i = 10$, $\beta = 200$, and $V_{CC} = 16$ V. Determine all element values, R_{in}, A_v, and the maximum output voltage swing.

3.22 Analyze the circuit shown in Figure P3.13 and determine the following when $\beta = 300$ and $V_{BE} = 0.6$ V.

a. I_{CQ} and V_{CEQ}
b. Undistorted output voltage swing
c. Power needed from power source
d. Maximum power output (ac undistorted)
e. Load lines

3.23 Design an EF amplifier as shown in Figure 3.11(a) to drive a 10-Ω load when $V_{CC} = 24$ V, $V_{BE} = 0.6$ V, $A_v = 1$, $R_{in} = 100\ \Omega$, and $\beta = 200$. Determine all element values, R_{in}, and maximum output voltage swing.

3.24 Analyze the circuit shown in Figure P3.14 when $\beta = 100$ and $V_{BE} = 0.7$ V and determine the following:

a. I_{CQ} and V_{CEQ}
b. Undistorted output voltage swing
c. Power supplied from the power source
d. Maximum power output (ac undistorted)
e. Current gain

3.25 Design a CB amplifer (see Figure 3.13) that has a voltage gain of 10 and a 4-kΩ load. Use $\beta = 100$, $V_{BE} = 0.7$ V, $V_{CC} = 18$ V, and $R_E = 500\ \Omega$. Determine values of I_{CQ}, R_1, R_2, R_B, and the voltage output swing. What is the voltage gain when R_1 is bypassed with a large capacitor?

Figure P3.14

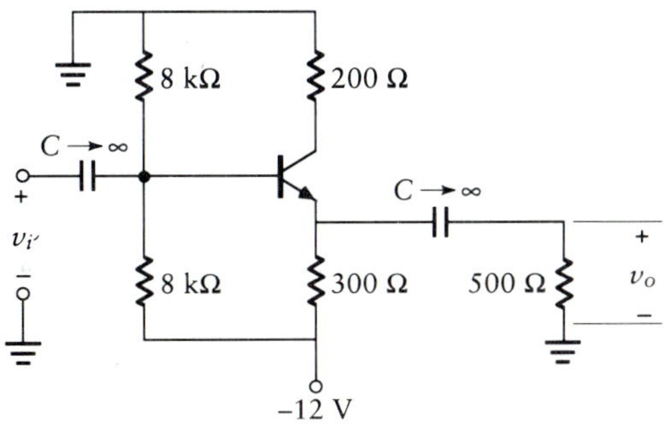

3.26 Design a CB amplifier using the values given in Problem 3.25 except that the voltage gain is 100. Determine value of R_1, R_2, I_{CQ}, R_B, and the maximum output voltage swing.

3.27 Design a CB amplifier for maximum voltage swing and at least 100 Ω input impedance, R_L = 8 kΩ, V_{CC} = 12 V, and R_E = 400 Ω. Use an *npn* transistor with a β = 200 and V_{BE} = 0.7 V. Determine the voltage gain and all resistor values.

3.28 Analyze a CB amplifier for R_{in}, A_v, and $V_o(p\text{-}p)$ that has the following values: V_{CC} = 16 V, R_1 = 2 kΩ, R_2 = 25 kΩ, R_E = 200 Ω, R_C = R_L = 4 kΩ, β = 200, and V_{BE} = 0.7 V. The base is ac grounded in Figure 3.13.

3.29 Determine the values of V_1, V_2, V_3, V_4, I_{C1}, and I_{C2} for the circuit shown in Figure P3.15. β is 300 for both transistors.

3.30 Directly couple a CE amplifier to an EF (see Figure 3.15(b)) for 4-V output swing with the following values: V_{CC} = 12 V, A_v = 10, Q_1 has β = 200 and V_{BE} = 0.7 V, Q_2 has β = 100 and V_{BE} = 0.7 V, and R_{E1} = 100 Ω. Let R_C = 4 kΩ, and find R_1, R_2, and R_E.

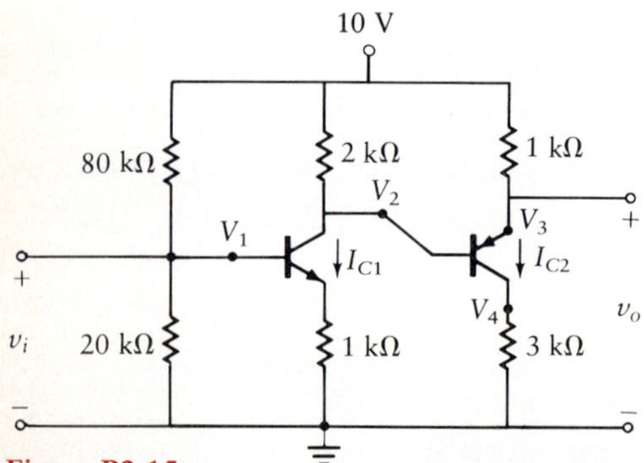

Figure P3.15

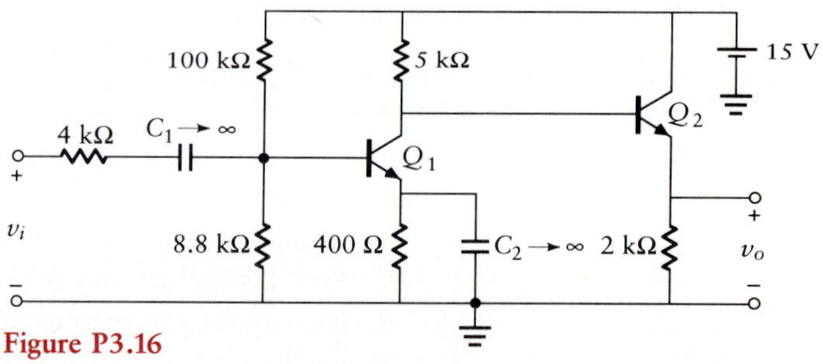

Figure P3.16

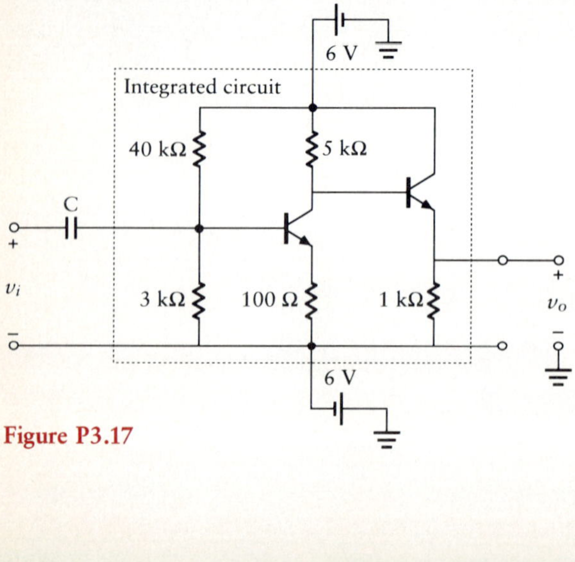

Figure P3.17

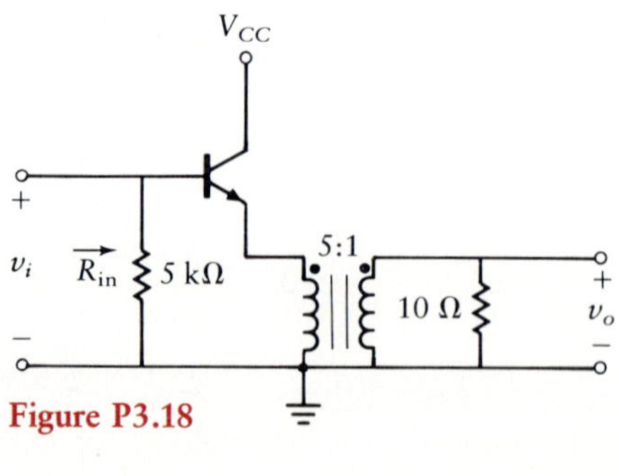

Figure P3.18

3.31 For the circuit in Figure P3.16, determine the following when $\beta = 400$ and $V_{BE} = 0.6$ V:

a. Q points for both amplifiers
b. Maximum undistorted output voltage swing
c. Sketch of the output signal
d. Voltage gain, v_o/v_i

3.32 For the circuit in Figure P3.17 when $v_i = 0.1 \sin 1000t$ V, determine the voltage output (assume $\beta = 200$ and $V_{BE} = 0.7$ V):

a. From $v_o(+)$ terminal to $v_o(-)$ terminal
b. From $v_o(+)$ terminal to ground

3.33 Determine A_v, A_i, and R_{in} for the EF amplifier shown in Figure P3.18 when $\beta = 200$ and $h_{ib} = 0$.

3.34 Determine the overall current and voltage gains and the input resistance for the transformer-coupled amplifier as shown in Figure P3.19. Use an *npn* transistor with $a = 4$, $R_1 = 2$ kΩ, $R_2 = 4$ kΩ, $V_{CC} = 15$ V, $\beta = 200$, and $R_L = 500$ Ω. Neglect h_{ie}.

3.35 Determine A_i and A_v for the two-stage amplifier shown in Figure P3.20. The transistors are silicon.

3.36 Determine A_i and A_v for the two-stage amplifier shown in Figure P3.21.

Figure P3.19

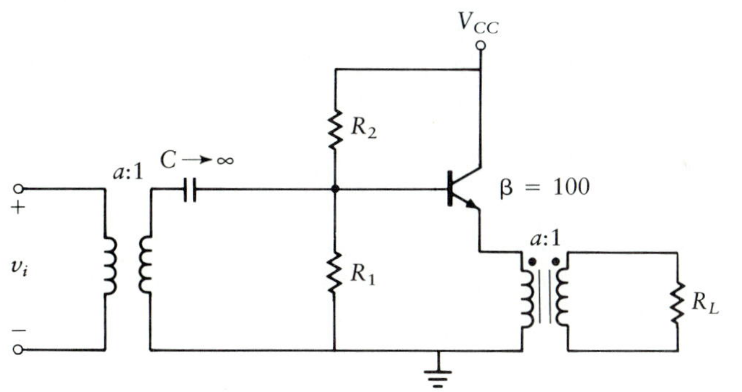

Figure P3.20

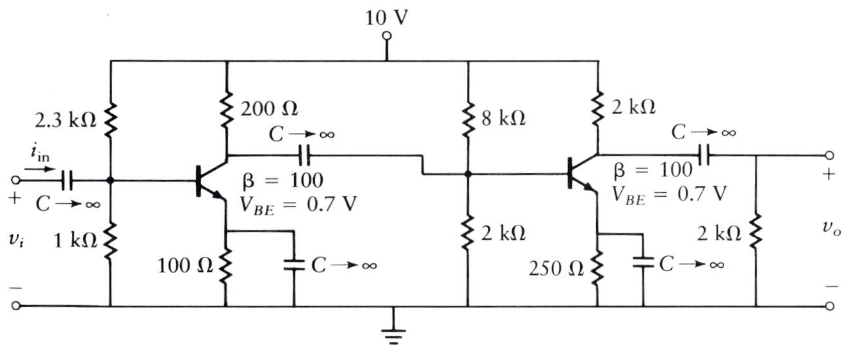

3.37 Determine A_v, A_i, and R_{in} for the two-stage amplifier shown in Figure P3.22.

3.38 Design a CE amplifier using an *npn* transistor for maximum voltage output with the following requirements: $A_v = -20$, $R_{in} = 4$ kΩ, $R_L = 5$ kΩ, $V_{CC} = 12$ V, $\beta = 300$, $V_{BE} = 0.7$ V. Determine all resistor values, undistorted peak-to-peak output voltage swing, and current gain.

3.39 Find R such that at dc, $V_o = 0$ for the circuit of Figure P3.23. Also find I_{CQ1}, I_{CQ2}, R_{in}, R_o, and A_v. Assume that $V_{BE} = 0.7$ V and $\beta = 100$ for both transistors.

Figure P3.21

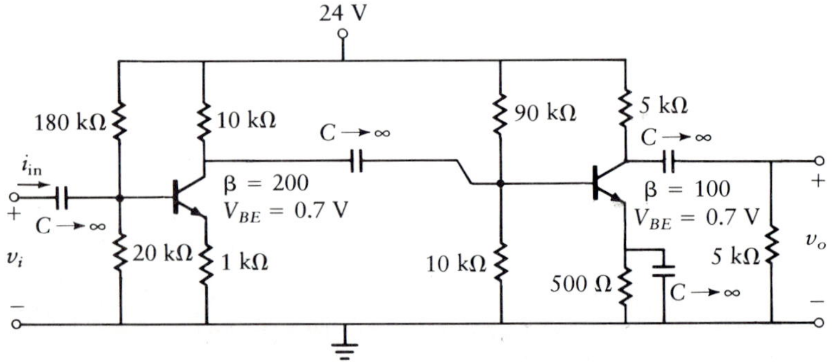

Figure P3.22

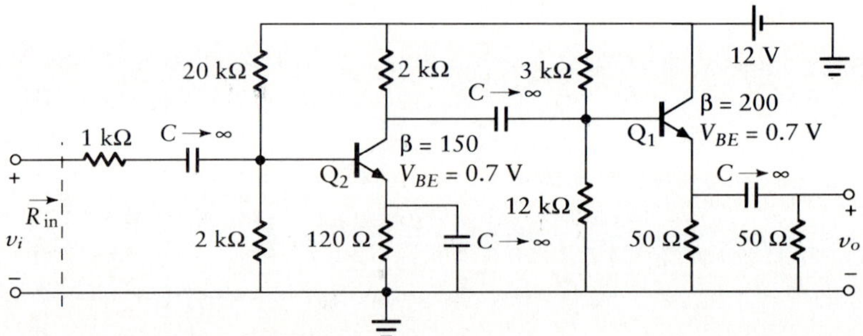

Figure P3.23

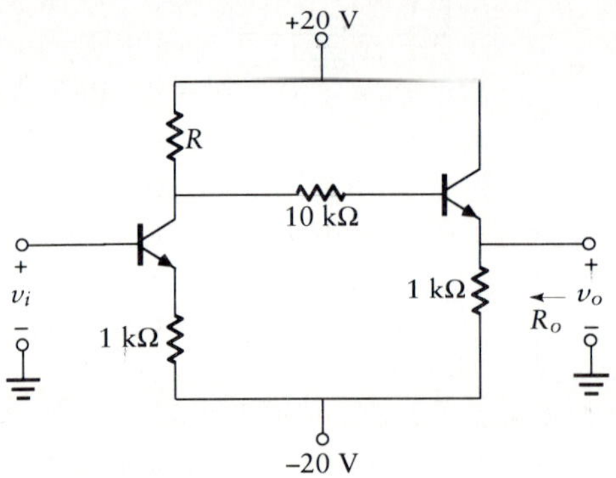

4

FIELD-EFFECT TRANSISTOR AMPLIFIERS

4.0 INTRODUCTION

The *field-effect transistor (FET)* was proposed by W. Shockley in 1952. Its performance differs from that of the BJT. The FET is a *majority carrier* device. Its operation depends on control of the majority carriers in a channel by an applied voltage. This voltage controls the current in the device by means of an electric field. Thus the FET is a voltage-controlled current source as opposed to a bipolar transistor, which is a current-controlled current source.

In comparing FETs and BJTs, we notice that the *drain* (*D*) is analogous to the collector and the *source* (*S*) is analogous to the emitter. A third contact, the *gate* (*G*), is analogous to the base. The source and drain of an FET can usually be interchanged without affecting transistor operation.

This chapter begins with a discussion of the FET characteristics and a comparison of these characteristics with those of the BJT. The construction and operation of both JFETs and MOSFETs are then described. We develop the biasing techniques for FETs, followed by ac analysis using equivalent circuits. We then derive the gain equations for the common-source (CS) amplifier. This is followed by development of a step-by-step design procedure, which is applied to several design examples.

Analysis and design of common-drain (CD) (source-follower (SF)) amplifiers are then presented. Step-by-step design procedures are developed, followed by application of these procedures to examples.

The chapter concludes with a brief discussion of other specialty devices.

4.1 ADVANTAGES AND DISADVANTAGES OF THE FET

The advantages of FETs can be summarized as follows:

1. FETs are voltage-sensitive devices with high input impedance (on the order of 10^7 to 10^{12} Ω). Since this input impedance is considerably higher than that of BJTs, FETs are preferred over BJTs for use as the input stage to a multistage amplifier.
2. One class of FETs (JFETs) generates lower noise than BJTs.
3. FETs are more temperature stable than BJTs.
4. FETs are generally easier to fabricate than BJTs. Greater numbers of devices can be fabricated on a single chip (i.e., increased *packing density* is possible).
5. FETs react like voltage-controlled variable resistors for small values of drain-to-source voltage.
6. The high input impedance of FETs permits them to store charge long enough to allow their use as storage elements.
7. Power FETs can dissipate high power and can switch large currents.
8. JFETs are not as sensitive to radiation as BJTs.

Several disadvantages limit the use of FETs in some applications:

1. FETs usually exhibit poor frequency response because of high input capacitance.
2. Some types of FETs exhibit poor linearity.
3. FETs can be damaged in handling due to static electricity.

4.2 TYPES OF FETs

We consider here three major types of FETs:

1. Junction FET (JFET)
2. Depletion-mode metal-oxide semiconductor FET (depletion MOSFET)
3. Enhancement-mode metal-oxide semiconductor FET (enhancement MOSFET)

The MOSFET is often called an insulated-gate FET (IGFET).

4.3 JFET OPERATION AND CONSTRUCTION

Like the BJT, the FET is a three-terminal device. It has basically only one *pn* junction between the gate and the drain-source channel, rather than two as in

(a) *n*-channel (b) *p*-channel

Figure 4.1 Physical structure of JFET.

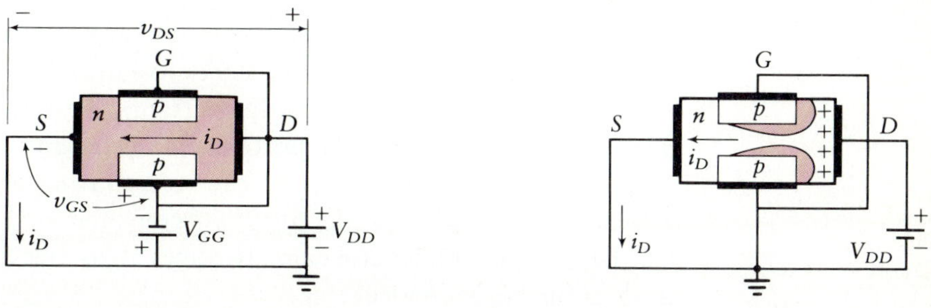

(a) (b)

Figure 4.2 Operation of the JFET.

the BJT. A schematic for the physical structure of the JFET is shown in Figure 4.1. The n-channel JFET, shown in Figure 4.1(a), is constructed using a strip of n-type material with two p-type materials diffused into the strip, one on each side. The p-channel JFET has a strip of p-type material with two n-type materials diffused into the strip, as shown in Figure 4.1(b).

To gain insight into the operation of the JFET, let us connect the n-channel JFET of Figure 4.1(a) to an external circuit. A supply voltage, V_{DD}, is applied to the drain (this is analogous to the V_{CC} supply voltage for a BJT) and the source is attached to common. A gate supply voltage, V_{GG}, is applied to the gate (this is analogous to V_{BB} for the BJT). This circuit configuration is shown in Figure 4.2(a).

V_{DD} provides a drain-source voltage, v_{DS}, that causes a drain current, i_D, from drain to source. The drain current, i_D, which is identical to the source current, exists in the channel surrounded by the p-type gate. The gate-to-source voltage, v_{GS}, which is equal to $-V_{GG}$ (see Figure 4.2(a)), creates a *depletion region* in the channel, which reduces the channel thickness and hence increases the resistance between drain and source. Since the gate-source junction is reverse-biased, zero gate current results.

We consider JFET operation with $v_{GS} = 0$, as shown in Figure 4.2(b). The drain current, i_D, through the n-channel from drain to source causes a voltage drop along the channel, with the higher potential at the drain-gate junction. This positive voltage at the drain-gate junction reverse-biases the pn junction and produces a depletion region, as shown by the shaded area in Figure 4.2(b). When we increase v_{DS}, the drain current, i_D, also increases, as shown in Figure 4.3. This action results in a larger depletion region and an increased channel resistance between drain and source. As v_{DS} is further increased, a

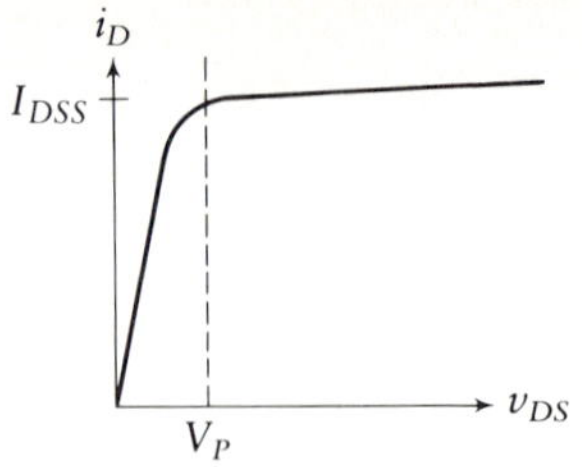

Figure 4.3
i_D-v_{DS} characteristic for n-channel JFET.

point is reached where the depletion region cuts off the entire channel at the drain edge and the drain current reaches its saturation point. If we increase v_{DS} beyond this point, i_D remains fairly constant. The value of the saturated drain current with $v_{GS} = 0$ is an important parameter and is denoted as the *drain-source saturation current,* I_{DSS}. As can be seen from Figure 4.3, increasing v_{DS} beyond this so-called channel *pinch-off* point causes no further increase in i_D, and the i_D-v_{DS} characteristic curve becomes fairly flat (i.e., i_D remains constant as v_{DS} is further increased).

4.3.1 JFET Gate-to-Source Voltage Variation

In the previous section, we developed the i_D-v_{DS} characteristic curve with $v_{GS} = 0$. In this section, we consider the complete i_D-v_{DS} characteristics for various values of the parameter v_{GS}. Note that in the case of the BJT, the characteristic curves (i_C-v_{CE}) have i_B as the parameter. The FET is a voltage-controlled device in which v_{GS} does the controlling. Figure 4.4 shows the i_D-v_{DS} characteristic curves for both the n-channel and p-channel JFET. Before proceeding with an analysis of these curves, note the symbols for an n-channel and a p-channel JFET, which are also shown on Figure 4.4. These two symbols are the same except for the direction of the arrow.

As v_{GS} is increased (more negative for an n-channel and more positive for a p-channel), the depletion region is formed and pinch-off is attained for lower values of i_D. Hence for the n-channel JFET of Figure 4.4(a), the maximum i_D decreases from I_{DSS} as v_{GS} is made more negative. If v_{GS} is further decreased (more negative), a value of v_{GS} is reached after which v_D will be zero regard-

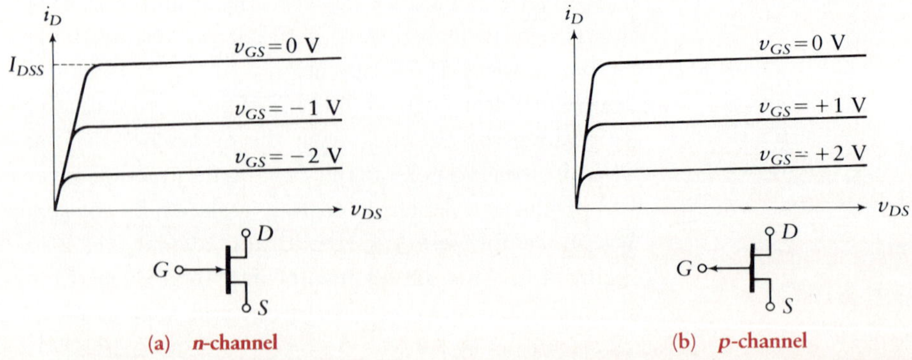

Figure 4.4
i_D-v_{DS} characteristic curves for JFET.

less of the value of v_{DS}. This value of v_{GS} is called $V_{GS(OFF)}$, or *pinch-off voltage* (V_p). The value of V_p is negative for an *n*-channel JFET and positive for a *p*-channel JFET.

4.3.2 JFET Transfer Characteristics

The transfer characteristic is of great value in JFET design. It is a plot of the drain current, i_D, as a function of gate-to-source voltage, v_{GS}. Although this is plotted with v_{DS} equal to a constant, the transfer characteristic is essentially independent of v_{DS} since, after the FET reaches pinch-off, i_D remains relatively constant for increasing values of v_{DS}. This can be seen from the i_D-v_{DS} curves of Figure 4.4, where each curve becomes flat for values of $v_{DS} > V_p$.

In Figure 4.5, we show the transfer characteristics and the i_D-v_{GS} characteristics for an *n*-channel JFET. We plot these with a common i_D axis. The transfer characteristics can be obtained from an extension of the i_D-v_{DS} curves. The most useful method of determining the transfer characteristic in the saturation region is with the following relationship (the Shockley equation):

$$\frac{i_D}{I_{DSS}} \approx \left(1 - \frac{v_{GS}}{V_p}\right)^2 \tag{4.1}$$

Hence, we need only know I_{DSS} and V_p to determine the entire characteristic. Manufacturers' data sheets often give these two parameters, so the transfer characteristic can be constructed or equation (4.1) can be used directly. V_p in the manufacturer's specification sheet is shown as $V_{GS(OFF)}$. Note that i_D saturates (i.e., becomes constant) as v_{DS} exceeds the voltage necessary for the channel to pinch off. This can be expressed as an equation for $v_{DS(sat)}$ for each curve, as follows:

$$v_{DS(sat)} = v_{GS} - V_p$$

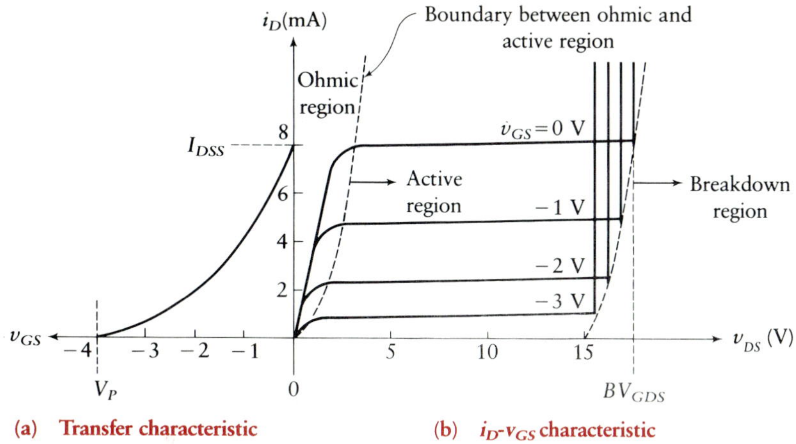

Figure 4.5 JFET transfer characteristics.

As v_{GS} becomes more negative, the pinch-off occurs at lower values of v_{DS} and the saturation current becomes smaller. The useful region for linear operation is above pinch-off and below the breakdown voltage. In this region, i_D is saturated and its value depends on v_{GS}, according to equation (4.1) or the transfer characteristic.

The transfer and i_D-v_{GS} characteristic curves for the JFET, which are shown in Figure 4.5, are different from the similar curves for a BJT: the FET is a voltage-controlled device, whereas the BJT is a current-controlled device.

There are two other distinct differences between the FET and BJT. First, the vertical spacing between pairs of parametric curves for the FET is not linearly related to the value of the controlling parameter. For example, the distance between the curves for $v_{GS} = 0$ and $v_{GS} = -1$ V in Figure 4.5 is not the same as that between the curves for $v_{GS} = -1$ V and $v_{GS} = -2$ V.

The second difference relates to the size and shape of the *ohmic* region of the characteristic curves. Recall that in using BJTs, we avoid nonlinear operation by not using the transistor at the lower 5% of values of v_{CE}, which is called the *saturation region.* We see that the width of the ohmic region for the JFET is a function of the gate-to-source voltage. The ohmic region is approximately linear until the knee occurs close to pinch-off. This region is called the ohmic region because, when the transistor is used in this region, it behaves like an ohmic resistor whose value is determined by the value of v_{GS}. We also note from Figure 4.5 that the breakdown voltage varies as the gate-to-source voltage is modified. In fact, to obtain reasonably linear signal amplification, we must utilize only a relatively small segment of these curves—the area of most linear operation is in the active region.

Note from Figure 4.5 that as v_{DS} increases from zero, a break point occurs on each curve, beyond which the drain current increases very little as v_{DS} continues to increase. At this value of drain-to-source voltage, pinch-off occurs. The pinch-off values in Figure 4.5 are connected with a dashed curve that separates the ohmic region from the active region. As v_{DS} continues to increase beyond pinch-off, a point is reached where the voltage between drain and source becomes so large that *avalanche breakdown* occurs. (This phenomenon also occurs in diodes and in BJTs.) At the breakdown point, i_D increases sharply with a negligible increase in v_{DS}. This breakdown occurs at the drain end of the gate-channel junction. Hence, when the drain-gate voltage, v_{DG}, exceeds the breakdown voltage, BV_{GDS}, for the *pn* junction, avalanche occurs (for $v_{GS} = 0$ V). At this point, the i_D-v_{DS} characteristic exhibits the peculiar shape shown at the right in Figure 4.5.

The region between pinch-off and avalanche breakdown is called the *active region, amplifier operating region, saturation region,* or *pinch-off region,* as shown in Figure 4.5. The ohmic region (before pinch-off) is sometimes called the *voltage-controlled region.* The FET is operated in this region when a variable resistor is desired.

The breakdown voltage is a function of v_{GS} as well as v_{DS}. As the magnitude of the voltage between gate and source is increased (more negative for *n*-channel and more positive for *p*-channel), the breakdown voltage decreases (see Figure 4.5). With $v_{GS} = V_p$, the drain current is zero (except for a small leakage current), and with $v_{GS} = 0$, the drain current saturates at a value

$$i_D = I_{DSS}$$

I_{DSS} is the *saturation drain-to-source current.*

Between pinch-off and breakdown, the drain current is saturated and does not change appreciably as a function of v_{DS}. After the FET passes the pinch-off operating point, the value of i_D can be obtained from the characteristic curves or from the equation

$$i_D \approx I_{DSS}\left(1 - \frac{v_{GS}}{V_p}\right)^2$$

A more accurate version of this equation is as follows:

$$i_D = I_{DSS}\left(1 - \frac{v_{GS}}{V_p}\right)^2 (1 + \lambda v_{DS})$$

λ^{-1} is analogous to the *early voltage* for bipolar transistors. Remember that the slope of the characteristic curves at the *Q*-point defines $1/r_{DS}$, where r_{DS} is the resistance of the current source modeling the transistor. Since λ is small, we assume that λv_{DS} approaches zero. This justifies omitting the second factor in the equation.

The saturation drain-to-source current, I_{DSS}, is a function of temperature. The effects of temperature on V_p are not large. However, I_{DSS} decreases as temperature increases, the decrease being as much as 25% for a 100° increase in temperature. Even larger percentage variations occur in V_p and I_{DSS} because of slight variations in the manufacturing process. This can be seen by viewing Appendix D for the 2N3822, where the maximum I_{DSS} is 10 mA and the minimum is 2 mA.

The currents and voltages in this section are presented for an *n*-channel JFET. The values for a *p*-channel JFET are the reverse of those given for the *n*-channel.

4.3.3 Equivalent Circuit, g_m and r_{DS}

To obtain a measure of the amplification possible with a JFET, we introduce the parameter g_m, which is the *forward transconductance.* The value of g_m, which has units of siemens (S), is a measure of the change in drain current for a change in gate-source voltage. This can be expressed as

$$g_m = \frac{\partial i_D}{\partial v_{GS}} \approx \left.\frac{\Delta i_D}{\Delta v_{GS}}\right|_{V_{DS}=\text{constant}} \tag{4.2}$$

The transconductance, g_m, does not remain constant as the Q-point is changed. This can be seen by geometrically determining g_m from the transfer characteristic curves. As i_D changes, the slope of the transfer characteristic curve of Figure 4.5 varies, thereby changing g_m.

We can find the transconductance by differentiating equation (4.1), with the result

$$g_m = \frac{\partial i_D}{\partial v_{GS}} = \frac{2I_{DSS}(1 - v_{GS}/V_p)}{-V_p} \tag{4.3}$$

We define

$$g_{mo} = \frac{2I_{DSS}}{-V_p}$$

which is the transconductance at $v_{GS} = 0$. Using this definition, the transconductance is given by

$$g_m = g_{mo}\left(1 - \frac{v_{GS}}{V_p}\right) \tag{4.4}$$

An alternative form of equation (4.4) can be found by defining

$$k_n = \frac{I_{DSS}}{V_p^2}$$

and rearranging terms, as follows:

$$i_D = I_{DSS}\left(1 - \frac{v_{GS}}{V_p}\right)^2 = \frac{I_{DSS}}{V_p^2}(V_p - v_{GS})^2 = k_n(V_p - v_{GS})^2$$

(Note: k_n is represented as β when $v_{GS} = 0$ for SPICE-based computer programs (see Appendix A).) The value of k_n depends on the construction of the JFET and is primarily a function of the width and length of the channel. It is measured in mA/V^2.

We select the Q-point so that $i_D = I_{DQ}$ and $v_{GS} = V_{GSQ}$. Thus we obtain

$$V_p - V_{GSQ} = -\sqrt{\frac{I_{DQ}}{k_n}} \tag{4.5}$$

But from equation (4.3),

$$g_m = -\frac{2I_{DSS}}{V_p}\left(1 - \frac{V_{GSQ}}{V_p}\right) = -\frac{2I_{DSS}}{V_p^2}(V_p - V_{GSQ})$$

Combining this with equation (4.5) yields

$$g_m = \frac{2I_{DSS}}{V_p^2}\sqrt{\frac{I_{DQ}}{k_n}} = 2k_n\sqrt{\frac{I_{DQ}}{k_n}} = 2\sqrt{k_n I_{DQ}} \tag{4.6}$$

The *inverse dynamic resistance,* r_{DS}, is defined by the inverse of the slope of the i_D-v_{DS} curve in the saturated region.

$$\frac{1}{r_{DS}} = \frac{\partial i_D}{\partial v_{DS}} \approx \frac{\Delta i_D}{\Delta v_{DS}}\bigg|_{\Delta v_{GS}=\text{constant}} \tag{4.7}$$

Because the slope of this curve is so small in the active region (see Figure 4.4), r_{DS} is large. Alternatively, r_{DS} is small when we operate near the origin.

We develop the ac equivalent circuit for a JFET much as we did for the BJT, with the expression

$$\Delta i_D = \frac{\partial i_D}{\partial v_{GS}}\Delta v_{GS} + \frac{\partial i_D}{\partial v_{DS}}\Delta v_{DS} \tag{4.8}$$

Equation (4.8) can be rewritten using the defined parameters as follows:

$$\Delta i_D = g_m \Delta v_{GS} + \frac{\Delta v_{DS}}{r_{DS}} \tag{4.9}$$

which leads to the equivalent circuit shown in Figure 4.6(a). Because r_{DS} is so large, we can usually use the simplified equivalent circuit of Figure 4.6(b) to

Figure 4.6 FET equivalent circuit.

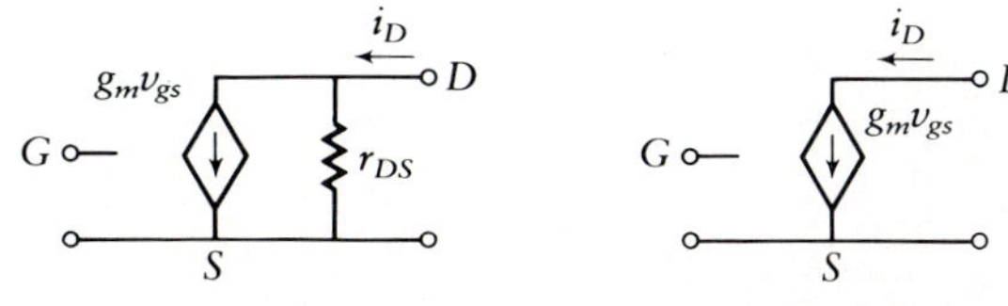

(a) **Including r_{DS}** (b) **Simplified**

determine the performance of a JFET in the active region. Equation (4.9) then reduces to

$$\Delta i_D = g_m \Delta v_{GS}$$

The performance of a JFET is specified by the values of g_m and r_{DS}. We now determine these parameters for an n-channel JFET using the characteristic curve shown in Figure 4.7. We select an operating region that is approximately in the middle of the curves, that is, between $v_{GS} = -0.8$ V with $i_D = 8.5$ mA and $v_{GS} = -1.2$ V with $i_D = 5.5$ mA. Now, from equation (4.2), we find

$$g_m = \left.\frac{\Delta i_D}{\Delta v_{GS}}\right|_{v_{DS}=\text{constant}} = 7.5 \text{ mS}$$

If the characteristic curves for a JFET are not available, g_m and v_{GS} can be obtained mathematically, provided I_{DSS} and V_p are known. These two parameters are usually given in the manufacturer's specifications. The quiescent drain current, I_{DQ}, can be selected to be between 30 and 70% of I_{DSS}, which locates the Q-point in the most linear region of the characteristic curves.

The relationship between i_D and v_{GS} can be plotted on a dimensionless graph (i.e., a normalized curve) as shown in Figure 4.8. The vertical axis of this graph is $i_D/|I_{DSS}|$ and the horizontal axis is $v_{GS}/|V_p|$. The slope of the curve is g_m.

A reasonable procedure for locating the quiescent value near the center of the linear operating region is as follows:

1. Select $I_{DQ} = I_{DSS}/2$ and, from the curve, $V_{GSQ} = 0.3V_p$. Note from Figure 4.8 that this is near the midpoint of the curve.

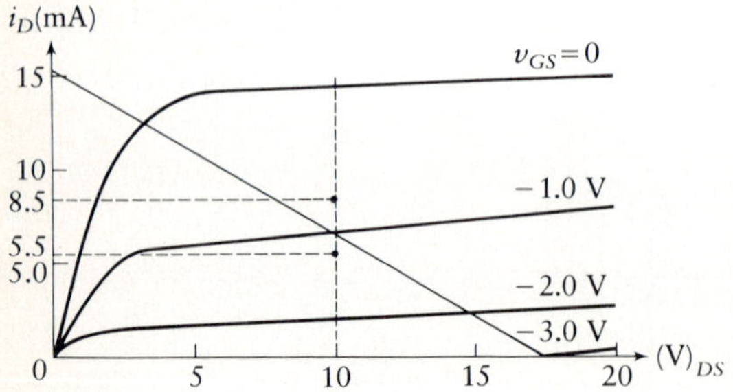

Figure 4.7 i_D-v_{DS} JFET characteristic curves.

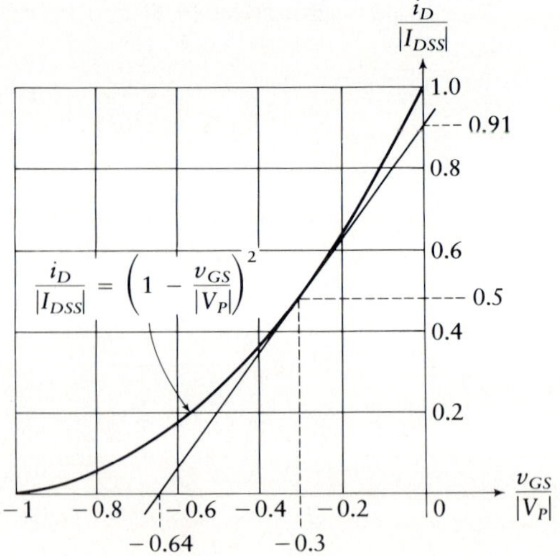

Figure 4.8 $i_D/|I_{DSS}|$ versus $v_{GS}/|V_p|$.

2. Select $V_{DSQ} = V_{DD}/2$.

We find the transconductance at the Q-point from the slope of the curve of Figure 4.8. This is given by

$$g_m = 1.42\frac{I_{DSS}}{|V_p|}$$

Remember that this value of g_m depends on the assumption that I_{DQ} is set at one-half I_{DSS}. These values usually represent a good starting point for setting the quiescent values for the JFET.

Example 4.1

Determine g_m for a JFET where $I_{DSS} = 7$ mA, $V_p = -3.5$ V, and $V_{DD} = 15$ V. Choose a reasonable location for the Q-point.

SOLUTION We start by referring to Figure 4.8 and selecting the Q-point as follows:

$$I_{DQ} = \frac{I_{DSS}}{2} = 3.5 \text{ mA}$$

$$V_{DSQ} = \frac{V_{DD}}{2} = 7.5 \text{ V}$$

$$V_{GSQ} = 0.3V_p = -1.05 \text{ V}$$

The transconductance, g_m, is found from the slope of the curve at the point $i_D/I_{DSS} = 0.5$ and $v_{GS}/V_p = 0.3$. Hence,

$$g_m = \frac{1.42I_{DSS}}{|V_p|} = 2840\ \mu\text{S}$$

Note that these results apply only for the case where I_{DQ} is set at $I_{DSS}/2$.

4.4 MOSFET OPERATION AND CONSTRUCTION

In this section, we consider the metal-oxide semiconductor FET (MOSFET). This FET is constructed with the gate terminal insulated from the channel with a silicon dioxide (SiO_2) dielectric and is constructed in either a *depletion* or an *enhancement* mode. We define and consider these two types in the following sections.

4.4.1 Depletion MOSFET

The construction of the n-channel and p-channel depletion MOSFETs is shown in Figures 4.9 and 4.10, respectively. Each of these figures illustrates the construction, the symbol, the transfer characteristic, and the i_D-v_{GS} characteristics. The depletion MOSFET is constructed with a physical channel inserted between the drain and the source. As a result, when a voltage, v_{DS}, is applied, i_D exists between drain and source.

The n-channel depletion MOSFET of Figure 4.9 is established on a p-substrate, which is p-doped silicon. The n-doped source and drain wells form low-resistance connections between the ends of the n-channel and the aluminum contacts of the source (S) and the drain (D). An SiO_2 layer, which is an insulator, is grown on the top surface of the n-channel, as shown in Figure 4.9(a). An aluminum layer is deposited on the SiO_2 insulator to form the gate (G) terminal. The performance of the depletion MOSFET is similar to that of the JFET. The JFET is controlled by the pn junction between the gate and the drain end of the channel. No such junction exists in the enhancement MOSFET, and the SiO_2 layer acts as the insulator. For the n-channel MOSFET, as shown in Figure 4.9, a negative v_{GS} pushes electrons out of the channel region, depleting the channel. When v_{GS} reaches V_p the channel is pinched off. Positive values of v_{GS} increase channel size, resulting in an increase of drain current. This is indicated on the characteristic curves of Figure 4.9(c).

Figure 4.9
The n-channel depletion MOSFET.

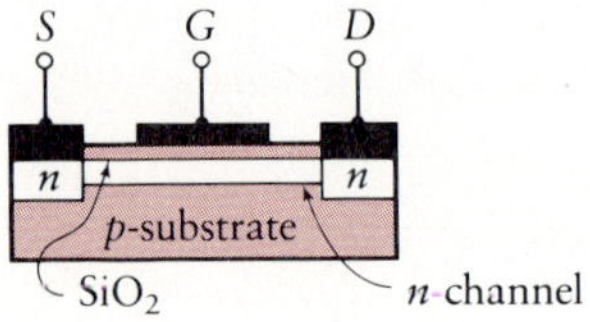

(a) Schematic of physical structure

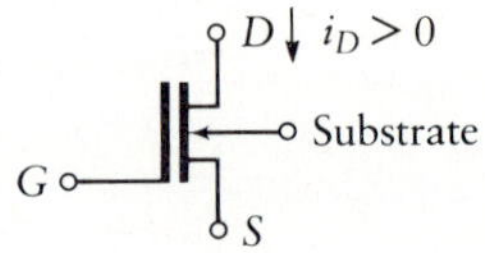

(b) Symbol

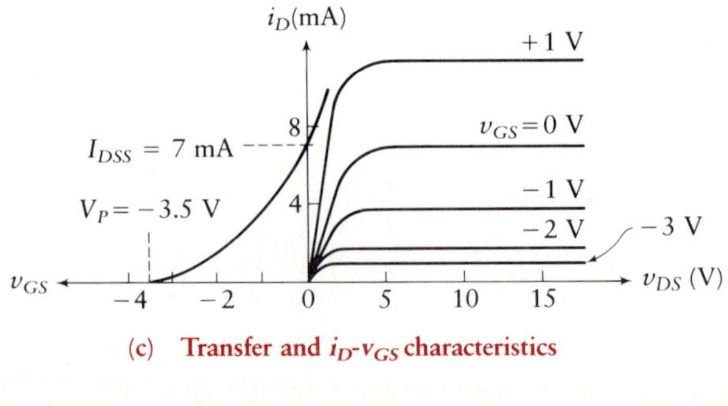

(c) Transfer and i_D-v_{GS} characteristics

Figure 4.10 The p-channel depletion MOSFET.

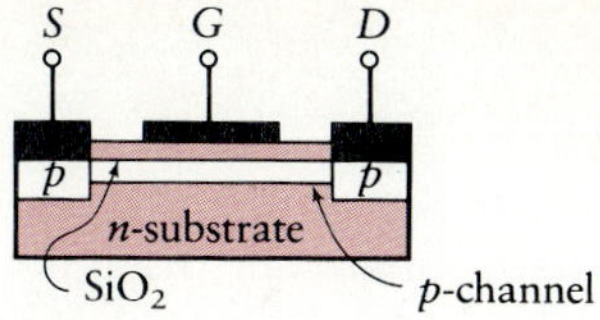

(a) Schematic of physical structure

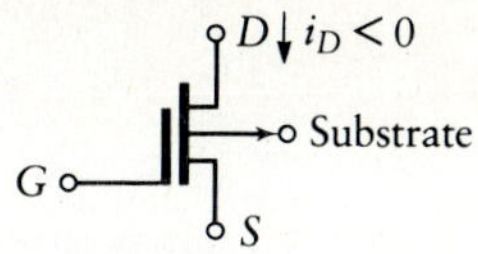

(b) Symbol

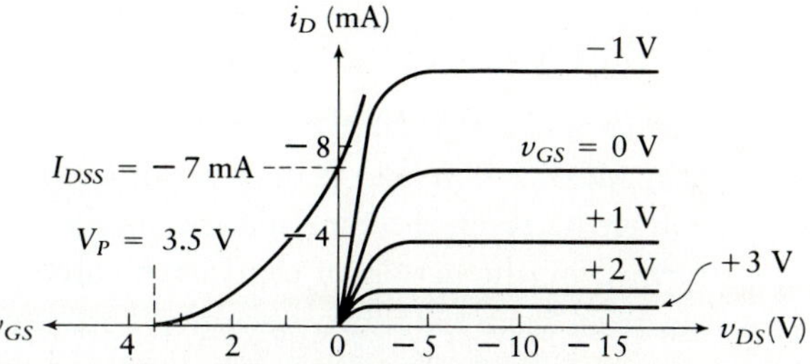

(c) Transfer and i_D-v_{GS} characteristics

Notice that the depletion MOSFET can operate with either positive or negative values of v_{GS}. We can use the same Shockley equation (equation 4.1)) to approximate the curves for negative v_{GS}. Notice, however, that the transfer characteristic continues on for positive values of v_{GS}. Since the gate is insulated from the channel, the gate current is negligibly small (10^{-12} A) and v_{GS} can be of either polarity.

The symbol for the MOSFET has a fourth terminal, the *substrate*. The arrowhead points in for an n-channel and out for a p-channel. The p-channel depletion MOSFET, which is shown in Figure 4.10, is the same as that of Figure 4.9, except that we reverse the n and p materials and reverse the polarity of the voltages and currents.

Example 4.2

Calculate the drain current, i_D, for the depletion MOSFET of Figure 4.9 for the following values of v_{GS}:

a. $v_{GS} = -1$ V c. $v_{GS} = -3$ V
b. $v_{GS} = -2$ V d. $v_{GS} = +0.5$ V

SOLUTION We use equation (4.1) for each case, where $I_{DSS} = 7$ mA and $V_p = -3.5$ V.

a. $$i_D = I_{DSS}\left(1 - \frac{v_{GS}}{V_p}\right)^2 = 3.57 \text{ mA}$$

b. $i_D = 1.29$ mA

c. $i_D = 0.14$ mA

d. $i_D = 9.14$ mA

It is interesting to note that i_D increases dramatically as v_{GS} becomes positive. The accuracy of the formula approximation decreases for positive values of v_{GS}.

4.4.2 Enhancement MOSFET

The enhancement MOSFET is shown in Figure 4.11. It differs from the depletion MOSFET in that it does not have the thin *n*-layer but requires a positive voltage between the gate and the source to establish a channel. This channel is formed by the action of a positive gate-to-source voltage, v_{GS}, which attracts electrons from the substrate region between the *n*-doped drain and source. Positive v_{GS} causes electrons to accumulate at the surface beneath the oxide layer. When the voltage reaches a threshold value, V_T, sufficient numbers of electrons are attracted to this region to make it act like a conducting *n*-channel. No appreciable current, i_D, will exist until v_{GS} exceeds V_T.

No value of I_{DSS} exists for the enhancement MOSFET, since the drain current is zero until the channel has been formed. I_{DSS} is zero at $v_{GS} = 0$. For values of

$$v_{GS} > V_T \quad \text{and} \quad (v_{GS} - V_T) < v_{DS}$$

the drain current can be calculated from the equation

$$i_D = k(v_{GS} - v_T)^2 \tag{4.10}$$

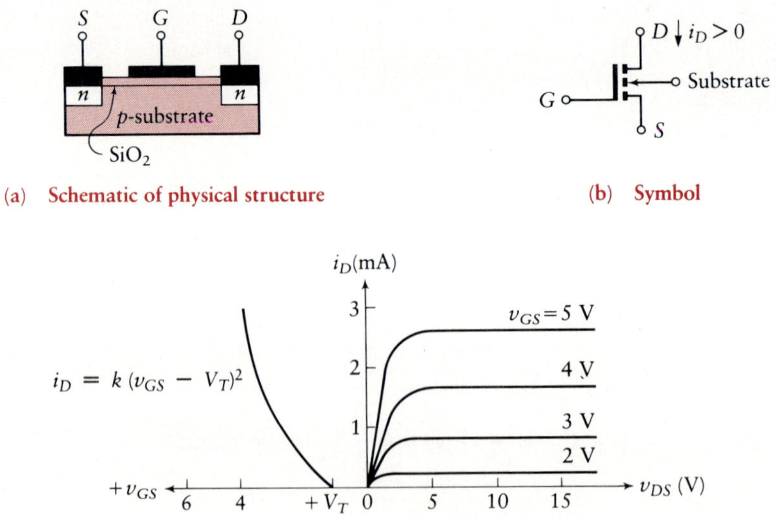

(a) Schematic of physical structure

(b) Symbol

(c) Transfer and i_D-v_{DS} characteristics

Figure 4.11 The *n*-channel enhancement MOSFET.

Note that this is not the same as equation (4.1). The value of k depends on the construction of the MOSFET and is primarily a function of the width and length of the channel. A typical value for k is 0.3 mA/V^2. The threshold voltage, V_T, is specified by the manufacturer. In SPICE-based computer simulations, the term KP is used as a parameter for MOSFETs. This parameter is related to k by

$$KP = 2k$$

We can find a value for g_m by differentiating equation (4.10), as we did with JFETs, with the result

$$g_m = \frac{\partial i_D}{\partial v_{GS}} = 2k(v_{GS} - V_T) \tag{4.11}$$

If

$$v_{GS} < V_T$$

then $i_D = 0$.

The p-channel enhancement MOSFET is shown in Figure 4.12. Its characteristics are similar but opposite to those of the n-channel enhancement MOSFET. Although it is more restricted in operating range than the depletion MOSFET, the enhancement MOSFET is useful in integrated circuit applications (see Chapter 15) because of its simple construction. It also has the advantage that it can be used in digital integrated circuits with single low-voltage supplies. The gate for both n- and p-channel MOSFETs is a metal deposit on a

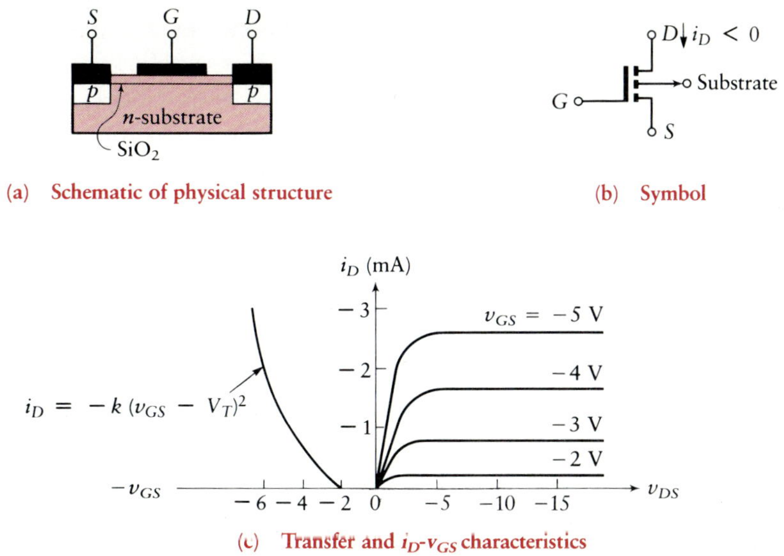

Figure 4.12 The p-channel enhancement MOSFET.

silicon oxide layer. The construction begins with a substrate material (p-type for n-channel; n-type for p-channel) on which the opposite type of material is diffused to form the source and drain. Notice that the symbol for an enhancement MOSFET, which is shown in Figures 4.11 and 4.12, shows a broken line between source and drain to indicate that no channel initially exists.

Example 4.3

Determine i_D for an n-channel enhancement MOSFET with $V_T = 3.0$ V when $k = 0.3$ mA/V^2 and v_{GS} given by the following values:

a. 3.0 V
b. 4.0 V
c. 5.0 V

SOLUTION We use equation (4.10).

a. $i_D = 0.3(v_{GS} - V_T)^2 = 0.3(3 - 3)^2 = 0$ mA

b. $i_D = 0.3(4 - 3)^2 = 0.3$ mA

c. $i_D = 0.3(5 - 3)^2 = 1.2$ mA

4.5 BIASING OF FETs

The same basic circuits that are used for biasing of BJTs can also be used for JFETs and depletion MOSFETs. However, for the active region of the JFET and the depletion-mode MOSFET, the polarity of v_{GS} can be opposite to that of the drain voltage source. In selecting the operating point, voltage of the opposite polarity may not be available from the source to meet the requirements of the circuit. It may be necessary to delete R_2 so that only voltage of correct polarity is acquired. It is not always possible to find resistor values to achieve a particular Q-point. In such instances, selecting a new Q-point can sometimes provide a solution to the problem.

We consider here the bias equations for the CS JFET amplifier as shown in Figure 4.13. The bias methods are similar for depletion MOSFETs. Including a resistor in the source circuit (R_S) provides increased linearity but also reduces the gain. In other words, the resistor provides a negative feedback, which is particularly useful in reducing the nonlinearity of the FET amplifier.

Figure 4.13(a) illustrates a FET amplifier using one dc voltage source for biasing. We form the Thevenin equivalent of the bias circuit to obtain

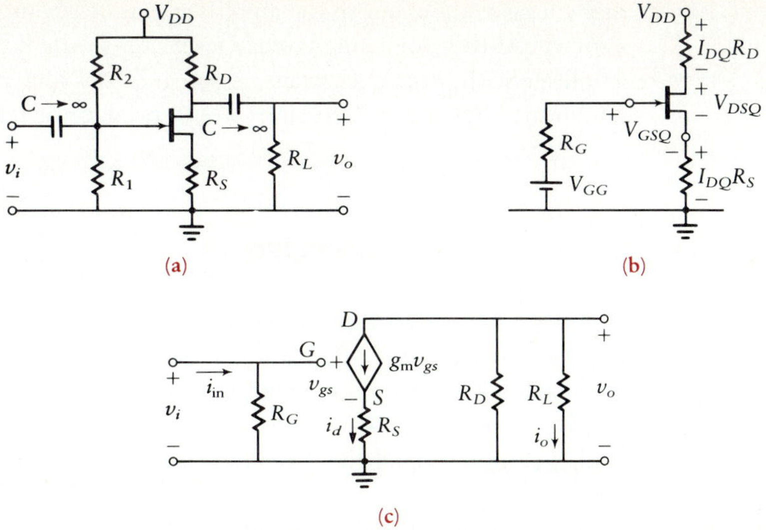

Figure 4.13
FET amplifier.

$$R_G = R_1 \parallel R_2 = \frac{R_1 R_2}{R_1 + R_2} \tag{4.12a}$$

$$V_{GG} = \frac{V_{DD} R_1}{R_1 + R_2} \tag{4.12b}$$

Since there are three unknown variables in these equations, I_{DQ}, V_{GSQ}, and V_{DSQ}, we need three dc equations. First, the dc equation around the gate-source loop is formed as follows:

$$V_{GG} = V_{GSQ} + I_{DQ} R_S \tag{4.13}$$

Notice that since the gate current is zero, a zero voltage drop exists across R_G. A second dc equation is found from the Kirchhoff's law equation in the drain-source loop, as follows:

$$V_{DD} = V_{DSQ} + I_{DQ}(R_S + R_D) \tag{4.14}$$

The third dc equation necessary to establish the bias point is found from equation (4.1), which is repeated here with $i_D = I_{DQ}$ and $v_{GS} = V_{GSQ}$.

$$\frac{I_{DQ}}{I_{DSS}} = \left(1 - \frac{V_{GSQ}}{V_p}\right)^2 \tag{4.15}$$

These three equations are sufficient to establish the bias for the JFET and depletion MOSFET.

Notice that we do not need to put the Q-point in the center of the ac load line as we often did for BJT biasing. This is because FET amplifiers are normally used at the input to the amplifier to take advantage of the high input resistance. At this point, the voltage levels are so small that we do not drive the amplifier with large excursions. Also, since the FET characteristic curves are nonlinear, large input excursions would produce distortion.

4.6 ANALYSIS OF CS AMPLIFIERS

Figure 4.13(c) shows the ac equivalent circuit for the FET. We assume r_{DS} is large compared to $R_D \parallel R_L$, so it can be neglected. If the source resistor, R_S, is bypassed by a capacitor, we need simply set the resistance equal to zero in the following equations. We do this at the conclusion of this derivation. Writing a KVL equation around the gate circuit, we find

$$v_{gs} = v_i - R_S i_d = v_i - R_S g_m v_{gs}$$

Solving for v_{GS} yields

$$v_{GS} = \frac{v_i}{1 + R_S g_m}$$

The output voltage, v_o, is given by

$$v_o = -i_D(R_D \parallel R_L) = \frac{-(R_D \parallel R_L) g_m v_i}{1 + R_S g_m}$$

The voltage gain, A_v, is

$$A_v = \frac{v_o}{v_i} = \frac{-g_m(R_D \parallel R_L)}{1 + R_S g_m}$$

$$= \frac{-(R_D \parallel R_L)}{R_S + 1/g_m} \tag{4.16a}$$

If the resistance, R_S, is bypassed by a capacitor, the voltage gain increases to

$$A_v = -g_m(R_D \parallel R_L)$$

The input resistance and current gain are given by

$$R_{\text{in}} = R_G = R_1 \parallel R_2 \tag{4.16b}$$

$$A_i = \frac{i_o}{i_{\text{in}}} = \frac{A_v R_{\text{in}}}{R_L} = \frac{-R_D \parallel R_L}{R_S + 1/g_m} \frac{R_{\text{in}}}{R_L} = \frac{-R_G}{R_S + 1/g_m} \frac{R_D}{R_D + R_L} \tag{4.16c}$$

Example 4.4 CS Amplifier (Analysis)

Find A_v for the JFET amplifier of Figure 4.14(a). The Q-point is at $V_{DSQ} = 12$ V and $I_{DQ} = 7$ mA. The FET parameters are given by

$$g_m = 3.0 \text{ mS}$$

$$r_{DS} = 200 \text{ k}\Omega$$

SOLUTION From the equivalent circuit of Figure 4.14(b), we obtain

$$A_v = \frac{v_o}{v_{GS}} = \frac{-i_D(R_D \parallel r_{DS})}{v_{GS}} = -g_m(R_D \parallel r_{DS}) = -52$$

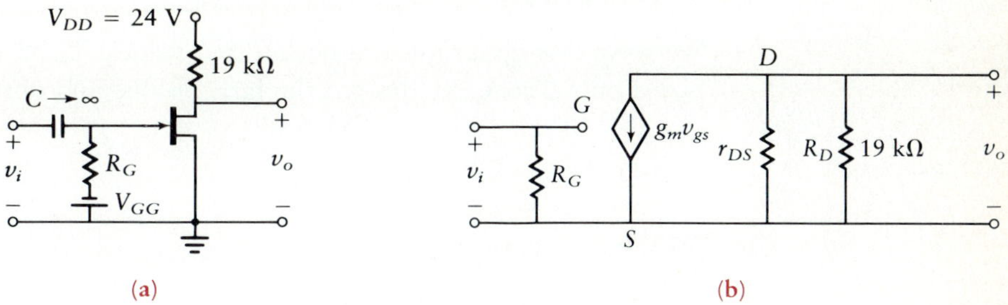

Figure 4.14
JFET circuit for Example 4.4.

Example 4.5 JFET CS Amplifier (Analysis)

Analyze the single-stage JFET CS amplifier shown in Figure 4.15 and determine A_v, A_i, and R_{in}. Assume $I_{DSS} = 2$ mA and $V_p = -2$ V.

SOLUTION We first determine g_m, since both A_v and A_i depend on this parameter. Remember that the value of g_m depends on the Q-point location. We need two equations in order to find I_{DQ} and V_{GSQ}, as follows:

$$V_{GSQ} = -R_S I_{DQ} = -0.4 I_{DQ}$$

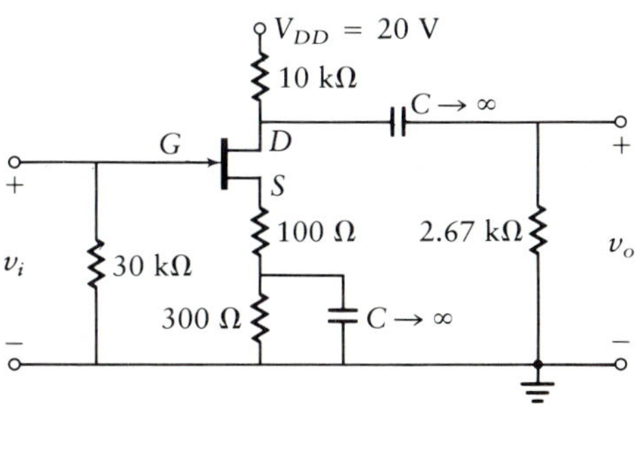

Figure 4.15
CS amplifier for Example 4.5.

and from equation (4.1)

$$\frac{I_{DQ}}{I_{DSS}} = \left(1 - \frac{V_{GSQ}}{V_p}\right)^2$$

When we solve these two equations, we obtain a quadratic in I_{DQ}.

$$\frac{I_{DQ}}{2} = \left(1 - \frac{-0.4I_{DQ}}{-2}\right)^2$$

where I_{DQ} is in mA.

This reduces to

$$I_{DQ}^2 - 22.5I_{DQ} + 25 = 0$$

We solve this quadratic and obtain two values: 21.33 mA and 1.17 mA. Since I_{DSS} is only 2 mA, we discard the larger value and obtain

$$I_{DQ} = 1.17 \text{ mA}$$

Hence

$$V_{GSQ} = (-0.4)\,(I_{DQ}) = -0.469 \text{ V}$$

We use equation (4.3) to find g_m.

$$g_m = \frac{2I_{DSS}}{V_p}\left(1 - \frac{V_{GSQ}}{V_p}\right) = 1.53 \text{ mS}$$

and

$$1/g_m = 653\ \Omega$$

We now find the voltage gain from equation (4.16a).

$$A_v = -\frac{R_D \parallel R_L}{R_{Sac} + 1/g_m} = -2.8$$

and

$$R_{\text{in}} = 30 \text{ k}\Omega$$

$$A_i = A_v R_{\text{in}}/R_L = -31.5$$

4.7 DESIGN OF CS AMPLIFIERS

The design procedure for a CS amplifier is presented in this section. The JFET and the depletion MOSFET amplifier design is presented as a step-by-step procedure. You should convince yourself that you understand the origin of each step, since several variations may subsequently be required.

Amplifiers are designed to meet gain requirements, assuming that the desired specifications are within the range of the transistor. The supply voltage, load resistance, voltage gain, and input resistance (or current gain) are usually specified. The designer's job is to select the resistance values R_1, R_2, R_D, and R_S. Refer to Figure 4.16 as you follow the steps in the procedure. This procedure assumes that a device has been selected and that its characteristics are known—at least V_p and I_{DSS}.

Step 1 Select a Q-point in the saturation region of the JFET characteristic curves. Refer to the curves of Figure 4.16(b) for an example. This identifies V_{DSQ}, V_{GSQ}, I_{DQ}, and g_m. If an i_D-v_{GS} characteristic curve is unavailable, use the dimensionless curve of Figure 4.8, with I_{DSS} and V_p given for the transistor type used. Alternatively, use equations (4.1), (4.3), and (4.4) to find V_{GSQ} and g_m.

Step 2 Write the dc KVL equation (equation (4.14)) around the drain-source loop:

$$V_{DD} = V_{DSQ} + (R_S + R_D)I_{DQ}$$

Solving for the sum of the two resistors yields

$$R_S + R_D = \frac{V_{DD} - V_{DSQ}}{I_{DQ}} = K_1 \tag{4.17}$$

Equation (4.17) represents one equation in two unknowns, R_S and R_D.

Figure 4.16
JFET CS amplifier.

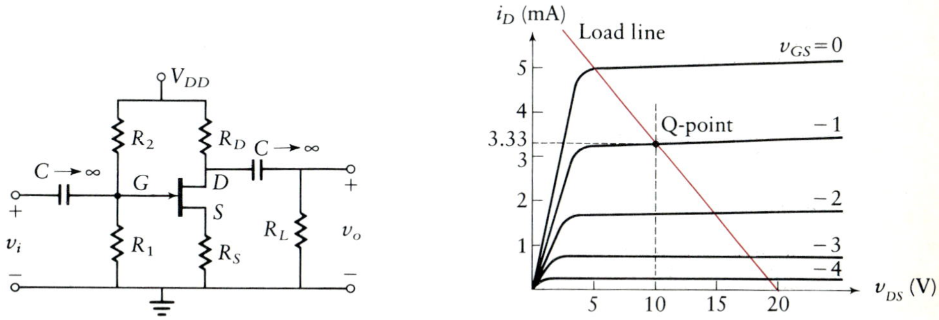

(a) CS circuit (b) Characteristic curve

Step 3 Use the voltage gain equations (equation (4.16a)) to yield a second equation in R_S and R_D. We can substitute equation (4.17) into equation (4.16a) to obtain

$$A_v = \frac{-R_L \parallel R_D}{R_S + 1/g_m} = \frac{-R_L \parallel R_D}{(K_1 - R_D) + 1/g_m} \tag{4.18}$$

The resistance, R_D, *is the only unknown in this equation.* Solving for R_D results in a quadratic equation having two solutions, one negative and one positive. If the positive solution results in $R_D > K_1$, thus implying a negative R_S, a new Q-point must be selected (i.e., restart the design at Step 1). If the positive solution yields $R_D < K_1$, we can proceed to Step 4.

Step 4 Solve for R_S using equation (4.17), the drain-to-source loop equation developed in Step 2.

$$R_S = \frac{V_{DD} - V_{DSQ}}{I_{DQ}} - R_D$$

With R_D and R_S known, we need only find R_1 and R_2.

Step 5 Write the KVL equation for the gate-source loop (equation (4.13)).

$$V_{GG} = V_{GSQ} + I_{DQ} R_S$$

The voltage, V_{GSQ}, is of opposite polarity to V_{DD}. Thus the term $I_{DQ} R_S$ must be greater than V_{GSQ} in magnitude. Otherwise, V_{GG} will have the opposite polarity to V_{DD}, which is not possible according to equation (4.12b).

Step 6 We now solve for R_1 and R_2, assuming that the V_{GG} found in Step 5 has the *same polarity* as V_{DD}. These resistor values are selected by finding the value of R_G from the current gain equation or from the input resistance. We solve equation (4.12) to find R_1 and R_2.

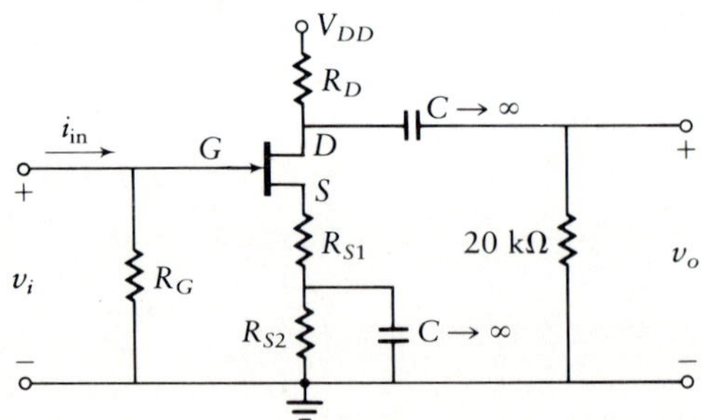

Figure 4.17
JFET design with bypassed source resistor.

$$R_1 = \frac{R_G}{1 - V_{GG}/V_{DD}}$$

$$R_2 = \frac{R_G V_{DD}}{V_{GG}}$$

Step 7 If V_{GG} has the *opposite polarity* to V_{DD}, it is not possible to solve for R_1 and R_2. The practical way to proceed is to let $V_{GG} = 0$ V. Thus, $R_2 \to \infty$. Since V_{GG} is specified by equation (4.13), the previously calculated value of R_S now needs to be modified. In Figure 4.17, where a capacitor is used to bypass a part of R_S, we develop the new value of R_S as follows:

$$V_{GG} = 0 = V_{GSQ} + I_{DQ} R_{Sdc}$$

Therefore, solving for R_{Sdc} yields

$$R_{Sdc} = \frac{-V_{GSQ}}{I_{DQ}}$$

The value of R_{Sdc} is $R_{S1} + R_{S2}$ and the value of R_{Sac} is R_{S1}.

Now that we have a new R_{Sdc}, we must repeat several steps.

Step 8 Determine R_D using KVL for the drain-to-source loop (repeat of Step 2).

$$R_D = K_1 - R_{Sdc}$$

The design problem now becomes one of calculating both R_{S1} and R_{S2} instead of finding only one source resistor.

With a new value for R_D of $K_1 - R_{Sdc}$, we go to the voltage gain expression of equation (4.18) with R_{Sac} used for this ac equation rather than R_S.

The following additional steps must be added to the design procedure.

Step 9 We find R_{Sac} (which is simply R_{S1}) from the voltage gain equation, equation (4.16a),

$$A_v = \frac{-(R_L \parallel R_D)}{R_{Sac} + 1/g_m}$$

R_{Sac} is the only unknown in this equation. Therefore,

$$R_{Sac} = -\frac{R_L \parallel R_D}{A_v} - \frac{1}{g_m}$$

Suppose now that R_{Sac} is found to be positive but less than R_{Sdc}. This is the desirable condition since

$$R_{Sdc} = R_{Sac} + R_{S2}$$

Then our design is complete and

$$R_1 = R_{\text{in}} = R_G$$

Step 10 Suppose that R_{Sac} is found to be positive but *greater* than R_{Sdc}. The amplifier cannot be designed with the voltage gain and Q-point as selected. A new Q-point must be selected, and we return to Step 1. If the voltage gain is too high, it may not be possible to effect the design with any Q-point. A different transistor may be needed or the use of two separate stages may be required.

4.8 SELECTION OF COMPONENTS

A design is not yet complete when the various component values are specified. It is still necessary to select the actual components to be used (e.g., choose the manufacturers' part numbers from a catalog). For example, if the design requires a resistor value, say 102.5 Ω, the designer will not be able to find this resistor in a standard parts catalog. Available nominal component values depend on tolerances. As an example, a 100-Ω resistor with a 5% tolerance can have any value between 95 and 105 Ω. It would not make much sense for the manufacturer to offer another *off-the-shelf* resistor with a nominal rating of 101 Ω, since that resistor could have a value between approximately 96 and 106 Ω. The distance between adjacent nominal component values is therefore related to the tolerance, with such distance decreasing as the tolerance decreases (i.e., higher-precision components). Standard values of resistors and capacitors are included in Appendix E.

Since component values are not readily available to arbitrarily high degrees of resolution, it would not make sense to carry out design calculations to an unreasonably large number of significant figures.

In our design examples, we specify values to at least three significant figures. This is important to ensure that we still maintain accuracy to two significant figures following arithmetic operations. For example, suppose we must add

$$0.274 + 0.474$$

If these two numbers are rounded to two significant figures, we obtain

$$0.27 + 0.47 = 0.74$$

If we first do the calculation, however, we obtain

$$0.274 + 0.474 = 0.748$$

which rounds to 0.75. Rounding the numbers prior to performing the addition results in an error in the second significant figure. Thus, to reduce accumulated errors and to increase the confidence that our answers are accurate to two significant figures, we maintain at least three figures throughout the calculations.

Example 4.6 JFET CS Amplifier (Design)

Design a CS JFET amplifier with a voltage gain of $A_v = -4$, $R_{in} = 100\ \text{k}\Omega$, $R_L = 20\ \text{k}\Omega$, $I_{DSS} = 6.67$ mA, $V_p = -3.33$ V, and $V_{DD} = 20$ V. Use the circuit configuration shown in Figure 4.16 and the dimensionless curves of Figure 4.8.

SOLUTION

Step 1 The Q-point is selected from Figure 4.8 as follows:

$$I_{DQ} = \frac{I_{DSS}}{2} = 3.33\ \text{mA}$$

$$V_{GSQ} = 0.3V_p = -1\ \text{V}$$

$$V_{DSQ} = \frac{V_{DD}}{2} = 10\ \text{V}$$

Then

$$g_m = 1.42\frac{I_{DSS}}{V_p} = 2.84 \times 10^{-3}\ \text{S}$$

and

$$\frac{1}{g_m} = 350\ \Omega$$

Note that these results apply only when $I_{DQ} = I_{DSS}/2$.

Step 2 From Step 2 in the design procedure, we have

$$R_D + R_S = \frac{20\ \text{V} - 10\ \text{V}}{3.33\ \text{mA}} = 3\ \text{k}\Omega = K_1$$

Step 3 Using the ac gain equation, we obtain

$$A_v = \frac{-(20\ \text{k}\Omega \parallel R_D)}{3\ \text{k}\Omega - R_D + 350\ \Omega} = -4$$

$$R_D^2 + (21.7\ \text{k}\Omega)R_D - 67\ \text{M}\Omega^2 = 0$$

From the quadratic for R_D we select the positive root,

$$R_D = 2.75\ \text{k}\Omega$$

Step 4 This quantity is less than K_1, so we proceed to Step 4. We find R_S using equation (4.15).

$$R_S = 3\ \text{k}\Omega - R_D = 253\ \Omega$$

Step 5 now yields

$$V_{GG} = -1 + 253(3.33 \times 10^{-3}) = -0.15\ \text{V}$$

Since this negative voltage cannot be obtained by dividing the source voltage using resistors, we skip to Step 7.

Step 7 Let

$$R_2 \rightarrow \infty$$

Then

$$V_{GG} = 0 = V_{GSQ} + I_{DQ}R_{Sdc}$$
$$= -1 + (3.33 \times 10^{-3})R_{Sdc}$$

Solving for R_{Sdc} yields

$$R_{Sdc} = 300\ \Omega$$

Step 8 then yields

$$R_D = 3\ \text{k}\Omega - R_{Sdc} = 2.7\ \text{k}\Omega$$

Step 9 R_{Sac} is determined from

$$R_{Sac} = -\frac{R_L \parallel R_D}{A_v} - \frac{1}{g_m} = 245\ \Omega$$

The final circuit is shown in Figure 4.17, where the component values are

$$R_D = 2.7\ \text{k}\Omega$$

$$R_{S1} = R_{Sac} = 245\ \Omega$$

$$R_{S2} = R_{Sdc} - R_{Sac} = 300\ \Omega - 245\ \Omega = 55\ \Omega$$

$$R_G = R_{\text{in}} = R_1 = 100\ \text{k}\Omega$$

Example 4.7 JFET CS Amplifier (Design)

Repeat Example 4.6, but select a Q-point that is not in the center of the linear region. Use a transistor with $I_{DSS} = 6.67$ ma and $V_p = -3.33$ V.

SOLUTION Let us arbitrarily select the new operating point as follows:

$$I_{DQ} = 3.5\ \text{mA}$$

$$V_{GSQ} = -0.8\ \text{V}$$

$$V_{DSQ} = 6\ \text{V}$$

Then using equation (4.3), we obtain

$$g_m = 3.04 \times 10^{-3}\ \text{S}$$

and

$$\frac{1}{g_m} = 329\ \Omega$$

From Step 2 in the design procedure we have

$$R_D + R_S = \frac{20 - 6}{3.5 \times 10^{-3}} = 4\ \text{k}\Omega = K_1$$

Then from Step 3 we find

$$A_v = -4 = \frac{-R_D \parallel 20\ \text{k}\Omega}{4\ \text{k}\Omega - R_D + 329\ \Omega}$$

from which we obtain, after solving the quadratic equation,

$$R_D = 3.571\ \text{k}\Omega$$

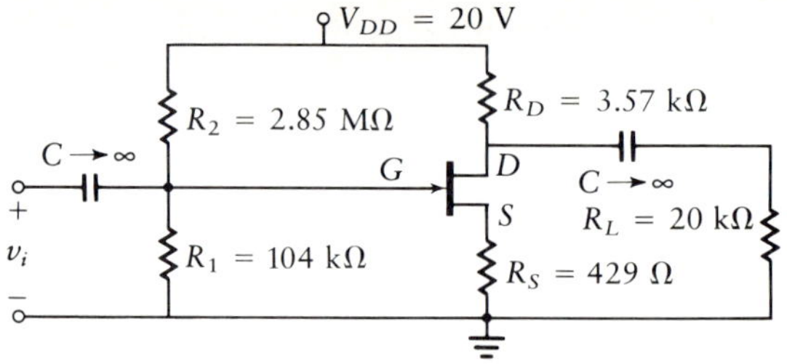

Figure 4.18 CS amplifier for Example 4.7.

This quantity is less than K_1, so we proceed to Step 4. From equation (4.17), we obtain

$$R_S = 4\ \text{k}\Omega - 3.571\ \text{k}\Omega = 429\ \Omega$$

Using the bias equation of Step 5, we find

$$V_{GG} = -0.8 + 429(3.5 \times 10^{-3}) = 0.702\ \text{V}$$

To determine R_1 and R_2 with $R_{in} = R_G$, we use Step 6, since V_{GG} is the same polarity as V_{DD} (this contrasts with the situation in the previous example). Thus,

$$R_1 = \frac{10^5}{1 - 0.702/20} = 104\ \text{k}\Omega$$

$$R_2 = 10^5\left(\frac{20}{0.702}\right) = 2.85\ \text{M}\Omega$$

The final circuit is shown in Figure 4.18. In practice, if $R_2 >> R_1$ and if R_2 is above 10 MΩ, we could remove R_2 completely. This causes V_{GG} to go to 0 V rather than a small positive voltage.

Drill Problems

D4.1 Design a CS JFET amplifier that has an R_L of 10 kΩ, $V_{DD} = 12$ V, $R_{in} = 500$ kΩ, and $A_v = -2$. Use the circuit of Figure 4.16(a). Select the Q-point as $V_{DSQ} = 6$ V, $V_{GSQ} = -1$ V, $I_{DQ} = 1$ mA, and $g_m = 2500\ \mu$S.

Ans: $R_S = 1.22$ kΩ; $R_D = 4.78$ kΩ;
$R_1 = 509$ kΩ; $R_2 = 27$ MΩ;
$A_i = -100$

D4.2 Redesign the amplifier of Drill Problem D4.1 for a transistor with $V_p = -4$ V and $I_{DSS} = 6$ mA.

Ans: $R_1 = 500$ kΩ; $R_2 \rightarrow \infty$;
$R_D = 1.61$ kΩ; $R_{Sdc} = 390$ Ω;
$R_{Sac} = 223$ Ω; $A_i = -100$

4.9 ANALYSIS OF CD (SF) AMPLIFIERS

The CD (SF) JFET amplifier is illustrated in Figure 4.19(a) and the equivalent circuit is shown in Figure 4.19(b). Our approach to the analysis of this amplifier parallels that of Section 4.7. The input resistance is $R_{\text{in}} = R_G$. Note in the equivalent circuit that we ignore r_{DS} since it is much larger than R_S. If r_{DS} is not much larger than R_S, we modify the following equations by replacing R_S with $r_{DS} \parallel R_S$.

Writing a KVL equation around the gate-to-source loop, we have

$$v_{gs} = v_i - g_m(R_S \parallel R_L)v_{gs}$$

from which we obtain

$$v_i = v_{gs}[1 + g_m(R_S \parallel R_L)]$$

The output voltage is

$$v_o = g_m(R_S \parallel R_L)v_{gs}$$

and the voltage gain is the ratio

$$A_v = \frac{v_o}{v_i} = \frac{g_m(R_S \parallel R_L)}{1 + g_m(R_S \parallel R_L)} = \frac{R_S \parallel R_L}{R_S \parallel R_L + 1/g_m} \tag{4.19a}$$

We find the current gain using the gain impedance formula as follows:

$$\begin{aligned} A_i &= A_v \frac{R_{\text{in}}}{R_L} \\ &= \frac{R_S \parallel R_L}{(R_S \parallel R_L) + 1/g_m} \frac{R_{\text{in}}}{R_L} = \frac{R_S}{(R_S \parallel R_L) + 1/g_m} \frac{R_G}{R_S + R_L} \end{aligned} \tag{4.19b}$$

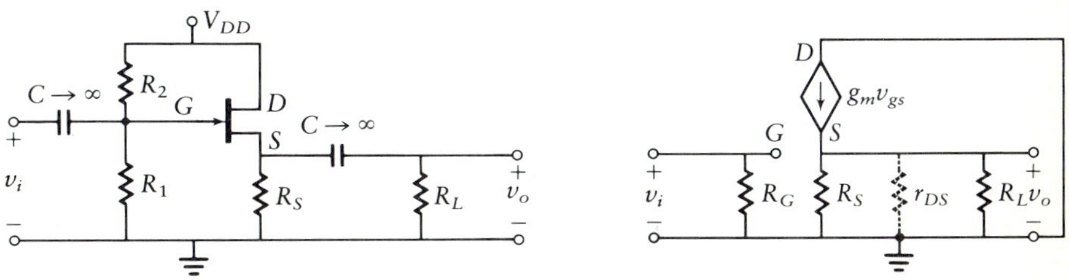

(a) Amplifier (b) Equivalent circuit

Figure 4.19
JFET SF amplifier.

4.10 DESIGN OF CD AMPLIFIERS

We now present the design procedure for the CD JFET amplifier. This procedure is the same as that used to design depletion MOSFET amplifiers. The following quantities are specified: current gain, load resistance, and V_{DD}. Input resistance may be specified, and if so, current gain may be omitted from the specifications. Refer to the circuit of Figure 4.19 as you study the following procedure.

Step 1 Select a Q-point in the center of the FET characteristic curves with the aid of Figure 4.8. This step determines V_{DSQ}, V_{GSQ}, I_{DQ}, and g_m.

Step 2 Write the dc KVL equation around the drain-to-source loop.

$$V_{DD} = V_{DSQ} + R_S I_{DQ}$$

from which we find the dc value of R_S,

$$R_{Sdc} = \frac{V_{DD} - V_{DSQ}}{I_{DQ}}$$

Step 3 We find R_{Sac} from the rearranged current gain equation, equation (4.19b), as follows:

$$R_{Sac} = \frac{R_L}{(R_G/A_i - R_L)g_m - 1} \tag{4.20}$$

where

$$R_G = R_{\text{in}}$$

If the input resistance is not specified, let $R_{Sac} = R_{Sdc}$ and calculate the input resistance from equation (4.20). If the input resistance is not high enough, it may be necessary to change the Q-point location.

If R_{in} is specified, it is necessary to calculate R_{Sac} from equation (4.20). In such cases, R_{Sac} is different from R_{Sdc}, so we bypass part of R_S with a capacitor.

Step 4 Determine V_{GG} using the equation

$$V_{GG} = V_{GSQ} + I_{DQ} R_S$$

No phase inversion is produced in a source-follower FET amplifier and V_{GG} is normally of the same polarity as the supply voltage.

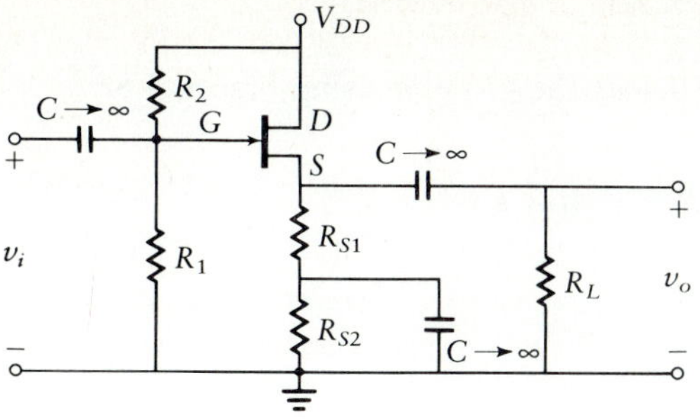

Figure 4.20
CD amplifier.

Step 5 Determine the values of R_1 and R_2 from equation (4.12).

$$R_1 = \frac{R_G}{1 - V_{GG}/V_{DD}}$$

$$R_2 = \frac{R_G V_{DD}}{V_{GG}}$$

There is usually enough drain current in an SF to develop the opposite-polarity voltage needed to offset the negative voltages required by the JFET gate. Therefore, normal voltage division biasing can be used.

We now return to the problem of specifying the input resistance. We can assume that part of R_S is bypassed, as in Figure 4.20, which leads to different values of R_{Sac} and R_{Sdc}. Step 2 is used to solve for R_{Sdc}. In Step 3, we let R_G equal the specified value of R_{in} and use equation (4.20) to solve for R_{Sac}.

If the R_{Sac} calculated above is smaller than R_{Sdc}, the design is accomplished by bypassing R_{S2} with a capacitor. Remember that $R_{Sac} = R_{S1}$ and $R_{Sdc} = R_{S1} + R_{S2}$. If, on the other hand, R_{Sac} is larger than R_{Sdc}, the Q-point must be moved to a different location. We select a smaller V_{DSQ}, thus causing increased voltage to be dropped across $R_{S1} + R_{S2}$, which makes R_{Sdc} larger. If V_{DSQ} cannot be reduced sufficiently to make R_{Sdc} larger than R_{Sac}, the amplifier cannot be designed with the given current gain, R_{in}, and FET type. One of these three specifications must be changed, or a second amplifier stage must be used to provide the required gain.

Example 4.8 CD Amplifier (Design)

Design a CD JFET amplifier (see Figure 4.19) with the following specifications: $A_i = 12$, $R_{\text{in}} > 9$ kΩ, $R_L = 400$ Ω, $I_{DSS} = 20$ mA, $V_p = -6.67$ V, and $V_{DD} = 12$ V.

SOLUTION

Step 1 We select the Q-point using Figure 4.8:

$$I_{DQ} = \frac{I_{DSS}}{2} = 10 \text{ mA}$$

$$V_{DSQ} = \frac{V_{DD}}{2} = 6 \text{ V}$$

Using the graphical technique, we obtain

$$V_{GSQ} = (0.3)(-6.67) = -2 \text{ V}$$

and

$$g_m = \frac{1.42 I_{DSS}}{V_p} = 4.26 \text{ mS}$$

and

$$\frac{1}{g_m} = 235\ \Omega$$

Step 2 Since $R_{\text{in}} > 9$ kΩ, we let $R_S = R_{Sac} = R_{Sdc}$ and

$$R_S = \frac{V_{DD} - V_{DSQ}}{I_{DQ}} = 600\ \Omega$$

Step 3 We use this step to find R_G by rearranging equation (4.19b).

$$R_G = A_i\left(R_L + \frac{1 + R_L/R_S}{g_m}\right) = 9.5 \text{ k}\Omega$$

Since R_G is greater than 9 kΩ, the design is acceptable and we can continue to Step 4.

Step 4

$$V_{GG} = -2 + (10 \times 10^{-3})(600) = 4 \text{ V}$$

Step 5 Finally, from this step we find

$$R_1 = \frac{9500}{1 - 4/12} = 14.25 \text{ k}\Omega$$

$$R_2 = \frac{9500 \times 12}{4} = 28.5 \text{ k}\Omega$$

$$A_v = A_i \frac{R_L}{R_{in}} = 0.51$$

Drill Problem

D4.3 Design a CD JFET amplifier (Figure 4.19) to provide a current gain of 15 to a load of $R_L = 20$ kΩ using $V_{DD} = 12$ V and $R_{in} = 400$ kΩ. Use an n-channel FET that has a $V_p = -3$ V and an $I_{DSS} = 6$ mA. Assume $V_{DSQ} = V_{DD}/2$ and $I_{DQ} = 0.4I_{DSS}$. Determine the resistor values and the voltage gain of the amplifier.

Ans: $R_{Sdc} = 2.5$ kΩ; $R_{Sac} = 1.25$ kΩ;
$R_1 = 676$ kΩ; $R_2 = 980$ kΩ
$A_v = 0.75$

4.11 SF BOOTSTRAP AMPLIFIERS

We now analyze an *SF (or CD) bootstrap FET amplifier.* This circuit is a special case of the SF and is illustrated in Figure 4.21. Here the bias is developed across only a part of the source resistor. This reduces the need for a capacitor bypass across part of the source resistor and thus attains a much larger input resistance than normally can be attained. This design allows us to take advantage of the high-impedance characteristics of the FET without using a high value of the gate resistor, R_G.

The equivalent circuit of Figure 4.22 is used to evaluate the circuit operation. If we assume that i_{in} is sufficiently small to approximate the current in R_{S2} as i_1, the output voltage is then found to be

$$v_o \approx g_m v_{GS}(R_S \parallel R_L) \tag{4.21}$$

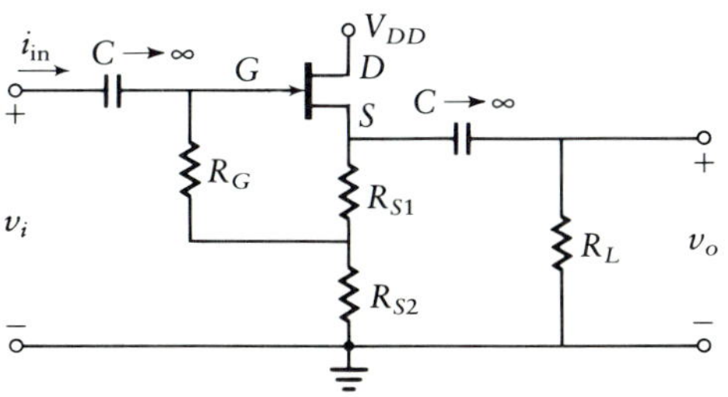

Figure 4.21 Bootstrap SF.

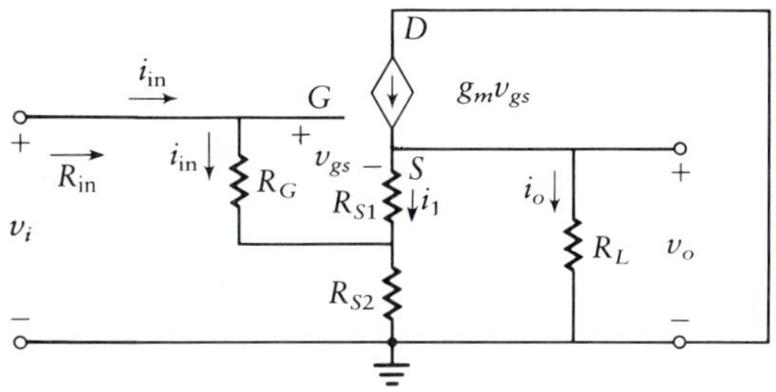

Figure 4.22 ac equivalent circuit for bootstrap SF.

where

$$R_S = R_{S1} + R_{S2}$$

If the assumption about i_{in} is not valid, R_S is replaced by the expression

$$R_{S1} = \frac{(i_{\text{in}} + i_1)R_{S2}}{i_1}$$

A KVL equation at the input yields v_i as follows:

$$v_i = v_{GS} + i_1 R_{S1} + (i_1 + i_{\text{in}})R_{S2} \tag{4.22}$$

The current, i_1, is found from a current-divider relationship,

$$i_1 = \frac{g_m v_{GS} R_L}{R_S + R_L}$$

Using this in equation (4.22), we obtain

$$v_i = v_{GS} + \frac{g_m v_{GS} R_L R_S}{R_L + R_S} + i_{\text{in}} R_{S2} \tag{4.23}$$

A second equation for v_i is developed around the loop through R_G and R_{S2} as follows:

$$\begin{aligned} v_i &= i_{\text{in}} R_G + \left(\frac{g_m v_{GS} R_L}{R_S + R_L} + i_{\text{in}}\right) R_{S2} \\ &= i_{\text{in}}(R_G + R_{S2}) + \frac{g_m v_{GS} R_L R_{S2}}{R_S + R_L} \end{aligned} \tag{4.24}$$

We eliminate v_i by setting equation (4.23) equal to equation (4.24) and solve for i_{in} to obtain

$$i_{\text{in}} = \frac{g_m v_{GS}}{R_G}\left[(R_L \parallel R_S) + \frac{1}{g_m} - \frac{R_L R_{S2}}{R_S + R_L}\right] \tag{4.25}$$

The input resistance, $R_{\text{in}} = v_i/i_{\text{in}}$, is found by dividing equation (4.24) by equation (4.25) with the result

$$R_{\text{in}} = \frac{R_G[1/g_m + (R_L \parallel R_S)]}{(R_L \parallel R_S) + 1/g_m - R_L R_{S2}/(R_S + R_L)} + R_{S2}$$

We solve for R_G as follows:

$$R_G = \frac{(R_{in} - R_{S2})[R_{S1}R_L + (R_S + R_L)/g_m]}{R_L R_S + (R_S + R_L)/g_m} \tag{4.26}$$

The current gain is

$$A_i = \frac{i_o}{i_{in}} = \frac{v_o}{R_L i_{in}}$$

Using equation (4.21) and equation (4.25) and noting that $R_S - R_{S2} = R_{S1}$, we find

$$A_i = \frac{R_G R_S}{R_L R_{S1} + (R_L + R_S)/g_m} \tag{4.27}$$

The voltage gain is

$$A_v = \frac{A_i R_L}{R_{in}}$$

$$= \frac{R_G R_S R_L}{R_{in}[R_{S1}R_L + (R_S + R_L)/g_m]} \tag{4.28}$$

Note that the denominator in equation (4.26) is larger than the bracketed part of the numerator, thus showing that $R_G < (R_{in} - R_{S2})$. This proves that a large input resistance can be attained without having the same order of size as R_G.

Example 4.9 Bootstrap SF (Design)

Design an SF amplifier circuit for the following conditions: $R_{in} = 100\ \text{k}\Omega$, $R_L = 10\ \text{k}\Omega$, and $V_{DD} = 20$ V. The circuit is connected as shown in Figure 4.21. The Q-point is selected at

$$V_{DSQ} = 10\ \text{V} \qquad I_{DQ} = 3.33\ \text{mA}$$

$$V_{GSQ} = -1\ \text{V} \qquad g_m = 2\ \text{mS}$$

SOLUTION Designing this circuit consists of choosing values for R_{S1}, R_{S2}, and R_G. The relationship $V_{GSQ} = -1$ V is used to find R_{S1}. We sum voltages around the gate-to-source loop, assuming that i_{in} is approximately equal to zero, as follows:

$$0 = V_{GSQ} + R_{S1}I_{DQ}$$

$$R_{S1} = \frac{1}{3.3 \times 10^{-3}} = 300\ \Omega$$

In order to find R_{S2}, we write a KVL equation around the source-to-drain loop.

$$V_{DD} = V_{DSQ} + (R_{S1} + R_{S2})I_{DQ}$$

Solving for R_{S2} yields

$$R_{S2} = 2.7\ \text{k}\Omega$$

We find R_G using equation (4.26) as follows:

$$R_G = \frac{R_{S1}R_L + (R_S + R_L)/g_m}{R_L R_S + (R_S + R_L)/g_m}(R_{\text{in}} - R_{S2}) = 25.3\ \text{k}\Omega$$

Drill Problem

D4.4 Determine the value of the resistors and the current gain for an SF JFET bootstrap amplifier that requires $R_{\text{in}} = 200\ \text{k}\Omega$, $R_L = 20\ \text{k}\Omega$, and $V_{DD} = 10$ V. The Q-point is selected as

$$V_{DSQ} = 5\ \text{V} \qquad V_{GSQ} = -1.5\ \text{V}$$

$$I_{DQ} = 0.5\ \text{mA} \qquad g_m = 4\ \text{mS}$$

Use the configuration of Figure 4.22.

Ans: $R_G = 62.8\ \text{k}\Omega$; $R_{S1} = 3\ \text{k}\Omega$; $R_{S2} = 7\ \text{k}\Omega$; $A_i = 9.3$

4.12 METAL SEMICONDUCTOR BARRIER JUNCTION TRANSISTOR

The *metal semiconductor barrier junction transistor* (MESFET) is similar to a FET, except that the junction is a metal semiconductor barrier, much as is the case with Schottky diodes (see Section 1.6.1). FETs made of silicon (Si) are constructed with diffused or ion-implanted gates. However, there are advantages to using a Schottky barrier metal gate when the channel is n-type and short channel widths are needed. Gallium arsenide (GaAs) is difficult to work with, yet it makes good Schottky barriers that are useful in high-frequency applications because electrons travel faster in GaAs than in Si. Using GaAs in MESFETs results in a transistor that exhibits good performance in microwave applications. In comparison with the silicon bipolar transistor, GaAs MESFETs have better performance at input frequencies above 4 GHz. These MESFETs exhibit high gain, low noise, high efficiency, high input impedance, and properties that prevent thermal runaway. They are used in microwave oscillators, amplifiers, mixers, and for high-speed switching.

4.13 OTHER DEVICES

Other devices that are an outgrowth of the normal two- and three-terminal devices are presented in this section.

4.13.1 VMOSFET (VMOS)

Considerable research effort has been applied to increasing the power capability of solid-state devices. An area that has shown much promise is the MOSFET in which the conduction channel is modified to form configurations other than the conventional source-to-drain straight line. One such modification employs a V-shaped channel, as shown in Figure 4.23. An additional semiconductor layer is added. The term *VMOS* is derived from the fact that the current between source and drain follows a vertical path due to the construction. The drain is now located on a piece of added semiconductor material. This allows the transistor drain area to be placed in contact with a heat sink to aid in dissipating the heat generated in the device. The V-shaped gate controls two vertical MOSFETs, one on each side of the notch. By paralleling the two *S* terminals, the current capacity can be doubled. VMOS is unsymmetrical so that the *S* and *D* terminals cannot be interchanged as is the case in low-power MOSFETs. Conventional FETs are limited to currents of the order of milliamperes, but VMOS FETs are available for operation in the 100-A current range. This provides a great improvement in power over the conventional FET.

The VMOS device can provide a solution to high-frequency, high-power applications. Ten-watt devices have been developed at frequencies in the lower ultrahigh-frequency (UHF) band. There are other important advantages of

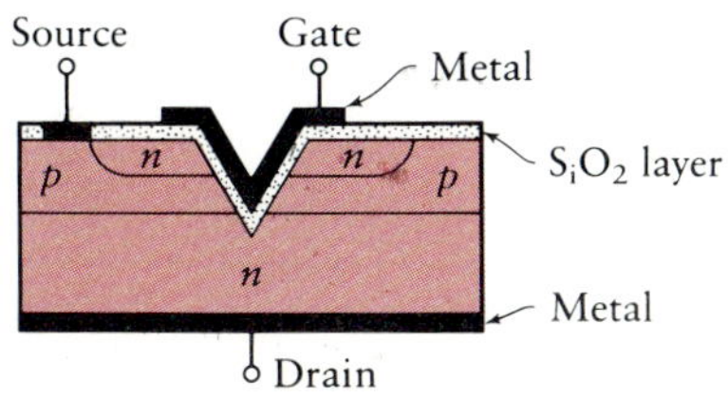

Figure 4.23
VMOS construction.

VMOS FETs. They have a negative temperature coefficient to prevent thermal runaway. They exhibit low leakage current. They are capable of achieving high switching speed. VMOS transistors can be made to have equal spacing of their characteristic curves for equal increments of gate voltage, so they can be used like bipolar junction transistors for high-power linear amplifiers [25].

4.13.2 Other MOS Devices

Another type of MOS device is a *double-diffused process fabricated FET,* sometimes called *DMOS.* This device has the advantage of providing excellent low-power dissipation and high-speed capability.

Fabrication of a FET on small silicon islands on a substrate of sapphire is sometimes referred to as *SOS.* The islands of silicon are formed by etching a thin layer of silicon grown on the sapphire substrate. This type of fabrication provides insulation between the islands of silicon, thus greatly reducing parasitic capacitance between devices.

MOS technology has the advantage that both capacitors and resistors (using MOSFETs) are made at the same time as the FET, although large-value capacitors are not feasible. Using an enhancement MOSFET, a two-terminal resistance is made and the MOSFET gate connected to the drain causes the FET to operate at pinch-off. The MOSFET gate is connected to the drain through a power source, causing the FET to be biased where it will operate in the voltage-controlled resistance region of the characteristics. In this way, drain-load resistors are replaced by a MOSFET rather than a deposited resistor, hence saving chip area.

PROBLEMS

4.1 The characteristic curves for the operating region of a specific n-channel FET transistor can be approximated by the equation

$$i_D = 0.5(4 + v_{GS})^2 \text{ mA}$$

when the following conditions hold: $R_S = 500\ \Omega$, $R_D = 2\ \text{k}\Omega$, $R_{\text{in}} = 100\ \text{k}\Omega$, $I_{DQ} = 5$ mA, and $V_{DD} = 20$ V. Determine the following parameters:

a. V_{GSQ}
b. V_D
c. V_{DSQ}
d. R_1 and R_2

Refer to Figure P4.1.

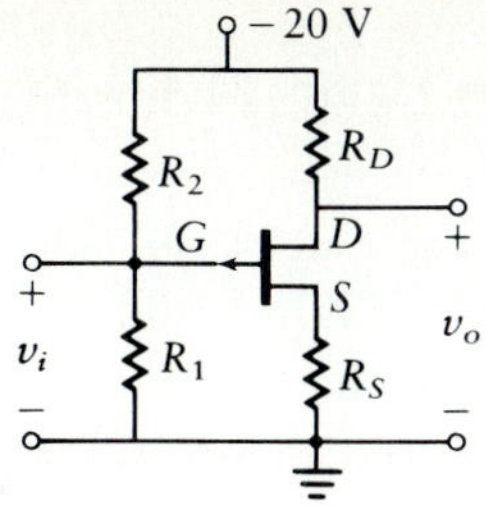

Figure P4.1

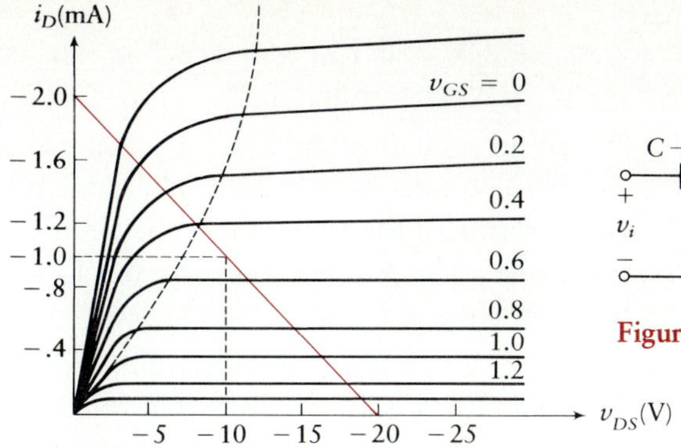

Figure P4.2

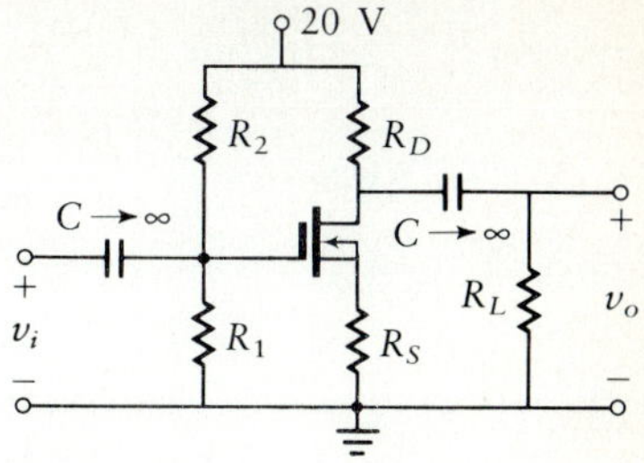

Figure P4.3

4.2 In the circuit of Figure 4.16(a), when $R_1 = 21\ \text{k}\Omega$, $R_2 = 450\ \text{k}\Omega$, $R_S = 500\ \Omega$, $R_D = 1.5\ \text{k}\Omega$, $R_L = 4\ \text{k}\Omega$, and $V_{DD} = 12$ V, find the following when $V_{DSQ} = 4$ V:

a. I_{DQ}
b. V_{GSQ}
c. R_{in}
d. A_v when $g_m = 3.16$ mS
e. A_i

4.3 In the circuit of Figure 4.16(a), $R_D = 2\ \text{k}\Omega$, $R_L = 5\ \text{k}\Omega$, $R_{\text{in}} = 100\ \text{k}\Omega$, $R_S = 300\ \Omega$, and $V_{DD} = 15$ V. Find the values of R_1 and R_2 needed for the transistor to operate at 4 mA when $V_p = -4$ V and $I_{DSS} = 8$ mA. Also determine the voltage and current gain of the amplifier.

4.4 a. Design a CS amplifier (Figure P4.1) using a *p*-channel JFET when the specification calls for $A_v = -10$ and $R_{\text{in}} = 20\ \text{k}\Omega$. Assume that the Q-point is chosen at $I_{DQ} = -1$ mA, $V_{DSQ} = -10$ V, $V_{GSQ} = 0.5$ V.

b. Calculate A_i, R_1, R_2, R_S, and R_D. (Refer to the characteristic curve in Figure P4.2. Note that you may have to split R_S and bypass part of it.)

4.5 Repeat Problem 4.4 when an R_L of 20 kΩ is coupled to the drain with a capacitor. Note that you may need to choose a different Q-point.

4.6 Design a CS amplifier using a MOSFET, as shown in Figure P4.3. Let $R_L = 1\ \text{k}\Omega$, $A_v = -1$, $R_{\text{in}} = 15\ \text{k}\Omega$. The Q-point is chosen at $V_{GSQ} = 3$ V, $I_{DQ} = 7$ mA, $V_{DSQ} = 10$ V, where $g_m = 2300\ \mu\text{S}$. Find values for all other elements.

4.7 Design a CS amplifier using an *n*-channel JFET for a circuit of the type shown in Figure P4.4 with $A_v = -1$, $V_{DD} = 12$ V, $R_L = 1\ \text{k}\Omega$, $R_{\text{in}} = 15\ \text{k}\Omega$, $I_{DSS} = 10$ mA, and $V_p = -4$ V. Use $I_{DQ} = I_{DSS}/2$.

4.8 Design a CS amplifier using an *n*-channel JFET when $R_L = 4\ \text{k}\Omega$, $A_v = -3$, and $R_{\text{in}} = 50\ \text{k}\Omega$. Assume that a transistor is used that has

$V_p = -4.2$ V and $I_{DSS} = 6$ mA. Use the circuit of Figure P4.4 with $V_{DD} = 20$ V. Determine A_i.

4.9 Design an n-channel JFET CS amplifier that has $A_v = -2$, $A_i = -20$, $V_{DD} = 12$ V, and $R_L = 5$ kΩ. Determine all component values and the power rating of the transistor. (The circuit may require changes to meet the design.) The selected transistor has $V_p = -5$ V and $I_{DSS} = 8$ mA. Use $I_{DQ} = 0.4 I_{DSS}$ and $V_{DSQ} = V_{DD}/2$. See Figure P4.4.

4.10 Design a CS p-channel JFET amplifier with $A_v = -4$, $A_i = -40$, $R_L = 8$ kΩ, and $V_{DD} = -16$ V. The selected transistor has $V_p = 3$ V and $I_{DSS} = -7$ mA. Use $I_{DQ} = 0.3 I_{DSS}$ and $V_{DSQ} = V_{DD}/2$. Use the circuit of Figure P4.4. Determine the power rating of the transistor.

4.11 Design a CS p-channel JFET amplifier with a 5-kΩ load using the circuit shown in Figure P4.4. Let $V_{DD} = -20$ V, $A_v = -2$, $A_i = -20$, $V_p = 6$ V, and $I_{DSS} = -5$ mA. Determine the power rating of the transistor.

4.12 Design a CS n-channel MOSFET amplifier using a 3N128 transistor (see Appendix D) for a 10-kΩ load with a voltage gain of $A_v = -10$. Use the circuit of Figure P4.3. Select the Q-point using the characteristic curves shown in the specifications when $R_{\text{in}} > 10$ kΩ.

4.13 Design a CS n-channel MOSFET amplifier using a 3N128 transistor (see Appendix D) for a 2-kΩ load with $A_v = -4$ and $R_{\text{in}} > 100$ kΩ. Assume that the Q-point is chosen as $V_{GSQ} = -0.6$ V, $V_{DSQ} = 10$ V, $I_{DQ} = 10$ mA, $V_{DD} = 20$ V. See Figure P4.3.

4.14 Analyze a CS n-channel JFET amplifier as shown in Figure P4.5 when the load is 20 kΩ, $R_D = 8$ kΩ, $V_{DD} = 24$ V, and $R_{\text{in}} = 50$ kΩ. Select the Q-point as $V_{GSQ} = -1.5$ V, $V_{DSQ} = 12$ V, $I_{DQ} = 1$ mA, and $g_m = 2.83$ mS. Find all component values, A_i, and A_v.

4.15 If R_S in Figure P4.4 is bypassed with a capacitor, what is the voltage gain? Assume that the Q-point is selected so that $g_m = 1.5$ mS, $R_D = 3.2$ kΩ, and $R_L = 5$ kΩ. Determine the current gain when $R_S = 500$ Ω, $R_1 = 200$ kΩ, and $R_2 = 800$ kΩ.

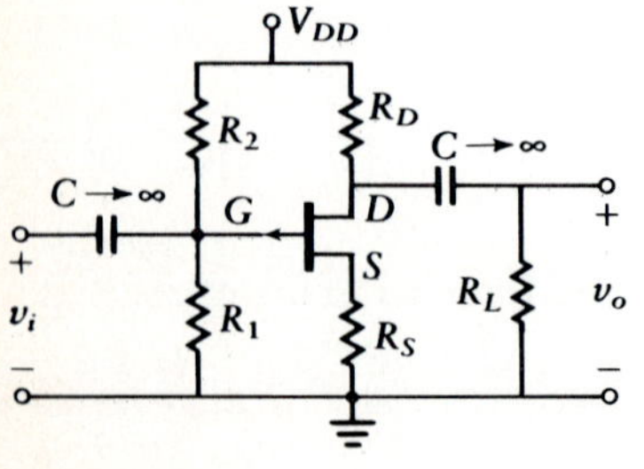

Figure P4.4

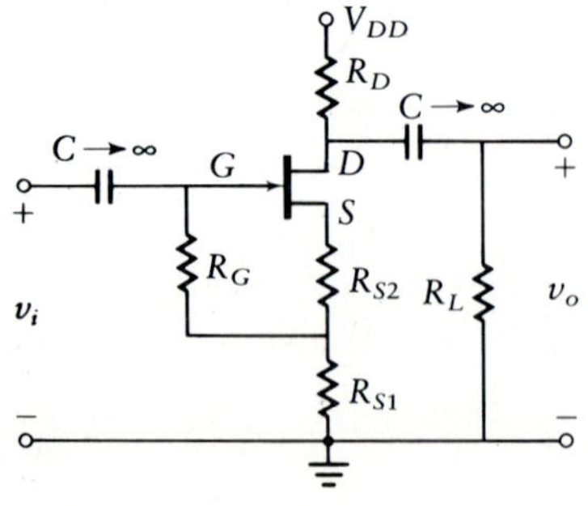

Figure P4.5

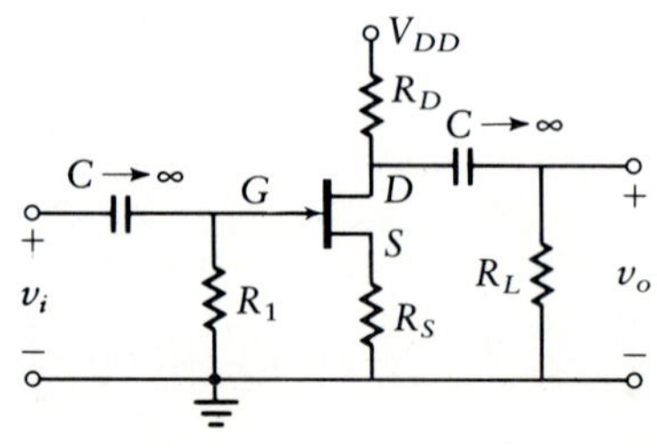

Figure P4.6

4.16 What is the voltage gain, A_v, of the circuit shown in Figure P4.4 if a signal is fed into the amplifier, which has a source voltage resistance, R_i, of 10 kΩ? Assume $R_D = 4$ kΩ, $R_L = 10$ kΩ, $R_S = 500$ Ω, $g_m = 2$ mS, $R_1 = 25$ kΩ, and $R_2 = 120$ kΩ.

4.17 In the circuit shown in Figure P4.6, assume that R_S is bypassed with a capacitor. $V_{DD} = 15$ V, $R_D = 2$ kΩ, $R_L = 3$ kΩ, $R_S = 200$ Ω, $R_1 = 500$ kΩ, $I_{DSS} = 8$ mA, and $V_p = -4$ V. Determine A_v, A_i, R_{in}, and the Q-point for the amplifier.

4.18 In the circuit shown in Figure P4.6, $V_{DD} = 20$ V, $R_D = 2$ kΩ, $R_L = 10$ kΩ, $R_S = 200$ Ω, $R_1 = 1$ MΩ, $I_{DSS} = 10$ mA, and $V_p = -5$ V. Determine the Q-point, A_v, A_i, and R_{in} for the amplifier.

4.19 For the amplifier shown in Figure P4.6, $V_{DD} = 20$ V, $R_D = 2$ kΩ, $R_L = 6$ kΩ, $R_S = 100$ Ω, $R_1 = 1$ MΩ, $I_{DSS} = 10$ mA, and $V_p = -5$ V. Determine the Q-point, A_v, A_i, and R_{in} for the amplifier.

4.20 A CS amplifier circuit as shown in Figure P4.1 uses a JFET for which $I_{DSS} = 2$ mA and $g_{mo} = 2000$ μS. If the value of $R_D = 10$ kΩ and $R_S = 200$ Ω, what is the voltage gain, A_v, for the following values of V_{GSQ}?

a. -1 V
b. -0.5 V
c. 0 V

4.21 A CS amplifier circuit as shown in Figure P4.6 has a transistor with $V_p = -4$ V, $I_{DSS} = 4$ mA, and $r_{DS} = 500$ Ω. If $R_D = 2$ kΩ, $R_L = 4$ kΩ, and $R_S = 200$ Ω, what is the voltage gain, A_v, of the circuit if $V_{GSQ} = -1$ V? What happens to A_v as r_{DS} approaches infinity?

4.22 Design a CS n-channel MOSFET amplifier as shown in Figure P4.3 when $R_L = 4$ kΩ, $A_v = -5$, and $A_i = -10$. Assume that the Q-point was selected as $V_{DSQ} = 10$ V, $V_{GSQ} = 4$ V, $I_{DQ} = 2$ mA, and $g_m = 4000$ μS.

4.23 Given the circuit shown in Figure P4.7 when $R_i = 50$ kΩ, $R_1 = 100$ kΩ, $R_2 = 800$ kΩ, $R_D = 4$ kΩ, $R_L = 6$ kΩ, $R_S = 200$ Ω, and $V_{DD} = 20$ V, determine the following when using a FET with $V_{DSQ} = 6$ V and $g_m = 2.5$ mS:

a. I_{DQ}, V_{GG}, and V_{GSQ}
b. A_v, R_{in}, and A_i.

4.24 For the circuit shown in Figure P4.7, if R_2 is removed, the FET transistor is operating at 2 mA. The component values are $R_i = 100$ kΩ, $R_1 = 400$ kΩ, $R_D = 3$ kΩ, $R_L = 5$ kΩ, and $V_{DD} = 12$ V. Determine the following when the transistor has an $I_{DSS} = 8$ mA, $R_1 = 400$ kΩ, and $V_p = -4$ V:

a. R_S
b. A_v, R_{in}, and A_i.

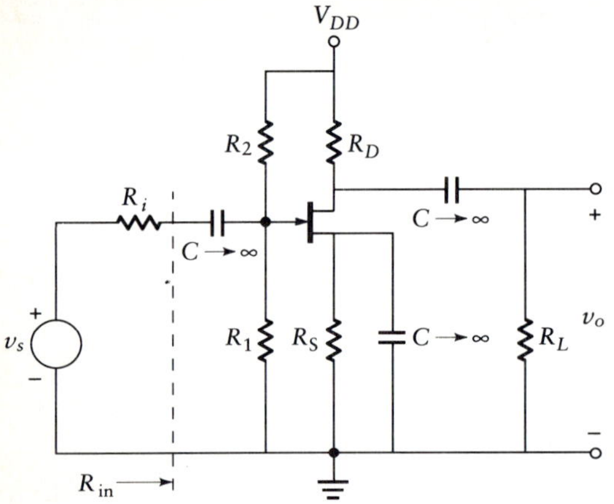

Figure P4.7

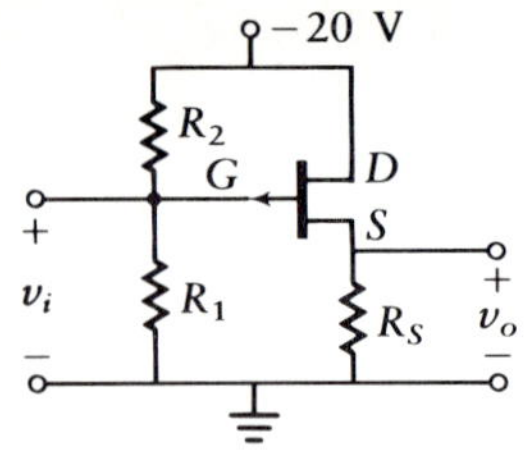

Figure P4.8

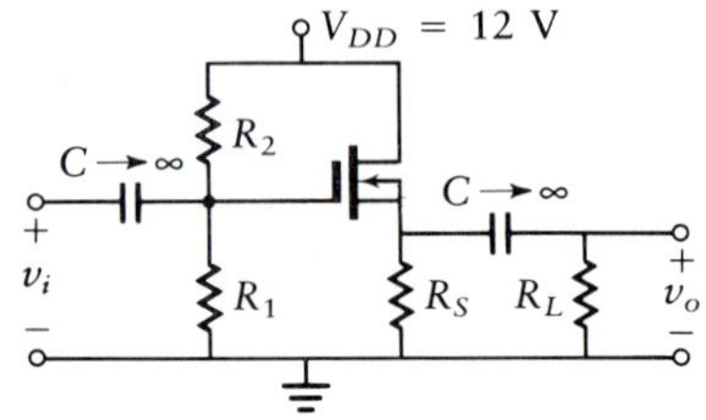

Figure P4.9

4.25 Design an SF amplifier using a p-channel JFET as shown in Figure P4.8 with $R_{in} = 20$ kΩ. Try to obtain a voltage gain, A_v, as close to unity as possible. Calculate A_i, R_1, R_2, and R_S. Use the characteristic curves as shown in Figure P4.2.

4.26 Repeat Problem 4.25 when a 20-kΩ load is capacitively coupled to the amplifier.

4.27 Design a CD MOSFET amplifier when $R_L = 100\ \Omega$, $A_i = 200$, and $R_{in} = 100$ kΩ. Use a transistor with $V_p = -6$ V and $I_{DSS} = 20$ mA. Determine A_v and all resistor values. Use the circuit of Figure P4.9.

4.28 Design a CD amplifier using an n-channel MOSFET where $R_{in} = 120$ kΩ, $A_i = 100$, $R_L = 500\ \Omega$, $V_{DD} = 20$ V, and the transistor selected has $V_p = -5$ V and $I_{DSS} = 15$ mA. Use the circuit of Figure P4.9 with $I_{DQ} = 0.6 I_{DSS}$ and $V_{DSQ} = V_{DD}/2$.

4.29 Design an SF amplifier using an n-channel JFET for a current gain of 100 and an input resistance of 500 kΩ. The load is 2 kΩ. Select the Q-point for the following parameters: $V_{DSQ} = 8$ V, $I_{DQ} = 5$ mA, $V_{GSQ} = -1$ V, and $g_m = 4$ mS. Determine the resistances and the voltage gain and draw the circuit when $V_{DD} = 10$ V.

4.30 Repeat Problem 4.29 except with a different transistor having the following parameter values: $V_p = -3$ V, $I_{DSS} = 10$ mA

4.31 Design the circuit shown in Figure 4.21 when $V_{DD} = 16$ V and $R_L = 8$ kΩ. Use a transistor that has $V_p = -3.33$ V and $I_{DSS} = 10$ mA. Determine all component values, A_i, and A_v when $R_{in} = 12$ kΩ.

4.32 In Problem 4.31, find all component values, A_i, and A_v when $R_{in} = 200$ kΩ.

5

BIAS STABILITY OF TRANSISTOR AMPLIFIERS

5.0 INTRODUCTION

Parameter variations, such as those due to temperature changes, aging, and device substitutions, cause the operating-point location of a transistor circuit to vary from the nominal value. Designers often expend considerable effort to reduce the effects of these parameter variations, since such changes can adversely affect performance. For example, a change in the location of the Q-point can reduce the maximum symmetrical undistorted output swing. In extreme cases, such changes can saturate or cut off the transistor without any input signal being present.

Parameters vary because of changes in supply voltage, changes in temperature, and tolerances in the transistor-manufacturing process. This chapter studies the effects of these changes. Once the effects are understood, we develop equations for several discrete amplifier configurations. Several design examples are given. We then introduce diode compensation as one method of reducing the effects of parameter variations on circuit operation. The chapter concludes with a summary of the design steps needed to ensure bias stability.

5.1 TYPES OF BIASING

In this chapter, we illustrate the various biasing techniques using BJTs. The same biasing techniques can be applied to FETs, although the motivation is somewhat different. For example, temperature does not affect FET operation in the same manner as it does BJT operation.

The voltage-divider biasing system is commonly used and is discussed in Chapter 2. This method of biasing allows the amplifier to operate close to the intended design point regardless of variations that occur in the transistor.

Feedback is defined as the return of part of the output back to the input, from which it is subtracted. It reduces the effect of parameter variations. Many systems use feedback around high-gain IC amplifiers (op-amps), and we discuss the concept of feedback as used with the various biasing techniques. The intent is to illustrate useful techniques of dc biasing of transistor circuits. Feedback systems are examined in more detail in Chapter 11.

5.1.1 Current Biasing

Figure 5.1 illustrates one form of biasing that exhibits moderate stability. This is known as *current feedback,* where the collector current through R_E develops a negative feedback voltage. The base resistor, R_B, is connected to the supply, V_{CC}. We begin by deriving the quiescent condition equations. The KVL equation for the bias current loop (assuming $I_{EQ} \approx I_{CQ}$) is given by

$$\begin{aligned} V_{CC} &= I_{BQ}R_B + R_E I_{CQ} + V_{BE} \\ &= V_{BE} + I_{CQ}\left(\frac{R_B}{\beta} + R_E\right) \end{aligned}$$

We solve for I_{CQ} to obtain

$$I_{CQ} = \frac{V_{CC} - V_{BE}}{R_B/\beta + R_E} \tag{5.1a}$$

Dividing this by β yields

$$I_{BQ} = \frac{V_{CC} - V_{BE}}{R_B + \beta R_E} \tag{5.1b}$$

Note that the value of I_{BQ} is affected by R_B, R_E, V_{CC}, V_{BE}, and β. The specific values of these parameters will then determine the quiescent operating point of the transistor.

We examine the reason for calling this configuration "current feedback." *Feedback* occurs when a circuit is configured in such a manner that the input is affected by the output. The emitter resistor in the circuit of Figure 5.1 provides a form of feedback. As the collector current increases, the voltage across the emitter resistor also increases. For a given input voltage, v_i, this increase in

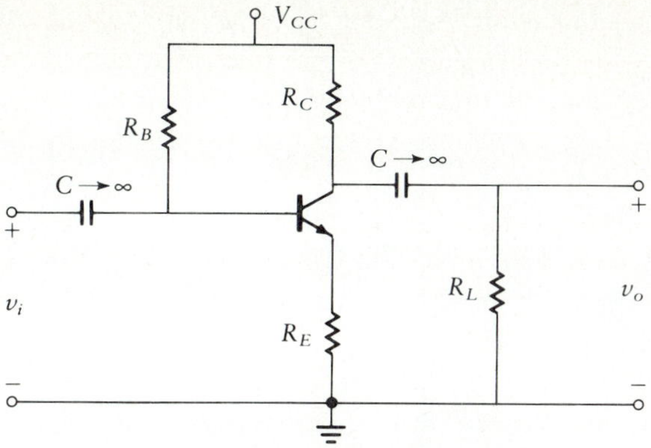

Figure 5.1
Current feedback.

voltage across R_E reduces the base-emitter voltage and therefore also reduces the base current. This, in turn, decreases the collector current to reduce the effect of the original change. Because this effect is "fighting" against the increase in collector current, the situation is known as *negative current feedback*. The emitter resistor can be bypassed with a capacitor, which eliminates the feedback for ac signals. The dc stability is still affected by the feedback. In fact, we cannot achieve bias stability relative to changes in β, as the transistor will be driven into saturation for any reasonable value of voltage gain. This is due to the large dc voltage at the base resulting in increased transistor current. With the value of R_B limited to $0.1\beta R_E$ for dc bias stability (see Chapter 2), there is not enough resistance in the base to restrict the dc bias current to a level low enough to keep the transistor from going into saturation.

5.1.2 Voltage and Current Biasing

A second type of feedback, shown in Figure 5.2, is *voltage feedback*. Note that current shunt feedback is still present due to R_E. Two forms of feedback are present in this circuit. We feed back a portion of the output voltage through R_F and also feed back a portion of the output from the voltage developed across R_E. Increased gain can be achieved without affecting the dc biasing of the circuit by bypassing R_E with a capacitor.

The feedback resistor, R_F, is connected between the collector and base. The base-to-ground voltage then consists of two components, one arising from the input voltage and one from the collector voltage. We analyze this circuit by writing the dc equations between base and collector. In the following, we assume that the quiescent base current is much less than the quiescent collector current so emitter current approximately equals collector current.

$$V_{CC} \approx I_{CQ}R_C + I_{BQ}R_F + V_{BE} + I_{CQ}R_E$$

$$= I_{CQ}R_C + \frac{I_{CQ}R_F}{\beta} + V_{BE} + I_{CQ}R_E$$

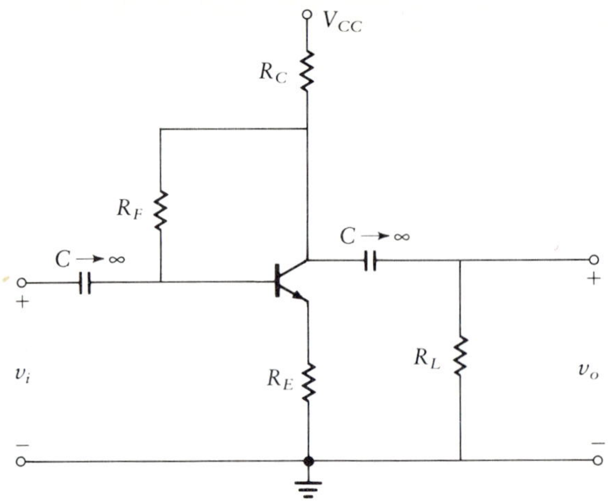

Figure 5.2
Voltage feedback.

Solving for I_{CQ} yields

$$I_{CQ} = \frac{V_{CC} - V_{BE}}{R_C + R_E + R_F/\beta} \tag{5.2a}$$

Dividing this by β, we get

$$I_{BQ} = \frac{V_{CC} - V_{BE}}{R_F + \beta(R_C + R_E)} \tag{5.2b}$$

Note that β appears in both equations (5.1) and (5.2). Variations in β therefore affect the Q-point location. It proves desirable to make R_B equal to $0.1\beta(R_E + R_C)$ when using voltage-division biasing circuits.

Drill Problems

D5.1 Design a CE amplifier for maximum output voltage swing with a voltage gain of −10 and an output resistance of 10 kΩ. Use the circuit shown in Figure 5.1 with V_{CC} = 16 V, V_{BE} = 0.7 V, and β = 100.

Ans: R_C = 10 kΩ; R_E = 474 Ω;
R_B = 1.48 MΩ

D5.2 Design a CE amplifier with A_v = −10. Use the circuit of Figure 5.2 with R_L = 10 kΩ, V_{CC} = 16 V, V_{BE} = 0.7 V, and β = 100.

Ans: R_C = 10 kΩ; R_E = 474 Ω;
R_F = 483 kΩ; $v_o(p\text{-}p)$ = 9.0 V

D5.3 Select a value of R_F in the amplifier of Drill Problem D5.2 to provide bias stability against changes in β. What is the maximum undistorted symmetrical output voltage swing?

Ans: $R_F = 105\ \text{k}\Omega$; $v_o(p\text{-}p) = 2.85\ \text{V}$

5.2 EFFECTS OF PARAMETER CHANGES—BIAS STABILITY

Temperature changes cause certain transistor parameters to change. In particular, the following parameters are temperature sensitive:

1. Collector leakage current between base and collector (I_{CBO}).
2. Base-emitter voltage (V_{BE}).

The supply voltage, V_{CC}, and the β of the transistor also vary, but these are usually not dependent on temperature. In many cases, the supply is sufficiently well regulated that we can ignore changes in V_{CC}. However, for completeness in our derivation, such changes are included [32].

Variations in β can be significant, but the largest variations are not caused by changes in temperature. The more likely cause is random variations from device to device that occur during the manufacture process.

We view temperature changes first. As the temperature increases, the changes in the parameters cause the Q-point to move up (i.e., an increase in I_{CQ}). If the temperature is reduced, the Q-point moves down (i.e., a decrease in I_{CQ}). Either condition causes the maximum possible peak-to-peak output voltage swing to be reduced, as shown in Figure 5.3.

We determine the amount of collector current change using partial derivatives. The collector current is a function of four variables,

$$I_C = f(V_{BE}, I_{CBO}, \beta, V_{CC})$$

For small parameter changes, the variation in I_C is approximately given by

$$\Delta I_C = \frac{\partial I_{CQ}}{\partial V_{BE}} \Delta V_{BE} + \frac{\partial I_{CQ}}{\partial I_{CBO}} \Delta I_{CBO} + \frac{\partial I_{CQ}}{\partial \beta} \Delta \beta + \frac{\partial I_{CQ}}{\partial V_{CC}} \Delta V_{CC}$$

We define four *variation constants* as the partial derivatives of I_C with respect to the four variables. These constants are designated δ_V, δ_I, δ_β, and δ_{VCC}. Then, in terms of these constants, we have

$$\Delta I_C = \delta_V \Delta V_{BE} + \delta_I \Delta I_{CBO} + \delta_\beta \Delta \beta + \delta_{VCC} \Delta V_{CC} \tag{5.3}$$

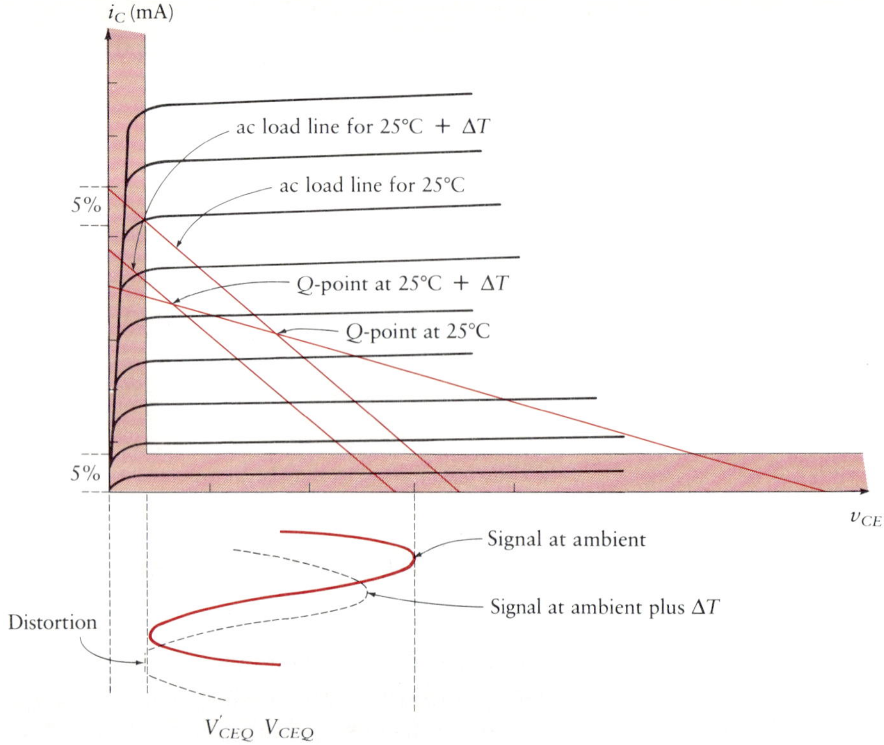

Figure 5.3
V_{CE} with changes in temperature.

We now examine the four variables in equation (5.3).

The variation ΔV_{BE} is the change of the junction voltage between the base and emitter. This behaves in the same way as does the voltage across a diode, as shown in Chapter 1. For a silicon transistor, the voltage varies linearly with temperature according to the following equation:

$$\Delta V_{BE} = -2(T_2 - T_1) \text{ mV} \tag{5.4}$$

In equation (5.4), T_2 and T_1 are in degrees Celsius.

The leakage current between the collector and base (sometimes referred to as the *reverse saturation current*) is also a function of temperature. The collector-to-base leakage current, I_{CBO}, approximately doubles for every 10°C temperature rise. This is given by the following formula, where I_{CBO1} is the reverse leakage current at room temperature (25°C).

$$I_{CBO2} = I_{CBO1} \times 2^{(T_2 - 25°\text{C})/10}$$

$$\Delta I_{CBO} = I_{CBO2} - I_{CBO1} = I_{CBO1}(2^{(T_2 - 25°\text{C})/10} - 1) \tag{5.5}$$

Since the primary variations of the other two parameters, V_{CC} and β, are due to factors other than temperature, information about the magnitude of these changes must be specified in the design statement.

In the following sections, we calculate values of the four variation constants for the common-emitter and common-collector configurations.

5.2.1 CE Configuration

We find the variation constant for V_{BE}, δ_V, by first writing the KVL equation around the base-emitter loop. Refer to the circuit shown in Figure 5.4, where, as in Chapter 2, V_{BB} and R_B are the Thevenin equivalent parameters for the base bias circuit. That is,

$$V_{BB} = \frac{V_{CC}R_1}{R_1 + R_2}$$

and

$$R_B = R_1 \parallel R_2$$

We then have

$$I_{CQ} = \frac{V_{BB} - V_{BE}}{R_E + R_B/\beta} \tag{5.6}$$

and

$$\delta_V = \frac{\partial I_{CQ}}{\partial V_{BE}} = -\frac{1}{R_E + R_B/\beta}$$

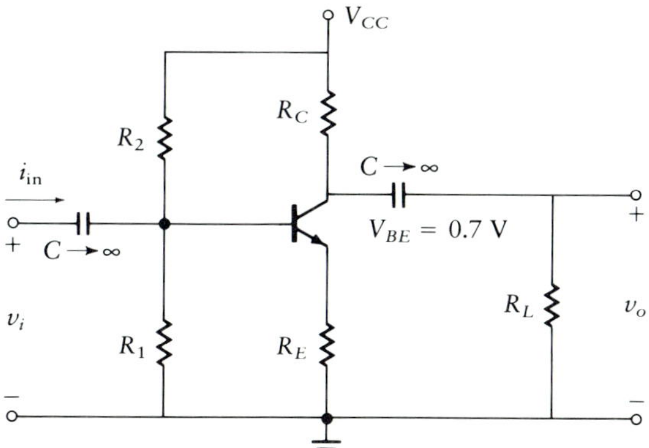

Figure 5.4 Common-emitter amplifier.

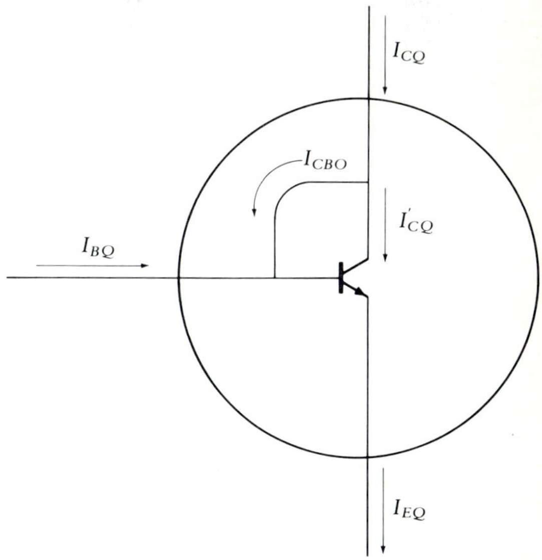

Figure 5.5 Current in the transistor.

The variation constant for I_{CBO}, δ_I, is developed by referring to Figure 5.5.

$$I_{CQ} = I'_{CQ} + I_{CBO}$$

and

$$I'_{CQ} = \beta(I_{BQ} + I_{CBO})$$

Thus,

$$I_{CQ} = \beta(I_{BQ} + I_{CBO}) + I_{CBO} = \beta I_{BQ} + (\beta + 1)I_{CBO} \tag{5.7}$$

If we assume that $1 + \beta \approx \beta$, then

$$I_{CQ} \approx \beta(I_{BQ} + I_{CBO})$$

so

$$I_{BQ} = \frac{I_{CQ}}{\beta} - I_{CBO} \tag{5.8}$$

Since we wish to solve for I_{CQ}, we need to eliminate I_{BQ} from equation (5.8). The base-emitter loop equation, equation (5.6), can be rewritten as follows:

$$V_{BB} - V_{BE} = I_{BQ}(R_B + R_E) + R_E I_{CQ}$$

Now substituting equation (5.8) for I_{BQ} yields

$$V_{BB} - V_{BE} = (R_B + R_E)\left(\frac{I_{CQ}}{\beta} - I_{CBO}\right) + I_{CQ}R_E$$

Solving for I_{CQ}, we have

$$I_{CQ} = \frac{V_{BB} - V_{BE} + (R_B + R_E)I_{CBO}}{(R_B + R_E)/\beta + R_E} \tag{5.9}$$

The variation constant is found by taking the partial derivative,

$$\delta_I = \frac{\partial I_{CQ}}{\partial I_{CBO}} = \frac{R_B + R_E}{(R_B + R_E)/\beta + R_E} = \frac{1}{1/\beta + R_E/(R_E + R_B)} \tag{5.10}$$

If $R_B = 0.1\beta R_E$ (for dc bias stability) and $R_B >> R_E$, then δ_I is approximately given by

$$\delta_I = \frac{\beta}{1 + \beta R_E/R_B} \tag{5.11}$$

We now evaluate the variation constant for β. Although β varies with temperature, the predominant variation is due to external factors. The range of values is usually given in the specification sheet. The value of β used in the following equations is normally the midvalue of the given range.

We start with the equation for collector current, equation (5.9), and differentiate this with respect to β to obtain

$$\frac{\partial I_{CQ}}{\partial \beta} = \frac{(R_B + R_E)[V_{BB} - V_{BE} + (R_B + R_E)I_{CBO}]}{(R_B + R_E + \beta R_E)^2}$$

If we can make the assumptions that $R_B >> R_E$, $\beta R_E >> R_B$, and $R_B I_{CBO} << (V_{BB} - V_{BE})$, the equation reduces to the following form:

$$\delta_\beta = \frac{\partial I_{CQ}}{\partial \beta} = \frac{R_B(V_{BB} - V_{BE})}{\beta^2 R_E^2} \tag{5.12}$$

Finally, the variation constant associated with changes in the voltage supply V_{CC} is found from the collector-to-emitter loop equation.

$$I_{CQ} = \frac{V_{CC} - V_{CE}}{R_E + R_C}$$

$$\delta_{VCC} = \frac{\partial I_C}{\partial V_{CC}} = \frac{1}{R_E + R_C} = \frac{1}{R_{dc}} \tag{5.13}$$

Putting all of these relationships together, the total change in I_{CQ} is found from equation (5.3) to be (assuming $\beta > 100$)

$$\Delta I_{CQ} \approx \frac{-1}{R_E + R_B/\beta}\Delta V_{BE} + \frac{\beta}{1 + \beta R_E/R_B}\Delta I_{CBO} + \frac{R_B(V_{BB} - V_{BE})}{\beta^2 R_E^2}\Delta\beta + \frac{\Delta V_{CC}}{R_{dc}} \tag{5.14}$$

We apply equation (5.14) to find the variation in I_{CQ} that occurs when the temperature rises or drops. Using this equation, we can determine which terms cause the largest changes in I_{CQ}. If the total variation is too large for the amplifier application, we concentrate on the term(s) that causes the largest variation in I_{CQ}. On viewing this equation, we determine whether a parameter value should be raised or lowered. For example, if the second term is the largest, it may be necessary to try to select a transistor that has a lower leakage current between collector and base (I_{CBO}). If this is not possible, the design must be changed to reduce R_B. For temperatures lower than ambient, all terms in equation (5.14) become negative. Thus, they cause I_{CQ} to become smaller.

Example 5.1 CE Amplifier

Design a CE amplifier with $A_v = -10$, $R_L = 1\ \text{k}\Omega$, and $V_{CC} = 12$ V. The temperature varies between $+25°$ and $+65°$C. The selected transistor has an $I_{CBO} = 1.5\ \mu$A at 25°C and a β that varies between 300 and 400. What is the maximum undistorted collector current swing?

SOLUTION Let us start the design by letting

$$R_C = R_L = 1\ \text{k}\Omega$$

Recall that setting R_C equal to R_L yields an optimum design. Now, using the gain equation for CE amplifiers,

$$A_v = -\frac{R_L \parallel R_C}{R'_E} = -10$$

so

$$R'_E = 50\ \Omega$$

$$R_{ac} = R_E + R_C \parallel R_L = 550\ \Omega$$

$$R_{dc} = R_C + R_E = 1050\ \Omega$$

$$I_{CQ} = \frac{V_{CC}}{R_{ac} + R_{dc}} = 7.5\ \text{mA}$$

$$h_{ib} = \frac{26\ \text{mV}}{|I_{CQ}|} = 3.47\ \Omega$$

$$R_E \approx 50 - 3 = 47\ \Omega$$

$$R_B = 0.1\beta R_E = 1.65\ \text{k}\Omega$$

$$V_{BB} = I_{CQ}(R_B/\beta + R_E) + V_{BE} = 1.09\ \text{V}$$

For the temperature change from 25° to 65°C, we first find the temperature-dependent variations in I_{CBO} and V_{BE}.

$$\Delta T = 40°\text{C}$$

$$\Delta I_{CBO} = I_{CBO1}(2^{40/10} - 1) = 22.5\ \mu\text{A}$$

$$\Delta V_{BE} = -2(40) = -80 \text{ mV}$$

The variation in β is given as

$$\Delta\beta = 100$$

Since the problem says nothing about variations in V_{CC}, we assume that $\Delta V_{CC} = 0$. The change in I_{CQ} is now found from equation (5.14).

$$\Delta I_{CQ} = 1.55 \text{ mA} + 0.72 \text{ mA} + 0.24 \text{ mA} = 2.51 \text{ mA}$$

We now find the new maximum symmetrical collector current swing. After movement in the positive direction, the new Q-point is at

$$I_{CQ} = (7.5 + 2.51) \text{ mA} = 10 \text{ mA}$$

If we avoid the upper 5% of the load line because of nonlinearities, the maximum amplitude of the current swing is

$$(12.7 \times 0.95 - 10) \text{ mA} = 2.07 \text{ mA}$$

The total swing of I_{CQ} has a peak-to-peak value of twice this amount, or 4.14 mA. This is the maximum peak-to-peak collector current for a distortion-free output and is illustrated in Figure 5.6. The maximum output voltage swing is reduced from 6.75 to 4.14 V.

In this design, the first term in the ΔI_{CQ} equation is the largest. Thus, the temperature effect on V_{BE} causes the largest change in I_{CQ}. If the amount of re-

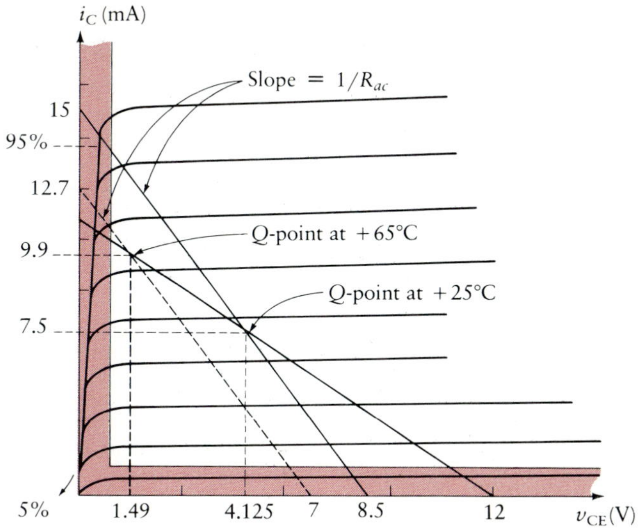

Figure 5.6 Load lines for Example 5.1.

duction in symmetrical output swing was not acceptable, one approach would be to reduce the temperature effect on V_{BE}. This could be accomplished by locating the amplifier in a cooler location or by providing for removal of heat from the transistor with a heat sink. Alternatively, we may alter the circuit values to reduce δ_V (e.g., increase the values of R_E and R_C).

Drill Problems

D5.4 Determine the variation of I_{CQ} for a CE amplifier that is designed using the following criteria: $A_v = -10$, $R_L = 4\ \text{k}\Omega$, $R_{\text{in}} = 5\ \text{k}\Omega$, $V_{CC} = 10$ V, and $I_{CBO} = 0.1\ \mu$A at 25°C. The value of β varies from 100 to 300 and the temperature ranges from 25° to 65°C (see Figure 5.4).

Ans: I_{CQ} starts at 1.56 mA and increases to 2.26 mA at 65°C.

D5.5 What is the maximum undistorted symmetrical voltage output swing for the amplifier of Drill Problem D5.4?

Ans: 0.5 V peak-to-peak

D5.6 If the amplifier in Drill Problem D5.4 is designed to operate from −25° to +25°C, what is the variation in I_{CQ}?

Ans: I_{CQ} decreases by 0.757 mA.

D5.7 What is the maximum undistorted symmetrical voltage output swing for the amplifier of Drill Problem D5.6?

Ans: 2.43 V peak-to-peak

5.2.2 EF Configuration

The EF amplifier is shown in Figure 5.7. The biasing technique for this amplifier is similar to that of the CE with the one exception that the collector resistor, R_C, is equal to zero. Thus, the derivation is based on the same bias equation,

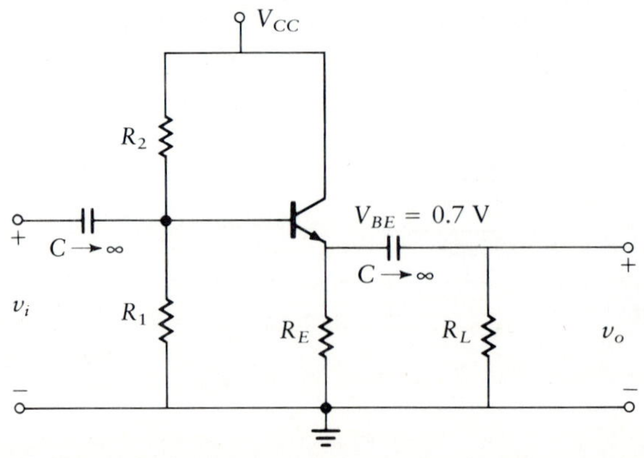

Figure 5.7
The emitter-follower amplifier.

$$I_{CQ} = \frac{V_{BB} - V_{BE}}{R_B/\beta + R_E}$$

The variation constants are then the same as those of the CE, except that R_{dc} is simply equal to R_E rather than $R_E + R_C$. Equation (5.14) is once again used to find the variation in quiescent collector current.

Example 5.2 EF Amplifier

Design an EF amplifier (see Figure 5.7) with $A_i = 10$, $I_{CBO1} = 10\ \mu A$ at 25°C, $V_{CC} = 18$ V, and $R_L = 200\ \Omega$. The temperature varies between 25° and 85°C and β varies between 80 and 120. The power supply voltage ranges between 17.5 and 18.5 V.

SOLUTION Refer to Chapter 3 and Table 3.1 for the necessary design equations. We start by setting

$$R_E = R_L = 200\ \Omega$$

$$I_{CQ} = \frac{V_{CC}}{R_{ac} + R_{dc}} = \frac{18}{100 + 200} = 60\ \text{mA}$$

Since this is an EF amplifier, we can ignore the effects of h_{ib} in the current gain equation and the current gain is given by

$$A_i = 10 = \frac{R_B}{R_B/\beta + R_E \parallel R_L} \times \frac{R_E}{R_E + R_L}$$

so

$$R_B = 2.5\ \text{k}\Omega$$

$$V_{BB} = V_{BE} + I_{CQ}\left(\frac{R_B}{\beta} + R_E\right) = 14.2\ \text{V}$$

Calculating the parameter variations based on the given conditions, we have

$$\Delta V_{BE} = -2\Delta T = -2(60) = -120\ \text{mV}$$

$$\Delta I_{CBO} = I_{CBO1}(2^{(T_2-25)/10} - 1) = 0.63\ \text{mA}$$

$$\Delta\beta = 40$$

$$\Delta V_{CC} = 1$$

We find ΔI_{CQ} using equation (5.14), where we use the average β value of 100.

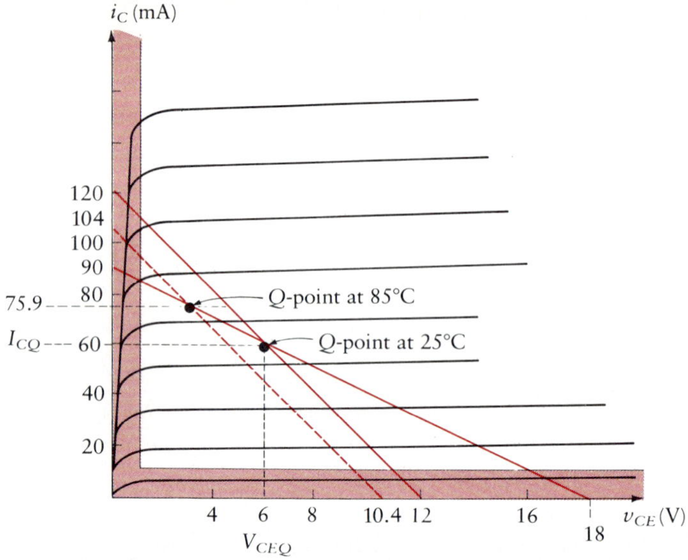

Figure 5.8 Load lines for Example 5.2.

$$\Delta I_{CQ} = 0.53 \text{ mA} + 7 \text{ mA} + 3.38 \text{ mA} + 5 \text{ mA} = 15.9 \text{ mA}$$

This change in I_{CQ} is shown in Figure 5.8. We find that the maximum current swing from zero to the peak is

$$i_{Cp} = 104 \times 0.95 - (60 + 15.9) = 22.9 \text{ mA}$$

and the maximum undistorted voltage swing at the output load is

$$2(22.9 \times 10^{-3})(100) = 4.58 \text{ V peak-to-peak}$$

This represents a reduction in output voltage swing from 10.8 to 4.58 V.

Drill Problems

D5.8 Design an EF amplifier for $R_L = 50\ \Omega$, $A_i = 15$, $V_{CC} = 15$ V, $V_{BE} = 0.7$ V, and $I_{CBO1}(25°\text{C}) = 2\ \mu\text{A}$. β ranges from 75 to 125 and the power supply voltage varies by ± 1 V. The amplifier is designed to operate at 100°C. What is I_{CQ} at 100°C and at 25°C?

Ans: $I_{CQ}(25°\text{C}) = 200$ mA;
$I_{CQ}(100°\text{C}) = 275$ mA

D5.9 What is the maximum undistorted symmetrical voltage swing for the amplitude of Drill Problem D5.8 if the temperature is 100°C?

Ans: 1.69 V peak-to-peak

5.3 DIODE COMPENSATION

The examples of the previous section show that changes in temperature can greatly affect the location of the Q-point. *Diode compensation* is a technique for reducing the effect of changes in temperature on I_{CQ}. A diode is selected that has temperature characteristics similar to those of the transistor. To ensure that the diode and transistor characteristic curves are identical, the same transistor type can be used in place of the diode junction. This is accomplished by shorting the collector to the base, thus using the transistor as a diode. This diode (transistor) is connected in the circuit as shown in Figure 5.9(a).

The addition of this diode in the base circuit compensates for changes arising from temperature variation, since V_γ varies in the same fashion as V_{BE}. R_f is the forward resistance of the diode. Assuming that the diode characteristic and the base-emitter junction characteristic are the same, then as the temperature changes both V_γ and V_{BE} change at the same rate, thus canceling out any variation in bias parameters. With proper diode selection, the effects of variations in V_{BE} are reduced. The new bias equation for the base-to-ground voltage is

$$V_B = V_\gamma + I_D R_D = V_{BE} + I_{CQ} R_E$$

Let us perform an analysis for the example shown in Figure 5.9(a). We begin the analysis by finding the Thevenin equivalent for the bias circuit. Figure 5.9(b) illustrates the circuitry connected to the base of the transistor. To find the open-circuit voltage from the base to ground, we first find the diode current.

$$I_D = \frac{V_{CC} - V_\gamma}{R_1 + R_2 + R_f}$$

The Thevenin voltage, V_{TH}, is then given by

$$V_{TH} = I_D R_1 + V_\gamma + I_D R_f$$

Figure 5.9
Diode compensation.

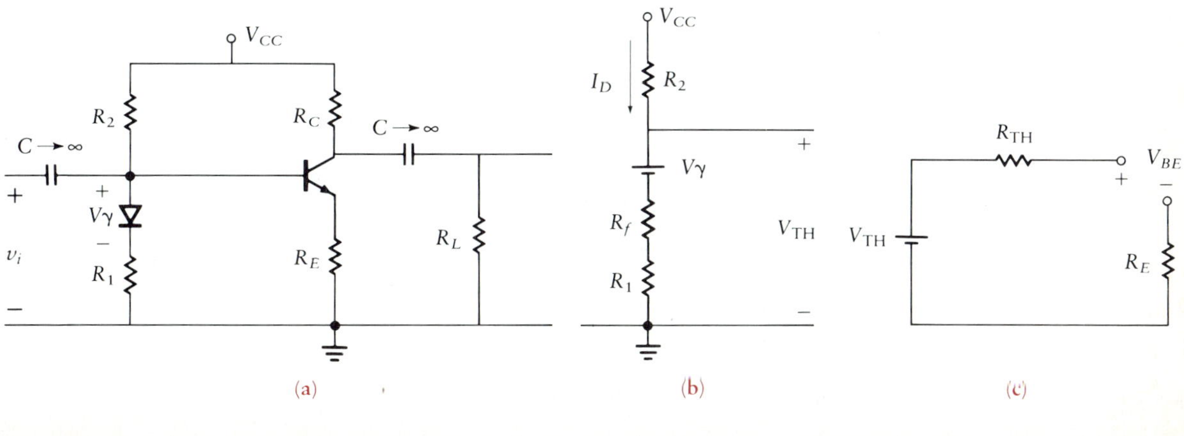

$$= \frac{V_{CC}(R_1 + R_f) + V_\gamma R_2}{R_1 + R_2 + R_f}$$

If $R_f << R_1$, then

$$V_{TH} = \frac{V_{CC} R_1 + V_\gamma R_2}{R_1 + R_2}$$

The Thevenin resistance is the parallel combination of R_2 with $R_1 + R_f$.

$$R_{TH} = R_2 \parallel (R_1 + R_f) \approx R_2 \parallel R_1$$

The equivalent of the bias circuitry is shown in Figure 5.9(c). We find the quiescent base current as follows:

$$I_{BQ} = \frac{V_{TH} - V_{BE}}{R_{TH} + \beta R_E}$$

The quiescent collector current is found by multiplying the base current by β. Following substitution of the earlier expressions, we obtain

$$I_{CQ} = \frac{V_{TH} - V_{BE}}{R_{TH}/\beta + R_E}$$

$$= \frac{(V_{CC} R_1 + V_\gamma R_2)/(R_1 + R_2) - V_{BE}}{R_{TH}/\beta + R_E} \tag{5.15}$$

The sensitivity of this circuit to variations of temperature is found by forming the partial derivative, $\partial I_{CQ}/\partial T$, as follows.

$$\frac{\partial I_{CQ}}{\partial T} = \frac{[R_2/(R_1 + R_2)]\partial V_\gamma/\partial T - \partial V_{BE}/\partial T}{R_{TH}/\beta + R_E}$$

Now if $R_2 >> R_1$, this is simplified to yield

$$\frac{\partial I_{CQ}}{\partial T} = \left(\frac{\partial V_\gamma}{\partial T} - \frac{\partial V_{BE}}{\partial T}\right)\frac{1}{R_{TH}/\beta + R_E} \tag{5.16}$$

This shows that if the diode temperature characteristic is matched to the base-emitter temperature characteristic, I_{CQ} can be made independent of changes in temperature. This applies if $R_1 << R_2$.

If $R_1 \approx R_2$, we need two diodes in series to obtain better compensation. That is, a different approximation is appropriate if two diodes are placed in series with R_1. In this case,

$$I_{CQ} = \frac{(V_{CC}R_1 + 2V_\gamma R_2)/(R_1 + R_2) - V_{BE}}{R_{TH}/\beta + R_E}$$

and

$$\frac{\partial I_{CQ}}{\partial T} = \frac{2[R_2/(R_1 + R_2)]\ \partial V_\gamma/\partial T - \partial V_{BE}/\partial T}{R_{TH}/\beta + R_E}$$

when $R_1 \approx R_2$, then

$$\frac{\partial I_{CQ}}{\partial T} = \left(\frac{\partial V_\gamma}{\partial T} - \frac{\partial V_{BE}}{\partial T}\right)\frac{1}{R_{TH}/\beta + R_E}$$

5.4 DESIGNING FOR BJT AMPLIFIER BIAS STABILITY

To reduce the effects of parameter changes on the Q-point location, we concentrate on reducing the large terms in equation (5.14). The design approach to accomplishing this goal is outlined by the following four steps.

1. Use diode compensation to reduce the changes that occur in V_{BE} from changes in temperature. The changes in V_{BE} are often significant and cause a large change in I_{CQ}. In using diode compensation, it is important that the characteristics of the diodes be the same as the characteristics of the transistor V_{BE}.
2. Select a transistor with low I_{CBO} so that the temperature change will not significantly affect I_{CQ}.
3. Ensure that the design uses some technique to reduce the effect of changes in β. For example, with voltage division biasing, R_B should be constrained to be less than $0.1\beta R_E$. This reduces the effect of changes in β, as can be seen from the third term in equation (5.14).
4. Use a well-regulated power supply to reduce the change in V_{CC} to such a small value that the Q-point location will not be affected.

5.5 FET TEMPERATURE EFFECTS

Temperature changes cause large variations in the bias point of BJTs. Fortunately, temperature instability is not as big a problem with FET amplifiers. However, the drain current is affected somewhat by temperature variations. This should be considered in designing FET amplifiers that are required to operate in a varying-temperature environment. Temperature increases also cause the gate-to-source leakage current of a JFET to increase.

Increasing the temperature of a FET amplifier tends to decrease the mobility of charge carriers in the channel of the FET. The effect of the smaller number of charge carriers is to reduce the drain current. However, the increased temperature also narrows the depletion region, which tends to increase the drain current. These two effects are in opposition, thus giving the FET its relatively low temperature coefficient. That is, the drain current changes are relatively small with large changes in temperature.

For some types of FET, the manufacturer specifies a quiescent drain current that will give a temperature coefficient near zero. This results from the decreased conductivity of the channel.

Since the temperature coefficient is not zero, it is good practice to consider bias stability in circuit design. One simple approach is to use a source resistor that results in negative feedback, as shown in Figure 5.10(a). When the drain current increases, v_S becomes more positive, causing v_{GS} to become more negative. Thus, as i_D increases, v_{GS} becomes more negative. This tends to decrease i_D. The result is that the amount of change in i_D is reduced. We illustrate this graphically in Figure 5.11. The manufacturer typically provides a range of V_p and I_{DSS}. The v_{GS}-i_D curve as shown in Figure 5.11 is used to account for the two extremes of these values. Note that as the value of the source resistor increases, the variation in i_{DS} decreases. This allows us to design the amplifier to reduce the effects of changes in drain current. A typical value of R_S that provides a reasonable I_{DQ} deviation is 10% of the value of R_D [25].

Figure 5.10
FET amplifier.

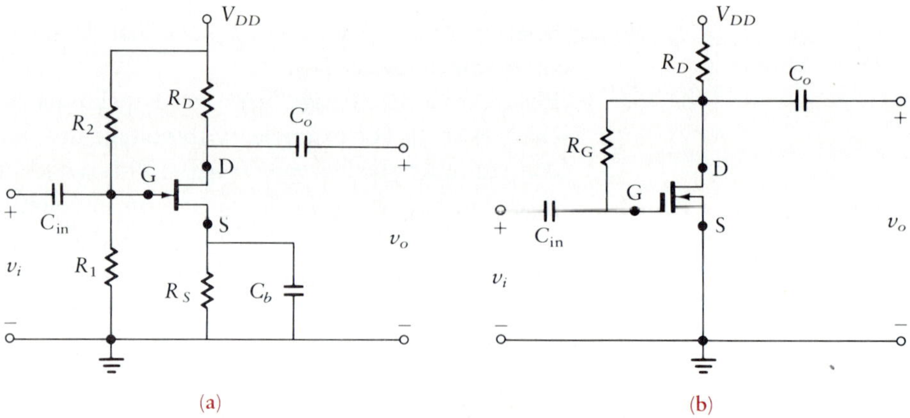

Bias stability is provided for MOSFETs in the same manner. However, a simpler circuit can be used, as shown in Figure 5.10(b). An increase in i_D causes v_{DS} to decrease. Reducing v_{DS} reduces v_{GS}, thereby tending to decrease the original increase in i_D. This is also negative feedback [32].

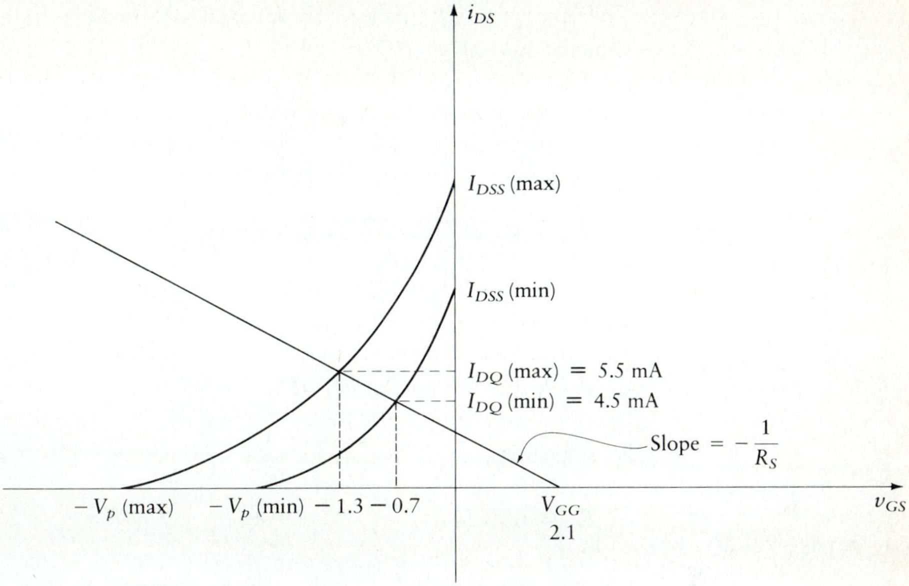

Figure 5.11
FET operating curves.

Example 5.3

Determine the value of the source and drain resistors for a JFET amplifier (see Figure 4.15(a)) that will allow for only a 10% variation in I_{DQ} for the following specifications:

- V_p ranges from -5 to -8 V (maximum and minimum values from the transistor specifications).
- I_{DSS} ranges from 7 to 10 mA (maximum and minimum values from the transistor specifications).
- The nominal value of I_{DQ} is 5 mA.
- $V_{DD} = 12$ V.
- $V_{DSQ} = 4$ V.

SOLUTION A 10% variation in I_{DQ} is a variation from a minimum of 4.5 mA to a maximum of 5.5 mA. The intercepts in Figure 5.11 show that

v_{GS} varies from -0.7 to -1.3 V

Then the source resistance is given by

$$R_S = \frac{\Delta v_{GS}}{I_{DQ\,\text{max}} - I_{DQ\,\text{min}}} = 600\ \Omega$$

The line passing through these points intersects the v_{GS} axis, resulting in a value of V_{GG} of +2.1 V.

$$V_{GSQ} = V_{GG} - 5\text{ mA }(600\ \Omega) = -0.9\text{ V}$$

$$V_{DD} = V_{DSQ} + (R_S + R_D)I_{DQ}$$

$$R_D + R_S = \frac{V_{DD} - V_{DSQ}}{I_{DQ}} = 1.6\text{ k}\Omega$$

Then

$$R_D = 1.6\text{ k}\Omega - R_S = 1\text{ k}\Omega$$

By selecting

$$R_S = 600\ \Omega$$

we can ensure that I_{DQ} will not vary more than 10% for any variation of the transistor parameters, V_p and I_{DSS}, within the specified range.

5.6 REDUCING TEMPERATURE VARIATIONS

Changing the temperature of any electronic device changes its operating characteristics. An increase in temperature may even cause the device to fail. Therefore it is important for the designer to consider the operating temperature of devices used in a system.

For BJTs, an increase in the temperature of the junction results in an increase of I_{CQ}, thus reducing the maximum output voltage swing.

Temperature increases can be caused either by external heat or by internal heat that is generated by operating the device at a high current level. Transistor capability is limited by the allowable transistor junction temperature that is specified by the manufacturer. It is your responsibility as a designer to ensure that your design does not allow the junction temperature to exceed the specified maximum value, or performance will deteriorate. It is also important that your design not allow the device to operate at or near the maximum allowable junction temperature, or the reliability of the device could be reduced.

Some transistors are rated at a certain power capability or current level, but these rated levels cannot be achieved without keeping the junction temperature within the allowable limits. There are two ways to keep the junction temperature from rising too high: active and passive cooling. *Active cooling* involves the use of fans or air conditioners. Such systems are expensive and bulky but they may be necessary when dissipation of large amounts of heat is required. The more inexpensive technique is to utilize *passive heat sinks* that employ metallic surfaces to conduct, and in some cases radiate, the heat to the surrounding media. To increase the dissipation of heat, cooling fins can be added to the metallic heat sink to increase the amount of surface area in contact with the surrounding media. A metal chassis supporting the electronic components is sometimes used as an effective and economical heat sink. If the problem is cold rather than heat, similar results can be achieved by using heaters.

Power transistors and other high-current devices require dissipation of large amounts of heat. These high-power devices are packaged in cases that permit contact between a metal surface and an external heat sink. In most cases, the metal surface of the device is electrically connected to one terminal. For example, power transistors have their cases connected to their collectors. For silicon transistors packaged in metal cases, the junction temperature is usually 200°C. When transistors are packaged in plastic cases, the junction temperature is usually 150°C.

To determine whether a particular heat sink is adequate for use with your design, some simple calculations are necessary. We use information from the specification sheets for the transistor and the heat sink selected. Examples of common heat sinks are shown in Figure 5.12. The thermal resistance, θ, which is defined as the heat rise divided by the power transferred, is a constant independent of temperature. It depends only on the properties of the mechanical joint. When a number of joints are in series, the total thermal resistance is the

Figure 5.12
Heat Sinks
Courtesy of International Electronic Research Corporation, Burbank, California

sum of the thermal resistances of the individual joints. Note that the lower the thermal resistance of these mechanical junctions, the better the heat transfer. When a transistor is mounted on a heat sink, the total thermal resistance of the system is the sum of the thermal resistance from the junction to the case of the transistor, θ_{jc}; the thermal resistance from the case to the heat sink, θ_{cs}; and the thermal resistance from the heat sink to the ambient environment, θ_{sa}. Thus, the temperature of the transistor junction can be determined from the following formula:

$$T_j = T_a + (\theta_{jc} + \theta_{cs} + \theta_{sa})P \tag{5.17}$$

where T_a is the temperature of the transistor environment in degrees Celsius and P is the power to be dissipated by the transistor. The heat sink should be selected to prevent the transistor from operating at or near the maximum junction operating temperature, since transistor life decreases rapidly when the operating temperatures are at or near the maximum.

Example 5.4

Determine whether a heat sink with a θ_{sa} of 3.3°C/W (from the specification of the heat sink) will maintain a temperature below the transistor maximum junction operating temperature when the transistor is operating in an environment with a temperature of 60°C. The transistor junction-to-case thermal resistance, θ_{jc}, is 1.5°C/W, maximum junction operating temperature is 200°C, and the transistor is contained in a TO-3 package (see the transistor manufacturer's specifications). The TO-3 package has a θ_{cs} of 0.3°C/W when the transistor is mounted with an insulated washer and a heat-conducting compound between the package and the heat sink (from the packaging specifications). The power of the transistor, $P_{(\text{transistor})}$, operating in this circuit is 21 W.

SOLUTION Using equation (5.17), we have

$$T_j = 60 + (1.5 + 0.3 + 3.3)21 = 167°\text{C}$$

Hence, this heat sink is adequate to maintain the temperature of the junction well below the maximum junction temperature allowed by the specification.

PROBLEMS

5.1 Prove that $V_{CC} = V_{BB}$ for the circuit shown in Figure 5.1. (Hint: Use the equations derived in Chapter 2.)

5.2 In the circuit of Figure 5.1, let $R_C = 1\ \text{k}\Omega$, $V_{CC} = 10$ V, and $A_v = -8$. Determine the values of R_B and R_E that would make the amplifier least susceptible to large changes in β. Assume $V_{BE} = 0.7$ V and $\beta = 150$.

5.3 For an amplifier of the type shown in Figure 5.2, choose values of R_F and R_E so that the amplifier operates consistently for large changes in β. Let $V_{CC} = 10$ V, $R_L = 1\ \text{k}\Omega$, $A_v = -10$, $V_{BE} = 0.7$ V, and $\beta = 100$.

5.4 Design a CE amplifier as shown in Figure P5.1 to obtain a voltage gain of -8. Do not exceed the power limit of the transistor, $P_{\max} = 50$ mW. If the temperature changes from 25° to 85°C and $I_{CBO}(25°\text{C}) = 0.4\ \mu\text{A}$, what is the peak-to-peak undistorted output voltage at 85°C when $V_{CC} = 12$ V?

5.5 For the design of Problem 5.4, determine the maximum peak-to-peak output voltage if β varies from 250 to 350 and the power supply varies from 11.5 to 12.5 V.

5.6 Design a CE amplifier to have a voltage gain of -8 and drive a 750-Ω load. Use the configuration of Figure P5.2 with a transistor that has $\beta = 300$ and $I_{CBO}(25°\text{C}) = 10\ \mu\text{A}$. Determine the maximum peak-to-peak output voltage swing when the temperature rises to 85°C, β varies from 250 to 350, and $V_{CC} = 24$ V.

5.7 The amplifier shown in Figure P5.2 is designed for operation where the temperature is -25°C. Find the maximum peak-to-peak output voltage swing at -25°C if all parameters are the same as in Problem 5.6 except that β varies from 200 to 350.

5.8 Design an EF amplifier to drive a 15-Ω load when $\beta = 60$, $V_{BE} = 0.7$ V, $I_{CBO}(25°\text{C}) = 1\ \mu\text{A}$, $V_{CC} = 20$ V, and $A_i = 8$. If the temperature now changes from 25° to 85°C, determine the peak-to-peak output voltage swing at 85°C. Use the circuit of Figure 5.7.

5.9 Design an EF amplifier to drive an 8-Ω load, as shown in Figure P5.3. Set the current gain $A_i = 10$. Determine the peak-to-peak output voltage swing if the temperature rises to 75°C. Assume that $I_{CBO}(25°\text{C}) = 0.5\ \mu\text{A}$, $V_{CC} = 24$ V, and β varies from 60 to 100.

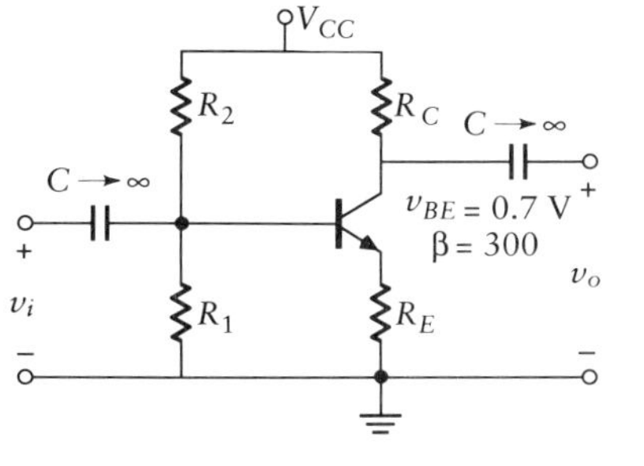

Figure P5.1

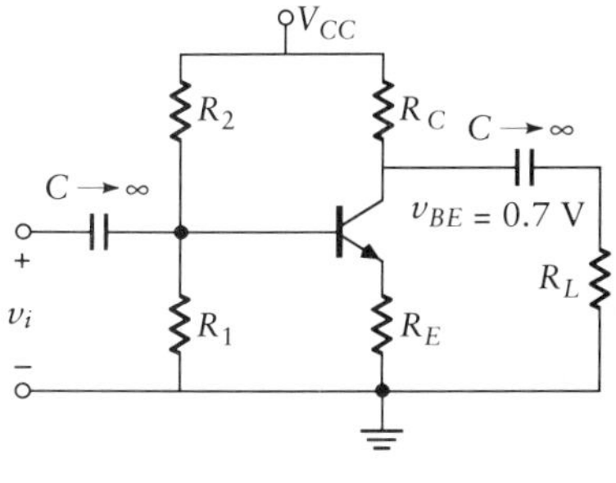

Figure P5.2

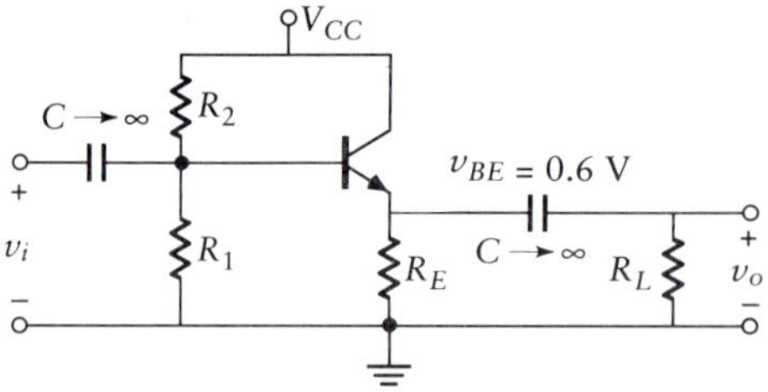

Figure P5.3

5.10 For the circuit of Figure P5.3, determine the maximum peak-to-peak output voltage swing if β varies from 40 to 80 and the voltage supply varies from 19 to 21 V with the temperature at 85°C. Assume $I_{CBO}(25°C) = 2\ \mu A$, $R_L = 200\ \Omega$, and $A_i = 8.8$.

5.11 An amplifier similar to that shown in Figure P5.3 is designed to have a current gain of 10 and to drive a 20-Ω load using a 24 V ± 2% regulated power supply. The transistor selected has a β variation of 60 to 100, $V_{BE} = 0.7$ V, and $I_{CBO}(25°C) = 1\ \mu A$. What is the maximum output voltage swing that can be obtained at −30°C and at 80°C? Let $V_{CC} = 24$ V.

5.12 Using the stability factor δ_B, find the value of R_E for an amplifier of the type shown in Figure P5.1. Use a silicon transistor designed so that the voltage across R_C will not vary more than ±0.5 V. Assume the supply voltage is 20 V, $V_{BE} = 0.7$ V, $I_{CQ} = 10$ mA, and β varies from 50 to 100.

5.13 An amplifier similar to the one shown in Figure P5.2 is being designed for use where the temperature ranges from 80° to −50°C. The battery source is 24 V, the transistor selected has a β variation of 200 to 300, and $I_{CBO}(25°C) = 2\ \mu A$. What is the maximum output voltage swing for a voltage gain of 10 if the load is 1 kΩ?

5.14 In the amplifier described in Problem 5.13, change the high temperature from 80° to 50°C. What is the maximum output voltage swing for the amplifier after this modification?

5.15 For the amplifier described in Problem 5.13, the transistor originally planned to be used went out of production and the only other transistor available that would meet the requirements had an $I_{CBO}(25°C)$ of 5 μA and a variation in β of 300 to 500. What is the maximum output voltage swing with the temperature changing from −50° to +50°C?

5.16 An amplifier as in Figure P5.3 requires a current gain of 10 into a 50-Ω load. The specification requires that the amplifier operate from −75° to 50°C. The transistor selected has an $I_{CBO}(25°C) = 5\ \mu A$ and a β variation of 200 to 300. Let $V_{CC} = 25$ V. Find the maximum output swing of the amplifier.

5.17 In Problem 5.16, change the high temperature to 100°C. With this change, what is the maximum output voltage swing?

5.18 Determine whether a heat sink with θ_{sa} of 3.3°C/W will maintain a temperature below the maximum transistor junction operating temperature when the transistor is operating in an environment of 80°C at a power rating of 15 W. Assume $\theta_{jc} = 20°C/W$ and $\theta_{cs} = 0.5°C/W$. The maximum transistor junction temperature is 180°C.

5.19 For Example 5.4, determine whether the transistor junction operating temperature would be exceeded if a heat sink with $\theta_{sa} = 4.5°C/W$ were used. How much would the output power have to be reduced in order to keep the transistor junction temperature at 10% below the maximum operating junction temperature?

6

POWER AMPLIFIERS AND POWER SUPPLIES

6.0 INTRODUCTION

We consider the design of power amplifiers in this chapter. The purpose of the power amplifier is to deliver a maximum undistorted symmetrical output voltage swing to a low-resistance load. In practice, a system may consist of several stages of amplification, the last of which is usually a power amplifier. The load fed by this power amplifier may be a loudspeaker, an actuator, a motor, or some other analog device. The input to the system is a low-level voltage, which is amplified through the voltage gain stages. The output of the voltage gain stages is of sufficient amplitude to drive the output power amplifier.

We begin the chapter with a discussion of the various biasing techniques leading to Class-A, -B, -AB, and -C operation. We then analyze the specific amplifier circuits and the effects of various coupling configurations. In particular, inductively coupled and transformer-coupled amplifiers are studied and several design examples are given.

The analysis of the zener-regulated supply of Chapter 1 is extended to include the use of power transistors. This allows regulation over a wider range of inputs and outputs. The integrated circuit regulator is also briefly discussed.

We conclude the chapter with a discussion of switching regulators.

6.1 CLASSES OF AMPLIFIERS

Power amplifiers are classified according to the percent of time that collector current is nonzero. There are four principal classifications: Class A, Class B, Class AB, and Class C. We discuss each of these in the following subsections.

6.1.1 Class-A Operation

In *Class-A* operation, which is considered in the amplifiers of Chapters 2 and 3, the amplifier reproduces the input signal in its entirety. The collector current is nonzero 100% of the time. This type of operation is inefficient, since even with zero input signal, I_{CQ} is nonzero. The transistor therefore dissipates power in the rest, or quiescent, condition.

Figure 6.1 illustrates typical characteristic curves for Class-A operation. The current, I_{CQ}, is usually set to be in the center of the ac load line. The figure shows an example of a sinusoidal input and the resulting collector current at the output. We covered this in detail in Chapter 3.

Figure 6.1
Class-A operation.

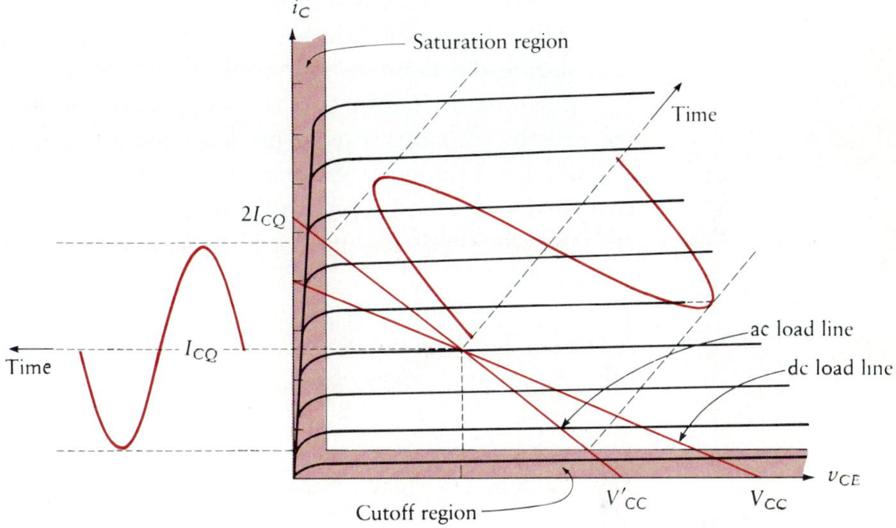

6.1.2 Class-B Operation

With *Class-B* operation, one amplifier is used to amplify the positive half-cycle of the input signal while a second amplifier is used to amplify the negative half-cycle. This amplifier configuration is known as *push-pull* or *complementary symmetry.*

Since a single transistor can respond only to a half-cycle, two transistors are required to produce the complete waveform. Each of these transistors is biased at cutoff rather than in the middle of the operating range, as is the case for Class-A operation. Each transistor operates half of the time, so the collector current of each is nonzero 50% of the time.

The advantage of Class-B operation is that the collector current is zero when the signal input to the amplifier is zero. Therefore the transistor dissipates no power in the quiescent condition.

Among the disadvantages of a Class-B amplifier is that the nonlinear cutoff region is included in the operating range. That is, unlike the Class-A situation, the 5% of the operating region shaded at the bottom of Figure 6.1 is included in the operating region. Therefore distortion occurring near the Q-point is included in the output signal.

Figure 6.2 illustrates a typical characteristic curve for a pair of transistors in the push-pull configuration. This figure is intended for conceptual purposes only, since we discuss the amplifier in more detail later. Since two transistors are connected back to back, we have repeated the set of transistor curves for the second transistor, but the sign of the collector current and the collector-to-emitter voltage have been reversed. That is, these two quantities increase in the downward and the left directions, respectively, for the second transistor characteristic. The upper left portion of the figure represents the first transistor, which conducts only during the positive half-cycle of the input. The lower right portion represents the second transistor, which is configured to conduct only during the negative half-cycle. A typical output waveform is shown in Figure 6.3. Note that the first transistor produces the positive part of the output and the second transistor produces the negative part. Also note that Figure 6.3(a) and (b) indicates some distortion near the point $i_C = 0$. When these two curves are added together, the output shown in Figure 6.3(c) results. This resembles the sinusoidal input, but the waveform is distorted near the zero axis.

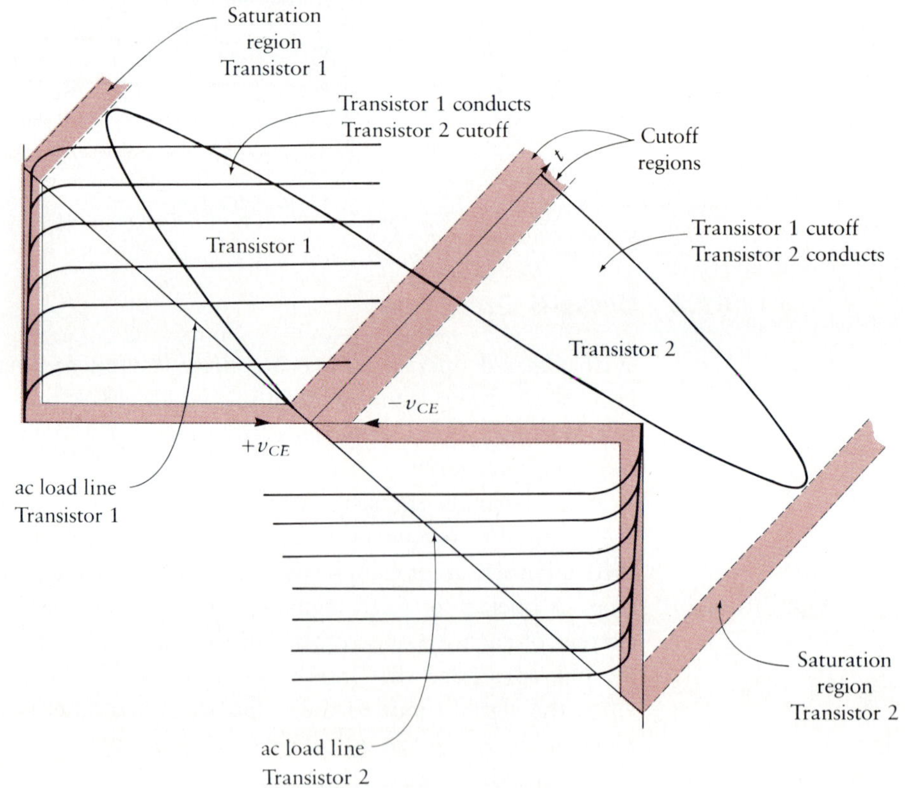

Figure 6.2
Class-B operation.

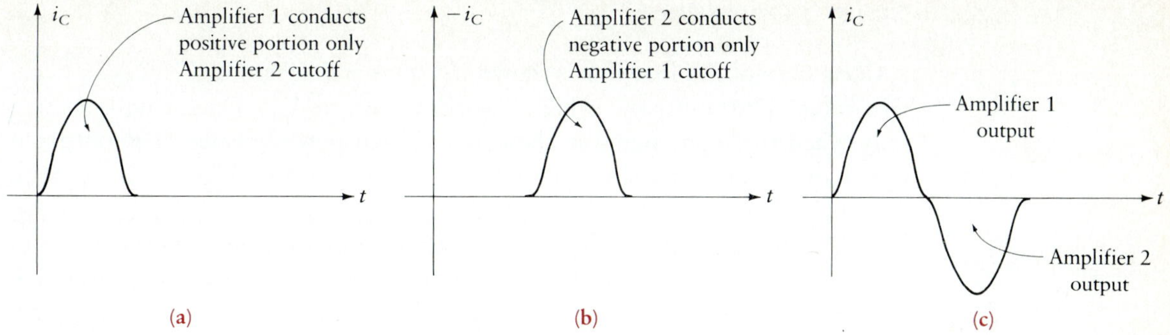

Figure 6.3
Push-pull output waveform.

It is important that the two transistors in a push-pull configuration be carefully matched. In this way, the positive and negative portions of the input are amplified by the same amount.

6.1.3 Class-AB Operation

Class-A operation has the advantage of little distortion, and Class B has the advantage of higher efficiency. Class-AB operation is a compromise between these two extremes. The *Q*-point is set slightly above the cutoff value, so it is at the lower boundary of the linear (no distortion) portion of the operating curves. The transistor therefore supports a nonzero collector current for slightly more than 50% of the time.

Figure 6.4 illustrates the positive portion of the operating curve for a sinusoidal input and Class-AB operation. Note that I_{CQ} is slightly above zero to reduce the distortion. The Class-AB amplifier is, of course, used in a push-pull type of arrangement.

Figure 6.4
Class-AB operation.

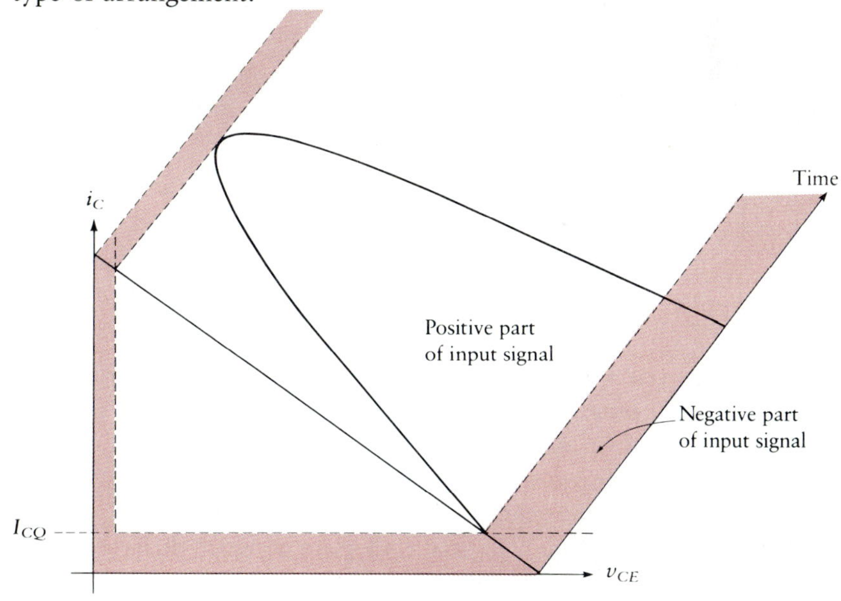

6.1.4 Class-C Operation

A *Class-C* amplifier load line is shown in Figure 6.5, where V_{BEQ} is set to a negative value. The transistor is biased with a negative V_{BB}. Thus it will conduct only when the input signal is above a specified positive value. The output is less than one-half of a sinusoid and the collector current is nonzero less than 50% of the time.

If a sinusoid forms the input to a Class-C amplifier, the output consists of "blips" at the frequency of the input. This is shown to the left in Figure 6.5. Since this is a periodic signal, it contains a fundamental frequency component plus higher-frequency harmonics. If this signal is passed through an inductor-capacitor (LC) circuit tuned to be resonant at the fundamental frequency, the output is approximately a sinusoidal signal at the same frequency as the input. This approach is often used if the signal to be amplified is either a pure sinusoid or a more general signal with a limited range of frequencies.

Class-C amplifiers are capable of providing large amounts of power. They are often used for transmitter power stages, where a tuned circuit is included to eliminate the higher harmonics in the output signal.

Figure 6.5
Class-C operation.

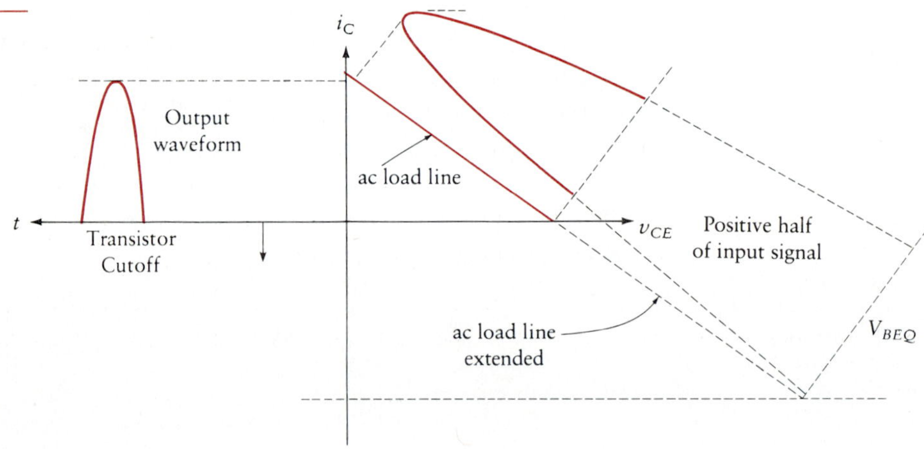

6.2 POWER AMPLIFIER CIRCUITS—CLASS-A OPERATION

Power amplifier circuits usually contain transistors capable of handling high power. These normally operate at higher voltages than do low-power transistors, and they therefore often require a separate power supply. For example, voltages of power transistors can exceed 450 V. Current ratings are also high, often in excess of 10 A of continuous current. Since these transistors need to dissipate high power, they are designed differently from low-power transistors. Protective circuits may be included to limit current. Additional effort is expended to dissipate heat that builds up during operation.

In this section, we discuss some useful circuit configurations for power amplifiers. These are categorized according to the type of coupling.

6.2.1 Inductively-Coupled Amplifier

High current gain is required to obtain power in the output load. The output voltage swing can be increased by using an inductor for the collector element instead of a resistor. We will see that this also increases the efficiency of the circuit. The inductor is selected so that it approximates an open circuit for the input frequency but a short circuit for dc. In other words,

$$\omega L >> R_L$$

at the lowest frequency and

$$R_{\text{coil}} << R_L \quad \text{and} \quad R_{\text{coil}} << R_E$$

Figure 6.6 illustrates the inductively coupled amplifier circuit and its load lines.

We choose the Q-point for maximum output swing. The current, I_{CQ}, is then given by (see Chapter 2)

$$I_{CQ} = \frac{V_{CC}}{R_{ac} + R_{dc}}$$

The ac resistance is simply R_L, since the inductor is an approximate open circuit for ac and the capacitors are short circuits. The dc resistance is R_E provided that we can neglect the resistance of the inductor. Therefore,

$$I_{CQ} = \frac{V_{CC}}{R_L + R_E} \tag{6.1a}$$

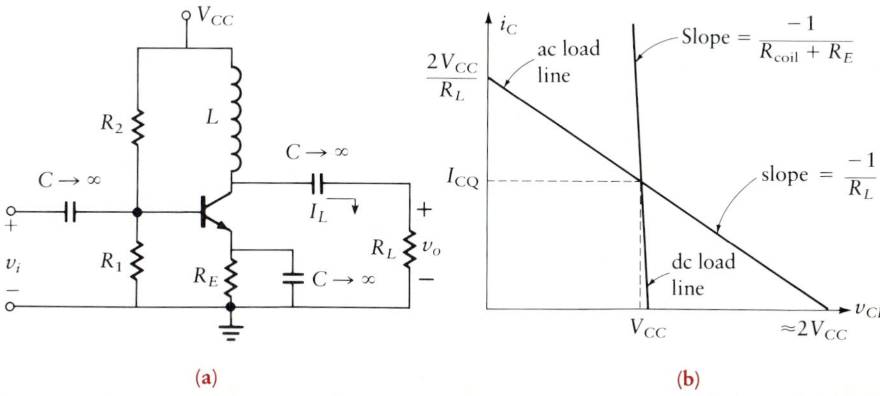

Figure 6.6 Inductively coupled amplifier.

Since both dc and ac load lines cross at the Q-point, the ac load line equation yields

$$I_{CQ} = \frac{V_{CEQ}}{R_L} \tag{6.1b}$$

We have assumed $R_E << R_L$, which is usually true. Then from equations (6.1a) and (6.1b) we see that $V_{CEQ} \approx V_{CC}$ and the ac load line intersects the v_{CE} axis at approximately $2V_{CC}$. The use of the storage device (inductor) results in a voltage swing that is effectively equivalent to doubling the supply voltage. The inductor field stores energy during the conducting cycle and thus acts like a second V_{CC} source in series with the dc supply.

The inductively coupled amplifier has higher efficiency than the amplifier that contains a collector resistance. To prove this, we calculate the efficiency of this amplifier, assuming sinusoidal input signals.

The power supplied by the voltage source is

$$P_{\text{supplied}} = V_{CC} I_{CQ} = \frac{V_{CC}^2}{R_L}$$

where we ignore R_E because we assume that $R_E << R_L$. The power delivered to the load, assuming that the input is sinusoidal with amplitude $I_{L\,\text{max}}$, is

$$P_{\text{load}} = I_{L\,\text{max}}^2 \frac{R_L}{2} = \frac{I_{CQ}^2 R_L}{2} = \frac{V_{CC}^2}{2R_L}$$

We define *conversion efficiency* as the ratio of ac load power to the power delivered by the source. This efficiency measure therefore depends on the power dissipated in the bias circuitry and in R_E. To derive a maximum value for efficiency, we assume that the power dissipated in the bias circuitry, in R_E, and in R_{coil} is negligible. The maximum conversion efficiency (with output swing a maximum) is then given by

$$\eta = \frac{V_{CC}^2/2R_L}{V_{CC}^2/R_L} = 50\%$$

The circuit with a collector resistor in place of the inductor is discussed in Section 2.6.1, where we derive the various power relationships. These can be used to show that the maximum efficiency of the amplifier with collector resistance is 25%, or one-half of the efficiency found for the inductively coupled amplifier. This is reasonable, since in the circuit with a collector resistor the load seen by the transistor is the parallel combination of the collector and load resistance. The maximum power delivered by the transistor is shared between these two resistors.

6.2.2 Transformer-Coupled Power Amplifier

The transformer-coupled amplifier is presented in Chapter 3, where basic relationships are developed. We study transformer coupling in the present section because it has useful application to power amplifiers. Figure 6.7(a) illustrates the CC (EF) transformer-coupled power amplifier, and Figure 6.7(b) shows the load lines. Note that the slope of the dc load line depends on the resistance of the primary coil of the transformer. This resistance is usually small. The slope of the ac load line depends on the reflected load resistance.

If the maximum output voltage swing is desired, we solve the design equation (equation (6.2)) in order to place the Q-point in the center of the load line. Note that the load resistance reflected by the transformer is

$$R_{ac} = a^2 R_L$$

The Q-point is then given by

$$I_{CQ} = \frac{V_{CC}}{R_{ac} + R_{dc}} = \frac{V_{CC}}{a^2 R_L} \tag{6.2}$$

In the last equality, we assume that the transformer primary resistance is negligible and therefore that $R_{dc} = 0$.

In the design of CE amplifiers, we select the base resistance, R_B, from the bias stability design equation

$$R_B = 0.1\beta R_E$$

In the design of CC amplifiers, a different criterion is used. The base resistance is constrained by the desired current gain, A_i, or by the specified input resistance, R_{in}. The voltage gain of this amplifier is near unity.

The maximum power transferred to the load, assuming a sinusoidal input and ignoring saturation effects, is

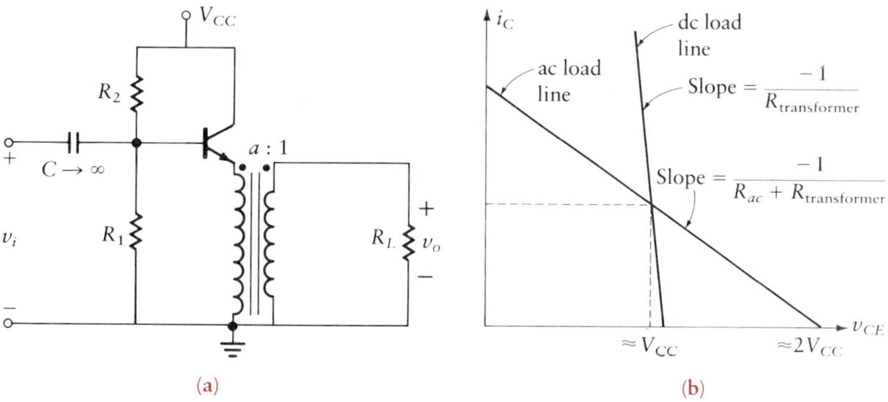

Figure 6.7 Transformer-coupled power amplifier.

$$P_{load} = \frac{V_{L\,max}^2}{2R_L} = \frac{V_{CC}^2}{2a^2R_L} \tag{6.3}$$

The maximum power conversion efficiency is then given by

$$\eta = \frac{V_{CC}^2/2a^2R_L}{V_{CC}^2/a^2R_L} = 50\%$$

Thus the transformer-coupled amplifier has characteristics similar to those of the inductively coupled amplifier. The maximum conversion efficiency of both circuits is 50%, and although the EF has a voltage gain near unity, the turns ratio of the transformer determines the voltage gain to the load. The transformer-coupled amplifier has the additional advantage of providing for impedance matching, as discussed in Chapter 3.

Example 6.1 Transformer-Coupled Amplifier

Design a transformer-coupled amplifier (see Figure 6.8) for a current gain of $A_i = 80$. Find the power supplied to the load and the power required from the supply.

SOLUTION We first use the design equation to find the location of the Q-point for maximum output swing.

$$I_{CQ} = \frac{V_{CC}}{R_{ac} + R_{dc}} = \frac{12}{a^2R_L} = 23.4 \text{ mA}$$

Since the problem statement requires a current gain of 80, the amplifier must have a current gain of 10 because the transformer provides an additional gain of 8. We use the equations from Table 3.1 to find the base resistance, R_B.

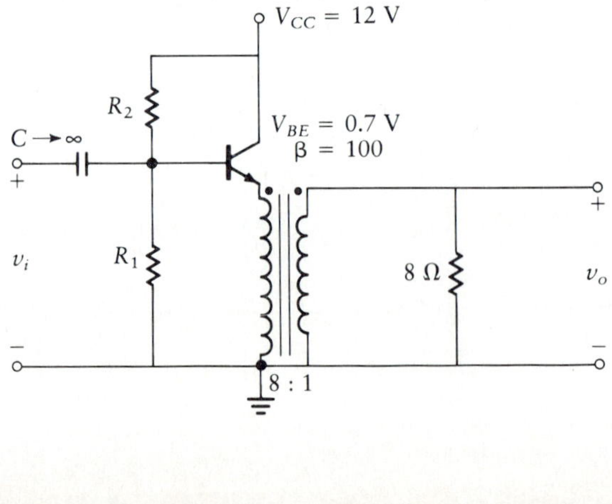

Figure 6.8 Transformer-coupled amplifier for Example 6.1.

$$A_i = \frac{R_B}{R_B/\beta + h_{ib} + R_E} = 10$$

where

$$R_E = a^2 R_L = 512\ \Omega$$

We note that h_{ib} is sufficiently small to be neglected. Then solving for R_B yields

$$R_B = 5.69\ \text{k}\Omega$$

$$V_{BB} = \frac{I_{CQ} R_B}{\beta} + V_{BE} = 2.03\ \text{V}$$

Now, solving for the bias resistors,

$$R_1 = \frac{R_B}{1 - V_{BB}/V_{CC}} = 6.85\ \text{k}\Omega$$

$$R_2 = \frac{V_{CC} R_B}{V_{BB}} = 33.6\ \text{k}\Omega$$

The design is now complete. The power delivered by the source is given by

$$P_{VCC} = V_{CC} I_{CQ} + \frac{V_{CC}^2}{R_1 + R_2} = 284\ \text{mW}$$

The power dissipated in the load is

$$P_L = \frac{(0.9 a I_{CQ})^2 R_L}{2} = 114\ \text{mW}$$

We have again restricted operation to the linear region by eliminating 5% of the maximum swing near cutoff and saturation. The efficiency is the ratio of load power to source power:

$$\eta = \frac{114}{284} = 0.4 \quad \text{or} \quad 40\%$$

Drill Problems

D6.1 Design an inductively coupled CE amplifier for $A_v = -10$, $R_{\text{in}} = 4\ \text{k}\Omega$, $R_L = 2\ \text{k}\Omega$, $V_{CC} = 12$ V, $\beta = 200$, and $V_{BE} = 0.7$ V. Determine A_i, the power delivered to the load, and the maximum undistorted symmetrical voltage output swing.

Ans: $R_E = 200\ \Omega$; $R_1 = 5.2\ \text{k}\Omega$;
$R_2 = 29.4\ \text{k}\Omega$; $A_i = -20$;
$P_o = 20.3$ mW; $v_o(p\text{-}p) = 18$ V

D6.2 Design a transformer-coupled EF amplifier to drive an 8-Ω load if $V_{CC} = 20$ V, $V_{BE} = 0.7$ V, $\beta = 100$, $R_{\text{in}} = 2\ \text{k}\Omega$, and the transformer has a turns ratio of 10:1. Determine the current gain, A_i, power output, and maximum undistorted voltage output swing.

Ans: $R_1 = 2.2\ \text{k}\Omega$; $R_2 = 33.8\ \text{k}\Omega$;
$A_i = 250$; $P_o = 203$ mW;
$v_o(p\text{-}p) = 3.6$ V

6.3 POWER-AMPLIFIER CIRCUITS—CLASS-B OPERATION

A Class-B audio amplifier uses one transistor to amplify the positive portion of the input signal and another transistor to amplify the negative portion of the input signal. As indicated earlier, the Class-B audio amplifier provides higher efficiency and lower output impedance to drive a typically low-impedance load. For example, a speaker load is normally 8 Ω.

6.3.1 Complementary-Symmetry Class-B Power Amplifier

A typical push-pull power amplifier can be designed with one *pnp* and one *npn* transistor with symmetrical characteristics as shown in Figure 6.9. This circuit is commonly called a *complementary-symmetry power amplifier.* We isolate the load with a capacitor and a single power supply is used. The capacitor, C_1, blocks the dc ($V_{CC}/2$) from the load. The capacitor also provides the power supply voltage to Q_2 when Q_1 is not conducting. That is, the capacitor charges to a dc value of $V_{CC}/2$ at the connection of the two emitters.

The dc load line is vertical, since the capacitor, C_1, acts as an open circuit for dc. Since the amplifier is to operate as Class B, I_{CQ} is set to zero.

As is the case for the transformer-coupled power amplifier, R_B is determined from the current gain or input resistance equations. The input resistance, R_{in}, is determined as follows ($h_{ib} = 0$):

$$R_{\text{in}} = R_B \parallel (\beta R_L) = \frac{R_B R_L}{R_B/\beta + R_L} \tag{6.4a}$$

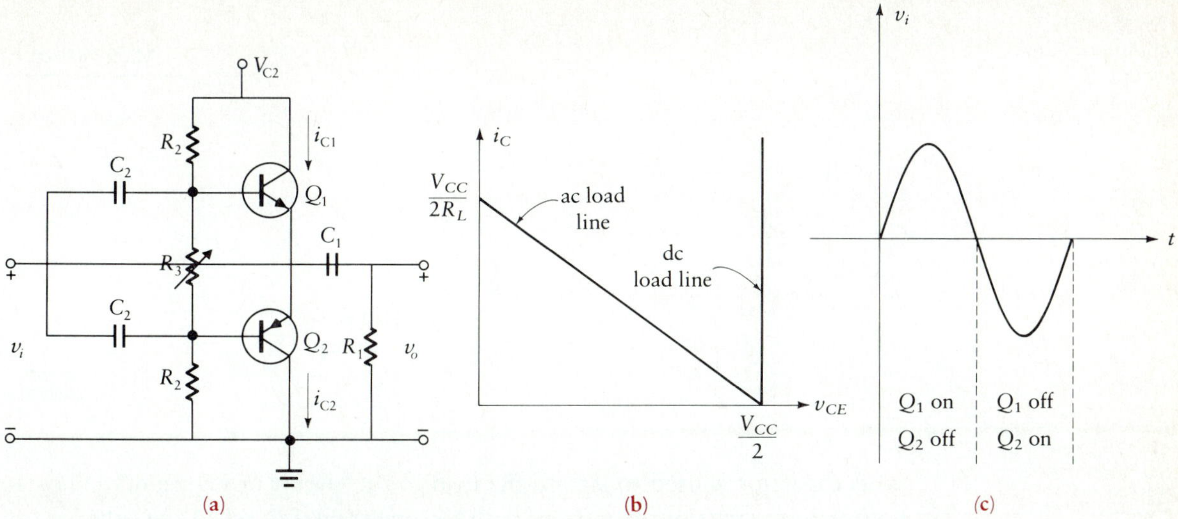

Figure 6.9
Complementary symmetry.

Each half of the circuit operates as an EF amplifier and the equivalent base resistance, R_B, is $R_2 \parallel R_2$. The current gain is found from the current divider equation as follows:

$$A_i = \frac{\beta R_B}{R_B + \beta R_L} = \frac{R_2}{R_2/\beta + 2R_L} \tag{6.4b}$$

From the gain-impedance formula, the voltage gain is

$$A_v = A_i \frac{R_L}{R_{\text{in}}} = 1$$

Notice that this reflects the EF nature of the circuit.

We find R_2 from equation (6.4b). R_1 is found from the following dc equation:

$$V_{BB} = \frac{R_1}{R_1 + R_2} \frac{V_{CC}}{2}$$

We select $V_{BB} = V_{BE}$, and

$$R_1 = \frac{2R_2 V_{BE}}{V_{CC} - 2V_{BE}} \tag{6.4c}$$

In order to avoid the nonlinear operating region near cutoff and thereby to obtain more symmetrical operation, the two R_1 resistors can be replaced by one adjustable resistor (larger than $2R_1$) in order to raise I_{CQ} above zero to compensate for the distortion. This accomplishes Class-AB operation.

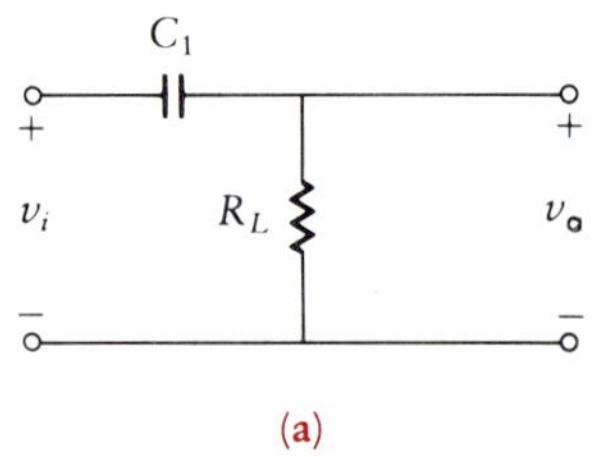

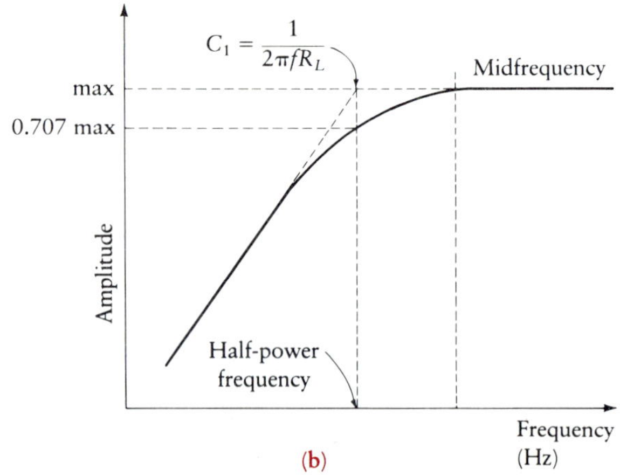

Figure 6.10
Input equivalent circuit.

A capacitor is used to isolate the load in the circuit of Figure 6.9. The capacitor forms part of the current path for one transistor when the other is cut off. Thus, the capacitor charges during conduction of Q_1 and discharges during conduction of Q_2.

With capacitance present, the circuit becomes frequency dependent. The low-frequency response of the stage is determined by the RC network shown in Figure 6.10(a). As the signal frequency decreases, the voltage across the series capacitor increases and the voltage across R_L decreases. This effect reduces the signal developed across R_L and hence decreases the gain of the amplifier.

The *half-power* or *3-dB** point specifies the lower-frequency cutoff. This is the frequency that causes a 3-dB drop ($1/\sqrt{2}$) in the output amplitude. At this point, the magnitude of the real part of the impedance (R) is equal to that of the imaginary part ($1/\omega C_1$) or

$$R_L = \frac{1}{\omega C_1}$$

or

$$\omega = \frac{1}{R_L C_1}$$

Figure 6.10(b) shows the amplitude response of the RC network. Note that at the half-power frequency, the amplitude drops by a factor of $1/\sqrt{2}$ from its peak value. This frequency is the lower radian cutoff frequency and normally represents the lowest frequency that can be effectively processed by the amplifier. That is, as frequency decreases, the output decreases. At some point, the

* The *decibel,* or dB, ratio is defined as 20 times the logarithm of the amplitude ratio (or 10 times the logarithm of the power ratio). Refer to Appendix B for a discussion of frequency response methods.

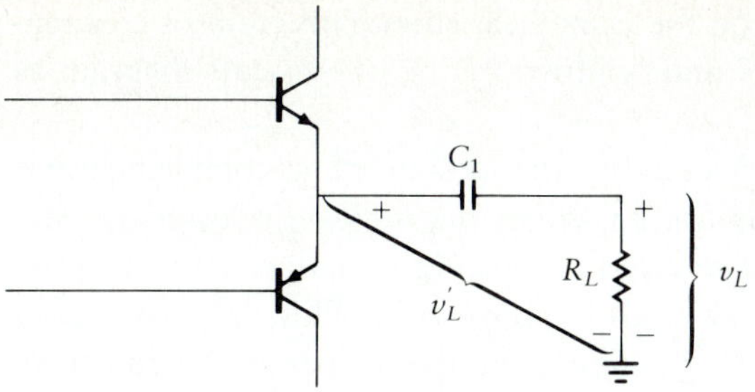

Figure 6.11
Output voltage divider at lowest frequency.

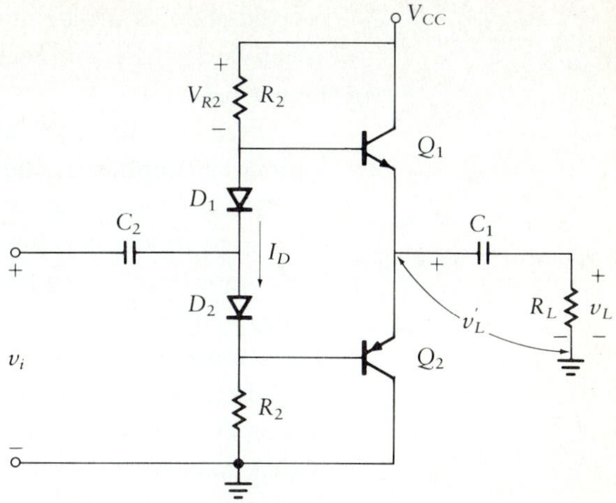

Figure 6.12 Complementary symmetry with diode compensation.

output amplitude is too small to be of use. The half-power point represents about a 30% drop in output voltage or current (i.e., $1/\sqrt{2} = 0.707$), and this may be more than can be permitted for the given application. Nonetheless, the half-power point is generally accepted as the limiting frequency.

Figure 6.11 shows that if C_1 is large the impedance is small and v_L is almost equal to v_L'. The output power is therefore approximately at its maximum value. Alternatively, if C_1 is relatively small, v_L is almost equal to zero and the output power is small. The impedance of C_1 determines the output amplitude, and this impedance depends on frequency. The impedance of a fixed capacitor decreases with increasing frequency, so the worst case occurs at the lowest operating frequency. Let us assume that the lowest frequency (i.e., the cutoff frequency) is f_{low} (hertz). Then the value of C_1 is found from the equation for the half-power point as follows:

$$C_1 = \frac{1}{2\pi f_{\text{low}} R_L} \tag{6.5}$$

For this value of C_1 the output voltage magnitude is given by

$$|V_L| = \frac{R_L|V_L'|}{\sqrt{X_{C1}^2 + R_L^2}} = \frac{|V_L'|}{\sqrt{2}} \tag{6.6}$$

Thus, as long as we operate above the cutoff frequency, $|V_L|$ is greater than the value shown in equation (6.6).

Further improvement in circuit operation is possible. The fluctuations of V_{BE} with temperature can be reduced by replacing the two R_1 resistors with diodes. These diodes should have characteristics similar to those of the transistor and they should be mounted on the same heat sink. This form of compensation is discussed in Section 5.3 and is illustrated in the circuit diagram of Figure 6.12.

There are three areas of concern in the design of a complementary-symmetry amplifier. The first is the *crossover distortion* discussed in Section 6.1.2. This distortion can be reduced by placing a small resistor in series with each diode to cause I_{CQ} to be slightly above zero. This, in turn, causes both amplifiers to amplify the ac input signal simultaneously in the cutoff region, thus compensating for the lower individual amplification in that region. The second area of concern is the possibility of *thermal runaway,* which can be caused by the two complementary transistors having different characteristics or by the value of V_{BE} decreasing at high temperatures. This can lead to a higher collector current, resulting in additional power dissipation and heating. This process continues until the transistor overheats and fails. Thermal runaway is prevented by placing a small resistor in series with each emitter to increase the bias level. With a load of 4 to 8 Ω, each resistor should be approximately 0.5 Ω.

The third area of concern is the distortion that results if the bias diodes, D_1 and D_2, stop conducting. One of the design requirements for the power amplifier of Figure 6.12 is to keep the diodes always turned on.

6.3.2 Design of Diode-Compensated Complementary-Symmetry Class-B Power Amplifier

The design of the power amplifier shown in Figure 6.12 requires knowledge of the diode forward resistance, R_f. We refer to the manufacturer's data sheet for an estimate of this value. For example, if we use the 1N4001 through 1N4007 diodes, we can estimate the value of R_f from Figure 1 on the data sheet (see Appendix D). The value of R_f varies widely with the value of instantaneous forward current. The estimated value of $1/R_f$ is found from the slope of the curve of Figure 1 of D1.1. The values of instantaneous voltage and current are read from the curve as follows:

Instantaneous forward voltage	Instantaneous forward current
0.8 V	90 mA
0.7 V	10 mA
0.6 V	1.5 mA

We calculate two values of R_f from these data by taking slopes.

$$R_{f1} = \frac{0.8 - 0.7}{90 - 10} \times 10^3 = 1.25\ \Omega$$

$$R_{f2} = \frac{0.7 - 0.6}{10 - 1.5} \times 10^3 = 11.8\ \Omega$$

The value of R_f clearly is a variable that depends on the diode forward current. Fortunately, the design of the amplifier is not highly dependent on the value of R_f and we use a fixed value for R_f in this design procedure. We will approximate R_f as 10 Ω.

It is important that the diode bias current be large enough to keep the diodes in their forward-biased region for all input voltages. The maximum negative peak current through the diode must be less than the direct-current bias. That is, the dc component of current must be larger than the ac component, so that when it adds to the ac component the resultant current does not go negative. If this were not true, the diode would be reverse-biased and distortion would result. This restriction is stated as

$$I_D > |i_{dp}| \tag{6.7}$$

where i_{dp} is the amplitude of the ac component of diode current.

The ac equivalent circuit is shown in Figure 6.13, where i_b is the transistor ac base current and v_L' is the ac voltage across the load, (R_L in series with C_1) at the low frequency. The quantity h_{ib} is omitted from Figure 6.13 because it is so small in value compared to the circuit resistors.

The direct current, I_D, through the diode is given by

$$I_D = \frac{V_{CC}/2 - 0.7}{R_2}$$

where we estimate the value of V_{BE} as 0.7 V. The *peak* signal current through the diode in the reverse direction, i_{dp}, is (refer to Figure 6.13) given by equa-

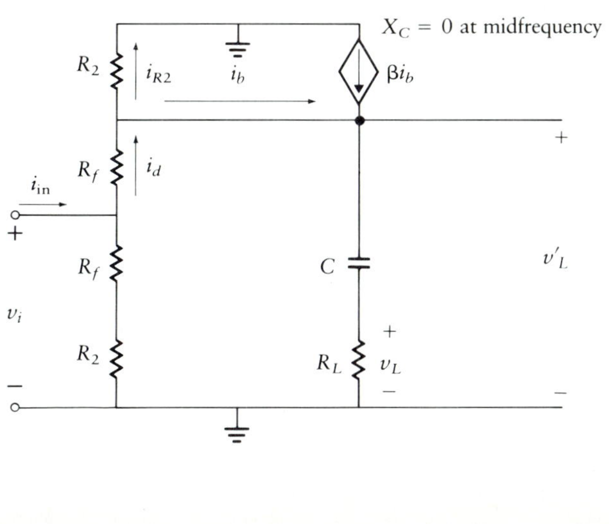

Figure 6.13 Base equivalent circuit.

tion (6.8). We use an additional subscript, p, to indicate that the peak value of the variable is being used.

$$\begin{aligned} i_{dp} &= i_{bp} + i_{R2p} \\ &= i_{bp} + \frac{v'_{Lp}}{R_2} \end{aligned} \tag{6.8}$$

Equation (6.8) is derived by assuming that the voltage gain is unity for the EF amplifier. That is, the ac voltage across R_2 is the same as the voltage from the emitter to ground.

By equating I_D to i_{dp}, we find the limiting condition for operating in the forward-biased diode condition (see equation (6.7)). From this, R_2 can be found as follows:

$$\frac{V_{CC}/2 - 0.7}{R_2} = i_{bp} + \frac{v'_{Lp}}{R_2}$$

so

$$R_2 = \frac{V_{CC}/2 - 0.7 - v'_{Lp}}{i_{bp}} \tag{6.9}$$

Since the amplifier is an emitter follower, $v_i \approx v'_L$. At midfrequency, the voltage across C_1 is zero, so the entire voltage, v_L, appears across R_L. Therefore, $v'_L = v_L$. At the lower 3-dB cutoff frequency, the output power drops to one-half of the power at the midrange frequency, and the magnitude of the voltage across R_L is equal to that of the voltage across C_1. Each of these voltage magnitudes is equal to $v_{Lp}/\sqrt{2}$. The peak magnitude of the voltage across the series combination of R_L and C_1 is

$$v'_{Lp} = \sqrt{\left(\frac{v_{Lp}}{\sqrt{2}}\right)^2 + \left(\frac{v_{Lp}}{\sqrt{2}}\right)^2} = v_{Lp}$$

Hence the value of v'_{Lp} in equation (6.9) can be written as

$$v'_{Lp} = R_L \beta i_{bp} = R_L i_{Cp} \tag{6.10}$$

The input resistance is determined from the equivalent circuit shown in Figure 6.14. We have assumed that $Z_L = R_L$ at the midfrequency of the amplifier where $X_{C1} = 0$. The capacitor is assumed to be a short circuit for midfrequency operation. Note that R_L reflects back as βR_L and the diode has a forward resistance of R_f.

The input resistance is found from Figure 6.14 as follows:

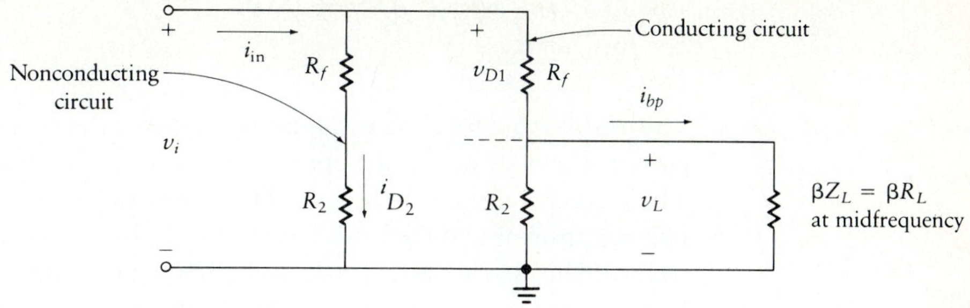

Figure 6.14 Input equivalent circuit.

$$R_{in} = (R_f + R_2) \parallel [R_f + (R_2 \parallel \beta R_L)] \tag{6.11}$$

The current gain is found using current division. The voltage across D_1 with Q_1 conducting is

$$v_{D1} = R_f\left(i_{bp} + \frac{v_L}{R_2}\right)$$

$$i_{D2} = \frac{v_{D1} + v_L}{R_f + R_2}$$

$$i_{in} = \frac{v_{D1} + v_L}{R_f + R_2} + i_{bp} + \frac{v_L}{R_2} \tag{6.12}$$

where $v_L = v'_{Lp}$ and

$$A_i = \beta\frac{i_{bp}}{i_{in}} \tag{6.13}$$

6.3.3 Power Calculations for Class-B Push-Pull Amplifier

The power delivered by the ac source is split between the transistor and the resistors in the bias circuitry. The ac signal source adds an insignificant additional amount of power, since base currents are small relative to collector currents. Part of the power to the transistor goes to the load, and the other part is dissipated by the transistor itself. The following equations specify the various power relationships in the circuit.

The input power is given by

$$P_{VCC} = V_{CC}I_{DC}$$

where I_{DC} is the average current drawn from the power supply for the transistor portion of the circuit. The current, I_{DC}, is determined by averaging across a full period. That is,

$$I_{DC} = \frac{1}{T}\int_0^{T/2} i_{C1}(t)\,dt = \frac{1}{\pi}I_{C\max}$$

This result requires close examination. During the first half-cycle of the input, current flows through the upper transistor into the capacitor and load resistor. The second transistor is cut off. The power during this half-cycle goes to the upper transistor and the load. During this half-cycle, energy is stored in the capacitor. During the second half-cycle, the upper transistor is cut off. Thus, the V_{CC} source *does not supply any power* during this half-cycle. Instead, stored energy in the capacitor is returned to the load and to the lower transistor.

$$P_{VCC}(\text{delivered to the transistor circuit}) = \frac{V_{CC}I_{C\max}}{\pi} \tag{6.14}$$

The maximum value of collector current is

$$I_{C\max} = \frac{V_{CC}}{2R_L}$$

The maximum power delivered to the transistor is

$$P_{VCC}(\text{max delivered to transistor}) = \frac{V_{CC}^2}{2\pi R_L}$$

The ac output power, assuming that the input is sinusoidal, is

$$P_o(\text{ac}) = \frac{I_{C\max}^2 R_L}{2} \tag{6.15}$$

The maximum ac output power is found by substituting $I_{C\max}$ to get

$$P_o(\text{ac max}) = \frac{1}{2}\left(\frac{V_{CC}}{2R_L}\right)^2 R_L = \frac{V_{CC}^2}{8R_L}$$

The total dc power supplied to the stage is the sum of the power to the transistor and the power to the bias and compensation circuitry.

$$P_{VCC} = \frac{V_{CC}I_{C\max}}{\pi} + \frac{V_{CC}^2}{2R_f + 2R_2} \tag{6.16}$$

If we subtract the power to the load from the power supplied to the transistor circuit, we find the power being dissipated in the transistors. Since this power is shared equally between the two transistors, the power dissipated by a single transistor is one-half of this value. Thus,

$$P_{\text{transistor}} = \frac{1}{2}\left(\frac{V_{CC}I_{C\max}}{\pi} - \frac{I_{C\max}^2 R_L}{2}\right) \tag{6.17}$$

We are assuming that the base current is negligible. The efficiency of the Class-B push-pull amplifier is the ratio of the output power to the power delivered to the transistor. Thus we neglect the power dissipated by the bias circuitry.

$$\eta = \frac{V_{CC}^2/8R_L}{V_{CC}^2/2\pi R_L} = \frac{\pi}{4} = 0.785 \quad \text{or} \quad 78.5\%$$

This amplifier is more efficient than a Class-A amplifier. It is often used in output circuits where efficiency is an important design requirement.

The amplifier designer must specify the power rating of the transistor. That is, it is important to know the maximum power dissipated by a single transistor. This parameter is found by differentiating equation (6.17) with respect to $I_{C\max}$ and setting the derivative to zero. Thus we will find the value of $I_{C\max}$ that results in maximum dissipated power as follows:

$$\frac{dP}{dI_{C\max}} = 0 = \frac{1}{2}\left(\frac{V_{CC}}{\pi} - I_{C\max}R_L\right) \tag{6.18}$$

and solving for $I_{C\max}$ yields

$$I_{C\max} = \frac{V_{CC}}{\pi R_L} \tag{6.19}$$

We now substitute this value back into equation (6.17) to find the maximum power:

$$P_{\max} = \frac{1}{2}\left(\frac{V_{CC}^2}{\pi^2 R_L} - \frac{V_{CC}^2}{2\pi^2 R_L}\right)$$

$$= \frac{V_{CC}^2}{4\pi^2 R_L} \tag{6.20}$$

Since equation (6.20) represents the maximum power, it is equivalent to the minimum transistor power rating. That is, in choosing a transistor, it is important for the power rating to equal or exceed this number.

Example 6.2 Class-B Push-Pull Amplifier (Design)

Design a diode-compensated complementary-symmetry (Figure 6.12) audio amplifier with a low-frequency cutoff of 60 Hz and a power output of $\frac{1}{2}$ W into an 8-Ω speaker. Use a 12-V power supply and silicon transistors with $\beta = 60$. The diodes have forward resistance of 8 Ω. Determine the current gain, power delivered to the amplifier, and power ratings of the transistors.

SOLUTION We first determine the value of $I_{C\max}$ needed to achieve the specified load power. From equation (6.15) we obtain

$$P_L = \frac{I_{C\max}^2 R_L}{2} = \tfrac{1}{2}\ \text{W}$$

$$I_{C\max} = \sqrt{\frac{1}{R_L}} = 0.354\ \text{A}$$

The maximum base current is

$$i_{bp} = \frac{I_{C\max}}{\beta} = 5.9\ \text{mA}$$

The lower frequency, 60 Hz, represents the half-power frequency used to find C_1. At this frequency,

$$C_1 = \frac{1}{\omega R_L} = 332\ \mu\text{F}$$

At 60 Hz, the impedance of the RC output circuit is $Z_L' = 8\sqrt{2} = 11.3\ \Omega$. Equation (6.10) can then be used to find the midfrequency value of load voltage.

$$v_{Lp} = R_L i_{Cp} = 2.83\ \text{V}$$

We use equation (6.9) to find R_2.

$$R_2 = \frac{V_{CC}/2 - 0.7 - v_{Lp}'}{i_{bp}} = 419\ \Omega$$

R_{in} and A_i are determined at midfrequency as follows:

$$V_{D1} = R_f\left(i_{bp} + \frac{v_L}{R_2}\right) = 0.1\ \text{V}$$

Equation (6.12) can now be used to find the input current, where $v_L = v_{Lp}'$.

$$i_{\text{in}} = \frac{v_{D1} + v_L}{R_f + R_2} + i_{bp} + \frac{v_L}{R_2} = 19.5\ \text{mA}$$

Now, using equation (6.11), we find R_{in}.

$$R_{\text{in}} = (R_f + R_2) \parallel [R_f + (R_2 \parallel \beta R_L)] = 150\ \Omega$$

Equation (6.13) is then used to evaluate A_i.

$$A_i = \beta \frac{i_{bp}}{i_{\text{in}}} = 18.2$$

The power to the amplifier, including the bias circuitry, is given by equation (6.16):

$$P_{VCC} = \frac{V_{CC} I_{C\max}}{\pi} + \frac{V_{CC}^2}{2R_f + 2R_2} = 1.52\ \text{W}$$

The power rating of each transistor is given by equation (6.20):

$$P_{\text{trans}} = \frac{V_{CC}^2}{4\pi^2 R_L} = 0.456\ \text{W}$$

Example 6.3 Class-B Push-Pull Amplifier (Design)

Design a complementary-symmetry push-pull diode-compensated Class-B amplifier (see Figure 6.15) to drive a 4-Ω load to ±3 V for a low-frequency 3-dB point of 50 Hz. Use *npn* and *pnp* transistors, each having a β of 100 and $V_{BE} = \pm 0.7$ V. The diodes have forward resistance $R_f = 10\ \Omega$. Determine all quies-

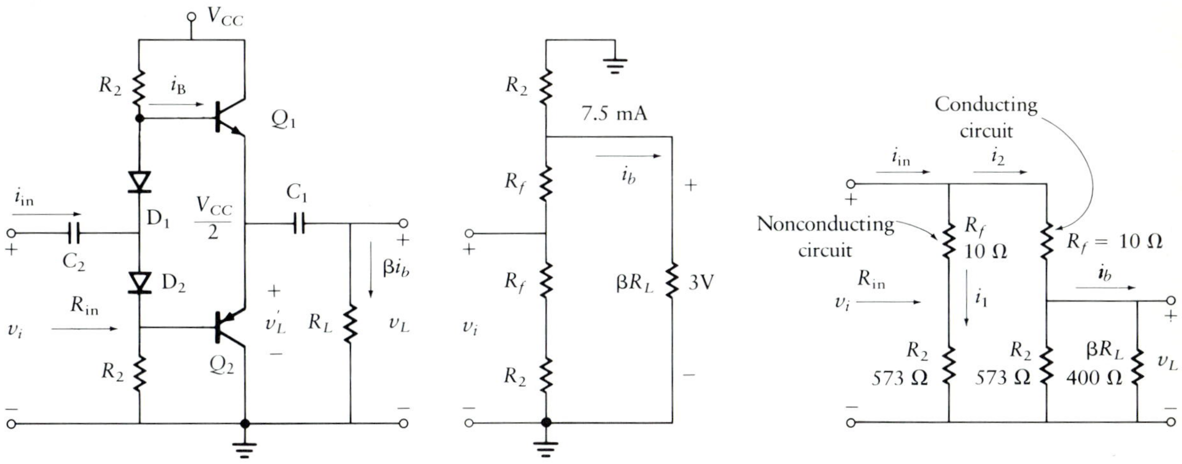

(a) Amplifier circuit (b) Equivalent circuit (midrange) (c) Equivalent circuit from input

Figure 6.15 Circuit for Example 6.3.

cent voltages and currents for $V_{CC} = 16$ V. Calculate the maximum power that is delivered from the power supply, the power delivered to the load, and the power rating of the transistors to be used.

SOLUTION Choose C_1 so that the half-power frequency is 50 Hz. Thus, at this frequency we have

$$C_1 = \frac{1}{2\pi f_L R_L} = 796\ \mu\text{F} \approx 800\ \mu\text{F}$$

We discuss component selection in Chapter 4. Some points from that chapter bear repetition here. In most designs in this text, we specify component values to three significant digits. In the majority of practical designs, this is much more accuracy than is necessary. For example, in dealing with nonprecision (everyday) resistors, the tolerance is normally ±10%. Thus, a resistor that is marked as 100 Ω can be expected to have a resistance value anywhere between 90 and 110 Ω. Therefore, if a 10% resistor is being used, it would not make much sense for the design engineer to specify a resistance value of 101.5936 Ω. The same is true of capacitors. The amount of rounding off that is permitted in a design depends on the precision of the components being used and the confidence the designer has in device parameters (e.g., β). The more precision, the more expensive are the components. Since we have not specified any particular level of precision, we continue to carry most of our answers to three significant digits. Rounding off numbers in intermediate computational steps can propagate errors. We feel justified in rounding off the value of C_1 found above to 800 μF. This represents a change of only $\frac{1}{2}$%.

The maximum collector and maximum base currents are related by β. Thus,

$$i_{bp} = \frac{i_{Cp}}{\beta}$$

but since $V_{L\,\max}$ is given as 3, we find

$$i_{Cp} = \tfrac{3}{4} = 750\ \text{mA}$$

Therefore,

$$i_{bp} = 7.5\ \text{mA}$$

and from equation (6.9), where $v_{Lp} = 3$ V,

$$R_2 = \frac{V_{CC}/2 - 0.7 - v_{Lp}}{i_{bp}} = 573\ \Omega$$

At midfrequency, the input resistance is found from equation (6.11) to be

$$R_{\text{in}} = (R_f + R_2) \parallel [R_f + (R_2 \parallel \beta R_L)] = 173\ \Omega$$

The supply power is given by equation (6.16).

$$P_{\text{sup}} = \frac{V_{CC} I_{C\max}}{\pi} + \frac{V_{CC}^2}{2R_f + 2R_2} = 4.04\ \text{W}$$

Equation (6.15) yields the output power.

$$P_o = \frac{I_{C\max}^2 R_L}{2} = 1.13\ \text{W}$$

Equation (6.20) is used to find the required power rating of each transistor.

$$P_{\text{trans}} = \frac{V_{CC}^2}{4\pi^2 R_L} = 1.62\ \text{W}$$

The current gain, A_i, is found by referring to Figure 6.15(c):

$$i_b = \frac{R_2 i_2}{R_2 + \beta R_L} = \frac{573 i_2}{973}$$

Therefore,

$$i_2 = \frac{973 i_{bp}}{573} = 12.7\ \text{mA}$$

i_2 and i_{in} are related by

$$i_2 = \frac{(R_f + R_2) i_{\text{in}}}{R_f + R_2 + R_f + R_2 \parallel \beta R_L} = \frac{583 i_{\text{in}}}{828}$$

Hence,

$$i_{\text{in}} = 18\ \text{mA}$$

Finally,

$$A_i = \frac{\beta i_b}{i_{\text{in}}} = 41.6$$

Drill Problems

D6.3 Design a complementary-symmetry diode-compensated Class-B amplifier to drive a 4-Ω load with 1 W of power for a low-frequency cutoff of 20 Hz. Use *npn* and *pnp* matched transistors, each having $\beta = 100$ and $V_{BE} = \pm 0.7$ V with equivalent characteristic diodes having $R_f = 50\ \Omega$. Let $V_{CC} = 12$ V. Determine R_2, R_{in}, C_1, P_{trans}, and $A_i = i_o/i_{in}$.

Ans: $R_1 = 350\ \Omega$; $C_1 = 2000\ \mu F$;
$P_{trans} = 0.91$ W;
$A_i = 29$; $R_{in} = 149\ \Omega$

D6.4 Design a complementary-symmetry diode-compensated Class-B power amplifier to deliver 2 W to a 10-Ω load with a low-frequency 3-dB point of 30 Hz. Use a matched pair of *npn* and *pnp* transistors, each having $\beta = 100$ and $V_{BE} = \pm 0.7$ V with equivalent characteristic diodes having $R_f = 5\ \Omega$. Determine R_2, C_1, R_{in}, and A_i when $V_{CC} = 16$ V.

Ans: $C_1 = 530\ \mu F$; $R_2 = 154\ \Omega$;
$R_{in} = 74\ \Omega$; $A_i = 7.1$

6.4 DARLINGTON CIRCUIT

Figure 6.16 illustrates a *Darlington circuit.* This configuration is composed of two cascaded transistors. This transistor combination possesses desirable characteristics that make it more useful than a single transistor in certain applications. For example, the circuit has high input impedance, low output impedance, and high current gain. One disadvantage of the Darlington transistor pair is that the leakage current of the first transistor is amplified by the second transistor.

If the two transistors are connected in the manner shown in Figure 6.17, the betas of the two transistors multiply together, forming a combination that looks like a single high-β transistor. The Darlington transistor pair can be used in either a CE or EF amplifier configuration. The h_{ie} of both transistors is not the same, since the quiescent point of the first transistor is different from

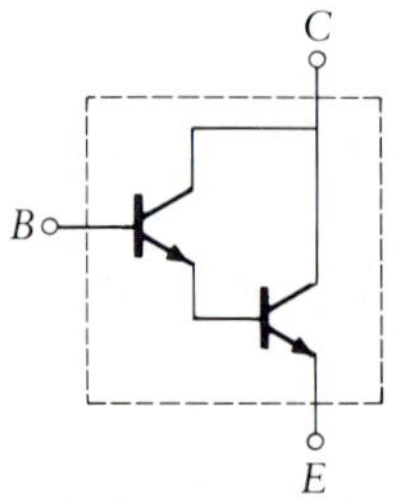

Figure 6.16 Darlington transistor pair.

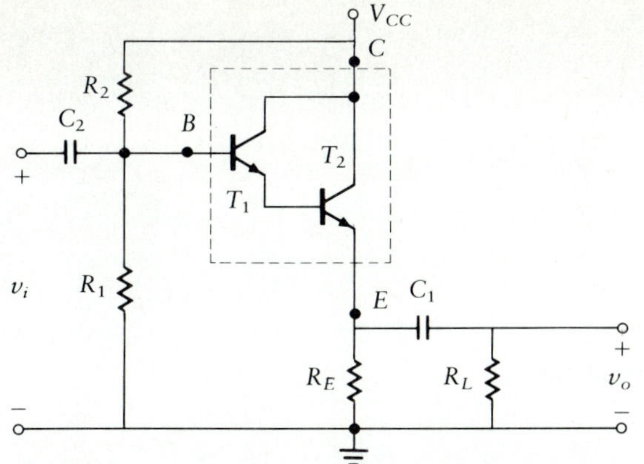

Figure 6.17 EF amplifier using Darlington pair.

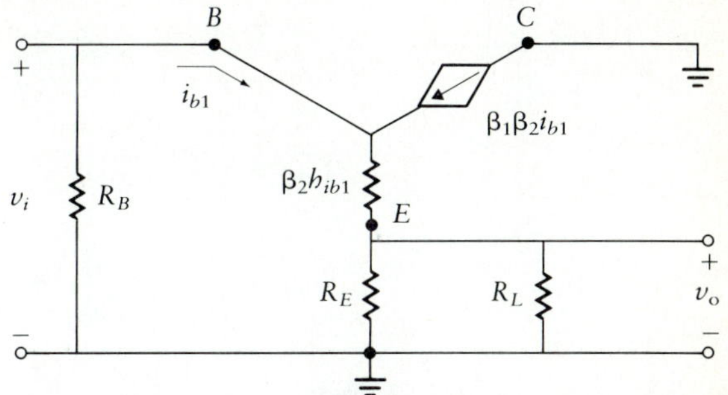

Figure 6.18 The ac equivalent circuit of Darlington pair.

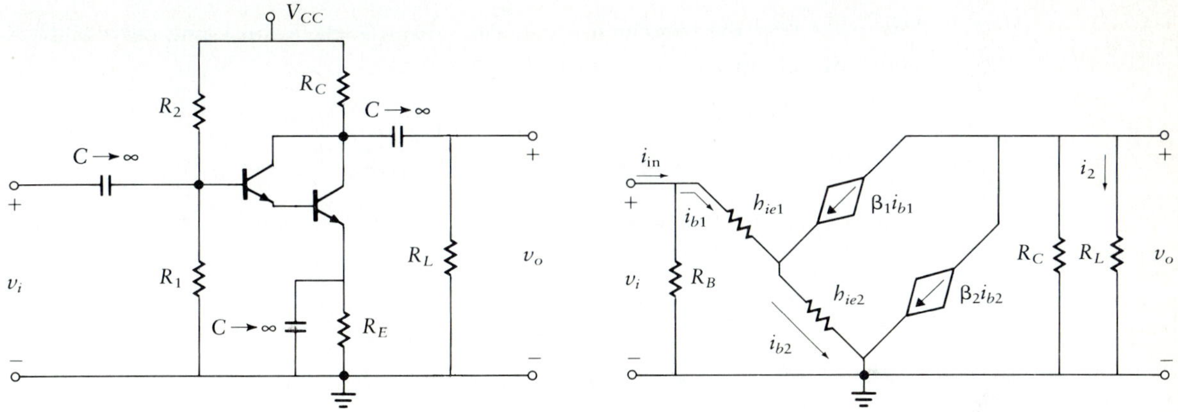

(a) (b)

Figure 6.19 Darlington CE amplifier.

that of the second. As can be seen from the equivalent circuit of Figure 6.18, the equivalent load on the first transistor is $\beta_2(R_L \parallel R_E)$, while the load on the second transistor is only $R_L \parallel R_E$. In practice, the first transistor can be of lower power rating than the second. The input resistance of the second transistor constitutes the emitter load for the first transistor.

To determine the ac parameters for a Darlington CE amplifier, we look at the equivalent circuit, as shown in Figure 6.19. We write the equation for R_{in} as follows:

$$R_{\text{in}} = R_B \parallel (h_{ie1} + \beta_1 h_{ie2})$$

and

$$h_{ie1} = \beta_1 h_{ib1}$$

$$h_{ie2} = \beta_2 h_{ib2}$$

The input resistance is given by

$$R_{\text{in}} = R_B \parallel (\beta_1 h_{ib1} + \beta_1\beta_2 h_{ib2})$$

where

$$h_{ib2} = \frac{V_T}{I_{CQ2}}$$

$$h_{ib1} = \frac{V_T}{I_{CQ1}} = \frac{V_T}{I_{BQ2}} = \frac{\beta_2 V_T}{I_{CQ2}} = \beta_2 h_{ib2}$$

Then

$$R_{\text{in}} = R_B \parallel (\beta_1\beta_2 h_{ib2} + \beta_1\beta_2 h_{ib2}) = R_B \parallel 2\beta_1\beta_2 h_{ib2}$$

We must find the current in order to determine the gain. Using current division, we obtain

$$i_{b1} = \frac{R_B}{R_B + h_{ie1} + \beta_1 h_{ie2}} i_{\text{in}} = \frac{R_B}{R_B + 2\beta_1\beta_2 h_{ib2}} i_{\text{in}}$$

$$i_{\text{in}} = \frac{R_B + 2\beta_1\beta_2 h_{ib2}}{R_B} i_{b1}$$

The current gain is then

$$A_i = \frac{i_2}{i_{\text{in}}} = \frac{-\beta_1\beta_2 i_{b1} R_C/(R_C + R_L)}{(R_B + 2\beta_1\beta_2 h_{ib2}) i_{b1}/R_B}$$

$$= \frac{-R_B}{R_B/\beta_1\beta_2 + 2h_{ib2}} \frac{R_C}{R_C + R_L}$$

Using the gain-impedance formula, we have

$$A_v = \frac{A_i R_L}{R_{\text{in}}}$$

$$= \frac{-\beta_1\beta_2 (R_C \parallel R_L)}{2\beta_1\beta_2 h_{ib2}} = \frac{-R_C \parallel R_L}{2h_{ib2}}$$

Although this is normally considered a voltage gain amplifier, it can provide high current gains due to the high input resistance. Most amplifiers with a bypassed emitter resistance have excellent voltage gain but have low input resistance, resulting in low current gain. The Darlington CE amplifier provides not only good voltage gain but also excellent current gain.

Some manufacturers package the Darlington transistor pair in a single package with only three external leads. Darlington pair transistors packaged in an integrated circuit are available with betas as high as 30,000.

Although the Darlington circuit may be viewed as if it were a single transistor, there are some important differences. One of these is speed of operation. Changing the voltage across a transistor junction requires a finite amount of time, since electrons must be moved. In fact, as capacitance increases, the time constant of any RC combination increases and the speed of operation decreases. Since the Darlington circuit has two base-emitter junctions in series with each other, the combination operates more slowly than a single transistor [32]. To speed up the operation, a resistor is placed between the emitter of the first transistor and the base of the second transistor. These resistors have typical values of several hundred ohms for power transistors and several thousand ohms for transistors used for signal amplification. In addition, since two base-emitter junctions exist, the overall V_{BE} is 1.4 V instead of 0.7 V.

The equivalent circuit of the Darlington pair used in EF amplifiers can be simplified as shown in Figure 6.18. The values of R_{in} and A_i are determined as follows:

$$R_{\text{in}} = R_B \parallel [\beta_1\beta_2(2h_{ib2} + R_E \parallel R_L)]$$

The current gain is given by

$$A_i = \frac{\beta_1\beta_2 i_{b1}}{i_{\text{in}}} \frac{R_E}{R_E + R_L}$$

where

$$i_{b1} = \frac{R_B}{R_B + (2h_{ib2} + R_E \parallel R_L)\beta_1\beta_2} i_{\text{in}}$$

and

$$i_{\text{in}} = \frac{R_B + (2h_{ib2} + R_E \parallel R_L)\beta_1\beta_2}{R_B} i_{b1}$$

We then obtain

$$A_i = \frac{R_B}{R_B/\beta_1\beta_2 + 2h_{ib2} + R_E \parallel R_L} \frac{R_E}{R_E + R_L}$$

We see from these equations that R_B can be made much larger than in the case of a single transistor. As a result, the input resistance and the current gain are both much larger for the Darlington pair.

Example 6.4 Darlington Pair in Class-A Amplifier (Design)

Design an EF amplifier for maximum symmetrical swing using a Darlington transistor pair (Figure 6.17) that has a combined β of 10,000 and $V_{BE} = 1.4$ V. The amplifier must drive a load of 20 Ω with $R_{\text{in}} = 3$ kΩ, $V_{CC} = 12$ V, and $f_L = 20$ Hz. Determine A_i and P_o.

SOLUTION Set $R_E = R_L$ since there is one less equation than unknowns. We calculate R_B from knowledge of R_{in} as follows.

$$R_{\text{in}} = R_B \parallel [\beta_1\beta_2(R_E \parallel R_L)]$$

$$3\text{ k}\Omega = \frac{R_B(10^5)}{R_B + 10^5}$$

Solving for R_B yields

$$R_B = 3.09\text{ k}\Omega$$

where we have ignored h_{ib}. The Q-point is at

$$I_{CQ} = \frac{V_{CC}}{R_{ac} + R_{dc}} = 400\text{ mA}$$

We use the bias equation to find V_{BB}.

$$V_{BB} = V_{BE} + I_{CQ}\left(\frac{R_B}{\beta_1\beta_2} + R_E\right) = 9.52\text{ V}$$

The bias resistors are given by

$$R_1 = \frac{R_B}{1 - V_{BB}/V_{CC}} = 15.0\text{ k}\Omega$$

$$R_2 = \frac{R_B V_{CC}}{V_{BB}} = 3.89\text{ k}\Omega$$

The current gain is

$$A_i = \frac{R_B}{R_B/\beta_1\beta_2 + R_E \parallel R_L}\,\frac{R_E}{R_E + R_L} = 150$$

The output power is

$$P_o = \frac{1}{2}\left(\frac{I_{CQ}}{2}\right)^2 R_L = 0.4\text{ W}$$

We calculate C_1 by noting that the total resistance in the discharge path of the capacitor is

$$R_L + R_E \parallel (h_{ib2} + R_B/\beta)$$

assuming R_i is large, so

$$C_1 = \frac{1}{2\pi f_L[R_L + R_E \parallel (h_{ib2} + R_B/\beta)]} = 395\ \mu\text{F}$$

The Darlington pair provides a large increase in current and power gain over the single-transistor amplifier. It also provides a higher input resistance than can be obtained using a single-transistor amplifier.

Drill Problems

D6.5 Design an EF amplifier for maximum symmetrical swing using a Darlington transistor pair (Figure 6.17) to drive an 8-Ω load with a combined β of 20,000, $V_{BE} = 1.4$ V, $V_{CC} = 20$ V, and $A_i = 500$. Design the amplifier with $f_L = 40$ Hz. Find R_E, R_1, R_2, C_1, R_{in}, and P_o.

Ans: $R_E = 8\ \Omega$; $R_1 = 17\ \text{k}\Omega$; $R_2 = 5.6\ \text{k}\Omega$;
$C_1 = 498\ \mu\text{F}$ (3 dB down);
$R_{\text{in}} = 4\ \text{k}\Omega$; $P_o = 2.3$ W

D6.6 Design a CE amplifier for maximum symmetrical swing using a Darlington transistor pair (Figure 6.19) with combined $\beta = 25{,}000$, $V_{BE} = 1.2$ V, $V_{CC} = 20$ V, $A_v = -120$, $A_i = 200$, $R_E = 200\ \Omega$, and $R_L = 5\ \text{k}\Omega$. Determine the following:

a. R_1, R_2, and R_{in}.
b. Maximum undistorted output voltage swing.
c. Input dc power required.

Ans: a. $R_1 = 9.24\ \text{k}\Omega$, $R_2 = 99.5\ \text{k}\Omega$, $R_{\text{in}} = 8.33\ \text{k}\Omega$;
b. $V_o = 11.2$ V;
c. $P_{\text{in}} = 53$ mW

6.5 POWER SUPPLY USING POWER TRANSISTORS

6.5.1 Power Supply Using Discrete Components

In Chapter 1 we analyze the regulated power supply using a zener diode as the voltage-controlling device. As the zener diode maintains a fixed voltage, the current through the diode changes. That is, as the input voltage increases, the current in the zener also increases. Since the diode has a nonzero resistance, the voltage across the diode is a function of the current through it. This leads to poor regulation and power is dissipated in the zener.

To obtain better regulation, the zener diode can be connected to the base circuit of a power transistor as shown in Figure 6.20. This configuration reduces the current flow in the diode. The power transistor used in this configuration is known as a *pass* transistor. The purpose of C_L is to reduce the effect of high-frequency variations.

Because of the current-amplifying property of the transistor, the current in the zener diode is small. Hence there is little voltage drop across the diode resistance, and the zener diode approximates an ideal constant-voltage source.

The current in the resistor, R_i, is the zener diode current plus the base current in the transistor, I_L/β. We take the two conditions

1. $V_{S\max}$, $I_{Z\max}$, $I_{L\min}/\beta$
2. $V_{S\min}$, $I_{Z\min}$, $I_{L\max}/\beta$

Refer to Chapter 1 for a definition of these terms. We calculate R_i for both conditions, and since R_i is constant we equate these two expressions as in equation (6.21).

$$R_i = \frac{V_{S\max} - V_Z}{I_{Z\max} + I_{L\min}/\beta} = \frac{V_{S\min} - V_Z}{I_{Z\min} + I_{L\max}/\beta} \tag{6.21}$$

As in Chapter 1, we let $I_{Z\min} = 0.1 I_{Z\max}$. Then we equate the expressions for equation (6.21) to obtain

$$(V_{S\max} - V_Z)\left(0.1 I_{Z\max} + \frac{I_{L\max}}{\beta}\right) = (V_{S\min} - V_Z)\left(I_{Z\max} + \frac{I_{L\min}}{\beta}\right)$$

Solving for $I_{Z\max}$, we obtain

$$I_{Z\max} = \frac{I_{L\min}(V_Z - V_{S\min}) + I_{L\max}(V_{S\max} - V_Z)}{\beta(V_{S\min} - 0.1V_{S\max} - 0.9V_Z)} \tag{6.22}$$

Equation (6.22) is the same as equation (1.48) except that $I_{Z\max}$ is reduced by the β of the transistor. The design is accomplished as in Chapter 1, with the exception that the value of $I_{Z\max}$ is reduced. We determine the equivalent load as seen from the capacitor, C_F, by looking at the circuit of Figure 6.20. We estimate the load resistance by taking the ratio of the minimum source voltage to the maximum load current.

$$R_L(\text{equivalent}) = R_L(\text{worst case}) = \frac{V_{S\min}}{I_{L\max}} \tag{6.23}$$

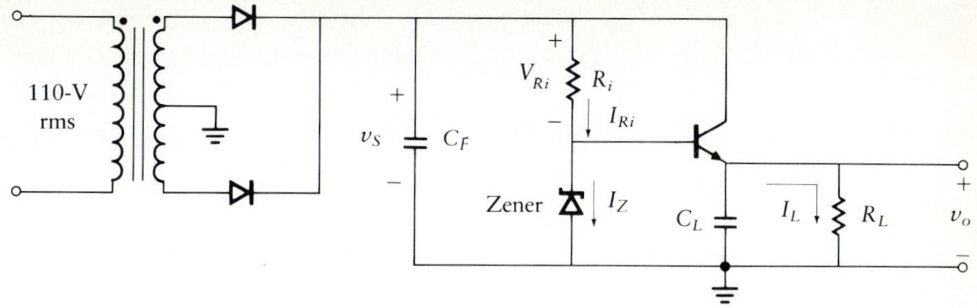

Figure 6.20 Regulated power supply.

We substitute equation (6.23) into equation (1.41) to estimate the capacitor size.

$$C_F = \frac{V_{S\max}}{\Delta V f_p R_L(\text{eq})} \tag{6.24}$$

Since the voltage gain of an EF amplifier is unity, the output voltage of the regulated power supply is

$$V_L = V_Z - V_{BE} \tag{6.25}$$

The percent regulation of the power supply is given by

$$\begin{aligned}\%\ \text{reg} &= \frac{V_{Z\max} - V_{Z\min}}{V_Z} \times 100 \\ &= \frac{R_Z(I_{Z\max} - I_{Z\min})}{V_Z} = 100\end{aligned} \tag{6.26}$$

The percent regulation is reduced since $I_{Z\max}$ is much smaller because of the division by β.

Example 6.5 Regulated Power Supply Using Discrete Components (Design)

Design an 11.3-V regulated power supply (see Figure 6.20) for a load current that varies between 400 and 500 mA. Assume an input of 120 V (rms) at 60 Hz into a 3 : 1 center-tapped transformer. Use a 12-V zener with $R_Z = 2\ \Omega$. The transistor has $V_{BE} = 0.7$ V and $\beta = 100$. Set C so that $\Delta v_S = 10\%$.

SOLUTION The design consists of choosing values for R_i and C_F. We first find $V_{S\max}$ by multiplying by $\sqrt{2}$ to obtain 170 V. The transformer output on either side of the center tap is one-sixth of the input, so $V_{S\max}$ is 28.3 V. Since $\Delta v_S = 10\%$,

$$V_{S\min} = (0.9)\,(V_{S\max}) = 25.45 \text{ V}$$

$$I_{L\,\max} = 500 \text{ mA}$$

$$I_{L\,\min} = 400 \text{ mA}$$

and

$$V_Z = 12 \text{ V}$$

Now, using equation (6.22), we obtain

$$I_{Z\,\max} = \frac{I_{L\,\min}(V_Z - V_{S\,\min}) + I_{L\,\max}(V_{S\,\max} - V_Z)}{\beta(V_{S\,\min} - 0.1V_{S\,\max} - 0.9V_Z)} = 2.32 \text{ mA}$$

Notice that the transistor keeps the value of $I_{Z\,\max}$ quite small, since β appears in the denominator of equation (6.22). We calculate the value of R_i from equation (6.21).

$$R_i = \frac{V_{S\,\max} - V_Z}{I_{Z\,\max} + I_{L\,\min}/\beta} = 2.58 \text{ k}\Omega$$

The capacitor size is estimated from equation (6.24):

$$C_F = \frac{V_{S\,\max}}{(V_{S\,\max} - V_{S\,\min})f_p R_L(\text{eq})} = 1630\ \mu\text{F}$$

Note that $R_L(\text{eq}) = V_{S\,\min}/I_{L\,\max} = 51\ \Omega$. Equation (6.26) can be used to evaluate the percent regulation at the load.

$$\%\text{ reg} = \frac{R_Z(I_{Z\,\max} - I_{Z\,\min})}{V_Z} \times 100 = 0.035\%$$

6.5.2 Power Supply Using IC Regulator (Three-Terminal Regulator)

Monolithic integrated circuits have greatly simplified the design of a wide variety of power supplies. Using a single IC regulator and a few external components, we can obtain better regulation (0.01%) with good stability and reliability and with overload protection. IC regulators are produced by a number of manufacturers.

The IC regulator improves on the performance of the zener diode regulator. It does this by incorporating an operational amplifier (see Chapter 8 for more information on operational amplifiers). In this section, we present basic design considerations for IC regulators. These techniques are useful in the design of power supplies for a variety of low-power applications. We consider the inter-

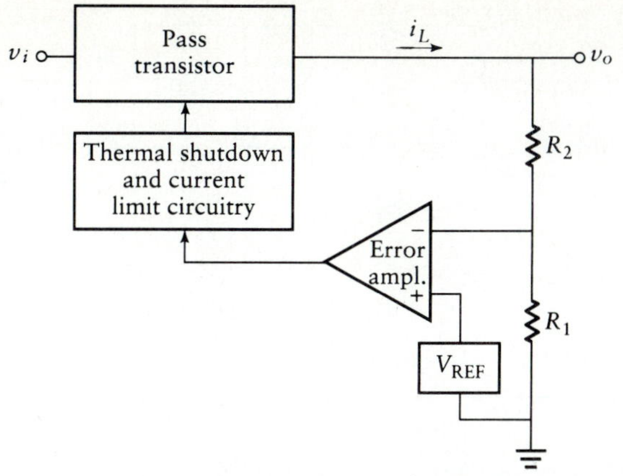

Figure 6.21 Functional block diagram of IC regulator.

nal theory of operation of these and other three-terminal voltage regulators in the current section. These products vary in the amount of current output. The most common range of output current is 0.75 to 1.5 A (depending on whether a heat sink is used).

The functional block diagram of Figure 6.21 illustrates the method of voltage regulation using this *series regulator*. The name series regulator is based on the use of a *pass transistor* (a power transistor), which develops a variable voltage that is in series with the output voltage. The voltage across the pass transistor is varied in such a manner as to keep the output voltage constant.

A reference voltage, V_{REF}, which is often developed by a zener diode, is compared with the voltage-divided output, v_o. The resulting error voltage is given by

$$v_\epsilon = \text{error voltage} = V_{REF} - \frac{R_1}{R_1 + R_2} v_o \tag{6.27}$$

v_ϵ is amplified through a discrete amplifier or an operational amplifier and used to change the voltage drop across the pass transistor. This is a feedback system that generates a variable voltage across the pass transistor in order to force the error voltage to zero. When the error voltage is zero, we obtain the desired equation (see equation (6.27)):

$$v_o = \left(1 + \frac{R_2}{R_1}\right) V_{REF}$$

Note that since R_1, R_2, and V_{REF} are constant, v_o is also a constant, independent of load current variations and input voltage variations.

Thermal shutdown and current limit circuitry exists between the error amplifier and the pass transistor. This circuitry protects the regulator in case the temperature becomes too high or an inadvertent short circuit exists at the output of the regulator.

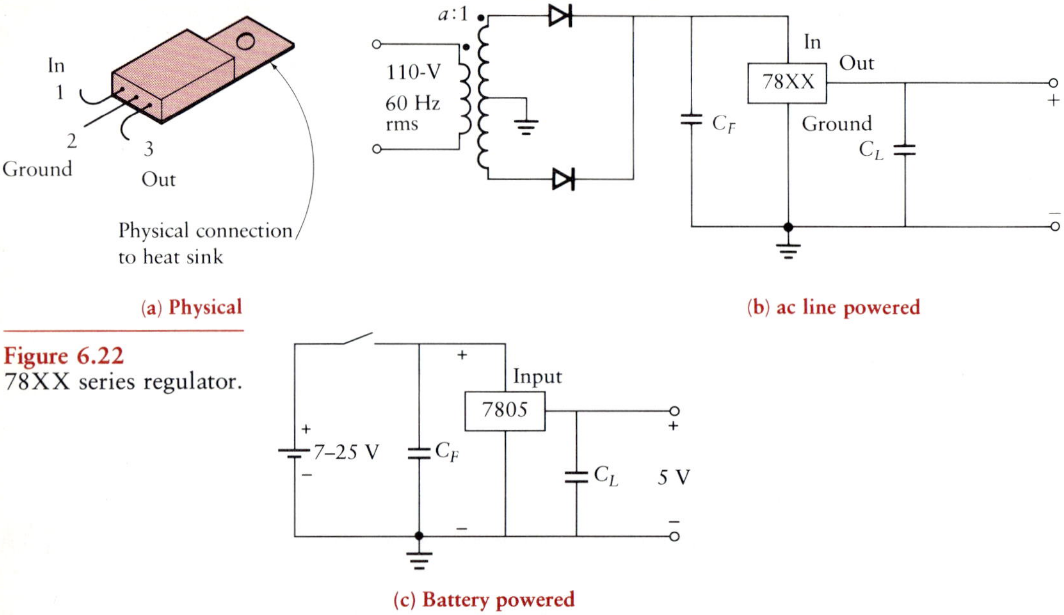

Figure 6.22
78XX series regulator.

The maximum power dissipated in this type of series regulator is the power dissipated in the internal pass transistor, which is approximately $(V_{S\max} - v_o)I_{L\max}$. Hence, as the load current increases, the power dissipated in the internal pass transistor increases. If I_L exceeds 0.75 A, the IC package should be secured to a heat sink. When this is done, I_L can increase to about 1.5 A.

We now focus our attention on the 78XX series of regulators. The last two digits of the IC part number denote the output voltage of the device. For example, a 7808 IC package produces an 8-V regulated output. These packages, although internally complex, are inexpensive and easy to use.

Specification sheets appear in Appendix D, and these should be referred to during the following discussion. A number of different voltages can be obtained from the 78XX series ICs; they are 5, 6, 8, 8.5, 10, 12, 15, 18, and 24 V. To design a regulator around one of these ICs, we need only select a transformer, diodes, and a filter. The physical configuration is shown in Figure 6.22(a). The ground lead and the metal tab are connected together. This permits direct attachment to a heat sink for cooling purposes. A typical circuit application is shown in Figure 6.22(b).

The specification sheet for this IC indicates that there must be a common ground between the input and output, and the minimum voltage at the IC input must be above the regulated output. To ensure this last condition, it is necessary to filter the output from the rectifier. The C_F in Figure 6.22(b) performs this filtering when combined with the input resistance to the IC. We use an a:1 step-down transformer, with the secondary winding center-tapped, to drive a full-wave rectifier.

Table 6.1 Minimum and Maximum Input Voltages for 78XX Regulator

Regulator	Min	Max
7805	7	25
7806	8	25
7808	10.5	25
7885	10.5	25
7810	12.5	28
7812	14.5	30
7815	17.5	30
7818	21	33
7824	27	38

The minimum and maximum input voltages for the 78XX family of regulators are shown in Table 6.1. We use Table 6.1 to select the turns ratio, n, for a 78XX regulator. As a design guide, we will take the average of V_{max} and V_{min} of the particular IC regulator (see Table 6.1) to calculate a. For example, using a 7805 regulator, we obtain

$$\frac{V_{max} + V_{min}}{2} = \frac{7 + 25}{2} = 16$$

The center tap provides division by 2, so the peak voltage out of the rectifier with 115 V applied is $115\sqrt{2}/2n = 16$. Therefore, $a = 5$. This is a conservative method of selecting the transformer ratio.

The filter capacitor, C_F, is chosen to maintain the voltage input range to the regulator as specified in Table 6.1.

The output capacitor, C_L, aids in isolating the effect of the transients that may appear on the regulated supply line. C_L should be a high-quality tantalum capacitor with a capacitance of 1.0 μF. It should be connected close to the 78XX regulator using short leads in order to improve the stability performance.

This family of regulators can also be used for battery-powered systems. Figure 6.22(c) shows a battery-powered application. The value of C_F is chosen in the same manner as for the standard filter.

The 79XX series regulator is identical to the 78XX series except that it provides negative regulated voltages instead of positive ones.

Example 6.6

Design an IC regulator to generate a 12-V output into a load whose current varies from 100 to 500 mA. The input is 115 V at 60 Hz.

SOLUTION We use the circuit of Figure 6.22(b) with a 7812 regulator. The center-tapped transformer and full-wave rectifier must produce a minimum voltage of at least 14.5 V and a maximum voltage of no more than 30 V. This information is obtained from Table 6.1. The input peak voltage is $115\sqrt{2}$ or 163 V. The center-tapped secondary divides this by 2 to yield 81.5 V. Let us choose the midpoint between 14.5 and 30, or 22.25 V, to select the transformer ratio. This yields a transformer ratio of 81.5/22.25 or 3.66.

Since $V_{S\min} = 14.5$ V (from Table 6.1) and $V_{S\max} = 22.3$ V, we have

$$C_F = \frac{V_{S\max}}{\Delta V f_p R_L} = 822\ \mu\text{F}$$

We have used the fact that

$$\Delta V = 22.3 - 14.5 = 7.8\text{ V}$$

and

$$R_L(\text{worst case}) = 29\ \Omega$$

The value of C_L is determined by the types of variations that occur in the load. A 1.0-μF high-quality tantalum capacitor will usually suffice for this application.

Drill Problems

D6.7 Design a 7.3-V regulated power supply for an 800 ±100 mA load. The input is 110 V rms at 60 Hz and a 4:1 center-tapped transformer is used. $V_Z = 8$ V and $\Delta V = 20\%$. The transistor has $V_{BE} = 0.7$ V and $\beta = 100$. Assume a full-wave rectifier and $R_Z = 5\ \Omega$. Find R_i, C_F, and the percentage of regulation.

Ans: $R_i = 777\ \Omega$; $C_F = 2380\ \mu$F; % regulation = 0.439%

D6.8 A 7815 IC is used as the regulator with a center-tapped transformer to full-wave rectify a 200-V peak-to-peak input voltage. What turns ratio and filter capacitor size are required to provide an output of 15 V at 400 mA?

Ans: $a = 4.20$; $C_F = 840\ \mu$F

6.5.3 Power Supply Using Three-Terminal Adjustable Regulator

The LM317 is an adjustable three-terminal positive-voltage regulator capable of supplying more than 1.5 A over an output range of 1.2 to 37 V. It is easy to use and requires only two external resistors to set the output voltage. Both line regulation and load regulation are better than in standard fixed-voltage regula-

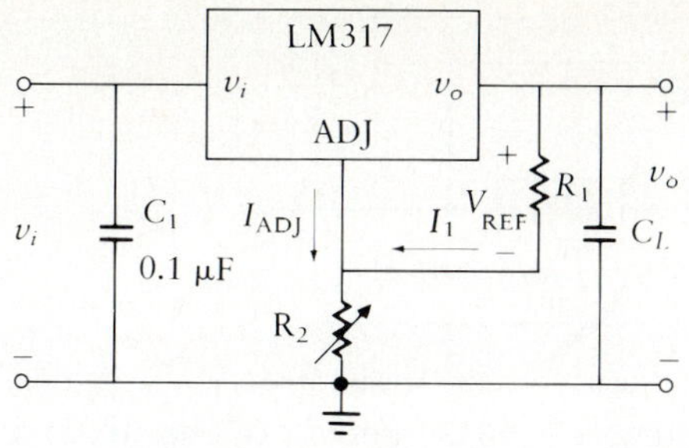

Figure 6.23
The LM317 regulator.

tors. It is packaged in a standard transistor package and provides overload protection, including current-limiting and thermal-overload protection. Figure 6.23 shows a connection diagram for the LM317. The capacitor, C_L, is optional, and when it is included the transient response improves. Output capacitors in the range of 1 to 10 μF (tantalum electrolytic) are used to provide improved output impedance and rejection of transients. The capacitor C_1 is needed if the device is physically located far from filter capacitors.

In operation, the LM317 has a precision internal voltage reference that develops a nominal 1.25-V voltage, V_{REF}, between the output and the adjustment terminal. The reference voltage appears across the *program resistor,* R_1. Since V_{REF} is constant, there is a constant current, I_1, through the program resistor. The output voltage is then given by

$$
\begin{aligned}
V_o &= V_{REF} + (I_1 + I_{ADJ})R_2 \\
&= V_{REF} + \frac{V_{REF}R_2}{R_1} + I_{ADJ}R_2 \\
&= V_{REF}\left(1 + \frac{R_2}{R_1}\right) + I_{ADJ}R_2
\end{aligned}
$$

Note that if V_{REF}, R_1, R_2, and I_{ADJ} are all constants, V_o is also a constant.

An input-bypass capacitor is usually used. A 0.1-μF disk or 1-μF solid tantalum capacitor is suitable for input bypassing for most applications. This capacitor shorts out high-frequency variations that occur in adjoining circuitry.

6.5.4 Higher-Current Regulator

Most IC regulators are limited to an output current of 1.5 A. This is because of the large power dissipated in the internal pass transistor.

The circuit of Figure 6.24 allows us to increase the output current to 5 A while still preserving the thermal shutdown and the short-circuit protection of the IC. The concept here is that an additional power transistor, Q_1, passes $0.8i_L$ while the regulator carries only $0.2i_L$. The inclusion of the safety features to the external pass transistor, Q_1, is based on the current sharing provided by

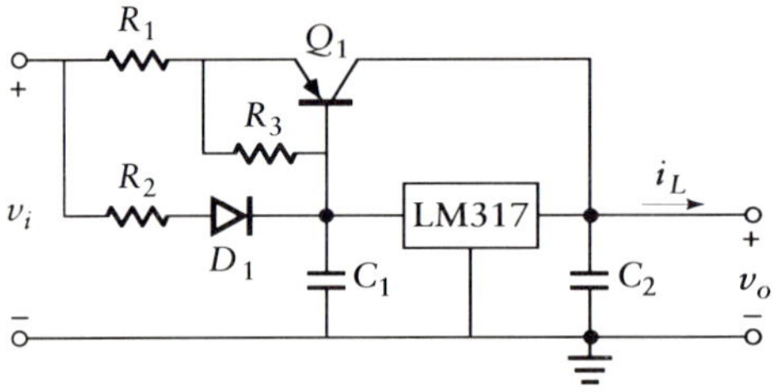

Figure 6.24 Circuit to increase output current.

R_1, R_2, and D_1. Since V_{BE} of Q_1 and V_γ of the diode are equal by design, the voltages across R_1 and R_2 are equal. Hence the current through R_1 will be four times the current through R_2 (since $R_2 = 4R_1$). The current through Q_1 is four times the current through the LM317. Hence the thermal shutdown and the short-circuit protection are provided to Q_1. The heat sink for Q_1 must, of course, have four times the capacity of the heat sink for the LM317.

6.6 SWITCHING REGULATORS

Discrete or integrated circuit-type series regulators were discussed in the previous sections. Although these series or power regulators are useful, they suffer from several limitations. The most serious difficulty lies in the area of the power generated within the regulator. These series regulators rely on the variable voltage dropped across the series resistor, R_i. The input voltage must be greater than the regulated output voltage. The difference is dropped across an internal component. The greater the difference, the greater the power dissipated within the regulator.

Because of the inefficiency of the series regulator, power supplies that generate high current output (in excess of 2 A) rely on a switching regulator. The block diagram of a switching regulator is shown in Figure 6.25. This "feedback" system compares a reference voltage, V_{REF}, with the regulated output voltage, V_{REG}. V_{REF} is obtained from a low-current output series regulator. Since little current (perhaps 20 mA) is required from this reference series regulator,

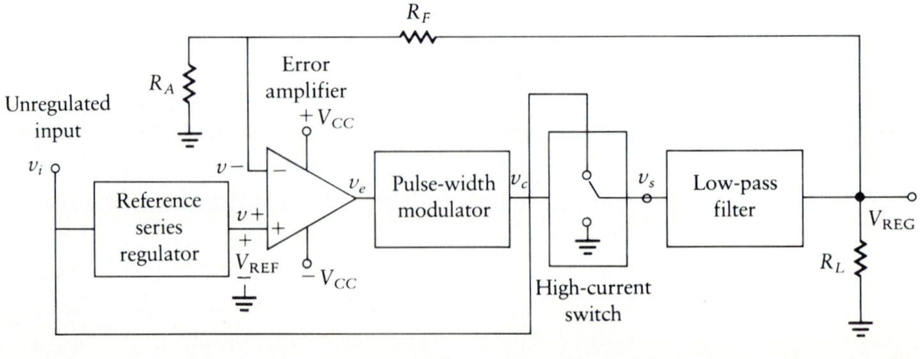

Figure 6.25 Switching regulator.

Figure 6.26
Low-pass filter for switching regulator.

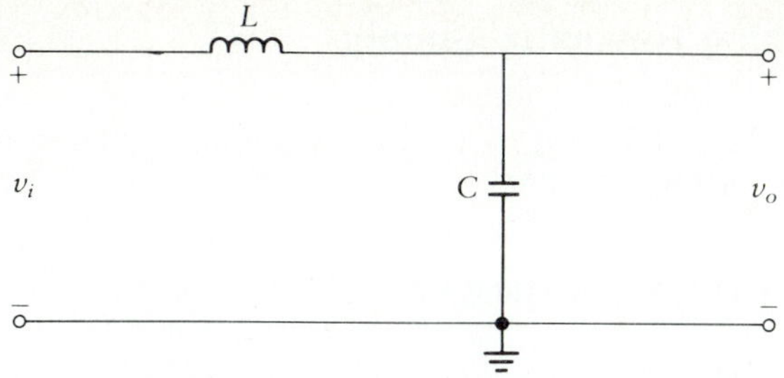

it dissipates little internal power and provides an accurate reference voltage. The regulated output voltage, V_{REG}, is compared to a fraction, $R_A/(R_A + R_F)$, of V_{REF}. The error amplifier is an operational amplifier, which is discussed more completely in Chapters 8 and 9. For our immediate purposes, the output of the error amplifier, v_e, is applied to a pulse-width modulator. The output of the pulse-width modulator, v_c, is used to control the high-current switch. This control voltage, v_c, is a square wave of period T whose duty cycle, σ, is given by

$$\sigma = kv \tag{6.28}$$

The high-current switch alternately supplies either unregulated voltage or zero volts to the filter. Hence, the voltage at the filter input, v_s, is a square wave with a period T and duty cycle, σ (just as the control voltage, v_c), but its low value is zero and its high value is v_i. The voltage, v_s, is passed through the low-pass filter to reduce the fundamental frequency, $2\pi/T$, and higher harmonics in V_{REG}. The low-pass filter is shown in Figure 6.26, where $\sqrt{LC}$ must be much larger than $T/2\pi$. The value of V_{REG} is then a constant and is equal to the average value of v_s.

An analysis of this operational amplifier circuit yields the following equations:

$$v_- = \frac{V_{REG} R_A}{R_A + R_F}$$

and

$$v_+ = V_{REF}$$

v_+ must equal v_- for an operational amplifier (see Chapter 8). Therefore,

$$V_{REG} = V_{REF}\left(1 + \frac{R_F}{R_A}\right) \tag{6.29}$$

Hence, assuming good filtering, the regulated voltage output depends on the accuracy of V_{REF} and the resistor ratio, R_F/R_A. The output, V_{REG}, does not vary as a function of either input voltage, v_i, or load current.

Example 6.7 Switching Regulator Supply

Select values of R_F and R_A for the switching regulator of Figure 6.25. V_{REG} is to be 12 V and V_{REF} is 5 V.

SOLUTION We use equation (6.29) and solve for the ratio R_F/R_A as follows:

$$\frac{R_F}{R_A} = \frac{V_{REG}}{V_{REF}} - 1 = 1.4$$

Now that the ratio is known, we can simply choose a value for R_A and solve for R_F. If, for example, R_A is 5 kΩ, then R_F would be equal to 14 kΩ.

We must be certain that $v_i > V_{REG}$, since V_{REG} is the average value of a chopped signal whose maximum value is v_i.

Some portions of the circuit of Figure 6.25 are consolidated onto one IC chip. These include the reference series regulator, the error amplifier, and the pulse-width modulator. As a result, the design of a switching regulator concentrates on the output circuitry, especially the high-current switch and the low-pass filter.

6.6.1 Efficiency of Switching Regulators

The output current comes directly from the unregulated input voltage, v_i, through the high-current switch and the filter inductor, L. Thus, if we use a transistor switch with low "ON" voltage drop and a low-loss inductor, conversion efficiency can be high (greater than 90%). The significant saving in this regulator is based on switching, or *modulating,* the input voltage with the high-current switch. In contrast, other regulators dissipate power across a series resistor or a pass transistor.

PROBLEMS

6.1 Determine the output power of an inductively coupled amplifier as shown in Figure 6.6(a). Let $V_{CC} = 12$ V, $V_{BE} = 0.7$ V, $R_E = 100$ Ω, $R_L = 1$ kΩ, and $\beta = 60$. Also find R_1 and R_2, the power provided by the power supply, the power dissipated in the transistor, and the maximum output voltage swing.

6.2 An inductively coupled amplifier is designed for a current gain of $A_i = -15$. Determine the output power when $R_L = 2$ kΩ, $V_{CC} = 12$ V,

$\beta = 200$, and $V_{BE} = 0.7$ V. Also determine the power supplied by the power supply, power rating required for the transistor, R_1, R_2, and R_E. Use the circuit of Figure 6.6(a), but delete the emitter bypass capacitor.

6.3 Use the circuit of Figure 6.6(a) to obtain a current gain of -60. If the power supply is 15 V, $R_L = 1$ kΩ, $V_{BE} = 0.7$ V, and $\beta = 100$, can the circuit provide the required gain when $R_E = 100$ Ω? Find the values of the circuit elements, power required from the power source, power rating of the transistor, R_1, R_2, and maximum power dissipated in the load resistor.

6.4 Design a transformer-coupled EF follower power amplifier to drive a 10-Ω load with $A_i = 100$ if $V_{CC} = 12$ V, $V_{BE} = 0.7$ V, the step-down transformer turns ratio is 10:1, and $\beta = 50$. Determine R_1, R_2, the power rating of the transistor, and the power dissipated in the load. Refer to the circuit of Figure 6.7(a).

6.5 What changes are required in the amplifier parameters of Problem 6.4 if the primary resistance of the transformer is 200 Ω?

6.6 A Class-A transformer-coupled EF power amplifier must deliver an output of $\frac{1}{2}$ W to an 8-Ω speaker. What transformer ratio is needed to provide this power if $V_{CC} = 18$ V? The transistor has $\beta = 100$ and $V_{BE} = 0.7$ V. Assume zero resistance in the transformer. What transistor power rating is needed?

6.7 Design a complementary-symmetry Class-B amplifier to drive a 12-Ω load. Refer to the circuit shown in Figure 6.9. Assume $V_{CC} = 18$ V, $V_{BE} = 0.7$ V, and $\beta = 60$. Calculate the total power dissipated in the load, the input resistance, and the power rating of the transistor. Select values of C_1, R_1 and R_2 for a 20 Hz lower corner frequency and a current gain of $A_i = 20$.

6.8 Design a complementary-symmetry Class-B amplifier to drive an 8-Ω load using $\beta = 60$, $V_{BE} = \pm 0.7$ V, $V_{CC} = 12$ V, and a lower corner frequency of 100 Hz. Use the circuit of Figure 6.9 and a current gain of $A_i = 20$.

a. Find all quiescent voltages and currents.
b. Find the maximum power delivered to the load.
c. Select values for R_1, R_2, and C_1.
d. Determine R_{in}.

6.9 Design a complementary-symmetry Class-B amplifier to drive an 8-Ω load using $\beta = 80$, $V_{BE} = \pm 0.7$ V, $V_{CC} = 16$ V, $f_L = 20$ Hz, and $R_f = 100$ Ω. Use the circuit of Figure 6.12 and 8 V peak-to-peak output.

a. Find all quiescent voltages and currents.
b. Determine R_2 and C_1.
c. Find the maximum power delivered to the load.
d. Determine R_{in} and the power rating of the transistors.

6.10 Design a complementary-symmetry Class-B diode-compensated amplifier to drive an 8-Ω load. Assume $\beta = 80$, a peak-to-peak output voltage of 6 V, $R_f = 10\ \Omega$, $V_{CC} = 12$ V, and $V_{BE} = \pm 0.7$ V. Let C_1 and $C_2 \to \infty$. Use the circuit of Figure 6.12.
a. Find all quiescent voltages and currents at midfrequency.
b. Find the maximum power delivered to the load.
c. Determine the values of R_2 and R_{in}.

6.11 Design a 5-W complementary-symmetry Class-B diode-compensated amplifier to drive an 8-Ω load. Assume $\beta = 80$, $R_f = 4\ \Omega$, $V_{CC} = 24$ V, $V_{BE} = \pm 0.7$ V, and $f_L = 50$ Hz. Use the circuit of Figure 6.12.
a. Find all quiescent voltages and currents at the midfrequency range.
b. Find the maximum power delivered by the power supply.
c. Determine R_2, C_1, C_2, A_i, and R_{in}.

6.12 Design a 8-W complementary-symmetry Class-B diode-compensated amplifier to drive an 8-Ω load. Assume $\beta = 100$, $R_f = 20\ \Omega$, $V_{CC} = 32$ V, $V_{BE} = \pm 0.7$ V, and $f_L = 20$ Hz. Use the circuit of Figure 6.12.
a. Find all quiescent voltages and currents at the midfrequency range.
b. Find the maximum power delivered by the power supply.
c. Find the power rating of the transistors.
d. Determine R_2, R_{in}, A_i, C_1, and C_2.

6.13 Design a 80-W complementary-symmetry Class-B audio amplifier to drive an 8-Ω speaker. Use $\beta = 200$, $V_{BE} = \pm 0.7$ V, $V_{CC} = 80$ V, and $f_L = 100$ Hz. Use the circuit of Figure 6.12. The diodes have forward resistance of 50 Ω and use a capacitor with the 3-dB point at 100 Hz.
a. Determine the power rating of the transistors at low frequency.
b. Determine R_2, R_{in}, C_1, and C_2.
c. Determine the current gain.

6.14 Design an EF Class-A amplifier using a Darlington transistor pair that has a combined β of 8000 and $V_{BE} = 1.4$ V to drive an 8-Ω load with $R_{\text{in}} = 5\ \text{k}\Omega$. Determine all the component values for $f_L = 20$ Hz, and find A_i and P_o when $V_{CC} = 24$ V. Use the circuit of Figure 6.17.

6.15 Design an EF Class-A amplifier using a Darlington transistor pair that has a combined β of 6000 and $V_{BE} = 1.4$ V to drive a 10-Ω load. The system requires $A_i = 500$ for $f_L = 500$ Hz. Use the circuit of Figure 6.17 with $V_{CC} = 12$ V. Determine all the component values for the amplifier and P_o.

6.16 Repeat Problem 6.13 but use a Darlington transistor pair with $\beta = 6000$ and $V_{BE} = 1.4$ V.

6.17 Design a Darlington pair CE amplifier as shown in Figure 6.19 to provide an A_i of -4000 to a 1-kΩ load. Design the amplifier for maximum output voltage swing and determine the value of the required maximum input voltage. $\beta_1 = 100$, $\beta_2 = 200$, V_{BE} for each transistor is 0.6 V, and $V_{CC} = 12$ V.

6.18 Design a dc bias stable Darlington pair CE amplifier to provide high input impedance with $A_v = -20$ to a 5-kΩ load. Determine all the values of the resistors, R_{in}, A_i, and maximum output voltage swing. Assume a combined $\beta = 30{,}000$, that V_{BE} for each transistor is 0.6 V, and that $V_{CC} = 24$ V.

6.19 Design a CE amplifier using Darlington transistor pairs for maximum output voltage swing. The requirements are $A_v = -50$, $A_i = -80$, $R_L = 5$ kΩ, and $V_{CC} = 15$ V. Assume a combined $\beta = 25{,}000$ and that the overall $V_{BE} = 1.2$ V. Determine:
a. R_E, R_1, R_2, and R_{in}.
b. Undistorted output power.
c. Maximum undistorted output voltage swing.
d. Input dc power required.

6.20 A power supply is to provide an amplifier with 300 ±50 mA of regulated power. Design a power source (see Figure 6.20) that will provide the power to the amplifier if the transformer has a turns ratio of 8:1 to each secondary and a 12-V zener diode is used. Assume the input voltage is 115 V at 60 Hz and the forward resistance of the diodes is zero. Use $\beta = 50$ and $V_{BE} = 0.7$ V. Select C_F so that $\Delta V = 10\%$. Determine C_1, R_i, and the power rating of the diodes and of the power transistor.

6.21 A power supply is to provide an audio amplifier with a 500 ±75 mA regulated current. Design a power source (see Figure 6.20) that uses a 20-V zener and a 5:1 transformer ratio to each secondary. The input voltage is 110 V, 60 Hz, $\beta = 100$, $V_{BE} = 0.7$ V, and $\Delta V = 10\%$.
a. Determine C_1 and R_i assuming no diode forward resistance.
b. Determine the power rating of the zener diode.
c. Determine the voltage variation at the load if $R_Z = 2\ \Omega$.

6.22 Design a regulated power supply to obtain an 11.3-V output using a power transistor having $V_{BE} = 0.7$ V and $\beta = 100$. The power supply will have an input of 120 V rms, 60 Hz, and a 4:1 center-tapped transformer and will be required to provide a load current varying from 500 mA to 1 A. Use a 12-V, $\frac{1}{4}$-W zener.

6.23 Design a regulated power supply to obtain an output of 5.5 V at 500 ±100 mA. Use a 1N753 zener diode. What are the values of R_i, C_F, and the power rating of the transistor, which has $V_{BE} = 0.7$ V and $\beta = 100$. The transformer has a ratio of 5:1 and an input voltage of 110 V at 60 Hz.

6.24 Determine the required filter capacitor size for a regulated power supply using an MC7812 full-wave rectifier with a 3:1 center-tapped transformer and 1 A output. The input is 110 to 120 V rms at 60 Hz.

6.25 Design a 10-V regulated power supply using an MC7810 voltage regulator. Select the transformer turns ratio, a, and the capacitor values using an input of 120 V, 60 Hz and 1 A output.

6.26 Determine the values of R_F and R_A of the switching regulator shown in Figure 6.25 when $V_{REG} = 20$ V and $V_{REF} = 5$ V.

7

FREQUENCY RESPONSE CHARACTERISTICS

7.0 INTRODUCTION

On receiving your electrical engineering degree, you were employed by the XYZ Audio Corporation, a company that designs and manufactures stereo equipment. Since your supervisor knew you had a solid background in electronic design, you were assigned the task of designing an amplifier as part of a compact disc music system. You mustered all your knowledge of electronics, made a number of trade-off decisions, and chose a common-emitter amplifier. The amplifier was designed using the techniques of the earlier chapters, and in order to achieve the necessary gain using only one stage, you bypassed the emitter resistor with a capacitor.

A prototype of the system was constructed, and a disc was played as a test. The result was a disaster! The sound was far less than full and, in fact, it was determined that the amplifier was not responding well to the low frequencies in the audio signal. The output was distorted! What went wrong?

You completely forgot that capacitors (and inductors) are frequency-dependent elements. As soon as they are present in a circuit, the response becomes a function of frequency. You failed to perform a frequency analysis of your amplifier! The current chapter teaches you how to correct this design error.

Our earlier work treated amplifiers as though their behavior was independent of the frequency of the input signal. This is not true in the practical case. The current chapter concentrates on the frequency-dependent aspects of various amplifiers.

Capacitance within an amplifier, which causes the response to be frequency dependent, may be present by design or may be present unintentionally. *Coupling* and *bypass* capacitors are often designed into the system, the desired condition being a short circuit for all signal frequencies. That is, these capacitors are intended to be open circuit for the dc bias voltages but to allow all signal components to pass through them without attenuation. Capacitors do not suddenly change from a short-circuit condition to an open circuit when the frequency reaches zero. Instead, they approach an open circuit as the frequency gets smaller, so the performance is degraded as the input frequencies decrease.

The second type of capacitance found in an amplifier is present unintentionally. Whenever two conductors are separated by any nonconducting material, capacitance exists. Internal capacitance exists within semiconductors, between contacts, and between the conductors of the circuit configuration. As frequencies increase, these capacitances tend to "short out" the signal and thereby decrease the gain.

Figure 7.1 illustrates a typical amplitude frequency response for an RC-coupled amplifier. Note that the maximum gain occurs for a *midrange* of frequencies and that the gain decreases at both low and high frequencies. The low and high limits of the midrange, f_L and f_H, are known as the *corner frequencies.* They are defined as the points at which the gain drops to 0.707 of its midrange value. This figure is the square root of 0.5 and therefore represents the frequency at which the output power decreases to one-half of its midrange value. It is known as the *half-power point.*

Amplifiers can be analyzed as linear systems—the frequency response can be described by a complex function yielding the magnitude and phase shift response for each input frequency. In our analysis, we concentrate on a simple method, known as the *Bode plot,* which permits drawing the graphs almost by inspection. If you have not previously been exposed to this method of developing asymptotic response curves, we refer you to Appendix B, where the theory of Bode plots is reviewed.

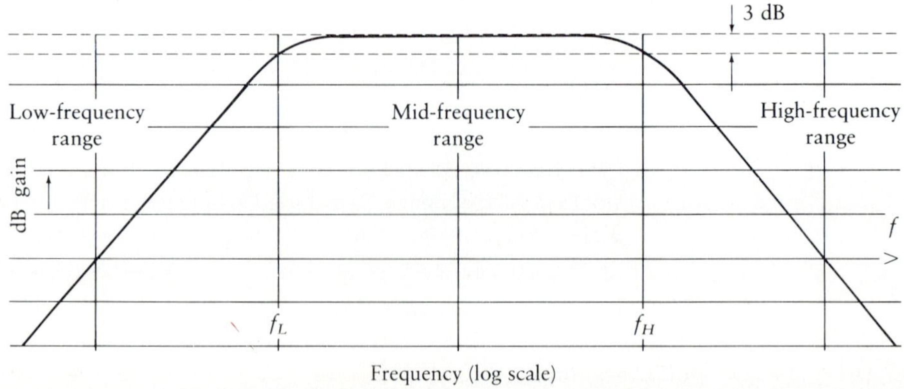

Figure 7.1
Gain versus frequency for RC-coupled amplifier.

7.1 LOW-FREQUENCY RESPONSE—BJT AMPLIFIER

The BJT amplifier often contains capacitors in order to couple the output to the load and to bypass the emitter resistor. These capacitors are intended to be open circuits for dc bias conditions but short circuits for the signal frequencies of interest. In practice, capacitors deviate from the short-circuit condition as the frequency decreases. This causes a degradation in response—the amplifier is now frequency dependent. In this section, we study the low-frequency response of the amplifier. We examine the effects of the coupling and bypass capacitors separately.

7.1.1 Low-Frequency Response—CE Amplifier

Figure 7.2(a) shows a CE amplifier with coupling capacitors C_1 and C_2. The equivalent circuit is shown in Figure 7.2(b). As the frequency decreases toward zero, both of the coupling capacitors approach open-circuit conditions and the input signal is attenuated. We shall find the voltage and current gains. These expressions include the effects of the capacitors. The input impedance is given by (see Figure 7.2(b))

$$Z_{\text{in}} = R_i + R_{\text{in}} + \frac{1}{sC_2} = R_i + [R_B \parallel (h_{ie} + \beta R_E)] + \frac{1}{sC_2}$$

where

$$R_B = R_1 \parallel R_2$$

The input current is the ratio of voltage to impedance,

$$I_{\text{in}} = \frac{V_i}{R_{\text{in}} + R_i + 1/sC_2}$$

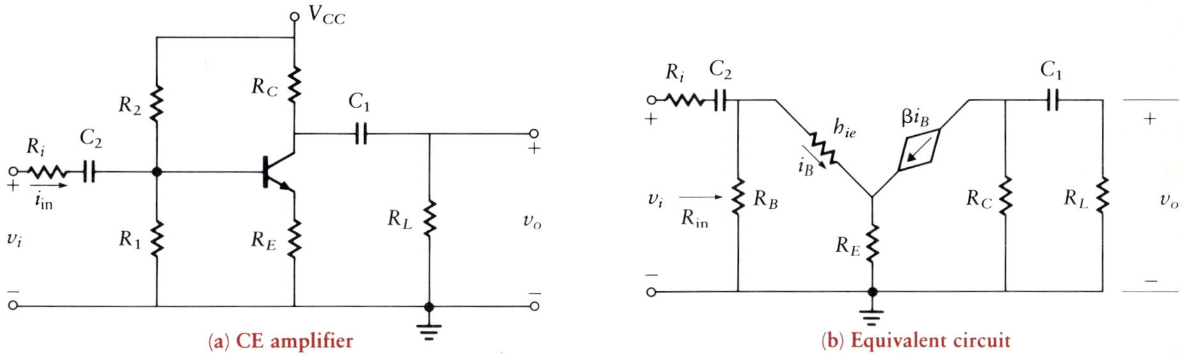

Figure 7.2
CE amplifier.

and the base current is

$$I_B = \frac{V_i R_B}{R_B + h_{ie} + \beta R_E} \frac{sC_2}{sC_2(R_{\text{in}} + R_i) + 1}$$

We can rewrite this expression by noting that

$$R_{\text{in}} = R_B \parallel (h_{ie} + \beta R_E) = \frac{R_B(h_{ie} + \beta R_E)}{R_B + h_{ie} + \beta R_E}$$

We then get

$$I_B = \frac{V_i}{h_{ie} + \beta R_E} \frac{R_{\text{in}} sC_2}{sC_2(R_{\text{in}} + R_i) + 1} \tag{7.1}$$

We define τ_2 as the time constant of the base resistor-capacitor combination.

$$\tau_2 = (R_{\text{in}} + R_i)C_2 \tag{7.2}$$

The base current is given by

$$I_B = \frac{V_i}{h_{ie} + \beta R_E} \frac{R_{\text{in}} C_2 s}{\tau_2 s + 1} = \frac{V_i}{h_{ie} + \beta R_E} \frac{R_{\text{in}} C_2}{\tau_2} \frac{s}{s + 1/\tau_2} \tag{7.3}$$

The output voltage is found from the circuit of Figure 7.2(b) using current division at the output.

$$V_o = \frac{-\beta I_B R_C R_L}{R_C + R_L + 1/sC_1} = -\beta I_B (R_C \parallel R_L) \frac{s}{s + 1/\tau_1} \tag{7.4}$$

where τ_1 is the time constant associated with the output RC loop. That is,

$$\tau_1 = C_1(R_C + R_L) \tag{7.5}$$

The voltage gain can now be found from equations (7.1), (7.2), and (7.4).

$$A_v = \frac{V_o}{V_i} = \frac{-R_C \parallel R_L}{h_{ib} + R_E} \frac{R_{\text{in}} C_2 s^2}{\tau_2 (s + 1/\tau_1)(s + 1/\tau_2)} \tag{7.6}$$

The current gain is found from two current dividers. In the input circuit, refer to Figure 7.2(b) to find

$$I_B = \frac{R_B I_{\text{in}}}{R_B + h_{ie} + \beta R_E}$$

and at the output circuit,

$$I_o = -\beta I_B \frac{R_C}{R_C + R_L + 1/sC_1}$$

$$= -\frac{\beta R_B I_B}{R_B + h_{ie} + \beta R_E} \frac{R_C}{R_C + R_L} \frac{s}{s + 1/[C_1(R_1 + R_2)]}$$

and the current gain is

$$A_i = -\frac{R_B}{R_B/\beta + h_{ib} + R_E} \frac{s}{s + 1/\tau_1} \frac{R_C}{R_C + R_L} \tag{7.7}$$

We plot these results for a particular example following a discussion of one simplification. Equations (7.6) and (7.7) can be simplified by defining a *midrange gain* as the value of the gain at frequencies to the right of the corner frequency. This is equivalent to taking the limit of the expressions in the two equations as s approaches infinity. We substitute for τ_1 and τ_2 and find, from equation (7.6),

$$A_{vm} = A_v|_{s\to\infty} = \frac{-(R_C \parallel R_L)R_{\text{in}}C_2}{(R_E + h_{ib})\tau_2} = \frac{-(R_C \parallel R_L)R_{\text{in}}}{(R_E + h_{ib})(R_{\text{in}} + R_i)} \tag{7.8a}$$

Similarly, equation (7.7) becomes

$$A_{im} = A_i|_{s\to\infty} = \frac{-R_B}{R_B/\beta + h_{ib} + R_E} \frac{R_C}{R_C + R_L} \tag{7.8b}$$

We can now use these midrange gains to normalize the expressions of equations (7.6) and (7.7). That is, we divide by the midrange values to obtain

$$\frac{A_v}{A_{vm}} = \frac{s^2}{(s + 1/\tau_1)(s + 1/\tau_2)} \tag{7.9a}$$

$$\frac{A_i}{A_{im}} = \frac{s}{s + 1/\tau_1} \tag{7.9b}$$

Suppose we wish to draw the Bode plot of the current gain in equation (7.9b). First note that the expression has a first-order zero at the origin and a first-order pole whose corner frequency is at $1/\tau_1$. For frequencies below the corner frequency, the zero causes the plot to rise at +20 dB/decade. At the corner frequency, $1/\tau_1$, we add a negative-going line of slope −20 dB/decade, so the sum is constant.

Example 7.1

Draw a Bode plot of the low-frequency voltage gain for the amplifier shown in Figure 7.3. Assume zero source resistance, R_i.

SOLUTION We need to calculate the midrange gains and the corner frequencies. In order to find the corner frequencies, we calculate the two time constants, τ_1 and τ_2.

$$V_{BB} = \frac{2200(10)}{2.2 \times 10^4} = 0.991 \text{ V}$$

$$R_B = 2.2 \text{ k}\Omega \parallel 20 \text{ k}\Omega = 1.98 \text{ k}\Omega$$

$$I_{CQ} = \frac{0.991 - 0.7}{1980/200 + 100} = 2.65 \text{ mA}$$

$$h_{iB} = \frac{26 \text{ mV}}{2.65 \text{ mA}} = 9.8 \ \Omega$$

$$R_{\text{in}} = 1.98 \text{ k}\Omega \parallel [(9.8 + 100)200] \ \Omega = 1.82 \text{ k}\Omega$$

$$\tau_1 = (5 \times 10^{-6})(2 \text{ k}\Omega) = 0.01 \text{ s} \quad \text{so } \omega_1 = 100 \text{ rad/s}$$

$$\tau_2 = (50 \times 10^{-6})(1.82 \text{ k}\Omega) = 0.091 \text{ s} \quad \text{and } \omega_2 = 11 \text{ rad/s}$$

The midrange gains are given by equations (7.8a) and (7.8b).

$$A_{vm} = \frac{500}{100 + 9.8} = -4.55$$

$$A_{im} = \frac{-1980}{9.9 + 9.8 + 100} \frac{1000}{2000} = -8.27$$

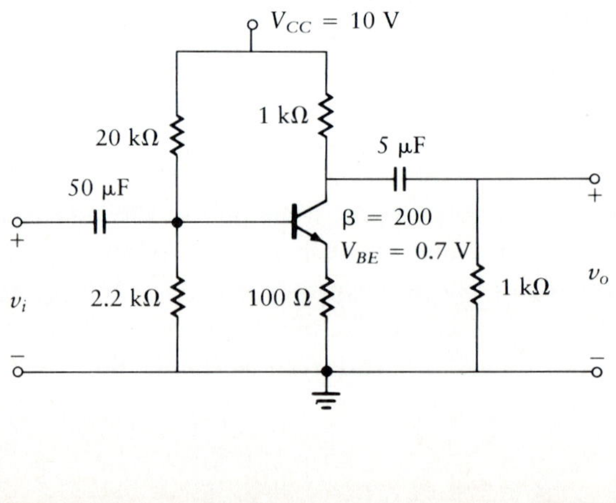

Figure 7.3
Amplifier for Example 7.1.

Figure 7.4 Bode plot for Example 7.1.

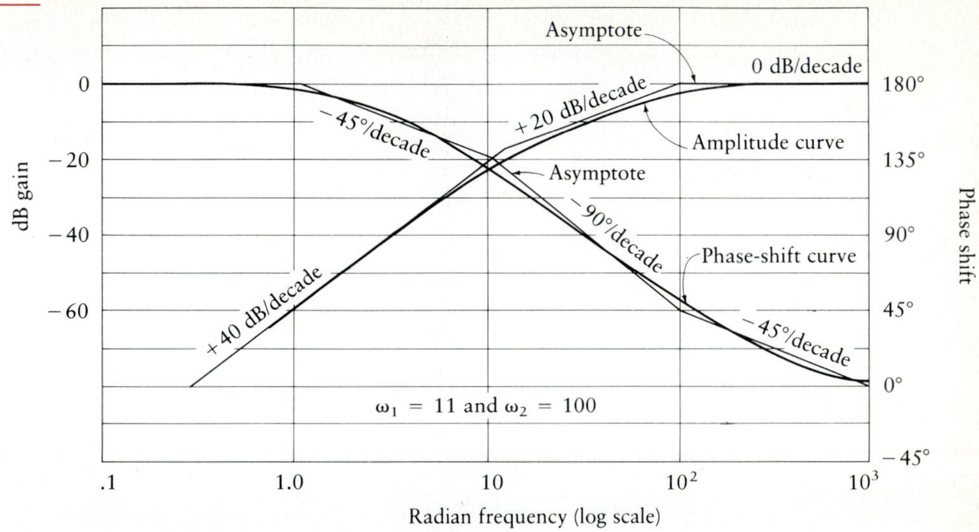

The Bode plots of the normalized gains are drawn from the following equations:

$$\frac{A_v}{A_{vm}} = 0.91 \times 10^{-3}\frac{s^2}{(0.01s + 1)(0.091s + 1)}$$

$$\frac{A_i}{A_{vm}} = 10^{-2}\frac{s}{0.01s + 1}$$

The voltage amplitude plot starts at a slope of +40 dB/decade. At the first corner frequency, $\omega = 11$, the slope changes to 20 dB/decade. The actual curve deviates from the straight-line asymptotic approximation by 3 dB at the corner frequency. At the second corner frequency, $\omega = 100$, the slope changes to 0 dB/decade. To the right of this point, the amplitude remains at 0 dB. This is shown as Figure 7.4.

The phase curve starts at 180°. The corner frequencies are located at $\omega = 11$ and $\omega = 100$ rad/s. Starting at one-tenth of the first corner frequency, or 1.1 rad/s, the curve follows a −45°/decade asymptote until one-tenth of the second corner frequency, or 10 rad/s. At $\omega = 100$ rad/s, the effect of the first pole disappears and the asymptotic curve has a slope of −45°/decade until a frequency of 1000. Thereafter, the slope of the phase curve is 0°/decade.

The corrections to the asymptotic curves for both phase and gain are shown in Figure 7.4.

The current amplitude plot starts at a slope of 20 dB/decade. It continues at this slope until the corner frequency at 100 rad/s. At this point the slope decreases to 0 dB/decade and the amplitude remains constant at 0 dB. The phase shift curve starts at $-90°$ at low frequency. At 10 rad/s, the slope becomes $-45°$/decade until 100 rad/s, at which point the phase shift goes to zero.

7.1.2 Design for a Given Frequency Characteristic

Suppose we wish to design an amplifier with a specific low-frequency cutoff. That is, the gain falls off at low frequency, and the point at which the response falls by 3 dB below the midrange value is specified. If there were only one corner frequency, we would set that corner frequency equal to the given frequency.

Since we usually have two corner frequencies that interact with each other, the design is more complex. We present two design approaches. The specific approach used in a particular design depends on the degree of separation of the two corner frequencies.

Approach 1 We let one pole reflect the total 3-dB drop and set the other pole one decade lower in frequency. Thus, as frequency is reduced toward zero, the 3-dB design requirement is achieved before the second pole takes effect. This approach usually results in two capacitors of roughly the same size, since the output load resistance is much smaller than the input resistance.

Approach 2 If the input and output resistances are approximately equal, we set the two pole corner frequencies equal. We allow each of the two poles to decrease the magnitude by 1.5 dB at the specified frequency. The total decrease is therefore 3 dB, as required by the design. The actual pole corner frequencies are below the specified (3-dB) design frequency. If we set both poles at the design corner frequency, the drop at that frequency would be 6 dB instead of the required 3 dB. We use equation (7.9a), with $1/\tau_1 = 1/\tau_2 = \omega_c$. Therefore, the normalized voltage gain is

$$\frac{A_v}{A_{vm}} = \left|\frac{s^2}{(s + \omega_c)^2}\right|$$

The 3-dB point occurs when this normalized magnitude drops to 0.707. Thus, we let $s = j\omega_o$, where ω_o is the specified 3-dB frequency,

$$\frac{\omega_o^2}{\omega_o^2 + \omega_c^2} = 0.707$$

Solving for ω_o yields

$$\omega_o = 1.55\omega_c \tag{7.10}$$

Example 7.2

A CE amplifier is designed with $R_C = R_L = 2\ \text{k}\Omega$, $R_{in} = R_B = 5\ \text{k}\Omega$, and $R_i << R_{in}$. Determine the capacitor sizes that will yield a low-frequency cutoff at 20 Hz. Verify your result using a computer simulation program.

SOLUTION We illustrate both approaches.

Approach 1 We place one pole at the given frequency and the second pole one decade below this point. Thus,

$$C_1 = \frac{1}{2\pi f(R_L + R_C)} = 2\ \mu\text{F}$$

$$C_2 = \frac{1}{2\pi(f/10)\,(R_{in} + R_i)} = 15.9\ \mu\text{F}$$

We now perform the computer simulation to verify the results of Approach 1. The result is representative of both approaches, so we will not present a separate analysis for Approach 2. Our first job is to configure the circuit from the given parameters. This involves choosing the resistor values not specified in the problem statement. We modify the bias circuitry used in Example 7.1 (see Figure 7.3) and scale the base resistor network by a factor of 2.5 to achieve the specified input resistance. We also select the 2N3904 transistor, since this is a general-purpose *npn* transistor that is in our computer library. The resulting circuit is shown in Figure 7.5(a). We now perform an ac analysis using a range of frequencies surrounding the expected break point of 20 Hz. The result is shown in Figure 7.5(b). Note that the gain reaches a midfrequency value of about 5.3 dB. The curve crosses the 3-dB point, 2.3 dB, at 20 Hz, thus verifying the design.

Approach 2 We set both corner frequencies to the same value. From equation (7.10) we have

$$f_o = 1.55 f_c$$

or

$$f_c = \frac{20}{1.55} = 12.9\ \text{Hz}$$

Then the capacitors are calculated to be

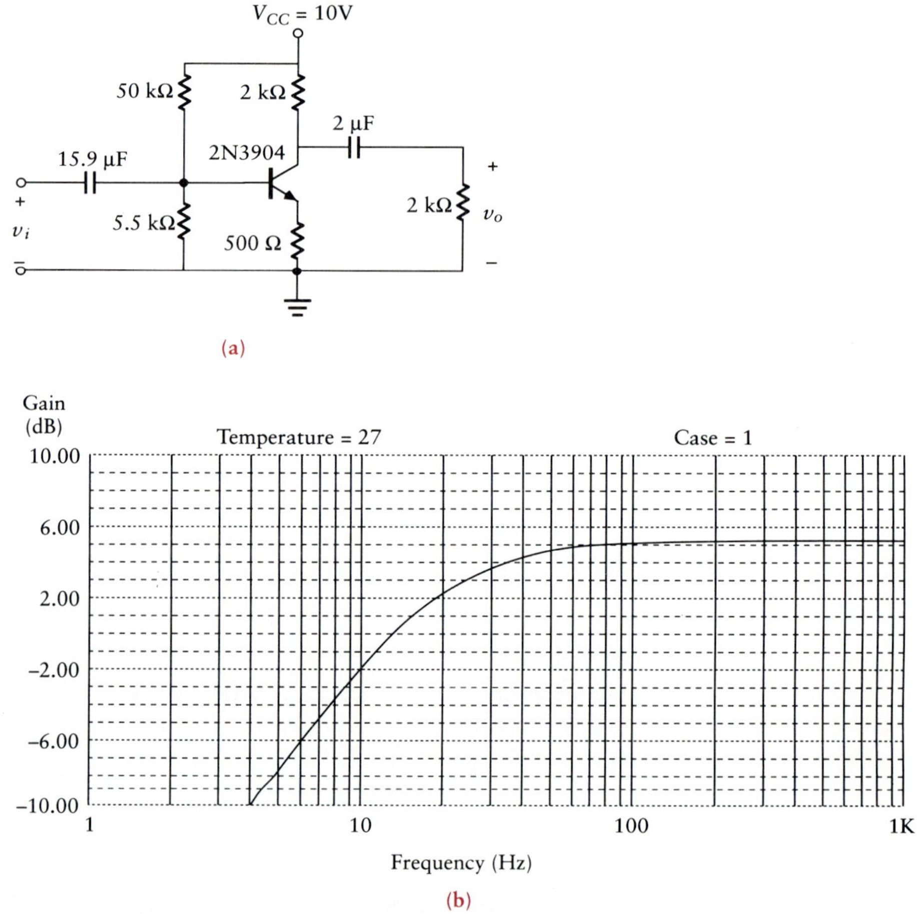

Figure 7.5
Circuit and result for Example 7.2.

$$C_1 = \frac{1}{2\pi f(R_L + R_C)} = 3.08\ \mu\text{F}$$

$$C_2 = \frac{1}{2\pi f(R_{\text{in}} + R_i)} = 2.46\ \mu\text{F}$$

Drill Problem

D7.1 Determine the low-frequency response for the amplifier of Figure 7.2 when $C_2 = 2\ \mu$F, $R_1 = 10$ kΩ, $R_2 = 90$ kΩ, $R_C = 1$ kΩ, $R_E = 200$ Ω, $C_1 = 4\ \mu$F, $R_L = 2$ kΩ, $\beta = 100$, $V_{BE} = 0.7$ V, $V_{CC} = 20$ V, and $R_i << R_{\text{in}}$.

Ans: $\tau_1 = C_1(R_L + R_C) = 0.012$ s

$\tau_2 = C_2 R_{\text{in}} = 0.0125$ s

$$f_1 = \frac{1}{2\pi\tau_1} = 13.3 \text{ Hz}$$

$$f_2 = \frac{1}{2\pi\tau_2} = 12.7 \text{ Hz}$$

The low-frequency decibel gain curve starts at a slope of 40 dB/decade. At f_1, the slope changes to 20 dB/decade until f_2 is reached. Finally, the curve is horizontal to the right of f_2.

7.1.3 Emitter-Resistor Bypass Capacitor

This section explores the addition of a bypass capacitor in parallel with the emitter-resistor for a single-stage CE amplifier. This is shown in Figure 7.6(a). Figure 7.6(b) is the equivalent circuit of this amplifier. We will let C_1 and C_2 be replaced by short circuits for this analysis (i.e., $C \to \infty$) so we can concentrate on the effects of the bypass capacitor. To evaluate the frequency-dependent effects of the bypass capacitor, we solve for the voltage gain. The base current is given by

$$I_b = \frac{V_i}{\beta(h_{ib} + R_E \parallel 1/sC_b)}$$

The output voltage is

$$V_o = -\beta I_b(R_L \parallel R_C)$$

We solve for the gain to find

$$A_v = \frac{V_o}{V_i} = \frac{-(R_L \parallel R_C)(1 + 1/sR_E C_b)}{h_{ib}(1 + 1/sC_b R_E) + R_E/sC_b R_E}$$

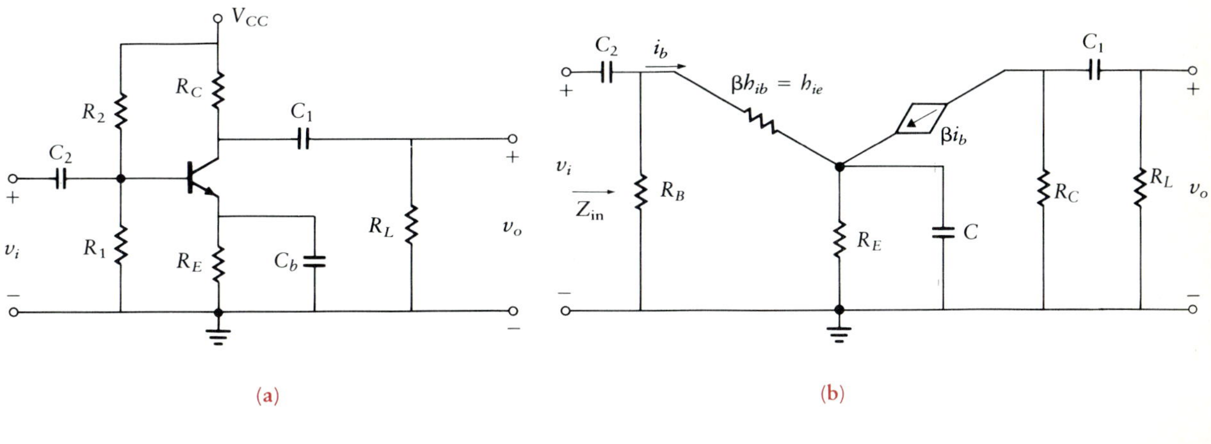

Figure 7.6
CE amplifier with bypass capacitor.

The midband gain is the value of A_v with the capacitor shorted. Thus,

$$A_{vm} = \frac{-(R_L \parallel R_C)}{h_{ib}}$$

The normalized gain is then

$$\frac{A_v}{A_{vm}} = \frac{s + 1/R_E C_b}{s + (1 + R_E/h_{ib})C_b R_E} = \frac{s + 1/\tau_a}{s + 1/\tau_b}$$

The time constants are given by

$$\tau_a = R_E C_b$$

$$\tau_b = \frac{R_E C_b}{1 + R_E/h_{ib}} = (R_E \parallel h_{ib})C_b$$

Note that τ_a is greater than τ_b. The Bode plot of the normalized gain contains two corner frequencies. The lower of these is due to the zero of the function. The plot is illustrated in Figure 7.7. It starts at low frequency with a horizontal line. At the corner frequency of the zero, $1/\tau_a$, the curve turns upward at a slope of +20 dB/decade. When the corner frequency due to the pole is reached, the curve once again turns horizontal.

Before proceeding to an analysis of the combined effects of coupling and bypass capacitors, we present a simplification that allows us to determine the time constants associated with poles. A resistor-capacitor combination develops the time constants found in electronic systems. A time constant is associated with each capacitor in the circuit. To determine the time constant associated with a particular capacitor, we need only find the equivalent (Thevenin) resistance

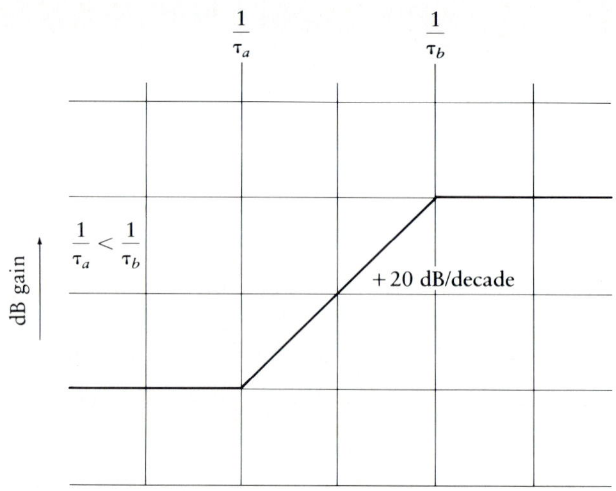

Figure 7.7
Amplitude plot for CE amplifier.

seen by that capacitor. That is, we determine the resistances through which the capacitor discharges. The product of the capacitor value and the value of the total resistance that the capacitor discharges through is the time constant for that particular capacitor.

As an example, consider the coupling capacitors of the previous section. We now find the time constants associated with the capacitors of Figure 7.2(b). For C_1, the discharge path is through $R_C + R_L$ since the impedance of the dependent current source (βI_b) is infinite. Hence the time constant is

$$\tau_1 = C_1(R_C + R_L)$$

For C_2 the discharge path is through R_i, since the ideal voltage source has zero impedance, plus R_{in}, where

$$R_{\text{in}} = R_B \parallel (h_{ie} + \beta R_E)$$

Hence, the time constant is

$$\tau_2 = C_2(R_i + R_{\text{in}})$$

These equations agree with those of the previous section. We will use a similar procedure to find the time constant for the bypass capacitor.

We again refer to Figure 7.6 and determine the discharge path for the bypass capacitor, C_b. For $C_b >> C_2$ (this is usually the case), the discharge path for C_b is

$$R_E \parallel h_{ib}$$

and the time constant is

$$\tau_b = C_b(R_E \parallel h_{ib})$$

The bypass capacitor then adds one more low-frequency roll off. The combined effect of the three capacitors, C_1, C_2, and C_b, is three corner frequencies. Since one end of C_b is at ground potential, we can use a chemical capacitor (tantalum or electrolytic) which can be quite large in value. Therefore we can place this pole one decade below the desired corner frequency.

7.1.4 Low-Frequency Response—EF Amplifier

Figure 7.8(a) shows an EF (CC) amplifier and Figure 7.8(b) shows the equivalent circuit. The low-frequency characteristic of the EF amplifier is similar to that of the CE amplifier. Again, there are two capacitors, C_1 and C_2. Hence

Figure 7.8
EF amplifier.

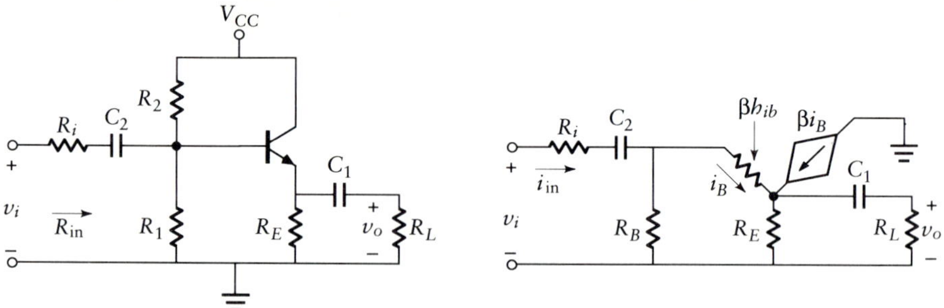

there are two time constants, which we find from the discharge paths as follows.

For C_1, the discharge path is through

$$R_L + R_E \left\| \left(h_{ib} + \frac{R_B \| R_i}{\beta} \right) \right.$$

so the time constant is

$$\tau_1 = C_1 \left[R_L + R_E \left\| \left(h_{ib} + \frac{R_B \| R_i}{\beta} \right) \right. \right] \tag{7.11a}$$

For C_2, the discharge path is through $R_i + R_{in}$, so the time constant is

$$\tau_2 = C_2(R_i + R_{in}) \tag{7.11b}$$

For an EF amplifier, R_{in} is larger than the output resistance. The input time constant is therefore larger than the output time constant, so we can allow the 3-dB loss to occur at the output circuit. That is, the break frequency due to the input circuit occurs at a lower frequency than that of the output circuit. The design limitation (i.e., lowest usable frequency) is then set by the output circuitry. This approach results in capacitor sizes that are as low as possible. We design the input capacitor-resistor combination to reflect no loss at the corner frequency. We do this by setting the input corner frequency one decade lower than the output circuit corner frequency.

7.1.5 Low-Frequency Response—CB Amplifier

Figure 7.9(a) illustrates a CB amplifier and Figure 7.9(b) shows the equivalent circuit. The design of the CB amplifier is similar to that of the EF amplifier.

We find the two time constants by finding the discharge capacitor discharge paths as follows: At C_1, the discharge path is through $R_C + R_L$, so the time constant is

Figure 7.9
CB amplifier.

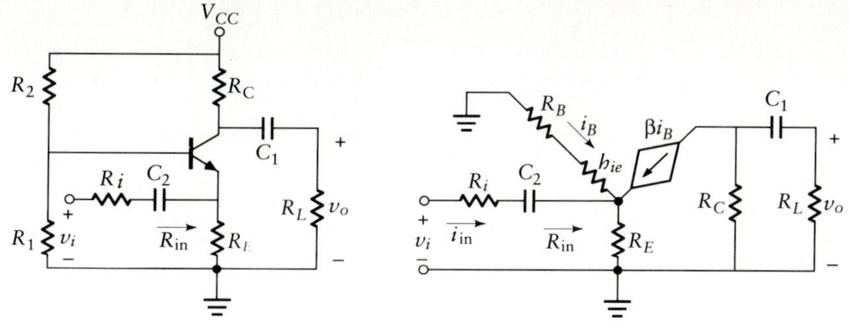

(a) CB amplifier

(b) Low-frequency equivalent circuit

$$\tau_1 = C_1(R_C + R_L) \tag{7.12a}$$

At C_2, the discharge path is through

$$R_i + R_E \left\| \left(h_{ib} + \frac{R_B}{\beta}\right)\right.$$

and the time constant is

$$\tau_2 = C_2\left[R_i + R_E \left\| \left(h_{ib} + \frac{R_B}{\beta}\right)\right.\right] \tag{7.12b}$$

Since τ_1 is usually larger than τ_2 to keep the capacitor sizes reasonable, we allow the 3-dB point to occur at the input circuit. We set the corner frequency of the output circuit to one decade below the specified lower frequency limit. When R_B is bypassed, R_B drops out of equation (7.12b).

7.2 LOW-FREQUENCY RESPONSE—FET AMPLIFIERS

7.2.1 Low-Frequency Response—CS Amplifier

Figure 7.10(a) illustrates a single-stage FET amplifier with coupling and bypass capacitors. The equivalent circuit is shown in Figure 7.10(b).

Since the input impedance of an FET amplifier is high, the input time constant is also high. This results in a low corner frequency. We can usually ignore the effects of the input coupling capacitor on the low-frequency response of the amplifier. The low-frequency performance is thus determined by the source resistor bypass capacitor, C_2, and the output coupling capacitor, C_1.

Figure 7.10
Single-stage FET amplifier.

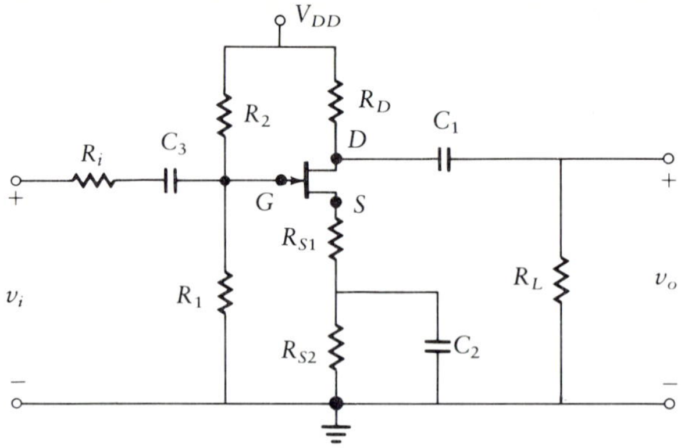

(a)

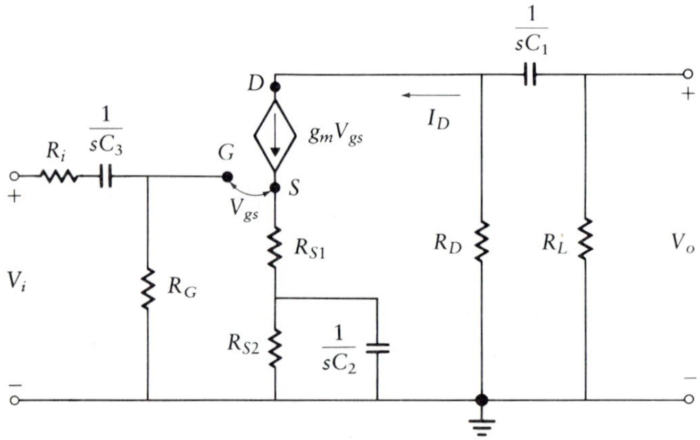

(b)

We first calculate the midrange voltage gain by assuming that the capacitors are short circuits, with the result

$$A_{vm} = \frac{-g_m(R_D \parallel R_L)}{1 + g_m R_{S1}}$$

where we have assumed $R_i << R_{\text{in}}$ to simplify the calculations. If this assumption is not correct, we can deal with the input circuit separately.

When the capacitors are not short circuits, the output voltage is found from a current-divider relationship as follows:

$$A_v = \frac{V_o}{V_i} = \frac{R_L R_D}{R_L + R_D + 1/sC_1} \frac{-V_{gs}g_m}{V_i}$$

Writing a loop equation around the gate-to-source loop yields

$$V_{gs} = V_i - g_m V_{gs}(R_{S1} + Z_{S2})$$

$$= \frac{V_i}{1 + g_m(R_{S1} + Z_{S2})}$$

where Z_{S2} is the parallel combination of R_{S2} and C_2, i.e.,

$$Z_{S2} = \frac{R_{S2}}{sC_2R_{S2} + 1}$$

Substituting the expressions for V_{gs} and Z_{S2} into the gain equation yields the final form for the voltage gain,

$$A_v = \frac{-R_LR_Dg_m}{(R_L + R_D + 1/sC_1)[1 + g_m(R_{S1} + Z_{S2})]}$$

$$= \frac{-R_LR_DsC_1}{sC_1(R_L + R_D) + 1} \frac{g_m}{1 + g_mR_{S1} + \dfrac{g_mR_{S2}}{sC_2R_{S2} + 1}}$$

This is simplified by defining three time constants,

$$\tau_1 = C_1(R_L + R_D) \tag{7.13a}$$

$$\tau_2 = R_{S2}C_2 \tag{7.13b}$$

$$\tau_3 = \frac{C_2R_{S2}}{1 + g_mR_{S2}/(1 + g_mR_{S1})} = C_2\left(R_{S2} \,\Big\|\, \left[R_{S1} + \frac{1}{g_m}\right]\right) \tag{7.13c}$$

Using these time constants, the voltage gain reduces to

$$A_v = \frac{-g_mR_LR_DC_1}{(1 + g_mR_{S1})[1 + g_mR_{S2}/(1 + g_mR_{S1})]} \frac{s(\tau_2s + 1)}{(\tau_1s + 1)(\tau_3s + 1)} \tag{7.14}$$

The normalized voltage gain is found by dividing equation (7.14) by the midrange gain

$$\frac{A_v}{A_{vm}} = \frac{\tau_1\tau_3}{\tau_2} \frac{s}{\tau_1s + 1} \frac{\tau_2s + 1}{\tau_3s + 1} \tag{7.15}$$

As before, we can simplify the derivation of poles (i.e., τ_1 and τ_3) by finding the discharge paths for each of the capacitors, as follows.

At C_1, the discharge path is through $R_D + R_L$, and the time constant is

$$\tau_1 = C_1(R_D + R_L)$$

The discharge path for C_2 is more difficult to establish due to the controlled current source in series with R_{S1}. One way to find the Thevenin equivalent resistance between two points is to impose a voltage across these points, measure the resulting current, and take the ratio. We therefore replace the capacitor, C_2, by a Thevenin voltage source, V_{TH}. With no other sources, the gate is at ground potential since there is no current in R_G. A loop equation then yields

$$V_{gs} + g_m V_{gs} R_{S1} + V_{TH} = 0$$

The current supplied by the Thevenin voltage source is

$$I_{TH} = \frac{V_{TH}}{R_{S2}} - g_m V_{gs}$$

Solving these equations yields

$$\frac{V_{TH}}{I_{TH}} = R_{TH} = \frac{R_{S2}}{1 + g_m R_{S2}/(1 + g_m R_{S1})}$$

The time constant is then given by

$$\tau_3 = \frac{C_2 R_{S2}}{1 + g_m R_{S2}/(1 + g_m R_{S1})}$$

If R_{S2} is bypassed, the amplifier has only one time constant, τ_1. The input circuit has such a large resistance $(R_i + R_{in})$, C_3 is chosen to set the corner frequency one decade below the design frequency so as not to affect the response.

Example 7.3

Given the FET amplifier of Figure 7.10 with the following parameter values: $R_{S2} = 200\ \Omega$, $R_{S1} = 100\ \Omega$, $R_D = 3\ \text{k}\Omega$, $R_L = 40\ \text{k}\Omega$, $R_2 \to \infty$, $R_1 = 10^6\ \Omega$, $I_{DQ} = 3.33$ mA, $V_{GSQ} = -1$ V, $V_{DSQ} = 10$ V, and $g_m = 2 \times 10^{-3}$ S, select values of C_1, C_2, and C_3 so that the low-frequency 3-dB point is at 20 Hz. Assume $R_i << R_{in}$.

SOLUTION Equation (7.13) yields the following time constants:

$$\tau_1 = C_1(R_L + R_D) = C_1(43\ \text{k}\Omega)$$

$$\tau_2 = R_{S2} C_2 = C_2(200\ \Omega)$$

$$\tau_3 = \frac{R_{S2} C_2}{1 + g_m R_{S2}/(1 + g_m R_{S1})} = C_2(150\ \Omega)$$

Recall that there are two zeros and two poles in the response characteristic for this amplifier. We let one of the zeros cancel one of the poles. That is, we let

$$\tau_1 = \tau_2$$

The normalized voltage gain now reduces to

$$\frac{A_v}{A_{vm}} = \frac{\tau_3 s}{\tau_3 s + 1}$$

By setting $\tau_1 = \tau_2$, we force

$$C_2(200\ \Omega) = C_1(43\ \text{k}\Omega)$$

This is one equation in the two unknown capacitors. The second equation comes from setting the remaining corner frequency to 20 Hz. Thus, we let

$$\frac{1}{\tau_3} = 2\pi \times 20$$

and

$$C_2 = \frac{1}{(150)\,(2\pi)\,(20)} = 53\ \mu\text{F}$$

From the first equation, we then have

$$C_1 = \frac{C_2 \times 200\ \Omega}{43\ \text{k}\Omega} = 0.25\ \mu\text{F}$$

Looking now at the input coupling capacitor, we have

$$R_G = R_1 = 10^6\ \Omega$$

$$C_3 R_G = C_3 10^6$$

We set this corner frequency at 1 decade below 20 Hz so it does not interact with the corner frequency of the pole. Thus,

$$10^6 C_3 = \frac{1}{4\pi}$$

$$C_3 = \frac{10^{-6}}{4\pi} = 0.08\ \mu\text{F}$$

7.2.2 Low-Frequency Response—CD Amplifier

In the case of the CD amplifier of Figure 7.11, we effectively have only one time constant since R_G is so large. We find the time constant at the input as

$$\tau_2 = (R_G + R_i)C_2 \tag{7.16a}$$

since the resistance in the discharge path is $R_G + R_i$.

At C_1, the equivalent resistance is found using the technique of the previous section. That is, since the controlled current source is connected to R_S, we cannot directly combine resistors in either parallel or series. We again replace the capacitor by a Thevenin voltage and solve for the ratio of Thevenin voltage to Thevenin current.

The loop equation for the input loop is

$$V_{gs} + (g_m V_{gs} + I_{TH})R_S = 0$$

The loop equation for the output loop is

$$V_{TH} - I_{TH}R_L - (g_m V_{gs} + I_{TH})R_S = 0$$

We solve these for the ratio of voltage to current to find

$$\frac{V_{TH}}{I_{TH}} = R_{TH} = R_L + \frac{R_S}{1 + g_m R_S}$$

so the time constant is

$$\tau_1 = \left(R_L + \frac{R_S}{1 + g_m R_S}\right)C_1 \tag{7.16b}$$

Figure 7.11
CD amplifier.

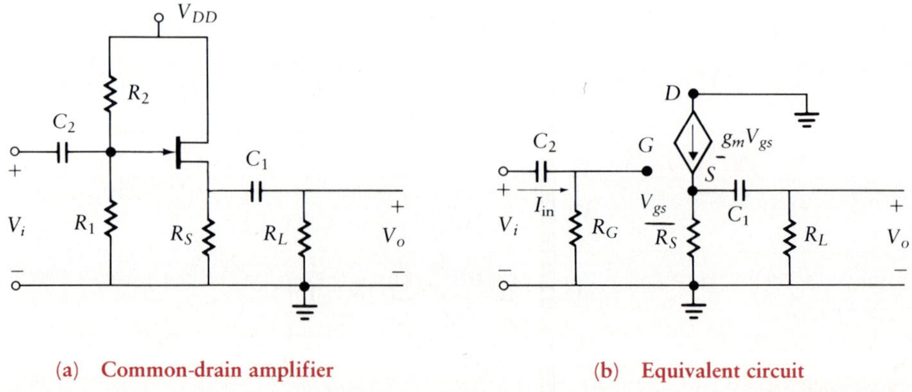

(a) Common-drain amplifier (b) Equivalent circuit

Example 7.4

A CD amplifier, as shown in Figure 7.11, has the following specifications: $R_i = 100\ \text{k}\Omega$, $R_L = 1\ \text{k}\Omega$, $R_S = 800\ \Omega$, $g_m = 2\ \text{mS}$, and $R_G = 500\ \text{k}\Omega$. Determine the size of the coupling capacitors for a low-frequency cutoff of 300 Hz.

SOLUTION The input circuit is designed to provide no loss to the signal at the low-frequency cutoff. Thus, we let $f_L = 30$ Hz.

$$\frac{1}{\tau_2} = 2\pi 30 = \frac{1}{(R_C + R_i)C_2} = \frac{1}{6 \times 10^5 C_2}$$

and solving for the capacitance, we obtain $C_2 = 0.0088\ \mu\text{F}$. Then

$$\frac{1}{\tau_1} = 2\pi 300 = \frac{1}{\left(R_L + \dfrac{R_S}{1 + g_m R_S}\right) C_1}$$

Solving for the capacitor value yields

$$C_1 = 0.41\ \mu\text{F}$$

7.3 HIGH-FREQUENCY RESPONSE—BJT AMPLIFIER

The low-frequency response of transistor circuits depends on the external capacitors used for coupling and for bypass. The high-frequency response depends on the *internal* capacitance of the wiring and of the transistor. The simplified equivalent circuit is expanded to include the effects of these internal capacitors.

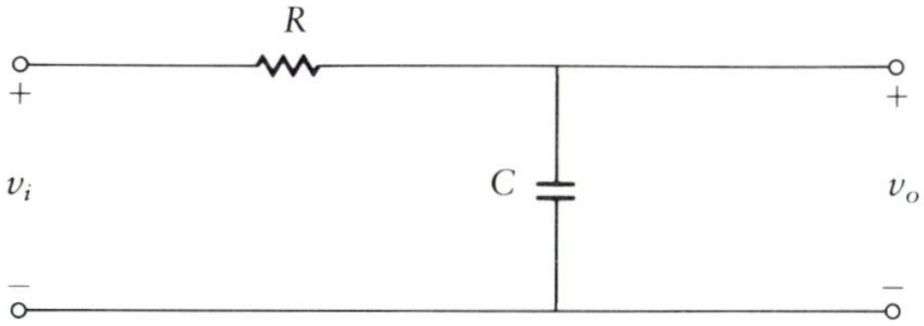

Figure 7.12
RC equivalent circuit.

These capacitors, which affect the high-frequency response, exist between the terminals of the device. We can view each of the capacitors as being in series with the equivalent (Thevenin) resistance of the associated circuitry. We begin by examining the simple RC circuit of Figure 7.12. As the frequency of the input increases, the output signal decreases in amplitude at −20 dB/decade according to the expression

$$\frac{V_o}{V_i} = \frac{1}{j\omega RC + 1} = \frac{1}{j\omega\tau + 1} \tag{7.17}$$

where the time constant $\tau = RC$.

Equation (7.17) specifies the high-frequency performance. The coupling and bypass capacitors are not considered in the discussion, since they are assumed to be short circuits at high frequencies.

7.3.1 High-Frequency Response—CE Amplifier

We use the *hybrid-π* model to estimate the high-frequency response of BJT amplifiers. The hybrid-π is similar to the *h*-parameter model studied in Chapter 2 of this text. The parameters of the two different models are related to each other. Figure 7.13(a) shows the hybrid-π model at low frequencies. The symbol B' represents an internal (intrinsic) base terminal that does not physically exist. It is separated from the actual physical base terminal, B, by a distributed resistance, $r_{bb'}$, the *base-spreading resistance.*

We define the resistances in the model and relate them to the *h*-parameters.

Base-Spreading Resistance The base-spreading resistance, $r_{bb'}$, is related to the *h*-parameter, h_{ie}, which is the input resistance with the output shorted. We apply a short circuit at the output of the circuit of Figure 7.13(a) in order to find h_{ie}. This yields

$$h_{ie} = r_{bb'} + r_{b'e} \parallel r_{b'c} \approx r_{bb'} + r_{b'e} \tag{7.18}$$

The last approximation is made since $r_{b'c} >> r_{b'e}$. Note that the base-to-collector junction is reverse-biased; thus the resistance is large. Looking from the base, $r_{b'c}$ typically on the order of $10\beta r_o$. The value of $r_{bb'}$ depends on current, a typical value being 100 Ω.

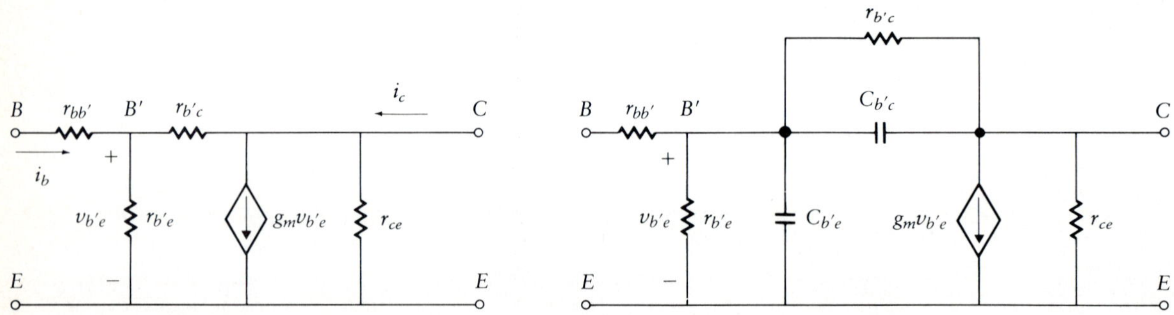

(a) Hybrid-π model of BJT at low frequencies

(b) Hybrid-π model for BJT at high frequencies

Figure 7.13 Hybrid-π transistor model.

Input Resistance The input resistance, $r_{b'e}$, is approximated by the ratio

$$r_{b'e} \approx \frac{v_{b'e}}{i_b}$$

The short-circuit collector current, i_c, is found from Figure 7.13(a) by short-circuiting the output and obtaining

$$i_c = g_m v_{b'e} \approx g_m r_{b'e} i_b$$

This quantity is related to the base current according to the h-parameter, h_{fe}; that is,

$$i_c = h_{fe} i_b$$

We therefore obtain

$$r_{b'e} = \frac{i_c}{g_m i_b} = \frac{h_{fe}}{g_m}$$

We can estimate the value of g_m from the equation

$$g_m \approx \frac{I_{CQ}}{26 \times 10^{-3}} = \frac{1}{h_{ib}} \tag{7.19}$$

where I_{CQ} is in amps. Thus

$$r_{b'e} = \frac{h_{fe} \times 26 \times 10^{-3}}{|I_{CQ}|}$$

Recall from Chapter 3 that

$$h_{ib} = \frac{26 \times 10^{-3}}{|I_{CQ}|} \tag{7.20}$$

Therefore,

$$r_{b'e} = h_{fe} h_{ib} = h_{ie}$$

Feedback Resistance The feedback resistance, $r_{b'c}$, is related to h_{re}, the reverse voltage gain, as follows:

$$h_{re} = \frac{v_{b'e}}{v_{ce}}$$

We now use a voltage-divider relationship to find

$$h_{re} = \frac{r_{b'e}}{r_{b'e} + r_{b'c}} \approx \frac{r_{b'e}}{r_{b'c}} \tag{7.21}$$

The last approximation results since $r_{b'c} >> r_{b'e}$. Finally,

$$r_{b'e} = h_{re} r_{b'c}$$

$r_{b'c}$ is typically in the megohm range.

Output Resistance The output conductance is given by h_{oe} in h-parameters, and it is determined with the input open-circuited. We see from the circuit of Figure 7.13(a) that

$$h_{oe} = \frac{i_c}{v_{ce}}$$

An output node equation yields

$$i_c = \frac{v_{ce}}{r_{ce}} + \frac{v_{ce}}{r_{b'c} + r_{b'e}} + g_m v_{b'e}$$

Therefore,

$$h_{oe} = \frac{1}{r_{ce}} + \frac{1}{r_{b'c}} + g_m h_{re} \tag{7.22}$$

Remember that

$$v_{b'e} = h_{re} v_{ce}$$

The hybrid-π model can be used to evaluate the high-frequency performance of BJTs if two capacitances are added, as shown in Figure 7.13(b). The capacitor, $C_{b'c}$, is small (typically in the range of 0.5 to 5 pF). The collector-junction capacitance is the capacitance of the collector-base junction. Although this is a varying capacitance, we consider it as constant for the particular transistor operating region. The value of this capacitance appears in the manufacturer's data sheets as C_{oB}. The capacitance $C_{b'c}$ varies depending on the voltage that appears across the junction. Most specification sheets have graphs of the capacitance versus the voltage across the junction. An example of this is shown in Figure 16 of the specification for the 2N3904 in Appendix D.

The other capacitor in Figure 7.13(b) is $C_{b'e}$, which is the base-emitter capacitor. The value of this capacitor appears in data sheets as C_{ib}. This capacitance is the sum of the emitter diffusion capacitance and the emitter-junction capacitance. Because the former capacitor is the larger of the two, $C_{b'e}$ is approximately equal to the diffusion capacitance (also known as the base-charging capacitance). The value of $C_{b'e}$ is small, typically in the range of 1 to 200 pF.

We can estimate the value of $C_{b'e}$ as

$$C_{b'e} \approx \frac{g_m}{2\pi f_T} = \frac{1}{2\pi f_T h_{ib}} = \frac{|I_{CQ}|}{52\pi f_T \times 10^{-3}} \tag{7.23}$$

In equation (7.23), f_T is the frequency at which the CE short-circuit current gain is 0 dB. The *gain-bandwidth product* is approximately constant as the input bandwidth varies. When the gain is unity, the gain-bandwidth product is simply the bandwidth. Therefore, we estimate the value of f_T to be the same as the gain-bandwidth product.

Miller Theorem We now show that the equivalent capacitance of a circuit can be considerably higher than the actual capacitor values present in the circuit. This is called the *Miller effect,* and in order to derive it, we start with a simple circuit transformation. Consider the circuits shown in Figure 7.14.

By appropriately selecting the values of the impedances, Z_1 and Z_2, in the circuit of Figure 7.14(b), the two circuits can be made identical. This is Miller's theorem, and it is most useful in high-frequency amplifier analysis.

Let us first write the equation for I_1 in the circuit of Figure 7.14(a).

$$I_1 = \frac{V_1 - V_2}{Z} = \frac{V_1}{Z}\left(1 - \frac{V_2}{V_1}\right) = \frac{V_1(1 - A_v)}{Z}$$

where we let V_2/V_1 be the voltage gain of the stage, A_v, which must be known by some other means. We cannot change the circuit configuration after we establish the gain, A_v. Because of this, the Miller effect is limited in its application. However, we can use it to determine amplifier forward gain. It cannot be used to determine output resistance.

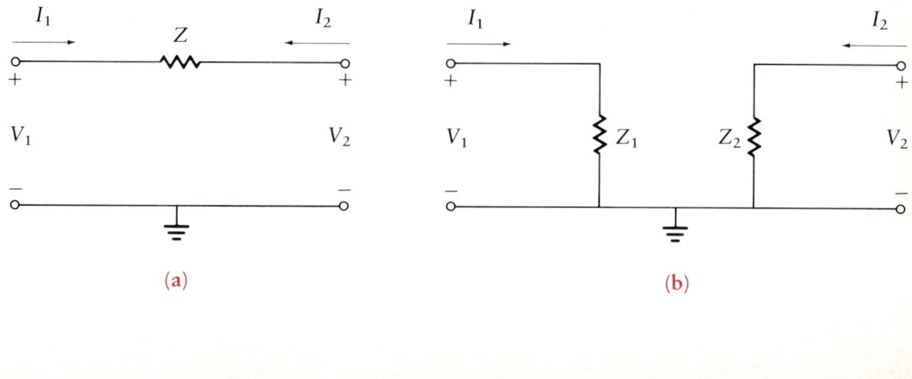

Figure 7.14 Miller's theorem.

For the circuit of Figure 7.14(b) to be identical to that of Figure 7.14(a), we must set the currents equal. Thus,

$$I_1 = \frac{V_1}{Z_1} = \frac{V_1(1 - A_v)}{Z}$$

and

$$Z_1 = \frac{Z}{1 - A_v} \tag{7.24}$$

In a similar manner, for the first circuit,

$$I_2 = \frac{V_2 - V_1}{Z} = \frac{V_2}{Z}\left(1 - \frac{V_1}{V_2}\right) = \frac{V_2}{Z}\left(1 - \frac{1}{A_v}\right)$$

where, again, $A_v = V_2/V_1$. We set the currents in the two circuits equal and solve for Z_2 with the result

$$Z_2 = \frac{Z}{1 - 1/A_v} \tag{7.25}$$

Let us now apply this result to the CE amplifier stage of Figure 7.15(a). If the

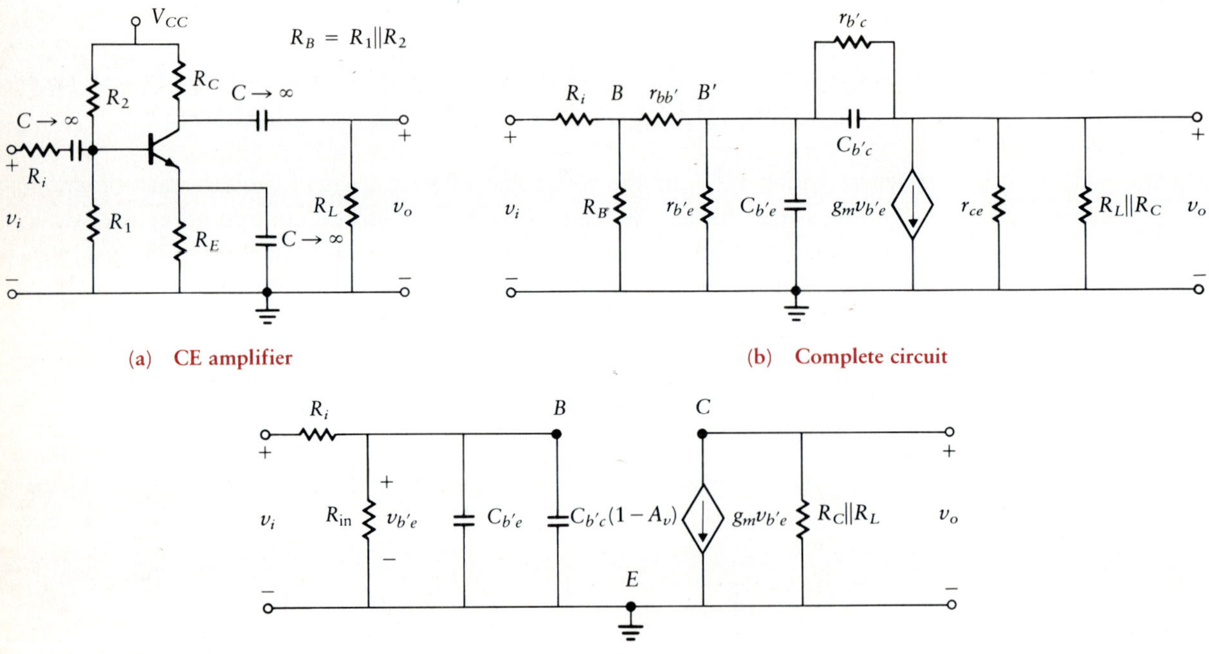

Figure 7.15
CE amplifier.

(a) CE amplifier (b) Complete circuit (c) Simplified equivalent circuit

hybrid-π model of Figure 7.13(b) is substituted for the transistor, the result is as shown in Figure 7.15(b). We make several simplifications as follows:

$r_{ce} >> R_L \parallel R_c$, so r_{ce} is deleted.

$r_{b'c} >>$ other resistors, so $r_{b'c}$ is deleted.

$r_{bb'} << r_{b'e}$, so $r_{bb'}$ is deleted.

We use Miller's theorem to reflect $C_{b'c}$ to the input and output. We further simplify the circuit by combining R_B and $r_{b'e}$ into a single resistance as follows: Since $r_{b'e} \approx h_{ie}$,

$$R_{\text{in}} = R_B \parallel h_{ie}$$

We find that there are two RC time constants in the transfer function: one in the base circuit and the other in the collector circuit. The base circuit dominates because the input capacitance is so large (it is multiplied by A_v). This results from the fact that A_v is usually greater than unity and the output capacitance is approximately equal to $C_{b'c}$, which is ignored to develop the equivalent circuit as shown in Figure 7.15(c).

We combine the two capacitors into a total capacitance, C_t, as follows:

$$C_t = C_{b'e} + C_{b'c}(1 - A_v) \tag{7.26}$$

The Miller capacitance is defined as

$$C_M = C_{b'c}(1 - A_v) \tag{7.27}$$

The high-frequency cutoff, f_H, is determined from the input resistance and capacitance, where the resistance is $R_i \parallel R_{\text{in}}$ and the capacitance is

$$C_t = C_{b'e} + C_M = C_{b'e} + C_{b'c}(1 - A_v)$$

The high-frequency cutoff is then given by the reciprocal of the RC time constant.

$$f_H = \frac{1}{2\pi(R_i \parallel R_{\text{in}})[C_{b'e} + C_{b'c}(1 - A_v)]} \tag{7.28}$$

Since we generalized equation (7.28) by using A_v and R_{in}, we can use this equation for any CE amplifier configuration. R_{in} and A_v will change depending on whether R_E is bypassed or not. Not bypassing R_E results in higher bandwidth and reduced gain. We can use C_{oB} in place of $C_{b'c}$, since the former notation is usually found on the data sheets and C_{oB} is approximately equal to $C_{b'c}$. However, when R_i becomes small, the dominant time constant changes from the input circuit to the output circuit [C_M (output) $= C_{b'c}(1 - 1/A_v)$]. This then raises the frequency corresponding to the dominant time constant, resulting in a higher-frequency capability of the amplifier. The Bode plot for equation (7.28) is a simple -20 dB/decade roll off starting from the high-frequency cutoff frequency, f_H.

Example 7.5

Calculate the high-frequency cutoff, f_H, for the 2N3904 transistor. Assume that a CE amplifier, with R_E bypassed, is used and that $\beta = 200$, $I_{CQ} = 10$ mA, $R_B = 5$ kΩ, $R_i = 1$ kΩ, and $R_L = R_C = 1$ kΩ. Further assume that you have checked the data sheets and found that $f_T = 0.25 \times 10^9$ Hz and $C_{oB} = 4.5$ pF. Verify your result using a computer simulation program.

SOLUTION In order to use equation (7.28), we need to find R_{in}, $C_{b'e}$, and C_M. These are found from the equations derived in this section. We first find h_{ie} as follows:

$$h_{ie} = 26 \times 10^{-3} \frac{\beta}{I_{CQ}} = 520\ \Omega$$

R_{in} is then given by

$$R_{\text{in}} = R_B \parallel h_{ie} = 5\text{ k}\Omega \parallel 520\ \Omega = 471\ \Omega$$

$C_{b'e}$ is found from equation (7.23).

$$C_{b'e} = \frac{|I_{CQ}|}{52\pi f_T \times 10^{-3}} = 245\text{ pF}$$

We need A_v to find C_M. This is given by

$$A_v = \frac{-R_L \parallel R_C}{h_{ie}/\beta} = -192$$

The Miller capacitance is then

$$C_M = C_{b'c}(1 - A_v) = 869\text{ pF}$$

where $C_{b'c} = C_{oB}$. Finally, from equation (7.28) we have

$$f_H = \frac{10^{12}}{(2\pi)(320)(245 + 869)} = 446\text{ kHz}$$

Equation (7.28) indicates that the high-frequency cutoff can be increased by lowering R_i and R_{in}, by lowering the voltage gain of the stage, $|A_v|$, or by selecting a high-frequency transistor with lower capacitance values. Suppose we recalculate f_H for Example 7.5, where the source impedance, R_i, is now reduced from 1 kΩ to 250 Ω. We calculate

$$R_i \parallel R_{\text{in}} = 250\ \Omega \parallel 471\ \Omega = 163\ \Omega$$

and

$$f_H = 874\ \text{kHz}$$

Hence, by reducing R_i, we nearly double the high-frequency cutoff.

We now perform the computer simulation. The circuit of Figure 7.5 was entered into the computer when we solved Example 7.2. We therefore call the circuit from memory (you should always save your circuits in meaningfully labeled files), modify the resistor values, and add a bypass capacitor across R_E as given in the problem. We must also set the bias circuitry and source voltage to achieve the required bias conditions. Note that the supply voltage must be increased beyond 10 V since the specified value of I_{CQ} is 10 mA, and the quiescent voltage drop across R_C is 10 V. We have performed this type of design many times, so the details are left to the student. The resulting ac analysis plot is shown as Figure 7.16. The midfrequency gain is approximately 36.3 dB, so the high-frequency cutoff occurs at a gain of 33.3 dB. This corresponds to a frequency of about 530 kHz. The theoretical solution is 446 kHz, so we shall explain this discrepancy. If we view the equation for f_H, we see that this

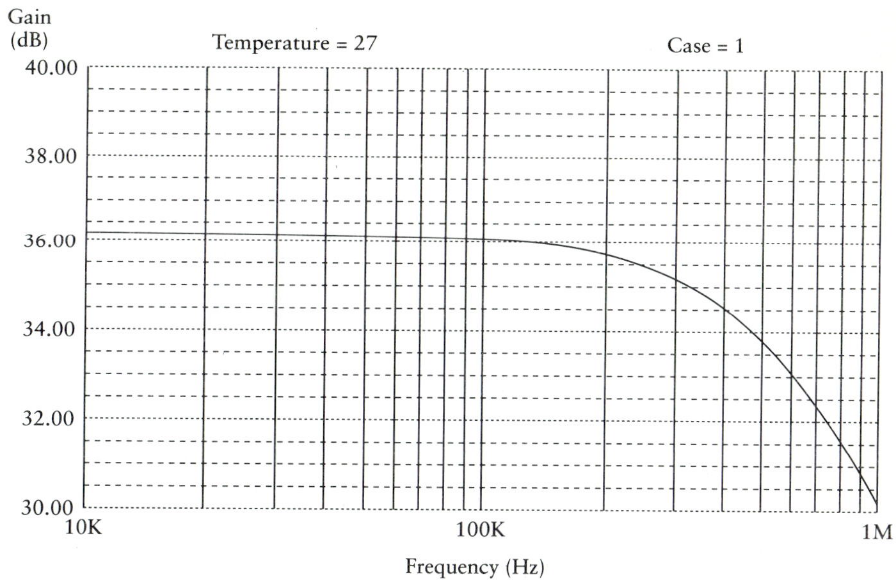

Figure 7.16 Computer simulation result for Example 7.5.

parameter increases beyond the theoretical value of 446 kHz if one or more of the following four changes occur:

1. $R_i \parallel R_{\text{in}} < 320\ \Omega$. This quantity is approximately equal to R_B, and R_B is fixed when the resistor values in the base bias circuitry are specified. This parameter will vary if the resistor model in the computer contains a tolerance other than 0%.
2. $C_{b'e} < 245$ pF. $C_{b'e}$ depends on g_m and f_T. Both of these parameters vary from those assumed in our paper design as the current changes.
3. $C_{oB} < 4.5$ pF. C_{oB} is a transistor parameter that varies from the value assumed in our paper design as the current increases.
4. $|A_v| < 192$. The midrange gain, A_v, deviates from the calculated value of -192 due to variation in h_{ie} and β. The computer varies these parameters away from their nominal values as a function of operating point. (Note that the simulated midband gain is 192 rather than 62 (36.3 dB). The 36.3 dB calculated earlier represents overall gain and includes the input resistor, R_i. If we take the voltage-divider ratio of R_i with R_{in} into account, the simulated gain of 192 reduces by a factor of 471 Ω/1471 Ω, and the two answers match very closely.)

Drill Problem

D7.2 Determine the high-frequency cutoff for the CE amplifier of Figure 7.15. The transistor parameters are $f_T = 0.4 \times 10^9$ Hz, $C_{oB} = 2$ pF, $\beta = 200$, $h_{ie} = 400\ \Omega$, $I_{CQ} = 13$ mA, $R_i = 270\ \Omega$, $R_B = 10$ kΩ, $R_E = 65\ \Omega$, and $R_C = R_L = 670\ \Omega$.

Ans: $R_{\text{in}} = 385\ \Omega$; $C_{b'e} = 199$ pF;
$C_M = 672$ pF; $f_H = 1.15$ MHz

7.3.2 High-Frequency Response—EF Amplifier

Figure 7.17 shows an EF amplifier, its equivalent circuit, and a modified equivalent circuit illustrating the Miller effect. The EF amplifier has a voltage gain approximately equal to unity. The load for EF amplifiers is frequently capacitive, so we include the load capacitor, C_L, in our models. Since the amplifier gain is close to unity, both $C_{b'e}(1 - A_v)$ and $C_{b'e}(1 - 1/A_v)$ are close to zero. We again find the time constants by determining the equivalent resistance in the discharge paths of the two capacitors.

For C_L, the discharge path is through $R_L' = R_L \parallel R_E$, so the output time constant is

$$\tau_{\text{out}} = C_L R_L' = C_L(R_L \parallel R_E) \tag{7.29a}$$

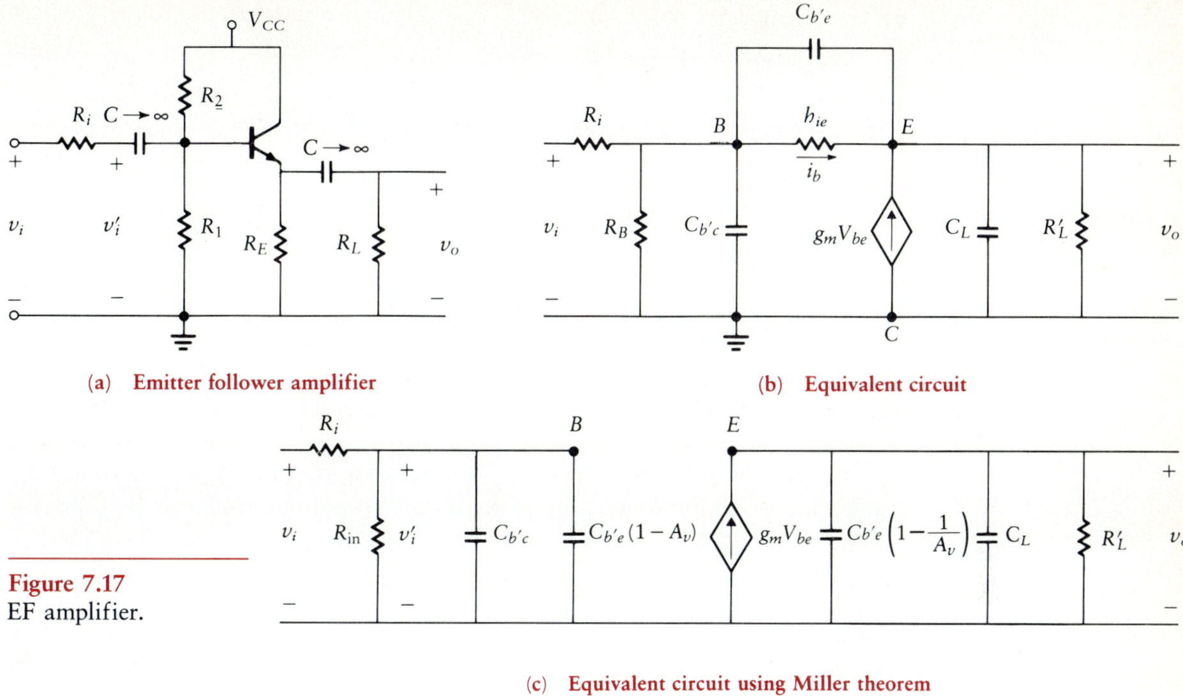

Figure 7.17
EF amplifier.

For $C_{b'c}$ (or C_{oB}), the discharge path is through $R_{in} \parallel R_i$, so the input time constant is

$$\tau_{in} = C_{oB}(R_{in} \parallel R_i) = C_{oB}[R_B \parallel \beta(R_E \parallel R_L)] \parallel R_i \tag{7.29b}$$

If C_L is approximately equal to C_{oB}, then $\tau_{in} >> \tau_{out}$ and the high-frequency cutoff is limited by τ_{in}. Hence, the high-frequency cutoff frequency is

$$f_H = \frac{1}{2\pi C_{oB}[R_B \parallel \beta(R_E \parallel R_L)] \parallel R_i} \tag{7.30}$$

7.3.3 High-Frequency Response—CB Amplifier

For the CE amplifier, the Miller effect causes the input capacitance effectively to increase by a factor of $(1 - A_v)\, C_{b'c}$. When the gain is large, the input capacitance becomes so large that the amplifier is essentially a single time constant amplifier with the input circuit providing the dominant time constant. In the case of the CB amplifier, the Miller effect does not exist, and there is no multiplying factor for $C_{b'c}$. This is shown in Figure 7.18, where the CB amplifier and its equivalent circuit are illustrated. Notice that the capacitor, C_b, effectively shorts the base to ground, as shown in Figure 7.18(b).

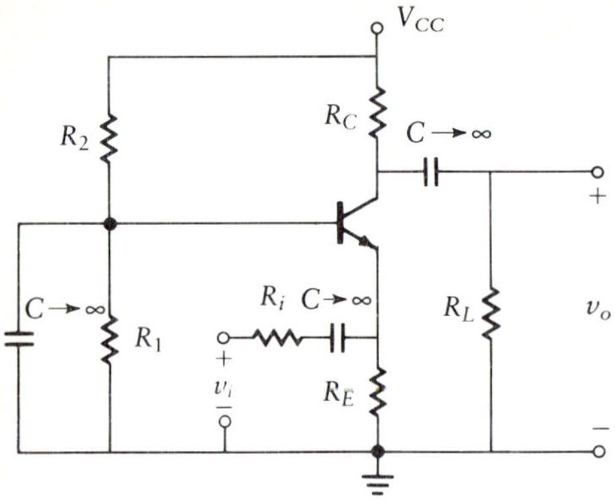

(a) **CB amplifier**

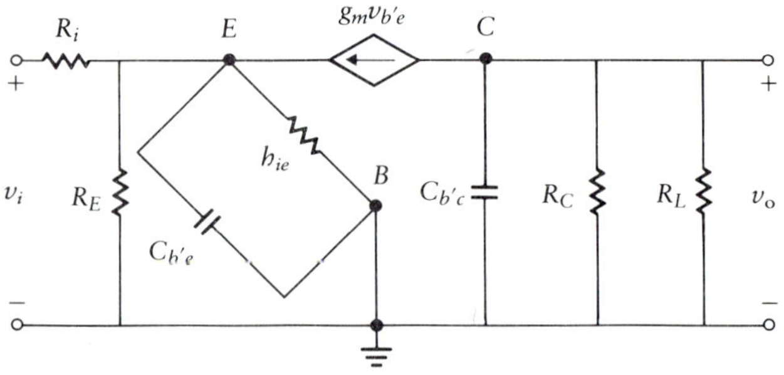

(b) **Equivalent circuit**

Figure 7.18
CB amplifier.

As with the CE amplifier, the CB amplifier response has two time constants. These can be found, as before, by observing the discharge path for the two capacitors. For the output circuit, the discharge path is through $R_C \parallel R_L$, so

$$\tau_{\text{out}} = C_{b'c}(R_C \parallel R_L) \tag{7.31a}$$

At the input, the time constant is

$$\tau_{\text{in}} = C_{b'e}\left(\frac{h_{ie}}{\beta} \middle\| R_E \parallel R_i\right) = C_{b'e}(h_{ib} \parallel R_E \parallel R_i) \tag{7.31b}$$

We use h_{ie}/β instead of h_{ie} in equation (7.31b) due to the effects of the controlled current source. The resulting pole in the input circuit is given by

$$f_H = \frac{1}{2\pi C_{b'e}(h_{ib} \parallel R_E \parallel R_i)} \tag{7.32}$$

Note that the high-frequency cutoff would be considerably higher if R_B were not bypassed (i.e., $C_b \to 0$).

Example 7.6

Determine the frequency response of the CB amplifier shown in Figure 7.18, with the following specifications: V_{CC} = 12 V, V_{BE} = 0.7 V, β = 100, R_i =

600 Ω, $R_E = 200$ Ω, $R_L = R_C = 2$ kΩ, $C_{oB} = 2$ pF, and $f_T = 10^8$ Hz. The amplifier is biased for maximum output swing.

SOLUTION First determine the Q-point for maximum output swing.

$$I_{CQ} = \frac{12}{2200 + 1200} = 3.53 \text{ mA}$$

Then

$$h_{ib} = \frac{26}{3.53} = 7.37 \ \Omega$$

Since $C_{b'c} = C_{oB} = 2$ pF, we have

$$\tau_{\text{out}} = (R_L \parallel R_C)(2 \text{ pF}) = 2 \text{ ns}$$

We find $C_{b'e}$ from equation (7.23) as follows:

$$C_{b'e} = \frac{I_{CQ}}{52 \times 10^{-3}\pi f_T} = 216 \text{ pF}$$

Then

$$\tau_{\text{in}} = C_{b'e}(h_{ib} \parallel R_E \parallel R_i) = 1.52 \text{ ns}$$

The high-frequency breakpoints are at

$$f_H \text{ (input)} = \frac{1}{2\pi \times 1.52 \times 10^{-9}} = 105 \text{ MHz}$$

and

$$f_H \text{ (output)} = \frac{1}{2\pi \times 2 \times 10^{-9}} = 80 \text{ MHz}$$

7.4 HIGH-FREQUENCY RESPONSE—FET AMPLIFIER

7.4.1 High-Frequency Response—CS Amplifier

The high-frequency equivalent circuit for an FET is obtained by adding three capacitors to the model developed in Chapter 4, as shown in Figure 7.19. The

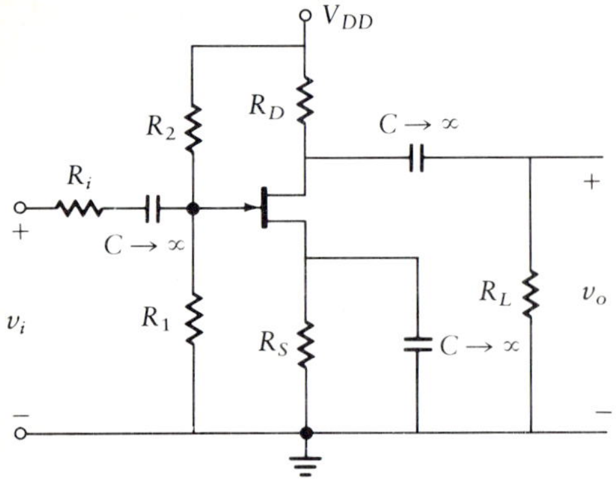

(a) CS amplifier circuit

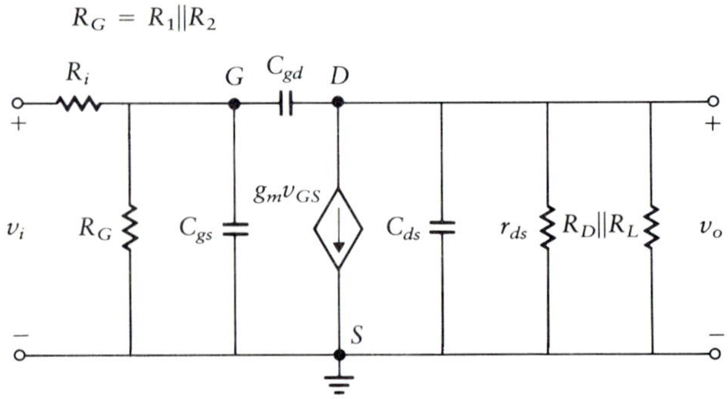

(b) Equivalent circuit

Figure 7.19 CS amplifier, high-frequency model.

capacitors are defined as follows:

C_{gs} is the capacitance between gate and source.

C_{gd} is the capacitance between gate and drain.

C_{ds} is the capacitance between drain and source.

The CS amplifier of Figure 7.19(a) is modeled at high frequencies with the small-signal equivalent circuit of Figure 7.19(b).

We note several simplifications in Figure 7.19(b). C_{ds} is so small that it is neglected, and r_{ds} is neglected because it is much larger than $R_D \| R_L$.

Now we apply the Miller theorem to the circuit of Figure 7.19(b) in a manner similar to that in the BJT analysis. The input capacitance becomes

$$C_{gs} + C_{gd}(1 - A_v) \tag{7.33}$$

and the output capacitance is so small because

$$C_{ds} + C_{gd}\left(1 - \frac{1}{A_v}\right) \approx C_{ds} + C_{gd} \tag{7.34}$$

The simplified circuit is shown in Figure 7.20, where C_{ds} and C_{gd} are neglected because of their small sizes.

The low-frequency voltage gain is

$$\frac{v_o}{v_i} \approx -g_m(R_L \| R_D)$$

The largest capacitor is the Miller capacitor because of the amplification of the stage. The dominant time constant is normally formed by the input characteristics of the circuit.

As in the case of BJT amplifiers, the high-frequency cutoff, f_H, is deter-

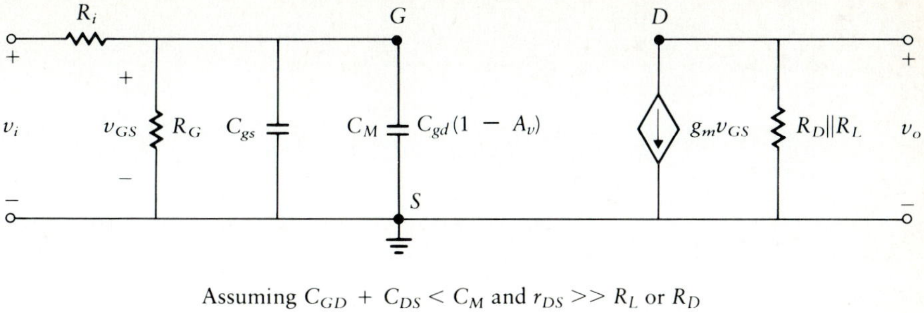

Figure 7.20 Simplified high-frequency equivalent circuit.

mined by the RC combination made up of the resistance, $R_G \| R_i$, and the capacitance,

$$C_{gs} + C_{gd}(1 - A_v) = C_{gs} + C_M$$

The high-frequency cutoff is then given by the reciprocal of the RC time constant.

$$f_H = \frac{1}{2\pi(R_G \| R_i)[C_{gs} + C_{gd}(1 - A_v)]} \tag{7.35}$$

The equivalent input capacitance, as shown in the data books, is

$$C_{is} = C_{gs} + C_{gd} \tag{7.36}$$

The equivalent forward transfer capacitance, as shown in the data books, is

$$C_{fs} = C_{gd} \tag{7.37}$$

The reverse transfer capacitance is

$$C_{rs} = C_{gd} \tag{7.38}$$

Finally, the output capacitance is

$$C_{os} = C_{gd} + C_{ds} \tag{7.39}$$

The input and reverse transfer capacitors, C_{is} and C_{rs}, are usually shown in the manufacturers' specification sheets as C_{iss} and C_{rss}. The second "s" is used to indicate that one of the ports is short-circuited.

The capacitors of Figure 7.19(b) can be approximately related to C_{rss} and C_{iss} by

$$C_{gd} = C_{rss}$$

and

$$C_{gs} = C_{iss} - C_{rss}$$

As noted previously, C_{ds} is so small that it can be neglected.

Example 7.7

Determine the high-frequency cutoff for the CS amplifier of Figure 7.19(a). Let $R_L = 20\ \text{k}\Omega$, $R_D = 4\ \text{k}\Omega$, $R_G = 100\ \text{k}\Omega$, $R_i = 100\ \text{k}\Omega$, $g_m = 2000\ \mu\text{S}$, $C_{gs} = C_{gd} = 16$ pF, and $C_{ds} = 0$.

SOLUTION We use equation (7.35) to find f_H as follows:

$$f_H = \frac{1}{(2\pi)50\ \text{k}\Omega\{16 \times 10^{-12} + 16 \times 10^{-12}[1 + (0.002)(20\ \text{k}\Omega \parallel 4\ \text{k}\Omega)]\}}$$

$$= 22.9\ \text{kHz}$$

This is a relatively low cutoff frequency because of the large values of the capacitors, C_{gs} and C_{cd}. We can increase this frequency by lowering C_{gs} and C_{gd} and by lowering the input resistor, or by decreasing the voltage gain ($|A_v|$) of the stage.

Drill Problem

D7.3 Determine the high-frequency cutoff, f_H, for the FET amplifier of Figure 7.19 with

$$C_{gs} = C_{gd} = 3\ \text{pF}, \qquad C_{ds} = 0$$

$$R_L = 20\ \text{k}\Omega, \qquad R_D = 4\ \text{k}\Omega, \qquad R_G = R_i = 10\ \text{k}\Omega$$

$$g_m = 2000\ \mu\text{S}$$

Ans: $f_H = 1.22$ MHz

7.4.2 High-Frequency Response—CD Amplifier

Figure 7.21 shows a CD amplifier, its high-frequency equivalent circuit, and a modification of the equivalent circuit that illustrates the Miller effect. Using the same transistor, the high-frequency cutoff of the CD amplifier is much

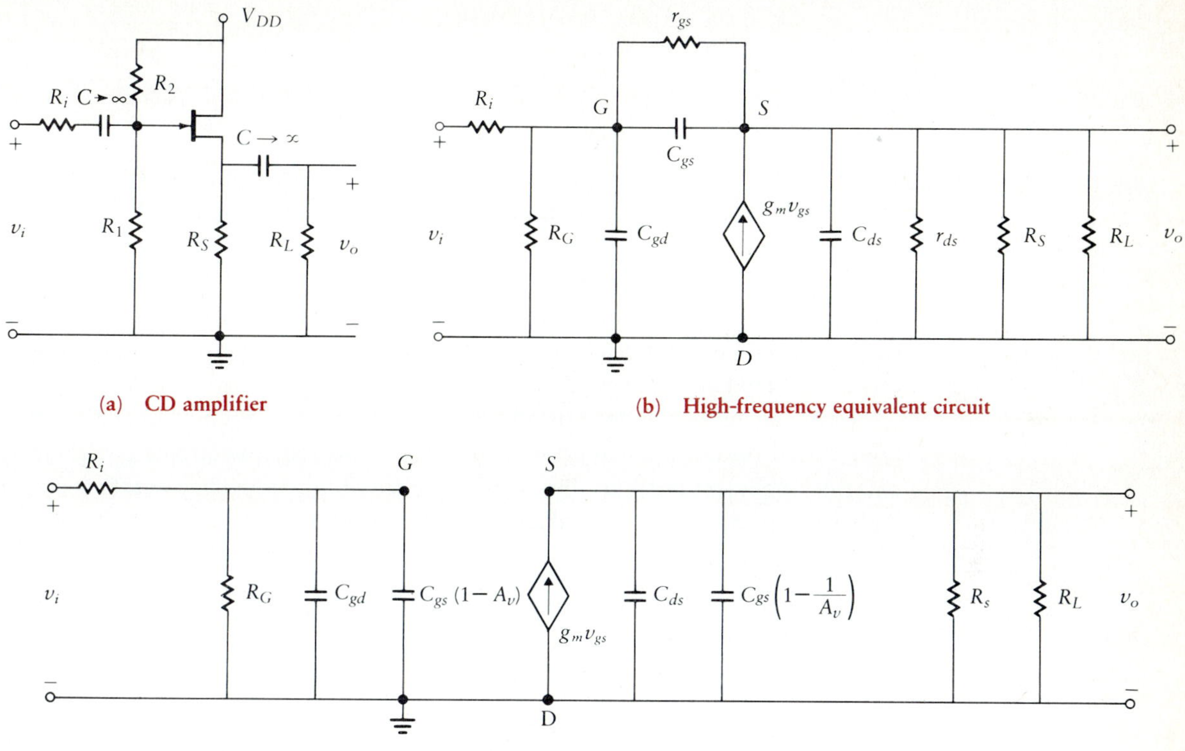

Figure 7.21
CD amplifier.

higher than that of CS amplifiers, as shown in Figure 7.21(c). The CD amplifier has a gain that is positive and less than unity. This allows the Miller effect to *reduce* the input capacitance. The gate-to-source and drain-to-source resistances, r_{gs} and r_{ds}, are ignored in the model of Figure 7.21(c). The reduction of the gate-to-source capacitance by the Miller effect reduces the effect of the input time constant. The input time constant is found by observing the discharge path for the input capacitors, with the result

$$\tau_{\text{in}} = (R_G \parallel R_i)[C_{gd} + C_{gs}(1 - A_v)] \tag{7.40}$$

The output time constant is found in a similar manner.

$$\tau_{\text{out}} = C_{ds}(R_S \parallel R_L) \tag{7.41}$$

The high-frequency cutoff, f_H, is found by inverting the time constant. The dominant pole is in the input since

$$R_G \parallel R_i >> R_S \parallel R_L$$

The high-frequency cutoff is given by

$$f_H = \frac{1}{2\pi(R_G \parallel R_i)[C_{gd} + C_{gs}(1 - A_v)]} \tag{7.42}$$

7.5 HIGH-FREQUENCY AMPLIFIER DESIGN

In the previous sections, we learned to analyze specific amplifier configurations to determine the high-frequency cutoff point. For any new amplifier configuration, the equivalent circuit is developed and reduced (taking into account the Miller effect). The time constants are then determined by finding the equivalent resistance through which each capacitor discharges. The series capacitors are shorted at the high frequency and affect only the low-frequency cutoff.

When an amplifier is constructed, it will attain a high-frequency cutoff point based on the particular transistor type selected, the circuit parameters, and the amplifier configuration. If the resulting high-frequency cutoff is not as high as desired, the designer does have some options:

1. Reduce the gain per stage to reduce the Miller effect.
2. Select a transistor that has a higher-frequency capability.
3. Use an amplifier circuit configuration that is not as frequency sensitive, such as the common-base configuration.

Changing any of the above parameters requires that a new high-frequency analysis be performed. Since this is time-consuming, it is an ideal application for computer simulation programs. Computer simulations not only model transistors more accurately than noncomputer analysis but also permit iterative analysis. Using the iteration feature present in many of the SPICE-based programs, you can view a family of output curves representing different temperature and/or different component values chosen from within their tolerance limits. You can then select the best designs for your application.

7.6 CONCLUDING REMARKS

Let us return to the XYZ Audio Corporation discussion presented in the introduction to this chapter. Recall that, as a newly graduated engineer, you designed an audio amplifier to be used in a compact disc player but, alas, that amplifier was found to have poor low-frequency response. Since the problem was identified as being at low frequencies rather than high frequencies, it can be traced to the presence of coupling and bypass capacitors. The techniques of this chapter are used to analyze your design. It is, of course, preferable to use the bandwidth criteria as part of the design process, but we assume that in this case the design was completed and we are now in the *troubleshooting* phase. The corner frequencies are found and compared to the design specifications.

Since the amplifier must handle audio signals, it is reasonable that it respond down to 15 or 20 Hz. When the corner frequency calculations are made for your initial design, they will probably indicate a cutoff frequency that is too high. To lower these frequencies, we must raise the time constants, which requires raising either the resistances or the capacitances. Raising the resistances requires complete redesign in order to maintain the same gain. Therefore, if your original design is close to specification, you would probably want to work with the capacitances.

It is not our intent to minimize the practical problems associated with a real-life design. Our study of electronics attempts to model the real world closely, but you should realize that we are simply working with models. There is a difference between real-world behavior and mathematical model performance.

PROBLEMS

7.1 Construct Bode plots for V_o/V_i for the circuits of Figure P7.1 and determine the lower and upper 3-dB frequencies.

7.2 A 2N3903 transistor is incorporated in the circuit of Figure 7.2. Calculate the lower-corner 3-dB frequency and draw the Bode plot for this amplifier. Let $R_L = R_C = 1\ \text{k}\Omega$, $C_1 = 0.1\ \mu\text{F}$, $R_E = 100\ \Omega$, $R_i = 2\ \text{k}\Omega$, $R_1 = 1.5\ \text{k}\Omega$, $R_2 = 10\ \text{k}\Omega$, $C_2 = 0.01\ \mu\text{F}$, and $V_{CC} = 12$ V.

7.3 Given the circuit of Figure 7.2 with $V_{CC} = 12$ V, $\beta = 250$, $R_L = 1\ \text{k}\Omega$, $R_i = 2\ \text{k}\Omega$, and $A_v = -5$, select C_1 and C_2 so that the amplifier has a lower 3-dB frequency of 20 Hz. I_{CQ} is to be in the middle of the ac load line.

7.4 Use a 2N3903 transistor in the circuit of Figure 7.2. Draw the Bode plot for this amplifier when $R_1 = 1\ \text{k}\Omega$, $R_2 = 9\ \text{k}\Omega$, $R_L = R_C = 1\ \text{k}\Omega$, $R_E = 100\ \Omega$, $R_i = 1\ \text{k}\Omega$, $C_2 = 0.01\ \mu\text{F}$, and $C_1 = 0.1\ \mu\text{F}$.

7.5 Design a CE amplifier with $R_{\text{in}} = 15\ \text{k}\Omega$, $A_v = -10$, $R_L = 10\ \text{k}\Omega$, $R_i = 5\ \text{k}\Omega$, and a low-frequency cutoff of 40 Hz. Use a transistor with $\beta = 200$, $V_{BE} = 0.7$ V, and $V_{CC} = 12$ V. What are the values of C_1 and C_2 in the following cases?

a. The time constants are one decade apart.
b. Both time constants are located at the same point.

7.6 The emitter resistor is bypassed in the amplifier of Problem 7.5. What value is required for the bypass capacitor if the cutoff frequency is 40 Hz?

7.7 Design a CE amplifier with $R_{\text{in}} = 4\ \text{k}\Omega$, $A_v = -10$, $R_L = 5\ \text{k}\Omega$, $R_i = 2\ \text{k}\Omega$, $V_{CC} = 16$ V, $\beta = 200$, $V_T = 26$ mV, and $V_{BE} = 0.7$ V. Use the circuit of Figure 7.2. Assume $f_c = 20$ Hz.

a. Determine all resistor values

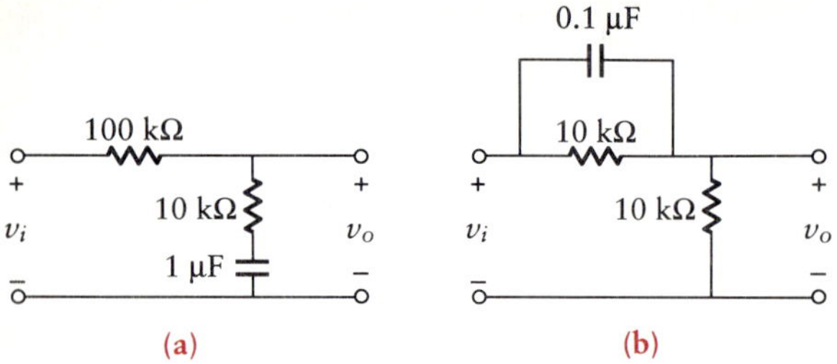

Figure P7.1

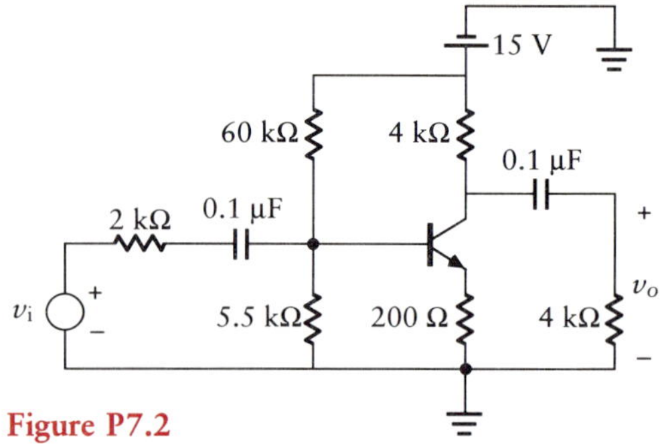

Figure P7.2

b. Find the values of C_1 and C_2 assuming that the time constants are one decade apart.
c. Find the values of C_1 and C_2 assuming that the time constants are identically located.

7.8 The EF amplifier of Figure 7.8 has $R_1 = 20$ kΩ, $R_2 = 2$ kΩ, $R_L = 100$ Ω, $R_E = 50$ Ω, $R_i = 1$ kΩ, $C_1 = 100$ μF, and $C_2 = 3.3$ μF. Draw the Bode plots for this amplifier if $V_{CC} = 10$ V, $\beta = 200$, and $V_{BE} = 0.7$ V.

7.9 Design an EF amplifier with $R_i = 1$ kΩ, $A_i = 10$, $R_L = 20$ Ω, and a low-frequency cutoff of 20 Hz. Use $V_{CC} = 10$ V and a transistor with $\beta = 80$ and $V_{BE} = 0.7$ V. What are the values of C_1 and C_2 in the following cases?
a. The time constants are one decade apart.
b. Both time constants are located at the same point.

7.10 Design a CB amplifier (Figure 7.9) for maximum output voltage swing. The amplifier must have a voltage gain of 10 into a 4-kΩ load. Let $\beta = 100$, $V_{BE} = 0.7$ V, $V_{CC} = 18$ V, $R_i = 100$ Ω, and $R_E = 500$ Ω. Choose the capacitors for a 3-dB corner frequency of 50 Hz.

7.11 Given the FET amplifier of Figure 7.10 with a 3N128 MOSFET, $V_{DD} = 12$ V, and $R_{in} = 50$ kΩ.
a. Design the amplifier to have a midrange voltage gain of -2 when $R_L = 3$ kΩ. Use $g_m = 5000$ μS, $V_{DSQ} = 7.5$ V, and $I_{DQ} = 2$ mA.
b. Select C_1, C_2, and C_3 so that the lower 3-dB frequency is 10 Hz.

7.12 Design a JFET amplifier for $A_v = -10$, $R_{in} = 50$ kΩ, and $R_L = 20$ kΩ, using $V_{DD} = 20$ V. The Q-point is selected at $V_{DSQ} = 7$ V, $V_{GSQ} = -0.35$ V, $I_{DQ} = 1.3$ mA, and $g_m = 1600$ μS. Select C_1 and C_2 for a lower 3-dB frequency of 20 Hz.

7.13 In the amplifier of Problem 7.12, the transistor is replaced by a MOSFET that has $g_m = 3333$ μS at a Q-point of $V_{DSQ} = 6$ V, $I_{DQ} = 1$ mA, and $V_{GSQ} = 2$ V.
a. Design the amplifier.
b. Select C_1 and C_2.

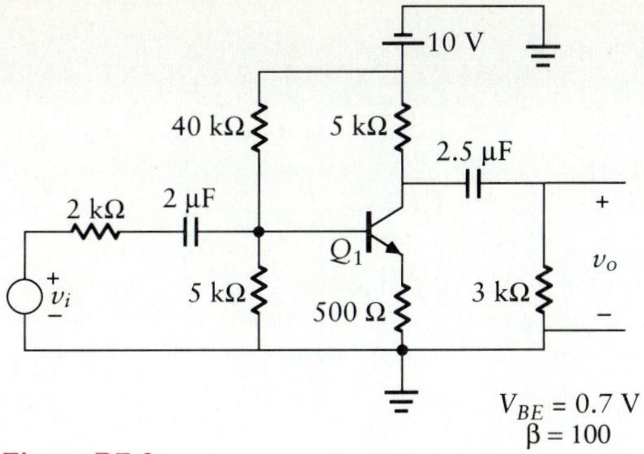

Figure P7.3

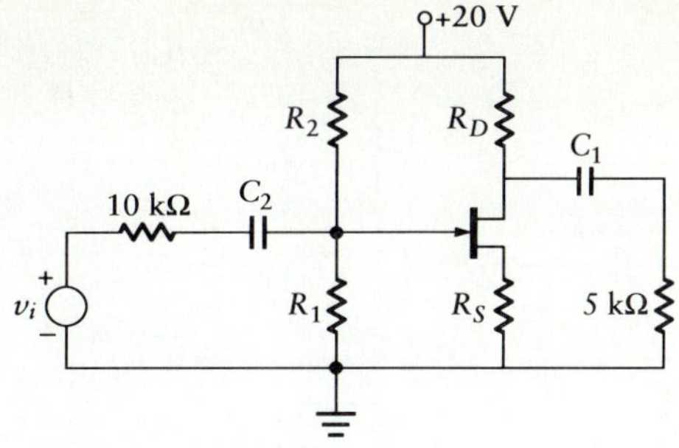

Figure P7.4

7.14 Design a JFET CS amplifier for a 3-dB lower-frequency cutoff of 40 Hz for $A_v = -2$ and $R_{in} = 100$ kΩ. The Q-point is selected at $V_{DSQ} = 6$ V, $V_{GSQ} = -1$ V, $I_{DQ} = 0.5$ mA, and $g_m = 2000$ μS. Determine each of the following.

a. All resistor values when $V_{CC} = 12$ V and $R_L = 10$ kΩ.

b. τ values.

c. Values of C_1 and C_2.

7.15 Design a CE amplifier for a low-frequency response of 50 Hz and an overall stage voltage gain of −7. Assume $V_{BE} = 0.7$ V, $\beta = 200$, $V_T = 26$ mV, $C_{oB} = 2$ pF, and $f_T = 2 \times 10^8$. The specifications for the circuit are $R_{in} = R_i = 4$ kΩ, $R_L = 5$ kΩ, and $V_{CC} = 15$ V. Design the amplifier such that the time constants are identical. Determine all resistor and capacitor values and the high-frequency cutoff.

7.16 Determine the high and low 3-dB cutoff frequencies of the amplifier shown in Figure P7.2 when the C_{oB} is 3 pF, $f_T = 4 \times 10^8$, $\beta = 100$, $V_{BE} = 0.7$ V, and $V_T = 26$ mV.

7.17 Using a 2N3904 transistor with the circuit of Figure 7.2, determine the upper 3-dB frequency and construct the Bode plot. Let $R_L = R_C = 10$ kΩ, $C_1 = 0.1$ μF, $R_E = 100$ Ω, $R_1 = 1.5$ kΩ, $R_2 = 21$ kΩ, $R_i = 500$ Ω, $C_2 = 0.01$ μF, $V_{CC} = 12$ V, $C_{oB} = 4.5$ pF, $f_T = 200$ MHz, and $\beta = 100$.

7.18 Determine the 3-dB high-frequency cutoff for the circuit shown in Figure P7.3 and sketch the Bode magnitude plot of the response. The transistor has $C_{oB} = 2$ pF, $V_{BE} = 0.7$ V, $\beta = 100$, $f_T = 2 \times 10^8$, and $V_T = 26$ mV.

7.19 Calculate the f_H for a junction transistor amplifier with the following values: $f_T = 500$ MHz, $h_{fe} = 400$, $C_{oB} = 0.5$ pF, $I_{CQ} = 5$ mA, $R_L = R_C = 7$ kΩ, $R_B = 20$ kΩ, and $R_i = 2$ kΩ.

7.20 Design a CB amplifier for maximum output swing and determine its high-frequency response when $f_T = 2 \times 10^7$, $\beta = 200$, $V_{BE} = 0.7$ V,

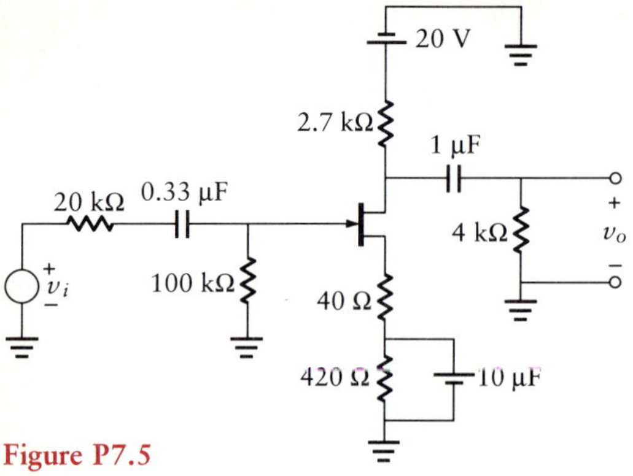

Figure P7.5

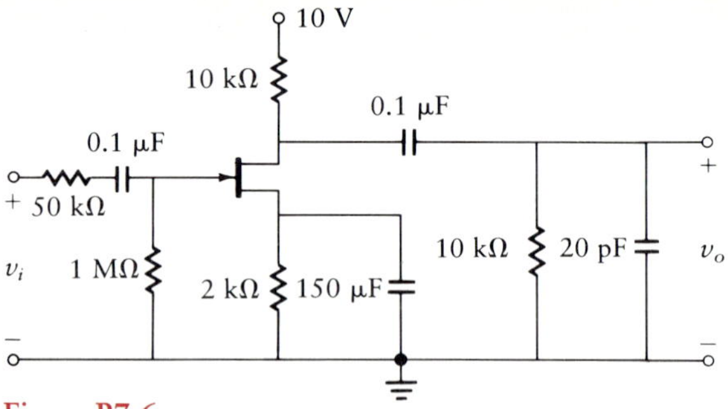

Figure P7.6

C_{oB} = 4 pF, R_i = 50 Ω, R_E = 400 Ω, R_L = 5 kΩ, and V_{CC} = 12 V. The low-frequency cutoff must be at 20 Hz. Use the circuit shown in Figure 7.18.

7.21 Determine the parameters of the model of Figure 7.19(b) from the information for the 3N128 given in the appendix. Use a frequency of 200 MHz at V_{DS} = 15 V and I_D = 5 mA. Assume the capacitor values are the same for 1 MHz as for 200 MHz and ignore the series resistors in the gate and source.

7.22 Design a CS JFET amplifier and find the high-frequency cutoff of the amplifier when $A_v = -2$, R_{in} = 100 kΩ, R_L = 5 kΩ, R_i = 10 kΩ, and V_{DD} = 20 V. The transistor used has I_{DSS} = 10 mA, $V_p = -6$ V, C_{gs} = 5 pF, C_{gd} = 2 pF, and C_{ds} = 1 pF. Use $I_{DQ} = 0.4 I_{DSS}$ and V_{DSQ} = 10 V.

7.23 Design a CS FET amplifier and find the high-frequency cutoff of the amplifier for $A_v = -3$, R_i = 10 kΩ, and R_{in} = 100 kΩ. The transistor used has I_{DSS} = 10 mA, $V_p = -5$ V, C_{gs} = 5 pF, C_{gd} = 2 pF, and C_{ds} = 1 pF. Use the circuit of Figure P7.4.

7.24 Analyze the circuit shown in Figure P7.5. The transistor used is a JFET with $V_p = -4$ V, I_{DSS} = 8 mA, C_{gd} = 3 pF, C_{ds} = 0, C_{gs} = 10 pF, and $1/r_{ds} = 0$. The transistor is operating at 3.2 mA. Determine all the time constants and break frequencies of the circuit.

7.25 For the circuit of Figure 7.19, develop the high-frequency portion of the Bode plot for an amplifier when R_i = 2 kΩ, R_L = 4 kΩ, R_D = 3 kΩ, R_G = 20 kΩ, r_{ds} = 10 kΩ, C_{gs} = 5 pF, C_{ds} = 1.1 pF, C_{gd} = 2 pF, and g_m = 0.002 S.

7.26 Determine the frequency response for the FET amplifier shown in Figure P7.6, where r_{DS} = 100 kΩ, g_m = 2.5 mS, C_{dg} = 2 pF, C_{gs} = 10 pF, and C_{ds} = 2 pF.

8

IDEAL OPERATIONAL AMPLIFIERS

8.0 INTRODUCTION

The rapid expansion of requirements for smaller, lighter, and more complex circuits resulted in the need to place not one but *hundreds* of transistors on a single silicon chip. Whenever more than one element is placed on a single chip, the resulting device is known as an *integrated circuit (IC)*.

The unmodified term, integrated circuit, is used to describe chips composed of less than about 60 elements. If a chip contains over 60 but less than about 300 elements, the term *medium-scale integration (MSI)* is used. If the number of elements is over 300 but less than about 1000, *large-scale integration (LSI)* is used. *Very large scale integration (VLSI)* refers to chips with more than 1000 elements.

Integrated circuits can be analog or digital, depending on the relationship between output and input waveforms. *Linear integrated circuits (LICs)* are designed to replace standard circuits, and these LICs are then used as building blocks for more complex systems. One of the most utilized analog circuits is the *operational amplifier (op-amp)*. Ideally, this amplifier has infinite gain, infinite input impedance, and zero output impedance. Practical operational amplifiers have performance characteristics that closely approach those of ideal operational amplifiers.

This chapter is the first of several devoted exclusively to the design of electronic systems that contain linear integrated circuits, or op-amps. The electronic engineer uses the op-amp as a building block in the design of larger systems. Our emphasis in these chapters is on the performance of op-amps and how they are incorporated in the overall system.

In this chapter, we consider the ideal op-amp and explore its use in design. The inverting amplifier and the noninverting amplifier are studied. A procedure is presented which permits a general approach to designing an amplifier that forms a weighted sum of any number of input voltages. We then explore a variety of useful op-amp applications, including negative-resistance circuits, integrators, analog computers, and instrumentation amplifiers.

8.1 IDEAL OP-AMPS

This section uses a *systems* approach to present the fundamentals of ideal operational amplifiers. We consider the op-amp as a block with input and output terminals. We are not currently concerned with the electronic devices within the op-amp.

An op-amp is a high-gain direct-coupled amplifier that is often powered by both a positive and a negative supply voltage. This allows the output voltage to swing both above and below ground. The op-amp finds wide application in many linear electronic systems.

The name operational amplifier is derived from one of the original uses of op-amp circuits: to perform mathematical operations in analog computers. This traditional application is discussed later in this chapter. Early op-amps used a single inverting input. A positive voltage change at the input caused a negative change at the output.

Figure 8.1 presents the symbol for the operational amplifier, and Figure 8.2 shows its equivalent circuit. The model contains a dependent voltage source whose voltage depends on the input voltage. The output impedance is represented in the figure as a resistance of value R_o. The amplifier is driven by two input voltages, v_+ and v_-. The two input terminals are known as the *noninverting* and *inverting* inputs, respectively. Ideally, the output of the amplifier does not depend on the magnitudes of the two input voltages but only on the difference between them. We define the *differential input voltage*, v_d, as the difference

$$v_d = v_+ - v_-$$

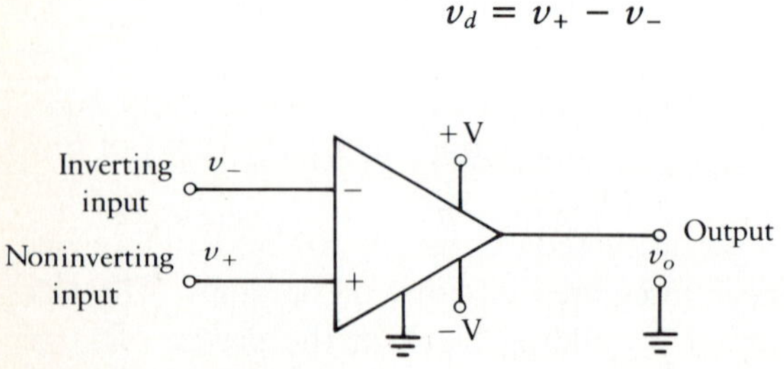

Figure 8.1 Ideal op-amp.

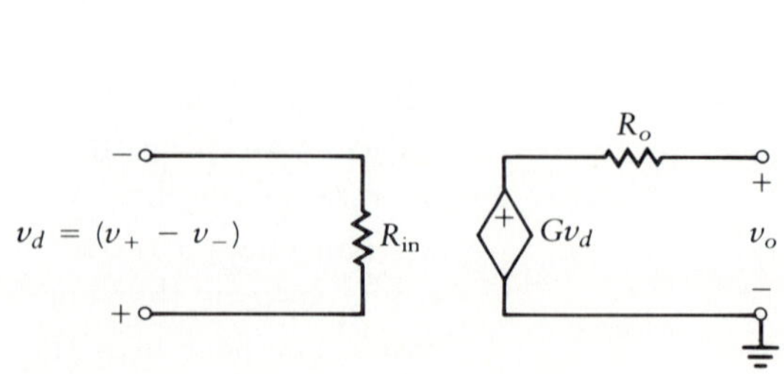

Figure 8.2 Equivalent circuit of op-amp.

The input impedance of the op-amp is represented as a resistance in Figure 8.2. The output voltage is proportional to the input voltage, and we designate the ratio as the open-loop gain, G. Thus, the output voltage is

$$v_o = G(v_+ - v_-) = Gv_d \tag{8.1}$$

As an example, an input of $E \sin \omega t$ (E is usually a small amplitude) applied to the inverting input with the noninverting terminal grounded produces $-G(E \sin \omega t)$ at the output. When the same source signal is applied to the noninverting input with the inverting terminal grounded, the output is $+G(E \sin \omega t)$.

The ideal operational amplifier is characterized as follows:

1. Input resistance, $R_{in} \to \infty$.
2. Output resistance, $R_o = 0$.
3. Open-loop voltage gain, $G \to \infty$.
4. Bandwidth $\to \infty$.
5. $v_o = 0$ when $v_+ = v_-$.

Let us explore the implication of the open-loop gain being infinite. If we rewrite equation (8.1) as follows:

$$v_+ - v_- = \frac{v_o}{G}$$

and let G approach infinity, we see that

$$v_+ - v_- = 0$$

Hence, *the voltage between the two input terminals is zero,* or

$$v_+ = v_-$$

Since the input resistance, R_{in}, is infinite, *the current into each input, inverting and noninverting, is zero.*

Practical op-amps have high voltage gain (typically 10^5 at low frequency), but this gain varies with frequency. For this reason, an op-amp is not normally used in the form shown in Figure 8.1. This configuration is known as *open loop* because there is no feedback from output to input. We see later that, while the open-loop configuration is useful for *comparator* applications, the more common configuration for linear applications is the *closed-loop* circuit with *feedback*.

External elements are used to "feed back" a portion of the output signal to the input. If the feedback elements are placed between the output and the inverting input, the *closed-loop gain* is decreased, since a portion of the output

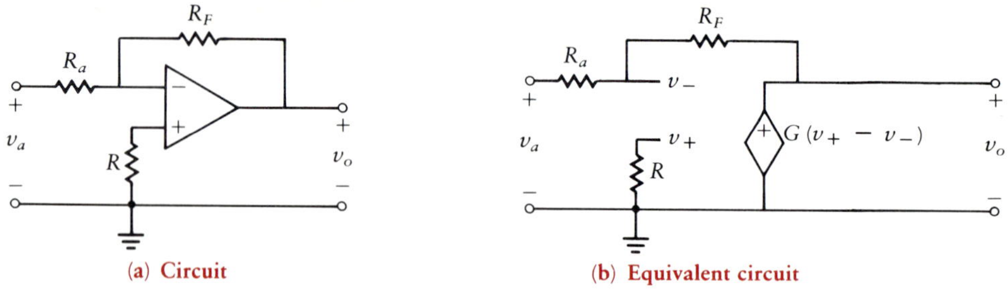

Figure 8.3
Inverting op-amp.

subtracts from the input. We will see later in this chapter that feedback not only decreases the overall gain but also makes that gain less sensitive to the value of G. With feedback, the closed-loop gain depends more on the feedback circuit elements and less on the basic op-amp voltage gain, G. Thus, the closed-loop gain is essentially independent of the value of G and depends only on values of the external circuit elements.

Figure 8.3 illustrates a single negative-feedback op-amp circuit. We will analyze this circuit later. For now, note that a single resistor, R_F, is used to connect the output voltage, v_o, to the inverting input, v_-. Another resistor is connected from the inverting input, v_-, to the input voltage.

Circuits using op-amps, resistors, and capacitors can be configured to perform many useful operations such as summing, subtracting, integrating, filtering, comparing, and amplifying.

8.1.1 Analysis Method

We use the two important ideal op-amp properties in order to analyze circuits:

1. The voltage between v_+ and v_- is zero, or $v_+ = v_-$.
2. The current into both the v_+ and v_- terminals is zero.

These simple observations lead to a procedure for analyzing any ideal op-amp circuit as follows:

1. Write the Kirchhoff node equation at the noninverting terminal, v_+.
2. Write the Kirchhoff node equation at the inverting terminal, v_-.
3. Set $v_+ = v_-$ and solve for the desired closed-loop gains.

When performing the first two steps, remember that the current into both the v_+ and v_- terminals is zero.

8.2 THE INVERTING AMPLIFIER

Figure 8.3(a) illustrates an inverting amplifier with feedback, and Figure 8.3(b) shows the equivalent-circuit form of this ideal inverting op-amp circuit. We have used the properties of the ideal op-amp to model the op-amp input as an open circuit. The controlled source is Gv_d. We wish to solve for the output voltage, v_o, in terms of the input voltage, v_a. Let us follow the procedure of Section 8.1.1.

1. Kirchhoff's node equation at v_+ yields

$$\frac{v_+}{R} = 0$$

2. Kirchhoff's node equation at v_- yields

$$\frac{v_a - v_-}{R_a} + \frac{v_o - v_-}{R_F} = 0$$

3. Setting $v_+ = v_-$ yields

$$v_+ = v_- = 0$$

We now solve for the closed-loop gain as

$$\frac{v_o}{v_a} = \frac{-R_F}{R_a}$$

Notice that the closed-loop gain, v_o/v_a, is negative (inverted) and is dependent on the ratio of two resistors, R_F/R_a, and independent of the open-loop gain, G. This desirable result is caused by the use of feedback of a portion of the output voltage to subtract from the input voltage. The feedback from output to input through R_F serves to drive the differential voltage, $v_d = v_+ - v_-$, to zero. In this example, since the noninverting input voltage, v_+, is zero, the feedback has the effect of driving v_- to zero. Hence, at the input of the op-amp,

$$v_+ = v_- = 0$$

and there is a *virtual ground* at v_-. The term *virtual* means that the voltage, v_-, is zero (ground potential) but no current actually flows into this apparent ground connection (since no current can flow into either the inverting or noninverting op-amp terminal).

No matter how complex the ideal op-amp circuit may be, by following this procedure the engineer can quickly analyze (and soon design) op-amp systems.

We now expand this result to the case of multiple inputs. The amplifier shown in Figure 8.4 produces an output that is a negative summation of several input voltages. The node equation at v_+ yields $v_+ = 0$. The node equation at the inverting input is given by

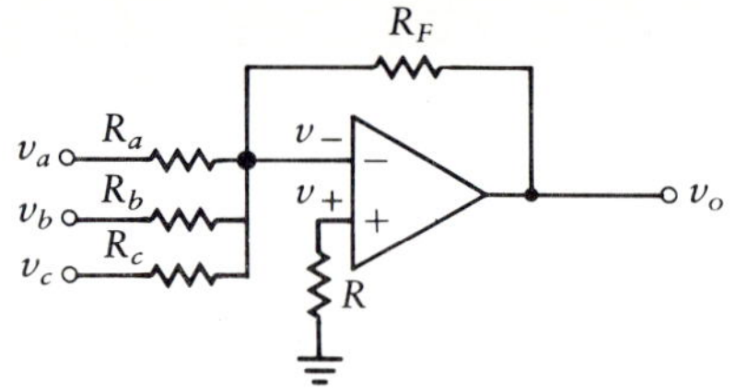

Figure 8.4
Op-amp circuit.

$$\frac{v_- - v_a}{R_a} + \frac{v_- - v_b}{R_b} + \frac{v_- - v_c}{R_c} + \frac{v_- - v_o}{R_F} = 0$$

Since $v_+ = v_-$, then $v_+ = 0 = v_-$ and we find v_o in terms of the inputs.

$$v_o = -R_F\left(\frac{v_a}{R_a} + \frac{v_b}{R_b} + \frac{v_c}{R_c}\right) = -R_F\sum_{j=a}^{c}\frac{v_j}{R_j} \tag{8.2}$$

The extension to n inputs should be obvious.

Drill Problems

Using the procedure of Section 8.1.1, determine v_o in terms of the input voltages for the following circuits. (The answers are shown directly on the figures.)

D8.1 Single inverting amplifier

D8.2 Voltage-divider amplifier

D8.3 Inverter

D8.4 Summing inverter

D8.5 Equal-gain summing inverter

D8.6 Dual inverted summer with gain

D8.7 Dual inverted weighted summer

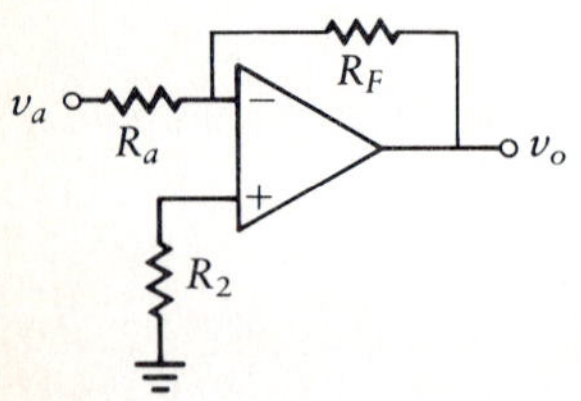

Ans: $\dfrac{-R_F}{R_a}v_a$

Figure D8.1

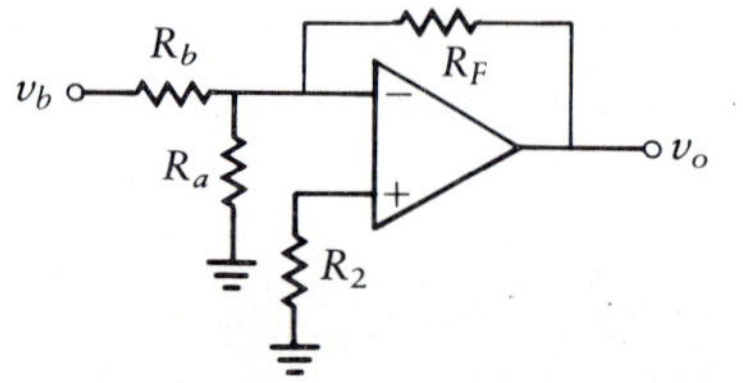

Ans: $v_o = \dfrac{-R_F}{R_b}v_b$

Figure D8.2

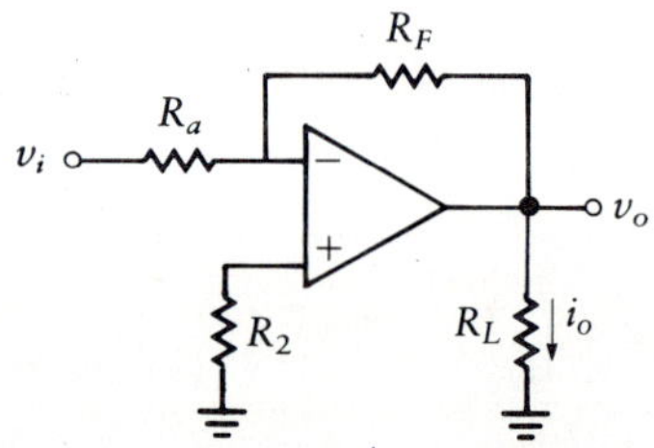

Ans: $v_o = -\dfrac{R_F}{R_a}v_i$

Figure D8.3

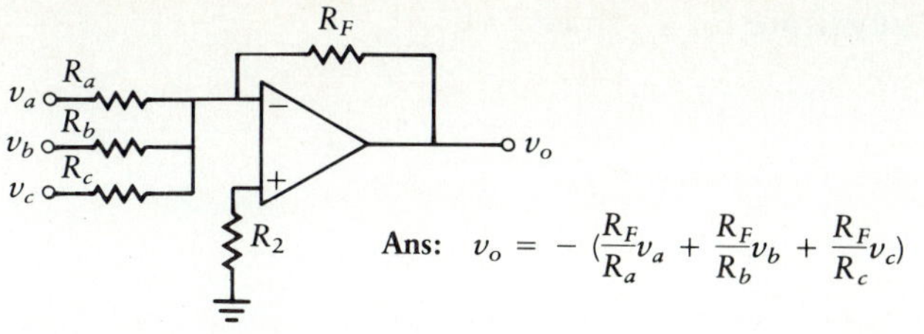

Figure D8.4

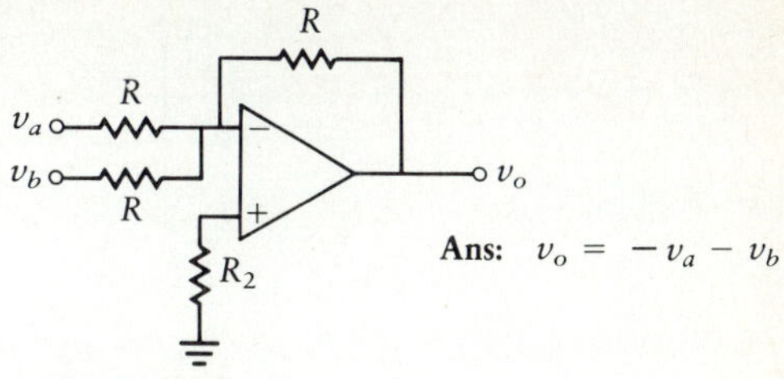

Figure D8.5

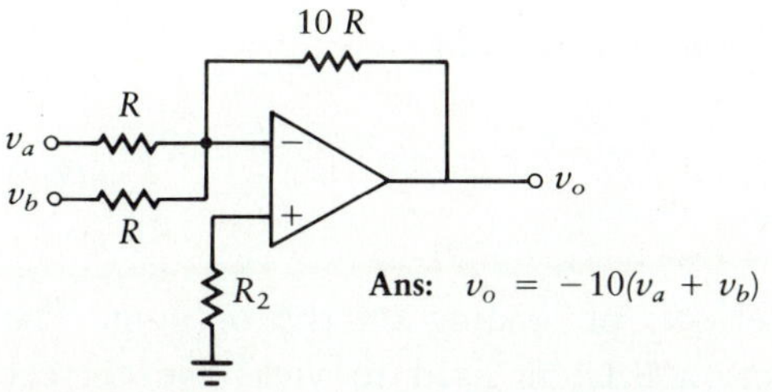

Figure D8.6

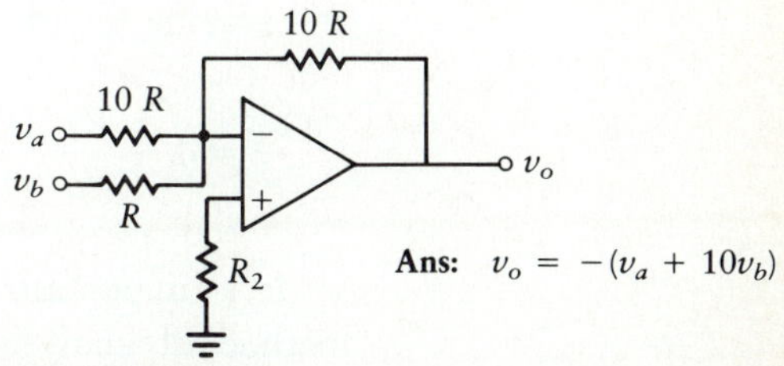

Figure D8.7

8.3 THE NONINVERTING AMPLIFIER

The operational amplifier can be configured to produce either an inverted or a noninverted output. In the previous section we analyzed the inverting amplifier, and in this section we repeat the analysis for the noninverting amplifier, as shown in Figure 8.5. We again follow the analysis procedure of Section 8.1.1 as follows:

1. Write a node equation at the v_+ node to get

$$v_+ = v_i$$

2. Write a node equation at the v_- node to get

$$\frac{v_-}{R_a} + \frac{v_- - v_o}{R_F} = 0$$

Figure 8.5
Noninverting amplifier.

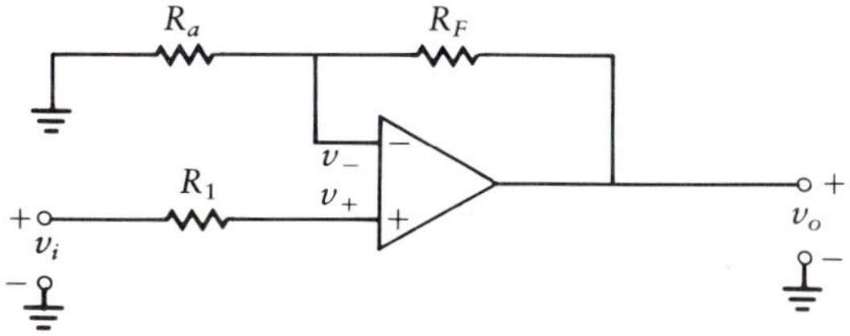

3. Set $v_+ = v_-$, and substitute for v_- since

$$v_+ = v_i = v_-$$

Then

$$\frac{v_i}{R_a} + \frac{v_i - v_o}{R_F} = 0$$

and solving for the gain, we obtain a noninverted gain

$$\frac{v_o}{v_i} = 1 + \frac{R_F}{R_a} \tag{8.3}$$

It is important to consider the effects of loading on the op-amp. The method of analysis presented in Section 8.1.1 is used to yield the correct voltage gain, v_o/v_i. The results are based on the assumption that the current required from the op-amp is within the range set forth by the manufacturer's data sheet. The next two examples illustrate this concept.

Example 8.1

Find v_o and i_o for the circuit of Figure D8.3.

SOLUTION We write the node equations, as set forth in Section 8.1.1, as follows:

$$v_+ = 0$$

$$\frac{v_- - v_i}{R_a} + \frac{v_- - v_o}{R_F} = 0$$

$$v_+ = v_-$$

Solving these three equations, we obtain

$$v_o = -\frac{R_F}{R_a} v_i$$

We must now find i_o. This is found from v_o by applying Ohm's law to the load resistor, R_L.

$$i_o = \frac{v_o}{R_L} = -\frac{R_F}{R_a R_L} v_i \tag{8.4}$$

Notice that the value of R_L does not appear in the expression for the gain, v_o/v_i. However, R_L cannot be so low as to demand an i_o that exceeds the op-amp specifications.

Example 8.2

Find v_o and i_o for the circuit shown in Figure 8.6(a). The input voltage is sinusoidal as indicated on the diagram. Check your result using a computer simulation.

SOLUTION We begin by writing the KCL equations at both the + and − terminals of the op-amp.

For the negative terminal,

$$\frac{v_o - v_-}{9.8 \times 10^4} + \frac{0 - v_-}{7000} = 0$$

Therefore,

$$15v_- = v_o$$

For the positive terminal

$$\frac{v_i - v_+}{10^4} + \frac{0 - v_+}{2 \times 10^4} = 0$$

Therefore,

$$\tfrac{2}{3}v_i = v_+$$

Substituting for v_-, since $v_+ = v_-$, we have

$$15(\tfrac{2}{3}v_i) = v_o$$
$$v_o = 10v_i$$

If v_i is now given by

$$v_i = 0.5 \sin \omega t \text{ V}$$

then

$$v_o = 5 \sin \omega t \text{ V}$$

Since the 2-kΩ resistor forms the load of the op-amp, by Ohm's law we have

$$i_o = \frac{v_o}{R_L} = 2.5 \sin \omega t \text{ mA}$$

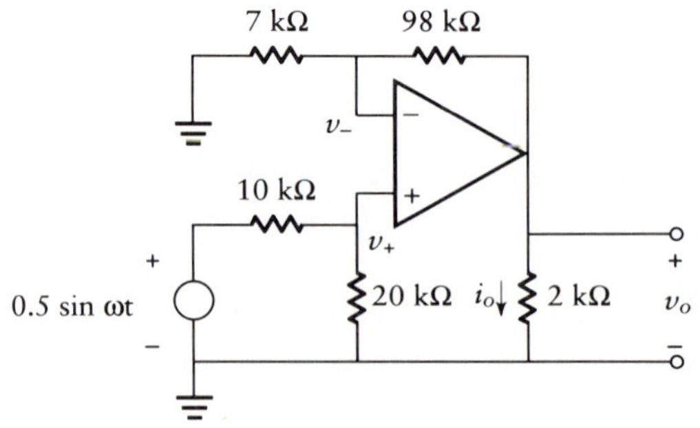

(a)

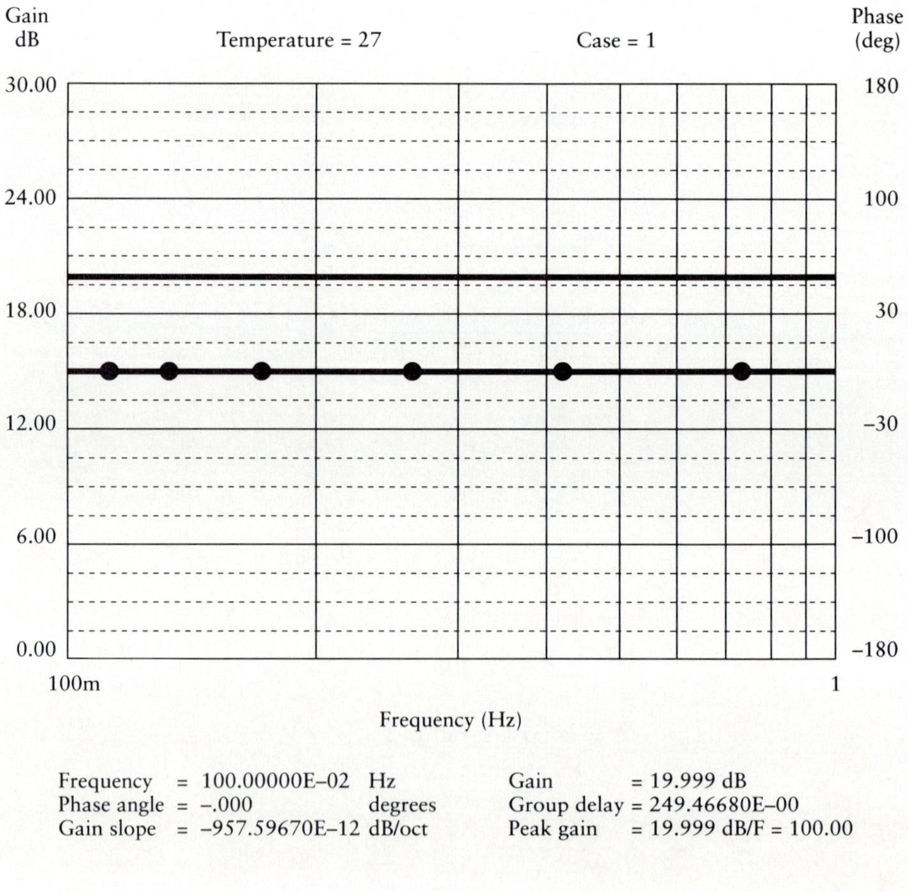

(b)

Figure 8.6
Circuit and result for Example 8.2.

We now verify this result using the computer simulation program. We first need to input the circuit. This is done in a straightforward manner whether using SPICE or a derivative program such as MICRO-CAP II. The only questions that may come up during the input process are related to the op-amp and to the source. The current chapter deals with ideal op-amps, yet the simulation programs model practical op-amps, which we cover in Chapter 10. However, we shall find that practical op-amps behave very much like their ideal counterparts as long as input frequencies are relatively low. The specified input frequency for this example is only 1 rad/s ($1/2\pi$ Hz), which is extremely low. We therefore can choose a practical op-amp and expect results very close to ideal. Let us choose the LM741 op-amp. We can then input the source by creating a library entry with that particular waveform.

There is, however, a much simpler way to solve this problem. Since the specified input is sinusoidal, we can perform an ac analysis and apply the results directly. Performing such an analysis (using a frequency range between 0.1 and 1 Hz) yields a gain of 19.999 dB and a phase shift of 0.00°. This response curve is shown in Figure 8.6(b). A gain of 10 would be 20 dB, so we see that the simulation results in virtually the identical answer for output voltage.

Drill Problems

Using the ideal op-amp approximations, determine v_o in terms of the input voltages for the following four circuits. (The answers are shown directly on the figures.)

D8.8 Noninverting amplifier

D8.9 Noninverting buffer

D8.10 Noninverting input with voltage divider

D8.11 Less than unity gain

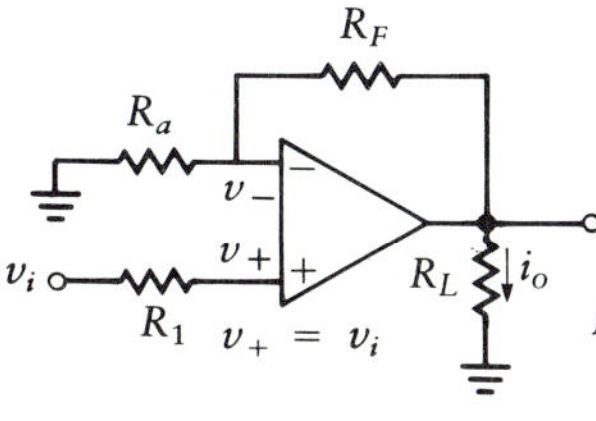

Ans: $v_o = (1 + \frac{R_F}{R_a}) v_i$

Figure D8.8

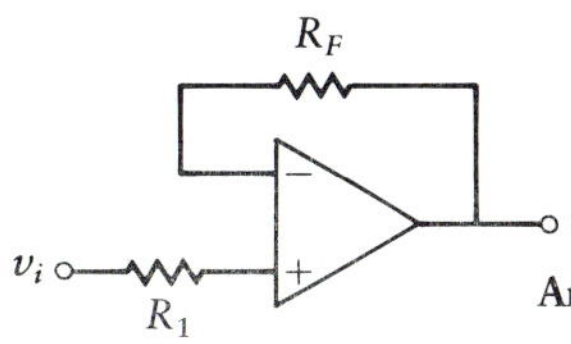

Ans: $v_o = v_i$

Figure D8.9

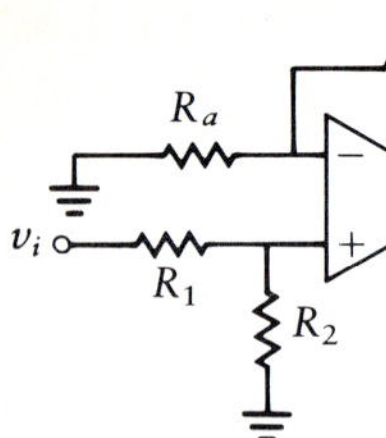

Ans: $v_o = (1 + \frac{R_F}{R_a})(\frac{R_2}{R_1 + R_2}) v_i$

Figure D8.10

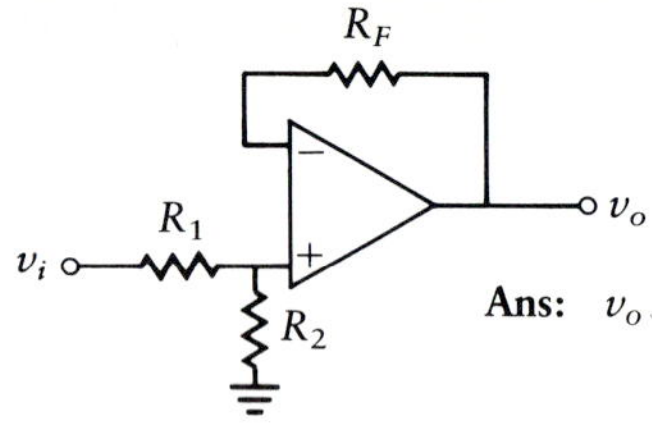

Ans: $v_o = \frac{R_2}{R_1 + R_2} v_i$

Figure D8.11

Let us now analyze noninverting op-amps with multiple inputs. Figure 8.7 illustrates a circuit with two input voltages, which are applied to the noninverting input of the operational amplifier. The analysis of this circuit follows the procedure of Section 8.1.1. In order to find v_+, we apply Kirchhoff's current law to the noninverting input terminal to yield (recall that the input current to the op-amp is zero)

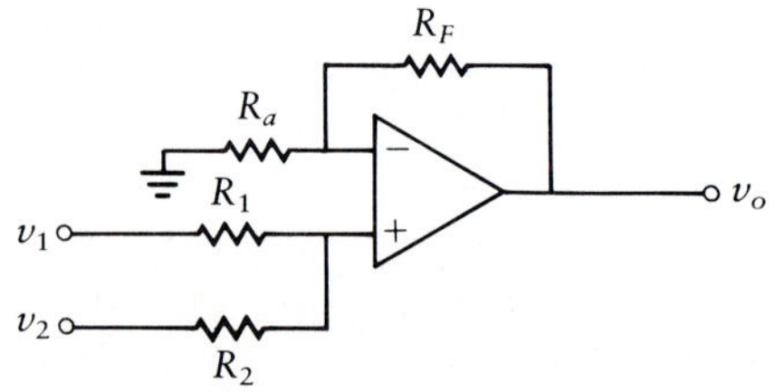

Figure 8.7
Two noninverting inputs.

$$\frac{v_1 - v_+}{R_1} + \frac{v_2 - v_+}{R_2} = 0$$

Solving for v_+, we obtain

$$v_+ = (R_1 \parallel R_2)\left(\frac{v_1}{R_1} + \frac{v_2}{R_2}\right)$$

The inverting voltage, v_-, is found from the node equation at v_- with the result

$$v_- = \frac{R_a v_o}{R_a + R_F}$$

Setting v_+ equal to v_-, we obtain

$$v_o = (R_1 \parallel R_2)\left(\frac{v_1}{R_1} + \frac{v_2}{R_2}\right)\left(1 + \frac{R_F}{R_a}\right) \tag{8.5}$$

Drill Problems

Determine v_o in terms of the input voltages for the following circuits. (The answers are shown directly on the figures.)

D8.12 Noninverting summer with gain

D8.13 Unity-gain summer

D8.14 Sum of two inputs with load resistor

D8.15 Weighted sum of two inputs

D8.16 Sum of three inputs

D8.17 Voltage divider weighted sum of two inputs

When analyzing both the inverting and the noninverting configurations, we must be certain that the op-amp is capable of driving the load resistor. Example 8.3 illustrates this concept.

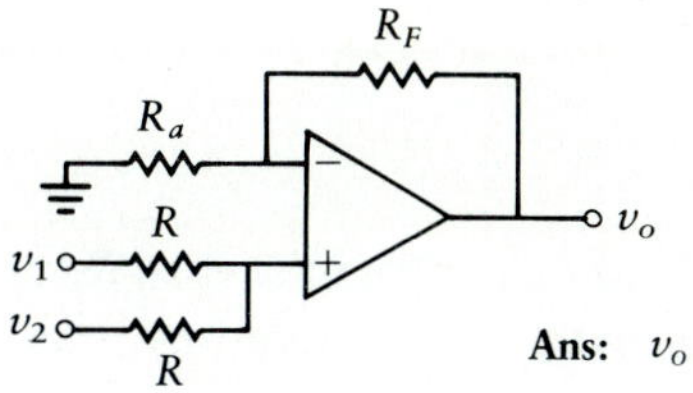

Ans: $v_o = \frac{1}{2}(1 + \frac{R_F}{R_a})(v_1 + v_2)$

Figure D8.12

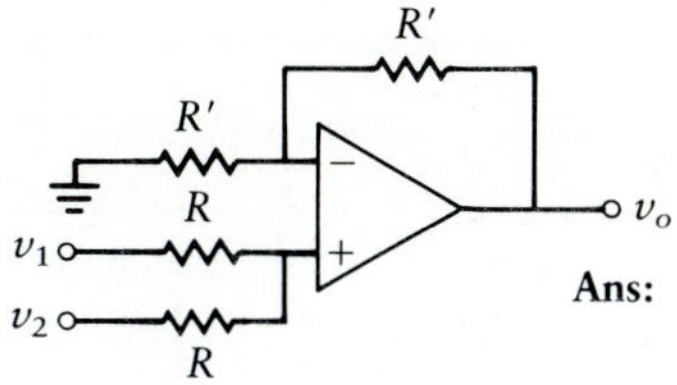

Ans: $v_o = v_1 + v_2$

Figure D8.13

Ans: $v_o = (1 + \frac{R_F}{R_a})(\frac{1}{2})(v_1 + v_2)$;

$i_o = \frac{v_o}{R_L}$

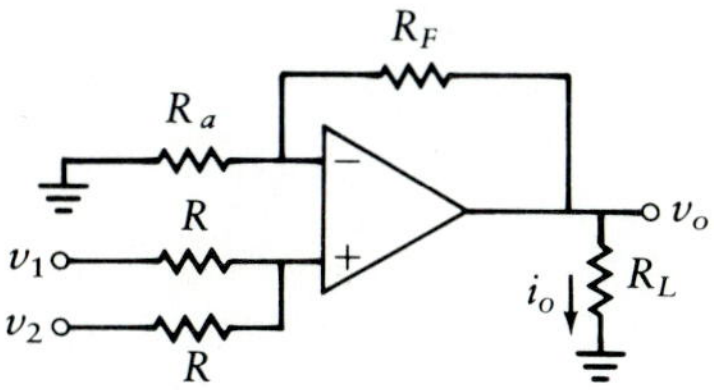

Figure D8.14

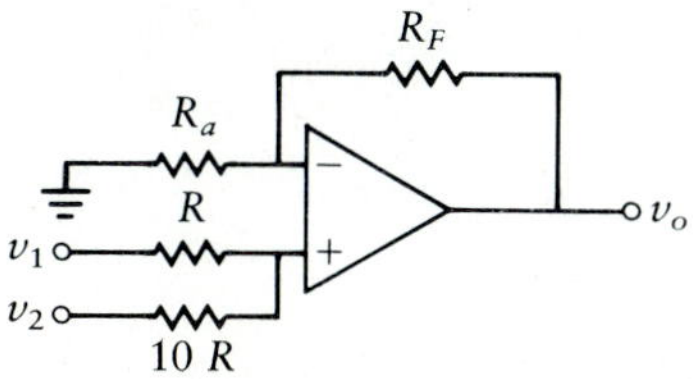

Ans: $v_o = (1 + \frac{R_F}{R_a})(\frac{1}{11})(10v_1 + v_2)$

Figure D8.15

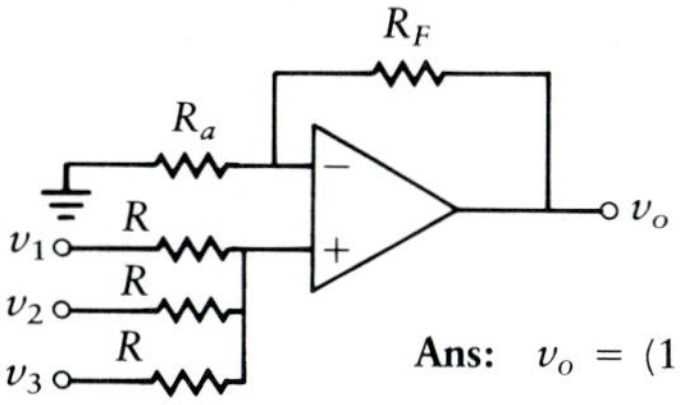

Ans: $v_o = (1 + \frac{R_F}{R_a})(\frac{1}{3})(v_1 + v_2 + v_3)$

Figure D8.16

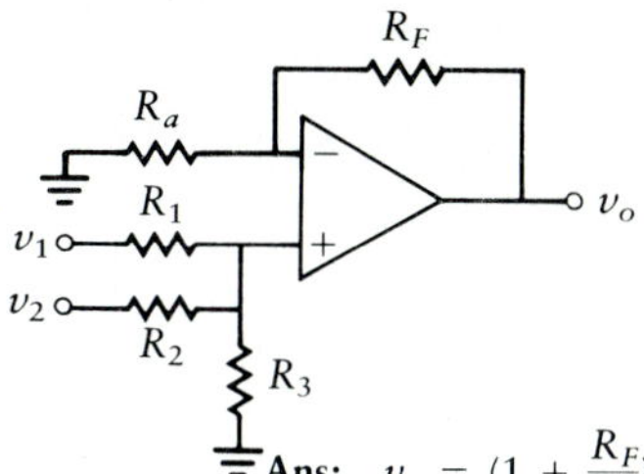

Ans: $v_o = (1 + \frac{R_F}{R_a})(R_1 \parallel R_2 \parallel R_3)(\frac{v_1}{R_1} + \frac{v_2}{R_2})$

Figure D8.17

Example 8.3

Find v_o and i_o for the circuit of Figure D8.8.

SOLUTION We write node equations, as set forth in Section 8.1.1.

$$v_+ = v_i$$

$$v_- = \frac{R_a}{R_a + R_F} v_o$$

$$v_- = v_+$$

Solving these three equations, we obtain

$$v_o = \left(1 + \frac{R_F}{R_a}\right) v_i$$

The current through the load resistor, R_L, is then

$$i_o = \frac{v_o}{R_L} = \frac{1}{R_L}\left(1 + \frac{R_F}{R_a}\right) v_i$$

Again, R_L does not affect the equations used to find v_o/v_i provided that R_L is not so low as to overload the op-amp. However, if the design requires more current than can be provided by the op-amp, it may be necessary to design a power amplifier, as discussed in Chapter 6, between the op-amp and the load, R_L.

8.4 INPUT RESISTANCE OF OP-AMP CIRCUITS

The input resistance of the ideal op-amp is infinite. However, the input resistance to a circuit composed of an ideal op-amp that is connected to external components is not infinite. In fact, it depends on the form of the op-amp circuit. In this section, we calculate the input resistance for an op-amp with external components. We first consider the inverting op-amp. The equivalent circuit for an inverting op-amp is shown in Figure 8.8(a). Figure 8.8(b) shows the same circuit rearranged for simplicity of analysis. Note that we have attached a voltage source to the input in order to calculate the equivalent resistance. Since the circuit contains a dependent voltage source, we find the input resistance by assuming a voltage, v, and finding the ratio of that voltage to the resulting current, i. The loop equation is given by

$$(R_a + R_F)i = v - Gv_d$$

but since

$$v_d = -(v - iR_a)$$

Figure 8.8
Input resistance, inverting amplifier.

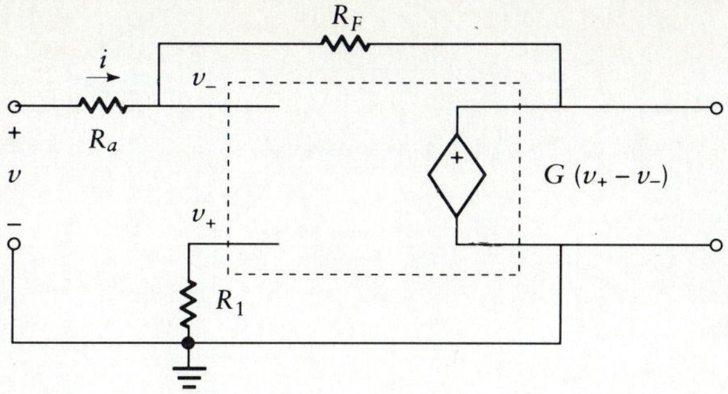

(a)

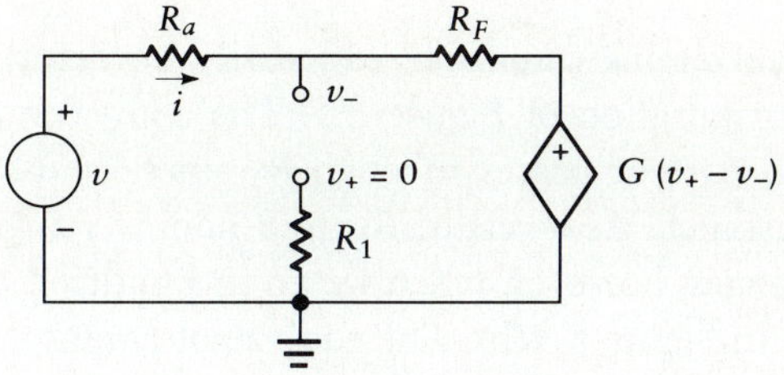

(b)

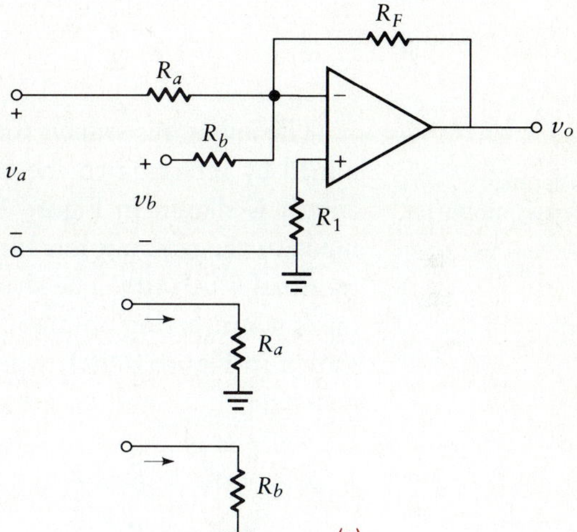

(c)

then

$$(R_a + R_F)i = (1 + G)v - GiR_a$$

The equivalent input resistance is then

$$R_{\text{in}} = \frac{v}{i} = \frac{R_a + R_F}{1 + G} + \frac{GR_a}{1 + G} \tag{8.6}$$

As the loop gain, G, approaches infinity, the first term in equation (8.6) approaches zero and the input resistance tends toward R_a. Thus, the input resistance seen by the source is equal to the value of the external resistance, R_a.

Figure 8.8(c) shows an inverting amplifier with two inputs, each of which is applied through a corresponding input resistance. This is a special case of the circuit of Figure 8.4. Since the voltage at the inverting input to the op-amp is zero (virtual ground), the input resistance seen by v_a is R_a, and that seen by v_b is R_b. The "grounded" inverting input also serves to isolate the two inputs from each other. That is, a variation in v_a does not affect the input v_b, and vice versa. The term virtual ground is used in this situation since the v_- terminal is virtually grounded for ac signals.

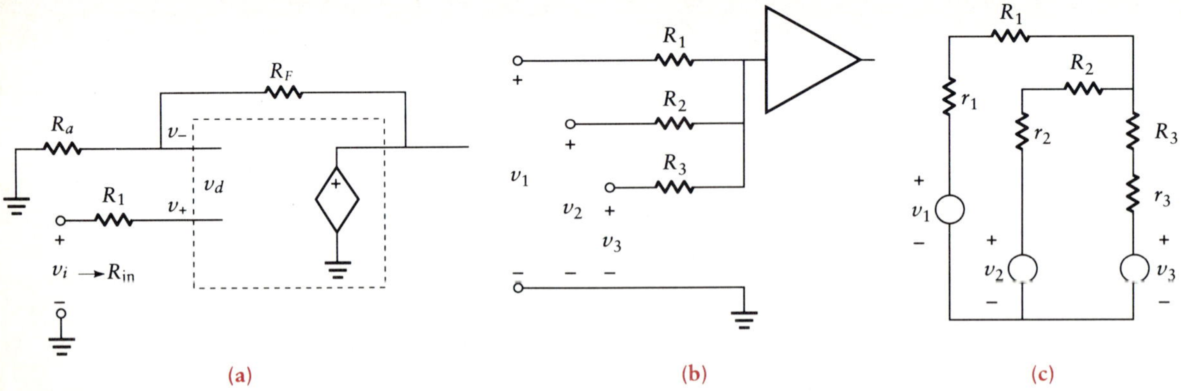

Figure 8.9
Input resistance, noninverting amplifier.

The input resistance for the noninverting amplifier can easily be determined by referring to the circuit configuration of Figure 8.5. The equivalent circuit is shown in Figure 8.9(a). No current passes through R_1 since the v_+ input to the op-amp has infinite resistance. As a result, R_{in} to a noninverting terminal is infinity. The situation changes, however, when we go to a multiple-input noninverting op-amp, as shown in Figure 8.9(b). The equivalent circuit is shown in Figure 8.9(c), where for simplicity we have assumed that the source resistors, r_1, r_2, and r_3, are all zero. The input resistance for v_1 is

$$R_{in1} = R_1 + R_2 \parallel R_3$$

$$R_{in2} = R_2 + R_1 \parallel R_3$$

$$R_{in3} = R_3 + R_1 \parallel R_2$$

This concept can easily be extended to n inputs.

It becomes clear from the discussion that if your design needs a large input resistance, you should use a single-input noninverting op-amp. Such a configuration is called a *noninverting buffer* and is shown in Drill Problem D8.9.

8.5 COMBINED INVERTING AND NONINVERTING INPUTS

The most general case of input configuration is a combination of the previous two cases. That is, we allow for both inverting and noninverting inputs. The general configuration is shown in Figure 8.10. The previous circuits can be considered as special cases of this general problem. The output relationship is found by applying superposition as follows. Equation (8.7) is derived by combining equation (8.2) with equation (8.5). (You should not just accept the result but should perform the simple derivation.)

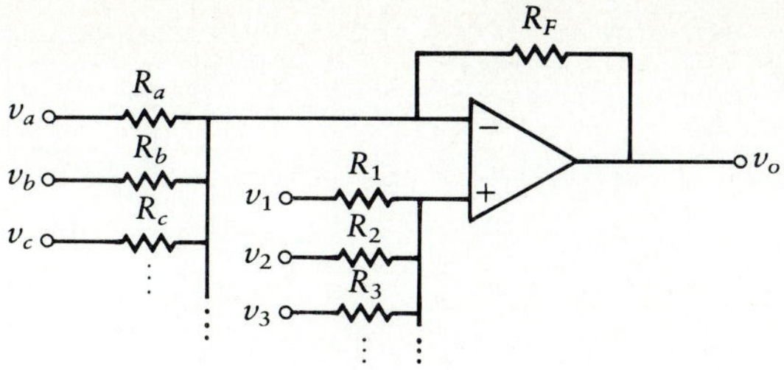

Figure 8.10 Inverting and noninverting inputs.

$$v_o = \left[1 + \frac{R_F}{R_a \parallel R_b \parallel R_c \parallel \cdots}\right](R_1 \parallel R_2 \parallel R_3 \parallel \cdots)\left[\frac{v_1}{R_1} + \frac{v_2}{R_2} + \frac{v_3}{R_3} + \cdots\right] - \left[\frac{R_F}{R_a}v_a + \frac{R_F}{R_b}v_b + \frac{R_F}{R_c}v_c + \cdots\right] \quad (8.7)$$

Equation (8.7) represents a general result that will prove useful in analyzing a wide variety of circuits.

Drill Problems

Determine v_o in terms of all input voltages for the following configurations. (The answers are shown directly on the figures.)

D8.18 Positive and negative gain configuration

D8.19 Differencing amplifier

D8.20 Weighted differencing amplifier

D8.21 Differencing amplifier with gain

D8.22 Sign switcher

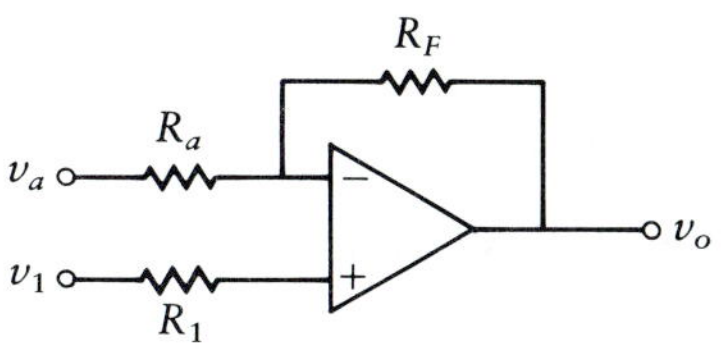

Ans: $v_o = (1 + \frac{R_F}{R_a})v_1 - \frac{R_F}{R_a}v_a$

Figure D8.18

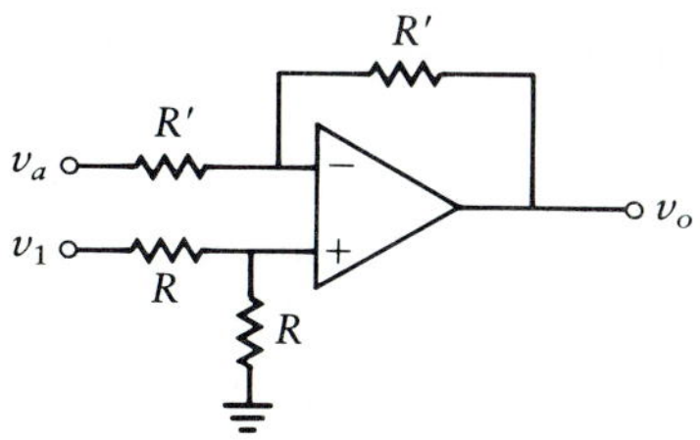

Ans: $v_o = v_1 - v_a$

Figure D8.19

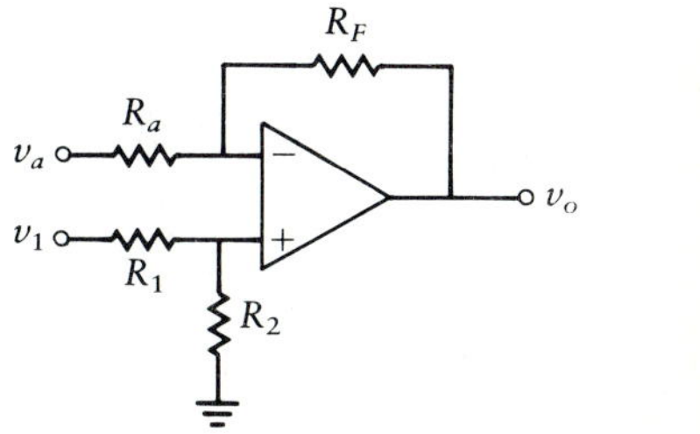

Ans: $v_o = (1 + \frac{R_F}{R_a})(\frac{R_2}{R_1 + R_2})v_1 - \frac{R_F}{R_a}v_a$

Figure D8.20

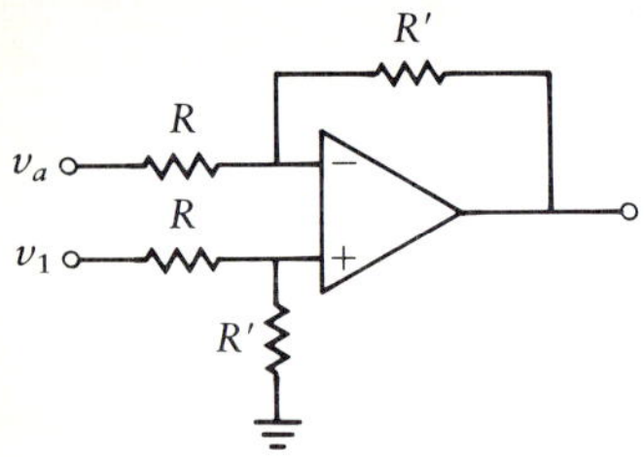

Ans: $v_o = \frac{R'}{R}(v_1 - v_a)$

Figure D8.21

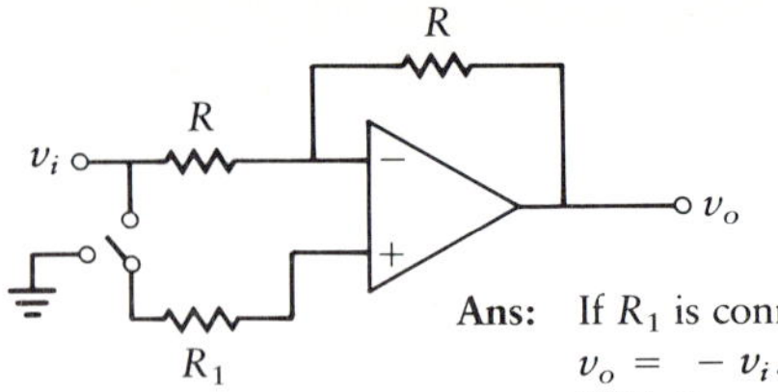

Ans: If R_1 is connected to ground, $v_o = -v_i$.
With R_1 connected to v_i, $v_o = (1 + 1)v_i - v_i = v_i$.

Figure D8.22

8.6 DESIGN OF OP-AMP CIRCUITS

When the configuration of an op-amp system is given, we *analyze* that system to determine the output in terms of the inputs, using the procedure of Section 8.1.1.

If you wish to *design* a circuit that combines both inverting and noninverting inputs, the problem is more complex. We present in this section a practical design technique. This technique allows us to design an op-amp summing circuit without elaborate solution of simultaneous equations.*

We first rearrange the circuit equations to put them into a different format. The design method will be applied to several example designs.

In a design problem, a desired linear equation is given and the op-amp circuit must be designed. The desired output of the operational amplifier summer can be expressed as a linear combination of inputs,

$$v_o = X_1v_1 + \cdots X_nv_n - Y_av_a - \cdots Y_mv_m \tag{8.8}$$

where $X_1, X_2, \ldots X_n$ are the desired gains at the noninverting inputs and $Y_a, Y_b, \ldots Y_m$ are the desired gains at the inverting inputs. Equation (8.8) is implemented with thc circuit of Figure 8.11.

Equation (8.7) shows that the values of the resistors, $R_a, R_b, \ldots R_m$ and $R_1, R_2, \ldots R_n$ are inversely proportional to the desired gains associated with the respective input voltages. In other words, if a large gain is desired at a particular input terminal, the resistance at that terminal is small.

When the open-loop gain of the operational amplifier, G, is large, the output voltage may be written in terms of the resistors connected to the operational amplifier as in equation (8.7). Equation (8.9) repeats this expression with slight simplification.

---

* This technique was devised by Phil Vrbancic, a student at California State University, Long Beach, and presented in a paper submitted to the IEEE Region VI 1982 Prize Paper Contest.

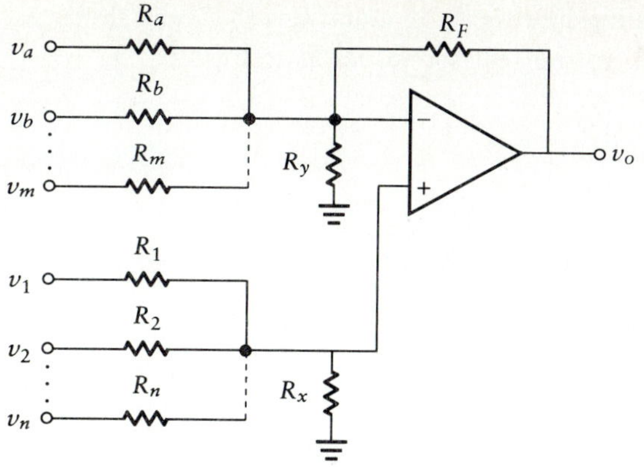

Figure 8.11 Multiple-input summer.

$$v_o = R_{eq}\left[\frac{v_1}{R_1} + \frac{v_2}{R_2} + \cdots + \frac{v_n}{R_n}\right] - R_F\left[\frac{v_a}{R_a} + \frac{v_b}{R_b} + \cdots + \frac{v_m}{R_m}\right] \tag{8.9}$$

where

$$R_A = R_a \parallel R_b \parallel \cdots \parallel R_m \parallel R_y \tag{8.10}$$

$$R_{eq} = \left(1 + \frac{R_F}{R_A}\right)(R_1 \parallel R_2 \parallel \cdots \parallel R_n \parallel R_x) \tag{8.11}$$

We see that the output voltage is a linear combination of inputs where each input is divided by its associated resistance and multiplied by another resistance. The multiplying resistance is R_F for inverting inputs and R_{eq} for noninverting inputs.

Although not required for design using ideal op-amps, we will use a design constraint that is important for nonideal op-amps. For the noninverting op-amp, the Thevenin resistance looking back from the inverting input is usually made equal to that looking back from the noninverting input. We show the importance of this design constraint later in this text. For the configuration shown in Figure 8.11, this constraint can be expressed as follows:

$$R_1 \parallel R_2 \parallel \cdots \parallel R_n \parallel R_x = R_a \parallel R_b \parallel \cdots \parallel R_m \parallel R_y \parallel R_F \tag{8.12}$$

Substituting equations (8.10) and (8.11) into equation (8.12) yields

$$\frac{R_{eq}}{1 + R_F/R_A} = R_A \parallel R_F \tag{8.13}$$

from which we obtain

$$R_{eq} = R_F \tag{8.14}$$

By equating like terms in equations (8.8) and (8.9), we can relate the values of X_i and Y_j to the resistor values as follows:

$$X_i = \frac{R_{\text{eq}}}{R_i} = \frac{R_F}{R_i} \tag{8.15}$$

and

$$Y_j = \frac{R_F}{R_j} \tag{8.16}$$

Equation (8.12) can now be rewritten as follows:

$$\frac{1}{1/R_x + \Sigma_{i=1}^{n} 1/R_i} = \frac{1}{1/R_F + 1/R_y + \Sigma_{j=a}^{m} 1/R_j} \tag{8.17}$$

or

$$\frac{1}{R_x} + \sum_{i=1}^{n} \frac{1}{R_i} = \frac{1}{R_F} + \frac{1}{R_y} + \sum_{j=a}^{m} \frac{1}{R_j} \tag{8.18}$$

Substituting equations (8.15) and (8.16) into equation (8.18), we obtain

$$\frac{1}{R_x} + \sum_{i=1}^{n} \frac{X_i}{R_F} = \frac{1}{R_F} + \frac{1}{R_y} + \sum_{j=a}^{m} \frac{Y_j}{R_F}$$

$$\frac{1}{R_x} + \frac{1}{R_F} \sum_{i=1}^{n} X_i = \frac{1}{R_F} + \frac{1}{R_y} + \frac{1}{R_F} \sum_{j=a}^{m} Y_j \tag{8.19}$$

Recall that our goal is to solve for the resistors in terms of the X_i and Y_j. Let us define

$$X = \sum_{i=1}^{n} X_i \tag{8.20}$$

and

$$Y = \sum_{j=a}^{m} Y_j$$

We can then rewrite equation (8.19) as follows:

$$\frac{1}{R_x} + \frac{1}{R_F} X = \frac{1}{R_F} + \frac{1}{R_y} + \frac{1}{R_F} Y \tag{8.21}$$

This is a starting point for our design procedure. Recall that R_x and R_y are the resistors between ground and the noninverting and inverting inputs, respectively. The feedback resistor is denoted by R_F.

We can eliminate either or both of the resistors, R_x and R_y, from the circuit of Figure 8.11. That is, either or both of these resistors can be set to infinity (i.e., open-circuited). This yields three design possibilities, which we denote as Case I, Case II, and Case III. Depending on the desired multiplying factors relating output to input, one of these cases will yield the appropriate design. The following results are summarized in Table 8.1.

Table 8.1 Summary of Summing Amplifier Design

<table>
<tr><th>Case</th><th>Z</th><th>R_y</th><th>R_x</th><th>R_i</th><th>R_j</th></tr>
<tr><td>I</td><td>>0</td><td>$\frac{R_F}{Z}$</td><td>∞</td><td rowspan="3">$\frac{R_F}{X_i}$</td><td rowspan="3">$\frac{R_F}{Y_j}$</td></tr>
<tr><td>II</td><td><0</td><td>∞</td><td>$\frac{R_F}{-Z}$</td></tr>
<tr><td>III</td><td>0</td><td>∞</td><td>∞</td></tr>
</table>

Case I If $R_x \to \infty$, equation (8.21) becomes

$$\frac{X}{R_F} = \frac{1}{R_F} + \frac{1}{R_y} + \frac{Y}{R_F} \tag{8.22}$$

When we choose an R_F, R_y is the only unknown in this equation. Solving for this, we obtain

$$\begin{aligned}\frac{1}{R_y} &= \frac{X}{R_F} - \frac{Y}{R_F} - \frac{1}{R_F}\\ &= \frac{X - Y - 1}{R_F}\end{aligned} \tag{8.23}$$

Let us now define

$$Z = X - Y - 1$$

Then equation (8.23) becomes

$$\frac{1}{R_y} = \frac{Z}{R_F}$$

Z must be positive for R_y to be physically realizable. If Z is negative, Case I does not apply.

Equations (8.15) and (8.16) yield the resistance values

$$R_i = \frac{R_F}{X_i}, \qquad R_j = \frac{R_F}{Y_j} \tag{8.24}$$

Case II When $R_y \to \infty$, equation (8.21) becomes

$$\frac{1}{R_x} + \frac{X}{R_F} = \frac{1}{R_F} + \frac{Y}{R_F} \tag{8.25}$$

and

$$\frac{1}{R_x} = -\frac{X - Y - 1}{R_F} = \frac{-Z}{R_F} \tag{8.26}$$

Hence, Z must be negative.

Case III When $R_x \to \infty$ and $R_y \to \infty$, equation (8.21) becomes

$$\frac{X}{R_F} = \frac{1}{R_F} + \frac{Y}{R_F} \tag{8.27}$$

and

$$0 = \frac{X - Y - 1}{R_F} = \frac{Z}{R_F} \tag{8.28}$$

Therefore, $Z = 0$ and $R_x = R_y \to \infty$.

The results of all three cases are combined in Table 8.1. Note that $Z = X - Y - 1$, where

$$X = \sum_{i=1}^{n} X_i$$

$$Y = \sum_{j=a}^{m} Y_j$$

Example 8.4 Op-Amp Summer (Design)

Design an op-amp summer to yield the following input/output relationship.

$$v_o = 10v_1 + 6v_2 + 4v_3 - 5v_a - 2v_b$$

SOLUTION The values of X, Y, and Z are calculated as follows:

$$X = \sum_{i=1}^{3} X_i = 10 + 6 + 4 = 20$$

$$Y = \sum_{j=a}^{b} Y_j = 5 + 2 = 7$$

$$Z = X - Y - 1 = 20 - 7 - 1 = 12$$

In this example, Z is greater than zero, so we are dealing with Case I, where R_x is open-circuited. A suitable value of R_F must first be chosen. When R_F has been determined, all other resistor values are easily found. Because there is one more unknown variable than the number of equations, R_F can be selected to be any value we wish. We choose a value that will enhance the final design. We offer one method of selecting a value of R_F. Suppose we want the minimum resistance, R_{min}, at any of the inputs to be 10 kΩ. Then the multiplying factor, K, would be the largest of any X_i, Y_j, or Z. Thus, $K = 12$ and $R_F = 10\ \text{k}\Omega \times 12 = 120\ \text{k}\Omega$. We do not want to choose too small a value, or the circuit will load previous circuitry. We also do not use an exceptionally large value for R_F, since this would increase the noise generated in the circuit. All resistors used in the op-amp circuit should be between 1 kΩ and 1 MΩ. Having determined R_F, we find the resistors from equation (8.24) as follows:

$$R_1 = \frac{R_F}{X_1} = \frac{120\ \text{k}\Omega}{10} = 12\ \text{k}\Omega$$

$$R_2 = \frac{R_F}{X_2} = \frac{120\ \text{k}\Omega}{6} = 20\ \text{k}\Omega$$

$$R_3 = \frac{R_F}{X_3} = \frac{120\ \text{k}\Omega}{4} = 30\ \text{k}\Omega$$

$$R_a = \frac{R_F}{Y_a} = \frac{120\ \text{k}\Omega}{5} = 24\ \text{k}\Omega$$

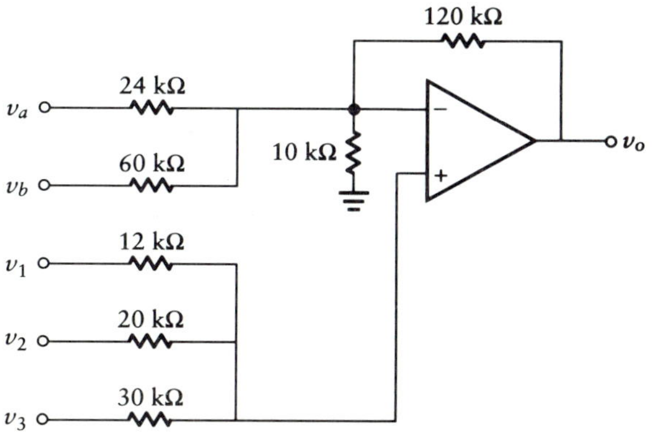

Figure 8.12 Amplifier for Example 8.4.

$$R_b = \frac{R_F}{Y_b} = \frac{120\text{ k}\Omega}{2} = 60\text{ k}\Omega$$

$$R_y = \frac{R_F}{Z} = \frac{120\text{ k}\Omega}{12} = 10\text{ k}\Omega$$

The resulting amplifier is shown in Figure 8.12.

Example 8.5 Op-Amp Summer (Design)

Design an op-amp circuit to implement the following equation:

$$v_o = 4v_1 + v_2 - 8v_a - 6v_b$$

SOLUTION We first calculate the values of X, Y, and Z.

$$X = 4 + 1 = 5$$

$$Y = 8 + 6 = 14$$

$$Z = 5 - 14 - 1 = -10$$

Since Z is less than zero, R_y is open-circuited and we are dealing with an example of Case II. Suppose that in this case we want the equivalent resistance at the + and − terminals to be 10 kΩ. Then the multiplying factor would be the larger of X or $Y + 1$. This would make $K = 15$ and $R_F = 15 \times 10\text{ k}\Omega = 150\text{ k}\Omega$.

$$R_x = \frac{R_F}{-Z} = \frac{150\text{ k}\Omega}{10} = 15\text{ k}\Omega$$

$$R_a = \frac{R_F}{Y_a} = \frac{150\text{ k}\Omega}{8} = 18.75\text{ k}\Omega$$

$$R_b = \frac{R_F}{Y_b} = \frac{150\text{ k}\Omega}{6} = 25\text{ k}\Omega$$

$$R_1 = \frac{R_F}{X_1} = \frac{150\text{ k}\Omega}{4} = 37.5\text{ k}\Omega$$

$$R_2 = \frac{R_F}{X_2} = \frac{150\text{ k}\Omega}{1} = 150\text{ k}\Omega$$

The complete circuit is shown in Figure 8.13. Note that at each input terminal, the equivalent resistance is 10 kΩ, calculated as follows:

$$37.5\text{ k}\Omega \parallel 150\text{ k}\Omega \parallel 15\text{ k}\Omega = 25\text{ k}\Omega \parallel 18.75\text{ k}\Omega \parallel 150\text{ k}\Omega = 10\text{ k}\Omega$$

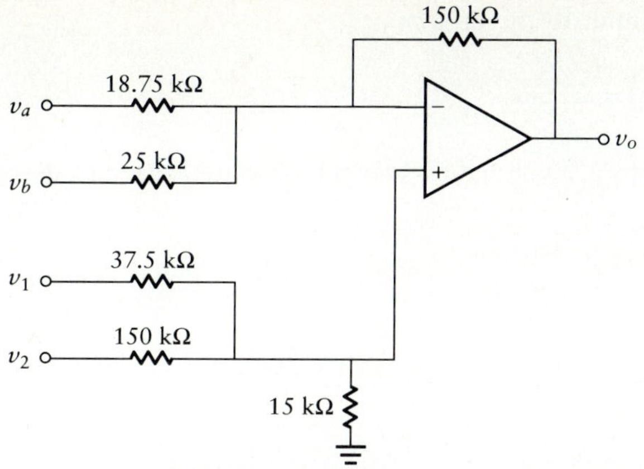

Figure 8.13 Amplifier for Example 8.5.

Recall that this design forced these to be equal to minimize the dc bias current offset. The initial selection of R_F is not critical, since if we are not happy with the resulting resistor values, they can all be multiplied by the same constant without changing the voltage relationships.

8.7 OTHER OP-AMP APPLICATIONS

We have seen that the op-amp can be used as an amplifier or as a means of combining a number of inputs in a linear manner. In the current section, we investigate several additional important applications of this versatile linear IC.

8.7.1 Negative-Impedance Circuit

The circuit shown in Figure 8.14 produces a negative input resistance (impedance in the general case). This circuit can be used to cancel an unwanted positive resistance. Numerous oscillator applications depend on a negative-resistance op-amp circuit. The input resistance, R_{in}, is defined as

$$R_{in} = \frac{v}{i}$$

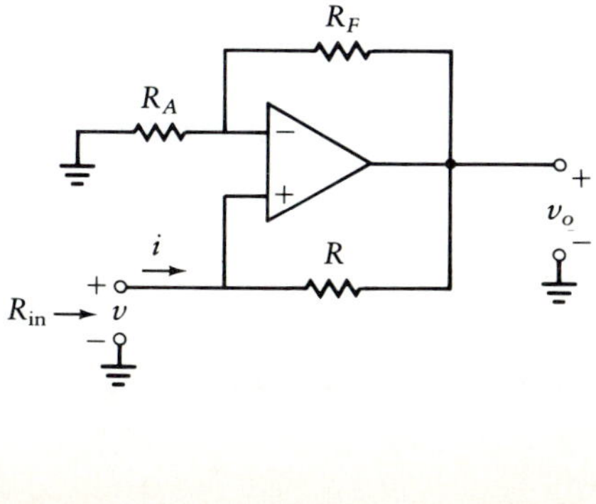

Figure 8.14 Negative impedance circuit.

The input to the op-amp, v_+, is

$$v_+ = v$$

As before, a voltage-divider relationship is used to derive the expression for v_-.

$$v_- = \frac{R_A v_o}{R_A + R_F}$$

We let $v_+ = v_-$ and solve for v_o in terms of v, which yields

$$v_o = v\left(1 + \frac{R_F}{R_A}\right)$$

Since the input impedance to the v_+ terminal is infinite, the current in R is equal to i and can be found as follows:

$$i = \frac{v - v_o}{R} = \frac{v - v(1 + R_F/R_A)}{R} = \frac{-R_F v}{R_A R}$$

The input resistance, R_{in}, is given by

$$R_{in} = \frac{v}{i} = \frac{-R_A R}{R_F} \tag{8.29}$$

Equation (8.29) shows that the circuit of Figure 8.14 develops a negative resistance. If R is replaced by an impedance, Z, the circuit develops a negative impedance.

8.7.2 Dependent-Current Generator

A dependent-current generator produces a load current that is proportional to an applied voltage, v_i, and is independent of the load resistance. It can be designed with a slight modification of the negative-impedance circuit. The circuit is shown in Figure 8.15(a). Suppose we let $R_F = R_A$. Equation (8.29) then indicates that the input resistance to the op-amp circuit (enclosed in the dashed box) is $-R$. The input circuit can then be simplified as shown in Figure 8.15(b). We wish to calculate i_L, the current in the load resistor, R_L. Although the resistance is negative, the normal Kirchhoff laws still apply, since nothing in their derivations assumed positive resistors. The input current, i_{in}, is then found by combining the resistances into a single resistor, R_{in}.

$$i_{in} = \frac{v_i}{R_{in}} = \frac{v_i}{R - RR_L/(R_L - R)} = \frac{v_i(R_L - R)}{R_L R - R^2 - RR_L} = \frac{v_i(R_L - R)}{-R^2}$$

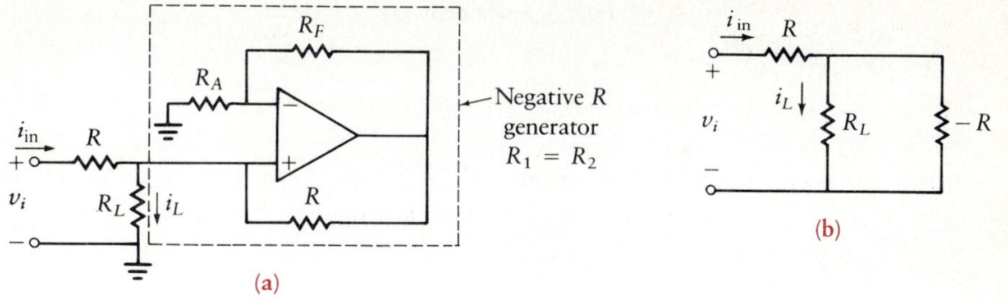

Figure 8.15 Dependent-current generator.

We then apply a current-divider ratio to the current split between R_L and $-R$ to obtain

$$i_L = \frac{-i_{in}R}{R_L - R} = \frac{-v_i(R_L - R)}{R^2} \frac{-R}{R_L - R}$$

$$= \frac{v_i}{R} \tag{8.30}$$

Thus the effect of the addition of the op-amp circuit is to make the current in the load proportional to the input voltage. It does not depend on the value of the load resistance, R_L. The current is therefore independent of changes in the load resistance. The op-amp circuit is effectively canceling out the load resistance. Since the current is independent of the load and depends only on the input voltage, we call this a *current generator* (or voltage-to-current converter).

Among the numerous applications of this circuit is a dc regulated current source. If we let $v_i = E$ (a constant), the current through R_L is constant independent of variations of R_L.

8.7.3 Current-to-Voltage Converter

The circuit of Figure 8.16 produces an output voltage that is proportional to the input current. We analyze this circuit using the properties of ideal op-amps.

$$v_+ = 0$$

$$v_- = v_o + i_i R$$

$$v_+ = v_- = v_o + i_i R$$

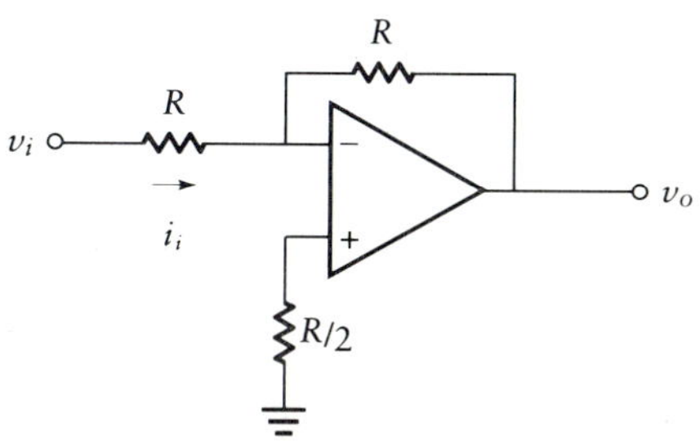

Figure 8.16 Current-to-voltage converter.

Hence, the output voltage, $v_o = -i_i R$, is proportional to the input current, i_i.

8.7.4 Voltage-to-Current Converter

With the circuit of Figure 8.17, we obtain a voltage-to-current converter. We analyze this circuit as follows:

$$v_+ = v_i$$

$$v_- = i_L R_1$$

$$v_+ = v_-$$

yielding

$$i_L = \frac{v_i}{R_1}$$

Therefore, the load current is independent of the load resistor, R_L, and is proportional to the applied voltage, v_i. This circuit develops a voltage-controlled current source. However, a practical shortcoming of this circuit is that neither end of the load resistor can be grounded.

As an alternative, the circuit of Figure 8.18 provides a voltage-to-current converter with one end of the load resistance grounded. We analyze this circuit by writing node equations as follows.

$$\frac{v_+ - v_i}{R_1} + \frac{v_+ - v_o}{R_2} + i_L = 0 \tag{8.31}$$

$$\frac{v_- - v}{R_1} + \frac{v_- - v_o}{R_2} = 0 \tag{8.32}$$

We wish to solve for i_L in terms of v_i. We begin by rewriting equations (8.31) and (8.32) as follows:

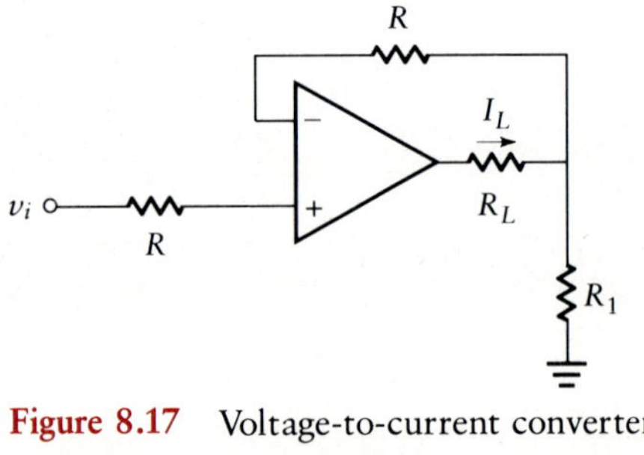

Figure 8.17 Voltage-to-current converter.

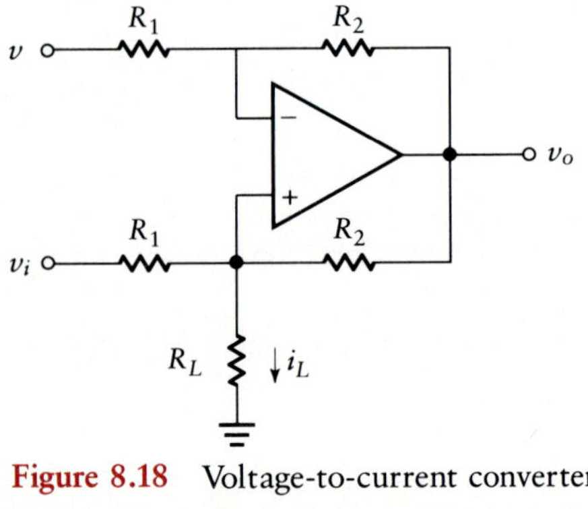

Figure 8.18 Voltage-to-current converter.

$$v_+\left(\frac{1}{R_1}+\frac{1}{R_2}\right)=\frac{v_i}{R_1}+\frac{v_o}{R_2}-i_L$$

$$v_-\left(\frac{1}{R_1}+\frac{1}{R_2}\right)=\frac{v}{R_1}+\frac{v_o}{R_2}$$

Now since $v_+ = v_-$, we equate these two expressions to obtain

$$\frac{v_i}{R_1}+\frac{v_o}{R_2}-i_L=\frac{v}{R_1}+\frac{v_o}{R_2}$$

We solve for i_L with the result

$$i_L=\frac{1}{R_1}(v_i-v)$$

The load current, i_L, is independent of the load, R_L, and is a function of only the voltage difference, $v_i - v$.

8.7.5 Inverting Amplifier with Impedances

The relationship of equation (8.2) is easily extended to include nonresistive components if R_j is replaced by an impedance, Z_j, and R_F is replaced by Z_F. For a single input, as shown in Figure 8.19(a), the output reduces to

$$V_o=\frac{-Z_F V_i}{Z_A} \tag{8.33}$$

Since we are dealing with the frequency domain, we use uppercase letters for the voltages and currents, thus representing the *complex variables.*

Figure 8.19
Use of impedance rather than resistance.

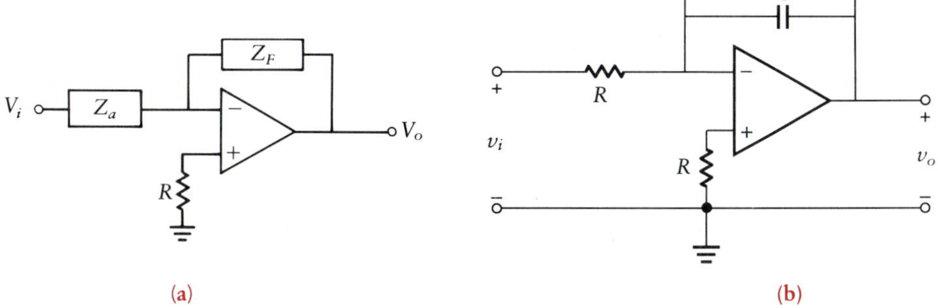

One useful circuit based on equation (8.33) is the *Miller integrator,* as shown in Figure 8.19(b). In this application, the feedback component is a capacitor, C, and the input component is a resistor, R, so

$$Z_F = \frac{1}{sC}$$

where s is the Laplace transform operator. For sinusoidal signals, $s = j\omega$. When we substitute these impedances into equation (8.33), we obtain

$$\frac{V_o}{V_i} = \frac{-1}{RC}\frac{1}{s}$$

In the complex frequency domain, $1/s$ corresponds to integration in the time domain. This is an *inverting integrator* because the expression contains a negative sign. Hence the output voltage is

$$v_o(t) = \frac{-1}{RC}\int_0^t v_i(\tau)\, d\tau + v_o$$

where v_o is the initial condition. The value of v_o is developed as the voltage across the capacitor, C, at time $t = 0$. The switch is closed to charge the capacitor to the voltage v_o and then at $t = 0$ the switch is open. We use electronic switches, which we discuss more fully in Chapter 15. In the event that the initial condition is zero, the switch is still used to reset the integrator to zero output voltage at time $t = 0$.

If the feedback element is a resistor and the input element is a capacitor, the input/output relationship becomes

$$\frac{V_o}{V_i} = -RCs$$

and in the time domain this becomes

$$v_o(t) = -RC\frac{dv_i}{dt}$$

The circuit is operating as an *inverting differentiator.* We should note that this is not a practical circuit because the input capacitor, $Z_a = 1/sC$, does not provide a path for dc.

8.7.6 Analog Computer Applications

In this section we present the use of interconnected op-amp circuits, such as summers and integrators, to form an analog computer that is used to solve dif-

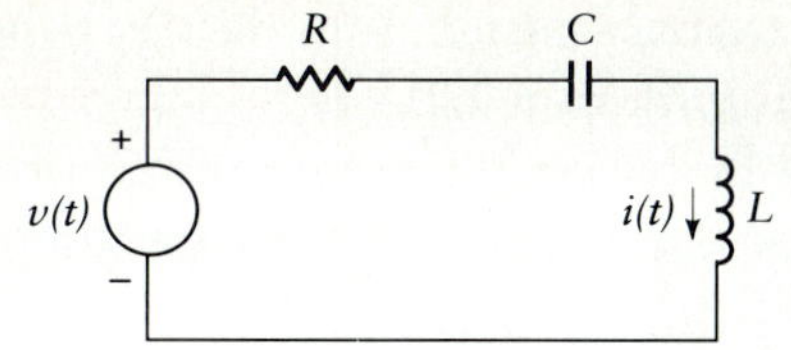

Figure 8.20 Analog computer application.

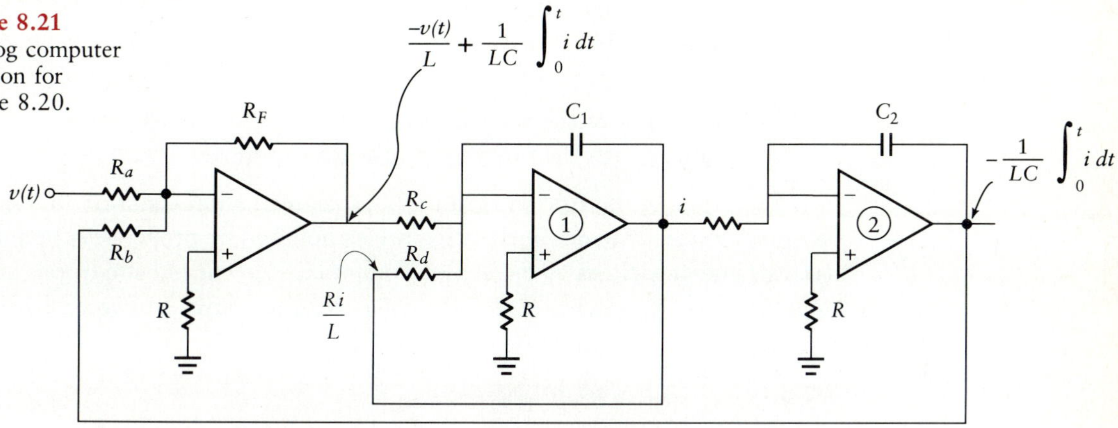

Figure 8.21 Analog computer solution for Figure 8.20.

ferential equations. Many physical systems are described by linear differential equations, and the systems can therefore be analyzed with the aid of an analog computer.

Let us solve for the current, $i(t)$, in the electric circuit of Figure 8.20. The input voltage is the driving function and the initial conditions are zero. We write the differential equation for the circuit of Figure 8.20 as follows:

$$v(t) = L\frac{di}{dt} + Ri + \frac{1}{C}\int_0^t i(\tau)\,d\tau$$

We now solve for di/dt, obtaining

$$-\frac{di}{dt} = \frac{-v(t)}{L} + \frac{Ri}{L} + \frac{1}{LC}\int_0^t i(\tau)\,d\tau \tag{8.34}$$

We know that for $t > 0$,

$$i = -\int_0^t -\frac{di}{d\tau}d\tau + i_o \tag{8.35}$$

From equation (8.34) we see that $-di/dt$ is formed by summing three terms, which are found in Figure 8.21 at the input to the first integrating amplifier, as follows:

1. The driving function, $-v(t)/L$, is formed by passing $v(t)$ through an inverting summer, with gain $1/L$.
2. Ri/L is formed by taking the output of the first integrating amplifier and adding it at the input of the first integrator to the output of the summing amplifier.
3. The term

$$-\frac{1}{LC}\int_0^t i(\tau)\,d\tau$$

is the output of the second integrator. Since the sign must be changed, we sum it with the unity-gain inverting summer.

The output of the first integrator is $+i$, as seen from equation (8.35). The constants in the differential equation are established by proper selection of the resistors and capacitors of the analog computer. Zero initial conditions are accomplished with switches across the capacitors, as shown in Figure 8.19(b).

8.7.7 Noninverting Miller Integrator

We use a modification of the dependent-current generator of the previous section to develop a noninverting integrator. The circuit is configured as shown in Figure 8.22. This is similar to the circuit of Figure 8.18, but the load resistance has been replaced by a capacitance. The current, I_L, is found by writing the node equation at V_+ as follows:

$$\frac{V_+ - V_i}{R} + \frac{V_+ - V_o}{R} + I_L = 0$$

The inverting voltage, V_-, is found from the voltage division between V_o and V_- as follows:

$$V_- = \frac{R_1 V_o}{R_1 + R_1} = \frac{V_o}{2}$$

Since $V_+ = V_-$, we solve and find $I_L = V_i/R$. Note that

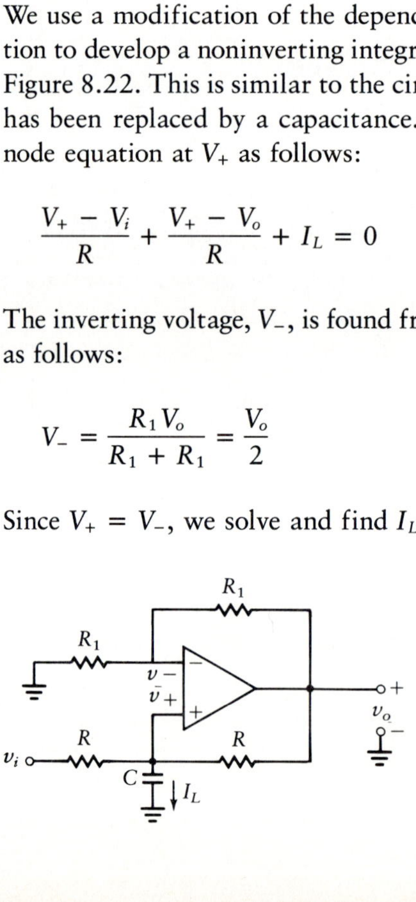

Figure 8.22 Noninverting integrator.

$$V_+ = \frac{I_L}{sC} = \frac{V_i}{sRC} = V_- = \frac{V_o}{2}$$

where s is the Laplace transform operator. The V_o/V_i function is then

$$\frac{V_o}{V_i} = \frac{2}{RC}\frac{1}{s}$$

Thus, in the time domain we have

$$v_o(t) = \frac{2}{RC}\int_0^t v_i(\tau)\,d\tau \qquad (8.36)$$

The circuit is therefore a noninverting integrator.

8.7.8 Op-Amps and Diodes

Numerous examples exist for the use of op-amps with diodes. We consider a wide range of these circuits in Chapter 13 and present several examples in this section.

Consider the ideal op-amp circuit shown in Figure 8.23 with

$$v_i = V \sin \omega t$$

The diode is conducting when v_+ attempts to go negative and nonconducting when v_+ is positive. In this presentation, we shall let $V_\gamma = 0$ rather than 0.7 V. With the diode *on*, when v_+ attempts to go negative, the diode shorts and

$$v_+ = 0$$

The inverting input voltage is found using the following voltage-divider relationship:

$$v_- = \frac{R_a v_o}{R_a + R_F}$$

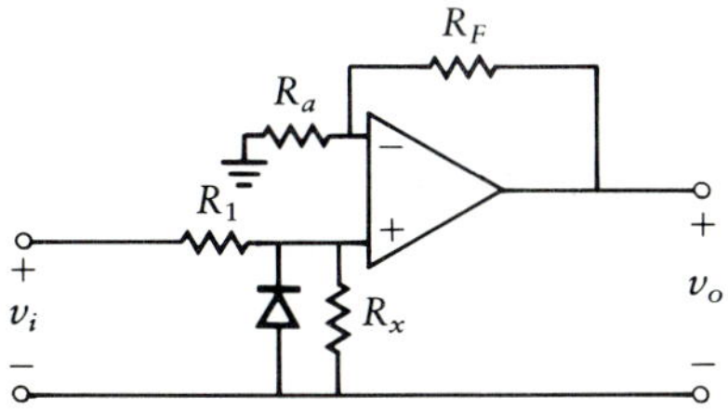

Figure 8.23
Half-wave rectifier circuit.

Now, since $v_+ = v_- = 0$, we see that v_o must also equal zero.

With the diode *off,* when v_+ goes positive, the diode is open-circuited and a voltage-divider relationship yields

$$v_+ = \frac{R_x v_i}{R_1 + R_x}$$

As before, the inverting voltage input is given by

$$v_- = \frac{R_a v_o}{R_a + R_F}$$

Now since v_+ is equal to v_-, we equate these two expressions to obtain

$$v_o = \frac{(1 + R_F/R_a)R_x v_i}{R_1 + R_x}$$

$$= v_i \frac{1 + R_F/R_a}{1 + R_1/R_x} = Av_i \tag{8.37}$$

where

$$A = \frac{1 + R_F/R_a}{1 + R_1/R_x}$$

The output waveform is plotted in Figure 8.24 for $v_i = V \sin \omega t$. This shows that the circuit is operating as a half-wave rectifier.

As another example of the use of diodes with op-amps, consider the *electronic thermometer* shown in Figure 8.25. Recall (see Chapter 1) that the voltage across a diode varies with temperature according to the expression

$$\Delta V_\gamma = -2(T_2 - T_1) \text{ mV}$$

At room temperature ($T_1 = 25°\text{C}$) the voltage across the diode is 700 mV. The

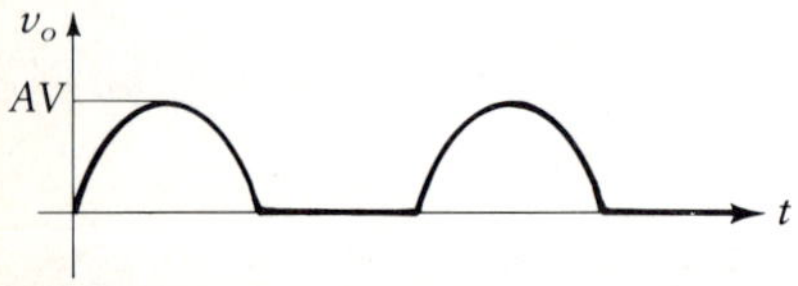

Figure 8.24 Output of half-wave rectifier.

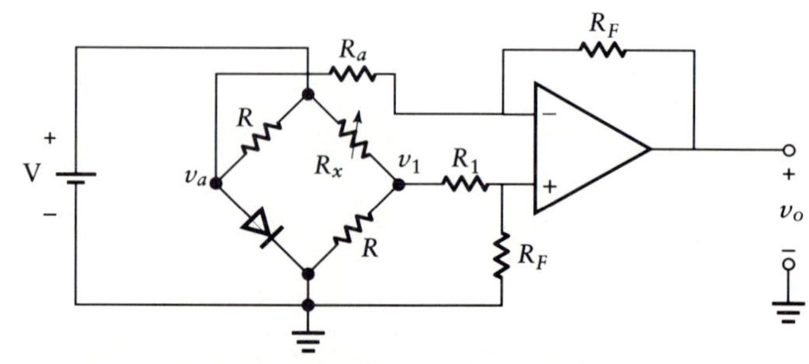

Figure 8.25 Electronic thermometer.

diode voltage decreases as the temperature increases. For example, at $T_2 =$ 125°C, the decrease in diode voltage is

$$\Delta V_\gamma = -2(125 - 25) = -200 \text{ mV}$$

As a result, the diode voltage drops to 500 mV. This voltage variation can be used as the basis for an inexpensive thermometer. We choose the resistance, R, so the diode is conducting and the diode voltage, which is the v_a input to the op-amp, is

$$\begin{aligned} V_a = V_\gamma &= 700 \text{ mV} - 2(T_2 - 25°\text{C}) \text{ mV} \\ &= (750 - 2T_2) \text{ mV} \end{aligned} \tag{8.38}$$

If we let $V = 10$ V and $R_1 >> R$, $R_1 >> R_x$, the bias voltage, which is the v_1 input to the op-amp, is

$$v_1 = 10^4 \frac{R}{R + R_x} \text{ mV} \tag{8.39}$$

and the op-amp equation is

$$v_o = -\frac{R_F}{R_a} v_a + \left[\frac{R_F}{R_1 + R_F} \frac{R_F}{R_a \parallel R_F} \right] v_1 \tag{8.40}$$

If we let $R_a = R_1$ and substitute equations (8.38) and (8.39) into equation (8.40), we obtain

$$\begin{aligned} v_o &= -\frac{R_F}{R_1}(750 - 2T_2) + \frac{R_F}{R_1} 10^4 \frac{R}{R + R_x} \\ &= \frac{R_F}{R_1}\left[10^4 \frac{R}{R + R_x} - 750 + 2T_2 \right] \text{ mV} \end{aligned}$$

We wish to cancel the dc components, so we let

$$750 = 10^4 \frac{R}{R + R_x} \tag{8.41}$$

and we obtain

$$v_o = \frac{2R_F T_2}{R_1} \text{ mV} \tag{8.42}$$

Proper selection of resistors yields the desired output voltage from equation (8.42). The design details are left to Problem 8.30.

PROBLEMS

In Problems 8.1 through 8.10 find the output voltage, v_o, in terms of the input voltage for each of the circuits shown.

8.11 Construct an ideal op-amp mathematical model and solve for the gain ratio, v_o/v_i, for each of the configurations of Figure P8.11.

8.12 Find the output voltage, v_o, in terms of the input signals for the circuits of Figure P8.12. Use the ideal op-amp mathematical model.

In Problems 8.13 through 8.16, design an op-amp amplifier circuit to yield the relationships shown in each equation. Use the procedure of Examples 8.4 and 8.5 and compare the results.

8.13 $v_o = v_1 + 10v_2 - 30v_a - 100v_b$

8.14 $v_o = 8v_1 + 8v_2 - 4v_a - 9v_b$

8.15 $v_o = 6v_1 + 8v_2 - 3v_a - 12v_b$

8.16 $v_o = 3v_1 + v_2 + 6v_3 - 4v_a - 5v_b$

8.17 For the circuit of Figure P8.13, what value of v_2 is required to produce $v_o = 500$ mV when $v_1 = 40$ mV, $R_1 = 50$ kΩ, and $R_2 = 150$ kΩ? What is the value of the output current, i_L, for the above conditions and $R_L = 4$ kΩ?

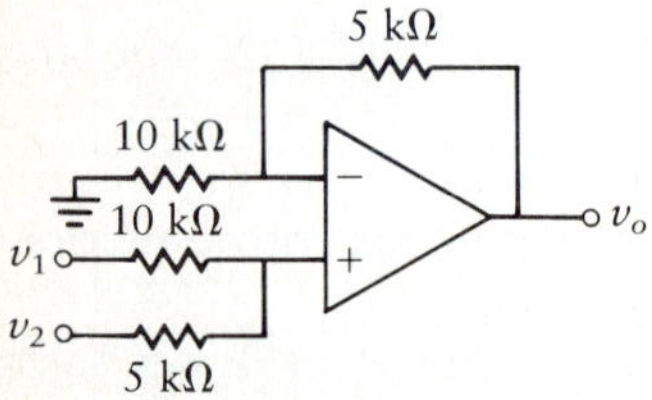

Figure P8.1

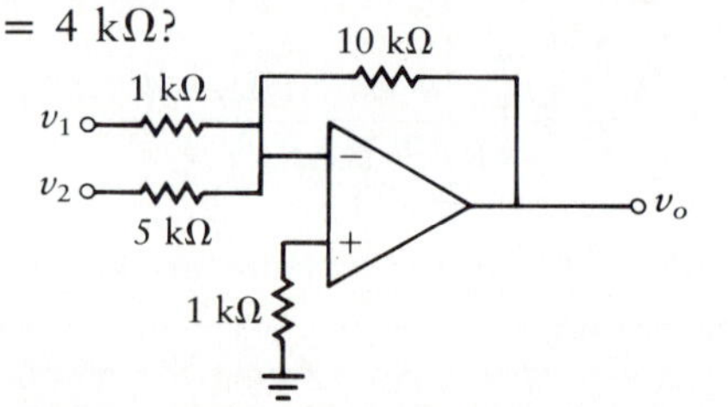

Figure P8.2

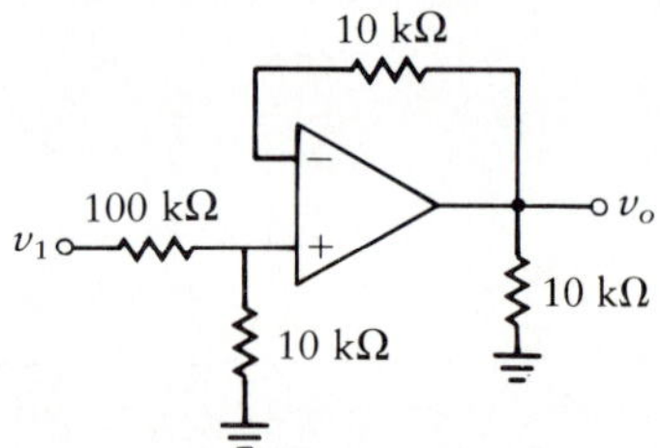

Figure P8.3

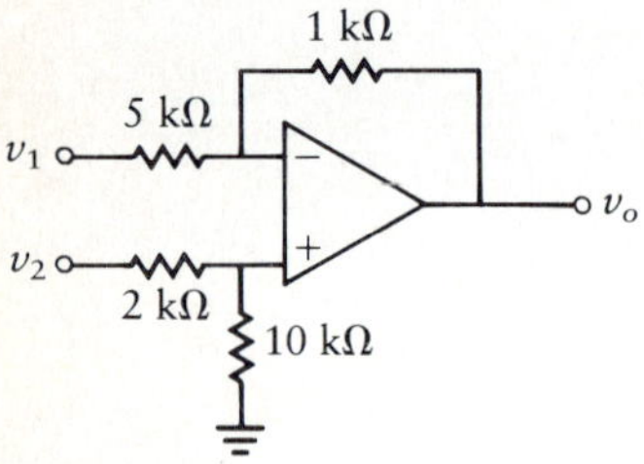

Figure P8.4

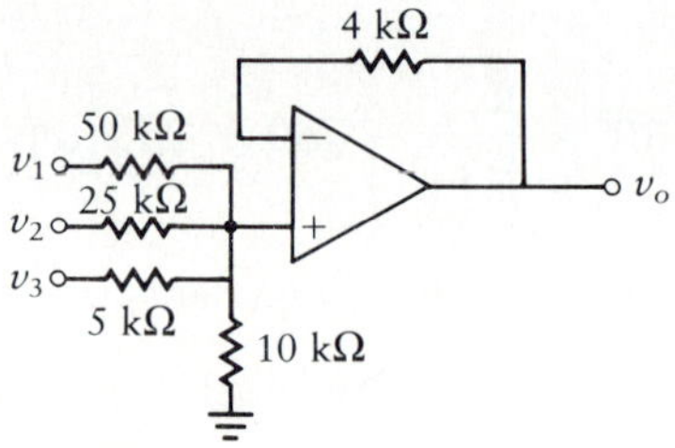

Figure P8.5

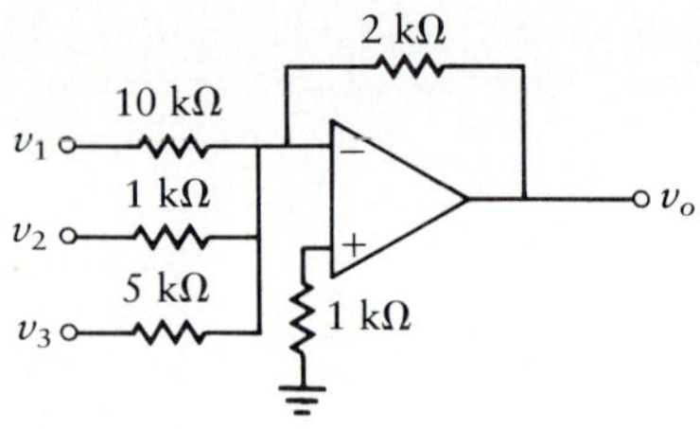

Figure P8.6

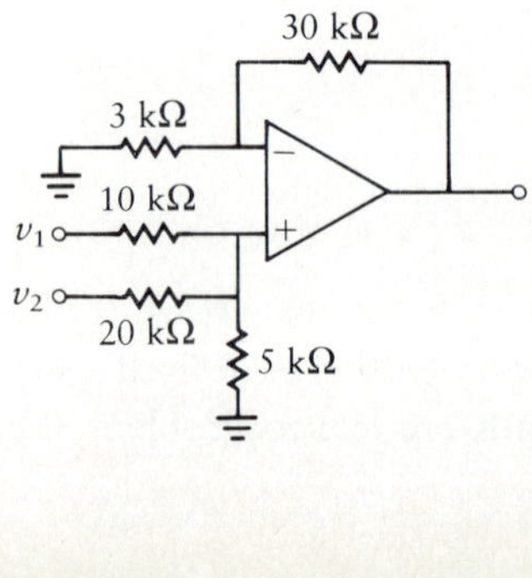

Figure P8.7

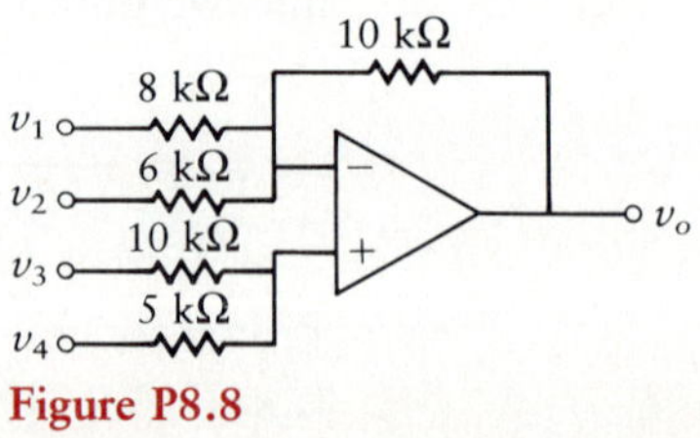

Figure P8.8

Figure P8.9

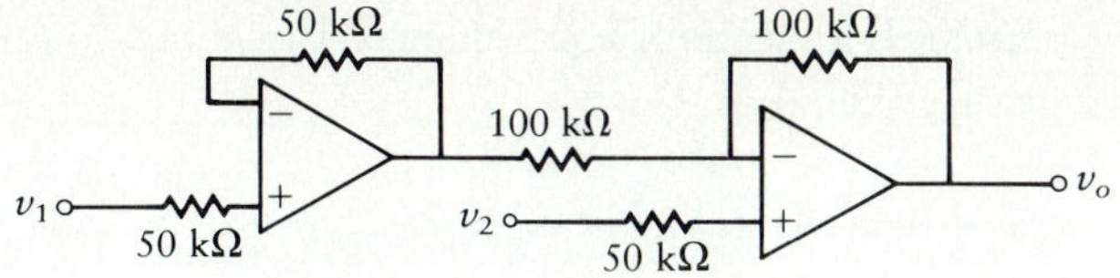

Figure P8.10

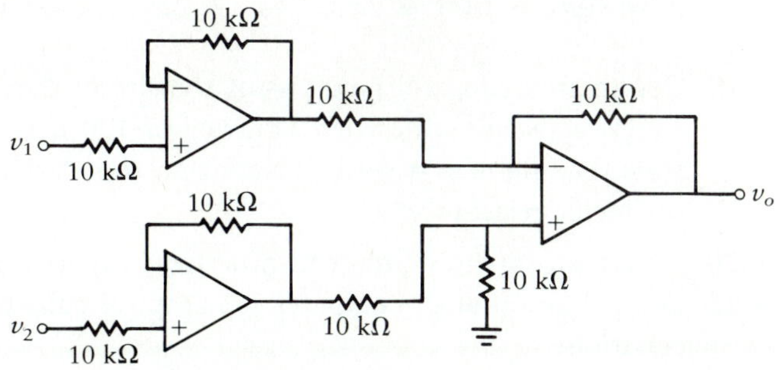

Figure P8.11

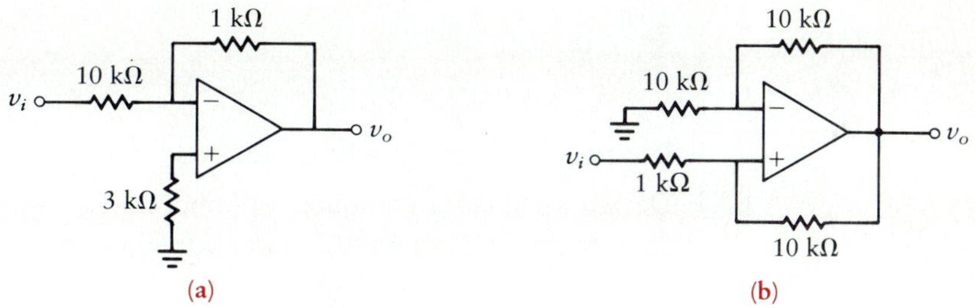

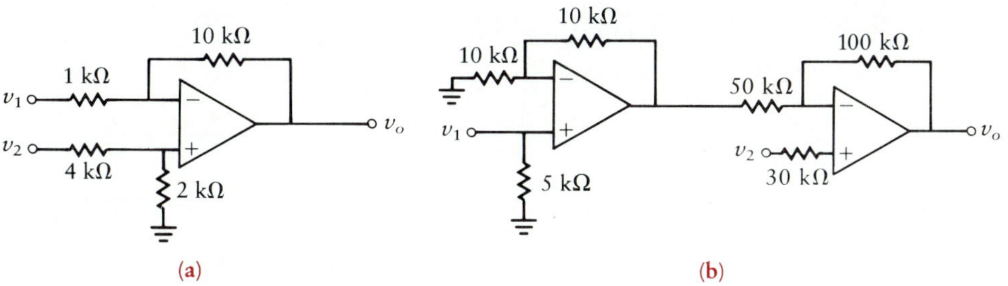

Figure P8.12

Figure P8.13

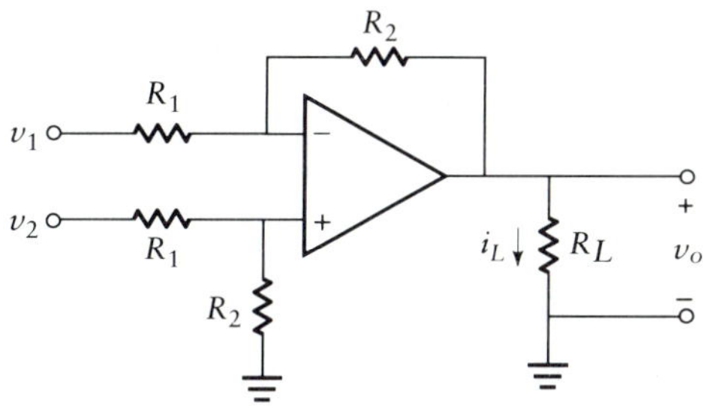

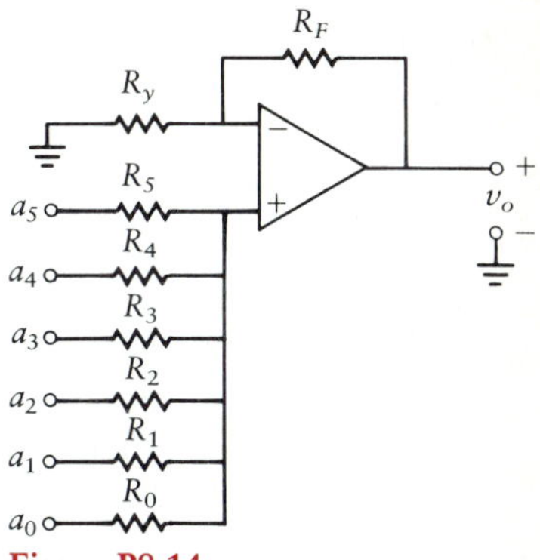

Figure P8.14

8.18 A digital-to-analog (D/A) converter can be designed using an op-amp as shown in Figure P8.14. Use the method of Section 8.6 to design a 6-bit D/A converter with $R_{\text{min}} = 10\ \text{k}\Omega$. What is a good choice for V_{CC} if logic 1 corresponds to 0.2 V?

Hint: The decimal equivalent of the binary number $a_5a_4a_3a_2a_1$, where a_i is either 0 or 1, is

$$\begin{aligned} N &= a_5 2^5 + a_4 2^4 + a_3 2^3 + a_2 2^2 + a_1 2^1 + a_0 2^0 \\ &= 32a_5 + 16a_4 + 8a_3 + 4a_2 + 2a_1 + a_0 \end{aligned}$$

8.19 Design an analog voltmeter using the circuit shown in Figure P8.15. The meter reads full scale with a current of 100 μA. Find R so that the full-scale reading is $v = +10$ V. Note that this design is independent of R_m, the meter resistance.

8.20 Design an op-amp circuit to produce a negative resistance of -10 kΩ.

8.21 Design an analog computer op-amp circuit to solve the following equation:

$$\frac{dx}{dt} + 9x = 3$$

Use two summing amplifiers and one integrator in your design.

8.22 Design an analog computer op-amp circuit to solve the following two first-order simultaneous differential equations:

$$\frac{dx}{dt} + a_1 x + b_1 y = f_1(t)$$

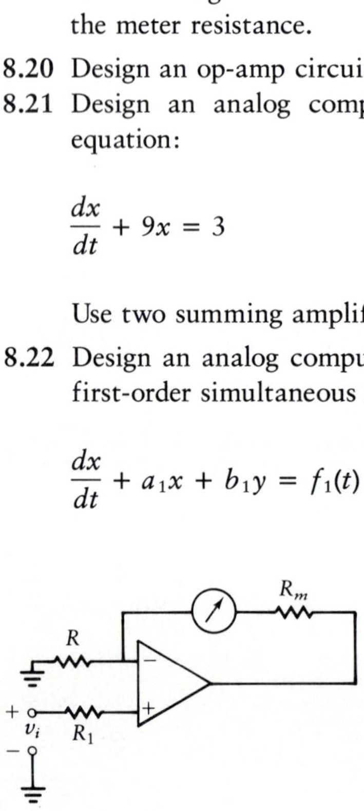

Figure P8.15

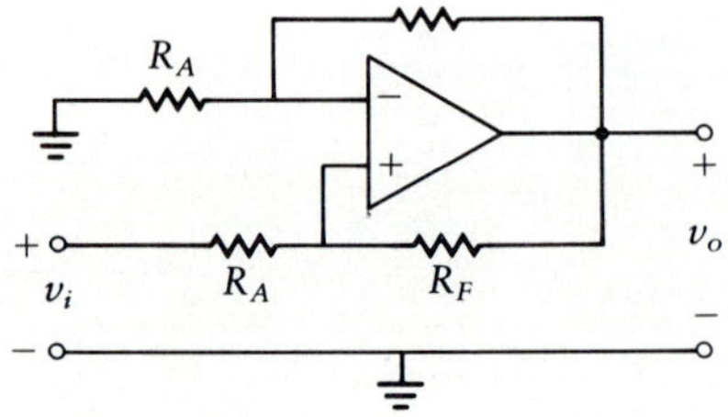

Figure P8.16

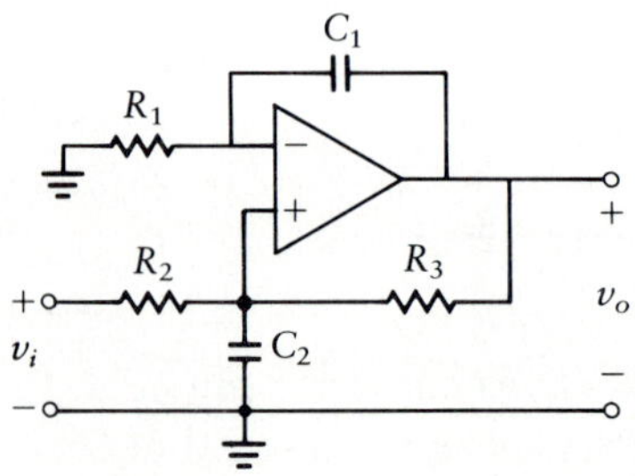

Figure P8.17

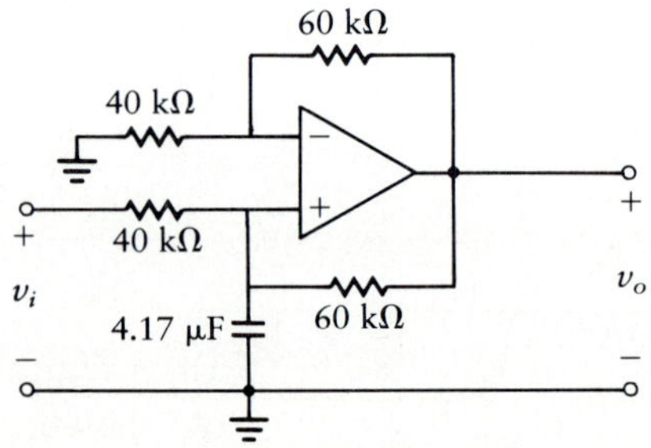

Figure P8.18

$$\frac{dy}{dt} + b_2y + a_2x = f_2(t)$$

Use two integrating op-amps and two summing op-amps in your design.

8.23 Determine the gain, v_o/v_i, for the ideal op-amp circuit shown in ure P8.16. Why does this circuit yield such a peculiar answer?

8.24 Determine the transfer ratio, V_o/V_i, for the circuit shown in Figure P8.17. The op-amp is ideal.

8.25 Determine the transfer ratio, V_o/V_i, for the circuit shown in Figure P8.18. The op-amp is ideal.

8.26 Find the gain, v_o/v_i, for the ideal op-amp circuits shown in Figure P8.19.

8.27 Determine the voltage gain, v_o/v_i, for the ideal op-amp circuit shown in Figure P8.20. This will be a function of the variable resistor. The maximum value of the resistor is R when $x = 1$, and x ranges from 1 to 0. Plot the gain, v_o/v_i, as a function of x.

8.28 Plot the output voltage, v_o, as function of R/R_x for the ideal op-amp circuit of Figure P8.21. Assume that $V = 15$ V.

8.29 Plot the output voltage for the op-amp system shown in Figure P8.22 when the input voltage is given by

$$v_i = 10 \sin 20t$$

Assume that $V_\gamma = 0.7$ V and that $R_f = 0$.

Figure P8.19

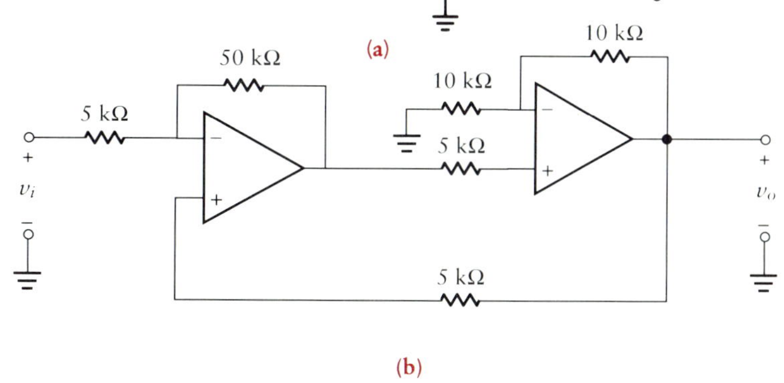

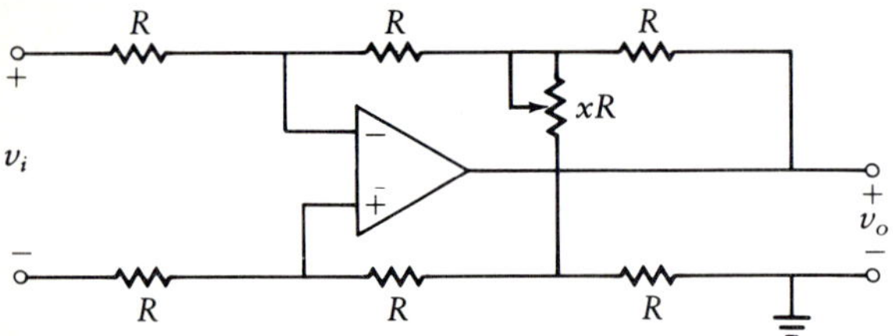

Figure P8.20

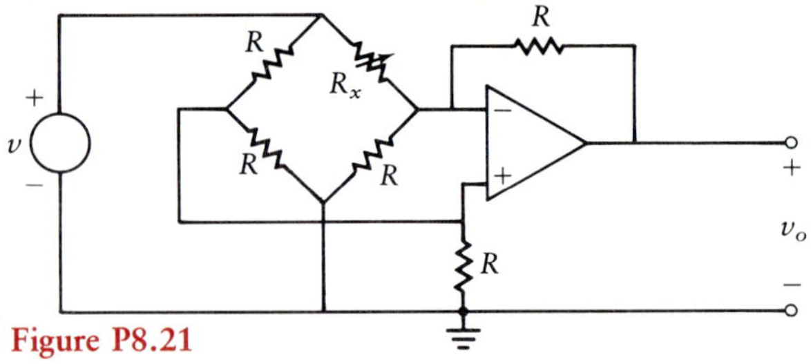

Figure P8.21

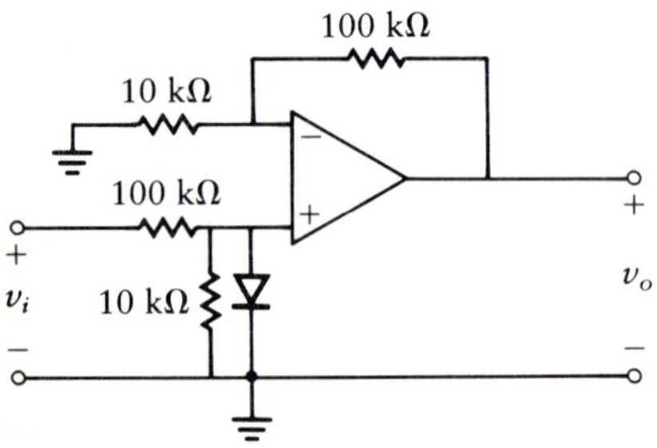

Figure P8.22

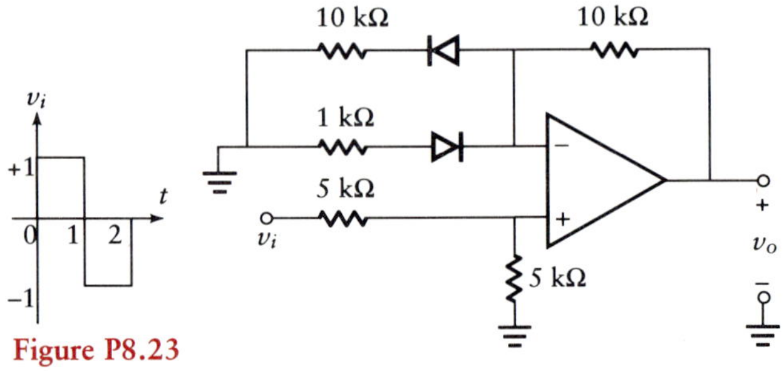

Figure P8.23

8.30 Figure 8.25 is an electronic thermometer. Plot the output voltage, V_o, as the temperature varies from room temperature (25°C) to 125°C. The voltage across the diode varies according to the relationship

$$\Delta V_\gamma = -2(T_2 - T_1) \text{ mV}$$

Select the resistor values so that the output, v_o, given by equation (8.42) must be 5 V when the temperature is 125°C.

8.31 Determine the output waveform of the circuit shown in Figure P8.23. Assume $V_\gamma = 0 = R_f$.

8.32 Design an instrumentation system as shown in Figure P8.24. Select R_F so that when $\Delta R = 30\ \Omega$, $v_o = 3$ V. Set $R_A = 30$ kΩ and $R = 1$ kΩ. The battery voltage is $V = 12$ V, and $0 < \Delta R < 30\ \Omega$. Plot v_o as a function of ΔR.

8.33 Determine the gain, v_o/v_i, for the op-amp system shown in Figure P8.25. Select a value for R_1 so the resistance looking out v_+ equals the resistance looking out v_-.

8.34 Plot the output voltage, v_o, as a function of the input voltage, v_i, for the circuit shown in Figure P8.26. The diode is ideal, so $V_\gamma = 0 = R_f$.

8.35 The input voltage to the circuit of Figure P8.27 is

$$v_i = 10 \sin 2\pi 60t$$

Plot the output voltage, $v_o(t)$, when $R = 10$ kΩ and $V_\gamma = 0 = R_f$. Specify the maximum value of the output voltage.

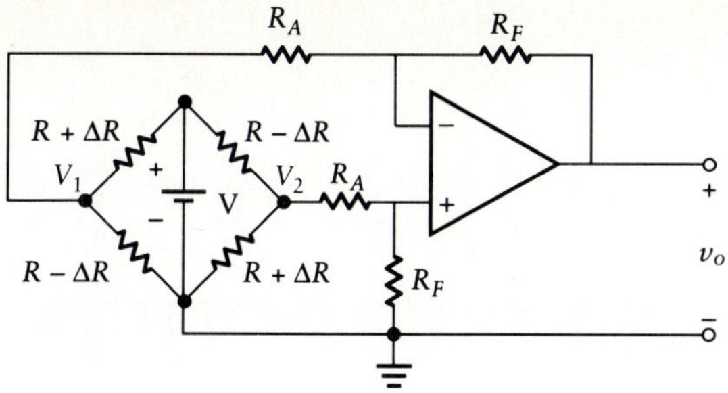

Figure P8.24

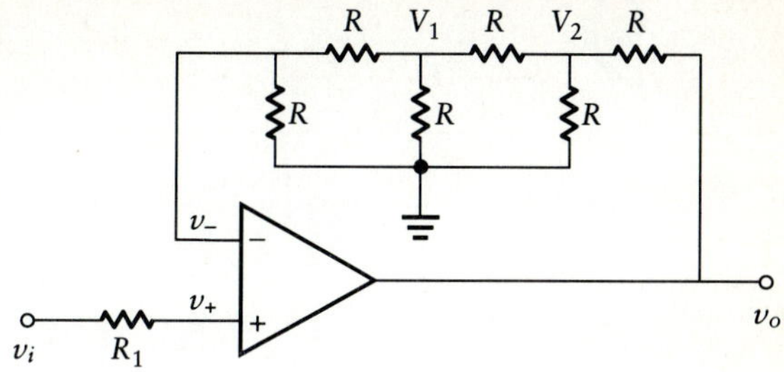

Figure P8.25

Figure P8.26

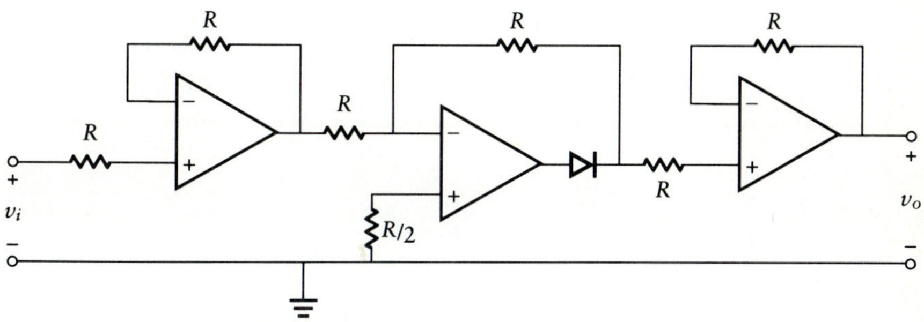

Figure P8.27

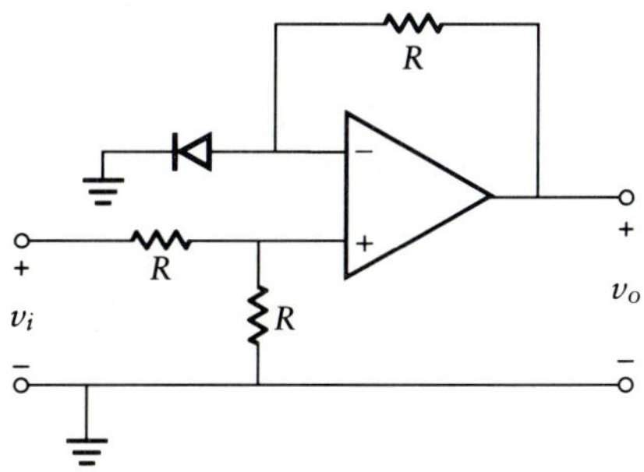

9

OPERATIONAL AMPLIFIERS AS INTEGRATED CIRCUITS

9.0 INTRODUCTION

We introduced ideal op-amps in the previous chapter. We presented these from the perspective of a systems engineer who would analyze and design an electronic system. In this chapter, we study a variety of discrete circuits that make up a typical op-amp.

We discuss the internal characteristics of the op-amp. We begin with a discussion of techniques for fabricating transistors, resistors, and capacitors on integrated circuits (ICs). A summary of the IC fabrication techniques is then presented. The basic building block, the difference or differential amplifier, is then analyzed. Several other internal circuit groups, including the constant-current source and single-ended input and output, are discussed. Various types of current sources are presented as examples of active sources used as circuit elements on a chip. The chapter concludes with information about the forms of packaging used for op-amps. A discussion of the 741 general-purpose op-amp is included.

In Chapter 10, we will modify the ideal op-amp mathematical model by recognizing the changes required to make the model more closely coincide with the real op-amp. Our design method will be refined to account for the necessary modifications from the ideal op-amp.

9.1 INTEGRATED CIRCUIT FABRICATION

A complex circuit can often be fabricated on a single silicon chip. The circuit can be composed of transistors, resistors, and capacitors, all of which are small enough to be located on the chip. We briefly discuss the fabrication of each of these elements in the following sections [25].

9.1.1 Transistors and Diodes

IC transistors and diodes are usually fabricated on a silicon substrate material, although gallium arsenide is more commonly used for high-frequency applications. The physical properties of silicon make it convenient for fabrication of active and passive devices on a single chip. Silicon also has the characteristic that it is easily oxidized to form insulating layers, which provide isolation between devices on the chip. This property also allows fabrication of different integrated circuit components at the same time on a single chip with the interconnections to provide the complete circuit. To fully understand how ICs are fabricated, you should become acquainted with the different processes used for developing the different types of ICs. We will mention only some of these processes and leave the details to books on IC fabrication. Some of the processes used are oxidation, diffusion, ion implantation, chemical vapor deposition, metallization, epitaxial growth, and photolithography. A simple explanation of the development of an *npn* transistor follows.

A substrate of *p*-type material is subjected to an oxidizing atmosphere at an elevated temperature. This forms a thin layer of silicon oxide, or glass, on the surface of the substrate. The thin layer of silicon oxide is removed by a photoengraving process, exposing the *p*-type material of the substrate. The silicon chip is then placed in an *n*-type atmosphere at an elevated temperature, which causes the *n*-type atoms to diffuse into the exposed *p*-type material. The original process of obtaining the SiO_2 layer is then repeated, and the photoengraving process is repeated over a smaller area, thus forming the base region. The chip is subjected to a *p*-type atmosphere at an elevated temperature, which causes the *p*-type atoms to diffuse into the *n*-type silicon. The oxidation process is then repeated and a smaller area is photoengraved, forming the emitter. This time, the chip is subjected to an *n*-type atmosphere to get the *n*-type diffusion as shown in Figure 9.1. Oxidizing and photoengraving are redone to expose each of the three layers of the transistor for the connections to the outside terminals. The terminals are formed by depositing metal in those areas. If the junction between the substrate and the collector is maintained in a reverse-biased condition during the processing, the leakage current will be small and the substrate and newly developed transistor are effectively insulated. As you can visualize from Figure 9.1, a number of devices can be made at the same

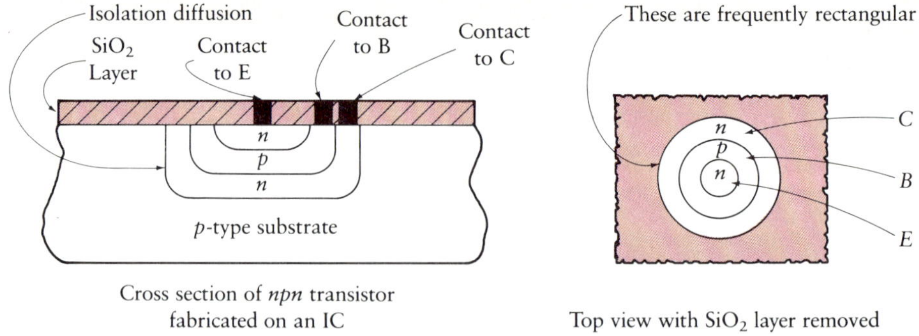

Figure 9.1
npn transistor on IC.

time. Diodes can be fabricated on the chip if the exposure to the final *n*-type atmosphere is eliminated. To fabricate a *pnp* transistor, an *n*-type silicon substrate material is used.

9.1.2 Resistors

Resistors are produced on an IC at the same time as the transistors. For example, a thin channel of *p*-type material can be used to produce the resistor. This channel is isolated by use of a layer of *n*-type material. The *p*-type material is diffused during the same step in production as the base of the transistor. The amount of resistance is controlled by varying the thickness and impurity level of the material. The channel width and its length affect the value of the resistance. In fact, the resistance is directly proportional to the length of the channel (assuming a constant width). If this fabrication technique does not provide a sufficiently high resistance, the channel can be overlaid upon itself several times, thus increasing the effective length and resistance. Resistances above about 100 kΩ are usually avoided on ICs because of physical size limitations.

9.1.3 Capacitors

Two types of capacitors are used on an IC chip: a diffused capacitor and a MOS capacitor. The *diffused capacitor* takes advantage of the incremental junction capacitance of a reverse-biased *pn* junction.

The *MOS capacitor* is made similarly to a conventional capacitor. The lower conductor is a highly doped *n* material, the insulator is a film of silicon dioxide, and the other plate is a metallized layer that also forms one contact point. This type of capacitor provides about 400 to 600 pF per square millimeter. Because of the relatively large amount of space needed for capacitors, the number used in IC design is usually kept small.

9.1.4 Lateral Transistors

Lateral pnp transistors can be fabricated on ICs using planar technology (25). Small *p*-type material is used as the emitter and is placed on a slab of *n* mate-

Figure 9.2
Cross section of lateral transistor.

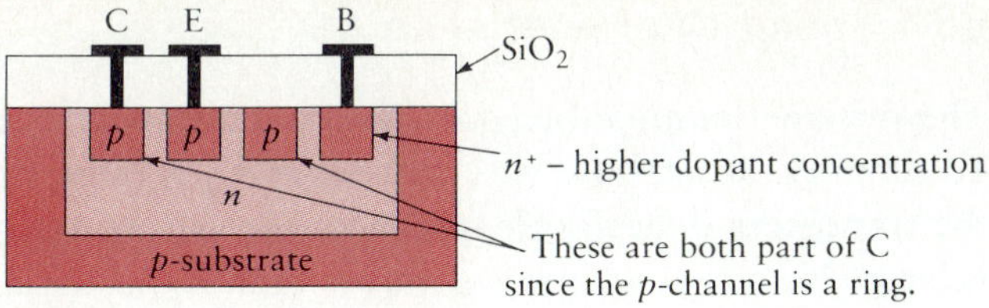

rial, which is used as the base. A ring of p-channel material is used as the collector. The emitter p-type material is placed within this ring, with separating n-material used as the base. Holes are injected from the emitter and flow parallel to the surface across the n-type base region. The holes are collected by the p-type collector before reaching the base contact. The emitter to collector current is lateral rather than vertical, as illustrated in Figure 9.2.

Lateral *pnp* transistors have relatively low current gains (on the order of 10 to 50) because the base is much wider than the standard *pnp*. Because they are easy to fabricate on ICs, these transistors are used for active loads, current sources, and level shifters (these are discussed later in this chapter). An advantage of the lateral transistor is that the base-emitter breakdown voltage (on the order of 50 V) is much larger than that of a standard *npn* or *pnp* transistor. These configurations are so easy to fabricate in integrated circuit design that parasitic lateral transistors may be formed unintentionally and cause problems with IC performance.

9.2 DIFFERENTIAL AMPLIFIERS

Most operational amplifiers consist of a series of transistors, resistors, and capacitors forming a complete system on a single chip. The amplifiers available today are reliable, small in size, and consume a small amount of power [9].

The input stage of most op-amps is a *differential amplifier,* shown in its simplest form in Figure 9.3. The differential amplifier is composed of two emitter-coupled common-emitter dc amplifiers with two inputs, v_1 and v_2, and three outputs, v_{o1}, v_{o2}, and v_o. The third output, v_o, is the difference between v_{o1} and v_{o2}.

Figure 9.3
Differential amplifier.

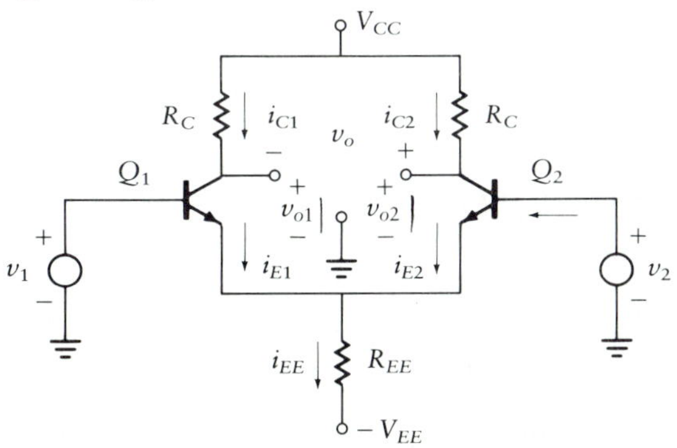

9.2.1 dc Transfer Characteristics

The differential amplifier does not operate linearly with large signal inputs. To simplify the analysis, we assume that R_{EE} is large, that the base resistance of the transistors is negligible, and that the output resistance of the transistors is large. Note that we use R_{EE} rather than R_E in the differential amplifier because the resistor used here is large and may be the equivalent resistance of a current source. The large value of R_{EE} keeps the emitter-resistor voltage drop nearly constant.

We wish to solve this circuit for the output voltage. We begin by writing a KVL equation around the base junction loop for the circuit of Figure 9.3.

$$v_1 = v_{BE1} - v_{BE2} + v_2 \tag{9.1}$$

We solve the circuit for the collector currents, i_{C1} and i_{C2}. The base-emitter voltages are given by the equation presented in Section 3.2.

$$v_{BE1} = V_T \ln\left(\frac{i_{C1}}{\beta I_{o1}}\right) \tag{9.2}$$

$$v_{BE2} = V_T \ln\left(\frac{i_{C2}}{\beta I_{o2}}\right) \tag{9.3}$$

In equations (9.2) and (9.3), I_{o1} and I_{o2} are the reverse saturation currents for Q_1 and Q_2, respectively. The transistors are assumed to be identical. Combining equations (9.1), (9.2), and (9.3) yields

$$v_1 - V_T \ln\left(\frac{i_{C1}}{\beta I_{o1}}\right) + V_T \ln\left(\frac{i_{C2}}{\beta I_{o2}}\right) - v_2 = 0$$

and

$$\frac{i_{C1}}{i_{C2}} = \exp\left(\frac{v_1 - v_2}{V_T}\right) \tag{9.4}$$

We assume that i_C is approximately equal to i_E. Therefore,

$$i_{EE} = i_{C1} + i_{C2} \tag{9.5}$$

Combining equations (9.4) and (9.5), we have

$$i_{C1} = \frac{i_{EE}}{1 + \exp[-(v_1 - v_2)/V_T]} \tag{9.6}$$

$$i_{C2} = \frac{i_{EE}}{1 + \exp[(v_1 - v_2)/V_T]} \tag{9.7}$$

Note that

$$v_o = (i_{C1} - i_{C2})R_C$$

An important observation can be made by viewing equations (9.6) and (9.7). If $v_1 - v_2$ becomes greater than several hundred millivolts, the collector current in transistor 2 becomes small and the transistor is essentially cut off. The collector current in transistor 1 is approximately equal to i_{EE}, and this transistor is saturated. The collector currents, and therefore the output voltage v_o, become independent of the difference between the two input voltages. Linear amplification occurs only for input voltage differences less than approximately 100 mV. In order to increase the linear range of the input voltage, small emitter resistors can be added.

9.2.2 Common-Mode and Differential-Mode Gains

The differential amplifier is intended to respond only to the difference between the two input voltages, v_1 and v_2. However, in a practical op-amp the output depends to some degree on the sum of these inputs. For example, if both inputs are equal, the output voltage should be zero but in a practical amplifier it is not. When the circuit responds to the difference, we say it is in the *differential mode.* If the two inputs are made equal, we say the circuit is in its *common mode.* Ideally, we would expect the circuit to produce an output only in the differential mode.

Any two input voltages can be resolved into a common and a differential part. That is, we define two new input voltages as follows:

$$\begin{aligned} v_{di} &= v_1 - v_2 \\ v_{ci} &= \frac{v_1 + v_2}{2} \end{aligned} \tag{9.8}$$

The voltage v_{di}, the differential-mode input voltage, is simply the difference between the two input voltages. The voltage v_{ci}, the common-mode input voltage, is the average of the two input voltages. The original input voltages can be expressed in terms of these new quantities as follows:

$$\begin{aligned} v_1 &= \frac{v_{di} + 2v_{ci}}{2} \\ v_2 &= \frac{2v_{ci} - v_{di}}{2} \end{aligned} \tag{9.9}$$

If we set the two input voltages equal, we have

$$v_1 = v_2 = v_{ci}$$

and

$$v_{di} = 0$$

Since the two inputs are equal, the emitter-base junction voltages are equal (if the transistors are identical). Thus, the collector currents must also be identical.

We now view the equivalent circuit for the differential-mode input voltage as shown in Figure 9.4(a). Note that when the current in the Q_1 circuit is increasing, the current in the Q_2 circuit is decreasing at the same rate and amplitude. This is true because the input to Q_2 is equal to that of Q_1 but 180° out of phase. Thus the voltage change across R_{EE} is zero. Since the ac signal voltage across R_{EE} is zero, it can be replaced by a short circuit in the ac equivalent circuit. Note that when we place voltages at each transistor base that are equal in amplitude but 180° out of phase, it is equivalent to placing a voltage between the two transistor bases of twice the amplitude. The voltages at v_{o1} and v_{o2} will be of equal amplitude but opposite phase and the differential gain is

$$A_d = \frac{v_{o1}}{v_{di}} = \frac{-R_c}{2h_{ib}} \tag{9.10a}$$

This differential-mode gain is a *single-ended output* because it is taken between one collector to ground. If the output is taken between v_{o1} and v_{o2}, the differential-mode gain is termed a *double-ended output* and is

$$A_d' = \frac{v_{o1} - v_{o2}}{v_{di}} = \frac{v_{od}}{v_{di}} = \frac{-R_c}{h_{ib}} \tag{9.10b}$$

A similar analysis can be applied to the common-mode equivalent circuit in Figure 9.4(b). If we divide the resistor R_{EE} into two parallel resistors that each have double the original resistance, we can find the output by analyzing only half of the circuit. Since the transistors are identical and the common-mode input voltages are equal and in phase, the voltages across the $2R_{EE}$ resistors are the same. Thus, the current between the two parallel resistors shown for R_{EE} is zero, and we need only look at one side of the circuit. The common-mode voltage gain is then

$$A_c = \frac{v_{oc}}{v_{ci}} = \frac{-R_c}{2R_{EE}} \tag{9.10c}$$

since R_{EE} is large and $h_{ib} << R_{EE}$.

We find the double-ended output voltage in terms of the common-mode and differential-mode gains as follows:

$$v_o = A_d v_{di} + A_c v_{ci}$$

It is desirable for the differential-mode gain to be much larger than the common-mode gain so that the amplifier will react only to the difference between the input voltages. The *common-mode rejection ratio (CMRR)* is defined as the ratio of the differential-mode gain to the common-mode gain (usually expressed in dB).

$$\text{CMRR} = 20\log\left(\frac{|-R_C/2h_{ib}|}{|-R_C/2R_{EE}|}\right)\text{ dB} = 20\log\left(\frac{R_{EE}}{h_{ib}}\right)\text{ dB} \tag{9.11}$$

Figure 9.4
Differential amplifier equivalent circuits.

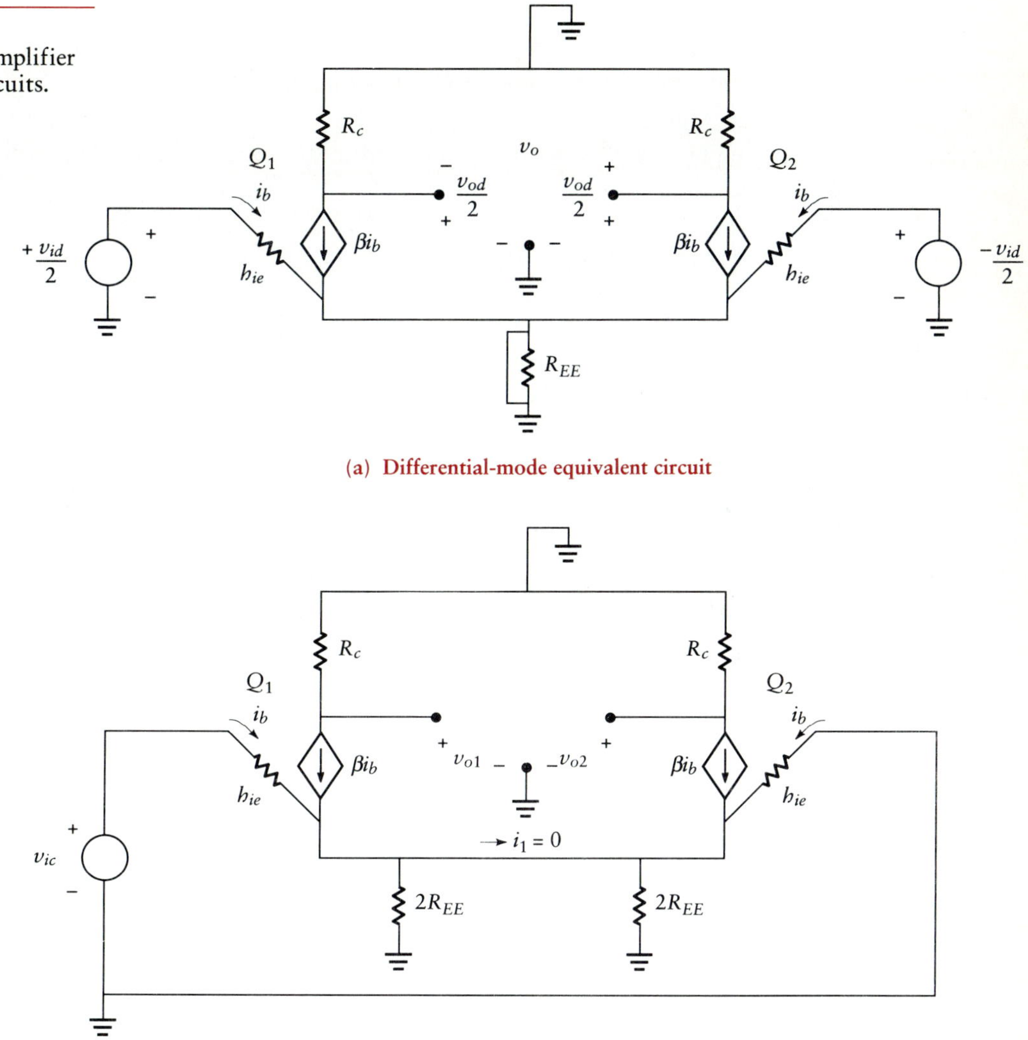

(a) Differential-mode equivalent circuit

(b) Common-mode equivalent circuit

Let us determine the input resistance of the amplifier in both the differential mode and the common mode. For the differential mode, we look into the amplifier at the base of both transistors. This results in a complete circuit through the emitter of both transistors, and the input resistance is

$$R_i(\text{differential mode}) = 2h_{ie}$$

For the common-mode input, we look into the amplifier in Figure 9.4(b). Thus, the input resistance is

$$R_i(\text{common mode}) = 2\beta R_{EE}$$

These results indicate that the input resistance of the common mode is much higher than that of the differential mode.

Our differential-amplifier analysis is based on BJTs as the transistor building blocks. JFETs can also be used in differential amplifiers, with the resulting advantages of reduced input bias current and higher input impedance. The analysis of the differential amplifier using JFETs is accomplished in the same way as the analysis for BJTs.

Differential amplifiers need matched transistors to ensure that the circuit operates correctly. If the differential amplifier is on an IC, this additional requirement is not a problem because the two transistors are fabricated at the same time using the same material.

9.2.3 Differential Amplifier with Constant-Current Source

It is desirable to make R_{EE} as large as possible in order to reduce the common-mode output. Equation (9.11) shows that to make the CMRR large, we must make R_{EE} large. Since large resistances are hard to fabricate on IC chips, we seek an alternative approach. This is accomplished by replacing R_{EE} with a constant-current generator. An ideal constant-current source has infinite impedance, so we investigate the possibility of replacing R_{EE} with such a current source. Figure 9.5 illustrates a differential amplifier in which the resistor, R_{EE}, is replaced with a constant-current source.

The closer the source is to the ideal constant-current source, the higher the common-mode rejection ratio. We illustrate a diode-compensated current source. The compensation, as discussed in Chapter 5, makes the operation of the circuit less dependent on temperature variations. Diode D_1 and transistor Q_3 are selected so that they have nearly identical characteristics over the range of operating temperatures.

To analyze the circuit of Figure 9.5(a) and find the CMRR, we determine the equivalent resistance, R_{TH}, the Thevenin equivalent of the constant-current source circuit. We see from Figure 9.5(b) that the equivalent resistance is given by

Figure 9.5
Differential amplifier with constant-current source.

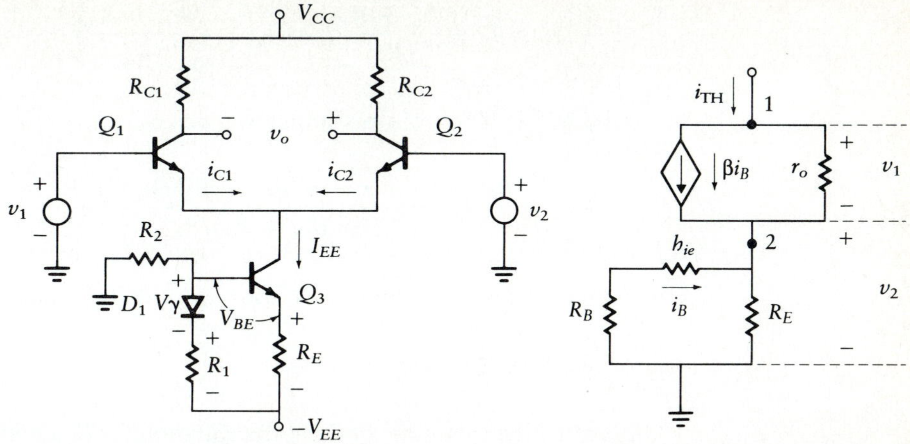

(a) Differential amplifier with constant-current source

(b) Equivalent circuit of current source

$$R_{TH} = \frac{v_1 + v_2}{i_{TH}}$$

Writing a KCL equation at node 1, we have

$$i_{TH} = \beta i_B + \frac{v_1}{r_o}$$

where r_o is the internal resistance of the transistor at the specified operating point and is given by

$$r_o = \frac{V_A}{I_{CQ}}$$

V_A is the early voltage (see Section 2.4.2).

A KCL equation at node 2 yields

$$\beta i_B + \frac{v_1}{r_o} + i_B - \frac{v_2}{R_E} = 0$$

but

$$v_1 = (i_{TH} - \beta i_B)r_o$$

and

$$v_2 = -i_B(h_{ie} + R_B)$$

Substituting v_1 and v_2 into the equation at node 2, we have

$$-i_{\mathrm{TH}} = i_B + \frac{h_{ie} + R_B}{R_E} i_B = i_B\left(1 + \frac{h_{ie} + R_B}{R_E}\right)$$

Finally, the Thevenin resistance is given by

$$R_{\mathrm{TH}} = \frac{h_{ie} + R_B + r_o[1 + (h_{ie} + R_B)/R_E] + \beta r_o}{1 + (h_{ie} + R_B)/R_E}$$

To maintain bias stability, we use the guideline that

$$R_B = 0.1\beta R_E$$

Substituting this value of R_B in the equation for R_{TH} and dividing by β, we have

$$R_{\mathrm{TH}} = \frac{h_{ib} + 0.1R_E + r_o[1/\beta + (h_{ib} + 0.1R_E)/R_E + 1]}{1/\beta + (h_{ib} + 0.1R_E)/R_E}$$

We can simplify this expression by noting that

$$1 >> \frac{1}{\beta}$$

and

$$\frac{h_{ib} + 0.1R_E}{R_E} >> \frac{1}{\beta}$$

We then have

$$R_{\mathrm{TH}} = R_E + r_o\left(1 + \frac{R_E}{h_{ib} + 0.1R_E}\right)$$

Since the second term in this equation is much greater than the first, we can ignore R_E to obtain

$$R_{\mathrm{TH}} = r_o\left(1 + \frac{R_E}{h_{ib} + 0.1R_E}\right)$$

This equation can be further simplified if the following condition exists:

$$0.1R_E >> h_{ib}$$

In that case, we have the simple result

$$R_{TH} \approx 11r_o \tag{9.12}$$

Hence, if all the approximations are valid, R_{TH} is independent of β and its value is quite large.

9.2.4 Differential Amplifier with Single-Ended Input and Output

Figure 9.6 shows a differential amplifier in which the second input, v_2, is set equal to zero and the output is taken as v_{o1}.

We use a constant-current source in place of R_{EE}, as discussed in the previous section. This is known as a *single-ended input and output amplifier with phase reversal.* The amplifier is analyzed by setting $v_2 = 0$ in the earlier equations. The differential input is then simply

$$v_{di} = v_1 - v_2 = v_1$$

so the output is

$$v_o = v_{o1} = A_d v_{di} = \frac{-R_C v_i}{2h_{ib}}$$

The minus sign indicates that this amplifier exhibits a 180° phase shift between the output and input. A typical sinusoidal input and output are illustrated in Figure 9.7.

If an output signal is to be referenced to ground but a phase reversal is not desired, the output can be taken from transistor Q_2.

Figure 9.6
Single-ended input with phase reversal.

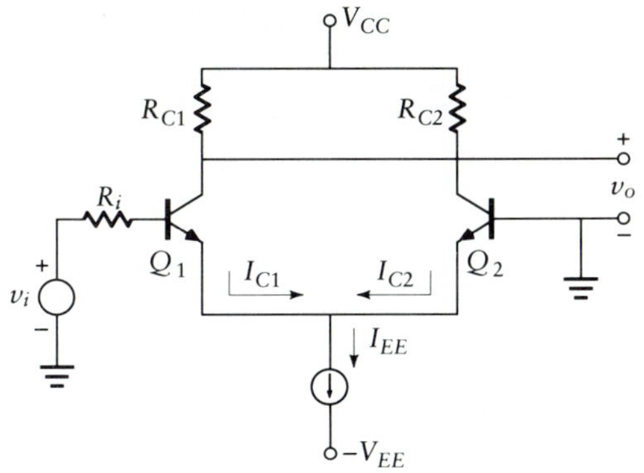

Figure 9.7
Single-ended input with in-phase output.

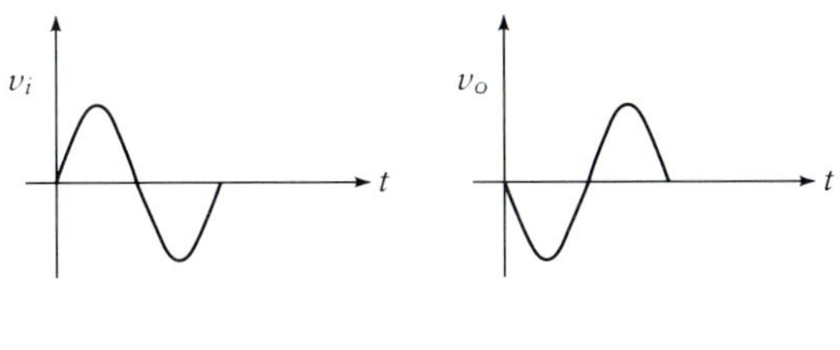

Example 9.1 **Differential Amplifier (Analysis)**

Find the differential voltage gain, the common-mode voltage gain, and the CMRR for the circuit shown in Figure 9.3. Assume that $R_i = 0$, $R_C = 5\ \text{k}\Omega$, $V_{EE} = 15$ V, $V_{BE} = 0.7$ V, $V_T = 26$ mV, and $R_{EE} = 25\ \text{k}\Omega$. Let $v_2 = 0$ and take the output from v_{o2}.

SOLUTION The current through R_{EE} is found at the quiescent condition. Since the base of Q_2 is grounded, the emitter voltage is $V_{BE} = 0.7$ V, and

$$I_{EE} = \frac{V_{EE} - V_{BE}}{R_{EE}} = 0.57\ \text{mA}$$

The quiescent current in each transistor is one-half of this amount.

$$I_{C1} = I_{C2} = \frac{I_{EE}}{2} = 0.29\ \text{mA}$$

Since

$$h_{ib} = \frac{V_T}{I_C} = 90\ \Omega$$

the differential voltage gain in each transistor is

$$A_d = \frac{-R_C}{2h_{ib}} = -28$$

The common-mode voltage gain is

$$A_c = \frac{-R_C}{2R_{EE}} = -0.1$$

The common-mode rejection ratio is then given by

$$\text{CMRR} = 20 \log\left(\frac{|A_d|}{|A_c|}\right) = 49\ \text{dB}$$

Example 9.2

For the differential amplifier described in Example 9.1, design a current source to replace R_{EE} and determine the new CMRR for the differential amplifier, with $r_o = 105\ \text{k}\Omega$, $V_{BE} = 0.7$ V, and $\beta = 100$.

SOLUTION We place the transistor operating point in the middle of the dc load line.

$$V_{CEQ} = \frac{V_{EE} - V_{BE}}{2} = 7.15 \text{ V}$$

Then, referring to the current source of Figure 9.5(a),

$$R_E = \frac{7.15 \text{ V}}{0.57 \text{ mA}} = 12.5 \text{ k}\Omega$$

For bias stability,

$$R_B = 0.1\beta R_E = 125 \text{ k}\Omega$$

Since $0.1R_E >> h_{ib}$ (i.e., $1.25 \text{ k}\Omega >> 26/0.57 \ \Omega$), from equation (9.12) we have

$$R_{TH} \approx 11r_o = 11(105 \text{ k}\Omega) = 1.16 \text{ M}\Omega$$

The CMRR is given by

$$\text{CMRR} = 20 \log\left(\frac{R_{TH}}{h_{ib}}\right) = 82.2 \text{ dB}$$

Drill Problems

D9.1 What are the differential- and common-mode gains of the amplifiers of Figure 9.3 if $R_{C1} = R_{C2} = 5 \text{ k}\Omega$ and $R_{EE} = 20 \text{ k}\Omega$? Assume that $V_{CC} = 12$ V, $V_{BE} = 0.7$ V, $V_T = 26$ mV, and $V_{EE} = 12$ V.

Ans: $A_c = -0.125$; $A_d = -53.8$

D9.2 What are the differential- and common-mode gains of the amplifier of Figure 9.6 if $R_{C1} = R_{C2} = 5 \text{ k}\Omega$, $V_{CC} = 15$ V, $V_{EE} = 15$ V, $V_{BE} = 0.7$ V, and $V_T = 26$ mV? Assume that V_{EE}/R_{TH} is a constant-current source with $R_{TH} = 10 \text{ k}\Omega$.

Ans: $A_c = -0.25$; $A_d = -69$

D9.3 For the constant-current source shown in Figure 9.5, determine R_E, R_1, and R_2, where $V_{EE} = 10$ V, $I_{EE} = 2$ mA, and $V_{BE} = 0.7$ V. Also assume that $V_T = 26$ mV and $\beta = 200$.

Ans: $R_E = 2.33 \text{ k}\Omega$; $R_1 = 93 \text{ k}\Omega$; $R_2 = 93 \text{ k}\Omega$

9.3 CURRENT SOURCES, ACTIVE LOADS, AND LEVEL SHIFTERS

In this section we explore alternative methods of simulating a constant-current source to replace R_{EE} in a differential amplifier. One type of source is discussed in Section 9.2.3, where a compensated fixed-bias transistor current source is used.

9.3.1 A Simple Current Source

A commonly used constant current source on ICs is a simple two-transistor current source as shown in Figure 9.8. The current source is achieved by using a reference current to a transistor connected as a diode and then using the voltage across this transistor to drive the second transistor where $R_E = 0$. Since the circuit has only one resistor, it can easily be fabricated on an IC chip. The disadvantage of this circuit is that the reference current is approximately equal to the current source. In this circuit, Q_2 is in the linear mode, since the collector voltage (output) is higher than the base voltage. The transistors, Q_1 and Q_2, are identical devices fabricated on the same IC chip. The emitter currents are equal since the emitters and bases are in parallel. If we sum the currents of Q_2, we obtain

$$I_B + I_C = I_E$$

so

$$I_o = I_C = I_E - \frac{I_E}{\beta + 1} = \frac{\beta I_E}{\beta + 1}$$

Summing currents at the collector of Q_1, we obtain

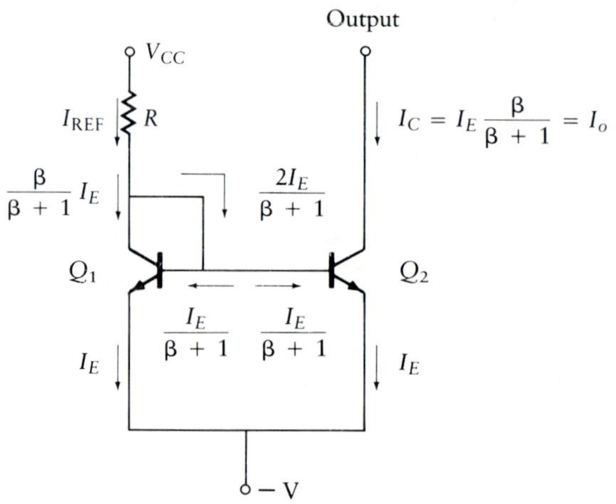

Figure 9.8
Current source.

$$I_{REF} = \left(\frac{\beta}{\beta + 1} + \frac{2}{\beta + 1}\right)I_E = \frac{\beta + 2}{\beta + 1}I_E \approx I_o$$

The current gain of the mirror is given by

$$A_i = \frac{I_o}{I_{in}} = \frac{\beta I_E}{\beta + 1}\frac{\beta + 1}{(\beta + 2)I_E} = \frac{1}{1 + 2/\beta} \approx 1$$

If β is large, this current gain is approximately equal to unity, and the current mirror has reproduced the input current.

A disadvantage of this current source is that R_{TH} of the circuit is limited by the r_o of the transistor.

$$R_{TH} = r_o = \frac{V_A}{I_o} \approx \frac{V_A}{I_{REF}}$$

9.3.2 Widlar Current Source

Because of the high gain of an operational amplifier, the bias currents must be small. Typical collector currents are in the range of 5 μA. Large resistors are often required to maintain small currents, and these large resistors occupy correspondingly large areas on the IC chip. It is therefore often desirable to replace these large resistors with current sources. One such device is the *Widlar current source* [41], as illustrated in Figure 9.9. This constant current source is not limited by the early voltage of the transistor. The two transistors, Q_1 and Q_2, are identical. We sum the voltages around the base loop of the two transistors to obtain

$$V_{BE1} - V_{BE2} - I_{C2}R_2 = 0 \qquad (9.13)$$

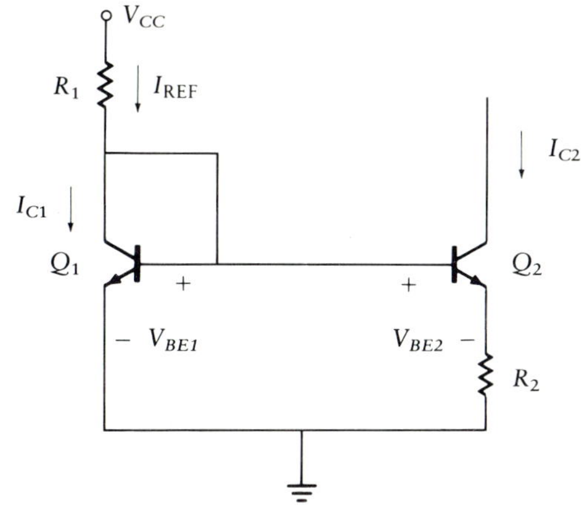

Figure 9.9 Widlar current source.

$$V_{BE} = V_T \ln\left(\frac{I_C}{\beta I_o}\right) \tag{9.14}$$

Substituting the expression of equation (9.14) into equation (9.13), we obtain

$$V_T \ln\left(\frac{I_{C1}}{\beta I_o}\right) - V_T \ln\left(\frac{I_{C2}}{\beta I_o}\right) - I_{C2}R_2 = 0 \tag{9.15}$$

We assume that the transistors are matched so that I_o, β, and V_T are the same for both transistors. Thus

$$V_T \ln\left(\frac{I_{C1}}{I_{C2}}\right) = I_{C2}R_2 \tag{9.16}$$

For design purposes, I_{C1} is usually known because it is used as the reference and I_{C2} is the desired output current. This allows us to solve equation (9.16) for the required value of R_2.

The Widlar circuit can be used to simulate a high resistance. The small-signal equivalent circuit is shown in Figure 9.10. Since this circuit has no independent sources, it is equivalent to a single Thevenin resistance. To find the value of this resistance, we assume a current, i_{TH}, find the resulting voltage, and take the ratio. We start by summing currents at node 1,

$$i_{TH} = \beta i_{B2} + \frac{v_2}{r_o} \tag{9.17}$$

and at node 2,

$$\beta i_{B2} + \frac{v_2}{r_o} - \frac{v_1}{R_2} + i_{B2} = 0 \tag{9.18}$$

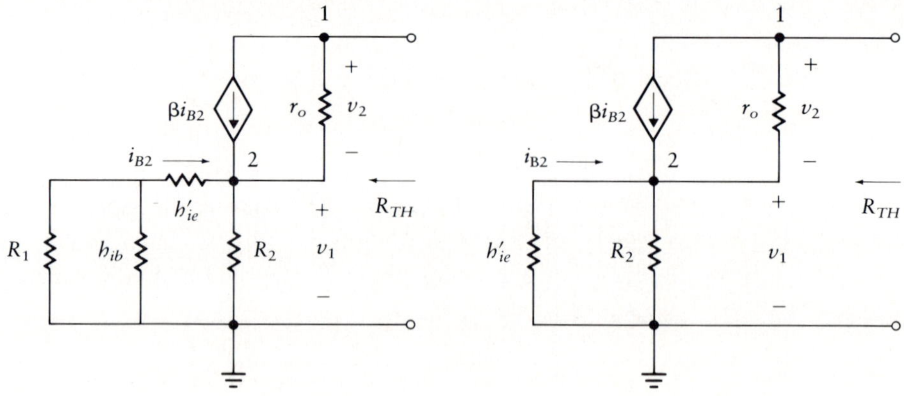

Figure 9.10 Widlar current source equivalent circuit.

(a) Actual equivalent circuit (b) Approximate equivalent circuit

From equation (9.17) we have

$$v_2 = (i_{TH} - \beta i_{B2})r_o \tag{9.19}$$

We define h'_{ie} for Q_2 and h_{ib} for Q_1.

$$v_1 = -i_{B2}h'_{ie} \tag{9.20}$$

Substituting equations (9.19) and (9.20) into equation (9.18) yields

$$-i_{TH} = i_{B2}\left(1 + \frac{h'_{ie}}{R_2}\right) \tag{9.21}$$

We now find R_{TH} as the ratio of voltage to current.

$$R_{TH} = \frac{v_{TH}}{i_{TH}} = \frac{r_o(1 + \beta + h'_{ie}/R_2) + h'_{ie}}{1 + h'_{ie}/R_2}$$

To simplify this result, assume that $h'_{ie} = \beta V_T/I_{C2}$ and

$$\frac{h'_{ie}}{R_2} >> 1, \qquad r_o >> h'_{ie}$$

Then

$$R_{TH} = r_o\left(1 + \frac{\beta}{h'_{ie}/R_2}\right)$$

$$= r_o\left(1 + \frac{\beta R_2 I_{C2}}{\beta V_T}\right)$$

Finally,

$$R_{TH} = r_o\left(1 + \frac{I_{C2}R_2}{V_T}\right) \tag{9.22}$$

Equation (9.22) shows that R_{TH} depends on $I_{C2}R_2$, which is the dc voltage drop across R_2. The larger the voltage drop, the higher the output resistance. Generally, the output resistance of a simple current source is increased by the addition of resistances (R_2 in Figures 9.9 and 9.10) in the emitters of the current source transistors.

The following example illustrates how a current source can be designed to provide a small constant current while using resistors that are easily fabricated on an IC chip.

Example 9.3 Widlar Current Source (Design)

Design a Widlar current source to provide a constant current of 3 μA and determine the R_{TH} with $V_{CC} = 12$ V, $R_1 = 50$ kΩ, $V_A = 50$ V, and $V_{BE} = 0.7$ V.

SOLUTION Use the circuit of Figure 9.9. Apply KVL to the Q_1 transistor to obtain

$$I_{C1} \approx I_{REF} = \frac{12 - 0.7}{5 \times 10^5} = 0.226 \text{ mA}$$

From equation (9.16) we have

$$0.026 \ln\left(\frac{2.26 \times 10^{-4}}{3 \times 10^{-6}}\right) = 3 \times 10^{-6} R_2$$

and

$$R_2 = 37.5 \text{ k}\Omega$$

$$R_{TH} = \frac{V_A}{I_{C2}}\left(1 + \frac{I_{C2} R_2}{V_T}\right) = 88.8 \text{ M}\Omega$$

Since R_2 is less than 50 kΩ, it is practical to fabricate it on an IC.

9.3.3 Wilson Current Source

One of the important characteristics of current sources is that they need to achieve a high Thevenin equivalent resistance—much higher than can be attained with the simple two-transistor current source. The *Wilson current source* [43] uses three transistors and provides this capability, and the output is almost independent of the internal transistor characteristics. The Wilson current source, as shown in Figure 9.11, uses the negative feedback provided by Q_3 to raise the output impedance. The difference between the reference current and I_{C1} is the base current of Q_2. This base current is multiplied by $(\beta + 1)$ and becomes the collector current of Q_3. Since the base of Q_1 is connected to the base of Q_3, the magnitude of the current in Q_1 is approximately independent of the voltage of the collector of Q_2. Thus, the collector current of Q_2 remains almost constant, providing a high output impedance.

We now illustrate that I_{C2} is approximately equal to I_{REF}. Applying Kirchhoff's current law at the emitter of Q_2 yields

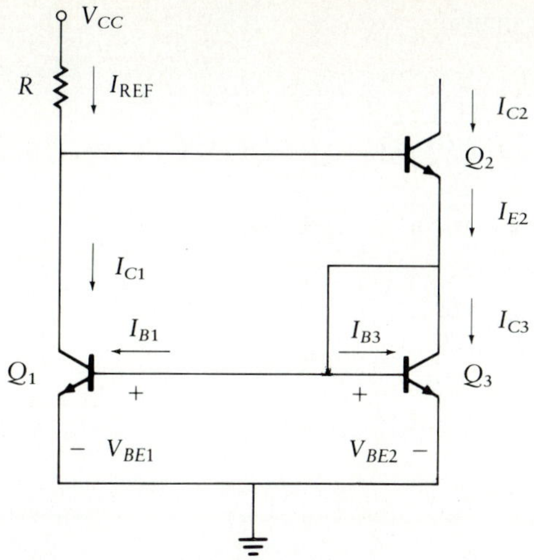

Figure 9.11
Wilson current source.

$$I_{E2} = I_{C3} + I_{B3} + I_{B1} \tag{9.23}$$

Using the relationship between collector and base current yields

$$I_{E2} = I_{C3}\left(1 + \frac{1}{\beta}\right) + \frac{I_{C1}}{\beta} \tag{9.24}$$

Since all three transistors are designed to be identical,

$$V_{BE1} = V_{BE2} = V_{BE3}$$

and

$$\beta_1 = \beta_2 = \beta_3$$

With identical transistors, current in the feedback path splits equally between the bases of Q_1 and Q_3, leading to the result that

$$I_{C1} = I_{C3}$$

Thus, equation (9.24) becomes

$$I_{E2} = I_{C3}\left(1 + \frac{2}{\beta}\right)$$

The collector current of Q_2 is

$$I_{C2} = \frac{I_{E2}\beta}{\beta + 1} = \frac{I_{C3}(1 + 2/\beta)\beta}{\beta + 1}$$

Solving for I_{C3} yields

$$I_{C3} = \frac{I_{C2}}{(1 + 2/\beta)\beta/(1 + \beta)} = I_{C2}\frac{\beta + 1}{\beta + 2} \tag{9.25}$$

Summing currents at the base of Q_2, we find

$$I_{C1} = I_{REF} - \frac{I_{C2}}{\beta}$$

or

$$I_{C2} = \beta(I_{REF} - I_{C1}) \tag{9.26}$$

Since $I_{C1} = I_{C3}$, we substitute I_{C3} from equation (9.25) for I_{C1} in equation (9.26) to obtain

$$I_{C2} = \beta I_{REF} - \frac{\beta(\beta + 1)}{\beta + 2}I_{C2}$$

and solving for I_{C2},

$$I_{C2} = \frac{\beta^2 + 2\beta}{\beta^2 + 2\beta + 2}I_{REF}$$

$$= 1 - \frac{2}{\beta^2 + 2\beta + 2}I_{REF} \tag{9.27}$$

Equation (9.27) shows that β has little effect on I_{C2}, since for reasonable values of β,

$$\frac{2}{\beta^2 + 2\beta + 2} << 1$$

We perform a similar analysis for the Wilson current source, and we simply state the result here. The equivalent resistance is given by

$$R_{TH} = \frac{\beta r_o}{2} = \frac{\beta V_A}{2I_{C2}} \tag{9.28}$$

The equivalent resistance of the Wilson source is much greater than that of the other current sources.

9.3.4 Current Mirrors

The circuit shown in Figure 9.12 is known as a *current mirror.* It is used in op-amp circuit design to reproduce a current from one location to one or more other locations.

The errors in base current accumulate when multiple outputs are used and the current gain tends to deviate from unity. Let us look at Q_4, since it has an emitter resistor. In this case, we have a Widlar current source using the same reference current. We can change the mirrored current to whatever value is needed in the circuit by sizing the emitter resistor as discussed in Section 9.3.2. This approach has the great advantage that we can use a single reference current to mirror different current values throughout a chip as needed.

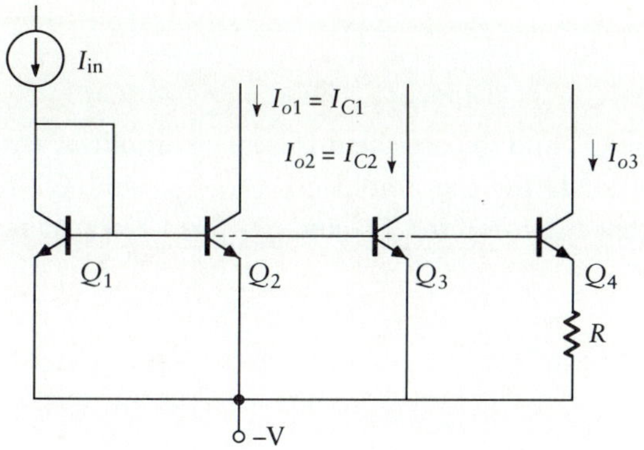

Figure 9.12
Current mirror.

9.3.5 Current Sources as Active Loads

In the conventional differential amplifiers of Section 9.2, the collector loads consist of resistors. The differential gain is

$$A_v = \frac{v_o}{v_i} = \frac{-R_C}{h_{ib}} = \frac{-R_C I_{CQ}}{V_T}$$

To achieve a large voltage gain, the $R_C I_{CQ}$ product must be large. This requires large values of either resistance or power supply voltage. Since the power supply size is usually fixed, large resistors are required. The preceding sections illustrate several alternative forms of almost ideal constant-current sources. These current sources can be substituted for the large resistors needed on an IC chip.

As an example of the use of constant-current circuits, consider Figure 9.13, which shows a differential amplifier using active loads, two single two-transistor current sources, and a current mirror. Circuits of this type are found in many linear ICs, since large resistors are avoided in the fabrication process.

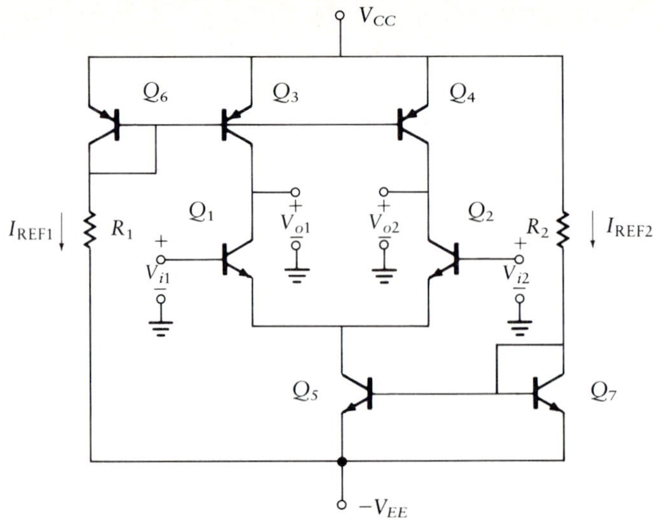

Figure 9.13 Common-emitter amplifier with active load.

Transistors Q_1 and Q_2 are the differential amplifier transistors; Q_3 and Q_4 are the active loads for the amplifier; Q_3, Q_4, and Q_6 form the current source with current mirrors; and Q_5 and Q_7 form a transistor current source.

Example 9.4

Design a circuit to attain the conditions as specified in Figure 9.14 for maximum output voltage swing. The five transistors Q_1 to Q_5 each have $\beta = 100$, while for transistor Q_6, $\beta = 200$. $V_{BE} = 0.6$ V for all transistors, $V_T = 26$ mV, and $V_A = 80$ V. Assume all transistors are identical. Determine:

a. R_C, R_1, R_2, and CMRR.
b. Common-mode output voltage.
c. Differential-mode output voltage.
d. Differential-mode input voltage for maximum output.

SOLUTION We shall treat the circuit in three sections:

1. Darlington amplifier

$$I_{CQ} = \frac{12}{4400 + 4000} = 1.43 \text{ mA for maximum swing}$$

$$h_{ib} = \frac{26}{1.43} = 18.2\ \Omega$$

$$R_{\text{in}}(\text{Darlington}) = R_L(\text{differential}) = 2h_{ib2}\beta_1\beta_2 = 728\ \text{k}\Omega$$

$$V_{od} = 0.9 \times 0.00143 \times 4000 = 5.15 \text{ V 0-p (zero-to-peak)}$$

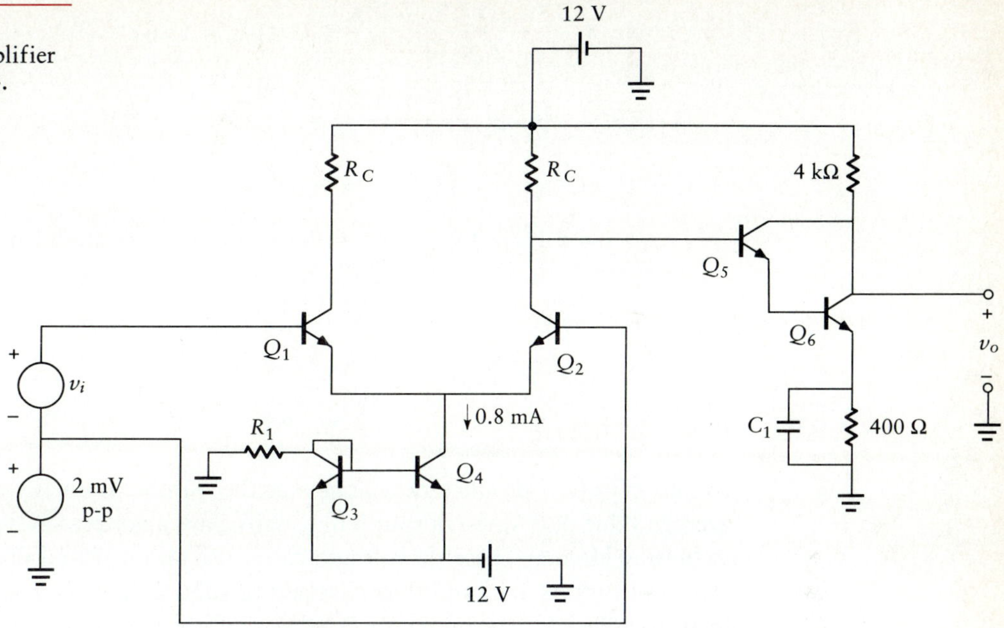

Figure 9.14
Differential amplifier for Example 9.4.

$$A_v = -\frac{4000}{2(18.2)} = -110$$

2. Differential amplifier

$$I_{C1} = I_{C2} = \frac{I_{EE}}{2} = \frac{0.0008}{2} = 0.4 \text{ mA}$$

$$V_C = 12 - R_C(0.0004) = V_B = 1.2 + 0.00143 \times 400$$

$$R_C = 25.6 \text{ k}\Omega$$

$$h_{ib}(\text{diff}) = \frac{26}{0.4} = 65 \ \Omega$$

3. Current source

$$R_1 = \frac{12 - 0.6}{0.8 \times 10^{-3}} = 14.25 \text{ k}\Omega$$

$$R_{\text{TH}} = \frac{80}{0.8 \times 10^{-3}} = 100 \text{ k}\Omega$$

Now, for the total system we have

$$A_{vd} = \frac{728 \text{ k}\Omega \parallel 25.6 \text{ k}\Omega}{2 \times 65} \times (-110) = -20.9 \times 10^3$$

$$A_{vc} = \frac{728\ \text{k}\Omega \parallel 25.6\ \text{k}\Omega}{2 \times 10^5} \times (-110) = 13.6$$

$$\text{CMRR} = 20 \log\left(\frac{20.9 \times 10^3}{13.6}\right) = 63.7\ \text{dB}$$

$$V_{oc} = 0.002 \times 13.6 = 28.2\ \text{mV p-p}$$

$$V_{id} = 5.14/(20.9 \times 10^3) = 0.246\ \text{mV 0-p} \quad \text{or } 0.492\ \text{mV p-p}$$

9.3.6 Level Shifters

Amplifiers often produce dc voltages at the output. Even if the input has an average value of 0 V, the output will usually have a nonzero average voltage due to biasing effects. These dc voltages can cause an undesired offset.

Level shifters are amplifiers that add or subtract a known voltage from the input in order to compensate for dc offset voltages. Since the op-amp is a multistage dc amplifier with high gain, unwanted dc voltages can be a source of concern. A small offset in an early stage can saturate a later stage. For this reason, op-amps have level shifters included in their design.

Figure 9.15 illustrates a level shifter. We show that this shifter acts as a unity-gain amplifier for ac while providing an adjustable dc output. We begin the analysis by using KVL in Figure 9.15(a) and letting $v_i = 0$ to obtain

$$V_{BB} = I_B R_B + V_{BE} + I_C R_E + V_o$$

Now since

$$I_B = \frac{I_C}{\beta}$$

we solve for the dc value of output voltage, V_o.

$$V_o = V_{BB} - \frac{R_B I_C}{\beta} - I_C R_E - V_{BE} \tag{9.29}$$

This equation shows that, by selecting a value of R_E, V_o can be set to any desired dc level (less than $V_{BB} - V_{BE}$). A complete circuit with an active current source is shown in Figure 9.15(b). Since V_{BB} is the dc level acquired from the previous stage, this amplifier is used to shift the level *downward* (to a lower value). If *upward* shifting is required, a similar circuit is used but *pnp* transistors are substituted for the *npn* transistors.

Figure 9.15
Level shifter.

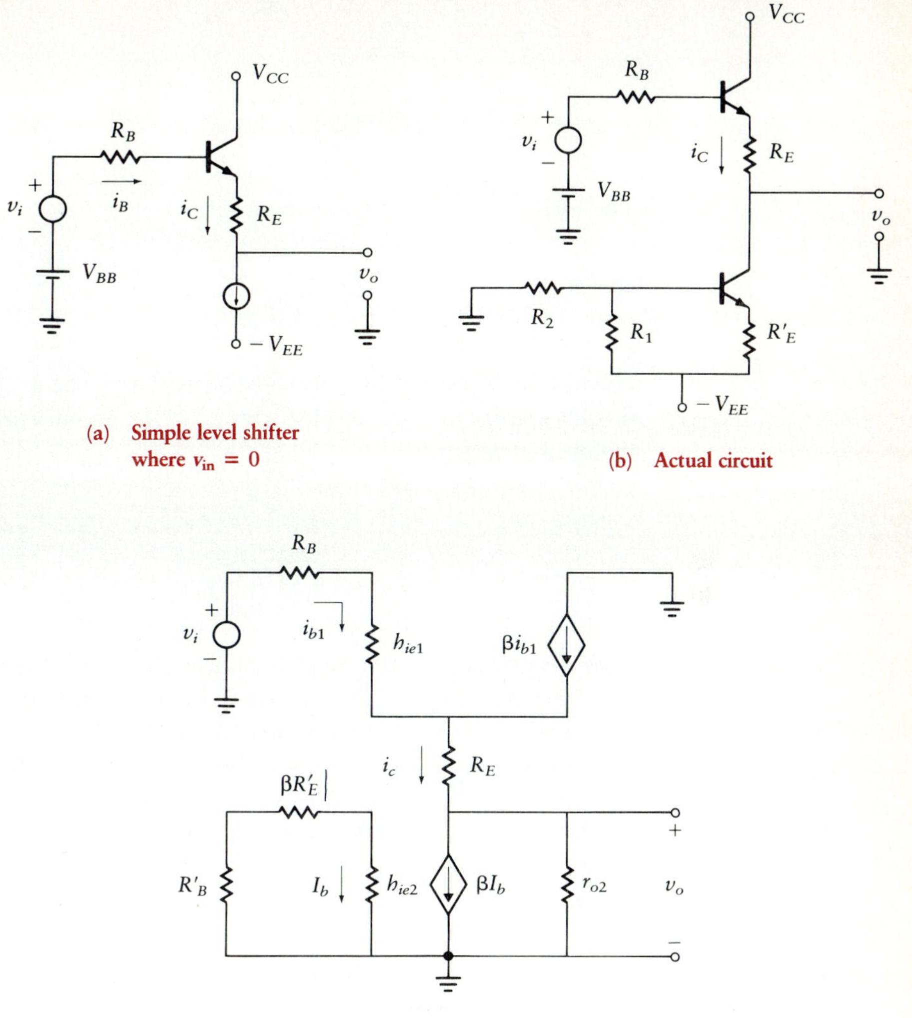

(a) **Simple level shifter where $v_{in} = 0$**

(b) **Actual circuit**

(c) **Small-signal ac equivalent circuit**

We now examine the circuit with ac signals applied. Figure 9.15(c) illustrates the ac equivalent circuit. Note that βI_B is the collector current in the active current source, and we assume it to be a constant. Because the ac value of the current is zero, this current source is replaced by an open circuit. We write the ac equations using KVL.

$$v_i = i_{b1} R_B + i_{b1} h_{ie1} + i_c R_E + v_o \tag{9.30}$$

and

$$v_o = i_c r_{o2}$$

$$i_c = \frac{v_o}{r_{o2}}$$

$$v_i = \frac{v_o R_B}{\beta r_{o2}} + \frac{v_o h_{ie1}}{\beta r_{o2}} + \frac{v_o R_E}{r_{o2}} + v_o$$

The ratio of ac output to ac input is

$$\frac{v_o}{v_i} = \frac{1}{1 + (R_B/\beta + h_{ie}/\beta + R_E)/r_{o2}} \tag{9.31}$$

Equation (9.31) shows that as r_{o2} becomes large, the ratio of output to input approaches unity and the level shifter acts like an emitter follower to ac. This is the desired result.

Example 9.5

Two direct-coupled CE amplifiers are placed in series to achieve the desired voltage gain. Design a level shifter to be placed between the two CE amplifiers to provide a dc voltage sufficiently low to prevent the second CE amplifier from saturating. Do this by providing a 1-V bias to the second stage. The collector voltage, V_C, of the first amplifier is 4 V, and the R_C of that amplifier is 1 kΩ. Design the level shifter to have an I_{CQ} of 1 mA using a 10-V power supply. Use a current source of the type shown in Figure 9.5 with transistors having $\beta = 100$, $V_{BE} = 0.7$ V, and $V_\gamma = 0.7$ V.

SOLUTION The level shifter is shown in Figure 9.15(b). We need to find the values of R_E, R_1, R_2, and R_E'. Since the first amplifier has a V_C of 4 V, the value of V_{BB} for equation (9.29) is 4 V, whereas the R_B of that formula is 1 kΩ. Note that this is using the Thevenin equivalent circuit of the previous amplifier. Equation (9.29) then yields

$$R_E = 2.29 \text{ k}\Omega$$

Setting the current-source transistor operating point in the middle of the dc load line, we have

$$V_{CEQ} = \frac{10 + 1}{2} = 5.5 \text{ V}$$

$$R_E' = \frac{5.5}{10^{-3}} = 5.5 \text{ k}\Omega$$

and

$$R'_B = R_1 \parallel R_2 = 0.1\beta R'_E = 55 \text{ k}\Omega$$

The voltage across R'_E is 5.5 V.

Then

$$V_{R2} = 10 - 5.5 - 0.7 = 3.8 \text{ V}$$

We now know the voltages across R_1 and R_2 and the parallel resistance. This yields two equations, where we assume that the base current in the lower transistor of Figure 9.15(b) is negligible.

$$R_B = \frac{(5.5/I)\,(3.8/I)}{(5.5/I) + (3.8/I)} = 55 \text{ k}\Omega$$

$$I = 0.041 \text{ mA}$$

and

$$R_1 = 134 \text{ k}\Omega$$

$$R_2 = 93 \text{ k}\Omega$$

The design is therefore complete.

Drill Problems

D9.4 Design a Widlar current source to obtain a constant current of 50 μA. The battery voltage is 20 V, and a 4-mA reference current is used. Determine the resistor value required and R_{TH} of the current source when r_o(transistor) = 50 kΩ, $V_{BE} = 0.7$ V, and $V_T = 26$ mV.

Ans: $R_1 = 4.83$ kΩ; $R_2 = 2.28$ kΩ; $R_{\text{TH}} = 269$ kΩ

D9.5 Design a Wilson current source to provide 100 μA when $V_{CC} = 15$ V. Use transistors with $\beta = 100$, $V_{BE} = 0.7$ V, and $V_T = 26$ mV. Determine R_{TH} when r_o of the transistor is 15 kΩ.

Ans: $I_{\text{REF}} = 100$ μA; $R_1 = 136$ kΩ; $R_{\text{TH}} = 750$ kΩ

D9.6 Design a level shifter to change V_C from 6 V to 2 V so that the voltage is compatible with a following CE amplifier. Use a CE amplifier that has $R_C = 5$ kΩ and a current source of 4 mA with $V_{CC} = 15$ V and $V_{EE} = 15$ V. The transistors have $\beta = 200$ and $V_{BE} = 0.7$ V.

Ans: $R_E = 805\ \Omega$; $R_E' = 2.13\ \text{k}\Omega$;
$R_1 = 110\ \text{k}\Omega$; $R_2 = 69\ \text{k}\Omega$

9.4 TYPICAL OP-AMPS

We have discussed numerous circuits in this text for use as analog amplifiers. In this chapter, we completed the review of the circuits necessary to fabricate an operational amplifier.

Most operational amplifiers are designed and constructed in accordance with the block diagram shown in Figure 9.16. The differential amplifier and the voltage gain stage are the only stages that provide the voltage gain. The differential amplifier also provides a large value for CMRR, which is important for good op-amp performance. The output of the differential amplifier is often connected to an emitter follower with a large emitter resistor to provide a high-impedance load to the differential amplifier. Linear op-amps are direct-coupled to provide dc gain. This also eliminates the need for a coupling capacitor that is too large to be placed on an IC chip. Level shifters are required to ensure that the output signal does not have any dc offset.

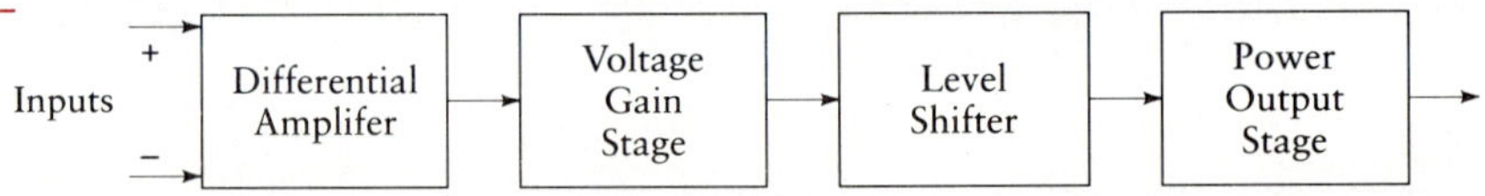

Figure 9.16
Typical configuration of an op-amp.

9.5 OP-AMP PACKAGING

Op-amp circuits are packaged in standard IC packages, including cans, dual-in-line packages (DIPs), and flat packs. Each of these packages has at least eight pins or connections. They are illustrated in Figures 9.17, 9.18, and 9.19. In a practical design, it is important to identify the various leads correctly (they are usually not numbered). The figures illustrate the manner in which pin 1 is identified from the physical configuration. In the *can package* of Figure 9.17, pin 1 is identified as the first pin to the left of the tab, and the pins are numbered consecutively counterclockwise looking from the top. In the *dual-in-line package* of Figure 9.18, the top of the package has an indentation to locate pin 1, and the pins are numbered down on the left and up on the right. Note that more than one op-amp (typically two or four) is packaged in one DIP.

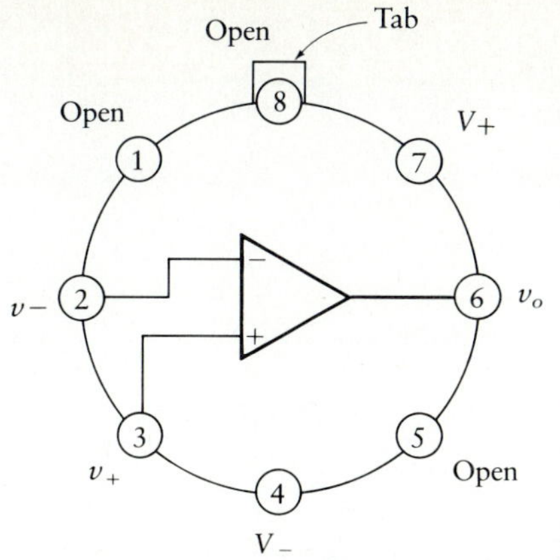

Figure 9.17 Op-amp connection for can package.

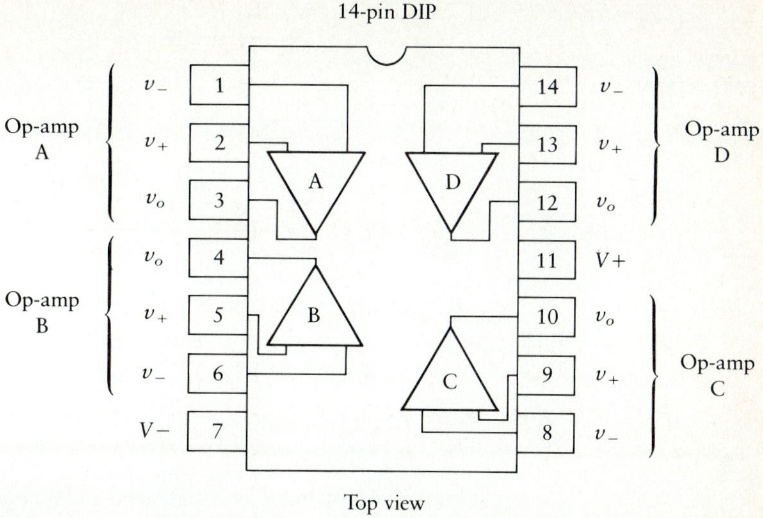

Figure 9.18 Op-amp connection for 14-pin DIP.

Figure 9.19 Op-amp connection for 10-pin flat pack.

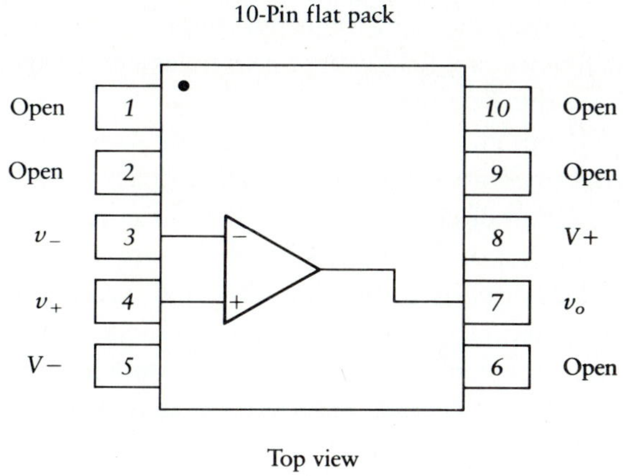

In the *flat pack* of Figure 9.19, pin 1 is identified by a dot and the pins are numbered as in the DIP.

9.5.1 Power Requirements

Many op-amps require both a negative and a positive voltage source. Typical voltage sources range from ±5 V to ±22 V. Figure 9.20 shows typical power supply connections to the op-amp.

The maximum output voltage swing is limited by the dc voltage supplied to the op-amp. Some operational amplifiers can be operated from a single voltage source. The manufacturer's specifications define the limits of operation in cases where the op-amp uses only one power supply.

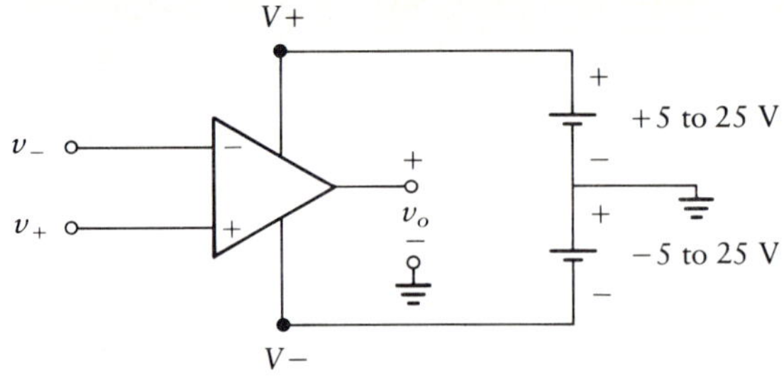

Figure 9.20 Power supply connections.

9.6 THE 741 OP-AMP

The μA741 op-amp is illustrated in the equivalent circuit of Figure 9.21. It has been produced since 1966 by most IC manufacturers, and although there have been many advances since its introduction, the 741 is still widely used.

The 741 op-amp has *internal compensation,* which refers to the RC network that causes the high-frequency amplitude response to fall off. Two cascaded difference amplifiers drive a complementary-symmetry power amplifier through another voltage amplifier.

The 741 op-amp consists of three major stages: an input differential amplifier, an intermediate single-ended high-gain amplifier, and an output buffering amplifier. Other circuitry important for its operation includes a level shifter to shift the dc level of the signal so that the output can swing both positive and negative, bias circuits to provide reference currents to the various amplifiers, and circuits that protect the op-amp from short circuits at the output. The 741 is internally compensated by means of an on-chip capacitor-resistor network.

The op-amp is further improved by adding more stages of amplification, isolating the input circuits, and adding more emitter followers at the output to decrease the output impedance. Other improvements result in increased CMRR, higher input impedance, wider frequency response, decreased output impedance, and increased power.

9.6.1 Bias Circuits

Several constant sources can be seen in the 741 op-amp of Figure 9.21. Transistors Q_8 and Q_9 are the current source for I_{EE} of the differential amplifier formed by Q_1, Q_2, Q_3, and Q_4. Transistors Q_5, Q_6, and Q_7 are the active loads substituting for the R_C resistors of the differential amplifier. Transistors Q_{10}, Q_{11}, and Q_{12} form the bias network for the differential amplifier current sources. Transistors Q_{10} and Q_{11} form a Widlar current source for this bias network, with the other transistor acting as current mirror.

Figure 9.21
The 741 op-amp.
Courtesy of Fairchild Semiconductor Corp.

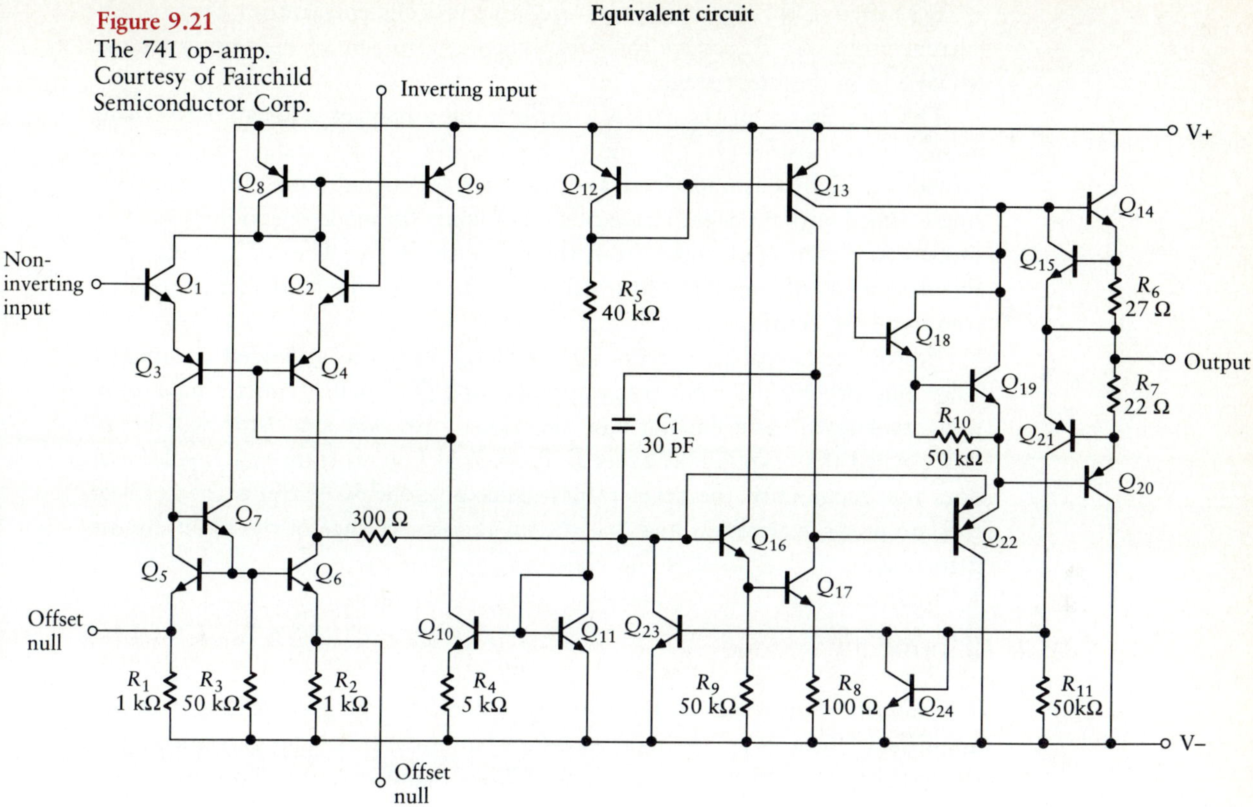

9.6.2 Short-Circuit Protection

The 741 circuit includes a number of transistors that are normally cut off and conduct only when a large current exists at the output. The bias on the output transistors is then changed to reduce this current to an acceptable level. In the circuit of Figure 9.21, this short-circuit protection network consists of transistors Q_{15} and Q_{22} and resistor R_{11}.

9.6.3 Input Stage

The input stage of the 741 op-amp is required to provide voltage gain, level shifting, and a single-ended differential amplifier output. The complexity of the circuitry causes a large offset-voltage error. In contrast to this, the standard resistor-loaded differential amplifier causes less offset-voltage error. However, the standard amplifier has limited gain, which means that more stages would be required to achieve the desired amplification. The resistor-loaded differential amplifiers are used in op-amps that have less voltage drift than the 741.

BJTs used in the input stage require large bias currents, introducing offset-current problems. To reduce the offset-current error, other op-amp types use MOSFETs in the input stage.

The input stage of the 741 is a differential amplifier with an active load formed by transistors Q_5, Q_6, and Q_7 and resistors R_1, R_2, and R_3. This circuit provides a high-resistance load and converts the signal from differential to single-ended with no degradation of gain or common-mode rejection ratio. The single-ended output is taken from the collector of Q_6. The input-stage level shifter consists of lateral *pnp* transistors, Q_3 and Q_4, which are connected in a common-base configuration.

Use of the lateral transistors, Q_3 and Q_4, results in an added advantage. They help protect the input transistors, Q_1 and Q_2, against emitter-base junction breakdown. The emitter-base junction of an *npn* transistor will break down when the reverse bias exceeds about 7 V. Lateral transistor breakdown does not occur until the reverse bias exceeds about 50 V. Since the transistors are in series with Q_1 and Q_2, the breakdown voltage of the input circuit is increased.

9.6.4 Intermediate Stage

The intermediate stages in most op-amps provide high gain through several amplifiers. In the 741, the single-ended output of the first stage is connected to the base of Q_{16}, which is in an emitter-follower configuration. This provides a high input impedance to the input stage, which minimizes loading. The intermediate stage also consists of transistor Q_{17} and resistors R_8 and R_9. The output of the intermediate stage is taken from the collector of Q_{17} and provided to Q_{14} through a phase splitter. The capacitor in the 741 is used for frequency compensation, which is discussed in subsequent chapters of this text.

9.6.5 Output Stage

The output stage of an op-amp is required to provide high current gain to a low output impedance. Most op-amps use a complementary-symmetry output stage to gain more efficiency without sacrificing current gain. In Chapter 6 we show that the maximum achievable efficiency for the complementary-symmetry, Class-B amplifier is 78%. The single-ended output amplifier has a maximum efficiency of only 25%. Some op-amps use Darlington pair complementary symmetry to increase their output capability. The complementary-symmetry output stage in the 741 consists of Q_{14} and Q_{20}.

The small resistors, R_6 and R_7, provide current limiting at the output. The Darlington pair, Q_{18} and Q_{19}, are used in place of the diode in the diode-compensated complementary-symmetry output stage as described in Chapter 6. The Darlington pair arrangement is favored over the two transistors connected as a diode, since it can be fabricated in a smaller area. The current source substituting for the bias resistor in the complementary-symmetry circuit is realized by one part of transistor Q_{13}. Transistors Q_{22}, Q_{23}, and Q_{24} are part of a level-shifter arrangement which ensures that the output voltage is centered around the zero axis.

9.7 MANUFACTURERS' SPECIFICATIONS

Manufacturers' specifications provide the characteristics of the op-amp under various operational conditions. The major parameters are shown in either tabular or graphical form. There may also be manufacturer-recommended typical applications for the op-amp. Other items shown in the specification may include examples of the external circuits required to balance the op-amp or alter the frequency response. This would be an appropriate time to familiarize yourself with Appendix D, where we illustrate examples of specification sheets. The μA741 is included and should be viewed as a typical example. We examine these op-amp parameters in more detail in Chapter 10.

Op-amps are versatile building blocks for use by the designer. As we explore the various applications of these building blocks (in the next four chapters), it would be most useful to obtain one of the latest copies of a manufacturer's *linear databook*. This contains the specification sheets and other valuable information for a variety of op-amps and related linear ICs.

PROBLEMS

9.1 What are the differential- and common-mode gains for the amplifier of Figure 9.3 if $R_C = 10\text{ k}\Omega$, $R_{EE} = 10\text{ k}\Omega$, and $V_{CC} = 15$ V? Assume that V_{EE}/R_{EE} is a constant-current source and that $V_{BE} = 0.7$ V, $V_T = 26$ mV, and $V_{EE} = 15$ V.

9.2 What are the differential- and common-mode gains of the amplifier of Figure 9.3 if $R_{EE} = 5\text{ k}\Omega$, $R_C = 5\text{ k}\Omega$, $V_{CC} = V_{EE} = 15$ V? Assume that $V_{BE} = 0.7$ V, $V_T = 26$ mV, and $\beta = 100$.

9.3 Find the differential voltage for the circuit of Figure 9.3 if $v_1 = 0.6$ V and $v_2 = 0.55$ V. Let $R_C = 5\text{ k}\Omega$, $R_{EE} = 5\text{ k}\Omega$, and $I_{EE} = 2.6$ mA.

9.4 A manufacturer lists the voltage gain of a differential amplifier as 200 and the CMRR as 80 dB. If a differential signal of 2 mV is applied to the input along with an unwanted common-mode signal of 10 mV, what is the amplitude of each signal at the output?

9.5 The differential amplifier of Figure 9.5 has $A_d = 200$, a differential voltage input of 3 mV, a common-mode voltage of 15 mV, and a CMRR of 95 dB. Calculate the differential voltage and the common-mode voltage at the output.

9.6 Calculate v_{o1}, v_{o2}, and v_{o3} in terms of the input v_1 and v_2 for the circuit of Figure P9.1. Assume that the differential voltage gain, A_d (double-ended output to double-ended input), is 100 and that the common-mode voltage gain, A_c, is -0.5 for each stage. All transistors can be assumed to be identical.

9.7 Calculate v_{o3} in terms of the inputs, v_1 and v_2, for the circuit of Figure P9.2. The differential- and common-mode voltage gains for the first stage are the same as those in Problem 9.6. For transistor Q_3, $\beta = 100$, $R_C = 10\ \text{k}\Omega$, $h_{ib} = 50\ \Omega$, and $R_E = 200\ \Omega$.

9.8 Calculate v_{o1} and v_{o2} for the input voltages as shown in Figure P9.3. A_d and A_c are the same as those in Problem 9.6.

9.9 Find A_c, A_d, and the CMRR for the circuit of Figure 9.5 if $R_{C1} = R_{C2} = 3\ \text{k}\Omega$, $I_{EE} = 4$ mA, $R_E = 2\ \text{k}\Omega$, $V_A = 60$ V, $V_{CC} = 12$ V, and $V_{EE} = 12$ V.

Figure P9.1

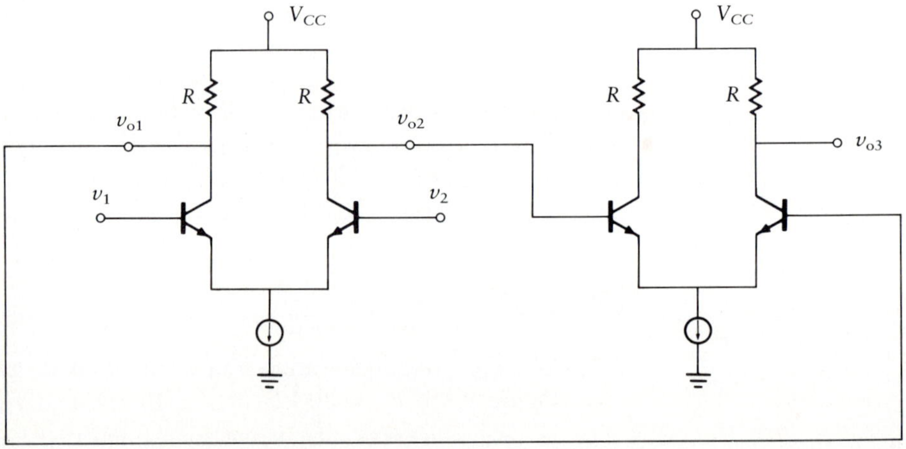

Figure P9.2

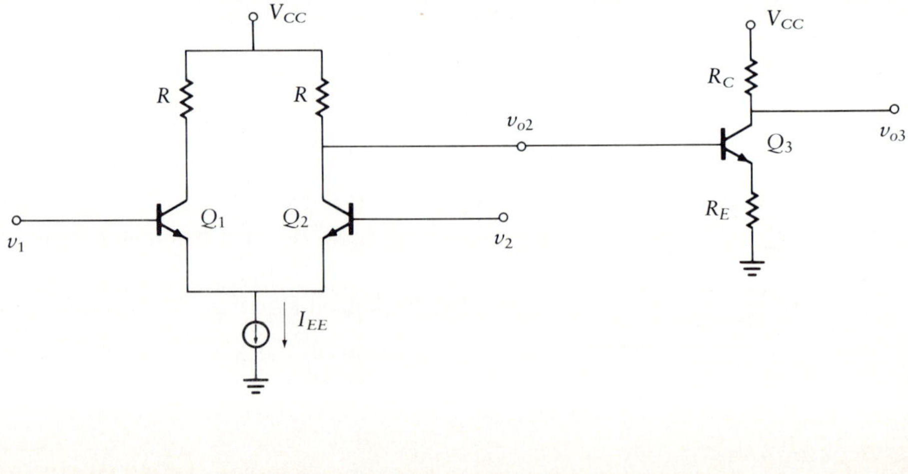

9.10 Find A_c, A_d, and the CMRR of the circuit of Figure 9.3 if $R_C = 4\ \text{k}\Omega$, $I_{EE} = 3$ mA, $V_{CC} = V_{EE} = 12$ V, and $R_{EE} = 40\ \text{k}\Omega$.

9.11 Find A_c, A_d, and the CMRR for the circuit of Figure P9.2 when R(current source) $= 50\ \text{k}\Omega$, $R = 6\ \text{k}\Omega$, V_2 is grounded, $I_{EE} = 2$ mA, $V_{CC} = 8$ V, $R_C = 2\ \text{k}\Omega$, $R_E = 500\ \Omega$, and $\beta = 200$.

9.12 In the current source shown in Figure 9.5, determine the values of R_1 and R_2 if $V_{EE} = 10$ V, $I_{EE} = 2$ mA, and $\beta = 200$. Make sure V_γ can be used to balance out the temperature variations of V_{BE}.

9.13 If the current source shown in Figure 9.5 has $R_1 = 20\ \text{k}\Omega$, $R_2 = 18\ \text{k}\Omega$, $V_{EE} = 15$ V, $R_E = 10\ \text{k}\Omega$, and $\beta = 200$, what is the value of I_{EE} for the circuit? What are the values of I_{C1} and I_{C2} if the differential amplifier is balanced? Assume that $V_{BE} = V_\gamma$.

9.14 For the circuit shown in Figure P9.4, determine R_{C1}, R_{C2}, I_{EE}, A_c, A_d, v_o, and the CMRR. Assume $V_{BE} = 0.7$ V and $\beta = 200$.

9.15 If the CE bypass capacitor was removed from the circuit in Problem 9.14, would A_c, A_d, and CMRR change? If so, by how much?

9.16 If R_{EE} in Problem 9.14 is replaced with a simple current source ($V_A = 80$ V), what is the new CMRR? Determine the value for R_{REF}.

9.17 Determine the differential-mode and common-mode output voltages and CMRR of the circuit shown in Figure P9.5. Assume that $V_{BE} = 0.7$ V, $V_T = 26$ mV, $V_A = 60$ V, and the transistors are identical with $\beta = 200$. Also determine R_{in}(DM) and R_{in}(CM).

9.18 For the amplifier shown in Figure P9.6, determine R_1, R_2, R_3, A_c, A_d, R_{C1}, R_{C2}, and CMRR when $\beta = 100$ and $V_{BE} = 0.7$ V for the single tran-

Figure P9.3

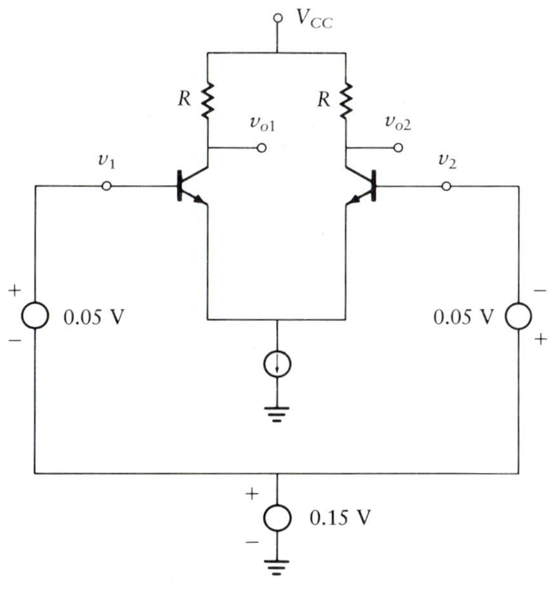

sistors and $\beta = 10{,}000$ and $V_{BE} = 1.4$ V for the Darlington pair.

9.19 Design a Widlar current source to provide $I_{C2} = 10\ \mu$A. Assume $V_{CC} = 30$ V, $R_1 = 40$ kΩ, and $V_{BE} = 0.7$ V. Refer to Figure 9.9.

9.20 Design a Widlar current source (as in Figure 9.9) to provide $I_{C2} = 100\ \mu$A for a reference current of 1 mA. Assume that $V_{CC} = 20$ V and $V_{BE} = 0.7$ V.

9.21 The Widlar current source of Problem 9.20 is used as an active load. What is the equivalent resistance of the current source when it is used in this manner? Assume that 2N3903 transistors are employed (see Appendix D1).

9.22 For the Wilson current source of Figure 9.11, calculate the output resistance if $V_{CC} = 15$ V, $R = 12$ kΩ, $V_T = 26$ mV, $V_{BE} = 0.7$ V, and

Figure P9.4

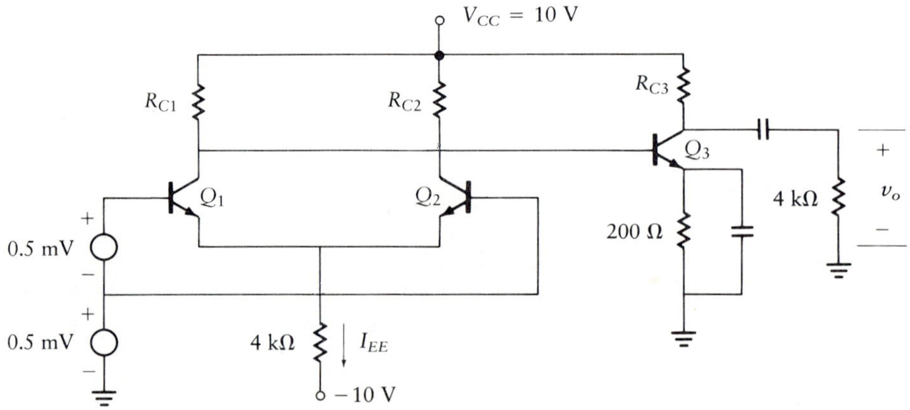

Figure P9.5

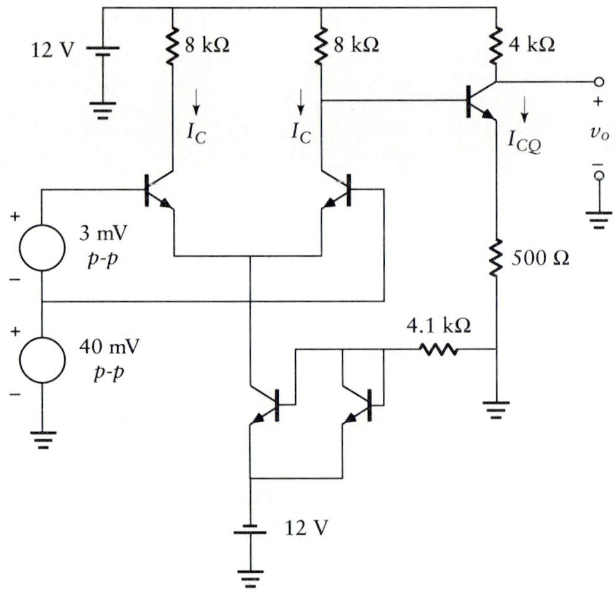

$\beta = 100$. Assume that 2N3903 transistors are used in the 500-μA range (see Appendix D1).

9.23 In the circuit of Problem 9.22, what are the values of I_{REF} and I_{C2}?

9.24 Design a current mirror of the type shown in Figure P9.7 where $I_1 = 10$ mA, $I_2 = 1$ mA, $I_3 = 0.1$ mA, $I_4 = 100\ \mu$A, and $I_5 = 50\ \mu$A. Assume the transistors are identical and have $\beta = 100$.

9.25 Find the dc offset (when $v_i = 0$), A_d, A_c, and the CMRR of the circuit shown in Figure P9.8. Assume that $\beta = 300$ for all transistors, $V_T = 26$ mV, $V_A = 80$ V, and $V_{BE} = 0.7$ V. Determine the value of R_{REF} and R_2 for $I_{EE} = 2$ mA and $I_{REF} = 50$ mA.

9.26 What is the output of the level shifter of Figure 9.15(a) if $V_{BB} = 8$ V, $R_B = 5$ kΩ, $V_{CC} = 10$ V, the current source provides 4 mA, $R_E = 800\ \Omega$, $V_{BE} = 0.7$ V, and $\beta = 100$?

Figure P9.6

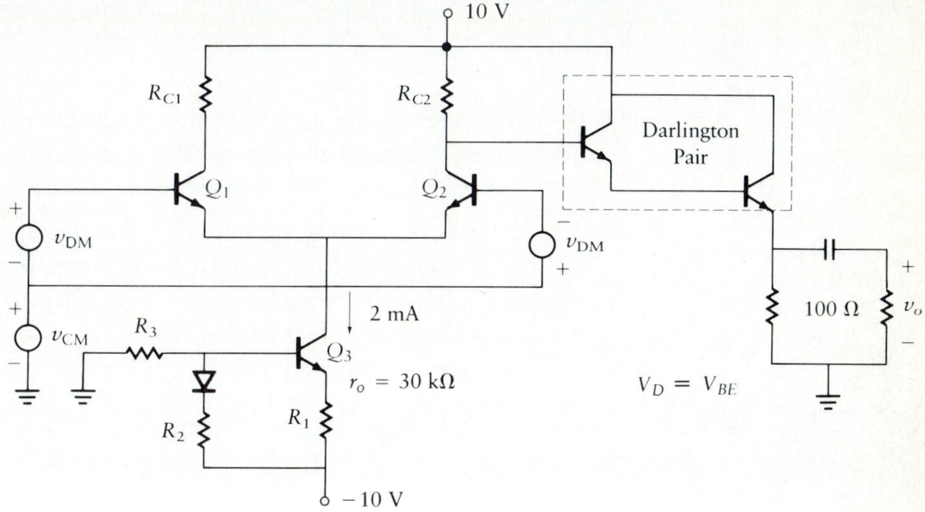

Figure P9.7

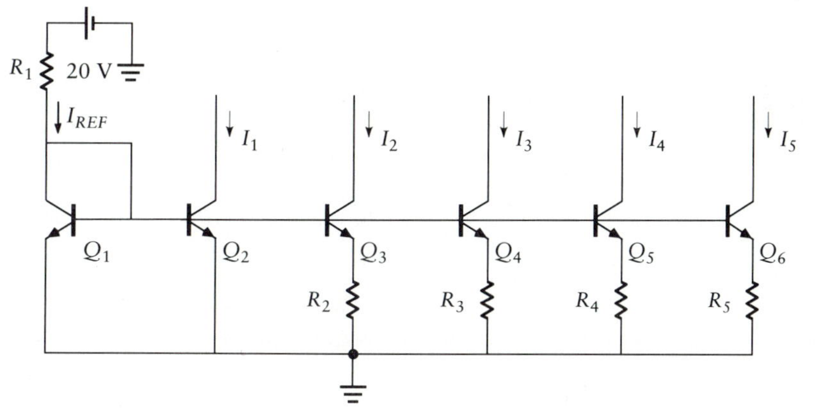

Figure P9.8

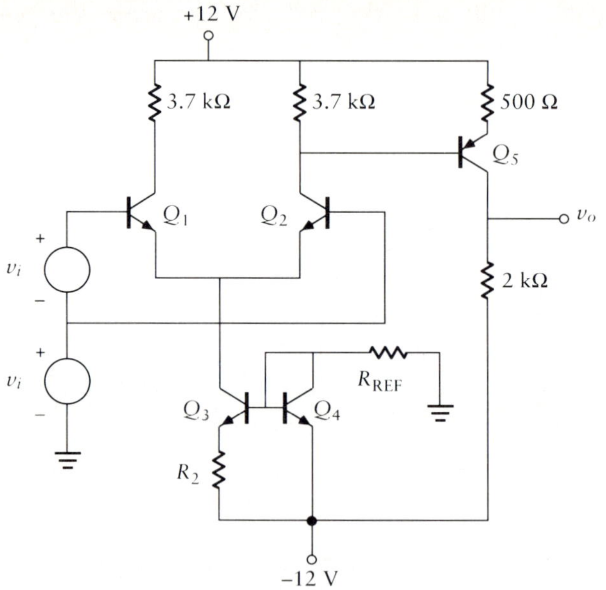

9.27 Design a level shifter to attain a 4-V offset using the circuit of Figure 9.15(b). Assume that $R_B = 4\ \text{k}\Omega$, $V_{BB} = 8$ V, $V_{CC} = 12$ V, $V_{EE} = 12$ V, $\beta = 100$, and $V_{BE} = 0.7$ V. Use a current source that provides 6 mA.

9.28 A direct-coupled multistage amplifier has a common-emitter amplifier for its last stage. The V_C for the CE amplifier is 6 V with an R_C of 5 kΩ. Design a level shifter to follow the CE amplifier and provide an ac output. The parameter values are $V_{CC} = 10$ V, $V_{EE} = 10$ V, $V_{BE} = 0.7$ V, and $\beta = 100$. Design the current source to have a 5-mA output.

10

PRACTICAL OPERATIONAL AMPLIFIERS

10.0 INTRODUCTION

Chapter 8 introduced the ideal operational amplifier and explored some of its applications. In that chapter, we assumed that the operational amplifier closely follows the ideal characteristics. Various internal circuits that make up the operational amplifier were explored in Chapter 9. The 741 op-amp was introduced as a useful general-purpose amplifier.

The current chapter develops a more complex op-amp model that takes into account several of the more important practical operational aspects. After examining these deviations from the ideal operating characteristics, we study the following important practical design applications:

1. Multiple op-amp design applications where bandwidth, gain per stage, and input resistance are part of the specification.
2. Design applications that require a balanced input or output.
3. Methods of eliminating cross-coupling between multiple inputs.

The chapter concludes with a consideration of the 101 operational amplifier and an introduction to power audio op-amps.

10.1 OP-AMP FREQUENCY RESPONSE

We now apply Bode plot concepts to the frequency analysis of op-amp circuits. (We review Bode plot construction in Appendix B.) Typical open-loop op-amp amplitude versus frequency characteristics are shown in Figure 10.1, where we illustrate a family of curves, each corresponding to a different closed-loop gain, A_v. The closed-loop gain for each curve is indicated on the figure. The numbers are typical of a general-purpose single roll-off op-amp such as the 741. The 741 op-amp includes *fixed compensation,* which is a technique of modifying the open-loop frequency response characteristic for the purpose of improving performance. Compensation is used in numerous feedback systems. One of the simplest forms of frequency compensation is provided by the introduction of an RC network as shown in Figure 10.2. The transfer function for this network is

$$\frac{V_o(j\omega)}{V_i(j\omega)} = \frac{1}{j\omega RC + 1} = \frac{1}{j\omega\tau + 1}$$

where the time constant, τ, is RC. This network has a single pole at a frequency, ω, of $1/\tau$. When the RC network is inserted into an op-amp, the gain decreases at a slope of -20 dB/decade for frequencies greater than the corner frequency, $\omega_o = 1/\tau$ (refer to Appendix B for details). The 741 op-amp has a fixed compensation network built into the device. This *internal compensation* network causes the 741 op-amp response to follow the amplitude-frequency re-

Figure 10.1
Op-amp open-loop gain characteristics.

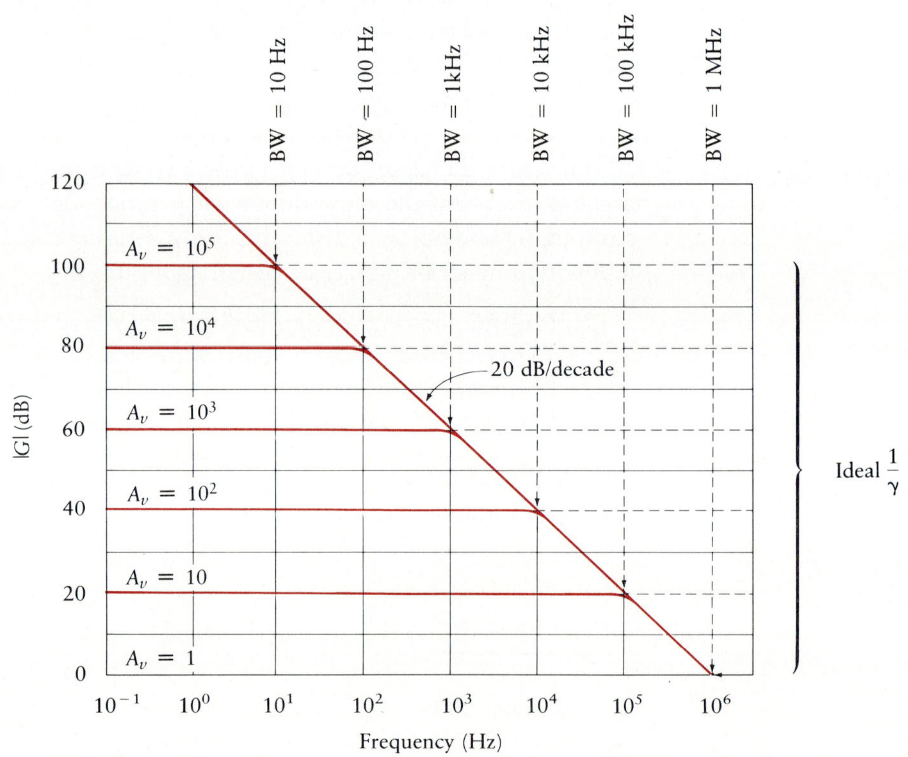

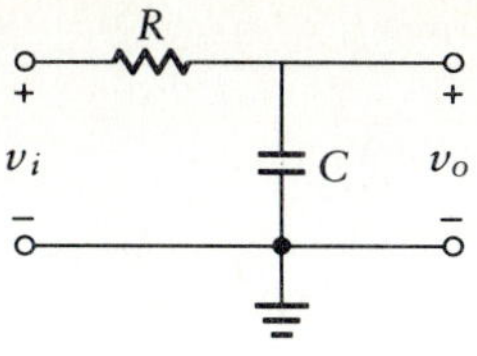

Figure 10.2 RC compensating network.

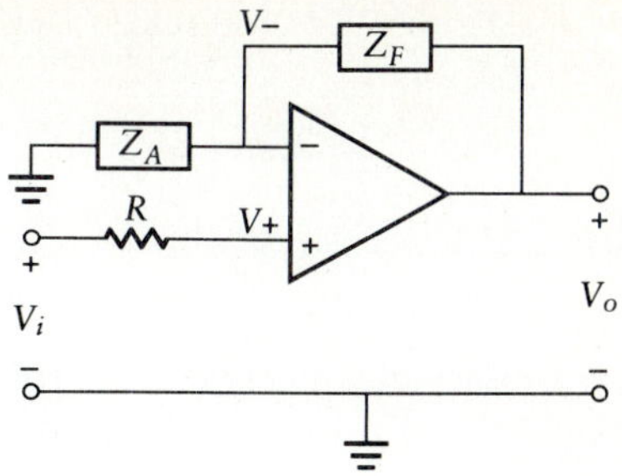

Figure 10.3 Circuit configuration for noninverting op-amp.

sponse shown in Figure 10.1. We can approximate this family of curves with the following expression:

$$G(s) = \frac{G_o}{1 + s/\omega_o} \tag{10.1}$$

The constant G_o is the dc open-loop gain of the op-amp and ω_o is the op-amp corner frequency. This amplitude-frequency characteristic results from the internal compensation built into the op-amp. The curve shown is for single-pole internal compensation. The manufacturer's specification sheets usually show both amplitude and phase response of the op-amp. This information is shown in Appendix D for the 741 op-amp.

Examination of the curve labeled $A_v = 10^5$ in Figure 10.1 shows that the gain is 100 dB at low frequency. Then, starting at $f = 10$ Hz, the gain falls off at a slope of -20 dB/decade, reaching 0 dB at a frequency of 10^6 Hz. The product of the gain with the bandwidth, which is defined as the *gain-bandwidth product (GBP),* is a constant and is equal to 10^6 Hz.

We verify that the gain-bandwidth product is constant by referring to the curves in Figure 10.1. The bandwidth is the corner frequency (the frequency beyond which the amplitude response falls at a slope of -20 dB/decade). We tabulate the gain (A_v), the bandwidth (BW), and the gain-bandwidth product (GBP) in Table 10.1. Notice that for this first-order compensation network, the product of gain and bandwidth is a constant, 10^6.

Figure 10.3 shows a typical noninverting amplifier configuration. We now find the closed-loop gain for this noninverting amplifier. The closed-loop gain is found by writing the nodal equations at V_+ and V_- as follows:

$$V_+ = V_i$$

$$\frac{V_-}{Z_A} = \frac{V_o - V_-}{Z_F}$$

Solving for the inverting voltage input, we obtain

Table 10.1

A_v	BW	GBP
10^5	10^1	10^6
10^4	10^2	10^6
10^3	10^3	10^6
10^2	10^4	10^6
10^1	10^5	10^6
10^0	10^6	10^6

$$V_- = V_o \frac{Z_A}{Z_A + Z_F}$$

We use the open-loop gain relationship to solve for the output voltage.

$$V_o = G(V_+ - V_-) = G\left(V_i - V_o \frac{Z_A}{Z_A + Z_F}\right)$$

$$= \frac{GV_i}{1 + GZ_A/(Z_A + Z_F)}$$

When we let

$$\gamma = \frac{Z_A}{Z_A + Z_F}$$

the voltage gain reduces to

$$A_v = \frac{V_o}{V_i} = \frac{G}{1 + G\gamma} \tag{10.2}$$

If the open-loop op-amp gain, G, is large, the closed-loop amplifier gain approaches

$$\frac{V_o}{V_i} = \frac{1}{\gamma}$$

The quantity $1/\gamma$ is the closed-loop gain of the noninverting op-amp system at low frequencies. We substitute G, which is the function of frequency given by equation (10.1), into equation (10.2) to obtain

$$A_v = \frac{G_o}{1 + G_o\gamma + s/\omega_o}$$

If the dc gain is large, we can assume that $\gamma G_o >> 1$. The equation for A_v then reduces to

$$A_v = \frac{1}{\gamma} \frac{1}{1 + s/\gamma G_o \omega_o} \tag{10.3}$$

If the impedances are purely resistive, $Z_A = R_A$ and $Z_F = R_F$, and

$$\frac{1}{\gamma} = 1 + \frac{R_F}{R_A}$$

The closed-loop gain is then

$$A_v = \frac{1}{1/(1 + R_F/R_A) + s/G_o\omega_o}$$

This gain equation contains a single pole. Therefore, the Bode plot starts at $20 \log(1 + R_F/R_A)$ and has a corner frequency at

$$\omega_c = \frac{G_o\omega_o}{1 + R_F/R_A}$$

Note that if we form the product of A_{vo} (the closed-loop noninverting gain at low frequency) with the bandwidth (the corner frequency ω_c), the resulting gain-bandwidth product is

$$A_{vo}\omega_c = \left(1 + \frac{R_F}{R_A}\right) \frac{G_o\omega_o}{1 + R_F/R_A} = G_o\omega_o$$

Hence we again see that for single-pole compensation, the gain-bandwidth product is constant for the inverting configuration. As we show later in this chapter (equations (10.20) and (10.29)), the bandwidths for the noninverting and inverting configurations are identical.

As the closed-loop gain increases, the corner frequency (i.e., bandwidth) must decrease so that their product remains constant. This is also verified in Table 10.1. The inverse relationship between gain and bandwidth is an important trade-off in design considerations.

10.1.1 Open-Loop and Closed-Loop Op-Amp Response

To determine the stability of closed-loop systems using the frequency response method (see Appendix B for more information), we need to know the open-loop (GH) function. We usually know the closed-loop function. In this section, we consider a generalized approach to evaluating the open-loop frequency response from the closed-loop function for op-amp systems. We start with a noninverting op-amp system as shown in Figure 10.3. The closed-loop response is found as before, by writing the node equations for V_- and V_+. For convenience, we repeat equation (10.2) below.

$$\frac{V_o}{V_i} = \frac{G}{1 + GZ_A/(Z_A + Z_F)}$$

The block diagram of Figure 10.4 represents equation (10.2).

This can be verified by finding the closed-loop response for the general block diagram for a feedback system, as shown in Figure 10.5. The *input signal* is $R(s)$, the *output signal* is $C(s)$, $G(s)$ is the *forward transmittance,* and $H(s)$ is the *feedback transmittance.* The *actuating error,* $\epsilon(s)$, is the difference

$$\epsilon(s) = R(s) - H(s)C(s)$$

In this diagram the closed-loop function is $C(s)/R(s)$, and it can be obtained from

$$C(s) = \epsilon(s)G(s) = [R(s) - H(s)C(s)]G(s)$$

Therefore,

$$\frac{C(s)}{R(s)} = \frac{G(s)}{1 + G(s)H(s)}$$

When we substitute the values from Figure 10.4, that is, $G(s) = G$, $H(s) = Z_A/(Z_A + Z_F)$, $C(s) = V_o$, and $R(s) = V_i$, we obtain

$$\frac{V_o}{V_i} = \frac{G}{1 + GZ_A/(Z_A + Z_F)}$$

which is identical to equation (10.2). The *open-loop response,* $G(s)H(s)$, is the product of all the transmittance functions around the loop. To find the open-loop response for equation (10.2), we compare it to the general equation and obtain

$$G(s)H(s) = G\frac{Z_A}{Z_A + Z_F} \tag{10.4}$$

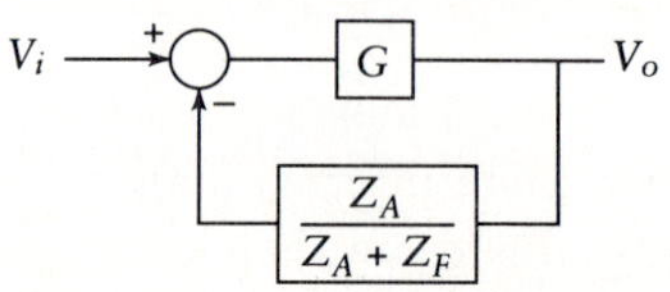

Figure 10.4 Block diagram for noninverting op-amp.

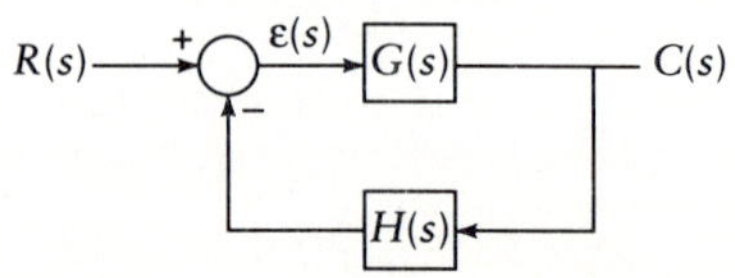

Figure 10.5 Block diagram of feedback system.

The Bode plot is useful for determining stability of electronic systems. For this application we construct a Bode plot of the open-loop response, not the closed-loop response. Equation (10.4) is useful in determining stability. (Stability is covered in Chapter 11.)

It is important to notice that the open-loop response for the *inverting* op-amp is also given by equation (10.4).

Example 10.1

Find the open-loop function for the op-amp system of Figure 10.6. The component values are $R = 1\ \text{k}\Omega$, $C = 1\ \mu\text{F}$, and $L = 100$ mH. The op-amp gain is

$$G(s) = \frac{G_o}{0.001s + 1}$$

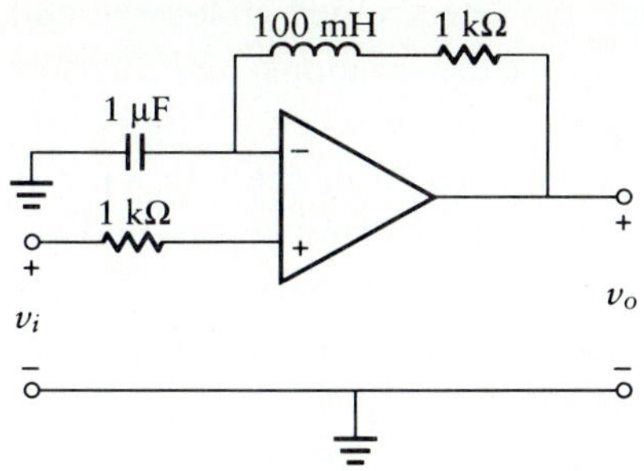

Figure 10.6 Op-amp system for Example 10.1.

SOLUTION In this system, $Z_A = 1/sC$ and $Z_F = sL + R$. The open-loop function is given by equation (10.4).

$$\frac{GZ_A}{Z_A + Z_F} = \frac{G(1/LC)}{s^2 + (R/L)s + (1/LC)}$$

We insert values for R, L, and C to obtain

$$\frac{GZ_A}{Z_A + Z_F} = \frac{G}{[s/(1.13 \times 10^3) + 1][s/(8.9 \times 10^3) + 1]}$$

We substitute for G to find the open-loop function,

$$G(s)H(s) = \frac{G_o}{[s/10^3 + 1][s/(1.13 \times 10^3) + 1][s/(8.9 \times 10^3) + 1]}$$

The open-loop function is used to construct the Bode plot and hence to determine the system stability.

10.2 PRACTICAL OP-AMPS

Practical op-amps approximate their *ideal* counterparts but differ in some important respects. It is important for the circuit designer to understand the differences between actual op-amps and ideal op-amps, since these differences can adversely affect circuit performance.

Our goal is to develop a detailed model of the practical op-amp—a model that takes into account the most significant characteristics of the nonideal device. We begin by defining the parameters used to describe practical op-amps. These parameters are specified in listings on data sheets supplied by the op-amp manufacturer.

Table 10.2 lists the parameter values for three particular op-amps, one of the three being the μA741. We use μA741 operational amplifiers in many of the examples and end-of-chapter problems for the following reasons: (1) they have been fabricated by many IC manufacturers, (2) they are found in great quantities throughout the electronics industry, and (3) they are general-purpose internally compensated op-amps, and their properties can be used as a reference for comparison purposes when dealing with other op-amp types. As the various parameters are defined in the following sections, reference should be made to Table 10.2 in order to find typical values.

The most significant difference between ideal and actual op-amps is in the voltage gain. The ideal op-amp has a voltage gain that approaches infinity. The actual op-amp has a finite voltage gain that decreases as the frequency increases. Actual op-amps provide a predictable voltage gain versus frequency characteristic, as shown in Figure 10.1. Some op-amps, such as the 741, are internally compensated with an RC network. Other op-amps, like the 101, provide for external addition of a capacitor for the purpose of altering the frequency characteristic.

Table 10.2 Parameter Values for Op-Amps.

	Ideal	General-Purpose 741	High Speed 715	Low Noise 5534
Voltage gain, G_o	∞	1×10^5*	3×10^4	10^5
Output impedance, Z_o	0	75 Ω	75 Ω	0.3 Ω
Input impedance, Z_{in} (open loop)	∞	2 MΩ	1 MΩ	100 kΩ
Offset current, I_{io}	0	20 nA	250 nA	300 nA
Offset voltage, V_{io}	0	2 mV	10 mV	5 mV
Bandwidth, BW	∞	1 MHz	65 MHz	10 MHz
Slew rate, SR	∞	0.7 V/ms	100 V/ms	13 V/ms

* The 741 op-amp has typical values of approximately 2 to 3×10^5; however, in this text we use 10^5.

10.2.1 Open-Loop Voltage Gain (G)

The *open-loop voltage gain* of an op-amp is the ratio of the change in output voltage to a change in the input voltage without feedback. Voltage gain is a dimensionless quantity. The symbol G is used to indicate the open-loop voltage gain. Op-amps have high voltage gain for low-frequency inputs. The op-amp specification lists the voltage gain in volts per millivolt or in decibels (dB) (defined as $20 \log_{10}(v_o/v_i)$).

The open-loop voltage gain is frequency dependent. Figure 10.1 illustrates this gain as a function of frequency for a typical op-amp. Note that the gain decreases with increasing frequency. This characteristic voltage gain versus frequency curve matches the performance of the μA741 op-amp (see Appendix D).

10.2.2 Modified Op-Amp Model

Figure 10.7 shows a modified version of the idealized op-amp model. We have altered the idealized model by adding input resistance (R_i), output resistance (R_o), and common-mode resistance (R_{cm}).

Typical values of these parameters (for the 741 op-amp) are

Input resistance: $R_i = 2\ \text{M}\Omega$

Common-mode resistance: $2R_{cm} = 400\ \text{M}\Omega$

Output resistance: $R_o = 75\ \Omega$

Open-loop gain at dc: $G_o = 10^5$

We now consider the circuit of Figure 10.8 in order to examine op-amp performance. The inverting and noninverting inputs of the op-amp are driven by sources that have series resistance. The output of the op-amp is fed back to the input through a resistor, R_F.

The sources driving the two inputs are denoted v_A and v_1, and the associated series resistances are R_A and R_1. If the input circuitry is more complex, these resistances can be considered as Thevenin equivalents of that circuitry.

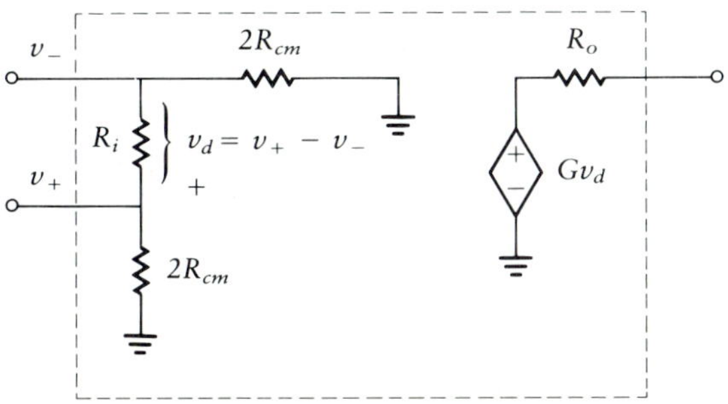

Figure 10.7 Modified op-amp model.

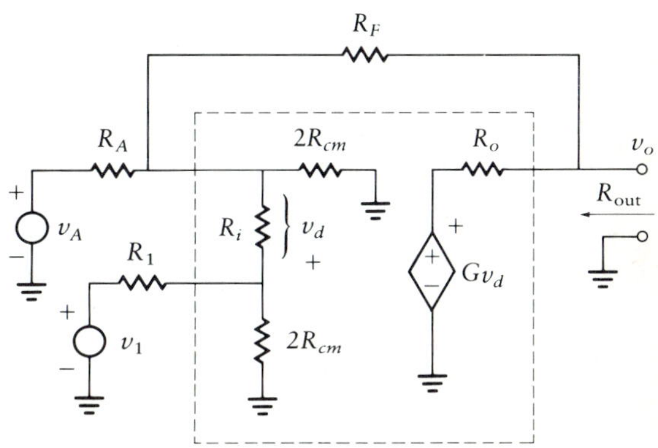

Figure 10.8 Op-amp circuit.

10.2.3 Input Offset Voltage (V_{io})

When the input voltage to an ideal op-amp is zero, the output voltage is also zero. This is not true for an actual op-amp. The *input offset voltage,* V_{io}, is defined as the differential input voltage required to make the output voltage equal to zero. V_{io} is zero for the ideal op-amp. A typical value of V_{io} for the 741 op-amp is 2 mV. A nonzero value of V_{io} is undesirable because the op-amp amplifies any input offset, thus causing a larger output dc error.

The following technique can be used to measure the input offset voltage. Rather than vary the input voltage in order to force the output to zero, the input is set equal to zero, as shown in Figure 10.9, and the output voltage is measured. The output voltage resulting from a zero input voltage is known as the *output dc offset voltage.* The input offset voltage is obtained by dividing this quantity by the open-loop gain of the op-amp.

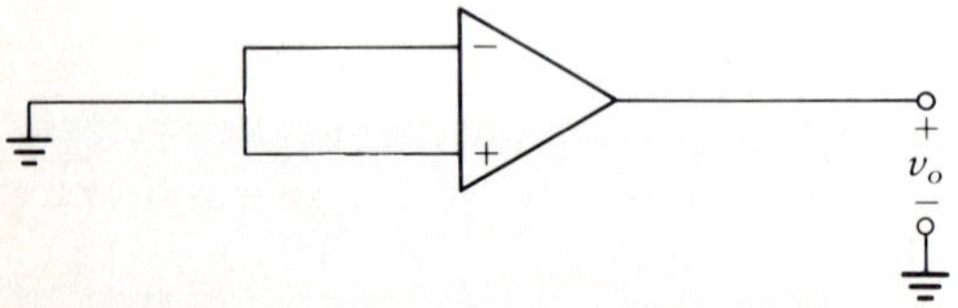

Figure 10.9 Technique for measuring V_{io}.

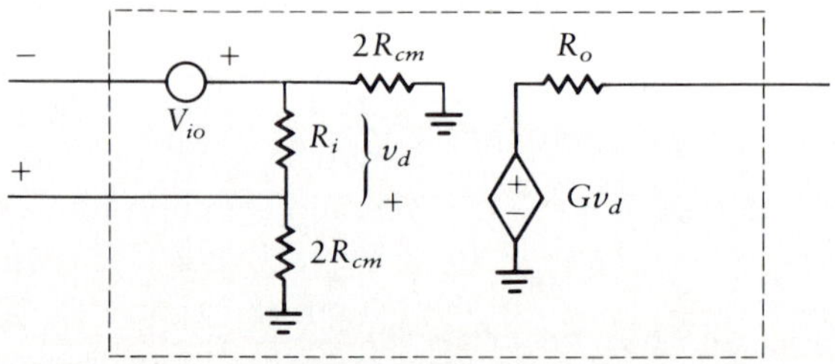

Figure 10.10 Input offset voltage model.

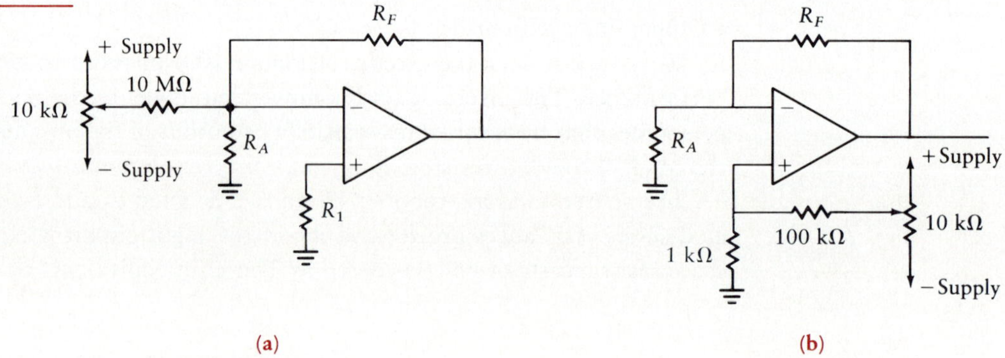

Figure 10.11 Offset voltage balancing.

The effects of input offset voltage can be incorporated into the op-amp model as shown in Figure 10.10. In addition to including input offset voltage, the ideal op-amp model has been further modified with the addition of four resistances. R_o is the *output resistance.* The *input resistance* of the op-amp, R_i, is measured between the inverting and noninverting terminals. The model also contains a resistor connecting each of the two inputs to ground. These are the *common-mode resistances,* and each is equal to $2R_{cm}$. If the inputs are connected together as in Figure 10.9, these two resistors are in parallel, and the combined Thevenin resistance to ground is R_{cm}. If the op-amp is ideal, R_i and R_{cm} approach infinity (i.e., open circuit) and R_o is zero (i.e., short circuit).

The external configuration shown in Figure 10.11(a) can be used to negate the effects of offset voltage. A variable voltage is applied to the inverting input terminal. Proper choice of this voltage cancels the input offset. Similarly, Figure 10.11(b) illustrates this balancing circuit applied to the noninverting terminal.

10.2.4 Input Bias Current (I_{Bias})

Although ideal op-amp inputs draw no current, actual op-amps allow some bias current to enter each input terminal. I_{Bias} is the current into the input transistor, and a typical value is 2 μA. When the source impedance is low, I_{Bias} has little effect, since it causes a relatively small change in input voltage. However, with high-impedance driving circuits, a small current can lead to a large voltage.

The bias current can be modeled as two current sinks, as shown in Figure 10.12. The values of these sinks are independent of the source impedance. The *bias current* is defined as the average value of the two current sinks. Thus

$$I_{\text{Bias}} = \frac{I_{B+} + I_{B-}}{2} \tag{10.5}$$

The difference between the two sink values is known as the *input offset current*, I_{io}, and is given by

$$I_{io} = I_{B+} - I_{B-} \tag{10.6}$$

Both the input bias current and the input offset current are temperature dependent. The *input bias current temperature coefficient* is defined as the ratio of change in bias current to change in temperature. A typical value is 10 nA/°C. The *input offset current temperature coefficient* is defined as the ratio of the change in magnitude of the offset current to the change in temperature. A typical value is -2 nA/°C.

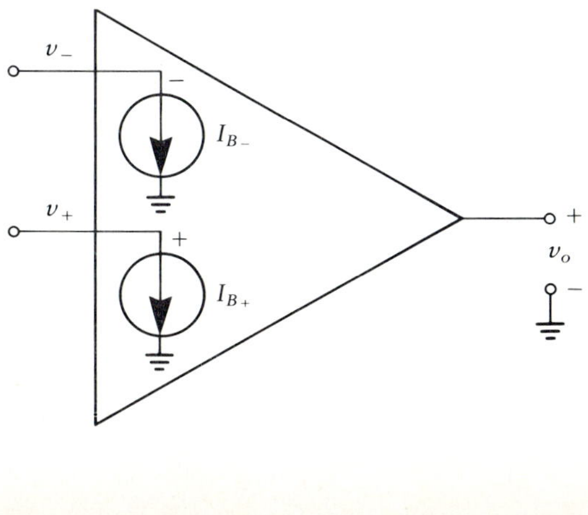

Figure 10.12
Input bias current.

The input bias currents are incorporated into the op-amp model of Figure 10.13, where we assume that the input offset current is negligible. That is,

$$I_{B+} = I_{B-} = I_B$$

We analyze this model in order to find the output voltage caused by the input bias currents. Figure 10.14(a) shows an op-amp circuit where the inverting and noninverting inputs are connected to ground through resistances. The circuit is replaced by its equivalent in Figure 10.14(b), where we have neglected V_{io}.

We further simplify the circuit in Figure 10.14(c) by neglecting R_o and R_L. That is, we assume $R_F >> R_o$ and $R_L >> R_o$. Output loading requirements usually ensure that these inequalities are met.

The circuit is further simplified in Figure 10.14(d), where the series combination of the dependent voltage source and resistor is replaced by a parallel combination of a dependent current source and resistor.

Finally, we combine resistances and change both current sources back to voltage sources to obtain the simplified equivalent of Figure 10.14(e). We use a loop equation to find the output voltage.

$$\begin{aligned} V_o &= GV_d \\ &= \frac{GR_1'(R_A' \parallel R_F - R_1')I_B(R_A' + R_F)}{(R_A' + R_F)\,(R_i + R_A') \parallel (R_F + R_1') + GR_1'R_A'} \end{aligned} \tag{10.7}$$

where

$$R_A' = R_A \parallel 2R_{cm}$$

$$R_1' = R_1 \parallel 2R_{cm}$$

$$R_F >> R_o$$

$$R_L >> R_o$$

The common-mode resistance, $2R_{cm}$, is in the range of several hundred megohms for most op-amps. Therefore

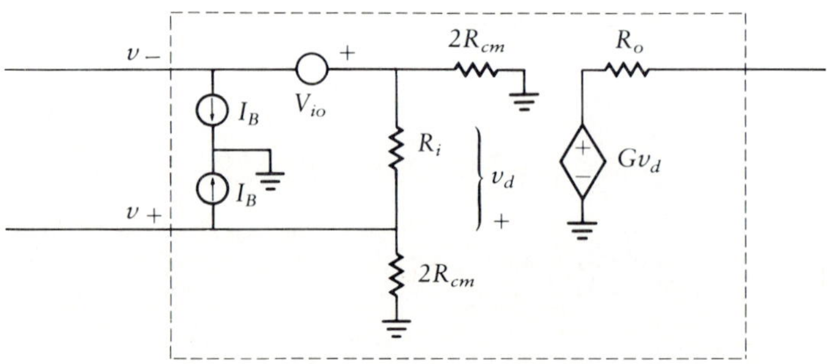

Figure 10.13 Input bias current model.

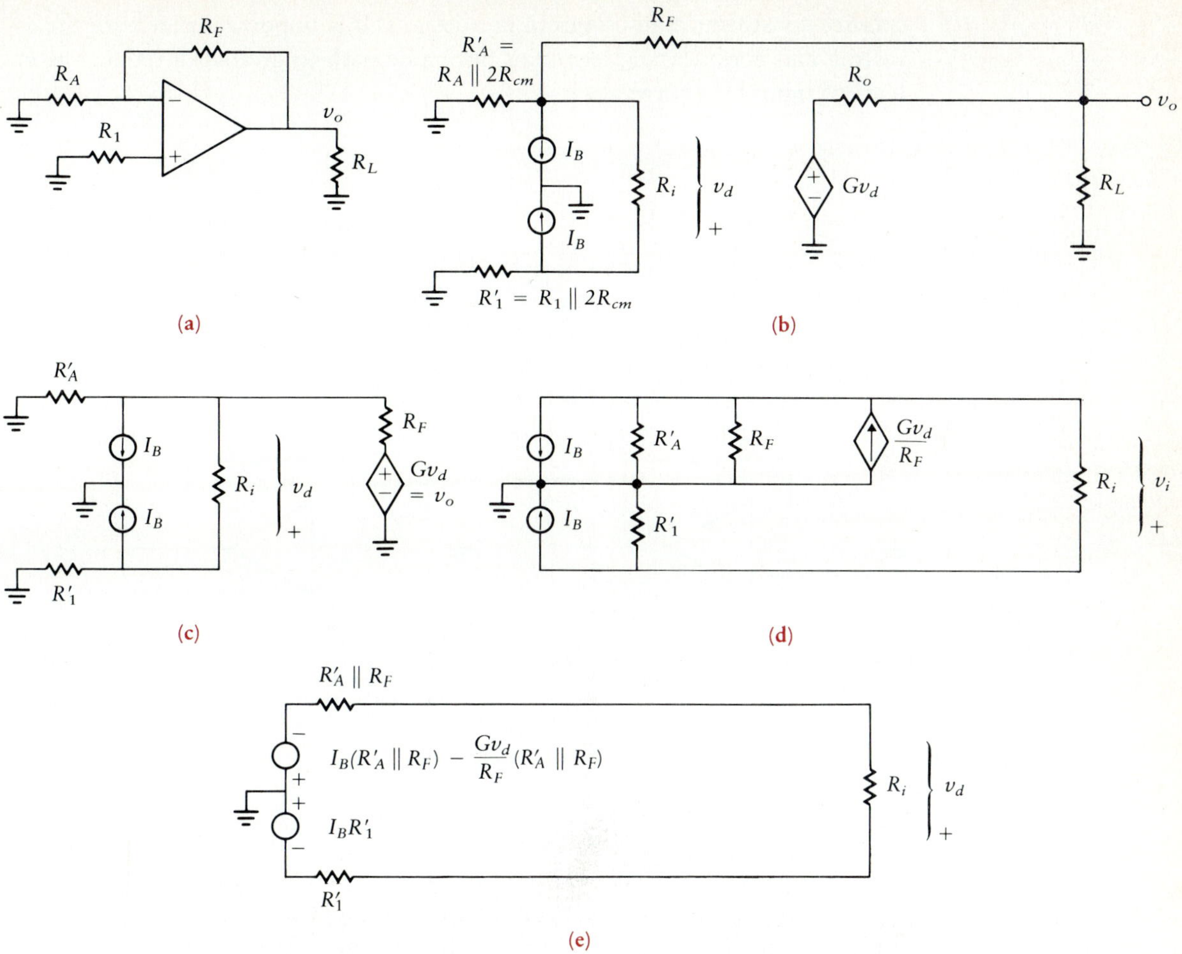

Figure 10.14 Input bias current effects.

$$R'_A \approx R_A$$

and

$$R'_1 \approx R_1$$

If we further assume that G is large, equation (10.7) becomes equation (10.8).

$$V_o = \left(1 + \frac{R_F}{R_A}\right) I_B(R_A \parallel R_F - R_1) \tag{10.8}$$

Note that if the value of R_1 is selected to be equal to $R_A \parallel R_F$, then the output voltage is zero. We conclude from this analysis that the dc resistance from v_+ to ground should equal the dc resistance from v_- to ground. We use this *bias balance* constraint many times in our designs. It is important that both the inverting and noninverting terminals have a dc path to ground to reduce the effects of input bias current.

Example 10.2

Find the output voltage for the configurations of Figure 10.15 where

$$I_B = 80 \text{ nA} = 8 \times 10^{-8} \text{ A}$$

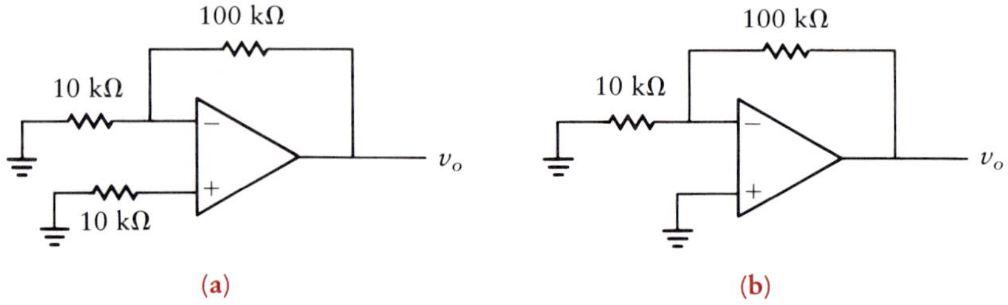

Figure 10.15 Configurations for Example 10.2.

SOLUTION We use the simplified form of equation (10.8) to find the output voltages for the circuit of Figure 10.15(a).

$$V_o = \left(1 + \frac{10^5}{10^4}\right)(8 \times 10^{-8})(9100 - 10^4) = -0.79 \text{ mV}$$

For the circuit of Figure 10.15(b), we obtain

$$V_o = \left(1 + \frac{10^5}{10^4}\right)(8 \times 10^{-8})(9100) = 8 \text{ mV}$$

By selecting $R_1 = 10\ \text{k}\Omega$ rather than $0\ \Omega$, we reduce the output voltage due to I_B by a factor of 10. We therefore balance the bias-current effect by equating the resistances connected between the positive terminal and ground with those connected between the negative terminal and ground.

10.2.5 Common-Mode Rejection

The op-amp is normally used to amplify the difference between two input voltages. It therefore operates in the *differential mode*. A constant voltage added to each of these two inputs should not affect the difference and should

Figure 10.16
Common mode.

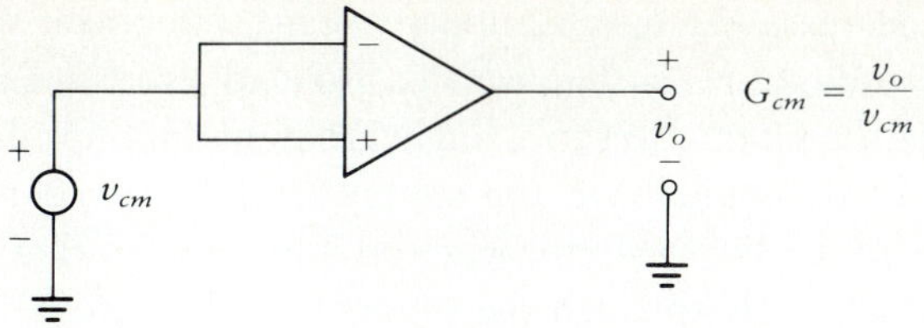

therefore not be transferred to the output. In the practical case, this constant, or average value of the inputs, *does* affect the output voltage. If we consider only the equal parts of the two inputs, we are considering what is known as the *common mode.*

Let us assume that the two input terminals of an actual op-amp are connected together and then to a common source voltage. This is illustrated in Figure 10.16. The output voltage would be zero in the ideal case. In the practical case, this output is nonzero. As discussed in Chapter 9, the ratio of the nonzero output voltage to the applied input voltage is the *common-mode voltage gain,* G_{cm}. The *common-mode rejection ratio* (CMRR) is defined as the ratio of the dc open-loop gain, G_o, to the common-mode gain. Thus,

$$\text{CMRR} = \frac{|G_o|}{|G_{cm}|} \tag{10.9a}$$

or, in decibels,

$$\text{CMRR} = 20 \log_{10}\left(\frac{|G_o|}{|G_{cm}|}\right) \tag{10.9b}$$

Typical values of the CMRR range from 80 to 100 dB. It is desirable to have the CMRR as high as possible.

10.2.6 Power Supply Rejection Ratio

Power supply rejection ratio is a measure of the ability of the op-amp to ignore changes in the power supply voltage. If the output stage of a system draws a variable amount of current, the supply voltage could vary. This load-induced change in supply voltage could then cause changes in the operation of other amplifiers sharing the same supply. This is a form of *crosstalk,* and it can lead to instability.

The *power supply rejection ratio* (PSRR) is the ratio of the change in v_o to the total change in power supply voltage. For example, if the positive and negative supplies vary from ± 5 V to ± 5.5 V, the total change is $11 - 10 = 1$ V. The PSRR is usually specified in microvolts per volt or sometimes in decibels. Typical op-amps have a PSRR of about 30 μV/V.

To decrease changes in supply voltage, the power supply for each group of op-amps should be *decoupled* (i.e., isolated) from those of other groups. This confines the interaction to a single group of op-amps. In practice, each printed circuit card should have the supply lines bypassed to ground via a 0.1-μF ceramic or 1-μF tantalum capacitor. This ensures that load variations will not feed significantly through the supply to other cards.

10.2.7 Phase Shift

If a sinusoidal signal forms the inverting input to an *ideal* op-amp, the output is 180° out of phase with this input. However, with *practical* op-amps, the phase shift between input and output signals is not 180°. The angle decreases as the frequency of the input signal increases. At high frequencies, the phase difference approaches zero and a portion of the output signal is fed back in phase. This changes the feedback from *negative* to *positive,* and the amplifier can exhibit behavior characteristics of an oscillator. (See Chapter 11 for a discussion of stability of op-amps.)

Op-amp manufacturers often prevent this situation by including an internal filter. This is done in a manner such that the op-amp has a gain of less than unity at frequencies where the phase difference between output and input approaches zero. Oscillation will not occur with positive feedback as long as the gain is less than unity. This modification of the op-amp is known as *internal frequency compensation.* If the manufacturer does not provide for this internal compensation, an external capacitor can be added to accomplish the same result.

10.2.8 Slew Rate (SR)

Because a practical op-amp has a response that is frequency dependent, the output due to a step input is not a perfect step function. This characteristic of op-amps, which is termed *slew rate limiting,* causes distortion of the output signal. We analyze slew rate limiting by referring to equation (10.3) for the closed-loop gain, A_v, of the noninverting op-amp systems (see Figure 10.17(a)). For convenience, we repeat this equation as follows:

$$A_v = \frac{1}{\gamma} \frac{1}{1 + s/\gamma G_o \omega_o}$$

The circuit includes resistors, R_A and R_F, so the low-frequency gain, $1/\gamma$, is $1 + R_F/R_A$. The gain-bandwidth product is $G_o \omega_o$, which for the 741 op-amp is 10^6 Hz.

We apply the step function shown in Figure 10.17(b) at the input to the op-amp system of Figure 10.17(a). We can use Laplace transform methods (reviewed in Appendix C) to find the output.

$$V_o(s) = \frac{V}{s}\left(\frac{1}{\gamma}\frac{1}{1 + s/\gamma G_o\omega_o}\right)$$

where the Laplace transform of the input step is V/s. The output time response is found using partial fraction expansions (see Section C.3 of the appendix) with the result

$$v_o(t) = \frac{V}{\gamma}[1 - e^{-t\gamma G_o\omega_o}]$$

Figure 10.17
Example of slew rate limiting.

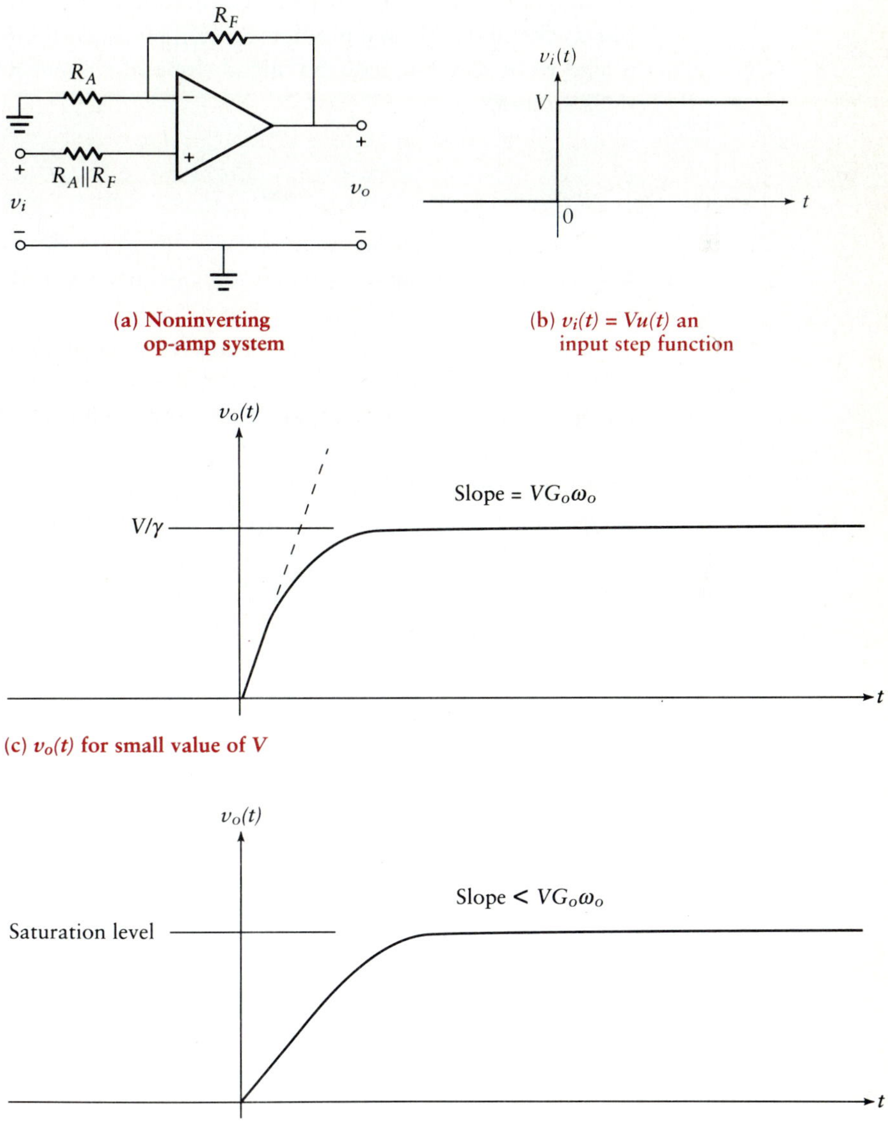

(a) Noninverting op-amp system

(b) $v_i(t) = Vu(t)$ an input step function

(c) $v_o(t)$ for small value of V

(d) $v_o(t)$ for large value of V

We sketch this output waveform as Figure 10.17(c). Notice that the initial slope of the waveform is

$$\left.\frac{dv_o}{dt}\right|_{t=0} = \left.\frac{V\gamma G_o\omega_o}{\gamma}e^{-t\gamma G_o\omega_o}\right|_{t=0} = VG_o\omega_o$$

These results are for linear operation, which means that V must be sufficiently small that the op-amp does not saturate. For large values of V, the amplifier saturates, and the output response resembles that shown in Figure 10.17(d). Notice that the initial slope of the output response curve is lower than that predicted by the linear theory. This inability of the op-amp output to rise as fast as the linear theory predicts is defined as *slew rate limiting.* The op-amp is said to be slewing, and the initial slope of the output response is defined as the *slew rate:*

$$\text{Slew rate} = \text{SR} = \text{maximum}\,\frac{dv_o}{dt} \tag{10.10}$$

An op-amp begins slewing when the initial slope of the $v_o(t)$ response is less than $VG_o\omega_o$.

The value of SR is usually specified in the manufacturer's data sheets in volts per microsecond at unity gain (see Appendix D). The equations derived for Figure 10.17 can be applied to a unity-gain buffer by letting $R_A \to \infty$, since this makes $1/\gamma$ equal to unity.

Slew rate is related to the *power bandwidth, f_p,* which is defined as the frequency at which a sine wave output, at rated output voltage, begins to exhibit distortion. For example, if the ouput signal is

$$v_o = V \sin 2\pi f_p t$$

the rate of change of this curve is

$$\frac{dv_o}{dt} = V2\pi f_p \cos 2\pi f_p t$$

and the maximum value of this slope occurs at $t = 0$:

$$\left.\frac{dv}{dt}\right|_{\text{max}} = 2\pi f_p V$$

Now if $2\pi f_p V > \text{SR}$, the output response is distorted. The power bandwidth is found from the slew rate at the point where the distortion begins. That is,

$$f_p = \frac{SR}{2\pi V_r}$$

where V_r is the rated op-amp output voltage. Note that if the output voltage is lower than V_r, slew rate distortion begins at a frequency that is higher than f_p.

Drill Problems

D10.1 If the slew rate for a unity-gain inverting 741 op-amp is 0.5 V/μs, how long will it take for the output to change by 5 V?

Ans: 10 μs

D10.2 For an op-amp with SR = 0.6 V/μs, find f_p for a peak undistorted output voltage of (a) 1 V and (b) 10 V.

Ans: (a) 95.5 kHz; (b) 9.6 kHz

D10.3 For a 741 op-amp with SR = 0.5 V/μs, find f_p with a peak output voltage of 13 V.

Ans: 6.1 kHz

10.2.9 Output Resistance

As a first step in determining the output resistance, R_{out}, we find the Thevenin equivalent for the portion of the op-amp circuit shown in the box enclosed in dashed lines in Figure 10.18. Note that we are ignoring the offset current and voltage in this analysis. Since the circuit contains no independent sources, the Thevenin equivalent voltage is zero, so the circuit is equivalent to a single resistor. The value of the resistor cannot be found using resistor combinations. To find the equivalent resistance, assume that a voltage source, v, is applied to the output leads. We then calculate the resulting current, i, and take the ratio v/i. This yields the Thevenin resistance.

Figure 10.19(a) illustrates the applied voltage source. The circuit is simplified to that shown in Figure 10.19(b) and further reduced to the circuit shown in Figure 10.19(c), where we define two new resistances as follows:

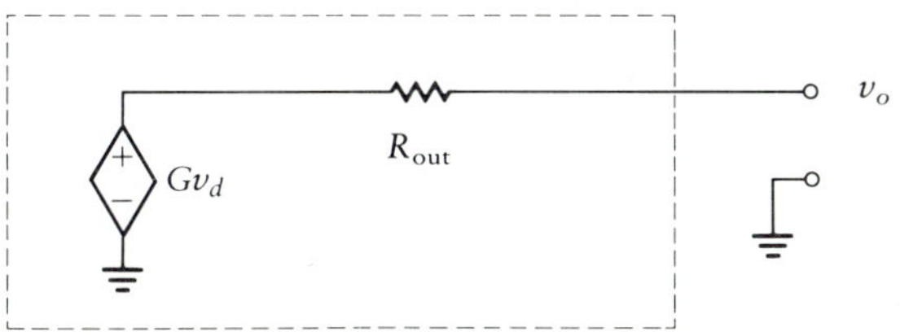

Figure 10.18 Thevenin equivalent for output resistance.

$$R_A' = R_A \parallel 2R_{cm}$$

$$R_1' = R_1 \parallel 2R_{cm}$$

We make the assumption that $R_A' << (R_1' + R_i)$ and $R_i >> R_1'$. The simplified circuit of Figure 10.19(d) results. The input differential voltage, v_d, is found from this simplified circuit using a voltage-divider ratio.

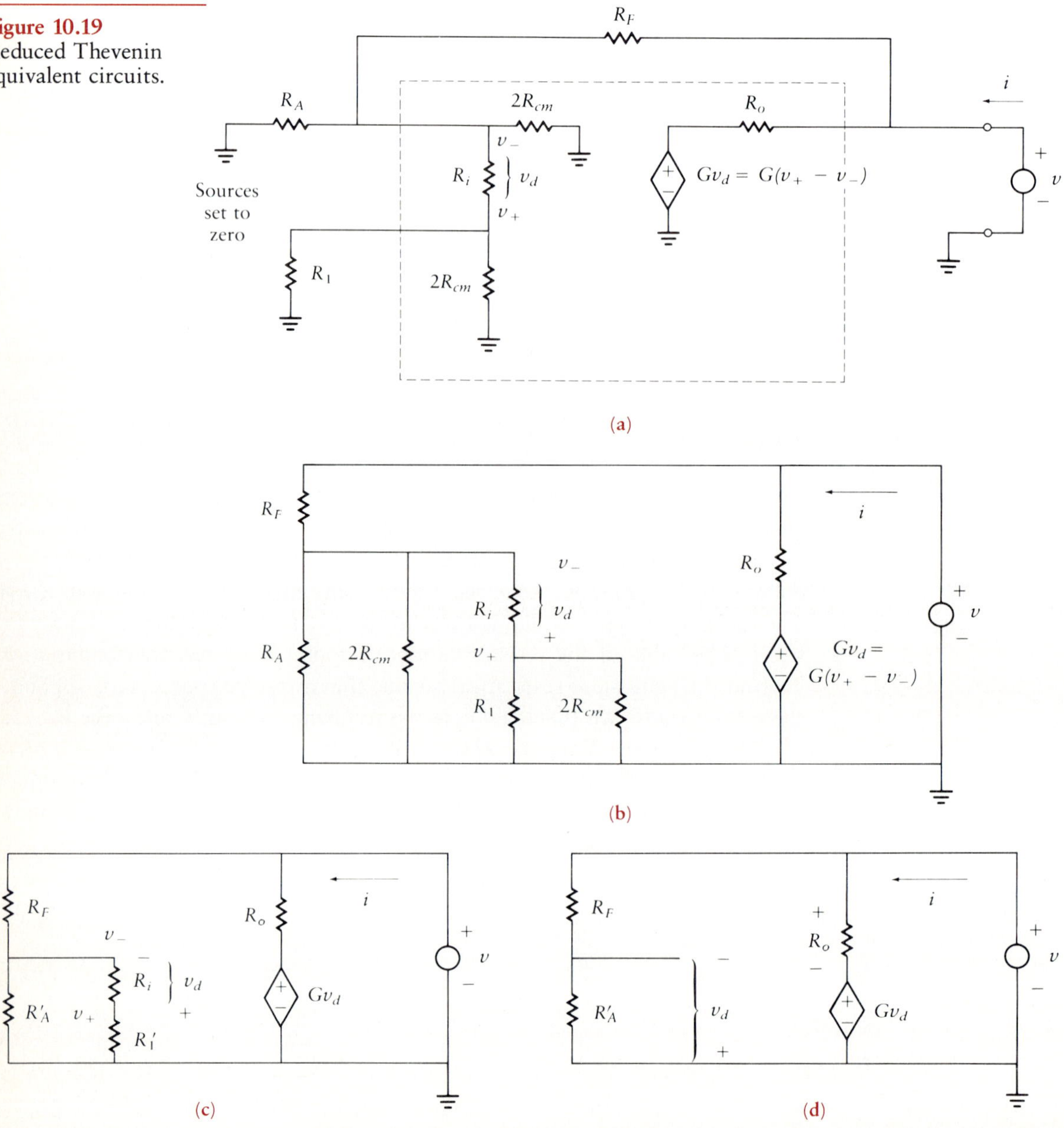

Figure 10.19
Reduced Thevenin equivalent circuits.

$$v_d = \frac{-R_A'v}{R_A' + R_F}$$

To find the output resistance, we begin by writing the output loop equation.

$$iR_o = v - Gv_d = v\left(1 + \frac{GR_A'}{R_F + R_A'}\right)$$

The output resistance is then given by equation (10.11).

$$R_{\text{out}} = \frac{v}{i} = \frac{R_o}{1 + GR_A'/(R_F + R_A')} \tag{10.11}$$

In most cases, R_{cm} is so large that $R_A' \approx R_A$ and $R_1' \approx R_1$. Equation (10.11) can be simplified using the zero-frequency voltage gain, G_o. The result is equation (10.12).

$$R_{\text{out}} = \frac{R_o}{G_o}\left(1 + \frac{R_F}{R_A}\right) \tag{10.12}$$

Example 10.3

Find the output impedance of a unity-gain buffer as shown in Figure 10.20.

SOLUTION When the circuit of Figure 10.20 is compared to the feedback circuit of Figure 10.19, we find that

$$R_A \to \infty$$

Therefore,

$$R_A' = \infty \parallel 2R_{cm} = 2R_{cm}$$

Equation (10.12) cannot be used, since we are not sure that the inequalities leading to the simplification of Figure 10.19(c) apply in this case. That is, the simplification requires that

$$2R_{cm} << R_1 \parallel (2R_{cm} + R_i)$$

Without this simplification, the circuit takes the form shown in Figure 10.21. This circuit is analyzed to find the following relations:

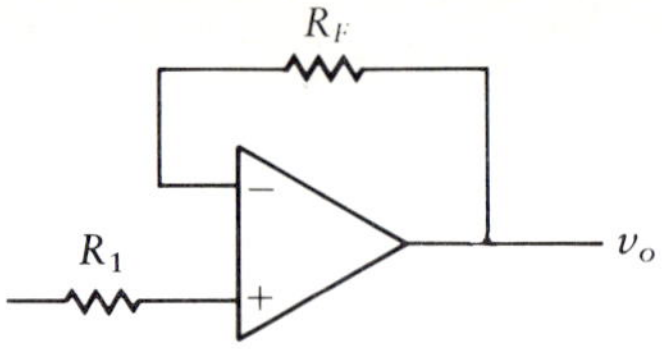

Figure 10.20 Unity-gain buffer.

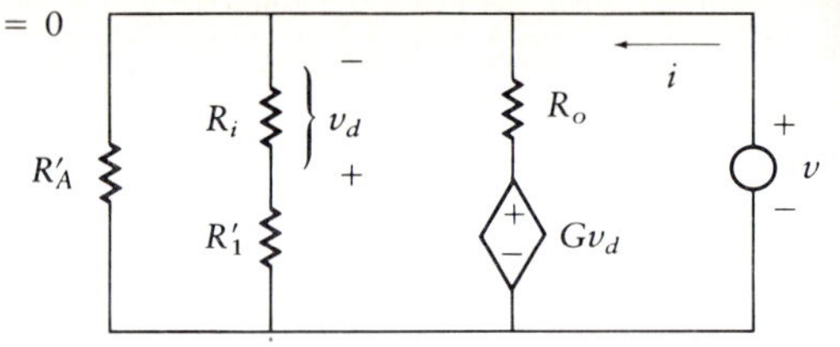

Figure 10.21 Equivalent circuit—unity-gain buffer.

$$v_d = \frac{-R_i v}{R'_1 + R_i}$$

$$R_o i = v - Gv_d = \left(1 + \frac{R_i}{R'_1 + R_i}\right)v$$

In the first of these equations, we have assumed that $R_o << (R'_1 + R_i) << 2R_{cm}$. The output resistance is then given by

$$R_{\text{out}} = \frac{v}{i} = \frac{R_o}{1 + R_i G_o/(R'_1 + R_i)} \tag{10.13}$$

where we again use the zero-frequency voltage gain, G_o.

10.3 NONINVERTING AMPLIFIER

Figure 10.22(a) illustrates the *noninverting amplifier,* and Figure 10.22(b) shows the equivalent circuit. The input voltage is applied through $R_1 = R_A \parallel R_F$ into the noninverting terminal.

10.3.1 Input and Output Resistances

The *input resistance* of this amplifier is found by determining the Thevenin equivalent of the input circuit. The load resistance is normally such that $R_L >> R_o$. If this were not true, the effective gain would be reduced and the effective value of R_o would be the parallel combination of R_o with R_L. Let us again define $R'_A = R_A \parallel 2R_{cm}$ and $R'_F = R_F + R_o$. We shall neglect R_1, since it is so much less than R_{in}. Now since $R_L >> R_o$, we can reduce Figure 10.22(a) to the simplified form of Figure 10.23(a). We find the Thevenin equivalent of the circuit surrounded by the dashed curve, resulting in Figure 10.23(b). In Figure 10.23(c), the resistance to the right of $2R_{cm}$ is given by v/i'. In order to evaluate this, we write a loop equation to obtain

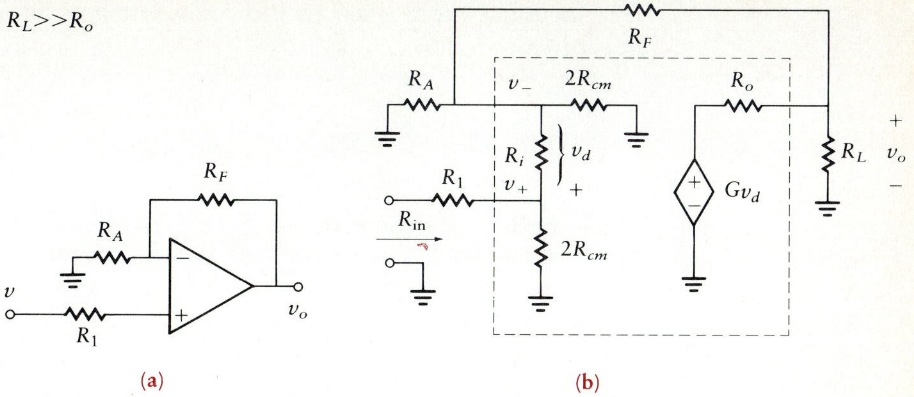

Figure 10.22 The noninverting amplifier.

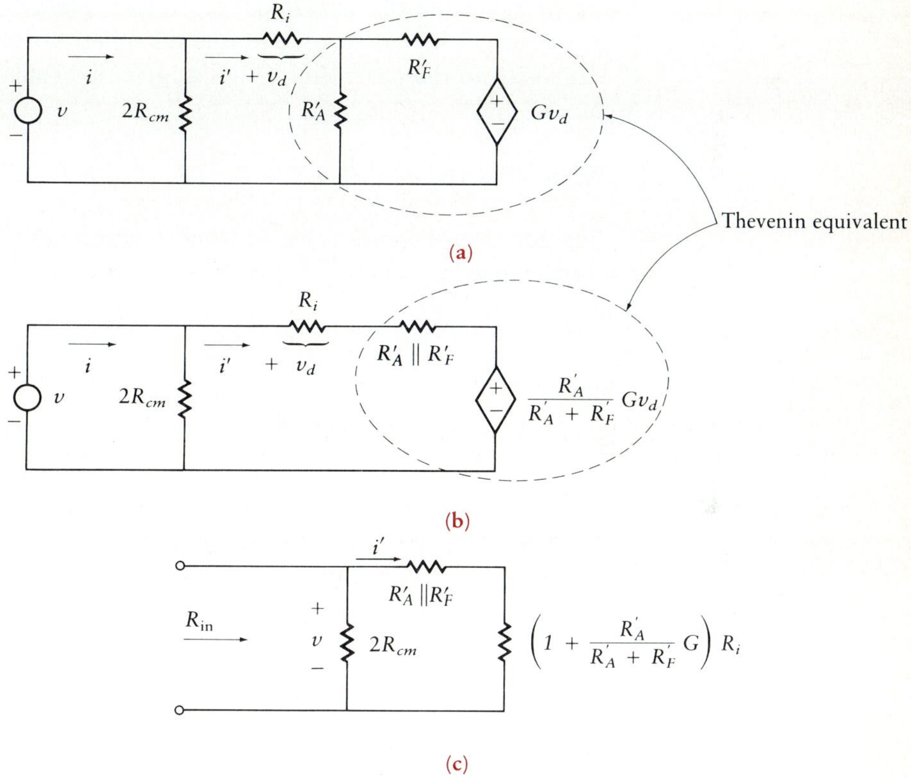

Figure 10.23 Reduced circuits for input resistance.

$$(R_i + R_A' \parallel R_F')i' = v - \frac{R_A' G R_i i'}{R_A' + R_F'}$$

Therefore,

$$\frac{v}{i'} = R_A' \parallel R_F' + \left(1 + \frac{R_A' G}{R_A' + R_F'}\right) R_i$$

The input resistance is the parallel combination of this quantity with $2R_{cm}$ and is written as follows:

$$R_{\text{in}} = 2R_{cm} \left\| \left[R'_A \| R'_F + \left(1 + \frac{R'_A G}{R'_A + R'_F}\right) R_i \right] \right. \tag{10.14}$$

Recall that $R'_A = R_A \| 2R_{cm}$, $R'_F = R_F + R_o$, and $R_L >> R_o$. If we retain only the most significant terms and note that R_{cm} is large, equation (10.14) reduces to

$$R_{\text{in}} = 2R_{cm} \left\| \frac{G_o R_i}{1 + R'_F/R'_A} \right. \tag{10.15}$$

where we again use the zero-frequency voltage gain, G_o.

Equation (10.15) can be used to find the input resistance of the 741 op-amp. If we substitute the parameter values as given in Table 10.2, equation (10.15) becomes

$$R_{\text{in}} \approx 400 \text{ M}\Omega \tag{10.16}$$

The *output resistance* is found from equation (10.12). We again use the assumptions that R_{cm} is large, that is, $R'_F \approx R_F$ and $R'_A \approx R_A$. Then the output resistance is given by

$$R_{\text{out}} = \frac{75(1 + R_F/R_A)}{10^5} \tag{10.17}$$

Example 10.4

Calculate the input resistance for the unity-gain follower shown in Figure 10.24(a).

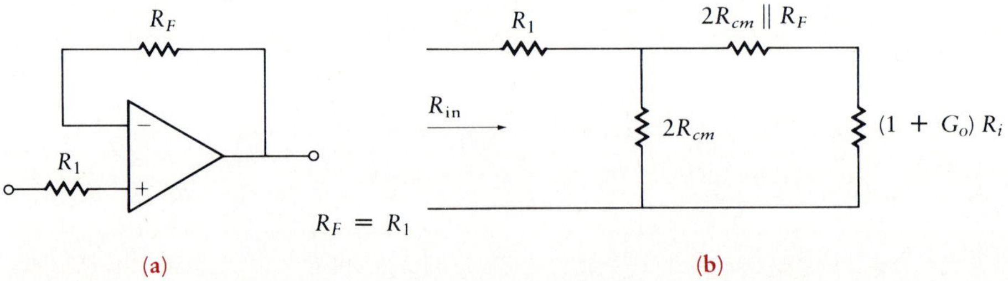

Figure 10.24 Unity-gain follower—Example 10.4.

SOLUTION The equivalent circuit is shown in Figure 10.24(b). Since we assume the zero-frequency gain, G_o, and the common-mode resistance, R_{cm}, are high, we can neglect the term $2R_{cm} \parallel R_F$ compared to $(1 + G_o)R_i$. Equation (10.14) cannot be used since $R_A \to \infty$. The input resistance is then given by

$$R_{\text{in}} = 2R_{cm} \parallel (1 + G_o)R_i = 2R_{cm}$$

This is typically equal to 200 MΩ or more, so we can neglect R_1 (i.e., set $R_1 = 0$).

10.3.2 Voltage Gain

We wish to determine the voltage gain, A_+, for the noninverting amplifier of Figure 10.25(a). This gain is defined by

$$A_+ = \frac{v_o}{v_i} = \frac{v_o}{v_+}$$

The equivalent circuit is shown in Figure 10.25(b). If we assume $R_F >> R_o$, $R_L >> R_o$, and $R'_A = R_A \parallel 2R_{cm}$, the circuit can be reduced to that shown in Figure 10.25(c). If we further define $R'_1 = R_1 \parallel 2R_{cm}$, then Figure 10.25(d) results.

The assumed conditions are desirable in order to prevent reduction of the effective gain. The operation of taking Thevenin equivalents modifies the dependent voltage source and the driving voltage source as in Figure 10.25(d). Note that

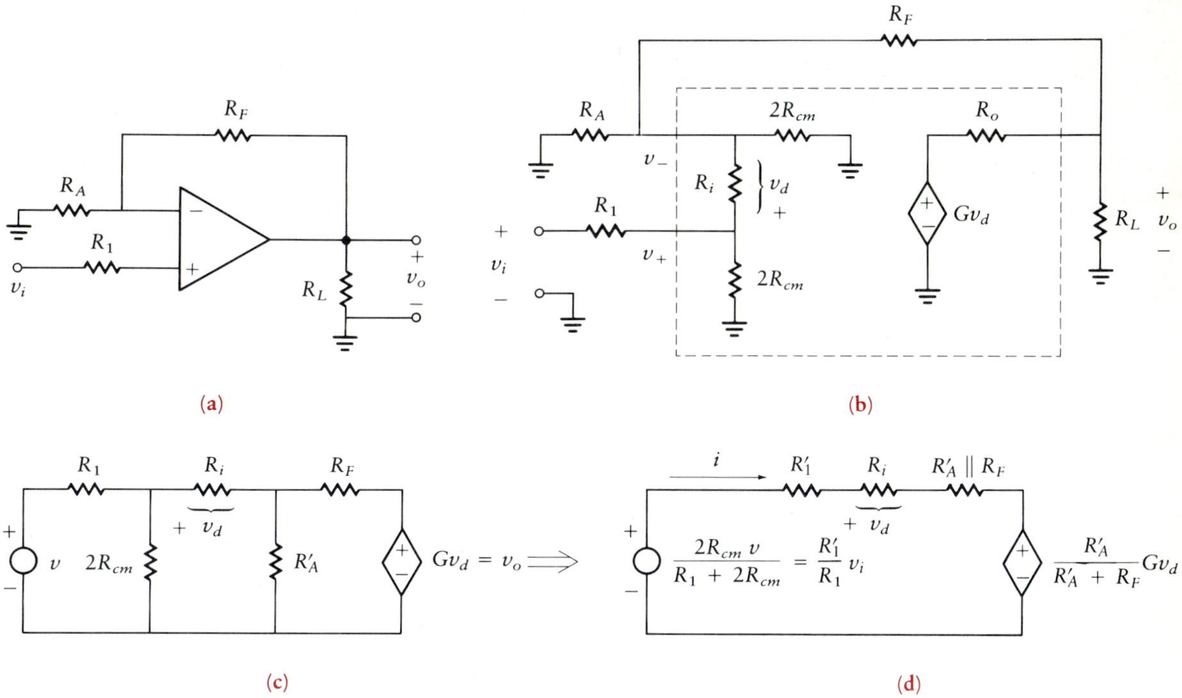

Figure 10.25 Noninverting amplifier.

$$\frac{R_1'}{R_1} \approx \frac{2R_{cm}}{R_1 + 2R_{cm}}$$

The output voltage is given by

$$v_o = Gv_d = R_i iG$$

We can find i by applying KVL to the circuit of Figure 10.25(d) to obtain

$$(R_1' + R_i + R_A' \parallel R_F)i = \frac{R_1' v_i}{R_1} - \frac{R_A' G v_d}{R_A' + R_F}$$

where

$$v_d = R_i i$$

Solving for the current, i, we obtain

$$i = \frac{R_1' v_i / R_1}{(R_A' \parallel R_F) + R_1' + [1 + R_A' G/(R_A' + R_F)]R_i}$$

The voltage gain is given by the ratio

$$A_+ = \frac{v_o}{v_i} = \frac{R_1' G R_i / R_1}{R_A' \parallel R_F + R_1' + [1 + R_A' G/(R_A' + R_F)]R_i} \tag{10.18}$$

As a check of this result, we can reduce the model to that of the ideal op-amp. We use the zero-frequency gain, G_o, in place of G in equation (10.18) and also the following equalities.

$$R_i G_o >> R_F, \quad R_1 = R_1', \quad R_i G_o >> R_1, \quad R_A = R_A', \quad R_i G_o >> R_L$$

When we let $G_o \to \infty$, equation (10.18) becomes

$$A_+ = \frac{R_A + R_F}{R_A}$$

which agrees with the result for the idealized model presented in Chapter 8.

Example 10.5

Find the gain of the unity-gain follower shown in Figure 10.26.

Figure 10.26
Unity-gain follower.

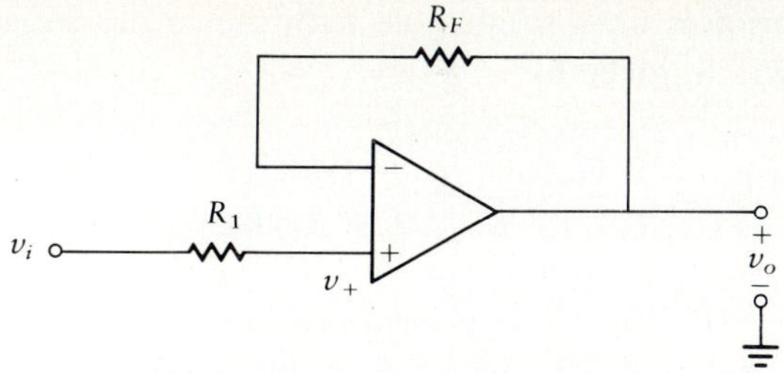

SOLUTION In this circuit, $R_A \to \infty$, $R'_A = 2R_{cm}$, and $R_F << R'_A$. We assume that G_o is large, $R'_1 \approx R_1$, and we set $R_1 = R_F$. Equation (10.18) then reduces to

$$A_+ = \frac{R'_A + R_F}{R'_A} = 1$$

so $v_o = v_i$ as expected.

10.3.3 Bandwidth

We will use the noninverting gain function given in equation (10.18) to determine the op-amp bandwidth. We note that $R'_A = R_A$ and $R'_1 = R_1$, with the result

$$A_+ = \frac{R_1 G R_i / R_1}{R_A \parallel R_F + R_1 + [1 + R_A G/(R_A + R_F)]R_i}$$

We now substitute

$$G = \frac{G_o}{1 + s/\omega_o}$$

to obtain

$$A_+ = \frac{G_o R_i}{(R_A \parallel R_F + R_1 + R_i)(1 + s/\omega_o) + R_i G_o/(1 + R_F/R_A)}$$

In this equation, we note that $R_i >> R_A \parallel R_F$ and $R_i >> R_1$. Taking these inequalities into account, the expression simplifies to equation (10.19).

$$A_+ = \left(1 + \frac{R_F}{R_A}\right)\frac{1}{s(1 + R_F/R_A)/(G_o\omega_o) + 1} \tag{10.19}$$

This equation is for a single pole with the corner frequency ω_c, which is also the bandwidth, given by

$$\omega_c = \text{BW} = \frac{G_o\omega_o}{1 + R_F/R_A} = \frac{\text{GBP}}{1 + R_F/R_A} \tag{10.20}$$

In equation (10.20), $G_o\omega_o$ is the open-loop gain-bandwidth product (GBP) of the basic single-pole compensated op-amp, and $1 + R_F/R_A$ is the gain factor. A typical value of GBP for the 741 op-amp is 1 MHz. Using this value, the bandwidth becomes

$$\text{BW} = \frac{10^6}{1 + R_F/R_A}\ \text{Hz} \tag{10.21}$$

The variables for the noninverting op-amp are summarized in Table 10.3 for both the nonideal and the 741 op-amp. Also given are values for the 741 op-amp. The inverting op-amp values are derived in Section 10.4.

Table 10.3 Summary of Equations for Nonideal Op-Amp.

	Noninverting Op-Amp		**Inverting Op-Amp**	
Variable	*Nonideal*	*741*	*Nonideal*	*741*
R_{in}	$2R_{cm} \Big\Vert \dfrac{GR_i}{1 + R_F/R_A}$	$(4 \times 10^8) \Big\Vert \dfrac{2 \times 10^{11}}{1 + R_F/R_A}$	R_A	R_A
R_{out}	$\dfrac{R_o}{1 + \dfrac{R_A G}{R_F + R_A}}$	$75\left(1 + \dfrac{R_F}{R_A}\right)10^{-5}$	$\dfrac{R_o}{1 + \dfrac{R_A G}{R_F + R_A}}$	$75\left(1 + \dfrac{R_F}{R_A}\right)10^{-5}$
BW	$\dfrac{\text{GBP}}{1 + R_F/R_A}$	$\dfrac{10^6}{1 + R_F/R_A}$	$\dfrac{\text{GBP}}{1 + R_F/R_A}$	$\dfrac{10^6}{1 + R_F/R_A}$ *
Voltage gain	$A_+ = \left(1 + \dfrac{R_F}{R_A}\right)$		$A_- = -\dfrac{R_F}{R_A}$	

*See Section 10.3.3 for a discussion of this equation.

Example 10.6

Design a system using μA741 op-amps to yield the following output voltage function:

$$v_o = 400v_1$$

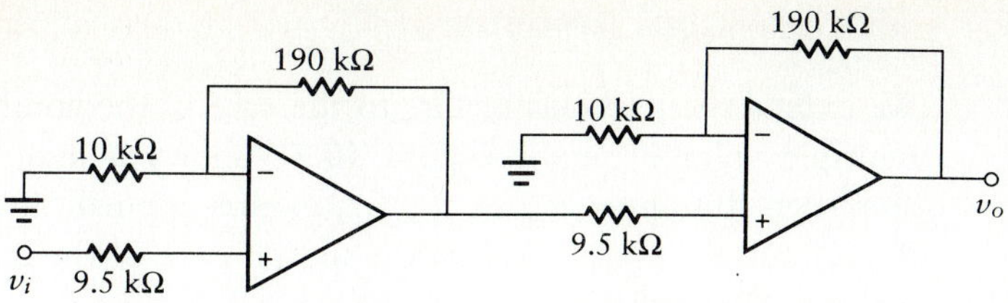

Figure 10.27
Op-amp circuit for Example 10.6.

The input resistance must be greater than 100 MΩ and the bandwidth must be greater than 20 kHz. Calculate the output resistance of the system.

SOLUTION We use the equation for bandwidth for the noninverting 741 op-amp (see Table 10.3).

$$BW = \frac{10^6}{1 + R_F/R_A}$$

The maximum gain for any stage is

$$1 + \frac{R_F}{R_A} = A_{\max} = \frac{10^6}{BW} = \frac{10^6}{20 \text{ kHz}} = 50$$

Since we must achieve an overall gain of 400, at least two op-amps are required. Let us select a gain of 20 for each stage. The stages will then be identical.

$$A = 20 = 1 + \frac{R_F}{R_A}$$

If we select $R_A = 10$ kΩ, then $R_F = 190$ kΩ.

$$R_i = R_A \parallel R_F = 9.5 \text{ k}\Omega$$

Using the equations from Table 10.3, we find

$$R_{\text{in}} \approx 400 \text{ M}\Omega$$

$$R_o = \frac{(75)(20)}{10^5} = 15 \text{ m}\Omega$$

The bandwidth is given by

$$BW = \frac{10^6}{20} = 50 \text{ kHz}$$

The configuration is as shown in Figure 10.27.

10.3.4 Multiple-Input Amplifiers

We extend the previous results to the case of the noninverting amplifier with multiple voltage inputs. Figure 10.28 shows a multiple-input noninverting amplifier. If inputs $v_1, v_2, v_3, \ldots, v_n$ are applied through input resistances $R_1, R_2, R_3, \ldots, R_n$, we obtain a special case of the general result derived in Chapter 8, as follows:

$$v_o = \left(1 + \frac{R_F}{R_A}\right)(R_1 \| R_2 \| \cdots \| R_n) \sum_{k=1}^{n} \frac{v_k}{R_k} \tag{10.22}$$

We choose

$$R_A \| R_F = R_1 \| R_2 \| \cdots \| R_n$$

in order to achieve bias balance. The input resistance, output resistance, and bandwidth are found from equations (10.15), (10.17), and (10.21), respectively, and these are summarized in Table 10.3.

Let us determine the output voltage of the two-input summer of Figure 10.29. The output voltage is found from equation (10.22), as follows:

$$v_o = \left(1 + \frac{R_F}{R_A}\right)\left(\frac{R_2 v_1}{R_1 + R_2} + \frac{R_1 v_2}{R_1 + R_2}\right) \tag{10.23}$$

We choose $R_A \| R_F = R_1 \| R_2$ to achieve bias balance. If we assume $R_F = R_1 = R_2 = R_A$, then equation (10.23) reduces to $v_o = v_1 + v_2$, which is a unity-gain two-input summer.

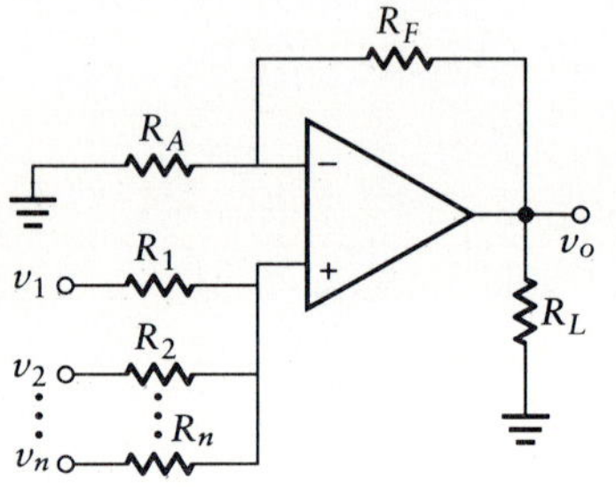

Figure 10.28 Multiple-input noninverting amplifier.

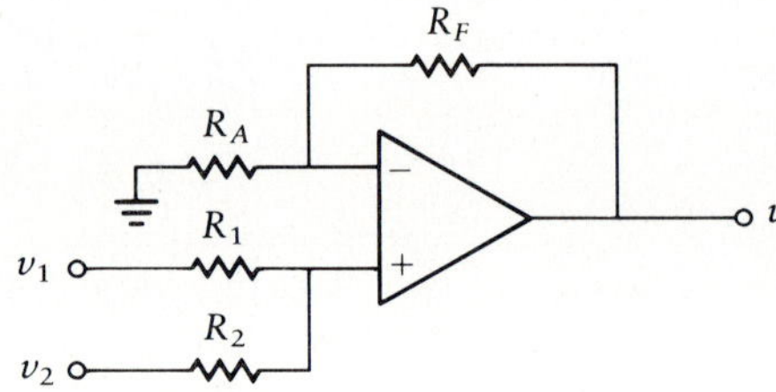

Figure 10.29 Two-input summer.

Example 10.7

Use μA741 op-amps in the design of a system to yield the following output voltage function:

$$v_o = 500v_1 + v_2$$

The input resistance for both v_1 and v_2 must be greater than 100 MΩ and the bandwidth must be greater than 20 kHz. Calculate the output resistance of the system.

SOLUTION The maximum gain per stage is $(10^6$ Hz$)/(2 \times 10^4)$ Hz or 50. We therefore need two noninverting amplifiers for the v_1 stage. If we set the gain of the first stage to 20, we can modify the output voltage equation to be

$$v_o = 25v_1' + v_2$$

where v_1' is the output of the first op-amp in the v_1 channel, or

$$v_1' = 20v_1$$

We solve the equation

$$v_o = 25v_1' + v_2$$

using the methods of Chapter 8. Doing so yields

$$X = 26, \qquad Y = 0, \qquad Z = 25$$

Let us choose $R_F = 250$ kΩ. Then

$$R_1 = 2.5 \times 10^5/1 = 250 \text{ k}\Omega$$

$$R_2 = 2.5 \times 10^5/25 = 10 \text{ k}\Omega$$

$$R_y = 2.5 \times 10^5/25 = 10 \text{ k}\Omega$$

$$R_x \to \infty$$

The resulting configuration is shown in Figure 10.30. Notice that op-amp 1 solves the equation

$$v_o = 25v_1' + v_2$$

and op-amp 2 solves the equation

$$v_1' = 20v_1$$

(This was derived in Example 10.6.) These two operations are necessary to develop the original equation,

$$v_o = 25(20v_1) + v_2 = 500v_1 + v_2$$

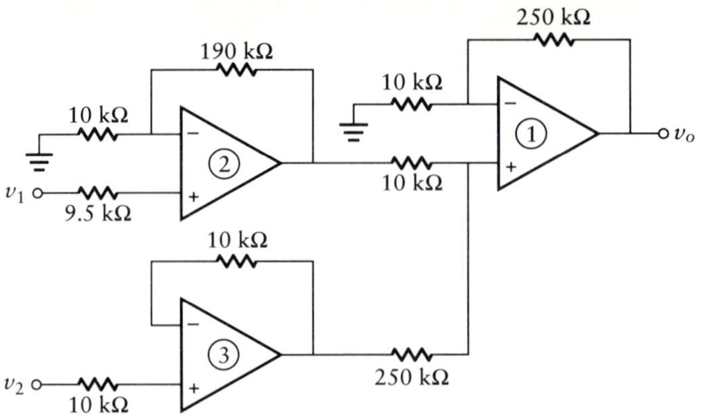

Figure 10.30 Op-amp circuit for Example 10.7.

Op-amp 3 is a noninverting buffer that is used to guarantee a high input resistance and reduce crosstalk between the two channels. More will be said about crosstalk later in this chapter.

The input resistance for both inputs is approximately 400 MΩ. The output resistance is given by equation (10.12).

$$R_{\text{out}} = \frac{R_o(1 + R_F/R_A)}{G} = 19.5\ \text{m}\Omega$$

The bandwidth is given by equation (10.21).

$$\text{BW} = \frac{10^6}{1 + R_F/R_A} = 38.5\ \text{kHz}$$

Drill Problems

D10.4 Find the input resistance and bandwidth of the 741 op-amp circuit of Figure D10.1. Let $R_A = R_F = R_L = 10$ kΩ.

Ans: $R_{\text{in}} = 400$ MΩ; BW = 0.5 MHz

D10.5 Calculate the output resistance, R_{out}, for the amplifier of Figure D10.1. Use a 741 op-amp with $R_A = R_F = R_L = 10$ kΩ.

Ans: 1.5 mΩ

D10.6 Find the voltage gain and bandwidth of the 741 op-amp circuit of Figure D10.1. Let $R_F = 10$ kΩ, $R_1 = R_A = 1$ kΩ.

Ans: $A_v = 11$; BW = 91 kHz

Figure D10.1

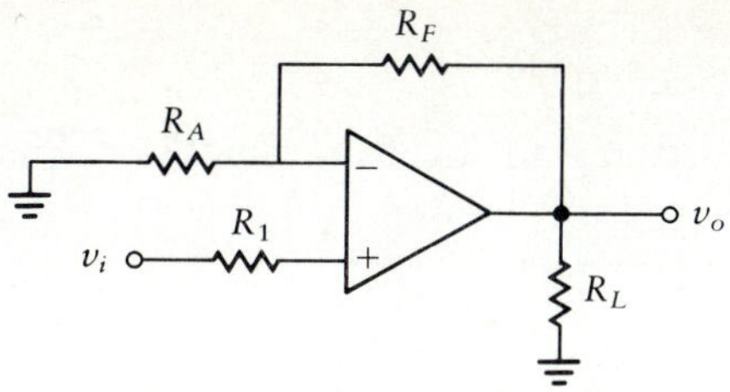

Example 10.8

Design and analyze a three-input summer using a 741 amplifier where

$$v_o = 3v_1 + 2v_2 + 5v_3$$

Select $R_F = 100$ kΩ.

SOLUTION We use the design method of Section 8.6, as follows: $X = 10$, $Y = 0$, $Z = 10 - 0 - 1 = 9$,

$$R_1 = 100 \text{ k}\Omega/3 = 33.3 \text{ k}\Omega$$

$$R_2 = 100 \text{ k}\Omega/2 = 50 \text{ k}\Omega$$

$$R_3 = 100 \text{ k}\Omega/5 = 20 \text{ k}\Omega$$

$$R_A = 100 \text{ k}\Omega/9 = 11.1 \text{ k}\Omega$$

The gain of the amplifier is $(1 + R_F/R_A) = 10$. The bandwidth of the amplifier is 1 MHz/10 = 100 kHz. The output resistance is $75(10)/10^5 = 7.5$ mΩ. The design method of Chapter 8 assures us that

$$R_1 \parallel R_2 \parallel R_3 = R_F \parallel R_A = 10 \text{ k}\Omega$$

or

$$33.3 \text{ k}\Omega \parallel 50 \text{ k}\Omega \parallel 20 \text{ k}\Omega = 100 \text{ k}\Omega \parallel 11.1 \text{ k}\Omega = 10 \text{ k}\Omega$$

The input resistances are given by

$$R_{\text{in}}(v_1) = R_1 + R_2 \parallel R_3 = 47.6 \text{ k}\Omega$$

$$R_{\text{in}}(v_2) = R_2 + R_1 \parallel R_3 = 62.5 \text{ k}\Omega$$

$$R_{\text{in}}(v_3) = R_3 + R_1 \parallel R_2 = 40 \text{ k}\Omega$$

10.4 INVERTING AMPLIFIER

Figure 10.31(a) illustrates an inverting amplifier. Figure 10.31(b) shows the equivalent circuit using the op-amp model developed earlier in this chapter.

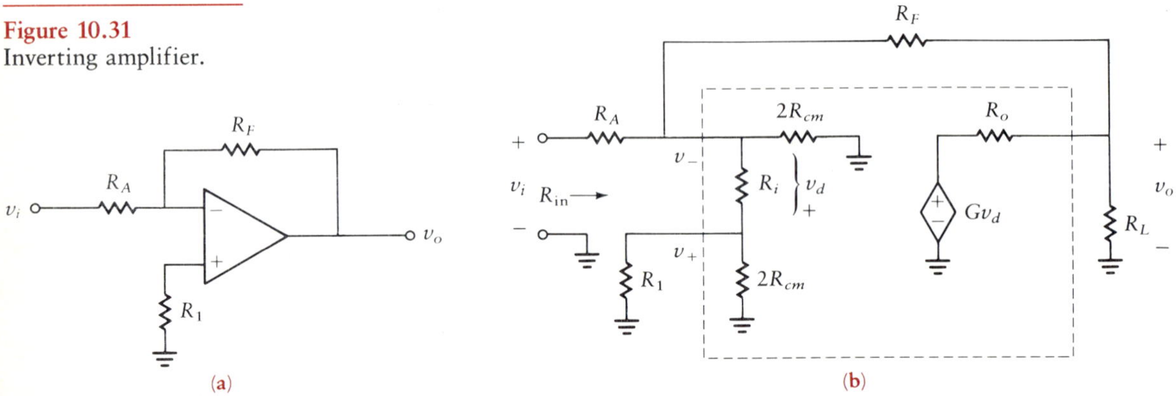

Figure 10.31
Inverting amplifier.

10.4.1 Input and Output Resistance

Figure 10.31(b) is reduced to Figure 10.32(a) if we let $R_1' = R_1 \parallel 2R_{cm}$, $R_F >> R_o$, and $R_L >> R_o$. It is reasonable to assume that these inequalities apply because, were they not true, the output would load the input and the gain would be reduced.

A voltage-divider relationship can be used to yield

$$v_d = \frac{-R_i v_-}{R_i + R_1'}$$

and a loop equation yields

$$R_F i'' = v_- - Gv_d = 1 + \frac{R_i G}{R_i + R_1'} v_- = \frac{R_1' + (1 + G)R_i}{R_i + R_1'} v_- \tag{10.24}$$

The input resistance, R_{in}, is obtained from Figure 10.32(b), where we have replaced the dependent source with an equivalent resistance. The value of this resistor is v_-/i'', which is found from equation (10.24). For large G (i.e., $G \rightarrow G_o$), the rightmost resistance in Figure 10.32(b) is approximately zero, and $R_{in} \approx R_A$. The output resistance of the inverting amplifier is the same as that of the noninverting amplifier. Thus,

$$R_{out} = \frac{R_o(1 + R_F/R_A)}{G_o} \tag{10.25}$$

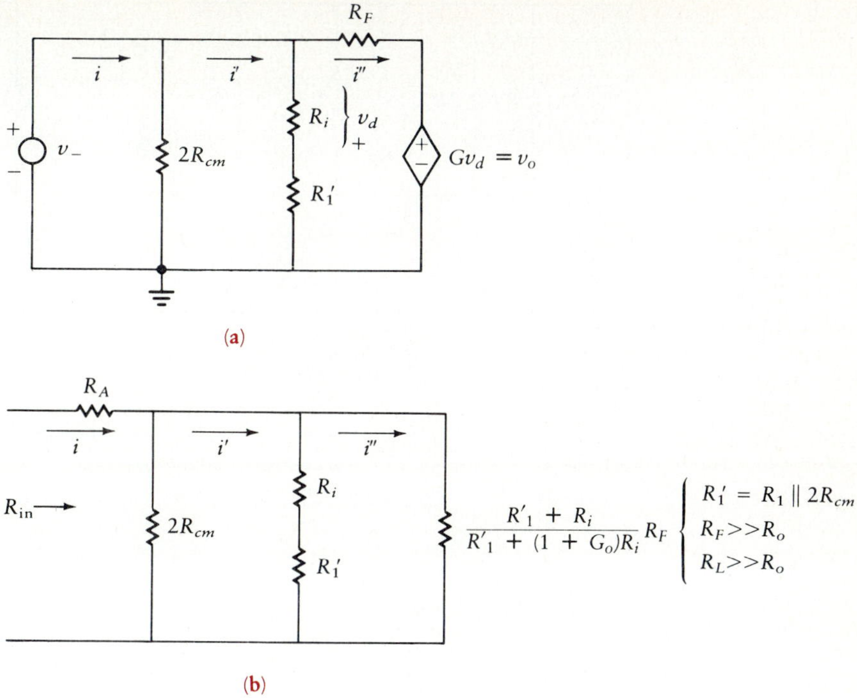

Figure 10.32 Simplified inverting amplifier model.

10.4.2 Voltage Gain

We use the equivalent circuits of Figure 10.31(b) and Figure 10.32(a) to determine the voltage gain. The inverting input gain, $A_- = v_o/v_i$, is obtained from the circuit of Figure 10.32(a) by again making the same assumptions that we made in finding the output resistance. These assumptions reduce the circuit to that shown in Figure 10.33(a), where we have changed the voltage source in series with a resistance to a current source in parallel with a resistance. The resistors can then be combined to yield the circuit of Figure 10.33(b). Finally, the current source is converted back to the voltage source to yield the simplified circuit of Figure 10.33(c). The loop equation for this circuit is given by

$$v_d = \frac{-R_i}{R_i + R_1' + R_A' \| R_F}\left[(R_A' \| R_F)\frac{v_i}{R_A} + \frac{R_A'G}{R_A' + R_F}v_d\right] \tag{10.26}$$

Since $v_o = Gv_d$, the inverting voltage gain is

$$A_- = \frac{v_o}{v_i} = -\frac{GR_A'R_iR_F/R_A}{(R_i + R_1')(R_A' + R_F) + R_A'R_F + R_iR_A'G} \tag{10.27}$$

We can verify this result relative to the gain of the ideal op-amp by making the approximations: $R_A << 2R_{cm}$ and $G >> 1$. Then

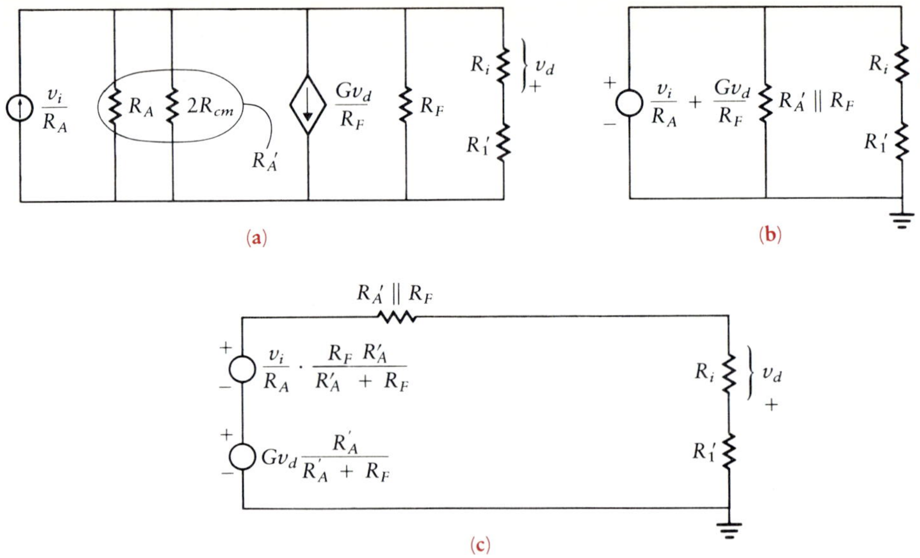

Figure 10.33
Inverting input gain.

$$R_A' = R_A \parallel 2R_{cm} = R_A$$

and

$$A_- = -\frac{R_F}{R_A}$$

This is the same as the result found earlier for the simplified model. The variables for the inverting op-amp are summarized in Table 10.3.

10.4.3 Bandwidth

We find the bandwidth by making a simplification of equation (10.27). Note that $R_A' = R_A$, $R_i >> R_1'$, and $R_i >> R_F$. We let

$$G = \frac{G_o}{1 + s/\omega_o}$$

with the result

$$A_- = -\frac{R_F}{R_A} \frac{G_o/(1 + s/\omega_o)}{(1 + R_F/R_A) + G_o/(1 + s/\omega_o)}$$

$$= -\frac{R_F}{R_A} \frac{G_o}{(1 + R_F/R_A)(1 + s/\omega_o) + G_o}$$

Now since $G_o >> (1 + R_F/R_A)$, we obtain equation (10.28).

$$A_- \approx -\frac{R_F}{R_A}\frac{1}{(1 + R_F/R_A)s/G_o\omega_o + 1} \tag{10.28}$$

We now obtain the bandwidth from this expression.

$$\omega_c = \text{BW} = \frac{G_o\omega_o}{1 + R_F/R_A} = \frac{\text{GBP}}{1 + R_F/R_A} \tag{10.29}$$

Notice that the bandwidth for the inverting amplifier (equation (10.29)) is identical to that of the noninverting op-amp (equation (10.20)). Since the closed-loop gain for the inverting op-amp configuration is $-R_F/R_A$, we must use the gain factor, $1 + R_F/R_A$ rather than $-R_F/R_A$ to calculate the bandwidth for the inverting configuration. This becomes obvious when we let $|R_F/R_A| < 1$, which would indicate a bandwidth greater than $G_o\omega_o$.

Example 10.9

Use μA741 op-amps in the design of a system to yield the following output voltage function:

$$v_o = -400v_1$$

The input resistance must be greater than 100 MΩ and the bandwidth must be greater than 20 kHz. Calculate the output resistance of the system.

SOLUTION Referring to Table 10.3, we select the equation for the bandwidth of an inverting 741 op-amp.

$$\text{BW} = \frac{10^6}{1 + R_F/R_A}$$

Since the GBP is 10^6, the maximum gain for any stage is

$$A_{\max} = \frac{10^6}{\text{BW}} = 50$$

Therefore, at least two op-amps are required. Let us select a gain of 20 for each stage. The first stage should consist of a noninverting amplifier with a gain of 20. We use a noninverting amplifier for the first stage (rather than the second) to obtain a high value for input resistance. The second stage consists of

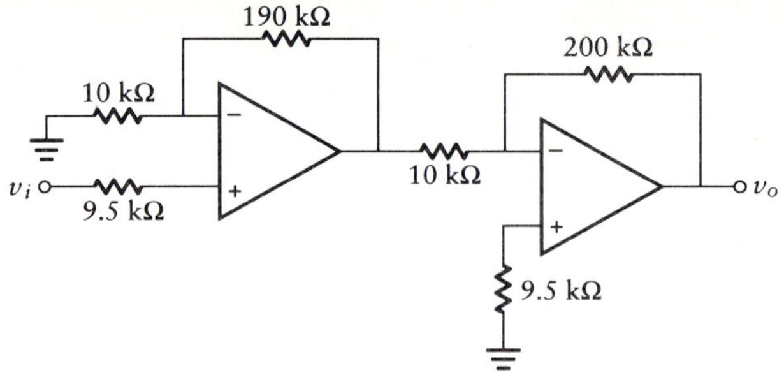

Figure 10.34 Amplifier configuration for Example 10.9.

an inverting amplifier with a gain of −20. Thus we provide the overall negative value of gain. The configuration is shown in Figure 10.34.

First Stage

$$A = 20 = 1 + \frac{R_F}{R_A}$$

If we select $R_A = 10\ \text{k}\Omega$, then the feedback resistor is given by

$$R_F = 190\ \text{k}\Omega$$

$$R_1 = R_A \parallel R_F = 9.5\ \text{k}\Omega$$

Second Stage Let us again set $R_A = 10\ \text{k}\Omega$. We find the value of feedback resistance from $-R_F/R_A = -20$, so $R_F = 200\ \text{k}\Omega$.

$$R_1 = R_A \parallel R_F = 9.5\ \text{k}\Omega$$

The input resistance is then approximately 400 MΩ and the output resistance is given by

$$R_o = \frac{(75)\,(21)}{10^5} = 15.8\ \text{m}\Omega$$

The bandwidth is given by

$$\text{BW} = \frac{10^6}{21} = 47.6\ \text{kHz}$$

thereby meeting the specifications.

10.4.4 Multiple-Input Amplifiers

If the voltages $v_a, v_b, \ldots, v_m$ are applied to the summing junction (inverting input to op-amp) through resistors $R_a, R_b, \ldots, R_m$, respectively, as shown in Figure 10.35, the output voltage is

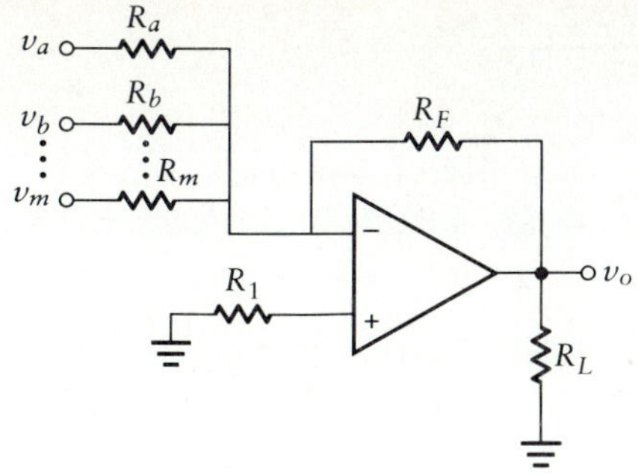

Figure 10.35 Multiple-input inverting amplifier.

$$v_o = -\frac{R_F v_a}{R_a} - \frac{R_F v_b}{R_b} - \cdots - \frac{R_F v_m}{R_m} = -R_F \sum_{j=a}^{m} \frac{v_j}{R_j} \tag{10.30}$$

To achieve bias balance, we choose

$$R_1 = R_F \parallel R_a \parallel R_b \parallel \cdots \parallel R_m$$

Let us define

$$R_A = R_a \parallel R_b \parallel \cdots \parallel R_m$$

The output resistance is then given by

$$R_{\text{out}} = \frac{R_o(1 + R_F/R_A)}{G_o}$$

and the bandwidth is

$$\text{BW} = \frac{\text{GBP}}{1 + R_F/R_A}$$

Suppose now that only two inputs are used. The output voltage is then

$$v_o = -\frac{R_F v_a}{R_a} - \frac{R_F v_b}{R_b} \tag{10.31}$$

The input resistance at v_a is approximately equal to R_a, and the input resistance at v_b is approximately R_b. We can make this circuit a unity-gain two-input summer with an output voltage of

$$v_o = -v_a - v_b$$

by setting $R_F = R_a = R_b$. The resistance from the noninverting input terminal to ground is chosen to achieve bias balance. Thus, $R_1 = R_F/3$, and we have

$$1 + \frac{R_F}{R_A} = 1 + \frac{R_F}{(R_a \parallel R_b)} = 3$$

An equal-gain (i.e., not unity) two-input summer is obtained by setting $R_a = R_b = R$ and $R_1 = (R_F \parallel R)/2$. In this case, the output voltage is

$$v_o = -\left(\frac{R_F}{R}\right)(v_a + v_b)$$

The input resistance is approximately R. Since $R_A = R/2$,

$$1 + \frac{R_F}{R_A} = 1 + \frac{2R_F}{R}$$

If m inputs are summed through equal resistors (R), the output voltage is

$$v_o = -\frac{R_F}{R}\sum_{j=a}^{m} v_j \tag{10.32}$$

For this equal-gain multiple-input inverting summer, the input resistance to each input is approximately R. Since $R_A = R/m$,

$$1 + \frac{R_F}{R_A} = 1 + \frac{mR_F}{R}$$

and

$$R_1 = R_F \left\| \frac{R}{m} \right.$$

The output resistance is

$$R_{\text{out}} = \frac{R_o}{G_o}\left(1 + \frac{mR_F}{R}\right) \tag{10.33}$$

Example 10.10

Design and analyze a three-input inverting amplifier using a 741 op-amp where

$$v_o = -4v_a - 2v_b - 3v_c$$

and the input resistance is $R_{\text{min}} = 8\ \text{k}\Omega$.

SOLUTION We use the design method of Chapter 8 to find

$$X = 0, \qquad Y = 9, \qquad Z = 10$$

Then

$$R_F = 10 \times 8\ \text{k}\Omega = 80\ \text{k}\Omega$$
$$R_a = 80\ \text{k}\Omega/4 = 20\ \text{k}\Omega$$
$$R_b = 80\ \text{k}\Omega/2 = 40\ \text{k}\Omega$$
$$R_c = 80\ \text{k}\Omega/3 = 26.7\ \text{k}\Omega$$
$$R_x = 80\ \text{k}\Omega/10 = 8\ \text{k}\Omega$$
$$R_A = R_a \parallel R_b \parallel R_c = 8.9\ \text{k}\Omega$$

The gain multiplier of the amplifier is $1 + R_F/R_A = 10$ and the bandwidth of the amplifier is 1 MHz/10 = 100 kHz. We find the input resistance as follows:

$$R_{\text{in}}(v_a) = 20\ \text{k}\Omega$$
$$R_{\text{in}}(v_b) = 40\ \text{k}\Omega$$
$$R_{\text{in}}(v_c) = 26.7\ \text{k}\Omega$$

The output resistance is approximately $75(10)/10^5 = 7.5\ \text{m}\Omega$. To achieve bias balance, we set

$$R_x = R_a \parallel R_b \parallel R_c \parallel R_F = 8\ \text{k}\Omega$$

10.5 DIFFERENTIAL SUMMING

We have seen that an op-amp can be configured to produce an output that is a weighted sum of multiple inputs. If the sum includes both positive and negative signs, *differential summing* results. The op-amp configuration of Figure 10.36 produces an output voltage, v_o, given by

$$v_o = \left(1 + \frac{R_F}{R_A}\right)(R_1 \parallel R_2 \parallel R_3 \parallel \cdots \parallel R_n \parallel R_x) \sum_{i=1}^{n} \frac{v_i}{R_i} - R_F \sum_{j=a}^{m} \frac{v_j}{R_j} \tag{10.34}$$

where

$$R_A = R_a \parallel R_b \parallel R_c \parallel \cdots \parallel R_m \parallel R_y$$

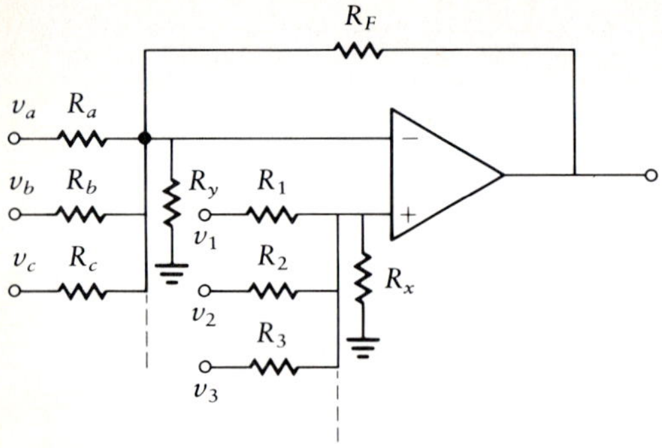

Figure 10.36 Differential summing.

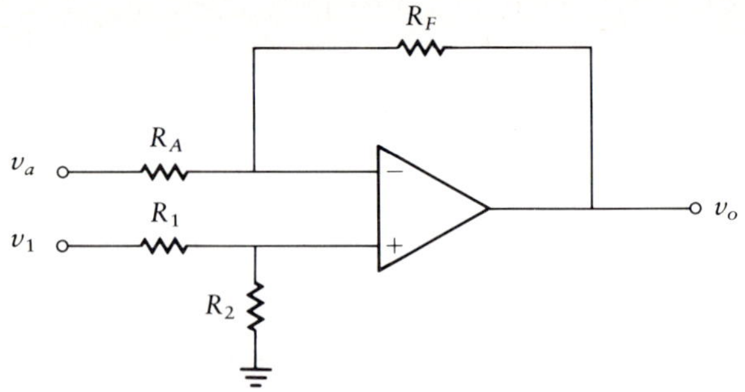

Figure 10.37 Differencing amplifier.

We choose the resistors to achieve bias balance as follows:

$$R_1 \parallel R_2 \parallel R_3 \parallel \cdots \parallel R_x = R_F \parallel R_a \parallel R_b \parallel R_c \parallel \cdots \parallel R_y$$

The input resistance for each inverting input, v_j, is R_j.

If the inverting and noninverting terminals each have only one input, the result is a *differencing amplifier*. This is illustrated in Figure 10.37. The output voltage for this configuration is

$$v_o = \left(1 + \frac{R_F}{R_A}\right)\frac{R_2}{R_1 + R_2}v_i - \frac{R_F v_a}{R_A} \tag{10.35}$$

In order to achieve bias balance, we choose

$$R_1 \parallel R_2 = R_A \parallel R_F$$

The input resistance for the v_a terminal is R_A. The input resistance for the v_1 terminal is $R_1 + (R_2 \parallel R_{\text{in}})$, where R_{in} is found from equation (10.15) to be

$$R_{\text{in}} = 2R_{cm} \left\| \frac{GR_i}{1 + R_F/R_A} \approx 2R_{cm} \right. \tag{10.36}$$

R_{out} is found in equation (10.12) and is equal to

$$R_{\text{out}} = \frac{R_o(1 + R_F/R_A)}{G_o} \tag{10.37}$$

The bandwidth is given by

$$\text{BW} = \frac{\text{GBP}}{1 + R_F/R_A} \tag{10.38}$$

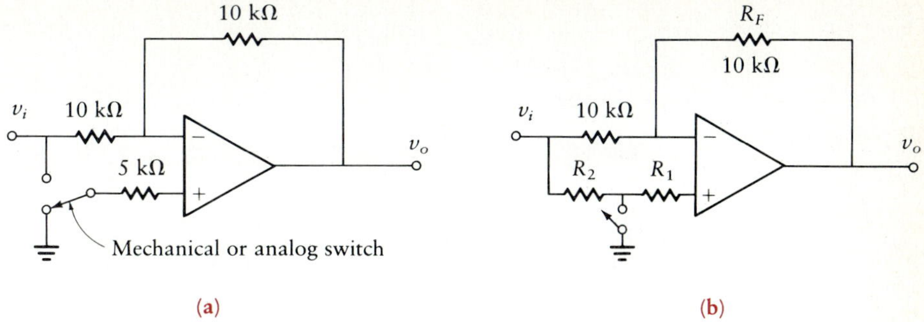

Figure 10.38 Sign switcher.

To achieve unit-gain differencing, where the output is given by $v_o = v_1 - v_a$, we set $R_A = R_F = R_1 = R_2$. If a 741 op-amp is used, a typical value for these four resistors is 10 kΩ. The bandwidth is approximately 500 kHz. The input resistance into v_a is then 10 kΩ, and into the v_1 terminal the resistance is 20 kΩ.

Suppose that equal-gain differencing is desired but that the gains need not be unity. We then set $R_1 = R_A$ and $R_2 = R_F$. The output voltage is then

$$v_o = \frac{R_F(v_1 - v_a)}{R_A} \tag{10.39}$$

The input resistance into the v_a terminal is R_A, and into the noninverting terminal it is $R_A + (R_F \parallel R_{in})$, which is approximately $R_A + R_F$ since $R_{in} >> R_F$. Values for input resistance, R_{in}, output resistance, R_{out}, and the BW are easily determined with the use of equations (10.36), (10.37), and (10.38).

A useful modification of the configuration of Figure 10.37 is the *sign switcher* as shown in Figure 10.38. With the switch in the position shown in Figure 10.38(a), $v_o = -v_i$, and for the opposite switch position, $v_o = 2v_i - v_i = v_i$. The input resistance is 10 kΩ for each position, and the bias is balanced in each position. Figure 10.38(a) shows a single-pole double-throw switch. Sign switching can also be accomplished with a single-throw switch as shown in Figure 10.38(b). However, for this implementation, the bias and input resistances are not equal in the two switch positions.

Drill Problems

D10.7 Find the gain and bandwidth for the 741 op-amp of Figure 10.31 with $R_F = 100$ kΩ, $R_A = 10$ kΩ, and $R_1 = 9$ kΩ.

Ans: $A_v = -10$; BW = 91 kHz

D10.8 Find the bandwidth and input resistance for the differencing amplifier of Figure 10.37 using the 741 op-amp. $R_F = R_2 = 100$ kΩ and $R_A = R_1 = 10$ kΩ.

Ans: $R_{in}(v_1) = 110\ \text{k}\Omega$; BW = 91 kHz;
$R_{in}(v_a) = 10\ \text{k}\Omega$

10.6 DESIGNING AMPLIFIERS USING MULTIPLE OP-AMPS

We now have knowledge of the external electrical characteristics of the op-amp and are able to use this knowledge as a building block in the design of amplifiers requiring more than one op-amp. The specified input resistance at each terminal is important to the designer, as are the characteristics of bandwidth and output resistance. Other characteristics that aid in selecting the correct op-amp type include CMRR, PSRR, and the slew rate.

The first step in designing multiple op-amp circuits is to determine the amount of gain per stage. This is accomplished by dividing the gain-bandwidth product of the selected op-amp by the required bandwidth.

The next step is to determine which inputs are negative and which are positive. This determines whether connections should be made to the negative or positive terminal of the op-amp.

If the required input resistance is above the value of the coupling resistor, the input voltage must be fed to the noninverting terminal of the op-amp. Recall that the input resistance to the inverting terminal is equal to the value of the coupling resistance used, and this is considerably less than 1 MΩ. There are other considerations in the design, such as isolation of various sources from each other and phase relationships. Having previously presented some simple examples, we now tie these concepts together through the following two design examples.

Example 10.11 Multiple Op-Amp (Design)

Design an amplifier that is composed of 741 op-amps in order to obtain a gain of 800 with an input impedance of at least 20 MΩ. The amplifier must respond to a frequency of 40 kHz. Determine all resistor values, the output resistance, the input resistance, and the bandwidth. Check the gain of your design using a computer simulation program.

SOLUTION The maximum gain per stage is determined by the gain-bandwidth product for the 741 op-amp and the specified bandwidth. Thus,

$$\text{Maximum gain/stage} = \frac{1\ \text{MHz}}{40\ \text{kHz}} = 25$$

In order to achieve an overall gain of 800, we therefore need at least three stages of amplification. To achieve the high input impedance, the input to the first op-amp must be connected to the noninverting input. The overall gain must be split among the three stages, so one possible choice is to design two of the stages with gains of 10 each and the third with a gain of 8. All stages are noninverting.

When selecting the gain per stage, it is normally preferable to balance the stages—that is, to have all of the individual stage gains approximately equal in magnitude. This is true since the bandwidth is limited by the highest stage gain.

In designing single op-amp stages in earlier sections of this text, we found that usually not enough information is given to specify all of the design parameters. We must choose the value of one or more resistors before proceeding with the design. The same is true in this case, and if the procedures are followed, one possible set of resistor values for the op-amps with gains of 10 is

$$R_F = 90 \text{ k}\Omega, \qquad R_A = 10 \text{ k}\Omega, \qquad R_1 = 9 \text{ k}\Omega$$

For the op-amp requiring a gain of 8, we can use

$$R_F = 70 \text{ k}\Omega, \qquad R_A = 10 \text{ k}\Omega, \qquad R_1 = 8.75 \text{ k}\Omega$$

The resulting design is shown in Figure 10.39. The bandwidth of the amplifier is the smallest of the three individual bandwidths of the various stages. For the stage with gain of 10 we have

$$1 + \frac{R_F}{R_A} = 1 + \frac{90 \text{ k}\Omega}{10 \text{ k}\Omega} = 10$$

The bandwidth is then given by

$$\text{BW} = \frac{1 \text{ MHz}}{10} = 100 \text{ kHz}$$

which exceeds the requirement. The input resistance is approximately 400 MΩ (see equation (10.15)), which also exceeds the requirement. The output resistance is found from equation (10.12) to be

$$R_{\text{out}} = \frac{75(1 + R_F/R_A)}{10^5} = 7.5 \text{ m}\Omega$$

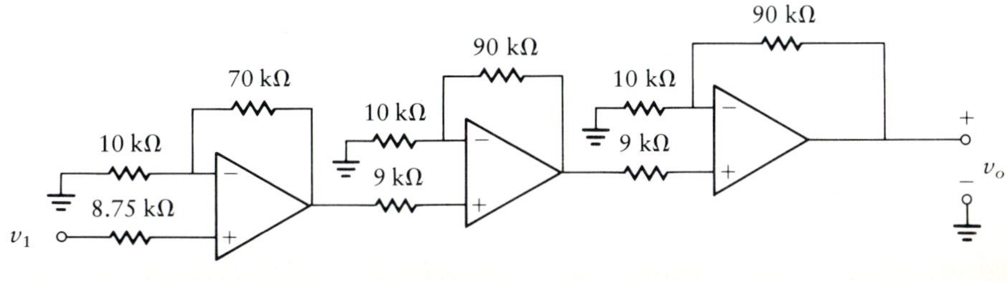

Figure 10.39
Solution for Example 10.11.

We now verify our design using a computer simulation program. We have used MICRO-CAP II, but any SPICE-based program can be used. The circuit of Figure 10.39 is entered, and the LM741 op-amp is used as it appears in the program library. We can verify the overall gain as well as the required upper frequency response by running an ac analysis from a low frequency (say 1 Hz) to 20 kHz.

Running the program produces a plot of gain as a function of frequency. The curve starts at a gain of 58.06 dB for low frequency and decreases to 57.7 dB for a frequency of 20 kHz. A gain factor of 800 is equivalent to about 58 dB, so our system is yielding a gain of almost exactly 800 at low frequency. At high frequency, the gain is changing by −0.3 dB, which is equivalent to 0.966. We thus see a decrease of less than 3.5% in the gain at the high frequencies. This verifies our design.

Example 10.12 Multiple Op-Amp (Design)

Design an amplifier that is composed of 741 op-amps to obtain an output of

$$v_o = 20v_1 - 40v_2 - 45v_3$$

The amplifier must respond to frequencies up to 50 kHz. Determine all resistor values, the output resistance, the input resistance, and the bandwidth of the amplifier.

SOLUTION The maximum gain in any one stage is checked using the GBP and the bandwidth requirements as follows:

$$\frac{\text{maximum gain}}{\text{stage}} = \frac{\text{GBP}}{\text{BW}} = \frac{1\text{ MHz}}{50\text{ kHz}} = 20$$

Thus, we can achieve the required gain of the v_1 channel with one amplification stage. If the gain of any channel is greater than 20, we obtain that gain with more than one amplifier in series.

We must use two amplifiers to achieve the required gain for v_2 and v_3. Suppose the first of each set of amplifiers multiplies v_2 by 4 and v_3 by 5. The main amplifier stage must then solve the modified equation

$$v_o = 20v_1 - 10v_2' - 9v_3'$$

We design the main amplifier stage using the procedure of Chapter 8.

$$X = 20, \qquad Y = 19, \qquad Z = 0$$

so

$$R_x = R_y \to \infty$$

We choose $R_F = 180$ kΩ and then solve for the other resistor values:

$$R_1 = 180 \text{ k}\Omega/20 = 9 \text{ k}\Omega$$

$$R_2 = 180 \text{ k}\Omega/10 = 18 \text{ k}\Omega$$

$$R_3 = 180 \text{ k}\Omega/9 = 20 \text{ k}\Omega$$

The first voltage, v_1, is connected directly to the noninverting terminal, yielding an input resistance of approximately 400 MΩ. The amplification of v_2 by 4 is achieved with a stage using

$$R_A = 10 \text{ k}\Omega$$

$$R_F = 30 \text{ k}\Omega$$

$$R_1 = 7.5 \text{ k}\Omega$$

The amplification of v_3 by 5 is achieved with a stage using

$$R_A = 10 \text{ k}\Omega$$

$$R_F = 40 \text{ k}\Omega$$

$$R_1 = 8 \text{ k}\Omega$$

The output resistance is given by

$$R_{\text{out}} = \frac{75[1 + R_F/(R_2 \parallel R_3)]}{10^5} = 15 \text{ m}\Omega$$

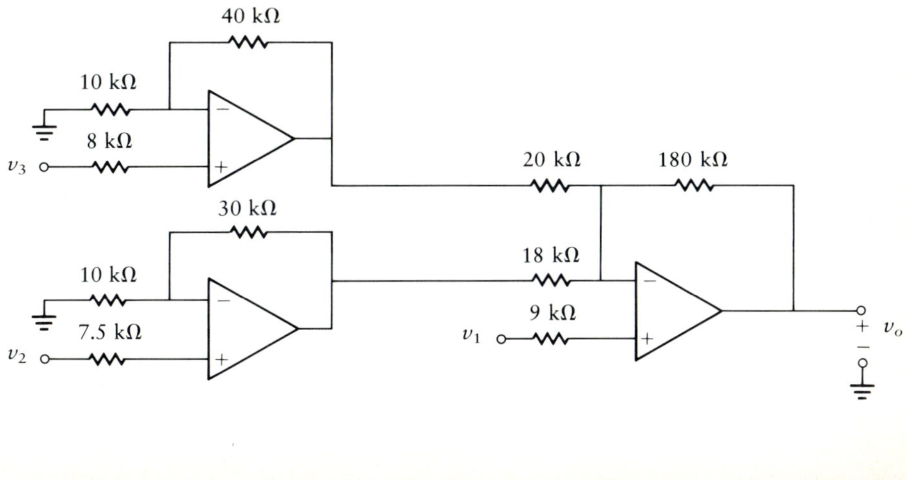

Figure 10.40 Solution for Example 10.12.

The bandwidth for the system is determined by the amplifier with the largest gain ($20v_1$),

$$BW = \frac{1 \text{ MHz}}{1 + R_F/(R_2 \parallel R_3)} = 50 \text{ kHz}$$

The resulting amplifier is shown in Figure 10.40.

10.7 101 AMPLIFIER

As we learned in Section 10.1, the 741 is internally compensated to enhance stability. The 101 op-amp has characteristics similar to those of the 741. However, as an additional feature the 101 amplifier provides a capability for external compensation, which can be used to increase the bandwidth. The 741 op-amp contains an internal RC network that provides stable performance for most feedback conditions. Alternatively, stabilization of the 101 op-amp is accomplished with an externally added RC network. If a 30-pF external capacitor is used, the 101 op-amp has characteristics similar to those of the 741 op-amp. If, however, a 3-pF capacitor is externally added, the break point is moved 10 times higher in frequency, providing a 10 times wider bandwidth. With this change, the amplifier is stable only for a low-frequency gain of 10 or greater.

The principal characteristics of the 101 op-amp are

$$R_i = 2 \text{ M}\Omega, \qquad R_o = 100\ \Omega$$

$$R_{cm} = 200 \text{ M}\Omega, \qquad G_o = 10^5 \text{ (externally compensated)}$$

Offset adjustment for the 101 op-amp is accomplished with either arrangement shown in Figure 10.41.

Single-pole compensation is accomplished by placing a capacitor, C, between pins 1 and 8 of the op-amp package (see Figure 10.42(a)). If a value of $C = 30$ pF is used, the gain-bandwidth product is

$$\text{GBP} = 10^6 \text{ Hz}$$

The amplitude response as a function of frequency is shown in Figure 10.43(a). If the capacitor value is changed to $C = 3$ pF, the gain-bandwidth product increases to

$$\text{GBP} = 10^7 \text{ Hz}$$

Figure 10.41
Input offset adjustment.

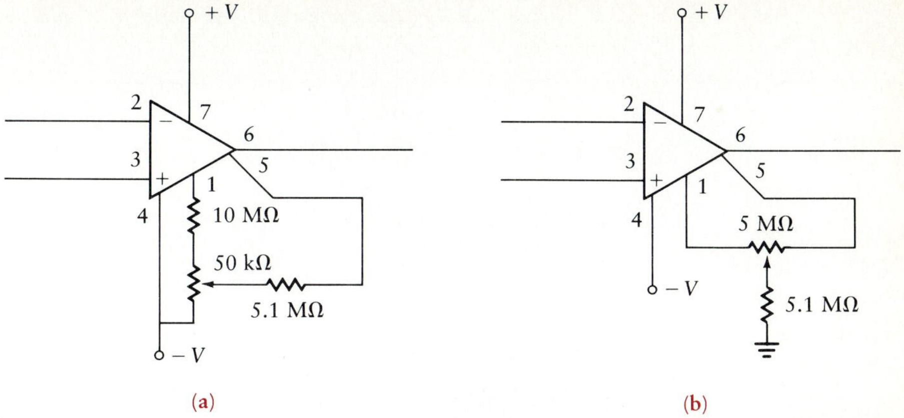

Figure 10.42
101 compensation.

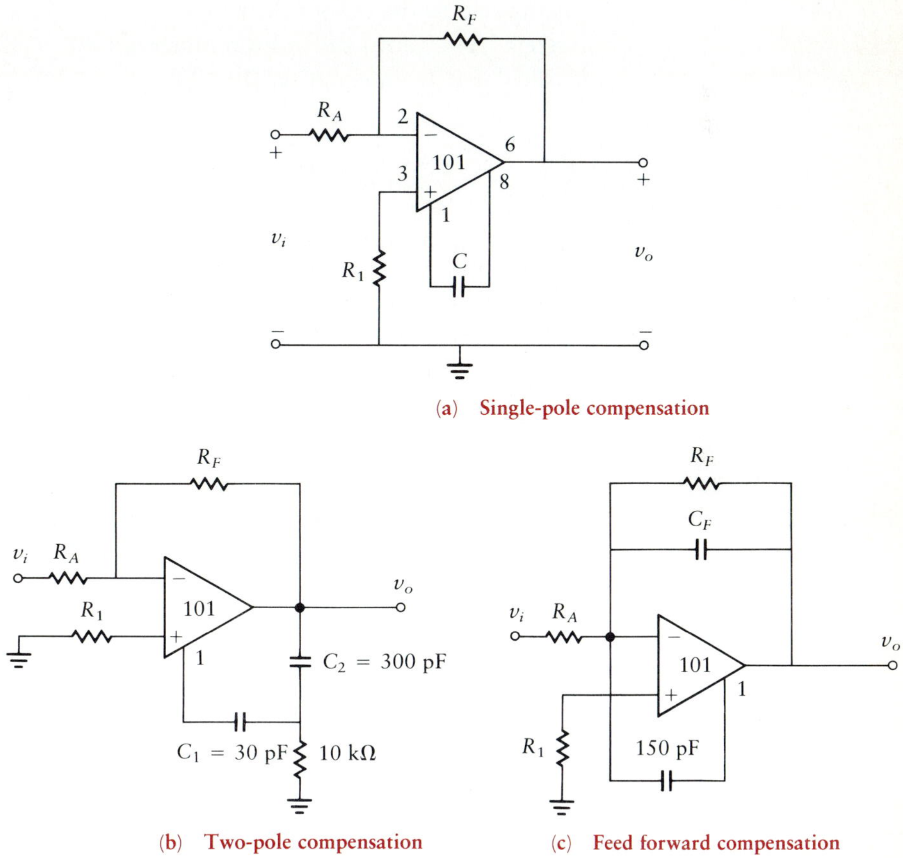

The resulting frequency response is shown in Figure 10.43(b). The figures illustrate straight-line *Bode plot* approximations. If the amplifier has a gain less than +20 dB, 3-pF compensation should not be used.

The circuit of Figure 10.42(b) is used if the compensation is *two-pole* in-

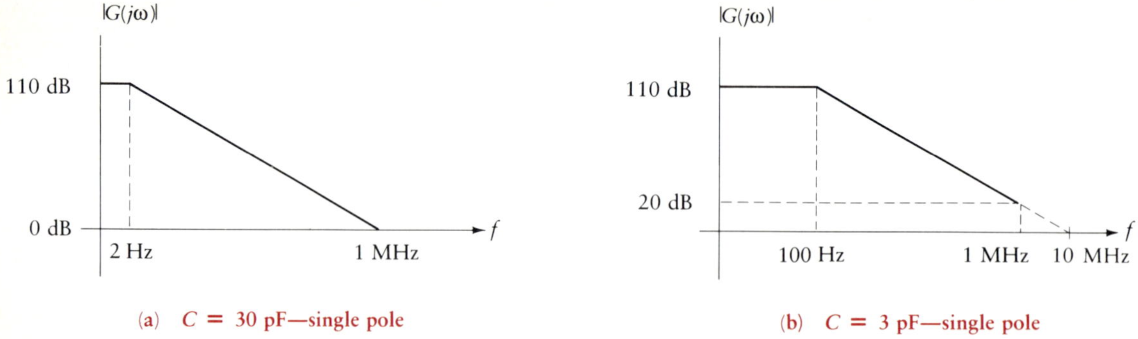

Figure 10.43 101 compensation—single pole.

stead of *one-pole.* The frequency response drops with frequency at twice the slope (−40 dB per decade), as is the case with one-pole compensation.

Figure 10.42(c) illustrates *feedforward compensation.* That is, instead of pin 1 being tied to the output through a capacitor, it is connected to the input. This system has a gain-bandwidth product of GBP = 10^7 Hz. The capacitance, C_F, is calculated from equation (10.40).

$$C_F = \frac{1}{2\pi \times 3 \times 10^6 \times R_F} \tag{10.40}$$

R_F is selected to obtain the desired gain.

10.7.1 Noninverting 101 Amplifiers

Figure 10.44 shows a noninverting amplifier. The output voltage is

$$v_o = \left(1 + \frac{R_F}{R_A}\right)v_i$$

The input resistance and output resistance of this amplifier are found by applying equations (10.15) and (10.12) as follows:

$$R_{\text{in}} = 2R_{cm} \,\Bigg\|\, \frac{R_A G_o R_i}{R_A + R_F} \approx 400 \text{ M}\Omega$$

$$R_{\text{out}} = \frac{R_o}{R_A G_o/(R_F + R_A)} = \frac{100(1 + R_F/R_A)}{10^5}$$

The bandwidth is the range of frequencies over which the gain is within 0.707 (3 dB) of its maximum value. This is derived in equation (10.20). If a 30-pF capacitor is used in single-pole compensation, the bandwidth is given by

$$\text{BW}_{30} = \frac{\text{GBP}_{30}}{1 + R_F/R_A} = \frac{10^6}{1 + R_F/R_A} \tag{10.41}$$

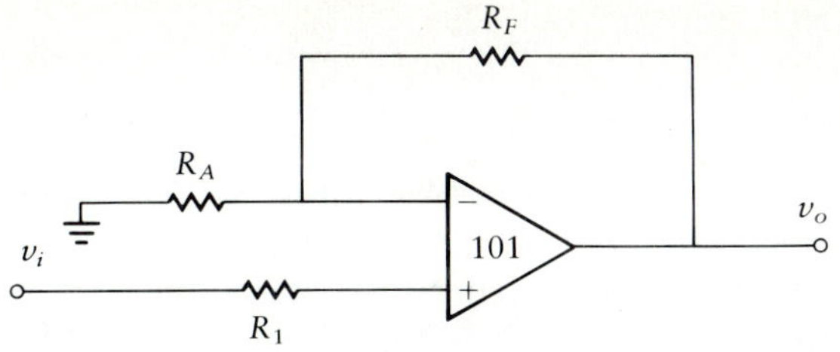

Figure 10.44 Noninverting amplifier.

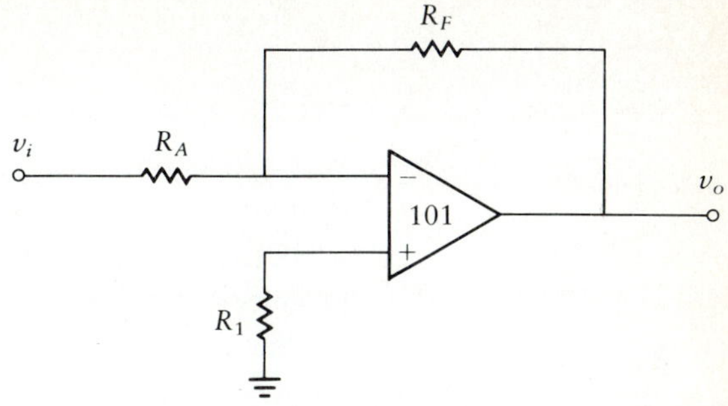

Figure 10.45 101 inverting amplifier.

If a 3-pF capacitor is used in single-pole compensation (or with feedforward compensation) the bandwidth is

$$BW_3 = \frac{GBP_3}{1 + R_F/R_A} = \frac{10^7}{1 + R_F/R_A} \tag{10.42}$$

To achieve bias balance, we set $R_1 = R_A \parallel R_F$.

10.7.2 Inverting 101 Amplifiers

The output voltage for the inverting 101 amplifier of Figure 10.45 is

$$v_o = \frac{-R_F v_i}{R_A}$$

Once again, to achieve bias balance, we choose $R_1 = R_F \parallel R_A$. R_{in} and R_{out} are found as in Section 10.4.1 and are given by

$$R_{in} = R_A$$

$$R_{out} = \frac{R_o(1 + R_F/R_A)}{G_o} = \frac{100(1 + R_F/R_A)}{10^5}$$

With 30-pF single-pole compensation, the bandwidth is given by

$$BW_{30} = \frac{GBP_{30}}{1 + R_F/R_A} = \frac{10^6}{1 + R_F/R_A}$$

This is the same as for the noninverting amplifier of the previous section. With 3-pF single-pole compensation,

$$BW_3 = \frac{10^7}{1 + R_F/R_A}$$

10.8 AMPLIFIERS WITH BALANCED INPUTS OR OUTPUTS

The op-amp systems presented so far in this text are driven by voltages for which one side is grounded. Numerous op-amp applications require that we deal with voltages that are balanced; that is, neither side of the voltage can be grounded. Op-amps are most useful for converting a balanced input to a grounded (or unbalanced) output.

Various configurations of balanced inputs and outputs are illustrated in Figure 10.46. These provide for balancing by cascading two op-amps.

For example, in the configuration shown in Figure 10.46(b), if v_i is supplied from a high-impedance source, the 5-kΩ resistors are no longer useful. Bias current may not be balanced in each amplifier. However, the effects of equal offsets due to unbalanced bias currents in the two input amplifiers should cancel one another. The alternative arrangement shown in the figure avoids this problem.

Figure 10.46 Balanced inputs and outputs.

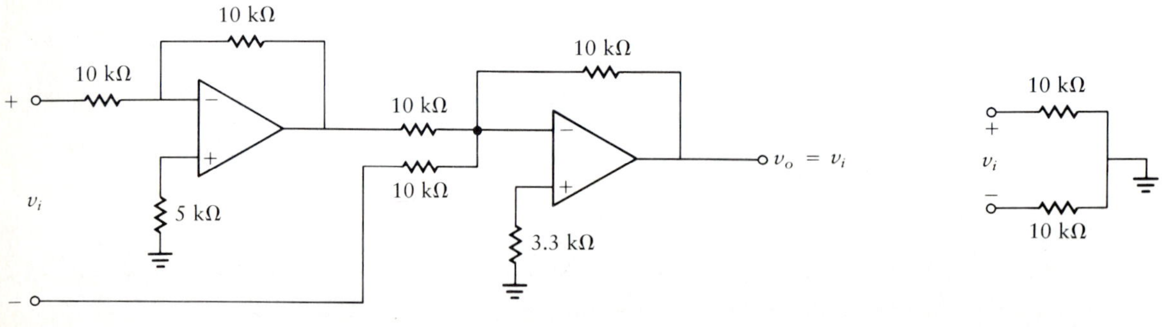

(a) Balanced input, unbalanced output

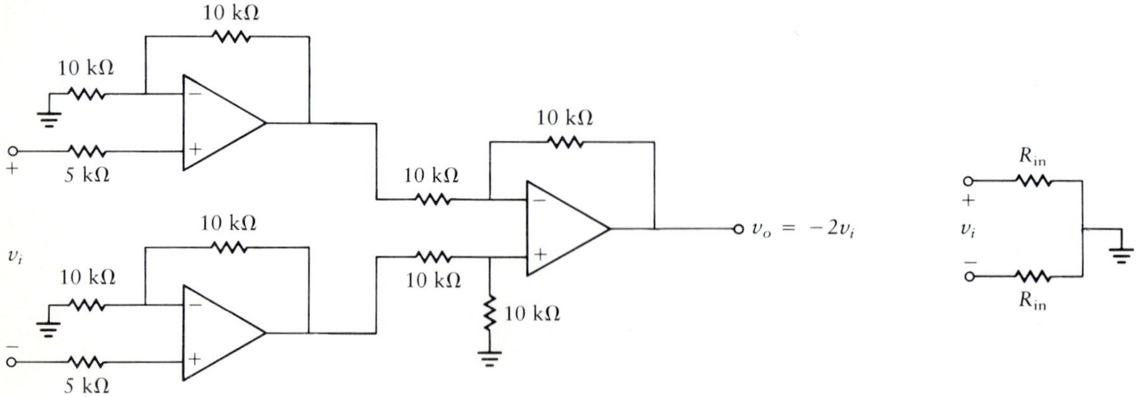

(b) Balanced high-impedance input, unbalanced output

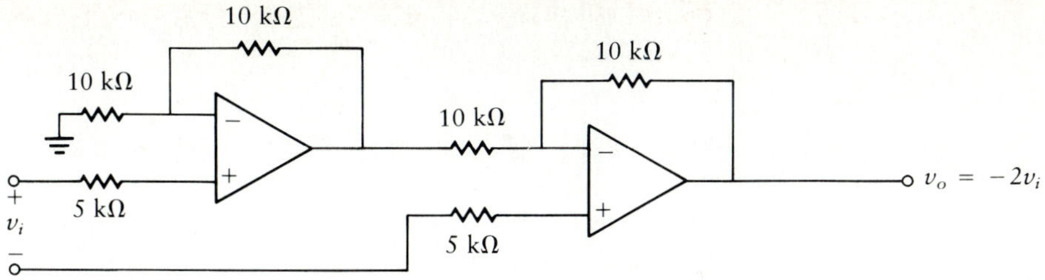

Alternate arrangement for (b)

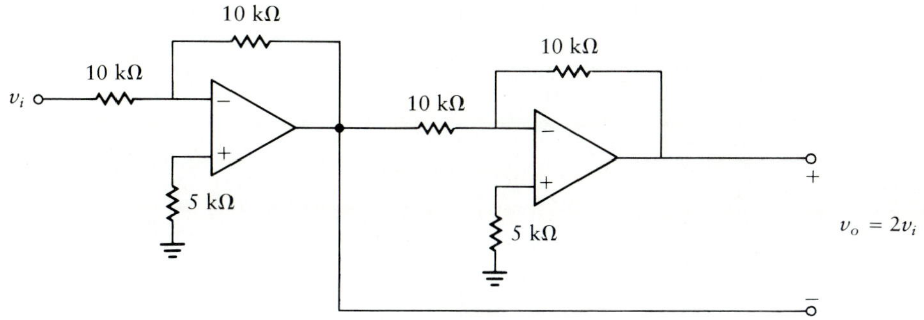

(c) **Unbalanced input, balanced output**

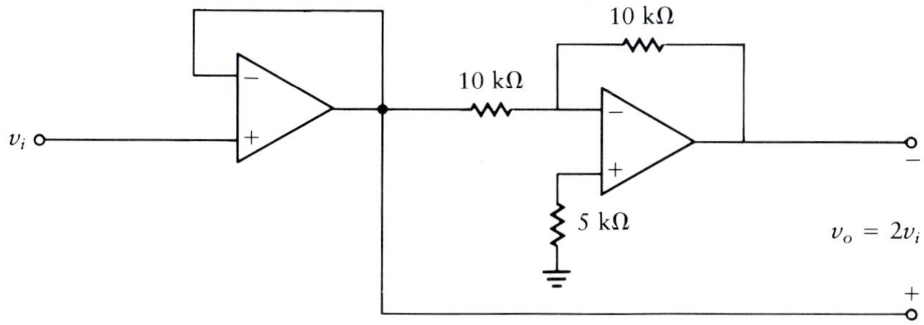

(d) **Unbalanced high-impedance input, balanced output**

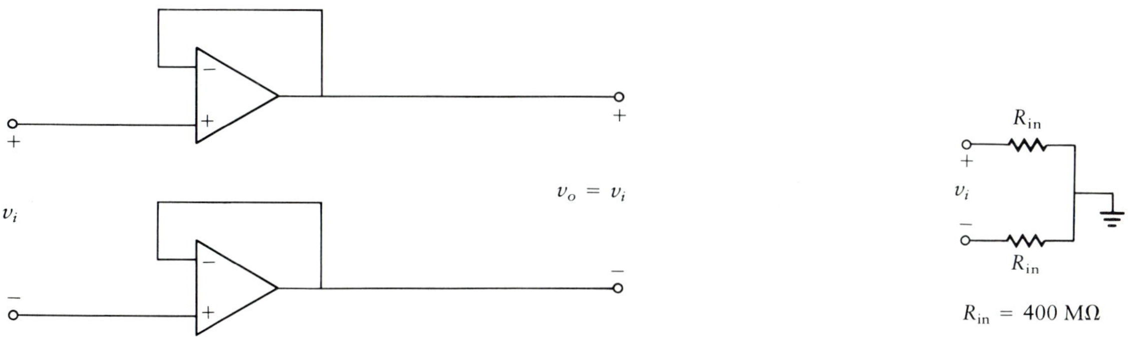

(e) **Balanced high-impedance input, balanced output**

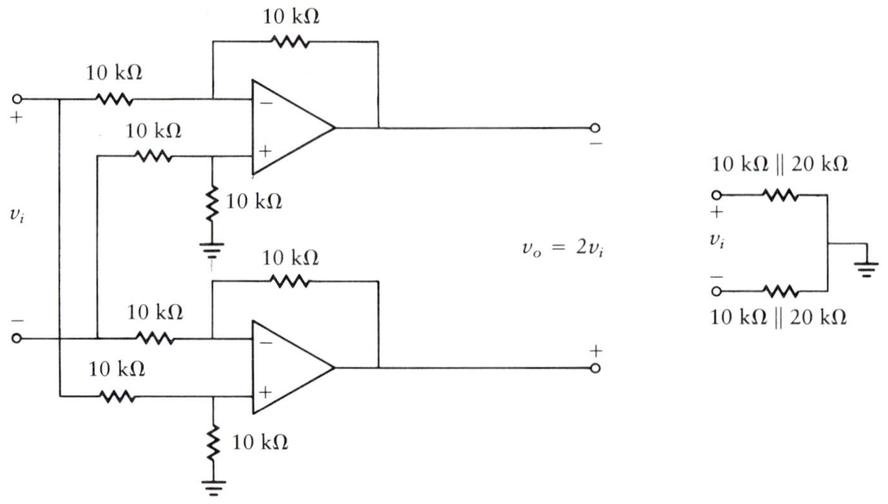

(f) **Balanced input, balanced output**

Example 10.13

Consider the design of an op-amp system to monitor one of the line-to-line voltages on the three-phase Δ-power system that operates at 60 Hz, as shown in Figure 10.47. The voltages

$$|V_{21}| = |V_{32}| = |V_{13}| = 190 \text{ V rms}$$

Let us design an op-amp system to monitor V_{21} (defined as $V_2 - V_1$) and deliver an unbalanced output given by

$$V_o = 0.1V_{21} \text{ rms}$$

SOLUTION We cannot use an op-amp system of the type shown earlier because, in those systems, one terminal is grounded. In this example it is clearly impossible to ground any of the three-phase power lines. Let us use 741 op-amps and be sure that the input impedances to the instrumentation system are greater than 50 MΩ. We connect a noninverting buffer to each input, v_1 and v_2, with the buffered outputs feeding into a differential op-amp system with a gain of 0.1. We use the design method of Section 8.6.

$$v_o = 0.1v_{21} = 0.1(v_2 - v_1)$$

$$X = 0.1, \qquad Y = 0.1, \qquad Z = X - Y - 1 = -1$$

Let us choose $R_F = 10 \text{ k}\Omega$.

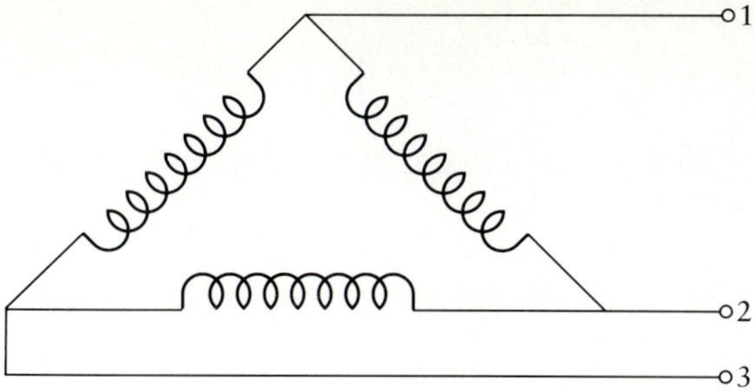

Figure 10.47 Three-phase power system.

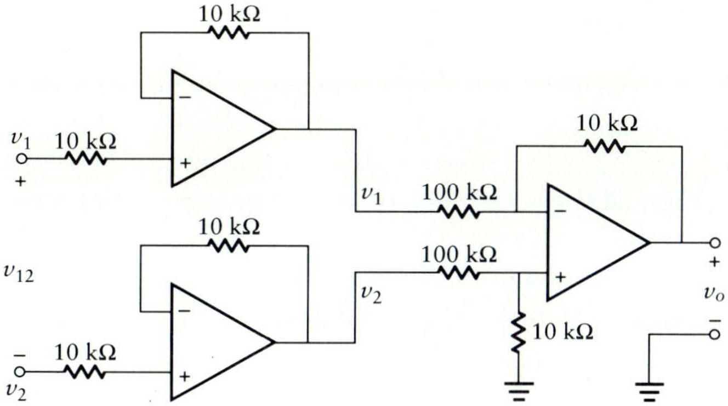

Figure 10.48 System for Example 10.13.

$$R_y \to \infty$$

$$R_x = \frac{R_F}{1} = 10\ \text{k}\Omega$$

$$R_2 = \frac{R_F}{0.1} = 100\ \text{k}\Omega$$

$$R_1 = \frac{R_F}{0.1} = 100\ \text{k}\Omega$$

The system is as shown in Figure 10.48. As a practical consideration, it is necessary to use an optocoupler (see Chapter 3) to isolate the instrumentation from the power lines.

10.9 COUPLING BETWEEN MULTIPLE INPUTS

When more than one input signal is connected to either the inverting or the noninverting input to the op-amp, coupling between the inputs can result. This is frequently a disturbing problem, since a variation in one channel can pro-

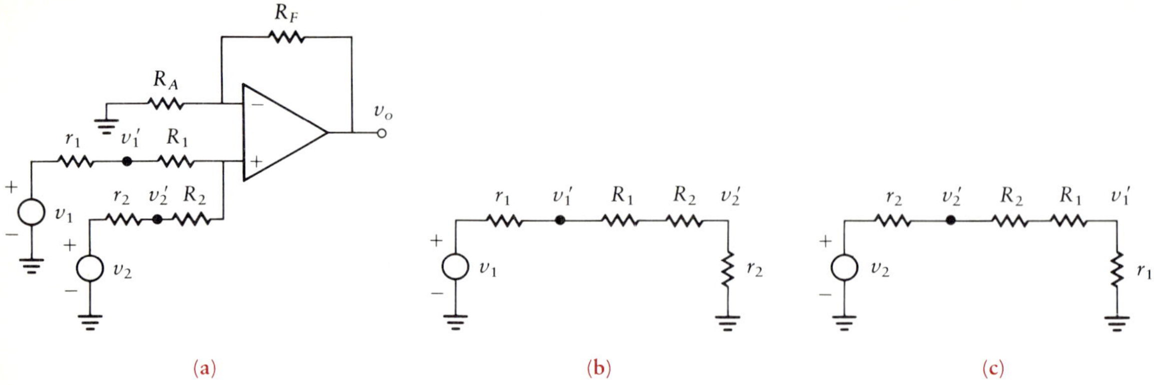

Figure 10.49
Coupling between multiple inputs.

duce an input into an adjoining channel. Consider the dual-input, noninverting op-amp of Figure 10.49(a), where each channel is driven with a voltage source in series with a source resistance (the internal resistance of the voltage source). The ideal source voltages, v_1 and v_2, have series resistances r_1 and r_2, respectively. Let us write the equations for the effective voltages, v_1' and v_2', into the summing amplifier. We assume that no current enters the op-amp. With v_2 set equal to zero, as shown in Figure 10.49(b), the voltage into the op-amp, v_2', is

$$v_2' = v_1 \frac{r_2}{r_1 + r_2 + R_1 + R_2}$$

When $v_1 = 0$, the v_1' voltage is (see Figure 10.49(c)),

$$v_1' = v_2 \frac{r_1}{r_1 + r_2 + R_1 + R_2}$$

Notice that the v_2' voltage comes from v_1 and the v_1' voltage comes from v_2. This coupling effect produces an undesirable *crosstalk* between the two inputs. The effect can be eliminated by designing a system with r_1 and r_2 approaching zero. Hence, to eliminate coupling, each noninverting multiple input should be driven with an op-amp that has zero (or very low) output impedance.

10.10 POWER AUDIO OP-AMPS

A common use for linear amplifiers is to provide gain for audio systems. An *audio amplifier* receives an input signal from a microphone, phonograph cartridge, tape deck, or AM/FM tuner. The output of the amplifier drives a speaker system, headphones, or a tape recorder. The input devices named above usually can be modeled by a voltage source with a low output voltage and high source impedance. Therefore the input impedance of the amplifier

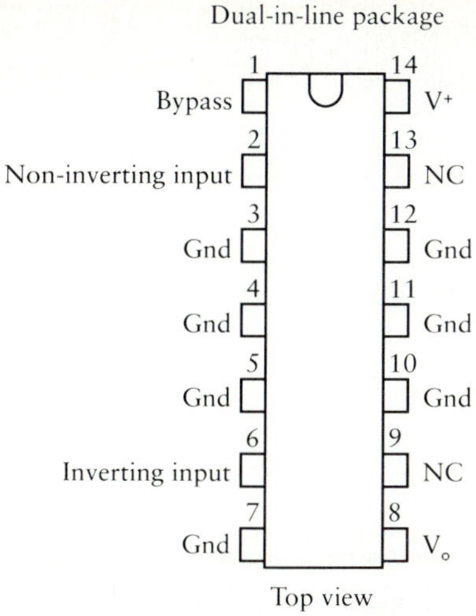

(a) Connection diagram

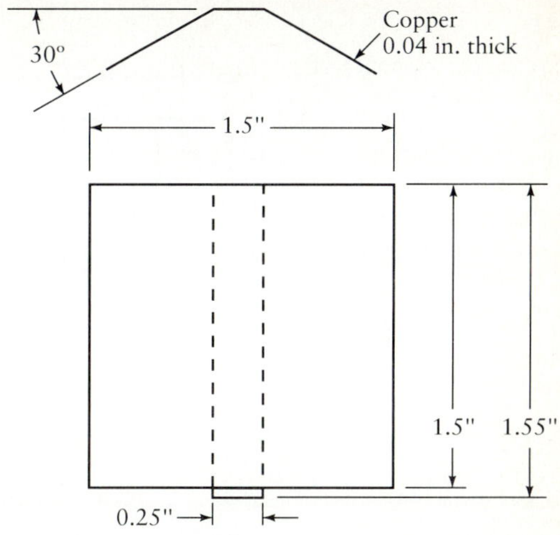

(b) Heat sink

Figure 10.50
The LM380 audio power amplifier.

following this device must be high (much greater than the source impedance of the input device). In this manner, the amplifier does not significantly load the input device and the gain is not decreased.

The devices that are driven by the amplifier usually have low impedance. For example, the impedance of a single speaker is normally 8 Ω. These devices may require powers on the order of 1 to 10 W.

A variety of integrated circuit audio power op-amps, with different output powers, are available to the electronic design engineer. As an example, we present the LM380 audio power amplifier* which is used in such consumer applications as phono and tape deck amplifiers, intercoms, line drivers, alarms, TV sound systems, AM/FM radios, small servo drivers, and power converters. It has an internally fixed gain of 50 (34 dB) and an output that centers itself around one half of the supply voltage. Inputs can be either referenced to ground or balanced. The output stage is protected with both short-circuit current-limiting and thermal shutdown circuitry. The amplifier is packaged in a 14-pin DIP package as shown in Figure 10.50(a). The output current is rated at 1.3 A peak. Since the device shuts down at junction temperatures above 150°C, a heat sink (see Figure 10.50(b)) should be soldered to pins 3, 4, 5 and 10, 11, 12. Maximum output power (with a heat sink) is 3.7 W. The Bode frequency response plot is shown in Figure 10.51. The device is internally biased and compensated.

Figure 10.52 shows the circuit configuration of a complete phono amplifier. A volume and tone control has been included in this circuit.

---

* The data and circuits are printed with the permission of the manufacturer, National Semiconductor Corp. The student is urged to use the data books when designing equipment with power op-amps.

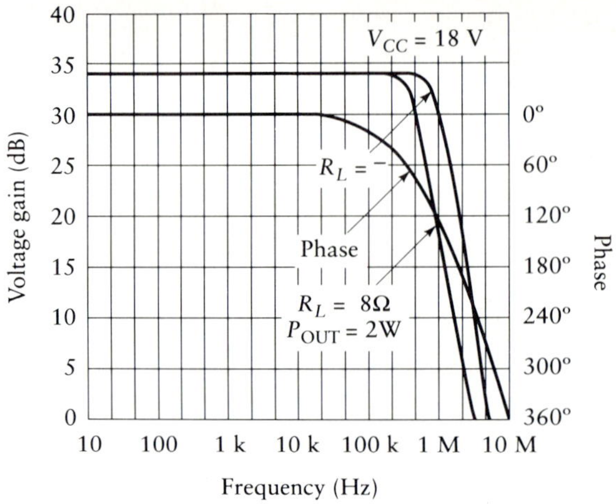

Figure 10.51 Frequency response of the LM380.

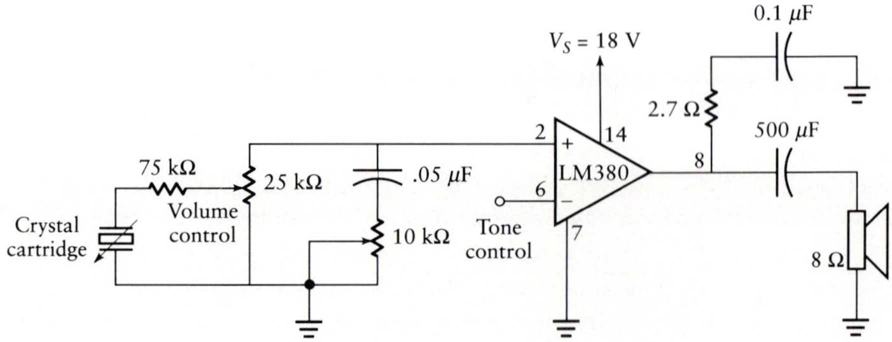

Figure 10.52 Phono amplifier using LM380.

10.10.1 Bridge Power Op-Amp

If a particular application requires more power than can be obtained from a single power op-amp, we can use the bridge configuration of Figure 10.53. Since this system provides twice the voltage swing across the load as the single-device system, the power capability is theoretically increased by a factor of 4 over the single amplifier (for a given power supply voltage). Since heat dissipation is the limiting concern in this design, we usually design the system conservatively and only double the output power.

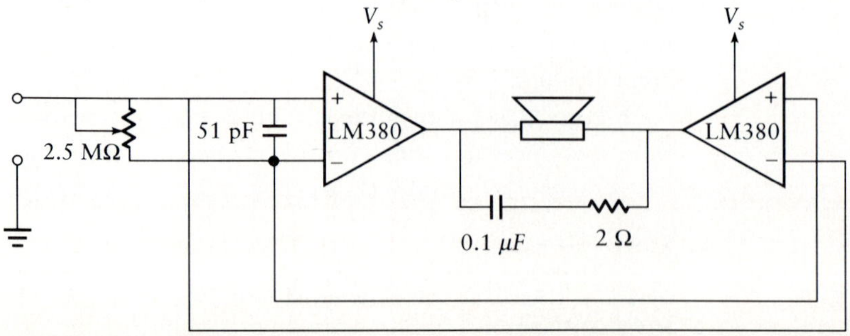

Figure 10.53 Bridge configuration for high power.

10.10.2 Intercom

Figure 10.54 shows an intercom incorporating a power op-amp and a few external components. With the dual two-position switch in the talk position (as shown in the figure), the speaker of the master station performs the function of a microphone, driving the power op-amp through a step-up transformer. The remote speaker is driven from the output of the power op-amp.

Switching to the *listen* position reverses the role of master and remote. Now the *remote* speaker plays the role of the microphone, and it drives the power amplifier through a step-up transformer. The master speaker is now driven from the output of the power op-amp. The student should trace the wiring with the switches in the listen position to verify this. A step-up transformer with a turns ratio of 1:25 can be used, and the potentiometer, R_v, acts as the volume control.

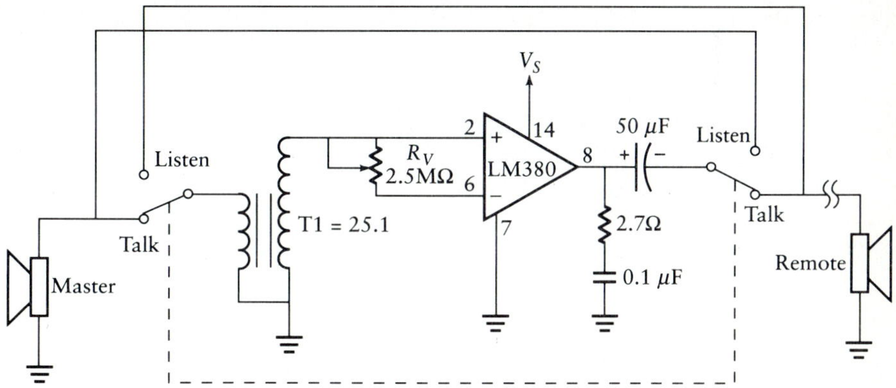

Figure 10.54 Intercom.

PROBLEMS

10.1 Consider the modified op-amp model with

$$G = 10^5$$

$$R_i = 1\ \text{M}\Omega$$

$$R_{cm} = 200\ \text{m}\Omega$$

Solve the network of Figure P10.1 for the resistance, R_o.

10.2 In each of the op-amp circuits of Figure P10.2, $V_{io} = 10$ mV, the input offset voltage temperature coefficient is 10 μV/°C, the temperature is 50°C, and $R_i = 1$ MΩ. Find the largest possible offset in v_o due to V_{io}.

Figure P10.1

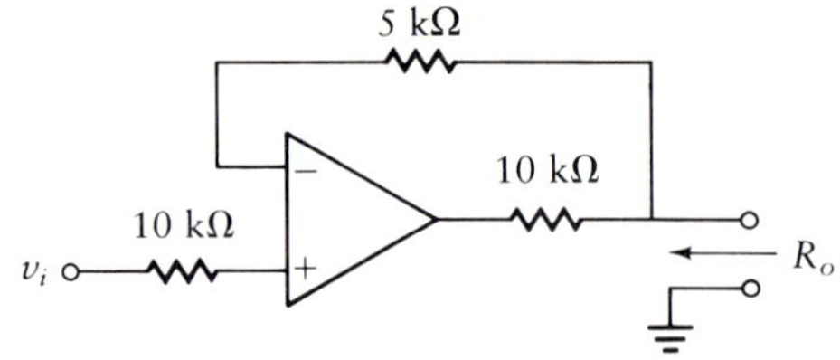

Figure P10.2

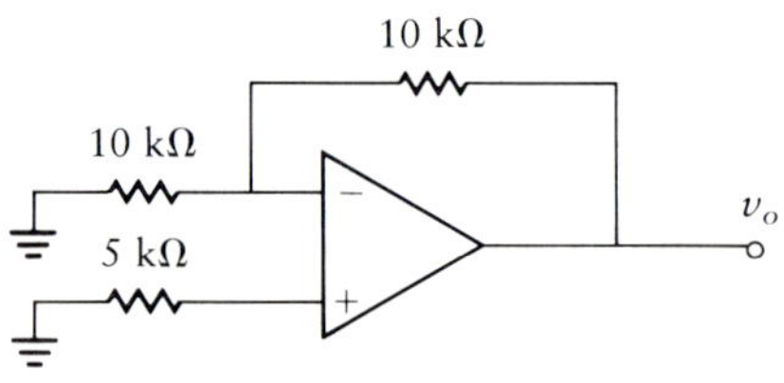

(a)

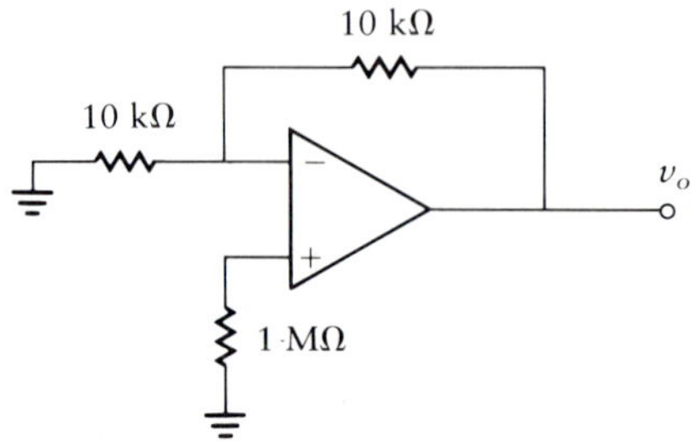

(b)

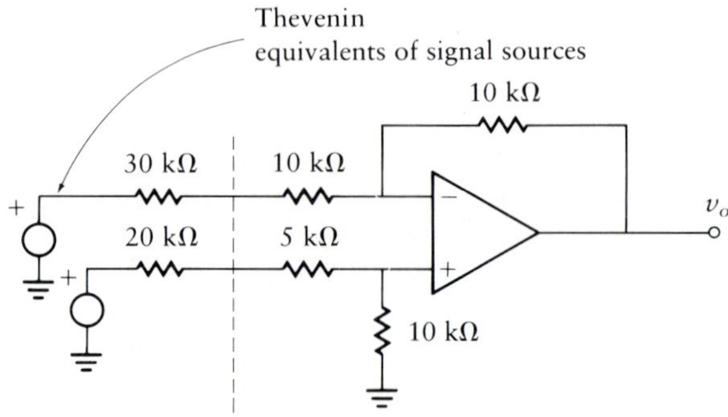

(c)

10.3 For each of the circuits of Figure P10.2, the op-amp input bias current is 800 nA, the bias offset is 20 nA, the input bias current temperature coefficient is −10 nA/°C, and the input offset bias current temperature coefficient is −2 nA/°C. Find:

a. The largest possible offset in v_o due to average bias current effects.
b. The largest possible offset in v_o due to bias offset effects.
c. The largest possible offset in v_o due to voltage offset bias current and bias offset combined.

10.4 Design a single 741 op-amp amplifier that will yield an output given by the equation

$$v_o = 10v_1 + 6v_2 + 4v_3$$

The equivalent resistance at the negative and positive terminals is 10 kΩ. Determine each resistor value, the bandwidth, the input resistance of each amplifier input, and the output resistance.

10.5 Design a single 741 op-amp amplifier that will yield an output given by the equation

$$v_o = -10v_1 - 5v_2 - 4v_3$$

The equivalent resistance at the negative and positive terminals is 10 kΩ. Determine each resistor value, the bandwidth, the input resistance at each amplifier input, and the output resistance.

10.6 Design a single 741 op-amp amplifier that will yield an output given by the equation

$$v_o = 10v_1 + 6v_2 - 3v_3 - 4v_4$$

The equivalent resistance at the negative and positive terminals is 10 kΩ. Determine each resistor value, the bandwidth, the input resistance at each amplifier input, and the output resistance.

In Problems 10.7–10.16, design 741 op-amp systems that will generate the indicated output voltage, v_o, from the input voltages, v_1, v_2, and v_3. Be sure to balance the bias currents in each design. For each design, find the input resistance for each input, the output resistance, and the bandwidth. The input resistance must be at least 100 MΩ and the bandwidth must be greater than 20 kHz. It may be necessary to use more than one op-amp in some of these designs.

10.7 $v_o = 700v_1$

10.8 $v_o = v_1/700$

10.9 $v_o = -700v_1$

10.10 $v_o = -v_1/700$

10.11 $v_o = v_1 - v_2$

10.12 $v_o = 10v_1 - v_2$

10.13 $v_o = v_1 - 10v_2$

10.14 $v_o = v_1 + 700v_2$

10.15 $v_o = -(v_1 + 700v_2)$

10.16 $v_o = v_1 - 2v_2 + 3v_3$

In Problems 10.17–10.22, design multiple 741 op-amp systems to develop the indicated output voltage, v_o, from the input voltages, v_1, v_2, and v_3. The input resistance to each input must be 100 MΩ or greater, and the inputs should not be directly coupled to one another. For each design, find R_{in}, R_{out}, and the bandwidth. A minimum bandwidth of 50 kHz must be achieved.

10.17 $v_o = 3(v_1 + v_2)$

10.18 $v_o = 3(v_1 - v_2)$

10.19 $v_o = 1000v_1 - 300v_2$

10.20 $v_o = 500v_1 - 50v_2$

10.21 $v_o = 70v_1 - \frac{1}{7}v_2$

10.22 $v_o = 100v_1 + 50v_2$

In Problems 10.23–10.26, design single-amplifier 101 circuits that will generate the indicated output voltage, v_o, from the input voltages, v_1 and v_2. Use a 3-pF compensating capacitor wherever possible. Be sure to balance the bias currents in each input. Calculate R_{in} and R_{out}. The minimum bandwidth is 20 kHz.

10.23 $v_o = 500v_1$

10.24 $v_o = 10v_1 - v_2$

10.25 $v_o = 20v_1 + 30v_2$

10.26 $v_o = v_1 - 15v_2$

In Problems 10.27–10.29, design multiple 101 operational amplifier systems to develop the voltage, v_o, from the input voltages, v_1 and v_2. The input resistance into each input must be at least 100 MΩ and the inputs should not be directly coupled to one another. The bandwidth must be greater than 20 kHz using 30-pF or 3-pF 101 compensation.

10.27 $v_o = 10(v_1 + v_2)$

10.28 $v_o = 10v_1 - v_2$

10.29 $v_o = 1000v_1 - 300v_2$

10.30 Design a system using 741 op-amps that, from the *differential* voltages, v_1 and v_2, develops an output voltage

$$v_o = 100v_1 + 50v_2$$

Input resistances should be balanced to ground and should be in excess of 100 MΩ. The minimum bandwidth must be 50 kHz.

10.31 Design a system using 741 op-amps that, from voltages v_1 and v_2, develops a *balanced* output voltage

$$v_o = 100v_1 - 50v_2$$

The output resistances should be less than 1 Ω.

10.32 Design a system using 741 op-amps that has a balanced input, a balanced output, a gain of 100, input impedance greater than 100 MΩ, and output impedance less than 1 Ω.

10.33 Two differential voltages, v_A and v_B, each of which is balanced with respect to ground, are available as inputs. The v_A source has a Thevenin resistance between 10 and 210 kΩ, whereas the v_B source has a Thevenin resistance between 50 and 150 kΩ. Design a multiple 741 system to generate the voltage

$$v_o = 10(v_A - v_B)$$

with respect to ground without coupling between the two input sources. The bias current balance should be as good as is possible.

10.34 Design an op-amp system to measure the output current of a power supply, as shown in Figure P10.3. The voltage across a precision 1-Ω resistor is used to measure the current out of the power supply. This output current varies from 75 to 125 mA. The digital voltmeter (DVM) input must be in the range of 0.75 to 1.25 V.

10.35 A balanced three-phase wye power system operates at 50 Hz. The magnitude of the line-to-line voltages is $V_{21} = V_{32} = V_{13} = 110$ V rms. Design an instrumentation system to monitor the voltage, V_{21}, as shown in Figure P10.4 and to yield an output voltage, V_o, given by the equation

$$V_o = 0.1V_{21}$$

Use 101 op-amps and select the proper value of compensating capacitor (30 pF or 3 pF). R_{in} must be greater than 100 MΩ. Select the proper value of V_{CC} for the 101 op-amps.

Figure P10.3

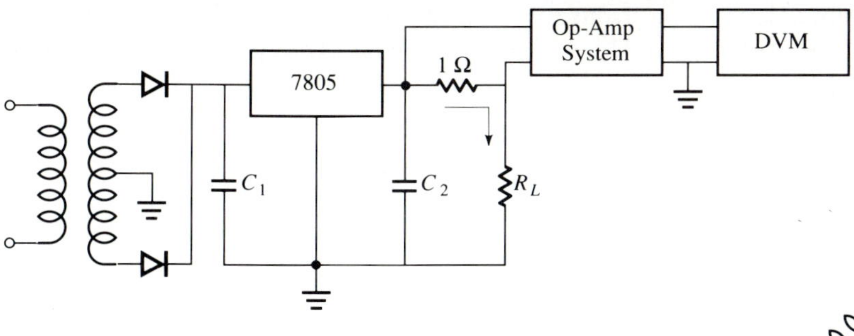

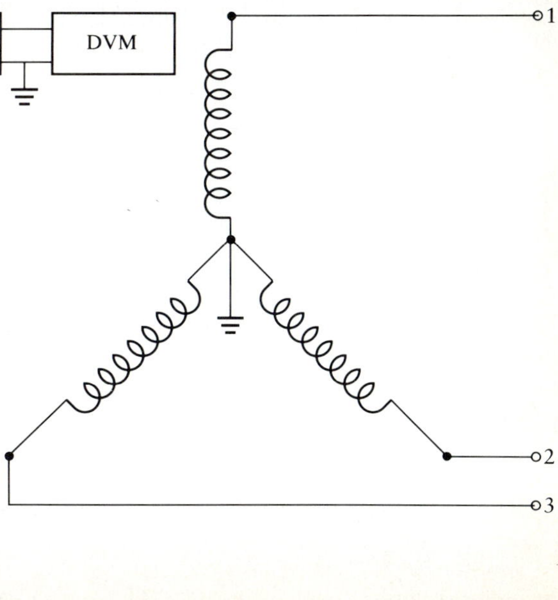

Figure P10.4

11

FEEDBACK AND STABILITY

11.0 INTRODUCTION

You have studied hard, learned well, and can now design amplifiers using BJTs, FETs, and ICs. You have designed many such amplifiers for the XYZ Audio Company and have been promoted to group leader. A new engineer has just joined your group, and you gave that person the assignment of designing a simple audio amplifier. The new engineer was knowledgeable regarding frequency response, and the resulting amplifier produced distortionless sound. However, one day during a demonstration for a prospective customer, a loud whistling sound came from the speaker. Instead of simply amplifying the audio signal, the system had become an oscillator! Pity! Both you and your young charge had failed to perform a stability analysis of the amplifier system. Heed well what follows in this chapter, and you can avoid such embarrassing situations!

Feedback exists when the output of a system is connected to the input in such a way that waveforms appearing at the output affect the input signal.

We have already seen several forms of feedback. Transistor amplifiers often include an emitter resistor. If that emitter resistor is left unbypassed, a form of feedback occurs since a part of the output voltage is subtracted from the input voltage and thereby affects the base-to-emitter voltage. The same current flows through the load resistor, R_L, as flows through the emitter resistor, R_E, thus causing the feedback.

Previous chapters dealt with op-amp circuits, most of which include feedback between output and input. This feedback has the desirable effect of reducing the sensitivity of the overall amplifier to changes in the op-amp parameters. Most notably, the gain of the open-loop op-amp (without feedback) varies significantly with changes in frequency, whereas the overall gain of the feedback amplifier is much less sensitive to frequency changes. The advantages of feedback are summarized as follows:

1. Closed-loop gain, although usually lower than that without feedback, is relatively independent of the variation of device parameters.
2. Input and output resistances of the closed-loop system are controlled.
3. Amplifier bandwidth is increased.
4. Nonlinearities and distortion are reduced.
5. Unwanted internally generated noise signals are reduced. (See Appendix E for a discussion of noise and noise generators.)

In this discussion, we consider *negative feedback*. This occurs when the portion of the output that is fed back to the input is subtracted from that input. *Positive feedback,* in which the signal fed back from the output is *added* rather than *subtracted* from the input, is frequently used in electronic circuits. Such feedback is intentionally introduced in the design of *oscillators*. The amplifier becomes unstable and, even in the absence of an input signal, begins to oscillate, producing a periodic output signal. This chapter develops the mathematical tools for analyzing the effects of feedback on electronic systems. We consider both discrete and linear ICs. We use the Bode method to determine the stability of the feedback system.

11.1 FEEDBACK AMPLIFIER CONSIDERATIONS

Figure 11.1 shows a model of a feedback system. This model can be used to represent an amplifier or any other feedback system. The circle denotes a summing operation. The signal leaving the circle, ϵ, is equal to the difference between the two signals entering. Thus,

$$\epsilon = R + (-Y) = R - Y$$

Each of the parameters in the above equation can represent a time function, a Laplace transform, or a complex phasor as used in the case of sinusoidal inputs. For the present analysis, we use Laplace transform notation (a review of the Laplace transform method is presented in Appendix C).

The system of Figure 11.1 is a *negative-feedback* system, since the signal driving the forward loop block, $G(s)$, is equal to the input signal, $R(s)$, *minus* some function of the output signal, $C(s)$. These quantities are functions of the Laplace transform variable, s, so are frequency dependent. In this case,

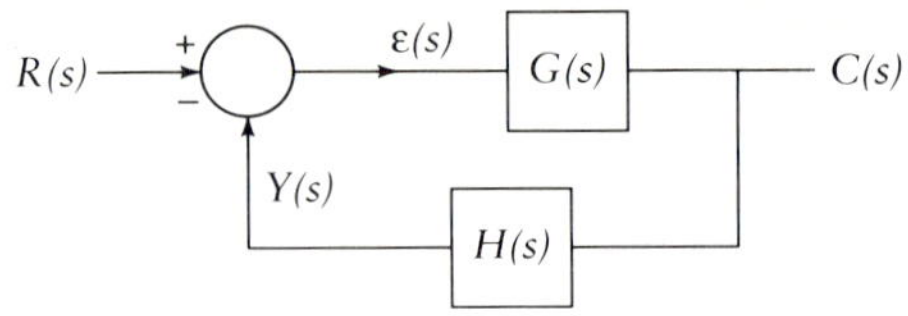

Figure 11.1
Typical closed-loop system.

$$Y(s) = H(s)C(s)$$

and

$$C(s) = G(s)\epsilon(s)$$

$H(s)$ and $G(s)$ are transfer functions, which represent the ratio of the Laplace transform of the output of a given block to the Laplace transform of the input to the block. The *closed-loop* transfer function, $C(s)/R(s)$, is derived as follows:

$$\epsilon(s) = R(s) - H(s)C(s)$$

and

$$\begin{aligned} C(s) &= G(s)\epsilon(s) \\ &= G(s)R(s) - G(s)H(s)C(s) \end{aligned}$$

Therefore,

$$\frac{C(s)}{R(s)} = \frac{G(s)}{1 + G(s)H(s)} \tag{11.1}$$

11.2 TYPES OF FEEDBACK

Feedback exists when a portion of the output is connected to the input through a circuit. Since both the output and input can be characterized by either a voltage or a current, there are four possible forms of feedback. For example, the output voltage can be fed back to the input in the form of a voltage signal that then subtracts from the input voltage signal. Similarly, either or both of these voltage parameters can be replaced with current parameters. This gives rise to the following four forms of feedback, where we include impedance characteristics.

1. *Voltage feedback–voltage subtraction:*
   Input is a voltage.
   Output is a voltage.
   Input impedance is high.
   Output impedance is low.
   Voltage is fed back as in Figure 11.2(a).
2. *Current feedback–current subtraction:*
   Input is a current.
   Output is a current.
   Input impedance is low.
   Output impedance is high.
   Current is fed back as in Figure 11.2(b).
3. *Voltage feedback–current subtraction:*
   Input is a current.
   Output is a voltage.
   Input impedance is low.
   Output impedance is low.
   Current is fed back as in Figure 11.2(c).
4. *Current feedback–voltage subtraction:*
   Input is a voltage.
   Output is a current.
   Input impedance is high.
   Output impedance is high.
   Voltage is fed back as in Figure 11.2(d).

Figure 11.2 Four types of feedback.

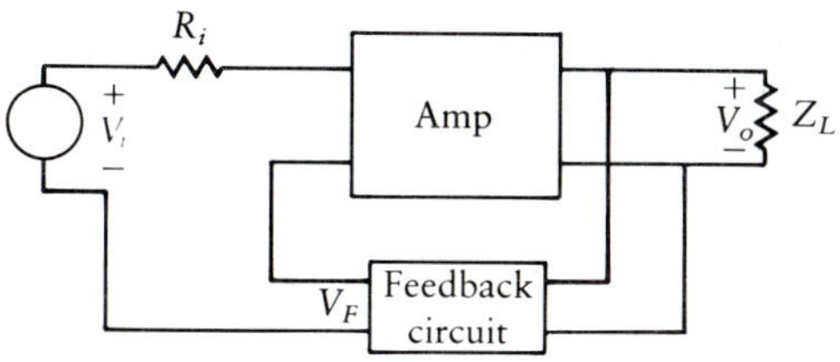

(a) Voltage feedback—voltage subtraction

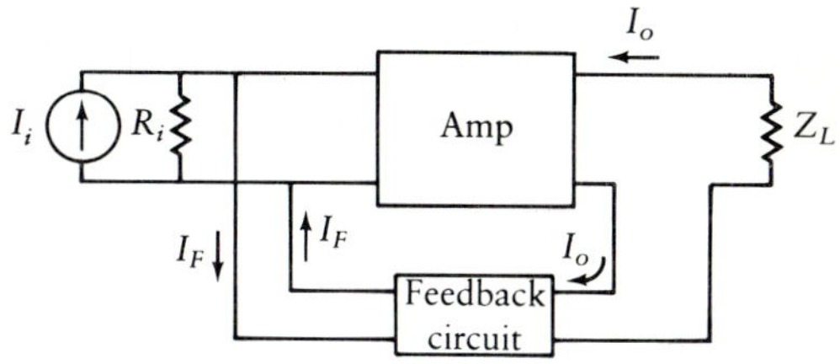

(b) Current feedback—current subtraction

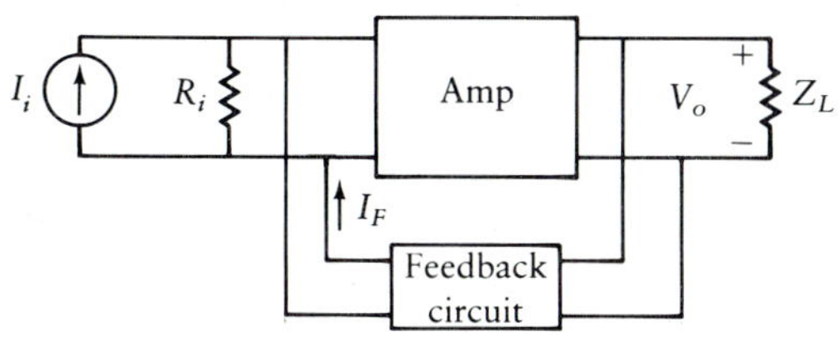

(c) Voltage feedback—current subtraction

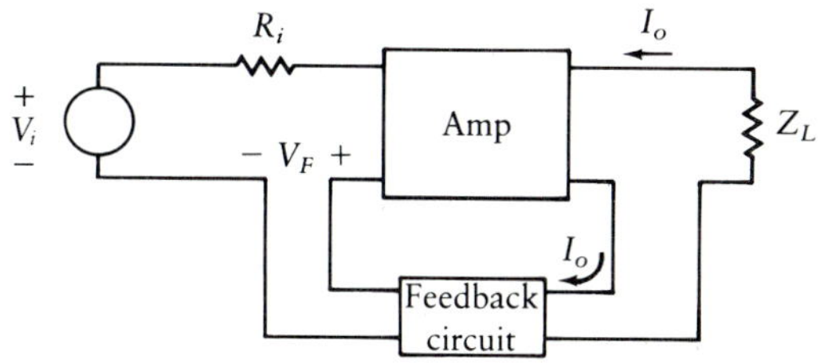

(d) Current feedback—voltage subtraction

11.3 FEEDBACK AMPLIFIERS

In this section, we discuss examples of both voltage and current feedback. In most cases, such as shown in Figure 11.2, the amplifier networks consist of active elements and the feedback networks consist of passive elements. The feedback networks can be either resistive or frequency sensitive. In the frequency-sensitive feedback networks, the amount of feedback is determined by the frequency of the input signals. In other words, as the frequency changes, the amount of feedback changes.

As indicated in the previous summary, negative feedback either increases or decreases the input impedance of an amplifier, depending on whether the feedback network is in series or in parallel with the input signal path.

11.3.1 Current Feedback—Voltage Subtraction for Discrete Amplifiers

Consider a common-emitter BJT amplifier with an emitter resistor that is not bypassed. Therefore, a voltage proportional to the output current is subtracted from, or fed back to, the input signal voltage. This form of feedback is illustrated in Figure 11.2(d).

Figure 11.3 shows the common-emitter circuit with a bypass capacitor. We can use this circuit to consider both feedback and no feedback simply by varying the value of the bypass capacitance. That is, as the capacitor value approaches zero, the capacitor approaches an open circuit and we have the feedback situation. As the capacitance approaches infinity, the capacitor becomes a short circuit and the amplifier has no feedback. Figure 11.4 shows the ac equivalent circuit of the amplifier.

The voltage gain for the CE amplifier is calculated from this equivalent circuit as in Chapter 3. We present the result of the derivation in equation (11.2).

$$A_v = \frac{V_o}{V_i} = \frac{-\beta(R_C \parallel R_L)}{h_{ie} + \beta(R_E \parallel 1/sC)} = \frac{-(R_C \parallel R_L)}{h_{ib} + (R_E \parallel 1/sC)} \tag{11.2}$$

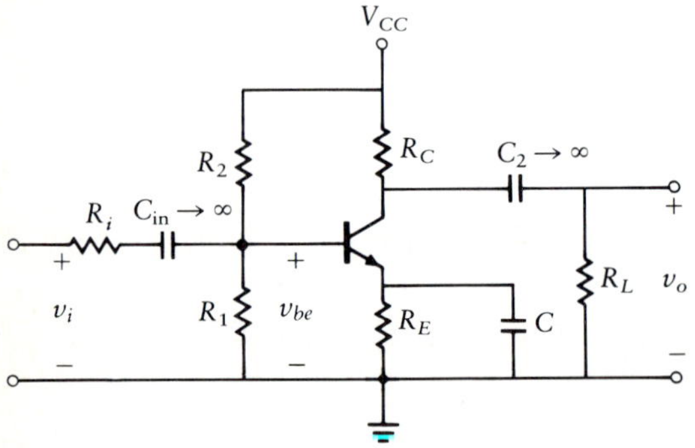

Figure 11.3 Common-emitter stage.

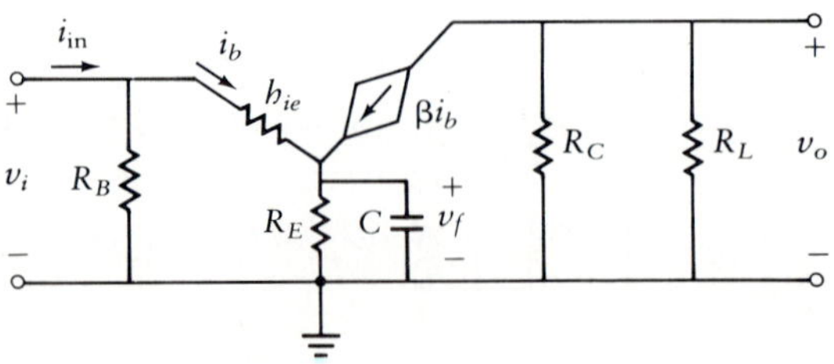

Figure 11.4 Equivalent circuit for CE stage.

The voltage gain with feedback is known as the *closed-loop gain* and is found by letting C approach zero in equation (11.2). Also assume that $R_E >> h_{ib}$. Equation (11.2) for the closed-loop gain then reduces to

$$A_v = \frac{-(R_C \parallel R_L)}{R_E}$$

The voltage gain without feedback is known as the *open-loop gain*, A_{vo}, and is found by letting C approach infinity in equation (11.2).

$$A_{vo} = \frac{-(R_C \parallel R_L)}{h_{ib}} = \frac{-\beta(R_C \parallel R_L)}{h_{ie}}$$

As an example, we calculate the open-loop and closed-loop gain for the CE amplifier with the following component values: $R_C = R_L = 1\ \text{k}\Omega$, $R_E = 100\ \Omega$, $h_{ie} = 600\ \Omega$, and $\beta = 300$. The gain with no feedback (open loop) is given by

$$A_{vo} = \frac{-\beta(R_C \parallel R_L)}{h_{ie}} = -250$$

The gain with feedback (closed loop) is

$$A_v = \frac{-(R_C \parallel R_L)}{R_E} = -5$$

As can be seen, A_v is much smaller than A_{vo} since the feedback considerably reduces the gain. Note that the equation for gain with feedback does not include any transistor parameter values. Thus, the gain with feedback depends only on the ratio of resistance values. This is the desired case. The higher gain, obtained when the capacitance approaches infinity, is dependent on the transistor parameters β and h_{ie}. In Chapter 5 we showed that these parameters can vary considerably.

To highlight the negative feedback, let us write the voltage around the base-emitter loop of Figure 11.4. The voltage, v_{be}, is given by

$$v_{be} = v_i - i_c R_E \tag{11.3}$$

Equation (11.3) shows that the voltage driving the transistor, v_{be}, is reduced by a signal proportional to the output current, i_c. We now return to equation (11.2) and rewrite it in a form that emphasizes the feedback, that is,

$$A_v = \frac{-R_C \parallel R_L}{h_{ib}} \frac{1}{1 + (R_E \parallel 1/sC)/h_{ib}} \tag{11.4}$$

We define γ as the ratio of feedback voltage, V_f, to output voltage, V_o, where V_f is defined in Figure 11.4 as the voltage across ($R_E \parallel 1/sC$). This *feedback attenuation factor* corresponds to $H(s)$ in Figure 11.1. Note that the symbol β is often used for the feedback factor. We have chosen γ so as not to confuse this with the current gain of a transistor. Therefore,

$$\gamma = \frac{V_f}{V_o} = \frac{-R_E \parallel (1/sC)}{R_C \parallel R_L}$$

Recall that the gain without feedback, or the open-loop gain, is

$$A_{vo} = \frac{-(R_C \parallel R_L)}{h_{ib}}$$

Using the equations for A_{vo} and for γ, we rewrite equation (11.4) in the simplified form

$$A_v = \frac{A_{vo}}{1 + \gamma A_{vo}} \tag{11.5}$$

This equation is now in the same form as the general feedback equation, equation (11.1). Keep in mind that both A_{vo} and γ are negative, so the denominator in equation (11.5) contains positive quantities.

Example 11.1

In a given feedback amplifier, $A_{vo} = -80$ and the ratio of feedback voltage to output voltage is $\gamma = -0.1$. What is the closed-loop voltage gain of the amplifier? Now suppose that the gain without feedback increases by 25%. Find the new overall gain.

SOLUTION The closed-loop gain is given by equation (11.5).

$$A_v = \frac{-80}{1 + 0.1 \times 80} = -8.89$$

We now let A_{vo} change from -80 to -100. The new closed-loop gain is

$$A_v = \frac{-100}{1 + 0.1 \times 100} = -9.09$$

Thus, a 25% change in the open-loop gain results in only a 1.1% change in the closed-loop gain.

Example 11.1 considers a specific example of the *sensitivity* of an amplifier to changes in gain. This sensitivity is determined in general by differentiating equation (11.5) with respect to the open-loop gain. Thus,

$$\frac{dA_v}{dA_{vo}} = \frac{1}{(1 + \gamma A_{vo})^2}$$

If we now divide this derivative by A_v as in equation (11.5), we have

$$\frac{dA_v}{A_v} = \frac{1}{1 + \gamma A_{vo}} \frac{dA_{vo}}{A_{vo}} \approx \frac{1}{\gamma A_{vo}} \frac{dA_{vo}}{A_{vo}} \tag{11.6}$$

since $\gamma A_{vo} >> 1$.

The expression on the left side of equation (11.6) is the fractional change in closed-loop gain and is therefore approximately equal to the fractional change in open-loop gain divided by γA_{vo}. Of course, since γA_{vo} is usually a large number, the variation in closed-loop gain is greatly reduced for a given variation in open-loop gain.

Drill Problem

D11.1 Calculate γ for the amplifier stage of Example 11.1.

Ans: $\gamma = -0.2$

11.3.2 Voltage Feedback—Current Subtraction for a Discrete Amplifier

Figure 11.5(a) illustrates an amplifier in which the output voltage is fed back to the input through a resistor, R_F. Since we are dealing with voltage feedback–current subtraction, the source should be a current source. The figure shows a voltage, v_i, in series with a source resistance, R_i. This represents the Thevenin equivalent of the current source, where $v_i = i_{in} R_i$. The equivalent circuit is shown in Figure 11.5(b). Here we bypass R_E with a large capacitor to eliminate the type of feedback presented in Section 11.3.

We develop equations for the gain of this circuit in the form of equation (11.5). We find the output voltage by assuming that the current through R_F is negligible with respect to βi_b. Thus,

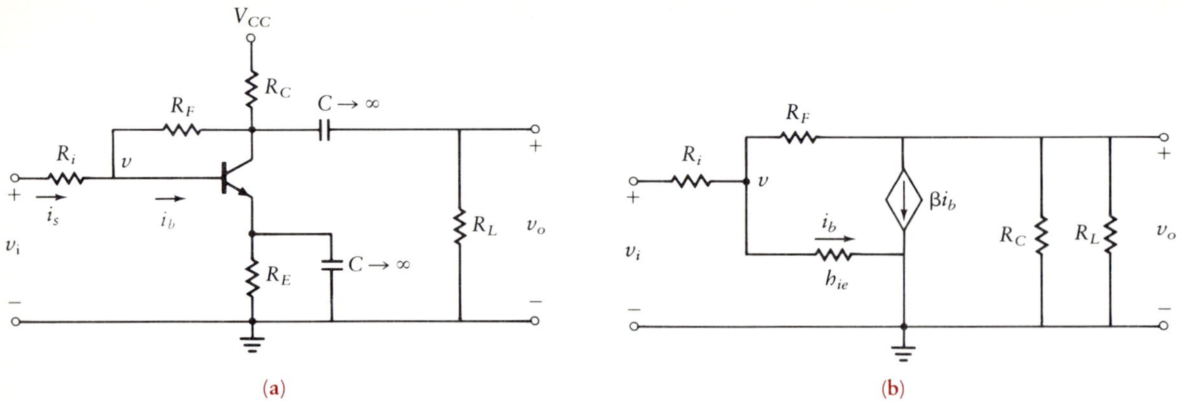

Figure 11.5 Voltage feedback–current subtraction.

$$v_o = -(R_C \parallel R_L)\beta i_b \tag{11.7}$$

The current, i_b, is found by writing a nodal equation at the node labeled v in Figure 11.5(b).

$$i_b = \frac{v_i - v}{R_i} + \frac{v_o - v}{R_F} = \frac{v_i}{R_i} + \frac{v_o}{R_F} - \frac{v}{R_i \parallel R_F}$$

Now since $i_b = v/h_{ie}$, we can multiply through by h_{ie} and solve for v as follows:

$$v\left(1 + \frac{h_{ie}}{R_i \parallel R_F}\right) = \left(\frac{v_i}{R_i} + \frac{v_o}{R_F}\right)h_{ie}$$

This simplifies to

$$\frac{v}{h_{ie}} = i_b = \frac{v_i/R_i + v_o/R_F}{\alpha} \tag{11.8}$$

where, to simplify the algebra, we define

$$\alpha = 1 + \frac{h_{ie}}{R_i \parallel R_F}$$

We now substitute equation (11.8) into equation (11.7) to obtain the output voltage:

$$v_o = \frac{-\beta(R_C \parallel R_L)}{\alpha}\left(\frac{v_i}{R_i} + \frac{v_o}{R_F}\right) \tag{11.9}$$

The voltage gain is found by dividing v_o in equation (11.9) by v_i to obtain

$$A_v = \frac{v_o}{v_i} = \frac{-\beta(R_C \parallel R_L)/\alpha R_i}{1 + \beta(R_C \parallel R_L)/R_F\alpha} \tag{11.10}$$

If R_F approaches infinity, we have the situation with no feedback, and A_v of equation (11.10) becomes

$$A_{vo} = \frac{-\beta(R_C \parallel R_L)}{\alpha_\infty R_i} \tag{11.11}$$

The value of α_∞ in equation (11.11) is the value that applies when R_F approaches infinity. That is,

$$\alpha_\infty = 1 + \frac{h_{ie}}{R_i}$$

Suppose now that R_F does not approach infinity (i.e., we have feedback) but, as is often the case, we can assume $R_i << R_F$. Under this assumption, α is still approximated by the above equation, and the gain of equation (11.10) reduces to

$$A_v = \frac{A_{vo}}{1 - R_i A_{vo}/R_F} = \frac{A_{vo}}{1 + \gamma A_{vo}} \tag{11.12}$$

where $\gamma = -R_i/R_F$. We have again reduced the result to the form of equation (11.2), the general feedback relationship. As A_{vo} becomes large, A_v approaches $1/\gamma = -R_F/R_i$, which is the closed-loop gain of the amplifier.

Example 11.2

In the single-stage amplifier of Figure 11.5(a), let $R_C = R_L = 1\ \text{k}\Omega$, $R_E = 100\ \Omega$, $h_{ie} = 800\ \Omega$, $\beta = 300$, $R_i = 1\ \text{k}\Omega$, and $R_F = 10\ \text{k}\Omega$. Determine the open- and closed-loop voltage gain of this amplifier.

SOLUTION We first solve for α.

$$\alpha = 1 + \frac{h_{ie}}{R_i} = 1.8$$

Now from equation (11.11) we have $A_{vo} = -83.3$. This is the voltage gain for no feedback (open loop). The gain with feedback (closed loop) is given by equation (11.12).

$$A_v = \frac{-83.3}{1 - (1000/10^4)(-83.3)} = -8.93$$

Although the gain, A_v, is less, the sensitivity to a change in A_{vo} is reduced by 0.11. We can use the sensitivity expressions of equation (11.6) to find

$$\frac{dA_v}{A_v} = \frac{1}{1 + (10^3/10^4)83.3}\frac{dA_{vo}}{A_{vo}} = 0.11\frac{dA_{vo}}{A_{vo}}$$

11.4 MULTISTAGE FEEDBACK AMPLIFIERS

The performance of an amplifier can be altered by the use of feedback whether the amplifier consists of one stage or many stages. Since we desire negative feedback, the number of negative-gain amplifier stages must be odd, so there is an odd number of polarity changes.

An example of a feedback amplifier with an odd number of polarity changes is shown in Figure 11.6(a). Increasing the number of stages before feedback

Figure 11.6 Multistage feedback amplifier.

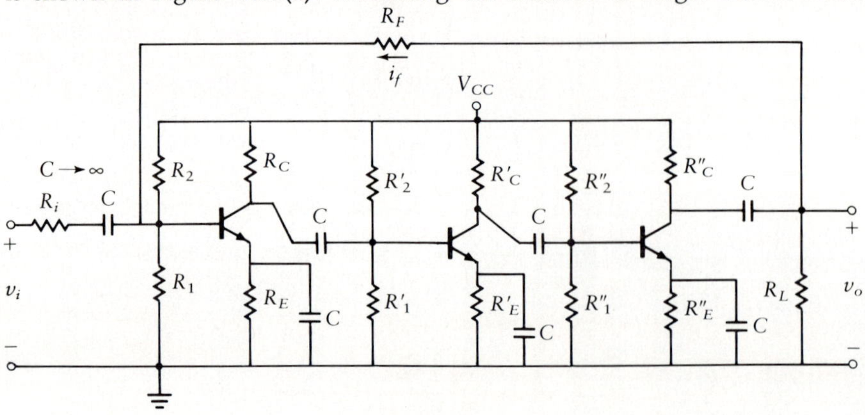

(a) Circuit

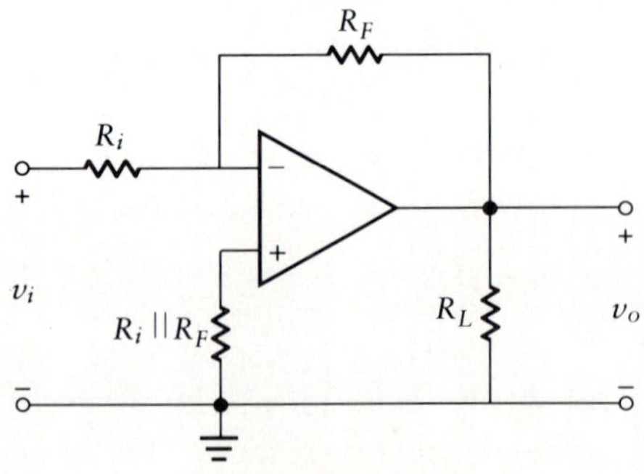

(b) Equivalent circuit

does not pose a major problem, since we can represent the amplifier of many stages as a single amplifier as long as we carefully monitor the number of phase reversals in the overall amplifier.

The open-loop gain of the circuit in Figure 11.6(a) is large (on the order of 10^4). From the equivalent circuit of Figure 11.6(b), we can solve the circuit of Figure 11.6(a) using the same theory as that developed for op-amps.

11.5 FEEDBACK IN OPERATIONAL AMPLIFIERS

In the previous chapter, we saw that the gain of the practical op-amp is high at dc but decreases as frequency increases. This frequency dependence is built into the op-amp by the manufacturer. Numerous op-amps permit the engineer to select *external compensation networks,* which change the shape of the Bode plot to improve performance. Figure 11.7 shows the curve of the open-loop gain for a μA741 op-amp (see Appendix D for a complete data sheet for this op-amp) as a function of frequency.

It is useful to express the open-loop frequency response of Figure 11.7 as an analytic expression. The figure represents a simple pole, which was studied in the previous chapter. For this case, the corner frequency, ω_o, is 20π radians per second. Note that $\omega_o = 2\pi f_o$, where f_o is in hertz. The zero-frequency gain, G_o, is 100 dB, or 10^5. The analytic expression representing a single-pole op-amp that decays at the slope of -20 dB/decade is

$$G(s) = \frac{G_o}{1 + s/\omega_o} \tag{11.13}$$

For a μA741 op-amp, we simply substitute $\omega_o = 20\pi$ radians per second and $G_o = 10^5$ (or 100 dB).

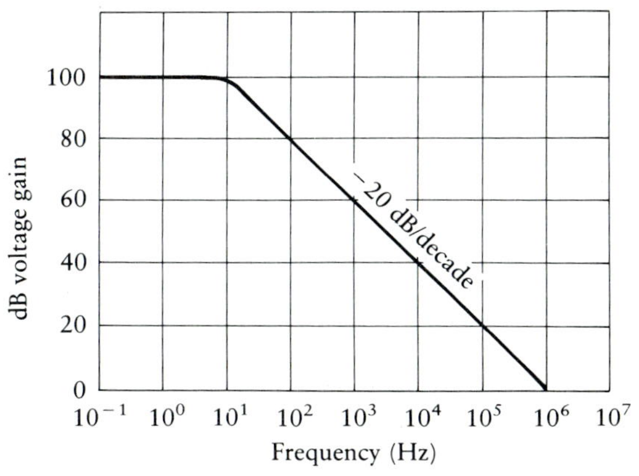

Figure 11.7
Typical op-amp open-loop gain response.

We find that, just as for discrete amplifiers, op-amps exhibit less sensitivity when feedback is employed. To calculate the error due to gain variations, we assume that all characteristics of the op-amp are ideal except for the gain variation with frequency. We use the inverting amplifier circuit of Figure 11.8(a) for the analysis and begin by finding the effect of the output voltage on the op-amp input voltage. We assume that $v_i = 0$, since we are finding only that portion of v_- due to v_o. In order to find v_-, we write the nodal equation at v_- of Figure 11.8(a) and solve for v_- to obtain

$$v_- = v_o \frac{R_A}{R_A + R_F} = v_o \frac{1}{1 + R_F/R_A} = \gamma v_o \tag{11.14}$$

where γ is the feedback attenuation factor. To find the closed-loop gain, we write a nodal equation at the inverting terminal of the amplifier of Figure 11.8(a) as follows (Note—$v_i \neq 0$):

$$\frac{v_- - v_i}{R_A} + \frac{v_- - v_o}{R_F} = 0$$

Now since $v_+ = 0$,

$$v_o = -Gv_- \qquad \text{and} \qquad v_- = \frac{-v_o}{G}$$

Solving for the closed-loop gain, we obtain

$$A_v = \frac{v_o}{v_i} = \frac{-R_F}{R_A} \frac{1}{1 + (1 + R_F/R_A)/G} \tag{11.15}$$

We again use the feedback attenuation factor of equation (11.14) to obtain the expression in standard feedback form,

$$\frac{v_o}{v_i} = -\frac{R_F}{R_A} \frac{G\gamma}{1 + G\gamma} \tag{11.16}$$

Figure 11.8
Inverting and noninverting amplifier.

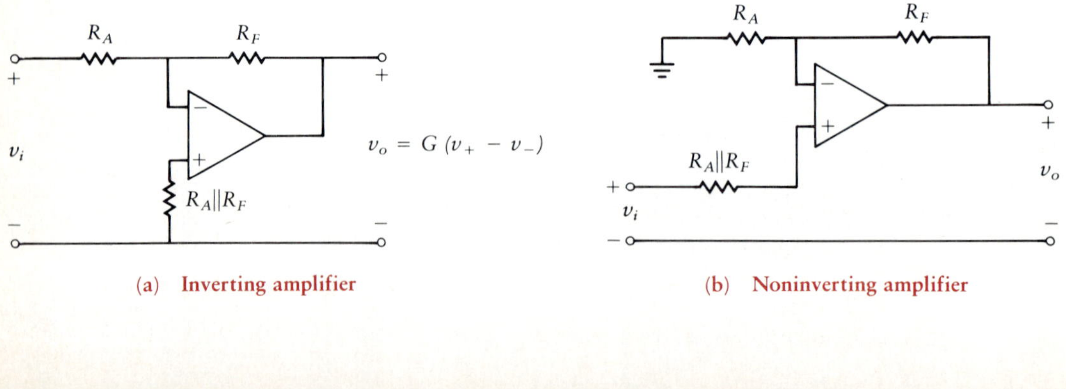

(a) **Inverting amplifier** (b) **Noninverting amplifier**

As the op-amp gain, G, increases, equation (11.16) approaches

$$A_{v\infty} = \frac{-R_F}{R_A}$$

Note that as the op-amp gain approaches infinity, A_v becomes independent of the specific value of G and is a function only of the two resistor values, R_F and R_A. This is the same result as obtained in Chapter 8 for an ideal op-amp.

We next consider a *noninverting* op-amp, shown in Figure 11.8(b). We assume that the op-amp is ideal except that the gain, G, depends on frequency, as in Figure 11.7. We find the closed-loop gain in a manner similar to that above. We write the node voltages for the inverting and the noninverting inputs as follows:

$$v_- = v_o \frac{R_A}{R_A + R_F}$$

and

$$v_+ = v_i$$

The output voltage is

$$v_o = G(v_+ - v_-) = G\left(v_i - v_o \frac{R_A}{R_A + R_F}\right)$$

We solve for v_o using equation (11.14) for γ as follows:

$$v_o\left(1 + G\frac{1}{1 + R_F/R_A}\right) = Gv_i$$

and the closed-loop voltage gain is

$$A_v = \frac{v_o}{v_i} = \frac{G}{1 + \gamma G} = \left(1 + \frac{R_F}{R_A}\right)\frac{\gamma G}{1 + \gamma G} \tag{11.17}$$

For very large G, equation (11.16) reduces to the ideal op-amp expression,

$$A_{v\infty} = \frac{1}{\gamma} = 1 + \frac{R_F}{R_A}$$

The two gain expressions, equations (11.15) and (11.17), can be *normalized* by dividing by $A_{v\infty}$, the gain for infinite G. The same expression results for both the inverting and the noninverting amplifier, as follows:

$$\frac{A_v}{A_{v\infty}} = \frac{G\gamma}{1 + G\gamma} \tag{11.18}$$

The expression $G\gamma$ is the *loop gain.* This is the gain obtained by tracing a loop through the amplifier and feedback path back to the starting point. Note that the loop gain of both amplifiers is identical.

To determine the sensitivity of the closed-loop gain, A_v, to changes in the open-loop gain, $G\gamma$, we follow a procedure similar to that used in developing equation (11.6). We differentiate equation (11.18) as follows:

$$\frac{dA_v}{dG\gamma} = \frac{A_{v\infty}}{(1 + G\gamma)^2}$$

We divide by A_v to obtain the sensitivity. This expression is simplified to obtain

$$\frac{dA_v}{A_v} = \frac{1}{1 + G\gamma} \frac{d(G\gamma)}{G\gamma} \tag{11.19}$$

Equation (11.19) applies to both the inverting and the noninverting amplifier. It expresses the effect of feedback. A variation in open-loop gain, $G\gamma$, is divided by $1 + G\gamma$ to yield a much smaller variation in closed-loop gain, A_v. As an example, if the open-loop gain, $G\gamma$, starts at 1000 (60 dB), a 10% variation in this gain will result in only a 0.01% variation in A_v.

Drill Problem

D11.2 Find the voltage gain, A_v, for the op-amp circuit of Figure D11.1. $R_F = 200\ \text{k}\Omega$ and $R_i = 10\ \text{k}\Omega$. Also find the sensitivity, dA_v/A_v, at zero frequency. Note that

$$G = \frac{10^5}{1 + s/20\pi}$$

$$\frac{d(G\gamma)}{G\gamma} = 50\%$$

Figure D11.1

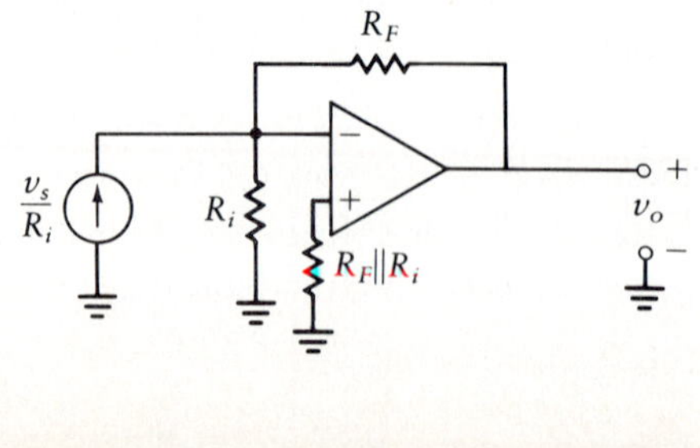

Ans: $$A_v = \frac{v_o}{v_i} = \frac{-20}{1 + 0.334 \times 10^{-5}s}$$

$$\frac{dA_v}{A_v} = 0.01\%$$

11.6 STABILITY OF FEEDBACK AMPLIFIERS

Negative feedback is advantageous, since the operation of the amplifier is less sensitive to parameter variations. Negative feedback requires that a portion of the output signal be subtracted from the input signal. Although we might design a system to provide such perfect subtraction, as soon as frequencies vary out of the midrange, phase shifts may make the subtraction less than perfect. In particular, at certain frequencies a phase shift of 180° could occur. This would change the subtraction to an addition and change *negative* feedback into *positive* feedback, yielding an unstable system.

In a practical unstable system, the output does not reach infinity even though the feedback is positive. Although the output may be bounded, it no longer depends on the input value. It is therefore important in amplifier design to make sure that the circuit remains stable for all operating frequencies.

Amplifier stability depends only on the properties of the system and not on the driving function. Hence, if a system is unstable, *any* excitation causes the system to operate in an unstable manner. Alternatively, if a system is stable, *any* bounded excitation will cause a bounded response.

If the transient time response of a system to an impulse input decays as time approaches infinity, the system is stable. For example, suppose the following time response, $h(t)$, is the result of applying an impulse function to the input of some system:

$$h(t) = Ae^{-a_1 t} + e^{-a_2 t}(B \cos \omega_r t + C \sin \omega_r t)$$

As t approaches infinity, $h(t)$ approaches zero and the system is stable. While stability analysis can be performed by examining $h(t)$, it is often easier if we use Laplace transforms.

We begin with a review of the basic concepts. (Appendix C presents a summary of important concepts from Laplace transform theory.) We let $f(t)$ be the time function associated with the Laplace transform, $F(s)$. If $F_1(s) = E_1/(s + a)$, then $f_1(t) = E_1e^{-at}$. If $F_2(s) = E_2/(s - b)$, the time response is $f_2(t) = E_2e^{bt}$. When the root of the denominator (pole) of the transfer function (known as the *characteristic function*) lies in the left half-plane, the exponent is negative. For example, the root of $F_1(s)$ lies at $s_i = -a$, and if $a > 0$, the system is stable. The root of $F_2(s)$ lies at $s_i = b$, and if $b > 0$, the system is unstable.

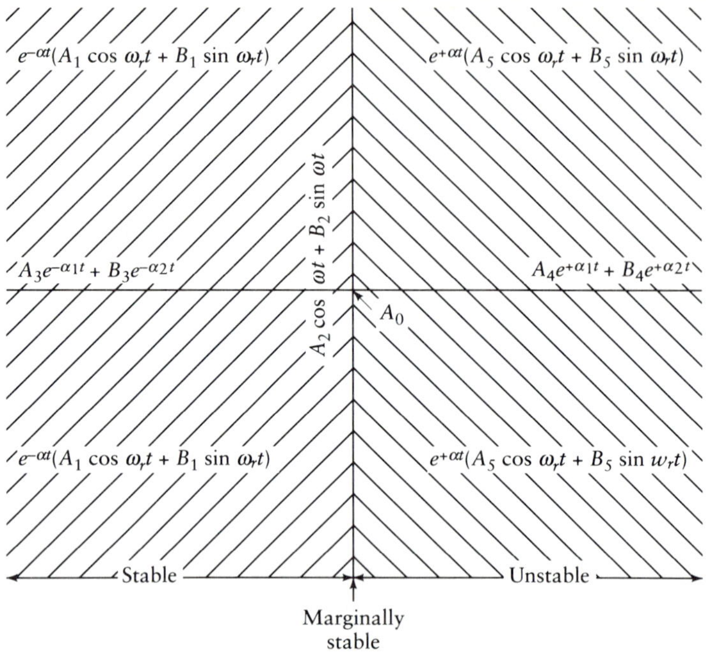

Figure 11.9 Root plane and location of roots.

The location of typical roots in the s plane and the corresponding time functions are shown in Figure 11.9. Table 11.1 illustrates the relationship between the location of the roots and the time function.

The nature of stability can now be stated in terms of the location of the roots of the characteristic equation $1 + G(s)H(s) = 0$. This is the denominator of equation (11.1). *A system is stable* if *all* roots lie in the left half-plane. *A system is unstable* if

1. any of the roots lie in the right half-plane, or
2. any multiple (double, triple, etc.) complex pairs of roots lie along the j axis, or
3. any multiple real roots lie at the origin.

A system is marginally stable if

1. any single pair of conjugate roots lies along the j axis or
2. a single root lies at the origin and all other roots lie in the left half-plane.

A system is conditionally stable if all roots lie in the left half-plane only for some particular condition of the system parameters. The system is often stable for a limited range of some parameters (e.g., loop gain constant).

Roots that lie in the right half-plane result in responses that increase with time, and such systems are unsuitable for practical use. Single conjugate pairs of roots that lie on the imaginary axis (other than at the origin) result in an undamped sinusoidal (oscillatory) response term. If all other roots lie in the left half-plane (except possibly a single root at the origin), this system may

Table 11.1 Root Location and Form of Solution*

Root Location	Form of Solution
Single roots along the negative real axis, such as $-\alpha_1$ and $-\alpha_2$, where α_1 and $\alpha_2 > 0$	$A_1e^{-\alpha_1 t} + B_1e^{-\alpha_2 t}$
Single roots along the positive real axis, α_1 and α_2, where α_1 and $\alpha_2 > 0$	$A_2e^{+\alpha_1 t} = B_2e^{+\alpha_2 t}$
A single complex conjugate pair of roots, $\pm j\omega$, along the imaginary or j axis	$A_3 \cos \omega t + B_3 \sin \omega t$
One simple pair of complex roots, $-\alpha \pm j\omega$, in the left half-plane ($\alpha > 0$)	$e^{-\alpha t}(A_4 \cos \omega_r t + B_4 \sin \omega_r t)$
A simple pair of complex roots, $\alpha \pm j\omega$, in the right half-plane ($\alpha > 0$)	$e^{+\alpha t}(A_5 \cos \omega_r t + B_5 \sin \omega_r t)$
Double roots of the characteristic equation occurring at one point on the real axis in the left half of the s plane assuming both roots at $-\alpha$	$(A_6 + B_6 t)e^{-\alpha t}$
A single root at the origin	A_0
A double root at the origin	$A_7 + B_7 t$

*In writing these expressions, the constants A_i and B_i used in each expression are arbitrary. Their values for a particular system depend upon the initial conditions.

be considered an oscillator. This is the limiting case between stable and unstable systems.

11.6.1 System Stability and Frequency Response

A designer is interested in knowing whether any system roots lie in the right half-plane, since this indicates an unstable system. Amplitude and phase plots are often readily available, whereas considerable analytical effort would be required to determine the location of system roots. This is particularly true if SPICE-based computer simulations are used. These programs can generate frequency curves directly from the circuit. It is therefore desirable to be able to make stability determinations directly from frequency response plots.

In order to extend stability criteria from the s plane to the frequency response curves, we need to invoke concepts from advanced calculus—in particular, contour integration. We will present this material in a manner that does not require that you have an extensive background in advanced calculus.

To determine whether any roots of $1 + G(s)H(s) = 0$ are in the right half-plane, we examine the contour shown in Figure 11.10. This contour encloses the entire right half of the s plane. The radius of the circular arc portion of the contour is made so large that all possible roots in the right half-plane are included. This is usually done by considering the limiting case of R approaching infinity. Stability depends on whether any roots fall within the closed contour.

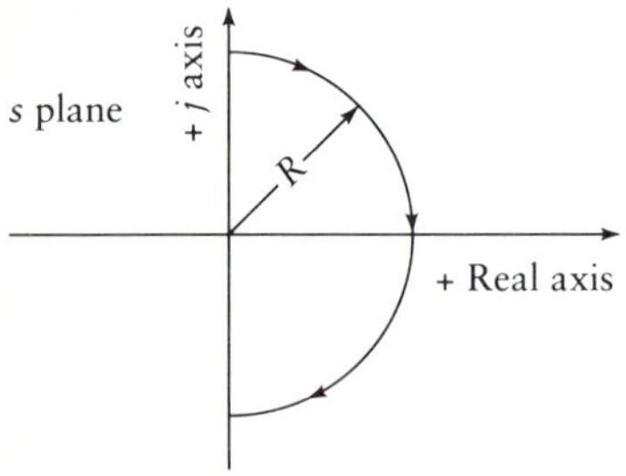

Figure 11.10 Contour enclosing the right half-plane.

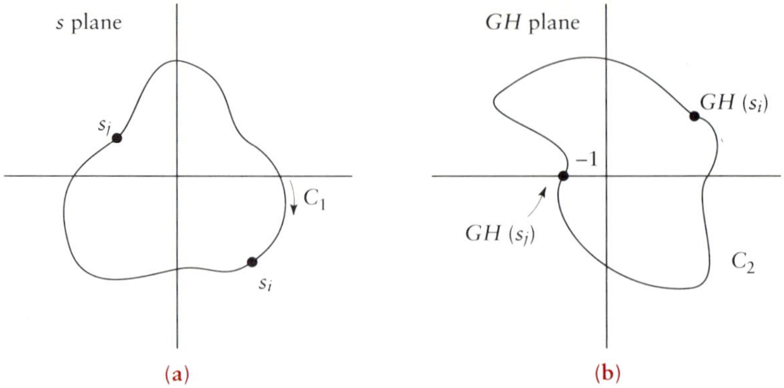

Figure 11.11 Contour in s plane mapped into contour in GH plane.

The plot of $G(s)H(s)$ as s runs the contour of Figure 11.10 yields the *mapping* of this contour located in the s plane into a contour in the GH plane. A closed curve, C_1, such as that shown in the s plane of Figure 11.11(a), maps into a closed curve, C_2, in the GH plane, as shown in Figure 11.11(b). For single-valued functions, a one-to-one correspondence exists between a point on C_1, the curve in the s plane, and a point on C_2, the map in the GH plane. If a point is moved along C_1 in the direction of the arrow (clockwise), the mapped point moves along C_2 in a direction that depends on the GH function.

Let s_j denote a root of $1 + G(s)H(s) = 0$. Now suppose the contour C_1 passes through the point s_j in the s plane, as shown in Figure 11.11(a). If s_j is a root, then $G(s_j)H(s_j) + 1 = 0$; thus for $s = s_j$, $G(s_j)H(s_j) = -1$. That is, the contour C_2 in the GH plane passes through the point $GH = -1$.

The characteristic function for a closed-loop system can be written in factored form, as shown in equation (11.20).

$$1 + G(s)H(s) = K_1 \frac{(s + s_1)(s + s_2)(s + s_3)}{(s + s_a)(s + s_b)(s + s_c)} \tag{11.20}$$

$-s_1, -s_2, \ldots$ are the roots and $-s_a, -s_b, \ldots$ are the poles of $1 + G(s)H(s)$. For simplicity, we will be examining $G(s)H(s)$ instead of $1 + G(s)H(s)$. Poles of $G(s)H(s)$ are also poles of $1 + G(s)H(s)$, since substitution of the pole (e.g., $s = -s_a$) in either equation makes the expression approach infinity.

Once a value is assumed for s, each factor in equation (11.20) is a complex number and hence can be represented by a vector, as shown in Figure 11.12. The vectors extend from the fixed points, s_1, s_2, s_3, and poles s_a, s_b, s_c, to the variable point s. Suppose the variable point moves in a clockwise direction on a contour so as to make one complete revolution about s_2 (see Figure 11.12). Then the vector $s + s_2$ makes one complete clockwise revolution because the contour encircles this root. Since all the other roots and poles are external to the contour, each of the remaining vectors makes no complete revolution. Since the term $s + s_2$ in equation (11.20) changes phase by 360° (corresponding to one complete revolution about s_2), $1 + G(s)H(s)$ experiences a change in

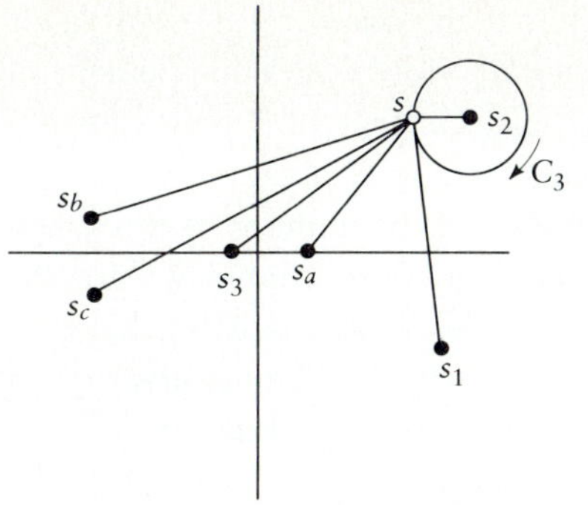

Figure 11.12 Encirclement of a root in the s plane.

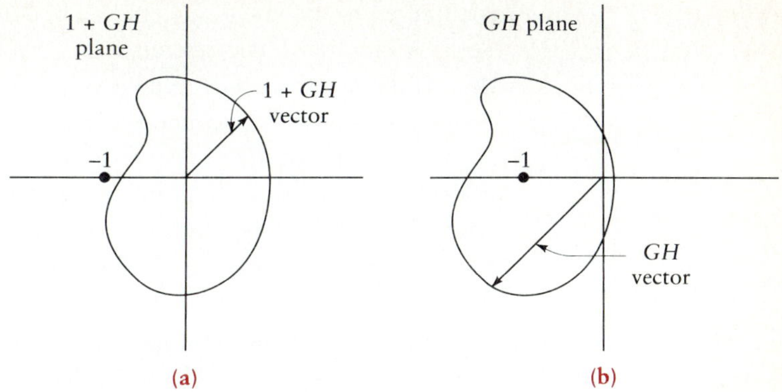

Figure 11.13 (1 + GH) plane and GH plane.

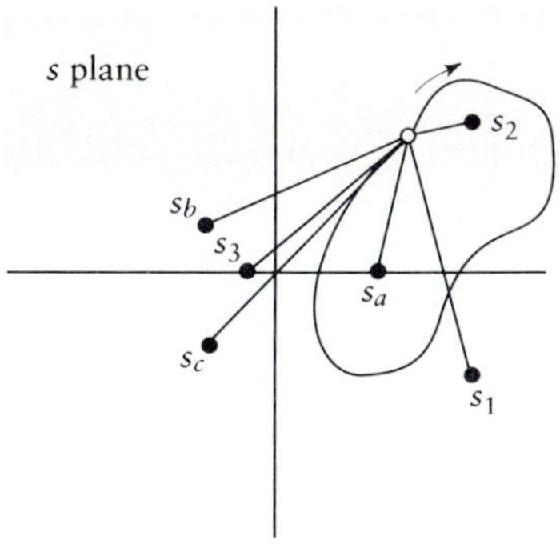

Figure 11.14 Enclosing a root and a pole.

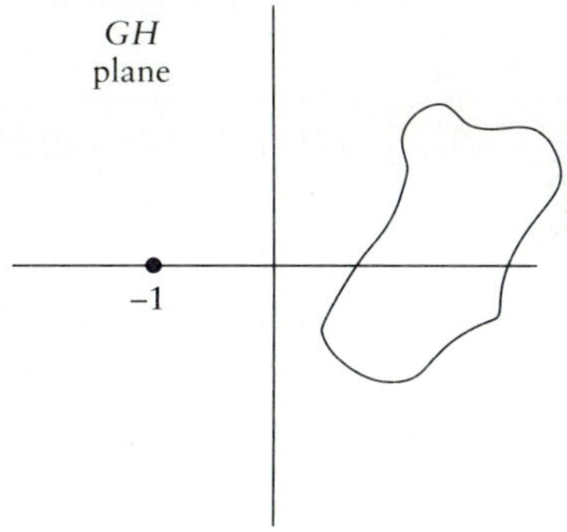

Figure 11.15 No encirclement of the −1 point in the GH plane.

phase of 360°. Hence, a vector representing $1 + G(s)H(s)$ (in the $(1 + GH)$ plane) would make one clockwise encirclement of the origin. This is shown in Figure 11.13(a). The remaining roots and poles contribute no net change to the phase of $1 + G(s)H(s)$.

Because the roots are in the numerator of equation (11.20), one clockwise rotation about s_2 results in one clockwise encirclement of the origin in the $(1 + GH)$ plane, which is related to the GH plane as shown in Figure 11.13(b). In this figure, one clockwise encirclement of the origin of the $(1 + GH)$ plane corresponds to one clockwise encirclement of the $-1 + j0$ point in the GH plane.

Suppose the closed contour in the s plane is made to encircle both a root and a pole in the clockwise direction, as shown in Figure 11.14(a). In this case both the vector from s_2 to s and that from s_a to s rotate through one complete clockwise revolution, or 360°. The factor, $s + s_2$, corresponding to the root s_2 contributes +360° to the change in phase of $1 + G(s)H(s)$ since it is in the numerator of equation (11.20). The factor $s + s_a$ contributes −360° to the change in phase since it is in the denominator. Hence the net change in phase of $1 + G(s)H(s)$ is zero, and the resulting map in the GH plane *does not* encircle the −1 point, as shown in Figure 11.15. If the closed contour in the s plane is enlarged to include s_1, s_2, and s_a, the net change in phase of $1 + G(s)H(s)$ is 360°. A clockwise encirclement of a root causes a clockwise encirclement of the -1 point in the GH plane. A clockwise encirclement of a pole causes a counterclockwise encirclement of the -1 point in the GH plane.

We now summarize these results. We define n_N as the number of clockwise encirclements of the -1 point, n_R as the number of roots located within the contour in the s plane, and n_P as the number of poles located within this contour. Then a clockwise encirclement of a region in the s plane causes $n_N = n_R - n_P$ clockwise encirclements of the -1 point in the GH plane. n_N is positive if $n_R > n_P$ in the right half-plane. In this case the -1 point is encircled in the same direction (clockwise) as the contour in the s plane. n_N is zero if $n_R = n_P$ and the -1 point is not encircled. n_N is negative if $n_R < n_P$ and the -1 point is encircled in the direction opposite (counterclockwise) to the contour in the s plane (clockwise).

11.6.2 Bode Plots and System Stability

A closed-loop system is unstable if roots of the characteristic equation, $1 + G(s)H(s) = 0$, lie in the right half of the s plane. Suppose the contour of Figure 11.10 is made large enough to include the entire right half of the s plane ($R \to \infty$). If this contour is mapped on the GH plane, the number of clockwise encirclements of the -1 point of the resulting map yields information regarding the stability of the closed-loop system.

The stability criterion can be stated as follows: the open-loop transfer function, $G(s)H(s)$, is expressed as the ratio of two factored polynomials in the variable s and written in the form

$$G(s)H(s) = \frac{K(s\tau_1 + 1)(s\tau_3 + 1)}{s^n(s\tau_2 + 1)(s\tau_4 + 1)} \tag{11.21}$$

Now as s travels a closed contour comprising the imaginary axis from $-j\infty$ to $+j\infty$ and then the right-hand semicircle from $s = Re^{j\pi/2}$ to $s = Re^{-j\pi/2}$ as $R \to \infty$, the polar plot of $G(s)H(s)$ encircles the $-1 + j0$ point in a clockwise direction n_N times. n_N is given by

$$n_N = n_R - n_P \tag{11.22}$$

Recall that n_N is the number of clockwise encirclements of the point -1 (a negative n_N corresponds to counterclockwise encirclements), n_P is the number of poles of $G(s)H(s)$ in the right half-plane, and n_R is the number of roots of $1 + G(s)H(s)$ that lie in the right half-plane. $G(s)$ is the forward-loop transfer function, and $H(s)$ is the feedback-loop transfer function. These functions are indicated in Figure 11.1. As noted earlier, the poles of $G(s)H(s)$ are also poles of $1 + G(s)H(s)$.

A somewhat simpler contour than that shown in Figure 11.10 is used in practice. Since systems considered in this text have characteristic equations with constant real coefficients, the roots of the characteristic function must either be real or occur in complex conjugate pairs. As a result, the map of the upper half of the $j\omega$ axis and the upper half of the infinite semicircle is the mirror image of the lower half of the $j\omega$ axis and the lower half of the infinite semicircle. Further, the pole and zero structures in these two areas are mirror images. Hence, it is necessary only to plot the simple closed contour comprising the upper portion of the imaginary axis and the upper portion of the semi-infinite circle to determine whether any roots of $1 + G(s)H(s) = 0$ lie in the right half of the s plane.

So that all the right half of the s plane is encircled, the semicircular contour must have an infinitely large radius, R. Physical systems usually have zero response to an infinite frequency; hence, the large semicircle in the s plane usually maps onto the GH plane as a point at the origin. We therefore arrive at the important result that only real, positive frequencies from zero to infinity need to be mapped (i.e., the positive imaginary axis). This plot is obtained by applying sinusoidal inputs and varying the frequency of the signals from zero to infinity. We have succeeded in relating the s-plane analysis to a frequency response analysis.

The stability criterion gives information regarding the difference between the number of roots and the number of poles of $1 + G(s)H(s)$. $G(s)H(s)$ can be expressed as the ratio of factored polynomials (and often it occurs naturally in this form). Since the poles of $1 + G(s)H(s)$ are identical to the poles of $G(s)H(s)$, the number of poles of $G(s)H(s)$ in the right half of the s plane can be found by inspection of equation (11.21). Hence, under the conditions just stated, the number of poles with positive real parts can easily be determined. The number of clockwise encirclements n_N of the point $-1 + j0$ in the GH plane is found from the GH-plane plot. Then both n_N and n_P in equation (11.22) are known. The number of roots of $1 + G(s)H(s)$ in the right half of the s plane can then be found from equation (11.22), with the result

$$n_R = n_N + n_P \tag{11.23}$$

Appendix B reviews techniques for generating Bode plots for a wide variety of $G(s)H(s)$ functions. The stability criterion can be applied to the Bode plot of decibel gain and phase shift. It is necessary only to find the -1 point on this Bode format. This is easy, since in complex form

$$-1 = 1e^{-j180}$$

so the -1 point occurs when the gain is 0 dB at $-180°$ phase shift.

We can determine system stability as follows: We look at the *perfected* Bode plot* and determine the frequency at which the phase shift crosses $-180°$. If the dB gain at this frequency is

1. Less than 0 dB, the system is stable.
2. Equal to 0 dB, the system is marginally stable; that is, the response neither builds up nor dies out.
3. Greater than 0 dB, the system is unstable.

Example 11.3

A three-stage amplifier is shown in Figure 11.16(a). Each identical stage has a gain, $-K$, and an input RC network which approximates the *RC* time constant within each stage. Plot the Bode diagram for $RC = 10^{-5}$ s and determine the gain, *K*, that will produce *marginal stability* (this is the point where the phase shift is $-180°$ and the gain is 0 dB).

Figure 11.16 Three-stage amplifier.

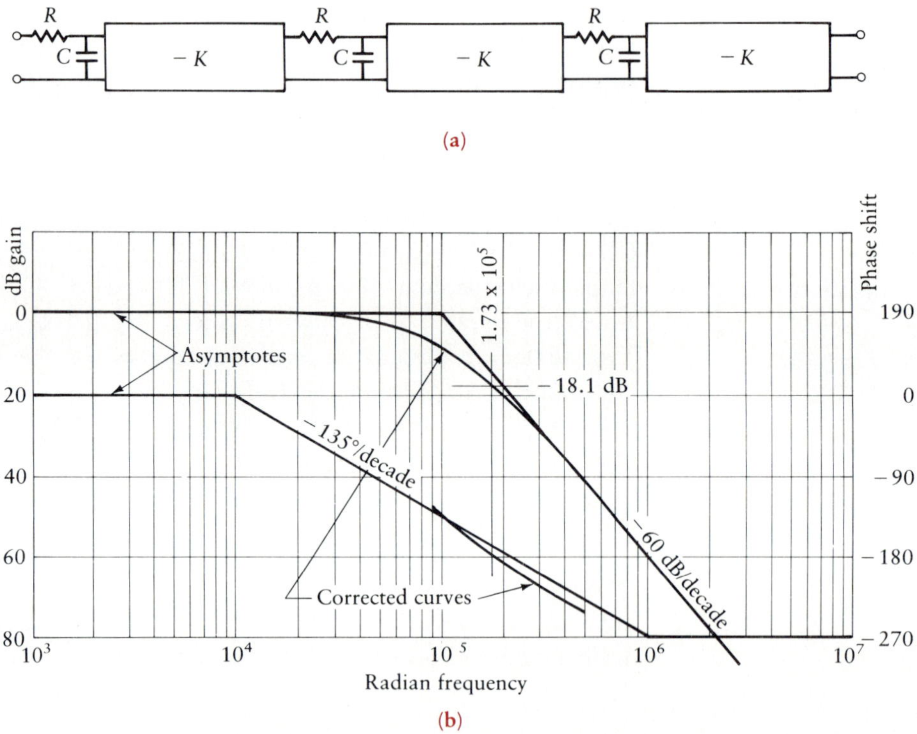

* A perfected Bode plot is one in which the corrections have been made to the straight-line asymptotes, as shown in Appendix B.

SOLUTION The Bode diagram is plotted in Figure 11.16(b) for $K = 1$. The frequency at which the phase shift crosses $-180°$ is 1.73×10^5 rad/s, or 27.53 kHz. At this frequency, the gain is -18.1 dB, which indicates that we can raise the combined gain, K^3, by 18.1 dB or

$$K^3 = 8.04$$

so

$$K = 2.$$

Example 11.4

With $\tau = 0.1$ s, examine the stability of the system with the following $G(s)H(s)$ function:

$$G(s)H(s) = \frac{A}{s(\tau s + 1)}$$

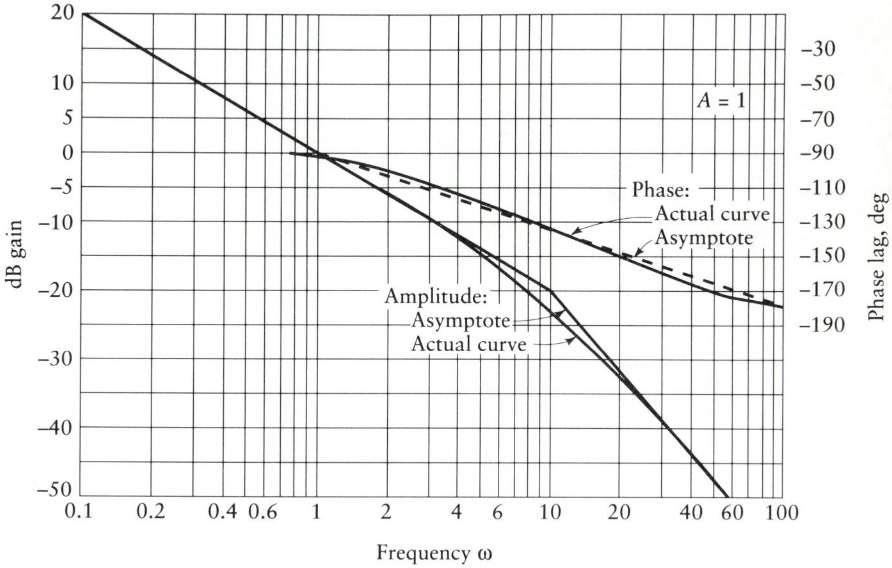

Figure 11.17 Bode plot for Example 11.4.

SOLUTION The Bode plot is constructed and is shown in Figure 11.17. The resulting amplitude plot consists of a straight-line segment with slope of -20 dB/decade for frequencies below 10 rad/s. For $\omega > 10$, the plot is a straight-line segment with slope of -40 dB/decade. The curve is plotted for

$A = 1$. A different value of A would result in a vertical translation of the amplitude curve. The phase curve starts at $-90°$, corresponding to the term $1/j\omega$ resulting from the pole at the origin, and continues to decrease as frequency increases, to a limit of $-180°$. The phase shift does not cross the $-180°$ line, so this system is stable for any value of A.

Example 11.5

Examine the stability of a system with the following $G(s)H(s)$ function:

$$G(s)H(s) = \frac{A}{s(s+1)(0.5s+1)}$$

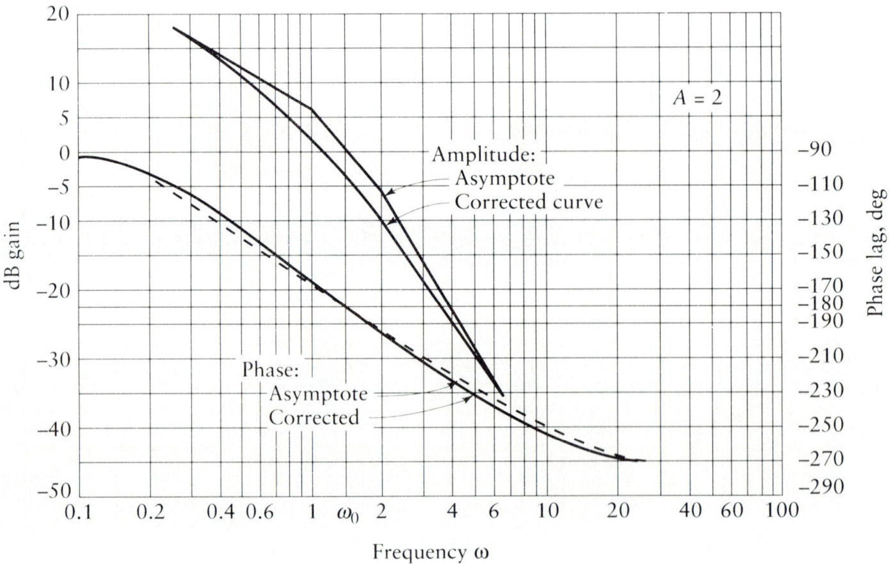

Figure 11.18 Bode plot for Example 11.5.

SOLUTION The Bode plot is constructed with the asymptotic approximations as shown in Figure 11.18. This Bode plot is constructed for $A = 2$. To determine stability, we look for the point where the phase shift curve crosses the $-180°$ line. At this frequency, we examine the corrected dB gain curve. We note that at the frequency where the phase is $-180°$, the gain curve is below 0 dB (approximately -4 dB). Hence the system is stable with a gain of $A = 2$. However, if the gain increases by 4 dB (A increases to 3.16), the amplitude curve shifts upward and reaches 0 dB at the $-180°$ phase shift point. Further increases in A move the system into instability.

11.7 FREQUENCY RESPONSE—FEEDBACK AMPLIFIER

The stability of a negative-feedback system is dependent on the loop gain being less than 0 dB when the phase shift crosses −180°. If the system has −180° of phase shift when the gain is 0 dB, the system is unstable and can oscillate. That is, an output will exist when no input signal is applied.

11.7.1 Single-Pole Amplifier

Let us return to the basic feedback system of Figure 11.1 and assume that $G(s)$ is a polynomial that has one pole and $H(s)$ is a constant. That is,

$$G(s) = \frac{-G_o}{1 + s/\omega_o}$$

and

$$H(s) = -\gamma$$

Since we are interested in the stability of the amplifier, we wish to examine the phase shift of the loop gain expression to ensure that the stability condition is met. Figure 11.19 illustrates the Bode plot of the normalized amplitude and phase shift. Since the Bode plot uses the $G(s)H(s)$ function, we obtain

Figure 11.19
Response for single-pole amplifier.

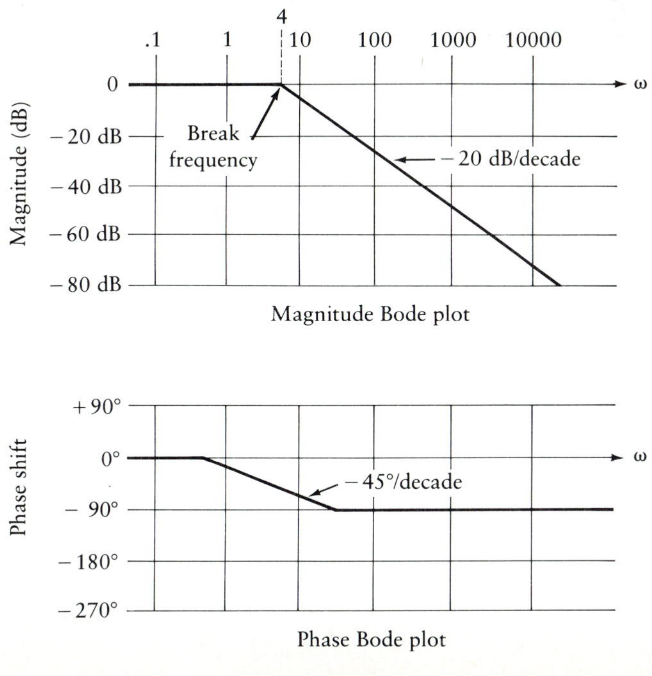

$$G(s)H(s) = \frac{\gamma G_o}{1 + s/\omega_o}$$

For purposes of illustration, the break frequency is selected to be $\omega_o = 4$. Since the maximum phase shift is $-90°$, this amplifier can never be unstable. A necessary condition for instability is that the phase shift reach $-180°$.

The single-pole amplifier has a closed-loop gain, $T(s)$, given by

$$T(s) = \frac{G(s)}{1 + G(s)H(s)}$$

$$= \frac{-G_o}{1 + G_o\gamma} \frac{1}{1 + s/(1 + G_o\gamma)\omega_o} \quad (11.24)$$

The 3-dB point for the closed-loop gain characteristic is at a frequency of

$$\omega = (1 + \gamma G_o)\omega_o$$

Since $\gamma G_o >> 1$, the 3-dB frequency is approximately $\gamma G_o \omega_o$, and the closed-loop gain becomes

$$T(s) = \frac{-1}{\gamma} \frac{1}{1 + s/(G_o\gamma\omega_o)} \quad (11.25)$$

11.7.2 Two-Pole Amplifier

Let us now assume that $G(s)$ has two poles at the same location. Thus,

$$G(s) = \frac{-G_o}{(1 + s/\omega_o)^2} \quad (11.26)$$

We again assume that $H(s)$ is a constant,

$$H(s) = -\gamma$$

To determine the stability, we plot the $G(s)H(s)$ function on a Bode diagram.

$$G(s)H(s) = \frac{\gamma G_o}{(1 + s/\omega_o)^2}$$

Let us set $\omega_o = 60$ rad/s, $\gamma = 0.1$, and $G_o = 10^5$. The plot for this function is shown in Figure 11.20.

$$G(s)H(s) = \frac{10^4}{(1 + s/60)^2} \quad (11.27)$$

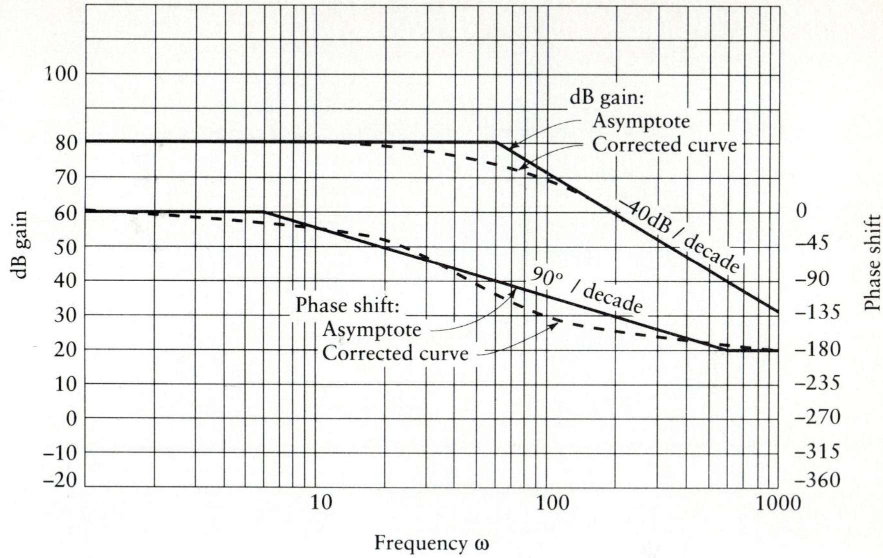

Figure 11.20
Bode plot for two-pole amplifier.

Since the phase shift curve does not cross the −180° point, the system is stable for any value of gain. We really did not need to plot this function, since any two-pole amplifier must be stable. The closed-loop gain function, $T(s)$, is given by

$$T(s) = \frac{-G_o/(1 + s/\omega_o)^2}{1 + G_o\gamma/(1 + s/\omega_o)^2}$$

$$= \frac{-G_o}{1 + \gamma G_o} \frac{1}{s^2/[\omega_o^2(1 + \gamma G_o)] + 2s/[\omega_o(1 + \gamma G_o)] + 1} \tag{11.28}$$

This can be written in a more useful form using the following two definitions:

$$\omega_n = \omega_o\sqrt{1 + \gamma G_o} \tag{11.29}$$

and

$$\xi = \frac{1}{\sqrt{1 + \gamma G_o}} \tag{11.30}$$

We define ω_n as the undamped natural radian frequency and ξ as the damping ratio. When these new terms are used in equation (11.28), the gain function can be written as

$$T(s) = \frac{-G_o}{1 + \gamma G_o} \frac{1}{s^2/\omega_n^2 + 2\xi s/\omega_n + 1} \tag{11.31}$$

Example 11.6

Determine the stability of the op-amp circuit shown in Figure 11.21(a) by plotting the Bode diagram for the $G(s)H(s)$ function. The 741 op-amp has a gain function given by

$$G(s) = \frac{G_o}{1 + s/\omega_o}$$

with $G_o = 10^5$ and $\omega_o = 20\pi$.

SOLUTION We write the node equation at V_-.

$$\frac{V_-}{R} + \frac{V_- - V_o}{1/sC} = 0$$

so, solving for V_-,

$$V_- = \frac{sRCV_o}{1 + sRC}$$

At V_+, the node equation yields $V_+ = V_i$. But since

Figure 11.21 Amplifier for Example 11.6.

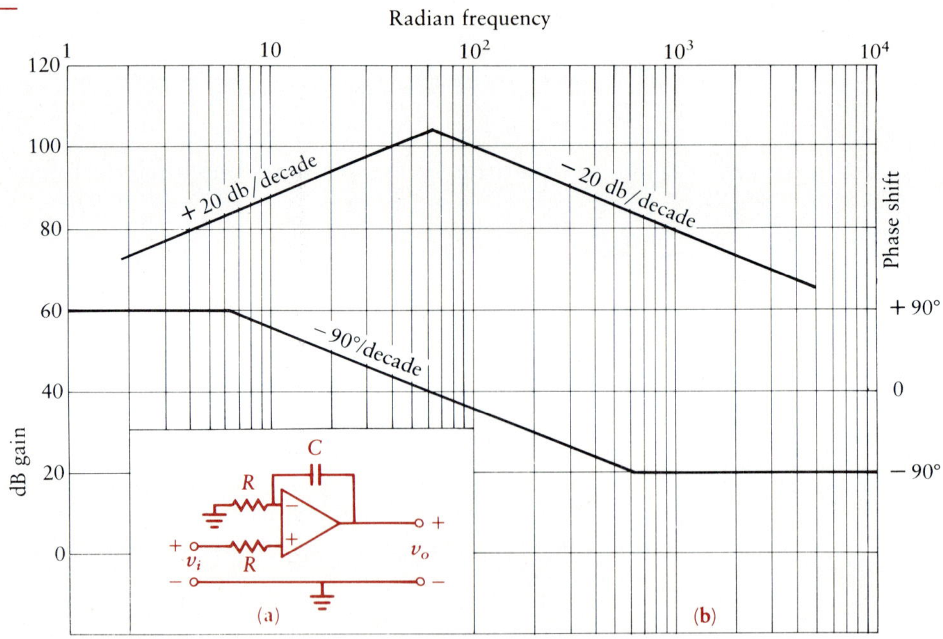

$$V_o = G(V_+ - V_-) = G\left(V_i - \frac{sRCV_o}{1 + sRC}\right)$$

we solve for the closed-loop gain,

$$\frac{V_o}{V_i} = \frac{G}{1 + GsRC/(1 + sRC)} = \frac{G_o/(1 + s/\omega_o)}{1 + G_osRC/[(1 + s/\omega_o)(1 + sRC)]} \tag{11.32}$$

In the last equality, we have substituted the given function for G. This is now in the form of

$$\frac{G(s)}{1 + G(s)H(s)} \tag{11.33}$$

The Bode diagram is a plot of the $G(s)H(s)$ function, and by comparison of equation (11.32) with (11.33), we obtain

$$G(s)H(s) = \frac{G_osRC}{(1 + s/\omega_o)(1 + sRC)}$$

where $G_o = 10^5$ and $\omega_o = 20\pi = 62.8$. In order to arrive at a numerical result, let us assume $1/RC = \omega_o$, so $RC = 0.016$. Then

$$G(s)H(s) = \frac{1.6 \times 10^3\, s}{(1 + s/62.8)^2}$$

We have chosen the RC time constant to yield two identical poles. This function is plotted in the Bode diagram of Figure 11.21(b). Only the straight-line asymptotes are shown. We see that the phase shift never reaches $-180°$, so the system is always stable. That is why the straight-line approximation proves sufficient.

Example 11.7

Determine the stability of the op-amp system of Figure 11.22(a) using a Bode plot. The op-amp gain function is given by

$$G(s) = \frac{G_o}{(1 + s/20\pi)^2}$$

Let $R = 100\ \text{k}\Omega$ and $C = 0.1\ \mu\text{F}$.

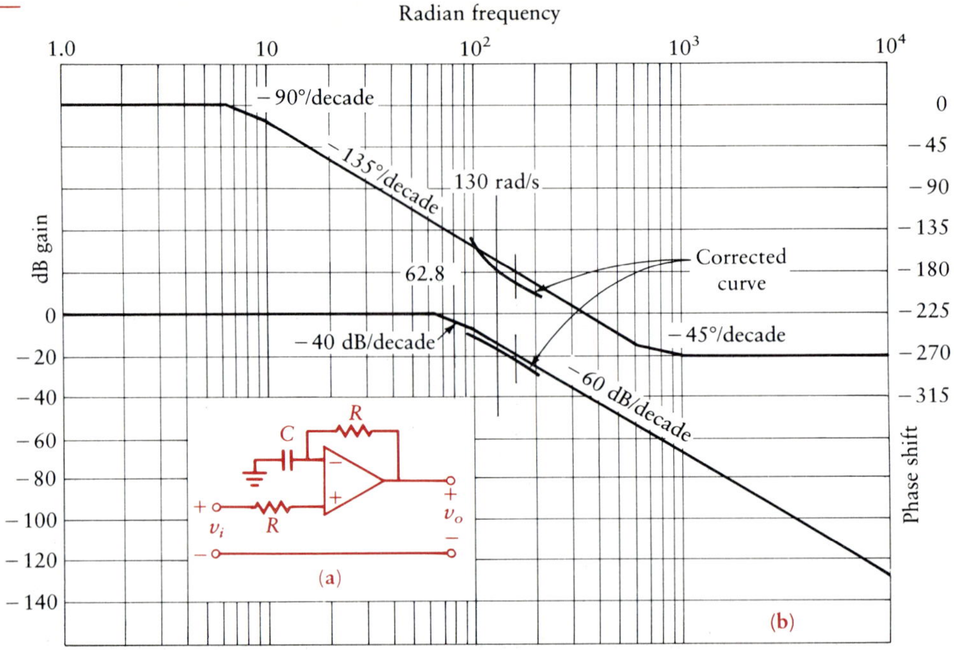

Figure 11.22 Amplifier for Example 11.7.

SOLUTION We again write the node equation at V_- and V_+ as follows:

$$V_- = V_o \frac{1}{1 + sRC}$$

and $V_+ = V_i$. The output voltage is $V_o = G(V_+ - V_-)$, so combining terms we obtain

$$V_o = G\left(V_i - \frac{V_o}{1 + sRC}\right)$$

The closed-loop gain is

$$\frac{V_o}{V_i} = \frac{G(s)}{1 + G(s)/(1 + sRC)} = \frac{G(s)}{1 + G(s)H(s)}$$

and the $G(s)H(s)$ function is given by

$$G(s)H(s) = \frac{G_o}{(1 + sRC)\,(1 + s/20\pi)^2}$$

where we have substituted in the expression for $G(s)$. Let us substitute the given component values and start with a gain, $G_o = 1$. The resulting function is

$$G(s)H(s) = \frac{1}{(1 + s/10^2)(1 + s/62.8)^2}$$

This is shown in Figure 11.22(b). The straight-line asymptotes are first plotted, and the curves are corrected in the vicinity of the frequency where the phase shift is −180°. From this corrected plot, we can see that the phase shift curve intersects the −180° line at a frequency of 130 rad/s, and the gain at this frequency is −18.7 dB (or 8.61). Thus, 18.7 dB is the gain for marginal stability. This would be a poor design since the gain is so low. The next section considers one method for improving performance.

11.8 DESIGN OF A THREE-POLE AMPLIFIER WITH A LEAD EQUALIZER

Consider an amplifier with three poles located at the same point, that is,

$$G(s) = \frac{-G_o}{(1 + s/\omega_o)^3}$$

Again, let $H(s)$ be a constant,

$$H(s) = -\gamma$$

so the loop function is

$$G(s)H(s) = \frac{\gamma G_o}{(1 + s/\omega_o)^3}$$

Figure 11.23 illustrates the Bode plot for the loop transfer function, $G(s)H(s)$, where we have let γG_o be normalized to 1. For purposes of illustration, we have selected a corner frequency of $\omega_o = 10$ rad/s.

As before, we determine the stability of the amplifier system with the loop transfer function rather than with the closed-loop gain of the amplifier system. We check for instability by seeing whether the $G(s)H(s)$ function has an amplitude of unity when its phase shift is −180°. We note that the phase shift does reach −180° at a frequency of 18 rad/s. The amplitude at this point is −18 dB, but recall that we have normalized the function by dividing it by γG_o. Thus, if the loop gain, γG_o, reaches +18 dB, the overall gain will reach 0 dB and the system will become unstable. A gain of +18 dB is relatively low, so this particular amplifier would probably be of little use in a practical application.

This leaves us with two options. We must either abandon the amplifier, since it becomes unstable at low gain, or change it so that the instability does

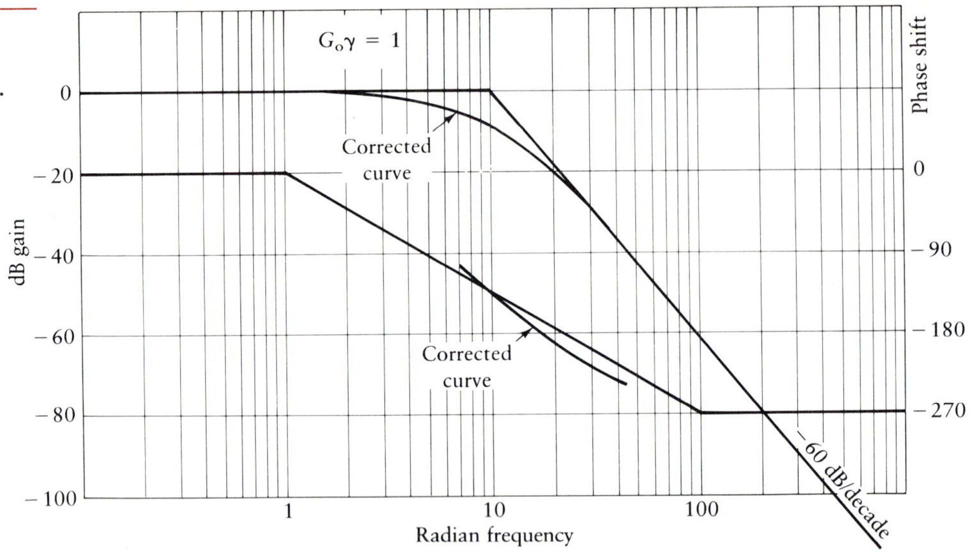

Figure 11.23 Bode plot for three-pole amplifier.

not occur. Fortunately, it is not difficult to alter the amplifier to reduce the chances of instability. *Compensation networks* (or *equalizers*) are used to alter the gain-phase characteristics in order to prevent instabilities. One way to view these networks is that they change the phase shift characteristic so that the shift is no longer −180° at frequencies where the gain exceeds 0 dB.

In the following analysis, we consider a *lead network*. The word *lead* refers to the relationship between input and output phase shift. The circuit for a passive lead network, sometimes referred to as a *passive circuit differentiator,* is shown in Figure 11.24. We first solve for the output-input relationship,

$$G(s) = \frac{V_o}{V_i} = \frac{R_2}{R_1 + R_2} \frac{1 + sR_1C}{1 + s(R_1 \parallel R_2)C} \tag{11.34}$$

The phase shift of this expression is found by setting $s = j\omega$ and subtracting the phase shift of the denominator from that of the numerator. Note that the phase shift of the numerator is larger than that of the denominator since

$$R_1C > (R_1 \parallel R_2)C$$

Therefore the overall phase shift of this expression is positive, and the output phase shift leads the input phase.

Equation (11.34) is written in a different form by defining

$$\alpha = 1 + \frac{R_1}{R_2}$$

and

Figure 11.24
Passive circuit lead network.

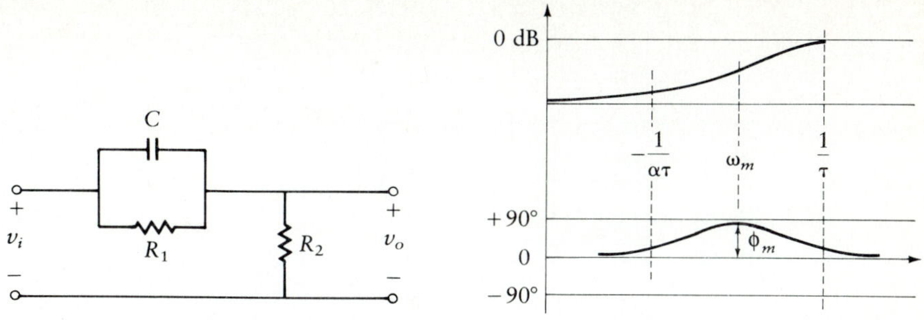

(a) Passive circuit lead network

(b) Bode plot of lead network

$$\tau = \frac{R_1 R_2 C}{R_1 + R_2}$$

With these two definitions, the transfer function becomes

$$G(s) = \frac{V_o}{V_i} = \frac{\alpha\tau s + 1}{\alpha(\tau s + 1)} \tag{11.35}$$

The Bode plot for this network is shown in Figure 11.24(b). Note that the maximum phase lead is a function of α. It approaches 90° as α approaches infinity.

The peak phase shift, ϕ_m, and the frequency at which it occurs, ω_m, can be related to the parameters α and τ by setting the derivative of the phase expression to zero. We obtain

$$\omega_m = \frac{1}{\sqrt{\alpha\tau}}$$

$$\phi_m = \sin^{-1}\left(\frac{\alpha - 1}{\alpha + 1}\right)$$

Example 11.8

Select a lead network with $\tau = 1/40$ and $\alpha = 4$ to adequately compensate a three-pole op-amp summing amplifier.

SOLUTION The transfer function for the lead network is

$$G(s) = \frac{0.1s + 1}{4(s/40 + 1)}$$

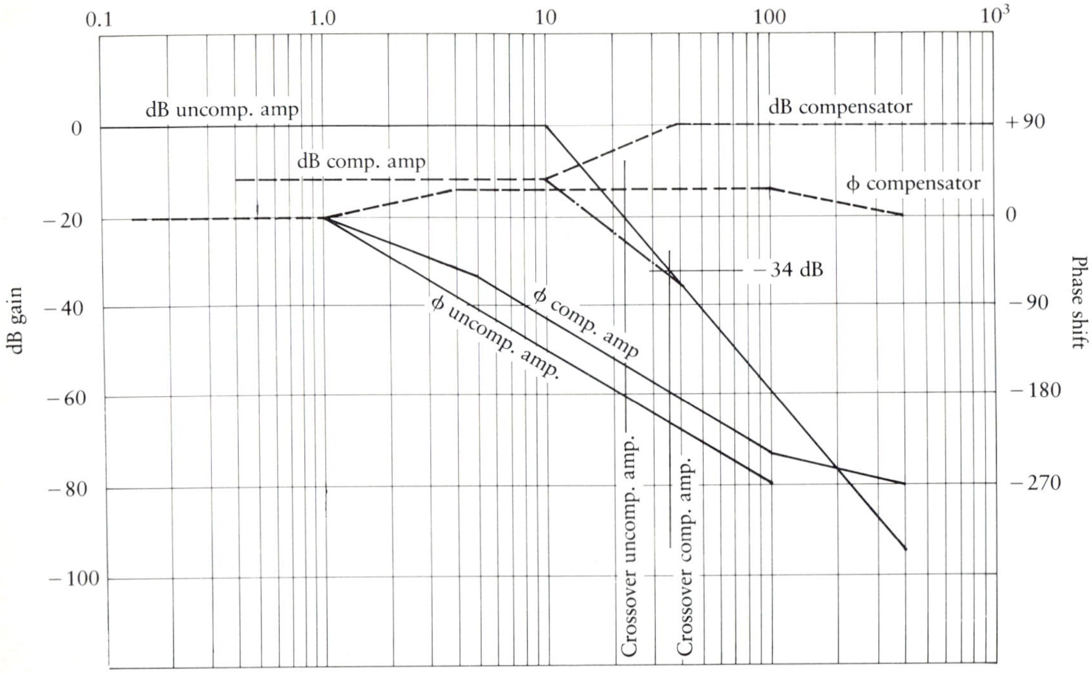

Figure 11.25
Bode plots for Example 11.8.

Figure 11.25 illustrates the gain and phase curves for the function that results when this is multiplied by the transfer function of the three-pole system. Shown in Figure 11.25 is a repeat of the three-pole amplitude and phase curves from Figure 11.23. When the compensation network is placed in cascade with this system, we add the Bode plots (both dB gain and phase) for the uncompensated amplifier and the compensation network. Thus, note that the amplitude curve has a slope that increases by 20 dB/decade between a frequency of 10 and 40 and then decreases by 20 dB/decade to return to a slope of −60 dB/decade. These two changes are due to the zero and pole of the compensation network function at frequencies of 10 and 40, respectively. The phase curve is similarly altered at the break frequencies of $\frac{1}{10}$ and 10 times the two corner frequencies, 10 and 40.

We now view the point at which the compensated phase curve crosses a phase shift of −180°. At this frequency (approximately 35 rad/s), the compensated gain is −34 dB. The system can therefore have any gain below 34 dB and still be stable. This is the overall gain including the compensation network. Since the compensation network has a gain (attenuation) of $\frac{1}{4}$ (−12 dB), the gain of the op-amp can be as high as

$$G_o = 34 \text{ dB} + 12 \text{ dB} = 46 \text{ dB}$$

Without compensation, the allowable gain is about 20 dB. Thus, an improvement of approximately 26 dB is realized. The system can be further improved if we vary the location of the zero and pole.

In the previous example, we found that the compensated gain is −34 dB at a frequency where the phase is −180°. The term *gain margin* is used for the difference between 0 dB and the actual gain when the phase is −180°. Thus, the gain margin is the amount that the gain can increase before the system is unstable.

Alternatively, we can view how much the phase can change at a frequency where the gain is 0 dB. This is known as the *phase margin.* (Note that since we did not assign an amplifier gain to the previous example, the concept of phase margin does not apply because the gain is always less than 0 dB.)

Example 11.9

Consider the negative-gain op-amp system of Figure 11.26(a). Investigate the stability of this three-pole amplifier.

SOLUTION The equivalent circuit of the amplifier is shown in Figure 11.26(b) where $v_d = v_+ - v_-$. The closed-loop gain, V_o/V_i, is found using equation (11.15), which is repeated here for convenience.

$$\frac{V_o}{V_i} = \frac{-G\gamma R_F/R_A}{1 + G\gamma}$$

where

$$\gamma = \frac{1}{1 + R_F/R_A}$$

The amplifier has the following third-order transfer function:

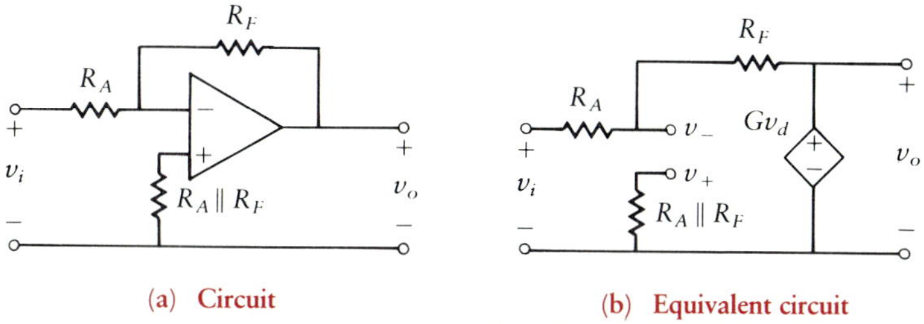

Figure 11.26 Amplifier for Example 11.9.

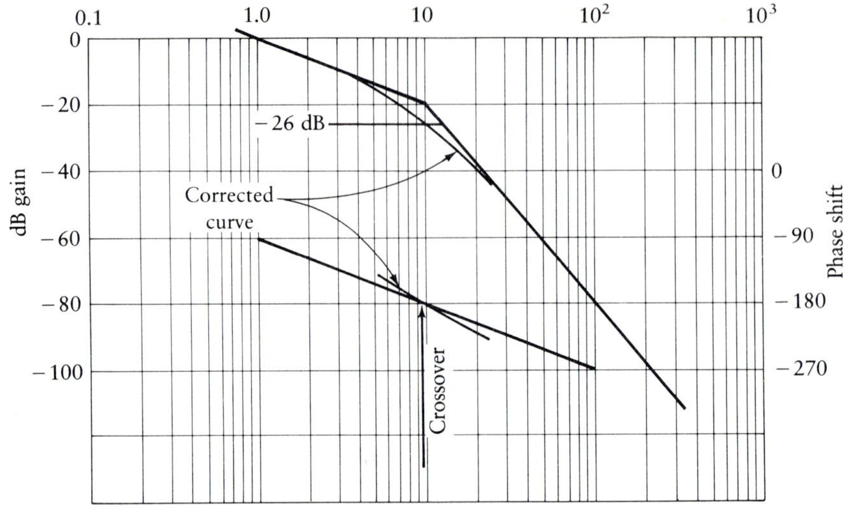

Figure 11.27 Bode plot for Example 11.9.

$$G = \frac{G_o}{s(1 + s/\omega_o)^2}$$

where

$$\omega_o = 10$$

The Bode plot for this system is shown in Figure 11.27. Notice that we construct the Bode plot for the loop gain, $G_o\gamma$, where we set $G_o\gamma = 1$. The phase shift reaches $-180°$ at a frequency of 10 rad/s, and the associated gain is -26 dB. The straight-line asymptotes account for -20 dB, and the additional -6 dB is due to the correction to the straight-line asymptote. Thus, the design allows for a gain of 26 dB before instability is experienced.

11.9 PHASE-LAG EQUALIZER

A *passive phase-lag network,* often termed an *integrator,* is shown in Figure 11.28(a). We obtain the output-input relationship,

$$\frac{V_o}{V_i} = \frac{CR_2s + 1}{C(R_1 + R_2)s + 1} \tag{11.36}$$

Since

$$C(R_1 + R_2) > CR_2$$

the phase shift of the output lags the phase shift of the input. That is, the phase of the denominator in equation (11.36) is larger than the phase of the numerator, so the overall phase shift is negative.

The transfer function is simplified by again defining two constants,

$$\alpha = 1 + \frac{R_1}{R_2}$$

$$\tau = R_2 C$$

The transfer function then becomes

$$\frac{V_o}{V_i} = G(s) = \frac{1}{\alpha}\frac{s + 1/\tau}{s + 1/\alpha\tau} \tag{11.37}$$

The Bode plot for this expression is shown in Figure 11.28(b). Hence we rely on the attenuation of the dB gain curve to improve the stability. Note that the maximum phase lag increases as α increases, and it approaches 90° as α approaches infinity. The frequency, ω_m, at which the maximum occurs decreases with increasing α. We again insert this network in the feedback path of the op-amp.

Lag networks possess certain disadvantages when contrasted with other forms of compensation. If the time constant of the lag network is too large, a large root occurs in the system function, which can lead to transients that die out very slowly.

Figure 11.28
Passive circuit lag network.

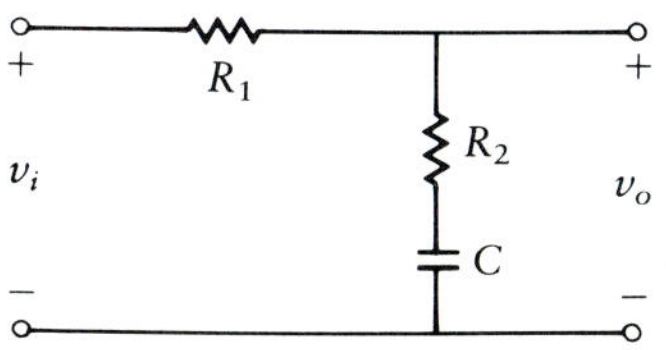

(a) Passive circuit lag network

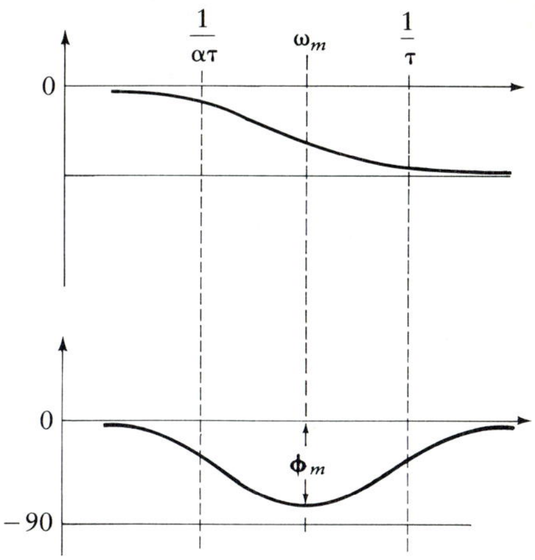

(b) Polar plot of phase-lag network

11.10 EFFECTS OF CAPACITIVE LOADING

The op-amp transfer function includes the effects of output capacitance and integrated circuit capacitance. A capacitive load lowers the corner frequency (i.e., a longer time constant leads to lower frequency) and is troublesome for capacitor values (C_L) of more than about 100 pF in a typical circuit configuration (see Figure 11.29(a)). The effect can be reduced by addition of a series resistance as shown in Figure 11.29(b). This resistance adds attenuation and modifies the gain-phase characteristic.

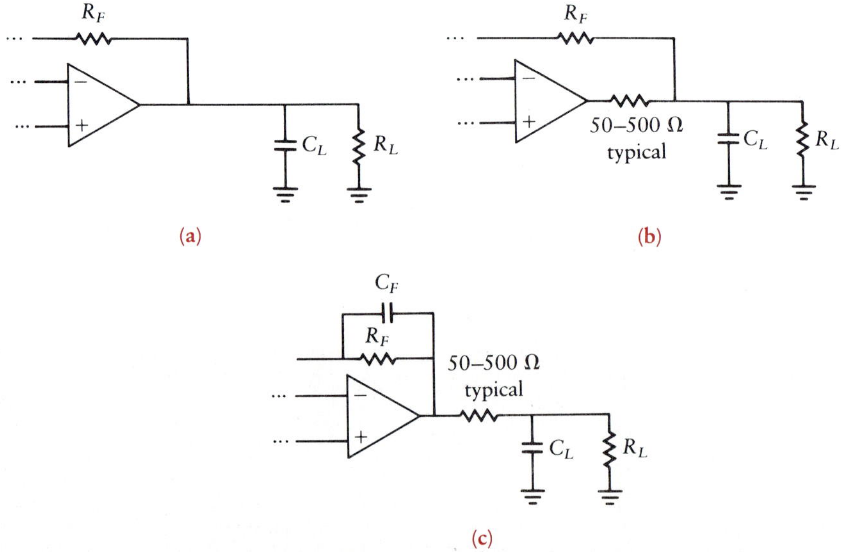

Figure 11.29 Capacitive loading.

Input impedance is frequently modeled as a capacitor, C_i, in parallel with a resistor, R_i. This R_iC_i network accounts for another of the poles of $G(s)$ and leads to a phase shift at high frequencies. Additional external capacitance across the input is troublesome when R_F is large because it lowers the corner frequency. Additional compensation in the form of a capacitor, C_F, in parallel with R_F is often used as a cure (see Figure 11.29(c)). This is known as *Miller-effect compensation.* The compensating circuit contributes one zero and one pole. The pole is far from the origin because of the Miller effect, which multiplies the capacitance value by the gain. The circuit parameters are chosen to place the zero coincident with the original circuit pole to effect cancellation. The result is a much higher corner frequency. This type of compensation is used extensively for internal compensation. Typical values for C_F are 3 to 10 pF.

11.11 OSCILLATORS

Sinusoidal driving sources are the building blocks of many systems. They are used extensively in communication systems, as well as in virtually every other application of electronics. A feedback system oscillates if the loop-transfer function reaches unity amplitude (0 dB) when the phase shift is −180°. Such a system produces an output without any input. In a sense, it "chases its tail," with the signal constantly recirculating and regenerating itself.

An equivalent technique exists for determining whether a feedback circuit will oscillate. We examine the loop gain of a negative-feedback system. If a phase shift of 180° occurs where the loop gain is 0 dB, the circuit is unstable and will oscillate.

The feedback model shown in Figure 11.1 is a good starting point for oscillator analysis. Instead of the negative feedback as shown, the feedback is positive. This occurs if the phase shift of the feedback network is 180°. The input signal can be removed without changing the output, since the output is fed back in phase with the input. If the circuit is to sustain oscillations, the loop gain must be unity at −180° phase shift.

While the basic concept of instability and positive feedback is common to all oscillators, there are a number of variations on the basic design. Common oscillators include the *Colpitts,* the *Hartley,* the *Wien bridge,* and the *phase shift* oscillators.

11.11.1 The Colpitts and Hartley Oscillator

We begin by analyzing the circuit shown in Figure 11.30. Several of the standard oscillators can be modeled in this manner. If Z_1 and Z_2 are capacitors and Z_3 is an inductor, the circuit is known as a *Colpitts oscillator.*

The impedances are given by $Z_1 = 1/sC_1$, $Z_2 = 1/sC_2$, and $Z_3 = sL$. Then

$$Z_L = sL \left\| \left(\frac{1}{sC_1} + \frac{1}{sC_2} \right) \right. \tag{11.38}$$

We expand equation (11.38) and find

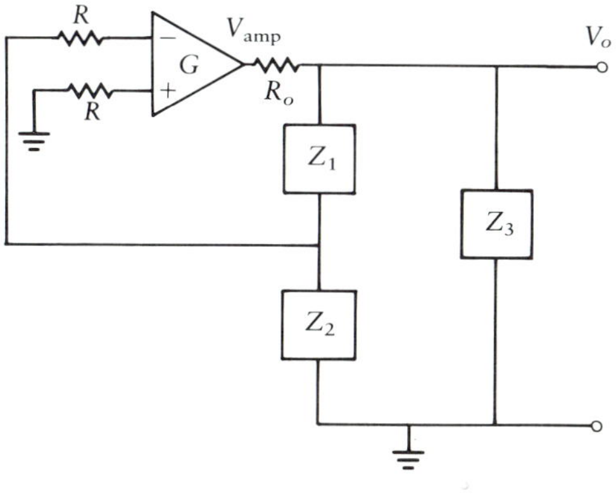

Figure 11.30
Oscillator circuit.

$$Z_L = \frac{sL(1/sC_1 + 1/sC_2)}{sL + 1/sC_1 + 1/sC_2} = \frac{sL}{s^2C_1C_2L/(C_1 + C_2) + 1} \tag{11.39}$$

The denominator of Z_L has poles on the j axis, which results in oscillation. When we substitute $s = j\omega$, the resonant frequency of the Colpitts oscillator is

$$\omega = \sqrt{\frac{C_1 + C_2}{C_1C_2L}} \tag{11.40}$$

Alternatively, in the circuit of Figure 11.30, if Z_1 and Z_2 are inductors and Z_3 is a capacitor, the circuit is a *Hartley oscillator.* The impedances are given by $Z_1 = sL_1$, $Z_2 = sL_2$, and $Z_3 = 1/sC$, and

$$Z_L = \frac{1}{sC} \,\Big\|\, (sL_1 + sL_2) = \frac{s(L_1 + sL_2)}{s^2C(L_1 + L_2) + 1} \tag{11.41}$$

Again, the denominator of Z_L has two poles on the imaginary axis, so the resonant frequency is

$$\omega = \frac{1}{\sqrt{C(L_1 + L_2)}} \tag{11.42}$$

11.11.2 Wien Bridge Oscillator

This relatively simple oscillator is shown in Figure 11.31. We analyze this circuit by writing equations for V_+ and V_- as follows:

$$V_+ = V_o\frac{Z_1}{Z_1 + Z_F} = \frac{RCsV_o}{s^2(RC)^2 + 3sRC + 1} \tag{11.43}$$

and

$$V_- = \frac{V_o}{1 + R_F/R_A} \tag{11.44}$$

We equate V_+ and V_- and let $s = j\omega$, $RC = \tau$,

$$\frac{1}{1 + R_F/R_A} = \frac{-\omega\tau}{j(1 - \omega^2\tau^2) - 3\omega\tau} \tag{11.45}$$

This complex equality is really two equations, since both the imaginary and real parts of the two expressions must be equal. Separating the two equations, we have

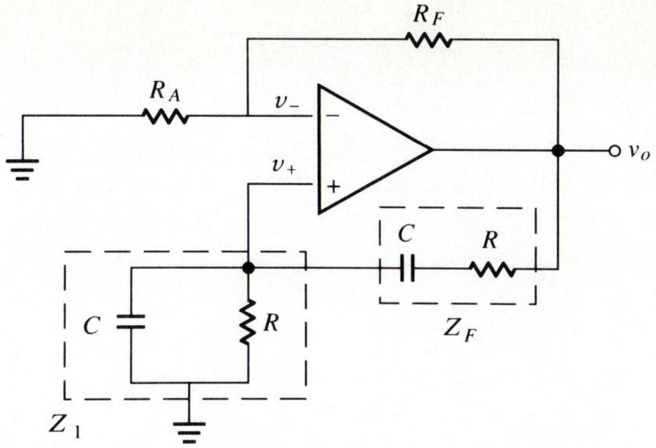

Figure 11.31
Wien bridge oscillator.

$$1 - \omega^2\tau^2 = 0$$

or

$$f = \frac{1}{2\pi RC} \tag{11.46}$$

and

$$\frac{1}{1 + R_F/R_A} = \frac{-\omega\tau}{-3\omega\tau} = \frac{1}{3}$$

with the result that

$$R_F = 2R_A \tag{11.47}$$

11.11.3 The Phase Shift Oscillator

The *phase shift oscillator* is shown in Figure 11.32. It includes three identical RC networks. Each provides a phase shift, resulting in the required $-180°$ total phase shift. To achieve oscillation, the loop gain (the product of the op-amp gain, A, and the transfer function, V_+/V_o) must satisfy the oscillation criteria. That is, the phase shift must be $-180°$ when the gain is 0 dB. This yields the design equation for this oscillator. We write three node equations for V_1, V_2, and V_+ as follows:

$$V_1\left(2sC + \frac{1}{R}\right) + V_2(-sC) = V_o sC \tag{11.48a}$$

$$V_1(-sC) + V_2\left(2sC + \frac{1}{R}\right) + V_+(-sC) = 0 \tag{11.48b}$$

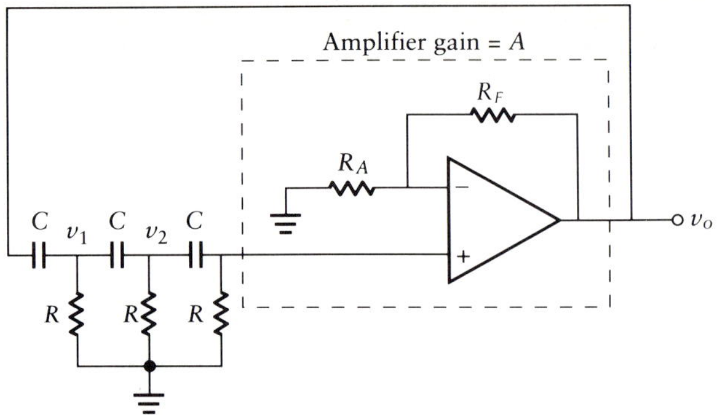

Figure 11.32
Phase-shift oscillator.

$$V_2(-sC) + V_+\left(sC + \frac{1}{R}\right) = 0 \tag{11.48c}$$

Equations (11.48) are solved for the transfer function, V_+/V_o, to obtain

$$\frac{V_+}{V_o}(s) = \frac{s^3\tau^3}{s^3\tau^3 + 6s^2\tau^2 + 5s\tau + 1} \tag{11.49}$$

where $\tau = RC$. We replace s with $j\omega$ to obtain

$$\frac{V_+}{V_o}(j\omega) = \frac{(\omega\tau)^3}{[(\omega\tau)^3 - 5\omega\tau] + j[1 - 6(\omega\tau)^2]} \tag{11.50}$$

To determine the frequency of oscillation, we set the phase shift of equation (11.50) to $-180°$, with the result

$$\text{Phase shift} = \tan^{-1}\left(\frac{1 - 6(\omega\tau)^2}{\omega\tau[(\omega\tau)^2 - 5]}\right) = -180° \tag{11.51}$$

Solving for ω (set the numerator of equation (11.51) to zero), we find

$$f = \frac{\omega}{2\pi} = \frac{1}{2\pi\sqrt{6}\,RC} \tag{11.52}$$

To determine the value of A, we find the magnitude of equation (11.50) at the frequency found in equation (11.52). The magnitude of equation (11.50) is

$$\left|\frac{V_+}{V_o}(j\omega)\right| = \frac{(\omega\tau)^3}{\{[(\omega\tau)^3 - 5\omega\tau]^2 + [1 - 6(\omega\tau)^2]^2\}^{1/2}} \tag{11.53}$$

We substitute the frequency from equation (11.52) to obtain

$$\left|\frac{V_+}{V_o}(j\omega)\right|_{\omega=1/\sqrt{6}RC} = \frac{1}{29} \tag{11.54}$$

For the loop gain to be 0 dB (unit amplitude), the amplifier gain must be at least 29.

This oscillator uses no inductors, is relatively easy to operate, and is inexpensive. However, the oscillator frequency is not accurately controlled, since it depends on the values of R and C. Similar accuracy problems exist for the Colpitts, Hartley, and Wien bridge oscillators. The next section presents a more accurate oscillator.

11.11.4 The Crystal Oscillator

When a voltage is applied across a piezoelectric crystal (usually quartz), the crystal oscillates in a stable and accurate manner. Electrodes are deposited on the opposite faces of the crystal so that the crystal can be excited. The frequency of oscillation is determined by the crystal dimensions. The sharpness of the frequency response curve, which is termed the Q of the resonant circuit (a high Q provides a sharp resonant curve), is much higher for a crystal than for other tuned circuits.

Since the crystal is a resonant circuit, it can be used in conjunction with discrete or integrated circuits. Crystal oscillators achieve accurate frequency control. Stabilities in the range of several parts per million in frequency variation can be achieved (over normal temperature ranges). Commercial quartz crystals are readily available with frequencies from a few kilohertz to a few hundred megahertz and with extremely high values of Q (10^4 to 10^5).

The equivalent circuit for a crystal is shown in Figure 11.33(a), where, since the Q is so high, we can neglect R. We write the equation for impedance as

$$Z(s) = \frac{1}{sC_p}\frac{s^2 + 1/LC}{s^2 + (1/C + 1/C_p)/L} \tag{11.55}$$

Figure 11.33 Crystal oscillator.

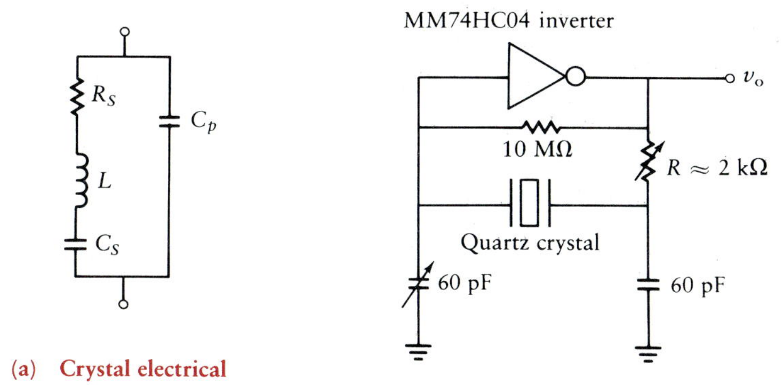

(a) Crystal electrical equivalent circuit

(b) Crystal oscillator circuit

Two frequencies exist, one for the zero

$$\omega_z = \frac{1}{\sqrt{LC}} \tag{11.56}$$

and one for the pole

$$\omega_p = \sqrt{\frac{1}{L}\left(\frac{1}{C} + \frac{1}{C_p}\right)} \tag{11.57}$$

However, since $C_p >> C$, $\omega_z \approx \omega_p$, so the frequency of equation (11.56) can be used for design.

A crystal can replace the tuned inductance-capacitance circuit in a wide variety of oscillators. The oscillator frequency is determined by the crystal rather than by the remaining part of the circuit.

A simple crystal oscillator can be designed using a CMOS inverter (CMOS inverters are discussed in Chapter 15) that is biased to operate as an ac amplifier. The square-wave oscillator, which is shown in Figure 11.33(b), provides a fixed frequency, which is determined by the crystal frequency. By placing a 10-MΩ resistor from the output to the input of the MM74HC04 CMOS inverter, we cause it to operate as an ac amplifier. This 10-MΩ negative-feedback resistor biases the inverter to operate at one-half of the supply voltage, with a gain of several thousand. The MM74HC04 inverter can produce a gain of 10^4. The loop gain can be adjusted with R, which is set to about 2 kΩ. The variable capacitor (approximately 60 pF) provides for small adjustments in frequency. The capacitors also prevent the crystal from oscillating at any frequency other than the fundamental. The higher harmonics are filtered out with the RC network around the crystal.*

11.11.5 Touch-Tone Generator

Touch-tone dialing permits a standard telephone keypad to be used for dialing a telephone number by generating a pair of frequencies or tones for each key pressed. The TP50-87DTMF tone generator, which is shown in Figure 11.34, supplies four row frequencies, R_i, and four column frequencies, C_i, as shown below.

R_1	697 Hz	C_1	1209 Hz
R_2	770 Hz	C_2	1336 Hz

---

* Additional information on this oscillator can be found in the CMOS application notes AN340, National Semiconductor Corp., 1984.

R_3	852 Hz	C_3	1477 Hz
R_4	941 Hz	C_4	1633 Hz

Figure 11.34 TP50-87DTMF tone generator. Courtesy of National Semiconductor Corp.

The block at the left of the figure indicates the tone combinations. For example, when the number 7 is pressed, the two sinusoidal frequencies of 852 Hz (R_3) and 1209 Hz (C_1) are simultaneously generated by the IC.

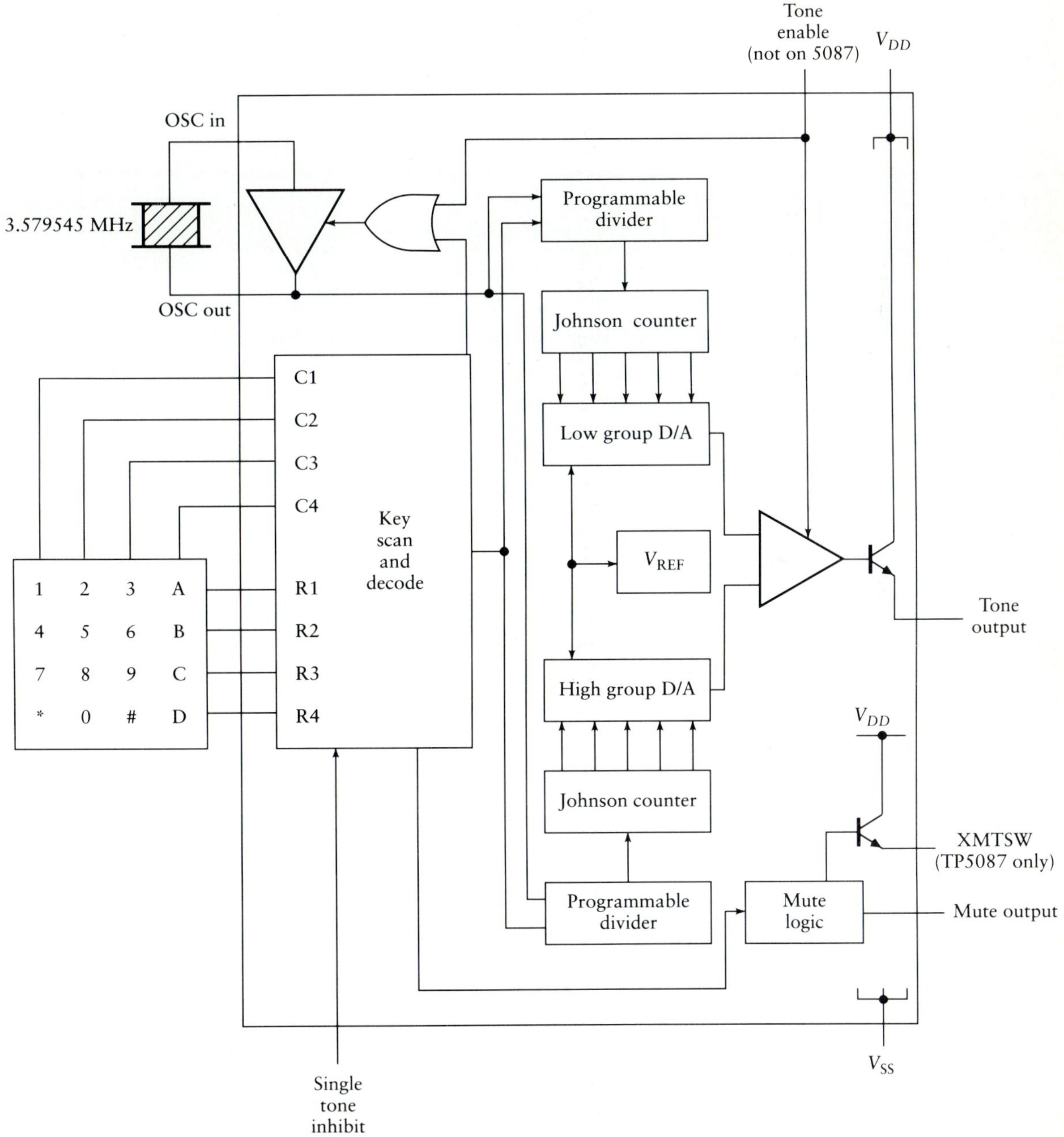

The basic frequency is generated from a low-cost 3.579545-MHz crystal (a standard NTSC TV color burst crystal), which feeds the IC. The row and column inputs, which are energized when a key is pressed, are decoded to develop the high-frequency tone and the low-frequency tone. The two tones are summed in a mixing amplifier and power amplified in an *npn* emitter follower. Some of the circuits shown in Figure 11.34 (i.e., dividers, counters, and D/A) are covered in more detail in subsequent chapters. For now, we note that the tones are derived by dividing the reference frequency by fixed numbers. For example, 697 Hz results from dividing the reference by 5136, and 1633 Hz results if we divide by 2192.

PROBLEMS

11.1 For the circuit of Figure P11.1, determine the voltage gain, input resistance, and current gain when $C_E \to \infty$.

11.2 Repeat Problem 11.1 when $C_E = 0$. Compare the values of A_v, R_{in}, and A_i found in each of these problems, and explain any differences.

11.3 For the amplifier of Figure P11.1, determine A_v, R_{in}, and A_i when $R_1 \to \infty$ and $R_2 = 200\ k\Omega$.

11.4 For the amplifier shown in Figure P11.1, determine the feedback attenuation factor when $C_E = 0$.

11.5 A feedback amplifier of the type shown in Figure 11.3(a) with $C = 0$ has a gain of -200 with no feedback, and the ratio of the feedback voltage to the output voltage is 0.2. What is the voltage gain of the amplifier? If the gain with no feedback increases to -300, what is the new voltage gain?

11.6 For the amplifier of Figure P11.2, determine A_v, R_{in}, and A_i when $R_F = 14\ k\Omega$.

11.7 A single-stage amplifier using voltage feedback, as shown in Figure P11.2, has a source voltage resistance of 2 kΩ and an R_F of 20 kΩ. Determine the open-loop and closed-loop gain of the amplifier.

Figure P11.1

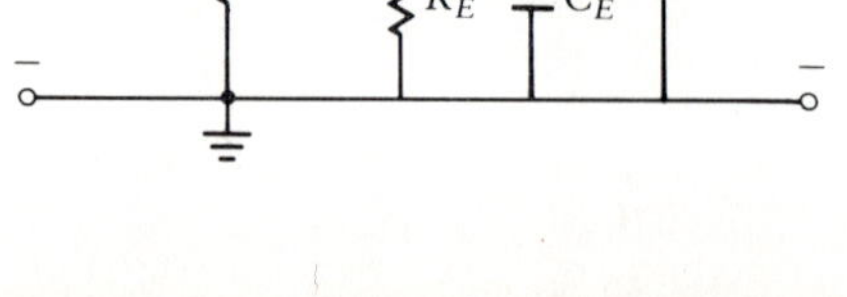

$R_C = R_L = 2\ k\Omega$

$R_E = 300\ \Omega$

$R_1 \| R_2 = 5\ k\Omega$

$\beta = 200$

$V_{CC} = 15\ V$

$h_{ie} = 400\ \Omega$

11.8 Using the same amplifier as in Problem 11.7, select the value of R_F to make the amplifier bias stable and determine the closed-loop gain.

11.9 Draw the amplitude and phase Bode plots for the operational amplifier of Figure P11.3. Let G_o = 120 dB and $\omega_c = 2\pi$ rad/s in the equation

$$G = \frac{G_o}{1 + s/\omega_c}$$

Use the following resistor values:

a. $\frac{R_F}{R_A} = 10^3$

b. $\frac{R_F}{R_A} = 10$

c. $\frac{R_F}{R_A} = 1$

11.10 Use the data sheets in Appendix D to determine the worst-case values of the following quantities for each of the three op-amps and conditions given below.

a. 741E op-amp at 25°C, ±15 V supplies, and a load resistance of 10 kΩ.

b. 741E op-amp at 50°C, ±20 V supplies, and a load resistance of 2 kΩ.

c. 101A op-amp at 50°C, ±15 V supplies, and a load resistance of 5 kΩ.

i. Gain (smallest is worst case).
ii. Input resistance (smallest is worst case).
iii. Input offset voltage (largest is worst case).
iv. Input bias current (largest is worst case).
v. Input offset current (largest is worst case).
vi. Output resistance (largest is worst case).
vii. Output voltage swing (smallest is worst case).

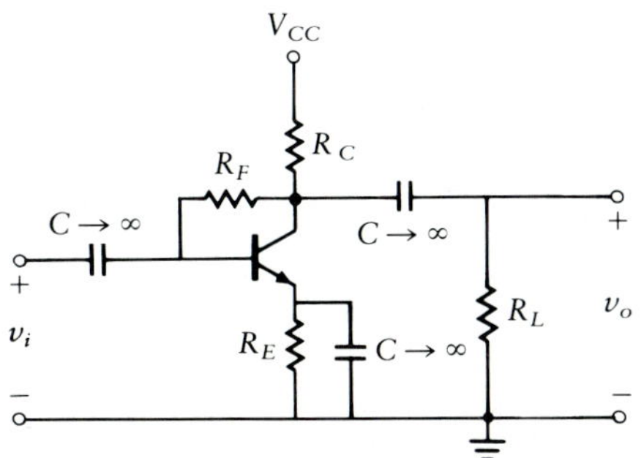

$R_C = R_L = 2$ kΩ

$R_E = 300$ Ω

$\beta = 200$

$V_{CC} = 15$ V

$h_{ie} = 400$ Ω

Figure P11.2

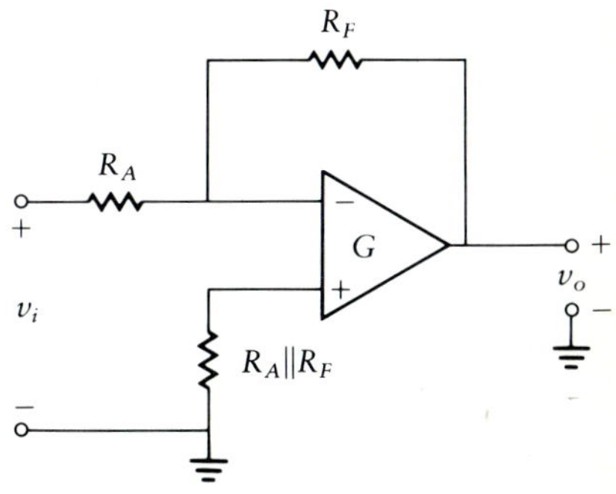

Figure P11.3

viii. Gain-bandwidth product (smallest is worst case).
ix. Power supply current (largest is worst case).

11.11 For the inverting op-amp amplifier shown in Figure 11.8(a), what is the percent error if the open-loop gain is only 50,000? Use $R_F = 200\ \text{k}\Omega$ and $R_A = 10\ \text{k}\Omega$.

In Problems 11.12 through 11.17, sketch decibel gain and phase curves for each function, and discuss the stability of each system as a function of K.

11.12 $\dfrac{K}{s(s+10)}$

11.13 $\dfrac{K}{s^2(s+10)}$

11.14 $\dfrac{K(s+1)}{s^2(s+10)}$

11.15 $\dfrac{K}{s(s+1)(s+100)}$

11.16 $\dfrac{K(s+1)}{s^3(s+100)}$

11.17 $\dfrac{K}{s(s^2+20s+100)}$

11.18 The op-amp system of Figure 11.22(a) has a gain function that is

$$G(s) = \frac{G_o}{1 + s/2\pi}$$

$R = 100\ \text{k}\Omega$ and $C = 0.1\ \mu\text{F}$. Determine the open- and closed-loop functions.

11.19 For the op-amp system of Problem 11.18, plot a Bode diagram with $G_o = 1$. Determine the stability of this system.

11.20 For the op-amp system of Figure 11.21(a), use

$$G(s) = \frac{G_o}{(1 + s/20\pi)^2}$$

$R = 160\ \text{k}\Omega$ and $C = 0.1\ \mu\text{F}$. Note the square in the op-amp gain function. Determine the open- and closed-loop functions.

11.21 For the op-amp system of Problem 11.20, plot the Bode diagram with $G_o = 1$. Determine the stability of the system.

For the expressions in Problems 11.22 through 11.24, sketch Bode plots and discuss the stability of each amplifier as a function of K.

11.22 $\dfrac{K}{s(s^2 + s + 25)}$

11.23 $\dfrac{K}{s^2(s^2 + s + 25)}$

11.24 $\dfrac{Ks}{s^2 + s + 25}$

11.25 Determine the value of gain, K, for which the system with

$$G(s)H(s) = \frac{K}{s(0.1s + 1)^2}$$

has phase shift of $-180°$ at a dB gain of 0 dB (i.e., marginal stability).

11.26 A zero is added to the system function of Problem 11.25 to improve performance. The new system open-loop function is

$$G(s)H(s) = \frac{K(0.125s + 1)}{s(0.1s + 1)^2}$$

Find the value of K for which the system is marginally stable.

11.27 Plot a Bode diagram for the op-amp system of Figure P11.4. For what values of the gain, K, is the system stable? Perfect the critical point before determining the value of K. Let $C = 1\ \mu\text{F}$, $L = 100$ mH, $R = 1\ \text{k}\Omega$. The op-amp gain function is given by

$$G(s) = \frac{K}{0.001s + 1}$$

11.28 Determine the stability of the integrator shown in Figure P11.5. The op-amp gain function is

$$G(s) = \frac{K}{(1 + s/20\pi)^3}$$

and $RC = 200$ ms. Find the value of K for marginal stability.

11.29 Determine the stability of the differentiator of Figure P11.6. The op-amp gain function is

$$G(s) = \frac{K}{(1 + s/40)^2}$$

and RC = 250 ms. Find K for marginal stability, i.e., 0-dB gain when the phase is $-180°$. Be sure to perfect the critical point.

11.30 Determine the ratio of R_F to R_A to meet the condition for oscillations of the Wien bridge oscillator shown in Figure P11.7 and determine the formula for frequency of oscillation.

11.31 Design a phase shift oscillator using a FET transistor as shown in Figure P11.8. Let g_m = 5000 μS, r_d = 50 kΩ, and the feedback resistors R = 20 kΩ. Select the value of C for the oscillator to operate at 10 kHz. Make sure the gain of the amplifier is at least -50 to compensate for loading.

11.32 Design a phase shift oscillator, as shown in Figure 11.32, to operate at 100 kHz. Let C = 0.1 pF and select appropriate values for R, R_F, and R_A.

11.33 In Problem 11.30, design the Wien bridge oscillator to oscillate at 100 MHz with R = 10 kΩ.

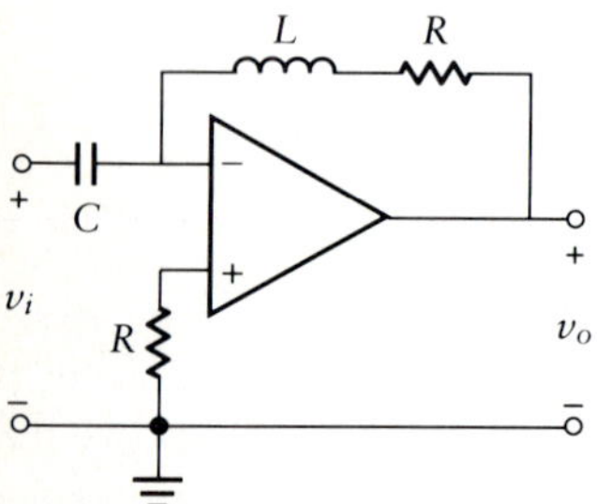

Figure P11.4

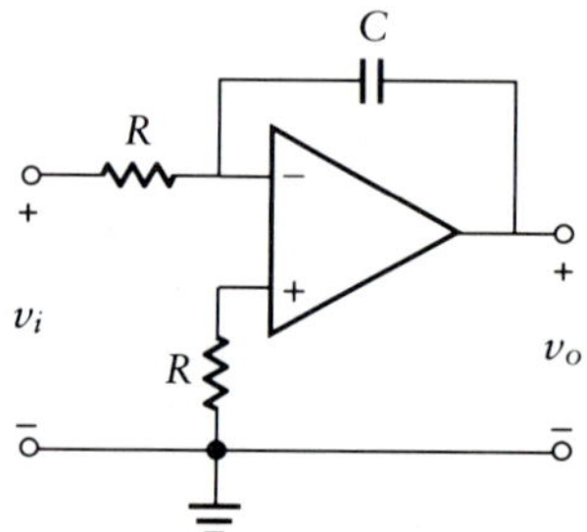

Figure P11.5

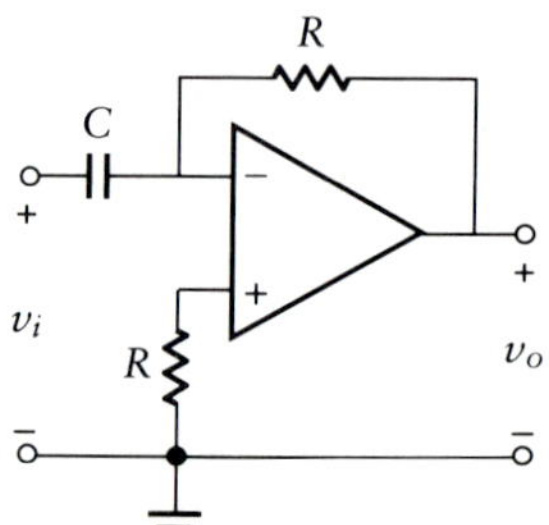

Figure P11.6

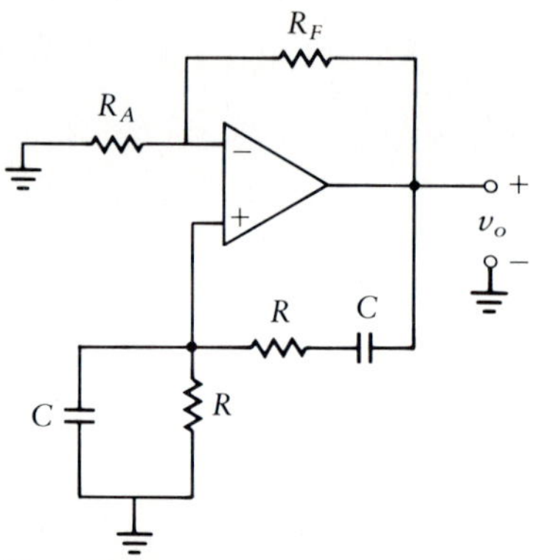

Figure P11.7

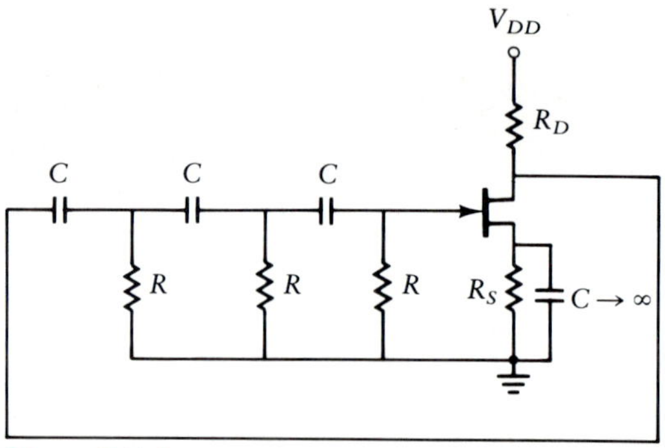

Figure P11.8

12

ACTIVE FILTERS

12.0 INTRODUCTION

Suppose you are employed by a company that designs and manufactures video games. The company has just embarked on manufacturing a new generation of games coupling compact disc technology with traditional microprocessor-controlled video games. The pioneer entry into this market is a game that has been named "The Colonization of Mars." The game couples actual footage of film taken on the surface of Mars with superimposed images of the player of the game. The Federal Communications Commission (FCC) must license the game because video signals from the player are transmitted by a small remote camera to the game module. Unfortunately, the FCC refused approval due to spurious signals. They claimed that your transmitter was interfering with signals in adjacent frequency bands. Your company has given you a small budget to eliminate the problem as quickly as possible, since firm orders for the game (at a fixed price) must be filled within 30 days.

Having taken undergraduate courses in network synthesis, you muster your knowledge and design a filter that meets the FCC specifications for reducing out-of-band transmissions. However, the specifications are so stiff that your filter requires a large number of components. Since components are inexpensive, this did not bother you until you became aware of the "real estate" problem. That is, the expense of the components is not so much the problem as is the space to house them. Alas, after searching your storehouse of experience and knowledge, you cannot come up with a solution. Another engineer working for the same company asks you if you considered active filters. Indeed, you have not. The current chapter will help you correct this deficiency in your background and will give you the tools to solve the problem described above.

A *filter* is a system designed to achieve a desired transfer characteristic. That is, it operates on an input signal (or signals) in a predetermined manner. *Passive linear filters* are normally considered as part of a study of circuits, networks, or linear systems. They are composed of combinations of resistors, inductors, and capacitors. Although it is possible to achieve a wide variety of transfer characteristics using these elements, a large number of components is often required. This leads us to explore alternatives to passive filters.

Active filters contain amplifiers, which permit the design of a wide range of transfer functions (within some broad restrictions related to the properties of the transfer function).

The practical details of active filters are presented in this chapter. Integrators and differentiators are first considered, followed by first- and second-order active filters. The important concepts of transfer function, impulse response, frequency response, and filter classification are developed.

The low-pass, high-pass, band-pass, and band-stop filters are then discussed. Transfer functions are analyzed prior to presentation of actual filter circuits. A general configuration is presented and then adapted to the various types of filters.

Section 12.5 explores classical analog filters. In particular, the Butterworth and Chebyshev filters are discussed. Practical design techniques are presented throughout the chapter. The chapter concludes with switched-capacitor filters, and an IC version of this useful filter.

12.1 INTEGRATORS AND DIFFERENTIATORS

A simple two-element passive circuit can approximate an integrator or differentiator. For example, a series RC circuit can perform this function.

Figure 12.1 illustrates an RC circuit with a voltage source applied. Since the current in a capacitor is proportional to the derivative of the voltage across the capacitor, the circuit acts as a differentiator provided that the resistance, R, is made small compared to the capacitive impedance, $1/\omega C$, at the frequencies of interest. The capacitor voltage is then almost equal to the input voltage, since

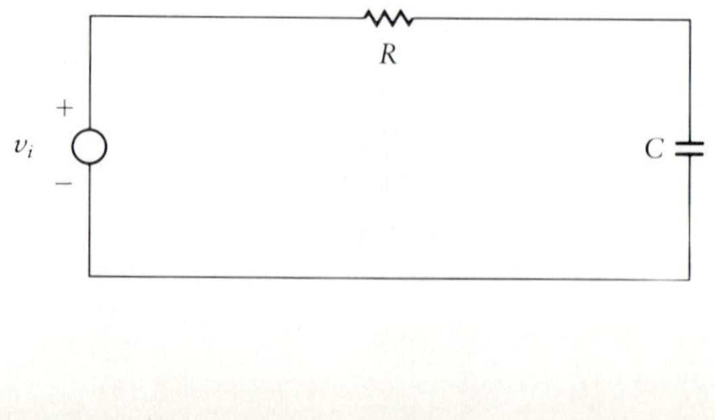

Figure 12.1
RC circuit.

the voltage across R is small. The current is therefore approximately equal to the derivative of the input voltage. The output, which is taken across the resistor, is approximately proportional to the derivative of the input voltage.

In order to approximate an integrator, the opposite approach is used. That is, we want the current in the circuit to be proportional to the input voltage. This is achieved by making the resistance predominate over the capacitive impedance. Since the voltage across a capacitor is proportional to the integral of the current through it, the output is taken across the capacitor.

This passive circuit approximates the operations of differentiation and integration, and the approximation gets better as the impedance of one of the elements is made much smaller than that of the other. Unfortunately, as this impedance gets smaller, the magnitude of the output voltage also gets smaller.

Considerable improvement of performance is possible if we switch from passive to active circuits. Let us begin our study with the basic op-amp of Figure 12.2(a). The gain of this ideal op-amp circuit is given by equation (12.1), where Z_F is the impedance in the feedback path and Z_A is the impedance between the source and the inverting input terminal.

$$\frac{V_o}{V_i} = -\frac{Z_F}{Z_A} \tag{12.1}$$

The Laplace transform (see Appendix C) operator notation for an integrator is $1/s$. Therefore, if we choose impedances that make the right side of equation (12.1) proportional to $1/s$, we have accomplished integration. For example, if Z_F is the impedance of a capacitor and Z_A is that of a resistor, the gain takes the desired form. Alternatively, Z_F can be the impedance of a resistor and Z_A that of an inductor. Figure 12.2(b) shows one form of the integrator. The behavior of this circuit is described by the following equation:

$$V_o(s) = -\left(\frac{1}{sRC}\right)V_i(s)$$

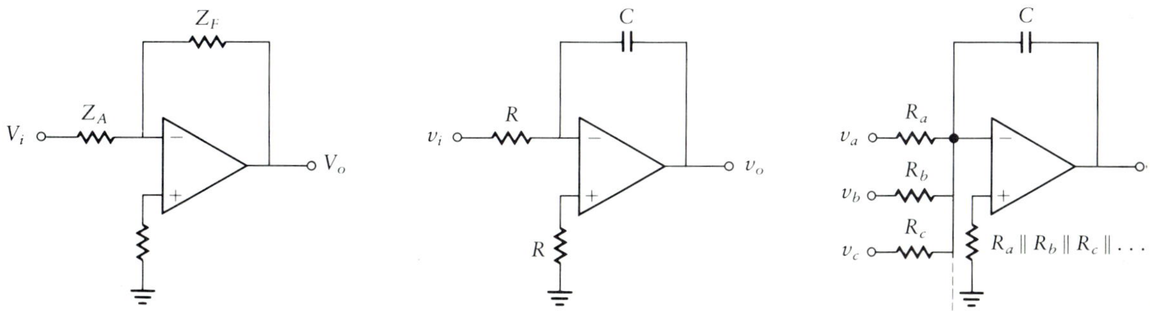

(a) (b) (c)

Figure 12.2 Op-amp circuits.

When this equation in operator notation is converted to the time domain, we obtain

$$v_o(t) = -\left(\frac{1}{RC}\right)\int_0^t v_i(\tau)\,d\tau \tag{12.2}$$

The above analysis makes the active filter appear to be a perfect integrator. Of course, we are using the ideal model for the op-amp, so this is still an approximation. However, this active circuit yields a much better approximation to an integrator than can be achieved with passive circuits.

Another advantage of using op-amps is that they can easily sum inputs. A *summing inverting integrator* is illustrated in Figure 12.2(c). This configuration yields the following equations:

$$V_o(s) = -\frac{1}{s}\left[\frac{V_a(s)}{R_a C} + \frac{V_b(s)}{R_b C} + \cdots\right] \tag{12.3}$$

In the time domain, this becomes

$$v_o(t) = -\int_0^t \left[\frac{v_a(\tau)}{R_a C} + \frac{v_b(\tau)}{R_b C} + \cdots\right] d\tau$$

It is easy to form differences between input signals when op-amps are used in the circuits. A *differencing integrator* is shown in Figure 12.3. The equations are obtained as follows:

$$V_- = \frac{(V_o - V_1)R}{R + 1/sC} + V_1$$

$$= \frac{RCsV_o + V_1}{RCs + 1}$$

$$V_+ = \frac{1/Cs}{R + 1/Cs}V_2 = \frac{RCsV_o + V_1}{RCs + 1}$$

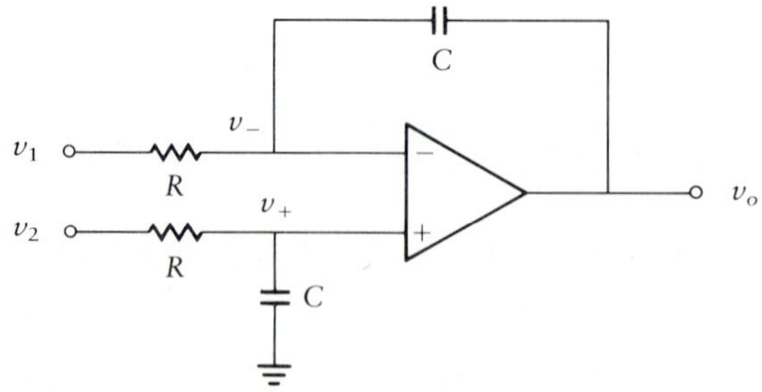

Figure 12.3 Differencing integrator.

Note that we are using uppercase notation because the voltages are functions of s. Since $G \rightarrow \infty$, $V_+ = V_-$, and

$$V_o = \frac{V_2 - V_1}{RCs} \tag{12.4}$$

In the time domain, this becomes

$$v_o(t) = \frac{1}{RC}\int_0^t [v_2(\tau) - v_1(\tau)]\,d\tau$$

The components R and C are chosen to achieve the required dc gain.

If the capacitor and resistor of Figure 12.2(b) are interchanged, the result is the *basic differentiator,* as shown in Figure 12.4(a). Since $V_+ = V_- = 0$, we write a node equation at the inverting terminal of the op-amp and solve for V_o, obtaining

$$V_o = -RCsV_i \tag{12.5}$$

In the time domain, equation (12.5) becomes

$$v_o(t) = -RC\frac{dv_i(t)}{dt}$$

Differentiation circuits are not often used. One of the reasons for this is the effect the circuit has on random noise. Integration is a smoothing process, whereas differentiation is the opposite. For example, if noise spikes are present, differentiation leads to higher spikes because of the presence of large slopes in the noise voltage waveform. A second reason for avoiding differentiation circuits is that, because the differentiator accentuates higher frequencies, these circuits are more seriously affected by op-amp bandwidths.

As with integrators, we can sum or take the difference between inputs. A

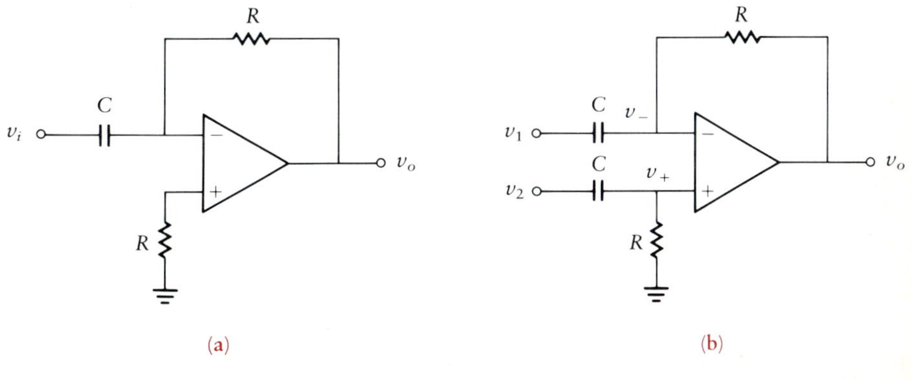

Figure 12.4
Differentiator.

differencing differentiator is shown in Figure 12.4(b). The equations describing its operation are derived from the basic op-amp relationships as follows:

$$V_+ = \frac{V_2 R}{R + 1/Cs} = \frac{V_2 RCs}{RCs + 1}$$

$$V_- = (V_o - V_1)\frac{1/Cs}{1/Cs + R} + V_1$$

$$= \frac{V_o + RCsV_1}{RCs + 1}$$

Since $G \to \infty$, $V_+ = V_-$, and we obtain

$$V_o = (V_2 - V_1)RCs \tag{12.6}$$

In the time domain, this becomes

$$v_o(t) = RC\frac{d[v_2(t) - v_1(t)]}{dt} \tag{12.7}$$

The process of differentiation accentuates high-frequency components. That is, taking the derivative of a sinusoid multiplies the amplitude by the frequency of the waveform. For this reason, if differentiators are used, they are often combined with filters whose transfer function attenuates high frequencies. In reality, the op-amp frequency limitations often provide this attenuation without the need for additional circuitry. Thus, at low frequency, the transfer function approaches a multiple of s, whereas at high frequencies the function approaches either a constant or zero. Examples of typical composite transfer functions are

$$H_1(s) = \frac{100s}{s + 100}$$

$$H_2(s) = \frac{100s}{s^2 + 20s + 100}$$

Both of these functions approximate $H(s) = s$ at low frequencies (i.e., $s = j\omega \to 0$).

12.2 ACTIVE-NETWORK DESIGN

Network synthesis is the study of techniques for going from a desired transfer function to a practical circuit implementation. The addition of an amplifier to

Figure 12.5
Operational amplifier used for active-network synthesis.

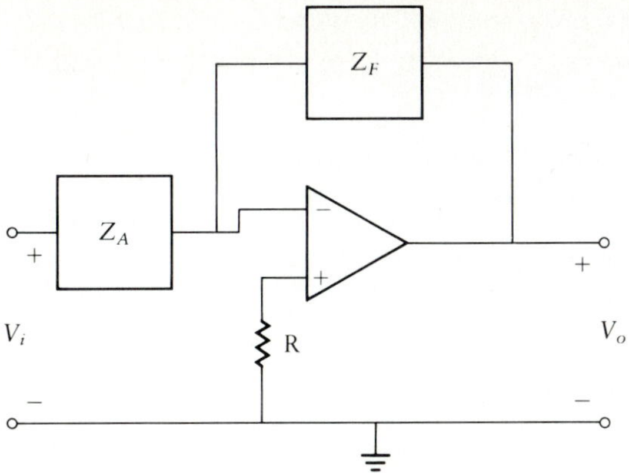

the passive circuit results in a transfer function that has the form of a ratio of two polynomials. The constant multiplier (overall gain) may be changed, but the form of the transfer function remains the same.

The general form of the active network is repeated as Figure 12.5.

A number of advantages are associated with using active networks as compared to passive networks. We list a few of these below.

- *Low cost* In low-frequency applications, inductors can become quite large and expensive. Active filters that use op-amps usually do not require inductors.
- *Cascadability* Because of good isolation, complex filters can be broken down into a series of simple sections. Cascadability allows each filter section to be designed separately and then cascaded so that the total transfer function becomes the product of the individual section functions.
- *Gain* Active filters can produce gain as needed to suit system or filter requirements.

Along with these advantages are some limitations.

- *Supply power* A power supply is needed for all active filters, whereas passive filters do not require any supply.
- *Signal limits* The op-amp inherently has definite signal limits beyond which nonlinear operation occurs.
- *Frequency limits* The op-amp cannot respond at high frequencies. It may have a cutoff frequency too low to allow it to be used in a particular application.

One of the most direct methods of active-network design is based on equation (12.1). The impedances, Z_F and Z_A, need not represent single elements. In fact, they can be any achievable two-terminal functions resulting from combinations of elements. Table 12.1 illustrates some common two-terminal configurations and their associated impedance functions.

Table 12.1 Two-Terminal Functions.

	Circuit	Transfer Function
(a)	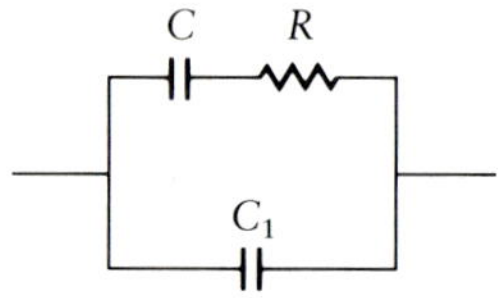	$\dfrac{s + \dfrac{1}{RC}}{C_1 s\left(s + \dfrac{1}{R}\dfrac{C_1 + C}{C_1 C}\right)}$
(b)	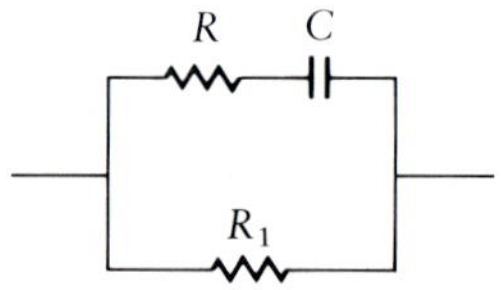	$\dfrac{RR_1\left(s + \dfrac{1}{RC}\right)}{(R + R_1)\left[s + \dfrac{1}{C(R + R_1)}\right]}$
(c)	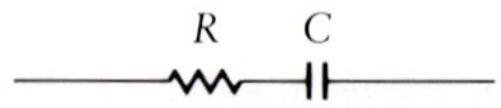	$\dfrac{RCs + 1}{Cs}$
(d)	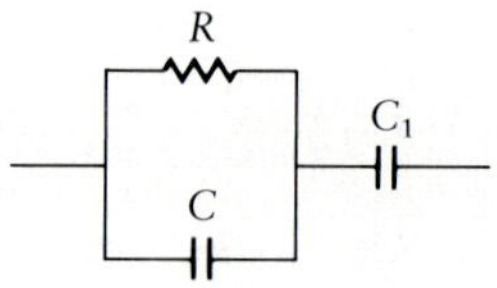	$\dfrac{(C + C_1)\left[s + \dfrac{1}{R(C + C_1)}\right]}{CC_1 s\left[s + \dfrac{1}{RC}\right]}$
(e)	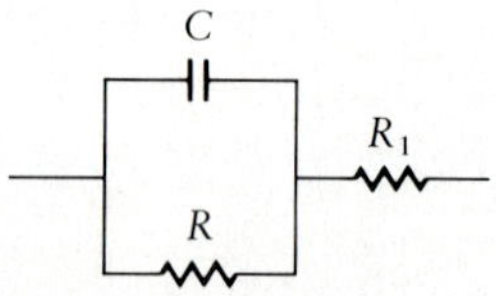	$\dfrac{R_1\left(s + \dfrac{1}{C}\dfrac{R + R_1}{R_1 R}\right)}{s + \dfrac{1}{RC}}$
(f)	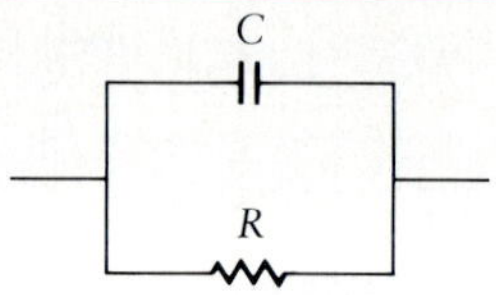	$\dfrac{1}{C\left(s + \dfrac{1}{RC}\right)}$

Example 12.1

Design a network to achieve the following transfer function:

$$\frac{V_o}{V_i} = \frac{-(s + 30)^2}{(s + 50)(s + 10)} = \frac{-(s + 30)}{s + 50}\,\frac{s + 30}{s + 10}$$

SOLUTION We use the circuit of Figure 12.5. Comparing equation (12.7) with equation (12.1) and solving for the two-terminal impedance functions, we find one possible implementation:

$$Z_F = \frac{s + 30}{s + 50}$$

and

$$Z_A = \frac{s + 10}{s + 30}$$

Reference to Table 12.1 indicates several two-terminal networks that will satisfy equation (12.8). In particular, entry (a) or (d) can be used.

Drill Problem

D12.1 Design an active network to generate the following transfer function.

$$\frac{V_o}{V_i} = -\frac{(s + 4)^2}{s(s + 20)^2}$$

Use no inductors and one op-amp.

Ans: Use circuit (b) from Table 12.1 for Z_A and circuit (d) for Z_F. The result is shown in Figure D12.1.

Figure D12.1

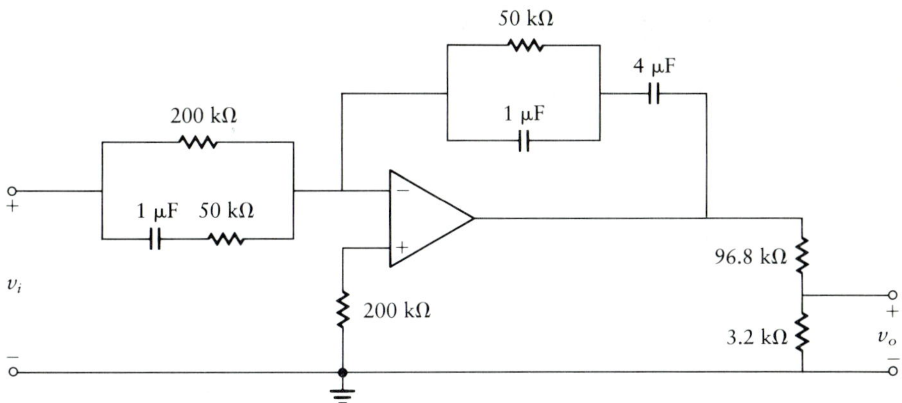

12.3 ACTIVE FILTERS

The word *filter* refers to the removal of an undesired portion of the frequency spectrum. It was originally applied to systems that eliminate undesired frequency components from a time waveform. The word is now used in a much more general way to include systems that simply weight the various frequency components of a function in a desired manner. Four of the most commonly encountered classes of filters are low-pass, high-pass, band-pass, and band-stop filters.

Ideal *low-pass* filters allow frequencies up to a given limit to pass and attenuate frequencies above that limit. Ideal *high-pass* filters are just the reverse of low-pass filters in that they pass frequencies above the limit and attenuate those below. Ideal *band-pass* filters allow only a particular band of frequencies to pass and attenuate the remaining frequencies. Ideal *band-stop* filters are just the reverse of band-pass filters in that they pass frequencies outside the particular band and reject frequencies within the band.

Active filters produce gain and usually consist only of resistors and capacitors along with integrated circuits. The op-amp, when combined with resistors and capacitors, can simulate the performance of passive inductive-capacitive filters. For high-order filters, the active configurations are much simpler than the passive ones.

When designing a circuit or system, constraints are usually imposed. Meeting the desired specifications is the heart of the design. Specifications can include the *roll-off* (the rate of attenuation of a signal outside the pass band), the *corner* (or *cutoff*) frequency, and the *peaking* (amount of gain produced at the resonant frequency of the circuit). These are frequency domain requirements. Time domain requirements are usually also important since they determine the transient response. These are commonly expressed in terms of *rise time, overshoot,* and *settling time* for prescribed inputs (usually step functions).

Often one constraint can be met only at the expense of another. In these situations, the engineer is forced to make a trade-off between the desired parameter and its unwanted counterpart.

12.3.1 Filter Properties and Classification

The transfer function, $H(s)$, for a single-input, single-output linear time-invariant system is the ratio of the Laplace-transformed output, $Y(s)$, to the Laplace-transformed input, $R(s)$, with all initial conditions set equal to zero. Note that since we are talking about general systems rather than specific circuits, we use the more general notation of $r(t)$ for the input and $y(t)$ for the output. For specific circuits, these become the voltages $v_i(t)$ and $v_o(t)$. The transfer function is the Laplace transform of the impulse response of the system, since the Laplace transform of an impulse is unity. Figure 12.6 illustrates

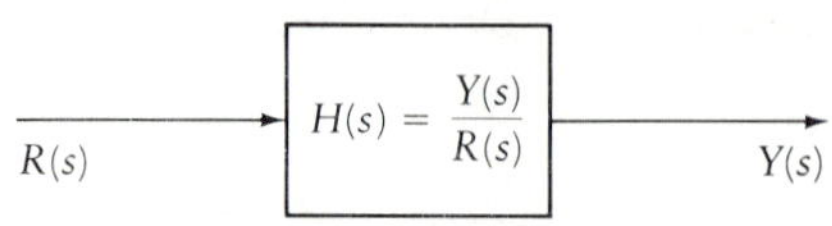

Figure 12.6
Definition of transfer function.

a general system and introduces the notation just described.

The output and input for a linear time-invariant system are related by a differential equation as follows:

$$\frac{d_n y}{dt^n} + a_{n-1}\frac{d_{n-1}y}{dt^{n-1}} + \cdots + a_1\frac{dy}{dt} + a_0 y = b_m\frac{d_m r}{dt^m} + b_{m-1}\frac{d_{m-1}r}{dt^{m-1}} + \cdots + b_0 r \tag{12.8}$$

The corresponding transfer function is then given by equation (12.9).

$$H(s) = \frac{b_m s^m + b_{m-1}s^{m-1} + \cdots + b_1 s + b_0}{s^n + a_{n-1}s^{n-1} + \cdots + a_1 s + a_0} \tag{12.9}$$

Example 12.2

Determine the transfer function of the system described by the following differential equation.

$$\frac{d^2y}{dt^2} + 2\frac{dy}{dt} + 10y = 6\frac{d^2r}{dt^2}$$

SOLUTION We take the Laplace transform of the equation with zero initial conditions to obtain

$$H(s) = \frac{Y(s)}{R(s)} = \frac{6s^2}{s^2 + 2s + 10}$$

The transfer function can be obtained from the impulse response, as indicated in Figure 12.7(a). If an impulse function is applied as the system input, the system output has a Laplace transform equal to the system transfer function. The system impulse response is thus the inverse Laplace transform of the transfer function.

The transfer function can also be determined from the sinusoidal steady-state response shown in Figure 12.7(b). If a sinusoidal signal forms the input to a system, the ratio of output magnitude to input magnitude is the magnitude

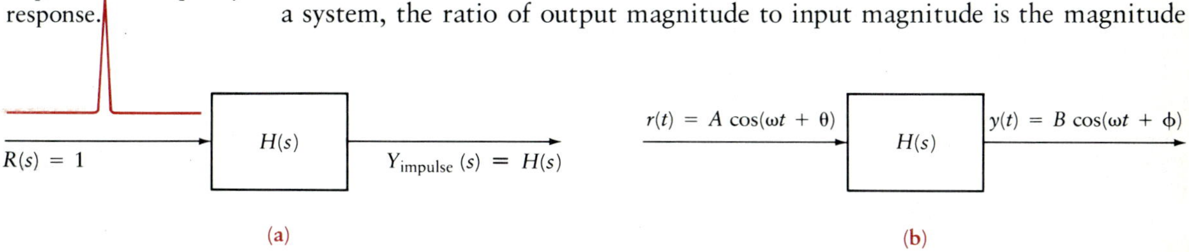

Figure 12.7
Impulse and frequency response.

of the transfer function evaluated at $s = j\omega$. The difference in phase between output and input is the phase shift of the transfer function at this frequency.

Example 12.3

A system has a transfer function given by

$$H(s) = \frac{10}{s + 4}$$

Determine the frequency response of this system.

SOLUTION The frequency response of this system has an amplitude function

$$|H(s = j\omega)| = \sqrt{\frac{10}{16 + \omega^2}}$$

and the phase shift function

$$\underline{/H(s = j\omega)} = \tan^{-1}\left(\frac{\omega}{4}\right)$$

Figure 12.8
Low-pass and high-pass filters.

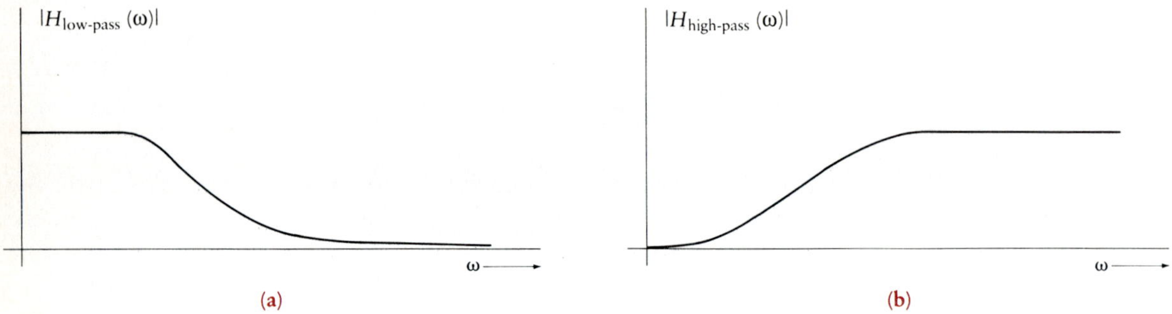

We now examine the frequency response of the general classifications of filters. A *low-pass* filter has a frequency response that is nonzero for $\omega = 0$ and approaches zero as $\omega \to \infty$, as shown in Figure 12.8(a). Note that we are considering amplitude only and ignoring phase shift.

Example 12.4

Describe the frequency response of a system with transfer function

$$H(s) = \frac{10}{s + 2} \quad \text{and} \quad H(j\omega) = \frac{10}{j\omega + 2}$$

SOLUTION The gain at $\omega = 0$ is 5 and declines at -20 dB/decade for frequencies larger than the corner frequency of $\omega = 2$ rad/s. This is a low-pass filter.

A *high-pass* filter has a magnitude characteristic that is zero for $\omega = 0$ and nonzero, approaching a constant, as $\omega \to \infty$, as shown in Figure 12.8(b).

Example 12.5

Describe the frequency response of a system with transfer function

$$H(s) = \frac{s}{s + 2} \quad \text{and} \quad H(j\omega) = \frac{j\omega}{j\omega + 2}$$

SOLUTION The zero at the origin yields zero magnitude for $\omega \to 0$. The magnitude approaches a constant, 1, for $\omega \to \infty$. This is a high-pass filter.

A *band-pass* filter has a magnitude characteristic that is zero for $\omega = 0$ *and* for $\omega \to \infty$ but has a nonzero magnitude for a band of frequencies between these extremes. This is illustrated in Figure 12.9(a).

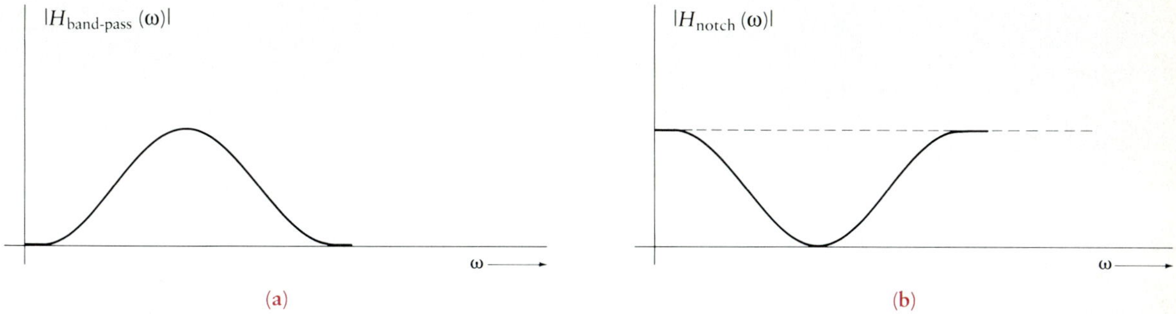

Figure 12.9
Amplitude response of band-pass and band-stop filters.

Example 12.6

Describe the frequency response of a system with transfer function

$$H(s) = \frac{s}{s^2 + s + 20} \quad \text{and} \quad H(j\omega) = \frac{j\omega}{(20 - \omega^2) + j\omega}$$

SOLUTION The zero at the origin leads to a magnitude of zero for $\omega = 0$. The higher-order denominator causes the magnitude to be zero for $\omega \to \infty$. For $\omega = \sqrt{20}$, the amplitude is unity. This is a band-pass filter.

A *band-stop* or *notch* filter has a magnitude characteristic that is nonzero for $\omega = 0$ and the same nonzero value for $\omega \to \infty$. The amplitude is zero (or nearly zero) over some range of values of frequency. This is shown in Figure 12.9(b).

Example 12.7

Describe the frequency response of a system with transfer function

$$H(s) = \frac{3s^2 + 60}{s^2 + s + 20} \quad \text{and} \quad H(j\omega) = \frac{60 - 3\omega^2}{(20 - \omega^2) + j\omega}$$

SOLUTION The numerator of this transfer function is zero for $\omega = \sqrt{20}$ (i.e., $s = j\sqrt{20}$) and the constant term and the highest-power terms have the same ratio, in this case 3:1. This is a notch filter.

A nontrivial filter for which the magnitude is constant (or nearly constant) is termed an *all-pass* filter or *delay equalizer.* A magnitude characteristic plot for such a filter is shown in Figure 12.10. Only the magnitude is shown in the figure.

The following two examples show typical forms of the transfer functions for an all-pass filter.

$$H(s) = \frac{s - 10}{s + 10} \quad \text{and} \quad H(j\omega) = \frac{j\omega - 10}{j\omega + 10}$$

$$H(s) = \frac{s^2 - 3s + 20}{s^2 + 3s + 20} \quad \text{and} \quad H(j\omega) = \frac{(20 - \omega^2) - 3j\omega}{(20 - \omega^2) + 3j\omega}$$

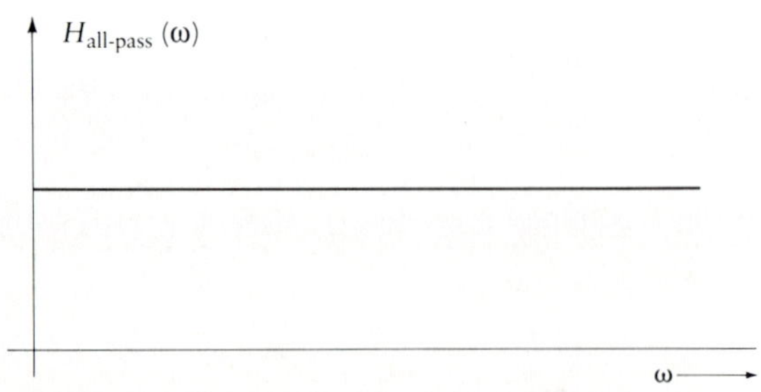

Figure 12.10
All-pass network.

It is not possible to achieve a phase that is identically zero, $\phi(\omega) = 0°$. In fact, if the phase were zero, the filter would be "trivial." That is, the resulting circuit would not produce any change in a signal. The closest approximation for many applications is to let the phase shift be proportional to frequency, as follows:

$$\phi(\omega) = -\tau\omega$$

where τ is a constant corresponding to a constant *time delay* of signals passing through the filter. The delay is τ seconds if ϕ is expressed in radians.

Drill Problems

Find the transfer function, $H(s)$, for the circuits shown in Drill Problems D12.2 to D12.4. Let all resistors, R_1, R_2, R_3, and R_4, be 100 kΩ. Also let $C_1 = 10\ \mu F$; $C_2 = 0.01\ \mu F$.

D12.2

Ans: $-\dfrac{(s + 2000)}{(s + 1000)}(s + 1)$

D12.3

Ans: $-\dfrac{s^2 + 2000s + 1000}{s(s + 1000)}$

D12.4

Ans: $\dfrac{10^3(s + 1)}{s(s + 2)}$

D12.5 What type of filter does the following transfer function describe?

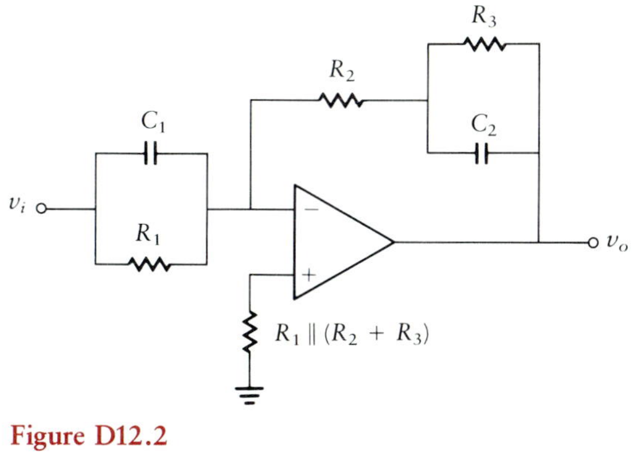

Figure D12.2

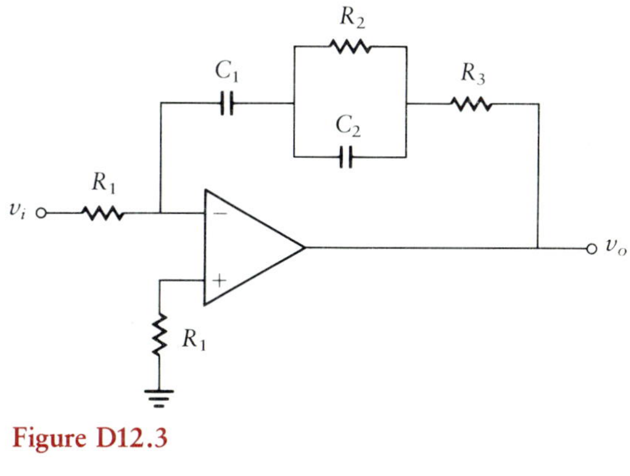

Figure D12.3

Figure D12.4

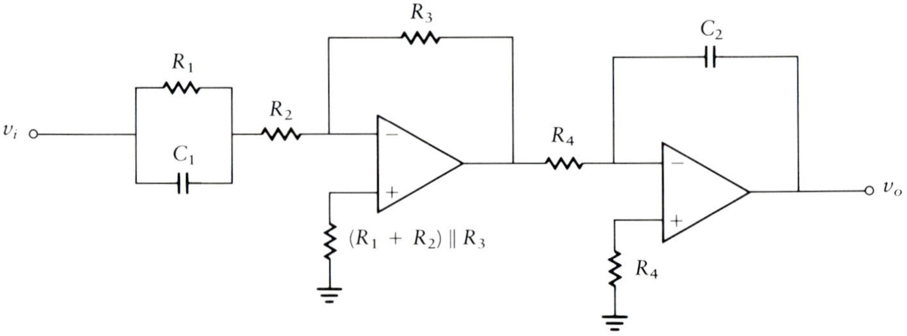

$$H(s) = \frac{5s - 10}{s^2 + s + 200}$$

Ans: Low-pass filter

12.3.2 First-Order Active Filters

An active RC low-pass filter is shown in Figure 12.11(a). The voltages for the two op-amp inputs are as follows:

$$V_+ = \frac{(1/sC)V_i}{R + 1/sC} = \frac{V_i}{RsC + 1}$$

Figure 12.11
The low-pass filter.

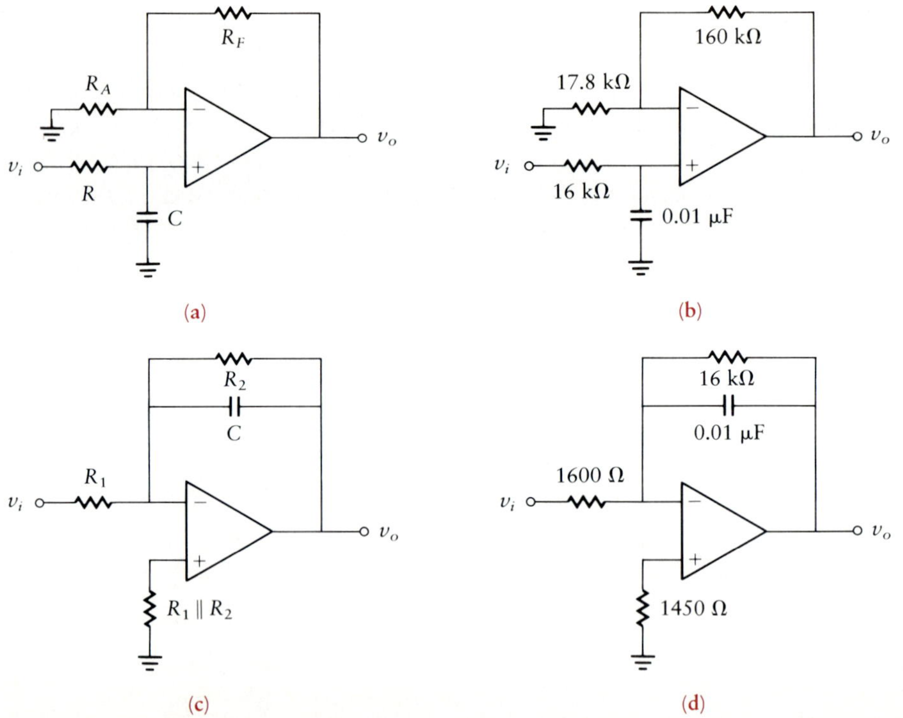

$$V_- = \frac{R_A V_o}{R_A + R_F}$$

Setting $V_+ = V_-$ (since we consider an ideal op-amp) and solving for the transfer function yields

$$H(s) = \frac{V_o}{V_i} = \frac{(1 + R_F/R_A)/RC}{s + 1/RC} \tag{12.10}$$

This is of the same form as the expression in Example 12.4 and represents a low-pass filter.

Notice that this first-order active filter has one pole (a first-order denominator polynomial). The corner frequency is at $1/RC$ and the dc gain (found by setting $s = 0$) is $1 + R_F/R_A$.

Example 12.8

Design a first-order active low-pass filter with a dc gain of 10 and a corner frequency of 1 kHz.

SOLUTION There are four unknown variables (R, R_A, R_F, and C) and only three equations: the dc gain, the corner frequency, and the equation for bias current balance. (Note that if we had specified the input impedance, we would have had four equations in four unknowns.) We therefore are free to choose one element value at will. In practice, we choose this to put all component values into reasonable ranges. (For example, it would be senseless to solve the equations and find that a 100-F capacitor is needed! It is preferable to keep component values in a range that permits use of off-the-shelf components.) Let us choose C as the specified component and use a 0.01-μF capacitor. We use the equation for corner frequency in order to solve for R; that is,

$$\omega = 2\pi f = 6280 = \frac{1}{RC}$$

Therefore,

$$R = 16 \text{ k}\Omega$$

If bias current balance is to be achieved,

$$R_A \parallel R_F = R = 16 \text{ k}\Omega$$

The dc gain gives a second equation in R_A and R_F.

$$H(0) = 1 + \frac{R_F}{R_A} = 10$$

Solving these two equations yields

$$R_F = 160 \text{ k}\Omega$$

$$R_A = 17.8 \text{ k}\Omega$$

and the design is complete. The resulting circuit is shown in Figure 12.11(b).

An alternative form of low-pass filter is shown in Figure 12.11(c). The transfer function is derived in the same manner as for the previous circuit. That is, we set $V_- = V_+$ to obtain

$$H(s) = \frac{-R_2}{R_1} \frac{1}{R_2 Cs + 1} \tag{12.11}$$

The dc gain is R_2/R_1 and the corner frequency is at $1/R_2 C$. Note that since we enter the inverting input to the op-amp, this low-pass filter function includes a negative sign.

Example 12.9

Design a first-order active filter with a dc gain of 10 and a corner frequency of 1 kHz. Use the circuit of Figure 12.11(c). Verify your design using a computer simulation program.

SOLUTION From the equation for corner frequency, we find

$$\omega = 2\pi f = 6280 = \frac{1}{R_2 C}$$

The dc gain equation yields

$$H(0) = -\frac{R_2}{R_1} = -10$$

There is one more unknown than the number of constraints. Suppose we choose $C = 0.01\ \mu\text{F}$. Then

$$R_2 = 16 \text{ k}\Omega$$

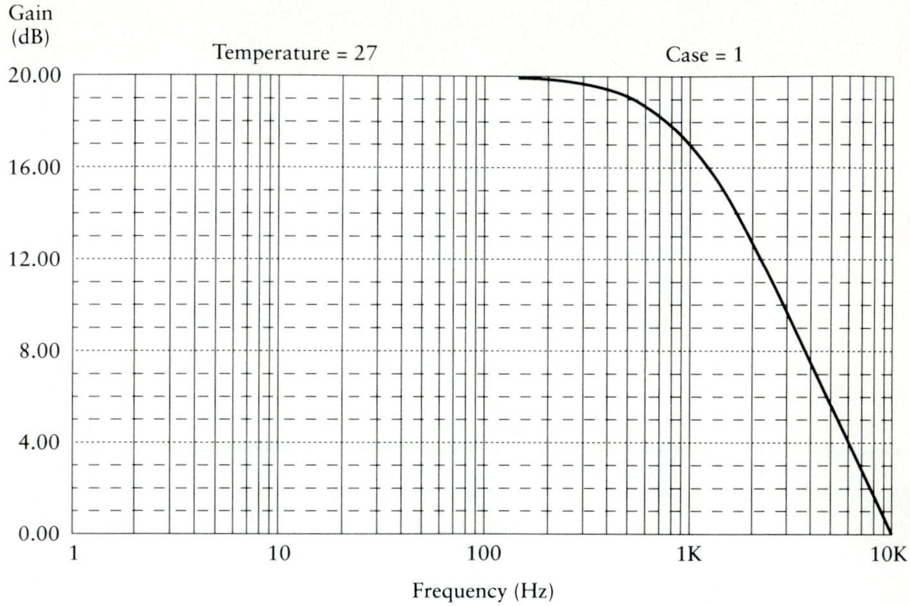

Figure 12.12 Computer simulation result for Example 12.9.

$$R_1 = 1.6 \text{ k}\Omega$$

The complete filter is shown in Figure 12.11(d).

We now enter the circuit of Figure 12.11(d) using MICRO-CAP II or any other SPICE-based computer simulation program. We chose a 741 op-amp for our simulation. However, since practical op-amps behave in a manner closely approximating ideal op-amps (for relatively low frequency), the result is not highly dependent on the choice of op-amp. The resulting ac response curve is shown in Figure 12.12. Note that the response starts at 20 dB, which corresponds to an amplitude factor of 10. The 3-dB point (17 dB gain on the curve) falls at 1 kHz, as expected. The design is therefore verified.

One advantage of active filters is that it is often quite simple to vary parameter values. As an example, a first-order low-pass filter with adjustable corner frequency is shown in Figure 12.13. The voltage at v_- is

$$v_- = \frac{v_1 R_A}{R_A + R_F}$$

The voltage at the noninverting input is found from the node equation at v_+.

$$v_+ = \frac{R_2 v_i}{R_1 + R_2} + \frac{R_1 v_o}{R_1 + R_2}$$

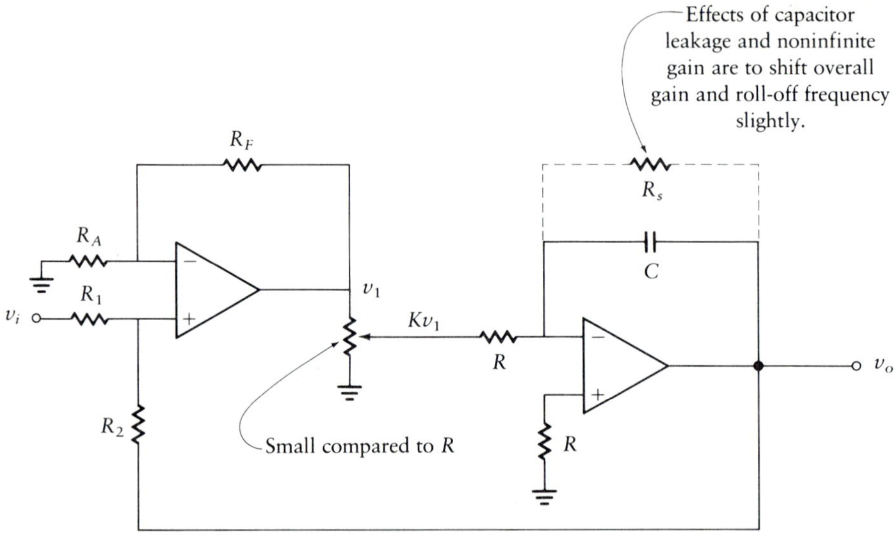

Figure 12.13 Low-pass filter with adjustable corner frequency.

$$\text{Where } K = \frac{\text{resistance from slider to ground}}{\text{total resistance}}$$

Setting $v_+ = v_-$, we obtain the voltage, v_1, as follows:

$$v_1 = \left(1 + \frac{R_F}{R_A}\right)\left(\frac{R_2}{R_1 + R_2}v_i + \frac{R_1}{R_1 + R_2}v_o\right)$$

$$= A_1 v_i + A_2 v_o$$

where

$$A_1 = \frac{(1 + R_F/R_A)R_2}{R_1 + R_2}$$

$$A_2 = \frac{(1 + R_F/R_A)R_1}{R_1 + R_2}$$

The second op-amp acts as an inverting integrator, and

$$V_o = -\frac{KV_1}{RCs}$$

Note that we use uppercase letters for the voltages, since these are functions of s. K is the fraction of V_1 sent to the integrator. That is, it is the potentiometer ratio, which is a number between 0 and 1.

$$V_o = \frac{-K(A_1 V_i + A_2 V_o)}{RCs}$$

The transfer function is given by

$$H(s) = \frac{V_o}{V_i} = \frac{-KA_1}{RCs + KA_2} \tag{12.12}$$

The dc gain is given by

$$-\frac{A_1}{A_2} = -\frac{R_2}{R_1}$$

The corner frequency is at KA_2/RC. Thus, the frequency is adjustable and is proportional to K. Without use of the op-amp, we would normally have a corner frequency that is inversely proportional to the resistor value. With a frequency proportional to K, we can use a *linear taper* potentiometer. The frequency is then linearly proportional to the setting of the potentiometer.

Example 12.10

Design a first-order adjustable low-pass filter with a dc gain of 10 and a corner frequency adjustable from near 0 to 1 kHz.

SOLUTION There are six unknowns in this problem (R_A, R_F, R_1, R_2, R, and C) and only three equations (gain, frequency, and bias balance). This leaves three parameters open to choice. Suppose we choose the following values:

$$C = 0.1\ \mu\text{F}$$

$$R = 10\ \text{k}\Omega$$

$$R_1 = 10\ \text{k}\Omega$$

The ratio of R_2 to R_1 is the dc gain, so with a given value of $R_1 = 10\ \text{k}\Omega$, R_2 must be 100 kΩ. We solve for A_1 and A_2 in order to find the ratio R_F/R_A.

The maximum corner frequency occurs at $K = 1$, so this frequency, A_2/RC, is set equal to $2\pi \times 1000$. Since R and C are known, we find $A_2 = 6.28$. Since A_2 and A_1 are related by the dc gain, we determine $A_1/A_2 = 10$ and $A_1 = 62.8$. Now, substituting the expression for A_2, we find

$$\frac{(1 + R_F/R_A)(R_1)}{R_1 + R_2} = A_2 = 6.28$$

and since

$$\frac{R_1}{R_1 + R_2} = \frac{1}{11}$$

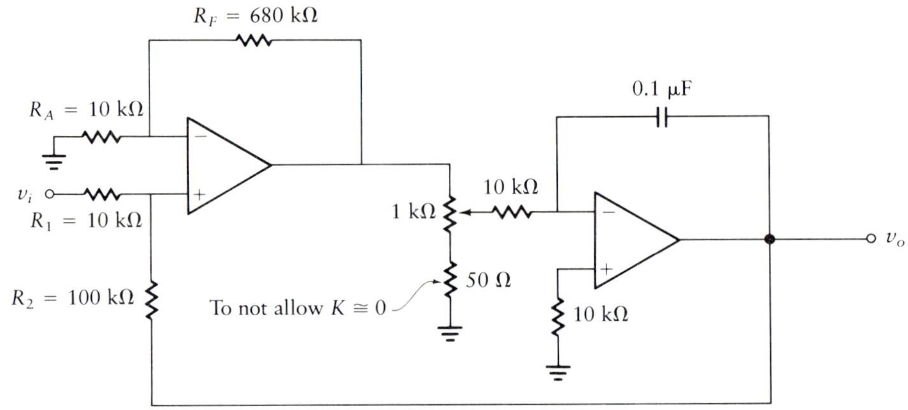

Figure 12.14
First-order filter for Example 12.10.

we find $R_F/R_A = 68$. R_A is chosen to achieve bias balance. The impedance attached to the noninverting input is

$$10\ \text{k}\Omega \parallel 100\ \text{k}\Omega \approx 10\ \text{k}\Omega$$

If we assume that R_F is large compared with R_A (we can check this assumption after solving for these resistors), the parallel combination will be close to the value of R_A. We therefore can choose $R_A = 10\ \text{k}\Omega$. With this choice of R_A, R_F is found to be 680 kΩ and bias balance is achieved. The complete filter is shown in Figure 12.14.

An RC *high-pass filter* is shown in Figure 12.15(a). Once again, we solve for V_+ and V_- and then set these equal to find the transfer function.

$$H(s) = \left(1 + \frac{R_F}{R_A}\right)\frac{s}{s + 1/RC}$$

The high-frequency gain is $1 + R_F/R_A$ and the corner frequency is at $1/RC$.

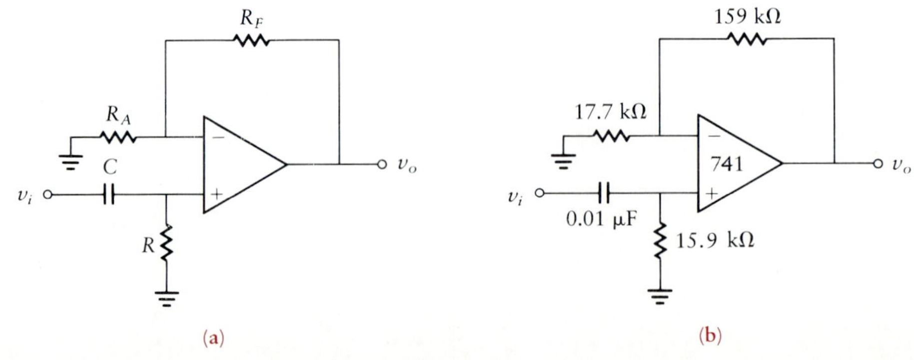

Figure 12.15
RC high-pass filter.

Example 12.11

Design an RC high-pass filter with a high-frequency gain of 10 and a corner frequency of 1 kHz.

SOLUTION In this problem we have three equations (high-frequency gain, corner frequency, and bias balance) and four unknowns (R, C, R_A, and R_F). Therefore, one parameter is arbitrary. Let us choose $C = 0.01\ \mu F$. We find R from the equation for corner frequency.

$$\omega = 2\pi f = \frac{1}{RC}$$

and

$$R = 15.9\ k\Omega$$

The high-frequency gain equation yields

$$1 + \frac{R_F}{R_A} = 10$$

This represents one equation in two unknowns, R_F and R_A. We derive a second equation from the bias balance relationship,

$$R_A \parallel R_F = R = 15.9\ k\Omega$$

Solving these two equations yields

$$R_F = 159\ k\Omega$$

$$R_A = 17.7\ k\Omega$$

The complete circuit is shown in Figure 12.15(b).

Drill Problem

D12.6 What are the dc gain and corner frequency of the circuit shown in Figure D12.5? What type of circuit is this?

Ans: dc gain = 11; corner frequency is 10^4 rad/s; circuit is a low-pass filter

Figure D12.5

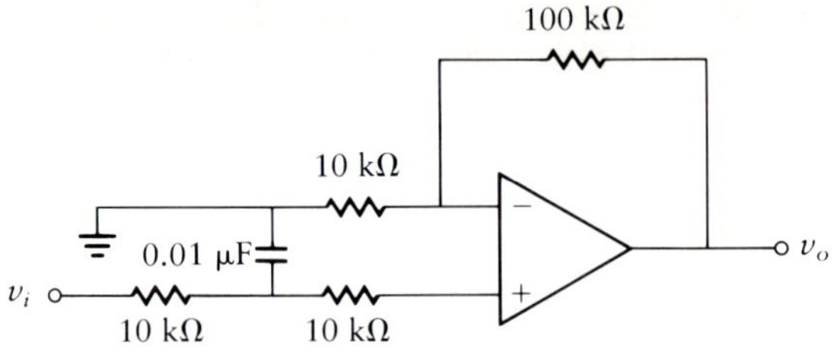

Figure 12.16
Adjustable high-pass filter.

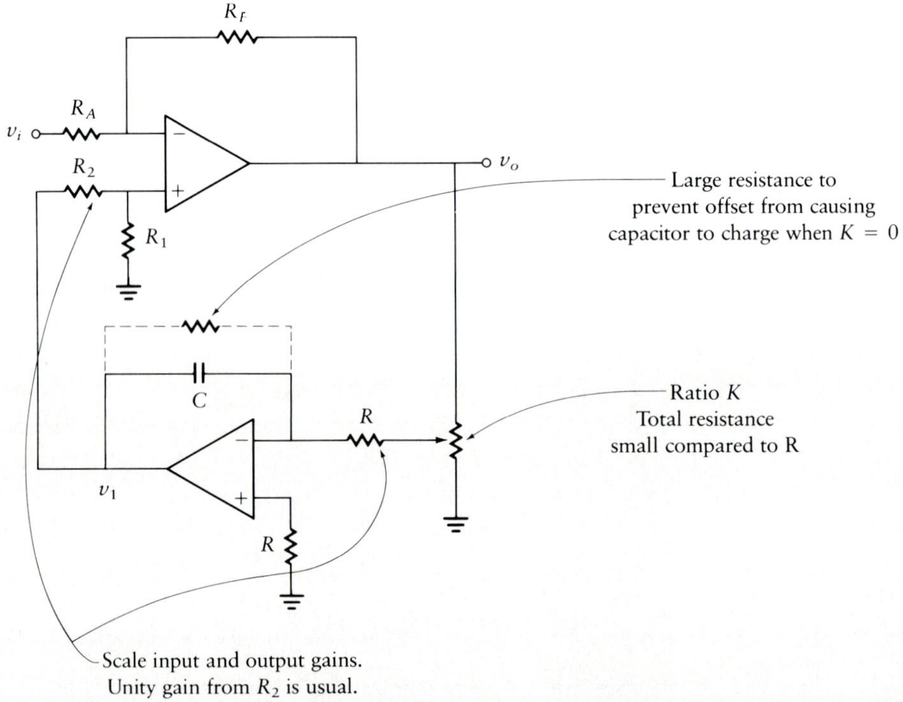

We now analyze an adjustable high-pass filter circuit. Figure 12.16 combines an amplifier for signal summing with a second amplifier for low-pass summing. Analysis yields the following equations:

$$V_o = \frac{(1 + R_F/R_A)(R_1)V_1}{R_1 + R_2} - \frac{R_F V_i}{R_A}$$

$$= A_1 V_1 - A_2 V_i$$

where

$$A_1 = \frac{(1 + R_F/R_A)(R_1)}{R_1 + R_2}$$

$$A_2 = \frac{R_F}{R_A}$$

If K is the potentiometer attenuation, then

$$V_1 = -\frac{KV_o}{RCs}$$

and

$$V_o = \frac{-KA_1V_o}{RCs} - A_2V_i$$

The transfer function is given by

$$H(s) = \frac{V_o}{V_i} = -A_2\frac{s}{s + KA_1/RC} \tag{12.13}$$

The result is a high-pass filter with adjustable corner frequency.

Example 12.12

Design a high-pass filter with a high-frequency gain of 10 and a corner frequency adjustable from 100 Hz to 400 Hz. Verify your design using a computer simulation program.

SOLUTION We have three equations and six unknowns, so we choose values for three variables: R_A, A_1, and C. We start with the high-frequency gain equation,

$$A_2 = \frac{R_F}{R_A} = 10$$

If we now choose $R_A = 10\ \text{k}\Omega$, we find $R_F = 100\ \text{k}\Omega$. We select $A_1 = 1$ to achieve unity gain for the noninverting input. Then

$$A_1 = \frac{(1 + R_F/R_A)(R_1)}{R_1 + R_2} = 1$$

The bias balance equation yields

$$R_1 \parallel R_2 = R_A \parallel R_F = 10\ \text{k}\Omega \parallel 100\ \text{k}\Omega$$

We thus have two equations in R_1 and R_2. Solving these yields

$$R_1 = 10\ \text{k}\Omega$$

$$R_2 = 100\ \text{k}\Omega$$

Since K cannot exceed unity, we design for the maximum corner frequency at $K = 1$ to yield

$$\frac{A_1}{RC} = \frac{1}{RC} = \omega = 2\pi f = 2512\ \text{rad/s}$$

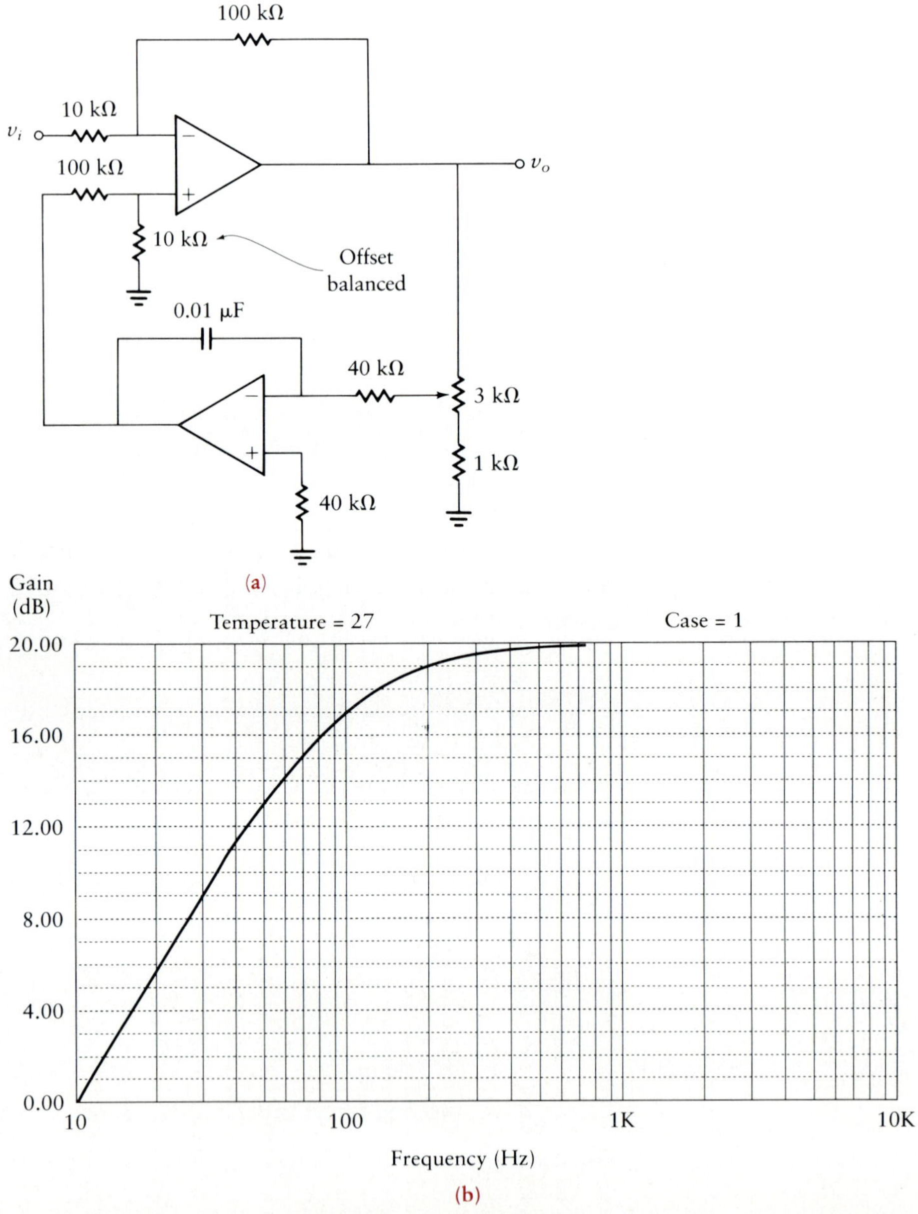

Figure 12.17 Circuit for Example 12.12.

If we now choose $C = 0.01\ \mu F$, we find

$$R = \frac{10^8}{2512} = 40\ k\Omega$$

Adjustment of K from 0.25 to 1 yields the desired frequency range of 100 to

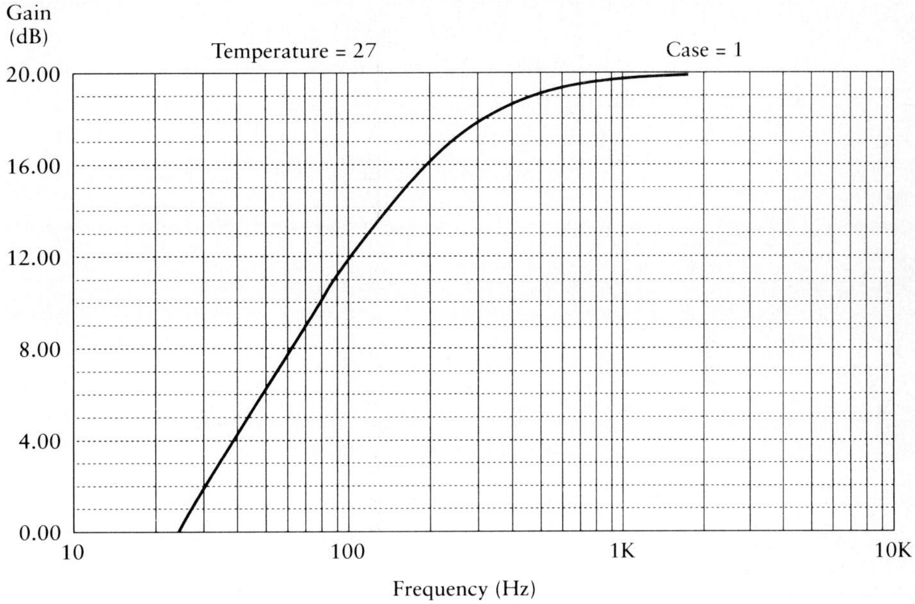

(c)

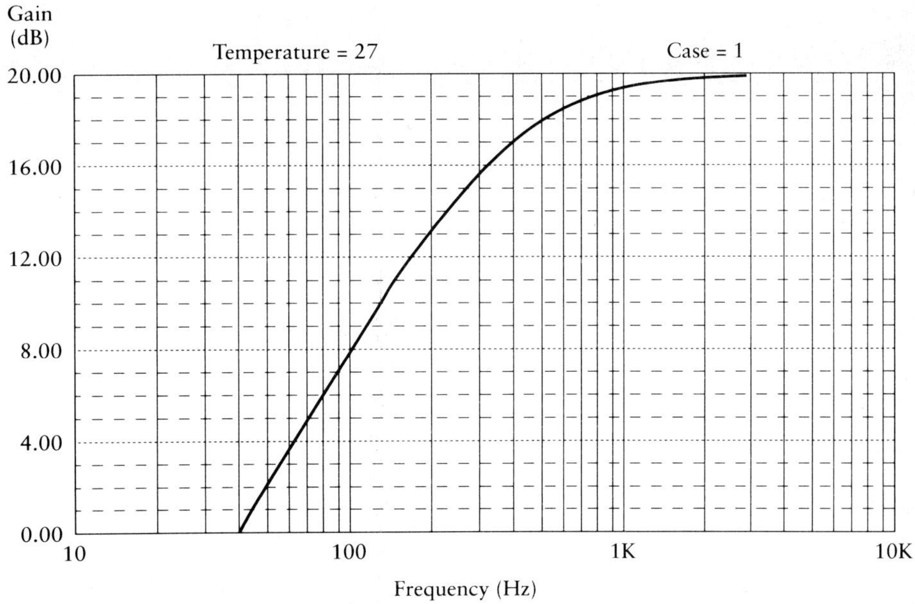

(d)

400 Hz. This is accomplished by using a 4-kΩ potentiometer with a limiting tab or by placing a 1-kΩ fixed resistor between ground and the 3-kΩ potentiometer. The latter approach is shown in the input of the feedback op-amp in the circuit of Figure 12.17.

We now use a computer simulation to check our result. Since the simulation we used does not provide for potentiometers, we broke the 3 kΩ into two 1.5 kΩ. In this manner, we were able to achieve values of $K = 0.25$, $K = 0.625$, and $K = 1$ by varying the point at which the right side of the 40 kΩ is connected. The resulting frequency plots are shown in Figure 12.17(b), (c), and (d) for the three values of K. Note that for $K = 1$, the 3-dB break point occurs at 400 Hz. For $K = 0.625$ the break is at 230 Hz, and for $K = 0.25$ it occurs at 100 Hz. We have therefore verified the design.

Drill Problem

D12.7 What are the gain and the adjustable corner frequency range for the circuit shown in Figure D12.6?

Ans: $A_1 = 1$; range of corner frequencies is 28.6 to 200 Hz

Figure D12.6

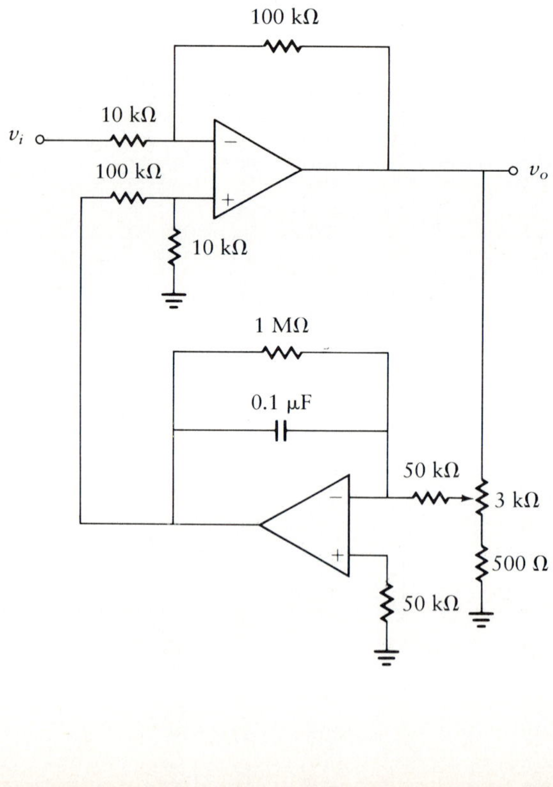

12.4 SINGLE AMPLIFIER—GENERAL TYPE

We considered the low-pass and high-pass filters in the previous sections. These filters are specific in form. In this section, we consider a "general-type" first-order filter. The circuit diagram for the single op-amp general filter is shown in Figure 12.18.

This one circuit can be configured to produce the following first-order filters:

1. High-pass, when $R_1R_4 = R_2R_3$
2. All-pass, when $R_2R_3 = 2R_1R_4$
3. Low-pass, when $R_3 \to \infty$

If we assume that the op-amp is ideal (i.e., $V_- = V_+$), the transfer function, $H(s)$, is found as in Chapter 8. The transfer function is developed as follows:

$$H(s) = \left(1 + \frac{R_2/R_1}{R_2Cs + 1}\right)\frac{R_4}{R_3 + R_4} - \frac{R_2/R_1}{R_2Cs + 1}$$

$$= \frac{R_4}{R_3 + R_4}\,\frac{s + (1/R_1C)(R_1/R_2 - R_3/R_4)}{s + 1/R_2C} \qquad (12.14)$$

In order to make this a *high-pass filter,* we want the numerator zero to be at the origin. We therefore let

$$R_1R_4 = R_2R_3$$

Then

$$H(s) = \frac{R_4}{R_3 + R_4}\,\frac{s}{s + 1/R_2C} \qquad (12.15)$$

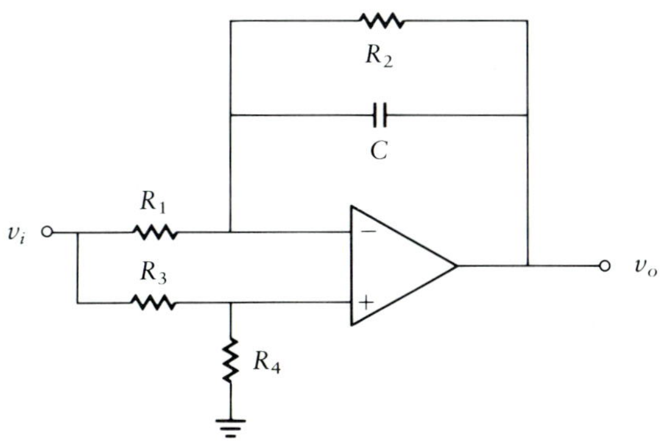

Figure 12.18
Single op-amp general filter.

The gain at zero frequency is zero, and as the frequency approaches infinity, the gain is

$$\frac{V_o}{V_i} = \frac{R_4}{R_3 + R_4} \tag{12.16}$$

If we wish to make this circuit into an *all-pass filter,* we set

$$R_2R_3 = 2R_1R_4$$

in equation (12.14). Then

$$H(s) = \frac{R_4}{R_3 + R_4}\frac{s - 1/R_2C}{s + 1/R_2C} \tag{12.17}$$

Finally, if we wish to construct a *low-pass filter,* we let R_3 approach infinity in equation (12.14). This yields

$$H(s) = \frac{-1/R_1C}{s + 1/R_2C}$$

Example 12.13

Design a high-pass filter with a high-frequency gain of $\frac{1}{2}$ and a corner frequency of 4 kHz.

SOLUTION The unknowns are R_1, R_2, R_3, R_4, and C. We have equations for gain, corner frequency, bias balance, and the high-pass constraint. This leaves one unknown, which we can select. Let us choose $C = 0.01\ \mu$F. Then from the corner frequency we have

$$\frac{10^8}{R_2} = 2\pi f = 25{,}120 \text{ rad/s}$$

Therefore, $R_2 = 4$ kΩ, and for a high-frequency gain of $\frac{1}{2}$ we find (see equation (12.16)) $R_3 = R_4$. The high-pass resistor constraint yields $R_1R_4 = R_2R_3$, and hence $R_1 = R_2 = 4$ kΩ. To achieve bias balance, we choose $R_3 = R_4 = 4$ kΩ.

Example 12.14

Design an all-pass filter with a gain of $\frac{1}{2}$ and a corner frequency of 4 kHz.

SOLUTION Let us choose $C = 0.01\ \mu F$. From the corner frequency equations, we have

$$\frac{1}{R_2 C} = \frac{10^8}{R_2} = 2\pi f = 25{,}120 \text{ rad/s}$$

so $R_2 = 4\ k\Omega$. To achieve a gain of $\frac{1}{2}$, $R_3 = R_4$. The all-pass constraint yields $R_2 R_3 = 2R_1 R_4$, but since $R_3 = R_4$ we have

$$R_1 = \frac{R_2}{2} = 2\ k\Omega$$

The parallel combination of R_1 and R_2 has a resistance of 1.33 kΩ, so to achieve bias balance we require

$$R_3 = R_4 = 2.67\ k\Omega$$

12.5 CLASSICAL ANALOG FILTERS

If the filter specification requires a roll-off of only −20 dB/decade, then the first-order single-amplifier filter, described in the previous section, is the proper choice. If a faster roll-off is required, a higher-order filter must be designed. The balance of this chapter is directed toward the design of higher-order filters (up to and including tenth order).

Before we can design more complex active filters, we must first become familiar with some results in passive filter theory.

There are many types of filters, each of which exhibits special characteristics. *Butterworth* filters produce no ripple in the pass band and attenuate unwanted frequencies outside this band. *Chebyshev* filters attenuate unwanted frequencies with fewer components than Butterworth filters, but they exhibit ripple in the pass band. Other important classical filters include the elliptic, parabolic, Bessel, Papoulis, and Gaussian. The current discussion is limited to Butterworth and Chebyshev filters, although similar design approaches can be used for the other types.

Specifications of filters are often presented by use of a modified amplitude versus frequency characteristic diagram. For example, Figure 12.19 is a diagram of the characteristics of a low-pass filter. The *pass band* of the filter extends from zero frequency to a maximum frequency of f_{pass}. The range of frequencies between f_{pass} and f_{stop} is known as the *transition band,* and the range above f_{stop} is the *stop band.* The magnitudes shown on the diagram are in decibels. Within the pass band, the unshaded region indicates that the magnitude must be above A_p throughout that band. Some filters, such as the Chebyshev, have a

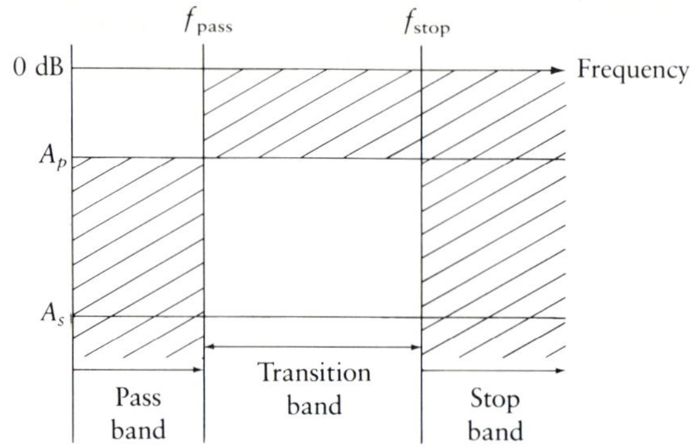

Figure 12.19 Low-pass filter characteristic.

ripple in the pass band. Nonetheless, the minimum magnitude is greater than A_p. Within the transition band, attenuation occurs as the frequency increases, with the magnitude at f_{stop} (the upper frequency) equal to A_s.

The standard form of the transfer function of a filter is often given as follows:

$$|H(j\omega)|^2 = \frac{1}{1 + e^2 s_n^2(\omega)} \tag{12.18}$$

In this equation, $s_n(\omega)$ is a classical nth-order polynomial and e is related to the ripple in the pass band. We explore several specific examples in the following sections.

12.5.1 Butterworth Filters

Butterworth filters are also known as *maximally flat magnitude (MFM)* filters, since the transfer function is chosen so that the magnitude response curve is as flat as possible within the pass band of the filter. This is accomplished by setting as many derivatives of the function as possible equal to zero at the center of the pass band (zero frequency for a low-pass filter).

Suppose we set $s_n = \omega^n$ and $e = 1$ in equation (12.18). This yields the Butterworth transfer function. The magnitude-squared transfer function of the nth-order Butterworth filter is given as follows:

$$|H(j\omega)|^2 = \frac{1}{1 + \omega^{2n}} \tag{12.19}$$

For example, the transfer function for a second-order Butterworth filter is given by

$$H(s) = \frac{1}{s^2 + \sqrt{2}\,s + 1} \tag{12.20}$$

If $s = j\omega$ is substituted into equation (12.20), the magnitude of the resulting complex function is that given in equation (12.19) with $n = 2$. The magnitude of the transfer function is unity within the pass band, and the break frequency (i.e., the frequency at which $|H(j\omega)| = 1/\sqrt{2}$) is also unity. We refer to this as a *normalized filter.*

Equation (12.19) is the general expression for the magnitude-squared transfer function of the Butterworth filter. As the order of the filter, n, increases, the roll-off in the transition region becomes steeper. (The roll-off is $20n$ dB/decade, where n is the filter order.)

When a roll-off is specified in decibels per decade, the required order of the filter is given by this quantity divided by 20.

Suppose that the specifications give the magnitude of the characteristic at two points; that is, the magnitude is $|H_1|$ at a frequency of ω_1 and $|H_2|$ at a frequency of ω_2. We find the required order of the filter from equation (12.19). Substituting the two values into this equation, we find

$$n \geqslant \frac{\log(|H_2|^{-2} - 1)^{0.5}/(|H_1|^{-2} - 1)^{0.5}}{\log(\omega_2/\omega_1)} \tag{12.21}$$

The transfer function of the Butterworth filter is expressed as the reciprocal of a polynomial in s. If $H(s)$ is the transfer function and $B_n(s)$ is the polynomial for the nth-order filter, we obtain

$$H(s) = \frac{1}{B_n(s)}$$

The transfer function for a normalized filter has a corner frequency at $\omega = 1$. The magnitude at this corner frequency is 0.707.

The normalized polynomials for Butterworth filters are as follows:

$$B_1(s) = s + 1$$

$$B_2(s) = s^2 + 1.414s + 1$$

$$B_3(s) = s^3 + 2s^2 + 2s + 1$$

$$B_4(s) = s^4 + 2.61s^3 + 3.41s^2 + 2.61s + 1$$

$$B_5(s) = s^5 + 3.24s^4 + 5.24s^3 + 5.24s^2 + 3.24s + 1$$

$$B_6(s) = s^6 + 3.86s^5 + 7.46s^4 + 9.14s^3 + 7.46s^2 + 3.86s + 1 \tag{12.22}$$

In equation (12.20) we used the order 2 polynomial, $B_2(s)$.

For each value of n, the poles of $H(s)$ are in the left-half s plane on the unit circle. The poles are symmetrically spaced as shown in Figure 12.20.

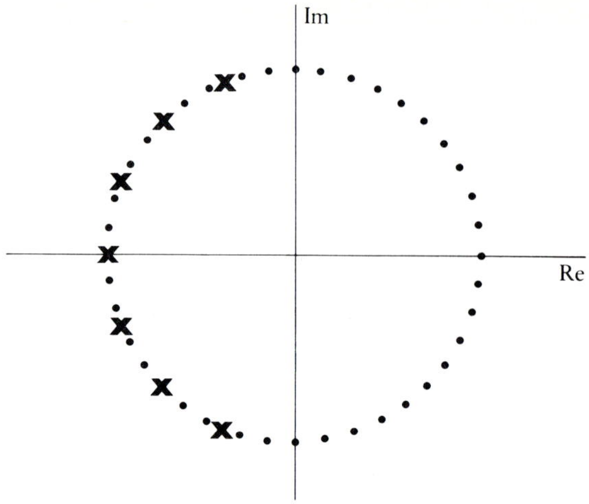

Figure 12.20 Poles for Butterworth filter.

Example 12.15

Determine the transfer function for a third-order Butterworth low-pass filter with a Bode plot as shown in Figure 12.21.

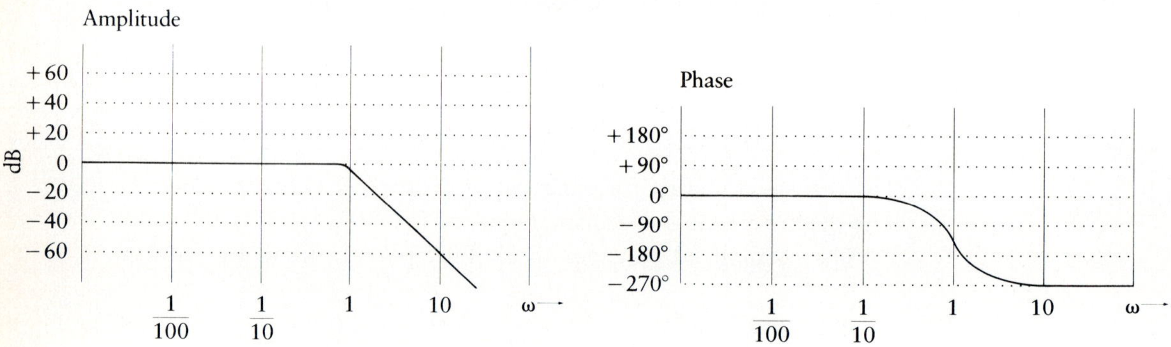

Figure 12.21 Bode plot for Example 12.15.

SOLUTION We begin with the transfer function of a normalized third-order Butterworth low-pass filter.

$$H(s) = \frac{1}{s^3 + 2s^2 + 2s + 1}$$

We now scale the function to account for the dc gain, and we scale s to provide for a cutoff frequency different from unity. The resulting third-order Butterworth low-pass filter with a dc gain of 10 and a cutoff frequency of 1000 Hz is given by

$$H(s) = \frac{10}{(s/2\pi \times 1000)^3 + 2(s/2\pi \times 1000)^2 + 2(s/2\pi \times 1000) + 1}$$

12.5.2 Chebyshev Filters

The Chebyshev low-pass filter has a transfer function built around a Chebyshev polynomial. The polynomials are sketched in Figure 12.22 and are given by the following equations:

$$\begin{aligned}
C_0 &= 1 \\
C_1(x) &= x \\
C_2(x) &= 2x^2 - 1 \\
C_3(x) &= 4x^3 - 3x \\
C_4(x) &= 8x^4 - 8x^2 + 1 \\
C_5(x) &= 16x^5 - 20x^3 + 5x \\
C_6(x) &= 32x^6 - 48x^4 + 18x^2 - 1
\end{aligned} \tag{12.23}$$

and in general

$$C_{n+1}(x) = 2xC_n(x) - C_{n-1}(x) \tag{12.24}$$

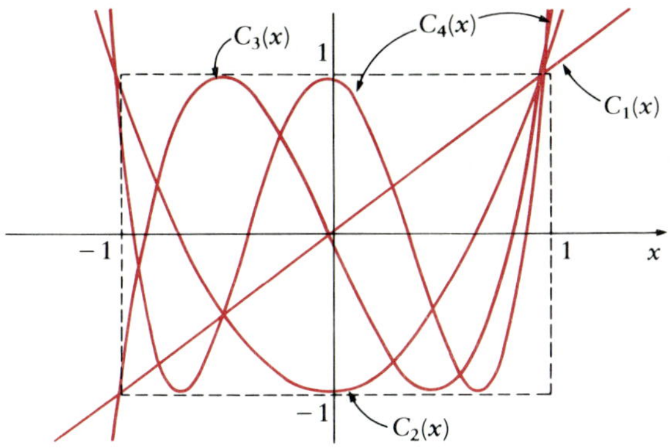

Figure 12.22 Chebyshev polynomials.

Chebyshev filters exhibit ripple in the pass band. However the same specification can frequently be met with a lower order Chebyshev filter than with a Butterworth filter. At the higher frequencies, both filter characteristics have the same asymptotic slope. The magnitude of the transfer function is

$$|H^2(\omega)| = \frac{1}{1 + \epsilon^2 C_n^2(\omega)} \tag{12.25}$$

where C_n is the nth-order polynomial. The constant, ϵ, determines the ripple magnitude and is less than 1.

The normalized second-order Chebyshev low-pass filter with $\epsilon = 0.5$ has a ripple bounded by

$$\text{Ripple} = \frac{1}{\sqrt{1 + \epsilon^2}} = 0.894$$

The transfer function is found by substituting the value of the Chebyshev polynomial into the general expression to find

$$|H^2(\omega)| = \frac{1}{1 + (0.5)^2 C_2^2(\omega)}$$

$$= \frac{1}{1 + 0.25(2\omega^2 - 1)^2}$$

$$= \frac{1}{\omega^4 - \omega^2 + 0.75}$$

Letting $s = j\omega$ yields

$$H^2(s) = \frac{1}{s^4 + s^2 + 0.75}$$

This can be rewritten by completing the square, with the result

$$H(s) = \frac{1}{(s^2 + 0.5)^2 + 0.5}$$

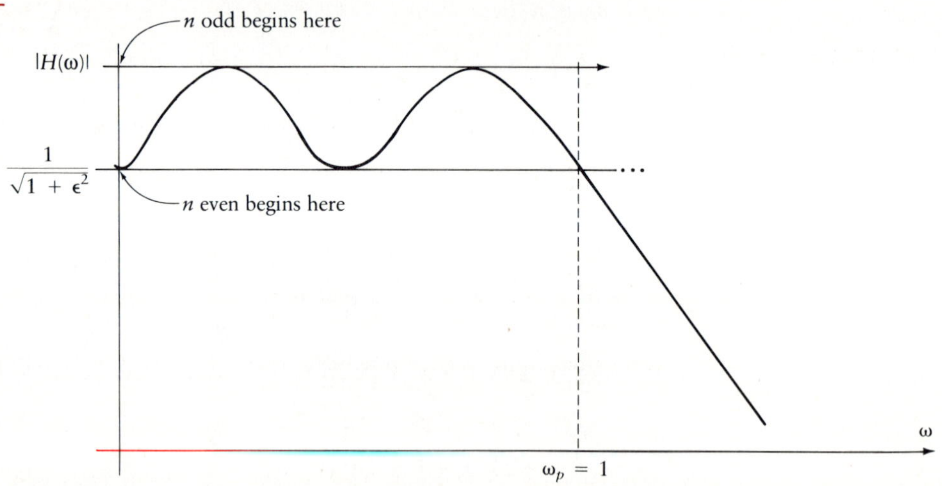

Figure 12.23 Example of a Chebyshev filter.

If the filter order is odd, the response curve starts at $|H(j\omega)| = 1$ or, in decibels, at 0 dB. If the filter order is even, the curve starts at the magnitude of the ripple. Excluding the point where the curve crosses at $\omega_p = 1$, the order of the filter equals the sum of the response maxima and minima in the pass band. The magnitude curve of Figure 12.23 would thus represent an even-order filter of order 4.

When designing higher-order filters, the most convenient approach is to break the function up into second- and third-order stages. These successive stages will then combine to yield the desired response. This cascading is possible with op-amps because of the isolation between adjacent stages. If one stage were to load another stage, cascading would not be possible.

12.6 TRANSFORMATIONS

There are close relationships between the various classifications of filters. Once a design is completed for one type, it is often easy to modify the design to change the filter into one of a different classification. In this manner, lengthy redesigns may be avoided. We examine two particular transformations from one filter format to another.

12.6.1 Low-Pass to High-Pass Transformation

Substituting s for $1/s$ converts a normalized low-pass transfer function to a high-pass transfer function. This transformation has the effect of interchanging the resistors with the capacitors.

Example 12.16

Transform a third-order Butterworth low-pass filter into a high-pass filter.

SOLUTION The low-pass transfer function is given by

$$H_{\text{LP}}(s) = \frac{1}{s^3 + 2s^2 + 2s + 1}$$

We now replace s by $1/s$ to yield

$$H_{\text{HP}}(s) = \frac{1}{(1/s)^3 + 2(1/s)^2 + 2(1/s) + 1}$$

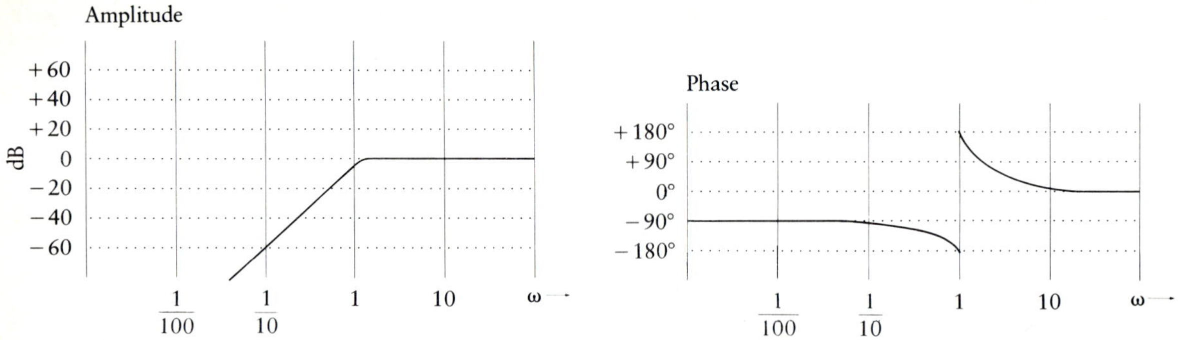

Figure 12.24
Bode plot of transformed Butterworth filter.

$$= \frac{s^3}{s^3 + 2s^2 + 2s + 1}$$

The amplitude and phase of this transformed Butterworth network are shown in Figure 12.24.

Example 12.17

Transform a second-order Chebyshev low-pass filter into a high-pass filter.

SOLUTION We start with the low-pass transfer function,

$$H_{\mathrm{LP}}(s) = \frac{1}{(s + 0.84)^2 + 0.87^2}$$

$$= \frac{1}{s^2 + 1.7s + 1.47}$$

We now replace s by $1/s$ to yield

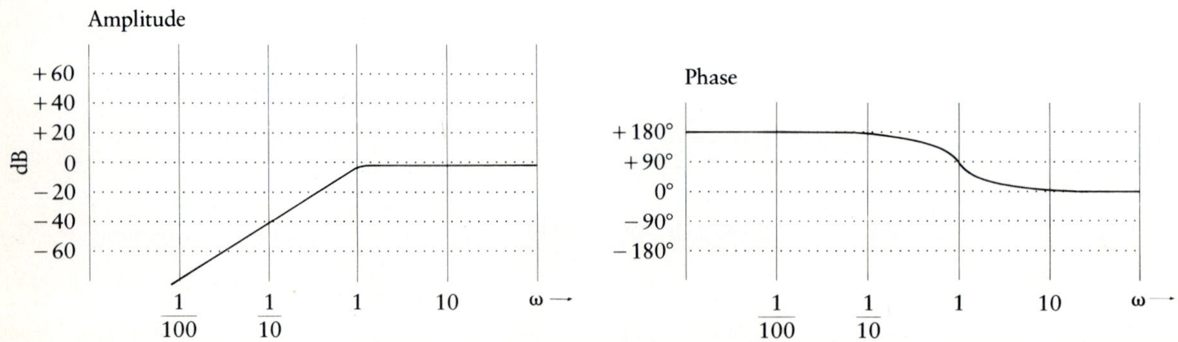

Figure 12.25
Bode plot of transformed Chebyshev filter.

$$H_{\mathrm{HP}}(s) = \frac{1}{(1/s)^2 + 1.7(1/s) + 1.47}$$

$$= \frac{s^2}{1.47s^2 + 1.7s + 1}$$

The Bode plot for this transformed Chebyshev network is shown in Figure 12.25.

12.6.2 Low-Pass to Band-Pass Transformation

We can convert a normalized low-pass transfer function into a normalized band-pass transfer function by making the substitution

$$\frac{s^2 + 1}{s} \quad \text{in place of} \quad s$$

The order of the band-pass filter is twice that of the original low-pass filter.

Example 12.18

Transform a third-order Butterworth low-pass filter into a band-pass filter.

SOLUTION We start with the low-pass transfer function,

$$H_{\mathrm{LP}}(s) = \frac{1}{s^3 + 2s^2 + 2s + 1}$$

Replacing s with $(s^2 + 1)/s$ yields

$$H_{\mathrm{BP}}(s) = \frac{1}{[(s^2 + 1)/s]^3 + 2[(s^2 + 1)/s]^2 + 2[(s^2 + 1)/s] + 1}$$

$$= \frac{s^3}{s^6 + 2s^5 + 4s^4 + 5s^3 + 4s^2 + 2s + 1}$$

In the next section, we present another method for quickly designing a band-pass and band-reject filter.

12.7 DESIGN PROCEDURE FOR BUTTERWORTH AND CHEBYSHEV FILTERS

In this section, we examine a constant-resistance or constant-capacitance approach toward the design of Butterworth and Chebyshev filters up to tenth order. Each order is formed by combining second- and third-order filters. The design approach set forth assumes 1-Ω values for the resistors in the two-pole and three-pole low-pass filters shown in Figure 12.26.

To design a high-pass Butterworth or Chebyshev filter, we interchange the capacitors and resistors from those of the low-pass design. In a high-pass filter, the capacitors are of equal value (instead of the resistors, as is the case in low-pass filters). We set the capacitances to 1 F. The two-pole and three-pole high-pass filters are shown in Figure 12.27.

The circuit configuration of the Chebyshev filter is identical to that of the Butterworth filter. Only the component values are different.

Figure 12.26 Unity-R active low-pass filter.

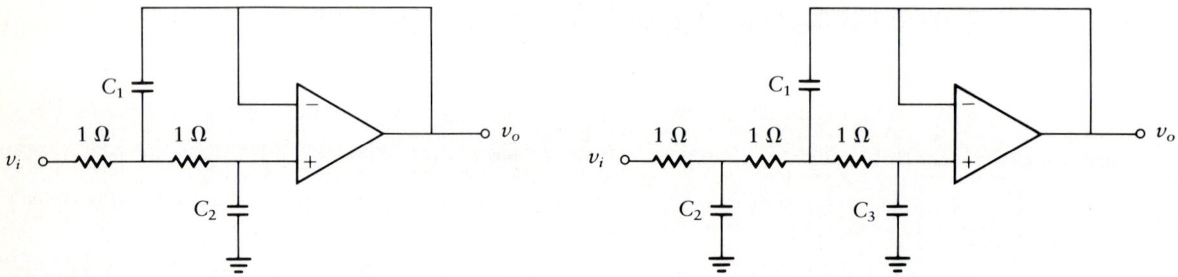

(a) Two-pole filter (b) Three-pole filter

Figure 12.27 Unity-C active high-pass filter.

(a) Two-pole filter (b) Three-pole filter

12.7.1 Low-Pass Filter Design

Four parameters must be specified in the design of Butterworth or Chebyshev filters. These four parameters can be stated in various ways. One set of choices

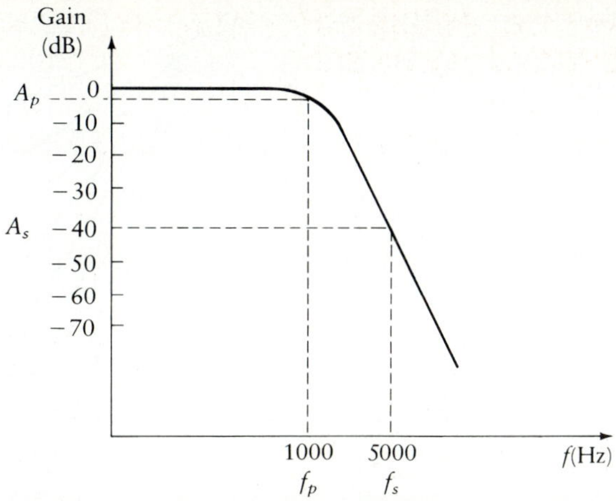

Figure 12.28 Frequency response requirement for low-pass filter.

is shown in Figure 12.28, and the parameters are

A_p = dB attenuation in the pass band

A_s = dB attenuation in the stop band

f_p = frequency at which A_p occurs

f_s = frequency at which A_s occurs

In the case of the Chebyshev filter, one additional parameter is specified. This parameter is the maximum ripple, ϵ, permitted in the pass band, given in decibels.

The design procedure is divided into two parts. It is necessary first to find the required *order* of the filter and then to find the *scale factor* that must be applied to the normalized parameter values.

12.7.2 Filter Order

We first determine what order of filter is needed to satisfy the specification for A_p at f_p and A_s at f_s. Notice that A_p and A_s are magnitudes, so we need not be concerned with the minus sign that appears in the expressions. We denote the filter order by n_B for the Butterworth filter and n_C for the Chebyshev filter. We use equation (12.21), where $|H_1|$ is A_p, ω_1 is ω_p, $|H_2|$ is A_s, and ω_2 is ω_s. Since A_p is the number of decibels *below* 0 dB, the value of $|H_1|$ is $10^{-A_p/20}$. Other quantities are similarly transformed. Thus,

$$|H_1|^{-2} = 10^{+0.1A_p}$$

Therefore, the equation for the Butterworth order, n_B, is

$$n_B = \frac{\log(\epsilon_2/\epsilon_1)}{\log(f_s/f_p)} \tag{12.26}$$

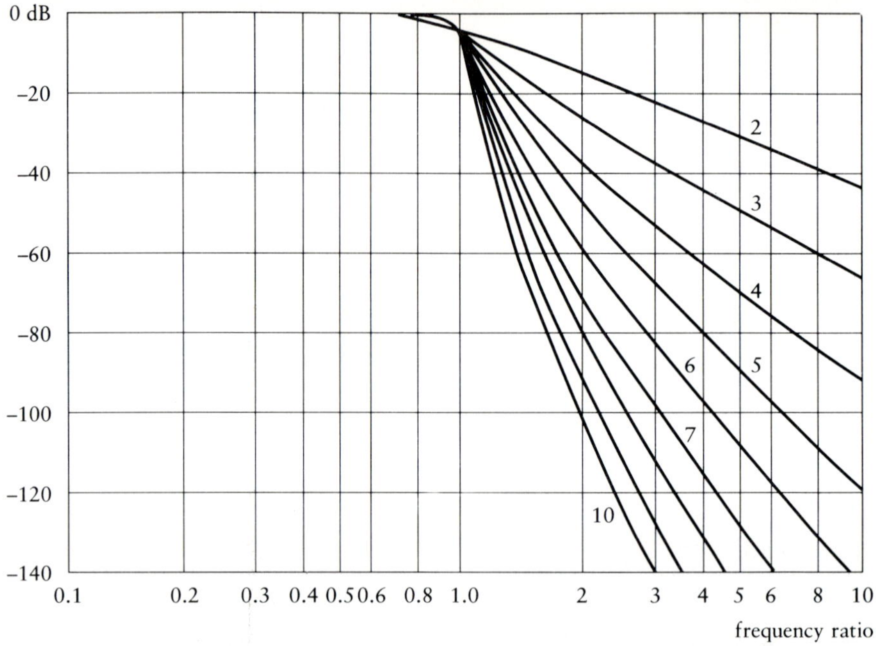

Figure 12.29 Normalized 1-dB Chebyshev low-pass filter.

where

$$\epsilon_1 = \sqrt{10^{0.1A_p} - 1}$$

$$\epsilon_2 = \sqrt{10^{0.1A_s} - 1} \tag{12.27}$$

Sets of equations exist for finding the required order of a Chebyshev filter. However, the application of these equations is somewhat difficult for the following reason. We have been using the 3-dB point to define the corner of the pass band. In the case of the Chebyshev filter, this leads to contradictions, since the maximum ripple within the pass band is specified. For example, a 0.1-dB Chebyshev filter has a maximum ripple of 0.1 dB within the pass band. The 3-dB point is not within the pass band of the filter. The *3-dB bandwidth* and the *filter pass band* are not the same, and caution must be exercised in applying any formula based on pass-band properties.

We shall avoid these ambiguities by using a normalized response curve to find the required filter order, n_C. Figure 12.29 illustrates normalized Chebyshev low-pass filter curves for a 1-dB filter. Other ripple limits would result in slight deviations from these curves, but the asymptotic slopes depend only on the order, regardless of the ripple values. Because of these slight variations, if a set of specifications places us close to the borderline between two choices of filter order, it may be wise to (1) choose the higher order or (2) perform a simulation of the resulting circuit to ensure that the specifications are met. We shall see how to apply the curves of Figure 12.29 in the following examples.

Example 12.19

Select the order for a low-pass filter that has the following specifications:

$$A_p = 3 \text{ dB} \qquad \text{at } f_p = 1 \text{ kHz}$$

$$A_s = 40 \text{ dB} \qquad \text{at } f_s = 5 \text{ kHz}$$

This filter characteristic is shown in Figure 12.28.

SOLUTION Using equations (12.26) and (12.27), we obtain

$$\epsilon_1 = \sqrt{10^{(0.1)(3)} - 1} = 1$$

$$\epsilon_2 = \sqrt{10^{(0.1)(40)} - 1} = 100$$

We calculate n_B as follows:

$$n_B = \frac{\log(\epsilon_2/\epsilon_1)}{\log(f_s/f_p)} = \frac{\log(100)}{\log(5000/1000)} = 2.86 \quad \text{so we use } n = 3$$

We refer to the normalized curves of Figure 12.29 to find the required order of a Chebyshev filter. The two frequencies specified in this example are 1 kHz and 5 kHz. It is first necessary to normalize these by dividing by 1000. We are therefore looking for a filter with a 3-dB point at 1 Hz and an attenuation of at least 40 dB at 5 Hz. Figure 12.29 indicates that a second-order filter has an attenuation of about 32 dB at 5 Hz, while a third-order filter has an attenuation of about 50 dB. We would therefore have to choose the third-order filter.

Even though the required orders of the Butterworth and Chebyshev filters turned out to be equal, we will find in the following examples that the Chebyshev provides more rejection in the stop band than does the Butterworth.

12.7.3 Parameter Scale Factor

Now that we know the required order for the filter, we need to choose the component values.

Various synthesis methods are available for the design of filters. The constant-resistance/constant-capacitance technique developed here provides a direct and suitable method for designing these filters easily ([42] provides additional information regarding the design process). Once we know the order of the filter, we can use Table 12.2 to select the capacitor coefficients for the Butterworth filter or Table 12.3, a through e, for design of the Chebyshev filter.

Table 12.2 lists capacitor ratios for normalized Butterworth low-pass filters and resistor ratios for high-pass filters. For the low-pass filter, the resistors are assumed to be 1 Ω in value. As an example, viewing the first row of the table yields a second-order Butterworth filter design. We use one second-order stage with $C_1/C = 1.41$ and $C_2/C = 0.707$. For orders greater than 3, the table contains more than one row. The reason is that we build these filters using second- and third-order stages, so for order greater than 3, more than one stage is required. The table is configured for unit corner frequency, that is, $\omega_p = 1$.

To design a practical Butterworth filter, we obtain the capacitor or resistor ratios from the table and scale the values to practical sizes. That is, if the corner frequency is anything other than 1 rad/s and/or the resistors are any value other than 1 Ω, the capacitor values from the table must be appropriately scaled. Since the product, *RC*, is inversely proportional to the frequency, *R* and *C* are themselves inversely related for a given frequency. If, for example, we

Table 12.2 Butterworth Active Low-Pass Values.*

Order *n*	C_1/C or R/R_1	C_2/C or R/R_2	C_3/C or R/R_3
2	1.414	0.7071	
3	3.546	1.392	0.2024
4	1.082	0.9241	
	2.613	0.3825	
5	1.753	1.354	0.4214
	3.235	0.3090	
6	1.035	0.9660	
	1.414	0.7071	
	3.863	0.2588	
7	1.531	1.336	0.4885
	1.604	0.6235	
	4.493	0.2225	
8	1.020	0.9809	
	1.202	0.8313	
	1.800	0.5557	
	5.125	0.1950	
9	1.455	1.327	0.5170
	1.305	0.7661	
	2.000	0.5000	
	5.758	0.1736	
10	1.012	0.9874	
	1.122	0.8908	
	1.414	0.7071	
	2.202	0.4540	
	6.390	0.1563	

*See [42].

Table 12.3a 0.01 dB Chebyshev Active Values.*

Order n	C_1/C or R/R_1	C_2/C or R/R_2
2	1.4826	0.7042
4	1.4874	1.1228
	3.5920	0.2985
6	1.8900	1.5249
	2.5820	0.5953
	7.0522	0.1486
8	2.3652	1.9493
	2.7894	0.8196
	4.1754	0.3197
	11.8920	0.08672

*See [42].

Table 12.3b 0.1 dB Chebyshev Active Values.*

Order n	C_1/C or R/R_1	C_2/C or R/R_2	C_3/C or R/R_3
2	1.638	0.6955	
3	6.653	1.825	0.1345
4	1.900	1.241	
	4.592	0.2410	
5	4.446	2.520	0.3804
	6.810	0.1580	
6	2.553	1.776	
	3.487	0.4917	
	9.531	0.1110	
7	5.175	3.322	0.5693
	4.546	0.3331	
	12.73	0.08194	
8	3.270	2.323	
	3.857	0.6890	
	5.773	0.2398	
	16.44	0.06292	
9	6.194	4.161	0.7483
	4.678	0.4655	
	7.170	0.1812	
	20.64	0.04980	
10	4.011	2.877	
	4.447	0.8756	
	5.603	0.3353	
	8.727	0.1419	
	25.32	0.04037	

*See [42].

Table 12.3c 0.25 dB Chebyshev Active Values.*

Order n	C_1/C or R/R_1	C_2/C or R/R_2	C_3/C or R/R_3
2	1.778	0.6789	
3	8.551	2.018	0.1109
4	2.221	1.285	
	5.363	0.2084	
5	5.543	2.898	0.3425
	8.061	0.1341	
6	3.044	1.875	
	4.159	0.4296	
	11.36	0.09323	
7	6.471	3.876	0.5223
	5.448	0.2839	
	15.26	0.06844	
8	3.932	2.474	
	4.638	0.6062	
	6.942	0.2019	
	19.76	0.05234	
9	7.766	4.891	0.6919
	5.637	0.3983	
	8.639	0.1514	
	24.87	0.04131	
10	4.843	3.075	
	5.368	0.7725	
	6.766	0.2830	
	10.53	0.1181	
	30.57	0.03344	

*See [42].

Table 12.3d 0.5 dB Chebyshev Active Values.*

Order n	C_1/C or R/R_1	C_2/C or R/R_2	C_3/C or R/R_3
2	1.950	0.6533	
3	11.23	2.250	0.0895
4	2.582	1.300	
	6.233	0.1802	
5	6.842	3.317	0.3033
	9.462	0.1144	
6	3.592	1.921	
	4.907	0.3743	
	13.40	0.07902	
7	7.973	4.483	0.4700
	6.446	0.2429	
	18.07	0.05778	

Table 12.3d 0.5 dB Chebyshev Active Values.* *(Continued)*

Order n	C_1/C or R/R_1	C_2/C or R/R_2	C_3/C or R/R_3
8	4.665	2.547	
	5.502	0.5303	
	8.237	0.1714	
	23.45	0.04409	
9	9.563	5.680	0.6260
	6.697	0.3419	
	10.26	0.1279	
	29.54	0.03475	
10	5.760	3.175	
	6.383	0.6773	
	8.048	0.2406	
	12.53	0.09952	
	36.36	0.02810	

*See [42].

Table 12.3e 1 dB Chebyshev Active Values.*

Order n	C_1/C or R/R_1	C_2/C or R/R_2	C_3/C or R/R_3
2	2.218	0.6061	
3	16.18	2.567	0.06428
4	3.125	1.269	
	7.546	0.1489	
5	8.884	3.935	0.2540
	11.55	0.09355	
6	4.410	1.904	
	6.024	0.3117	
	16.46	0.06425	
7	10.29	5.382	0.4012
	7.941	0.1993	
	22.25	0.04684	
8	5.756	2.538	
	6.792	0.4435	
	10.15	0.1395	
	28.94	0.03568	
9	12.33	6.853	0.5382
	8.281	0.2813	
	12.68	0.1038	
	36.51	0.02808	
10	7.125	3.170	
	7.897	0.5630	
	9.952	0.1962	
	15.50	0.08054	
	44.98	0.02269	

*See [42].

double R, we halve C. Thus, to use resistor values other than 1 Ω, the capacitor ratios from the table are scaled. Likewise, if a corner frequency other than 1 rad/s is used, the capacitor values are again scaled. The resistor values, which are shown as 1 Ω in Figures 12.26 and 12.27, are raised to a more practical value. The capacitor values are also raised to a practical value.

We first select a value for all resistors, that is, $R_1 = R_2 = R_3 = R$. The actual capacitor values are found by using the scaling equation,

$$C_n = \frac{C_i}{2\pi f_p R} \tag{12.28}$$

where R is the chosen resistor value and the C_i are read from the table.

Example 12.20

Find component values for the design of Example 12.19. In that example, we specify

$$A_p = 3 \text{ dB} \qquad \text{at } f_p = 1 \text{ kHz}$$

$$A_s = 40 \text{ dB} \qquad \text{at } f_s = 5 \text{ kHz}$$

Verify your result using a computer simulation program.

SOLUTION Refer to Table 12.2 for the Butterworth filter of order $n_B = 3$. The capacitor ratios are read as follows:

$$\frac{C_1}{C} = 3.546, \qquad \frac{C_2}{C} = 1.392, \qquad \frac{C_3}{C} = 0.2024$$

These values must be scaled inversely with the frequency and the selected resistor value. Suppose we choose resistors of value 1 kΩ. The scaling factor is then

$$\frac{1}{2\pi f_p R} = \frac{1}{2\pi \times 10^3 \times 10^3} = 0.16 \times 10^{-6}$$

This yields capacitors of value

$$C_1 = (0.16)(3.546)\ \mu\text{F} = 0.56\ \mu\text{F}$$

$$C_2 = (0.16)(1.392)\ \mu\text{F} = 0.22\ \mu\text{F}$$

$$C_3 = (0.16)(0.2024)\ \mu\text{F} = 0.03\ \mu\text{F}$$

Figure 12.30 Three-pole Butterworth filter for Example 12.20.

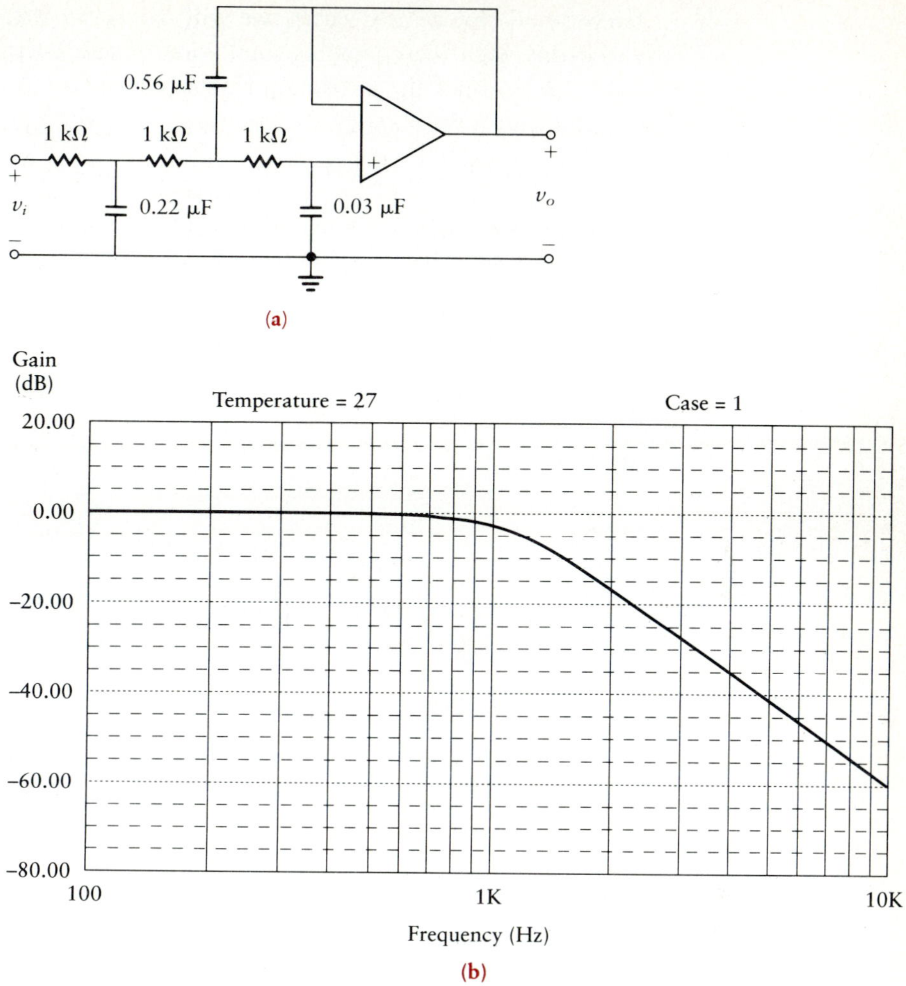

The complete filter is shown in Figure 12.30(a). All resistor values are equal to 1 kΩ. If the derived capacitor values were not practical because of size or availability, we could select a new value for R and recalculate the capacitor values.

The computer simulation yields the ac response of Figure 12.30(b). Note that the amplitude is down over 42 dB at a frequency of 5 kHz. Our design therefore meets the specifications.

Example 12.21

Derive the 0.1-dB Chebyshev filter for the specifications of Example 12.20.

SOLUTION Let us again select $R = 1$ kΩ and use the capacitor coefficients for a Chebyshev low-pass filter of order 3. Since Figure 12.29 indicates a relatively small change as ϵ varies, we will use $n_C = 3$ from Example 12.19 and check this with the computer simulation. These coefficients are taken from Table 12.3b, since the maximum ripple is specified as 0.1 dB. Therefore,

$$\frac{C_1}{C} = 6.653, \qquad \frac{C_2}{C} = 1.825, \qquad \frac{C_3}{C} = 0.1345$$

Using the scaling technique described above with $R = 1\ \text{k}\Omega$, we find the scale value to be

$$\frac{1}{2\pi f_p R} = 0.16 \times 10^{-6}$$

We thus obtain

Figure 12.31 Chebyshev filter for Example 12.21.

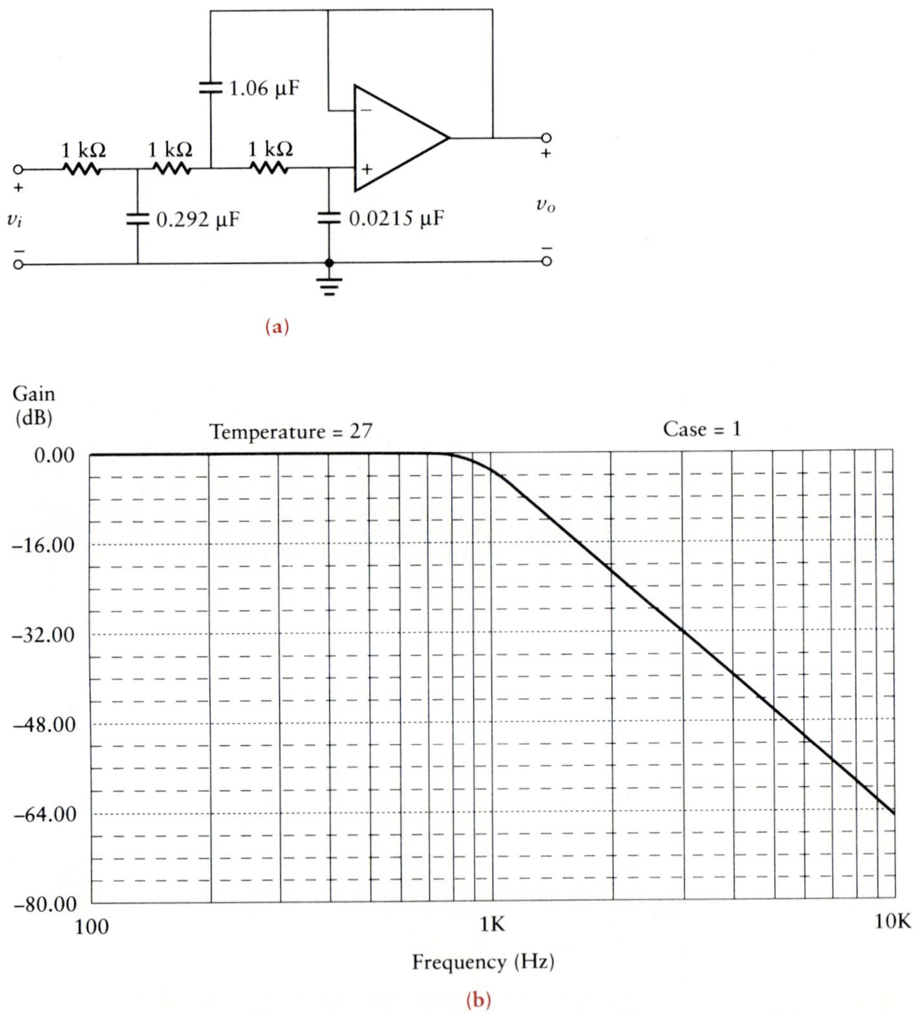

$$C_1 = (0.16)(6.653)\ \mu F = 1.06\ \mu F$$

$$C_2 = (0.16)(1.825)\ \mu F = 0.292\ \mu F$$

$$C_3 = (0.16)(0.1345)\ \mu F = 0.0215\ \mu F$$

The filter is shown in Figure 12.31(a).

Figure 12.31(b) illustrates the results of the computer simulation. Note that the attenuation at 5 kHz is approximately 46 dB. At 5 kHz, the Chebyshev filter attenuates further than the Butterworth filter and exceeds the specifications by a considerable amount.

12.7.4 High-Pass Filter

The circuits for a high-pass filter are shown in Figure 12.27(a) for a two-pole filter and Figure 12.27(b) for a three-pole filter. When calculating the order for a high-pass filter, f_p and f_s are interchanged and all of the capacitors are of the same value. We use the ratio R/R_i in the tables rather than C_i/C.

Example 12.22

Design an active high-pass filter to meet the following requirements:

$$A_p = 3\ \text{dB} \quad \text{at } f_p = 100\ \text{Hz}$$

$$A_s = 75\ \text{dB} \quad \text{at } f_s = 25\ \text{Hz}$$

Verify your design using a computer simulation program.

SOLUTION We first calculate the Butterworth filter order using equations (12.28) and (12.29).

$$\epsilon_1 = \sqrt{10^{(0.1)(3)} - 1} = 1$$

$$\epsilon_2 = \sqrt{10^{7.5} - 1} = 5623$$

The required order for the Butterworth filter is then obtained from equations (12.26) and (12.27).

$$n_B = \frac{\log 5623}{\log(100/25)} = 6.25 \quad \text{so use 7}$$

We now need to find the required order of a Chebyshev filter. Figure 12.29 applies to the low-pass filter, and our current application is a high-pass filter.

However, since roll-off depends on relative frequency (i.e., roll-off is stated in decibels per decade, which means a certain decibel decrease for each change in frequency by a factor of 10), we can simply reverse the direction and talk about division of frequency by 10. In this example, the attenuation is specified at a frequency that is one-fourth of the corner frequency, so we need to look at the point $f = 4$ in Figure 12.29. A fourth-order filter achieves about 65 dB of attenuation at $f = 4$, and a fifth-order filter achieves about 80 dB. Since the specification is for 75 dB, we must use a fifth-order Chebyshev filter.

Since the Chebyshev filter has a lower order than the Butterworth, we select this type of filter. We use the data for a 0.5-dB Chebyshev filter from Table 12.3d. Note that the ratios in this table are *resistor* ratios, R/R_i, for a high-pass filter. They are *capacitor* ratios for the low-pass filter. Hence, for this fifth-order filter we obtain

First Stage

$$\frac{R}{R_1} = 6.842, \qquad \frac{R}{R_2} = 3.317, \qquad \frac{R}{R_3} = 0.3033$$

Second Stage

$$\frac{R}{R_1'} = 9.462, \qquad \frac{R}{R_2'} = 0.1144$$

Let us choose $C = 0.15\ \mu\text{F}$. The scaling factor is calculated:

$$R = \frac{1}{2\pi f_p C} = \frac{1}{2\pi(100)\,(15 \times 10^{-9})} = 1.06 \times 10^5$$

We obtain the following resistor values:

$$R_1 = \frac{R}{6.842} = 15.5\ \text{k}\Omega$$

$$R_2 = \frac{R}{3.317} = 32\ \text{k}\Omega$$

$$R_3 = \frac{R}{0.3033} = 349\ \text{k}\Omega$$

$$R_1' = \frac{R}{9.462} = 11.2\ \text{k}\Omega$$

$$R_2' = \frac{R}{0.1144} = 927\ \text{k}\Omega$$

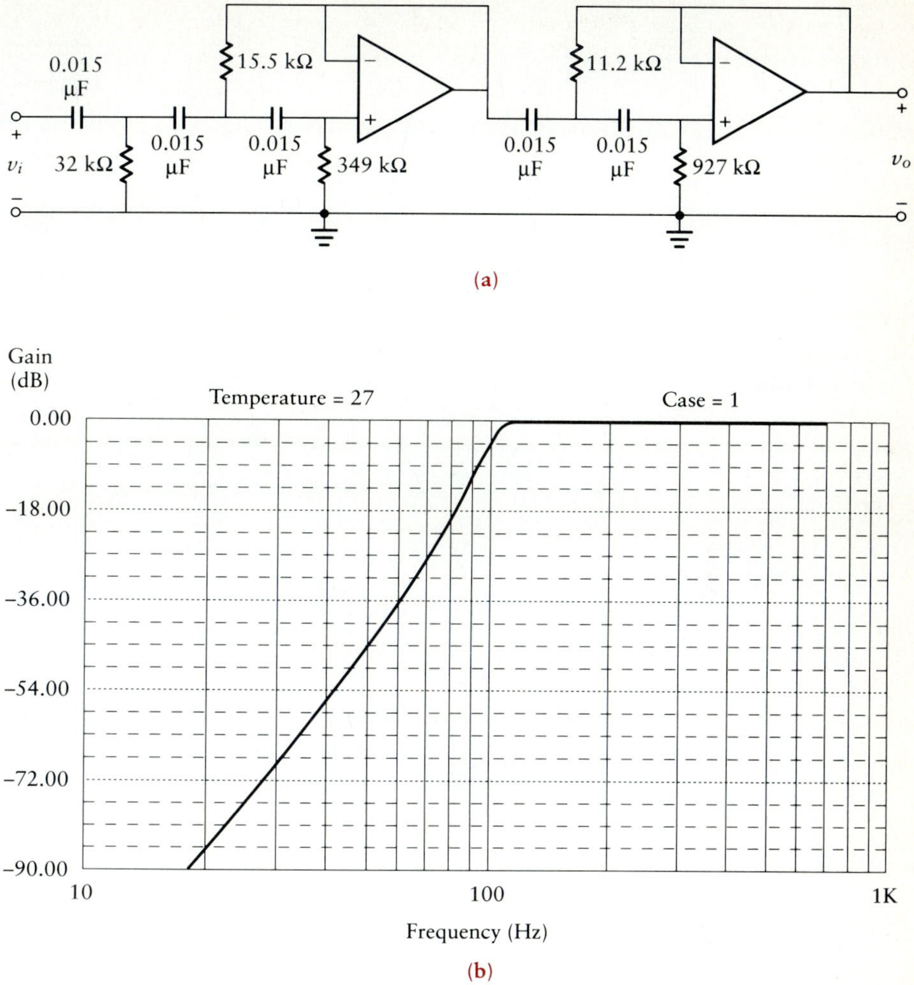

Figure 12.32 High-pass Chebyshev filter for Example 12.22.

The circuit is shown in Figure 12.32(a). We have placed the third-order filter in front of the second-order filter. We could just as easily reverse the order and obtain the same resulting response.

The computer simulation result is shown in Figure 12.32(b). Note that the simulation achieves an attenuation of about 77 dB at 25 Hz, thereby meeting the specifications.

12.7.5 Band-Pass and Band-Stop Filter Design

The band-pass and band-stop filters are each formed from low-pass and high-pass filters. The frequency response for the band-pass filter is shown in Figure 12.33(a). We form this type of filter by placing a low-pass filter in series with a high-pass filter, as shown in Figure 12.33(b). The frequency response

Figure 12.33
Band-pass filter.

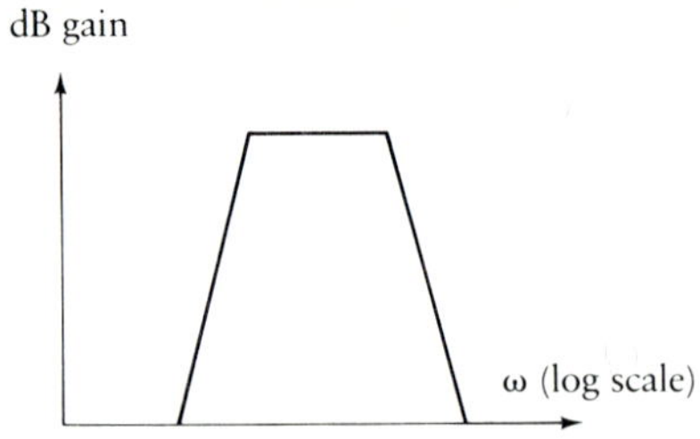

(a) Frequency response

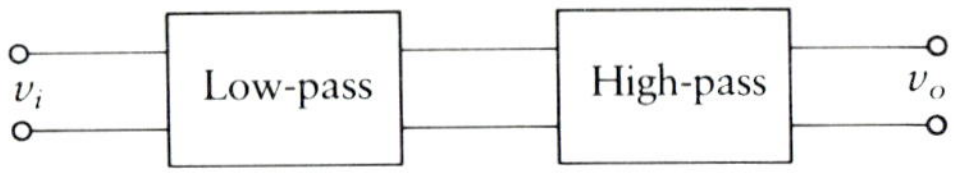

(b) Low-pass and high-pass filters in series

Figure 12.34
Band-stop filter.

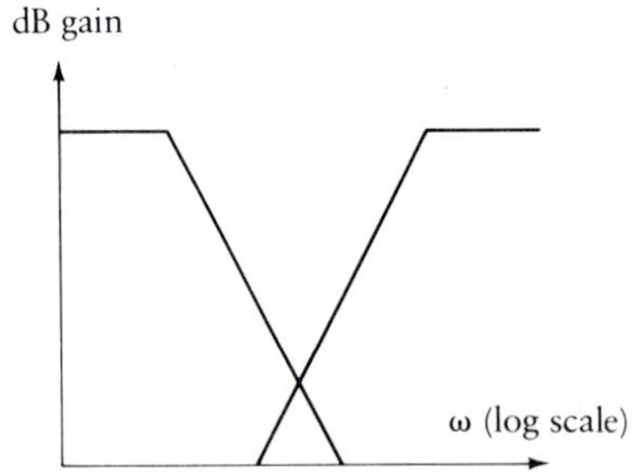

(a) Frequency response

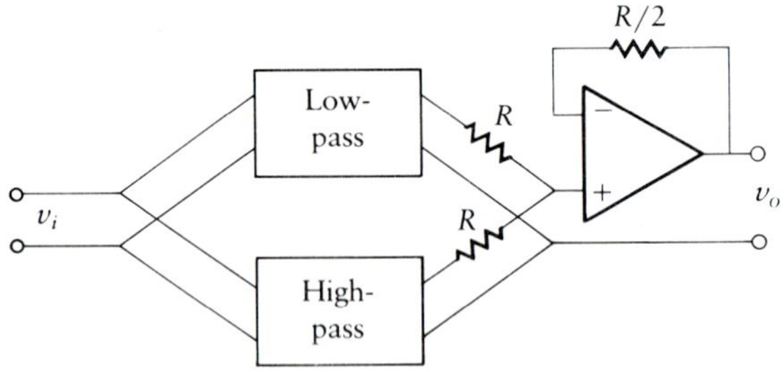

(b) Low-pass and high-pass filters in parallel

for the band-stop filter is shown in Figure 12.34(a). We form this filter by placing a low-pass filter in parallel with a high-pass filter, as shown in Figure 12.34(b). We use a summing amplifier to develop the output for the band-stop filter.

Example 12.23

Design a band-pass Butterworth filter that has a frequency response as shown in Figure 12.35. Verify your design using a computer simulation program.

Figure 12.35
Band-pass filter response for Example 12.23.

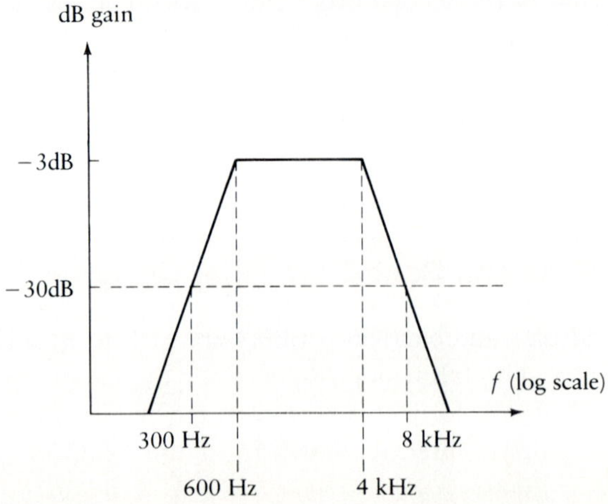

SOLUTION This filter is designed using a low-pass filter in series with a high-pass filter, as shown in Figure 12.33(b). We start the design of the low-pass filter by using equations (12.26) and (12.27) as follows:

$$\epsilon_1 = \sqrt{10^{(0.1)(3)} - 1} = 1$$

$$\epsilon_2 = \sqrt{10^{(0.1)(30)} - 1} = 31.6$$

$$n_B = \frac{\log(\epsilon_2/\epsilon_1)}{\log(f_s/f_p)} = \frac{\log(31.6)}{\log(8000/4000)} = 4.98 \quad \text{so we use 5}$$

We obtain the C_i/C coefficients from Table 12.2 as follows:

1.753	1.354	0.4214
3.235	0.309	

The scaling factor, C, is found, for a choice of $R = 10\ \text{k}\Omega$, to be

$$C = \text{scaling factor} = \frac{1}{2\pi f_p R} = \frac{1}{2\pi(4 \times 10^3)(10^4)} = 3.98 \times 10^{-9}\ \text{F}$$

The scale factor is multiplied by each of the C_i/C coefficients to yield the capacitor values. The final circuit for the low-pass section is shown in Figure 12.36(a). It makes no difference whether the third-order filter is before or after the second-order filter.

Figure 12.36 Low-pass filter for Example 12.23.

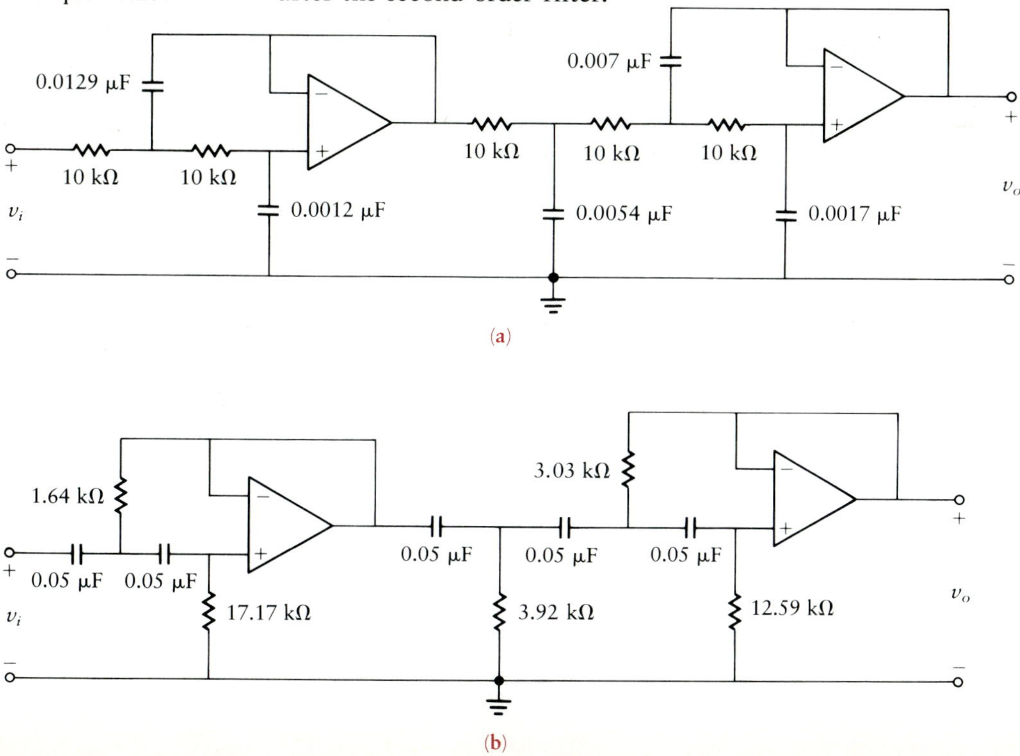

The second part of the filter comprises the design of the high-pass section. Equations (12.26) and (12.27) are used as follows:

$$\epsilon_1 = \sqrt{10^{(0.1)(3)} - 1} = 1$$

$$\epsilon_2 = \sqrt{10^{(0.1)(30)} - 1} = 31.6$$

and $n_B = 5$ just as for the low-pass filter. The coefficients for the high-pass filter are the ratios R/R_i but have the same numerical values as the C_i/C ratios. Thus, they are given as follows:

1.753	1.354	0.4214
3.235	0.309	

Let us choose $C = 0.05\ \mu\text{F}$, yielding the scaling factor, R, as

$$R = \text{scaling factor} = \frac{1}{2\pi(600)\,(0.5 \times 10^{-7})} = 5.305\ \text{k}\Omega$$

The resistor values, R_i, are found by dividing the coefficients into the scaling factor, R, as follows:

$$R_1 = \frac{5305}{1.753} = 3.03\ \text{k}\Omega$$

$$R_2 = \frac{5305}{1.354} = 3.92\ \text{k}\Omega$$

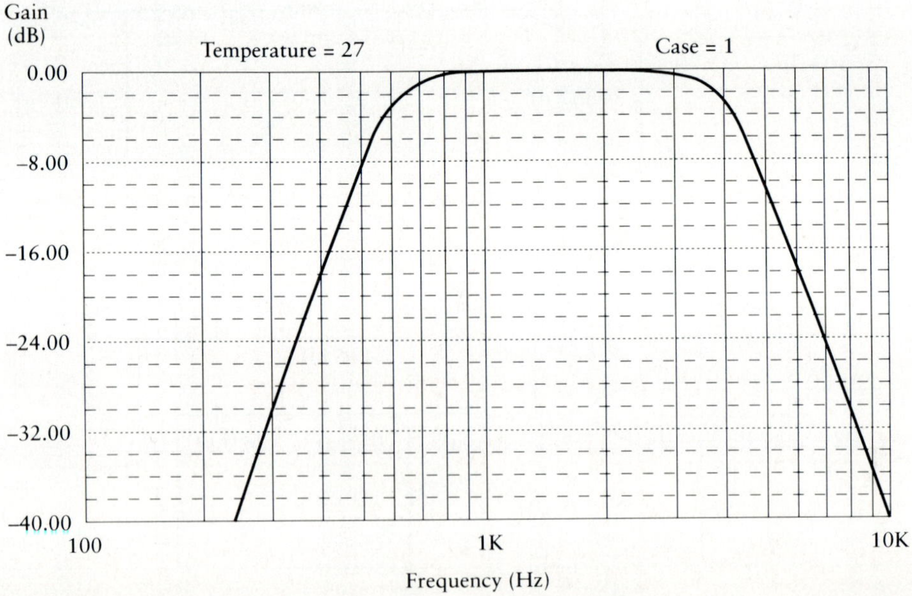

Figure 12.37 Computer simulation for Example 12.23.

$$R_3 = \frac{5305}{0.4214} = 12.59\ \text{k}\Omega$$

$$R_4 = \frac{5305}{3.235} = 1.64\ \text{k}\Omega$$

$$R_5 = \frac{5305}{0.309} = 17.17\ \text{k}\Omega$$

The final circuit for the high-pass section is shown in Figure 12.36(b). The complete circuit is formed by placing the low-pass section of Figure 12.36(a) in series with the high-pass section of Figure 12.36(b).

The results of the computer simulation (using MICRO-CAP II) are shown in Figure 12.37. This figure shows that the design meets the specifications given at the beginning of this example.

12.8 INTEGRATED CIRCUIT FILTERS

If we attempt to produce the filter designs presented in the first part of this chapter in monolithic integrated circuit form, we encounter the following problems:

1. The filter performance (especially the location of the corner frequency) is dependent on the accuracy of the component values. For example, with the high-pass filter of Example 12.22 (shown in Figure 12.32), it is necessary to select resistor values of 15.5, 32, 349, 11.2, and 927 kΩ. Reference to the standard component values in Appendix E indicates that approximations must be made in the implementation of this filter, since these resistor values are not generally available.
2. Large-valued capacitors are usually required. The examples set forth in the previous sections result in capacitor values in the microfarad range.

One technique that obviates these problems is use of the *switched-capacitor filter*. Several manufacturers produce IC filters based on this technique. We first consider the theory of switched-capacitor filters and then consider the method of designing IC switched-capacitor filters.

12.8.1 Switched-Capacitor Filters

A capacitor is switched between two circuit nodes at a high switching frequency. The result is that this circuit is equivalent to a resistor whose value can be accurately determined by the period of the switching operation.

An active RC integrator is shown in Figure 12.38(a). This network is discussed in Section 12.1. The input resistance to the integrator, R, is changed to

Figure 12.38 Switched-capacitor filter.

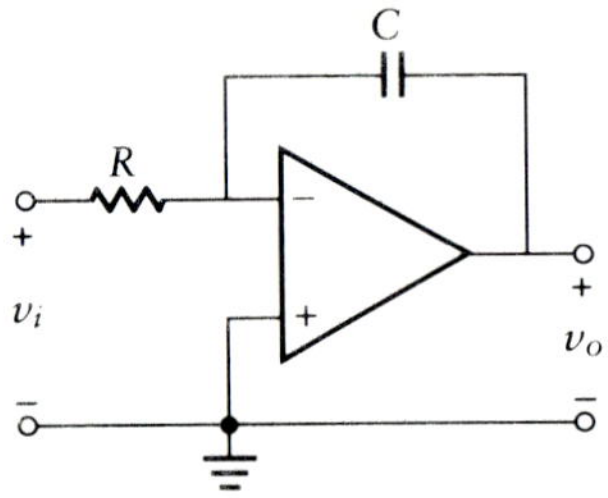

(a) Active integrator

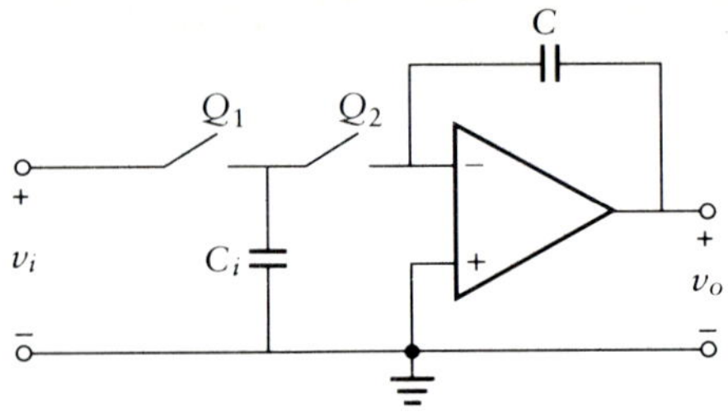

(b) Switched-capacitor integrator

Figure 12.39 Clock for switched-capacitor filter.

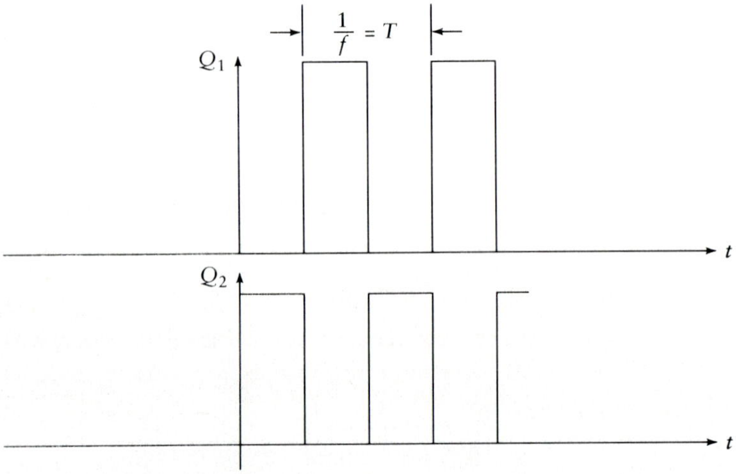

a switched capacitor in the circuit of Figure 12.38(b). The two CMOS switches,* Q_1 and Q_2, are driven by a nonoverlapping two-phase clock of period T, as shown in Figure 12.39. The frequency of the clock, $f = 1/T$, is chosen to be much higher than the highest frequency of the integrator input. The CMOS switch is closed when the applied clock signal is high and open when the applied clock signal is low. The two-phase clock signals must not overlap, since the two switches must never be closed or open at the same instant of time. When Q_1 is closed Q_2 must be open, and when Q_1 is open Q_2 must be closed.

Let us now consider the theory of operation of the switched-capacitor filter. The input capacitor, C_i, charges to v_i during the first half of the clock period. That is,

$$Q = C_i v_i$$

Since the frequency of the clock is so much higher than the frequency being filtered, v_i does not change while C_i is being charged. During the second half

---

* CMOS switches are discussed more fully in Section 15.10 of this text.

of the clock period, the charge, $C_i v_i$, is transferred to the feedback capacitor since $v_- = v_+ = 0$. The total transfer of charge in one clock cycle is

$$Q = C_i v_i \tag{12.29}$$

Current is defined as charge per unit time, so the average input current is

$$I_i = \frac{Q}{T} = \frac{C_i V_i}{T} = C_i V_i f \tag{12.30}$$

where f is the clock frequency. The equivalent input resistor (R) can be expressed as

$$R = \frac{v_i}{i_i} = \frac{1}{C_i f} \tag{12.31}$$

The equivalent RC time constant for the switched capacitor filter is

$$RC = \frac{C}{fC_i} \tag{12.32}$$

Hence the RC time constant that determines the integrator frequency response is determined by the clock frequency, f, and the capacitor ratio, C/C_i. The clock frequency can be set with an accurate crystal oscillator. The capacitor ratios are accurately fabricated on an IC chip (typical tolerances equal 0.1%). We need not accurately set an absolute value of C, since only the ratio C/C_i affects the time constant. We use small values of capacitance, such as 0.1 pF, to reduce the area on the IC devoted to the capacitors.

Example 12.24

A switched-capacitor integrator has a frequency $f = 50$ kHz. The capacitors are given by

$$\frac{C}{C_i} = \frac{1 \text{ pF}}{0.1 \text{ pF}} = 10$$

Find the resulting RC time constant of the integrator.

SOLUTION We apply equation (12.32) to find the time constant.

$$RC = \frac{10}{0.5 \times 10^5} = 2 \times 10^{-4} \text{ s} = 0.2 \text{ ms}$$

Note that we are using small capacitors in order to obtain a time constant of the order of 1 ms. Hence, we can obtain relatively large time constants, suitable for audio applications, with small areas on the IC.

12.8.2 A Sixth-Order Switched-Capacitor Butterworth Low-Pass Filter*

Of the numerous IC filters produced by National Semiconductor Corp., we select the MF6, a sixth-order switched-capacitor Butterworth low-pass filter, as a representative model. The block diagram of this filter is shown in Figure 12.40(a). The block diagram contains a Butterworth filter, a clock genera-

Figure 12.40
MF6 low-pass filter.

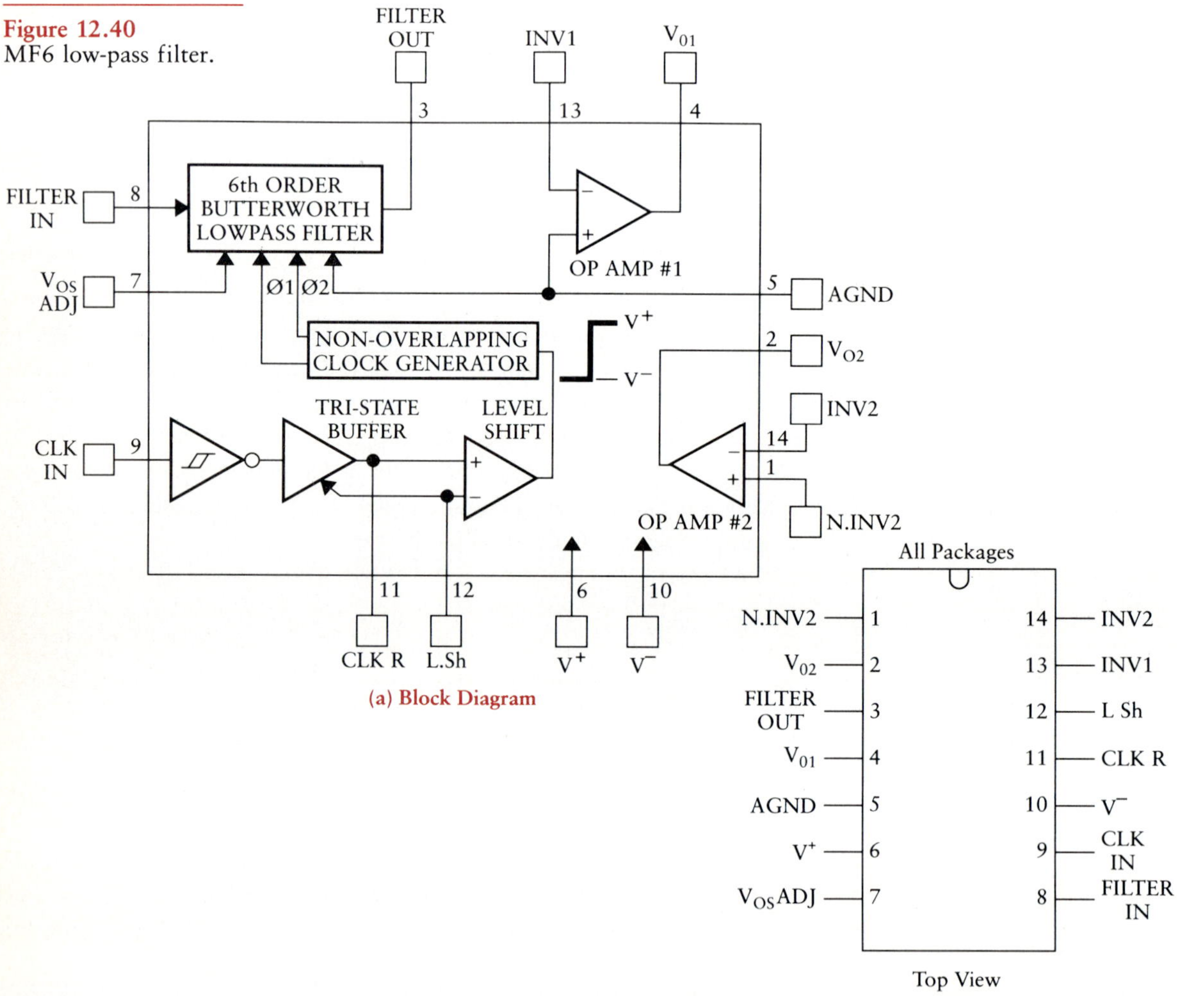

(a) Block Diagram

(b) Connection Diagram

* The information in this section is provided through the courtesy of National Semiconductor Corp. The reader is referred to Linear Data Book 2 published by National Semiconductor Corp.

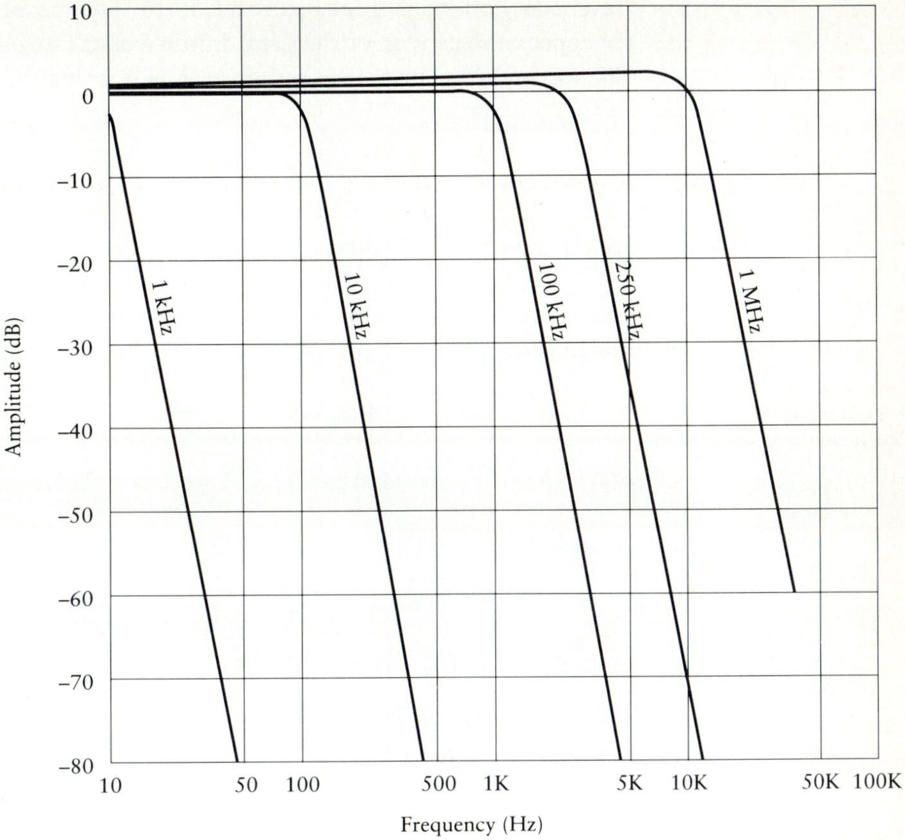

Figure 12.41 Response of MF6 low-pass filter.

tor, and a number of op-amps performing level shifting and buffer functions. The nonoverlapping clock generator is discussed in Chapter 14, and level shifters are discussed in Chapter 9. The buffer can be thought of as a power amplifier. The ratio of the clock frequency to the low-pass cutoff frequency is internally set to 100 to 1. For example, to obtain a cutoff frequency of 10 kHz, the clock frequency is 1 MHz. Two clock options are available:

1. With self-clocking, an external resistor and capacitor are used to set the clock frequency. This option is used for standard applications.
2. An external clock can be connected to the filter when more accurate frequency control is needed. This external clock must be compatible with the internal circuitry, so either a TTL- or CMOS-compatible clock is needed (we discuss TTL and CMOS in Chapter 15).

The filter is maximally flat (Butterworth) and produces a unity gain. Since the gain is unity, filters can be cascaded without worrying about exceeding proper input ranges.

The filter is packaged as a 14-pin DIP (dual in-line package) as shown in the connection diagram of Figure 12.40(b). The frequency response of this filter approximates that of the ideal Butterworth low-pass characteristic, as we can see from the amplitude response shown in Figure 12.41.

Example 12.25

Design an IC Butterworth filter to satisfy the following amplitude specifications:

$$A_p = 3\ \text{dB} \qquad f_p = 1\ \text{kHz}$$

$$A_s = 35\ \text{dB} \qquad f_s = 2\ \text{kHz}$$

SOLUTION From equation (12.27) we have

$$\epsilon_1 = \sqrt{10^{0.3} - 1} = 1$$

$$\epsilon_2 = \sqrt{10^{3.5} - 1} = 56.23$$

From equation (12.26) we obtain

$$n_B = \frac{\log(56.23)}{\log(2)} = 5.81$$

We therefore use a sixth-order filter.

We can use the MF6 to implement this. We select

$$f = 100 \times (1.0\ \text{kHz}) = 100\ \text{kHz}$$

12.9 CONCLUDING REMARKS

In the introduction to this chapter, we posed the problem of designing a filter to reduce out-of-band power output to the level required by the FCC. This design was to be accomplished quickly in order to bring the video game to market in a timely manner.

Because the filter was to be designed and manufactured quickly, it is probably not wise to attempt a custom design. The requirement for significant reduction outside the pass band leads us to consider the Butterworth or Chebyshev design. Since the selling price of our game was calculated without considering the filter, it is important that cost be held to an absolute minimum. Example 12.19 illustrates that the same level of out-of-band attenuation can be accomplished with fewer stages with a Chebyshev than with a Butterworth filter. For this reason, we would probably select the Chebyshev band-pass filter of the order necessary to meet the FCC specification. One precaution is necessary here. The ripple within the pass band of the Chebyshev filter normally would not cause great difficulty. In particular, if audio signals are involved, the listener would probably not hear any distortion. However, with video signals, this ripple could lead to ghost images on the TV screen. Given the tight time schedule, it would probably be best to build a prototype Chebyshev filter and test it in the game. If objectionable results occur, we could switch to a higher-order Butterworth filter.

PROBLEMS

In Problems 12.1–12.6, design active networks to provide the given V_o/V_i transfer function.

12.1 $H(s) = \dfrac{-10}{s}$

12.2 $H(s) = \dfrac{10}{s^2}$

12.3 $H(s) = \dfrac{-s}{s + 10}$

12.4 $H(s) = \dfrac{-10(s + 10)}{s + 1}$

12.5 $H(s) = \dfrac{10}{s(s + 1)}$

12.6 $H(s) = \dfrac{-(s + 1)^2}{s(s + 10)}$

In Problems 12.7 and 12.8, design a single-amplifier summing integrator to achieve the output voltage V_o related to input voltages V_1, V_2, and V_3 as indicated.

12.7 $V_o = \dfrac{-(V_1 + 10V_2 + 0.53V_3)}{s}$

12.8 $V_o = \dfrac{V_1 - V_2 - 2V_3}{s}$

12.9 Design a multiple-amplifier integrator using a single capacitor that achieves the input-output relationships of Problems 12.7 and 12.8.

12.10 Design a single-input integrator circuit with a switch that will reset the integrator output voltage to +10 V when thrown. When integrating, the output voltage should be the negative of the integral of the input voltage. A multiple-pole switch may be used if necessary.

12.11 Design a single-amplifier summing differentiator with the following relationship between output voltage, V_o, and input voltages, V_1, V_2, and V_3.

$$V_o = -s(V_1 + 10V_2 + 0.5V_3)$$

12.12 Design an active low-pass filter with a dc gain of $\frac{1}{2}$ and a roll-off frequency of 10 kHz.

12.13 Design an active low-pass filter with a dc gain of 1000 and a roll-off frequency of 0.4 Hz.

12.14 Design an active low-pass filter that is adjustable between 10 Hz and 100 Hz roll-off frequency. The frequency adjustment should be made with a potentiometer.

12.15 Design an active low-pass filter that has independently adjustable dc gain and roll-off frequency. Gain should be adjustable between 0 and 10 and frequency from near zero to 100 Hz. In addition, this system should have an input impedance greater than 50 MΩ.

12.16 Design an active high-pass filter with a high-frequency gain of 2 and a roll-off frequency of 20 Hz.

12.17 Design an active high-pass filter with a gain of 10 and a roll-off frequency of 200 Hz.

12.18 Repeat the design of Problem 12.14, making the roll-off frequency adjustable in the range between 500 and 1500 Hz.

12.19 Design an active filter with transfer function

$$H(s) = \frac{3s - 10}{s + 100}$$

12.20 Design an active filter with transfer function

$$H(s) = \frac{-10s + 5}{3s + 40}$$

12.21 Derive equation (12.17) from Figure 12.18.

12.22 Design an RC low-pass filter with a dc gain of 10 and a corner frequency of 2 kHz.

12.23 Design a first-order active filter with a dc gain of 20 and a corner frequency of 1 kHz.

12.24 Design a low-pass filter with adjustable roll-off frequency between 500 Hz and 1 kHz and a dc gain of 5.

12.25 Design a high-pass filter with a high-frequency gain of 20 and a corner frequency of 500 Hz.

12.26 Design an all-pass filter with a dc gain of $\frac{1}{2}$ and corner frequency at 100 Hz.

12.27 Design an adjustable high-pass filter with a high-frequency gain of 20 and a corner frequency adjustable from 200 to 600 Hz.

12.28 Design an adjustable high-pass filter with a high-frequency gain of 10 and a corner frequency adjustable from 100 to 200 Hz.

12.29 Design an all-pass filter with a gain of 10 and a corner frequency of 100 Hz.

12.30 Using the single amplifier of the general type, design a high-pass filter with a high-frequency gain of 1 and a corner frequency of 3 kHz.

12.31 Design a Butterworth low-pass filter with the following specifications:

$$f_p = 500 \text{ Hz} \qquad A_p = 3 \text{ dB}$$

$$f_s = 3 \text{ kHz} \qquad A_s = 50 \text{ dB}$$

12.32 Repeat Problem 12.31 using a Chebyshev filter with a 0.5-dB ripple.

12.33 Repeat Problem 12.32 but with a 1.0-dB ripple.

12.34 Design a Butterworth high-pass filter with the following specifications: 3 dB at 1 kHz

45 dB at 300 Hz

12.35 Repeat Problem 12.34 using a Chebyshev filter with a 0.25-dB ripple.

12.36 Design a "crossover" network to distribute the high frequencies of an audio signal into a high-frequency speaker and the low frequencies into a low-frequency speaker. This is shown in Figure P12.1. Choose either a Butterworth or a Chebyshev filter, whichever requires fewer components. If you choose a Chebyshev filter, choose a maximum ripple of 0.25 dB. Draw the filter diagrams.

Figure P12.1

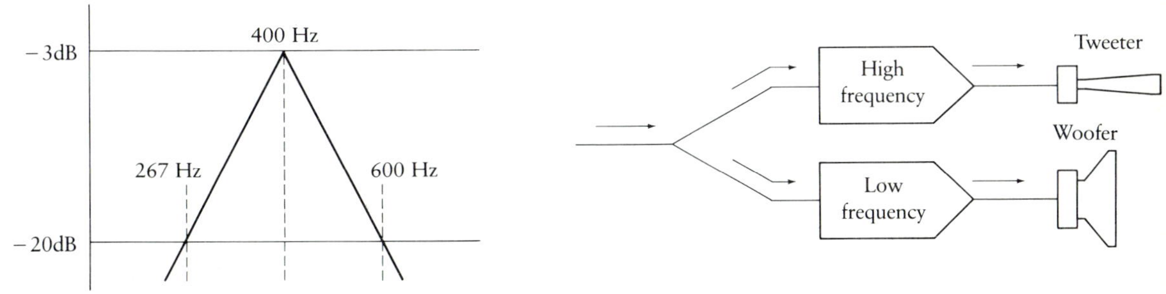

12.37 Determine the transfer function, V_o/V_i, for the ideal op-amp circuit shown in Figure P12.2. Since the op-amp is ideal, $V_+ = V_-$. The parameter values are

$$C_1 = 3.546C, \qquad C_2 = 1.392C, \qquad C_3 = 0.2024C, \qquad RC = 1.0$$

Solve this problem in two steps. First find the equations; then solve them using determinants. Construct a Bode plot for this network.

12.38 Design a Butterworth band-pass filter that has the amplitude characteristics shown in Figure P12.3.

12.39 Design a Chebyshev band-stop filter that has the amplitude characteristics shown in Figure P12.4. Set the maximum ripple at 1 dB.

12.40 Design two high-pass filters to satisfy the decibel gain plot of Figure P12.5. One filter is a Butterworth and the other is a 0.1-dB Chebyshev. Compare the two filters.

12.41 Design a telephone filter to develop the frequency response shown in Figure P12.6. Use a Chebyshev 1-dB filter with "reasonable" capacitor and resistor values.

12.42 Use the MF6-100 IC filter to satisfy the specification for the low-pass filter of Problem 12.31.

Figure P12.2

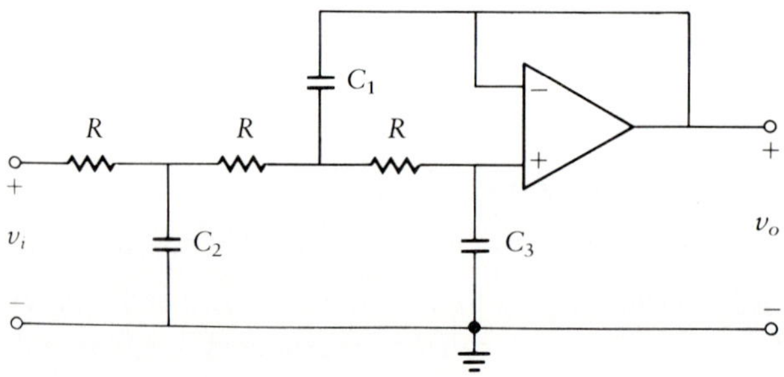

Figure P12.3

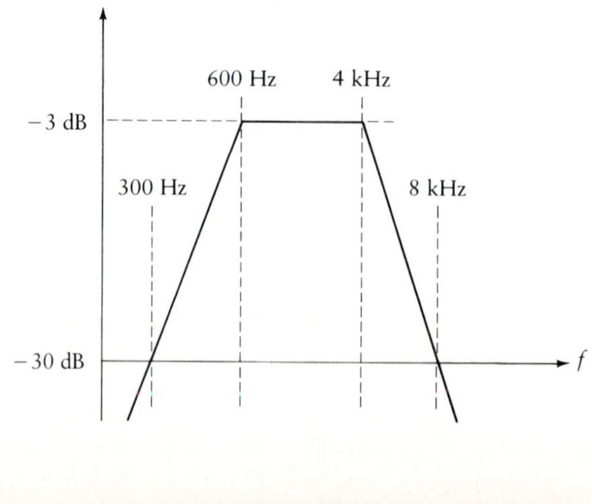

Figure P12.4

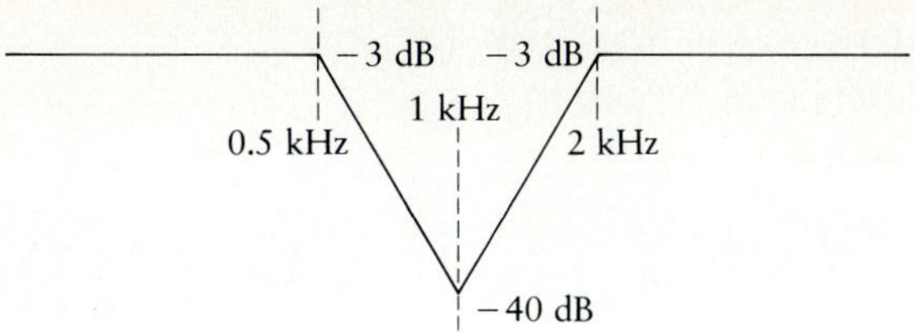

Figure P12.5

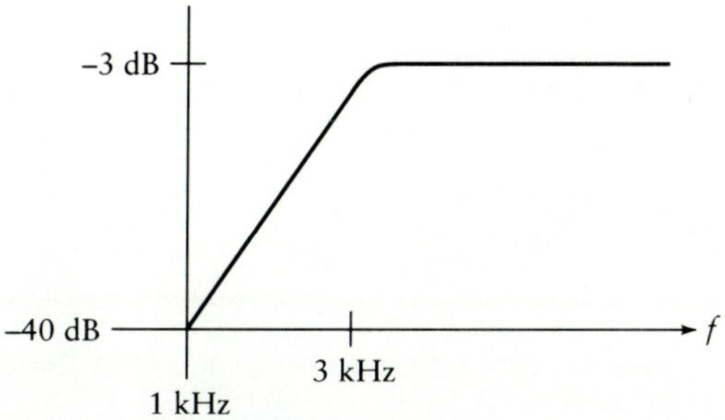

Figure P12.6

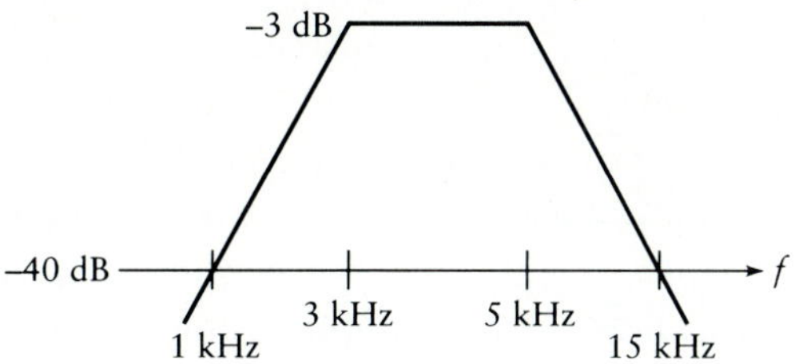

13

QUASI-LINEAR CIRCUITS

13.0 INTRODUCTION

Several diode circuits were considered in Chapter 1 of this text. One of the examples was the power supply circuit that converts an ac signal into dc. We also analyzed various clipping and clamping configurations that are used either to chop off a portion of a time-varying signal or to change its dc level. In the current chapter, we combine diodes with op-amps in various circuit configurations. The advantage of this approach is that the diodes are able to operate closer to their ideal characteristics. In many cases, the effects of the forward voltage drop across the diode are virtually eliminated because of the high open-loop gain of the op-amp. These diode circuits are frequently termed *quasi-linear* since we can break them into two linear circuits—one when the diode is on and another when the diode is off. We consider rectifiers, limiters, comparators, and Schmitt triggers. Our objective is to be able to design systems that achieve specified instantaneous output-to-input voltage transfer characteristics.

13.1 RECTIFIERS

The *rectifier* is one of the most basic and useful nonlinear circuits. Rectifiers operate on an input signal in a manner that depends on the sign of the instantaneous input voltage. They can be designed to chop off either the negative or positive part of the signal or to yield an output that is the mathematical absolute value of the input.

For illustration, we use the μA741 in the circuits presented in this chapter, and we include a data sheet for that op-amp in Appendix D. However, any internally compensated op-amp can be used in these circuits.

The circuit shown in Figure 13.1 is known as an *inverting half-wave rectifier.* Since a diode can operate in either of two states, we analyze the rectifier as two separate circuits. Which of the two separate circuits applies depends on the sign of v_1.

If we assume the op-amp to be ideal, then

$$v_- = v_+ = 0$$

For positive v_i, the amplifier output voltage, v_1, is less than zero. Diode D_2 *conducts,* and it can therefore be replaced by a small resistor, R_f. This small diode forward resistance acts as a feedback resistor, thus leading to lowered amplifier gain. Diode D_1 acts as an open circuit under this condition, so

$$v_o = v_- = 0$$

Alternatively, when v_1 is positive, D_2 is *nonconducting* (off) and D_1 is *conducting* (on). Figure 13.2(a) repeats the circuit of Figure 13.1 with the diodes replaced by their equivalents for the situation when v_1 is positive. That is, D_1 is replaced by a forward resistance, R_f, in series with the diode forward voltage drop, V_γ. D_2 is open-circuited.

The node equations for v_+ and v_- are found from Figure 13.2(a) as follows:

$$v_+ = 0$$

Figure 13.1
Inverting half-wave rectifier.

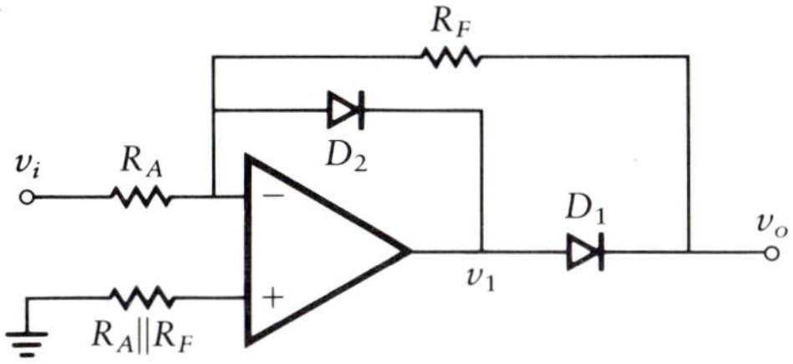

Figure 13.2
Inverting half-wave rectifier with positive input.

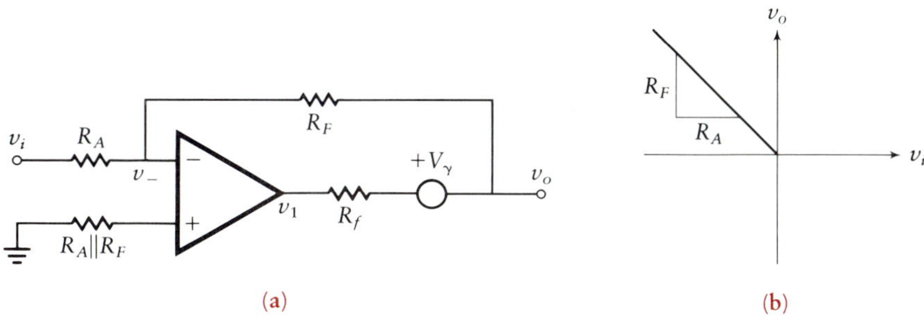

$$v_-\left(\frac{1}{R_A}+\frac{1}{R_F}\right)-\frac{v_i}{R_A}=\frac{v_o}{R_F} \tag{13.1}$$

Since $v_- = v_+ = 0$, we set the first term of equation (13.1) to zero and solve for the output voltage as follows:

$$v_o = -R_F\frac{v_i}{R_A} \tag{13.2}$$

The expression in equation (13.2) does not depend on the diode forward voltage, V_γ. Thus, because of the high open-loop gain of the op-amp, the feedback acts to cancel the diode turn-on (forward) voltage. This leads to improved performance, since the diode more closely approximates the ideal device. The transfer characteristic of equation (13.2) is shown in Figure 13.2(b).

The half-wave rectifier is one of the simplest nonlinear circuits. There are many variations on the basic circuit. Some of these alternative forms are illustrated in Table 13.1. We recommend that you not proceed beyond this point until you review (and understand) the operation of each circuit in the table.

The half-wave rectified output waveform can be shifted along the v_i axis by using a reference voltage added to the input voltage of the rectifier circuit, as shown in Figure 13.3. This is termed "axis shifting," and it adds to or sub-

Figure 13.3
Axis shifting.

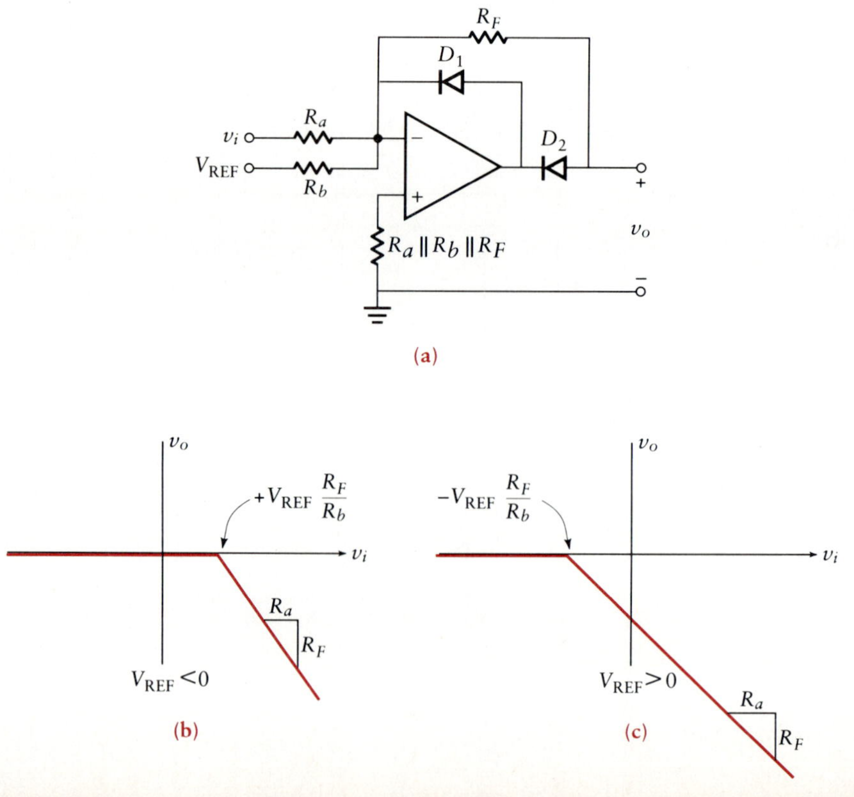

Table 13.1 Half-Wave Rectifier Configurations.

1. Basic positive output, inverting

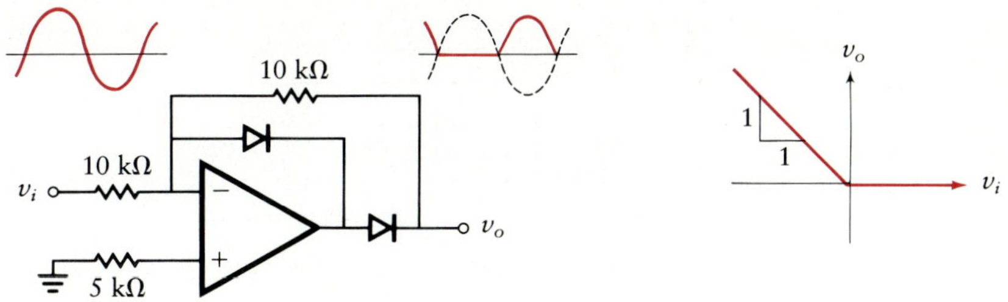

2. Positive output, inverting with gain

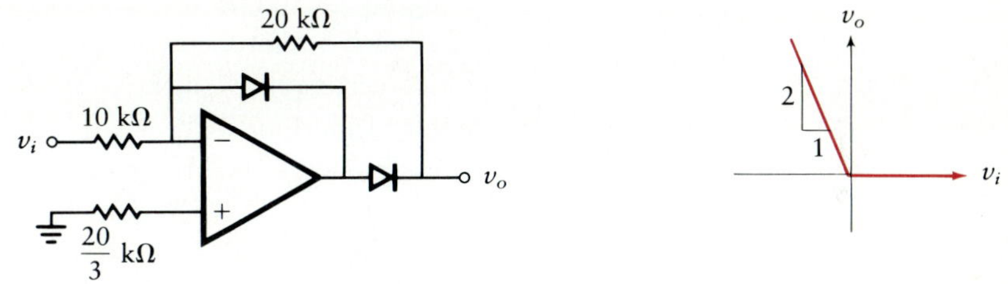

3. Positive output, inverting and summing

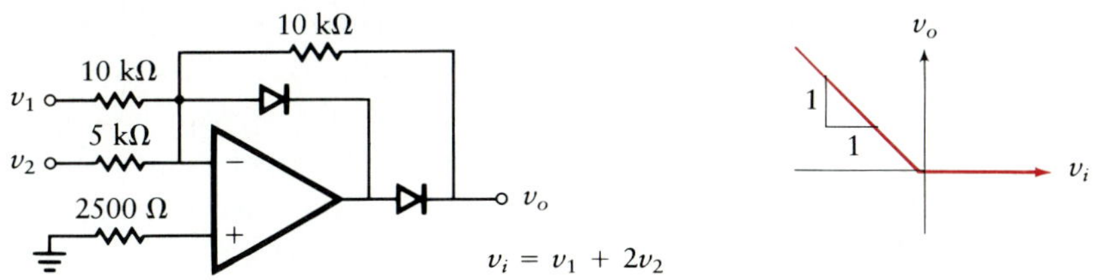

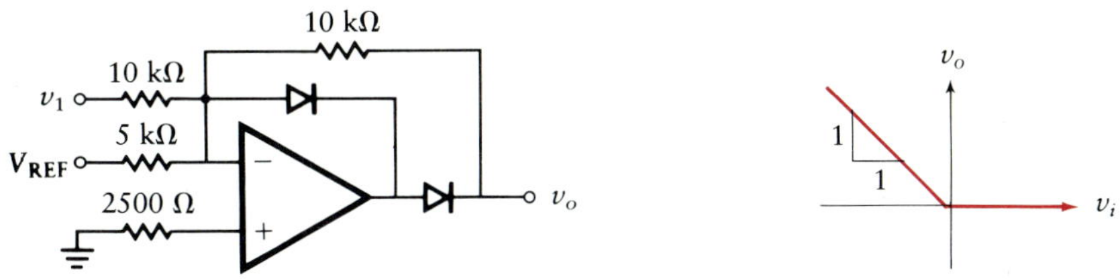

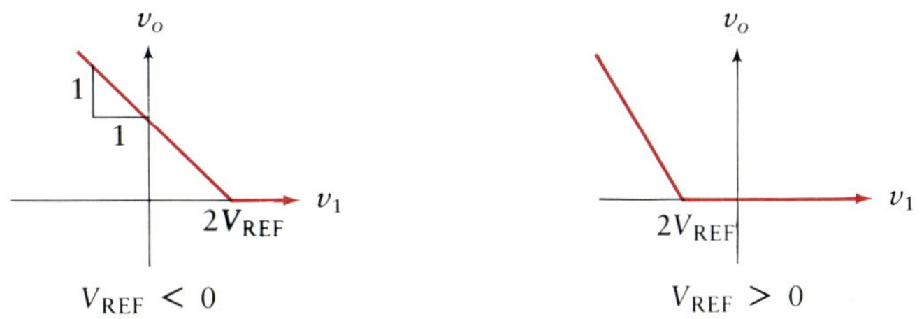

Table 13.1 Half-Wave Rectifier Configurations. *(Continued)*

4. Basic negative output, inverting

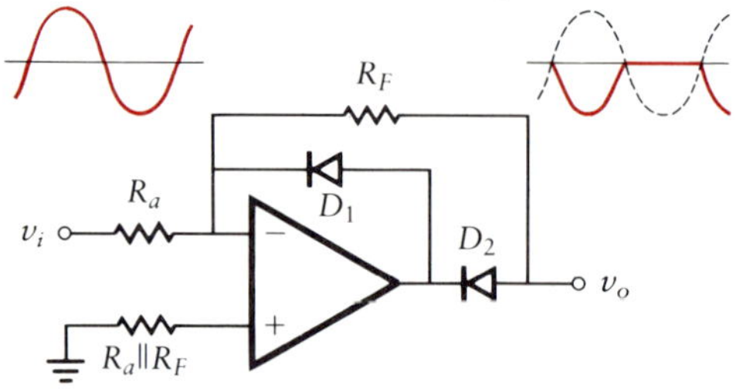

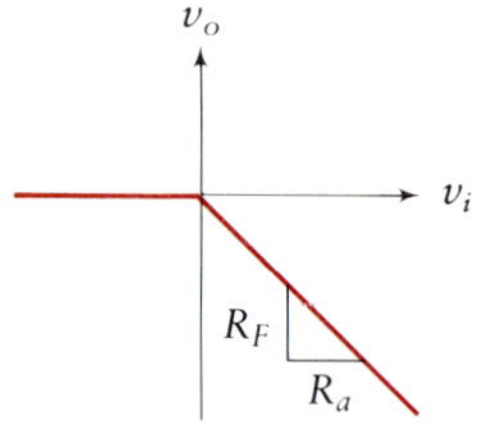

tracts from the input signal a fixed dc voltage, thereby shifting the diode turn-on voltage point. If a negative reference voltage, V_{REF}, is applied to the circuit of Figure 13.3, the diode turns on when the input voltage is still positive, thus shifting the v_o/v_i transfer function to the right. If a positive reference voltage is applied, the v_o/v_i transfer function shifts to the left. These shifted characteristic curves are shown in Figure 13.3. Accuracy is improved by making V_{REF} a well-regulated, low-current power supply from which almost no current is drawn.

The input-output voltage characteristics can also be shifted up or down. This is termed "level shifting" and is accomplished by adding a second op-amp with a reference voltage added to the negative input terminal, as shown in Figure 13.4. The reference voltage is used to vary the amount of voltage shift by adjusting the resistor ratio of the second op-amp, R_F'/R_b'. The resistors of the second op-amp should be made larger than those of the first op-amp to reduce attenuation (by voltage division) when D_2 is off.

A *full-wave rectifier,* or *magnitude operator,* produces an output that is the absolute value, or magnitude, of the input signal waveform. One method of accomplishing full-wave rectification is to use two half-wave rectifiers. One of these operates on the positive portion of the input and the second operates on the negative portion. The outputs are summed with the proper polarities. Figure 13.5 illustrates one such configuration. Note that the resistive network attached to the output summing op-amp is composed of resistors of higher value than those attached to the op-amp that generates v_1. This is necessary for the following reason. For negative v_i, v_2 follows the curve shown directly above the node labeled v_2. That is, as the input increases in a negative direction, v_2 increases in a positive direction. Since the input impedance to the + terminal of the summing op-amp is high, the voltage v_+ is simply one-half of v_2 (i.e., the two 100-kΩ resistors form a voltage divider). The voltage at the negative summing terminal, v_-, is the same as v_+ and therefore is equal to $v_2/2$. Now when v_i is negative, D_2 is open, and the node v_1 is connected to the inverting input of the first op-amp through a 5-kΩ resistor. The inverting input is a virtual

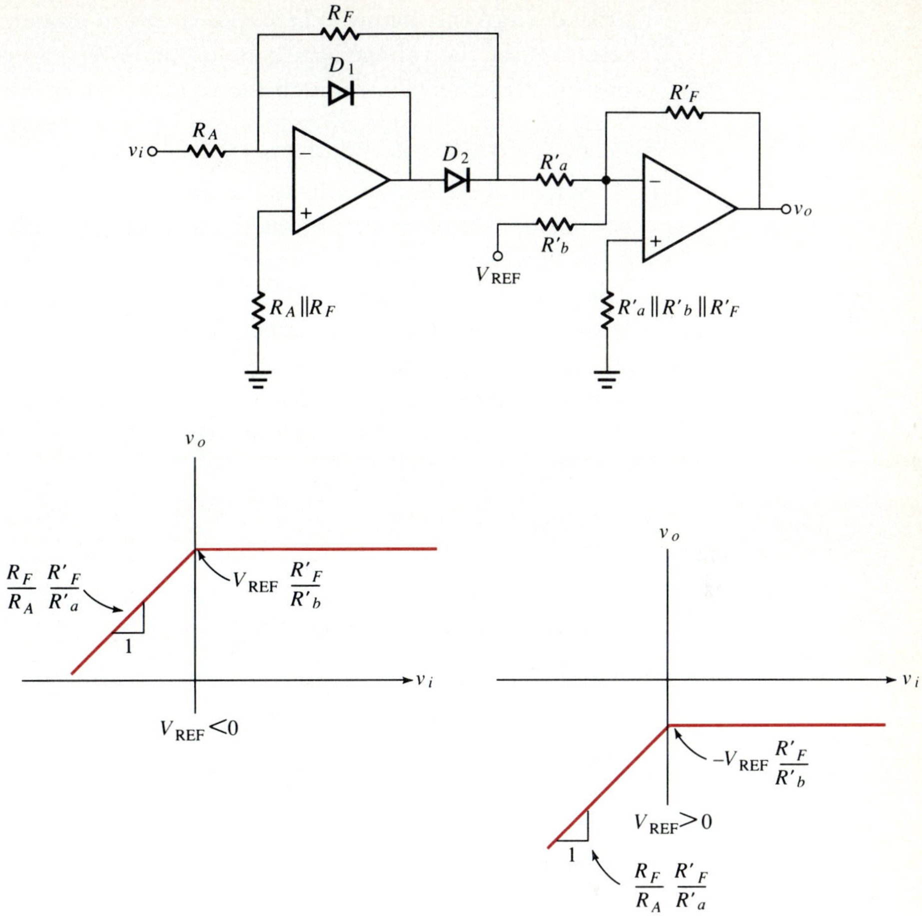

Figure 13.4
Level shifting.

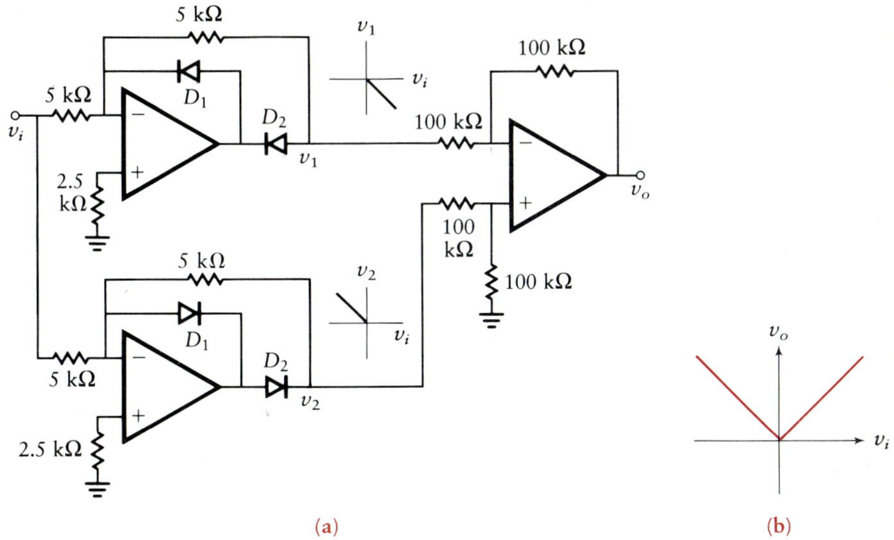

Figure 13.5
Full-wave rectifier.

ground, since the noninverting input is tied to ground through a resistor. The result is that the voltage appearing on the inverting terminal of the summing op-amp also affects v_1 through the voltage divider formed by the 100-kΩ and 5-kΩ resistors. In order to achieve a characteristic resembling that shown in the figure, this voltage divider must have a small ratio, on the order of 1 to 20.

Since this method of full-wave rectification requires three separate amplifiers, we now explore simpler methods. One approach follows from an arithmetic observation. Note that the mathematical operation of taking the absolute value is the same as that of reversing the sign of the negative part of the signal. If the half-wave rectified output is doubled and the original signal is subtracted from this, the result is the full-wave rectified waveform. This is easily proved by considering the two operating conditions separately. First, suppose that the input is positive. Then the half-wave output is equal to the input, and the difference described above becomes

$$2v_i(t) - v_i(t) = v_i(t)$$

Thus, the output is equal to the input. Now if the input is negative, the half-wave output is zero, and the difference becomes

$$2 \times 0 - v_i(t) = -v_i(t)$$

Thus, the output is equal to the negative of the input. The composite output is then the absolute value of the input. This form of full-wave rectifier is shown in Figure 13.6(a). The curve of Figure 13.6(b) shows v_1 as a function of v_i. In Figure 13.6(c), we show the output voltage due to v_i through the upper connec-

Figure 13.6
Full-wave rectifier.

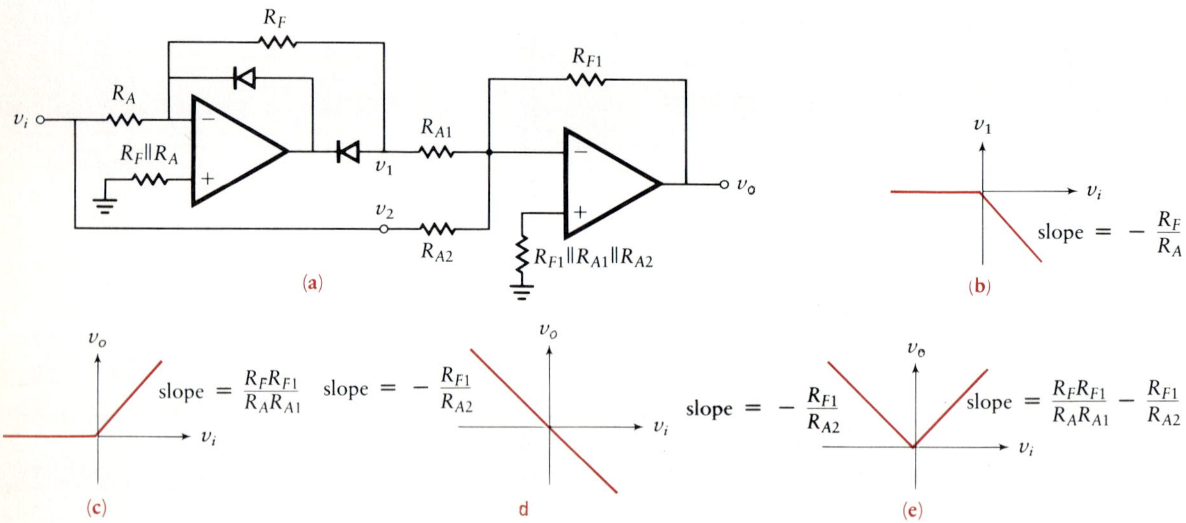

tion. Figure 13.6(d) shows the output voltage as a function of v_i through the lower connection. This is given by

$$v_o\big|_{v_1=0} = -\frac{R_{F1}}{R_{A2}} v_i$$

To obtain the total output voltage, we add Figure 13.6(c) and Figure 13.6(d) to obtain the curve shown in Figure 13.6(e). To make the full-wave rectifier symmetrical (i.e., magnitude of the slope is the same for both positive and negative v_i), we select the resistors so that

$$\frac{R_{F1}}{R_{A2}} = \frac{R_F R_{F1}}{R_A R_{A1}} - \frac{R_{F1}}{R_{A2}}$$

Table 13.2 shows some alternative methods of forming the full-wave rectified output. You should be able to verify the operation of each circuit.

Example 13.1

The circuit of Figure 13.7 has $V_{REF} = 5$ V at the negative terminal of the op-amp fed through a 20-kΩ resistor. Find the input-output voltage characteristic.

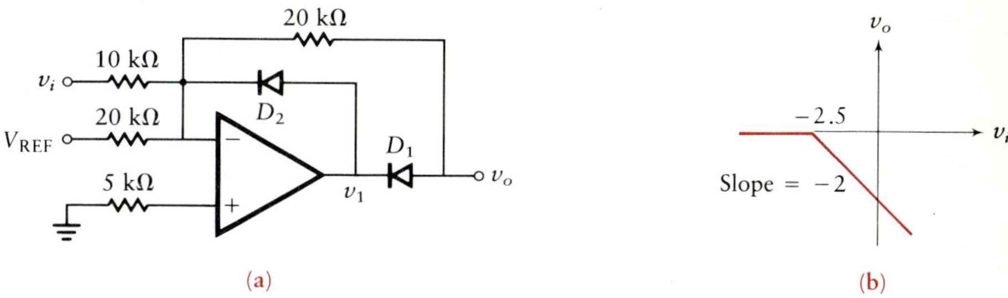

Figure 13.7 Circuit for Example 13.1.

SOLUTION If the reference voltage was zero, the output would be nonzero only in the fourth quadrant of the v_o versus v_i characteristic. This differs from the situation of Figure 13.1 by 180°, since the diodes are reversed. In that quadrant, the slope is $-R_F/R_a$ or -2, and the curve intercepts the origin. With a reference voltage of +5 V and an associated gain of unity (based on the given resistor values), the axis of the line is moved to where $v_i = -2.5$ when $v_o = 0$. That is, v_i must be -2.5 V for the summed input to equal zero. The 5-V reference causes the diode to conduct when v_i is -2.5 V. A gain of -2 is realized for all input voltages greater than or equal to -2.5 V. The resulting v_o versus v_i characteristic is shown in Figure 13.7(b).

Table 13.2 Full-Wave Rectifiers.

1. Standard full-wave rectifier with minimum op-amps

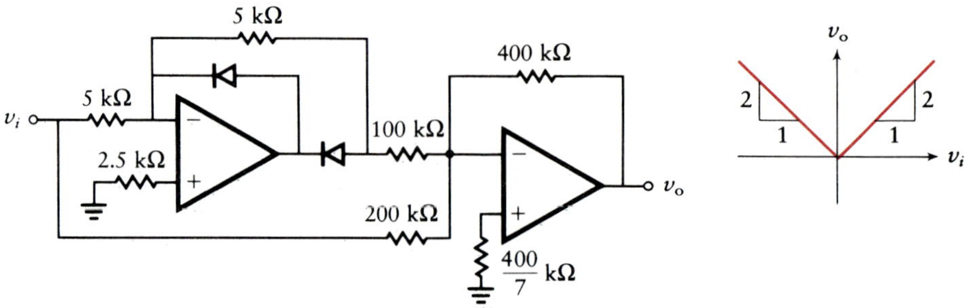

2. Full-wave rectifier with level shifting

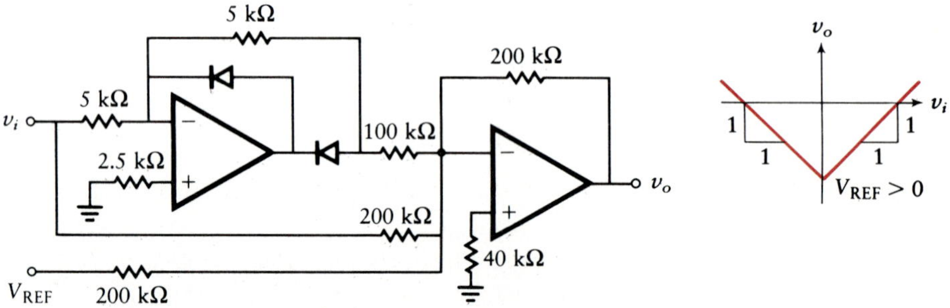

3. Full-wave rectifier with both level and axis shifting

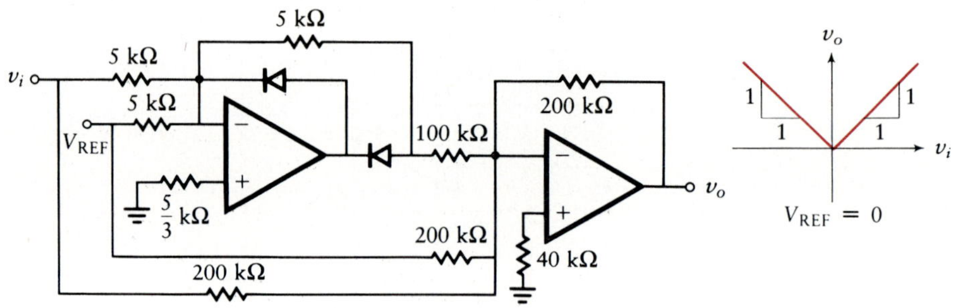

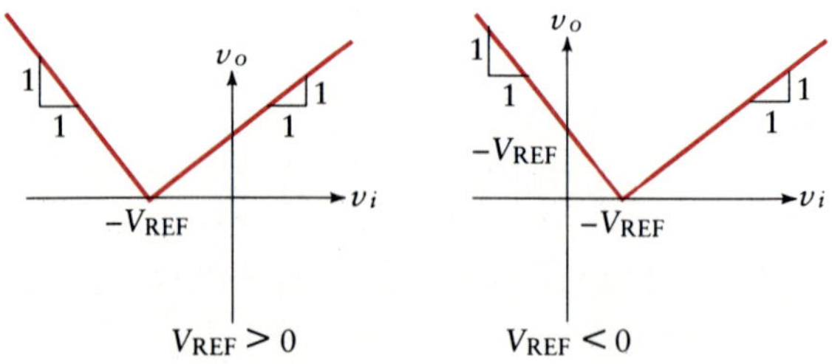

Example 13.2

Design an op-amp circuit to provide the transfer characteristic as shown in Figure 13.8(a). Use a minimum number of op-amps.

Figure 13.8 Characteristic and solution for Example 13.2.

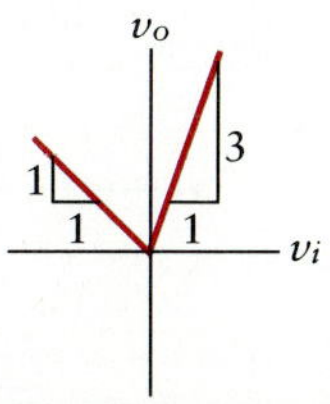

(a) Transfer characteristic

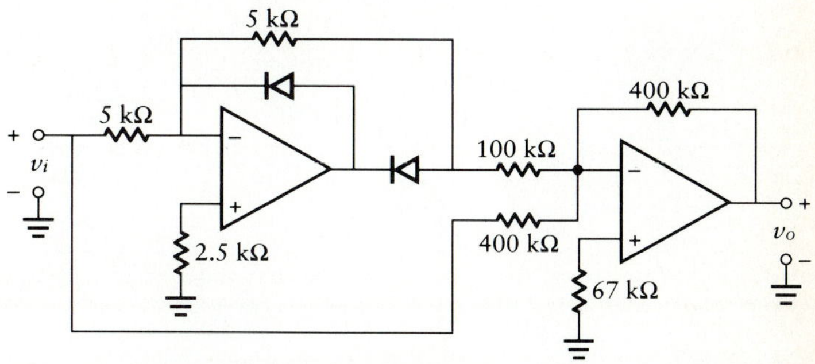

(b) Solution

SOLUTION The design requirement calls for a minimum number of op-amps and the output is in the form of a full-wave rectified version of the input. We choose the circuit of Figure 13.6, since it requires only two op-amps. We establish two equations from Figure 13.6(e) and the required slopes as follows:

$$\frac{-R_{F1}}{R_{A2}} = -1 \quad \text{so } R_{F1} = R_{A2}$$

and

$$\frac{R_F R_{F1}}{R_A R_{A1}} - \frac{R_{F1}}{R_{A2}} = +3 \tag{13.3}$$

We have five unknown variables, R_A, R_F, R_{A1}, R_{A2}, and R_{F1}, and only two equations, so we can choose three resistor values. Let us choose $R_F = R_A = 5\ \text{k}\Omega$ and $R_{A1} = 20$ times this amount (to reduce attenuation) or 100 kΩ. Then, from equation (13.3), we obtain $R_{F1} = R_{A2} = 400\ \text{k}\Omega$. The design is complete, and the circuit is as shown in Figure 13.8(b).

Drill Problems

Design op-amp circuits with transfer characteristics as shown in the figures. Assume that the reference supply available is ±10 V and that $V_\gamma = 0.7$ V and $R_f = 100\ \Omega$. The solutions are shown in Figures D13.1 and D13.2.

Find approximate relationships between the input voltages, v_1 and v_2, and the output voltage, v_o, for each of the circuits shown. Assume that $V_\gamma = 0.7$ V and $R_f = 100\ \Omega$. The solutions are shown in Figures D13.3 and D13.4.

D13.1

Figure D13.1

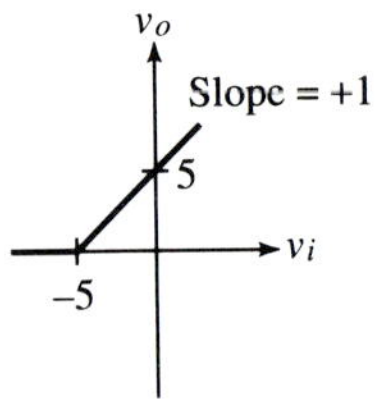

Ans:

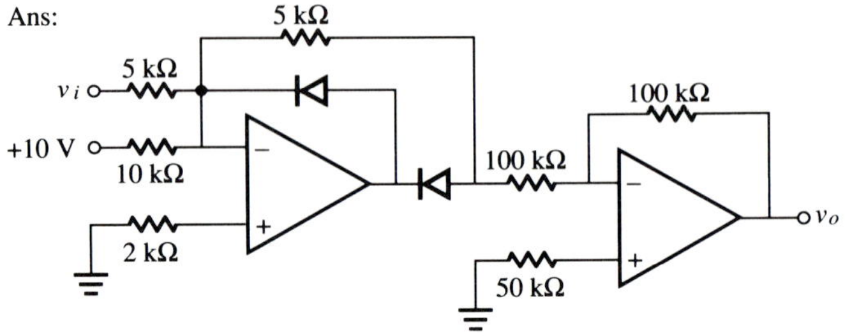

D13.2

Figure D13.2

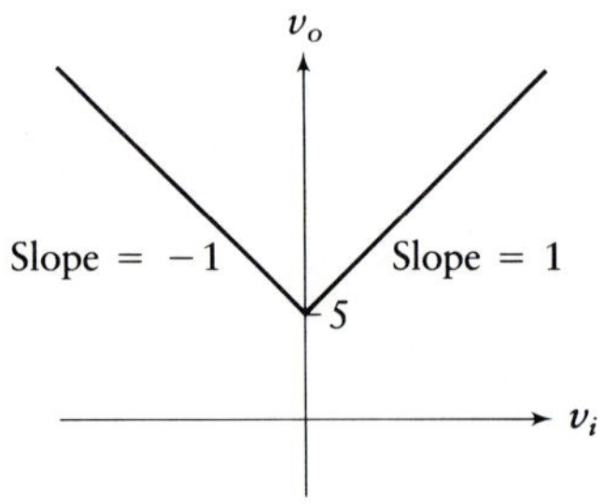

Ans:

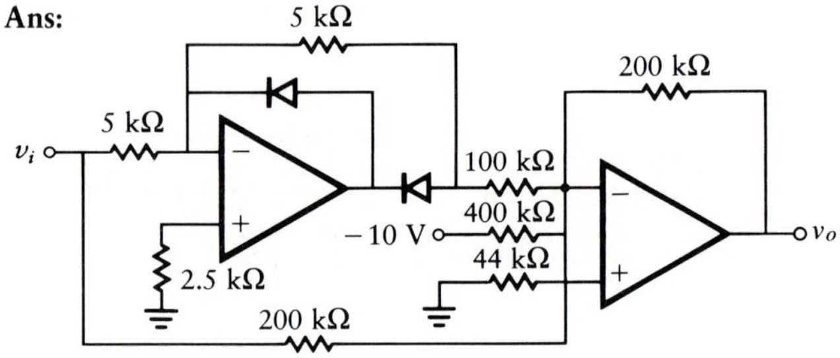

D13.3

Figure D13.3

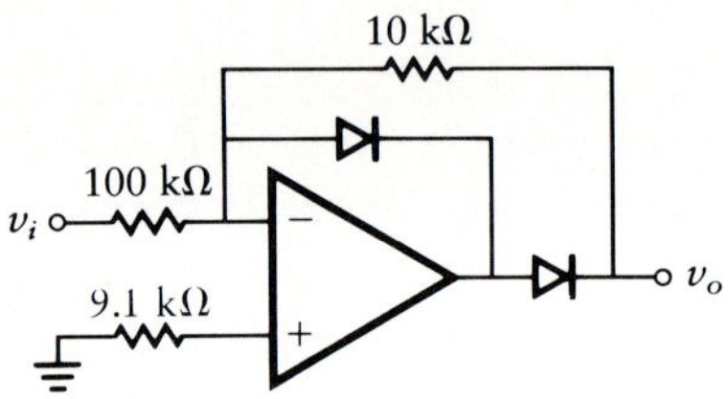

Ans:

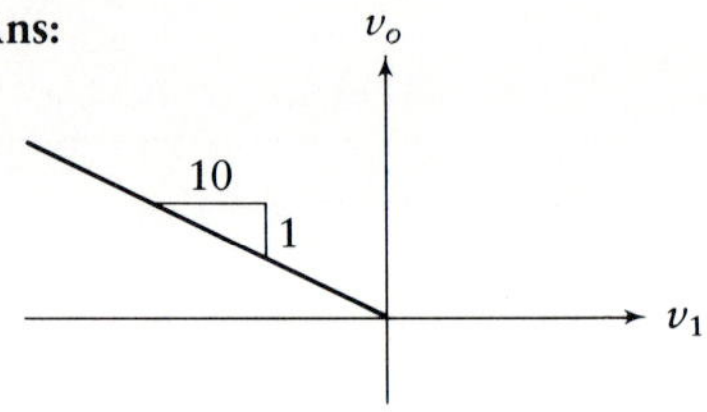

D13.4

Figure D13.4

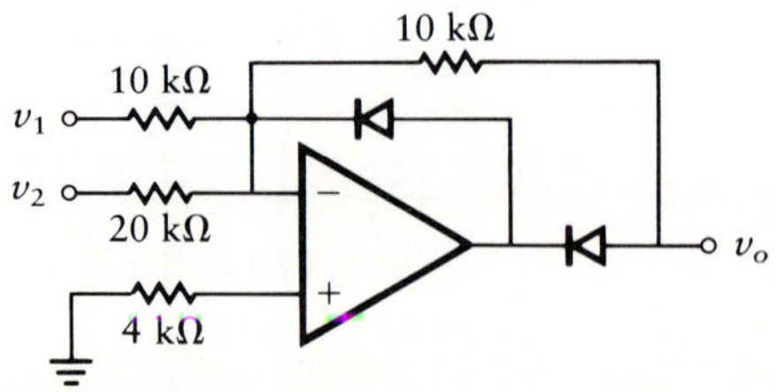

Ans:

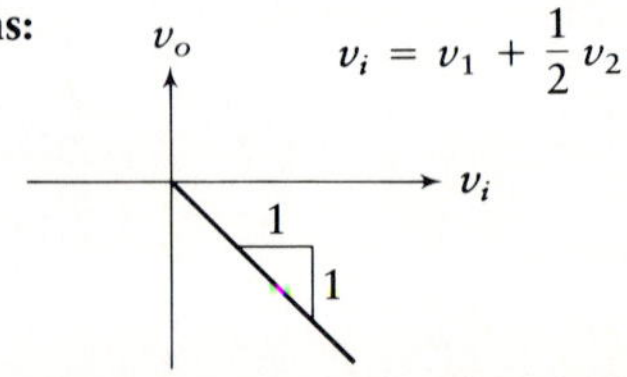

13.2 FEEDBACK LIMITERS

An ideal *limiter* constrains a signal to be below (or above) a particular specified value, the *break point*. The output signal is proportional to the input below (or above) this break point and is constant for inputs above (or below) this value.

There is a wide variety of configurations of the basic limiter circuit. In fact, any characteristic composed of two straight lines intersecting at a point can be considered as a form of limiter and can be realized by using a diode in the feedback path of an op-amp.

As an example, consider the system of Figure 13.9. Because of the presence of the diode, this circuit is analyzed by considering the two diode states separately. That is, the diode is first assumed to be an open circuit, and the circuit is solved. The circuit is again solved for the case when the diode is a short circuit. With the diode conducting, the gain of the op-amp is greatly reduced. Analysis simply requires that we find the location of the break point between the two regions.

When the diode of Figure 13.9 is not conducting, the circuit operates as an inverting amplifier with the output given by

$$v_o = -v_i \frac{R_F}{R_A}$$

and the gain is $-R_F/R_A$.

The break point occurs when the voltage across the diode reaches V_γ. To find the break point, we solve for v_1, assuming that the diode is not conducting, as follows:

$$v_1 = \frac{V_{\text{REF}} - v_o}{R_1 + R_2} R_2 + v_o = \frac{R_2 V_{\text{REF}} + R_1 v_o}{R_1 + R_2}$$

$$= (R_1 \parallel R_2)\left(\frac{V_{\text{REF}}}{R_1} + \frac{v_o}{R_2}\right)$$

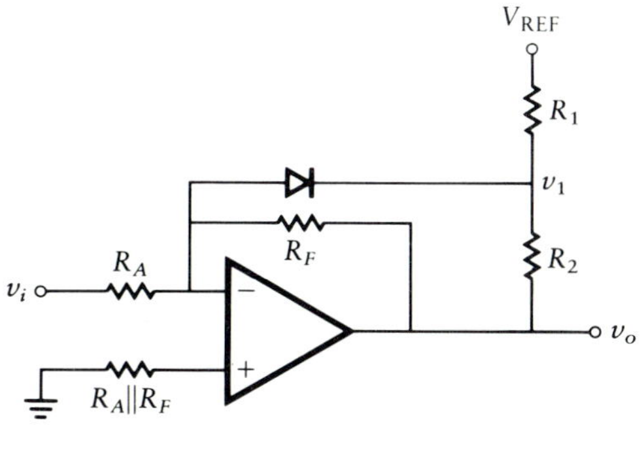

Figure 13.9
Feedback limiter.

At the break point, the diode current is zero. The diode conducts when v_1 tries to go below V_γ. (Note that $v_- = v_+ = 0$). That is,

$$v_1 = \frac{R_2 V_{REF} + R_1 v_o}{R_1 + R_2} < V_\gamma$$

We solve for v_o of the break point by setting $v_1 = -V_\gamma$. Then

$$v_o = \frac{-(R_1 + R_2)}{R_1} V_\gamma - \frac{R_2}{R_1} V_{REF} \tag{13.4}$$

If we can assume that V_γ is zero, this becomes

$$v_o = \frac{-R_2 V_{REF}}{R_1} \tag{13.5}$$

Equations (13.4) and (13.5) represent the break point between the two circuit conditions.

To analyze the condition when the diode is conducting, we find the Thevenin equivalent of the resistor divider network to the right of the diode as shown in Figure 13.10. The two voltage sources with series resistors of Figure 13.10(a) reduce to the Thevenin equivalent of Figure 13.10(b). The equivalent resistance is the parallel combination of R_1 with R_2, and the open-circuit voltage is found by either loop or nodal analysis.

When the diode is on, it is replaced by a "turn-on" voltage generator, V_γ, and a forward resistance, R_f. The circuit of Figure 13.9 then takes the form shown in Figure 13.11. The voltage, v_o, is determined from the circuit as follows:

$$v_o = \frac{R_F}{R_A} v_i - \frac{R_F}{R_1 \parallel R_2 + R_f}\left(\frac{R_2 V_{REF} + R_1 v_o}{R_1 + R_2} + V_\gamma\right)$$

and

$$\left(1 + \frac{R_F}{R_1 \parallel R_2 + R_f}\frac{R_1}{R_1 + R_2}\right) v_o = -\frac{R_F}{R_A} v_i - \frac{R_F}{R_1 \parallel R_2 + R_f}\left(\frac{R_2}{R_1 + R_2} V_{REF} + V_\gamma\right)$$

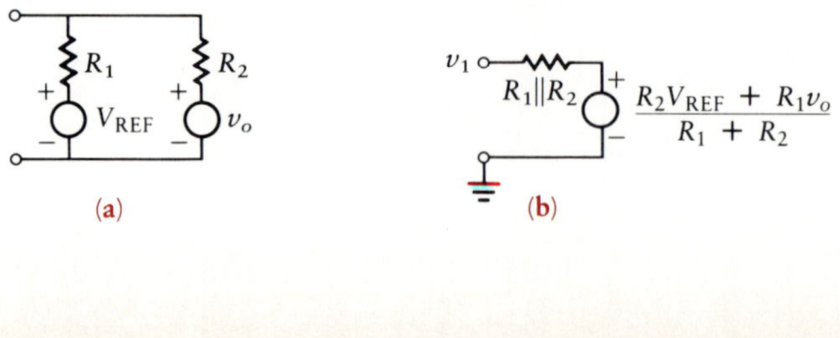

Figure 13.10 Thevenin equivalent for feedback limiter output circuit.

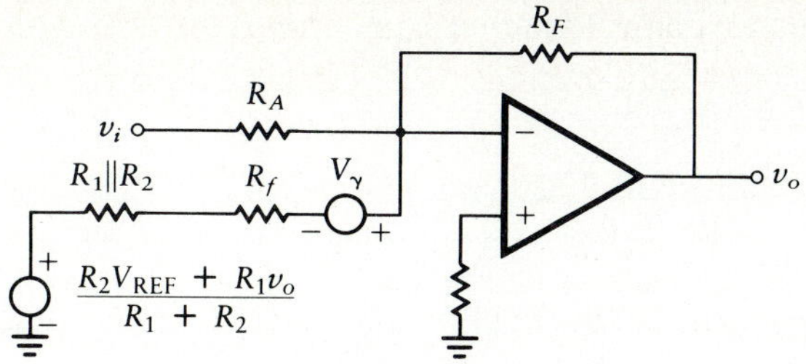

Figure 13.11
Feedback limiter with diode *on*.

Let us now assume that

$$\frac{R_F}{R_1 \parallel R_2 + R_f} >> 1$$

as is normally the case. The equation then reduces to

$$v_o = \frac{R_F v_i/R_A}{[R_F/(R_1 \parallel R_2 + R_f)]\,[R_1/(R_1 + R_2)]} - \frac{R_2}{R_1}V_{REF} - \frac{R_1 + R_2}{R_1}V_\gamma$$

$$= -\left(\frac{1}{R_A}\right)\left(1 + \frac{R_2}{R_1}\right)(R_1 \parallel R_2 + R_f)v_i - \frac{R_2}{R_1}V_{REF} - \left(1 + \frac{R_2}{R_1}\right)V_\gamma \qquad (13.6)$$

The gain is given by

$$\frac{dv_o}{dv_i} = -\left(\frac{1}{R_A}\right)\left(1 + \frac{R_2}{R_1}\right)(R_1 \parallel R_2 + R_f) \qquad (13.7)$$

If $R_f << R_1 \parallel R_2$, the expression for the gain in equation (13.7) reduces to

$$\frac{dv_o}{dv_i} = -\frac{R_2}{R_A}$$

The resulting characteristic curve is shown in Figure 13.12. The slope changes from $-R_F/R_A$ to approximately $-R_2/R_A$ as v_i increases beyond the break point. If V_{REF} is negative, the break in the characteristic curve occurs at a positive value of v_o.

Notice that R_2 must be much smaller than R_A if good limiting is to be achieved. That is, if the limiter is to have a slope near zero beyond the break point, then $R_2/R_A << 1$.

The values of v_o and v_i when the slope changes (we define these as V_{oc} and V_{ic}, respectively) are given by the following equations:

$$V_{oc} = \frac{-R_2 V_{REF}}{R_1} - \left(1 + \frac{R_2}{R_1}\right)V_\gamma \qquad (13.8)$$

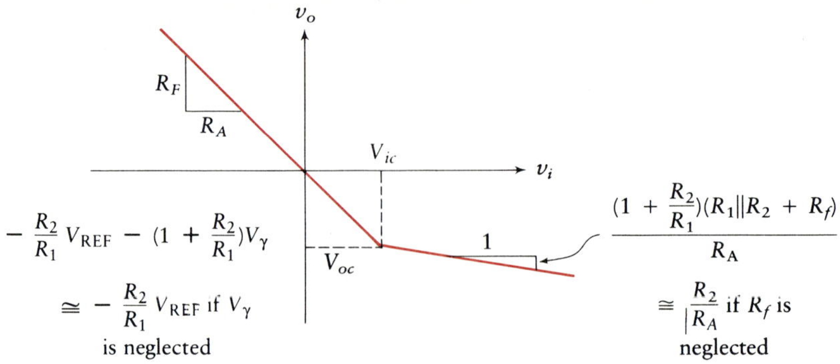

Figure 13.12
Limiter characteristics.

$$V_{ic} = \frac{R_A}{R_F}\left[\frac{R_2 V_{REF}}{R_1} + \left(1 + \frac{R_2}{R_1}\right)V_\gamma\right] \approx \frac{R_A R_2}{R_F R_1}V_{REF} \tag{13.9}$$

The last approximation is valid if V_γ is neglected.

The limiting effect is produced when the diode is on and the slope is

$$-\left(1 + \frac{R_2}{R_1}\right)(R_1 \parallel R_2 + R_f)$$

Note that the value of the output voltage when the slope changes, $|V_{oc}|$, must be less than the op-amp saturation voltage, $|E|$. The amplifier cannot produce a voltage greater than the saturation voltage, so if the break point is beyond this value, that point will never be reached. If $E = V_{REF}$, as is frequently the case, R_2 must be less than R_1. Recall that R_f is the forward resistance of the diode and R_F is the feedback resistance of the op-amp circuit.

Table 13.3 presents a variety of limiter configurations. You should study these before proceeding.

An important constraint in the design of limiters is that R_1 and R_2 be kept large enough to draw little current from the reference voltage source. When two inputs are provided to a limiter as shown in circuit 4 of Table 13.3, one input can be selected as the reference for the purpose of shifting the input-output characteristics. The input voltage, v_i, will then be equal to the one input plus a fraction of the second input (e.g., $v_i = v_1 + 0.5v_2$).

Figure 13.13 shows an application of a balanced limiter. Here we use a balanced limiter to provide amplitude control of a Wien bridge oscillator. The slope (R_2/R_A) in the range where limiting occurs is -20. The break voltage, V_{oc}, is found from equation (13.8).

$$V_{oc} = \frac{R_2}{R_1}10 + \left(1 + \frac{R_2}{R_1}\right)0.7 = 4.27 \text{ V}$$

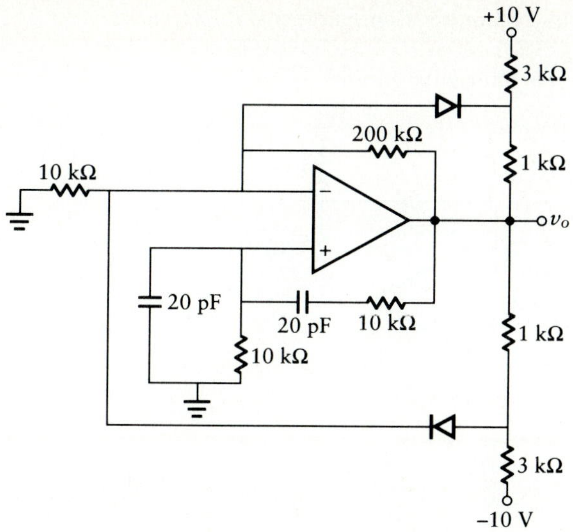

Figure 13.13
Oscillator amplitude limiting.

Table 13.3 Limiter Configurations. In These Examples, $R_f = 100\ \Omega$ and $V_\gamma = 0.7$ V.

1. Basic lower limit

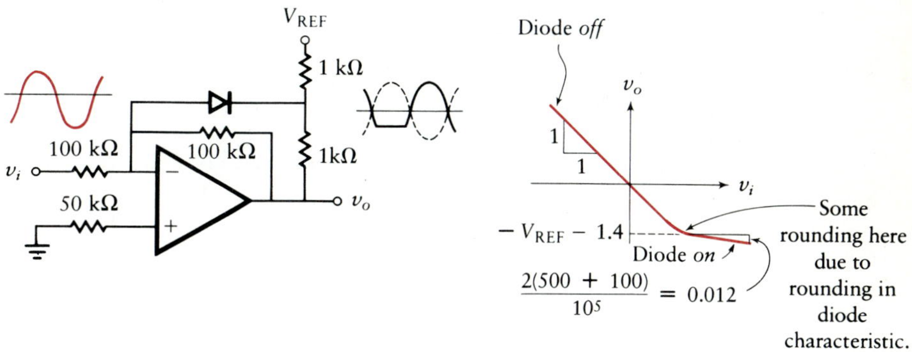

2. Negative reference voltage

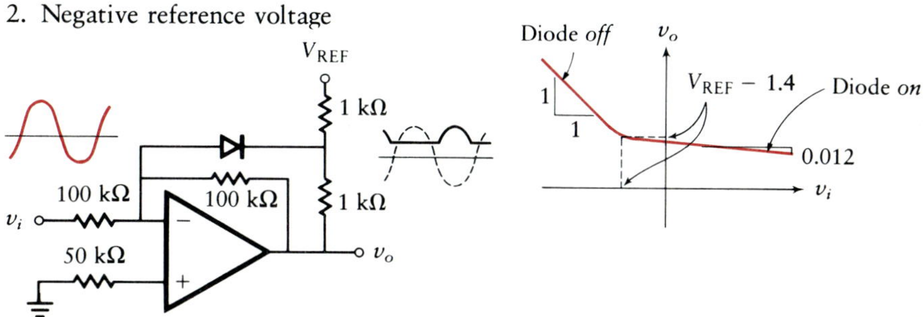

Table 13.3 Limiter Configurations. *(Continued)*

3. Lower limit with gain

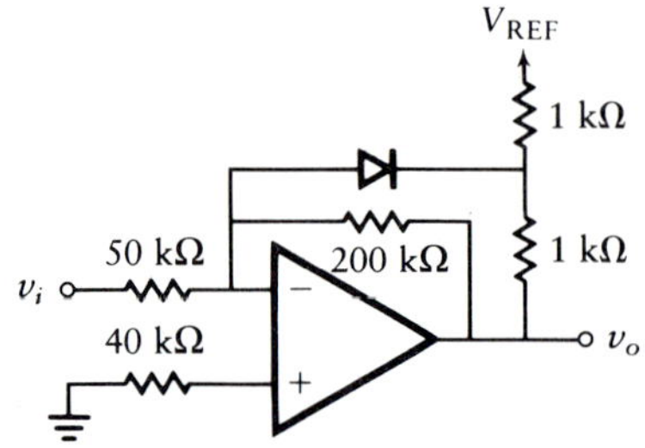

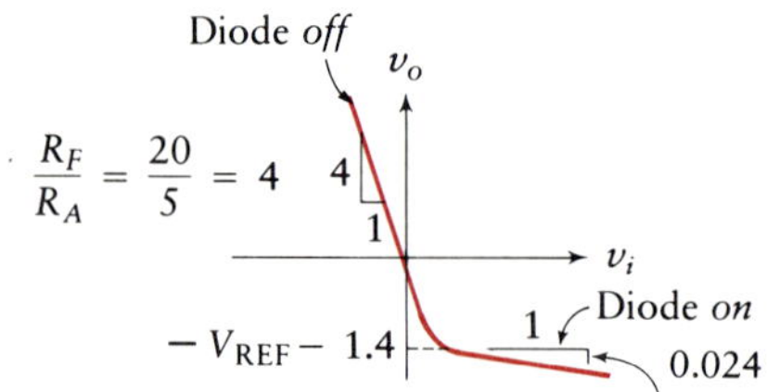

We have a negative reference voltage with the slopes the same as before, except now the output voltage is limited above zero.

4. Summation and lower limit

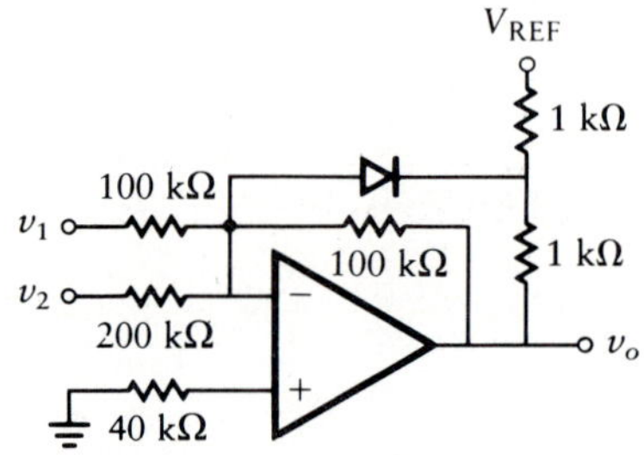

$v_i = v_1 + \frac{1}{2} v_2$

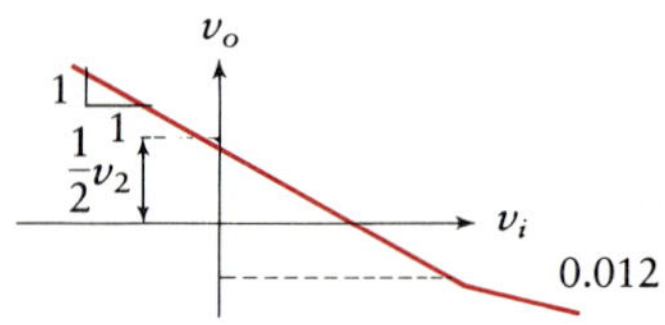

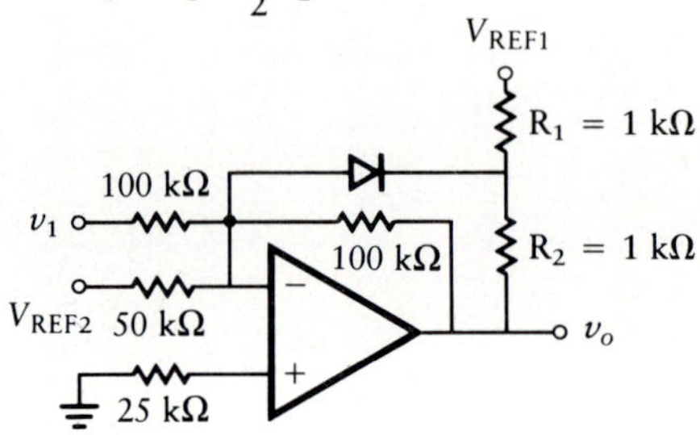

$v_i = v_1 + 2V_{REF2}$

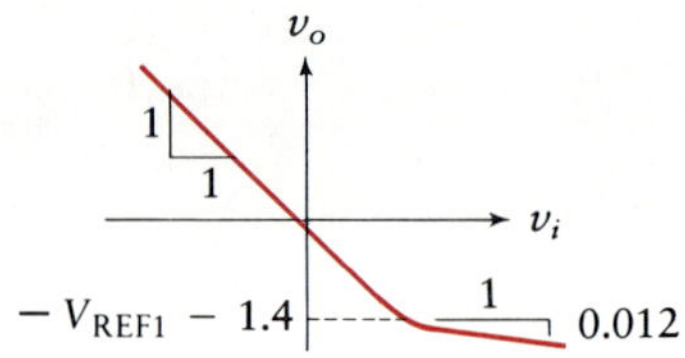

5. Unequal divider resistances

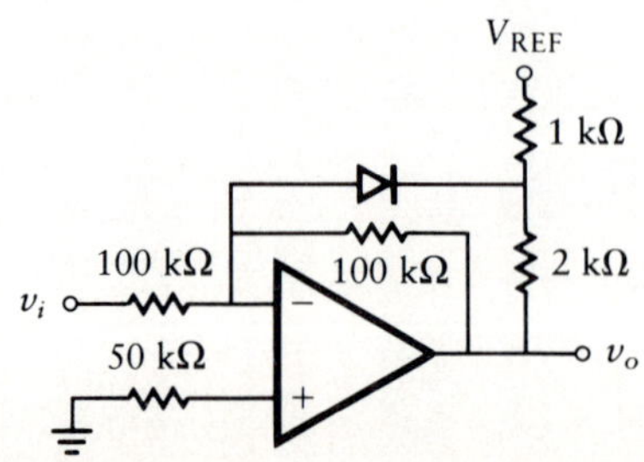

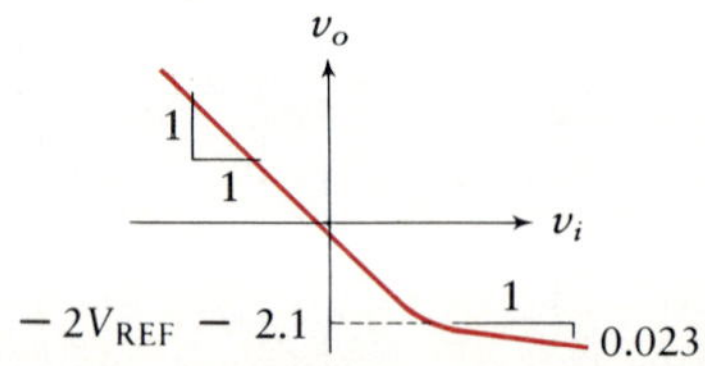

Table 13.3 Limiter Configurations. *(Continued)*

6. Basic upper limit

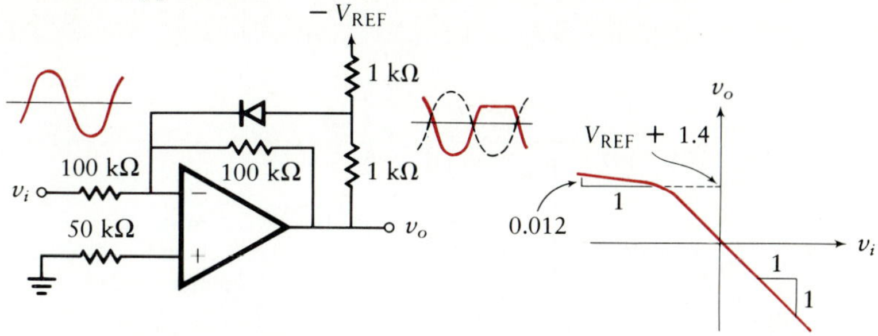

7. Positive reference voltage

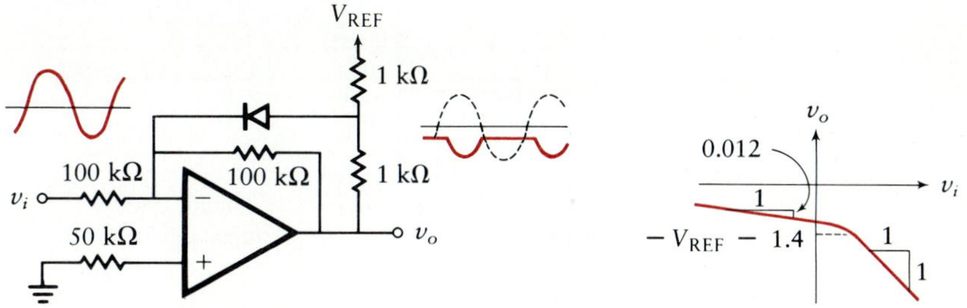

8. Upper and lower limiting

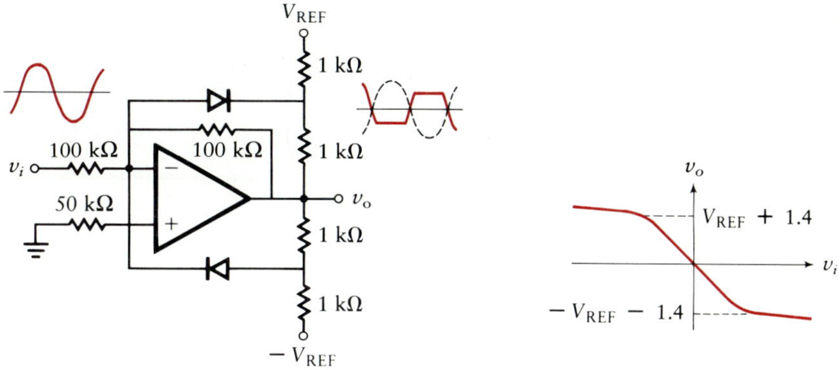

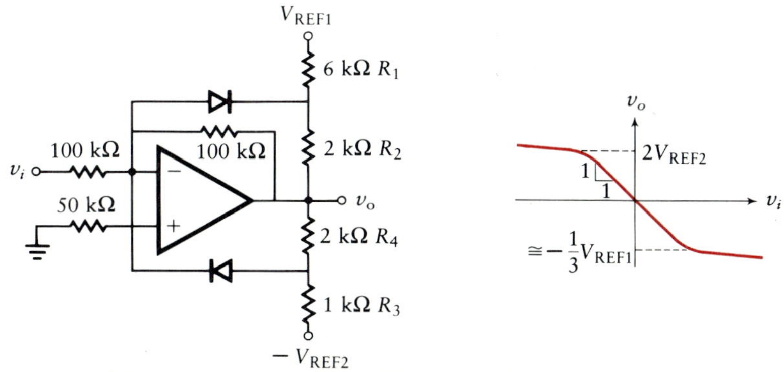

Table 13.3 Limiter Configurations. *(Continued)*

9. Adjustable limiting

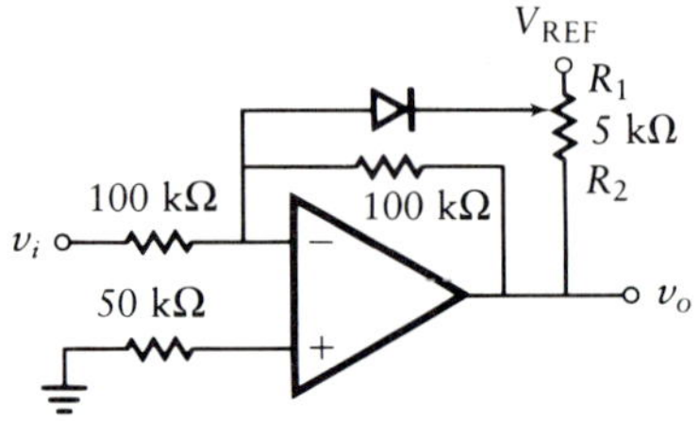

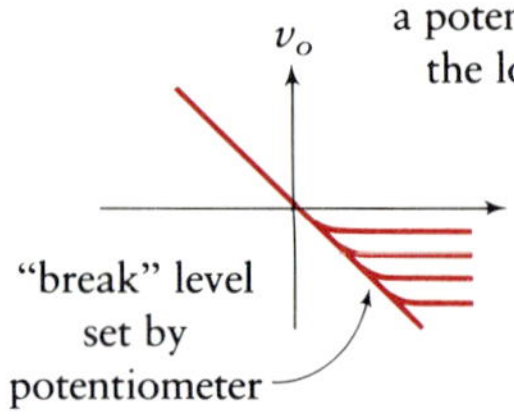

From the characteristic plot, we can see that by using a potentiometer for R_1 and R_2, the lower limit decreases as R_2 gets larger.

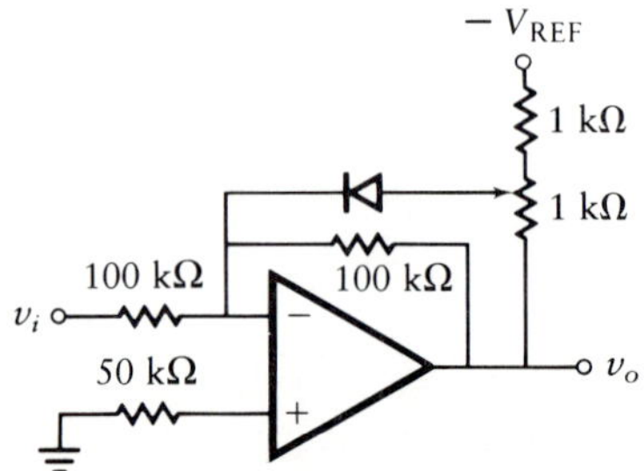

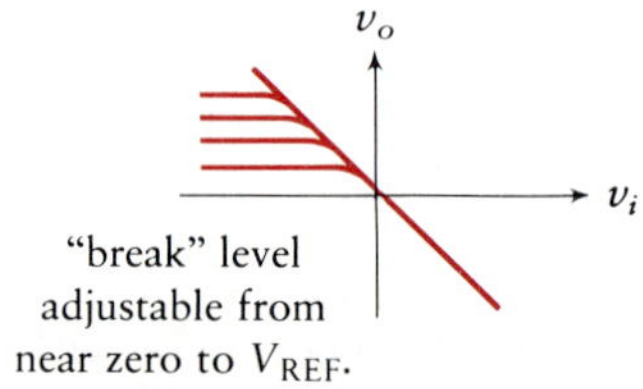

10. Upper and lower adjustable limiting

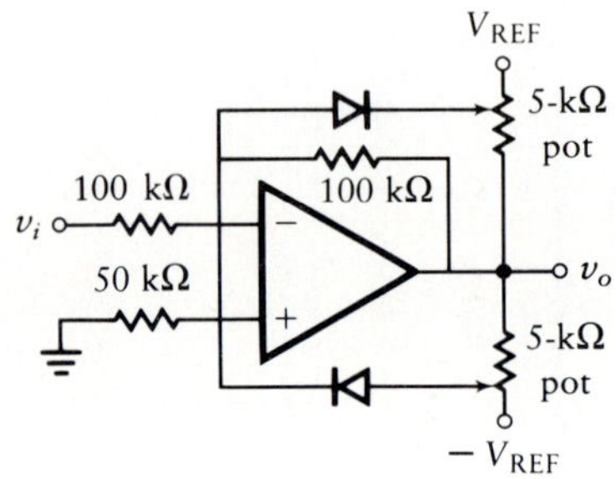

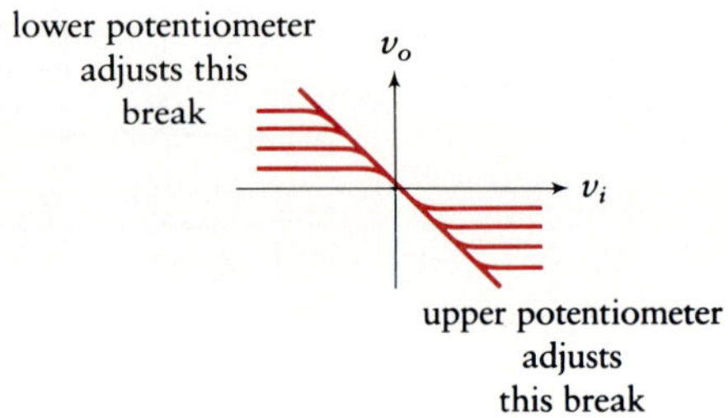

11. Voltage-controlled limiting

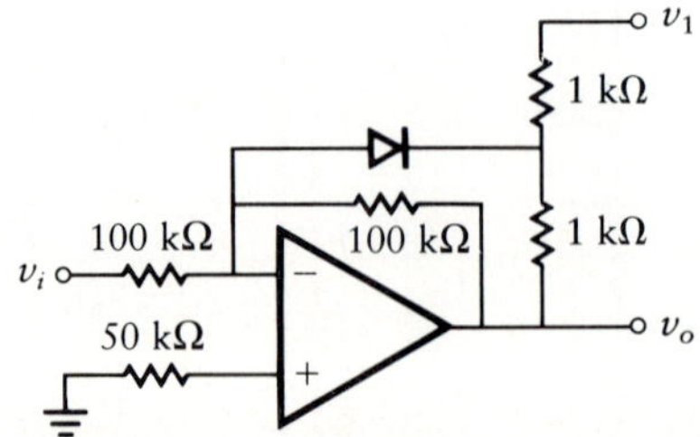

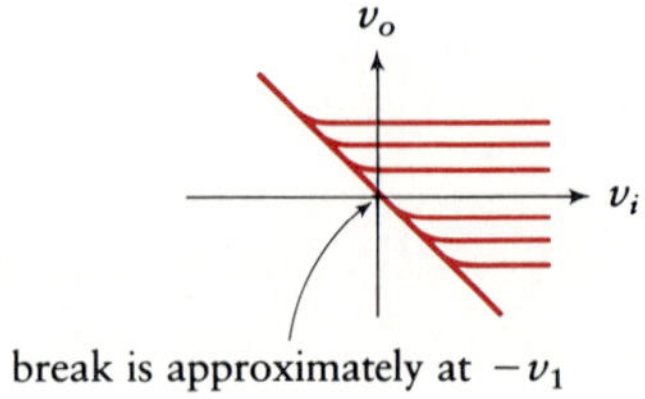

Table 13.3 Limiter Configurations. *(Continued)*

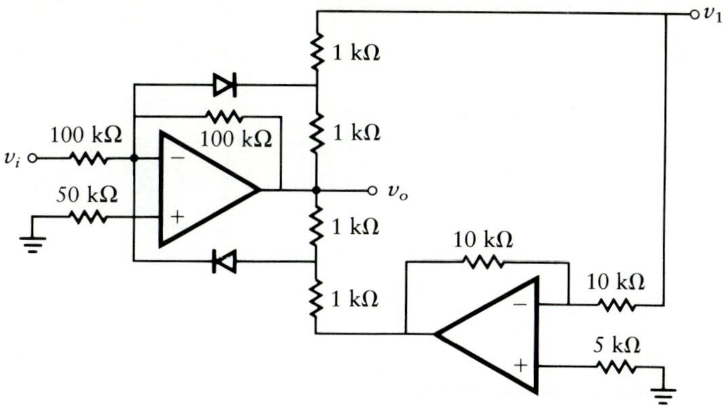

Example 13.3

A feedback limiter, as shown in Figure 13.9, has $R_A = 10$ kΩ, $R_F = 20$ kΩ, $R_1 = 4$ kΩ, $R_2 = 2$ kΩ, $V_{REF} = 10$ V, $V_\gamma = 0.7$ V, and $R_f = 50$ Ω. Determine where the characteristic of v_o versus v_i changes slope, and also find the slope in the saturation region (diode on).

SOLUTION From the equations for the break point, equations (13.8) and (13.9), we find $V_{oc} = -6.05$ and $V_{ic} = 3.02$. The slope is given by equation (13.7) to be -0.21.

The circuits and v_o versus v_i characteristics of selected additional limiter applications are shown in Table 13.4. We now comment on these circuits. The TTL interface (circuit 1) provides an output voltage of 4.23 V if v_i is negative and 0 V if v_i is positive. The break voltage at v_1 is calculated from equation (13.6) as follows:

$$v_1 = \frac{1\text{ k}\Omega}{3\text{ k}\Omega}(12) + 0.7\left(1 + \frac{1\text{ k}\Omega}{3\text{ k}\Omega}\right) = 4.93\text{ V}$$

Diode D_3 prevents the output voltage from going negative and reduces it to

$$v_o = v_1 - 0.7 = 4.23\text{ V}$$

Other values of R_1 and R_2 will accommodate other reference voltages.

Table 13.4 Limiter Applications.

1. TTL interface

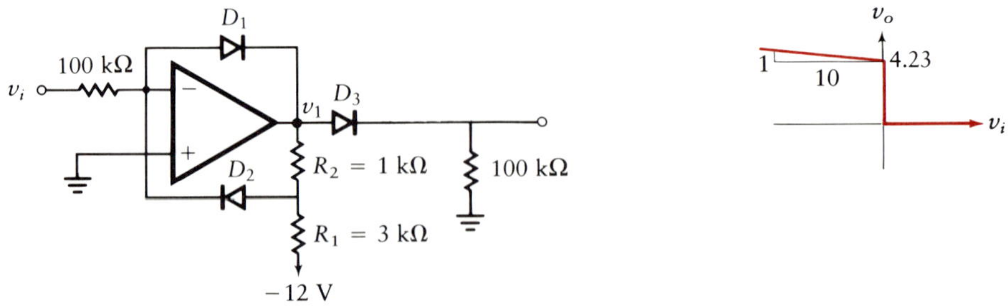

2. Fixed saturation levels

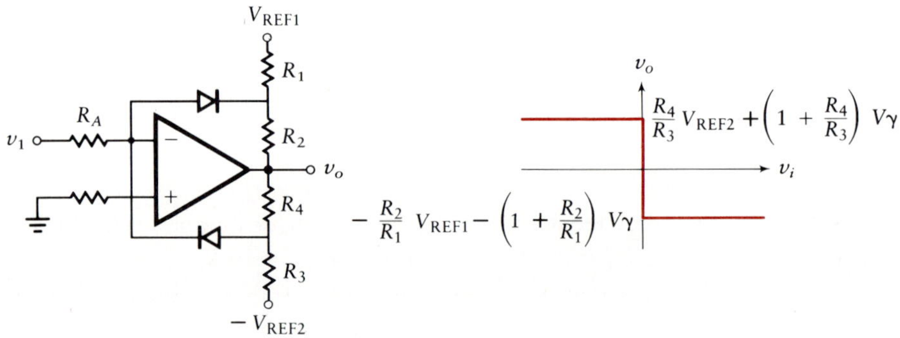

3. Adjustable saturation levels

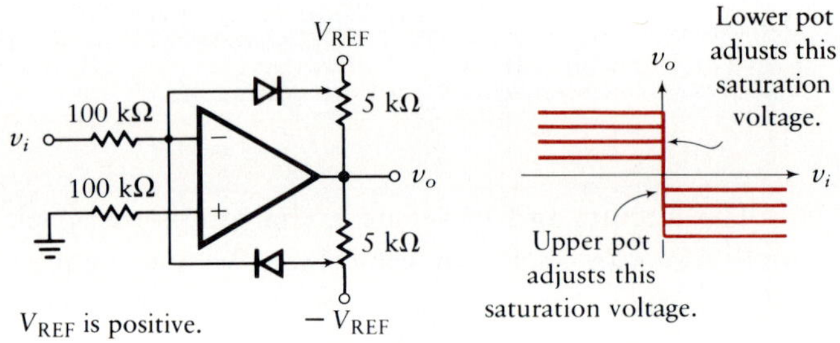

4. Voltage-controlled saturation levels

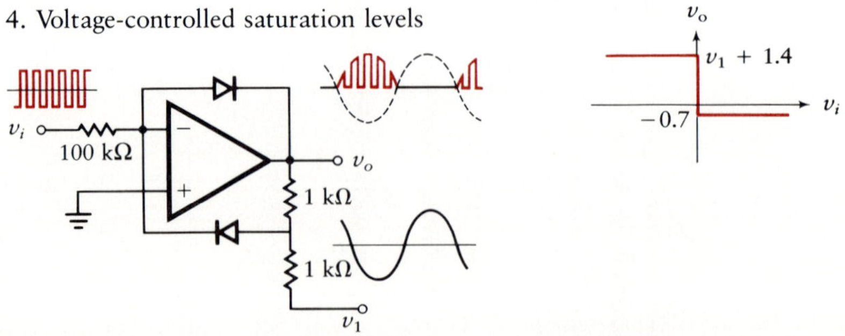

Table 13.4 Limiter Applications. *(Continued)*

5. Voltage-controlled saturation levels

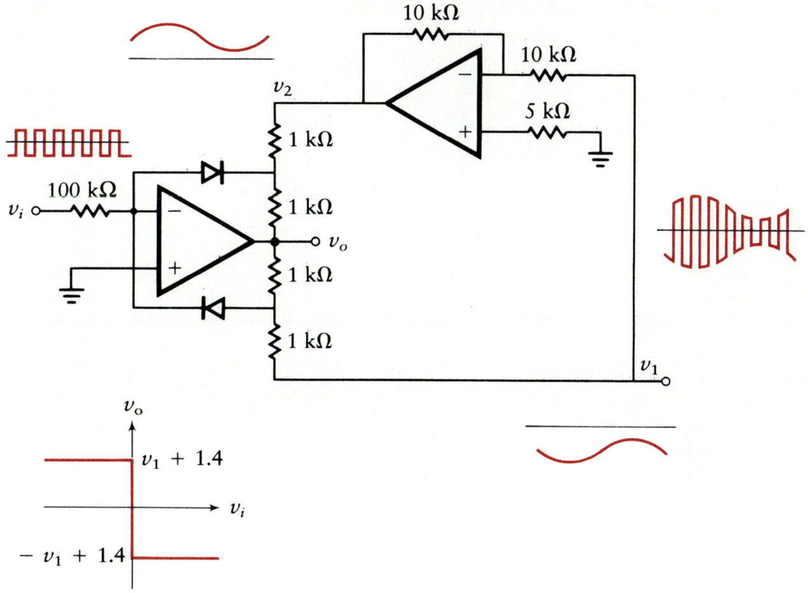

---

6. Input-axis shifting

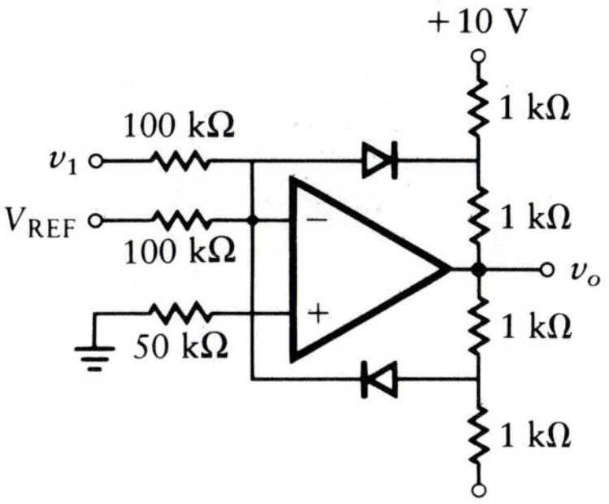

$v_i = v_1 + V_{REF}$

In this case, V_{REF} is added to v_1. This changes the value of v_1 that is required to swing from one saturation voltage to another.

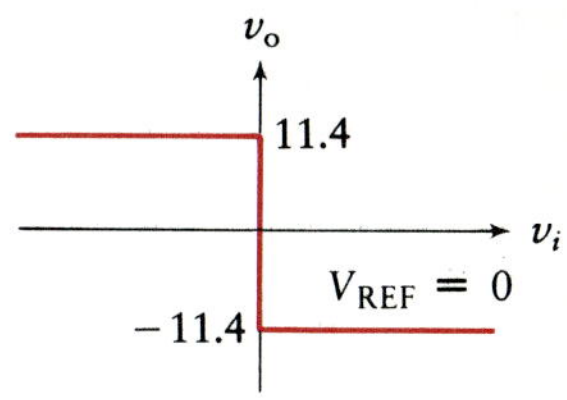

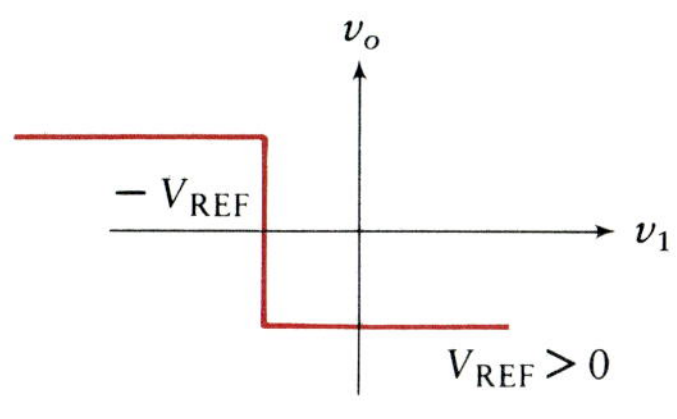

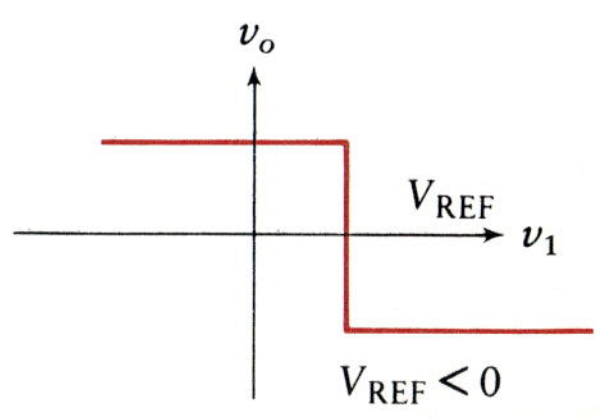

Table 13.4 Limiter Applications. *(Continued)*

7. Input-axis shifting

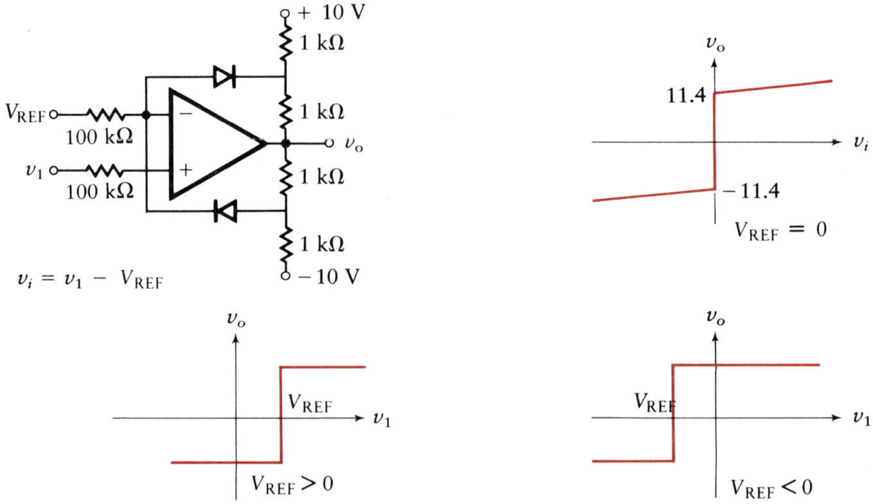

Circuit 2 provides fixed saturation levels (dependent on resistor and reference source values). Two different V_{REF} voltages are used and the values of the resistors are unequal, so the saturation levels for positive and negative v_i are different. Circuit 3 is a modification of circuit 2 using the same magnitude of reference voltage in both places but with 5-kΩ potentiometers in place of the resistors. The result is to make the upper and lower saturation voltages variable.

Circuit 4 provides a saturation level that can be adjusted with the reference voltage, v_1. As v_1 is varied, the output saturation level changes. The output voltage is determined by the ratio of the two 1-kΩ resistors. Circuit 5 behaves much like circuit 4 except that the second op-amp inverts the reference voltage, v_1, and feeds v_2 (which is $-v_1$) to the top half of the bias network. The result is output saturation for both positive and negative values of v_i.

Circuit 6 provides a method of shifting the saturation curve along the input axis. The two voltages, v_1 and V_{REF}, are combined into one input voltage, v_i, which equals $v_1 + V_{REF}$. The three curves show the shifting that takes place for $V_{REF} = 0$, $V_{REF} > 0$, and $V_{REF} < 0$. Circuit 7 provides a similar type of input-axis shifting. However, since v_1 is brought into the noninverting terminal of the op-amp, the saturation curves are the mirror images of those of circuit 6. The input signal is now $v_i = v_1 - V_{REF}$ and the saturation curves are shifted as shown for $V_{REF} = 0$, $V_{REF} > 0$, and $V_{REF} < 0$.

Example 13.4

Design a limiter that will provide limiting at $v_o = \pm 6$ V with a gain of -4 and a slope in the saturation region of $-1/25$. Assume that $V_{REF} = 10$ V, $V_\gamma = 0.7$ V, and $R_f = 100\ \Omega$.

SOLUTION The number of constraints is one less than the number of unknown component values. Let us select $R_A = 40$ kΩ. If this leads to unreasonable values of other components, we will return and revise our selection. Solving for the other parameters, since the gain is -4,

$$\frac{R_F}{R_A} = 4$$

so

$$R_F = 160\ \text{k}\Omega$$

$$V_{oc} = 6 = \frac{R_2 V_{REF}}{R_1} + \left(1 + \frac{R_2}{R_1}\right)V_\gamma$$

We solve this to find

$$\frac{R_2}{R_1} = 0.495$$

which represents one equation in R_1 and R_2. The slope is $-1/25$. We obtain (from Figure 13.12)

$$\frac{-1}{25} = -\left(1 + \frac{R_2}{R_1}\right)\frac{R_1 \parallel R_2 + R_f}{R_A}$$

This is a second equation in R_1 and R_2. We solve for the resistor values as follows:

$$R_2 = R_4 = 1.45\ \text{k}\Omega$$

$$R_1 = R_3 = 2.93\ \text{k}\Omega$$

The final circuit is similar to the type shown in Table 13.3 for "upper and lower limiting" (circuit 8).

Example 13.5

Design a feedback limiter to provide the transfer characteristics as shown in Figure 13.14(a). Assume $V_{REF} = 10$ V, $R_f = 100\ \Omega$, and $V_\gamma = 0.7$ V.

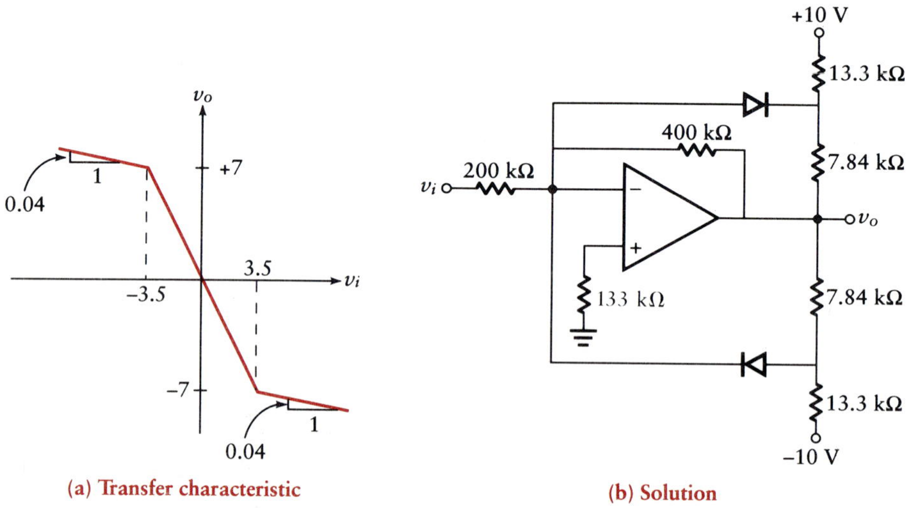

Figure 13.14 Characteristics for Example 13.5.

SOLUTION This is a balanced limiter, so we will solve for R_1 and R_2 and use the same values for R_3 and R_4.

When diodes are off, the gain is

$$\frac{R_F}{R_A} = 2 \quad \text{so let } R_A = 200\text{ k}\Omega \text{ and } R_F = 400\text{ k}\Omega$$

We substitute values into equation (13.9) to obtain

$$V_{ic} = -\frac{200\text{ k}\Omega}{400\text{ k}\Omega}\left[\frac{R_2}{R_1}(10) + \left(1 + \frac{R_2}{R_1}\right)0.7\right] = -3.5$$

and

$$\frac{R_2}{R_1} = 0.589$$

When diodes are on, the slope is given by equation (13.7), and we obtain

$$0.04 = \frac{1}{200\text{ k}\Omega}\left[1.589\left(\frac{R_2}{1 + 0.589} + 100\right)\right]$$

We solve this to find $R_2 = 7.84\ \text{k}\Omega$. Then the previous equation yields $R_1 = 13.3\ \text{k}\Omega$. The complete circuit is shown in Figure 13.14(b).

Drill Problems

Design op-amp circuits with the following approximate transfer characteristics. Available reference supply voltages are ±10 V. Op-amp supply voltages are ±15 V. Assume that $V_\gamma = 0.7$ V. The solutions are shown in Figures D13.5 and D13.6.

D13.5

Figure D13.5

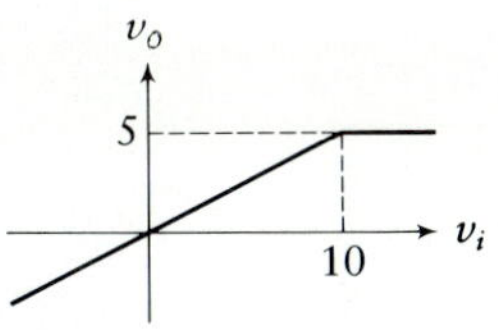

Ans:

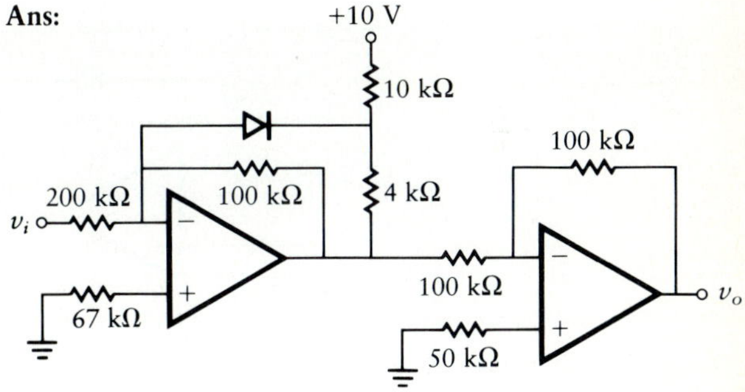

D13.6

Figure D13.6

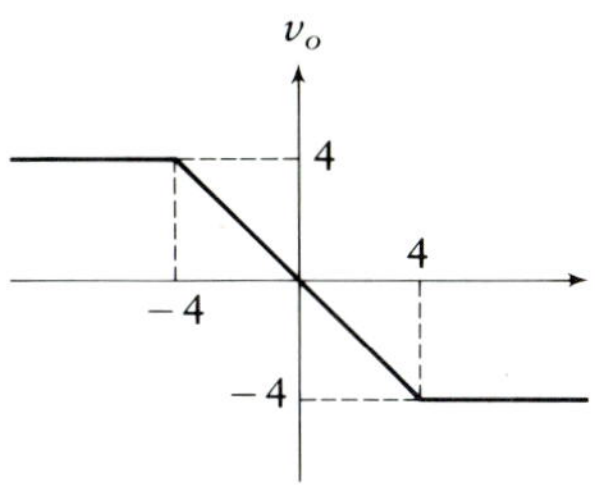

Ans:

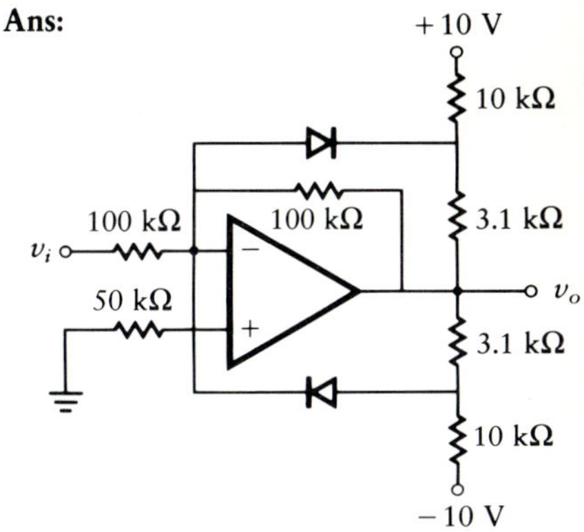

Find approximate relationships between the input voltage, v_i, and the output voltage, v_o, for the circuits shown. Assume that $V_\gamma = 0.7$ V. The solutions are shown in Figures D13.7 and D13.8.

D13.7

Figure D13.7

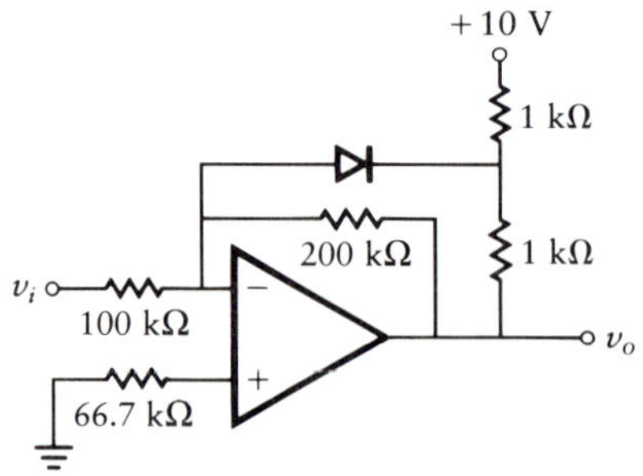

Ans:

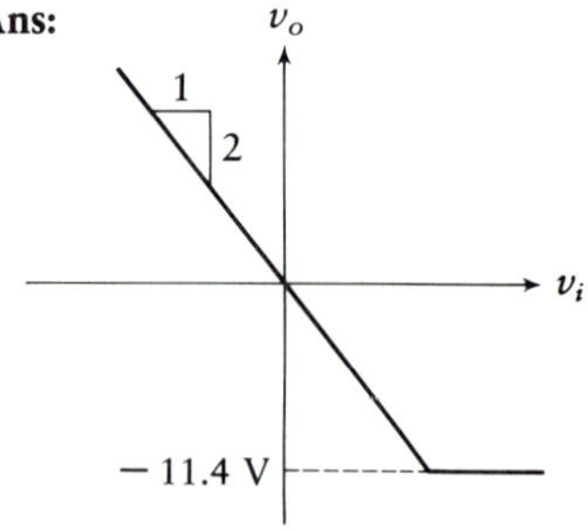

D13.8

Figure D13.8

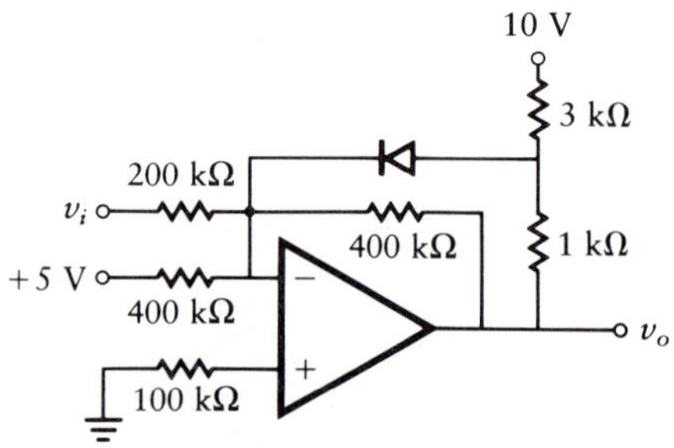

Ans:

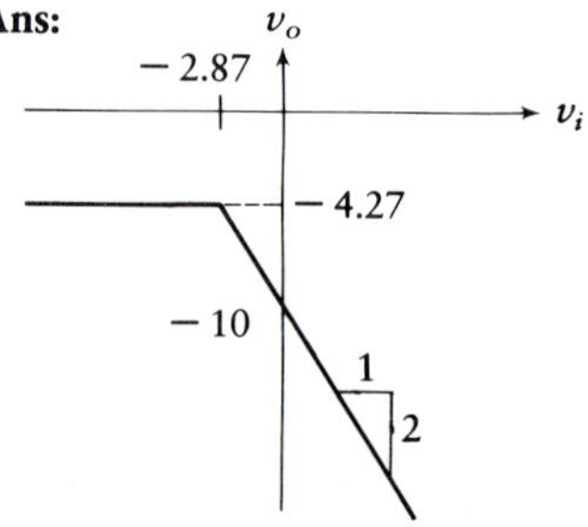

13.3 COMPARATORS

It is frequently important to compare two voltages to determine which is the larger. As a simple application, consider an electronic thermostat where the temperature is converted to a voltage. When the voltage corresponding to the room temperature is less than that corresponding to the desired temperature (setting of thermostat), the system should produce a signal that turns on the heater. As a more complex application, one form of digital communication (delta modulation) requires that a continuous time signal be replaced by a staircase approximation. At each sampling point, the decision of whether the approximation should step up or step down is based on a comparison of the staircase approximation with the original continuous function.

Since feedback control systems usually operate on the difference between two signals (inverting and noninverting), comparators are ideally suited to these applications.

The output of one type of *comparator* in Figure 13.15 is positive when the circuit voltage, v_i, is less than V_{REF} and negative when v_i is greater than V_{REF}. If the gain of the circuit is large, the output saturates. That is, as soon as v_i becomes slightly below V_{REF}, the output rapidly changes to the positive supply

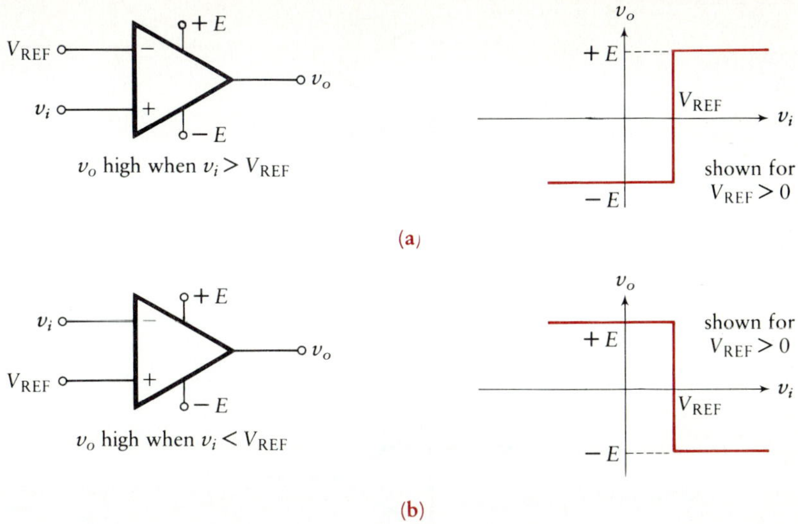

Figure 13.15
Saturation comparator.

voltage. Likewise, the voltage saturates at the negative supply voltage for any value of v_i greater than V_{REF}. Thus, the output takes on only one of two possible values: positive or negative.

When the input is varying around V_{REF}, transitions occur in the output whenever v_i crosses the V_{REF} axis. That is, at one instant v_i might be less than V_{REF}, while at the next instant the reverse is true. Ideally, the output should jump instantaneously from its positive saturation value to its negative value. In practice, a small amount of response time is required due to capacitive effects in the circuit. A typical value of this response time is a few microseconds. For example, the μA741 switches in about 40 μs.

The *accuracy* of a practical comparator is the voltage difference required between the input and reference to cause the output to change its state from one saturation value to the other.

Saturation comparators, which utilize the op-amp in the open-loop mode, depend on high open-loop gain to drive the op-amp into saturation. Figure 13.15 shows an op-amp used in two saturating comparator configurations. In both of these cases, the op-amp saturates at $\pm E$, the supply voltage for the op-amp. The reference voltage, V_{REF}, may be either positive or negative. The output-input curves are shown for positive V_{REF}. As V_{REF} is made more positive, the $v_o = 0$ crossover point moves to the right.

Figures 13.16 and 13.17 show other comparator configurations that can be used to vary the *crossover* voltage and the shape of the curves. Again, either polarity of V_{REF} may be used. The output-input curve of Figure 13.16 is for $V_{REF} < 0$, and for Figure 13.17, $V_{REF} > 0$. In Figure 13.16, v_o is high when

$$\frac{V_{REF}}{R_1} + \frac{v_i}{R_2} > 0$$

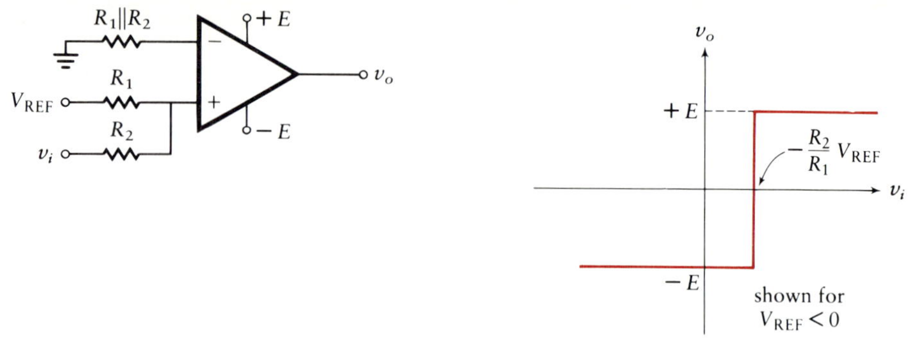

Figure 13.16
Variable crossover comparator.

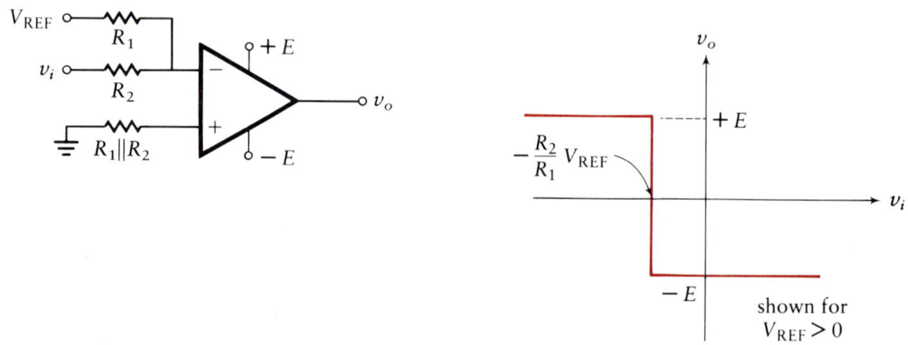

Figure 13.17
Variable crossover comparator.

or when

$$v_i > -\frac{R_2}{R_1}V_{REF} \tag{13.10}$$

In Figure 13.17, v_o is high when

$$\frac{V_{REF}}{R_1} + \frac{v_i}{R_2} < 0$$

or when

$$v_i < -\frac{R_2}{R_1}V_{REF} \tag{13.11}$$

A *limiting comparator* is formed with a diode as the feedback element, as shown in Figure 13.18(a). When the diode is open-circuited, the lack of feedback causes the op-amp to operate in the open-loop mode and to saturate for negative input voltages. Then

$$v_o = E = V_{REF}$$

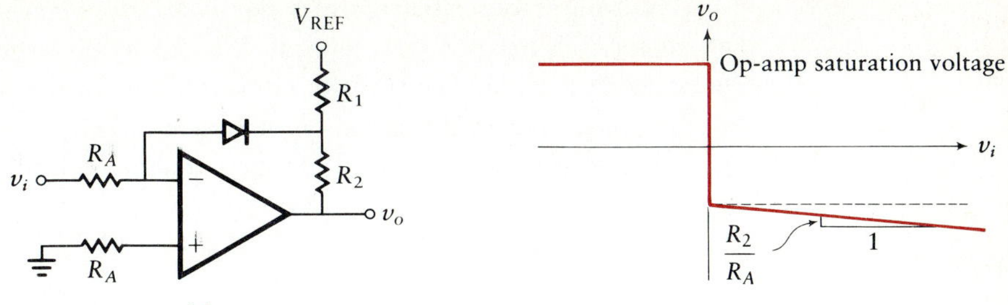

Figure 13.18 Limiting comparator.

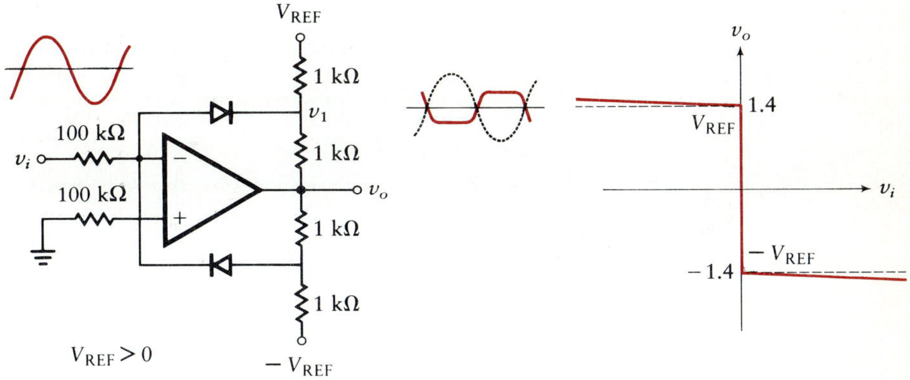

Figure 13.19 Balanced limiting comparator.

As v_i increases, the diode forward-biases (turns on). This happens when v_i exceeds $(V_{REF} + V_\gamma)$.

To find the equation for the saturation curve when the diode is conducting, we refer to Section 13.2, where we use equation (13.6) for a limiter. In this equation, we let $R_F \to \infty$, $R_f << (R_1 \parallel R_2)$, and $V_\gamma = 0$. This results in

$$v_o = -\frac{R_2}{R_A}v_i - \frac{R_2}{R_1}V_{REF} \tag{13.12}$$

This is the output voltage, v_o, as a function of the input voltage and the reference voltage.

The slope of the curve, for positive input voltage, is given by

$$\frac{v_o}{v_i} = -\frac{R_2}{R_A}$$

The transfer characteristics for equation (13.12) are shown in Figure 13.18(b).

Note that the output voltage for an input of $v_i = V_{REF}$ is given by

$$-R_2 V_{REF}\left(\frac{1}{R_A} + \frac{1}{R_1}\right)$$

The limiter described here is not *balanced,* since the switching transfer characteristics are not symmetrical. We can modify the circuit to provide for balanced operation. The transfer characteristic for the resulting balanced positive and negative limiter is shown in Figure 13.19. The analysis is performed in a manner similar to that used for the unbalanced limiter.

Example 13.6

Design a balanced limiting comparator of the type shown in Table 13.4.2 with limiting required at 8 V and the slope of the limited part of the characteristic to be $-\frac{1}{20}$. Assume that $V_{REF} = \pm 10$ V, $R_A = 50$ kΩ, $V_\gamma = 0.7$ V, and $R_f = 100$ Ω. The gain in the nonlimited region is −10.

SOLUTION From the slope requirement, we obtain

$$-\frac{R_F}{R_A} = -10 \quad \text{so } R_F = 0.5\ \text{M}\Omega$$

We use equation (13.8) to find

$$\frac{R_2}{R_1} = 0.682$$

We have one equation in two unknowns. The second equation comes from the given slope in the limited region ($-\frac{1}{20}$ in this example). From equation (13.7), we solve for the two resistor values with the result

$$R_2 = 2.33\ \text{k}\Omega, \qquad R_1 = 3.42\ \text{k}\Omega$$

The design is complete.

13.4 SCHMITT TRIGGERS

One class of comparator, known as the *Schmitt trigger,* uses positive feedback to speed up the switching cycle. With positive feedback, a small change in the input is amplified and fed back, in phase, to reinforce itself, thereby leading rapidly to larger changes. The feedback effectively increases the gain and therefore steepens the transition between the two output levels. Positive feedback

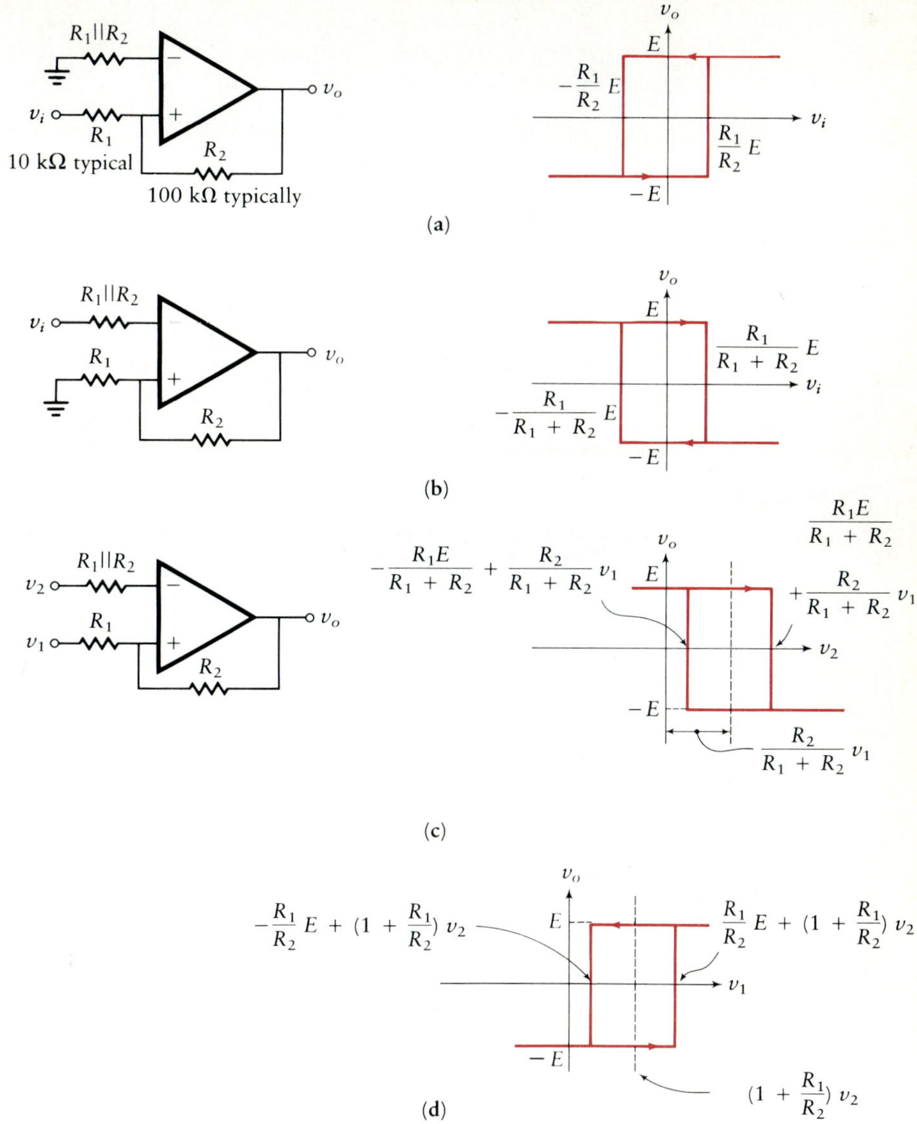

Figure 13.20
Schmitt trigger.

holds a comparator in one of the two saturation states unless a sufficiently large input is applied to overcome the feedback.

Figure 13.20(a) illustrates one form of Schmitt trigger where a reference voltage of 0 V is implied since $v_- = 0$. We use Figure 13.20(a) to develop the characteristic curves. We start with v_i as a large positive voltage. This causes the output voltage, v_o, to be at $+E$, the op-amp saturation voltage. The noninverting voltage, v_+, is calculated by writing a node equation at the v_+ node as follows:

$$\frac{v_+ - v_i}{R_1} + \frac{v_+ - v_o}{R_2} = 0$$

so

$$v_+\left(\frac{1}{R_1} + \frac{1}{R_2}\right) = \frac{v_i}{R_1} + \frac{v_o}{R_2} \tag{13.13}$$

We start reducing the magnitude of v_i to find the switching point. Since $v_- = 0$ and $v_+ = v_-$ (once the op-amp comes out of saturation), we set equation (13.13) to zero to obtain

$$v_i = \frac{-R_1 v_o}{R_2} = \frac{-R_1 E}{R_2} \tag{13.14}$$

As v_i is reduced from a large positive voltage, the output voltage, v_o, is switched from $+E$ to $-E$ at the point where v_+ goes to zero. This happens at the point where v_i reaches $-R_1 E/R_2$. As the input voltage, v_i, is reduced further, v_o remains at $-E$.

If we now increase the input voltage from a large negative value, the output voltage will switch to $+E$ when $v_+ = 0 = v_-$. Hence the switching takes place at

$$v_i = \frac{-R_1 v_o}{R_2} = \frac{-R_1(-E)}{R_2} = \frac{+R_1 E}{R_2} \tag{13.15}$$

v_o remains at $+E$ as v_i is further increased past $+R_1 E/R_2$.

The loop shown in the characteristic curve of Figure 13.20(a) is a form of *hysteresis.* This word is used to describe a situation in which the system has *memory.* That is, the output at any particular time depends not only on the present value of the input but also on past values. For example, for an input voltage of $v_i = 0$ there are two possible values of v_o, depending on the direction in which we approach $v_i = 0$.

This observation about hysteresis indicates one important application of the Schmitt trigger. This circuit can be used as a *binary memory device.* That is, since the output depends on past values of the input, we can apply a voltage to the input and then remove that voltage. The trigger circuit remembers whether the voltage was above or below the reference level. We can therefore "write" one of two possible values into this memory.

A second important application of the Schmitt trigger is as a square-wave generator. A continuous signal at the input (e.g., a sine wave) produces an output that rapidly jumps between two levels. The jump occurs as the input moves across the reference level. In this manner, a pulse-type waveform can be generated from a continuous sinusoidal input.

The inverting Schmitt trigger of Figure 13.20(b) interchanges the ground and the input voltages at the op-amp input. It is analyzed in a manner similar to that applied above. The switching point is found from the equations

$$v_- = v_i$$

$$v_+ = \frac{R_1 v_o}{R_1 + R_2}$$

The circuit switches state when the two voltages are equal, that is, when

$$v_i = \frac{R_1 v_o}{R_1 + R_2}$$

When $v_o = +E$ and v_i is increasing from a large negative voltage toward a positive voltage, the switching point occurs at

$$v_i = \frac{R_1 v_o}{R_1 + R_2} = \frac{R_1 E}{R_1 + R_2}$$

If $v_o = -E$ and v_i is decreasing from a large positive voltage toward a negative voltage, the switching point occurs at

$$v_i = \frac{-R_1 E}{R_1 + R_2}$$

The circuit of Figure 13.20(c) replaces the ground of the v_+ input of Figure 13.20(b) with a reference voltage, v_1. The second voltage, v_2, is the input voltage. The equations are derived in the same manner as those above and are included along with the hysteresis curve on the figure. Note that the entire characteristic is shifted to the right, so it is no longer symmetrical about the origin. You should verify these results before continuing.

The same circuit can be viewed in a different manner when v_2 is the reference and v_1 is the input. This represents a variation of the circuit of Figure 13.20(a), where the ground input to v_- is replaced by the v_2 reference. The curve of Figure 13.20(d) results.

The algebraic sign of the following expression determines the switching for the Schmitt triggers of Figure 13.20.

$$\frac{R_2}{R_1 + R_2} v_1 + \frac{R_1}{R_1 + R_2} v_o - v_2$$

This expression allows us to determine the $v_o = 0$ crossings shown on the output-input curves.

Example 13.7

Determine the output voltage for the Schmitt trigger of Figure 13.20(a) if the input voltage is

$$v_i = 20 \sin 200\pi t$$

$E = 5$ V, $R_1 = 20$ kΩ, and $R_2 = 100$ kΩ.

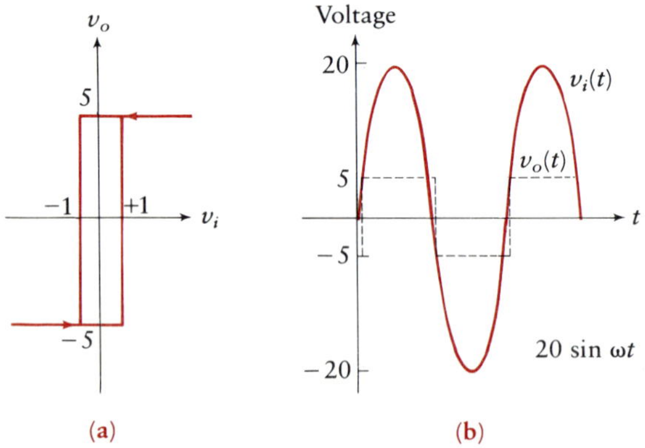

Figure 13.21
Schmitt trigger output for Example 13.7.

SOLUTION The hysteresis curve for these values is plotted in Figure 13.21(a). When an input sinusoid, $v_i = 20 \sin 200\pi t$, is applied to the Schmitt trigger of Figure 13.20(a), a square wave results. Figure 13.21(b) shows the input sinusoid (shown as a solid line) and the output pulse train (shown as a dashed line). The peak-to-peak voltage is 10 V and the zero crossing is offset by ±1 V because of the hysteresis loop of Figure 13.21(a).

13.4.1 Schmitt Triggers with Limiters

The limiting comparator, which is studied in Section 13.3, can be used in conjunction with any of the Schmitt triggers of Figure 13.20. In so doing, the op-amp saturation voltage, E, which is not very precise, is replaced with an accurate voltage, V_{REF}.

Figure 13.22(a) illustrates an unbalanced limiting comparator. Diode D_1 controls the lower saturation point and D_2 controls the upper saturation point. The resulting characteristic curve is shown in Figure 13.22(b).

Using this configuration, the saturation voltages, E, are replaced with the limiting voltages shown on the figure.

As an example of the use of a limiter with a Schmitt trigger, consider the circuit of Figure 13.23. This circuit uses positive feedback to achieve the faster switching action. The operation follows the hysteresis loop for a Schmitt trigger but with the limits of the voltage output curve determined by V_{REF}.

When a dc voltage is placed at one input of a Schmitt trigger, it causes the hysteresis loop to move along the v_i axis. This is shown by using Figure 13.20(c) with v_1 as a dc reference voltage (V_{dc}) and $v_2 = v_i$.

Figure 13.22 Unbalanced limiting comparator.

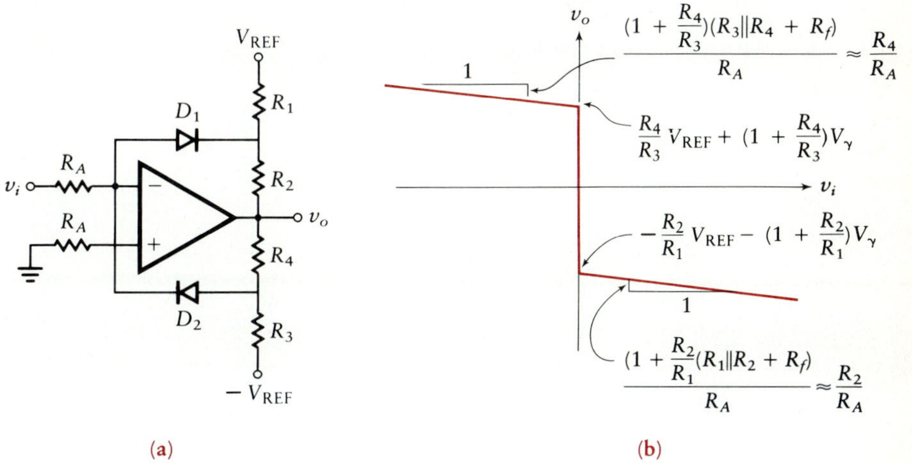

Figure 13.23 Schmitt trigger with limited values.

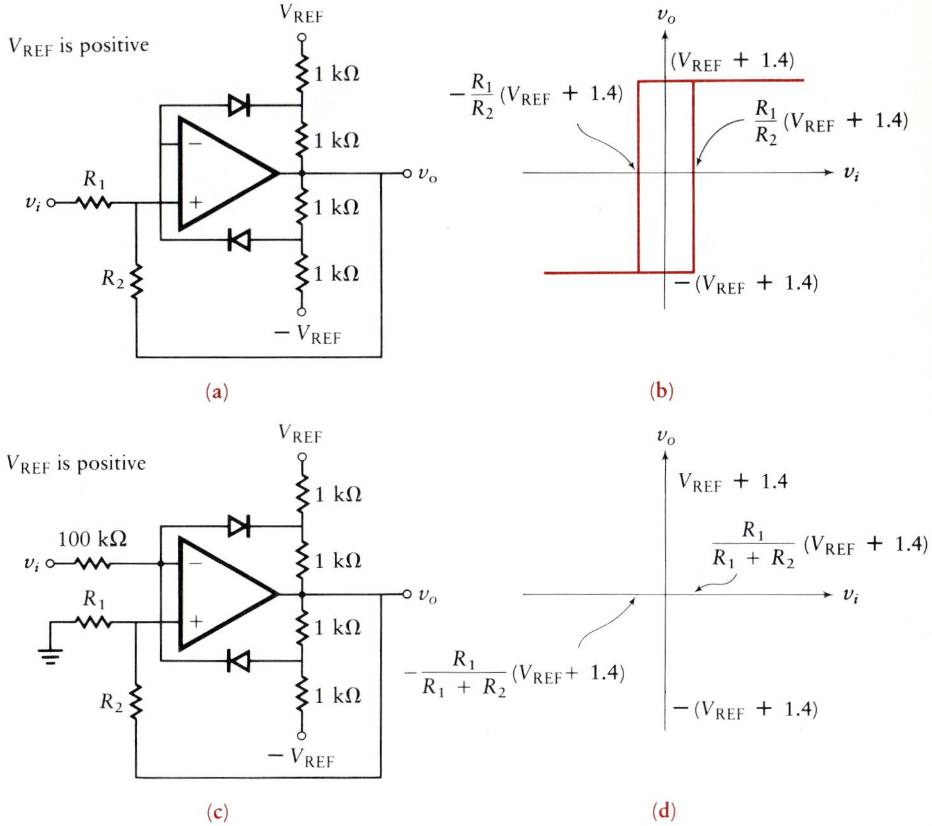

The $v_o = 0$ intersections are as follows:
On the right:

$$v_i = \frac{R_2 V_{dc}}{R_1 + R_2} + \frac{R_1 E}{R_1 + R_2}$$

On the left:

$$v_2 = \frac{R_2 V_{dc}}{R_1 + R_2} - \frac{R_1 E}{R_1 + R_2}$$

Example 13.8

Design an unbalanced limiter comparator of the type shown in Figure 13.22. Limiting should occur at +6 V and −4 V with the slope of the limiting characteristic as $-\frac{1}{25}$. Assume $V_{REF} = 10$ V, $R_A = 50$ kΩ, $V_\gamma = 0.7$ V, and $R_f = 100$ Ω.

SOLUTION We find R_2/R_1 from equation (13.8) with $V_{oc} = +6$ V.

$$\frac{R_2}{R_1} = 0.495$$

The slope requirement provides us with the second equation in R_1 and R_2. In equation (13.7), we equate the slope to $-\frac{1}{25}$ and solve these two equations for R_1 and R_2 to obtain

$$R_1 = 3.74 \text{ k}\Omega, \qquad R_2 = 1.85 \text{ k}\Omega$$

Note that this first part of the design parallels that of Example 13.4. In a similar manner, we solve for the remaining two resistors. Again, from equation (13.8), we let $V_{oc} = -4$ and find

$$\frac{R_4}{R_3} = 0.308$$

We equate the slope to $-\frac{1}{25}$ and, using equation (13.7), solve for the resistances.

$$R_3 = 6.07 \text{ k}\Omega, \qquad R_4 = 1.87 \text{ k}\Omega$$

Example 13.9

Design a Schmitt trigger that will operate according to the v_o/v_i characteristic curve shown in Figure 13.24. Assume that a ±10-V precision reference voltage is available.

SOLUTION We solve first for the hysteresis loop shown in Figure 13.25. This is a shifted version of Figure 13.24, which has symmetrical voltage swings. We use the Schmitt trigger shown in Figure 13.20(d) with $v_1 = E$ and $v_2 = v_i$. The values of R_1' and R_2' are calculated as follows (see Figure 13.20(d)). Setting the left edge of the hysteresis loop at $v_i = 0$, we have

$$\frac{-R_1'}{R_2'}E + \left(1 + \frac{R_1'}{R_2'}\right)v_2 = 0$$

Therefore,

$$\left(1 + \frac{R_1'}{R_2'}\right)v_2 = \frac{R_1'}{R_2'}E \tag{13.16}$$

The center of the hysteresis loop falls at $v_i = 1$, so from Figure 13.20(d) we have

$$\left(1 + \frac{R_1'}{R_2'}\right)v_2 = 1 \tag{13.17}$$

Combining equations (13.16) and (13.17) and recognizing that $E = 2.5$ V, we have

$$\frac{R_1'}{R_2'}E = 1 = \frac{R_1'}{R_2'}2.5$$

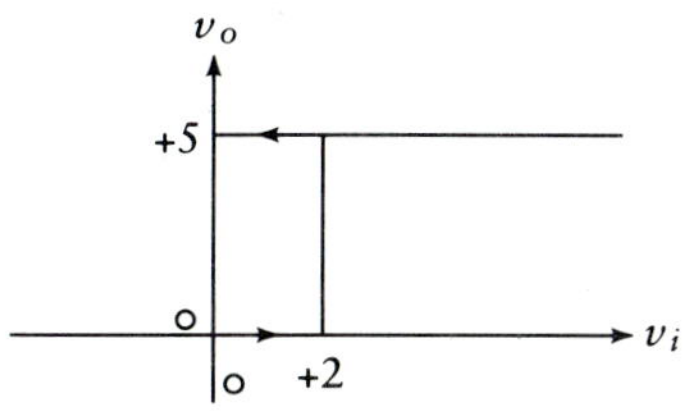

Figure 13.24 Characteristic curve for Example 13.9.

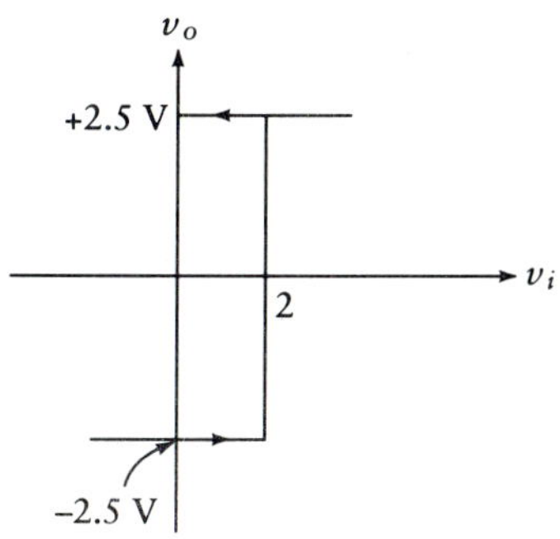

Figure 13.25 Hysteresis loop for Example 13.9.

so

$$R_2' = 2.5R_1'$$

Now from equation (13.16) we have

$$\left(1 + \frac{R_1'}{R_2'}\right)v_2 = 1$$

Therefore,

$$\left(1 + \frac{1}{2.5}\right)v_2 = 1$$

and

$$v_2 = 0.71 \text{ V}$$

If we set

$$R_1' = 100 \text{ k}\Omega$$

then

$$R_2' = 250 \text{ k}\Omega$$

The resulting design is shown in Figure 13.26. We now design the limiter, referring to Figure 13.22. To establish the horizontal portions of the hysteresis loop, we set the slope $R_4/R_1' \approx 0$. Then $R_4 << R_1'$.

We find the value of V_{REF} from the equation

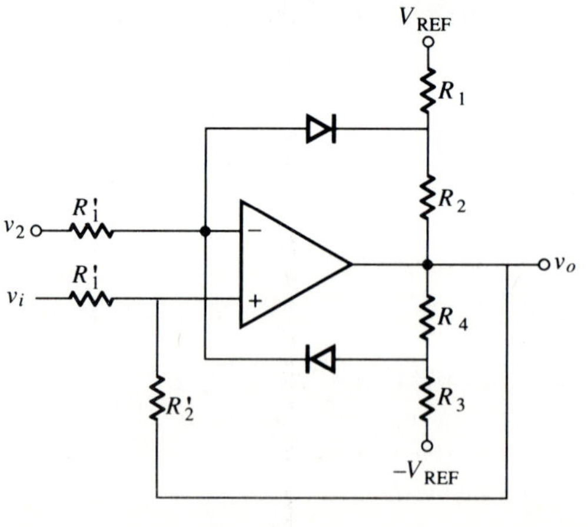

Figure 13.26 Circuit for Schmitt trigger.

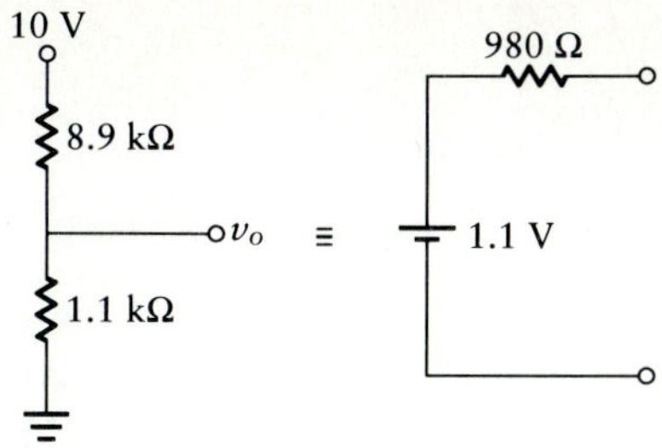

Figure 13.27 Voltage divider.

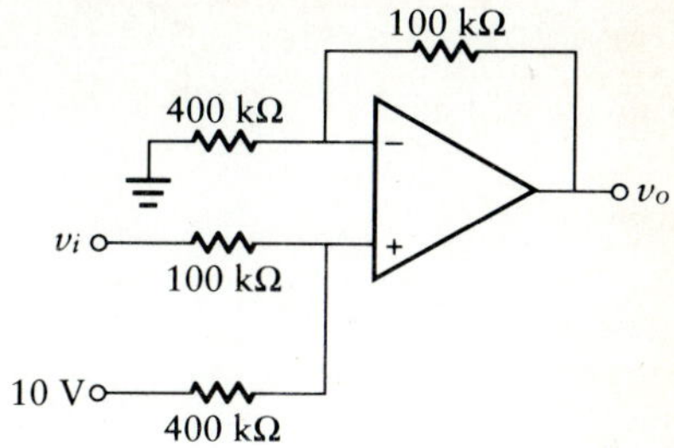

Figure 13.28 Addition of op-amp for level shift.

$$\frac{R_4}{R_3}V_{REF} + \left(1 + \frac{R_4}{R_3}\right)V_\gamma = 2.5 \text{ V} \tag{13.18}$$

Suppose we let $R_1 = R_2 = R_3 = R_4$. Let us choose 5 kΩ for these resistor values (we do not want to choose too high a value since this would require an excessively large R_1'). Then, from equation (13.18), we have

$$V_{REF} + 1.4 = 2.5$$

and

$$V_{REF} = 1.1 \text{ V}$$

Since we have a ±10 V precision reference source, we use a voltage divider to obtain V_{REF} as shown in Figure 13.27. Since the equivalent resistor of the voltage divider is in series with the sources, it adds to R_1 and R_3. Hence, we must reduce R_1 and R_3 from 5 kΩ to 5 kΩ − 980 Ω, so $R_1 = R_3 = 4020$ Ω.

We next make a level shift to move the v_o/v_i relationship up by +2.5 V, as specified in Figure 13.24. We add another op-amp at the output, as shown in Figure 13.28. This develops the final desired v_o/v_i curve of Figure 13.24.

Drill Problems

Design op-amp circuits with the following approximate transfer characteristics. Reference supplies available are ±10 V. Op-amp supply voltages are ±15 V. The solutions are shown in Figures D13.9 and D13.10. Assume V_γ = 0.7 V and R_f = 100 Ω.

D13.9

Figure D13.9

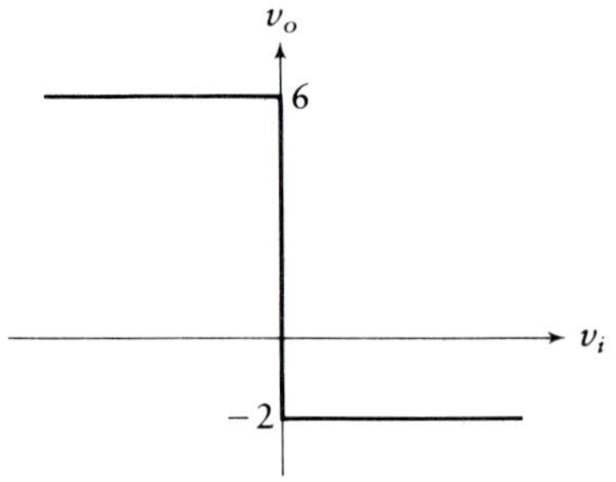

Ans:

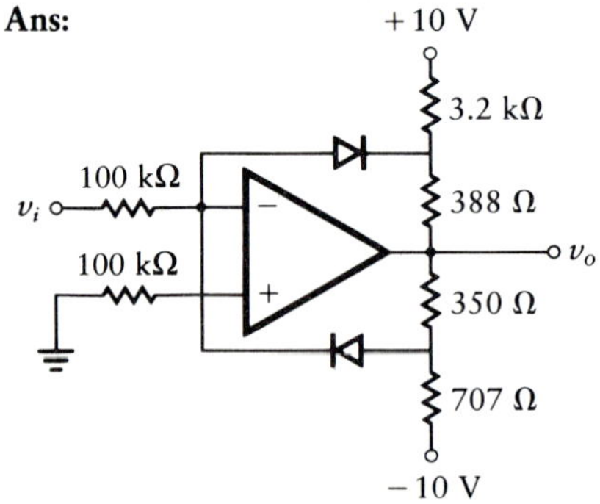

D13.10

Figure D13.10

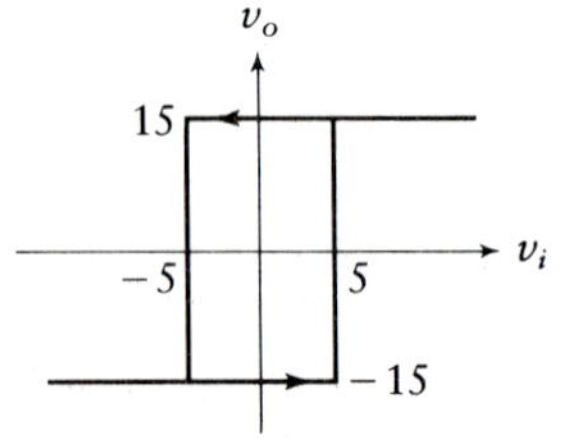

Ans:

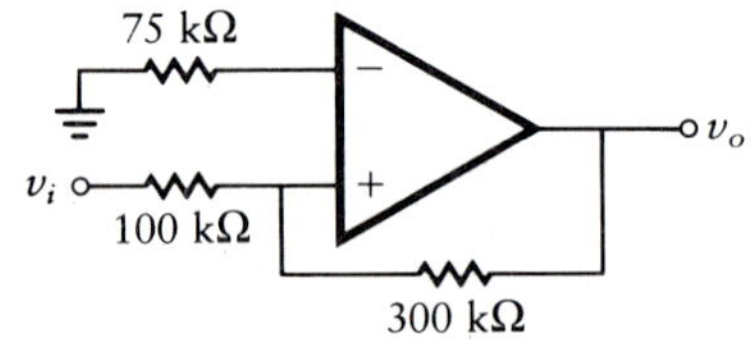

Find approximate relationships between the input voltage, v_i, and the output voltage, v_o, in each circuit shown. $V_\gamma = 0.7$ V and $R_f = 100\ \Omega$. The solutions are shown in Figures D13.11 and D13.12.

D13.11

Figure D13.11

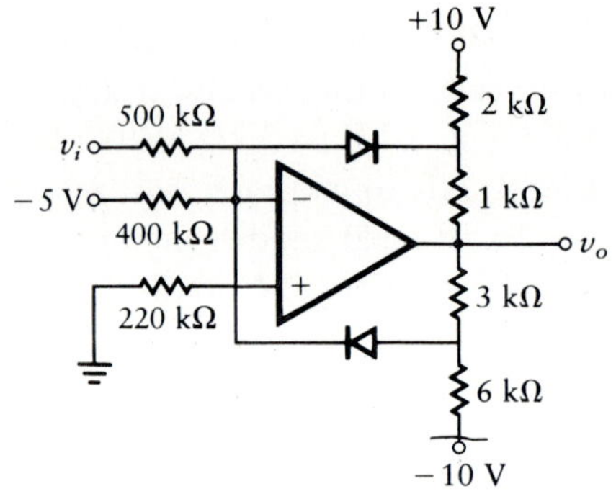

Ans:

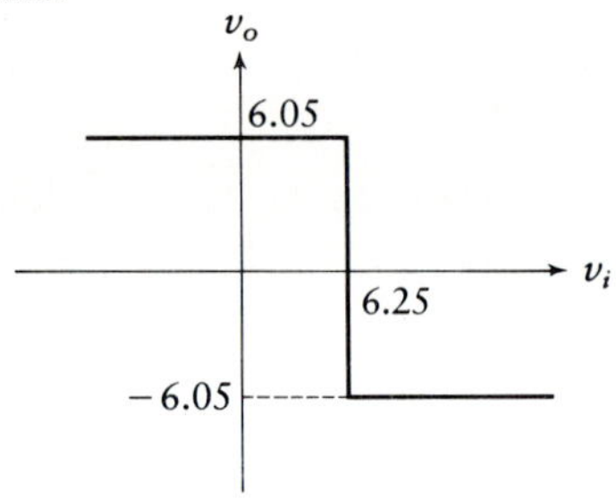

D13.12

Figure D13.12

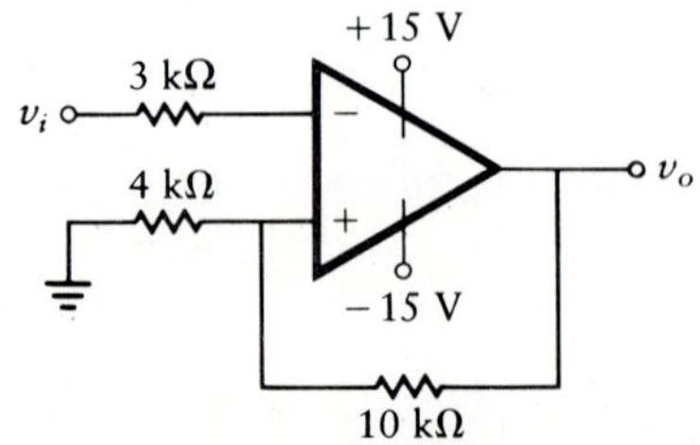

Ans:

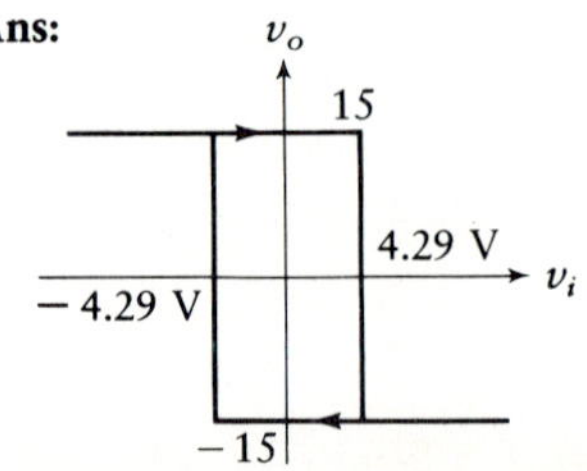

13.5 INTEGRATED CIRCUIT SCHMITT TRIGGER

We will usually not find it necessary to design a Schmitt trigger. We will, however, need to use Schmitt triggers in a wide assortment of design applications. A variety of Schmitt triggers are available for our designs. The hysteresis loop of the Schmitt trigger is incorporated into several digital integrated circuits (ICs). In this section, we consider the 7414 Hex Schmitt trigger inverter, which contains six independent inverters. Figure 13.29(a) shows the symbol of a Schmitt trigger inverter and Figure 13.29(b) shows the associated function table. When the input, A, is low (perhaps 0 V) the output, Y, is high (5 V). When the input is high, the output is low. We consider inverters and many other digital ICs in Chapter 15. For now we consider the effect of the Schmitt trigger on the inverter of Figure 13.29. Because of the Schmitt trigger action, each inverter has different input threshold levels for positive-going (V_{T+}) and for negative-going (V_{T-}) signals. The ICs produce clean and jitter-free signals even when triggered from slow input ramps. The package configuration is shown in Figure 13.29(c). The logic diagram and the function diagram are the same as for an ordinary inverter, except that the small hysteresis symbol is shown within the inverter. The six independent inverters are contained in a 14-lead molded package (known as a DIP, for dual in-line package), which is shown in Figure 15.4.

The IC Schmitt trigger finds wide application in signal conditioning, as in Example 13.7. When we explore the design of more complex electronic systems in Chapter 17, we will make frequent use of the IC Schmitt trigger.

Figure 13.29
7414 Hex Schmitt trigger inverter.

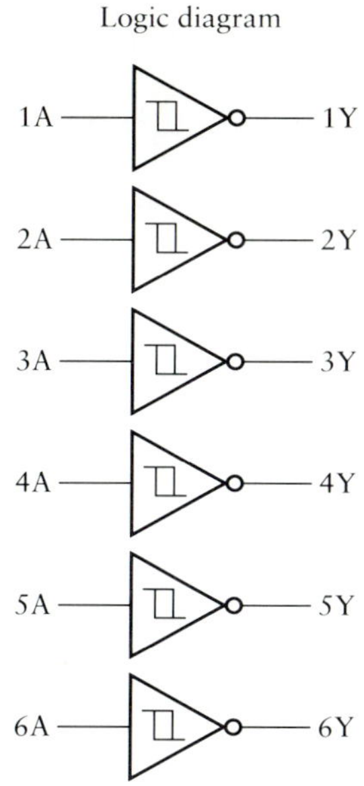

(a)

A	Y
0	1
1	0

(b)

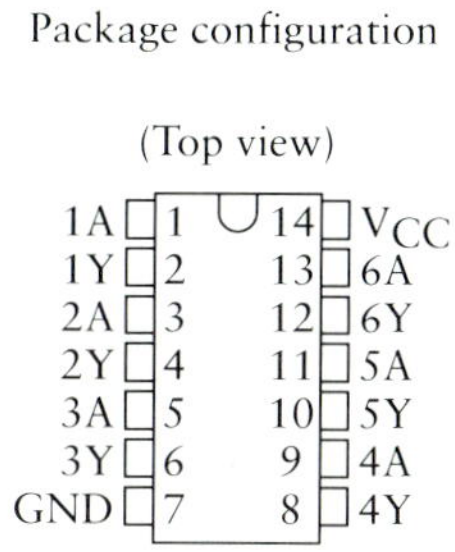

(c)

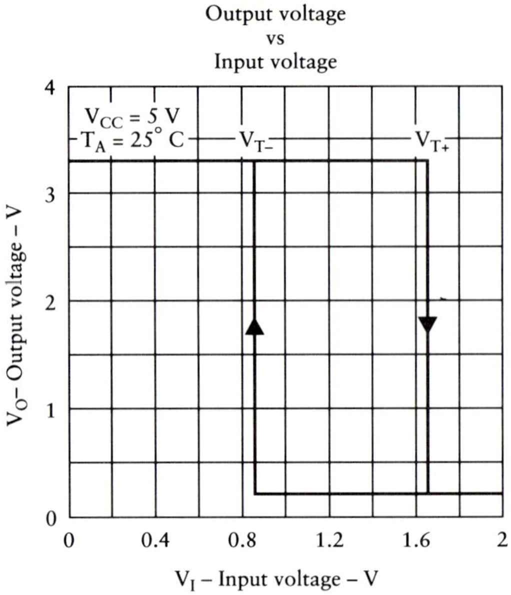

Figure 13.30 Hysteresis loop for 7414 Hex Schmitt trigger inverter.

The hysteresis loop for the 7414 Hex Schmitt trigger inverter is shown in Figure 13.30. As you can see from this figure, $V_{T-} = 0.85$ V and $V_{T+} = 1.66$ V. The high state occurs at a voltage of 3.3 V and the low state is at 0.2 V. In Chapter 15, we will discover that these voltages are the high and low values for the TTL logic family.

PROBLEMS

Find approximate relationships between the input voltages, v_1 and v_2, and the output voltage, v_o, in each of the circuits of Problems 13.1 to 13.5. Assume $V_\gamma = 0.7$ V and $R_f = 100$ Ω.

13.1

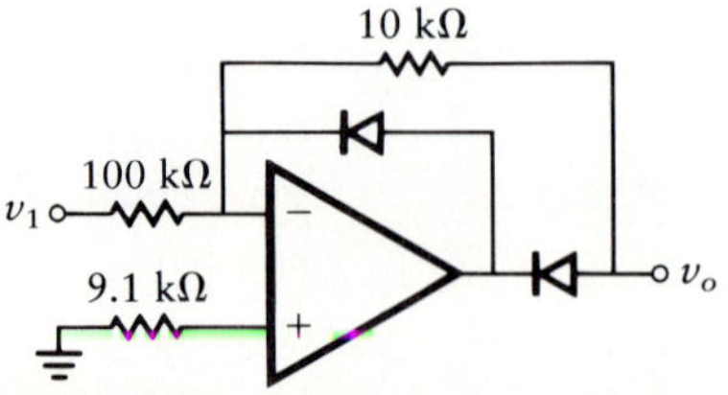

Figure P13.1

13.2

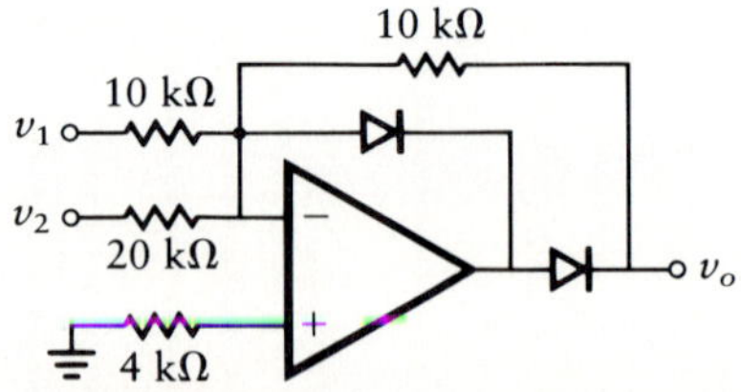

Figure P13.2

13.3

Figure P13.3

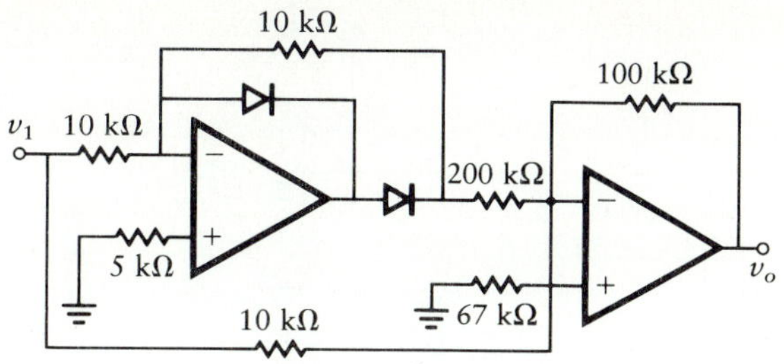

13.4

Figure P13.4

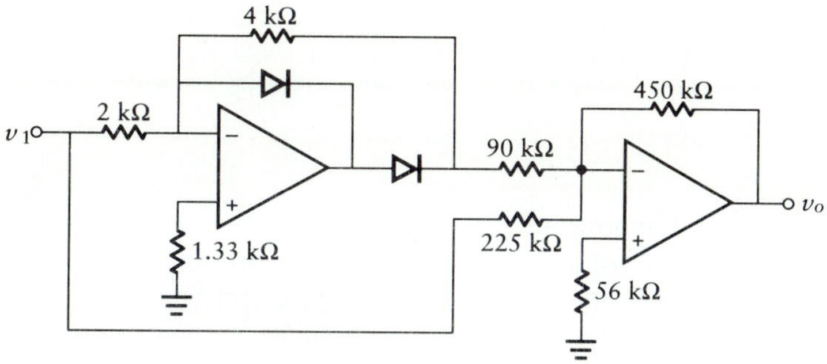

13.5

Figure P13.5

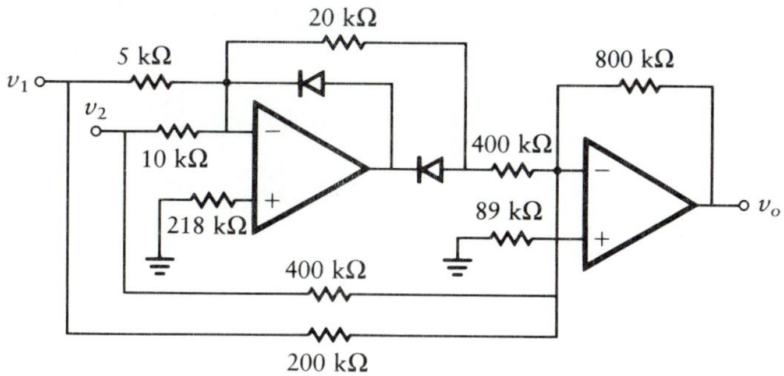

Design op-amp circuits with the approximate transfer characteristics shown in Problems 13.6 to 13.15. Reference supplies of +10 V and −10 V are available. Assume that $V_\gamma = 0.7$ V and $R_f = 100\ \Omega$.

13.6

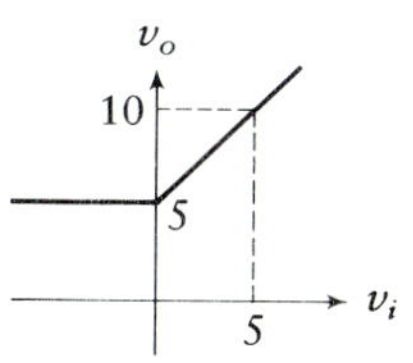

Figure P13.6

13.7

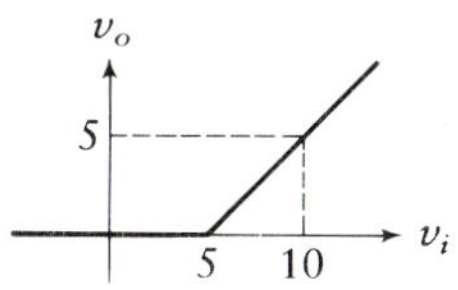

Figure P13.7

13.8

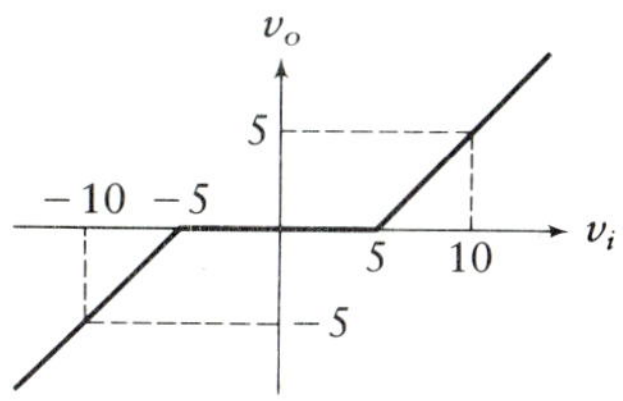

Figure P13.8

13.9

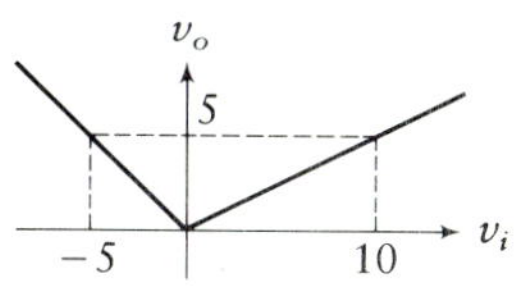

Figure P13.9

13.10

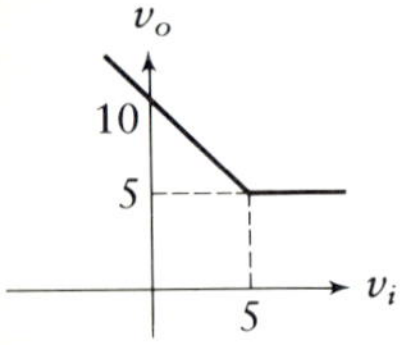

Figure P13.10

13.11

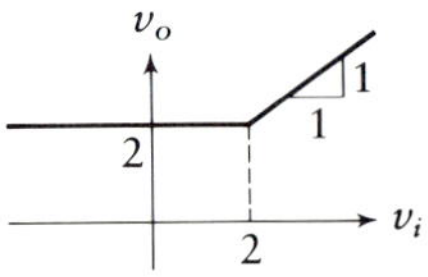

Figure P13.11

13.12

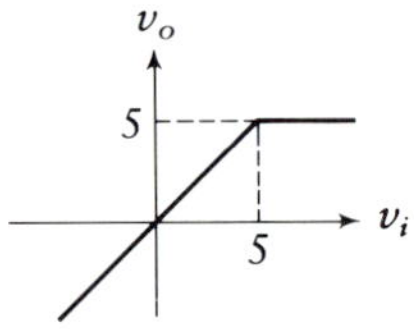

Figure P13.12

13.13

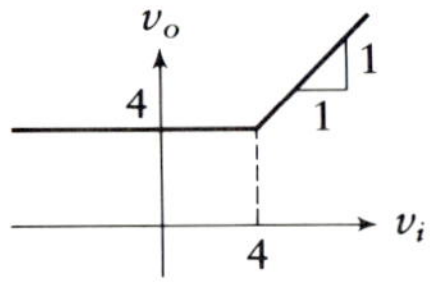

Figure P13.13

13.14

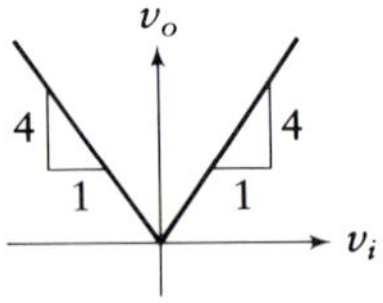

Figure P13.14

13.15

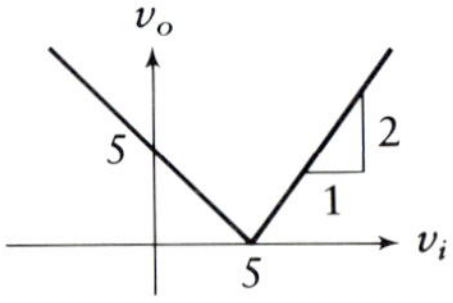

Figure P13.15

Find approximate relationships between the input voltages, v_1 and v_2, and the output voltage, v_o, in each of the circuits shown in Problems 13.16 to 13.20. Assume that $V_\gamma = 0.7$ V and $R_f = 100\ \Omega$.

13.16

Figure P13.16

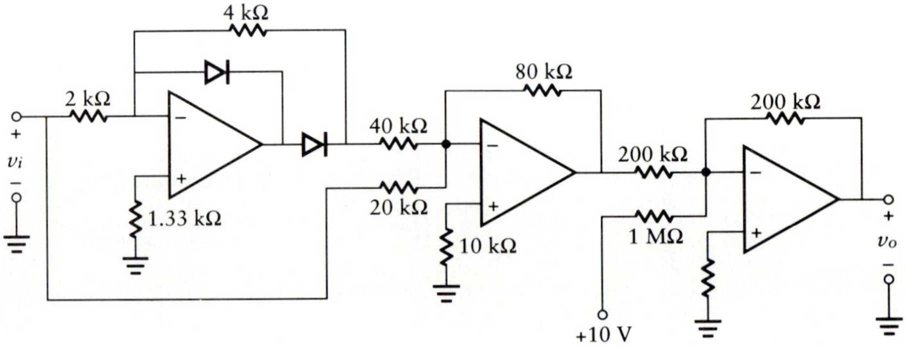

13.17

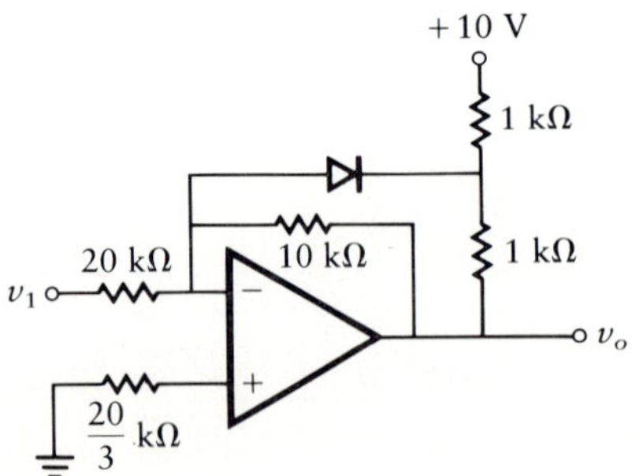

Figure P13.17

13.18

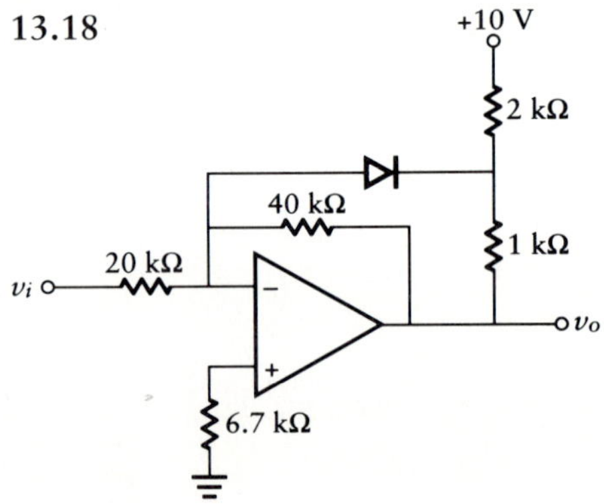

Figure P13.18

13.19

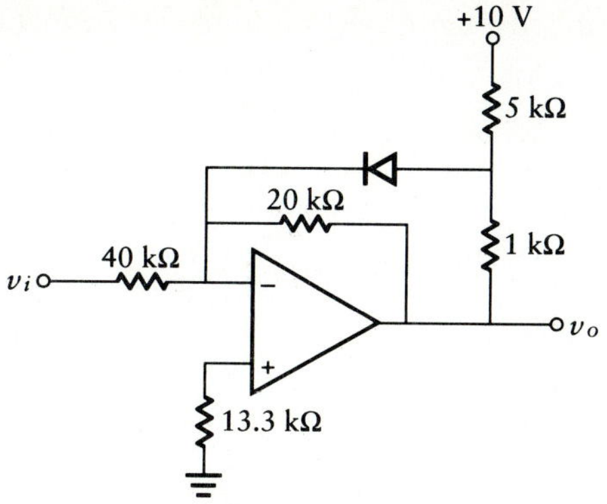

Figure P13.19

13.20

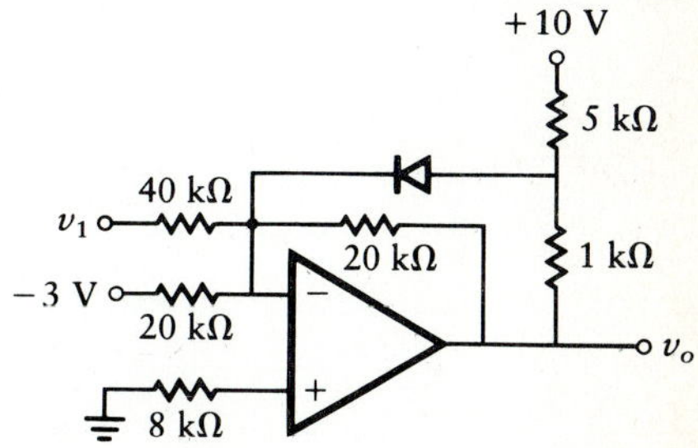

Figure P13.20

Design op-amp circuits with the approximate transfer characteristics shown in Problems 13.21 to 13.25. Reference supplies of ±10 V are available. Op-amp supply voltages are ±15 V. Assume that $V_\gamma = 0.7$ V and $R_f = 100\ \Omega$.

13.21

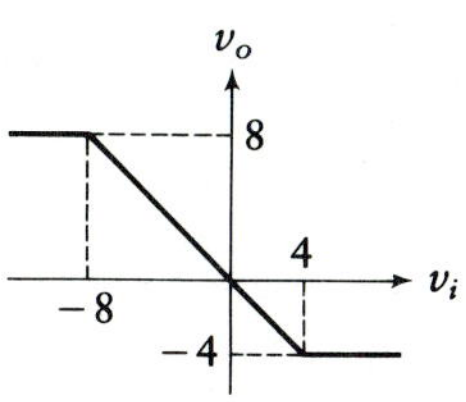

Figure P13.21

13.22

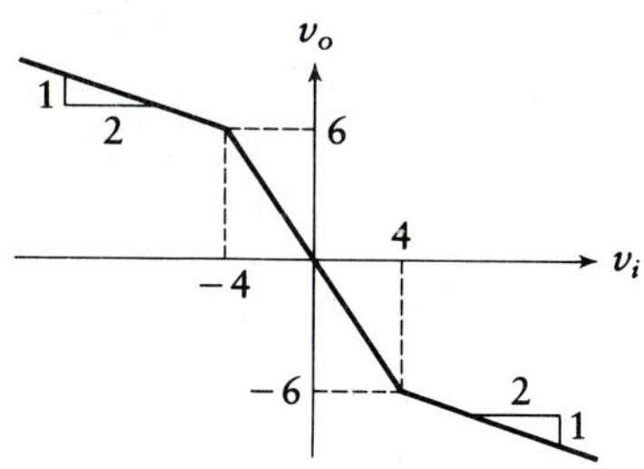

Figure P13.22

13.23

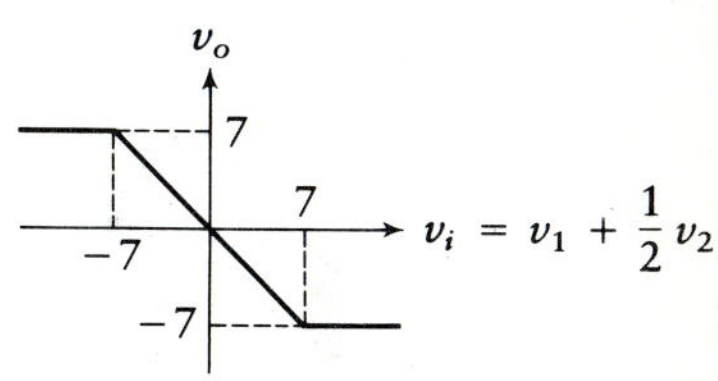

Figure P13.23

13.24

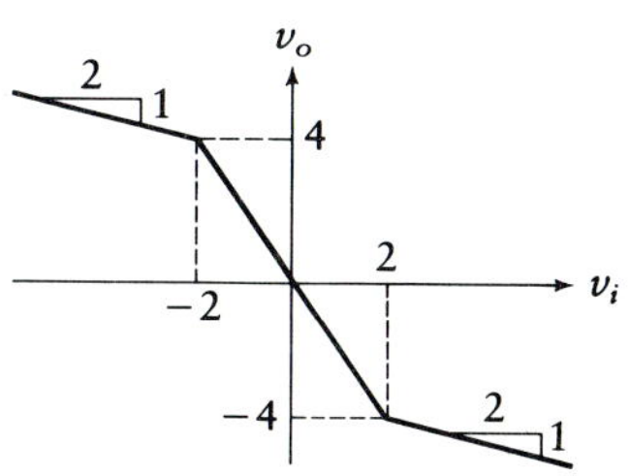

Figure P13.24

13.25

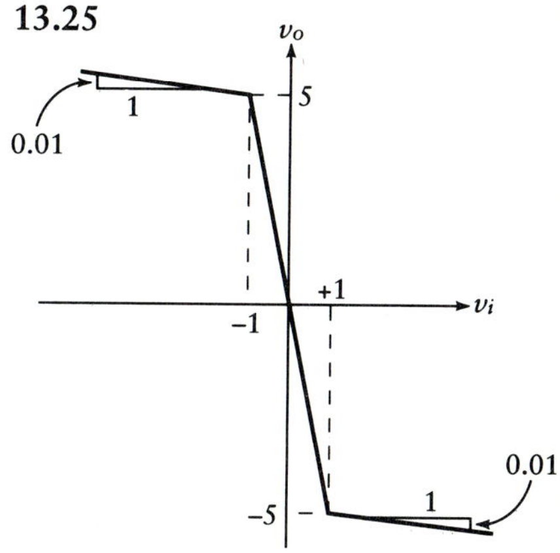

Figure P13.25

Find approximate relationships between the input voltages, v_1 and v_2, and the output voltage, v_o, in each of the circuits shown in Problems 13.26 to 13.31. Assume that $V_\gamma = 0.7$ V and $R_f = 100\ \Omega$.

13.26

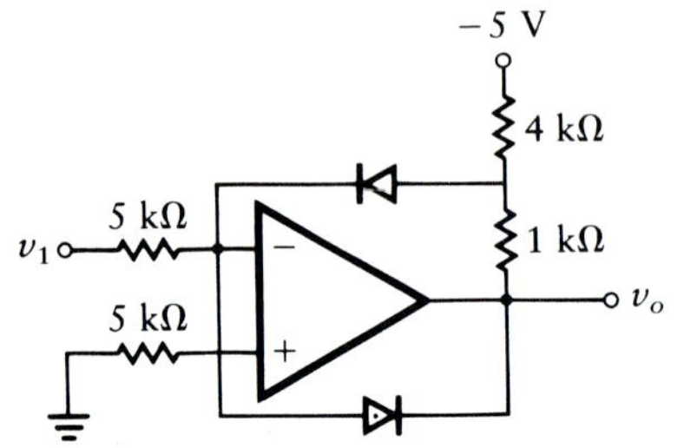

Figure P13.26

13.27

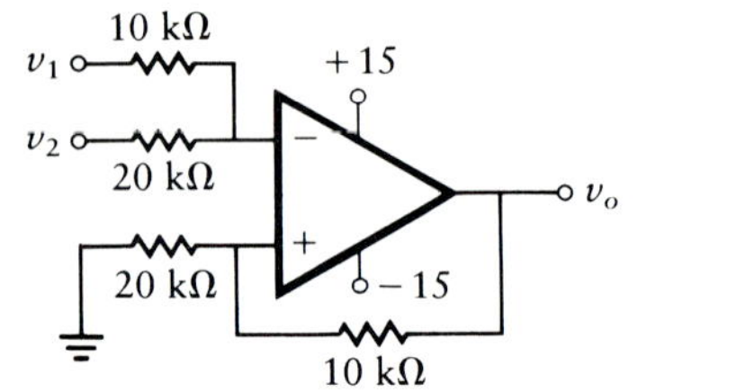

Figure P13.27

13.28

Figure P13.28

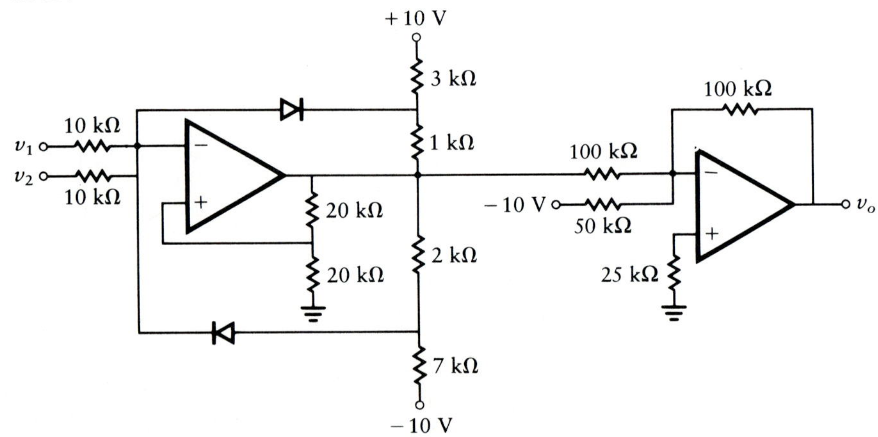

13.29

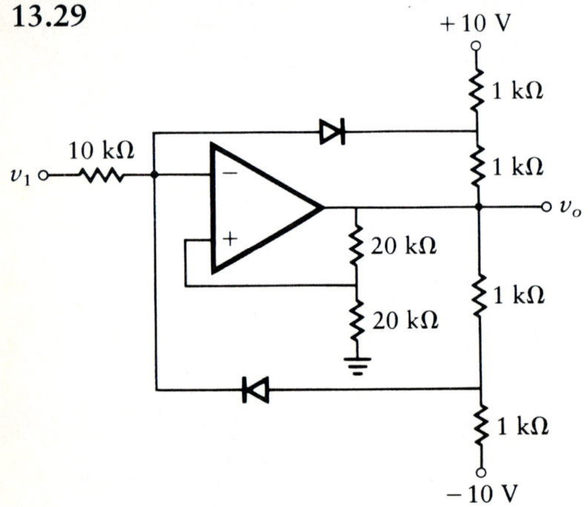

Figure P13.29

13.30

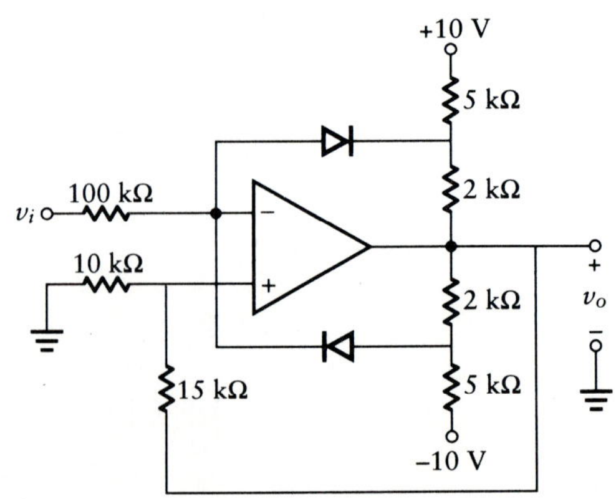

Figure P13.30

13.31

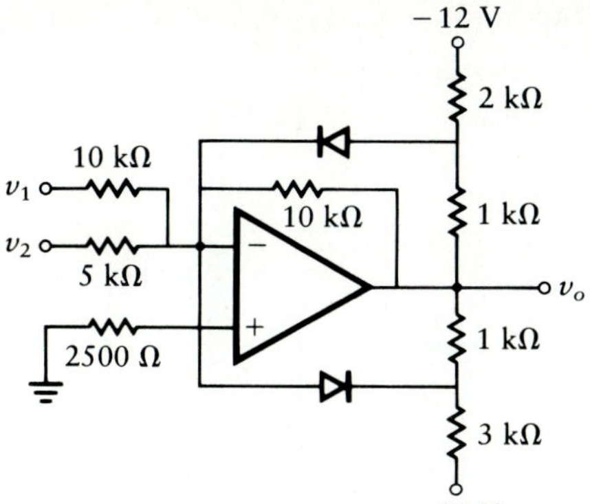

Figure P13.31

Design op-amp circuits with the approximate transfer characteristics shown in Problems 13.32 to 13.38. Reference supplies of ±10 V are available. Op-amp supply voltages are ±15 V. Assume $V_\gamma = 0.7$ V and $R_f = 100\ \Omega$.

13.32

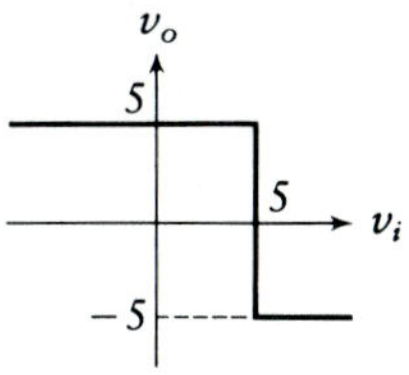

Figure P13.32

13.33

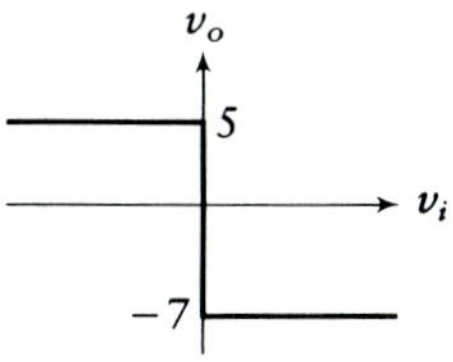

Figure P13.33

13.34

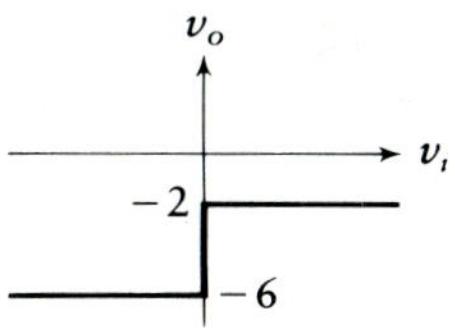

Figure P13.34

13.35

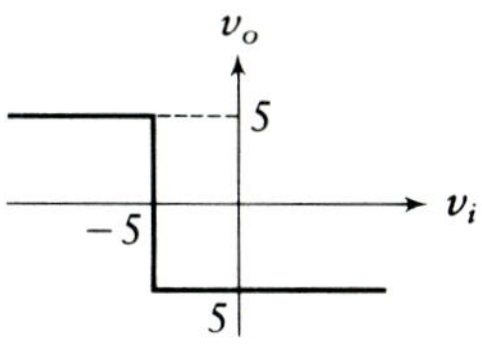

Figure P13.35

13.36

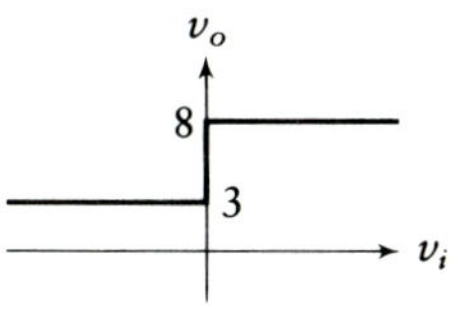

Figure P13.36

13.37

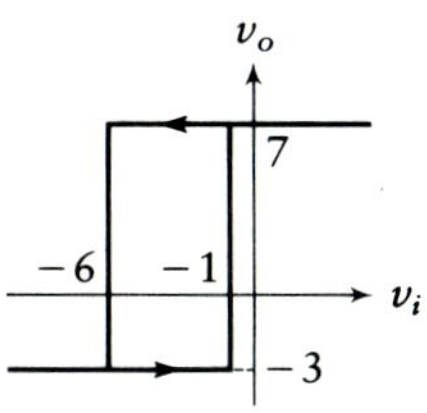

Figure P13.37

13.38

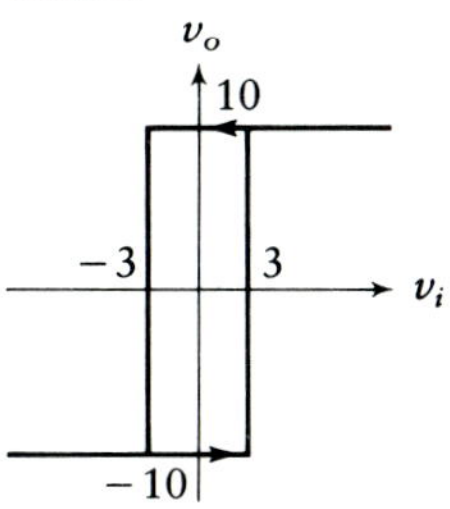

Figure P13.38

13.39 Analyze the precision absolute value circuit of Figure P13.39. The output, v_o, of this circuit equals $|v_i|$. Because of the high gain of the op-amps, the *on* voltage drop of the diode is effectively zero.

Figure P13.39

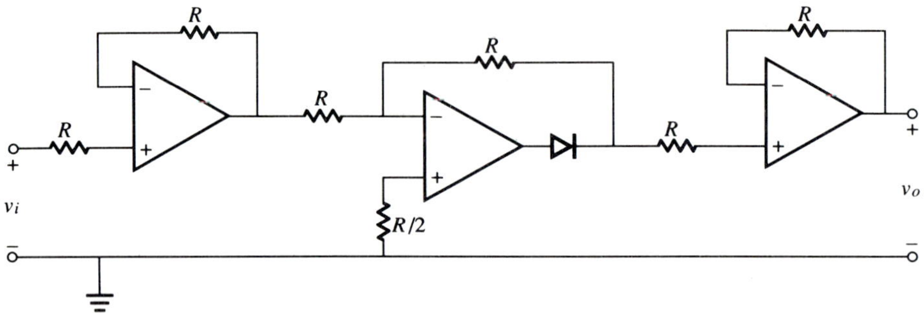

14

PULSED WAVEFORMS AND TIMING CIRCUITS

14.0 INTRODUCTION

The topics covered throughout the first 13 chapters of this text emphasize *analog* electronics. In such circuits, we are interested in voltage and current gain, maximum undistorted amplitude swings, and input and output resistance. Typical applications include the amplification of music or speech.

Chapters 14 through 17 are devoted to *digital* electronics. It has often been said that "the world is going digital." In the area of music and speech, the digital format of the *compact disc* has greatly improved the quality of sound (e.g., improved dynamic range and reduced distortion due to noise, scratching, and wow). Indeed, the continuing movement toward increased use of digital systems is exemplified by the transition from music systems using analog tape to digital compact discs and *digital audiotape.*

Digital electronics deals with signals that can take only one of two discrete values: high or low, on or off, 1 or 0. These signals are said to be *binary* and are the type found in present-day electronic digital systems. The bipolar junction transistors are either in the *cutoff* or *saturated* condition. We are no longer interested in the linear operating region of electronic devices. Instead, we are concerned with whether the device is on or off, that is, whether the output voltage is near zero or near the supply voltage value. By reducing the range of possible voltages to one of two values, the circuitry often becomes simpler and considerably more reliable. This last point deserves more emphasis. In an analog system, the output can take on any value within a continuum of values. Therefore, any disturbance, whether in the form of additive noise or distortion, results in an error. In a binary digital system, we are dealing with two possible signal values. Any disturbance must be large enough to make one value look like the other to cause a change. If the disturbance is not that large, the error is zero.

The signals that are processed through digital systems are a series of lows and highs, that is, a series of discrete values. For example, we could use +5 V for the high value and 0 V for the low value. A representative series of values is shown in Figure 14.1 as a train of pulses. Since digital systems respond to pulse trains such as that of Figure 14.1, we first study the behavior of these systems when a pulse train is applied. In particular, we begin our study of digital electronics by analyzing and designing RC networks with a pulse train as the input. Many of these networks include diodes. In particular, we are interested in the *steady-state* output waveforms of such networks. These types of networks are used to trigger (turn on) various ICs.

Most systems found in nature are analog and not digital. For example, the velocity of an automobile is a signal that can take on a continuum of values; the temperature of the human body is also an analog variable. For this reason, when we design digital systems we must first change the continuous variables into pulse trains that can be processed by the digital electronics. We must *digitize* or *quantize* the magnitude of samples of the continuous variable using an *analog-to-digital* converter. The resulting signal is a pulse train or sequence of binary numbers that represents the magnitude of the samples of the analog signal.

An electronic system frequently measures the continuously varying input variable, such as the velocity of a car. The signal is amplified and digitized to become a pulse train that is processed by the digital system. The output may be displayed on a digital readout (e.g., the digital speedometer in a car) or may be used to drive an analog actuator (e.g., as in a speed-control system). The design engineer must be proficient in both analog and digital electronics, as we will see in Chapter 17, where we concentrate on design of systems.

The heart of most digital systems is a clock. The clock produces a steady stream of pulses, which is used to synchronize the timing of the electronic system. We consider several clocks (square-wave pulse generators) and spend time analyzing the 555 pulse generator (oscillator). We use the 555 in a variety of applications.

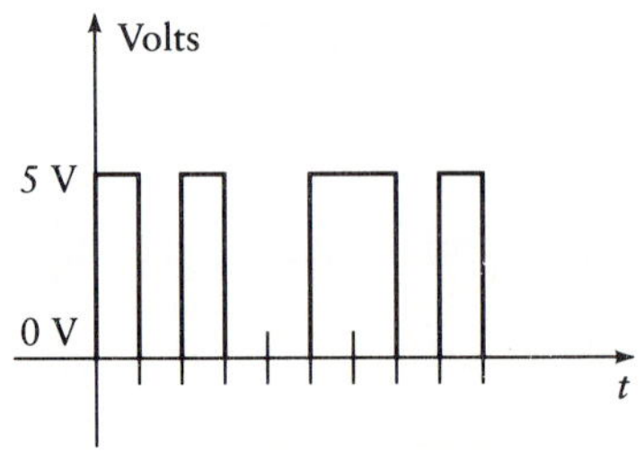

Figure 14.1
Pulse train.

14.1 HIGH-PASS RC NETWORK

A series capacitor and a shunt resistor form a simple high-pass network. The input voltage is applied across the combination and the output voltage is taken across the resistor, as shown in Figure 14.2(a). We analyze the first-order high-pass filter in Chapter 12, where we find the transfer function to be of the form

$$\frac{V_o}{V_i} = \frac{\tau s}{\tau s + 1}$$

The time constant, τ, is given by

$$\tau = RC$$

The sinusoidal steady-state amplitude plot of this response, which is covered in Appendix B, is shown in Figure 14.2(b). Notice that the corner frequency on this figure is f_c, which is equal to $1/(2\pi RC)$. Suppose the input to this system is a step function of amplitude V. That is,

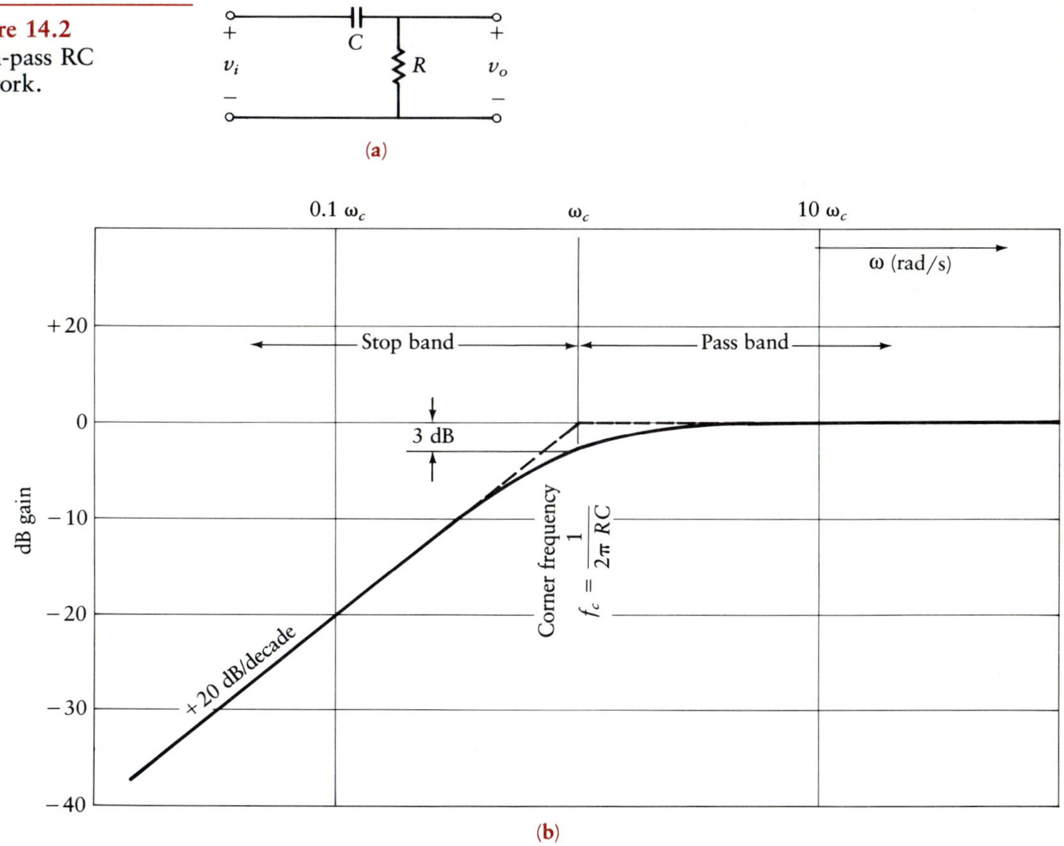

Figure 14.2 High-pass RC network.

$$v_i(t) = Vu(t)$$

where $u(t)$ is the unit step function. We can use Laplace transform analysis to find the output (see Appendix C for a review of the Laplace transform method). The Laplace transform of the output is found by multiplying the transfer function by V/s, the Laplace transform of the step input. If the initial conditions are zero, this becomes

$$V_o = \frac{\tau V}{\tau s + 1}$$

The corresponding time function is

$$v_o(t) = Ve^{-t/\tau}u(t) \tag{14.1}$$

The input step function and the output exponential response are shown in Figure 14.3 (for zero initial conditions). Since the equations for a single RC network are governed by a first-order differential equation, the form of the response to a step-function input must be exponential. For example, the voltage functions must be of the form

$$v(t) = A + Be^{-t/\tau} \tag{14.2}$$

The time constant, τ, is given by RC, and the constants A and B are chosen to match the initial and final values. If the initial and final values of $v(t)$ are given by V_i and V_f, respectively, we solve for A and B as follows:

$$v_o(0) = V_i = A + Be^0 = A + B$$

$$v_o(\infty) = V_f = A + Be^{-\infty} = A$$

with the result

$$A = V_f$$

$$B = V_i - V_f$$

The equation for the curve is then

$$v(t) = V_f + (V_i - V_f)e^{-t/\tau}, \qquad t > 0 \tag{14.3}$$

Equation (14.3) provides the engineer with a simple method for developing these time responses. As an example of the use of equation (14.3), let us once again find the solution for the high-pass network of Figure 14.1 when the input is the step function shown in Figure 14.3.

The initial value of the output voltage is V, since the capacitor voltage cannot change instantaneously when the input jumps from 0 to V volts. The final value is zero since the capacitor looks like an open circuit for a dc input. Therefore we have $V_i = V$ and $V_f = 0$. Substituting these values into equation (14.3) yields

$$v_o(t) = Ve^{-t/\tau}$$

as we found in equation (14.1). The $u(t)$ is understood.

We now complicate the system by forming the input from a composite of two steps. Consider the input to be a pulse of amplitude V and duration T. This is formed by adding a step of amplitude V at the origin to a step of amplitude $-V$ that is delayed by T seconds, as shown in Figure 14.4(a). We solve for the output waveform in two steps. We find the output for times prior to $t = T$ as done previously. That is, since the system cannot see into the future (it is *causal*), it does not know that another step occurs at $t = T$ until the input reaches that point. Therefore, the output prior to time T is given by

$$v_o(t) = Ve^{-t/\tau}$$

We use this equation to find the output just prior to time T, the time at which the second step is applied.

$$v_o(T^-) = Ve^{-T/\tau}$$

The output voltage of the high-pass network of Figure 14.1 *can* jump instantaneously, since the output voltage is measured across a resistor, R. Hence, we use the notation T^- to indicate the time just prior to the transition time. At time $t = T$, the input voltage jumps negatively by V volts. Since the capacitor

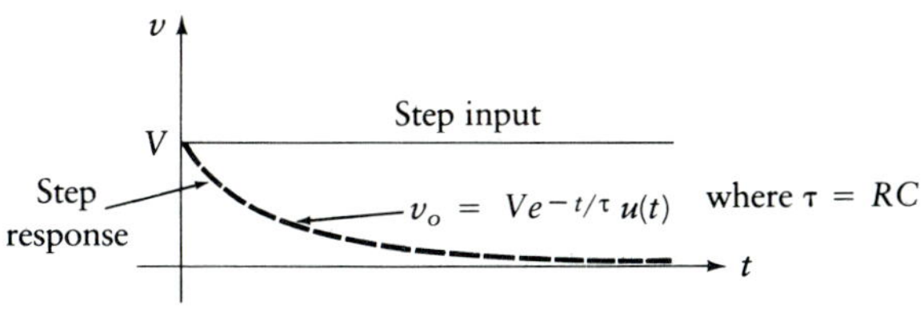

Figure 14.3 Step response of high-pass network.

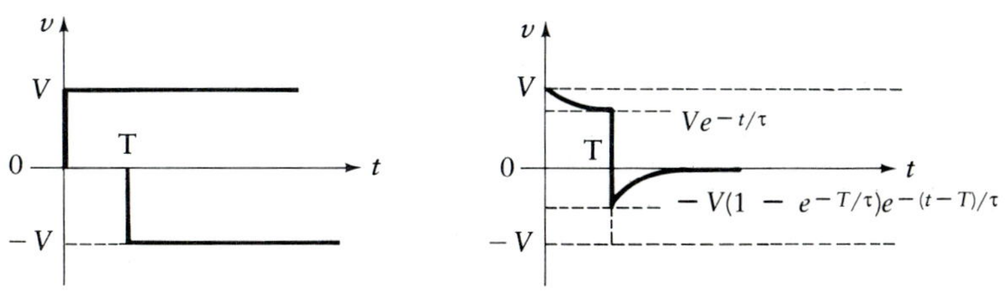

(a) Input (b) Output

Figure 14.4 Single pulse formed from two step functions.

voltage cannot change instantaneously and the sum of voltages around the loop is zero, the output must jump down by V volts to equal

$$v_o(T^+) = v_o(T^-) - V = V(e^{-T/\tau} - 1)$$

This forms the initial value for the second portion of the output. The final value is zero. Therefore, we use equation (14.3) in a shifted version, as follows:

$$v_o(t) = V_f + (V_i - V_f)e^{-(t-T)/\tau}$$

You should verify that this equation reduces to V_i at $t = T$ and to V_f as t approaches infinity. We now find initial and final values for this segment.

$$V_i = V(e^{-T/\tau} - 1)$$

$$V_f = 0$$

The equation for the second portion of the output is then given by

$$v_o(t) = -V(1 - e^{-T/\tau})e^{-(t-T)/\tau} \tag{14.4}$$

The composite output is plotted as Figure 14.4(b).

Before extending this single-pulse result to a pulse train input, we briefly consider the ramp input with slope α, as shown in Figure 14.5(a). We find the response of the RC network using Laplace transforms. The transform of the ramp is α/s^2. Therefore,

$$V_o(s) = \frac{\alpha}{s^2}\frac{s\tau}{s\tau + 1} = \alpha\frac{1}{s(s + 1/\tau)}$$

The inverse Laplace transform of this function is given by

$$v_o(t) = \alpha\tau(1 - e^{-t/\tau})u(t) \tag{14.5}$$

This is shown in Figure 14.5(b).

Figure 14.5
Ramp response of high-pass RC network.

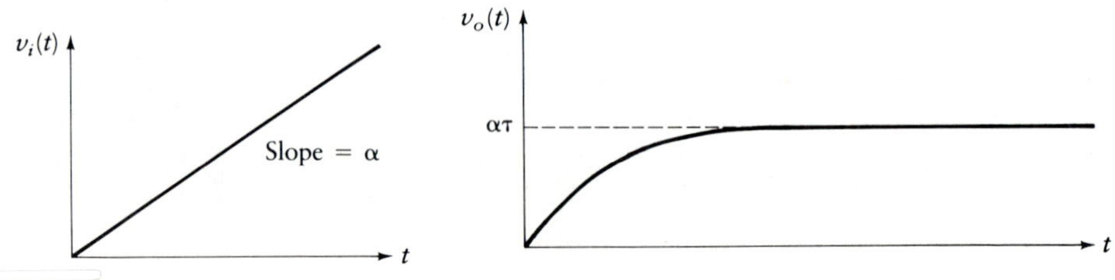

(a) Ramp input (b) Output

Example 14.1

The ramp of Figure 14.5(a) forms the input to a high-pass RC filter with a lower 3-dB frequency of 10^4 rad/s. The slope of the input ramp is 10 V/s.

a. Find the steady-state output voltage.
b. How long does it take the output to reach 90% of its steady-state value?
c. Sketch the output voltage waveform.

SOLUTION

a. Since the 3-dB frequency is the reciprocal of the time constant,

$$\tau = 10^{-4}$$

The steady-state output is found from equation (14.5) or Figure 14.5 to be $\alpha\tau$. Therefore

$$v_o\big|_{t\to\infty} = \alpha\tau = (10\ \text{V/s})\,(10^{-4}\ \text{s}) = 10^{-3}\ \text{V}$$

b. The output voltage will reach 90% of its steady-state value when

$$\alpha\tau(1 - e^{-t/\tau}) = 0.9\alpha\tau$$

$$t = -\tau\ \ln(0.1) = 0.23\ \text{ms}$$

Note that this result is independent of α.

c. The output voltage waveform is shown in Figure 14.6.

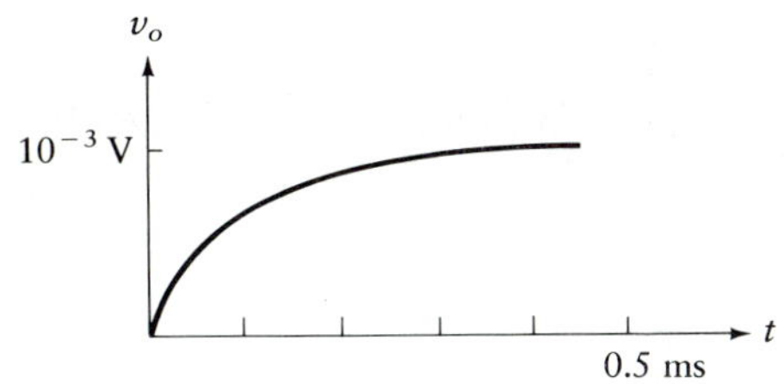

Figure 14.6
Output voltage for Example 14.1.

14.1.1 Steady-State Response—High-Pass Network

We now have the necessary background and tools to consider the periodic pulse input as shown in Figure 14.7(a). We apply this waveform to the high-pass network of Figure 14.2. Without making some logical assumptions, the solution of this circuit analysis problem is tedious. Without such assumptions, we would start with the first pulse and repeat the analysis of Figure 14.4. But then, when the second pulse comes along, we have a new initial value for the voltage and we must repeat the analysis for this second pulse. This analysis

Figure 14.7
High-pass RC filter response to periodic waveform.

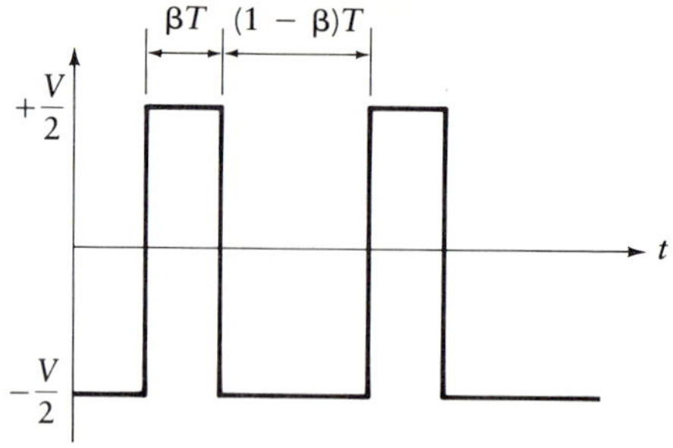

(a) Input

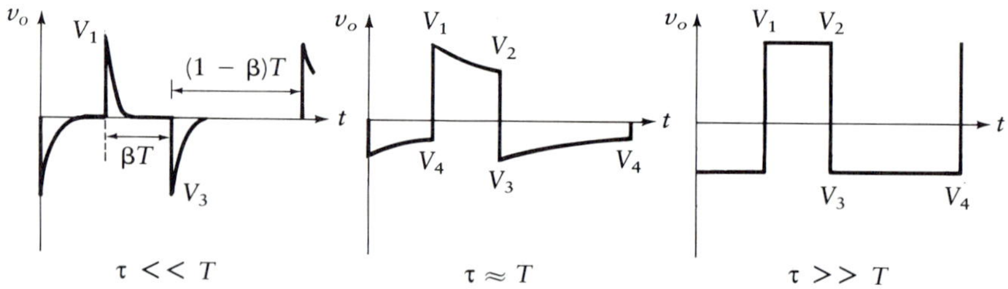

(b) Output

would be repeated for each input pulse. We would know that steady state is reached only if the initial values no longer change from those at the start of the previous cycle. That is, we must continue solving for the pulse response until the transient dies out. Of course, the transient is exponential, so it may take a long time to die out. We can find a better way to evaluate the steady-state response.

We assume that the output resembles the waveform shown in Figure 14.7(b). (We will soon see why three curves are shown, but for now concentrate on the center curve, labeled $\tau \approx T$.) We can make this assumption since the input remains constant over portions of the time axis, and during these periods the output must exponentially approach zero. We know that steady state has been reached when the values of the voltages repeat for each cycle. That is, we go from V_1 to V_2 and then to V_3 and exponentially decay to V_4. We have completed a cycle when we go from V_4 back to the beginning. If the curve returns to V_1, we have reached steady state. This condition is used to develop the steady-state solution.

Let us examine the case in which the time constant and the period of the input are of the same order of magnitude. This is represented by the second sketch in Figure 14.7(b). The output decays by a noticeable amount but does not reach zero between transitions. The relationship between V_1 and V_2 is found from equation (14.3). The voltage V_2 results from an exponential relationship starting at V_1 and decaying toward zero for βT seconds. This is written as follows:

$$V_2 = V_1 e^{-\beta T/\tau} \tag{14.6a}$$

When we arrive at V_2, the input voltage changes by $-V$ volts, and since the voltage across the capacitor cannot change instantaneously, V_2 and V_3 differ by V.

$$V_3 = V_2 - V \tag{14.6b}$$

We find V_4 from an exponential that starts at V_3 and decays toward zero for $(1 - \beta)T$ seconds.

$$V_4 = V_3 e^{-(1-\beta)T/\tau} \tag{14.6c}$$

When we reach V_4, the input voltage jumps by $+V$ volts and again, since the voltage across the capacitor cannot change instantaneously, the output voltage must also jump by $+V$ volts. We know that steady state is achieved if $V_4 + V$ is equal to V_1, the voltage where we started at the beginning of the cycle. Hence,

$$V_1 = V_4 + V \tag{14.6d}$$

We now have four equations (equations (14.6a) through (14.6d)) in four unknowns. These equations are easily solved for the four voltage values.

$$V_1 = \frac{V(1 - e^{-(1-\beta)T/\tau})}{1 - e^{-T/\tau}} \tag{14.7a}$$

$$V_2 = V_1 e^{-\beta T/\tau} = V\frac{e^{-\beta T/\tau} - e^{-T/\tau}}{1 - e^{-T/\tau}} \tag{14.7b}$$

$$V_3 = V_2 - V = V\frac{e^{-\beta T/\tau} - 1}{1 - e^{-T/\tau}} \tag{14.7c}$$

and

$$V_4 = V_3 e^{-(1-\beta)T/\tau} = \frac{Ve^{-T/\tau}(1 - e^{\beta T/\tau})}{1 - e^{-T/\tau}} \tag{14.7d}$$

Let us now find the shape of the waveform when we change the relative size of τ with respect to T.

If the time constant is much less than the period of the input ($\tau << T$), the exponential has time to decay almost to zero between transitions, and the output resembles the first sketch in Figure 14.7(b). Let $\tau << T$ in equation (14.7) to obtain

$$V_1 \to V$$

$$V_2 \to 0$$

$$V_3 \rightarrow -V$$

$$V_4 \rightarrow 0$$

Note that the peak-to-peak output voltage is now 2 times V.

If $\tau >> T$, note that equations (14.7a) and (14.7b) become equivalent, and $V_1 = V_2$. Similarly, equations (14.7c) and (14.7d) become equivalent, and $V_3 = V_4$, which is the shape of the right chart in Figure 14.7(b). We can find the values of all four voltages by noting that the average value of the output voltage must be zero. This is true since the series capacitor cannot pass direct current. Because of this, the area of the output curve above 0 V must equal the area of the output curve below 0 V. Hence, when the time constant is much larger than the input period, the output hardly has any time to decay between input transitions, and the output resembles the third sketch in Figure 14.7(b). Since $V_2 = V_1$ and $V_4 = V_3$, and since the height of the transitions must be V, we have

$$V_3 + V = V_1$$

Since the average value of the output is zero, we match negative and positive areas to find

$$\beta T(V_1) = -(1 - \beta)TV_3$$

Finally,

$$V_1 = V_2 = V(1 - \beta) \tag{14.8}$$

$$V_3 = V_4 = -V(\beta) \tag{14.9}$$

Note that, for this case, the output waveform is similar to that of the input but with a voltage shift.

In designing a system, the choice of time constant is often critical. If narrow pulses are required for a triggering operation, the output must look similar to the first sketch in Figure 14.7(b). We therefore design with a time constant much less than the period of the input. On the other hand, if we wish to reproduce the input at the output, we should choose a time constant much larger than the period of the input.

Example 14.2

A symmetrical square wave with a peak-to-peak voltage of 10 V is applied to a high-pass RC filter with $R = 1\ \text{k}\Omega$ and $C = 1\ \mu\text{F}$. Sketch the output waveform and identify key voltages. Assume that the frequency of the square wave is

a. 50 Hz
b. 500 Hz
c. 5 kHz

SOLUTION Note that since the input waveform is symmetrical, it spends equal portions of the period at positive and at negative values. Therefore $\beta = \frac{1}{2}$ and the previous equations are simplified.

We first find the time constant

$$\tau = RC = (1 \text{ k}\Omega)(1 \ \mu\text{F}) = 1 \text{ ms}$$

and then compare this time constant to the period of the square wave.

a. $T = 1/f = 20$ ms, which is much larger than the time constant,

$$\tau << T$$

Therefore,

$$V_1 = V = 10 \text{ V}$$

$$V_3 = -V = -10 \text{ V}$$

The result is shown in Figure 14.8(a). Note that the peak-to-peak voltage is 20 V.

b. For the second input condition,

$$T = \frac{1}{f} = \frac{1}{500} = 2 \text{ ms}$$

For this case, the input period and the time constant are of the same order of magnitude. Note that $V_1 = -V_3$ and $V_2 = -V_4$ due to symmetry. We solve for the four voltages from equations (14.7a) through (14.7d) with $\beta = \frac{1}{2}$ to find

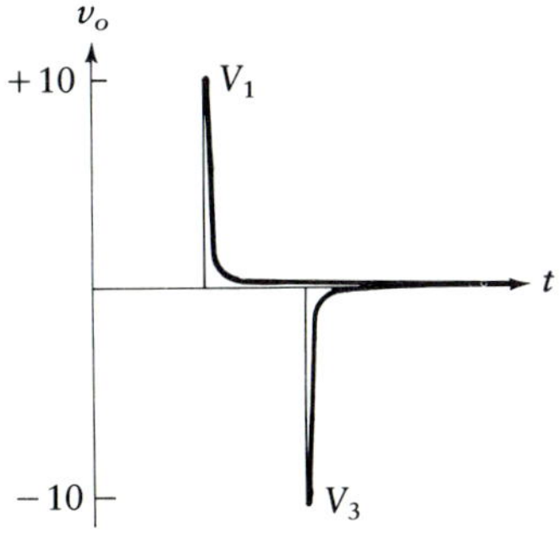

(a) $T >> \tau$

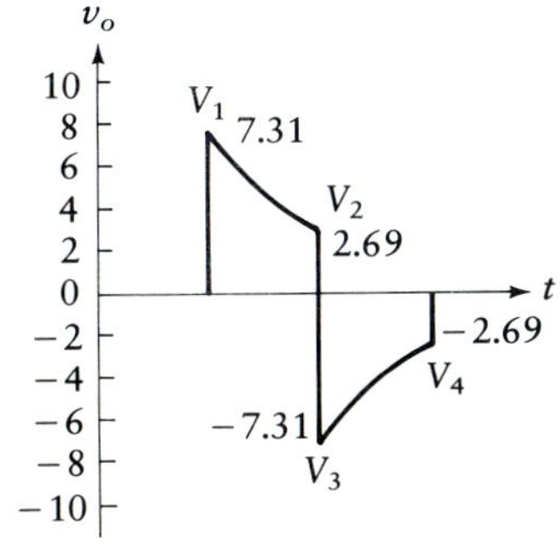

(b) $T \approx \tau$

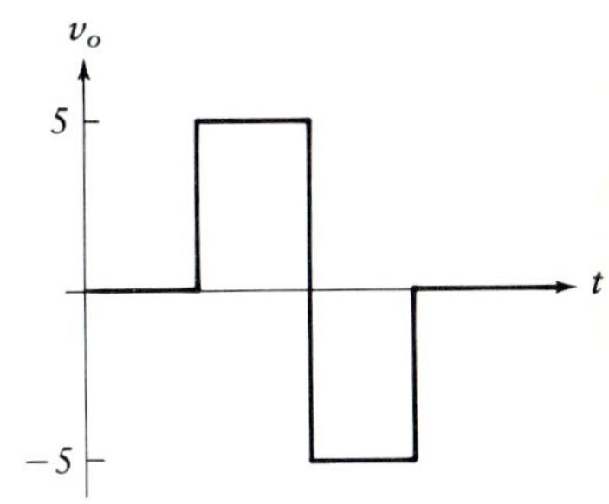

(c)

Figure 14.8
Output waveforms for Example 14.2.

$$V_1 = -V_3 = \frac{V(1 - e^{-T/2\tau})}{1 - e^{-T/\tau}} = 7.31 \text{ V}$$

$$V_2 = -V_4 = V_1 e^{-T/2\tau} = 2.69 \text{ V}$$

This waveform is shown in Figure 14.8(b). The peak-to-peak voltage is 14.62 V for this case.

c. For the third input condition,

$$T = \frac{1}{f} = \frac{1}{5000} = 0.2 \text{ ms}$$

This represents a case where the time constant is much larger than the period of the input. Therefore,

$$V_1 = V_2 = \frac{V}{2} = 5 \text{ V}$$

$$V_3 = V_4 = \frac{V}{2} = -5 \text{ V}$$

This waveform is plotted in Figure 14.8(c). Note that it resembles the input waveform. Here the peak-to-peak voltage is only 10 V.

Drill Problems

The asymmetric input signal of Figure 14.7(a) with $\beta = \frac{1}{3}$, $T = 10$ ms, and $V = 10$ V forms the input to the high-pass network of Figure 14.2. Calculate the four important voltages and sketch the output waveform for the conditions given in Drill Problems D14.1 through D14.3.

D14.1 Let $T/\tau = 100$.

Ans: $V_1 = 10$ V; $V_2 = 0$ V; $V_3 = -10$ V; $V_4 = 0$ V

D14.2 Let $T/\tau = 1$.

Ans: $V_1 = 7.7$ V; $V_2 = 5.52$ V; $V_3 = -4.48$ V; $V_4 = -2.3$ V

D14.3 Let $T/\tau = 0.01$.

Ans: $V_1 = 6.68$ V; $V_2 = 6.66$ V; $V_3 = -3.34$ V; $V_4 = -3.32$ V

14.2 LOW-PASS RC NETWORK

We now reverse the situation of the previous section by taking the output across the capacitor instead of the resistor. The result is the low-pass RC network shown in Figure 14.9. This network passes low frequencies and attenuates high frequencies. The sinusoidal steady-state amplitude response is shown in Figure 14.10, where the solid line is the actual curve and the dashed line is the Bode plot asymptotic approximation. The transfer function for the circuit of Figure 14.9 is given by

$$\frac{V_o}{V_i} = \frac{1}{1 + s\tau}$$

where the time constant, τ, is RC. Let us turn our attention to the pulse train input shown in Figure 14.11(a). Both the square-wave input and the output are shown on the same set of axes. The input square wave is no longer symmetrical around the zero axis. The positive amplitude is shown as αV and the negative as $-(1 - \alpha)V$. The total swing is therefore V volts.

The output is taken across the capacitor, and therefore it cannot change instantaneously. This "continuity" constraint on the output simplifies the solution, since there are only two unknown voltages, V_1 and V_2, as shown in Figure 14.11(a). We first examine the region between $t = 0$ and $t = \beta T$. The initial value of this exponential waveform is

Figure 14.9
Low-pass RC network.

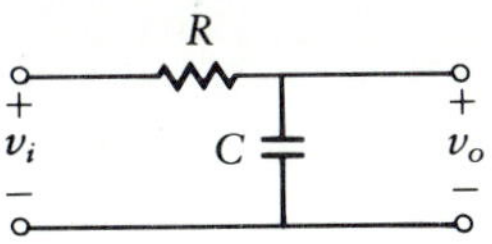

Figure 14.10
Amplitude response of low-pass network.

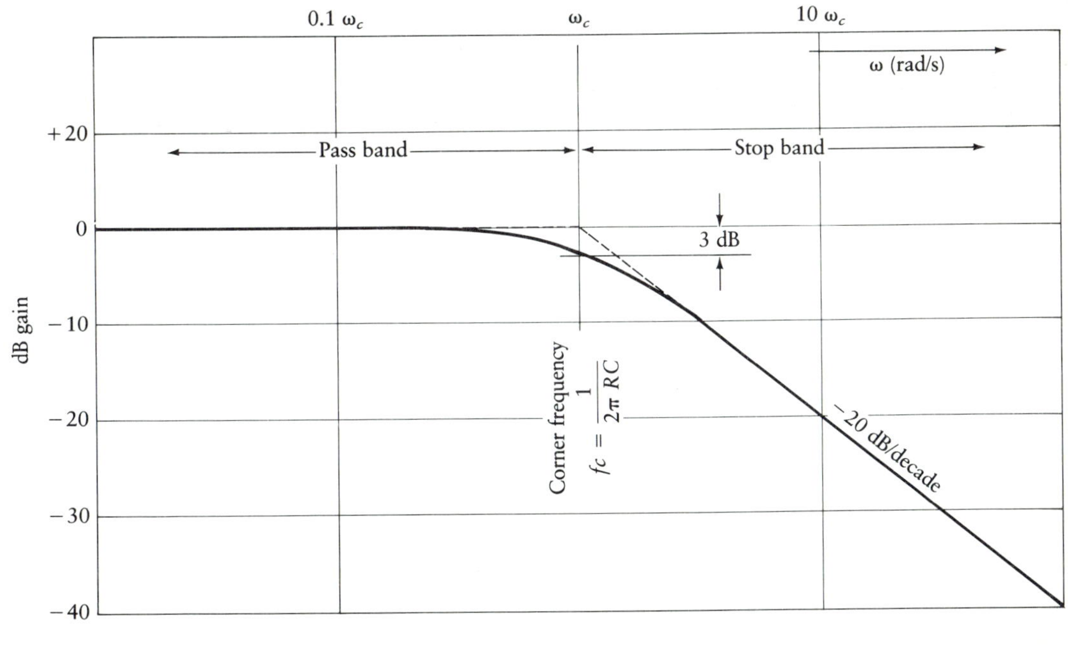

Figure 14.11
Input and output of low-pass filter.

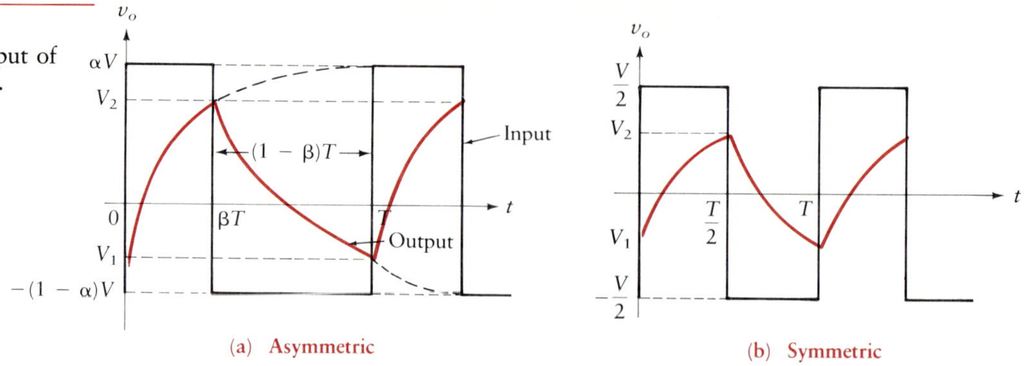

$$V_i = V_1$$

The final value is the value the voltage would attain if given infinite time. This is seen to be the value of the input, or

$$V_f = \alpha V$$

These two values are used in the general expression of equation (14.3) to find the waveform in the first interval:

$$v_o(t) = \alpha V + (V_1 - \alpha V)e^{-t/\tau}$$

Evaluating this expression at $t = \beta T$ yields an expression for V_2 as follows:

$$v_o(\beta T) = V_2 = \alpha V + (V_1 - \alpha V)e^{-\beta T/\tau} \tag{14.10}$$

For the next region between $\beta T < t < T$, we have

$$V_f = -(1 - \alpha)V$$

$$V_i = V_2$$

so

$$v_o(t) = -(1 - \alpha)V + (V_2 + V - \alpha V)e^{-(t-\beta T)/\tau}$$

Evaluating this expression at $t = T$ yields the second equation in the two unknowns.

$$v_o(T) = V_1 = -(1 - \alpha)V + (V_2 + V - \alpha V)e^{-(1-\beta)T/\tau} \tag{14.11}$$

We next solve equations (14.10) and (14.11) to obtain expressions for V_1 and V_2.

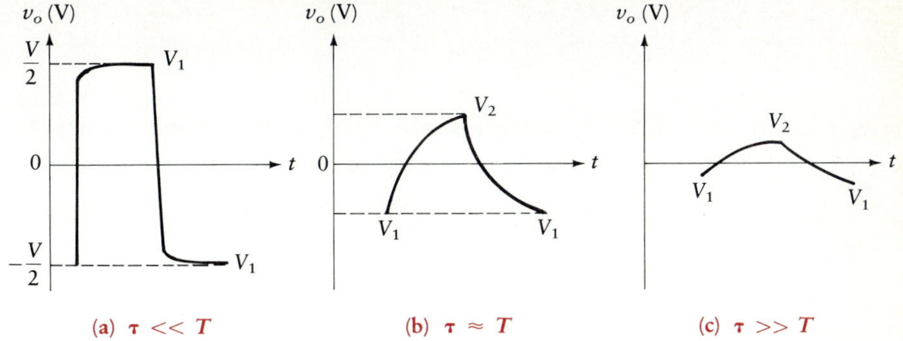

Figure 14.12 Output of low-pass RC filter with symmetrical square-wave input.

If the input waveform is symmetrical, several simplifications result, as follows:

$$\alpha = \beta = \tfrac{1}{2} \qquad \text{and} \qquad V_1 = -V_2$$

Under these conditions, we find

$$-V_1 = V_2 = \frac{V(1 - e^{-T/2\tau})}{2(1 + e^{-T/2\tau})} \tag{14.12}$$

This waveform is shown in Figure 14.11(b).

When a symmetrical square waveform is applied to the low-pass RC network of Figure 14.10, the output is as shown in Figure 14.12. Note the dependence of the output on the relationship between the time constant, τ, and the period of the input, T. The critical voltages in the output waveforms of Figure 14.12 are as follows:

1. For a time constant much less than the input period,

$$-V_1 = +V_2 = \frac{V}{2} \tag{14.13}$$

2. For a time constant of the same order as the input period, we use equation (14.12).
3. For a time constant much greater than the input period, the same result obtains as in case 2. It should, however, be realized that V_1 and V_2 will be small values.

If the input waveform is no longer symmetrical about the zero axis, α is not equal to $\frac{1}{2}$. It is not necessary to rederive all of the expressions. Since there is no capacitor blocking the dc level, the results of Figure 14.12 are simply shifted in the vertical direction by the average value of $v_i(t)$.

Example 14.3

A symmetric square wave, with a peak-to-peak voltage of 5 V, is applied to the low-pass RC filter of Figure 14.9 with $R = 10\ \text{k}\Omega$ and $C = 0.1\ \mu\text{F}$. Sketch the output voltage waveform and identify key voltages for the following three values of the square-wave frequency:

a. 50 Hz
b. 500 Hz
c. 5 kHz

SOLUTION The time constant of the circuit is

$$\tau = RC = (10\ \text{k}\Omega)(0.1\ \mu\text{F}) = 1\ \text{ms}$$

a. The period of the input is

$$T = \frac{1}{f} = 20\ \text{ms}$$

In this case, the time constant is much less than the period, so

$$-V_1 = +V_2 = \frac{V}{2} = 2.5\ \text{V}$$

The output is as shown in Figure 14.12(a).

b. The period of the input is

$$T = \frac{1}{f} = 2\ \text{ms}$$

In this case, the time constant is of the same order of magnitude as the input, so we solve for the two key voltages,

$$-V_1 = V_2 = \frac{V}{2}\frac{1 - e^{-T/2\tau}}{1 + e^{-T/2\tau}} = 2.5\frac{1 - e^{-1}}{1 + e^{-1}} = 1.155\ \text{V}$$

The resulting waveform is shown in Figure 14.12(b).

c. The period is given by

$$T = \frac{1}{f} = 0.2\ \text{ms}$$

In this case, the time constant is much greater than the period and the output is close to zero, as shown in Figure 14.12(c). The actual values of the key voltages are found as in part b of this example.

$$V_2 = \frac{V}{2}\frac{1 - e^{-T/2\tau}}{1 + e^{-T/2\tau}} = 2.5\frac{1 - e^{-0.1}}{1 + e^{-0.1}} = 0.125 \text{ V}$$

$$V_1 = -V_2 = -0.125 \text{ V}$$

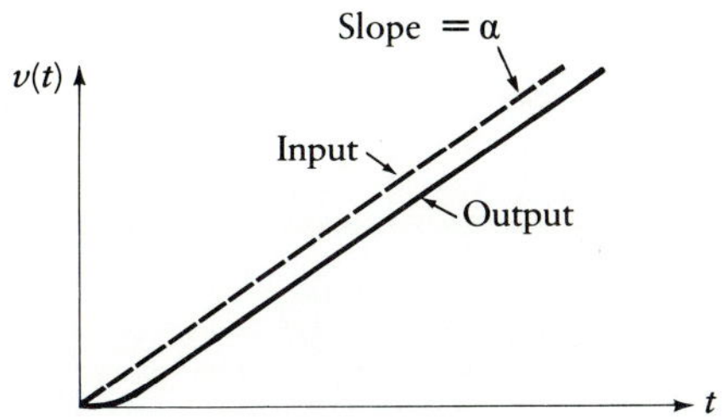

Figure 14.13
Low-pass RC filter ramp response.

Let us examine the response of the low-pass filter to the ramp input shown in Figure 14.13. The output is found in the same manner as for the high-pass filter and is given by

$$v_o(t) = (-\alpha\tau + \alpha t + \alpha\tau e^{-t/\tau})u(t) \tag{14.14}$$

This is also shown in Figure 14.13. Once again, α is the slope of the input ramp in volts per second.

Example 14.4

The ramp input of Figure 14.14 is applied to a low-pass RC filter with a corner frequency of 5000 rad/s. The slope of the input ramp is 10 V/s. Calculate and sketch the output voltage waveform.

SOLUTION The time constant is the inverse of the 3-dB frequency, so

$$\tau = \frac{1}{5000} \text{ s}$$

The output waveform is written directly from equation (14.14).

$$\begin{aligned} v_o(t) &= (-\alpha\tau + \alpha t + \alpha\tau e^{-t/\tau})u(t) \\ &= (-0.002 + 10t + 0.002e^{-5000t})u(t) \end{aligned}$$

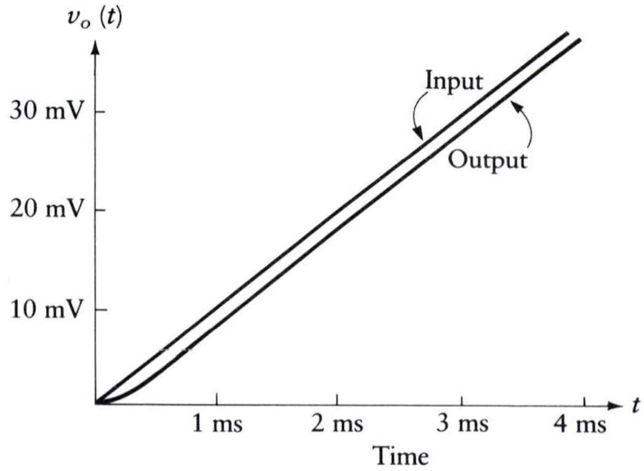

Figure 14.14 Response for Example 14.4.

This is also plotted in Figure 14.14.

14.3 DIODES

Recall from Chapter 1 that diodes are often coupled with RC networks in order to achieve desirable results. In analyzing such circuits, we use the piecewise linear diode model shown in Figure 14.15. When the diode is forward-biased and therefore conducting, it is represented by a small forward resistance, R_f. (We are neglecting the forward voltage, V_γ.) When the diode is reverse-biased, it is represented by a large reverse resistance, R_r.

14.3.1 Steady-State Response of Diode Circuits to Pulse Train

Circuits used in digital applications often combine resistors, capacitors, and diodes. One such configuration is shown in Figure 14.16(a).

If the diode is ideal, the circuit acts as a *clamper.* Clamping is a shifting operation, where the amount of the voltage shift depends on the actual waveform. The capacitor cannot discharge, so it charges to the peak value of the input. The output is therefore "clamped" to a zero voltage level. (Refer to Chapter 1

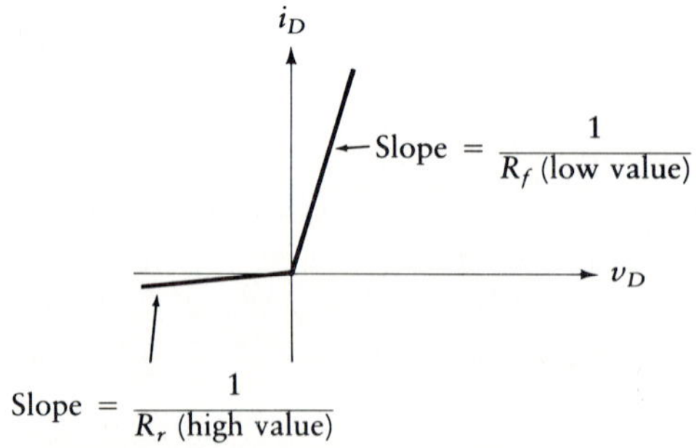

Figure 14.15 Diode model.

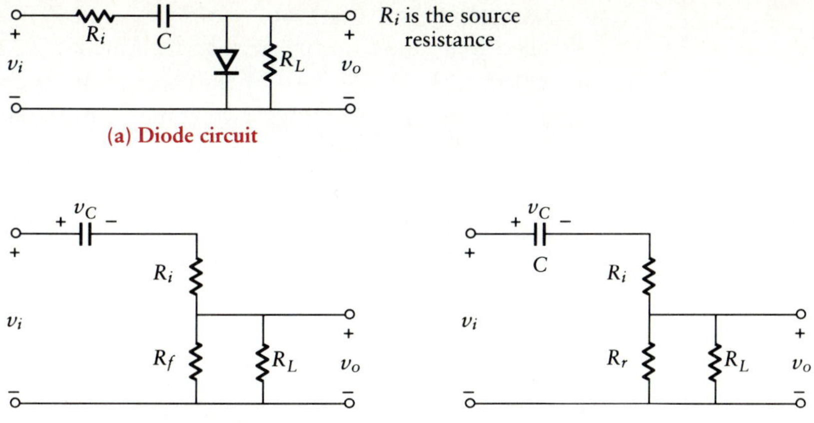

(a) Diode circuit (b) Diode conducting (c) Diode not conducting

Figure 14.16 Steady-state response to pulse train.

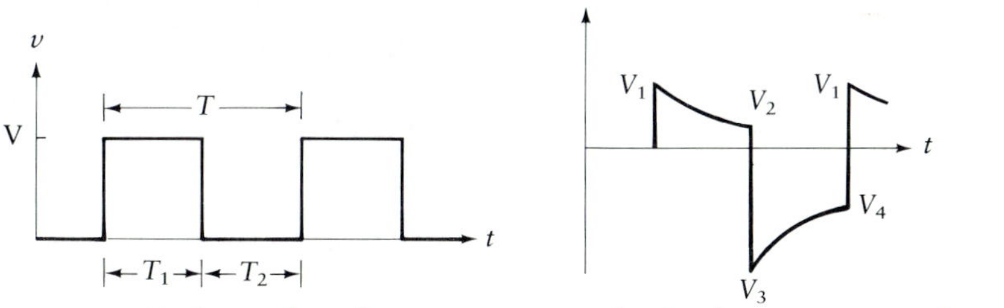

(a) Input pulse train (b) Steady-state output waveform

Figure 14.17 Steady-state response of diode circuit.

for a more complete discussion of clampers.) The circuit of Figure 14.16 is also known as a *voltage restorer.* The capacitor charges to the magnitude of the positive peak voltage of the input signal. Subsequently, when the diode turns off, the capacitor remains charged to this peak voltage.

We continue the analysis assuming that the diode is piecewise linear with $V_\gamma = 0$. The pulse train, as shown in Figure 14.17(a), forms the input to this circuit. The circuit behaves similarly to the RC circuit analyzed earlier in this chapter. The difference is that two different time constants apply, depending on the state of the diode. The steady-state output is shown in Figure 14.17(b).

Figure 14.16(b) shows the equivalent circuit under the condition that the diode is forward-biased. This situation occurs whenever the output is positive. Figure 14.16(c) shows the comparable situation for negative outputs, when the diode is reverse-biased. The two time constants, for the forward- and reverse-biased conditions, are therefore given by

$$\tau_f = (R_f + R_i)C$$

$$\tau_r = (R_r + R_i)C$$

Note that since $R_r >> R_f$, then $\tau_r >> \tau_f$.

We need four equations in order to find V_1, V_2, V_3, and V_4. As in the earlier circuit without the diode, two of these equations come from the exponential decay relationship and the other two come from the magnitude of the voltage transitions. The exponential decay relationships are

$$V_2 = V_1 \exp\left(\frac{-T_1}{\tau_f}\right) \tag{14.15a}$$

$$V_4 = V_3 \exp\left(\frac{-T_2}{\tau_r}\right) \tag{14.15b}$$

At the instant before the input voltage drops by V volts, the output is at V_2. When the input drops, the capacitor voltage cannot change instantaneously. Therefore, the voltage across the series combination of R_i and R_f must drop by V volts. If the resistor value did not change from R_f to R_r during this transition, we could simply use a voltage-divider formula to find the relationship between V_3 and V_2. Since the diode does indeed change state during this transition, we must find the new output by examining the capacitor voltage. The capacitor voltage just prior to this transition is found by writing a loop equation for the circuit of Figure 14.16 and realizing that

$$v_i = V$$

$$v_o = V_2$$

The capacitor voltage is

$$v_C = V - \frac{V_2(R_f + R_i)}{R_f}$$

At the instant after the input voltage drops to zero, the capacitor voltage is still at this value. The new value of the output is found by writing a loop equation for the circuit in Figure 14.16(c). This equivalent is used since the diode is now reverse-biased. The result is

$$V_3 = \frac{-R_r}{R_r + R_i}\left[V - V_2\left(\frac{R_f + R_i}{R_f}\right)\right] \tag{14.15c}$$

We use a similar approach to find the relationship between V_1 and V_4. The result is (you should verify this expression before continuing)

$$V_1 = \frac{R_f V + V_4 R_f(R_r + R_i)/R_r}{R_f + R_i} \tag{14.15d}$$

The four equations, (14.15a), (14.15b), (14.15c), and (14.15d), are sufficient to solve for the unknown reference points. They are rewritten as equations (14.16) for easy reference.

$$V = \frac{R_f + R_i}{R_f}V_1 - \frac{R_r + R_i}{R_r}V_4 \tag{14.16a}$$

$$V = \frac{R_f + R_i}{R_f}V_2 - \frac{R_r + R_i}{R_r}V_3 \tag{14.16b}$$

$$V_2 = V_1 \exp\left(\frac{-T_1}{\tau_f}\right) \tag{14.16c}$$

$$V_4 = V_3 \exp\left(\frac{-T_2}{\tau_r}\right) \tag{14.16d}$$

Example 14.5

Given the circuit of Figure 14.16(a) with $R_i = 100\ \Omega$, $R_f = 100\ \Omega$, $R_r = 1\ \text{M}\Omega$, $R_L = 10\ \text{k}\Omega$, $C = 1\ \mu\text{F}$, and $V_\gamma = 0$, find the critical voltages of the output waveform if the input is as shown in Figure 14.18. Use a piecewise linear diode model.

SOLUTION Since $R_L >> R_f$ and $R_r >> R_L$,

$$R_L \parallel R_f \approx R_f$$

$$R_r \parallel R_L \approx R_L$$

We now find the two time constants.

$$\tau_f = (R_f + R_i)C = 200\ \mu\text{s}$$

$$\tau_r = (R_L + R_i)C = 10\ \text{ms}$$

Equation (14.16) yields

$$10 = 2V_1 - 1.01V_4 \tag{14.17a}$$

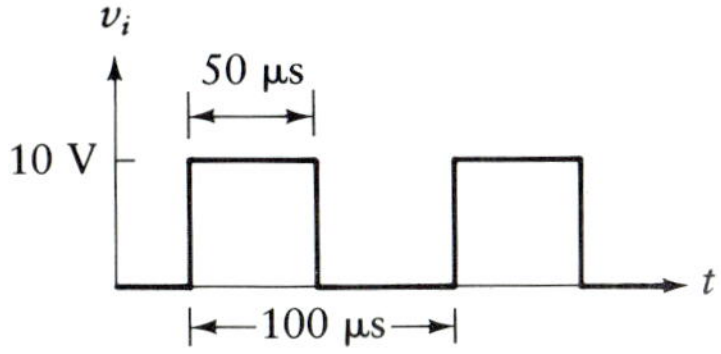

Figure 14.18 Input waveform for Example 14.5.

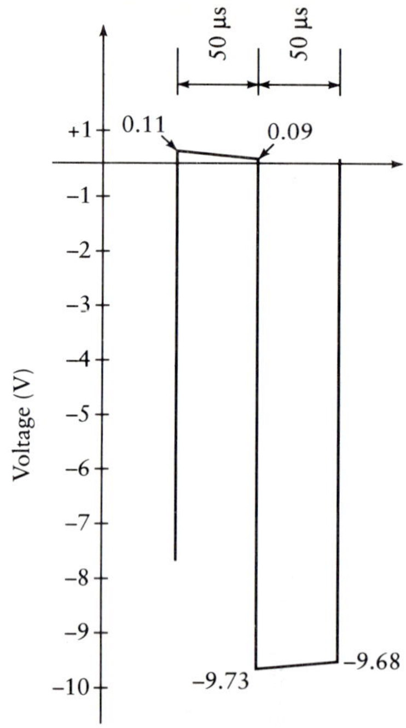

Figure 14.19 Solution to Example 14.5.

$$10 = 2V_2 - 1.01V_3 \tag{14.17b}$$

$$V_2 = V_1 \exp\left(\frac{-50\ \mu\text{s}}{200\ \mu\text{s}}\right) = V_1 e^{-1/4} = 0.78V_1 \tag{14.17c}$$

$$V_4 = V_3 \exp\left(\frac{-50\ \mu\text{s}}{10\ \text{ms}}\right) = V_3 e^{-0.005} = 0.995V_3 \tag{14.17d}$$

Solving these four equations for the four unknown voltages yields

$$V_1 = 0.11\ \text{V}$$

$$V_2 = 0.09\ \text{V}$$

$$V_3 = -9.73\ \text{V}$$

$$V_4 = -9.68\ \text{V}$$

The resulting waveform is plotted in Figure 14.19. Note that both time constants are large compared to the period of the input, so the exponentials do not decay appreciably between transitions. Further note that the derivation of this section neglected the diode forward voltage, V_γ. Since V_1 and V_2 turned out to be of the same order of magnitude as V_γ, the model used in this section is less accurate for this example.

14.4 TRIGGER CIRCUITS

Binary digital systems operate with discrete time signals. That is, since the information is contained in sequences of binary digits, each of these bits occurs at a specific time. *Timing* is therefore extremely important, and transitions must occur at carefully controlled points in time. Many digital circuits require a trigger pulse to be generated, and this trigger pulse is used to control the timing.

Trigger pulses are usually generated at either the leading or trailing edge of a pulse train. As one example, consider the *positive-edge trigger* of Figure 14.20(a). The input to the circuit is a pulse that has a transition between 0 and 5 V. This step function is applied to the high-pass RC network. Equation (14.3) can be used to find the output for $t > 0$.

$$v_o(t) = Ve^{-t/\tau}u(t)$$

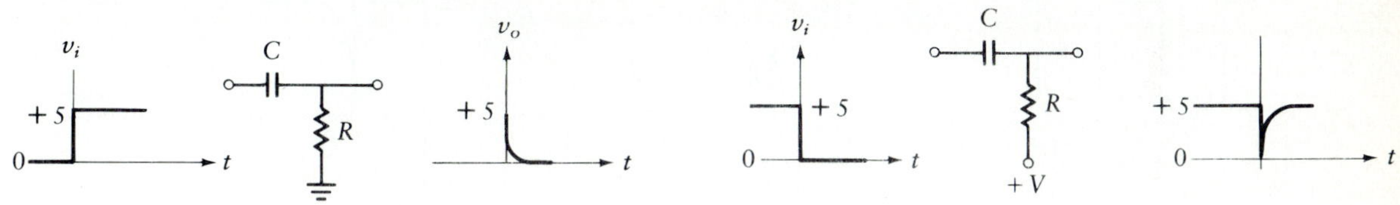

(a) Positive-edge detection

(b) Negative-edge detection

Figure 14.20
Trigger circuits.

In this case, $\tau = RC$ and $V = 5$ V, resulting in a positive-going trigger pulse. The trigger pulse is narrow, with the width dependent on the RC time constant. It can be used for accurate timing purposes.

The *negative-edge trigger* is shown in Figure 14.20(b). The output is also shown in the diagram. For $t > 0$, the output is found from equation (14.3) to be

$$v_o(t) = V(1 - e^{-t/\tau})$$

The exponential decay of both of these outputs can be shortened by reducing the time constant. We shall explore methods of making the output pulse even sharper. This is possible using inverters or buffers, both of which are discussed in the next chapter.

14.4.1 Pulse Train Response

The networks of the previous section detected both a positive and a negative edge of a single step function. We now study a pulse train as the input to an RC circuit. The input and corresponding output are shown in Figure 14.21(a). The output consists of a series of positive- and negative-going impulses of amplitude $\pm V$ volts. Recall from the discussion of Figure 14.7(b) that impulses of

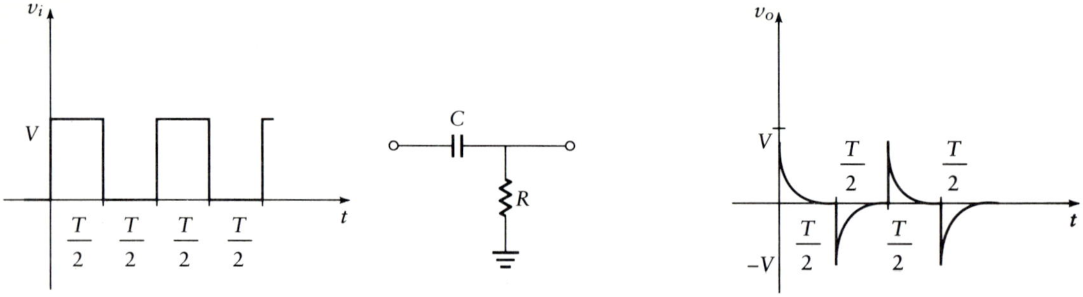

(a) No diode

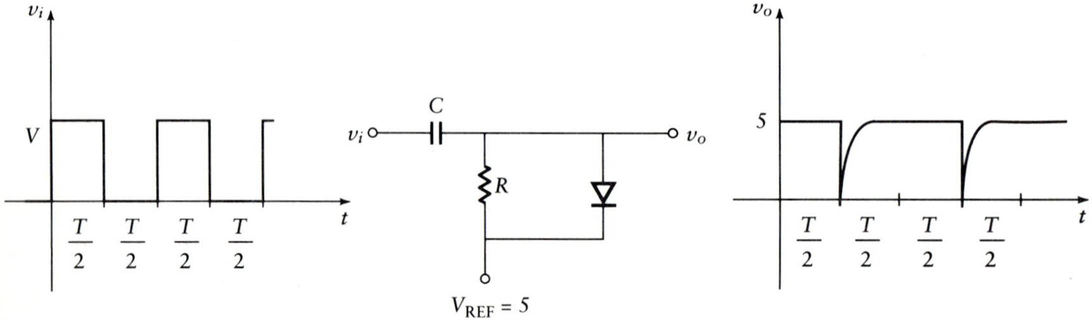

(b) With diode trailing edge

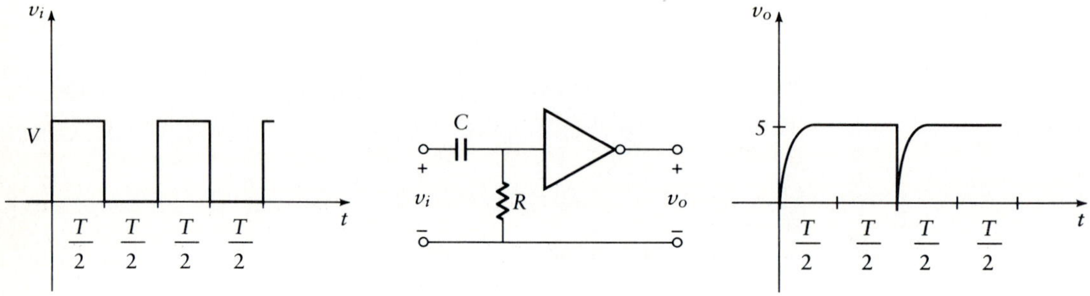

(c) With inverter leading edge

Figure 14.21
Response to a pulse train.

this type occur if the time constant is much smaller than the period of the input. Therefore, the *RC* product must be

$$\tau = RC << T$$

We often require only a negative-going pulse to act as a trigger, so a diode can be used to remove the positive-going pulses. This is shown in Figure 14.21(b). If the diode is reversed, the negative-going pulses are removed and the positive-going pulses remain. The circuit of Figure 14.21(c) will produce the same trigger as Figure 14.21(b), except that no V_{REF} is required. The 7404 inverter shown in this figure is discussed in Chapter 15.

Example 14.6

A symmetrical 60-Hz rectangular pulse train forms the input to a system. Design a circuit to develop a trigger pulse that must be less than 2 ms wide and exists only on the negative-going edges of the input clock pulse train. The waveforms are shown in Figure 14.22(a).

Figure 14.22
Waveforms and circuit for Example 14.6.

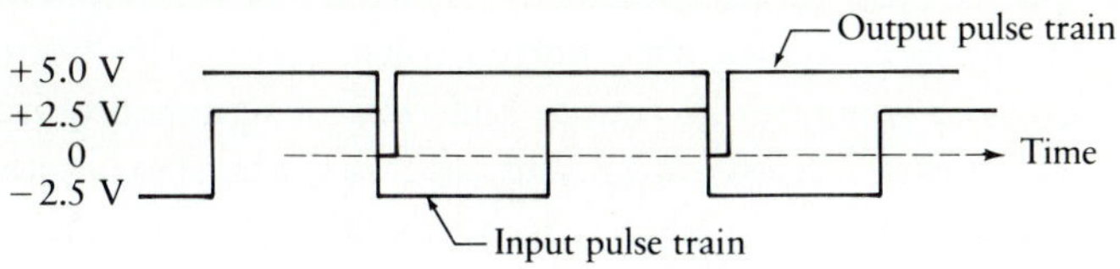

(a) Input and output signals

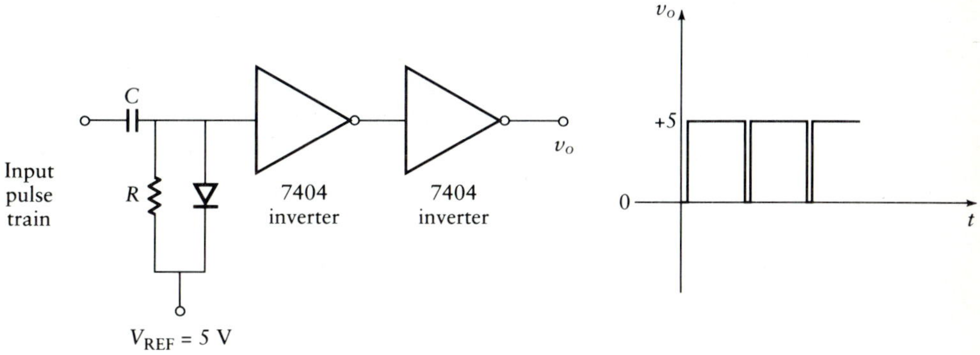

(b) Solution

SOLUTION The period of the input is 16.7 ms, so we want the RC time constant to be much less than this. The time constant must be sufficiently small that the output pulse has decayed almost to zero within 2 ms, as specified in the problem statement. Suppose we choose a time constant of 0.5 ms, which is obtained with a capacitor value of $C = 0.1\ \mu$F and a resistance value of $R = 5$ kΩ. The resulting circuit is shown in Figure 14.22(b).

A *buffer,* which can be formed from two inverters in series, can be used at the output of Figure 14.22(b). The buffer changes this exponential signal to a rectangular pulse and provides a low output impedance. The buffer is also discussed in Chapter 15.

14.5 THE 555 PULSE GENERATOR

Although this chapter is intended to introduce *digital* electronics, the circuits and signals we have been discussing are really analog. They are operating on and producing waveforms that take on a continuum of values rather than simply responding to one of two levels.

In between analog and digital circuits there exists a class of devices that possess the characteristics of both types. Within this class are *timers* and *waveform generators* that find wide use in both analog and digital circuitry. The major characteristic of these circuits is that the time (period) may be set either by an external voltage or by a resistor-capacitor combination. These devices often have external control lines so that the frequency or pulse width may be easily controlled by an external source. When these are implemented as integrated circuits, the internal construction uses both analog and digital circuits to generate the timing and control signals required for operation.

In this section we present several types of relaxation oscillator out of the many that exist. Others can be found in the IC data books published by the manufacturers.

14.5.1 The Relaxation Oscillator

We lead our way into timing circuits by first analyzing the principle of a *relaxation oscillator.* Numerous circuits are designed using this principle. A capacitor is alternately charged and discharged to produce a repetitive waveform. Several schemes are available to cause the capacitor to switch back and forth from the charging mode to the discharging mode.

One example of a relaxation oscillator, shown in Figure 14.23(a), uses an operational amplifier with both negative and positive feedback. This circuit is also called a *free-running* or *astable multivibrator* (most IC data books index this circuit under the label "multivibrator"). We now describe the performance of the astable multivibrator of Figure 14.23(a).

The voltage at the noninverting input of the op-amp is obtained from the output voltage fed back through a resistor divider composed of R_1 and R_2. This is positive feedback, so the amplifier has only two possible outputs: $+E$ or $-E$. If the differential input is positive, the op-amp output saturates at a value near the positive supply voltage. If the differential input is negative, the output saturates at a value near the negative of the supply voltage. The voltage at the inverting input is developed across the capacitor as part of an RC combination.

When the output is at the positive value, the capacitor exponentially charges toward this value with time constant RC. At some point, this increasing inverting input voltage causes the op-amp to switch to the other state, where the output voltage is negative. The capacitor then starts discharging toward this negative value until the differential input once again becomes posi-

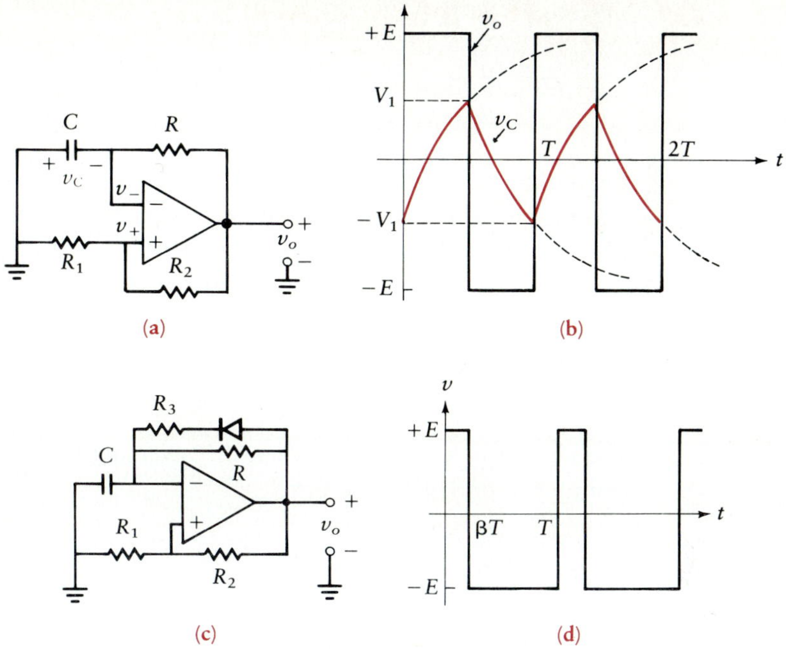

Figure 14.23
Relaxation oscillator.

tive. Notice that the capacitor is alternately charging toward $+E$ and discharging toward $-E$. The positive feedback causes the switching to occur as the output alternates from $+E$ to $-E$.

The output is therefore a square wave as shown in Figure 14.23(b). The exponential waveform in this figure represents the capacitor voltage. We find the equation for the exponential curve by using the initial and final values and the time constant as described earlier in this chapter (see equation (14.3)). For the first segment of the curve between times 0 and $T/2$, the initial value is $-V_1$, the final value is the supply voltage, $+E$, and the time constant is $\tau = RC$. The equation is therefore

$$v_-(t) = E + (-V_1 - E)e^{-t/\tau} \tag{14.18a}$$

For the next segment, between $T/2$ and T, the starting value is V_1 and the final value is $-E$. Note that we are assuming that, in the steady state, the waveform is symmetrical around the zero axis.

$$v_-(t) = -E + (V_1 + E)e^{-(t-T/2)/\tau} \tag{14.18b}$$

Equation (14.18a) and equation (14.18b) are evaluated at the transition time to obtain

$$v_-\left(\frac{T}{2}\right) = V_1 = E + (-V_1 - E)e^{-T/2\tau}$$

Therefore

$$e^{-T/2\tau} = \frac{E - V_1}{E + V_1} \tag{14.19}$$

The unknowns in this equation are T and V_1. We need a second equation to find the period of oscillation. This equation comes from the noninverting input since the transitions occur when the *differential* input goes through zero (i.e., $v_- = v_+$). The noninverting input during the first half-cycle is then given by

$$v_+ = \frac{R_1 E}{R_1 + R_2}$$

The change in state occurs when the inverting input reaches this value. Thus, from equation (14.19) we have

$$e^{-T/2\tau} = \frac{E - ER_1/(R_1 + R_2)}{E + ER_1/(R_1 + R_2)} = \frac{R_2}{R_2 + 2R_1}$$

For simplification, let us set $R_1 = R_2$. We then have

$$e^{-T/2\tau} = 1/3$$

Taking natural logs of both sides yields

$$\frac{-T}{2\tau} = \ln(\tfrac{1}{3}) = -1.1$$

and

$$T = 2.2RC$$

The frequency of oscillation is the reciprocal of the period.

$$f = \frac{1}{T} = \frac{0.455}{RC}\ \text{Hz}$$

This equation is based on ideal op-amp theory ($v_+ = v_-$). As we learned in previous chapters of this text, if the oscillation frequency is too high, the op-amp gain decreases with increasing frequency, thus causing this equation to be in error.

Suppose now that we desire an *asymmetric* pulse train at the output of the oscillator. The circuit of Figure 14.23(c) is used to produce the waveform of Figure 14.23(d). Note that a diode is added in order to permit two different time constants to occur. When the diode is *on*, the charging path for the capacitor is through a resistance made up of R in parallel with R_3 (we are neglecting the forward resistance of the diode). When the diode is *off*, the discharge path is through a resistor of value R.

Example 14.7

Find the output of the circuit of Figure 14.23(c) if $C = 0.1\ \mu\text{F}$, $R = 20\ \text{k}\Omega$, $R_1 = R_2$, and $R_3 = 1\ \text{k}\Omega$.

SOLUTION We first find the two time constants for charging and discharging the capacitor.

$$\tau_c = C(R \parallel R_3) = 0.095 \text{ ms}$$

$$\tau_d = RC = 2 \text{ ms}$$

The equation for the first portion of the waveform is

$$v_-(t) = E + (-V_1 - E)e^{-t/0.095 \text{ ms}}$$

The transition occurs when this voltage reaches $E/2$ since $R_1 = R_2$. Thus,

$$v_-(\beta T) = -E + (-V_1 - E)e^{-\beta T/0.095 \text{ ms}} = \frac{E}{2}$$

but setting $V_1 = E/2$ yields

$$\exp\left(\frac{-\beta T}{\tau_c}\right) = \frac{1}{3}$$

and

$$\beta T = 0.1 \text{ ms}$$

For the discharge region,

$$v_-(t) = -E + (V_1 + E) \exp\left[\frac{-(t - \beta T)}{\tau_d}\right]$$

Setting $t = T$ for the second transition yields

$$v_-(T) = \frac{-E}{2} = -E + \left(\frac{-E}{2} + E\right) \exp\left[\frac{-(T - \beta T)}{\tau_d}\right]$$

with the result that

$$(1 - \beta)T = 2.2 \text{ ms}$$

The period is then

$$T = \beta T + (1 - \beta)T = 2.3 \text{ ms}$$

The circuit of Figure 14.23(a) operates at low frequencies (less than 2 kHz) because of the bandwidth limitation of most op-amps. The output is low impedance.

We consider a 74124 voltage-controlled oscillator (VCO) in Chapter 16. When configured with a fixed applied voltage, the 74124 becomes an astable multivibrator that operates without the low-frequency restriction of the op-amp circuit. It can be used with frequencies into the megahertz region. We postpone further discussion of this circuit to Chapter 16.

The *monostable multivibrator* generates a single pulse at the output. When a monostable multivibrator is triggered with an appropriate input pulse, an output pulse of a predetermined length is produced. The duration of the output pulse is programmed through selection of an external capacitance. Some ICs permit the selection of both external resistances and capacitances. Some of these circuits provide Schmitt trigger outputs. Typical ICs that provide this function are the 74121, 74122, and 74123. You should take the time to locate details of these ICs in the TTL Data Book.

14.5.2 The 555 as an Oscillator

We now examine the useful *555 timer.* This inexpensive device can operate in two modes:

1. As a free-running oscillator producing a continuous square wave of variable frequency and duty cycle (which is defined as time high/time low). We term this mode the *astable mode* since the 555 is operating as a free-running oscillator.
2. As a single-pulse generator to produce a single-pulse output with an accurate duration time. This mode is termed the *monostable mode.* In this mode, the circuit is not free-running but produces a single pulse of predetermined duration each time a trigger pulse is applied at the input.

The 555 is also available in a dual version (the 556). The 555 consists of a mixture of digital and analog circuitry. A description of the internal operation of the 555 is presented in this section. Figure 14.24 shows a block dia-

Figure 14.24
Block diagram of the 555.

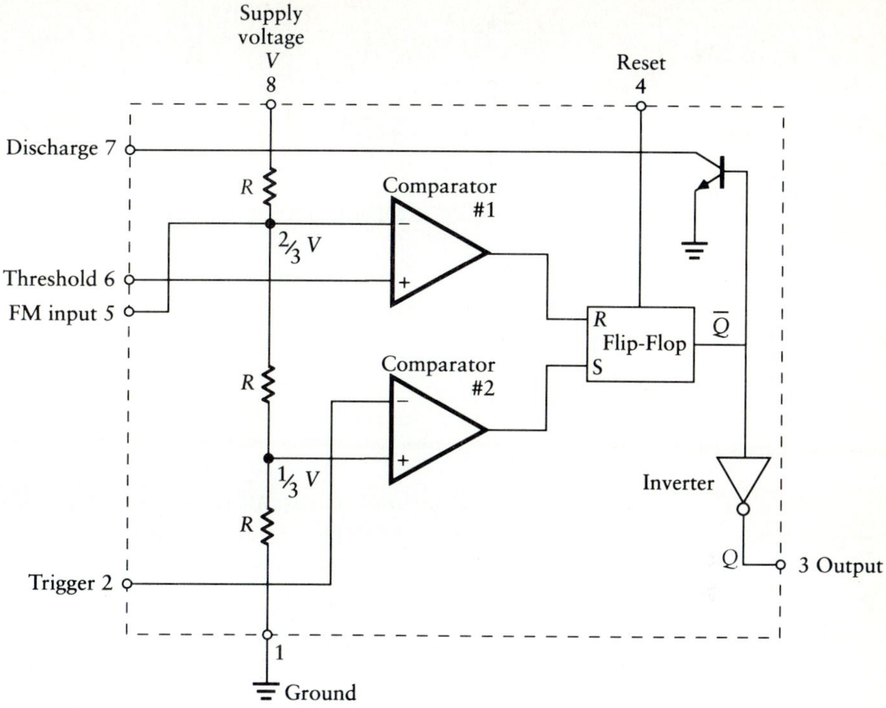

gram of the 555. The circuit contains analog comparators, amplifiers, buffers, and a flip-flop. Analog comparators are discussed in Chapter 13. Flip-flops, buffers, and inverters are considered in Chapters 15 and 16. For now, it is sufficient to think of the *set-reset flip-flop* as an electrically operated switch. Setting the switch causes the output, Q, to go high (approximately to the supply voltage), and resetting the flip-flop causes the output to go low (almost to zero). The three resistors, R, are used as a voltage divider to provide two-thirds of the supply voltage and one-third of the supply voltage levels to the analog comparators.

Pin 2 (trigger) and pin 6 (threshold) control the output of the 555 circuit. When the voltage at pin 6 goes above the $\frac{2}{3}$ supply level, the output of the comparator *sets* the flip-flop, causing its output to go high. The output of the flip-flop will then bias the discharge transistor *on,* causing its output (pin 7) to go low. A voltage at pin 2 of less than the $\frac{1}{3}$ supply level *resets* the flip-flop, causing the output to go low, and turns off the discharge transistor, allowing its output to float. Pin 4 is used to reset the flip-flop. By connecting pin 4 to common, the flip-flop will be reset to low. When not in use, this pin is connected to the positive supply. Pin 4 is used to disable the 555 in that the output becomes zero when pin 4 is set to 0 V. The output oscillates normally when pin 4 is at +5 V.

The 555 can be used as a pulse generator (astable mode) if configured as shown in Figure 14.25. In operation, the capacitor starts to charge through R_1

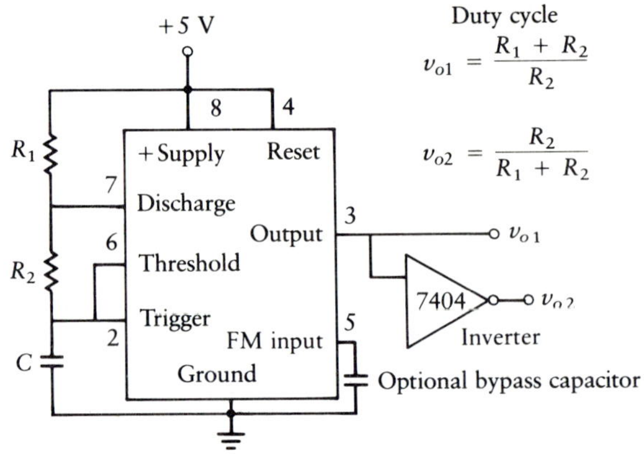

Figure 14.25 The 555 as a pulse generator.

and R_2. The flip-flop is reset (turned off) since pin 2 starts at a low voltage. With the flip-flop reset, the input to the inverter is low and the circuit output is high. When the capacitor voltage gets high enough for the voltage on pin 6 to reach $\frac{2}{3}$ of the supply, the flip-flop *sets* and pin 7 goes low. The capacitor then starts to discharge through R_2. When it has discharged to a value of $\frac{1}{3}$ of the supply, the second comparator, pin 2, causes the flip-flop to *reset* and the cycle repeats.

The capacitor voltage therefore goes exponentially between $\frac{1}{3}$ and $\frac{2}{3}$ of the supply voltage. During charging, the time constant is given by

$$\tau_c = (R_1 + R_2)C$$

During discharge, the time constant is

$$\tau_d = R_2 C$$

The switching occurs when the exponential has gone one-half of the way between initial and final values. For example, during charging, the capacitor voltage starts at $\frac{1}{3}$ of the supply voltage and is exponentially charging toward the supply voltage. Switching takes place when the voltage reaches $\frac{2}{3}$ of the supply. The reverse is true during discharge.

An exponential reaches $\frac{1}{2}$ of its total value after 0.693 of a time constant, that is, when

$$t = 0.693\tau$$

In the equation, $e^{-t/\tau}$, we have

$$e^{-0.693} = 0.5$$

Therefore, the charging time is given by

$$T_c = 0.693(R_1 + R_2)C$$

In a similar fashion, the discharge time is

$$T_d = 0.693R_2C$$

The output is high during charge and low during discharge. The total period of the square-wave output is given by the sum of the two times.

$$T = 0.693(R_1 + 2R_2)C$$

The *frequency* is the reciprocal of this, as follows:

$$f = \frac{1}{T} = \frac{1.44}{(R_1 + 2R_2)C} \tag{14.20}$$

When the 555 is connected to +5 V, the output pulse train goes from 0 to +5 V. As we will see in Chapter 15, this is compatible with TTL circuits. The circuit can operate with supply voltages in the range of 4 to 15 V. The output is capable of sinking more than 100 mA. This output is sufficient to drive a small loudspeaker for warning devices.

The *duty cycle* is defined here as the time high divided by the time low. For the 555, the equation is found as the ratio

$$\text{Duty cycle} = \frac{\text{time high}}{\text{time low}} = \frac{T_c}{T_d} = \frac{R_1 + R_2}{R_2} \tag{14.21a}$$

Since the duty cycle, given by equation (14.21a), has $R_1 + R_2$ in the numerator and R_2 in the denominator, we are limited to duty cycle values greater than one. If we desire a duty cycle less than unity, we can use an inverter (the 7404 inverter is discussed in the next chapter) to invert the output of the 555. Hence, if we include an inverter as shown in Figure 14.25, the duty cycle for the output, v_{o2}, is given by

$$\text{Duty cycle} = \frac{R_2}{R_1 + R_2} \tag{14.21b}$$

Example 14.8

Design an astable 555 square-wave generator to deliver a 1-kHz output signal with nearly unity duty cycle.

SOLUTION We take this opportunity to give some practical limitations of the 555 timer. The capacitance should be kept larger than 500 pF (5×10^{-10} F) to swamp out stray capacitance. Since one end of the capacitor is connected to ground, we can use electrolytics of high capacitance to produce low frequencies. Each of the resistors, R_1 and R_2, should be larger than 1 kΩ to limit the current. The sum, $R_1 + R_2$, should be no larger than 3.3 MΩ. It should be noted that, with these constraints, we cannot achieve a frequency of greater than about 1 MHz.

Let us select $C = 0.01\ \mu$F. We wish to have a duty cycle of nearly unity. That is, we desire the waveform to spend about as much time at the positive value as at the zero value. The discharge time should then be the same as the charging time. Equation (14.21a) indicates that these times cannot be exactly equal. If we choose R_2 to be much greater than R_1, the times are close to being equal. Let us assume that this is the case. The frequency is then given approximately by

$$f = 1 \text{ kHz} \approx \frac{1.44}{2R_2C}$$

Using the assumed value of C, this yields

$$R_2 = 72 \text{ k}\Omega$$

We can choose R_1 to be about 1 kΩ and still be within the design guidelines given above. The design is then complete.

The 555 is relatively insensitive to supply voltage variation because the comparator values are determined by the three equal resistors (R) shown in Figure 14.24. Since the external C, R_1, and R_2 are driven from the same V_{CC} as the internal resistors, changes in power supply voltages have little effect on the frequency of the output. If high accuracy (1.0%) is required, a multiturn potentiometer should be used for R_1 and R_2 to adjust the frequency. If extremely high accuracy (0.1%) is required, it may be necessary to use the ac power line or a crystal-controlled oscillator. The 555 astable is best suited to holding a particular frequency rather than operating at a certain frequency based on the nominal values of C, R_1, and R_2.

Example 14.9

Find the frequency and the duty cycle for the 555 circuit shown in Figure 14.26. Note that the 555 is driven from a 5-V power source and the timing resistors

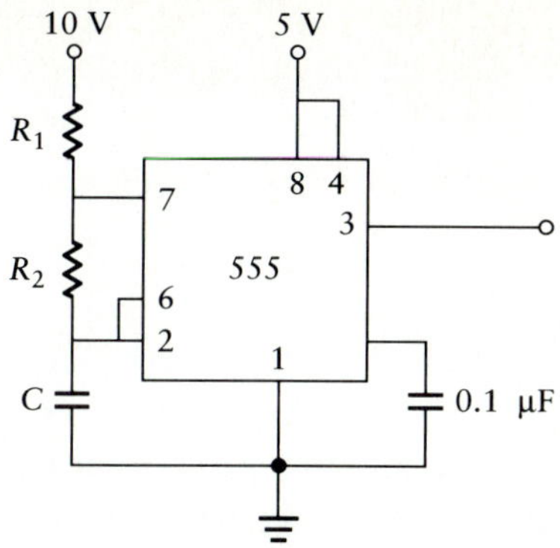

Figure 14.26
555 drive from two power sources—Example 14.8.

and capacitors are driven from a 10-V power source. Let $R_1 = 10\ \text{k}\Omega$, $R_2 = 66.7\ \text{k}\Omega$, and $C = 0.01\ \mu\text{F}$.

SOLUTION We will use equation (14.3), which is repeated here.

$$v(t) = V_f + (V_i - V_f)e^{-t/\tau_c}$$

$V_f = 10$ V and $V_i = 1.67$ V, since the 555 is driven from a 5-V power source, so

$$v(t) = 10 + (1.67 - 10)e^{-t/\tau_c} = 10 - 8.33e^{-t/\tau_c}$$

The time constants are

$$\text{During charge:} \quad \tau_c = (10^{-8})(76.7 \times 10^3) = 7.67 \times 10^{-4}$$

$$\text{During discharge:} \quad \tau_d = (10^{-8})(66.7 \times 10^3) = 6.67 \times 10^{-4}$$

The time, t_1, required to charge C to 3.33 V is found from

$$3.33 = v(t_1) = 10 - 8.33e^{-t_1/\tau_c}$$

Solving, we obtain $t_1 = 0.17$ ms. During discharge, $V_f = 0$ and $V_i = 3.33$ V. Then

$$v(t) = 3.33e^{-(t-t_1)/\tau_d}$$

At $t = t_2$, $v(t_2) = 1.667$ V and $t_2 - t_1 = 0.462$ ms. Hence the total time is

$$t_1 + (t_2 - t_1) = 0.17 + 0.46 = 0.63 \text{ ms}$$

and the frequency is

$$\frac{10^3}{0.63} = 1.587 \text{ kHz}$$

The duty cycle is

$$\frac{\text{time high}}{\text{time low}} = \frac{t_1}{t_2 - t_1} = \frac{0.17}{0.46} = 0.37$$

The 555 is a *user-friendly* device, but a few practical considerations must be observed. The supply voltage must be held constant with a regulated power supply and a decoupling capacitor from pin 8 to ground. This prevents power supply variations and associated timing errors. If large resistors are used, ac hum can be introduced into the output. To reduce this effect, timing resistors and capacitors should be physically located as close to the 555 IC as possible.

The voltage on pin 5 is normally $\frac{2}{3}$ of the supply voltage and is the inverting input to comparator 1. Pin 5, the FM input, can be used to frequency-modulate the output of the 555. A voltage applied to pin 5 produces a fairly linear variation in frequency. Shown in Figure 14.27 is a plot of the relative frequency, f/f_n, as a function of applied voltage on pin 5. To develop this curve, the 555 was operated with $V_{CC} = +8$ V and with nominal frequencies of 10 Hz, 100 Hz, 1 kHz, and 10 kHz. For each pin 5 voltage, f/f_n was measured and plotted. Although the curve is relatively linear, it must be modified with an

Figure 14.27
555 astable frequency variation.

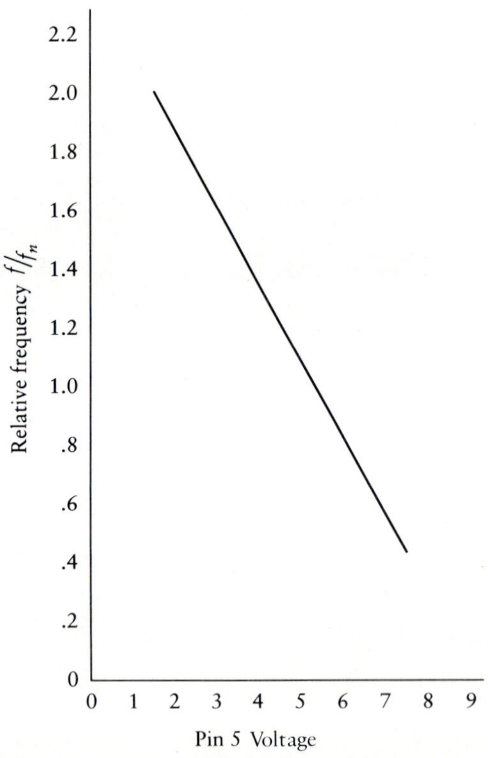

operational amplifier to change its slope. Figure 14.27 shows that, as we increase the pin 5 voltage, relative frequency decreases. If we are to use the 555 for a voltage-controlled oscillator, we must use an op-amp to provide an increase in frequency as the pin 5 voltage increases. This feature has some limited uses in electronic music systems. For other timing applications, it is good practice to bypass pin 5 to ground through a 0.1-μF capacitor.

14.5.3 The 555 as a Monostable Circuit

The 555 can be used to produce a single pulse, with an accurate pulse width. This is the *monostable mode* and requires eliminating R_2 and connecting C between pin 6, pin 7, and ground, as shown in Figure 14.28.

In the monostable mode, the circuit output consists of a single pulse with a specified duration each time a specially configured trigger pulse is delivered to pin 2, driving the comparator below $V_{CC}/3$. When pin 2 goes low, the flip-flop *resets* and the output (pin 3) goes high. Since the discharge transistor is turned off, pin 7 is allowed to float, and the capacitor starts to charge from its initial value of 0 V. When it has charged to $\frac{2}{3}$ of the supply voltage, the flip-flop *sets* and the output goes low. The trigger pulse, several of which are shown in Figure 14.21, must start at +5, drop to zero, and then return to +5, as shown in Figure 14.28. The trigger pulse must be of shorter duration than the desired output pulse. (The duration of the trigger pulse should be $\frac{1}{5}$ the duration of the desired output pulse). The *on-time* is the amount of time the output pulse is high. This is the time it takes the exponential charging voltage to go from 0 to $\frac{2}{3}$ of its final value. This will be 1.1 time constants, since

$$(1 - e^{-1.1}) = 2/3$$

The on-time is therefore

$$T = 1.1RC \tag{14.22}$$

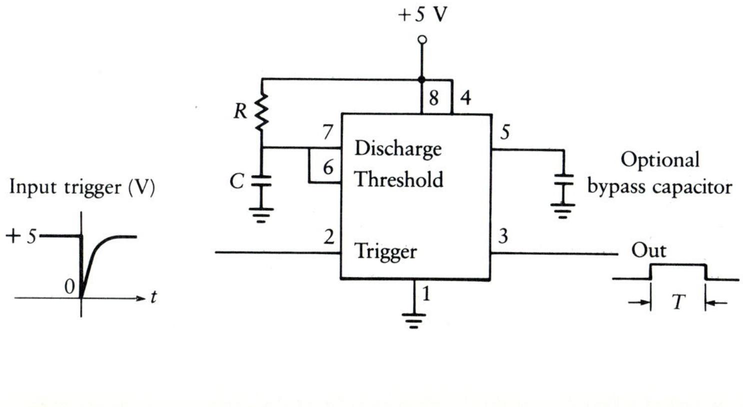

Figure 14.28 The 555 in a monostable mode.

Example 14.10

Design a 555 in the monostable mode to provide a 100-ms pulse each time an input trigger is applied.

SOLUTION Let us use a capacitor of 1 μF. Equation (14.22) then yields the value of R:

$$R = \frac{T}{1.1C} = \frac{100 \times 10^{-3}}{1.1 \times 10^{-6}} = 91 \text{ k}\Omega$$

The circuit is as shown in Figure 14.28 with the parameter values as calculated above.

The form of the trigger pulse necessary to initiate the 555 monostable is most important. As shown in Figure 14.28, the trigger pulse, applied to pin 2, is at +5 V. This voltage is briefly dropped to zero to initiate the timing sequence. At the instant when the trigger is brought to zero, an output pulse of duration 1.1RC is generated. In contrast to the astable mode of operation, the cycle does not repeat. Another trigger pulse is required to initiate another output pulse.

The method of obtaining these trigger pulses for the 555 in the monostable mode is shown in Figure 14.21. If a leading-edge trigger is desired, use the circuit shown in Figure 14.21(c). The inverter (the inverter is covered in Chapter 15) is used to invert the pulse into the form necessary to trigger the 555.

A trailing-edge trigger is obtained with the circuit of Figure 14.21(b). Note that to obtain the proper trigger the resistor is connected to +5 V rather than to ground. It may be desirable to use a noninverting buffer (discussed in Chapter 15) at the output of the circuit of Figure 14.21(b) to "square up" the trigger pulse and also to reduce the loading effect on the v_i source.

The reset pin (pin 4) can be used to disable or stop the timing cycle after it begins. When the reset pin is brought to ground, the operation is inhibited. When not required, pin 4 should be tied to the positive supply, as shown in Figure 14.28.

The FM input pin (pin 5) can be used to vary the charging time by applying a modulation voltage. We can produce a pulse-width-modulated (PWM) signal by changing the charging time using a voltage applied to pin 5. We drive the monostable 555 with a continuous pulse train of fixed frequency and with short-duration +5 V to 0 to +5 V pulses.

PROBLEMS

14.1 A 10-Hz symmetrical square wave has a peak-to-peak voltage of 5 V. It forms the input to a high-pass circuit with lower corner frequency of

5 Hz ($f_c = 1/(2\pi RC)$). Sketch the output waveform. What is the peak-to-peak output amplitude?

14.2 A 10-Hz symmetrical square wave forms the input to an amplifier that acts as a high-pass circuit. The peak-to-peak voltage is V. Plot the output waveform if the lower corner frequency is

a. 0.3 Hz
b. 3 Hz
c. 30 Hz

14.3 A square wave extends ±2 V with respect to ground. The duration of the positive section is 0.1 s and that of the negative section is 0.2 s. The waveform is applied to a high-pass network with time constant $RC = 0.2$ s. Plot the steady-state output waveform and calculate the important maximum and minimum voltages.

14.4 The limited ramp of Figure P14.1 is applied to a high-pass network. Let $V_m = 1$ V. Draw the output waveform for each value of T.

a. $T = 0.2(RC)$
b. $T = RC$
c. $T = 5(RC)$

14.5 A symmetrical square wave with an average value of zero has a peak-to-peak voltage of 10 V and a period of 2 μs. This waveform is the input to a low-pass circuit with a corner frequency of 0.16 MHz. Calculate and sketch the steady-state output waveform. What is the peak-to-peak output voltage?

14.6 The pulse train shown in Figure P14.2 is applied to the input of a diode-capacitor network. Plot the steady-state output voltage, $v_o(t)$, and calculate the four important voltages in the output. Assume that the diode voltage is zero when forward-biased (i.e., $V_\gamma = 0$), $R_f = 100\ \Omega$, $R_r = 100\ \text{k}\Omega$, and $R_i = 1\ \text{k}\Omega$.

14.7 A 10-kHz pulse train with peak value $V_{\text{peak}} = 10$ V with respect to ground is applied to the diode clamping circuit of Figure P14.3(b). The parameter values are $R = 10\ \text{k}\Omega$ and $C = 1\ \mu$F. The diode has $R_r = \infty$, $R_f = 0$, and $V_\gamma = 0$ and the source impedance, R_i, is zero.

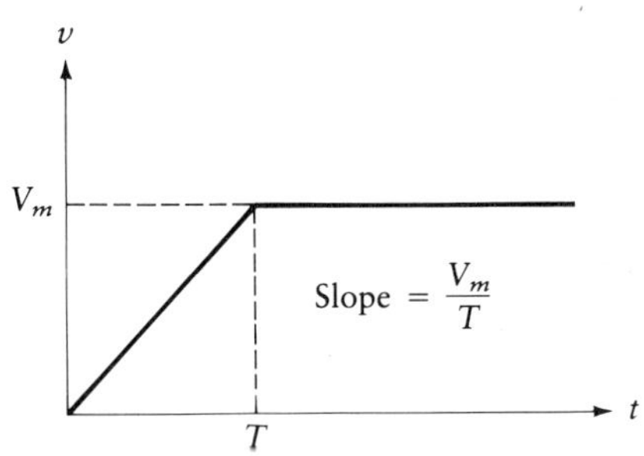

Figure P14.1

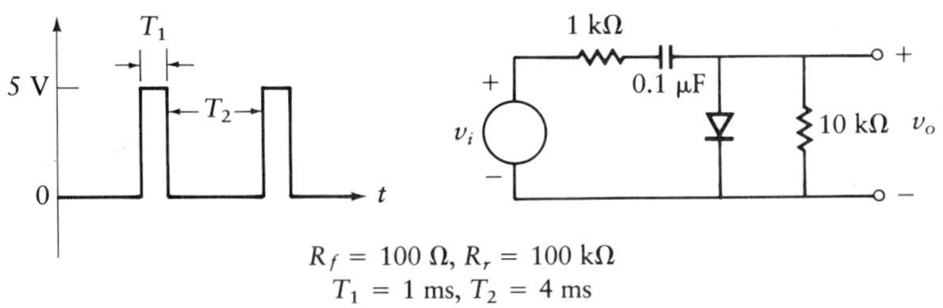

Figure P14.2

Figure P14.3

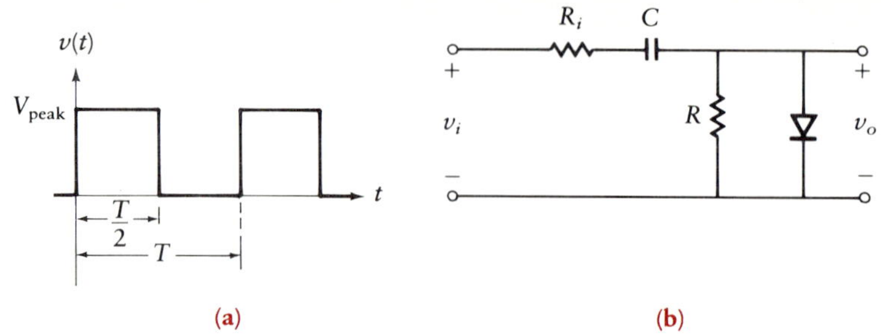

a. Sketch the output waveform.
b. If the diode forward resistance is 1 kΩ, sketch the output waveform. Calculate the maximum and minimum voltage with respect to ground.
c. Repeat part b if R_i = 1 kΩ.

14.8 The pulse train shown in Figure P14.3a with T = 50 μs is applied to the circuit of Figure P14.3(b). The circuit has R_i = 1 kΩ, R = 10 kΩ, C = 0.1 μF, R_f = 100 Ω, R_r = 100 kΩ, and V_γ = 0. Find the steady-state output waveform, $v_o(t)$, if V_{peak} = 10 V.

14.9 The limited ramp of Figure P14.1 is applied to a low-pass network with V_m = 1 V. Draw the output waveform for
a. $T = 0.2(RC)$
b. $T = RC$
c. $T = 5(RC)$

14.10 Design a signal generator using a 555 circuit in the astable mode to provide pulse rates of 100 kHz, 10 kHz, and 1 kHz. Keep the duty cycle constant at 1.15.

14.11 Design a trigger circuit and a monostable 555 to produce a train of 100-ms pulses. Assume that you have a symmetrical 0 to 5 V, 1 Hz pulse train input to this circuit.

14.12 Use a 555 in the astable mode and an inverter (needed to obtain a duty cycle less than 1) to produce a pulse generator for the following continuously variable frequency ranges:
a. 10 Hz to 100 Hz
b. 100 Hz to 1 kHz
c. 1 kHz to 10 kHz

Switch three capacitors into the circuit, one for each frequency range, and use a potentiometer to vary the frequency in each range. It is required that the duty cycle be no greater than 0.2. Select all resistor and capacitor values, and select the chip number for the inverter. Assume that you have available any power you need. Draw the circuit diagram for the design.

Figure P14.4

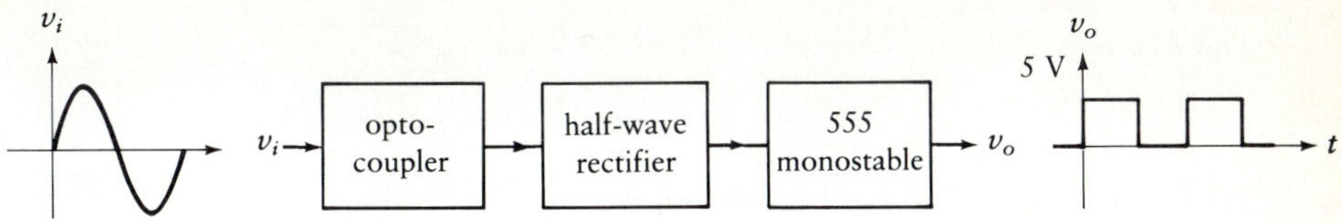

Figure P14.5

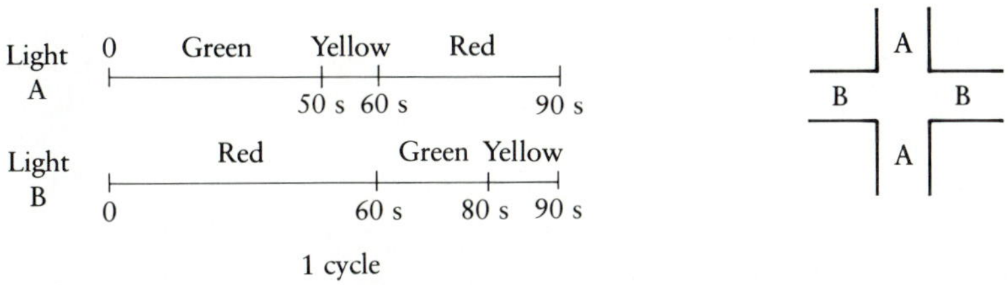

14.13 Develop a 60-Hz pulse train using the 60-Hz, 110-V power line voltage with the system shown in Figure P14.4. The output is a pulse train with a 100% duty cycle. Select all resistor and capacitor values. Draw the circuit diagram for your complete design.

14.14 Design a pulse generator that will have the following continuously variable ranges:

a. 0.1 kHz to 1 kHz
b. 1 kHz to 10 kHz
c. 10 kHz to 100 kHz

A duty cycle of 1.0 is required. Assume you have available any power needed in the design.

14.15 Design a traffic-light control system using several 555s. The traffic light is to be used at a four-way intersection, and each direction will have three lights: red, yellow, and green. Your system must provide an output of +5 V for each lamp that is to be turned on. If the output to a certain lamp circuit is zero, that lamp will be off. The required times and a map of the intersection are shown in Figure P14.5.

14.16 Design a pulse-delay network to provide a 15-ms delay for a pulse input whose duration is 1 ms as shown in Figure P14.6.

14.17 Design a pulse generator to provide the clock pulse for an electronic system. Use the 110 V, 60 Hz power line as the basic drive for the pulse generator. The duty cycle of the clock pulse output is 0.5. Calculate all resistor and capacitor values and specify the type number of the ICs used.

14.18 Analyze the circuit shown in Figure P14.7. Sketch the output signal as a function of time. Calculate all important frequencies and pulse widths. Switch SW is normally closed. You are to plot the output signal after the time that the switch is open.

14.19 Design a 1-kHz pulse generator with a 0 to 5 V output, using an astable 555 timer to provide the 1-kHz pulse train. The duty cycle of

Figure P14.6

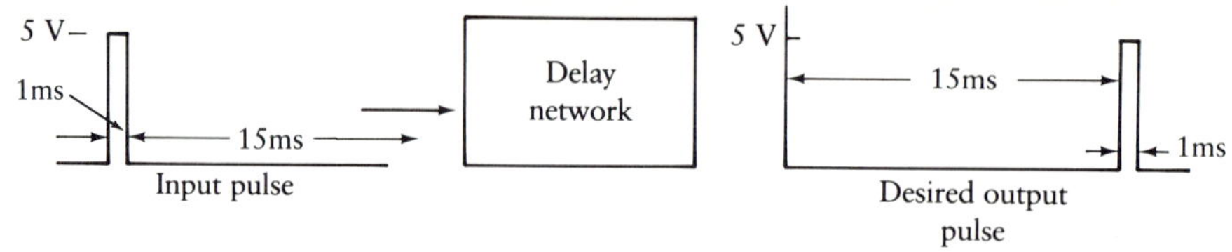

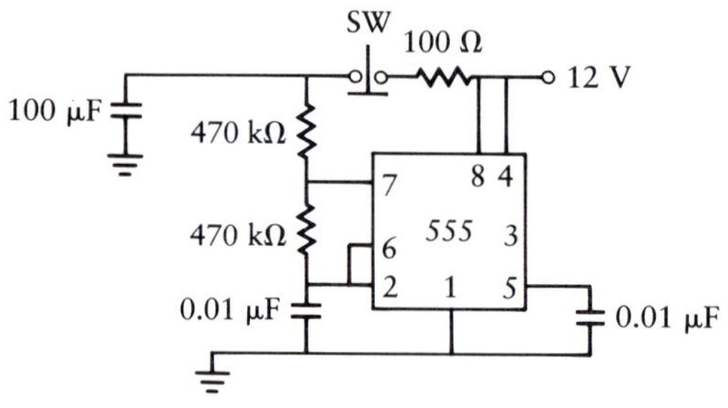

Figure P14.7

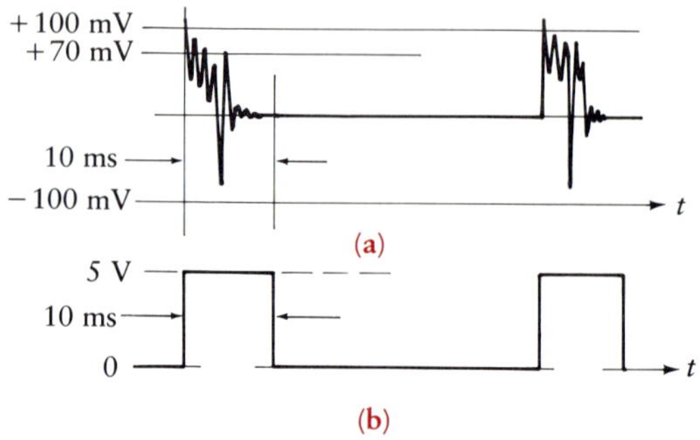

Figure P14.8

the output clock pulses must be variable from 0.2 to 10. Use one potentiometer, an astable 555, and one monostable 555 to generate the duty cycle. Calculate all resistor and capacitor values and specify the type numbers for the ICs used in the design.

14.20 Use op-amps and a 555 monostable in the design of a signal-conditioning system. The input for the system is obtained from the ignition system for an engine. One hundred turns of wire are wrapped around the supply line to one of the spark plugs. This produces a voltage pulse once for each two revolutions of the engine. The signal has the form shown in Figure P14.8(a). The transients are large and do not die out for 10 ms.

Design an electronic system to condition this input signal so it will be suitable to drive a TTL system. The suitable signal will have only one output pulse for each firing of the spark plug. That is, the signal-conditioning system should ignore the transients and produce only *one* pulse for each burst of energy. The conditioned pulse train should look like that shown in Figure P14.8(b). Each pulse of the output pulse train corresponds to two revolutions of the engine, and the maximum engine speed is 6000 rpm.

14.21 Design a warning alarm signal that has the form shown in Figure P14.9. The desired alarm signal is a 2-kHz square wave (developed by one 555 astable) that is turned on and off 30 times each minute (by another 555 astable). The time on is approximately equal to the time off.

14.22 An input clock signal has a variable pulse width, as shown in Figure P14.10. The pulse width varies from 1 ms to 3 ms, and the frequency

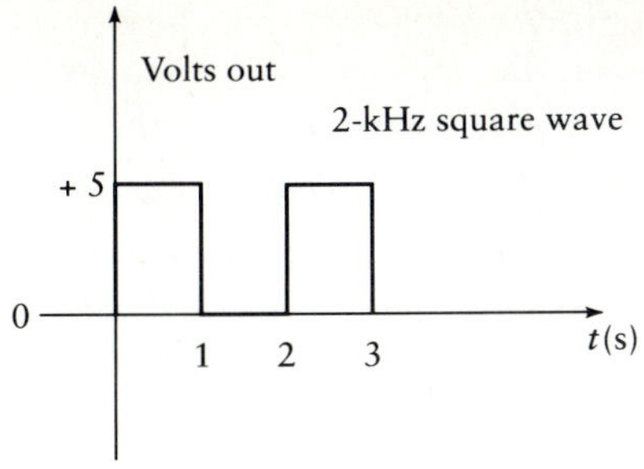

Figure P14.9

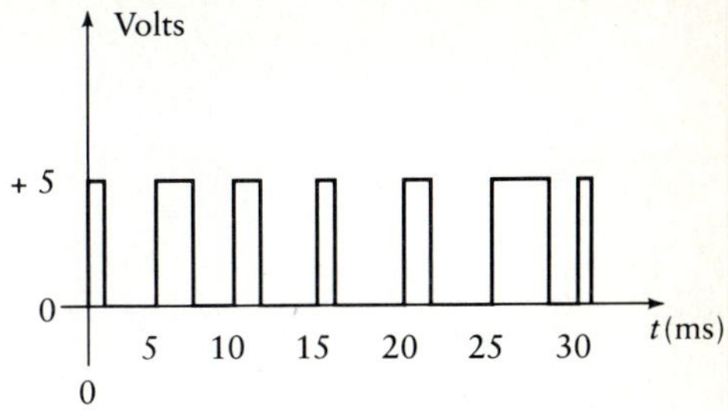

Figure P14.10

Figure P14.11

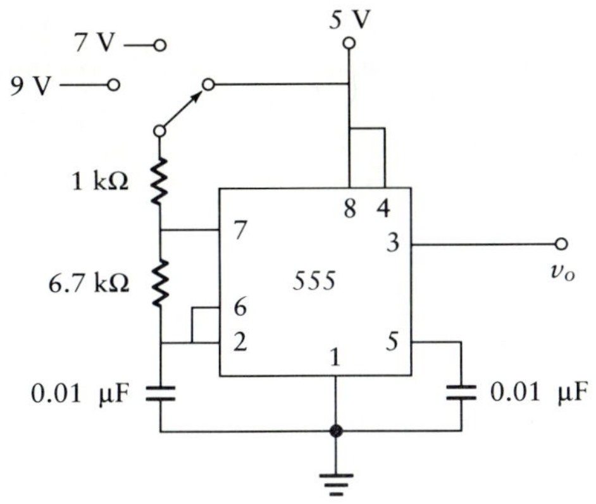

Figure P14.12

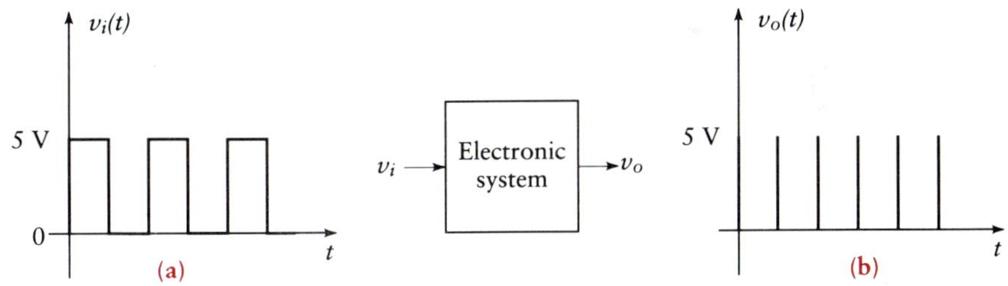

is constant at 200 Hz. Use a monostable 555 and a trigger circuit in the design of an electronic system to process this input signal. The output pulse train must have a frequency of 200 Hz with a constant duty cycle of 0.2.

14.23 The circuit shown in Figure P14.11 produces a frequency from the astable mode 555 that varies as the voltage applied to the timing resistors and capacitors changes. We refer to such a circuit as a *voltage-*

controlled oscillator. Calculate the output frequency for each of the three settings of the tap switch. Calculate the frequency as in Example 14.9 (use equation (14.3)). Plot the output frequency as a function of these three applied voltages.

14.24 The pulse train shown in Figure P14.12(a) forms the input to an electronic system. Use a trigger circuit and an op-amp in the design of a system that will produce a train of impulses, each of which will be no wider than 2 ms. The output, shown in Figure P14.12(b), must contain an impulse for every leading edge and a second impulse for every trailing edge of the incoming pulse train.

14.25 Use the output of the system of Problem 14.24 (Figure P14.12(b)) to produce a train of pulses with unity duty cycle but with twice the frequency of the input signal (Figure P14.12(a)). The results of Problems 14.24 and 14.25 constitute a frequency doubler circuit.

15

DIGITAL LOGIC FAMILIES

15.0 INTRODUCTION

You have secured an excellent engineering position working for a major auto corporation in their advanced design section. You have been assigned the task of designing an electronic control module for the safety features for next year's automobiles. The device specifications have been developed by the systems department of your corporation. They include requirements for a warning system that accomplishes a number of tasks. The system warns the driver if the headlights are off after dark. It also flashes a warning if the lights are left on after the car is parked and the passengers exit. Low fuel level, low tire pressure, low brake fluid, low oil pressure, and excessive engine temperature must also lead to warnings, as must driver intoxication, failure to turn on windshield wipers in the rain, and excessive speed. Finally, the device must also work as a burglar alarm. Each of these functions has override switches.

You carefully review all of the specifications and, with the help of various data books, select a variety of linear ICs to design what you think is a sophisticated system. Unfortunately, you significantly exceed the budget for this device. Even more serious than this, you install a prototype and find that false alarms frequently occur. You trace this to ignition noise generating undesired signals that trigger the comparators. If you had not already run beyond your budget, you would consider reducing the noise by providing shielding and by using resistive ignition wires. Alas, you seem to have run out of options, so you turn to your colleagues for help. One of them suggests that this is an ideal application for digital electronics rather than analog. You are told that this will not only reduce the probability of a false alarm but also lead to more compact, simpler, and less expensive circuitry. Since you studied digital electronics in school, you are in an ideal position to perform the design.

Digital electronics forms a subset of the broader field of digital engineering. A typical study of digital engineering includes procedures for reducing Boolean functions to their simplest form. A particular function is specified and then, prior to considering implementation, the function is reduced to its simplest form. It is only following that reduction that we consider filling in the blocks on the system diagram with electronic circuitry. In this text, we concentrate on the circuitry, while presenting a brief summary of some digital engineering results in Section 15.1. It is assumed that you either have had a course in digital engineering or will accept the specifications as being the result of logic operations.

The integrated circuits studied earlier in this text are *analog;* that is, they are able to accept inputs and produce outputs that range over a continuum of amplitudes. The op-amp is an example of a linear IC.

In this chapter, we consider digital integrated circuits. Binary digital ICs accept inputs consisting of either of two logic levels, usually denoted by a 0 or a 1. That is, the signals can take on only one of a set of specific amplitudes, with that set consisting of two entries in the case of binary circuits. Since each component of a (binary) digital integrated circuit produces a logic 0 or 1 at the output, depending on the value of the input(s), the components can be thought of as "logic gates." A *gate* will let either a 1 or a 0 through to the output.

The first section of this chapter briefly discusses function tables for the most important logic gates. The basics of Boolean algebra are then summarized. These results are needed for circuit simplification operations.

Three IC families are discussed: TTL, ECL, and CMOS. These families are compared to facilitate choice of the appropriate family for a given application. We stress operational properties so that you will be in a position to apply these circuits in your system designs.

15.1 BASIC CONCEPTS OF DIGITAL LOGIC

Logic has been a field of study since the days of the Greek philosophers. Many of the concepts that are used in digital logic were derived from earlier sources. Several of the basic functions were used by philosophers to prove statements and solve puzzles.

Two of these operations, known as AND and OR, form the basic building blocks of a logic system. These can be considered as analogous to addition and multiplication in the basic algebra system. We start with these two and then expand to more complex functions.

Prior to defining the functions, we need to explore ways to express a digital relationship. In algebra, a *function* is usually expressed in one of three ways. First, we might give an *equation* for the function, where that equation contains basic operations such as addition and multiplication. Alternatively, we can present a *graph* of the function in two dimensions, with one variable (independent) on the abscissa and the other (dependent) on the ordinate of the graph. A third technique for specifying the function is to give the value of the function for *every* possible value of the independent variable. In the analog continuous case, this would involve a table with an infinite number of entries.

Two of these analog functional representation techniques lend themselves nicely to describing digital functions. An *equation* can be written once we define some notation for the basic logic operations. Alternatively, a *table* can be presented that specifies the value of the function for every possible input. In contrast to the analog case, this table has only a finite number of entries. It is known as a *function table,* or *truth table,* and we use it to define the various functions.

15.1.1 State Definitions—Positive and Negative Logic

Since binary digital signals can take on only one of two possible values, the inputs to the system can also be only one of these values. Although they are manifested as voltages in a circuit, the values are usually referred to as 0 and 1. The two voltages developed by a widely used family (TTL family) of circuits consist of one close to zero and the other close to 5 V. We refer to these as LOW (0 or OFF) and HIGH (1 or ON), respectively. In order to convey digital information, two distinct signals are needed. There is nothing magic about 0 and 1. For example, the digital number 10110, when appearing in this text, could just as well be represented by XYXXY. As long as two distinct signals are used, the information is conveyed. We therefore have two choices in assigning the binary numbers 0 and 1 to the two voltages present in a circuit, as follows:

Positive logic: Associate HIGH voltage with 1 and LOW voltage with 0.

Negative logic: Associate HIGH voltage with 0 and LOW voltage with 1.

The manufacturer's data sheets define the logic levels in terms of H (high-voltage level) and L (low-voltage level). The selection of H equal to 1 and L equal to 0 results in positive logic. If we were to select H equal to 0 and L equal to 1, we would obtain negative logic.

The logic gates are usually named using positive logic. However, we can change the type of gate, and also its name, by changing from positive to negative logic. For example, a positive logic AND gate is the same circuit as a negative logic OR gate. We will show this when we consider the actual circuits in Section 15.1.4. Since the logic is independent of whether we choose positive or negative logic, we may economize on parts and hence power consumption by using a combination of positive and negative logic in the same electronic system. If your design results in several unused portions of logic packages, a change from positive to negative logic may simplify the parts requirements.

Throughout this text, we use positive logic unless specifically stated to the contrary. Because of the simplifications that can result by using negative logic, or a mixture of positive and negative logic, the other possibilities should be explored.

15.1.2 Time-Independent or Unclocked Logic

A *time-independent logic function* is one that has no memory. This form of logic responds only to the *present* inputs that are applied, and previous inputs have no effect on the present output. In essence, the circuit forgets any previous logic conditions. Function tables for such circuits are easy to construct, since there are no clock signals or previous conditions to be considered. The elementary logic functions considered in this chapter are all time independent.

Time-independent logic is also known as *unclocked* or *asynchronous logic.*

15.1.3 Time-Dependent or Clocked Logic

A *time-dependent logic function* is one that has a *memory.* The response of this form of logic depends not only on the present inputs that are applied to it but also on previous input and/or output conditions (i.e., the *state* of the system). A dependence on the previous output state represents a form of feedback in digital circuits. Time-dependent logic functions often have inputs with labels such as *clock, strobe, enable, set,* or *reset.* We encounter these terms in the following chapters. The output may change state at the rising edge of an input or at the falling edge of an input, or it may be a function of a high or low logic level of a critical input or combination of inputs. Time-dependent logic is also known as *clocked* or *synchronous logic.*

15.1.4 Elementary Logic Functions

The first digital ICs produced were simple gates and inverters containing one to six devices in a package. The variety of gates has increased dramatically.

The various levels of IC complexity and size are as follows. *Small-scale integration (SSI)* includes simple devices such as gates and flip-flops. Devices within this classification contain between one and ten equivalent gates. *Medium-scale integration (MSI)* consists of more complex devices such as counters, shift registers, encoders, decoders, and small memories. MSI devices contain from 10 to 100 equivalent gates. *Large-scale integration (LSI)* includes larger memories and microprocessors. LSI devices consist of between 100 and 1000 equivalent gates. *Very large scale integration (VLSI)* includes the largest memories and microprocessors. VLSI devices have more than 1000 equivalent gates. *Ultra large scale integration (ULSI)* devices have more than 10,000 equivalent gates.

We now briefly define the most basic and common logic gates. Even with the continuing development of MSI and LSI logic functions and microprocessors, these basic functions still play a major role in digital electronic design.

Eight functions are presented in Figure 15.1. All of these are described, but first we pick two for more detailed discussion. We do this to clarify the concept of the truth table.

Figure 15.1 Elementary gate summary with part numbers of various gates.

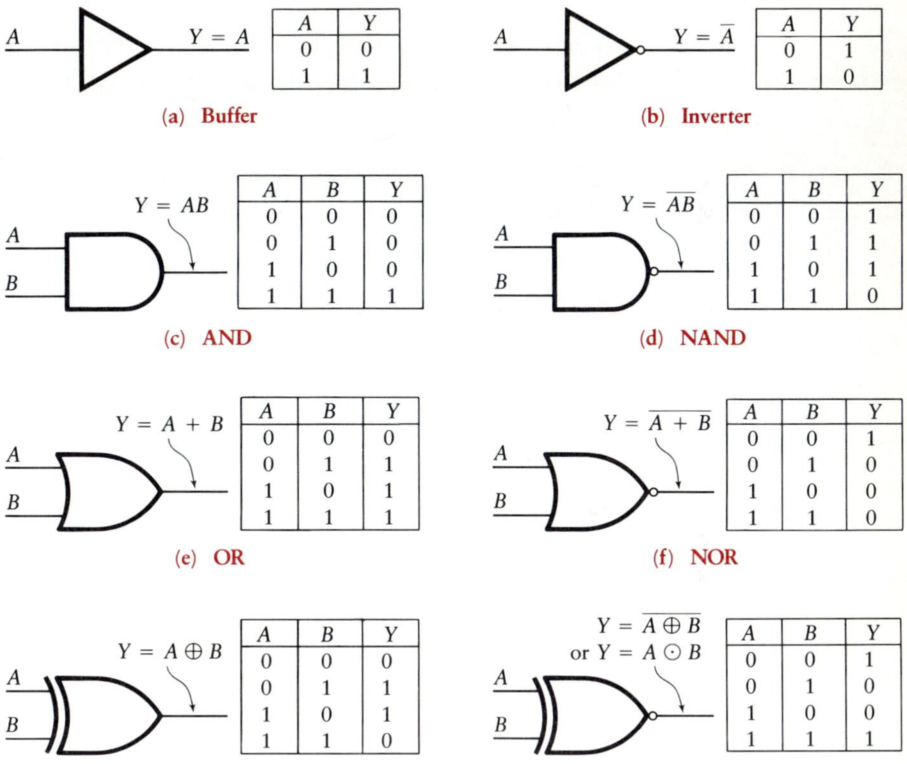

(a) Buffer

A	Y
0	0
1	1

(b) Inverter

A	Y
0	1
1	0

(c) AND

A	B	Y
0	0	0
0	1	0
1	0	0
1	1	1

(d) NAND

A	B	Y
0	0	1
0	1	1
1	0	1
1	1	0

(e) OR

A	B	Y
0	0	0
0	1	1
1	0	1
1	1	1

(f) NOR

A	B	Y
0	0	1
0	1	0
1	0	0
1	1	0

(g) XOR

A	B	Y
0	0	0
0	1	1
1	0	1
1	1	0

(h) XNOR (or equivalence)

A	B	Y
0	0	1
0	1	0
1	0	0
1	1	1

	BUFFER	INVERTER	AND	NAND
TTL —	7417	7404	7408	7400
ECL —	MC10188	MC10189	MC10104	MC10121
CMOS —	MM74HC125	MM74HC04	MM74HC08	MM74HC00

	OR	NOR	XOR	XNOR
TTL —	7432	7402	7486	74266
ELL —	MC10103	MC10102	MC10113	MC10107
CMOS —	MM74HC32	MM74HC02	MM74HC86	MM74HC266

Note: This summary is based on positive logic.

Begin with the *inverter,* which is a device for which the output is the *opposite* of the input. We used the inverter in Chapter 14 to change the duty cycle of the pulse train generated by the 555 clock. If the input is 0, the output is 1 and vice versa. The output is always the *complement* of the input. The truth table for the function is simple since there are only two possible inputs. We therefore enter each possible input value and write the corresponding output next to this value. The truth table is shown in Figure 15.1(b). The equation is given in the form

$$Y = \overline{A}$$

We have introduced the notation of a bar over the variable to indicate the logic operation of inversion. An alternative notation is to use a prime for the inversion operation, so in equation form this becomes

$$Y = A'$$

The notation used in the block diagram to denote inversion is a small circle at the output of the logic symbol. The inversion operation is sometimes referred to as a *NOT gate.* Thus, the output of the inverter with A as input is sometimes known as "NOT A."

The second operation we discuss is that of the *OR gate.* The OR gate has more than one input, and a two-input gate is illustrated in Figure 15.1(e). The gate provides a logic 1 at the output when *any one or more* of the inputs is in the logic 1 state. That is, the output, Y, is 1 if A and/or B is equal to 1.

Since this gate has two inputs, the truth table has three columns, one for each input and one for the output. Since each of the two inputs can take on either of two values, there are four possible input combinations. These form the four rows of the table. In general, the truth table will have 2^n rows, where n is the number of inputs.

It is a simple matter to develop a truth table with only four rows, but the process can become quite complex if the number of input variables increases. For example, with 6 inputs, the number of possible combinations is 2^6 or 64. In cataloging the 64 possible input combinations, it is usually helpful to arrange them in binary counting order. Thus, for two inputs, the rows are ordered by input combination as follows: 00, 01, 10, 11. This is helpful to avoid omitting any possible combination in more complex situations.

The formula notation for the operation of OR is the plus sign. The circuit symbol contains a curved line at the left edge, as shown in Figure 15.1(e).

We now briefly define the remaining six functions in Figure 15.1.

A *buffer* has a single input and a single output. The output is always in the same logic state as the input. The purpose of a buffer is to provide additional power to drive other logic inputs. A buffer can be built by cascading two inverters together. That is, taking the inverse twice returns us to the original value. The buffer is shown in Figure 15.1(a).

An *AND* gate has two (or more) inputs and one output. The output is at logic level 1 only when *all* of the inputs are in the logic 1 state. The AND gate is shown in Figure 15.1(c). Note that the notation of the AND operation used in the formula is to write the two variables next to each other as if we were talking about algebraic multiplication. We also sometimes use a dot between the variables. Thus,

$$AB$$

and

$$A \cdot B$$

are both read "A and B." Also note that the circuit symbol has a straight line on the input edge, as distinguished from the OR gate symbol, which has a curved line.

The *NAND* gate provides a logic 0 at the output only when *all* of the inputs are in the logic 1 state. The NAND gate can be viewed as an AND gate followed by an inverter. In words, the operation is *NOT AND.* This is illustrated in Figure 15.1(d). Note that the equation and the circuit diagram are both developed by combining the notation for the AND with that of the inversion operation. Note that *any* 0 at the input produces a 1 at the output.

A *NOR* gate provides a logic 0 output when *any one or more* of the inputs are in the logic 1 state. This is shown in Figure 15.1(f). Note that the NOR gate can be viewed as an OR gate followed by an inverter. Both the equation and the circuit diagram are developed by combining the two individual representations.

Normally, since the NAND and NOR gates are internally simpler than the AND and OR, and hence require less power, we attempt to use these gates rather than AND and OR gates.

The *XOR* gate, or *exclusive OR,* provides a logic 1 output when *any one, but only one,* of its inputs is in the logic 1 state. This is shown in Figure 15.1(g). Note that the formula notation is a plus sign (as used for the OR) with a circle around it. The circuit notation starts with the symbol for an OR gate and adds a second curved line at the input. This could be viewed as binary (modulo 2) addition where

$$0 \oplus 0 = 1 \oplus 1 = 0$$

$$1 \oplus 0 = 0 \oplus 1 = 1$$

The final logic operation shown is the *exclusive NOR* or *XNOR* gate. The

Figure 15.2
Positive logic to negative logic.

A	B	Y
1	1	1
1	0	1
0	1	1
0	0	0

(a)

A	B	Y
1	1	0
1	0	0
0	1	0
0	0	1

(b)

XNOR gate provides a logic 0 output when *any one, but only one,* of its inputs is in the logic 1 state. This gate is also called *equivalence,* since the output is logic 1 when the two inputs are equal. Note that the XNOR can be viewed as an exclusive OR gate followed by an inverter. This is shown in Figure 15.1(h), where two formula notations are shown.

Also included in Figure 15.1 are some IC part numbers for each of the gates in three different families. Later in this chapter, we consider these three families in greater detail.

Before continuing, let us consider the effect of changing from positive logic to negative logic. Let us take the positive-logic AND gate of Figure 15.1(c) and change the truth table to negative logic by changing zeros to ones and ones to zeros. This results in the truth table shown in Figure 15.2(a). Hence the same electronic hardware that is named an AND gate with positive logic is named an OR gate with negative logic.

Example 15.1

What is the name of the positive logic NAND gate of Figure 15.1(d) when we use negative logic?

SOLUTION We start with the truth table of Figure 15.1(d) and reverse the assignment of 0 and 1 by changing 0 to 1 and 1 to 0. This results in the truth table of Figure 15.2(b). If we look at the function tables of Figure 15.1, we see that this is now the function table for a NOR gate. Hence the same electronic hardware is used for a positive-logic NAND gate and a negative-logic NOR gate.

15.1.5 Boolean Algebra

Whenever binary variables are manipulated, equations result. We need rules for working with these equations. Boolean algebra gives us the necessary set of rules. In Boolean algebra, the two elementary algebraic operations of addition and multiplication are replaced by the elementary logic operations of OR and AND. Functions are defined either through an equation or by giving the value of the function for every possible input. This latter approach gives rise to the truth (or function) table, as discussed in the previous section.

The Boolean equation is a shorthand way of writing the truth table. *Boolean identities* can be proved by referring to the function tables for the gates as shown in Figure 15.1.

The following drill problems present an overview of several important Boolean identities.

Drill Problems

Prove each of the relationships given in Drill Problems D15.1 through D15.16.

D15.1 $A + (B + C) = (A + B) + C$ (associative law)

D15.2 $A(BC) = (AB)C$ (associative law)

D15.3 $\overline{A + B} = \overline{A}\,\overline{B}$ (DeMorgan's theorem)

D15.4 $\overline{ABC} = \overline{A} + \overline{B} + \overline{C}$ (DeMorgan's theorem)

D15.5 $A + 0 = A$

D15.6 $A + A = A$

D15.7 $A + \overline{A} = 1$

D15.8 $A \cdot 0 = 0$

D15.9 $A \cdot 1 = A$

D15.10 $A \cdot A = A$

D15.11 $A \cdot \overline{A} = 0$

D15.12 $A + AB = A$

D15.13 $AB + A\overline{B} = A$

D15.14 $\overline{(\overline{A})} = A$

D15.15 $\overline{(A + B)}\,\overline{(\overline{A} + \overline{B})} = 0$

D15.16 $B + A\overline{B} = A + B$

DeMorgan's theorem (see Drill Problems D15.3 and D15.4) can be used to develop alternative forms for implementing the NOR and NAND operations. These alternative forms are sometimes preferable for reasons related to practical considerations, as we shall see later.

Starting with the NOR operation, we can use the theorem to show

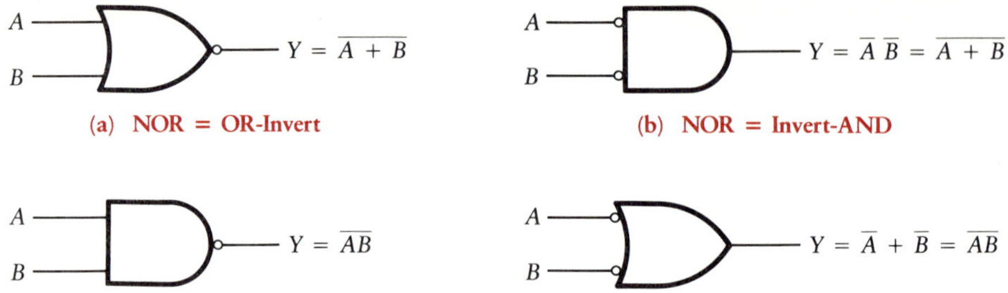

(a) NOR = OR-Invert (b) NOR = Invert-AND (c) NAND = AND-Invert (d) NAND = Invert-OR

Figure 15.3 Alternative forms for NOR and NAND gates.

$$\overline{A + B} = \bar{A}\bar{B} \tag{15.1}$$

For the case of three variables,

$$\overline{A + B + C} = \bar{A}\bar{B}\bar{C} \tag{15.2}$$

Alternatively, for the NAND operation we have

$$\overline{AB} = \bar{A} + \bar{B} \tag{15.3}$$

In the three-variable case,

$$\overline{ABC} = \bar{A} + \bar{B} + \bar{C} \tag{15.4}$$

The resulting alternative forms of the basic gates are shown in Figure 15.3.

15.2 IC CONSTRUCTION AND PACKAGING

Integrated circuits are fabricated using a sequence of processing steps that includes growing, slicing, and etching silicon wafers; masking and doping them with *n*- and *p*-type impurities; and depositing conductor patterns. This complex processing sequence produces complete functional circuits containing patterns of resistors, capacitors, diodes, and transistors.

A functional circuit pattern is repeated numerous times to fill the usable area of a silicon wafer. On completion of the processing steps, the wafer is sliced into small rectangles or *chips,* each containing one or more functional circuits. Each chip is tested and mounted in a package. A typical package is the 14-pin *dual in-line package* (known as a *DIP*). The package provides structural support, an arrangement of pins for external connections, and a sealed cover to keep out moisture and contamination. The complexity of VLSI chips, with their large numbers of inputs and outputs, has caused some decline in the use of the DIP. Newer package configurations provide for additional external connections and faster (therefore less expensive) manufacturing techniques. This is particularly true when these chips are incorporated into larger system modules. Several of the popular packages are illustrated in Figure 15.4.

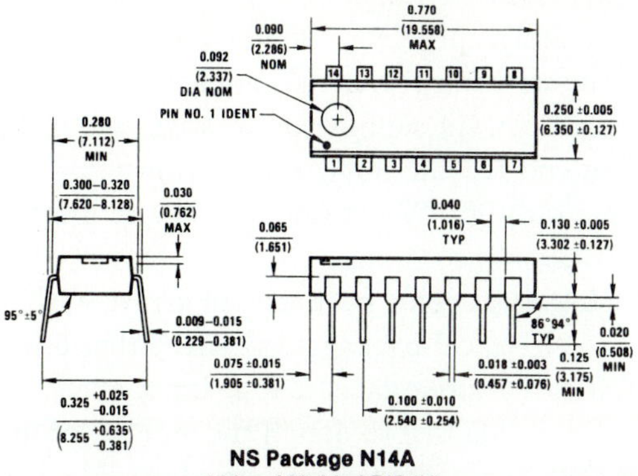

NS Package N14A
14-Lead Molded DIP (N)

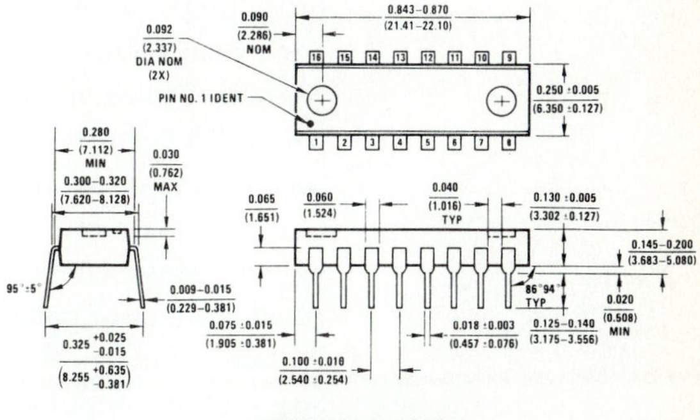

NS Package N16A
16-Lead Molded DIP (N)

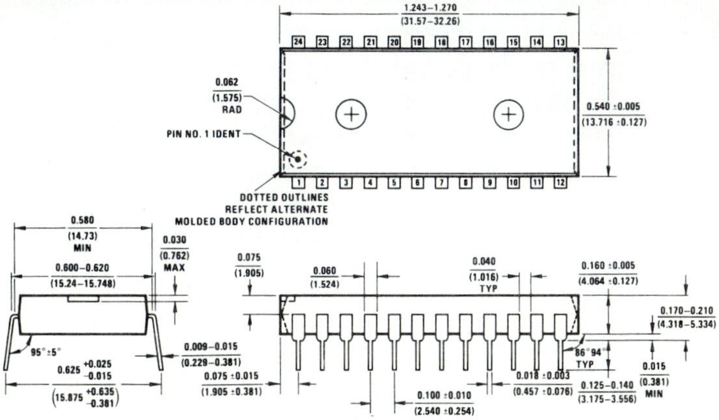

NS Package N24A
24-Lead Molded DIP (N)

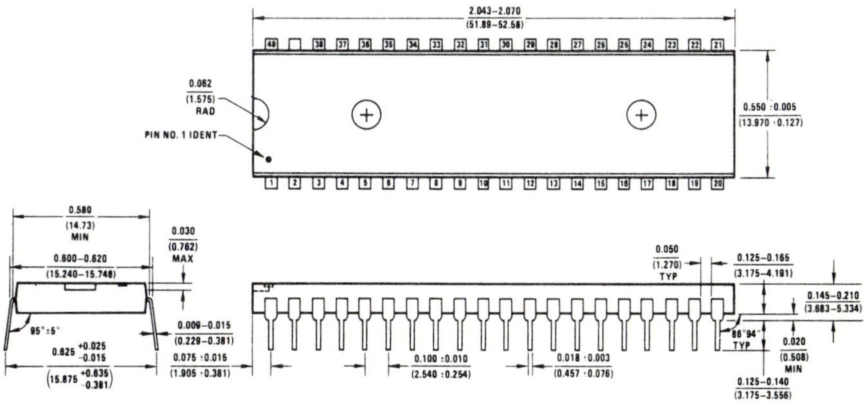

NS Package N40A
40-Lead Molded DIP (N)

Figure 15.4 IC package configurations. Courtesy of National Semiconductor Corp.

15.3 PRACTICAL CONSIDERATIONS IN DIGITAL DESIGN

A number of limitations or constraints must be taken into account when approaching a practical digital electronic design problem. We must be concerned with such factors as the number of inputs per device, the number of devices contained in each integrated circuit package, the number of packages required to implement a complete logic system, noise immunity, power consumption, speed, time delay, and loading of logic outputs. We examine the most important of these factors.

One of the first considerations is that gates come packaged in arrays with a fixed number of gates in each package. A typical package contains either 6 inverters or buffers, 4 two-input gates, 3 three-input gates, 2 four-input gates, or a single eight-input gate. These numbers are chosen to match the number of input and output leads available on a DIP. For example, one standard DIP has 14 pins. Two of these pins must be used to supply power to the chip, which leaves 12 for the gates. A two-input, one-output gate requires 3 leads, so 4 such gates can be packaged on a single chip.

Suppose, for example, a particular design required seven inverters. Since a typical package contains six inverters, you would need two packages, and this would leave five devices unused. On the other hand, consider the fact that a spare NAND or NOR gate may be used as an inverter. This can be seen from Figure 15.1. For example, suppose the first input to a NAND gate is set equal to 1 (tied to the HIGH of the supply). Then the relationship between the second input and the output is that of the inverter.

As another example, if you needed a five-input gate and only had an eight-input gate available, you could set the three unused inputs to the appropriate value and use the remaining five inputs to produce the desired function. In this manner, you may save purchase of additional chips and reduce the associated power and space requirements. Unused input pins should not be left open. Otherwise, they could assume an incorrect voltage level and affect the operation. They should be connected to a suitable logic LOW or logic HIGH level. Alternatively, input pins may be connected together in groups. The function table is used to determine the proper logic level to be applied to unused inputs. For OR and NOR gates, the unused inputs should be connected to a logic 0, or LOW voltage. For AND and NAND gates, the unused inputs should be connected to a logic 1, or HIGH voltage.

Noise immunity, power consumption, and *speed* are important considerations in the design of a logic circuit. The location or environment in which the logic circuit must function should be carefully considered. Locations in which electrical noise is prevalent (such as in factories with large electric motors and tools or near radio, television, or radar transmitters) may require a logic family that has *high noise immunity.* Digital ICs tend to ignore the noise inputs into the system more effectively than do linear ICs, where noise that enters the system propagates throughout the entire system. If a digital IC can distinguish between a HIGH and a LOW input and interpret it correctly, the noise has little effect on the proper operation of the circuit. Because digital families differ in the LOW and HIGH values that turn an IC gate off and on, noise immunity is one of the parameters that determines which digital IC family we choose. High noise immunity means that the logic family is insensitive to noise voltages that are radiated or conducted into the electronic system. That is, the probability of noise making one logic level look like the other is very small. If noise is still a problem, it may be necessary to use shielded enclosures, filtered power, and shielded logic wiring.

Power consumption, which is the power that must be supplied by the power supply, is usually of little consequence for equipment that is connected to a 110 V ac outlet. On the other hand, portable battery-powered equipment requires a logic family that consumes low power. Operating speed requirements also affect the choice of logic families.

Output loading should be carefully checked to make sure that each logic output is not being asked to drive an excessive number of logic inputs. This consideration also applies to any other type of load that exceeds the manufacturer's rating for the device. The number of logic inputs that a single IC can drive is defined as *fan-out*. For example, if all ICs require the same input current, fan out is equal to the amount of load current that the IC can drive divided by the current required into each logic input. The fan out specifies the number of loads that the output of a single gate can drive without its normal operation deteriorating.

Another performance characteristic of logic circuits is *time delay*, or *propagation delay*. This is the time between the application of a logic input and the appearance of the corresponding logic output. This delay can sometimes cause serious problems such as undesired transients, or *glitches*.

Consider the example shown in Figure 15.5(a), which demonstrates how a

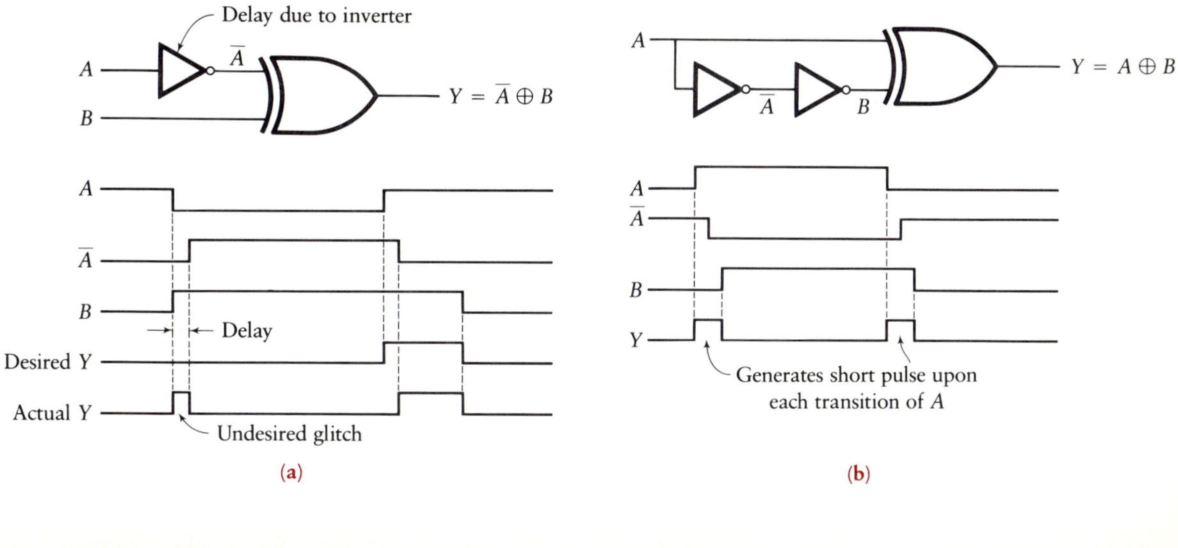

Figure 15.5
(a) Generation of undesired glitch.
(b) Example of useful glitches.

small delay of one logic input can cause a glitch. Underneath the circuit diagram is a series of five sketches, which constitute what is known as a *timing diagram.* We start by drawing an assumed shape for inputs *A* and *B*. In this case, we let *A* start HIGH, then go LOW, and finally go back to HIGH. The second input, *B*, starts LOW. It goes from LOW to HIGH at the same time that input *A* is changing from HIGH to LOW. Then *B* returns to LOW following the second transition in *A*.

The circuit performs an exclusive OR operation between *B* and the inverse of *A*. Thus the output, *Y*, should be HIGH if either *A* is low or *B* is high, but not both. The desired output is shown as "desired *Y*." However, now suppose that the inverter introduces a delay. The resulting output is as shown as "actual *Y*." Note that a narrow pulse has been generated at the first transition time. This undesired pulse is known as a glitch.

Computer simulation programs are available to produce timing diagrams. These programs use the actual propagation delays for the specified digital ICs. One example is *MICRO-LOGIC II* (software by Spectrum Software; student edition available from Addison-Wesley Publishing Company and Benjamin/Cummings Publishing Company).

Sometimes glitches can be useful, as shown in the example of Figure 15.5(b). This example demonstrates how two inverters (or any even number) may be used with an exclusive OR to generate a narrow pulse whenever a logic input changes from one logic state to the other (i.e., at both the rising transition and the falling transition).

Drill Problems

In Drill Problems D15.17 through D15.20, form a five-input AND gate using only the specified gates.

D15.17 Use two-input AND gates.

Ans: See Figure D15.1.

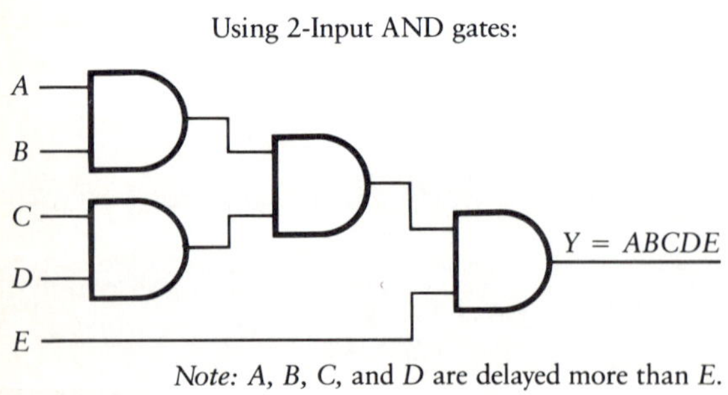

Figure D15.1

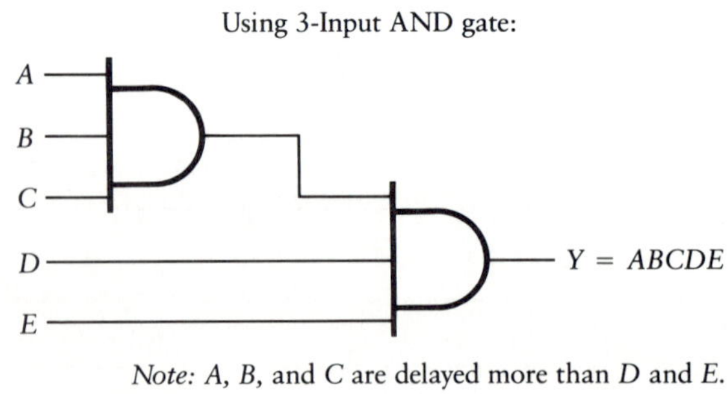

Figure D15.2

Using 4-Input AND gate:

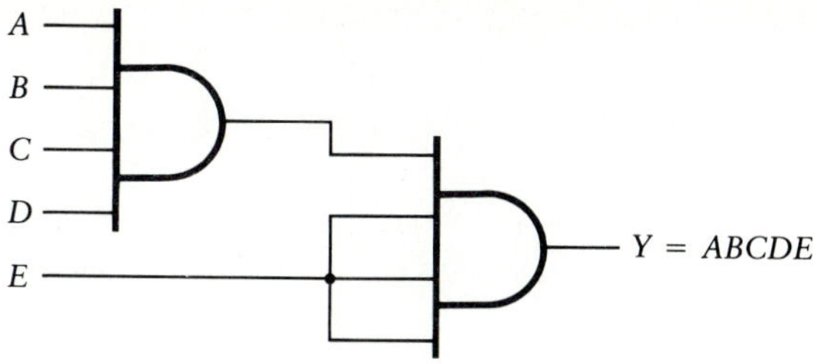

Note: A, B, C, and D are delayed more than E.

Figure D15.3

Using an 8-input AND gate:

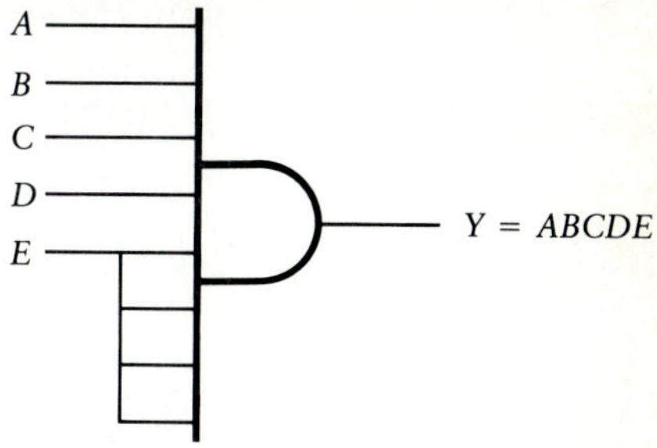

Note: All inputs are delayed the same amount.

Figure D15.4

D15.18 Use three-input AND gates.

Ans: See Figure D15.2.

D15.19 Use four-input AND gates.

Ans: See Figure D15.3.

D15.20 Use eight-input AND gates.

Ans: See Figure D15.4.

15.4 DIGITAL CIRCUIT CHARACTERISTICS OF BJTs

The operation of transistors was considered in the earlier chapters of this text. The design techniques presented in those chapters concentrate on ensuring operation in the linear or active region of the characteristic curves. In the case of digital circuits, we require that the transistor be either ON (in the saturated condition) or OFF (in the cutoff condition). In effect, we operate at either of the two extremes of the transistor characteristic curves, as shown in Figure 15.6(a). When the transistor is saturated, the voltage from the collector to the emitter, v_{CE}, is approximately 0.2 V. That is,

$$V_{CE\,\mathrm{sat}} = 0.2 \text{ V}$$

When the transistor is cut off, it acts like an open circuit between collector and emitter.

The i_B versus v_{BE} characteristic curve is shown in Figure 15.6(b). When v_{BE} is less than about 0.6 V, the transistor is cut off and

$$i_B = i_C = 0$$

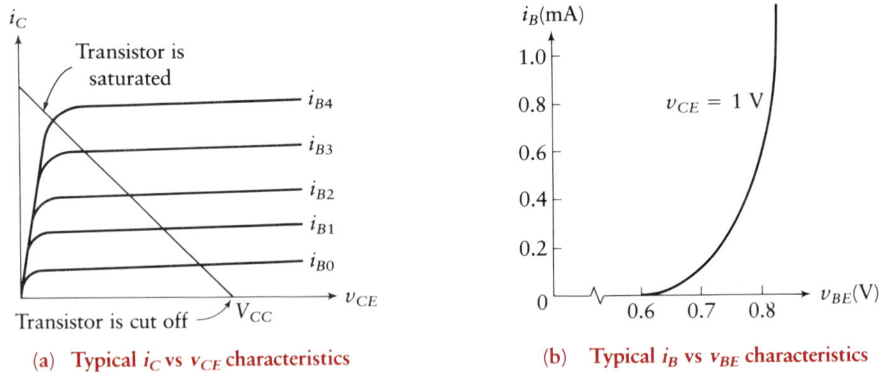

Figure 15.6
npn silicon transistor characteristic.

When v_{BE} increases, i_B and i_C increase rapidly. The saturation voltage between base and emitter, $V_{BE\,(sat)}$, is approximately 0.7 to 0.8 V.

The first step in the analysis of the BJT family of ICs is to prepare a function table that shows the status of each transistor (either ON or OFF) for all possible combinations of high and low levels of each of the inputs. The next step is to determine the numerical value of v_{BE} for each transistor. If v_{BE} is less than 0.7 V, the transistor is cut off and

$$i_B = i_C = 0$$

If v_{BE} is greater than 0.7 V, the transistor could be in the active region or it could be saturated. With digital ICs, we avoid the active region and we cause the transistors to be saturated when they are ON.

15.5 BIPOLAR LOGIC FAMILIES

There are six important families of bipolar logic. These are known as TTL, ECL, RTL, DTL, HTL, and HNIL. The first two of these are discussed in detail in Sections 15.6 and 15.7. The last four are of little current utility and are included in this section only to provide historical perspective.

Resistor-transistor logic (RTL) was the earliest integrated circuit family. It utilizes only transistors and resistors and provides logic gating by placing transistor collectors in parallel. The RTL family has disappeared, since other families dissipate less power and show improved noise immunity.

Diode-transistor logic (DTL) was a popular logic family for several years. It provided improvements in noise immunity and fan-out capabilities as compared to RTL. The gating of DTL is performed by diode OR gates at the input to each logic circuit.

Several manufacturers developed digital logic families that were functionally similar to DTL, but with some of the diodes replaced by zener diodes so that the transistors would not conduct until the input voltages reached a level of about 6 V. This improved the noise immunity, making this logic family more suitable for use in industrial environments (such as factories and refineries, where heavy electrical equipment produces spikes, transients, and other variations in the power line voltage). DTL has been essentially replaced by TTL, since the latter logic family is faster.

The *high-threshold logic* (HTL) and *high-noise-immunity logic* (HNIL) families are currently available only in limited SSI and MSI functions.

15.6 TRANSISTOR-TRANSISTOR LOGIC (TTL)

Transistor-transistor logic (TTL) began as a single, unique logic family. The operation of this family is presented beginning with Section 15.6.1. As the family grew, various requirements for higher speed, lower power, or higher output drive led to the development of several subgroups, as discussed below.

The original logic family is still in use, although most new logic functions are being implemented in newer subgroups of the family. TTL utilizes BJTs and provides moderate speed and power consumption.

The speed of the original TTL circuits is limited by the following conditions:

1. In digital ICs, we operate the transistors in the saturated mode, i.e., turned ON. For a saturated transistor to turn from ON to OFF, we must remove the base charge, much as we discharge a capacitor through a large resistor.
2. The wiring capacitance and the transistor capacitance must discharge through the circuit resistances.

The following subgroups deal with these two limitations to achieve an improved result. We can speed up the operation by:

a. Preventing the transistor from fully saturating when in the ON mode.
b. Reducing the resistor values in the circuit to reduce the time required to discharge the capacitors (i.e., reducing the RC time constant). This method increases the amount of power consumed by the IC.

High-power TTL (H-TTL) was developed for driver circuits and other applications requiring a high fan-out (number of circuits that the output must drive) and high speed. The H-TTL family consumes more power than other TTL subgroups. It is not widely used and is available only in a few logic functions. It utilizes lower resistor values than those used in TTL. This leads to lower time constants and higher operating speeds.

Low-power TTL (LP-TTL) was developed for applications requiring lower power consumption and where reduced operating speed can be tolerated. It utilizes saturating transistors and attains lower power consumption because resistor values are higher than those used in TTL. It has essentially been replaced by LS-TTL as described below.

Schottky TTL (S-TTL) was developed for high-speed applications. It attains high-speed operation by both preventing transistor saturation and reducing resistor values. We prevent transistor saturation by connecting a Schottky diode from base to collector (see Figure 15.7(b)) for each saturated transistor in the IC. The resulting transistor shunted with a Schottky diode is called a *Schottky transistor.* The Schottky diode, which is discussed more fully in Section 15.6.4, diverts some of the base driving current away from the base, thereby preventing the transistor from saturating. This technique of increasing speed is preferred to reducing resistor values, since it does not increase the required power. By reducing resistor values, we can speed up the IC; however, this does increase the power. Figure 15.7 shows a comparison between Schottky transistors and the standard BJT. The base-to-emitter voltage is typically 0.7 V, and the diode forward voltage drop is typically 0.4 V. Thus, when the collector voltage drops to 0.3 V, the excess base current is shunted through the diode to the collector, and the transistor is prevented from saturating. Since the transistor does not saturate, it does not suffer from the delay required to remove the excess base charge and turn the transistor OFF. Propagation delay is approximately 30% lower than that of standard TTL circuitry.

Low-power Schottky TTL (LS-TTL) is a subgroup that provides the speed of the original TTL family but with substantially reduced power consumption.

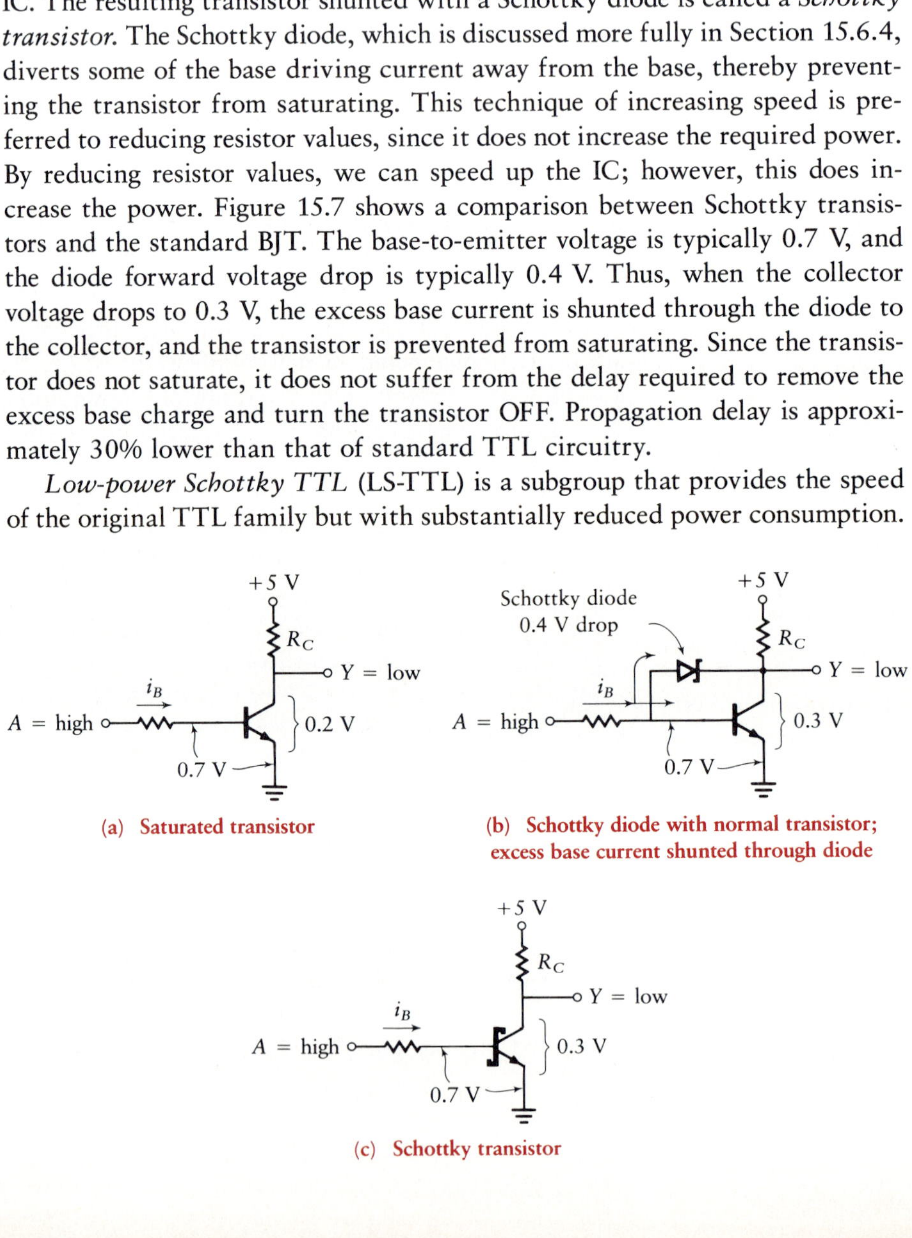

Figure 15.7 Evolution of Schottky transistor.

It can be described as low-power TTL using Schottky transistors instead of saturating transistors. It has been one of the most popular subgroups for many years, and most logic functions are available in LS-TTL.

Advanced Schottky TTL (AS-TTL) is a TTL subgroup that provides somewhat higher speed and lower power than S-TTL.

Advanced low-power Schottky TTL (ALS-TTL) provides an excellent combination of desirable speed and power consumption properties. It utilizes Schottky transistors but incorporates material developments and smaller circuit layouts with reduced capacitance.

15.6.1 Open Collector

We now examine the internal circuitry of a TTL IC. Figure 15.8 shows the schematic diagram and function table for a two-input TTL NAND gate with open collector. The operation of this basic TTL gate depends on the dual-emitter transistor, Q_1. The two inputs, A and B, are connected to the emitters. A TTL-compatible signal should be 0.2 V for the LOW level and between 2.4 and 5 V for the HIGH level. Since we are using positive logic in this text, 0.2 V corresponds to a binary 0 and a voltage between 2.4 and 5 V corresponds to a binary 1. The key to understanding the operation of TTL circuits depends on the input stage, Q_1.

We first let either or both inputs to the gate be brought to logic level 0 (0.2 V). Since the base-emitter junction of Q_1 is forward-biased, the current, I_B, goes through the base to the emitter, and Q_1 operates in the active mode. The large collector-to-emitter current, I_{CQ1}, draws current away from the base of Q_2, so Q_2 is driven into cutoff. As Q_2 turns OFF, the base voltage declines and Q_1 becomes saturated.

We now let both inputs to the gate be brought to logic level 1 (approximately V_{CC}). The current now flows from V_{CC} through R_1 (4 kΩ). This current forward-biases the base-collector junction of Q_1. The base-emitter junction of Q_1 is reverse-biased. Q_1 is operating in the reverse active mode, i.e., with emitter and collector interchanged. This is a nonconventional operation for a transistor in the active range. Q_1 has a low effective β. Although the current into the gate is small, the base current into Q_2 is large enough to drive Q_2 into

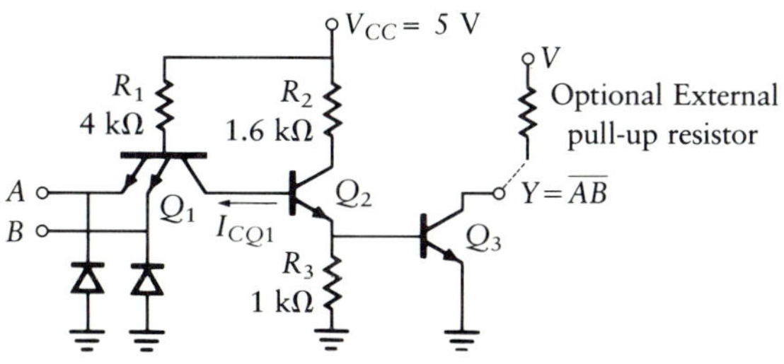

(a) Schematic

A	B	I_{CQ1}	Q_2	Q_3	Y
0	0	+	OFF	OFF	1
0	1	+	OFF	OFF	1
1	0	+	OFF	OFF	1
1	1	−	ON	ON	0

(b) Function table

Figure 15.8 Two-input TTL NAND gate (open collector).

saturation. With Q_2 ON, the saturation voltage, $V_{CE\,sat}$, decreases to approximately 0.2 V.

The output of this open collector gate is taken from the collector of Q_3. A "pull-up" resistor is externally inserted between the IC output and a voltage of level V, frequently 5 V. This pull-up resistor causes the output of the IC to "pull up" to V when Q_3 is OFF. Since the logic block is a NAND gate, the output should be 1 when either or both of the inputs are 0.

When either or both of the emitters of Q_1 are low, I_{CQ1} is large and positive (+ on the function table) and Q_2 turns OFF. With both emitters of Q_1 high, I_{CQ1} is small and negative (− on the function table); that is, a small current flows into the base of Q_2 and Q_2 turns ON. These conditions are shown in the function table of Figure 15.8(b).

With the status of Q_2 determined, we can establish the condition of Q_3. When Q_2 is saturated (ON), the voltage at the base of Q_3 is approximately given by

$$\frac{(V_{CC} - V_{CE})R_3}{R_2 + R_3} = \frac{V_{CC} - 0.2}{2.6 \times 10^3}10^3 = 1.85 \text{ V}$$

We have set $V_{CE} = 0.2$ V. This is sufficient to drive Q_3 into saturation (ON). Alternatively, when Q_2 is cut off (OFF), the voltage at the base of Q_3 is zero and Q_3 is cut off (OFF).

With Q_3 turned OFF, the output, Y, is brought to HIGH (the voltage V), and with Q_3 turned ON, the output, Y, is brought to LOW (0.2 V). This is summarized in the function table of Figure 15.8(b).

Let us now examine some of the major applications of the open collector gate. The gate can be used as a means of interfacing the TTL logic family to another logic family. For example, in a system that includes both TTL and CMOS ICs (discussed later in this chapter), the HIGH and LOW voltage levels may be different. In particular, the HIGH level for TTL is between 2.4 and 5 V, whereas for CMOS it can be 10 V. If the pull-up resistor of Figure 15.8(a) is connected to 10 V, the open collector gate acts as an interface between the TTL portion of the system and the CMOS portion.

A second application of the open collector gate occurs when several gates are attached to a common *bus.* In the operation of this common bus, each gate can control the logic level of the bus only by bringing it LOW, but not HIGH. With the collector left open, each gate that has Q_3 OFF exerts no control on the bus. Only when the output transistor is ON does the gate bring the bus to logic level 0.

A third application of the open collector occurs when a lamp or relay is driven from the output of a gate. The lamp or relay is placed between the output terminal of the open collector and the voltage source through an appropriate limiting resistor. When Q_3 is saturated, a path to ground exists and the lamp or relay is ON. When Q_3 is OFF, there is no path to ground, so the lamp or relay is OFF.

15.6.2 Active Pull-Up

Figure 15.9(a) shows a two-input TTL NAND gate with active pull-up on the output. This circuit configuration is known as a *totem-pole* output. The circuit is similar to that of Figure 15.8(a) except for the addition of D_3, Q_4, and R_4. This type of gate is used when increased operating speed is required. The *propagation delay* for an open collector gate is approximately 35 ns. The propagation delay for the active pull-up gate is only about 8 ns. The clear advantage is in increasing the speed of this totem-pole compared to that of the open collector output gate. One potential problem in the use of these totem-pole output TTL devices is that during the transition from ON to OFF and OFF to ON, large current spikes are drawn from the power supply, which can interfere with other ICs. The effects of this problem can be reduced by physically placing a bypass capacitor next to the IC and connected between the V_{CC} terminal of the IC and ground.

Suppose first that Q_2 and Q_3 are ON (the bottom entry in the table of Figures 15.8(b) and 15.9(b)). The voltage at the base of Q_4 is then given by

$$V_B(Q_4) = V_{CE}(Q_2) + V_{BE}(Q_3)$$
$$= 0.2 + 0.7 = 0.9 \text{ V}$$

The voltage necessary to turn Q_4 ON must be greater than 1.4 V. That is,

$$V_{BE}(Q_4) + V(D_3) + V_{CE}(Q_3) = 0.6 + 0.6 + 0.2 = 1.4 \text{ V}$$

Since the base voltage of Q_4 is only 0.9 V, Q_4 is OFF, as shown in the bottom entry of Figure 15.9(b). When Q_2 and Q_3 are OFF, Q_4 saturates. The output goes to logic 1. Since the load capacitance can now discharge through a resistor of about 150 Ω (130 Ω plus the resistances of Q_4 and D_3), the time to

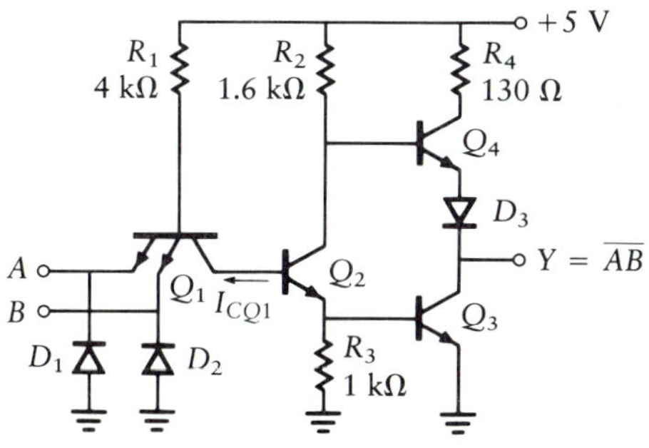

(a) Schematic

A	B	I_{CQ1}	Q_2	Q_3	Q_4	Y
0	0	+	OFF	OFF	ON	1
0	1	+	OFF	OFF	ON	1
1	0	+	OFF	OFF	ON	1
1	1	−	ON	ON	OFF	0

(b) Function table

Figure 15.9 Two-input TTL NAND gate (active pull-up).

change from 1 to 0 is less than for the open collector gate, which must discharge through a larger value pull-up resistor.

The totem-pole output gate cannot be used for the applications cited in the previous section. For example, suppose we tried to connect two totem-pole outputs to a common bus. If the output of one gate was HIGH and the output of the other was LOW, an excessive amount of current would result and this could overheat and damage the ICs.

Notice that we have added protective diodes D_1 and D_2 in the schematic diagrams of Figures 15.8 and 15.9. These prevent input A or B from going negative. This is done because the IC can be destroyed if the input is brought to more than 0.7 V below ground with a low-impedance source.

15.6.3 H-TTL and LP-TTL Gates

The two-input NAND gate for the H-TTL circuit is shown in Figure 15.10(a), and the function table is presented as Figure 15.10(b). Note that H-TTL has one more transistor than the TTL NAND gate with active pull-up. The function tables are identical through the column for Q_2. Starting with the next column, the state of Q_3 for the H-TTL gate is opposite to that of Q_2. In the H-TTL gate, transistors Q_3 and Q_5 are fed by the collector and emitter of Q_2. In the TTL NAND gate of Figure 15.9(a), transistors Q_4 and Q_3 are fed by the collector and emitter of Q_2. The results are therefore comparable (i.e., columns Q_3 and Q_5 of Figure 15.10(b) match columns Q_4 and Q_3 of Figure 15.9(b)).

Figure 15.10
Two-input H-TTL NAND gate.

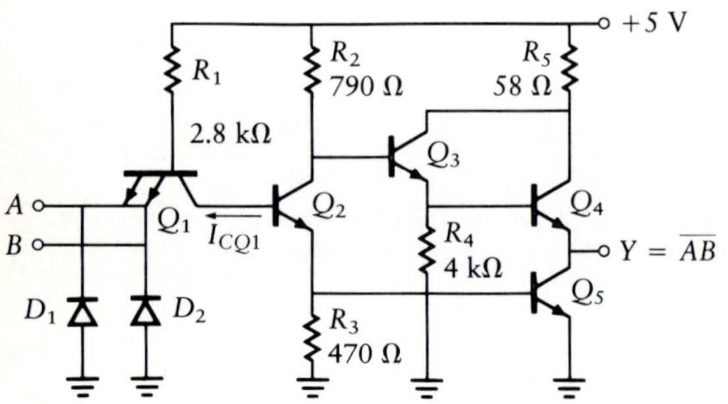

(a) Schematic

A	B	I_{CQ1}	Q_2	Q_3	Q_4	Q_5	Y
0	0	+	OFF	ON	ON	OFF	1
0	1	+	OFF	ON	ON	OFF	1
1	0	+	OFF	ON	ON	OFF	1
1	1	−	ON	OFF	OFF	ON	0

(b) Function table

15.6.4 Schottky TTL Gates

The Schottky subgroup of the TTL family is designed to reduce the propagation delay time of the standard TTL gates. The time needed for a transistor to switch from ON to OFF can be greatly reduced if the transistor is not permitted to go into saturation. Refer to Figure 15.7(b), where, by placing a Schottky diode between base and collector, we shunt base current through the diode.

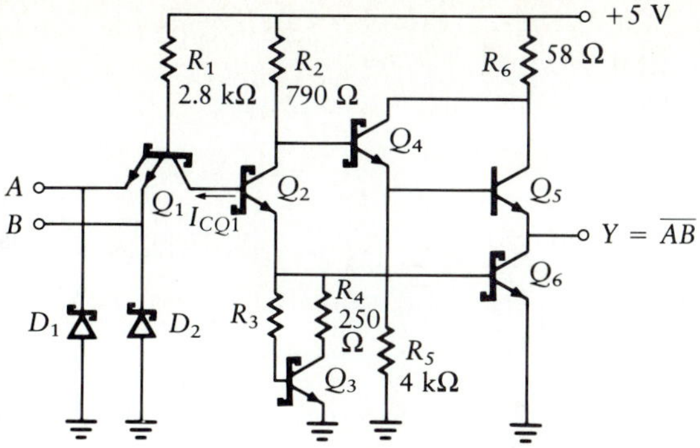

(a) Schematic

A	B	I_{CQ1}	Q_2	Q_4	Q_5	Q_6	Y
0	0	+	OFF	ON	ON	OFF	1
0	1	+	OFF	ON	ON	OFF	1
1	0	+	OFF	ON	ON	OFF	1
1	1	−	ON	OFF	OFF	ON	0

(b) **Function table** (Q_3 is used for base drive control of Q_4 and is not ON or OFF in a logic sense)

Figure 15.11 Two-input S-TTL NAND gate.

This reduces current in the base, thereby preventing the transistor from going into saturation. A Schottky diode differs from a conventional diode in that the Schottky diode is formed with a connection of a *metal* and a semiconductor, rather than a junction of p- and n-type semiconductor material. The voltage across a conducting Schottky diode used in this configuration is 0.4 V. The Schottky diode and the transistor are combined to form a Schottky transistor as in Figure 15.7(c).

The circuit schematic and the function diagram for the two-input NAND gate are shown in Figure 15.11. The transistors in the gate are of the Schottky type except for Q_5, which is of the standard type. The protection diodes are also Schottky. In this circuit, Q_5 and Q_4 form a *Darlington pair,* providing a high current gain and low output resistance (see Section 6.4 for a discussion of Darlington pair transistor configurations). The resistor values in Figure 15.11(a) are approximately one-half of those in the standard TTL circuit of Figure 15.9. This increases the speed of the gate; however, the power dissipation is approximately doubled.

The schematic for the S-TTL gate (Figure 15.11(a)) has an additional sixth transistor, Q_3, which does not appear in the H-TTL gate. The state of this sixth transistor is not carried in the table, since it has no direct effect on the states of the other transistors. In performance, Q_3 is analogous to the Q_4 transistor of the active pull-up (totem pole) discussed in Section 15.6.2. The fixed resistor of Figure 15.9(a) is replaced with an active circuit comprising R_3, R_4, and Q_3. Just as the totem pole speeds up the turn-on and turn-off time of the output stage, this active circuit speeds up the turn-on and turn-off time of Q_6. Hence the propagation delay of the gate is reduced. Verification of the entries in the function table is left to the student; the procedure parallels that of the previous sections.

15.6.5 Tri-State Gates

The *tri-state* family of gates combines the high-speed advantage of the totem-pole output with the advantages of an open collector output. However, because

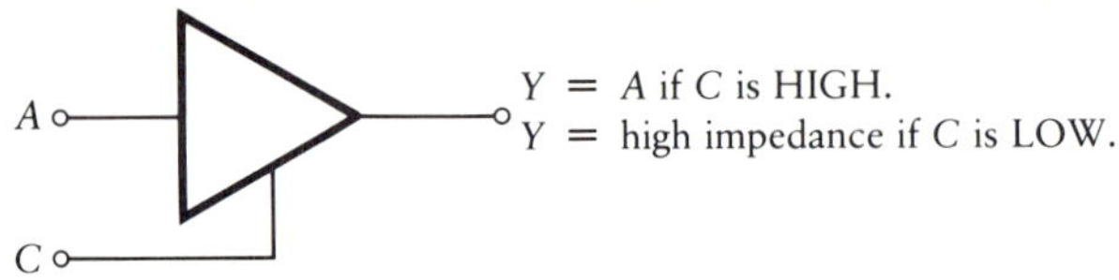

Figure 15.12
Tri-state bus driver.

of its design, it can be connected to a bus system (as an open collector gate can). The output transistors, i.e., Q_3 and Q_4 of Figure 15.9, can assume three states (hence the name tri-state) as follows:

1. $Y = 0$ when Q_3 is ON and Q_4 is OFF.
2. $Y = 1$ when Q_3 is OFF and Q_4 is ON.
3. Y is an open circuit when Q_3 is OFF and Q_4 is OFF.

The tri-state bus driver (buffer) is shown in the schematic of Figure 15.12. Tri-state operation is achieved with a control line, C, which permits the gate to operate normally if C is HIGH. The gate exhibits a high impedance when C is LOW. This permits several of these gates to be attached to a common bus, as with the open collector gate.

Drill Problems

Determine the state of each transistor and the output in the circuits shown in Drill Problems D15.21 and D15.22.

D15.21

Figure D15.5

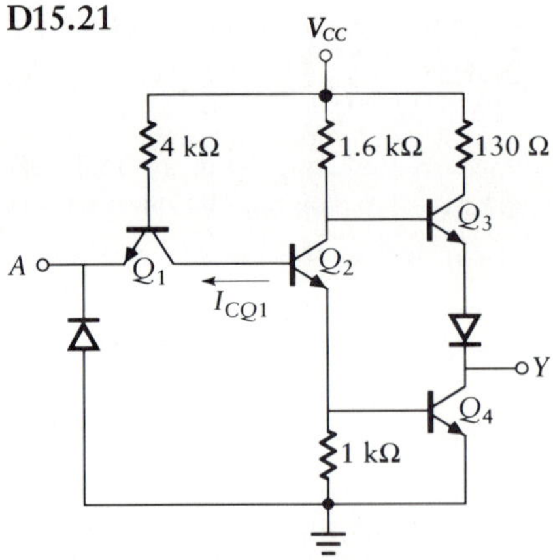

A	I_{CQ1}	Q_2	Q_3	Q_4	Y
0	+	F	N	F	1
1	–	N	F	N	0

N = ON
F = OFF

D15.22

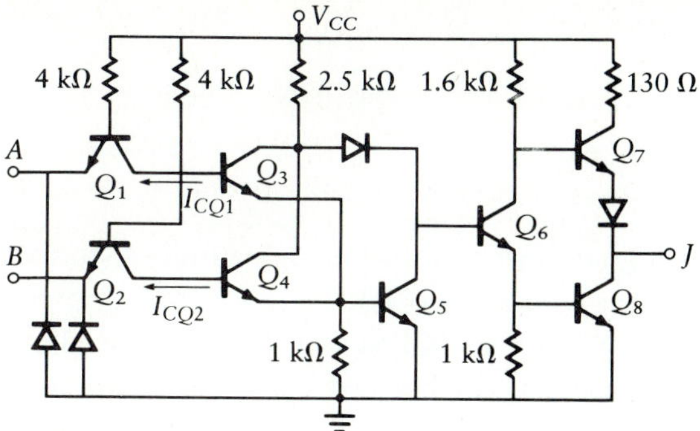

Figure D15.6

Ans:

A	B	I_{CQ1}	I_{CQ2}	Q_3	Q_4	Q_5	Q_6	Q_7	Q_8	J
0	0	+	+	F	F	F	N	F	N	0
0	1	+	−	F	N	N	F	N	F	1
1	0	−	+	N	F	N	F	N	F	1
1	1	−	−	N	N	N	F	N	F	1

N = ON
F = OFF

15.6.6 Device Listings

Table 15.1 shows a partial listing of the devices in the TTL family. At this point, we suggest that the student obtain one or more of the *data books* supplied by the various IC manufacturers. These are listed in the references at the

Table 15.1 Partial TTL Device Listing.

Type Number	Description
7400	Quad 2-input NAND
7401	Quad 2-input NAND, open collector
7402	Quad 2-input NOR
7403	Quad 2-input NOR, open collector
7404	Hex inverter
7405	Hex inverter, open collector
7406	Hex inverter, open collector to 30 V
7407	Hex buffer/driver, open collector to 30 V
7408	Quad 2-input AND
7409	Quad 2-input AND, open collector
7410	Triple 3-input NAND
7411	Triple 3-input AND
7414	Hex Schmitt-trigger inverters
7420	Dual 4-input NAND
7421	Dual 4-input AND
7427	Triple 3-input OR
7430	8-input NAND
7432	Quad 2-input OR
7486	Quad 2-input XOR

TTL Data Book, Vol. 2, Texas Instruments, Inc., Dallas, Texas.

end of this text. It is virtually impossible to complete digital electronic designs successfully without access to these data books.

15.7 EMITTER-COUPLED LOGIC (ECL)*

The ECL family operates at the highest speed of all the logic families studied in this text because none of the transistors are operated in saturation. Since propagation delays of 1 to 2 ns are achievable, ECL is useful in high-speed applications such as radar signal processors, high-speed computers, and data transmission.

We illustrate the MECL 10,000, one of several types of ECL gate, in Figure 15.13. This basic gate can be divided into three sections:

Figure 15.13 MECL 10,000 basic gate. Courtesy of Motorola Semiconductor Products, Inc.

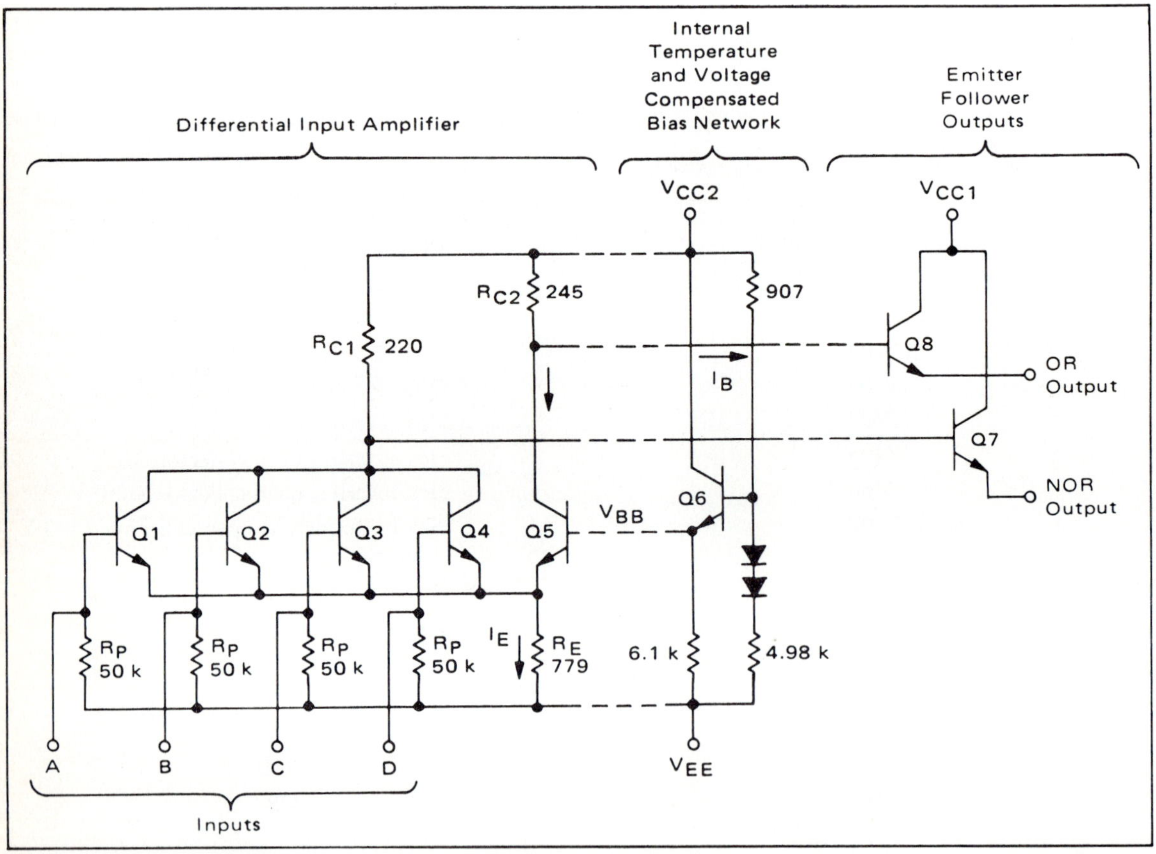

* Much of the material in this section is courtesy of Motorola Semiconductor Products, Inc.

1. The *differential input amplifier section,* which contains the differential amplifiers. This section provides the logic gating and voltage gain for the circuit.
2. An *internal temperature- and voltage-compensated bias network,* which supplies a reference voltage for the differential amplifier. The bias voltage, V_{BB}, is shown in Figure 15.13 and is set at −1.29 V, which is the midpoint between a 1 and a 0. V_{CC1} and V_{CC2} are set to zero, and V_{EE} is −5.2 V. The two diodes and Q_6 provide the required temperature compensation.
3. The *emitter-follower output devices,* which provide level shifting from the differential amplifier to the ECL output levels. The emitter followers also provide a low-impedance output for driving transmission lines.

The ECL family utilizes a differential amplifier circuit that allows current to flow in only one of the transistors at a time. This is known as *current steering.* One side of the differential pair (Q5) is connected to an internal reference voltage, while the other side is composed of two or more transistors connected in parallel (in Figure 15.13, four parallel transistors are shown). If the base voltage of one or more of these transistors is driven to a level higher than the internal reference voltage, the current is steered through the transistor with the highest input voltage. The collector voltages of the differential pair change by only about 0.8 V. The transistor voltages and currents are controlled to prevent transistor saturation without requiring special Schottky transistors. Emitter-coupled logic uses a negative supply voltage and consumes a large amount of power. Output voltage levels are −0.8 V for logic 1 and −1.8 V for logic 0. The devices are more susceptible to noise than are most other logic families. The fast rise and fall times of the output waveform often require use of special wiring techniques to prevent overshoot, ringing, or reflections of the wave returning from the other end of a cable.

Figure 15.14 shows an example of an ECL gate that provides both an OR and a NOR function. The Y output provides the OR, and Z provides NOR. The function table is shown in Figure 15.14(b). You should take the time to verify the entries in this table.

Figure 15.14 Two-input ECL OR/NOR gate.

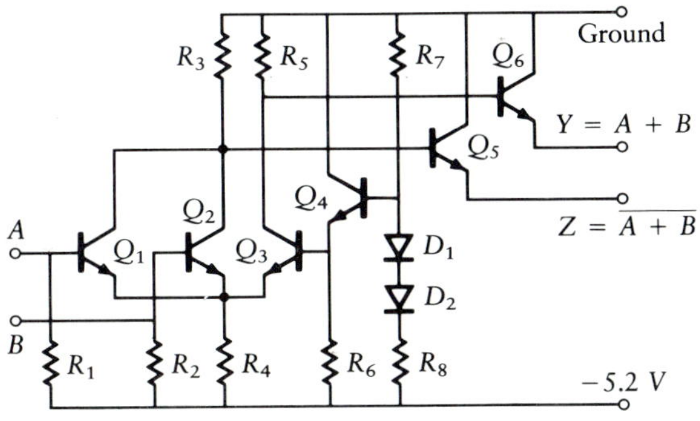

(a) Schematic

A	B	Q_1	Q_2	Q_3	Y	Z
0	0	OFF	OFF	ON	0	1
0	1	OFF	ON	OFF	1	0
1	0	ON	OFF	OFF	1	0
1	1	ON	ON	OFF	1	0

(b) Function table

15.7.1 Device Listings

Table 15.2 presents a partial listing of devices in the ECL family. Refer to an ECL data book for additional details.

Table 15.2 Partial ECL Device Listing.

Type Number	Description
10100/10500	Quad 2-input NOR with strobe
10101/10501	Quad OR/NOR
10102/10502	Quad 2-input NOR
10103/10503	Quad 2-input OR
10104/10504	Quad 2-input AND
10105/10505	Triple 2-3-2-input OR/NOR
10106/10506	Triple 4-3-3-input NOR
10107/10507	Triple 2-input exclusive OR/exclusive NOR
10109/10509	Dual 4-5-input OR/NOR
10110	Dual 3-input 3-output OR
10111	Dual 3-input 3-output NOR
10113/10513	Quad exclusive OR
10117/10517	Dual 2-wide 2-3-input OR-AND/OR-AND-invert
10118/10518	Dual 2-wide 3-input OR-AND
10119/10519	4-wide 4-3-3-input OR-AND
10121/10521	4-wide OR-AND/OR-AND-invert
10123	Triple 4-3-3-i bus driver

MECL Data Book, Motorola Semiconductor Products, Inc., Mesa, Arizona.

15.8 DIGITAL CIRCUIT CHARACTERISTICS OF FETs

The characteristics of the *n*-channel and *p*-channel enhancement MOSFETs for operation in the linear mode are presented in Section 4.4.2. In the current section we consider the characteristics of the *n*- and *p*-channel enhancement MOSFETs for use in digital applications, where the transistor is operating in an ON-OFF mode. The enhancement MOSFET is used in IC applications because of its small size and simple construction.

15.8.1 *n*-Channel Enhancement MOSFET

Refer to Figure 4.11, where we present the schematic of the physical structure, the symbol, and the characteristics for an *n*-channel enhancement MOSFET.

Note that this transistor has no channel between source and drain (a broken line is shown in the symbol). However, as the gate-to-source voltage, v_{GS}, becomes more positive, an *n*-channel region forms that extends from source to drain. This region provides a low resistance between source and drain. In order to turn the transistor ON, we apply a v_{GS} that is greater than the threshold voltage, V_T, as shown in Figure 4.11(c). This voltage, V_T, can vary upward from 1 V depending on the device. To turn the transistor OFF, we simply let $v_{GS} < V_T$ (we usually say $v_{GS} = 0$) so that no channel is formed. With the gate grounded, the resistance between source and drain is high, and the transistor is OFF. The enhancement MOSFET is normally OFF because no channel exists between source and drain. The gate-to-source voltage, v_{GS}, must be greater than V_T to form the channel and hence turn the transistor ON.

It is important to note that the input gate is an open circuit and resembles a small capacitor. No input current is required, except for the brief period when we charge or discharge this small input gate capacitor. When $v_{GS} = 0$, the resistance between source and drain is high and we can model this as an open switch. When v_{GS} is greater than V_T, the transistor turns ON and the channel between source and drain becomes equivalent to a low-value resistor (i.e., a closed switch). The only voltage drop is a small drop from source to drain across this resistance.

15.8.2 *p*-Channel Enhancement MOSFET

The *p*-channel enhancement MOSFET is shown in Figure 4.12, where we present the schematic of the physical structure, the symbol, and the characteristics. This device is the mirror image of the *n*-channel enhancement MOSFET. With a negative voltage applied to v_{GS} (i.e., when $v_{GS} < V_T$), the channel is formed and the transistor turns ON. When v_{GS} is zero, the transistor is an open circuit.

15.9 FET TRANSISTOR FAMILIES

In this section we consider *n*-channel and *p*-channel MOS integrated circuits. These are considered separately to aid in understanding the operation of the more useful CMOS devices. Recall from Chapter 4 that the input (or gate) of a FET is an open circuit. Since the gate of a FET leads only to an equivalent capacitor, no input current is required except for that needed to charge and discharge the small gate capacitor. This means that we are not concerned with the number of gates that can be driven from a single gate (fan-out).

15.9.1 *n*-Channel MOS

The n-*channel MOS* (NMOS) ICs are constructed with *n*-channel MOS transistors. An example of an NMOS inverter and its function table are shown in Figure 15.15.

We analyze the *n*-channel FET by noting that if the gate voltage and the substrate voltage are the same, the transistor is OFF. When the gate voltage is greater than the substrate voltage, the transistor is ON. With Q_n ON, the output voltage is near zero, depending on the value of R_L and the FET characteristics. We show it as zero in the function table. With Q_n OFF, the output is raised to $+V$ volts. The function table of Figure 15.15 shows that with the input HIGH, the output is LOW, and vice versa. This is the definition of an inverter.

Figure 15.15
NMOS inverter.

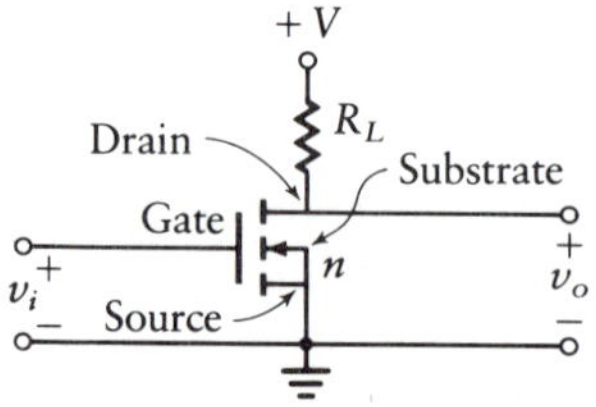

v_i	Q_n	v_o
0	OFF	$+V$
$+V$	ON	0

Although NMOS has some good qualities, it suffers from the need for large current from the power supply during the ON state. A large power supply would be needed to drive a system composed of multiple NMOS ICs.

15.9.2 *p*-Channel MOS

The p-*channel MOS* (PMOS) IC is constructed with *p*-channel MOS transistors. These were the first types of memory circuits to be developed. An example of a PMOS inverter and its function table are shown in Figure 15.16. We analyze this circuit by noting that if the voltage on the gate (the input voltage) is the same as the voltage on the substrate, the transistor is OFF. If the voltage on the gate is less than that on the substrate, the transistor is ON. With Q_p ON, the output voltage is raised to nearly $+V$, depending on R_L and the FET characteristics. We show it as $+V$ in the function table. With Q_p OFF, the output voltage is zero. Again, the function table of Figure 15.16 is that of an inverter. That is, a LOW input produces a HIGH output and vice versa. The

Figure 15.16
PMOS inverter.

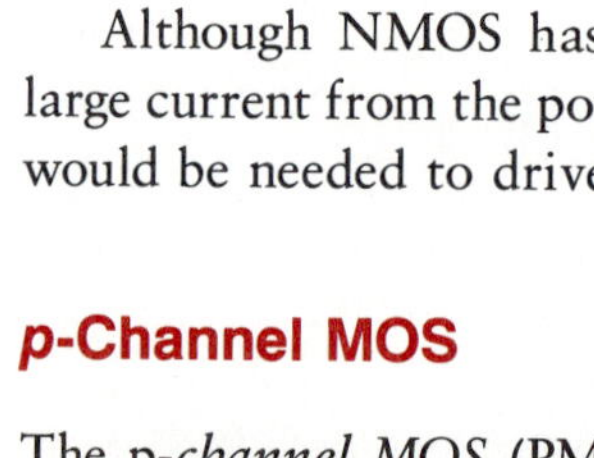

v_i	Q_p	v_o
0	ON	$+V$
$+V$	OFF	0

problems associated with PMOS are similar to those of NMOS. In fact, this brief presentation of both NMOS and PMOS is made only to lead to the exciting CMOS family of ICs. By combining NMOS and PMOS technology, we obtain the complementary MOS (CMOS) family, as discussed in the next section.

15.10 COMPLEMENTARY MOS (CMOS)

Complementary MOS (CMOS) integrated circuits are constructed with complementary pairs, or sets, of p-channel and n-channel MOS transistors. A CMOS inverter is formed from one p-channel and one n-channel MOSFET, as shown in Figure 15.17. Notice that no resistor (such as R_L in Figures 15.15 and 15.16) is needed in this circuit. The load resistor for the p-channel MOSFET is the n-channel MOSFET, and the load resistor for the n-channel MOSFET is the p-channel MOSFET.

The function table is formed by reference to the function tables of Figures 15.15 and 15.16. When v_i is zero, Q_p is ON and Q_n is OFF, so the output is connected to $+V$ through the on-resistance of Q_p. With no load resistor on the output, $v_o = +V$ as shown in the first line of the function table. When v_i is $+V$ volts, Q_p is OFF and Q_n is ON. Now the output is connected to ground through the on-resistance of Q_n. With no load resistor on the output, $v_o = 0$ V. Notice that the small supply power is needed only when the gate changes state. It takes a little energy to charge and discharge the gate capacitor and to supply the source-drain energy while the transistors are briefly on during transitions between states.

Figure 15.18 shows a CMOS two-input NAND gate. The circuit behaves like four switches as shown in Figure 15.18(b). Two of the switches are connected in parallel to V_{DD} and one is controlled by $\overline{A}$ and $\overline{B}$. Two are connected in series to ground and are controlled by A and B. When A is LOW, Q_1 is ON and Q_3 is OFF. When A is HIGH, Q_1 is OFF and Q_3 is ON. Likewise, when B is LOW, Q_2 is ON and Q_4 is OFF. When B is HIGH, Q_2 is OFF and Q_4 is ON. The function table is shown in Figure 15.18(c). Notice that in this table we use a 0 to indicate a low voltage near 0 and a 1 to indicate a voltage near V_{DD} (depending on the load on the gate output).

Figure 15.17
CMOS inverter.

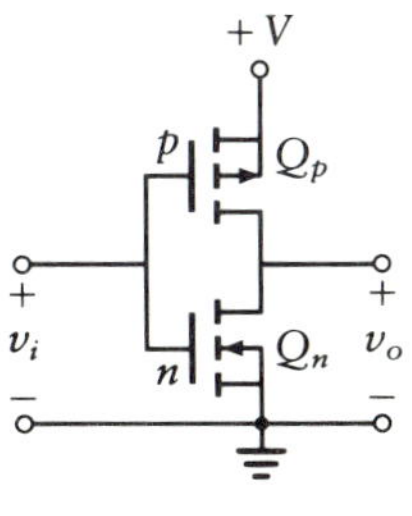

v_i	Q_p	Q_n	v_o
0	ON	OFF	$+V$
$+V$	OFF	ON	0

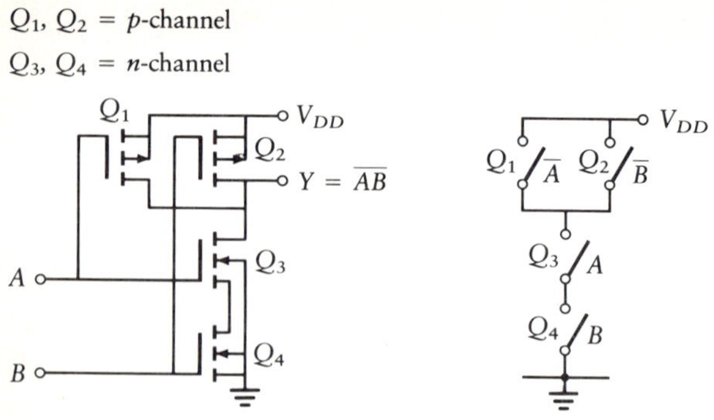

(a) (b)

A	B	Q_1	Q_2	Q_3	Q_4	Y
0	0	ON	ON	OFF	OFF	1
0	1	ON	OFF	OFF	ON	1
1	0	OFF	ON	ON	OFF	1
1	1	OFF	OFF	ON	ON	0

(c)

Figure 15.18
Two-input CMOS NAND gate.

CMOS logic can operate over a wide range of supply voltages, typically 3 to 18 V. Special circuits, such as digital watches, can operate at lower supply voltages. The primary form of power consumption for CMOS logic involves the charging and discharging of wiring and load capacitance as logic levels switch between ground and the supply voltage. As a result, CMOS operates with low power supply drain (especially at low frequencies, where capacitive impedance is high). CMOS is ideally suited for battery-powered portable devices.

CMOS is slower than most of the bipolar logic families. However, as the manufacturers succeed in reducing the physical size of the devices, the capacitance is reduced and the speed of operation is increased. CMOS ICs cannot drive as much power into output devices as other types of ICs. However, since the input to a CMOS gate is essentially an open circuit, a large number of CMOS inputs can be driven by one CMOS output, so a high-powered output is not so important. In addition, a buffered series of CMOSs is available for supplying a larger output power.

The CMOS family exhibits excellent noise immunity, as can be seen from the input-output transfer curve of Figure 15.19. The output voltage changes state at an input voltage of approximately $V/2$. Thus, noise must be of sufficient magnitude to break this threshold voltage in order for an error to occur. In addition, the CMOS family allows for large swings in voltage from HIGH to LOW logic states.

We summarize the advantages and characteristics of the CMOS logic family as follows:

The gate input is an open circuit, so is easy to drive.

Power-supply current is low, so CMOS is ideal for battery-powered devices.

Good noise immunity exists, since the logic changes from LOW to HIGH at $V/2$.

CMOS can operate over a wide range of power supply voltage.

CMOS creates little noise of its own.

We have considered the CMOS inverter and the CMOS NAND gate. We

Figure 15.19 CMOS input-output transfer curve.

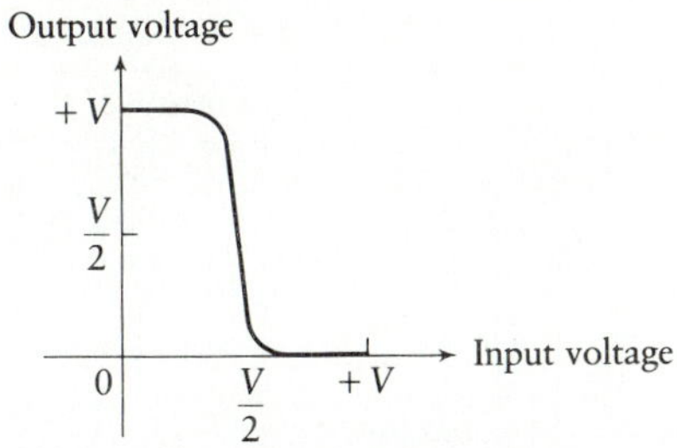

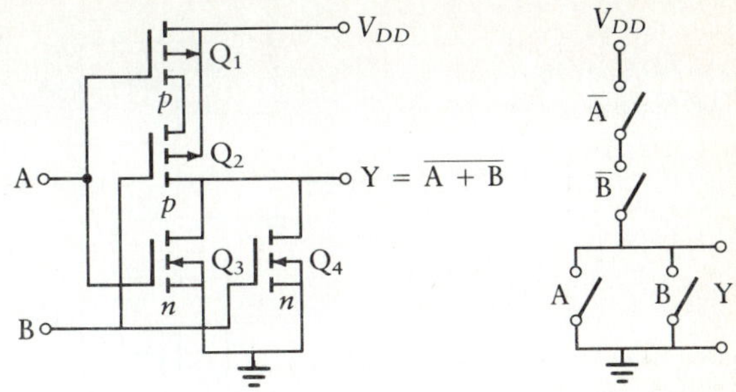

(a) **Circuit** (b) **Schematic**

A	B	Q_1	Q_2	Q_3	Q_4	Y
0	0	ON	ON	OFF	OFF	1
0	1	ON	OFF	OFF	ON	0
1	0	OFF	ON	ON	OFF	0
1	1	OFF	OFF	ON	ON	0

(c) **Function table**

Figure 15.20 Two-input CMOS NOR gate.

now examine the CMOS two-input NOR gate, as shown in Figure 15.20. This circuit can be represented by two switches in series with V_{DD} and two switches in parallel to ground, as shown in Figure 15.20(b). The two switches in series are controlled by A and B. The two switches in parallel are driven by $\overline{A}$ and $\overline{B}$. We use the rule that if the voltages on the gate and on the substrate are the same, the transistor is OFF. If the gate voltage is different from the substrate voltage, the transistor is ON. We obtain the function table of Figure 15.20(c) by applying these conditions.

15.10.1 CMOS Analog Switch

In this section, we discuss a useful circuit that is not found in the TTL or ECL families—the *analog switch* or *transmission gate*. This device can turn on or turn off an analog signal with a peak voltage below the supply voltage. It uses a digital control signal in the same range. When the control line is low the switch is open, and when the control line is high the switch is closed.

A circuit diagram for the MM74HC4016 (CD 4016) analog switch is shown in Figure 15.21. The switch consists of two sections. Q_1 and Q_2 form a CMOS inverter identical to that shown in Figure 15.17. The second section is composed of two parallel transistors, Q_3 (an n-channel FET) and Q_4 (a p-channel FET). The gate of Q_3 is driven with A and the gate of Q_4 is driven with $\overline{A}$. As a result, when $A = 0$ ($\overline{A} = 1$) the switch is open. When $A = 1$ ($\overline{A} = 0$) the switch is closed. The supply voltage should be between 3 and 15 V, and the maximum input/output voltage is limited to the supply voltage.

The switch is enabled in 20 ns, with an input voltage in the range of 0 to

Figure 15.21
MM74HC4016
CMOS analog switch.

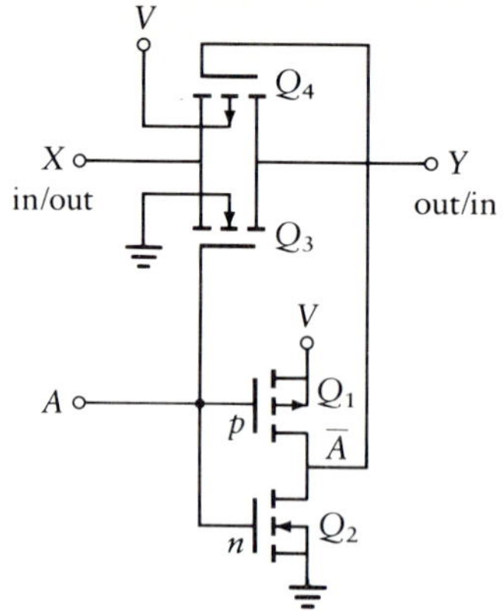

Ans:

A	$\bar{A}$	Q_1	Q_2	Q_3	Q_4	Condition
0	N	F	1	F	F	Open
1	F	N	0	N	N	X = Y

N = ON
F = OFF

12 V and V_{CC} = 15 V. At room temperature, the switch typically has 50-Ω on-resistance and low leakage current (+0.1 nA) when off. The switches are *bilateral*—either switch terminal can be used as the input. All analog inputs and outputs and the control lines are protected from electrostatic damage by diodes connected to V_{CC} and ground.

The analog switch provides the electronic engineer with a versatile tool to use in a number of design applications, including:

1. Analog signal switching and multiplexing
2. Modulation/demodulation
3. Commutation/chopping
4. Sampling for analog-to-digital conversion
5. Digital control of frequency, phase, and gain
6. Switching regulation
7. Switched capacitor filter design

Drill Problems

Determine the status of each transistor and the output in the circuits shown in Problems D15.23 and D15.24.

D15.23

Figure D15.7

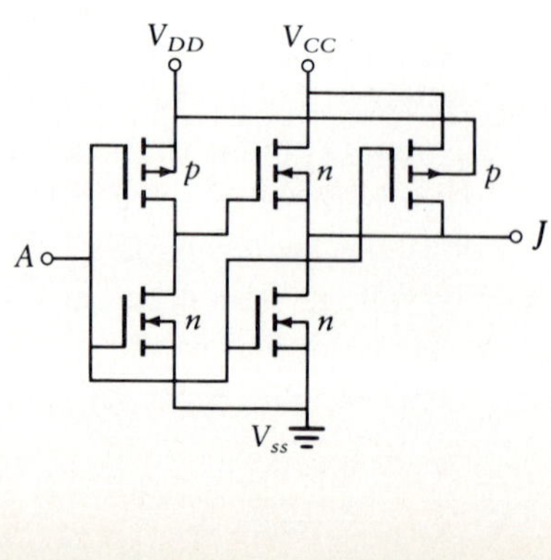

Ans:

A	Q_1	Q_2	Q_3	Q_4	Q_5	J
0	N	F	N	F	N	1
1	F	N	F	N	F	0

N = ON
F = OFF

D15.24

Figure D15.8

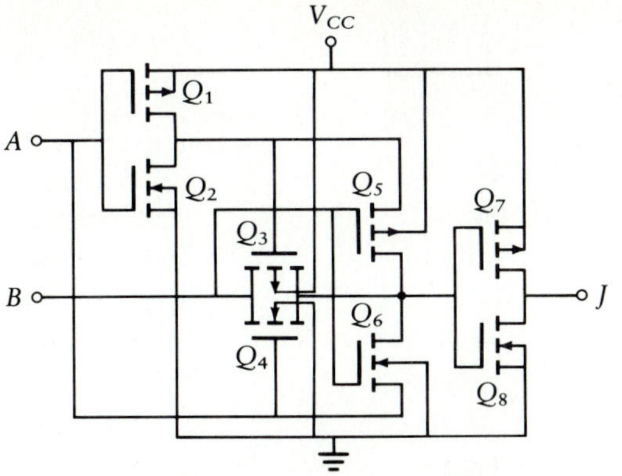

Ans:

A	B	Q_1	Q_2	Q_3	Q_4	Q_5	Q_6	Q_7	Q_8	J
0	0	N	F	F	F	N	F	F	N	0
0	1	N	F	F	F	F	N	N	F	1
1	0	F	N	N	N	N	F	N	F	1
1	1	F	N	N	N	F	N	F	N	0

N = ON
F = OFF

15.10.2 CMOS Device Listings and Usage Rules

Table 15.3 shows a partial listing of devices of the CMOS family. We again refer you to the appropriate data books for more details and an expanded listing.

We conclude with the following rules, which are important when using the CMOS family:

1. Attach every terminal to $+V$, ground, or an input signal.
2. Avoid exceeding the limits of the IC, thereby preventing the input diodes from conducting.
3. Avoid static electricity by storing CMOS devices on conductive foam or foil.

15.11 COMPARISON OF LOGIC FAMILIES

Figure 15.22 summarizes the parameters of the various logic families. It is presented to provide comparisons among the various groups and aid in selecting the proper family of circuit for a particular application. The most important families are TTL, ECL, and CMOS.

Table 15.3 Partial CMOS Device Listing.

Type Number	Description
MM54HC00/MM74HC00	Quad 2-input NAND gate
MM54HC02/MM74HC02	Quad 2-input NOR gate
MM54HC03/MM74HC03	Quad 2-input open drain NAND gate
MM54HC04/MM74HC04	Hex inverter
MM54HC08/MM74HC08	Quad 2-input AND gate
MM54HC10/MM74HC10	Triple 3-input NAND gate
MM54HC11/MM74HC11	Triple 3-input AND gate
MM54HC14/MM74HC14	Hex inverting Schmitt trigger
MM54HC20/MM74HC20	Dual 4-input NAND gate
MM54HC27/MM74HC27	Triple 3-input NOR gate
MM54HC30/MM74HC30	8-input NAND gate
MM54HC32/MM74HC32	Quad 2-input OR gate
MM54HC51/MM74HC51	Dual AND-OR-invert gate
MM54HC58/MM74HC58	Dual AND-OR gate
MM54HC86/MM74HC86	Quad 2-input exclusive OR gate
MM54HC132/MM74HC132	Quad 2-input NAND Schmitt trigger
MM54HC133/MM74HC133	13-input NAND gate
MM54HC266/MM74HC266	Quad 2-input exclusive NOR gate
MM54HC4002/MM74HC4002	Dual 4-input NOR gate
MM54HC4049/MM74HC4049	Hex inverting logic level down converter
MM54HC4050/MM74HC4050	Hex logic level down converter
MM54HC4075/MM74HC4075	Triple 3-input OR gate
MM54HC4078/MM74HC4078	8-input NOR/OR gate
MM54HCT00/MM74HCT00	Quad 2-input NAND gate
MM54HCT04/MM74HCT04	Hex inverter
MM54HCT05/MM74HCT05	Hex inverter (open drain)
MM54HCT34/MM74HCT34	Noninverter gate (TTL input)
MM54HCU04/MM74HCU04	Hex inverter

CMOS Databook, National Semiconductor, Inc., Santa Clara, California.

15.12 CONCLUDING REMARKS

You have been introduced to the topic of digital electonics and gained some familiarity with the basic time-independent logic families. Now you are in a position to execute the first level of digital electronic design and to formulate some useful circuits.

If we return to the auto safety control box described in the introduction to

Figure 15.22 Logic family comparison.

Logic Family	Supply Voltage	Power per Gate	Propagation Delay per Gate	Maximum Clock Frequency	Maximum Logic Zero Input	Minimum Logic One Input	Maximum Logic Zero Output	Minimum Logic One Output
RTL	3.6 V	20 mW	10 nS		.50 V	.88 V	0.3 V	
DTL	5 V	8 mW	30 nS	5 MHz				
HTL	15 V	40 mW	110 nS	0.5 MHz	6.5 V	8.5 V	1.0 V	14.4 V
TTL	5 V	10 mW	10 nS	35 MHz	0.8 V	2.0 V	0.4 V	2.4 V
HTTL	5 V	22 mW	6 nS	50 MHz	0.8 V	2.0 V	0.4 V	2.4 V
LPTTL	5 V	1 mW	33 nS	3 MHz	0.8 V	2.0 V	0.4 V	2.4 V
STTL	5 V	16 mW	4 nS	75 MHz	0.8 V	2.0 V	0.5 V	2.7 V
LSTTL	5 V	2 mW	10 nS	40 MHz	0.8 V	2.0 V	0.5 V	2.7 V
ALSTTL	5 V	1 mW	4 nS	50 MHz	0.8 V	2.0 V	0.5 V	2.5 V
ASTTL	5 V	8 mW	2.5 nS	100 MHz	0.8 V	2.0 V	0.5 V	3.0 V
ECL	− 5.2 V	25 mW	2 nS		−1.48 V	−1.13 V	−1.6 V	−0.98 V
PMOS	− 9 V − 5 V	≈ 1 mW	4 μS	100 kHz	−4.0 V	−1.2 V	−8.5 V	−1.0 V
NMOS	+ 5 V +12 V	≈ .1 mW	≈100 nS	3 MHz	0.8 V	2.4 V	0.4 V	2.4 V
CMOS	3–15 V	0.5 mW*	100 nS	3 MHz**	1.5 V	3.5 V	0.5 V	4.5 V
HCMOS	5 V	0.5 mW*	10 nS	30 MHz	1.0 V	3.5 V	0.05 V	4.95 V

*at 1 MHz
**at 5 V
(Some parameters vary with device type or manufacturer)

this chapter, we find that this device can be effectively designed using the basic blocks presented in this chapter.

Observe that the change from analog to digital electronics will go a long way toward reducing the effects of ignition noise on the control circuitry. You could choose the CMOS family for your design, since it provides excellent noise immunity and the operating frequencies in your system are low. However, since power is not of critical concern (an automotive battery looks like a virtually unlimited source of power when compared to the power required by ICs!), TTL could also be used.

Example 15.2

Design a theft alarm system for an automobile. Use two 555s (see Section 14.5), one operating in the monostable mode and one operating in the astable mode.

SOLUTION* A block diagram for the alarm is shown in Figure 15.23. Either a motion sensor, which we attach to the automobile, or a door-open detector, which turns the dome light on and off, will trigger the 555 monostable and develop a 60-s pulse. The motion detector momentarily closes a switch when the car is moved, and the door-open detector is the switch on the car door that turns the dome light on when the door is opened.

The alarm is activated by a lock switch located on the front fender of the car. When the alarm is on, an indicator LED starts to blink continuously. This LED alerts anyone approaching the car that the alarm is set.

When the lock switch is on, the astable 555 generates a continuous pulse

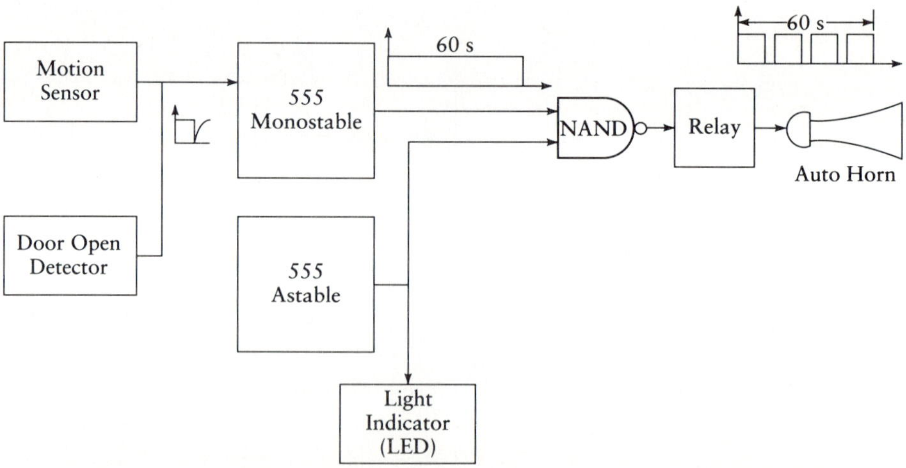

Figure 15.23 Block diagram for Example 15.2.

* This project was designed and built by James Mao and presented at the IEEE Design Competition at California State University, Los Angeles, in February 1989.

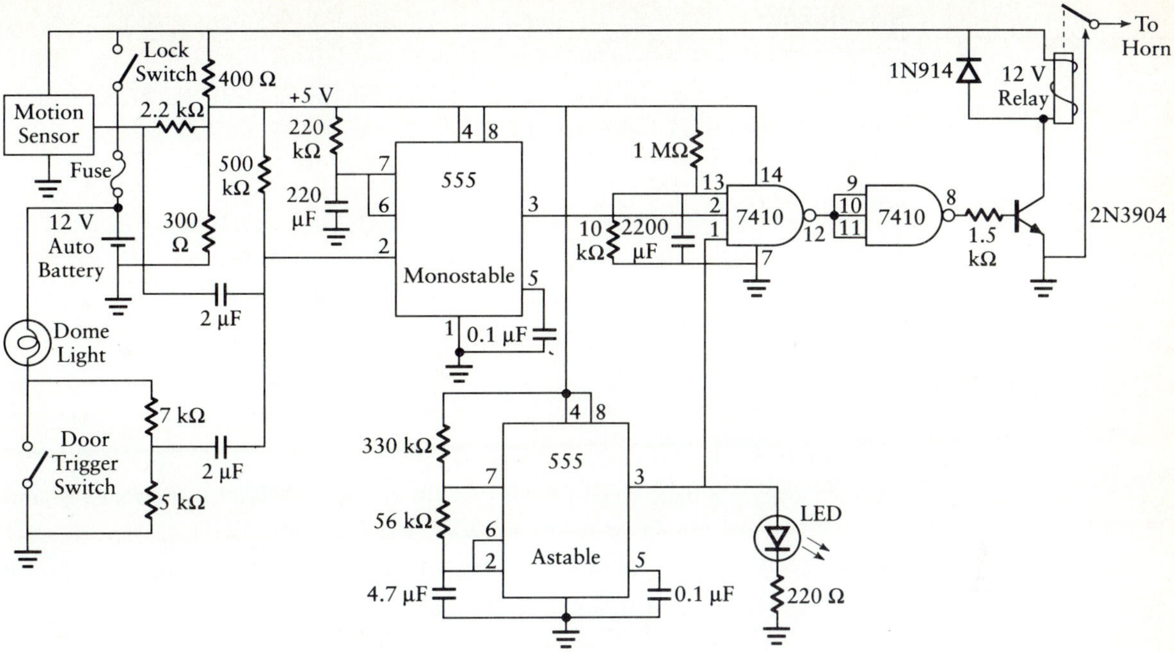

Figure 15.24
Wiring diagram for Example 15.2.

train for the LED to blink continuously and provides the signal for the car horn. We desire the horn "on period" to be longer than the "off period."

The duty cycle, $(R_1 + R_2)/R_2$, is variable. In this design, we set it to 6.89. The frequency of this astable 555 oscillator, $1.44/[(R_1 + 2R_2)C_1]$, can be adjusted with the selection of R_1, R_2, and C_1. In this design, we select $C_1 = 4.7\ \mu\text{F}$ and $f = 0.69$ Hz. The wiring diagram is shown in Figure 15.24.

When the motion sensor or the door-open detector triggers the 555 monostable, it develops a pulse. The pulse duration time, $1.1R_3C_2$, is set here to 53.2 s. If the value of C_2 was changed to 470 μF, the pulse duration time would increase to approximately 2 minutes.

The alarm uses the 12-V car battery, and the ICs are supplied with power from a voltage divider that produces 5.14 V. We decide to use the TTL family, since we have so much available power. We use a 7410 NAND gate, discussed in this chapter, instead of an AND gate because of lower cost and also because it can be used as an inverter. An *npn* transistor is used for the relay driver. A diode is connected in parallel with the relay coil to protect the transistor from relay inductance transients.

The motion sensor is readily available, as is the lock switch. The alarm can be further improved by using a remote control system, as developed for a home garage door opener, to replace the lock switch located on the fender.

PROBLEMS

Constuct logic diagrams and function tables for the Boolean algebraic expressions of Problems 15.1 through 15.7.

15.1 $Y = \overline{A} \cdot B \cdot \overline{C} \cdot D$

15.2 $Y = \overline{A} \cdot B \cdot \overline{C} + A \cdot B \cdot \overline{C}$

15.3 $Y = AB + A\overline{B}$

15.4 $Y = A\overline{B} + B$

15.5 $Y = A + B$

15.6 $Y = \overline{A + B + C}$

15.7 $Y = \overline{A \cdot B \cdot C}$

15.8 Develop a logic circuit such that the output is HIGH if inputs *A*, *B*, and *C* are all HIGH or if inputs *D*, *E*, and *F* are all HIGH.

15.9 Develop a logic circuit such that the output is HIGH if inputs *A* and *B* are HIGH or if inputs *C* and *B* are HIGH.

Develop logic networks to satisfy the functions given in Problems 15.10 through 15.19.

15.10 To win a prize, you must send in a receipt and also have the correct answer to at least one of the questions.

15.11 Today's luncheon special consists of a hamburger with either soup or salad, but not both.

15.12 You can paint the walls blue or green, but not both. But even if you do not paint the walls, you must paint the ceiling white.

15.13 A pair of kings or a pair of aces will win the hand.

15.14 They will rent the house to a couple or a single person, but not to both.

15.15 On this TV set, you can get the sound and picture separately or together on Channel 2, but only the sound is available on Channel 4.

15.16 To get in, you must have \$3 and a discount card, or you must pay an additional dollar.

15.17 You must attend at least one night session or one afternoon session of either day of the conference.

15.18 If you take a course in law or history or both, you must also take one in either English or speech, but not both.

15.19 You can play doubles at tennis, but if a player on either team fails to show up, the game is called off.

15.20 Design a four-passenger auto seat belt warning system that will sound an alarm if the ignition is on and there is a passenger with an unfastened seat belt. Normally open switches with one contact grounded are available to sense the presence of passengers and to sense the connection of each seat belt. This is shown in Figure P15.1(a). A normally open switch with one contact grounded senses whether the engine ignition is on. This circuit is shown in Figure P15.1(b). A 2-kHz signal should sound for any unsafe condition.

15.21 Design a safety reminder system for your automobile to sound a 2-kHz tone when the headlights are left on and the key is left in the switch. This system uses the following normally open switches:

a. Headlight switch is closed when lights are on and open when lights are off.
b. Ignition key switch is closed when key is in the lock and open when key is removed from the lock.
c. Door switch is closed when the door is open and open when the door is closed.

The alarm sounds only when the door is open and the headlights are on and/or the key is in the ignition switch.

15.22 Design a burglar alarm system that will sound a 500 Hz, 5 W sound. An alarm switch is built into the left front fender. It is a normally open switch. When the key is turned to the "armed" position, the switch is closed. A motion switch is mounted in the automobile such that when any motion occurs, a momentary contact is made. This is shown in Figure P15.2. When any disturbance to the automobile occurs, the output signal momentarily drops to zero (ground). The burglar alarm should sound the (500 ±10% Hz) signal *only* if the fender switch was

Figure P15.1

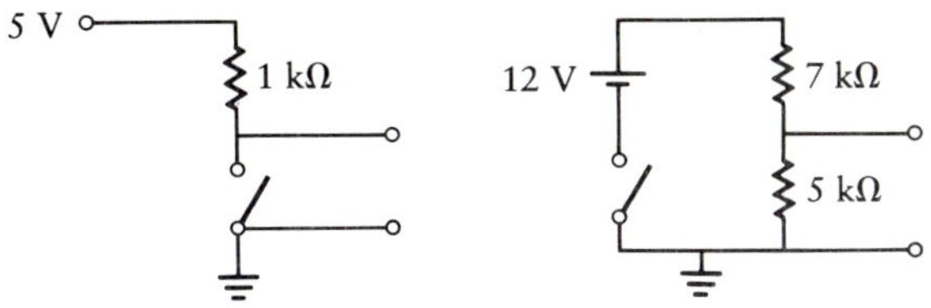

(a) Switch is closed when a passenger is sitting. Same for seat belt.

(b) When the ignition is turned on, the switch is closed.

Figure P15.2

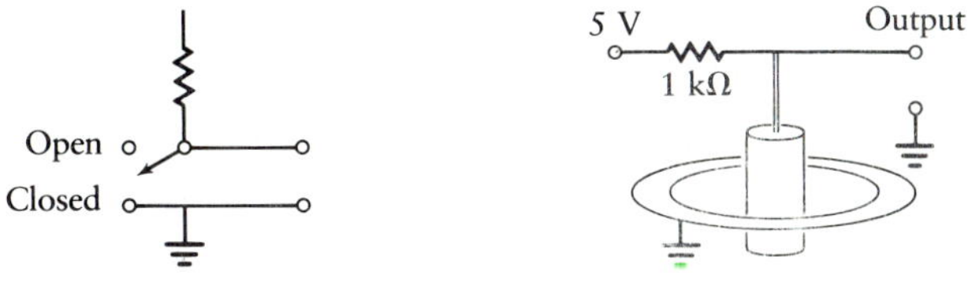

previously armed. The alarm should sound for 2 minutes and then shut off.

15.23 Analyze the auto burglar alarm system illustrated in Figure P15.3. Describe the operation of this electronic system.

15.24 Design the electronic circuitry to accomplish the following operation. A 3-ms-wide input pulse must be reproduced and sent on to another circuit. After a 10-ms delay, an echo pulse of 3 ms must follow, as shown in Figure P15.4.

15.25 Design a digital lock that operates in two steps. First, the code is set by activating 3 of the 6 two-position push-button switches. For this example, set the code to be: b d e. Second, the toggle switch connected to the output is thrown. If the correct code is selected, the output will go LOW and the lock will be released with a solenoid (a solenoid is an electromechanical device that produces a mechanical motion when a signal is applied to the solenoid coil). If the incorrect code is selected, a 1-kHz tone will sound for 1 minute. See Figure P15.5 for details.

15.26 Sketch the output of Figure 15.5(b) if an odd number of inverters is used in the circuit. Assume the delay time for one inverter is 35 ns.

Figure P15.3

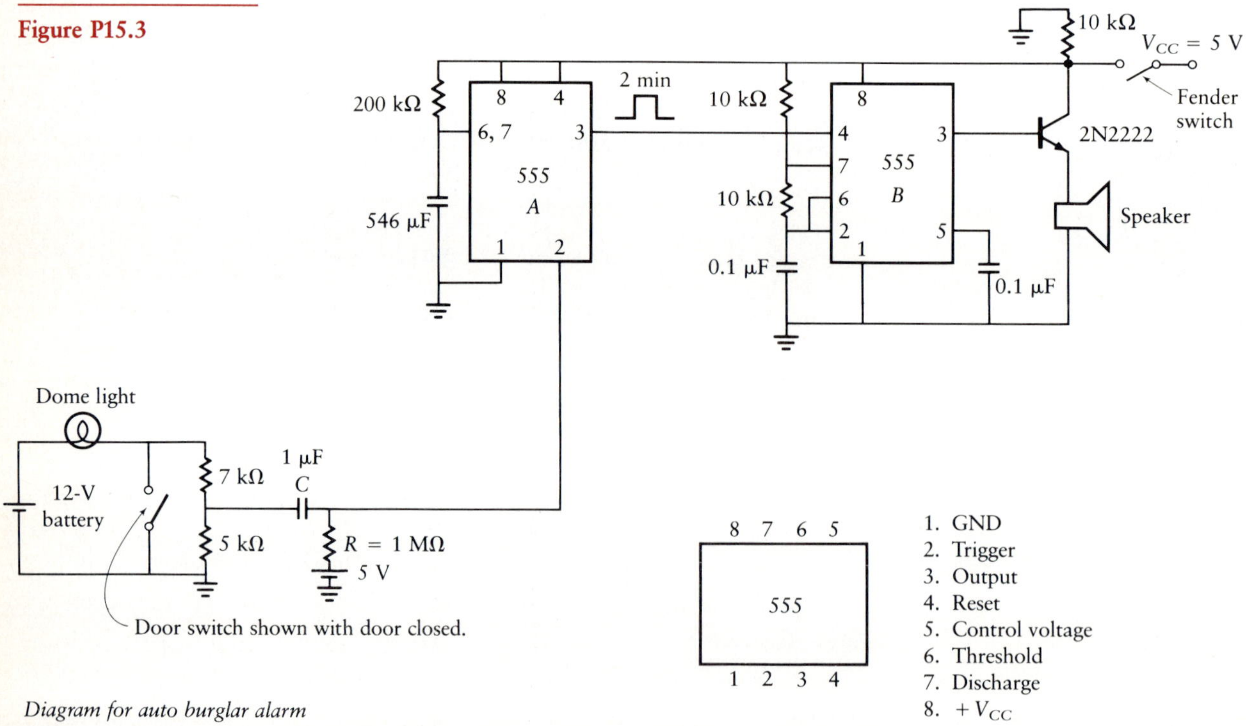

Diagram for auto burglar alarm

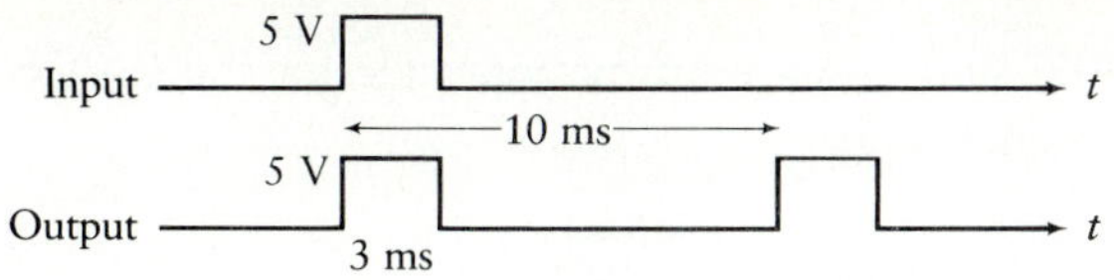

Figure P15.4

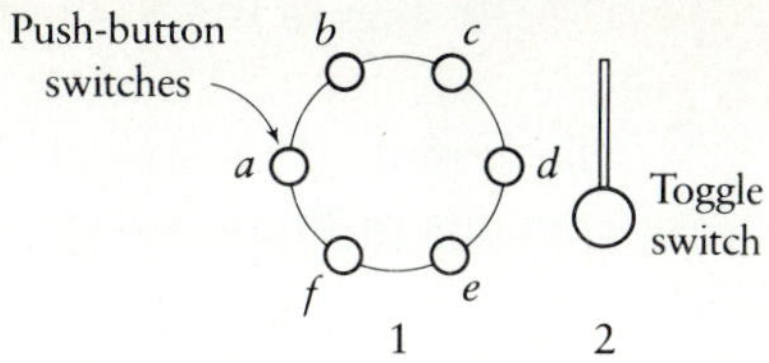

Figure P15.5

Figure P15.6

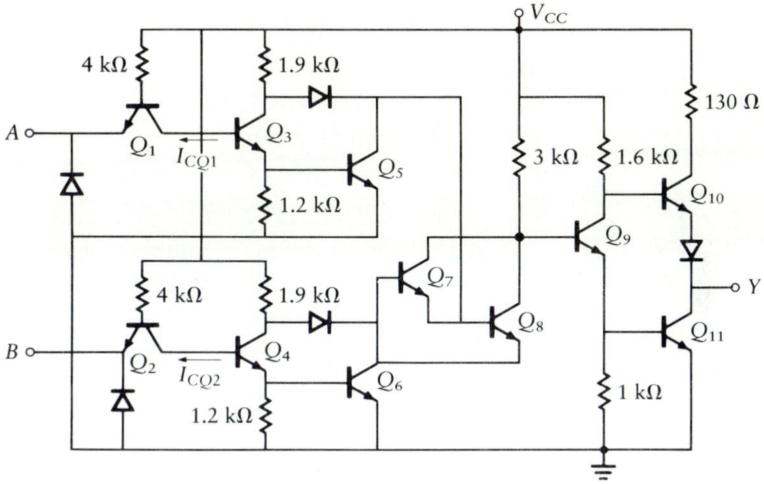

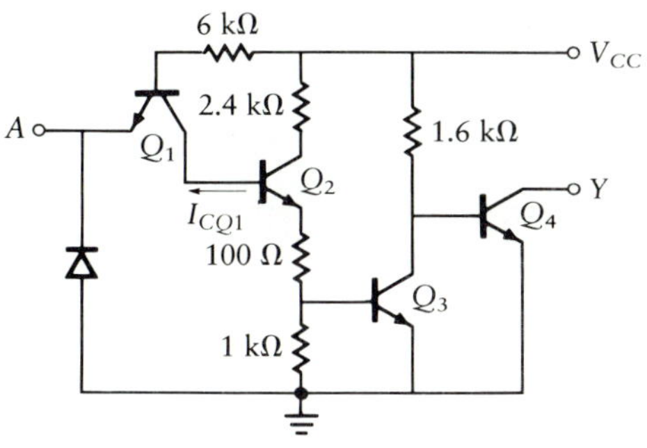

Figure P15.7

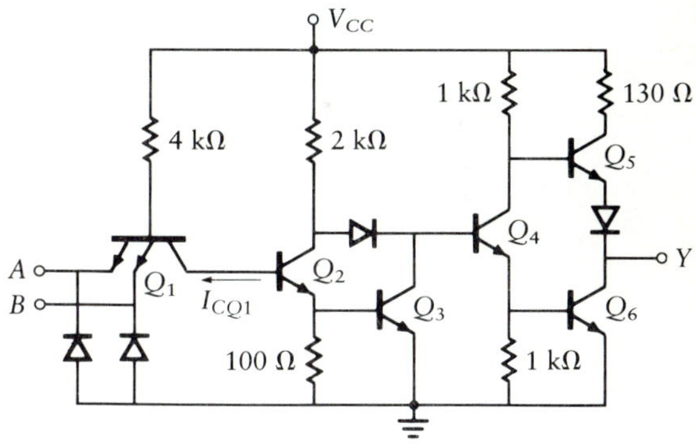

Figure P15.8

Determine the status of each transistor and prepare a function table for each of the TTL circuits given in Problems 15.27 through 15.30.

15.27 Figure P15.6—a 7486 quad two-input XOR, active collector output.

15.28 Figure P15.7—a 7407/17 hex driver, noninverting, open collector output.

15.29 Figure P15.8—a 7408 quad two-input AND, active collector output.

15.30 Figure P15.9—a 7400 quad two-input NAND, active collector output.

Determine the status of each FET and prepare a function table for the CMOS circuits of Problems 15.31 through 15.33.

15.31 Figure P15.10—a CD4001 quad two-input NOR.

15.32 Figure P15.11—a CD4-81 quad two-input AND.

15.33 Figure P15.12—a CD4001 quad two-input NAND.

15.34 Complete the function table for the TTL CMOS circuit shown in Figure P15.13.

15.35 We are using a 7447 BCD-to-seven-segment decoder/driver to drive a seven-segment LED display. Design an electronic system to vary the brightness of the display from almost off to maximum. Use a variable duty cycle astable 555 in your design.

15.36 Design a circuit to sample a function of time every 100 ms. The sampling time must be no greater than 5 ms. Draw the circuit diagram showing the complete pin-out of each IC used. The function of time, $f(t)$, and the sampled output are shown in Figure P15.14.

15.37 Design a system to provide an emergency warning signal. Use a 555 and a CMOS switch to provide a signal form as shown in Figure P15.15. The

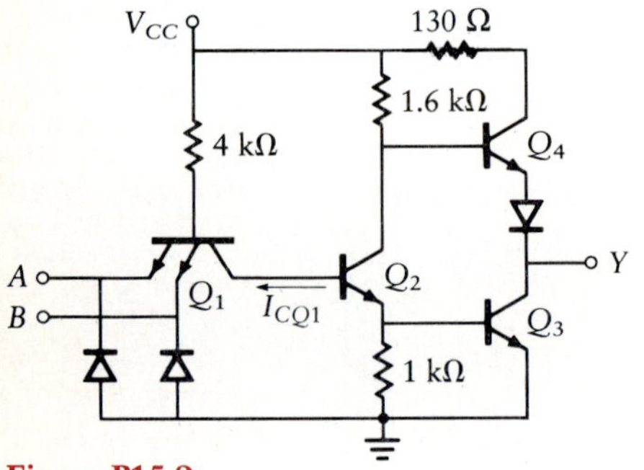

Figure P15.9

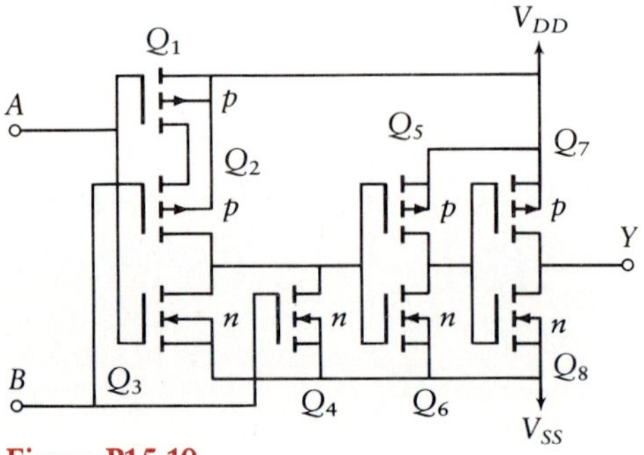

Figure P15.10

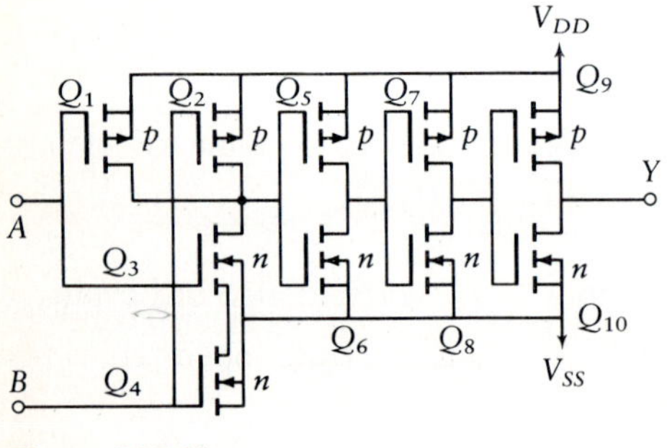

Figure P15.11

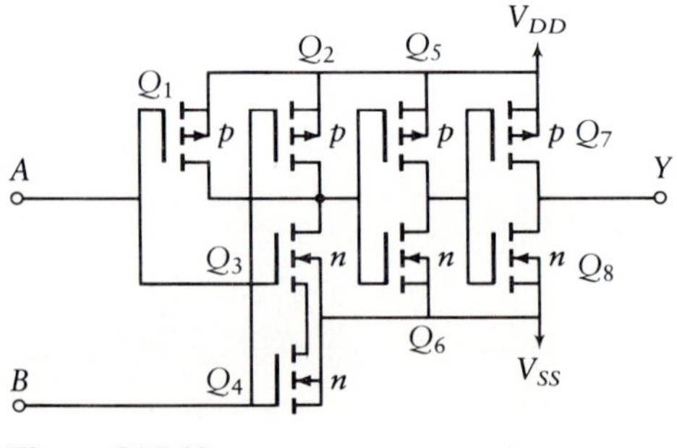

Figure P15.12

signal consists of a 3-kHz sinusoid with ±5 V amplitude that is turned on and off 30 times each minute. The time on is approximately equal to the time off. The 3-kHz sinusoidal voltage is provided to you from a sine wave generator.

Figure P15.13

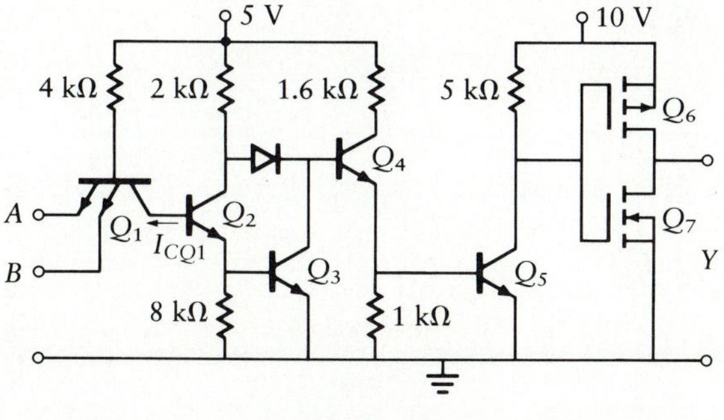

(a)

A	B	I_{CQ1}	Q_2	Q_3	Q_4	Q_5	Q_6	Q_7	Y
0	0								
0	1								
1	0								
1	1								

(b)

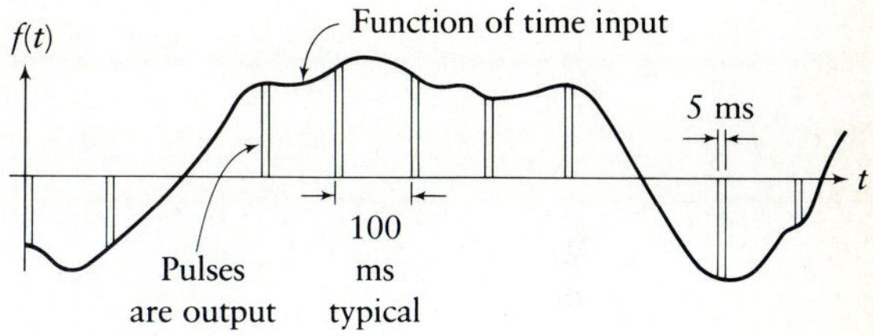

Figure P15.14

Figure P15.15

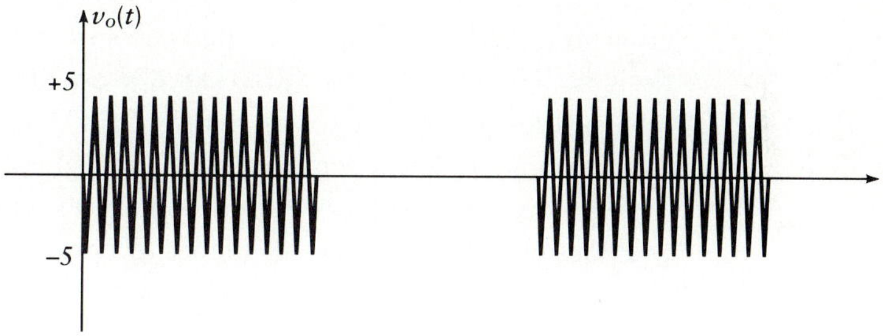

16

DIGITAL ICs

16.0 INTRODUCTION

In Chapter 15 we introduce the major families of digital integrated circuits. Included in that chapter is a discussion of the various basic circuits.

The intent of the present chapter is to expand this catalog of integrated circuits so that the engineer has the necessary tools to perform a wide variety of designs. There are many different ICs on the market. These can be divided into several major categories, which form the major sections of this chapter. Application examples are presented throughout.

The figures in the text often contain excerpts from data books. As in the previous chapter, we urge the reader to obtain several of these books. Although they are listed in the references, we repeat the major listings as follows:

1. *The TTL Data Book for Design Engineers,* Vol. 2, Texas Instruments Corp., Dallas, Texas.
2. *CMOS Data Book,* National Semiconductor Corp., Santa Clara, California.
3. *Linear and Data Acquisition Products,* Harris Semiconductor Products Division, Melbourne, Florida.
4. *CMOS Data Book,* Fairchild Camera and Instrument Corp., Mountain View, California.
5. *Master Selection Guide,* Motorola Semiconductor Products Inc., Phoenix, Arizona.
6. *Schottky and Low Power Schottky Data Book,* Advanced Micro Devices Inc., Sunnyvale, California.

We begin our study in Section 16.1 with decoders and encoders. The examples include binary-coded decimal to seven-segment displays, parity checkers, multiplexers, and arithmetic units.

Circuits with memory are examined in the later part of the chapter, beginning with Section 16.3. These include flip-flops, latches, shift registers, and counters. Since circuits with memory often require timing, we explore methods of accomplishing this. Particular attention is given to clocks, oscillators, phase-locked loops, and voltage-controlled oscillators (VCOs).

Section 16.7 considers the important topic of memories, with attention given to RAMs, ROMs, PROMs, and EPROMs.

The chapter concludes with a discussion of more complex devices, including arithmetic logic units, A/D and D/A converters, and programmable array logic (PAL).

Throughout this chapter, we present a systems approach. The circuit diagrams do not show transistors. Indeed, the basic building blocks are the integrated circuits of the previous chapters. Details of the internal construction of the various ICs can be found in the references.

16.1 DECODERS AND ENCODERS

A wide variety of multiple-input/multiple-output devices can be combined under the term *decoder.* Devices that fit into this category are given more specific names that accurately describe their function. Examples of these include *multiplexers* and *demultiplexers.* Whatever the specific name, decoders share a common trait with the elementary gates: these ICs have *no memory.* Regardless of the values of previous inputs and outputs, the device's present output depends only on its present input.

The first example we present is a 3- to 8-line decoder/demultiplexer as shown in Figure 16.1. We have selected a 74LS138 TTL circuit for purposes of illustration. Note that the pin-out diagram and function table apply to several device numbers. These ICs differ in speed of operation and in power requirements. These properties are shown elsewhere in the manufacturer's data sheets. The 3- to 8-line decoder/demultiplexer is used to enable (i.e., turn on) one among 8 possible outputs. It comprises a series of gates as shown in Figure 16.1(b). The 3- to 8-line decoder accepts 3-bit binary inputs representing a binary number between 000 and 111. These inputs correspond to the decimal numbers 0 through 7. Of the 8 output lines, only one is LOW at any one time, as shown in Figure 16.1(c). The output that is LOW is the one numbered to match the binary input (SELECT). For example, if the binary number 010 forms the input, output 2 (Y2) is LOW and all the others are HIGH. If we want

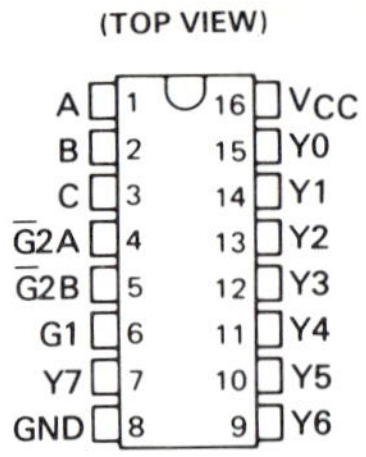

(a)

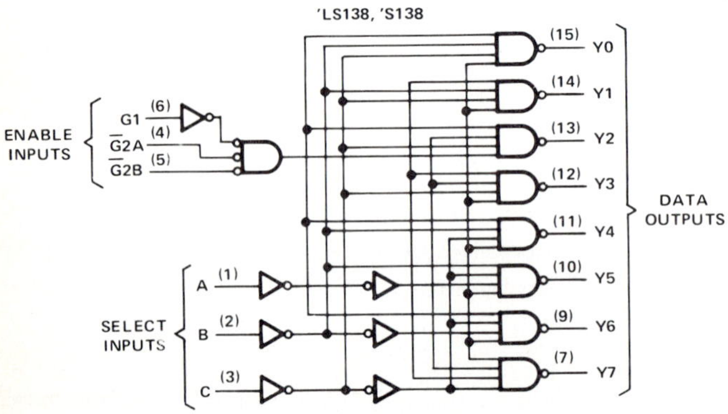

(b)

'LS138, 'S138
FUNCTION TABLE

INPUTS					OUTPUTS							
ENABLE		SELECT										
G1	$\overline{G2}$*	C	B	A	Y0	Y1	Y2	Y3	Y4	Y5	Y6	Y7
X	H	X	X	X	H	H	H	H	H	H	H	H
L	X	X	X	X	H	H	H	H	H	H	H	H
H	L	L	L	L	L	H	H	H	H	H	H	H
H	L	L	L	H	H	L	H	H	H	H	H	H
H	L	L	H	L	H	H	L	H	H	H	H	H
H	L	L	H	H	H	H	H	L	H	H	H	H
H	L	H	L	L	H	H	H	H	L	H	H	H
H	L	H	L	H	H	H	H	H	H	L	H	H
H	L	H	H	L	H	H	H	H	H	H	L	H
H	L	H	H	H	H	H	H	H	H	H	H	L

*$\overline{G2} = \overline{G2A} + \overline{G2B}$
H = high level, L = low level, X = irrelevant

(c)

Figure 16.1
74LS138 3- to 8-line decoder. Courtesy of Texas Instruments Inc.

the selected output to be HIGH instead of LOW, we can feed each of the outputs into a separate inverter.

The circuit diagram and the function table contain several *enable* lines. For the system to be operated as a decoder, enable input G1 must be HIGH and input G2 must be LOW. Note that $\overline{G2}$ is formed as an OR operation on $\overline{G2A}$ and $\overline{G2B}$. If enable input G1 is LOW, all of the outputs are high regardless of the condition of the other enable or of the select lines. This is also true if G2 is HIGH.

The 3- to 8-line decoder illustrates the concept that 3 bits of input data can control 2^3 lines of output. In more general terms, an n- to m-line decoder converts n input lines into a maximum of $m = 2^n$ output lines.

Before going on to the next topic, you should take the time to verify some of the entries in the function table. Do this by applying the appropriate logic HIGH and LOW inputs and tracing these through the logic block diagram.

In its most general form, the decoder accepts one binary word as input and produces a second binary word as output. The decoder circuit can thus be thought of as a *translator* that accepts one language and changes it into another. This leads to a large number of possible decoder configurations. Six of these configurations are used in many applications, and we discuss these below.

1. *Binary to single output.* An n-bit binary word selects a single output from one of 2^n possible outputs. The circuit of Figure 16.1 is an example of this configuration with $n = 3$.
2. *BCD to seven-segment display.* This decoder accepts a binary input and produces a 7-bit binary code as the output. The input is a 4-bit binary-coded decimal (BCD) number between 0000 and 1001, representing the decimal digits between 0 and 9. This is known as the BCD code. The output is matched to a seven-segment display. Seven straight-line segments (three horizontal and four vertical) are used to display any digit between 0 and 9. The decoder yields the appropriate output to light the necessary segments, thereby producing the decimal digit. We examine this decoder in more detail in Section 16.2.
3. *Multiple-input to multiple-output.* Many devices are used to convert numbers from one form to another. These devices are often used to perform simple mathematical operations or to multiplex several signals together.

   An important example of a multiple-input to multiple-output device is the *BCD-to-binary converter.* The BCD code represents each decimal digit between 0 and 9 by a 4-bit binary number. For example, the decimal number 64 becomes 01100100 in BCD, since each decimal digit is handled independently. Note that this differs from a simple conversion of 64 to binary; such a conversion would result in the binary number 1000000, which is only 7 bits in length instead of 8 bits as in the BCD code.

   The 74HC42 *BCD-to-decimal* decoder is illustrated in Figure 16.2. The input to this device is the 4-bit BCD code of a single decimal digit. There are 10 separate outputs, and only one of these is LOW for any valid input combination. The one that is LOW represents the decimal equivalent of the input BCD number. Note that the BCD code uses only 10 of the 16 possible 4-bit combinations. Therefore six input combinations are invalid and should not exist in the BCD code. If the input matches any of these invalid combinations, an error must have occurred prior to that point in the circuitry. The 74HC42 IC is designed to respond to these invalid inputs by not allowing any of the 10 outputs to go LOW.
4. *Testers.* This group of devices performs tests on coded information and produces an output containing information regarding the outcome of the test. The *comparator* is an example of a tester. It accepts two binary codes as the input, and the output indicates which of the two inputs is larger. *Parity checkers* represent another important application of testing ICs. These are discussed in Section 16.1.2.
5. *Arithmetic.* Arithmetic ICs perform simple mathematical operations such as addition and subtraction. A *full adder* is one example of such a

Connection Diagram

Dual-In-line Package

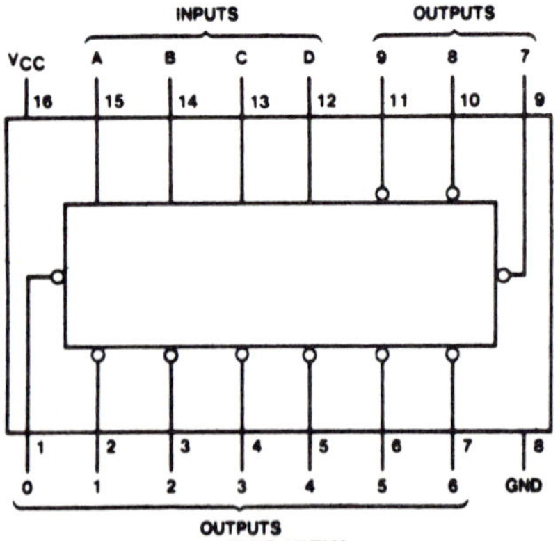

TOP VIEW
MM54HC42/MM74HC42

TL/F/5301-1

54HC42(J) 74HC42(J,N)

Truth Table

No.	Inputs				Outputs									
	D	C	B	A	0	1	2	3	4	5	6	7	8	9
0	L	L	L	L	L	H	H	H	H	H	H	H	H	H
1	L	L	L	H	H	L	H	H	H	H	H	H	H	H
2	L	L	H	L	H	H	L	H	H	H	H	H	H	H
3	L	L	H	H	H	H	H	L	H	H	H	H	H	H
4	L	H	L	L	H	H	H	H	L	H	H	H	H	H
5	L	H	L	H	H	H	H	H	H	L	H	H	H	H
6	L	H	H	L	H	H	H	H	H	H	L	H	H	H
7	L	H	H	H	H	H	H	H	H	H	H	L	H	H
8	H	L	L	L	H	H	H	H	H	H	H	H	L	H
9	H	L	L	H	H	H	H	H	H	H	H	H	H	L
INVALID	H	L	H	L	H	H	H	H	H	H	H	H	H	H
	H	L	H	H	H	H	H	H	H	H	H	H	H	H
	H	H	L	L	H	H	H	H	H	H	H	H	H	H
	H	H	L	H	H	H	H	H	H	H	H	H	H	H
	H	H	H	L	H	H	H	H	H	H	H	H	H	H
	H	H	H	H	H	H	H	H	H	H	H	H	H	H

H = High Level, L = Low Level

Logic Diagram

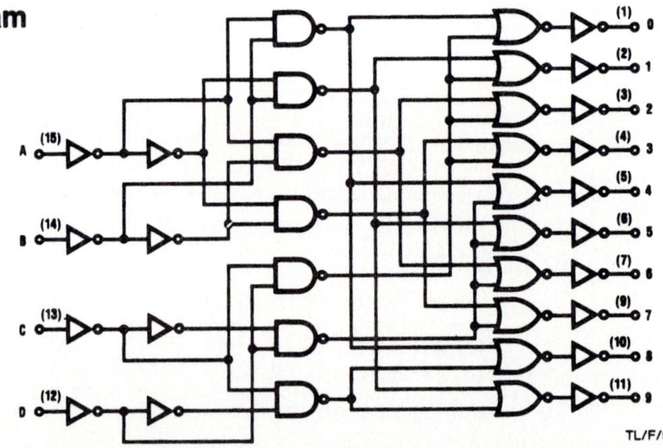

TL/F/5301-2

Figure 16.2
74HC42 BCD-to-decimal decoder. Courtesy of National Semiconductor Corp.

device. This adder accepts three input bits. Two of these are the bits to be added together, and the third is a *carry* bit from a previous, or lower-weight, column. Thus, the full adder is suited to applications where multiple-bit binary numbers must be added together. A full adder has two outputs: the *sum* and the *carry* bit. For example, the binary sum of 1 + 0 + 0 is 1 with a carry of 0. The sum of 1 + 1 + 1 is 1 with a carry of 1. We discuss full adders in more detail in Section 16.8.2.

6. *Multiplexers.* The multiplexer selects one of many inputs to be transferred to one output. This is the reverse of the *demultiplexer,* which routes one input to one of many outputs.

   Multiplexers are used to interleave data from several sources. For example, if only one transmission line were available to send four different signals, the bits would have to be *interleaved* (*commutated*) to form a *time-division-multiplexed* signal. The first bit of the first signal is followed by the first bit of the second, and so on. Finally, the first bit of the fourth signal is followed by the second bit of the first signal, and the cycle continues until the four signals are sent in their entirety.

At the receiving end, a *demultiplexer* (*decommutator*) is used to separate the various signals and sort them out onto different transmission paths.

We now examine the *data selector/multiplexer* in more detail. This circuit selects one input from many possibilities and transfers this input to the output. Figure 16.3 shows the SN74150 data selector/multiplexer.

There are four data-*select* inputs, labeled A, B, C, and D (pins 15, 14, 13, and 11, respectively). Depending on the value of these select inputs, the corresponding data input is transferred to the output. This particular IC yields an output (W) on pin 10 that is the inverse of the selected input. For example, if the data-select inputs are LHHL, representing the binary number 0110 (the decimal number 6), the output is the inverse of the signal connected to input pin 2, the sixth input bit (E6). This particular unit is *clocked,* and the appropriate input is transferred to the output only when a LOW signal is placed on the *strobe,* pin 9. If the strobe is HIGH, the output is HIGH, independent of the input levels.

It is important to supply proper timing to avoid transients in the output. The *data select,* which is a 4-bit binary number, changes whenever it is desired to transfer a different input to the output (pin 10). If, for example, the select is changed from 1010 to 0101, we desire that the input on pin 21 be transferred

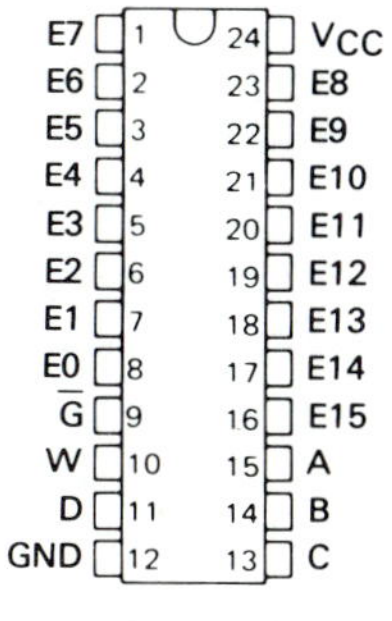

(a) Connection diagram

'150

FUNCTION TABLE

INPUTS					OUTPUT
SELECT				STROBE	W
D	C	B	A	$\overline{G}$	
X	X	X	X	H	H
L	L	L	L	L	$\overline{E0}$
L	L	L	H	L	$\overline{E1}$
L	L	H	L	L	$\overline{E2}$
L	L	H	H	L	$\overline{E3}$
L	H	L	L	L	$\overline{E4}$
L	H	L	H	L	$\overline{E5}$
L	H	H	L	L	$\overline{E6}$
L	H	H	H	L	$\overline{E7}$
H	L	L	L	L	$\overline{E8}$
H	L	L	H	L	$\overline{E9}$
H	L	H	L	L	$\overline{E10}$
H	L	H	H	L	$\overline{E11}$
H	H	L	L	L	$\overline{E12}$
H	H	L	H	L	$\overline{E13}$
H	H	H	L	L	$\overline{E14}$
H	H	H	H	L	$\overline{E15}$

(b) Function table

Figure 16.3 SN74150 data selector/multiplexer. Courtesy of Texas Instruments Inc.

to the output prior to the change and that the input on pin 3 be transferred after the change. However, the four binary select inputs may not change at exactly the same moment, so there may be intermediate select configurations prior to arrival at 0101. Without clocking, the output reflects each of these intermediate input values. The clock is used to ensure that the output changes only *after* the select has arrived at its final configuration.

Note that the IC has 24 pins in a DIP configuration. Two of these are for power (pin 24 for V_{CC} and pin 12 for ground), one is for strobe (pin 9), one for the output (pin 10), four for select, and the remaining 16 for the data inputs.

16.1.1 Keyboard Encoders/Decoders

A keyboard consists of a series of switches activated by keys. There is a wide variety of keyboard configurations, including that of a typewriter or of a numeric keypad. Prior to the development of digital electronics, each key was connected to the appropriate printing mechanism, often through a series of mechanical linkages. With the advent of digital electronics, it became preferable to code the input into an appropriate binary code, thus eliminating the necessity of connecting each key switch to the final output device.

The purpose of the encoder is to produce a binary output that contains information regarding which of the keys is being depressed. As a specific example, we consider the HD0165 keyboard encoder, as illustrated in Figure 16.4. This is a 16-line to 4-bit encoder intended for use with manual data-entry devices such as calculators or typewriter keyboards. Any 4-bit output code can be implemented.

Inputs are normally wired through the keyboard switches to the +5 V power supply. The 16 possible outputs represent all possible combinations of 4 bits. The outputs can be configured to yield the binary equivalent of the particular input that is HIGH. For example, if input 9 is HIGH (i.e., switch 9 is depressed), the output is LHHH, where, on Figure 16.4(b), we start with output 4 and work our way to the left to output 1. If L corresponds to 1 and H to 0, this is the binary number 1000 or, in decimal, 8. That is correct, since the input switches are labeled 1 through 16 and the output binary numbers range from 0 to 15. Thus, if we depress the input switches in order from number 1 to number 16, the output will count in binary numbers from 0 to 15. This one-to-one association between the ordered input switches and the output binary numbers is not the only way we could configure the output. That is, as input switches are depressed in order, we may wish the output to follow a sequence other than the standard counting sequence. Some thought will show that there are 16! or about 2.1×10^{13} ways to order the 16 output words. Particular configurations have been developed to yield desirable properties. The more common among these are the *Gray code* and the *1-2-4-2 BCD code.*

Figure 16.4
HD0165 keyboard encoder. Courtesy of Harris Semiconductor.

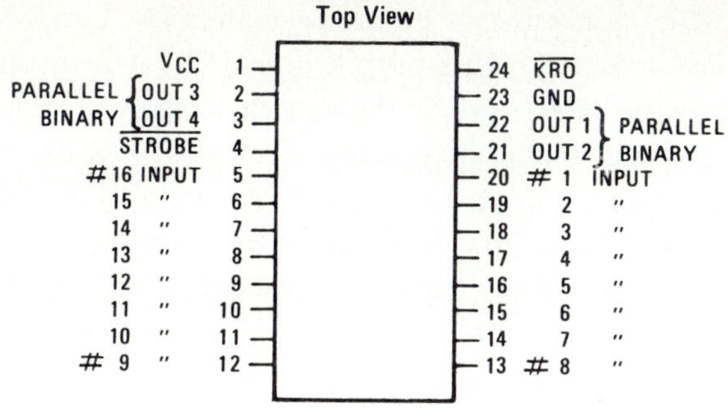

(a) Pinout

INPUTS																OUTPUTS					
1	2	3	4	5	6	7	8	9	10	11	12	13	14	15	16	1	2	3	4	$\overline{\text{St.}}$	$\overline{\text{KRO}}$
L	L	L	L	L	L	L	L	L	L	L	L	L	L	L	L	H	H	H	H	H	H
H	L	L	L	L	L	L	L	L	L	L	L	L	L	L	L	H	H	H	H	L	H
L	H	L	L	L	L	L	L	L	L	L	L	L	L	L	L	L	H	H	H	L	H
L	L	H	L	L	L	L	L	L	L	L	L	L	L	L	L	H	L	H	H	L	H
L	L	L	H	L	L	L	L	L	L	L	L	L	L	L	L	L	L	H	H	L	H
L	L	L	L	H	L	L	L	L	L	L	L	L	L	L	L	H	H	L	H	L	H
L	L	L	L	L	H	L	L	L	L	L	L	L	L	L	L	L	H	L	H	L	H
L	L	L	L	L	L	H	L	L	L	L	L	L	L	L	L	H	L	L	H	L	H
L	L	L	L	L	L	L	H	L	L	L	L	L	L	L	L	L	L	L	H	L	H
L	L	L	L	L	L	L	L	H	L	L	L	L	L	L	L	H	H	H	L	L	H
L	L	L	L	L	L	L	L	L	H	L	L	L	L	L	L	L	H	H	L	L	H
L	L	L	L	L	L	L	L	L	L	H	L	L	L	L	L	H	L	H	L	L	H
L	L	L	L	L	L	L	L	L	L	L	H	L	L	L	L	L	L	H	L	L	H
L	L	L	L	L	L	L	L	L	L	L	L	H	L	L	L	H	H	L	L	L	H
L	L	L	L	L	L	L	L	L	L	L	L	L	H	L	L	L	H	L	L	L	H
L	L	L	L	L	L	L	L	L	L	L	L	L	L	H	L	H	L	L	L	L	H
L	L	L	L	L	L	L	L	L	L	L	L	L	L	L	H	L	L	L	L	L	H
ANY TWO OR MORE HIGH																X	X	X	X	L	L

INPUTS: L = Open Circuit or < +1.0V H = > +4.5V Current Source
OUTPUTS: L = < +0.4V H = > +2.4V X = Erroneous Data

(b) Function table

As the mechanical switch is depressed, a *bounce* effect sometimes occurs. The mechanical switch makes momentary contact, then momentarily breaks contact, and finally makes a solid connection. The STROBE can be used to debounce the switch, eliminating this effect. Note from the function table that when any key is depressed, thus causing the associated encoder input to go HIGH, the STROBE output goes LOW. Debouncing can be accomplished by using the STROBE output to trigger a monostable, such as a 555 (see Section 14.5.3, Figure 14.28), with an on-time longer than the anticipated bounce time. Thus, the first contact of the switch causes the monostable output to go HIGH, and this output stays HIGH until the switch stops bouncing. When the monostable returns to the OFF state, a second strobe is activated, and it is this strobe that transfers the keyboard input to the switch. The debounce operation is a time-delay process where the keyboard input is transferred to the IC some time after the first switch closure.

The function table of Figure 16.4 illustrates an additional output labeled *key rollover (KRO)*. This output goes LOW when more than one key is depressed at the same time. It informs the circuit that the encoded output word is not valid and should be ignored. A more sophisticated approach involves using this output to set up a delayed strobe, much as is done to debounce the keyboard. We assume that depression of two keys is normally a temporary condition, so the delay may be sufficient to return to a single depressed key situation.

16.1.2 Parity Generators/Checkers

The *parity* of a binary number indicates whether the total number of 1's is odd or even. Thus, for example, the binary number 101 has *even parity* and the number 100 has *odd parity*. By adding one additional bit to a number, it is possible to force the enlarged number to have either even or odd parity. To create even parity, the added bit is a 1 if the original number had odd parity and is a 0 if the original number had even parity.

If a single bit in a word is changed, the parity of the word changes. Therefore, addition of a parity bit allows detection of single bit errors.

Suppose, for example, that a parity bit is added to each word to force the parity of the enlarged word to be always even. The system could then check parity at critical test points, and if it detects odd parity it knows a bit error occurred. In actuality, any odd number of bit errors causes a parity change but any even number of errors causes no change. For example, if two bit errors occur, the word still has even parity and the errors go undetected. The single parity bit is thus effective when we can rarely expect more than one bit error per word.

Discrete probability theory can be used to find the probability of an undetected error. For example, suppose that each bit of a 4-bit word has a probability of 10^{-5} of being in error (i.e., the *bit error rate*). Then the probability of no errors in a word is $(1 - 10^{-5})^4$. The probability of a single, and therefore detected, bit error is approximately 4×10^{-5}. The probability of two bit errors is approximately 6×10^{-10}, which is considerably less than the probability of one bit error. This represents the fraction of transmissions that will have undetected errors. The probability of three bit errors is approximately 4×10^{-15}, and the probability of four bit errors is 10^{-20}.

The problem associated with detecting multiple bit errors can prove serious in applications where errors frequently occur, such as in a high-noise environment where bit error rates are much higher than those assumed above. (In very high noise environments, the bit error rate can approach 0.5.) In such cases, undetected multiple errors can be expected. Some improvement is possible by using additional parity bits.

16.2 DRIVERS AND ASSOCIATED SYSTEMS

Driver ICs are used to power displays and other special-purpose devices. The driver circuits in this family are designed to operate over a wide range of voltages and currents, since the display devices often use nonstandard voltages and currents. Figure 16.5 illustrates a 7447 BCD-to-seven-segment decoder/driver. The function table for this device is shown in Figure 16.5(b). Four of the inputs represent a 4-bit binary number between 0 and 15. As can be seen from the "numerical designations and resultant displays" in Figure 16.5, the 4-bit binary numbers between 0 and 9 generate a display that is the deci-

Figure 16.5
7447 BCD-to-seven-segment decoder/driver. Courtesy of Texas Instruments Inc.

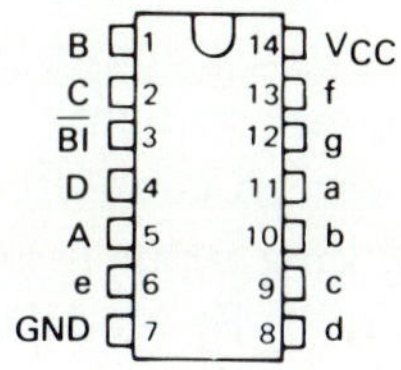

(a) Connection diagram

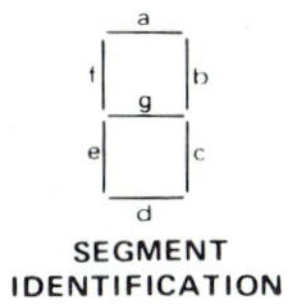

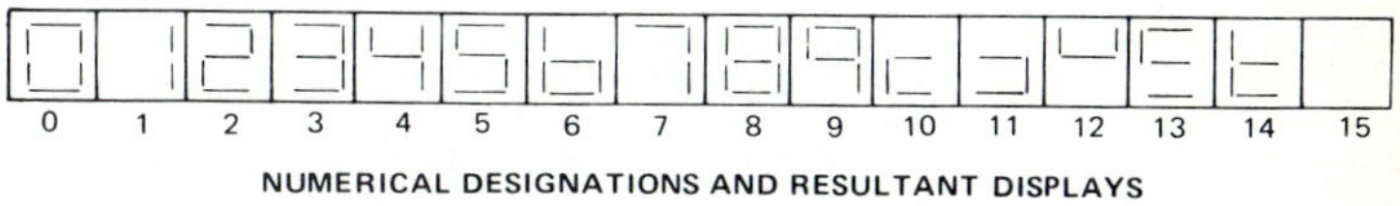

'46A, '47A, 'L46, 'L47, 'LS47 FUNCTION TABLE

DECIMAL OR FUNCTION	INPUTS						$\overline{BI}/\overline{RBO}$†	OUTPUTS						
	$\overline{LT}$	$\overline{RBI}$	D	C	B	A		a	b	c	d	e	f	g
0	H	H	L	L	L	L	H	ON	ON	ON	ON	ON	ON	OFF
1	H	X	L	L	L	H	H	OFF	ON	ON	OFF	OFF	OFF	OFF
2	H	X	L	L	H	L	H	ON	ON	OFF	ON	ON	OFF	ON
3	H	X	L	L	H	H	H	ON	ON	ON	ON	OFF	OFF	ON
4	H	X	L	H	L	L	H	OFF	ON	ON	OFF	OFF	ON	ON
5	H	X	L	H	L	H	H	ON	OFF	ON	ON	OFF	ON	ON
6	H	X	L	H	H	L	H	OFF	OFF	ON	ON	ON	ON	ON
7	H	X	L	H	H	H	H	ON	ON	ON	OFF	OFF	OFF	OFF
8	H	X	H	L	L	L	H	ON	ON	ON	ON	ON	ON	ON
9	H	X	H	L	L	H	H	ON	ON	ON	OFF	OFF	ON	ON
10	H	X	H	L	H	L	H	OFF	OFF	OFF	ON	ON	OFF	ON
11	H	X	H	L	H	H	H	OFF	OFF	ON	ON	OFF	OFF	ON
12	H	X	H	H	L	L	H	OFF	ON	OFF	OFF	OFF	ON	ON
13	H	X	H	H	L	H	H	ON	OFF	OFF	ON	OFF	ON	ON
14	H	X	H	H	H	L	H	OFF	OFF	OFF	ON	ON	ON	ON
15	H	X	H	H	H	H	H	OFF	OFF	OFF	OFF	OFF	OFF	OFF
BI	X	X	X	X	X	X	L	OFF	OFF	OFF	OFF	OFF	OFF	OFF
RBI	H	L	L	L	L	L	L	OFF	OFF	OFF	OFF	OFF	OFF	OFF
LT	L	X	X	X	X	X	H	ON	ON	ON	ON	ON	ON	ON

ON = 0 V
OFF = 5 V

(b) Function table

mal digit corresponding to the number. The remaining six input combinations generate symbols that may be used to convey various types of information (e.g., overload). Suppose, for example, the 4-bit input is HLLH representing the binary number 1001 and the decimal digit 9. Reference to the seven-segment LED display shows that we wish to light segments a, b, c, f, and g of the display to show the integer 9. The function table verifies that it is these specific five outputs that will be ON. The 7447 delivers active low outputs designed for driving common-anode LEDs or incandescent lamps. The driver portion of the circuit provides an open collector output, which is discussed in Chapter 15.

In addition to the 4-bit binary word, two additional binary inputs are provided. These are designated *lamp test* (LT) and *blanking input* (BI). The last line of the table shows that a LOW input on the lamp-test pin causes every display segment to light. This is used to test the circuit and the display. The display is turned completely off if the binary number 1111 forms the input or if the blanking input is LOW. This is used both to turn off the display and to *modulate* it. By modulating the display, or turning it on and off periodically, the perceived brightness can be adjusted.

Figure 16.6
7447 BCD decoder and display.

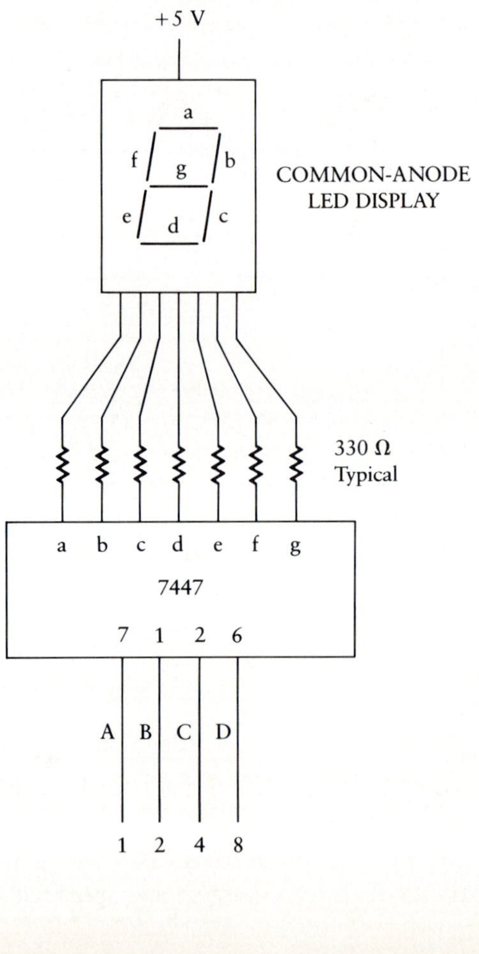

The 7447 is easily combined with a common-anode LED display, as shown in Figure 16.6. The typical forward voltage of the LED is 1.7 V with a current of 10 mA. We have used 330-Ω resistors to connect the open collector outputs of the 7447 to the V_{CC} = +5 V source through each segment of the LED. These resistors are necessary to limit the current through each LED segment.

Example 16.1

Design a circuit using a 555 oscillator to control the brightness of the seven-segment display.

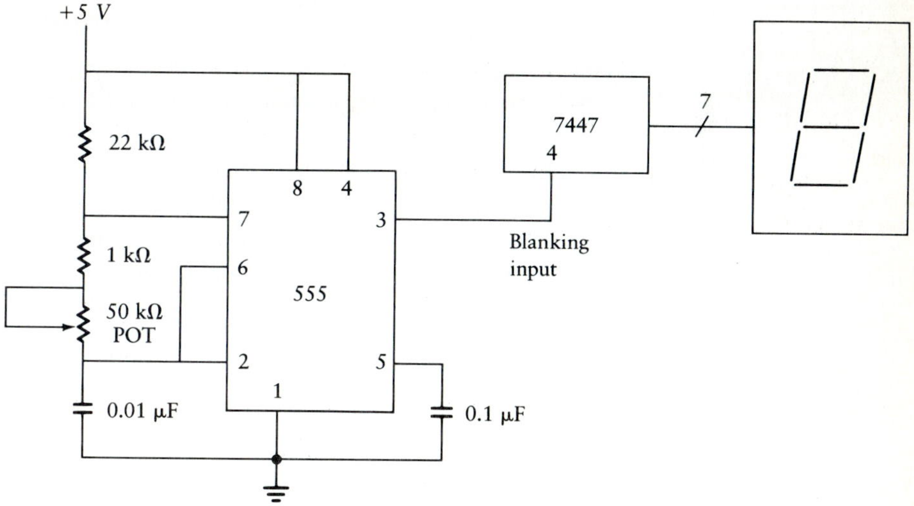

Figure 16.7
Variable brightness control for seven-segment display.

SOLUTION With the 7447 BCD-to-seven-segment decoder, we place a LOW input on the blanking in order to turn the display OFF. We use a 555 timer to construct a pulse generator to drive the decoder blanking input. The 555 pulse generator modulates the blanking input so that the brightness of the seven-segment display is adjustable. Variation of the duty cycle of the 555 changes the display brightness. The design is shown in Figure 16.7. The duty cycle is varied by adjusting the setting of the 50-kΩ potentiometer. The frequency must be at least 100 Hz to eliminate flicker in the display. A higher frequency is recommended.

16.3 FLIP-FLOPS, LATCHES, AND SHIFT REGISTERS

We now turn our attention to the study of circuits with memory. The outputs depend not only on the current value of the inputs but also on the past values.

This is known as *clocked logic, synchronous logic,* or *time-dependent logic.*

The function tables considered so far contain only two symbols, 0 and 1 or L and H. We now expand our vocabulary of function table symbols by introducing some new symbols in Figure 16.8. The first three entries in the figure represent symbols with which you are already familiar. The first entry is the data HIGH, or 1. The second is the data LOW, or 0. The third symbol, X, is used in function tables to represent values that do not matter — that is, they have no effect on the output. These are sometimes referred to as *don't care* conditions.

In some cases, the transitions are important in determining system output. In Chapter 14 we briefly discuss "edge triggering." A particular effect could occur on the leading (or trailing) edge of a clock input. In such a case, the transitions are important, and we use the fourth and fifth entries in Figure 16.8 to specify the direction of transition. The symbol *NC* is used to designate no change in an output.

More than one function table is often used to indicate the values of outputs and inputs before and after a clock pulse or transition. It is possible to combine these tables and to apply a name to variables before and after the change. In such cases, a variable Q is designated as Q_n and Q_{n+1} to represent the value before and after a given event, respectively. Some authors and data books have adopted variations of this terminology. However, the particular choice of symbols is usually self-explanatory.

We can start with circuits that have no memory and then change these to circuits with memory by feeding the output back to the input. A simple example is shown in Figure 16.9. The output of an OR gate is fed back to the input. The equivalent form uses a NAND gate with feedback. This configuration is known as an *always gate* for reasons that will shortly become clear.

Suppose we start with input and output equal to 0 (LOW). If we now apply a HIGH to input *A*, the output will go HIGH. If input *A* is now made LOW again, the output remains HIGH because the *B* input is now HIGH. Once *A* goes HIGH, the output will *always* be HIGH regardless of subsequent changes in *A*.

Figure 16.8
List of equivalent symbols.

Symbols	Common function table
1, H	Logical one or high
0, L	Logical zero or low
X	Don't care input; input can be any level or waveform
↑ or _┌	Low-to-high transition
↓ or ┐_	High-to-low transition
NC	No change in output
Q_{n+1}	Output after a given event

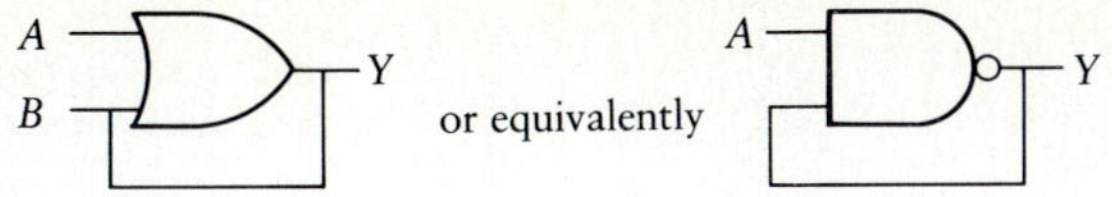

Figure 16.9
Always gate.

The always gate is an exceedingly simple circuit with limited applications. Nonetheless, it illustrates the fact that feedback can be used to create a circuit with memory capability from one that previously had none.

16.3.1 Flip-Flops

SR Flip-Flop We can combine two NOR gates in the manner shown in Figure 16.10 to obtain a useful configuration. This is known as the *set-reset,* or *SR, flip-flop,* which has two inputs (S and R) and two usually complementary outputs (Q and $\bar{Q}$). The two inputs are S, for *set,* and R, for *reset.* The flip-flop is set when Q is 1 and $\bar{Q}$ is 0, and it is reset when $Q = 0$ and $\bar{Q} = 1$. We can follow the performance of the *SR* flip-flop viewing the function diagram of Figure 16.10(b). Note that $Q(t)$ is the logic value of Q before a change of R or S, and $Q(t + 1)$ is the value after a change of R or S.

First assume both inputs, R and S, are set at 0. In this case, no change occurs from the previous state. You can verify this easily by considering the two separate cases ($Q = 0$ and $Q = 1$). Now suppose that S is brought to 1. In that case, $\bar{Q}$ becomes 0 and Q becomes 1. This is the set condition. Now suppose we again start with $R = S = 0$ but bring R to 1. Then Q becomes 0 and $\bar{Q}$ is 1. This is the reset condition. The output remains in either of these two stable states until the SET or RESET lines are changed. There are four possible combinations of inputs, and we have already dealt with three of these. The fourth occurs when both R and S are set to 1. The NOR gates would then force both Q and $\bar{Q}$ to be 0, so they are no longer a complementary pair. In fact, if the R and S inputs are not exactly synchronized, or if the two NOR gates are not identical (i.e., have different propagation delays), the output can oscillate. We shall see this when we perform the computer simulation. This condition, where we are instructing the flip-flop to both set and reset at the same time, is a disallowed state.

The function table of the *SR* flip-flop is shown in Figure 16.10(b). The device acts like a memory, since the outputs remain the same until a momentary change occurs at the input.

The *SR* flip-flop has several significant shortcomings. The first is the disallowed state that we have already discussed. A second shortcoming of this simple circuit is that transitions can occur at any time, depending on the state changes of the inputs. The circuit contains no clock. Careful control of the SET and RESET inputs is therefore required. For example, suppose we are making a transition from the RESET instruction ($S = 0$ and $R = 1$) to the

Figure 16.10
Set-reset flip-flop.

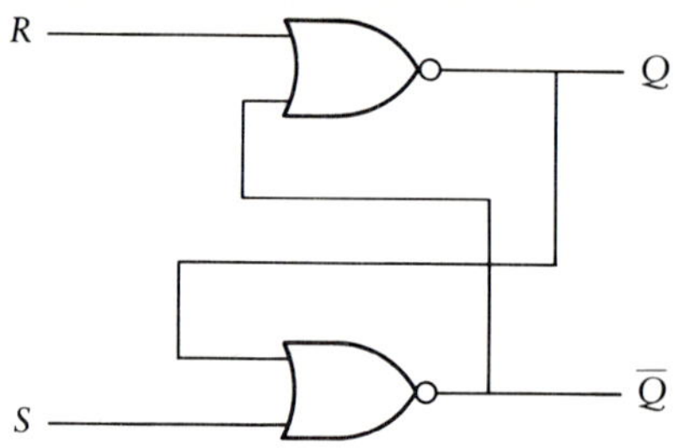

(a) Circuit

R	S	$Q(t+1)$	
1	1	0	Disallowed
0	1	1	Set
1	0	0	Reset
0	0	$Q(t)$	No change

(b) Function table

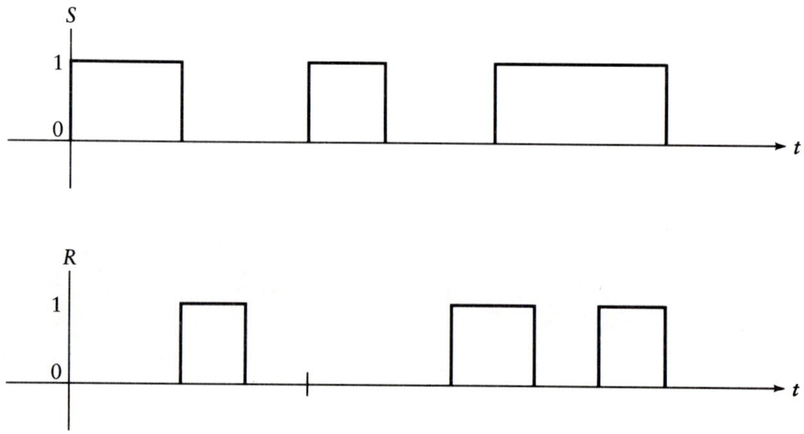

(c) S and R inputs

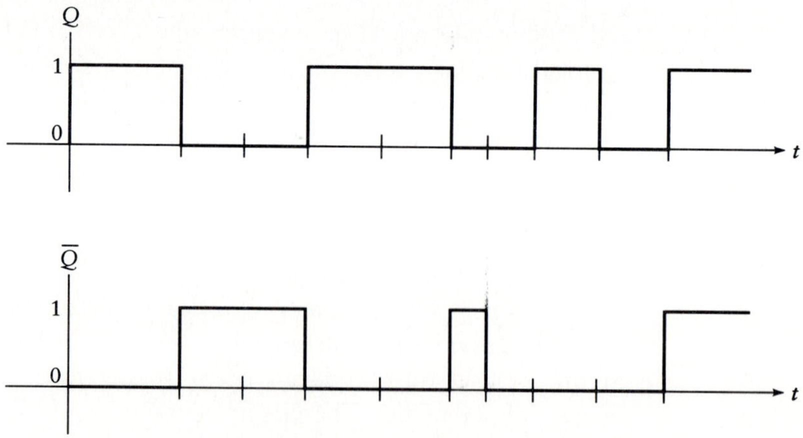

(d) Computer simulation output

SET instruction ($S = 1$ and $R = 0$). If during this transition both S and R are momentarily 1, then the disallowed output state occurs. The presence of the disallowed state and the lack of a clock represent severe shortcomings of the basic *SR* flip-flop. Because of these shortcomings, we concentrate on an improved circuit known as the *JK flip-flop*. However, before studying that type of flip-flop, we explore digital computer simulation.

Computer Simulation Digital circuits can be simulated using SPICE. However, digital analysis is not available on many of the simpler PC versions of SPICE. We choose to use MICRO-LOGIC II (from Spectrum Software; also available in a student version from Addison-Wesley Publishing Company). We input the circuit using a drawing board and a series of menus (accessed from either keyboard or mouse). To simulate the flip-flop of Figure 16.10(a), we first use the component menu to draw two NOR gates. Then we draw the connecting lines by designating the beginning and end of each line (when using the mouse, this is done by positioning the cursor on each of the two points and using two mouse buttons, one to designate the beginning and one for the end). Finally, we attach two data-input generators for the set and reset lines and label the two outputs, Q and $\bar{Q}$. We must then define a digital time function for each data input. Since we wish to simulate all four input combinations, we have chosen the data inputs shown in Figure 16.10(c).

We can set a variety of simulation parameters and display options. We can also generate netlists, node counts, and component counts. The simulation includes propagation delay, which can be set to minimum, maximum, or a random number within the range of the gate specifications. The program can also calculate propagation delay based on the number of devices connected to the output of the gates (i.e., the current supplied by the gate).

The simulation produces the timing diagram shown in Figure 16.10(d). This diagram verifies our earlier observations. You should note that in the disallowed state, both outputs are equal to 0. If we go from the disallowed state to the condition of $S = R = 0$, the output oscillates; in fact, the simulation appears as a shaded function, since the oscillation is at a high rate (dependent on the propagation delay).

***JK* Flip-Flop** The *JK* flip-flop does not suffer from the problems of the *SR* flip-flop. It features both time-dependent and time-independent signals and uses a mixture of the two signals to control the output signals, Q and $\bar{Q}$. Figure 16.11 presents a simplified circuit diagram and the function table for a *JK* flip-flop. In this table, the letter J is for *set* and the letter K is for *reset*. The clock is omitted to simplify the table.

By adding two gates to the input of the *SR* flip-flop, we obtain the circuit for Figure 16.11(a). The flip-flop output responds to changes in the input only when the clock pulse appears at the *CP* input. The output of this "clocked" flip-

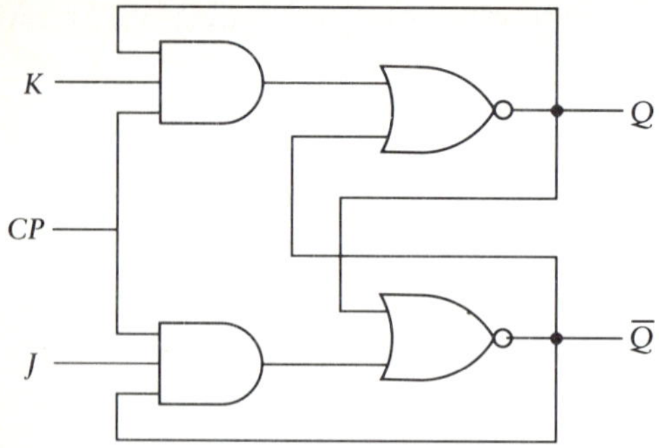

(a) Circuit

J	K	$Q(t+1)$
0	0	$Q(t)$
0	1	0
1	0	1
1	1	$\overline{Q(t)}$

(b) Function table

Figure 16.11 Function table for *JK* flip-flop.

flop responds to changes in the input only when the clock pulse is equal to 1.

In addition, feedback of the Q and $\overline{Q}$ outputs to the input of the *JK* flip-flop removes the indeterminate state from the *SR* flip-flop.

Since Q is ANDed with the K and CP inputs, the flip-flop is reset (cleared) when $K = Q = 1$ and the clock pulse occurs. Similarly, since $\overline{Q}$ is ANDed with the J and CP inputs, the flip-flop is set when $\overline{Q} = J = 1$ and the clock pulse occurs.

The table of Figure 16.11(b) should be contrasted with that of Figure 16.10(b) to see the differences between the *SR* and *JK* flip-flops. Note that if the J is considered as the SET input and the K as the RESET, two of the entries in the tables match. Thus a 0 on the J and a 1 on the K input RESETS the flip-flop regardless of what state it was in before the application of these signals. The opposite condition, $J = 1$ and $K = 0$, SETS the flip-flop. Now, if both inputs are 1, the *SR* flip-flop is in the disallowed state whereas the *JK* flip-flop changes state. That is, the output of the *JK* flip-flop under the condition $J = 1$ and $K = 1$ is given as

$$Q(t+1) = \overline{Q(t)}$$

which means that the new output is the inverse of the old. We say that the flip-flop *toggles.* For example, if the flip-flop is set (i.e., $Q = 1$) and both J and K are 1 when the clock pulse occurs, only the K signal "gets through" the associated AND gate. The J signal is ANDed with $\overline{Q}$, which is equal to 0.

If both J and K are 0, the *JK* flip-flop remains in the current state. That is,

$$Q(t+1) = Q(t)$$

Clocked Flip-Flops *Clocked flip-flops* can change state only when the clock signal appears at the input. No matter how many changes occur in J or K between clock signals, the state of the circuit will not change. This form of logic can hold an output constant while some of the inputs are changing. A clocked logic device can therefore be used as a memory device to store an output so that it can be referred to again and again. Another advantage of clocked logic is that all changes in a complex circuit can be forced to occur at exactly the same time. This restriction is used to prevent potentially severe problems.

There are two basic types of clocking—level and edge. With *level clocking* the input data cannot be changed except immediately after a clock pulse arrives. It is important that the input change only once during the period when the clock pulse is present. The state of the clock (either 0 or 1) determines whether changes in the output can occur. Alternatively, with *edge clocking* the input data can change at any time. Changes in the output occur only during *transitions* of the clock signal (HIGH to LOW or LOW to HIGH), so it is the value of the inputs at these times that matters.

Figure 16.12 illustrates the 74108 dual *JK* flip-flop connection diagram and function table. Note that this IC contains both a *preset* and a *clear* input. If the preset input is LOW and the CLEAR input is HIGH, the flip-flop SETS independent of the values of *J*, *K*, and the CLOCK. Thus, the output, *Q*, goes HIGH. If the CLEAR input is LOW and the PRESET input is HIGH, the flip-flop RESETS independent of the other inputs. Setting both PRESET and CLEAR to LOW is an illegal instruction, and both Q and $\overline{Q}$ temporarily go HIGH until the input signal is removed.

If both the PRESET and CLEAR are kept HIGH, the *J* and *K* inputs control the flip-flop whenever a CLOCK signal is present. The arrow notation in the CLOCK column of the table is used by this manufacturer to indicate that changes in state take place on the negative clock transition (that is, the pulse input goes from HIGH to LOW). The *J* input acts as a positive logic SET instruction, and the *K* acts as a RESET (or CLEAR). If neither *J* nor *K* is HIGH, the flip-flop remains in its previous state, denoted by Q_0. If both *J* and *K* are HIGH, the flip-flop changes state, or toggles.

***D* and *T* Flip-Flops** Two other major categories of flip-flops, in addition to the *JK* flip-flop, are the *D* and *T* flip-flops. The *D*, or *data,* flip-flop has only

SN54H108 . . . J OR W PACKAGE
SN74H108 . . . J OR N PACKAGE
(TOP VIEW)

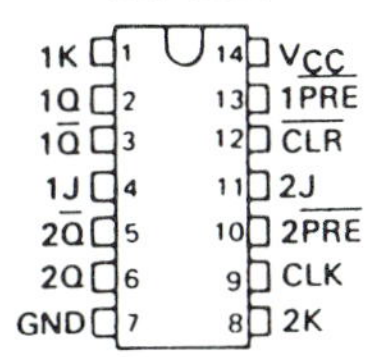

FUNCTION TABLE

INPUTS					OUTPUTS	
$\overline{PRE}$	$\overline{CLR}$	CLK	J	K	Q	$\overline{Q}$
L	H	X	X	X	H	L
H	L	X	X	X	L	H
L	L	X	X	X	H†	H†
H	H	↓	L	L	Q_0	$\overline{Q}_0$
H	H	↓	H	L	H	L
H	H	↓	L	H	L	H
H	H	↓	H	H	TOGGLE	
H	H	H	X	X	Q_0	$\overline{Q}_0$

†This configuration is nonstable; that is, it will not persist when preset and clear inputs return to their inactive (high) level.

Figure 16.12
74107 dual *JK* flip-flop. Courtesy of Texas Instruments Inc.

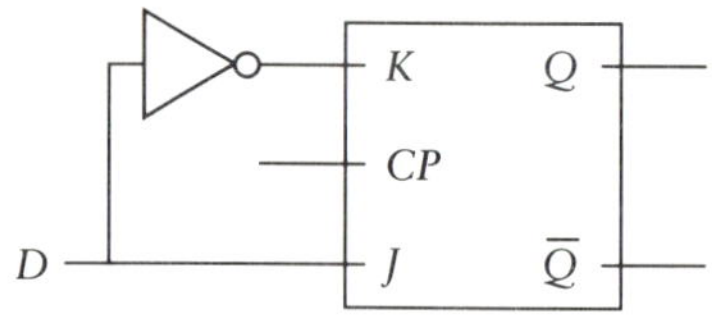

(a) Circuit

D	$Q(t + 1)$
0	0
1	1

(b) Function table

Figure 16.13
D flip-flop.

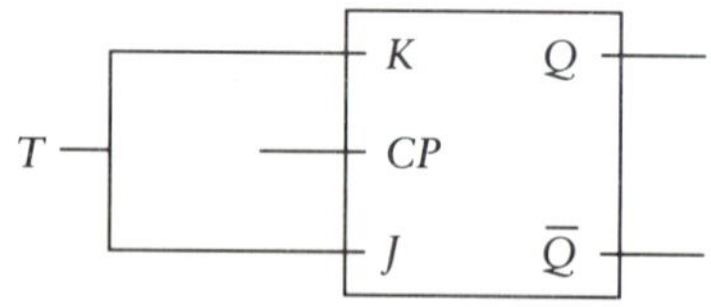

(a) Circuit

T	$Q(t + 1)$
0	$Q(t)$
1	$\overline{Q(t)}$

(b) Function table

Figure 16.14
T flip-flop.

one input instead of two (as is the case with the *SR* and *JK* flip-flops). Regardless of the input level, the *D* input is transferred to the output. Figure 16.13 illustrates the circuit configuration and function table for this type of flip-flop. Notice that the next state of the output is given by the current value of the input. A 1 at the input SETS the flip-flop, whereas a 0 at the input RESETS the flip-flop.

The *D* flip-flop can be constructed from a *JK* flip-flop by setting *J* equal to the *D* input and *K* equal to the complement of the *D* input. Thus, $D = 1$ for the *D* flip-flop is the same as $J = 1$, $K = 0$ for the *JK* flip-flop, and this SETS the device. The reverse is true for an input of $D = 0$. By forcing *K* to be the complement of *J*, we have eliminated two rows from the *JK* function table.

The *T*, or *toggle,* flip-flop also has only one input. If the input, *T*, is equal to 1, the flip-flop changes state. If the input is 0, the flip-flop remains in its current state. This is shown in the function table of Figure 16.14. A *JK* flip-flop can be turned into a *T* flip-flop by setting both *J* and *K* equal to *T*. That is, the two inputs to the *JK* are tied together to form a single input, *T*, as shown in Figure 16.14(a). Thus, if $T = 1$, this is the same as $J = 1$ and $K = 1$, which toggles the *JK* flip-flop. If $T = 0$, then $J = 0$ and $K = 0$, so the *JK* flip-flop does not change state.

16.3.2 Latches and Memories

The *latch memory* is a form of flip-flop that has the ability to remember a previous input and store it until the device is cleared or the data are called up to be read by another IC. Latch ICs come in sizes ranging from one latch to memory devices that store thousands of bits of information.

Figure 16.15 illustrates the pin diagram, function table, and single-bit circuit diagram for the 74373 tri-state octal *D*-type latch, which is an 8-bit storage element.

Figure 16.15
74373 tri-state octal *D*-type latch. Courtesy of Texas Instruments Inc.

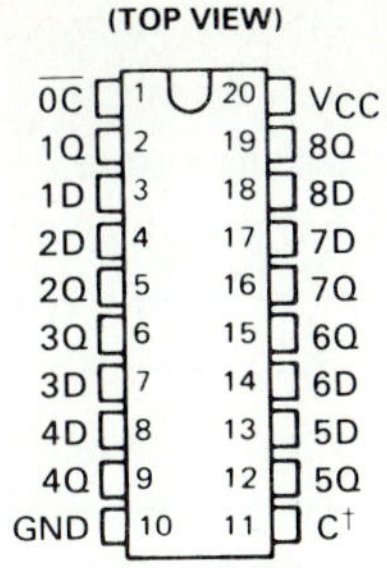

(a) Connection diagram

FUNCTION TABLE

OUTPUT ENABLE	ENABLE LATCH	D	OUTPUT
L	H	H	H
L	H	L	L
L	L	X	Q_0
H	X	X	Z

(b) Function table

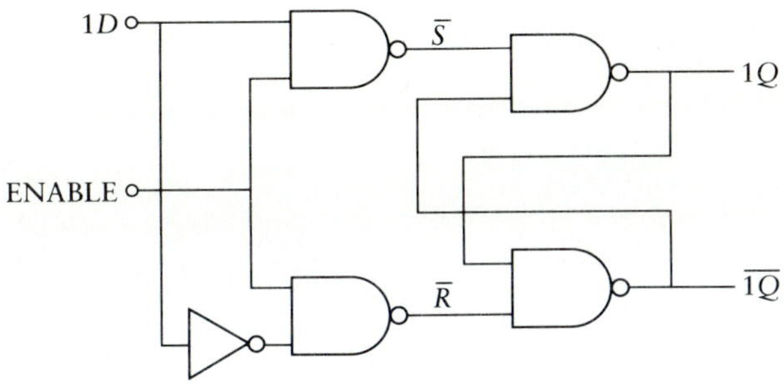

(c) Single-bit operation

There are eight data input lines, labeled 1*D* through 8*D*, and eight output lines, labeled 1*Q* through 8*Q*. The device is assembled in a 20-pin dual in-line (DIP) package. Eight of the pins are used for data input, eight for data output, and two for power, and there are two additional inputs as described below. The outputs are specially designed to drive high-capacitive loads such as are found in a system bus. An additional input is labeled ENABLE LATCH on the function table and C, for clock, on the connection diagram. When this is HIGH, the *Q* outputs follow the *D* inputs just as in the case of the *D*-type flip-flop. In this state the latch is said to be "transparent" since the outputs follow the input. When the ENABLE LATCH input is LOW, the outputs do not change. This is indicated in the manufacturer's function table by showing *Q* to equal Q_0. In this state, the data are latched at the level that existed when C was brought LOW.

An additional input is labeled OUTPUT DISABLE ($\overline{OC}$ on the connection diagram). When this input is HIGH, all of the outputs go to a high-impedance state regardless of the status of the other inputs. The "Z" in the last entry of the function table of Figure 16.15(b) signifies the high-impedance state.

The operation of the latch IC can be appreciated by viewing the circuit diagram of a single-bit operation. This is shown in Figure 16.15(c), where we use NAND gates as the building blocks. The right portion of this circuit is a variation of the *SR* flip-flop as presented in Figure 16.10. If 1*D* is high at the time the enable is high, $\overline{S}$ is low and $\overline{R}$ is high, so the flip-flop sets. If 1*D* is low, the flip-flop resets.

A *memory device* stores information until it is either cleared, set, or written over by another bit of information. The write operation can be triggered either by a rising or falling clock transition or by a steady logic state. Some devices will write the true value of the input data and others will invert it before storage. Likewise, the output may be either true or inverted. The output normally has its own triggering mechanism and can therefore be read by means of an enabling signal.

Many memory elements *multiplex* the inputs and outputs. That is, a single line is used to read or write more than one bit. It is therefore necessary to have a read and write signaling system that controls the read and write process. This contrasts with most flip-flops and simple latches, since for these circuits, outputs are always available for reading.

Example 16.2

Design a drive for one seven-segment LED display that will hold the output fixed while the input is changing. This is often required when we are measuring quantities such as velocity.

Figure 16.16 LED display with latch.

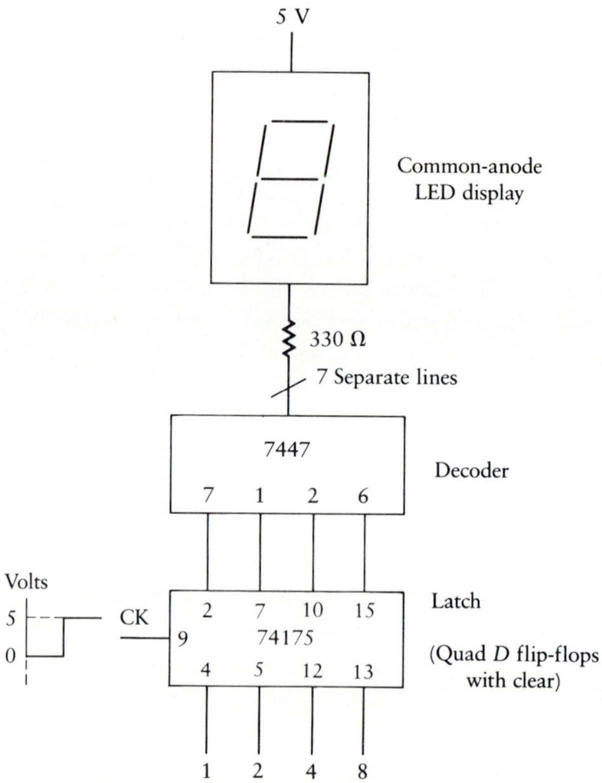

SOLUTION One possible solution for this design problem is shown in Figure 16.16. Since only four data lines are required, we have chosen a 74175 latch to hold the data on the display while the inputs are changing. Information on the four input lines (pins 2, 7, 10, and 15) is transferred to the output lines (pins 4, 5, 12, and 13) on the positive-going edge of the clock pulse on pin 9. When the clock input is at either the HIGH or LOW level, the data input has no effect on the output. The latch output feeds the 7447 BCD-to-seven-segment decoder/driver, which in turn feeds the seven inputs of the LED display through 330-Ω resistors. We have introduced the notation of a connection with a slash through it. This is shown in the figure as a slash with a 7 next to it. The slash-7 is a shorthand notation used to avoid drawing seven separate lines and represents seven wires with seven 330-Ω resistors connecting the SN7447 to the common-anode LED display.

16.3.3 Shift Registers

A *shift register* comprises a number of *JK* or *D* flip-flops cascaded in a string so that, on clocking, the contents of each stage are moved, or shifted, one stage to either the left or the right. A simple shift register is shown in Figure 16.17. We illustrate eight stages, Q_A through Q_H. The bits of data, either 0 or 1, are passed on in order, so the first bit in is the first bit out. The shifting takes place on the rising edge of the clock signal. As a result, the eight-stage register of Figure 16.17 will delay the input data for eight clock pulses.

A shift register can be thought of as a memory device that consists of N memory elements connected together in a chain. Each cell in the chain is capable of remembering one bit of information. That bit can be transferred to the adjacent cell, left or right, upon a proper control instruction.

Figure 16.18 illustrates one particular class of 4-bit shift register, the 74194. This is known as a "universal" shift register because it can be configured in a variety of ways. The circuit contains 46 equivalent gates. The shift register can be loaded in a parallel fashion by setting both MODE controls, $S1$ and $S0$, HIGH. Then the 4 bits of data are loaded into the associated flip-flops by us-

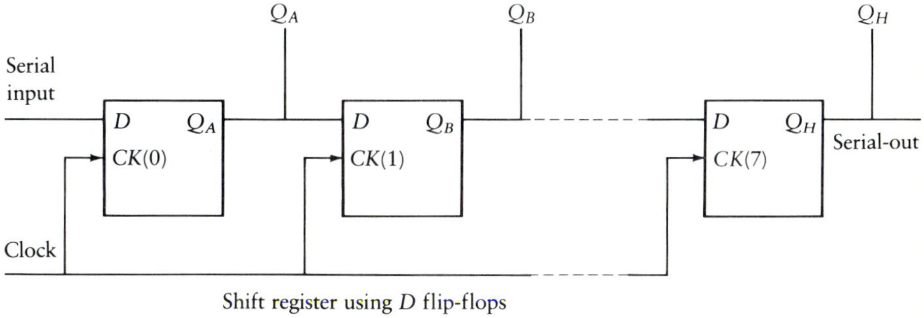

Figure 16.17 Shift register.

Figure 16.18
74194 4-bit bidirectional universal shift register. Courtesy of Texas Instruments Inc.

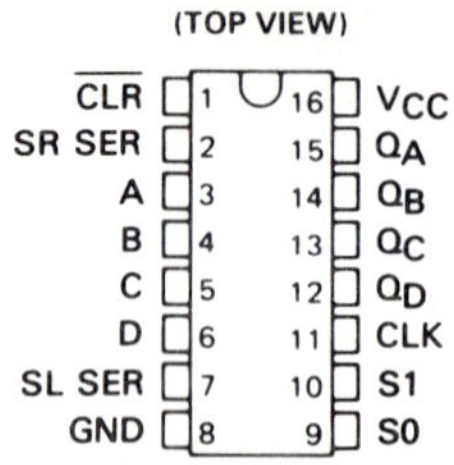

(a) Connection diagram

INPUTS										OUTPUTS			
CLEAR	MODE		CLOCK	SERIAL		PARALLEL				Q_A	Q_B	Q_C	Q_D
	S1	S0		LEFT	RIGHT	A	B	C	D				
L	X	X	X	X	X	X	X	X	X	L	L	L	L
H	X	X	L	X	X	X	X	X	X	Q_{A0}	Q_{B0}	Q_{C0}	Q_{D0}
H	H	H	↑	X	X	a	b	c	d	a	b	c	d
H	L	H	↑	X	H	X	X	X	X	H	Q_{An}	Q_{Bn}	Q_{Cn}
H	L	H	↑	X	L	X	X	X	X	L	Q_{An}	Q_{Bn}	Q_{Cn}
H	H	L	↑	H	X	X	X	X	X	Q_{Bn}	Q_{Cn}	Q_{Dn}	H
H	H	L	↑	L	X	X	X	X	X	Q_{Bn}	Q_{Cn}	Q_{Dn}	L
H	L	L	X	X	X	X	X	X	X	Q_{A0}	Q_{B0}	Q_{C0}	Q_{D0}

H = high level (steady state)
L = low level (steady state)
X = irrelevant (any input, including transitions)
↑ = transition from low to high level
a, b, c, d = the level of steady-state input at inputs A, B, C, or D, respectively.
Q_{A0}, Q_{B0}, Q_{C0}, Q_{D0} = the level of Q_A, Q_B, Q_C, or Q_D, respectively, before the indicated steady-state input conditions were established.
Q_{An}, Q_{Bn}, Q_{Cn}, Q_{Dn} = the level of Q_A, Q_B, Q_C, respectively, before the most-recent ↑ transition of the clock.

(b) Function table

ing inputs *A*, *B*, *C*, and *D*. The output is read in parallel as Q_A through Q_D. The *X* in the SERIAL columns of the table indicates that, during the parallel loading mode, serial data flow is inhibited. If we now set the mode controls to *S*0 HIGH and *S*1 LOW, the data shift to the right on the rising edge of the clock pulse (study the function table to convince yourself that the fourth and fifth rows confirm this observation). Alternatively, a LOW input to mode control *S*0 and a HIGH to *S*1 cause a shift to the left on the rising edge of the clock pulse. The sixth and seventh rows of the table confirm this. If both mode controls are LOW, the state of the register does not change, so clocking is inhibited. A CLEAR input is provided which sets all registers to LOW independent of the values of the other inputs or of the clock. The mode inputs must change only while the clock input is not changing.

Although there exist numerous shift register configurations, they can be divided into the following four broad categories, depending on whether inputs and outputs are handled serially (one after another) or in parallel.

1. *Serial-in serial-out (SISO).* Input data enter the shift register serially, and the data are taken from the output lead in a serial fashion, delayed by a number of clock pulses equal to the number of storage cells. Figure 16.17 is an example of an SISO shift register.
2. *Serial-in parallel-out (SIPO).* Input data enter the shift register serially, but the data are taken from the output leads in a parallel fashion. This

requires more than one output lead, since the bits are read in groups of multiple bits. For example, if the bits represent a BCD code, the output bits are read in groupings of four to represent one BCD word.

3. *Parallel-in serial-out (PISO).* This type of shift register has the capability of loading the data in parallel and shifting the data out serially. This register uses NAND gates and inverters with the flip-flops to properly sequence the input data.
4. *Parallel-in parallel-out (PIPO).* This type of parallel-access shift register is considerably more complex because of the additional gates that must be added. It can be thought of as a parallel combination of SISO shift registers.

16.4 COUNTERS

Counters can be either asynchronous (ripple) or synchronous. Figure 16.19 illustrates an *asynchronous counter.* Each of the blocks in this diagram is a *JK* flip-flop configured as a *T* flip-flop, since the *J* and *K* inputs are tied together. Note that the data input is used for the CLOCK. A clocked *T* flip-flop toggles only when its input goes HIGH and it receives a clock signal.

Suppose we feed a pulse train into the first flip-flop. The first time the input goes from LOW to HIGH, the flip-flop SETs. The second time the input goes from LOW to HIGH, the flip-flop CLEARs, and so on. Thus, the output of the first flip-flop, labeled 1, is a pulse waveform at one-half of the frequency of the input. The process repeats itself at the second flip-flop, and each device toggles once for every two toggles of the circuit to its left. The outputs are labeled according to the weight in a binary number. Thus, the third output corresponds to 2^2 or 4. The counter generates a 3-bit number that cycles as follows:

$$000, 001, 010, 011, 100, 101, 110, 111, 000, \ldots.$$

Note that the frequency at which each bit changes from 0 to 1 is one-half of that of the bit to its left. Thus, the circuit counts between 000 through 111 and

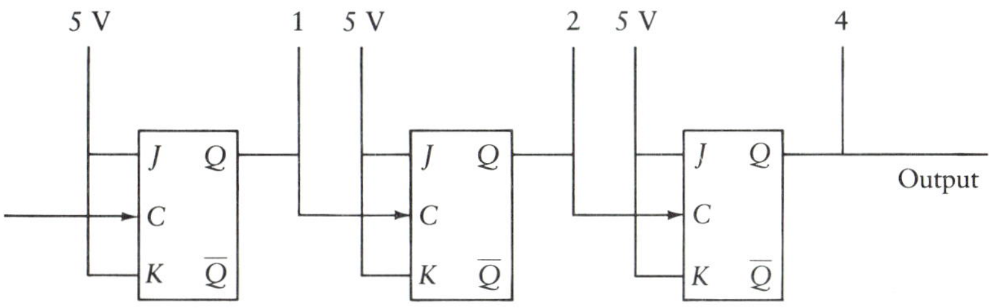

Figure 16.19
Asynchronous counter.

back to 000 as the input is pulsed eight times. The counter is asynchronous, since counts occur in the right flip-flop only after the clock pulse "ripples" from the left to the right flip-flop.

The circuit of Figure 16.20 is a *synchronous counter,* since the clock input feeds into all three flip-flops simultaneously. The second flip-flop toggles on a clock pulse only if the output of the first flip-flop is a 1. This represents a *carry* condition in the operation of adding 1 to the previous output. The third flip-flop toggles on a clock pulse only if both the first and second outputs are 1.

Since the output of each flip-flop is at a frequency that is one-half that of its input, the flip-flop is often known as a *divide-by-2* circuit. The counter is therefore often called a *divide-by-n counter,* where n is the number of input cycles required to produce one output cycle.

The majority of TTL and CMOS counters are *up-only* counters; that is, they count only in a direction of increasing binary numbers. Inverters can be used on the outputs to change an up-counter into a down-counter. This works only if the counter counts to a binary length that is a power of two. That is, if a 3-bit counter counts from 000 to 111 (2^3 counts), the inversion operation will change from up-counting to down-counting. However, if the counter is used to count only between 000 and 101, inversion will not have the desired effect.

An *up/down counter* either adds to, ignores, or subtracts from the current count at any time. Examples are the TTL 74190 and 74192 and the CMOS 74HC190 and 74HC192. Although these are the most flexible and versatile of counters, they are usually also more expensive and consume more power than the up-only counters.

Counters can be connected in sequence, with the output of the first forming the input to the second, thereby lengthening the count. It is possible to shorten the count sequence of a particular counter by presetting it to a nonzero number. An example is shown in Figure 16.21. This is a 74161 synchronous presettable binary counter to count from 8 to 14. The load inputs, L_1, L_2, L_4, and L_8, are used to preset the counter to any desired value. In this example, the connections preset the counter to 1000 or, in decimal, 8. These bits are

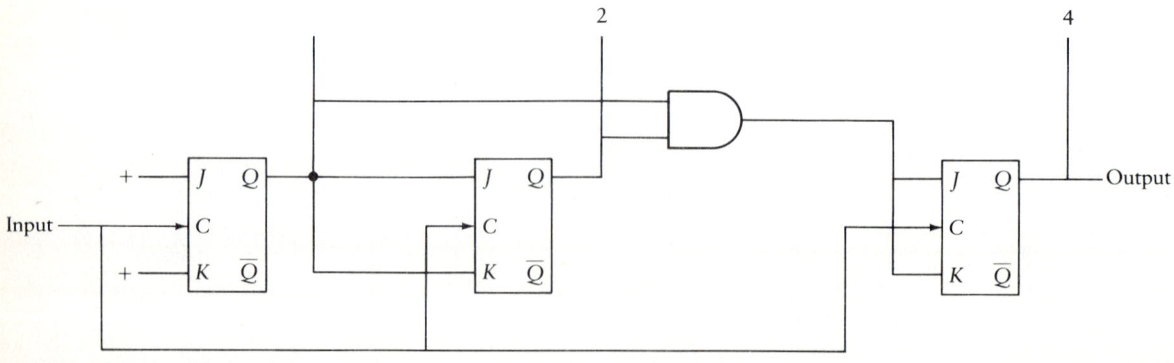

Figure 16.20
Synchronous counter.

loaded into the counter whenever the LOAD pin, 9, goes LOW. The CARRY OUT (also known as carry look-ahead) lead, pin 15, goes high when the count reaches 15 (1111). This is fed through the inverter so that the LOAD input goes LOW, thus resetting the counter to 1000.

Figure 16.22 shows another example of a truncated count sequence. In this case we set the load input to 0000, so the counter will start counting at 0000. When the count output reaches 0111, or 7, all three inputs to the NAND gate are HIGH and the NAND output goes LOW. This clears the counter to 0000. We see that the counter can be configured to start at any value and to end at any other value.

We now take a more detailed look at the 74161 IC, which is used in the previous two examples. Figure 16.23 presents some information regarding the 74160 through 74163 series of synchronous 4-bit counters. This figure is abstracted from the data sheets and represents only one piece of information available to the design engineer. The ICs are composed of flip-flops that are all clocked from the same signal. This class of counters is *programmable;* that is, the output can be preset to any desired combination. When signals are placed into the load inputs and the LOAD is enabled, the counter presets, regardless of the values of the other inputs.

We examine the decade counter timing diagram in Figure 16.23. Moving from left to right (increasing time), the first action is that the $\overline{\text{CLR}}$ input goes low. Both *asynchronous clear* and *synchronous clear* are available. Note that in the asynchronous clear case, the outputs, Q_A through Q_D, clear as soon as the $\overline{\text{CLR}}$ input goes LOW. In the synchronous clear case, clearing occurs at the first positive clock transition following the $\overline{\text{CLR}}$ input.

The next action shown in the diagram is the $\overline{\text{LOAD}}$ input going LOW. This

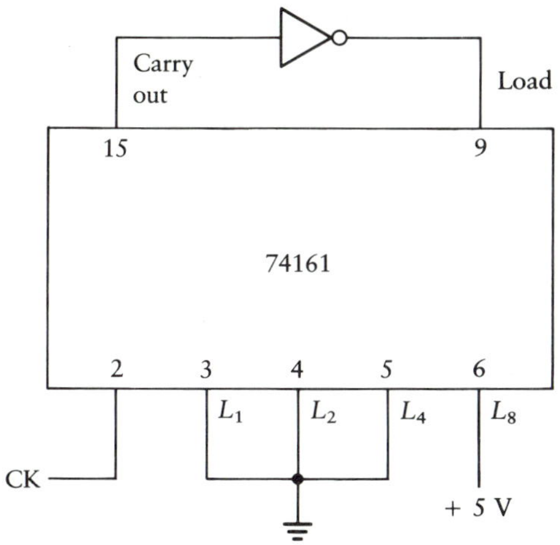

Figure 16.21 Connection diagram to count from 8 to 14.

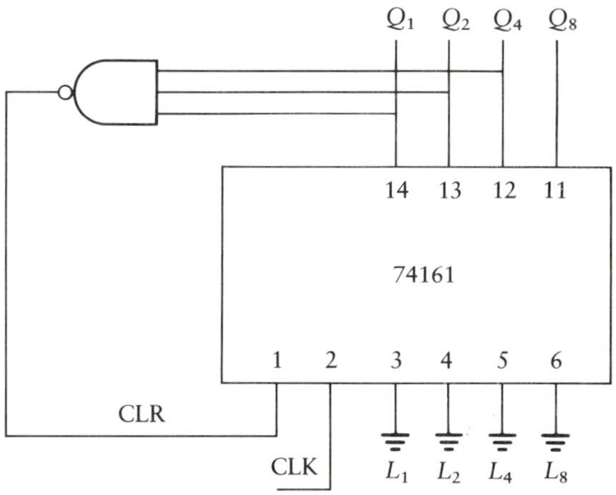

Figure 16.22 Connection diagram to cause counter to clear.

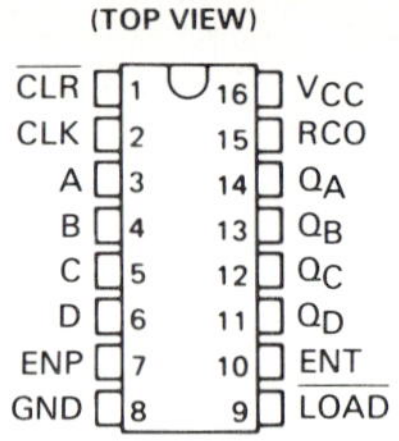

Figure 16.23
74160 through 74163 series synchronous 4-bit counters. Courtesy of Texas Instruments Inc.

typical clear, preset, count, and inhibit sequences

Illustrated below is the following sequence:

1. Clear outputs to zero ('160 and 'LS160A are asynchronous; '162, 'LS162A, and 'S162 are synchronous)
2. Preset to BCD seven
3. Count to eight, nine, zero, one, two, and three
4. Inhibit

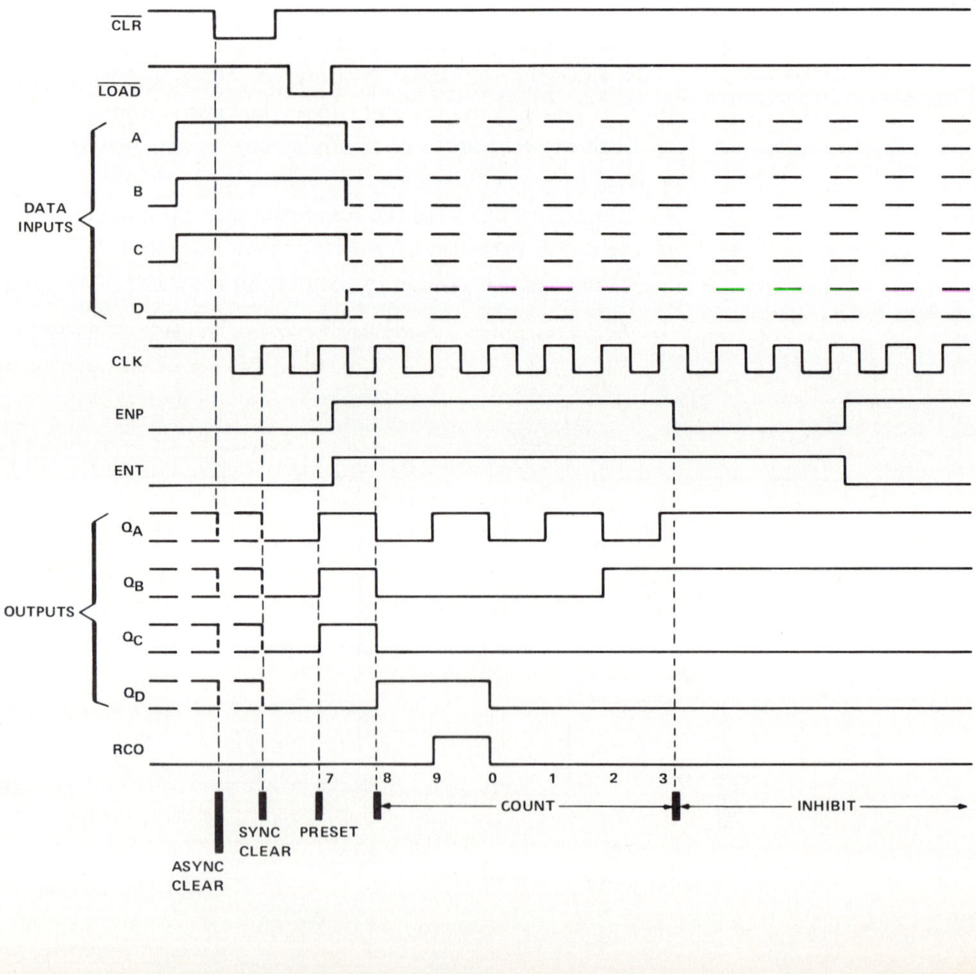

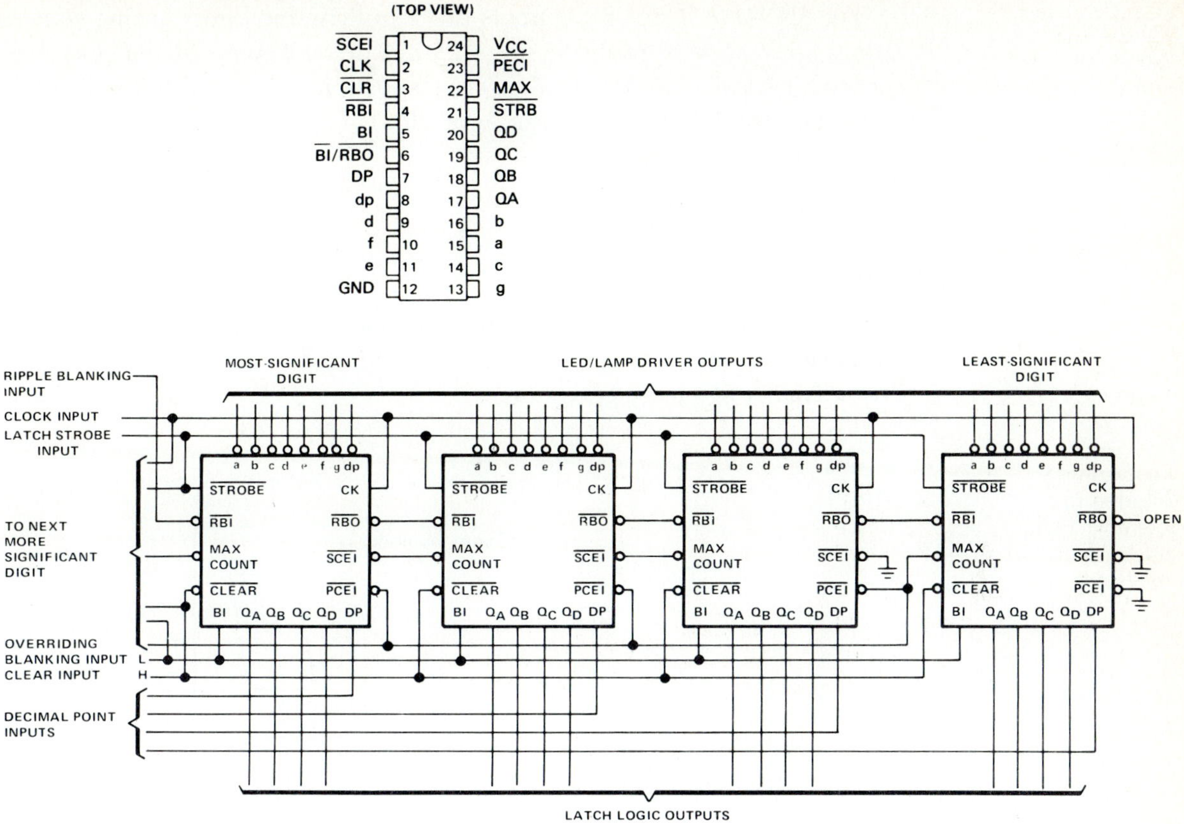

Figure 16.24
SN74143 and SN74144 series 4-bit counter/latch seven-segment LED/lamp drivers. Courtesy of Texas Instruments Inc.

causes the data input to be loaded at the next clock transition. Note that the example shows 0111, or 7, as the load sequence (*D* is the most significant bit and *A* is the least significant). The count is then enabled with the enable inputs going HIGH. The counter is shown incrementing through 0111, 1000, 1001, 0000 (the count reverts to zero since this is a *decade* counter), 0001, 0010, 0011. At that point, the ENABLE P (ENP on the diagram) goes low and the counter holds the last value.

Figure 16.24 illustrates the SN74143 MSI/TTL counter, which includes, in the same IC, the counter, latch, decoder, and display drivers. This useful IC combines the wide range of functions needed for LED/lamp drivers. Also shown in the figure is an example of cascading these devices to lengthen the count.

The 74143 and 74144 series TTL MSI circuits contain the equivalent of 86 gates on a single IC. They include relatively large resistors in series with the bases of the input transistors, which lowers the drive current requirements.

The *SN74143 driver* has outputs designed to maintain a constant current of approximately 15 mA into the loads for the seven segment outputs, *a* through *g*.

The *SN74144 driver* has outputs that can drive indicators having voltage ratings up to 15 V or requiring up to 25 mA drive. The maximum clock frequency is typically 18 MHz, and power dissipation is typically 280 mW. Additional data on this IC are given in Appendix D.

Example 16.3

Design a drive circuit for a single seven-segment LED display that will hold the output fixed while the input is changing. Use a 74162 decade counter driven from a 555 pulse generator.

Figure 16.25 Solution to Example 16.3.

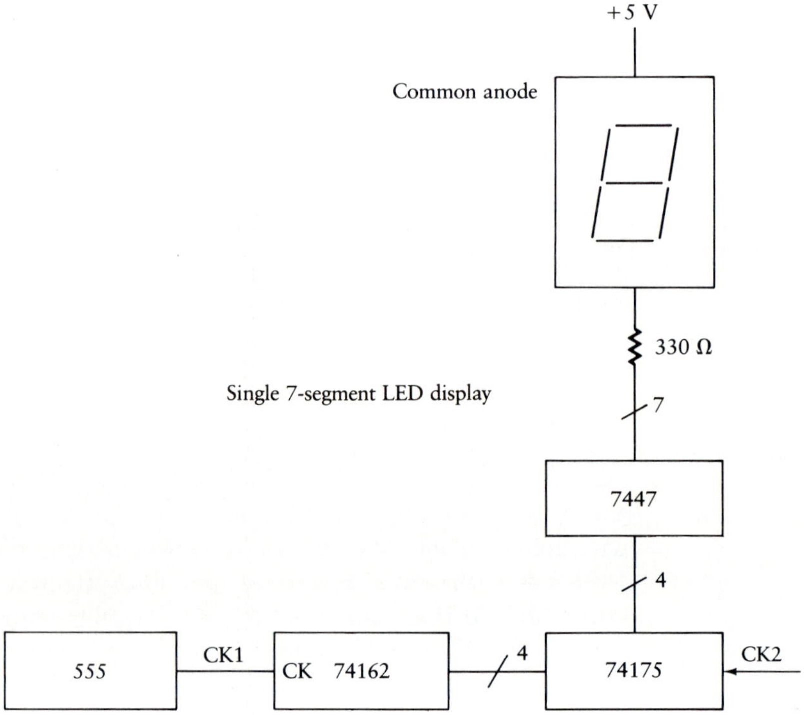

SOLUTION Refer to Figure 16.25 for a solution to this problem. The 555 clock (CK1) drives the 74162 synchronous 4-bit counter providing a signal to the 74175 latch. Whenever the second clock signal (CK2) goes from LOW to HIGH, the data at the input of the latch are transferred to the output, then to the 7447 decoder, and then to the seven-segment display. The latch output signal holds the display fixed even though the input to the latch is changing. The latch transfers the information into the display only when CK2 goes from LOW to HIGH.

Example 16.4

Repeat Example 16.3 using a 74143 4-bit counter/latch, seven-segment LED driver.

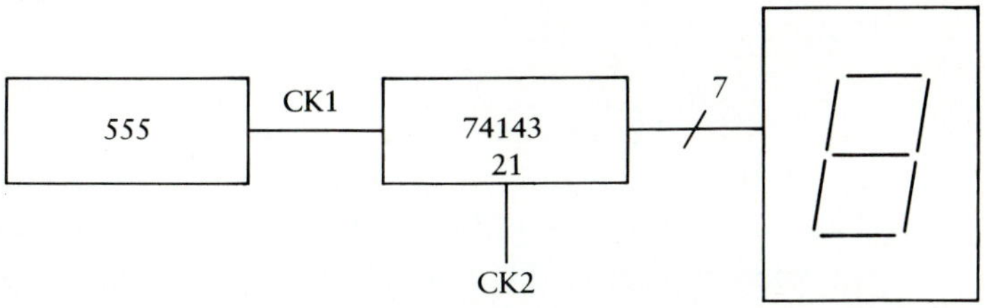

Figure 16.26 Solution to Example 16.4.

SOLUTION As can be seen from the block diagram of Figure 16.26, the single IC will solve the complete design since it contains a BCD counter, 4-bit latch, and decoder/driver. An important difference in using the 74143 is that we latch the count into the display with a LEVEL latch strobe voltage. We must hold pin 21 high to latch the data into the display. This is shown in Figure 16.26 as CK2.

16.4.1 Frequency Measurement

An important application of counters is in frequency measurement. A counter can be combined with a pulse generator to form a frequency counter. This is shown in Figure 16.27. The counter counts the number of pulses (or cycles) that occur during the time that the "window" pulse is HIGH. Since the width of the window pulse is known, the frequency of the input can be calculated. For example, if the window is exactly 1 s long, the counter yields an output in hertz.

Figure 16.27 Frequency counter.

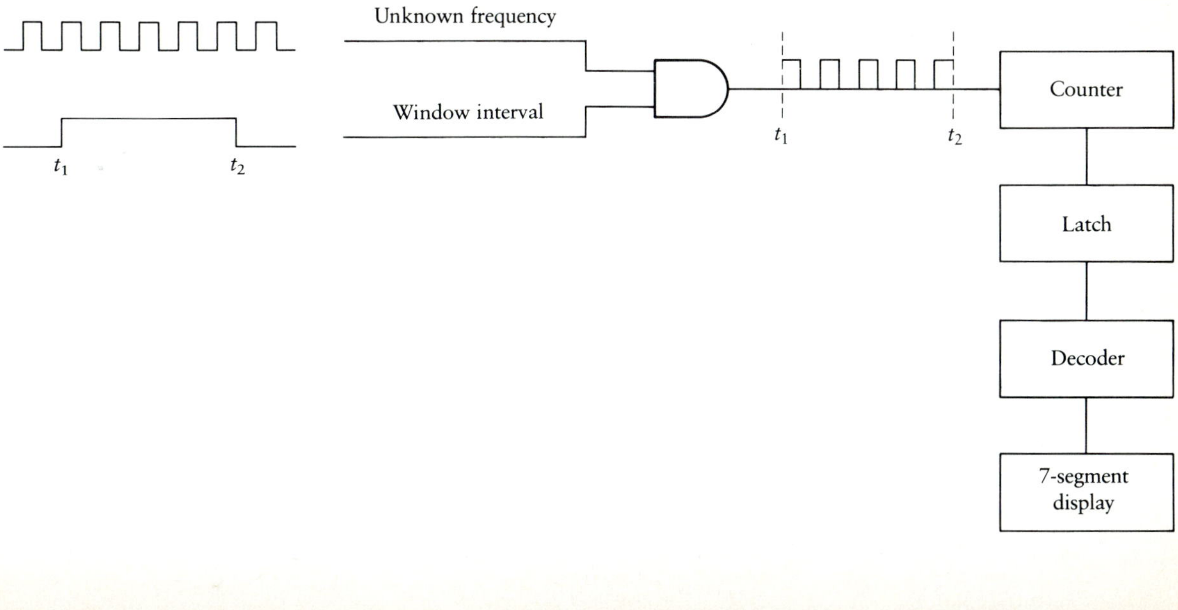

Example 16.5

Design a frequency counter to measure the frequency of a sinusoidal signal in the range between 5 and 15 kHz. The amplitude of the input signal is 10 V rms. Display the output on three 7-segment LEDs with three significant digits (i.e., *XX.X* kHz).

SOLUTION Let us choose the TTL family for this solution. We refer to the system shown in Figure 16.27, where we let the window period be 0.1 s long. At 5 kHz we obtain 500 pulses, and at 15 kHz we will obtain 1500 pulses within the window period. This meets the requirement for three significant figures. The sinusoidal input signal must first be changed into a pulse train so that the input is compatible with the digital circuitry. We refer to this process as *signal conditioning.* As we see from Figure 16.28, the sinusoidal input signal is first half-wave rectified and then processed through a 7414 Schmitt trigger to produce a 0 to 5 V pulse train that is compatible with TTL circuitry.

The window is developed with an astable 555 operating with a duty cycle of 10. We want the time low to be short so that little time is wasted before the window goes high and counting resumes. As a starting point, choose the time

Figure 16.28 Frequency measurement for Example 16.5.

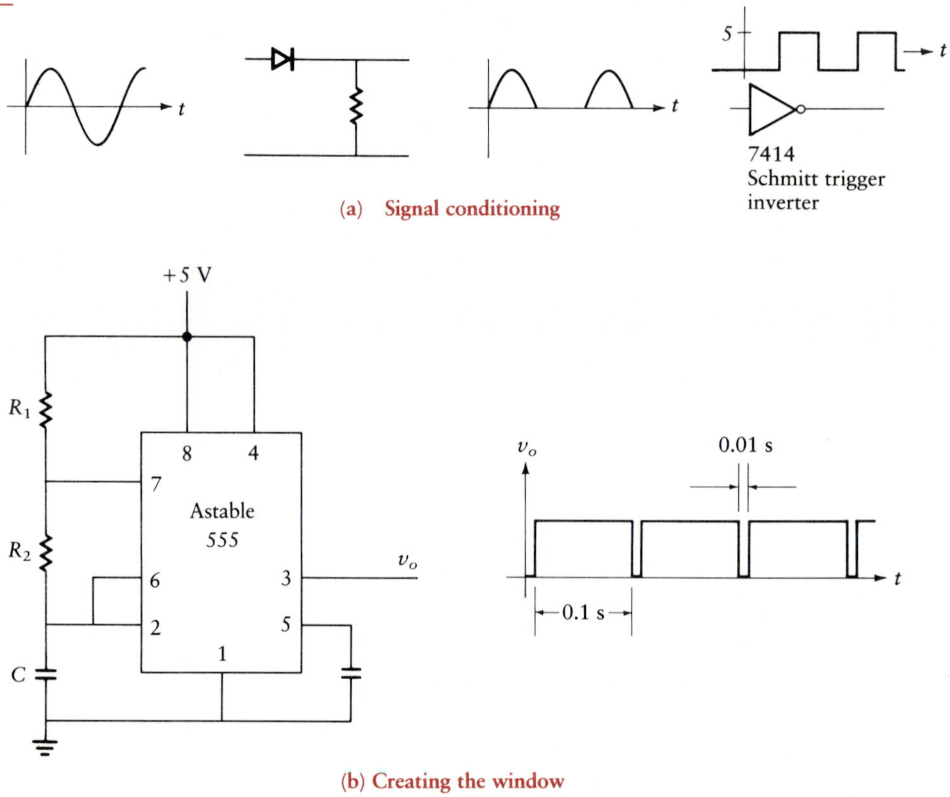

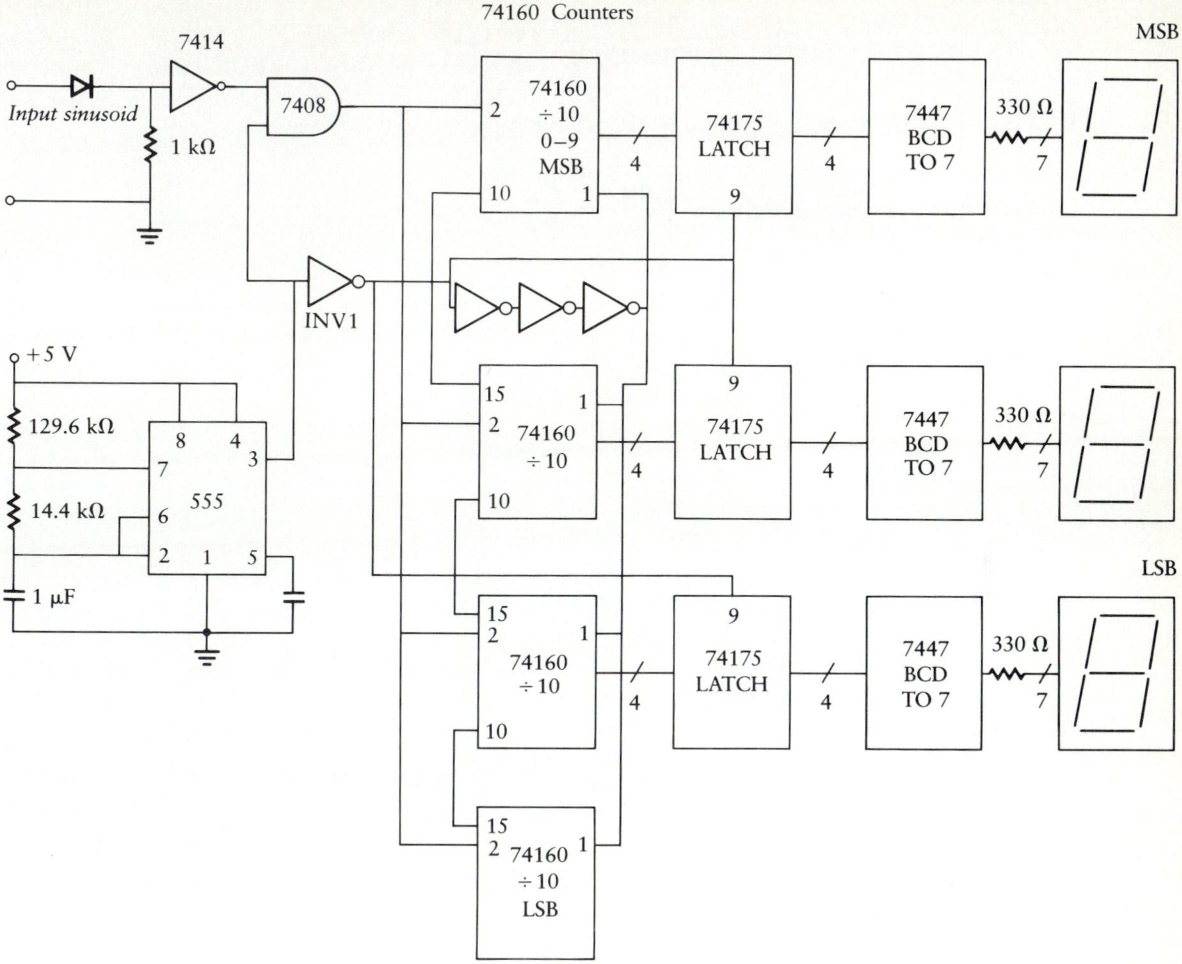

(c) **System block diagram**

low to be 10% of the time high (i.e., duty cycle of 10). This circuit is shown in Figure 16.28(b). The output of the astable 555 is a pulse train with a fixed frequency. Since we have chosen a window size of 0.1 s, the period of the astable 555 must be $T = 0.11$ s, so the frequency is $f = 1/T = 9.09$ Hz. We then use the techniques of Chapter 14 to design the timer, as follows. If we select $C = 1\ \mu$F, then

$$9.09 = \frac{1.44}{C(R_1 + 2R_2)}$$

from which we obtain

$$R_1 + 2R_2 = \frac{1.44}{9.09 \times 10^{-6}} = 158 \text{ k}\Omega$$

Since the duty cycle is 10, we have

$$\text{Duty cycle} = \frac{\text{time high}}{\text{time low}} = \frac{R_1}{R_2} + 1 = 10$$

Hence,

$$\frac{R_1}{R_2} = 9$$

and

$$R_1 + 2R_2 = 9R_2 + 2R_2 = 158 \text{ k}\Omega$$

We solve for R_2 and R_1, with the result

$$R_2 = 14.4 \text{ k}\Omega$$

$$R_1 = 129.6 \text{ k}\Omega$$

We now AND the output of the pulse train from Figure 16.28(a) with the window of Figure 16.28(b) to yield the number of pulses in 0.1 s. The complete block diagram is shown in Figure 16.28(c), where the pulses from the output of the 7408 AND gate are counted with four 74160 decade synchronous counters. Since we need only three 7-segment LED displays, we need only three 74175 latches and three 7447 BCD-to-7-segment decoder drivers.

The trailing edge of the window is used for two purposes:

1. To latch the ultimate count into the LED display and then, later,
2. To clear the counter back to 0000.

To ensure that the counter is cleared *after* the ultimate count is latched into the LED displays, the three extra inverters are included between the latch signal, after INV1, and the counter CLEAR signal. This is important, since if we do not delay the CLEAR signal, the display will show all zeros for any input count.

16.5 CLOCKS

Clocks are used to control the times at which changes occur in a digital circuit. One of the most popular clocks is formed using the 555 timer/oscillator, which is discussed in Chapter 14. A wide variety of other devices can be used for clocks, such as multivibrators, timers, and oscillator/dividers.

16.5.1 Voltage-Controlled Oscillator (VCO)

The output frequency of most oscillators depends on the setting of an *RC* time constant. We sometimes require a frequency that varies according to an input *voltage*. Examples of such situations include frequency modulators (FM), tone generators, analog-to-digital converters, and digital voltmeters. Oscillators of this type are termed *voltage-controlled oscillators (VCOs)*.

The frequency of oscillation of a relaxation oscillator, such as the 555 in the astable mode, depends on both the *RC* time constant and the voltage to which the capacitor charges. In the applications of the 555 discussed in Chapter 14, we drive the external charging circuit and the internal voltage dividers with the same voltage. We repeat the circuit diagram for a 555 in the

Figure 16.29
VCO.

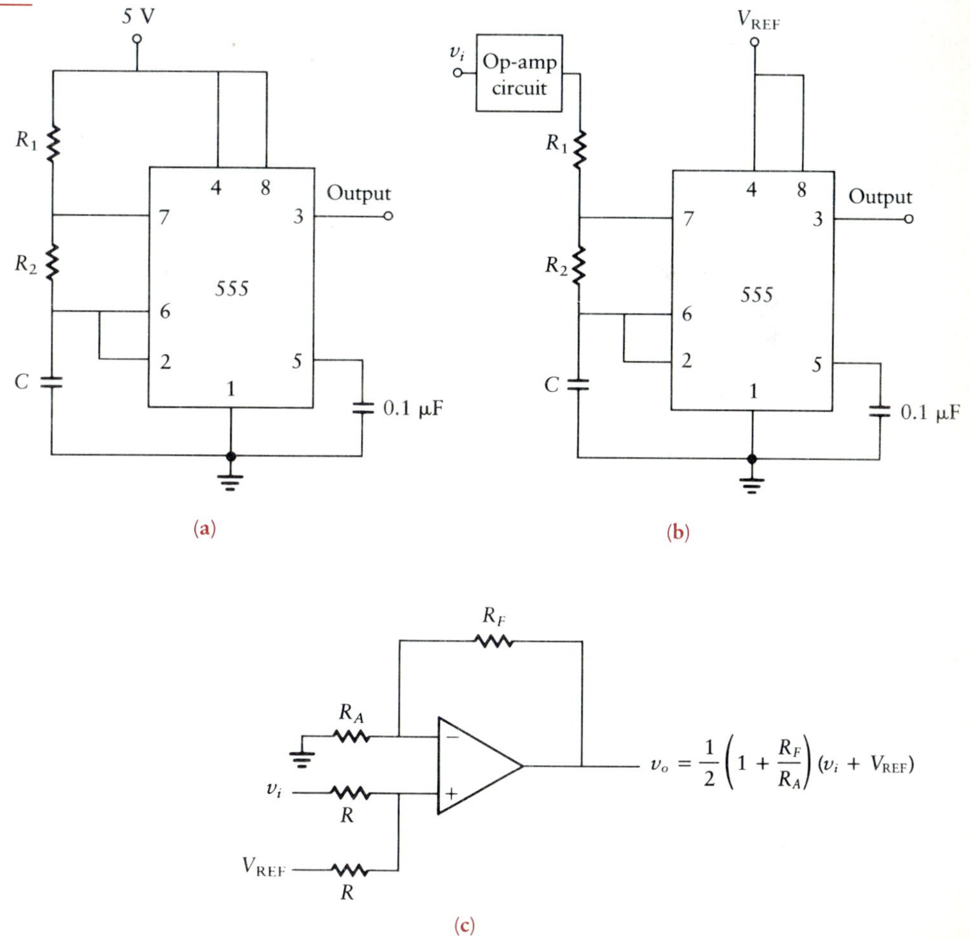

$$v_o = \frac{1}{2}\left(1 + \frac{R_F}{R_A}\right)(v_i + V_{REF})$$

astable mode as Figure 16.29(a). We use the same voltage drivers to ensure that the output frequency is independent of the supply voltage variations.

We now configure the 555 as shown in Figure 16.29(b), with the external charging circuit driven by an input voltage, v_i, through an op-amp circuit, such as shown in Figure 16.29(c). The internal voltage dividers are driven by a fixed reference voltage, V_{REF}. If we vary the voltage over a small region, the output frequency changes as a function of this voltage. Example 14.9 illustrates this mode of operation for the 555. The result is a simplified VCO with limitations of linearity and frequency range. These limitations cause us to consider a more sophisticated circuit.

Figure 16.30 shows the pin-out diagram and the typical characteristic curves for the 74LS124 dual voltage-controlled oscillator. This IC features two independent voltage-controlled oscillators in a single package. The output frequency of each VCO is established by a single external component, either a capacitor or a crystal, in combination with two voltage-sensitive inputs. One of the inputs controls the frequency range and the other controls the actual

Figure 16.30
74LS124 dual voltage-controlled oscillator. Courtesy of Texas Instruments Inc.

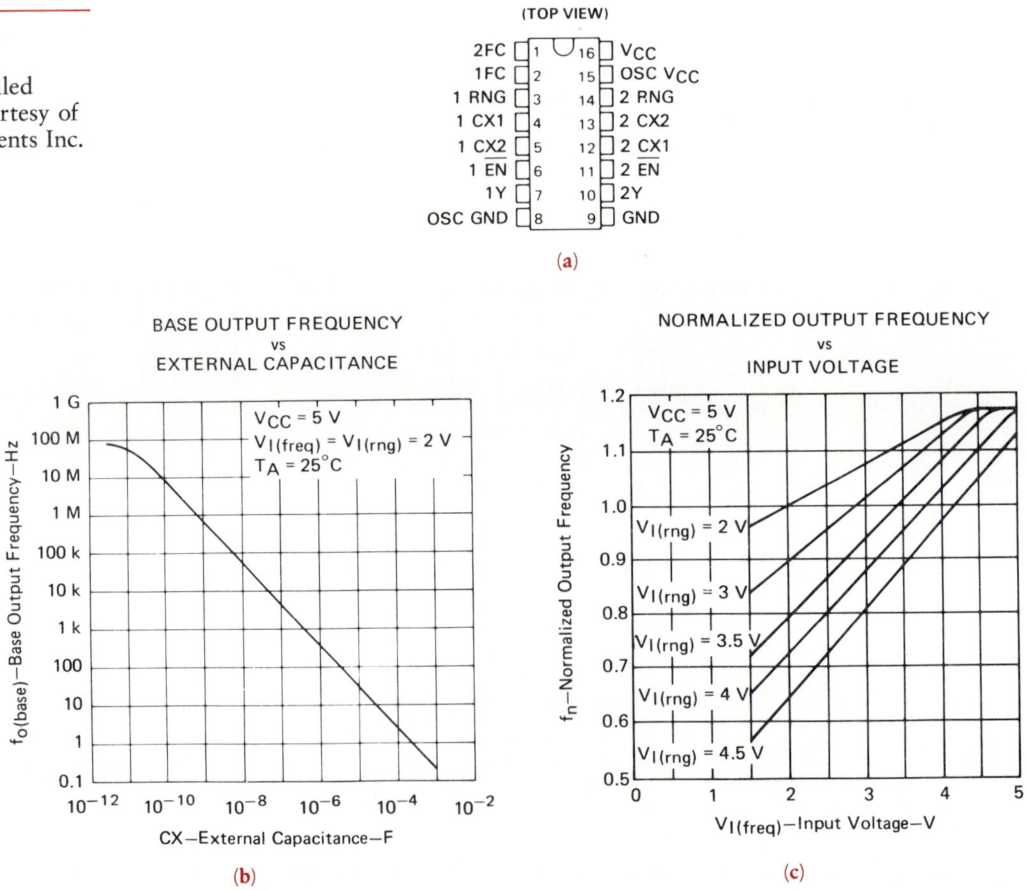

frequency. The relationship of these inputs to output frequency is shown in the curves of Figure 16.30. The curve of Figure 16.30(b) shows the center frequency as a function of the external capacitance which is attached between CX1 and CX2 in Figure 16.30(a). The curves of Figure 16.30(c) show the range of output frequency, expressed as a multiple of the center frequency, as the input voltage varies. There are a number of parametric curves on this graph. Each represents a different voltage on the range input. These stable oscillators can be set to operate at any frequency between about 0.12 Hz and 85 MHz. They can operate from a single 5 V supply. However, one set of supply voltage and ground pins (V_{CC} and GND) is provided for the enable, synchronization-gating, and + output sections and a separate set (OSC V_{CC} and OSC GND) is provided for the oscillator and associated frequency-control circuits. This is done so that effective isolation can be accomplished in the system.

Example 16.6

Use the 74S124 VCO to convert the voltage output of a temperature-sensitive bridge circuit to a frequency that is proportional to temperature. The center frequency should be 5 kHz. The temperature sensor provides the following outputs:

With 110°F, the output is 4 V.

With 90°F, the output is 2 V.

SOLUTION The design consists of the 74S124 with appropriate voltage inputs. We first set the center frequency using the frequency curves of Figure 16.30(b). To achieve a frequency of 5 kHz, the capacitance is approximately 10^{-7} F or 0.1 μF.

We must next decide what voltage value to use for the RANGE input. The set of curves in Figure 16.30(c) provides the necessary information. The input voltage ranges from 2 to 4 V, and this is read on the abscissa. We wish to choose a $V_{1(rng)}$ that will allow this voltage variation to cause a symmetrical swing around a normalized output frequency of unity, which corresponds to the center frequency, f_0. For example, if we choose a RANGE voltage of 4 V, the input of 2 V would create an output frequency of $0.72f_0$, where f_0 is the center frequency. A voltage of 4 V would create an output frequency of $1.03f_0$. These values are read from the $V_{1(rng)} = 4$ V curve in Figure 16.30(c). For an input of 4 V, the curve yields a normalized frequency of 1.13, so the output frequency is 1.13×5 kHz, or 5.65 kHz. For an input of 2 V, the curve yields a normalized frequency of 0.9, so the output frequency would be 0.9×5 kHz, or 4.5 kHz.

This is a poor choice of range voltage for this particular input voltage variation, since the frequencies are not symmetrical around f_0. Perhaps a better choice would be a range input of approximately 3 V. The output frequency range is then found from the figure.

We present a step-by-step procedure for scaling the SN74S124 voltage-controlled oscillator. We convert a specific voltage to a frequency that can then be measured using a window, as shown in Figure 16.27. We illustrate the procedure for the specifications of Example 16.6. We wish to display the number 90 when the VCO reads 2 V and display the number 110 when the VCO reads 4 V. The physical electronics and display are not affected by what the 90 and 110 represent. Breaking the problem solution into steps, we have the following:

1. Decide what the frequency output of the VCO, f_{01}, is going to be when the input voltage to the VCO is 2 V:

$$f_{01} = \frac{90}{t_2 - t_1}$$

   The window time, $t_2 - t_1$, is used later to read the number 90 from counter outputs. The frequency output, f_{02}, when the input voltage to the VCO is 4 V is given by

$$f_{02} = \frac{110}{t_2 - t_1}$$

   Hence the frequency range of the VCO output is from f_{01} to f_{02}. Note that for greater accuracy we might count 900 and 1100 pulses per window instead of 90 and 110 as indicated above.

2. Select the base frequency as the average

$$f_{\text{base}} = \frac{(f_{01} + f_{02})}{2}$$
$$= \frac{1}{2}\left(\frac{110}{T} + \frac{90}{T}\right)$$

   where we define

$$T = t_2 - t_1$$

   The effect is to allow a full frequency swing in the range of frequencies to be measured.

3. The external capacitor value is now calculated (or use specifications as in Figure 16.30(b)).

$$C_{ext} = \frac{5 \times 10^{-4}}{f_{base}}$$

4. Now that f_{base} and f_0 are known, we calculate the normalized frequencies

$$f_{n1} = \frac{f_{01}}{f_{base}}$$

$$f_{n2} = \frac{f_{02}}{f_{base}}$$

5. With this information, we look at the characteristic curves for the 74S124 and find the input voltages needed to provide the f_n calculated above. Pick a voltage range ($V_{1(rng)}$) that provides the greatest slope for the frequency range.
6. If the voltages corresponding to f_{n1} and f_{n2} are not equal to 2 and 4 V, we need to use an operational amplifier to perform the necessary conversion. Let us denote the voltages corresponding to f_{n1} and f_{n2} as V_{11} and V_{12}, respectively. The op-amp must accept an input with voltages ranging between 2 and 4 and convert this to an output with voltage ranges between V_{11} and V_{12}. The output versus input characteristic is a straight line with slope

$$m = \frac{V_{12} - V_{11}}{4 - 2} \tag{16.1}$$

and intercept

$$v_o = 2V_{11} - V_{12} \tag{16.2}$$

7. The op-amp configuration of Figure 16.29(c) can be used to develop an output voltage of

$$v_o = \left(\frac{1}{2} + \frac{R_F}{2R_A}\right)v_i + \left(\frac{1}{2} + \frac{R_F}{2R_A}\right)V_{REF} \tag{16.3}$$

The first expression in parentheses represents the op-amp gain, or the slope m. We equate terms and let

$$m = \frac{V_{12} - V_{11}}{2} = \frac{1}{2} + \frac{R_F}{2R_A} \tag{16.4}$$

The second term of equation (16.3) is the offset and can be found by equating terms, with the result

$$2V_{11} - V_{12} = \left(\frac{1}{2} + \frac{R_F}{2R_A}\right)V_{\text{REF}} \tag{16.5}$$

8. Select a value of R_F and solve for V_{REF} and R_A from equations (16.4) and (16.5).
9. Finally, connect the output of the op-amp to the voltage input of the VCO and the design is complete.

16.6 CONVERSION BETWEEN ANALOG AND DIGITAL

In Example 16.6, the VCO is used to develop a digital signal corresponding to a given analog signal. This electronic system is termed an *analog-to-digital (A/D) converter.* Analog-to-digital converters (and digital-to-analog converters) are not truly digital devices, but rather a combination of both analog and digital circuits. We include them in this chapter because they often form a critical starting (or ending) point in applying digital circuitry to analog systems.

16.6.1 Digital-to-Analog (D/A) Converter

Digital-to-analog (D/A) converters change a digital word into an analog voltage or current. Numerous techniques are used to accomplish this. Two methods are presented here.

The magnitude of the D/A output is generally proportional to the current flowing through weighted resistors or inversely proportional to the resistor values. An example of an 8-bit binary D/A converter with a current-to-voltage converter operational amplifier is shown in Figure 16.31. Each of the inputs is weighted according to the input-summing resistors so that the proper power of 2 is developed. An 8-bit signal at the input yields an analog output. You should analyze Figure 16.31 and verify that v_o is the analog equivalent of the digital input.

Another method is based on using a CMOS switch to change the resistors in a resistance ladder, as shown in Figure 16.32. This is called a *current-switching R-2R ladder* and involves a series of deposited silicon chromium resistors. These resistors, of value R or $2R$, are arranged in the ladder and constitute the National Semiconductor Corp. DAC0830 D/A converter. The circuit uses only two values of resistance, R and $2R$. Since the ladder network divides the current at each of the nodes, the accuracy of the ratio of the two resistors is more critical than their absolute values. Hence, the R-$2R$ ladder is relatively easy to fabricate. The ladder also presents a relatively constant resistance load to the V_{REF} source.

The digital input code applied to the input of the D/A converter controls the position of the current switches. In this manner, the available ladder cur-

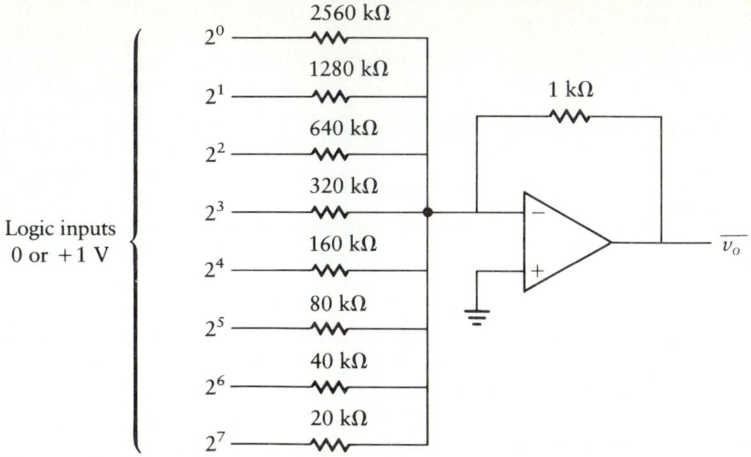

Figure 16.31
D/A converter.

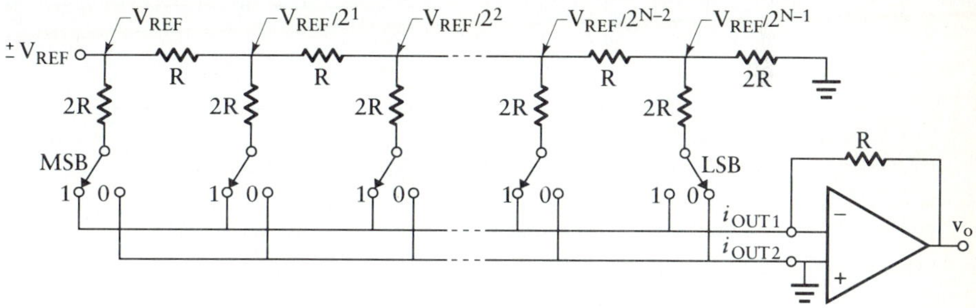

Figure 16.32
DAC0830 8-bit D/A converter.

rent is steered to either i_{OUT1} or i_{OUT2} as determined by the logic level (0 or 1, respectively). The CMOS switches are bilateral and so can switch currents of either polarity with only a small voltage drop.

With the use of the R-$2R$ ladder network, this D/A converter can produce 0.05% of full-scale maximum linearity error. Typical conversion time is 1 μs, and with an 8-bit input this circuit is capable of generating 256 distinguishable output current levels. The resolution is 8-bit. Additional technical details are given in the manufacturer's data sheets.

16.6.2 Analog-to-Digital (A/D) Converter

Analog-to-digital converters change an analog voltage level into a corresponding digital word. There are numerous methods of producing an A/D converter. We discuss several in this section.

One way of accomplishing the conversion is to increment a counter that feeds a D/A converter and stop the counter when the D/A converter's output exceeds the analog voltage in question. This method is illustrated in Figure 16.33. The D/A converter output is a staircase function. It can be thought of as a discrete ramp function. The number of counts it takes before the ramp

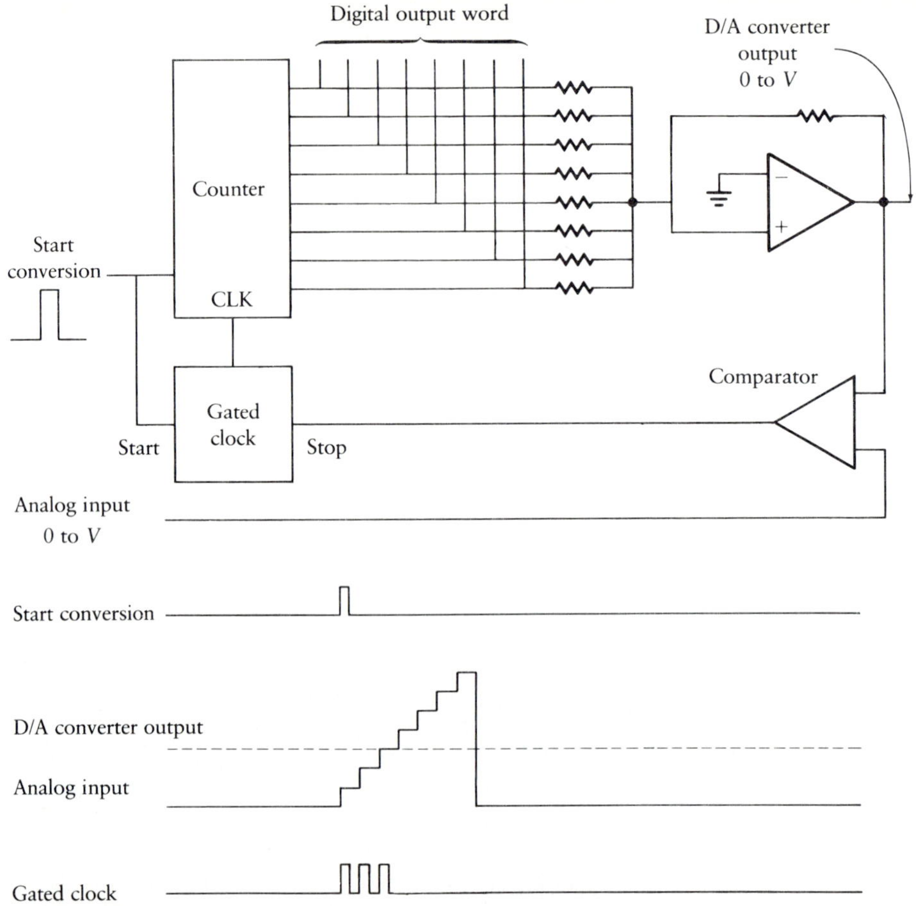

Figure 16.33
A/D converter.

crosses the analog value is proportional to that value. The digital output word is the counter output. An 8-bit counter resets and starts from zero for each measurement.

A second method of generating a digital word from an analog voltage is to use successive approximations. If we assign binary numbers to various voltage levels starting with the lowest voltage (all 0's) and counting toward the highest (all 1's), we can use the basic properties of binary sequences to simplify the conversion. The most significant bit in the binary number indicates whether the voltage is in the upper or lower half of the range. The next bit subdivides this range in half, and so on. This is equivalent to the observation that in a binary counter, each bit is oscillating at half the frequency of the previous bit. The conversion is then accomplished by a series of comparisons with the regional dividing points.

A specific example of this type of A/D converter is the ADC0801 IC, which contains a high-input-impedance comparator, 256 series resistors and analog switches, control logic, and output latches. Conversion is performed using a successive approximation technique, where the unknown analog voltage is compared to the voltage at the resistor tie points using analog switches. When the appropriate tie-point voltage matches the unknown voltage, conversion is complete. The digital outputs contain an 8-bit complementary binary word corresponding to the unknown voltage.

16.6.3 The $3\frac{1}{2}$-Digit A/D Converter

The $3\frac{1}{2}$-digit A/D converter is essentially a digital voltmeter. Incorporated in the design of this digital voltmeter IC is the *dual-slope* method of analog-to-digital conversion. We present the ICL7106/7107 CMOS A/D converter as an example. The pin-out diagram for this IC is shown in Figure 16.34. The ICL7106 drives a liquid crystal display (LCD), and the ICL7107 drives light-emitting diode (LED) display. Included are seven-segment decoders, display drivers, references, and a clock. The IC operates in three phases: (1) auto-zero, (2) signal integrate, and (3) reference integrate. In Phase 1 of the dual-slope conversion, the cycle is zeroed for a new start. This process is known as the auto-zero phase. The block diagram is shown in Figure 16.35(a). In Phase 2 of the dual-slope method, the signal is integrated for a fixed period of time with the slope depending on the RC combination of the integrating op-amp. Since the time period is fixed, the value of the integral is proportional to the signal amplitude. In Phase 3, the integrator input is switched from v_i to V_{REF}. The polarity is determined during Phase 2 so that the integrator discharges back toward zero. Since the slope of the discharge function is fixed, the time required to return to zero is proportional to the starting amplitude. We have therefore succeeded in translating the sample amplitude value into a time interval whose length is proportional to the original amplitude. We need now only convert the

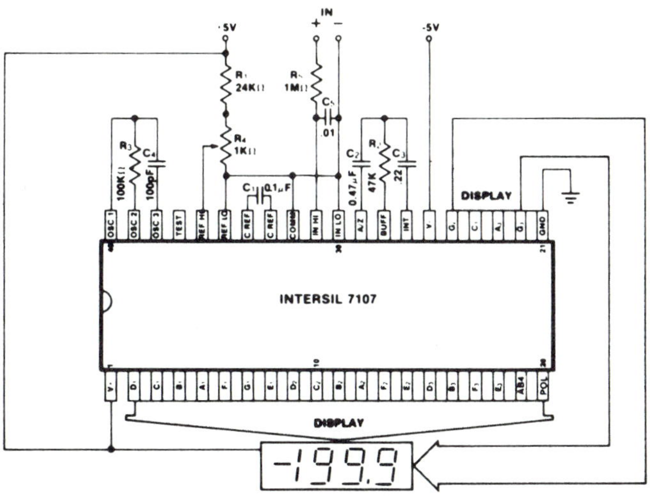

ICL7107 with LED Display

Figure 16.34 ICL7106/7107 digital voltmeter IC. Courtesy of Intersil Inc.

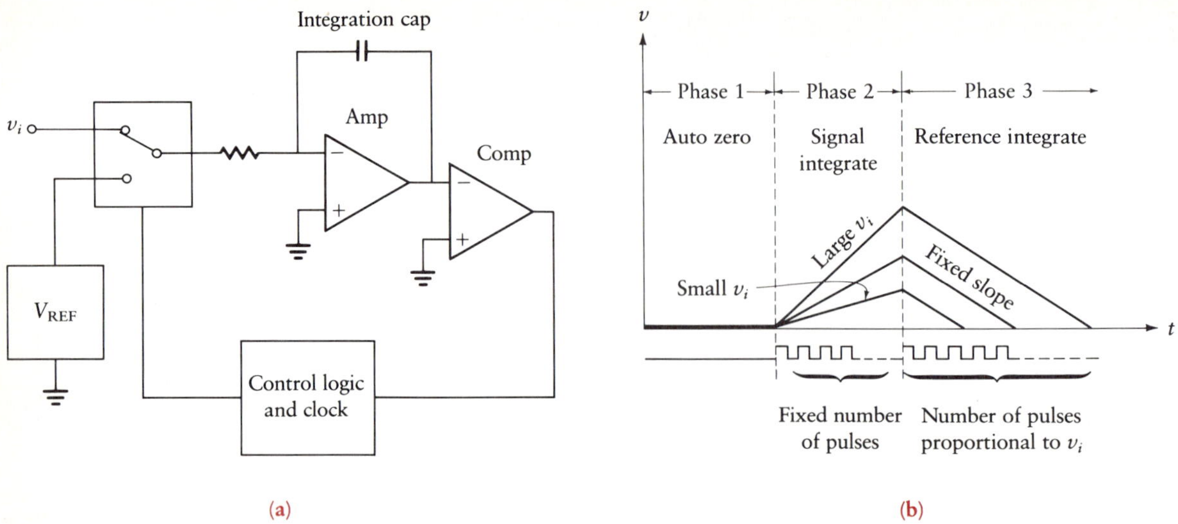

Figure 16.35
Phases in dual-slope conversion.

time interval to a digital number using a clock. The number of clock pulses counted between the beginning of this cycle (Phase 3) and the time when the integrator output passes through zero is a digital measure of the magnitude of v_i. This digital measurement is then fed to digital counters, which provide the digital output. The digital control logic synchronizes the display output for each cycle and begins the A/D conversion cycle again. In the dual-slope comparison method, the accuracy of the system is limited by the number of bits of the counter and by the accuracy of the reference voltages. This A/D converter depends only on the ratio of v_i to V_{REF}.

Drill Problem

D16.1 Analyze the input circuit for a thermometer that uses the $3\frac{1}{2}$-digit DVM (ICL7106) of Figure 16.34 for the digital voltmeter.

ANSWER The 1N914 diode provides the basic sensor to measure temperature. The bridge circuit provides the signal to the op-amp and also the zero adjust. This is shown in Figure D16.1(a). In Figure D16.1(b), we show the full-scale adjustment. The calibration is as follows.

a. The $3\frac{1}{2}$-digit DVM used as a digital centigrade thermometer: Calibration is achieved by placing the sensing diode in ice water and adjusting the zeroing potentiometer for a 000.0 reading. The sensor is then placed in boiling water and the scale factor potentiometer is adjusted for a reading of 100.
b. The $3\frac{1}{2}$-digit DVM used as a digital Fahrenheit thermometer: The procedure is the same as that for the digital centigrade thermometer except that

Figure D16.1

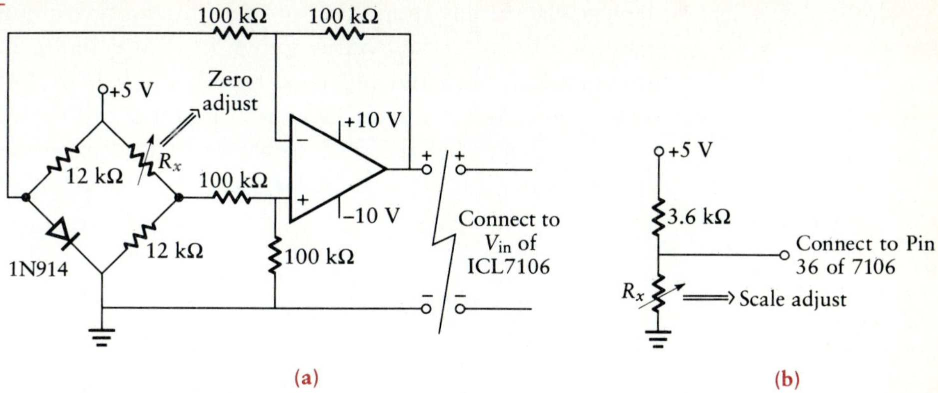

we adjust the zeroing potentiometer for a 032.0 reading and adjust the scale factor potentiometer for a 212.0 reading.

16.6.4 Liquid Crystal Display (LCD)

The *liquid crystal display* (LCD) requires low power and is ideally suited for battery-operated devices, such as digital watches. Liquid crystal displays are driven by applying a symmetrical square wave to the back plane (BP). To turn on a segment, a waveform 180° out of phase with BP (and of equal amplitude) is applied to that segment. Excessive dc voltages (>50 mV) will permanently

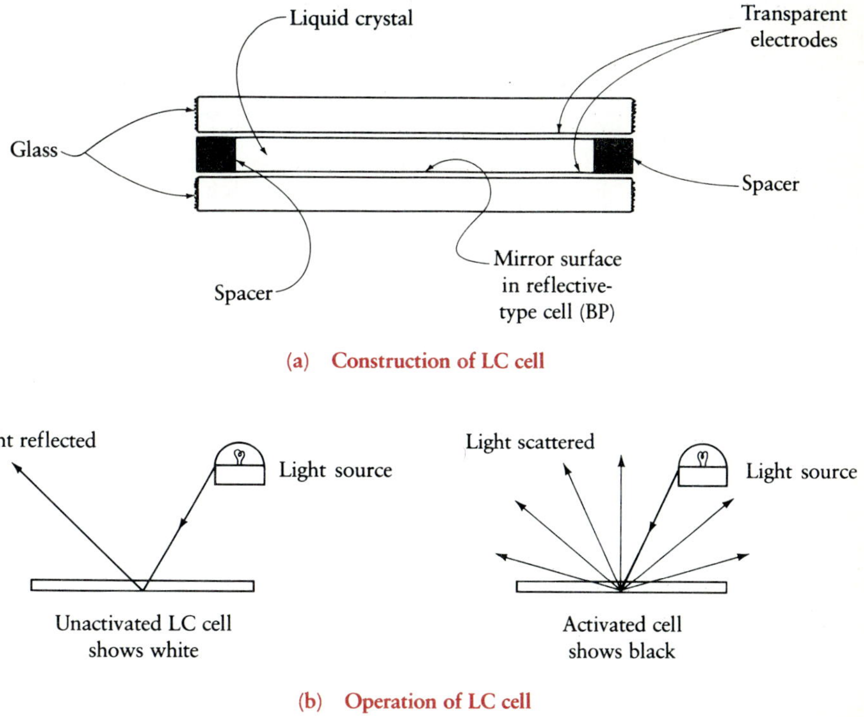

Figure 16.36
Liquid crystal display.

damage the display if applied for more than a few minutes. A schematic diagram of a liquid crystal cell is shown in Figure 16.36. As can be seen in Figure 16.36(b), when the cell is activated, the light is scattered so the display shows black. When the cell is unactivated, the cell shows white.

The segments of the seven-segment LCD display are driven to form the numbers from 0 to 9, as is done for the seven-segment LED display.

16.7 MEMORIES

We consider various types of memory devices, which are summarized in the block diagram of Figure 16.37.

16.7.1 Serial Memories

Data entered into the storage of serial memory devices are not immediately available for reading. Typically, each stored bit is transferred sequentially through 64 or more storage locations between the time it is written into memory and the time it first becomes available for reading.

Figure 16.37
Memory overview.

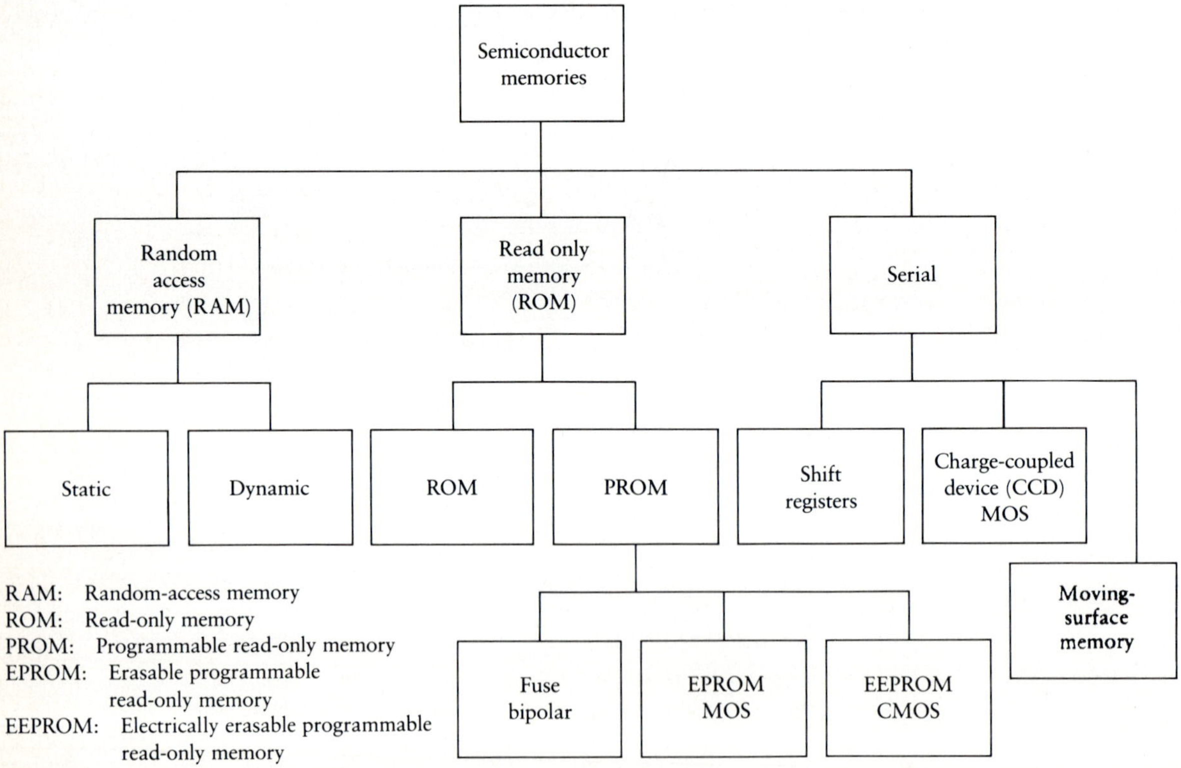

In Section 16.3.2 we studied the use of latches to hold the output lines (which drive the LED display) fixed while the inputs are being updated. This is a form of memory, since the latch remembers the most recent value and holds this value on the display while the data are being updated. With the 74175 latch, we transfer the most recent information to the display with the rising edge of the clock pulse. The *D* flip-flop in the 74175 remembers this number until another rising edge of a clock pulse is applied.

A shift register (presented in Section 16.3.3) is another form of memory, where the data are transferred from flip-flop to flip-flop. These stored data can be shifted right or left or manipulated in a parallel manner. Throughout these operations, the shift register remembers the data. Frequently, data at the output flip-flop can be fed back to the input flip-flop. In this manner, the data in the register are held unchanged but circulate in what is termed a *ring counter.* A variation of this is the *switch-tail* ring counter (or Johnson counter), in which we use a circular shift register. This form of shift register is shown in Figure 16.38(a). We take the *complement* of the output flip-flop and feed that back to the input flip-flop. When a 1 is injected into the input *D* flip-flop, at each clock pulse we generate the function table as shown in Figure 16.38(b).

Moving-surface memories are the slowest form of serial memory. The moving magnetic tape player is an example of this type. This form of memory is cheaper than electronic memories because there is no need to define individual

Figure 16.38 Switch-tail ring counter.

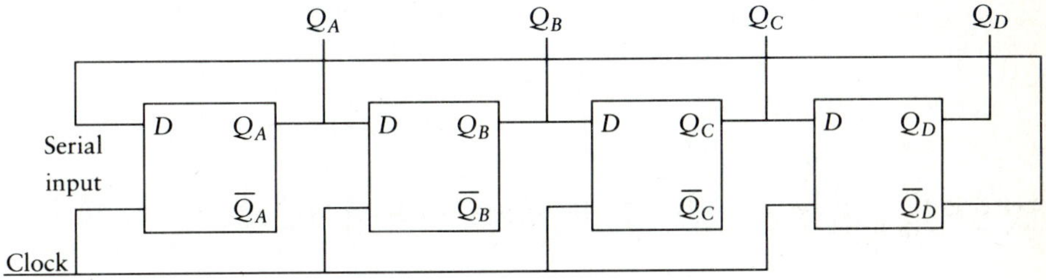

(a) Circuit diagram

Clock pulse	Q_A	Q_B	Q_C	Q_D
0	0	0	0	0
1	1	0	0	0
2	1	1	0	0
3	1	1	1	0
4	1	1	1	1
5	0	1	1	1
6	0	0	1	1
7	0	0	0	1
0	0	0	0	0

(b) Function table

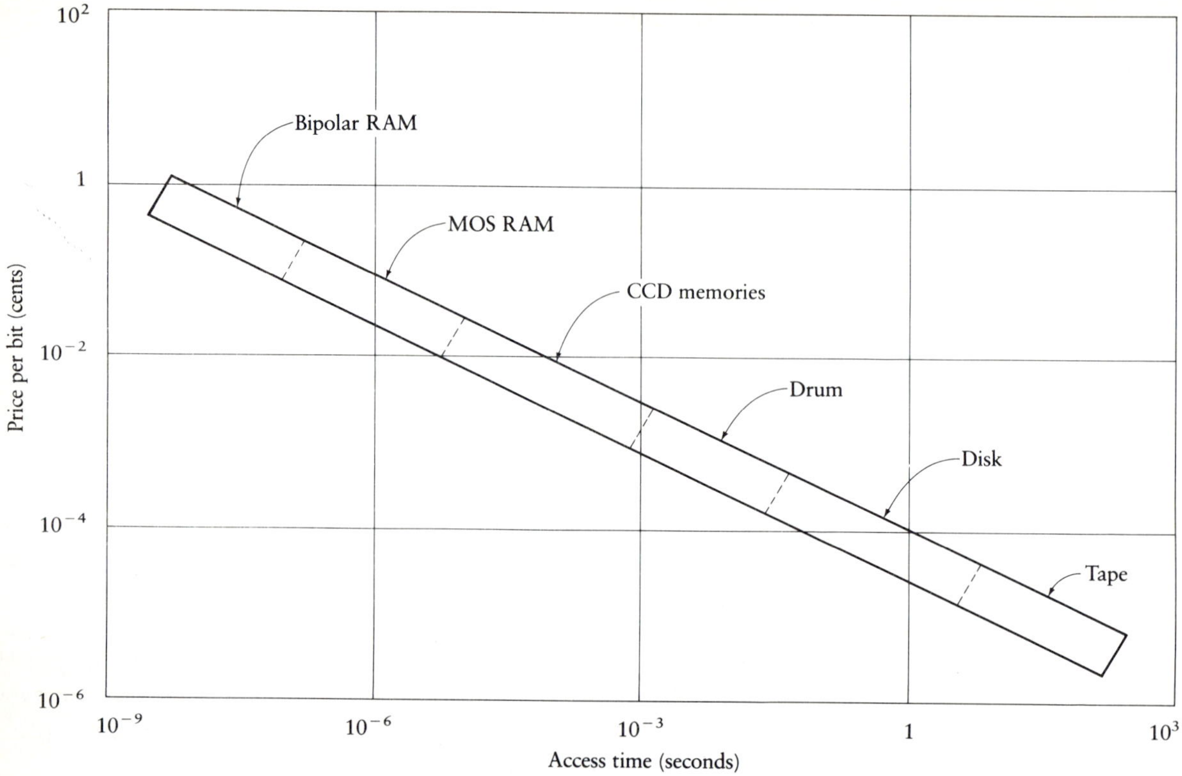

Figure 16.39 Access time versus price.

physical patterns or structures for each storage cell. However, precision mechanical components may be needed to transport the magnetic storage medium, leading to a high initial cost.

When selecting a particular memory, it is important to consider the price per bit of storage capacity. Price depends strongly on access time. In currently available systems, the access time ranges from about 10 ns for bipolar transistor memories to about 10 s or more for magnetic tape memories. The cost per bit varies over a very wide range. For bipolar memories it is of the order of 1 cent per bit, whereas for tape it is only about 10^{-5} cent per bit. The relationship between price per bit and access time is approximately logarithmic over a wide range of prices per bit. This relationship is shown in Figure 16.39.

The reliability of memory systems is a function of both fundamental and practical problems. The fundamental problems have to do with phenomena such as corrosion. The practical problems have to do with defective manufacturing, packaging, or testing and with mistakes in the use and maintenance of components and systems. Most of the failures in today's memory systems result from the practical problems.

16.7.2 Random-Access Memory (RAM)

Dynamic memories store data in the form of a charge on a small capacitor. A dynamic memory cell cannot store data indefinitely. The capacitor storage cell can lose stored information in two ways. First, the capacitor itself has an associated leakage current. Second, when the cell is selected for a read operation, the charge stored is shared between the cell capacitor and the large capacitance of the data line.

For the dynamic memory to retain valid data, the capacitor charge must be restored periodically. This restoration must occur at least once every 2 ms. Rewriting is accomplished internally without the need to reapply the original external data. This rewriting operation is called *refresh.*

Memories that do not require refresh operations are termed *static memories.* In spite of their higher cost per bit of storage, they are favored for small memory systems because they require a minimum of external support circuitry. At a further premium in cost, the power consumption of static memories can be significantly reduced. These memories are found in pocket calculators, where small batteries must provide sufficient power for operation over days and weeks.

A static RAM is an array of latches with a common addressing structure for both reading and writing. In the WRITE mode, the information at the data input is written into the latch selected by the ADDRESS. In the READ mode, the content of the selected latch is fed to the data output.

Semiconductor memories have *nondestructive readout.* That is, the memory can be read without destroying the data stored therein. Semiconductor read/write memories are *volatile.* That is, data can be stored only as long as the power is uninterrupted.

Memories are generally identified by specifying the number of words, number of bits per word, and function. For example, a 1024 × 16 RAM is a random-access read/write memory containing 1024 words of 16 bits each.

High-density RAM memories are usually organized in arrays of n words of 1 bit each. Only one input/output lead is needed in addition to the address leads, thus optimizing lead usage.

Addressing (word selection) in a semiconductor memory consists of two operations. First, a given device or group of devices must be selected; second, a given location in a device or group of devices must be specified. The device can be selected by supplying an input to the CHIP SELECT function of each device. The input is LOW for all but the desired device. The input can be derived from a binary-to-n decoder (see Section 16.1). The binary address of the device is fed in, and only the select output for that device is HIGH.

Figure 16.40 shows the block and connection diagrams for the NMC6164 8192 × 8-bit static RAM. This is an 8192-word by 8-bit random-access read/

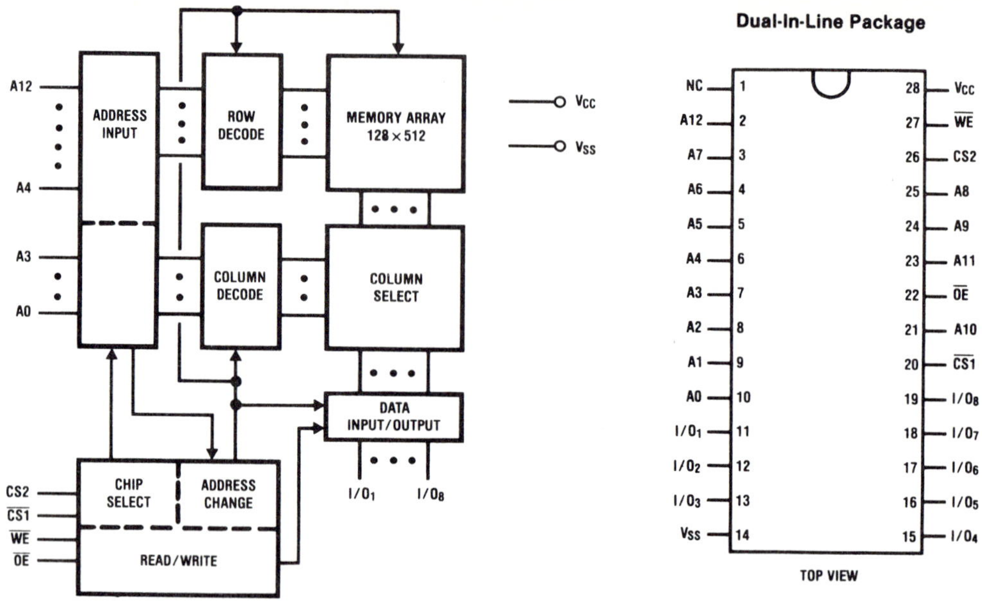

Figure 16.40
NMC 6164 static RAM.

write memory that uses silicon gate CMOS technology. The address is specified by a row and a column. Nine of the address bits specify one of 512 possible rows, and four address bits are used to specify one of 16 possible columns.

The actual 8-bit word is fed into or read out of pins 11–13 and 15–19. Data can be read when the write enable ($\overline{WE}$) and chip select 2 (CS2) inputs are HIGH and $\overline{CS1}$ and $\overline{OE}$ are LOW. Writing into memory occurs with the write enable LOW. This device has an access time on the order of 100 ns.

16.7.3 ROMs and PROMs

With a *read-only memory* (ROM) the stored information is fixed and nonvolatile. A semiconductor ROM is a circuit whose stored information is fixed during manufacture, whereas a *programmable ROM* (PROM) can be programmed after manufacture. A ROM is best suited for systems produced in large volume. Here the tooling charge for a unique *mask* is relatively small on a per-unit basis and is often counterbalanced by the economies of batch processing. PROMs are the best choice in low-volume production or in systems having limited useful life, in short procurement cycle situations for applications where some degree of system tailoring is required for each installation, and where there is a high probability that the stored information will be changed at some future date.

16.7.4 EPROMs

PROMs can be programmed with any desired array of binary numbers. The process of programming is sometimes known as *burning* the PROM, since the programming can be thought of as *burning out* appropriate fuses. Once this process is performed, the PROM can never be reprogrammed, although it is possible to make additional outputs high by burning additional fuses. That is, none of the output 1's can be changed to 0's, but the reverse is possible.

A class of PROMs exists where the programmed data can be cleared and the PROM reprogrammed with new data. *Erasable PROMs* (EPROMs) are available with MOS technology. An erasable PROM allows the programmed

Figure 16.41 NMC27C16 (2048 × 8) UV-erasable PROM. Courtesy of National Semiconductor Corp.

Pin Names

A0–A14	Addresses
$\overline{CE}$	Chip Enable
$\overline{OE}$	Output Enable
O_0–O_7	Outputs
$\overline{PGM}$	Program
NC	No Connect

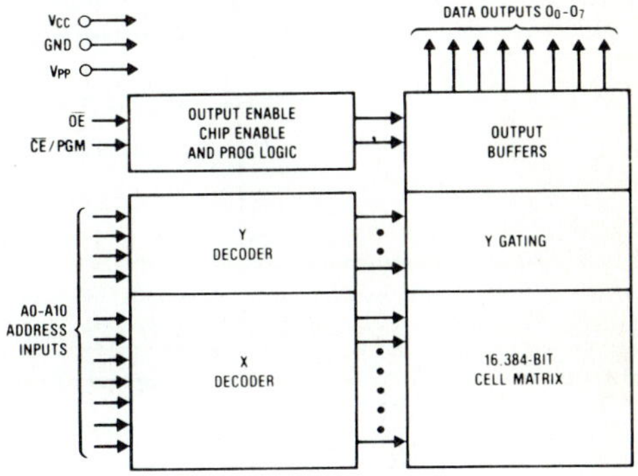

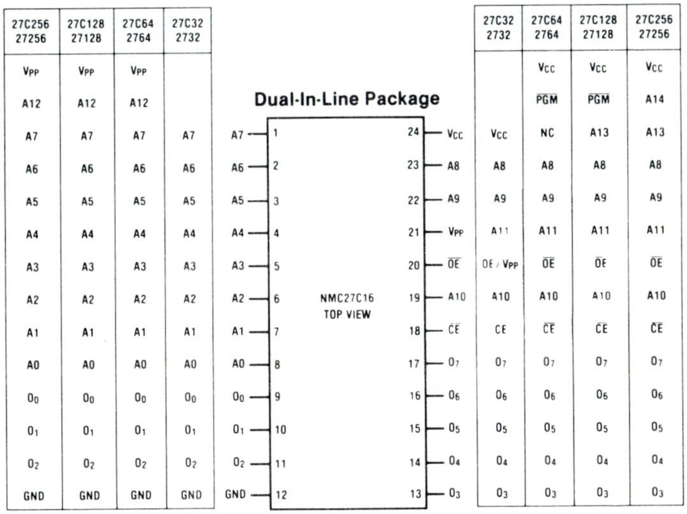

(a) **Block and connection diagrams**

Mode \ Pins	$\overline{CE}$/PGM (18)	$\overline{OE}$ (20)	V_{PP} (21)	V_{CC} (24)	Outputs (9–11, 13–17)
Read	V_{IL}	V_{IL}	V_{CC}	5	D_{OUT}
Standby	V_{IH}	Don't Care	V_{CC}	5	Hi-Z
Program	Pulsed V_{IL} to V_{IH}	V_{IH}	25	5	D_{IN}
Program Verify	V_{IL}	V_{IL}	25	5	D_{OUT}
Program Inhibit	V_{IL}	V_{IH}	25	5	Hi-Z

(b) **Mode selection**

information to be erased by exposure to ultraviolet light of the correct intensity and wavelength.

Figure 16.41 presents information for a 2048 × 8 EPROM. This 16k PROM is erased by applying an ultraviolet light to the window in the IC. The memory is reprogrammed electrically. The memory is packaged in a 24-pin dual in-line package (DIP) with a transparent window. This window permits the user to expose the device to ultraviolet light for the purpose of erasing the bit pattern. The IC uses a single +5 V power supply. The EPROM operates in five modes, which are shown in Figure 16.41(b) and are summarized as follows:

1. *READ mode.* Two control functions: chip enable ($\overline{CE}$) and output enable ($\overline{OE}$) are required to gate the addressed data to the output.
2. *STANDBY mode.* When in this mode, the outputs are in a high-impedance state, independent of $\overline{OE}$. In this standby mode, the power dissipation is reduced by 98%.
3. *PROGRAM mode.* After erasure, all bits in the memory are in the logic 1 state. Data are introduced by selectively programming logic 0 into the desired bit locations. To change a 0 to a 1, however, we must use ultraviolet light erasure. The memory is in a programming mode when the V_{PP} power supply is at 25 V and $\overline{OE}$ is at a high input voltage (V_{IH}).
4. *PROGRAM VERIFY mode.* To be certain that the bit pattern is correctly programmed, we can use this mode. We verify the program with V_{PP} = 25 V.
5. *PROGRAM INHIBIT mode.* When programming multiple memories in parallel, it is necessary to inhibit the memories that are not being programmed. A low-level $\overline{CE}$/PGM input inhibits the other parallel memories from being programmed.

An *electrically erasable PROM* (EEPROM) is useful when we wish to alter stored data. Erasing and programming are accomplished by applying electrical signals to the appropriate inputs of the IC.

16.8 MORE COMPLEX CIRCUITS

We now study some of the more complex digital integrated circuits. We concentrate on those that are used to perform mathematical operations.

16.8.1 Arithmetic Logic Unit (ALU)

The *arithmetic logic unit (ALU)* performs logic or arithmetic operations. We examine the SN74181 series as a typical example. The pin diagram and function table for these ICs are shown in Figure 16.42. The ALU has a complexity

Figure 16.42
Arithmetic logic unit/function generators. Courtesy of Texas Instruments Inc.

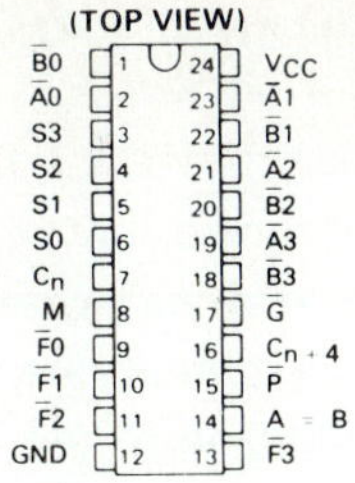

(a)

SELECTION				ACTIVE-LOW DATA		
				M = H	M = L; ARITHMETIC OPERATIONS	
S3	S2	S1	S0	LOGIC FUNCTIONS	Cn = L (no carry)	Cn = H (with carry)
L	L	L	L	F = $\overline{A}$	F = A MINUS 1	F = A
L	L	L	H	F = $\overline{AB}$	F = AB MINUS 1	F = AB
L	L	H	L	F = $\overline{A}$ + B	F = A$\overline{B}$ MINUS 1	F = A$\overline{B}$
L	L	H	H	F = 1	F = MINUS 1 (2's COMP)	F = ZERO
L	H	L	L	F = $\overline{A + B}$	F = A PLUS (A + $\overline{B}$)	F = A PLUS (A + $\overline{B}$) PLUS 1
L	H	L	H	F = $\overline{B}$	F = AB PLUS (A + $\overline{B}$)	F = AB PLUS (A + $\overline{B}$) PLUS 1
L	H	H	L	F = $\overline{A \oplus B}$	F = A MINUS B MINUS 1	F = A MINUS B
L	H	H	H	F = A + $\overline{B}$	F = A + $\overline{B}$	F = (A + $\overline{B}$) PLUS 1
H	L	L	L	F = $\overline{A}$B	F = A PLUS (A + B)	F = A PLUS (A + B) PLUS 1
H	L	L	H	F = A $\oplus$ B	F = A PLUS B	F = A PLUS B PLUS 1
H	L	H	L	F = B	F = A$\overline{B}$ PLUS (A + B)	F = A$\overline{B}$ PLUS (A + B) PLUS 1
H	L	H	H	F = A + B	F = (A + B)	F = (A + B) PLUS 1
H	H	L	L	F = 0	F = A	F = A PLUS A PLUS 1
H	H	L	H	F = A$\overline{B}$	F = AB PLUS A	F = AB PLUS A PLUS 1
H	H	H	L	F = AB	F = A$\overline{B}$ PLUS A	F = A$\overline{B}$ PLUS A PLUS 1
H	H	H	H	F = A	F = A	F = A PLUS 1

SELECTION				ACTIVE-HIGH DATA		
				M = H	M = L; ARITHMETIC OPERATIONS	
S3	S2	S1	S0	LOGIC FUNCTIONS	$\overline{C}_n$ = H (no carry)	$\overline{C}_n$ = L (with carry)
L	L	L	L	F = $\overline{A}$	F = A	F = A PLUS 1
L	L	L	H	F = $\overline{A + B}$	F = A + B	F = (A + B) PLUS 1
L	L	H	L	F = $\overline{A}$B	F = A + $\overline{B}$	F = (A + $\overline{B}$) PLUS 1
L	L	H	H	F = 0	F = MINUS 1 (2's COMPL)	F = ZERO
L	H	L	L	F = $\overline{AB}$	F = A PLUS A$\overline{B}$	F = A PLUS A$\overline{B}$ PLUS 1
L	H	L	H	F = $\overline{B}$	F = (A + B) PLUS A$\overline{B}$	F = (A + B) PLUS A$\overline{B}$ PLUS 1
L	H	H	L	F = A $\oplus$ B	F = A MINUS B MINUS 1	F = A MINUS B
L	H	H	H	F = A$\overline{B}$	F = A$\overline{B}$ MINUS 1	F = A$\overline{B}$
H	L	L	L	F = $\overline{A}$ + B	F = A PLUS AB	F = A PLUS AB PLUS 1
H	L	L	H	F = $\overline{A \oplus B}$	F = A PLUS B	F = A PLUS B PLUS 1
H	L	H	L	F = B	F = (A + $\overline{B}$) PLUS AB	F = (A + $\overline{B}$) PLUS AB PLUS 1
H	L	H	H	F = AB	F = AB MINUS 1	F = AB
H	H	L	L	F = 1	F = A	F = A PLUS A PLUS 1
H	H	L	H	F = A + $\overline{B}$	F = (A + B) PLUS A	F = (A + B) PLUS A PLUS 1
H	H	H	L	F = A + B	F = (A + $\overline{B}$) PLUS A	F = (A + $\overline{B}$) PLUS A PLUS 1
H	H	H	H	F = A	F = A MINUS 1	F = A

(b)

of 75 equivalent gates. These circuits perform 16 binary arithmetic operations on two 4-bit words as shown in the table of Figure 16.42(b). These operations are selected by the four function-select lines (S0, S1, S2, and S3). The 32 possible configurations of the select lines each lead to a form of *addition, subtraction, decrement by one,* and *straight transfer.* These functions are provided in various combinations, with and without carry bits.

In addition to use as an arithmetic processor, the ALU can be utilized as a *digital comparator* by placing it in the subtract mode so that one input is subtracted from the other. The IC is then configured to test whether this difference is positive, negative, or zero.

This circuit has been designed to incorporate most of the requirements that a design engineer may desire for arithmetic operations and also to provide 16 possible functions of two Boolean variables without the need for external circuitry. These Boolean functions are shown in Figure 16.42(b). The Boolean logic functions are selected by use of the four function-select inputs and with the mode-control input (M) at HIGH to disable the internal carry operation. The 16 logic functions include AND, OR, NAND, NOR, and exclusive OR, as shown in the tables of Figure 16.42(b).

16.8.2 Full Adders

A *full adder* is a circuit that forms the arithmetic sum of three input bits, as shown in the circuit of Figure 16.43. This circuit has three inputs and two outputs. Two of the input variables, denoted A_i and B_i, represent the two significant bits to be added. The third input, C_i, represents the carry from the previous lower significant-position addition operation. Two outputs are necessary because the arithmetic sum of three binary digits ranges in value from 0 to 3, thus requiring 2 bits. The two outputs are designated S (for sum) and C (for carry).

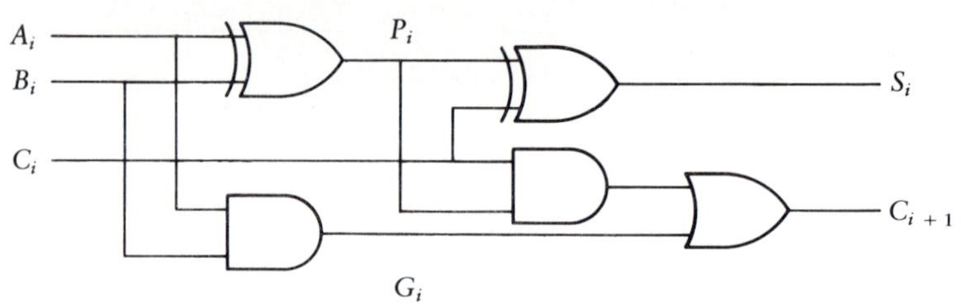

Figure 16.43 Full adder.

16.8.3 Look-Ahead Carry Generators

When two numbers are added together, we assume that the numbers are immediately available for computation at the same time. However, with digital circuits the signals must propagate through gates before the correct level is available at the output terminals. Rather than wait for the signals to propagate through a group of adders, we consider two new variables defined as follows. The *carry generate*, G_i, is defined by

$$G_i = A_i B_i$$

The *carry propagate*, P_i, is defined by

$$P_i = A_i \oplus B_i$$

The sum, S_i, and the carry, C_{i+1}, can be written as

$$S_i = A_i \oplus B_i \oplus C_i = P_i \oplus C_i$$

$$C_{i+1} = G_i + P_i C_i$$

Look-ahead carry generator circuits are used to speed up the operation of a group of adders. The 74S182 look-ahead carry generator IC is illustrated in Figure 16.44. This is a high-speed look-ahead carry generator capable of antici-

Figure 16.44
74182 look-ahead carry generator. Courtesy of Texas Instruments Inc.

PIN DESIGNATIONS

ALTERNATIVE	DESIGNATIONS†	PIN NOS.	FUNCTION
$\overline{G0}, \overline{G1}, \overline{G2}, \overline{G3}$	G0, G1, G2, G3	3, 1, 14, 5	CARRY GENERATE INPUTS
$\overline{P0}, \overline{P1}, \overline{P2}, \overline{P3}$	P0, P1, P2, P3	4, 2, 15, 6	CARRY PROPAGATE INPUTS
C_n	$\overline{C}_n$	13	CARRY INPUT
$C_{n+x}, C_{n+y}, C_{n+z}$	$\overline{C}_{n+x}, \overline{C}_{n+y}, \overline{C}_{n+z}$	12, 11, 9	CARRY OUTPUTS
$\overline{G}$	Y	10	CARRY GENERATE OUTPUT
$\overline{P}$	X	7	CARRY PROPAGATE OUTPUT
V_{CC}		16	SUPPLY VOLTAGE
GND		8	GROUND

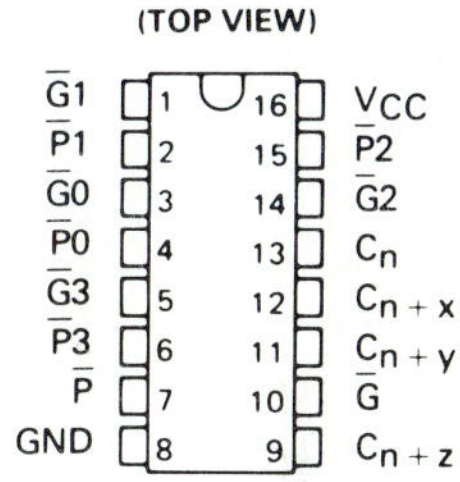

FUNCTION TABLE FOR $\overline{G}$ OUTPUT

INPUTS							OUTPUT
$\overline{G3}$	$\overline{G2}$	$\overline{G1}$	$\overline{G0}$	$\overline{P3}$	$\overline{P2}$	$\overline{P1}$	$\overline{G}$
L	X	X	X	X	X	X	L
X	L	X	X	L	X	X	L
X	X	L	X	L	L	X	L
X	X	X	L	L	L	L	L
All other combinations							H

FUNCTION TABLE FOR $\overline{P}$ OUTPUT

INPUTS				OUTPUT
$\overline{P3}$	$\overline{P2}$	$\overline{P1}$	$\overline{P0}$	$\overline{P}$
L	L	L	L	L
All other combinations				H

FUNCTION TABLE FOR C_{n+x} OUTPUT

INPUTS			OUTPUT
$\overline{G0}$	$\overline{P0}$	Cn	C_{n+x}
L	X	X	H
X	L	H	H
All other combinations			L

FUNCTION TABLE FOR C_{n+y} OUTPUT

INPUTS					OUTPUT
$\overline{G1}$	$\overline{G0}$	$\overline{P1}$	$\overline{P0}$	C_n	C_{n+y}
L	X	X	X	X	H
X	L	L	X	X	H
X	X	L	L	H	H
All other combinations					L

FUNCTION TABLE FOR C_{n+z} OUTPUT

INPUTS							OUTPUT
$\overline{G2}$	$\overline{G1}$	$\overline{G0}$	$\overline{P2}$	$\overline{P1}$	$\overline{P0}$	C_n	C_{n+z}
L	X	X	X	X	X	X	H
X	L	X	L	X	X	X	H
X	X	L	L	L	X	X	H
X	X	X	L	L	L	H	H
All other combinations							L

pating a carry across four binary adders or a group of adders. The IC looks across all four individual binary summation operations and generates an overall carry generate, G, and carry propagate, P. That is, rather than wait for the four individual binary operations to be completed before passing information on to the next IC in an arithmetic operation, the IC examines the four individual operations and develops the resulting carry information before the arithmetic is completed.

16.8.4 Magnitude Comparator

The comparison of two numbers is an operation that determines whether one number is greater than, less than, or equal to the other number. A *magnitude comparator* is a circuit that compares two numbers, A and B, to determine their relative magnitudes. The outcome of the comparison is specified by three binary variables that indicate whether $A > B$, $A = B$, or $A < B$.

The reduced circuit diagram for the comparator follows a bit-by-bit procedure to compare the two numbers. Suppose we are dealing with two 4-bit numbers, designated $A_3A_2A_1A_0$ and $B_3B_2B_1B_0$. The two numbers are equal if $A_i = B_i$ for all i between 0 and 3. To see if $A > B$, we first examine the most significant bits, A_3 and B_3. If these are unequal (one is 1 and the other is 0), the

Figure 16.45 74LS85 4-bit magnitude comparator. Courtesy of Texas Instruments Inc.

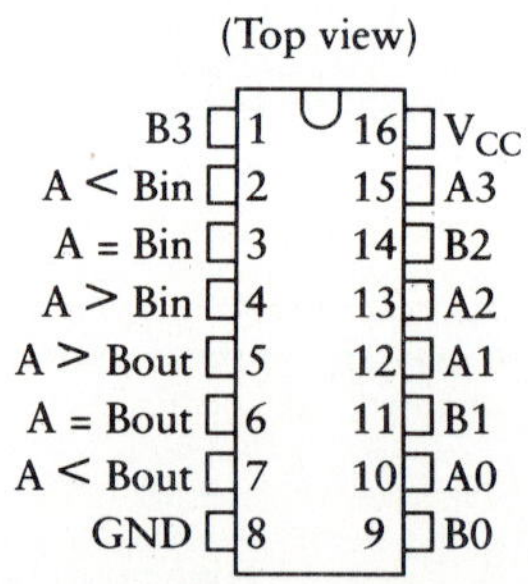

FUNCTION TABLE

COMPARING INPUTS				CASCADING INPUTS			OUTPUTS		
A3, B3	A2, B2	A1, B1	A0, B0	A > B	A < B	A = B	A > B	A < B	A = B
A3 > B3	X	X	X	X	X	X	H	L	L
A3 < B3	X	X	X	X	X	X	L	H	L
A3 = B3	A2 > B2	X	X	X	X	X	H	L	L
A3 = B3	A2 < B2	X	X	X	X	X	L	H	L
A3 = B2	A2 = B2	A1 > B1	X	X	X	X	H	L	L
A3 = B3	A2 = B2	A1 < B1	X	X	X	X	L	H	L
A3 = B3	A2 = B2	A1 = B1	A0 > B0	X	X	X	H	L	L
A3 = B3	A2 = B2	A1 = B1	A0 < B0	X	X	X	L	H	L
A3 = B3	A2 = B2	A1 = B1	A0 = B0	H	L	L	H	L	L
A3 = B3	A2 = B2	A1 = B1	A0 = B0	L	H	L	L	H	L
A3 = B3	A2 = B2	A1 = B1	A0 = B0	X	X	H	L	L	H
A3 = B3	A2 = B2	A1 = B1	A0 = B0	H	H	L	L	L	L
A3 = B3	A2 = B2	A1 = B1	A0 = B0	L	L	L	H	H	L

comparator need look no further. If they are equal, the comparator must look at the next set of bits. The 74LS85 is an example of a magnitude comparator and is illustrated in Figure 16.45. The function table in the figure expands on the discussion above. For example, the first row in the table indicates that if $A_3 > B_3$, we need look no further. The $A > B$ output goes high, and the other two go low.

If we wish to compare words of length greater than 4 bits, we can cascade these ICs together. For example for 8-bit comparisons, we cascade two ICs together. The inputs to the second (most significant) IC include the two 4-bit numbers plus the three outputs from the first IC. This is included in the function table. Note that as long as the four most significant bits are not the same (the first eight rows in the table), there is no need even to look at the outputs from the previous IC. Therefore, the cascading inputs are marked as X for don't care or irrelevant inputs.

16.9 PROGRAMMABLE ARRAY LOGIC (PAL)

A basic knowledge of digital engineering combined with the practical material presented so far in this chapter should be sufficient to design digital systems using the basic logic packages. It is not unusual for a design to require several hundred TTL logic circuits. For example, the early video games required about 150 ICs to control a simple simulated sport such as tennis. Although the cost of individual ICs is quite low, these systems become expensive because of manufacturing costs associated with wiring all the ICs together and also because the large number of circuits requires an unreasonably large amount of space.

As integrated circuit technology developed, circuits became smaller and versatility increased until the ultimate goal of full programmability was reached. The microprocessor represents a major breakthrough in versatility and reduction of manufacturing complexity.

This section deals with a user-programmable array of logic gates that can be used to replace a number of separate packages with a single IC. The array is known as *programmable array logic (PAL)*. It uses Schottky components, so it operates at high speed. With this IC, we can implement registers, flip-flops, and basic logic. The PAL typically is packaged in a single 20-pin DIP and can be used to replace between 4 and 12 SSI and MSI packages.

The basic logic structure of a PAL includes a programmable AND array that feeds a fixed OR array. The ICs are available in sizes ranging from 10 × 8 (10 input, 8 output) to 16 × 2. The wide variety of input/output formats allows the PAL to replace many different-sized blocks of combinational logic with a single package. Additional information can be obtained from manufacturer's data books.

PROBLEMS

16.1 Compare the 4028 and the 7442 decoders.

16.2 Contrast the 74150 TTL and 74C150 CMOS data selector/multiplexer.

16.3 Draw a diagram showing the connection of a CD4511 to drive a seven-segment LED display.

16.4 Design a system to drive four 7-segment LED displays with 74143 drivers.

16.5 Use the circuit of Figure 16.27 to design a device to measure the frequency of the ac power line and drive two 7-segment LEDs. Select $t_2 - t_1$ to be 10 s and use the 74160 decade counter.

16.6 Use a CD4047 as an astable multivibrator to operate at 50 kHz.

16.7 Design a VCO to operate at a base output frequency of 10 kHz. Use a 74LS124 and calculate the maximum obtainable output frequency variation.

16.8 Design a digital voltmeter using the ICL7106 IC (see Figure 16.34 and the manufacturer's data book) to measure voltages in the following ranges:

10 mV to 100 mV

100 mV to 1 V

1 V to 10 V

Use a voltage-divider network, since the maximum input to a $3\frac{1}{2}$-digit DVM is 199.9 mV.

16.9 Convert the 74161 binary counter into a decade counter.

16.10 Analyze the circuit of Figure P16.1, where the input voltage, v_i, varies linearly from 2 to 3 V. Determine the frequency of the output voltage as a function of v_i. Do this by calculating the frequency of the output for the following values of v_i: 2 V, 2.5 V, and 3 V. This circuit exhibits an output frequency that varies with input voltage.

16.11 Design an instrument to measure the frequency of a sinusoidal signal over the range 1.0 to 9.0 kHz. The voltage level of the input signal is 10 V rms. Display the output on two 7-segment LED displays, as shown in Figure P16.2.

16.12 Design a 0–5 V pulse train generator to develop the following continuously variable frequency ranges:

100 Hz to 1 kHz

1 kHz to 10 kHz

10 kHz to 100 kHz

Use a *JK* flip-flop to produce an output that is symmetrical; that is, the time high is the same as the time low.

Figure P16.1

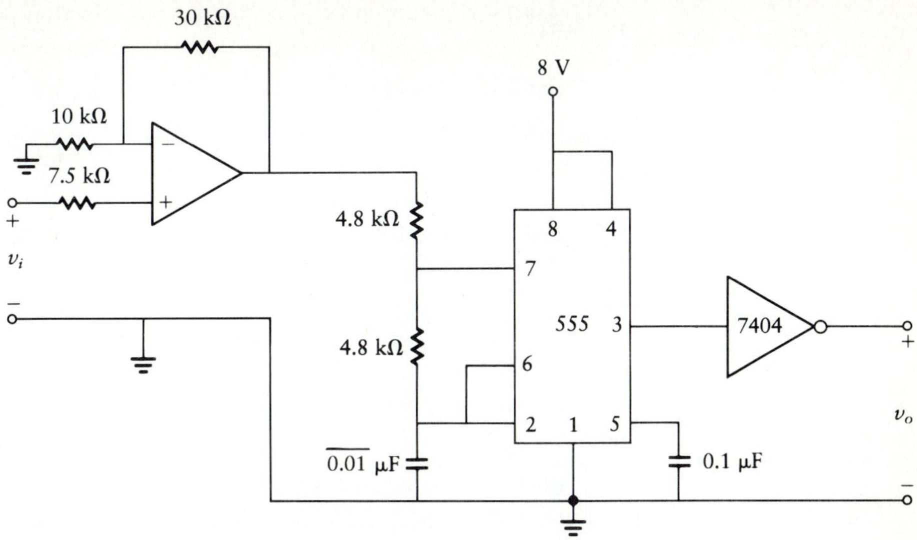

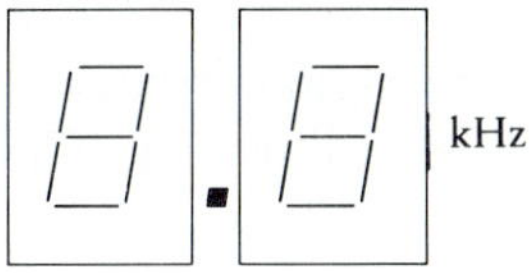

Figure P16.2

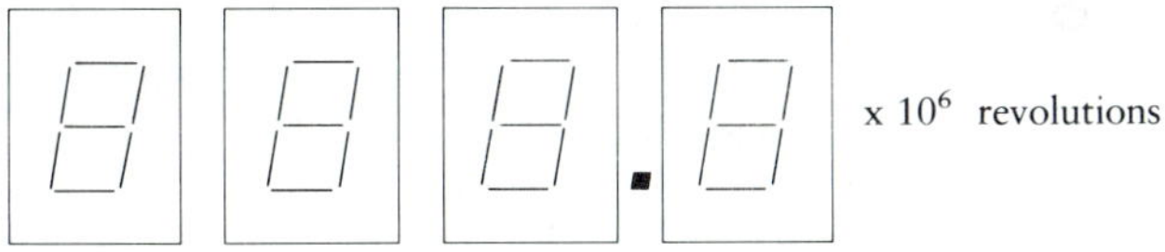

Figure P16.3

16.13 Design a 1-kHz clock that has a variable duty cycle. One potentiometer should be used to vary the duty cycle from 0.2 to 10. The frequency must not change from 1 kHz. Use a 555 in the astable mode to set the frequency. A 555 in the monostable mode is triggered from the astable 555 to provide the variable duty cycle. (Note that duty cycle is defined as time high divided by time low.)

16.14 Design a key chain equipped with an electronic system that will help locate your keys if they are lost. The device is to emit a 1-kHz tone for 30 s whenever you loudly clap your hands together. Use a duty cycle of 0.5 for the 1-kHz oscillator, and provide 0.25 W into the speaker. The crystal microphone outputs a 300-mV peak-to-peak signal when you clap your hands together within 20 feet of the key chain. Calculate all resistor and capacitor values and specify the type numbers for the ICs used in the design.

16.15 Design an electronic system to measure the total number of revolutions of an engine. This electronic system uses the conditioned pulse train from Problem 14.20 as the input, which is shown in Figure P14.8(b). The output of this electronic system is displayed on four 7-segment LED displays. This display is illustrated in Figure P16.3 and shows total revolutions in 10^6 revolutions. Remember that each pulse corresponds to two revolutions of the engine. Be sure to provide battery power to the critical parts of this system so that the total number of revolutions displayed is not lost during a power failure.

16.16 Design a pair of digital dice that uses the LED pattern shown in Figure P16.4. The dice are electronically "rolled" by pressing a button, and the digital dice box displays a random number between 1 and 6, as shown in the figure.

16.17 Design an rpm meter to display engine speed, which ranges in value from 0 to 6000 rpm. The input to this system is a pulse train from Problem 14.20, which is shown in Figure P14.8(b). This conditioned pulse train is TTL compatible, and each pulse corresponds to two revolutions of the engine. The output is shown on three 7-segment LED displays, as shown in Figure P16.5.

16.18 Design a system to measure the audience viewing interest in the various TV channels. To sample which channels the audience is viewing, the TV channel selector is instrumented. Eight families are selected to have their TV receivers fitted with potentiometers. The signals from the eight instrumented TV receivers are sent to a central station. The voltage on each of the lines is identical to the channel being viewed. For example, if Channel 4 is being viewed, the voltage on that line is 4 V. If the set is off, the voltage is 0 V. Each of the eight TV receivers is continuously sampled for 8 s each. While the receiver is being sampled, a display, consisting of three 7-segment LED displays, reads the TV family number and the channel being viewed. The display is shown in Figure P16.6. Use a CD4051 single 8 channel analog multiplexer.

16.19 Design a system that will count the number of cars in a parking lot. The maximum capacity of the lot is 99 cars. Each time a car enters the parking lot, an entrance gate opens. When the gate opens, a pulse appears on the IN line, as shown in Figure P16.7(a). Each time a car leaves, an exit gate opens and a pulse appears on the OUT line, as shown in Figure P16.7(b). These are two separate lines. Use two 7-segment LEDs to show continuously the number of cars in the parking lot.

Figure P16.4

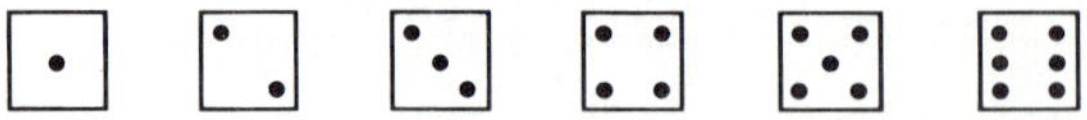

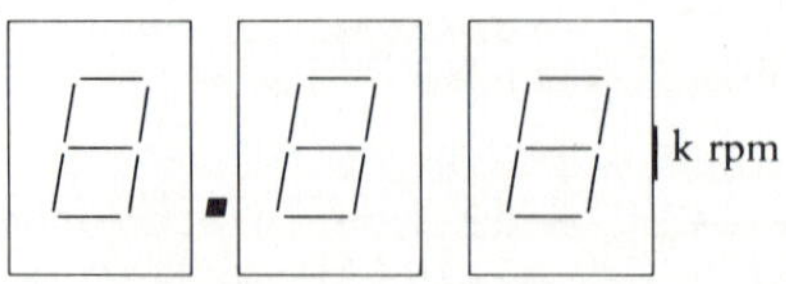

Figure P16.5

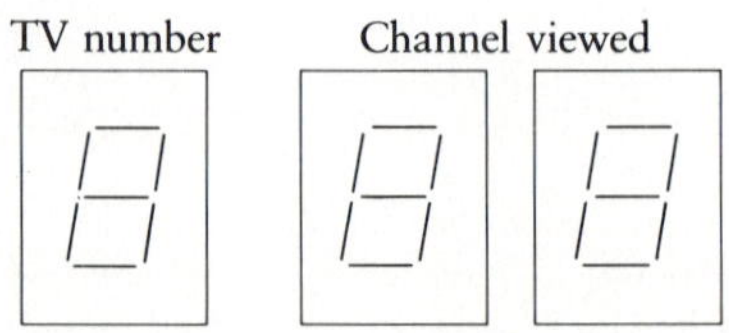

Figure P16.6

When the lot is full (99 cars in the lot), light four LEDs displaying the word FULL.

16.20 Design a tachometer for your automobile engine that operates in the range 0 to 5000 rpm. A Hall-effect generator provides 10 pulses for each revolution of the engine. The pulses are 100 mV in magnitude, so they must be signal conditioned. The output of this tachometer is displayed on two 7-segment LEDs with only the first two digits significant, as shown in Figure P16.8. Divide the design into three parts, as follows:

a. Design an electronic circuit to produce 10 pulses per engine revolution, where each pulse is 0 to 5 V in magnitude.
b. Design a window that will scale the tachometer so that when the engine is rotating at 5000 rpm, the two 7-segment LEDs will read 50. The LEDs are shown in Figure P16.8.
c. Design a counter and display for the tachometer. Do this by ANDing the input pulse train of part a with the window of part b and using the appropriate counters, latches, and decoders to display the rpm to two significant digits, as shown in Figure P16.8.

16.21 Design an electronic system to measure the frequency of a 0–5 V pulse generator that operates between 10 and 99 Hz. The output of this instrument is displayed on two 7-segment LEDs that read the pulse generator frequency to two significant digits. Conduct this design in two parts. In part a, you will design the window to obtain the scaling and division by time necessary to measure hertz (cycles per second). In part b, you will design the counter and the display.

a. Design a window circuit whose output will be ANDed with the input signal (frequency of 10 to 99 Hz) to be delivered to the counter and display system of part b. It is necessary to scale the window time so that when the input frequency is 25 Hz, the display shows the number 25. It is also necessary to deliver the appropriate CLEAR and LATCH signals to the circuit of part b.
b. Design a counter and a display system to display the frequency, as provided from part a, on two 7-segment LEDs. Be sure to include all CLEAR and LATCH signals.

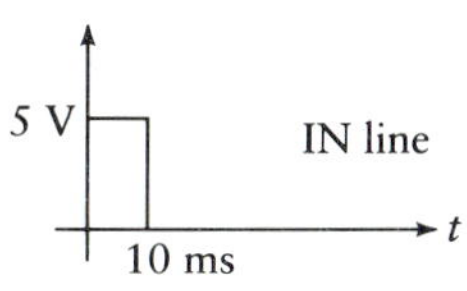

(a)

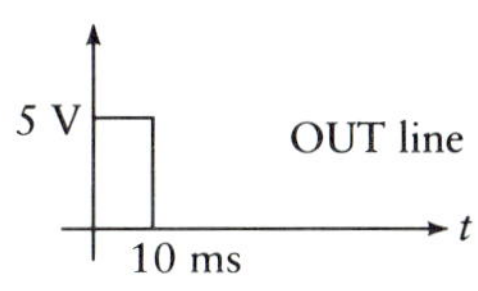

(b)

Figure P16.7

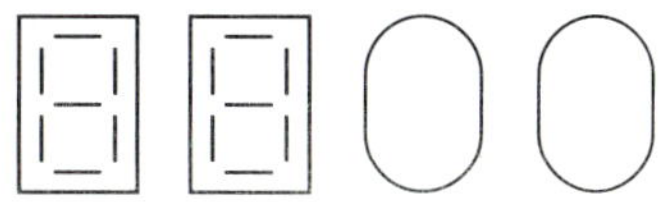

Figure P16.8

17

ELECTRONIC SYSTEM DESIGN

17.0 INTRODUCTION

In the previous three chapters, we studied the building blocks that make up digital electronic systems. We are now ready to put the pieces together in an orderly and rational fashion to form electronic systems. Problems facing the electronics engineer can involve material from every chapter in this text, including discrete components, linear ICs, and digital ICs. This chapter sets forth the necessary techniques to perform electronic system design. A number of examples are presented throughout the chapter. These require knowledge of the principles of discrete, linear integrated, and digital ICs. Electronic system design draws from a good portion of your past knowledge. This makes design extremely challenging—indeed, it has the potential to be discouraging. It is important to keep in mind that good design skills are acquired over a long period of time. You should not expect to be an expert electronic circuit designer having just taken your first course covering this material.

17.1 PRINCIPLES OF DESIGN

The orderly approach to problem solving consists of five major steps. We briefly state the steps here and expand on the discussion in the following sections.

1. *Define the problem.* State what your product is supposed to accomplish, including any special requirements and specifications.

2. *Subdivide the problem.* To simplify and speed the design process, break the main problem up into several smaller problems. It is difficult for even the most experienced engineer to solve a large, complex problem in one operation.
3. *Create documentation.* The essence of engineering is to generate drawings or plans so that the system can be manufactured and sold on the market. The best piece of engineering design work is useless unless others are aware of it. It would not be satisfying or profitable if all your work had to be repeated each time the product is produced.
4. *Build a prototype.* We like to think that our theory and equations represent good models of real-life behavior. In practice, this is not always the case. Until a prototype is built and tested, the designer cannot be sure that all contingencies have been considered and that the design specifications have been met. This step may include computer simulations.

   Since the prototype is not built until we have some confidence that the "paper" design is complete, we include a section entitled "Design Checklist." It is suggested that such a checklist be developed just prior to prototype construction.
5. *Finalize the design.* Once the prototype is working to your satisfaction, test it under the conditions in which it will be used. Then complete any documentation that may be required in addition to the drawings that have already been generated.

This concludes the design cycle. However, it is valuable to consider other options or methods of designing the product. A discussion with several of your engineering colleagues can often lead to valuable improvements. A flowchart of the process is shown in Figure 17.1. Once the design process is complete, the finished documents are sent to the appropriate departments and construction of the product begins. If your job has been done properly, you have generated a clear set of plans, instructions, and additional information needed to build, service, and update the design.

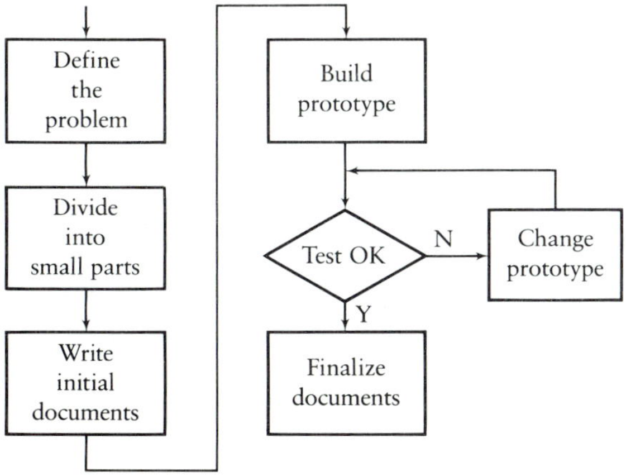

Figure 17.1 Design process flowchart.

17.2 PROBLEM DEFINITION

The first step in the design process is to *define the problem.* In this stage of the design, the engineer faces one of two possibilities. A fully specified design may be available, or the design may be a vague idea in the mind of a customer. The engineer will often be faced with a combination of these two, where part of the design must perform certain precise functions while the other part of the design is in the form of a hazy sketch.

In the case of the fully specified design, the engineer's job is to analyze and understand the specifications given by the customer. After reviewing the requirements, the engineer will often meet with the customer to make sure that they both understand and agree on what the product is supposed to do. It is now the engineer's task to design a product that will meet the specifications. If a product does not measure up to these requirements, the manufacturer will be forced to modify the design in order to make the product meet the standards or face loss of money, reputation, and time. The consequences of this course of action for the design engineer are open to speculation. It is, at best, an unpleasant situation.

In cases where the design is not fully specified, the customer may have only a general idea of what the product should accomplish. It is then the job of the engineer to define the product for the customer. This will usually require that the engineer and customer work together until both sides have agreed on the product definition. Once again, the engineer must exercise care in determining special customer requirements. If the product fails to perform to expectations, the customer may decide to take future business elsewhere.

17.3 SUBDIVIDING THE PROBLEM

Once the product has been defined and specified, the problem is usually broken up into a series of smaller designs. The number of smaller designs depends on the complexity of the product.

As an example, consider the design of an electronically controlled toaster. One design group could be working on the toaster mechanism while another group works on the electronic control portion. In this case, the product design is broken into two parts. Even with such a simple example, there are other ways to divide the problem. It is the project engineer's job to divide the problem into the best subsets so that the design can be rapidly and inexpensively implemented. Frequently, an engineering department is divided into groups that specialize in the various disciplines—for example, power supplies, electronic circuits, and control electronics.

17.4 DOCUMENTATION

A primary responsibility of an engineer is to tell someone else how to build, service, use, and update the product that has been designed. This information *must be precisely written* so that it will not be lost and there will be no room for misunderstandings. We use many terms and symbols in specifying electronic hardware design. In order to put these words and symbols on paper, engineers were forced to create their own language. Like any living language, the language of electronics has certain standard symbols that everyone understands and other symbols that have come to mean something else for a certain group of users.

The language of electronics can be broken down into three parts. The first is the set of universally recognized symbols and words—those that are used and accepted by a majority of the industry. Some of these words include the symbols for resistors and capacitors used in drawing schematics. Also included are universal measurements such as the ohm, volt, and henry. The engineer can use these symbols to communicate with others about electronics even if the readers do not speak the same conversational language.

The second part of the language consists of the symbol and word set that is recognized by most but is not standardized. For example, several different symbols are used for the field effect transistor (FET), all of which mean the same thing but look different from each other. This subset will be recognized and the meaning understood, but the people involved will continue to use their own symbols and words until the symbol becomes obsolete or nonstandard. Many "jargon" words such as *bus* and *interface* can be used in several ways and mean various things to different people.

The final portion of this language is composed of the special forms that each separate business uses to communicate internally. Many of these forms are specialized for use only by the company in question. Others are semistandard forms that are variants on industry standards. Every time an engineer moves to a new company, a new dialect must be learned so that documents can be generated in acceptable form. Examples of these forms include schedules, requisitions, and product change requests—information that the company must know but outsiders need not know.

17.4.1 The Schematic Diagram

The *schematic* or *circuit diagram* is a plan of an electronic device that is drawn using standard and nonstandard symbols. This diagram shows the interconnections and components used to build the circuit. A schematic usually shows only the electrical connections needed to build a circuit and not the physical layout and construction of the circuit.

The schematic is one of the most important documents that an engineer must draw. Most of the other documents needed for production and servicing are derived from the schematic drawing.

In order to draw a clear and understandable schematic of the entire system, it is usually desirable first to draw individual schematics for each of the blocks in a subdivided system. The complete schematic is then drawn using the smaller ones and combining them in appropriate ways.

The master schematic is arranged to put related modules next to each other so that the interconnecting signal lines traverse the shortest path. The master schematic is drawn so that each of the modules occupies a separate part of the plan. As a rule of good design, modules should be shown as a whole and not scattered over the drawing. Related stages that have inputs and outputs in common should be drawn next to each other.

17.4.2 The Parts List

Anyone who has ever used a list for grocery shopping understands the concept behind the parts list. This is a compendium of all of the parts needed to construct the product. The list shows the component tolerances, power ratings, voltage levels, and type of component (e.g., if a capacitor is electrolytic, paper, or mica). This list is often subdivided into modules so that it is easier to read. Components of the same type are listed together for ease of reference. Parts lists are used by both engineers and managers for cost evaluation and as a checklist for building the prototype. There is no standard way in which to make a parts list, since each manufacturer has its own ideas on what should be included. A sample list with some of the possibilities is shown in Figure 17.2. Such a parts list is often produced and updated using a computer simulation program.

17.4.3 Running Lists and Other Documentation

In addition to the schematic and the parts list, other lists are used at the prototype production stages. These lists are assigned various names and can be arranged in several different ways. The purposes of these lists are to keep track of wiring, maintain a list of signals, and ease the construction of the prototype. Typical lists, including running, wrapping, signal, and wiring lists, are illustrated in Figure 17.3. These lists, in combination with the parts lists and the assembly drawings, are used by engineers to build and test the prototype and generate the paperwork that the production department needs in order to build the unit. Computer simulation programs provide a great tool for the engineer to generate and update these lists.

Figure 17.2
Sample parts list.

Schematic symbol	Value in appropriate units	Additional descriptive information	Manufacturer's or in house part number	Quantity required
R_1, R_2	10 k 1/4 W	carbon film 5%	R10353	2
R_3	12 k 1/4 W	carbon composition 10%	R12310	1
C_1	0.01 μF	50 V disc ceramic	C18201	1

In addition, the following may also be listed:

1. Alternate parts
2. Parts manufacturer
3. Serial numbers if applicable
4. Other information deemed necessary by the engineering, marketing, and production departments.

Each parts list should also be identified as follows:

	Product name and number	Date issued	Which changes
XYZ	motor gauge parts list	Issued 6/5/89	Revision 4 Page 1 of 4

Quantity	Part	Description	Number
2	R_1, R_2	10 k 1/4 W carbon film 5%	R10353
1	R_3	12 k 1/4 W carbon composition 10%	R12310
1	C_1	0.01 μF disc ceramic 50 V	C18201
2	C_2, C_3	0.1 μF disc ceramic 50 V	C17201
1	C_4	10 μF 25 wVdc electrolytic	C15803
1	Q_1	100 μF 15 wVdc electrolytic	C14803

17.4.4 Using Documents

Different types of documents and paperwork are used throughout the engineering profession. Many of these documents are for company use in accounting, advertising, or management. We have covered a few types of documents used for building and planning a unit. To keep track of what changes have been made to a unit, schematics and lists must be constantly updated and changed. Whenever changes are made to the unit, all of the appropriate documents should be changed as soon as possible. Trusting to human memory is a sure way to create errors in the documents and waste time for everyone who must use these documents to test and build the unit.

Signal	Location
CE	10C12, 5B6, 3D2, 2D3, 1D12
D1	10C11, 9C11, 8C11
D2	10C9, 9C9, 8C9
D3	10C8, 9C8, 8C8

Signal name — Location of signal — Chip identifier — Pin number

OR

Signal	Location	X	Y
CE	10C12	1119	110
CE	5B6	509	90
CE	3D2	330	130
CE	2D3	220	120

Grid system used to specify exact pin location

Wire number	Location	X	Y	Location	X	Y	Length	Signal
1	10C12	1119	110	5B6	509	90	Y3	CE
2	3D2	330	130	2D3	220	120	Y3	CE
3	5B6	509	90	3D2	330	130	Y3	CE
4	2D3	220	120	1D12	100	96	Y4	CE

Order in which wires get wrapped — Location for both ends of wire — Wire color and length code

Location	X	Y	Signal	Wire number
10C9	1117	110	X	34 36
10C10	1118	110	Y	58
10C11	1119	110	CE	1

Chip number and pin — Location

Figure 17.3
Running lists.

17.5 DESIGN CHECKLIST

We are almost ready to build and test the prototype. However, since the transition from "paper design" to hardware represents a major step in the design process, we recommend a pause at this point to double-check the previous work.

Some of the problems that may occur include the following: outputs that are inadvertently tied together and assume different states, floating IC terminals, ground-loop problems, intercoupling between circuits, and false triggering.

Most engineers who design digital circuitry have a checklist that they run through in their minds while they are designing the circuits. Unfortunately, these lists are usually developed as a result of making design errors. They represent attempts to avoid repeating the same mistake twice.

A short version of a typical checklist might contain the following four steps.

1. *Input and output precautions.* Make sure that all unused inputs are tied to either an output, a ground, or a power supply. Inputs that are left floating can assume incorrect levels and "fool" the circuit into thinking that another logic level has been applied. Another input-related item is to check that the number of inputs connected to any one output does not exceed the specifications for the device. If the number is too large, additional buffers must be added so that the circuit can properly feed the load.
2. *Timing problems.* Many troubles occur in circuits when events are expected to happen simultaneously but do not. These problems can be traced back to propagation delays. Timing problems can cause a chip with several outputs to change state at different times, even though the design calls for the events to happen at the same time. Another problem can occur when several inputs require data at the same time and the inputs are delayed by first going through other circuitry. Although compensating time delays can be added, often the only sure way to cure time-related problems is to redesign the circuit.
3. *Special requirements.* This category includes any special quirks that must be taken into account. For example, certain memory ICs cause spikes to appear on their power inputs. For these pulses not to interfere with other devices, the power input pins must be bypassed with a capacitor that is located in close proximity to the power input pin. In addition, some TTL chips must have their outputs tied to the supply voltage through a resistor for proper operation.
4. *Power requirements.* The designer must make sure that the power supply can deliver the required current and voltage. In addition, the supply must be adequately filtered and there must be sufficient bypass capacitors on the supply leads. Provision must be made for the escape of excess heat. Some logic families such as TTL and ECL get quite warm when operated at maximum fan-out.

17.6 PROTOTYPING CIRCUITS

Although this text is not intended as a guide to circuit construction, a few words on the subject are in order. Engineers, experimenters, and hobbyists re-

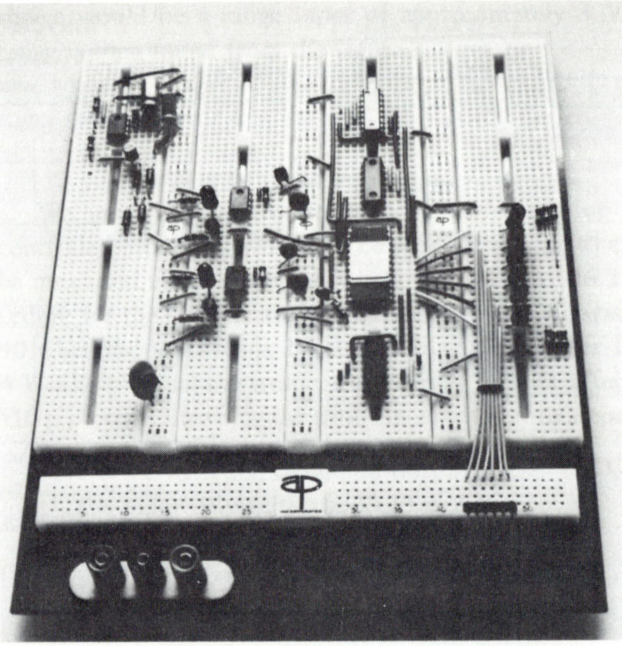

Figure 17.4
Proto-Board. Courtesy of AP Products

quire fast, economical means of constructing circuits. None of these people need to construct large numbers of their designs. Production methods of building circuits are not usually warranted because of cost and complexity. Mass-produced circuits use printed circuit boards (often double-sided). These boards are plated and processed at a printed circuit board fabricating facility. It is difficult to rectify mistakes on such a board or to make changes. When changes are required, traces are cut and lifted from the board and/or jumper wires are added. To avoid this problem, several methods have been developed that permit easy alteration of circuits.

One device that is commonly used for simple circuits goes under a variety of trade names including *Proto-Board* and *Circuit Board.* A representative board is shown in Figure 17.4. It consists of a plastic base with a grid of interconnected holes that can be used to connect components together. These boards are designed to accept standard integrated circuits and component leads so that any element may be used. They are fast to use, and components may be reused. There are some disadvantages associated with using these boards. Because of their layout, they suffer from high interlead capacitance and, after much use, develop intermittent lead contacts.

Another popular method for prototyping and small production runs is *wirewrap,* as illustrated in Figure 17.5. Connections are made using a special tool-and-socket combination that permits fast and reliable circuit interconnections. The special socket has a top portion that allows the insertion of an integrated circuit and a bottom portion that consists of square posts around which

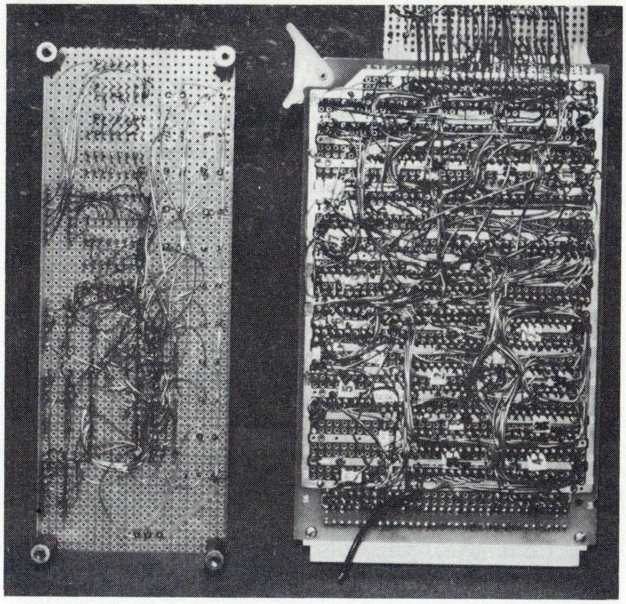

Figure 17.5
Wirewrap.

the wire is wrapped. The tool is designed to wrap neat, compact coils around the posts. These wires are run by the assembler from post to post for interconnections within the circuit. Wirewrap boards have several advantages. The connections are secure, circuits are easy to change, and the boards are inexpensive for small production runs. Among their disadvantages are that they are susceptible to noise from other circuits and crosstalk. They can be miswired, and they are bulkier than a printed circuit board.

No matter which method of prototyping is used, it is important to test the circuit as you go along. An inexperienced engineer often builds (or has built) an entire circuit only to find that it fails to operate properly when tested. In constructing a new circuit, build a portion of it (perhaps the counters) and immediately test this portion. Do not go further until the first section is working as you have designed it. Once this circuit works properly, build the next part and test it together with the first part (which you already know is working properly). If there is a problem in the design or the wiring, it is easier to check the smaller part of the circuit than to check out the entire circuit at one time.

17.7 DESIGN EXAMPLES

Although the design process follows a logical progression of specific steps, it is far from being simple. Decades ago, when design engineers were called on to design power supplies and other well-developed systems, they used various handbooks that contained examples and all of the necessary equations. Many of these people were affectionately referred to as "handbook engineers."

The diversity of applications of electronics and the myriad continuing developments in the field have required that the engineer become more creative. Although you now have the necessary skills to perform many designs, you must polish these skills and become a good designer. You will also have to expend a great deal of energy to keep abreast of new developments in electronics.

Since design skills are enhanced by practice, we present five design examples for your study. In addition, the problems at the back of this chapter contain many challenging situations to help you further develop your skills.

Example 17.1 Digital Tachometer

The assignment: The XYZ Motor Company has determined that the mechanical tachometers used on their engine test stands have become too expensive. In hopes of finding a cheaper electronic alternative, they contact ABC Engineering Company and tell a salesperson what they want. You are the ABC engineer assigned to this project. Trace the steps in this design process.

SOLUTION You contact the customer to refine the problem accurately and define the desired product. The customer and you then discuss and work out the requirements of the new digital tachometer. You then make a list of specifications for the new digital tachometer. Let us assume the following is a list of functions and specifications.

1. Read out rpm.
2. Have a range from 200 to 8000 rpm.
3. Have an accuracy of 1% and read out in tens of rpm.
4. Use the 60 Hz, 110–120 V power line for power supply and timing.
5. Use a timing transducer that produces one pulse for each revolution.

The problem has been defined, and you now understand what the customer wants. This completes the first step in the design cycle.

You now move to subdividing the problem. You study this problem and define a block diagram such as the one shown in Figure 17.6. The problem is broken into five parts. One part consists of the input to the tachometer and the processing required to make this input compatible with the remainder of the circuitry. Since the purpose of this circuit is to count the number of revolutions of an engine, the circuit needs a counter device. The count must last for a specified period of time because the specifications called out by the customer ask for revolutions per minute. Therefore, the circuit needs a clock to keep track of time. The rpm count must be displayed, so a display system is re-

Figure 17.6
Block diagram for Example 17.1.

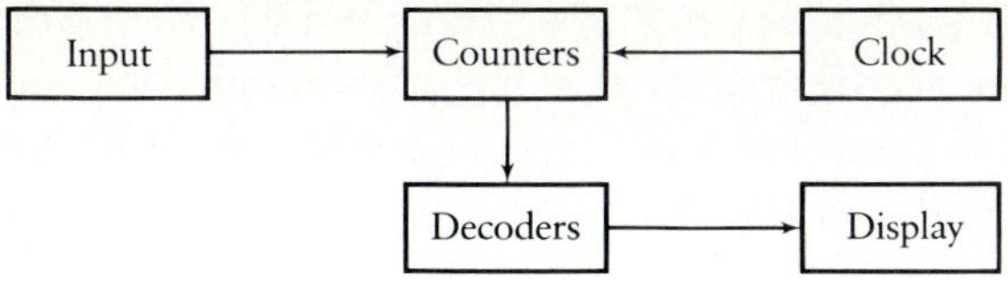

Figure 17.7
Input signal.

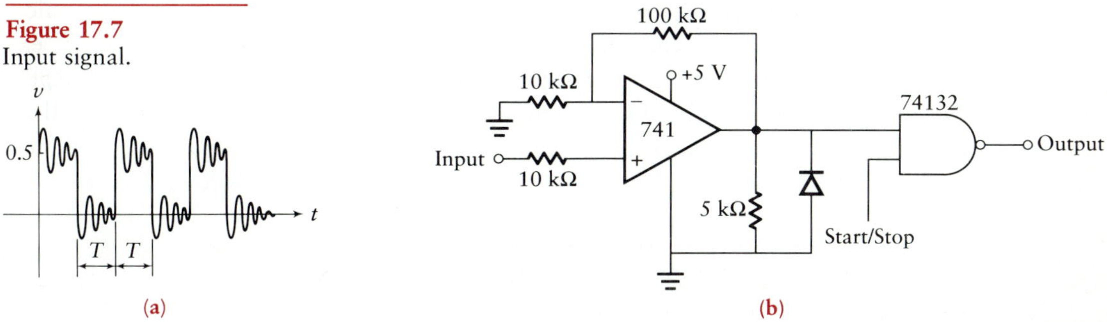

quired. From past experience, you know that the displays will not "understand" the raw information output from the counter. A decoder circuit is needed between the displays and the counters.

The five portions of the problem have now been defined, and a block diagram of the system is drawn showing the interconnections of the separate modules. This completes the second phase of the design process. Each of the modules can now be designed.

The input module is considered first. You must determine the type of signal the engine can generate for an input to the digital tachometer. The signal comes from a small variable-reluctance transducer (sometimes termed an "E-pickoff") mounted on the motor housing. This type of magnetic transducer is discussed in Section 17.8. The signal, shown in Figure 17.7(a), is contaminated with noise and transients. As a result, the signal must be amplified and cleaned up before it enters the remainder of the circuit. To perform this signal conditioning, you decide to use a 741 op-amp and a 74132 Schmitt trigger. Since ample power is available and the required speed is relatively low, you decide to use the inexpensive TTL family of ICs. This circuit exhibits high noise immunity because of the way it recognizes ones and zeros. The output of the op-amp and Schmitt trigger circuit is a clean, 0 to 5 V pulse that is usable by the rest of the system. You design circuitry for the input signal conditioning as shown in Figure 17.7(b).

The next stage to be considered is the counter module. This module must count the number of pulses that enter via the input circuit and output this number to the decoder circuitry. The counter is started, stopped, and reset by the clock, so provision must be made for a start, stop, and reset input. The display must be read out in decimal numbers, so it would be best if a decimal counter module is used. You scan the TTL Data Book looking for a divide-by-10 or decade counter that has a reset input and a start/stop input. A 74160 is selected, and you connect the counter circuit of Figure 17.8. The three ICs provide 1000 as the most significant digit (MSD) and 10 as the least significant digit (LSD).

The data that come from the counter must be processed. The decoder module, which is shown in Figure 17.9, performs the functions of storing and decoding the data so that the data may be shown on the displays.

The data lines are constantly changing or being reset, and there is only a short period of time during which they may be read. To store and use these data, you place a latch in the data lines. This stores the BCD signals while the input data are changing. Three 7475 4-bit latches are selected. The informa-

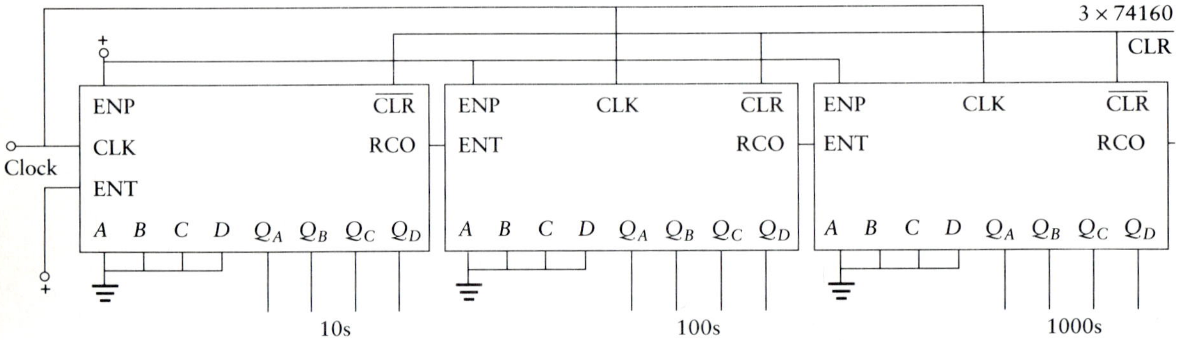

Figure 17.8 Counter circuit.

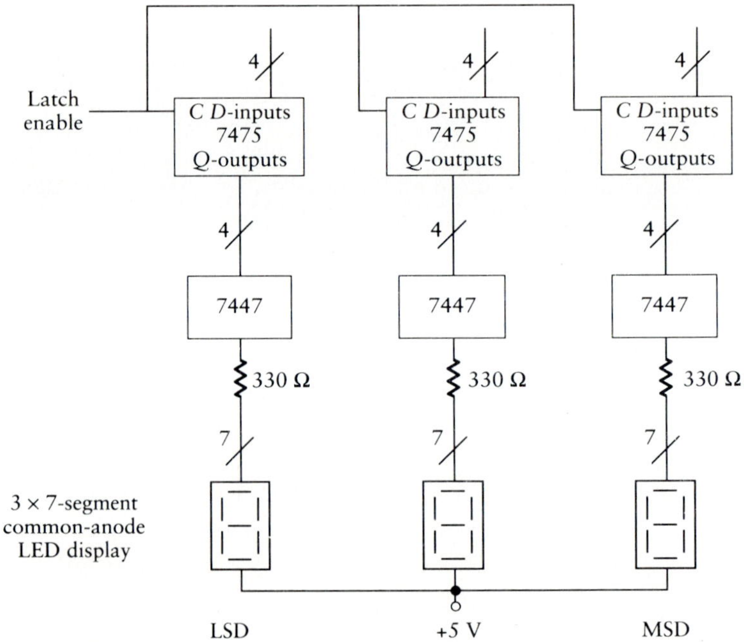

Figure 17.9 Decoder module.

tion present at the data (*D*) inputs is transferred to the *Q* outputs when the enable (*C*) is HIGH. The *Q* outputs will follow the data inputs as long as the enable is HIGH. When *C* goes LOW, the information that was present when the transition occurred is held at the *Q* output until *C* goes HIGH.

The signal must be translated from BCD to a form that the seven-segment LED display can understand. A decoder IC performs this function, and you choose the 7447, a BCD-to-seven-segment decoder/driver. This decoder features active-low outputs that are designed to drive common-anode LEDs. Each driver output provides up to 24 mA of current. As long as the displays need no more than 24 mA for operation, no transistors need be added for power gain. In fact, 330-Ω resistors are placed in series with each LED segment, as shown in Figure 17.9.

Before proceeding to the remainder of the design, you recall reading about a single IC that combines the functions of counter, latch, and decoder. Your data book shows that such an IC is available, the 74143. This is a 4-bit counter, latch, and seven-segment LED driver. The output is capable of driving 15-mA constant-current common-anode LEDs. (This IC is discussed in Chapter 16.) The block diagram for coupling three of these ICs for the present application is shown in Figure 17.10. (A complete wiring diagram appears in the Texas Instruments TTL Data Book.) You have simplified the system considerably! This could result in savings of power, heat, weight, and space while decreasing the probability of circuit failure.

You now turn your attention to the clock circuit. The 60-Hz input line is used for the clock, as shown in Figure 17.11.

You are concerned about a conductive connection to the power line, since a short circuit could cause the tachometer user to be injured. It is necessary to provide electrical isolation from the power line. We accomplish this with an *optoisolator*, as discussed in Chapter 3. This circuit combines a light-activated diode (LAD) with an LED to transmit only the 60-Hz signal. These devices commonly operate in the infrared region to eliminate effects due to room lighting. The output of the optoisolator, which provides only an optical path

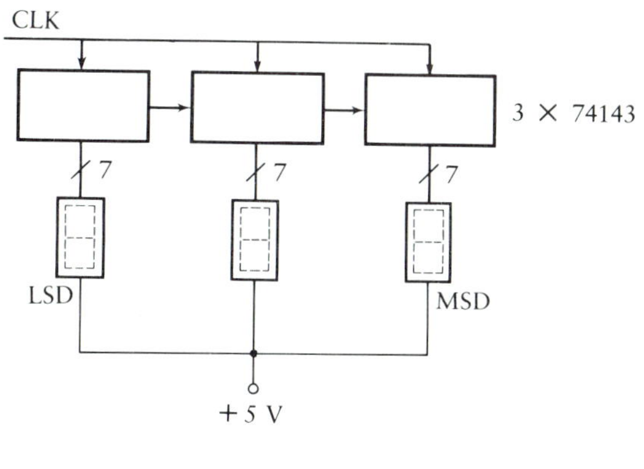

Figure 17.10 Counter latch, seven-segment LED driver.

Figure 17.11
Clock circuit.

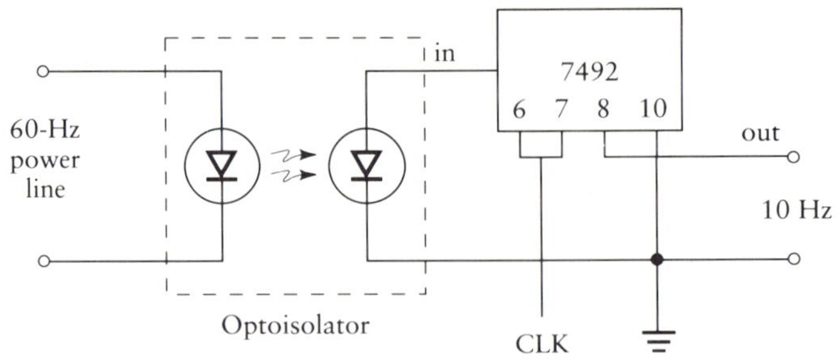

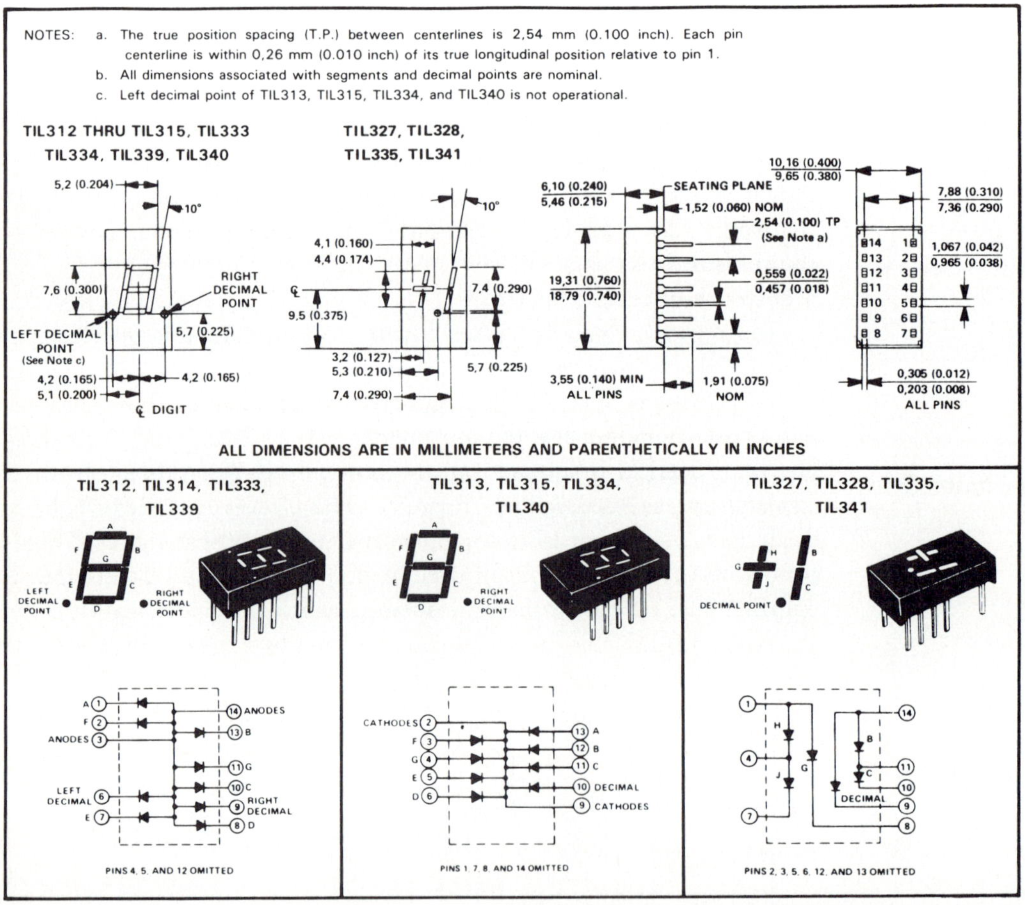

Figure 17.12
Display. Courtesy of
Texas Instruments Inc.

to the 60-Hz power line, is a 0 to 5 V pulse train at a 60-Hz rate. This signal is the input to the 7492 divide-by-6 counter, as shown in Figure 17.11. The output of the 7492 is a 10-Hz signal.

Several outputs are required from the timer circuit: a start/stop signal, a timing signal, a reset signal, and a data-ready signal. Since the output must only read tens of revolutions per minute, we decide to make the window pulse stay high for 0.1 min (6 s). During this time, the counting circuit will count the number of pulses entering the system. When the window time has passed, the trailing edge of the window pulse is used to latch the data into the display. We use a delayed version of the same signal to clear the counters and repeat the cycle. Once the window pulse ends, the data must be sent to the decoder circuits. A reversed or inverted start/stop signal is used to control the data latches in the three 74143 ICs. Remember that the 74143 is level-triggered.

You are now ready to turn your attention to the display. It is important because the customer will be viewing this part of your design more than the rest of the circuitry. The display must be bright enough to be seen under a variety of lighting conditions, and the numbers must be large enough to be read without eye strain. We decide to use $\frac{3}{4}$-inch seven-segment LED displays for the output because of their large size. Even though the $\frac{3}{4}$-inch LEDs are slightly more expensive, we find that they are more pleasing than the smaller displays. This is shown in Figure 17.12.

Now that the system has been defined, you must determine the power requirements. Both the total power and the voltages necessary should be set forth.

With preliminary design of the system finished, the prototype stage is entered. During this phase of the project, it is important to check out the performance of the system under varying temperature conditions. You should also determine how the design changes when component values vary. In cases where timing errors can degrade system performance (our specifications call for 1% accuracy), use of a calibrating potentiometer should be considered. This is important in setting the window pulse time. When you are satisfied that the prototype is functioning properly, demonstrate it to your colleagues and to your supervisor so that they can help you check performance. Then, when the prototype is passed, complete all of the documentation and schematics and send them to other departments in the ABC Engineering Company. You have apparently done a good job!

Example 17.2 Faulty-Lamp Indicator

The assignment: Design a faulty-lamp indicator that monitors 64 lamps at the end of an airport runway. When a lamp is operating, the voltage at that lamp's

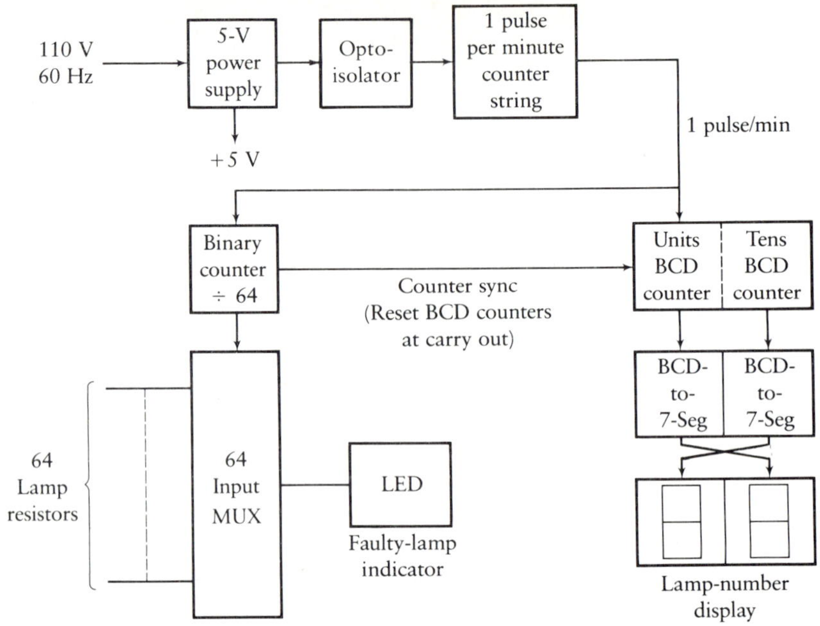

Figure 17.13 Block diagram for Example 17.2.

terminal is 5 V. When the lamp is faulty, the voltage is zero. The output of the indicator must have two 7-segment LEDs that indicate the number of the lamp being tested (0 to 63). If a fault is detected, a single LED must light to indicate that the lamp at this number position is faulty. Set the timing so that each lamp is tested for 1 minute once for each 64-minute cycle. Derive power from the 60 Hz, 110 V line.

SOLUTION Let us start with the block diagram as shown in Figure 17.13. A 5-V power supply converts 110 V, 60 Hz to 5 V dc, with a maximum current of 750 mA. The unit also provides a 60-Hz clock pulse, through an optoisolator, to a counter string that produces an output of one pulse per minute. This is done by dividing by 6, 10, 6, and 10 successively.

A divide-by-64 binary counter generates the addresses for a 64-input multiplexer. A pair of BCD decade counters is used to count from 0 to 63 and to drive BCD-to-seven-segment decoders and LED displays. These indicate the number of the particular runway lamp being tested.

The 64-input multiplexer contains four 16-input multiplexers (addressed simultaneously) whose outputs are combined by being fed to the inputs of a

4-input multiplexer. If the selected input (1 of 64) is LOW, the multiplexer output is LOW. This energizes an LED, which signifies that the runway lamp being tested is faulty. The complete schematic diagram is shown in Figure 17.14. The 7805 in the power supply must be able to handle the total current, so this is now checked. The current required by the various ICs is summarized below. These data are obtained from the TTL Data Book.

IC Number	Quantity	Current Required (mA)	Total Current (mA)
7492	2	31	62
7490	2	32	64
74161	2	34	68
74160	2	34	68
7447	2	43	86
74150	4	40	160
74151	1	29	29
		Total IC current requirement =	537 mA

Each segment of the LED requires 10 mA, and a maximum of 12 segments must be lighted (e.g., to display 28, 38, or 58).

The maximum display current is	120
The faulty lamp LED requires	10
The RC differentiator in the BCD reset line requires	2
The total current is then	669 mA

This is within the 750-mA rating of the 7805. The initial design is therefore complete, but the other steps of Section 17.1 must now be followed.

Example 17.3 Reactor Chamber Pressure Detector

The assignment: Design a system to display reactor chamber pressure, p_g. The pressure transducer outputs a 400-Hz signal with peak-to-peak voltage amplitude proportional to gauge pressure according to the equation

$$\text{Amplitude} = p_g \text{ mV/psig}$$

The pressure transducer has a range of 0 to 1000 psig. The system output is to be displayed on three 7-segment LEDs that read pressure from 0 to 999 psig. Derive power from the 60 Hz, 110 V line.

Figure 17.14
Schematic diagram for Example 17.2.

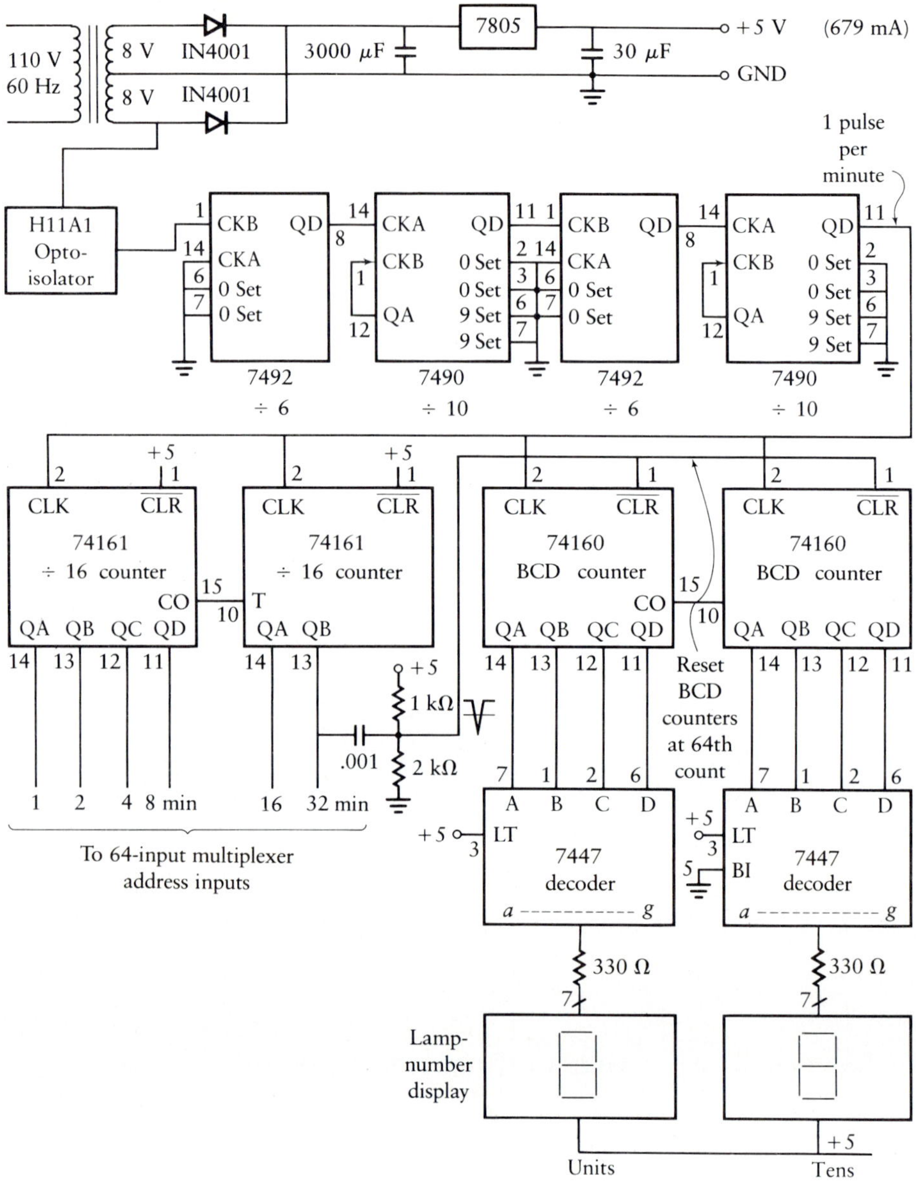

(a) **Power supply clock, counters, and lamp-number display**

SOLUTION Refer to the system block diagram of Figure 17.15. The amplitude of the input signal is proportional to pressure, but this signal is amplitude-modulated on a 400-Hz carrier, so it must first be demodulated. We now have

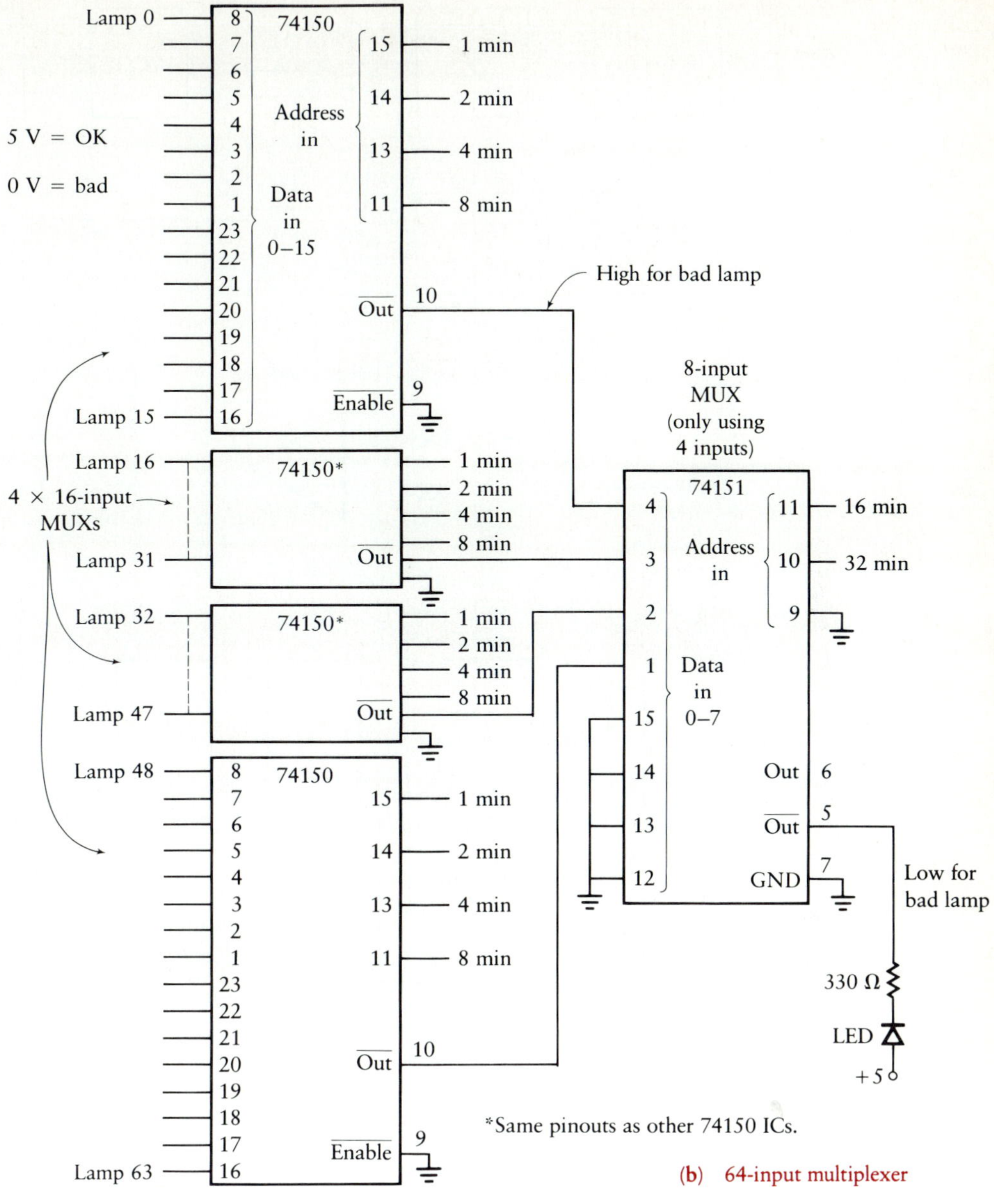

(b) 64-input multiplexer

a dc voltage that is proportional to pressure. At this point there are several possible approaches, including simply feeding this signal into a digital voltmeter ($3\frac{1}{2}$ DVM), as discussed in Chapter 16. After investigating several techniques and comparing costs, we have chosen to feed the voltage into a VCO and then

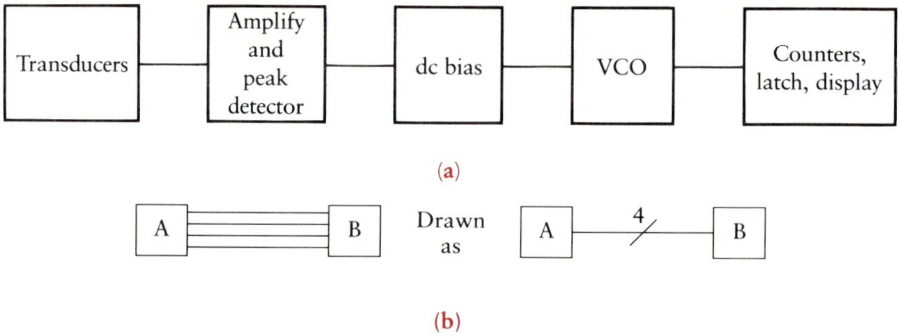

Figure 17.15 Block diagram for Example 17.3.

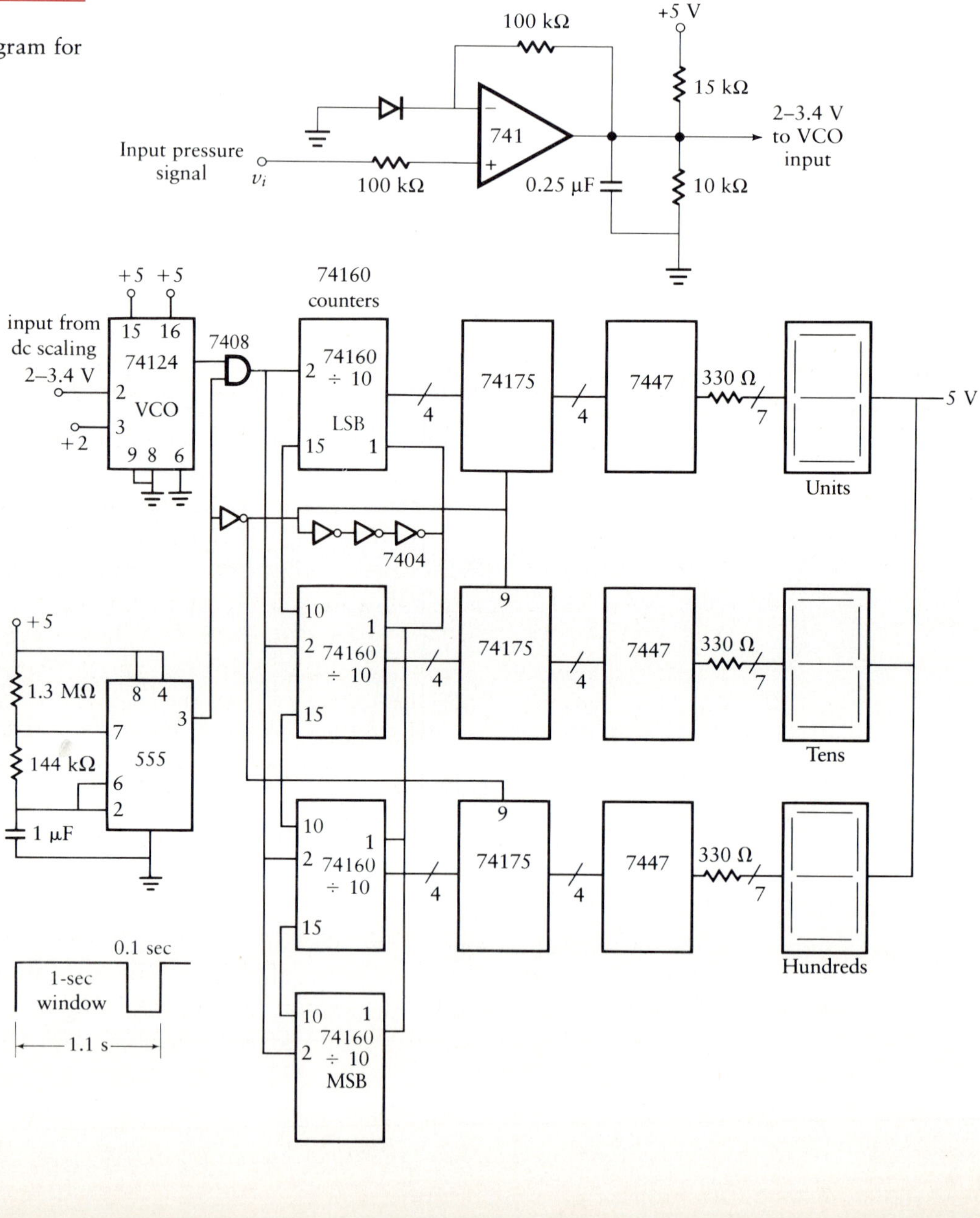

Figure 17.16 Schematic diagram for Example 17.3.

use a counter to measure the output frequency. Since VCOs do not operate well at 0-V input for zero frequency, we decide to shift the 0-mV input to 2 V and the 1000-mV input to 3.4 V.

The VCO is set to generate a frequency between 1000 and 1999 Hz. This frequency is counted during a 1-s window, and the count is displayed on the LEDs through a latch. The 1-s period is provided by a 555 timer operating in the astable mode. Only the last three digits are displayed (0–999, omitting the thousands digit). It is necessary to latch the data in the display before clearing the counters. The latch is therefore set approximately 30 ns before the counters are reset by feeding the reset signal through a series of inverters.

Note that in the diagrams, we simplify the drawing by using a single line with a slash and a number to indicate that more than one line exists. For example, in Figure 17.15(b), instead of drawing four lines between block A and block B, we draw one line with a slash-4 to represent the four wires connecting block A to block B. The schematic diagram for the system is shown in Figure 17.16.

Example 17.4 Warning Speedometer

The assignment: Design a warning speedometer to measure the speed of an automobile. The input to the system is taken from one spark plug. Each pulse from the plug is of $\frac{1}{2}$ V magnitude, and occurs once for each 2 revolutions of the engine. An engine speed of 2500 rpm results at a speed of 60 mph (in high gear). The speed is proportional to the engine rpm assuming that no slippage occurs and that a valid reading occurs when the transmission is in high gear.

Two outputs are required from the tachometer. These are in the form of a visual display of the speed and an audible tone warning of high speeds.

Two 7-segment LEDs display the speed of the automobile. An audible tone increases in frequency as the velocity increases above 60 mph. The tone is interrupted at a 10 Hz rate. The frequency of the tone changes as a function of the velocity. That frequency is 500 Hz at 60 mph, 2 kHz at 70 mph and 5 kHz at 80 mph. An 8-Ω speaker must be driven with $\frac{1}{2}$ watt of power.

SOLUTION A block diagram of our solution is shown in Figure 17.17, and the schematic diagram is shown in Figure 17.18. The input signal from the spark plug is conditioned to TTL compatibility with an op-amp and Schmitt trigger. Because we have so few input pulses, we must multiply and scale the input pulses (see Section 17.8 for more information on this technique).

$$\frac{2500 \text{ rev}}{60 \text{ s}} \times \frac{1}{2} = 20.83 \text{ pulses/s}$$

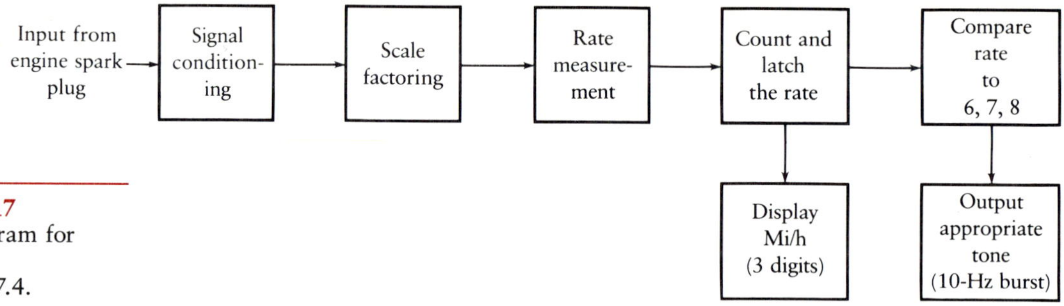

Figure 17.17
Block diagram for system of Example 17.4.

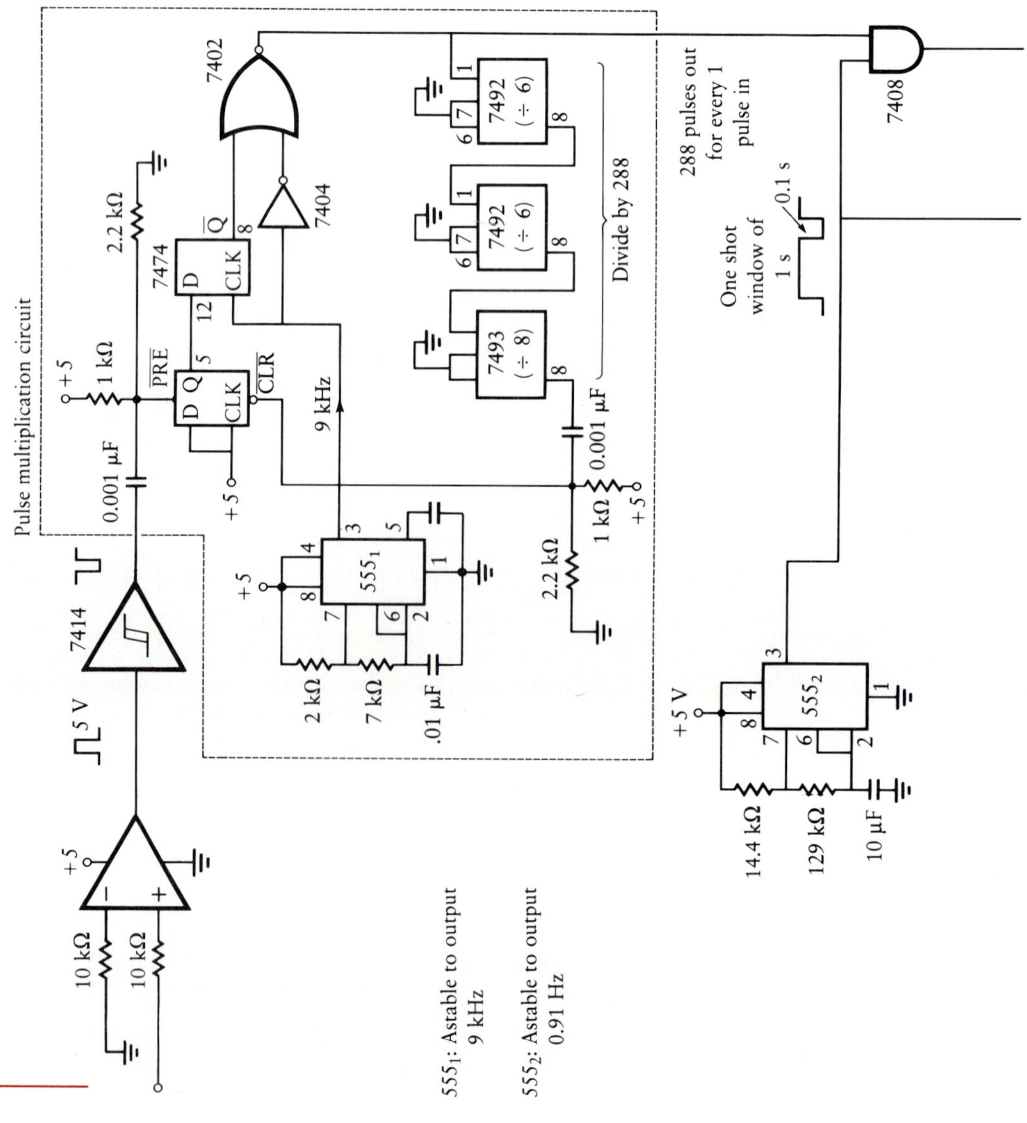

Figure 17.18
Schematic for Example 17.4.

This number of pulses per second corresponds to 60 mph. We multiply these pulses so that for each pulse coming into the pulse multiplication circuit, 288 pulses come out. As a result, 20.83 input pulses generate 6000 pulses/s. Therefore, we need simply divide the number of output pulses per second by 10^2 to obtain the speed.

The next step is to count the number of pulses that occur in a window that is 1.0 s long. We use a duty cycle of 10 and a frequency of 0.91 Hz and count

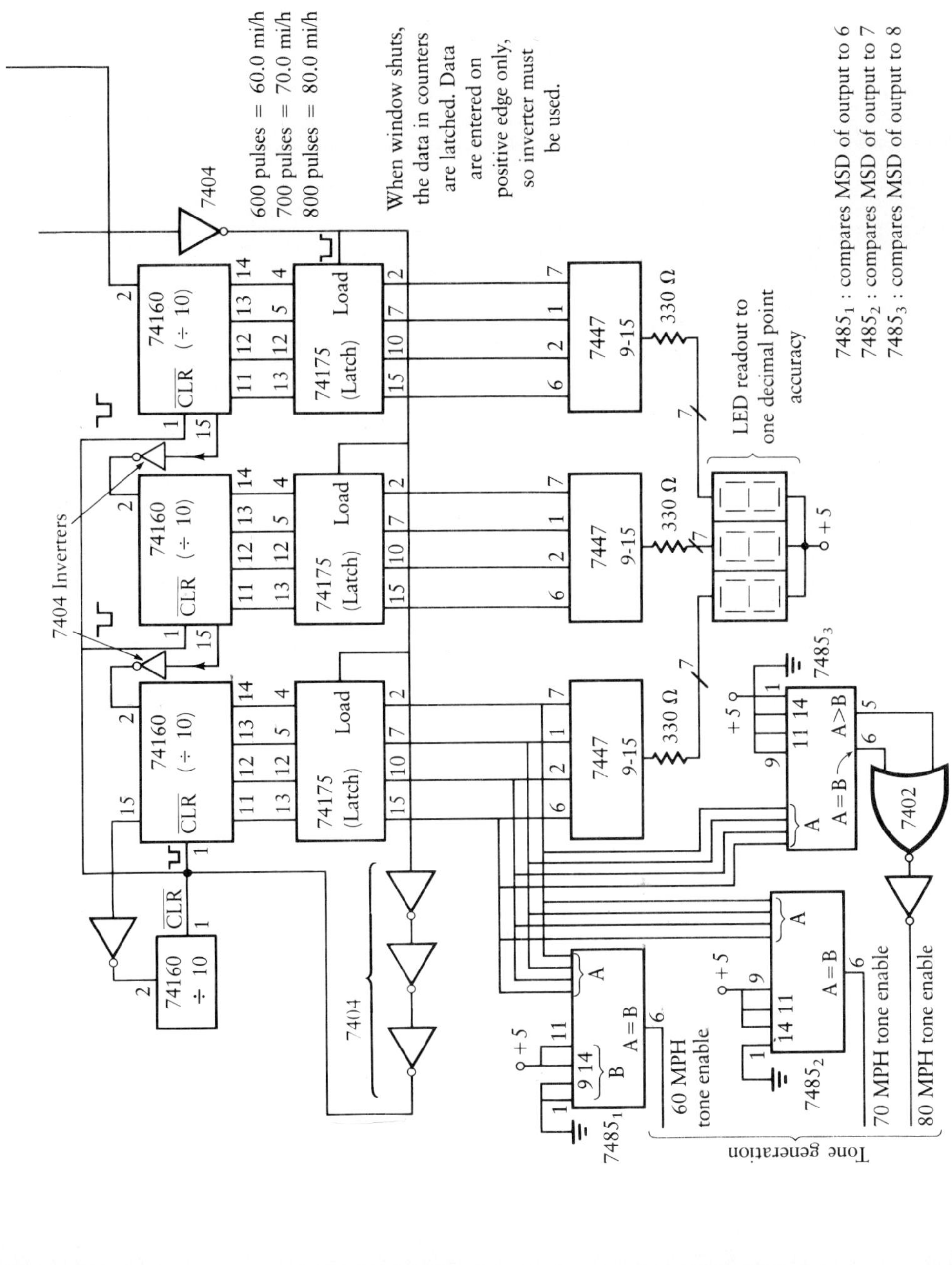

(a) Schematic diagram for warning tachometer

the number of pulses occurring during the window time. To update more often, we design the window signal so that it is low for only 0.1 s. While the window is open, we count the pulses with four decade counters. The MSD divide-by-10 counter is not latched or displayed because we wish to display only the most significant three digits. When the window shuts, the data in the counters are latched. Data are entered on the positive edge only, so 7404 inverters must be used.

Only three LED displays are used. As soon as the window closes, the count is latched into three 4-bit latches to produce an output to the decoder drivers.

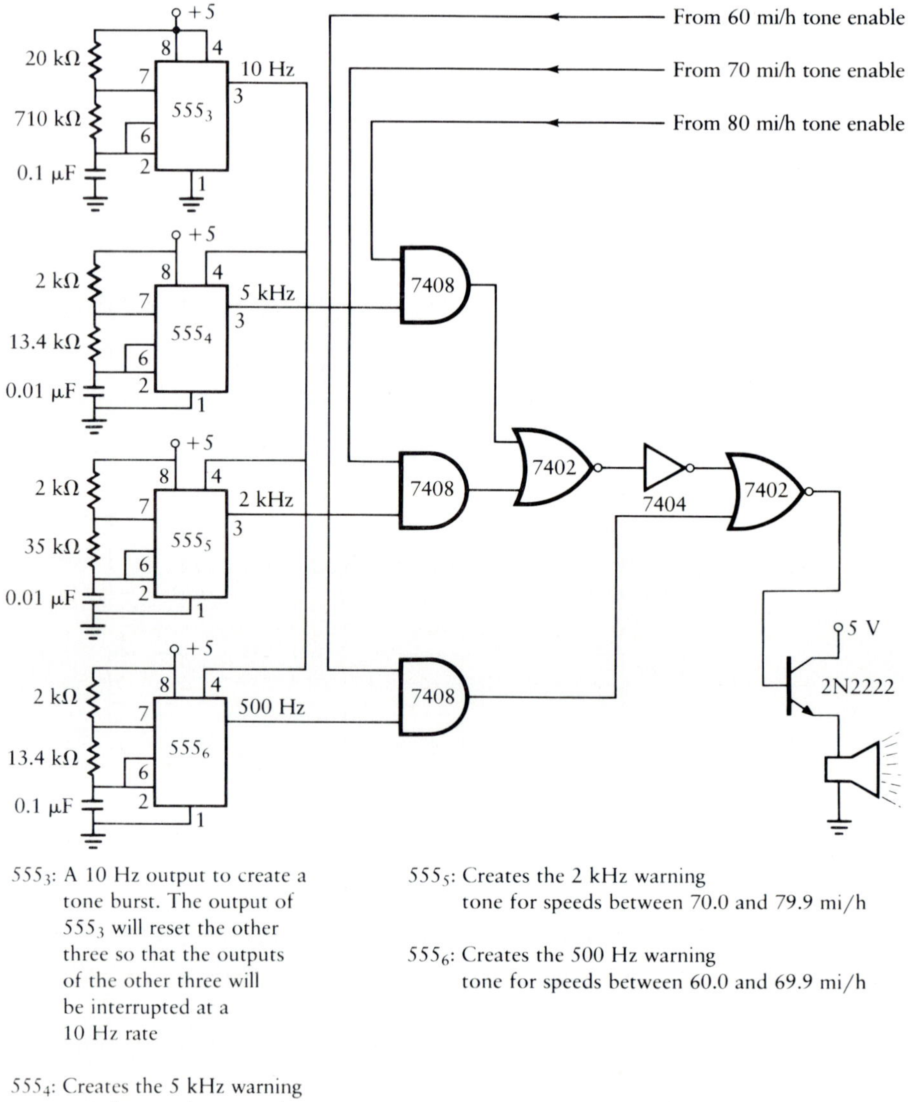

555_3: A 10 Hz output to create a tone burst. The output of 555_3 will reset the other three so that the outputs of the other three will be interrupted at a 10 Hz rate

555_4: Creates the 5 kHz warning tone for speeds greater than 80 mi/h

555_5: Creates the 2 kHz warning tone for speeds between 70.0 and 79.9 mi/h

555_6: Creates the 500 Hz warning tone for speeds between 60.0 and 69.9 mi/h

(b) Tone generation module for Example 17.4

Once the count is latched, the counters are cleared and the window can be opened again. In the meantime, the latched count is displayed on three LEDs. Displayed data from the MSD counter are sent to three 4-bit magnitude comparators. These comparators compare the MSD to 6, 7, and 8. The outputs of the comparators are used to enable the appropriate warning signal. The $\frac{1}{2}$ W of power is ensured by driving the speaker through a 2N2222 transistor. The schematic diagram is shown in Figure 17.18. A partial parts list is shown in Table 17.1.

Table 17.1 Partial Parts List for Example 17.4.

Number	Description	Quantity required
555	Timer	6
741	Op-amp	1
7402	Quad 2-input NOR gate	1
7404	Hex inverter	2
7408	Quad 2-input AND gate	1
7414	Hex Schmitt trigger inverter	1
7447	BCD-to-7-segment decoder/driver	3
7474	Dual *D* flip-flop	1
7485	4-bit magnitude comparator	3
7492	Divide-by-12 counter	2
7493	4-bit binary counter	1
74160	Synchronous decade counter	3
74175	Quad *D* flip-flop	3
	$2\frac{3}{4}$-in. 8-Ω speaker	1
	Red LED digital display	4
	Assorted resistors	
	Assorted capacitors	

Example 17.5 Tape Controller

The assignment: Design a digital tape-recorder controller that must have outputs and inputs as specified in Figure 17.19. There are no limits on the size, power supply, or type of ICs that may be used in this problem, but the circuit must supply 12-V signals for the motor controllers.

SOLUTION We first subdivide the problem into three parts: motor control, audio control, and switch-processing section. The power supply voltage for the CMOS family can be 3 to 15 V, and the outputs require a voltage level of 12 V.

Figure 17.19
Recorder control specifications.

Present state	Button pushed: Stop	Play	Fast forward	Rewind
Stop	NC	Wait 0 s. Start M1, M2. Turn audio on.	Wait 0 s. Start M1.	Wait 0 s. Start M3.
Play	Wait 0 s. Stop M1, M2. Turn audio off.	NC	Wait 0 s. Stop M2. Turn off audio.	Stop M1, M2. Turn off audio. Wait 1 s. Start M3.
Fast forward	Wait 0 s. Stop M1.	Wait 0 s. Start M2. Turn audio on.	NC	Stop M1. Wait 1 s. Start M3.
Rewind	Wait 0 s. Stop M3.	Stop M3. Wait 0 s. Start M1, M2. Turn audio on.	Stop M3. Wait 1 s. Start M1.	NC

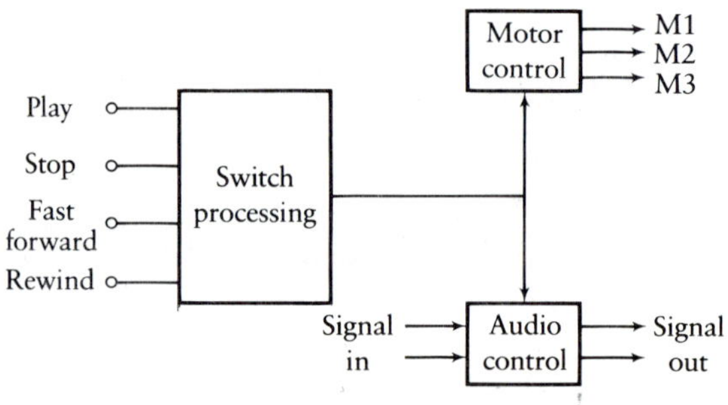

Figure 17.20
Block diagram for Example 17.5.

We decide to use CMOS ICs and set the power supply voltage to 12 V. A block diagram is drawn with indications of what outputs and inputs each module expects. This is shown in Figure 17.20. After the problem has been subdivided, the individual blocks are designed.

The motor control circuitry must generate delays depending on the order of switch operation. Since the amount of delay time is not critical (i.e., 1.05 s is as good as 1.00 s), a monostable multivibrator circuit is used. We find, by referring to the CMOS data book, that the MM74C221 has the necessary inputs and outputs for generating the delays. An external resistor, R_{EXT}, and an external capacitor, C_{EXT}, are used to set the time delay, t_w. Since t_w is equal to the product $R_{EXT}C_{EXT}$, a 1-s value for t_w can be obtained with $C = 1\ \mu F$ and $R = 1\ M\Omega$.

The audio portion of the system needs to switch low-level (2 V or less) audio signals. The CMOS family includes the MM74HC4066 quad bilateral switch, which can be switched using digital signals. For the switch-processing portion of the unit (see Figure 17.20), a memory device such as an MM74C76 *JK* flip-flop with clear and preset can be used to debounce and remember which switch was last pressed. The rest of the logic is then designed to generate the waveforms needed to interface the signals. A schematic is drawn, as shown in Figure 17.21. We use the clear (CLR) and preset (PR) as the inputs, since they are independent of the clock. The J, K, and clock (CLK) inputs are brought to V_{CC}.

Next, a prototype is built and tested. It is important to test the prototype under a wide variety of conditions. This testing, which is frequently termed "worst-case" testing, subjects the prototype to the environmental conditions of the specification, especially those of temperature and vibration. For example, the prototype is operated in a temperature chamber at the low and high extremes of temperature to determine the variation of system performance (sensitivity) due to temperature change.

Other worst-case testing is concerned with the variation of system performance as individual component values are changed. Because the system performance deviates by an amount that varies with the choice of component values, this testing helps the design engineer specify the required tolerance of the com-

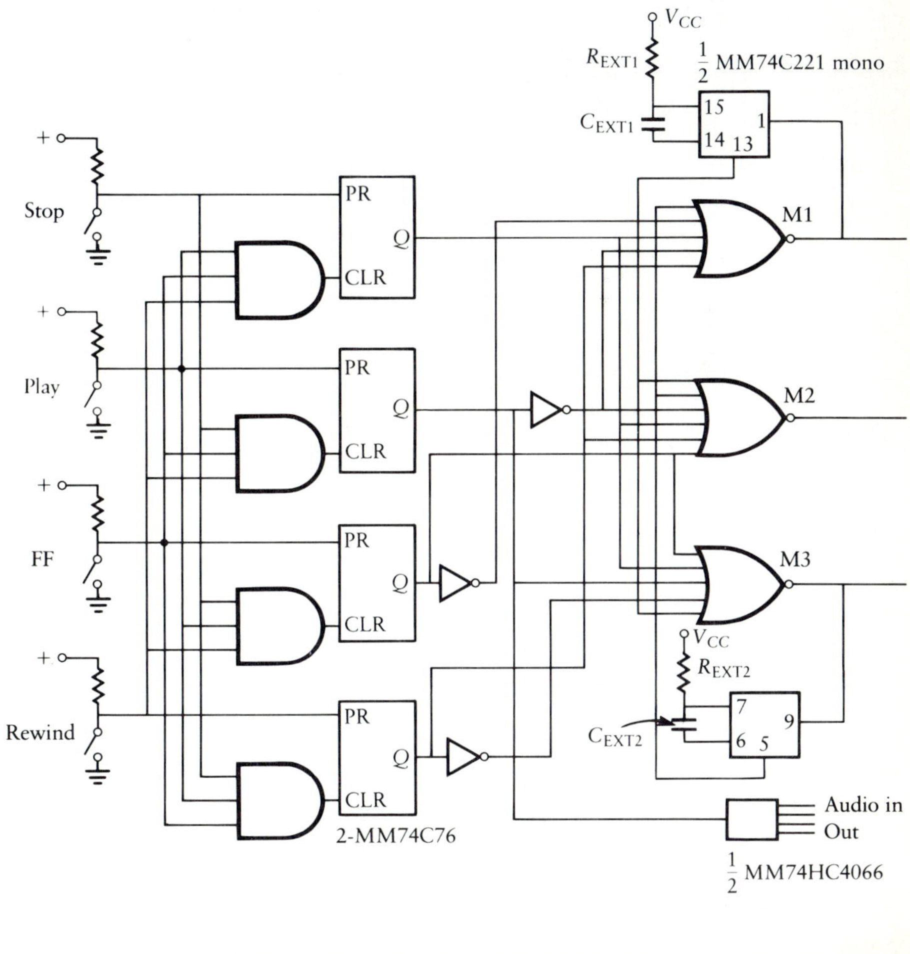

Figure 17.21
Schematic for Example 17.5.

ponents. For a large system, a computer simulation can be used to assist in the worst-case design.

If the unit is to be put into production, a parts list would be developed, printed circuit boards would be designed, and other documents needed for production would be written.

17.8 INTRODUCTION TO PROBLEMS

The problems that conclude this chapter are generally quite challenging, and each has no single correct solution. In fact, as time goes on and new and better devices are developed, the possible solutions to these problems will improve.

In formulating these problems, we have tried to pick situations for which you now have the necessary tools at least to make an attempt at the solution. There are, however, just a few loose ends we would like to tie up now to enlarge your repertoire of available tools.

In this section, we present several techniques that you will find useful in designing the systems at the end of this chapter. In particular, we consider the following:

1. Generating random numbers.
2. Measurement of mechanical angle or velocity.
3. The Hall-effect switch.
4. Use of timing windows.

17.8.1 Generating Random Numbers

In game-related situations, you will find it necessary to generate *random numbers.* Fortunately, this is a simple task with a number of possible approaches. One of the simplest ways to generate a random number is as shown in Figure 17.22. A 555 astable clock is run at a high frequency. The output of the 555 is used to select one output line of a decoder. When the signal from the 555 to the decoder is interrupted, the decoder stops and one output line is high. All output lines of the decoder have the same probability of going high. Since the frequency of the 555 is high compared to the other frequencies in the system (e.g., your reaction time if you are pushing a button), the instant at which the timer stops is random.

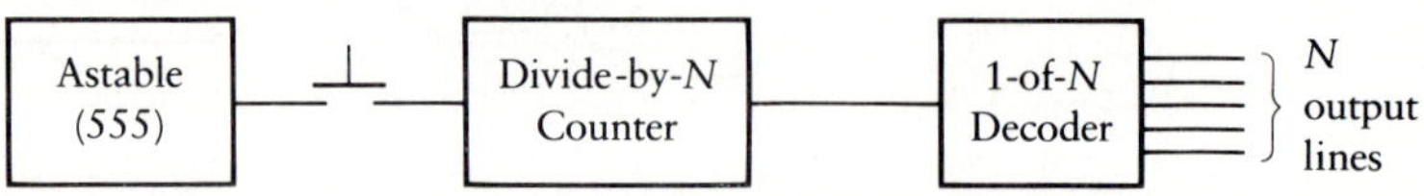

Figure 17.22 Random-number generator.

17.8.2 Measurement of Mechanical Angle or Velocity

In a number of physical situations, you will need ways to provide input to digital systems from mechanical devices, such as shaft angle or velocity. A *variable-reluctance pickup* is a device for providing such a signal to a digital system. Figure 17.23 illustrates one example of the use of this pickup device in conjunction with a rotating gear. As the gear tooth, which is made of magnetic material, moves past the magnet, a voltage is generated at the output of the coils wrapped about the magnetic circuit. The resulting output pulse train is noisy and the voltage levels are generally incompatible with digital circuits because they are so small. The signal must therefore be conditioned.

The conditioning of the signal consists of amplifying it to a level compatible with the particular digital logic being used and then applying it to a Schmitt trigger, as shown in Figure 17.24. The output is then a 0 to +5 V (in the case of TTL) signal that is compatible with the digital circuitry.

An optoelectronic detector used in an interrupted reflector module can count the number of revolutions of a shaft. Such a device is discussed in Chapter 3.

Figure 17.23
Variable-reluctance pickup.

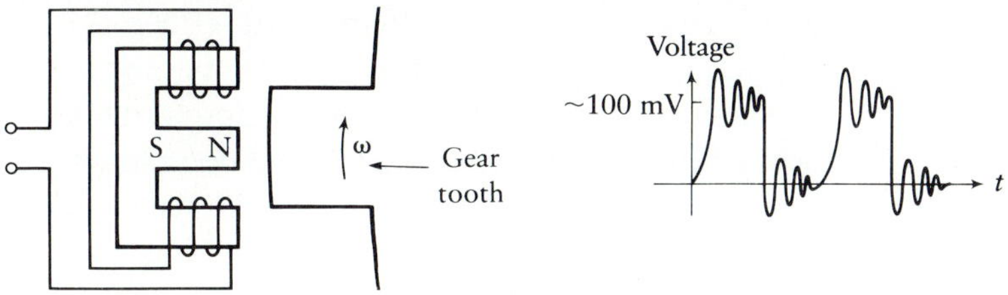

Figure 17.24
Signal conditioning.

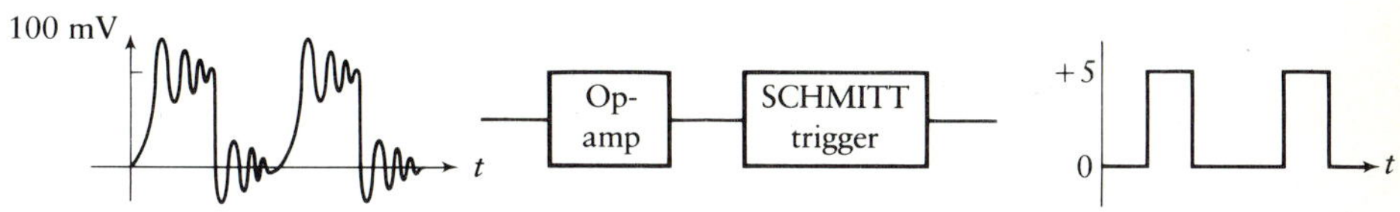

17.8.3 The Hall-Effect Switch

Electronic design for instrumentation systems requires measurement of numerous physical properties. We often deal with magnets in the process of position sensing, thickness determination, weight measurement, speed control, and pressure monitoring. We must be able to sense a magnetic field. This can be done with a *silicon Hall-effect switch*. The TL170C, as shown in Figure 17.25, is one example of such a switch. This is a low-cost magnetically operated switch that consists of a Hall-effect sensor, signal conditioning, hysteresis functions, and an output transistor. The outputs of these circuits are usually compatible with the digital ICs, so little, if any, signal conditioning is required.

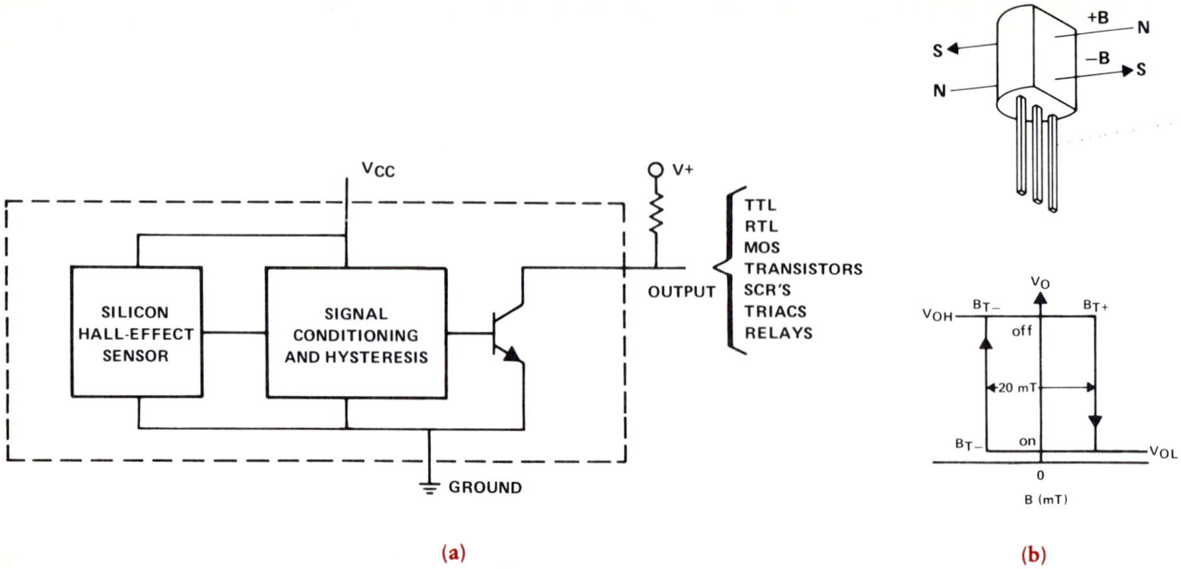

Figure 17.25
Hall-effect generator.

Hall-effect switches find wide application in security systems. For example, we can use a normally off Hall-effect switch as an open-door sensor. The switch is installed in the door frame and a permanent magnet is positioned in the door. While the door is closed, the magnet keeps the Hall-effect switch ON and there is current in the door-sensing circuit. When the door is opened, the magnet moves away from the sensor and the Hall-effect switch returns to the OFF condition, thus triggering an alarm. Wires are connected only to the door frame and not in the door. As an additional security feature, the alarm sounds if any connecting wire is cut open.

17.8.4 Use of Timing Windows

We have used a *window* in the solution of several design problems. Before moving to the design problems at the end of this chapter, we present three uses of a window for various electronic system applications.

1. *Use of a window to measure frequency or velocity.* If we hold a window high for a fixed time, t_w, we will count the number of pulses in the incoming signal with unknown frequency. We accomplish this by ANDing the window with the pulse train of unknown frequency, as shown in Figu 17.26(a). The output of the AND gate yields the number of pulses of the signal with unknown frequency that occur in the time period t_w. As shown in Figure 17.26(a), the ideal window is high for the time t_w and low for a short time, $0.1t_w$. The trailing edge of the window is used to set the latch and then clear the counter in anticipation of the next time the window goes high.

Figure 17.26
Use of windows.

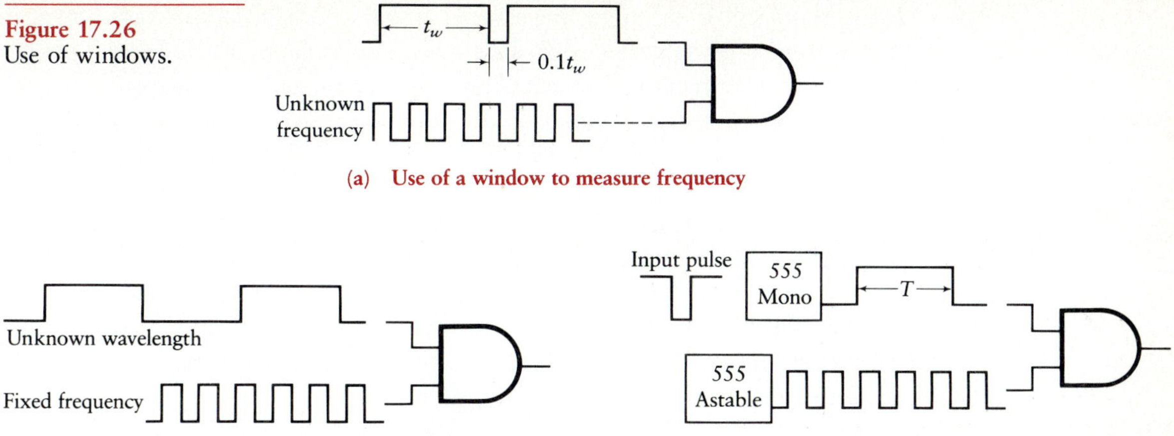

(a) **Use of a window to measure frequency**

(b) **Use of a window to measure wavelength**

(c) **Frequency multiplication**

2. *Use of a window to measure wavelength or 1/f.* This mode, which is shown in Figure 17.26(b), uses the signal with unknown wavelength as the window and a higher fixed-frequency signal as the input. These are ANDed with the result that the number of pulses of fixed frequency is proportional to the length of time the window is held open. Hence, the number of output pulses is proportional to the wavelength.
3. *Use of a window to produce multiplication of input pulses.* In applications where the number of pulses coming from the sensor is insufficient to drive the counter, we must multiply the number of pulses by a constant. This method is used in the solution of Example 17.4. Suppose, for example, we require 10 pulses each time we receive 1 pulse from the instrumentation. We set up a window so that with 1 pulse as input to the system, 10 pulses are generated at the output. This is accomplished by letting the input pulses trigger the 555 monostable of Figure 17.26(c) to provide a window of duration T. We design a 555 astable in such a manner that during the time T, precisely 10 pulses are produced at the output. The window of period T is ANDed with the pulse train of the 555 astable to produce the 10 desired pulses for each input pulse.

17.9 CONCLUDING REMARKS

The design process probably appears overwhelming if this is your first exposure to it. It is important to realize that this is an area where practice is extremely important. You should also know that help is available in the form of data books, other engineers who may be working in the same company with you, your professors, application notes, books, magazines, and technical publications. Your skill is an important asset in effective design, but it is also critical that you be knowledgeable regarding new developments in the field. This requires extensive reading and attendance at meetings of professional societies. By being organized in methods and willing to research information and by writing down the steps followed to produce a working prototype, you can solve many difficult problems.

With this assortment of tools combined with what you have learned in this book, you are now ready to approach the wide array of challenging electronic system design problems. After solving some of these problems, we hope you will be motivated to invent your own problems. You need only examine your daily environment to formulate a virtually unlimited list of projects.

Good luck in this exciting undertaking!

PROBLEMS

Use TTL or CMOS integrated circuits and/or discrete elements in each of the following problems. Your final design should include a written explanation of how the system operates and a schematic diagram. Include the component values used and the type numbers of the integrated circuits and discrete elements. Assume that any power supply voltages you need are available, so do not design the power supply for these problems. Use as few components as possible.

17.1 Design an oil well monitoring system to measure the output of a field of 16 oil wells. The output of *each* well is measured with a flow meter that outputs a binary digital number corresponding to one of the following four flow conditions. Logic 1 corresponds to 5 V.

00 no flow

01 33.3% flow

10 66.6% flow

11 100% flow

The information from all 16 wells is transmitted to a control center where two LEDs and two 7-segment LED displays are used in the format shown in Figure P17.1. The two 7-segment LED displays identify the oil well number (0 to 15) being measured. The two single LEDs indicate the flow information (status). Each oil well is to be sampled and read for 0.1 min once every 1.6 min. In addition, when the flow is 00 for any oil well, a 1-kHz tone is sounded when that oil well number is displayed.

17.2 Design a lamp test indicator for a commercial aircraft to test 16 important running and landing lights about the aircraft. Each lamp has a small resistor between the light and ground, as shown in Figure P17.2. When the lamp is operational, the voltage across the resistor is 5 V.

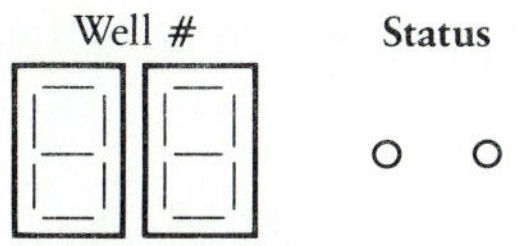

Figure P17.1

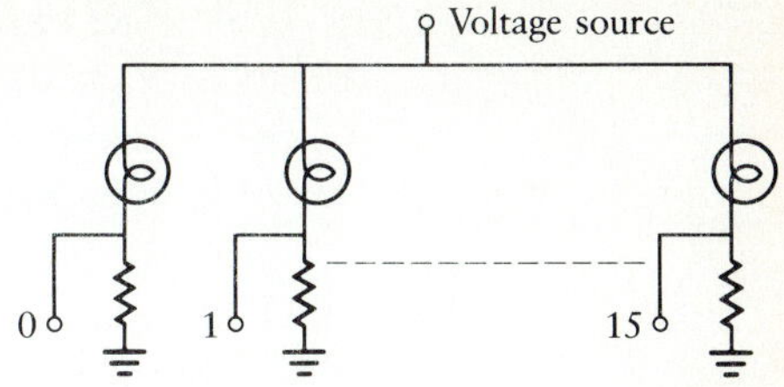

Figure P17.2

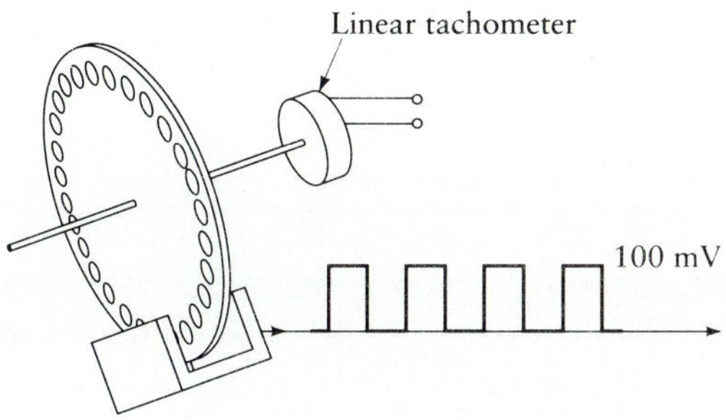

Figure P17.3

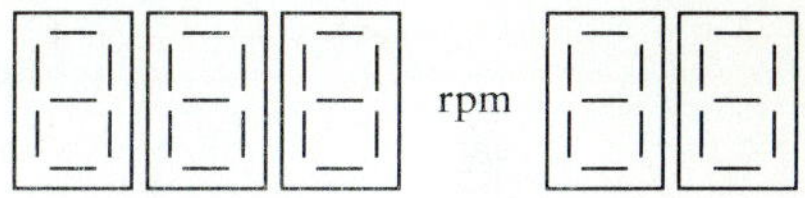

Figure P17.4

When the lamp is not operational, the voltage across the resistor drops to zero. Test each of the lamps for a period of 5 s and display the lamp number (0 to 15) on two 7-segment LED displays. Each time a lamp is faulty, sound a 1-kHz tone for the 5 s that the faulty lamp is being tested. Power of $\frac{1}{4}$ W is sufficient to drive the 8-Ω speaker.

17.3 Design a counter that will display the number of revolutions per minute (rpm) of a shaft. The mechanical system, shown in Figure P17.3, is used to determine the shaft rpm. Mounted on the shaft is a circular disk that has 60 holes around its edge. A light sensor is interrupted 60 times for each revolution of the disk. The output of the light sensor is a pulse train, as shown in Figure P17.3, with a 100-mV amplitude and with 60 pulses each revolution. The speed of the shaft is in the range of -999 to $+999$ rpm. Mounted on the same shaft is an analog tachometer that produces a $+5$ V signal for clockwise rotation and a -5 V signal for counterclockwise rotation. The output display comprises five 7-segment LEDs, as shown in Figure P17.4. The first three 7-segment LEDs show the rpm value, and the second two 7-segment LEDs show the letter C if the shaft is going clockwise and CC if the shaft is going counterclockwise.

a. Design an electronic system to count the rpm and display the speed in the range of -999 to $+999$ rpm with the first three 7-segment LEDs, as shown in Figure P17.4. Be sure to condition the 100-mV input.

b. Use the analog tachometer to drive the two 7-segment LEDs that show the direction of rotation.

17.4 Use two 7-segment LEDs to design a circuit that shows the most significant digit (MSD) on one 7-segment display and the "$\times 10^n$" range on another 7-segment display. The input signal is between 1 and 9000 Hz. The reading on this "*n*-display" would be as follows:
If frequency is 1 to 9, display 0, since the frequency is $\times 10^0$.

If frequency is 10 to 99, display 1, since the frequency is $\times 10^1$.

If frequency is 100 to 999, display 2, since the frequency is $\times 10^2$.

If frequency is 1000 to 9999, display 3, since the frequency is $\times 10^3$.

For example, if the frequency is 857, the MSD display would indicate 8 and the $\times 10^n$ display would show 2.

17.5 Design an electronic circuit that displays the amount of liquid in a large storage tank on one 7-segment display. The display shows the amount of liquid in the tank in percent full from 10% to 90% in 10% increments. Sound a 1-kHz alarm if the liquid level falls below 10% full or exceeds 90% full. The level is measured with a sonar device that provides pulses as shown in Figure P17.5.

17.6 Design a countdown sequencer to time the firing of a sounding rocket. It is desired to count down from 10 minutes before firing and then keep track of the time after the launch of the rocket per the following schedule:

10.0 minutes	Start countdown
7.0 minutes	Transfer to rocket power supply
6.0 minutes	Activate the on-board cooling
4.0 minutes	Transfer to on-board computer
3.0 minutes	Activate firing sequencer
2.0 minutes	Remove external cable
0.0 minutes	Fire

After firing, the counter will count up for at least 20 minutes. Use three 7-segment LED displays to show the time before and after firing to tenths of minutes, and use a digital CMOS switch (such as the 74HC4066) to activate each of the important times in the countdown. The three LED displays are to be arranged as shown in Figure P17.6.

Figure P17.5

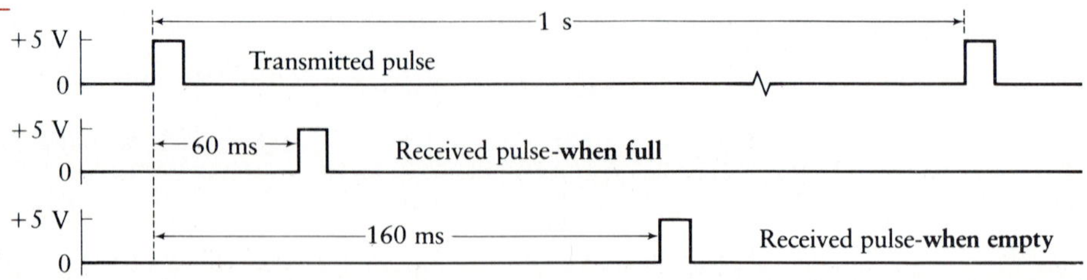

Preload a 74190 UP/DOWN decade counter with 10.0, and in the DOWN mode go through the prefire events. When the count reaches 00.0, change the counter to count UP to record time. As a suggestion, use three 74190 decade up/down counters. (A zero on pin 5 makes the counter count up and a one makes it count down.) An astable 555 operating at 100 Hz can be used for the clock.

17.7 Design a 24-hour clock and a system to control the turn-on and turn-off functions needed in an apartment complex.

If the clock reads 14.3, this means 2:18 P.M. since $\frac{3}{10}$ of an hour is $\frac{3}{10}$ of 60 min, which is 18 min. For clock timing accuracy, use the 60-Hz line frequency with an optoisolator. Design the system to activate the following:

1. Turn garden lights on at 18.0 h and off at 01.1 h by providing a 5-V signal to a relay.
2. Use a data distributor (74154) to turn on 10 possible sets of water sprinklers (one at a time). Turn on the first set of sprinklers by providing a +5 V signal to the voltage-controlled valve at 04.0 h. Provide a watering sequence by allowing the first set to flow for $\frac{1}{10}$ h (6 min). Then turn the first set off and turn the second set on, also for 6 min. Continue this sequence until all 10 sets have been turned on and off.
3. Turn the music system on (by providing a 5-V signal to a relay) at 09.0 h and off at 21.0 h.

17.8 Design a controller to activate a large incandescent advertising sign as shown in Figure P17.7. There are 64 lights around the circumference of the sign, and they are lighted 4 at a time in a sequence rotating as follows:

	Lights on
1st second	0, 16, 32, 48
2nd second	1, 17, 33, 49
3rd second	2, 18, 34, 50
⋮	

Figure P17.6

Figure P17.7

Each light is to remain on for 1 s. The letters are formed by an array of lights and are lighted in sequence as follows:

	Letter on
1st interval	C
2nd interval	CA
3rd interval	CAS
4th interval	CASI
5th interval	CASIN
6th interval	CASINO
7th interval	————
1st interval	Repeat sequence

Each interval is to be 4 s long. In this design, do not concern yourself with the output power requirements but only with the electronic design. The output of the electronic design is sufficient to drive one LED for each of the peripheral lights and only one LED to drive each letter C, A, S, I, N, and O.

17.9 Design a pressure-altitude hold system for a small commercial aircraft. The system is to operate over the range of 0 to 50,000 ft with a 200-ft resolution. The actual pressure altitude, A, which is measured in feet, is obtained from a digital pressure transducer that outputs an 8-bit binary signal with each binary number equivalent to 200 ft. So if the pressure transducer outputs 01101101, the pressure altitude is 109 × 200 ft = 21,800 ft.

The pilot sets the desired altitude, B, in feet, with a three-position switch on the control stick, as shown in Figure P17.8. This three-position switch allows the pilot to increase, by pushing the switch forward, or decrease, by pushing the switch back, the set altitude. The set altitude will remain the same if the pilot's finger is removed from the switch.

Three 7-segment LED displays indicate the set altitude (desired altitude) where the least significant digit corresponds to 200 ft. Compare the actual altitude with the set altitude, and provide a 5-V signal according to the following schedule:

Figure P17.8

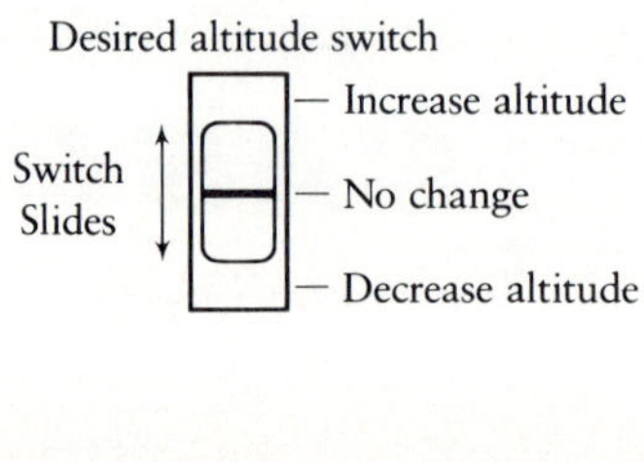

Terminal	$A > B$	$A = B$	$A < B$
Altitude decreases	1	0	0
No change	0	1	0
Altitude increases	0	0	1

17.10 Design a temperature-monitoring system for an aircraft engine. It is necessary to monitor eight points throughout the engine with the circuit shown in Figure P17.9. The voltage, V_γ, across the diode is 0.7 V at 25°C = T_1, and this voltage decreases as the temperature increases as follows:

$$\Delta V_\gamma = -2(T_2 - T_1)\text{ mV}$$

with T_1 and T_2 in degrees centigrade. The output is displayed on three 7-segment LEDs as shown in Figure P17.10. A single LED identifies the diode being read, and the three 7-segment LEDs read the temperature at the point in question. Each temperature is sampled sequentially and read for 6 s.

An analog multiplexer (74HC4051) is used to sample each of the eight values of voltage. The analog signal from the diode must be conditioned with the op-amp circuit of Figure P17.11.

The voltage V_γ is

$$V_\gamma = 700 - 2(T - 25) = (750 - 2T)\text{ mV}$$

where T is expressed in degrees centigrade. The voltage V_c is

Figure P17.9

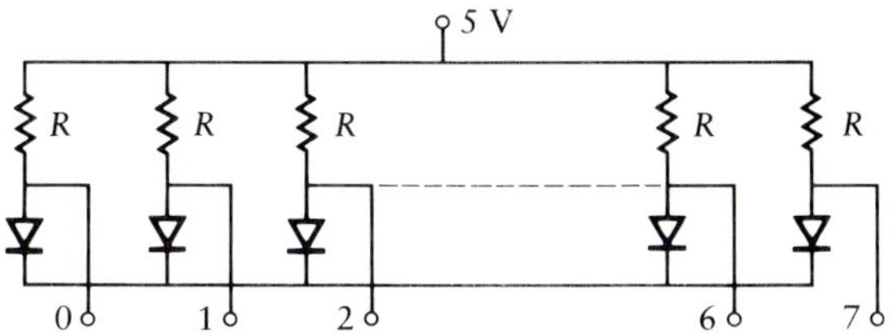

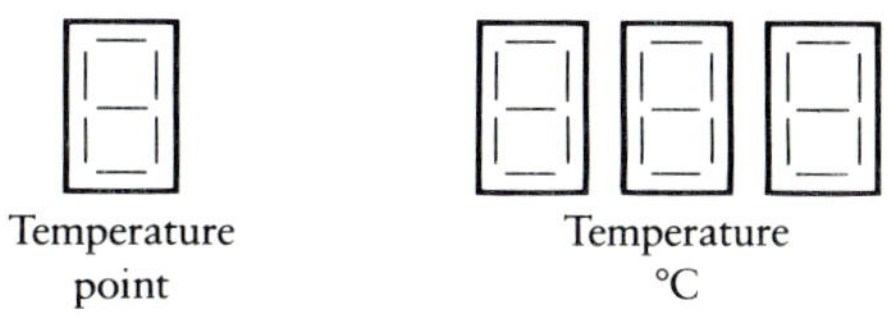

Figure P17.10

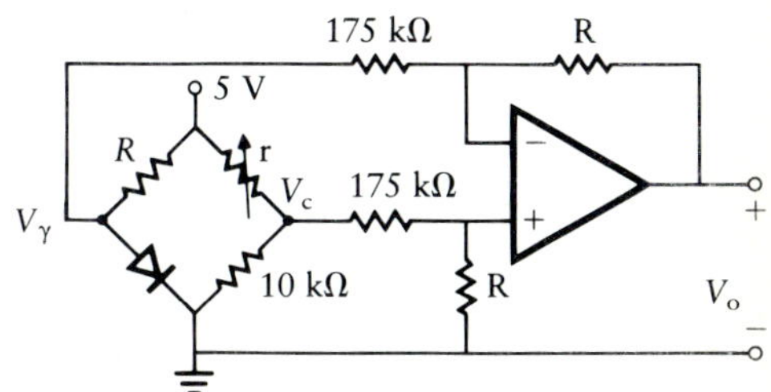

Figure P17.11

$$V_c = \frac{50}{r + 10}\text{ V}$$

where r is in kilohms. The output voltage, V_o, from the op-amp is

$$V_o = \frac{R}{175\text{ k}\Omega}(V_c - V_\gamma)$$

and substituting, we obtain

$$V_o = \frac{-R}{175\text{ k}\Omega}\left(750 - 2T - \frac{50 \times 10^3}{r + 10}\right)\text{ mV}$$

We choose r by letting

$$\frac{50 \times 10^3}{r + 10} - 750 = 0$$

We set the gain by selecting R from the equation

$$V_o = \frac{2R}{100\text{ k}\Omega}T\text{ mV}$$

Use a $3\frac{1}{2}$-digit DVM to output the signal to the LED display.

17.11 Design a temperature-monitoring system that will display seven different temperatures on three 7-segment displays. Seven diodes are used to measure the seven temperatures. The output voltage of each diode, which varies linearly with temperature, is fed into an op-amp. When the op-amp output is properly calibrated and amplified, a dc signal is obtained which ranges from 0.25 V for 25°C to 1.25 V for 125°C. Your design should monitor each of these outputs, in sequence, and display the temperature for a period of 1 minute. Seven single LEDs will indicate which temperature point is being displayed, as shown in Figure P17.12. Two modes of operation are required, as follows:

a. Automatic mode. Each temperature is displayed for 1 minute.
b. Manual mode. The temperature displayed remains the same until the user presses the advance button to see the next temperature.

17.12 Energy is to be saved in a large museum by automatically turning the lights on and off in an infrequently visited exhibition room, as shown in Figure P17.13. Design an electronic circuit that will turn the lights on when the first person enters and turns the lights off when the last person leaves the room. The electronic system should be able to accom-

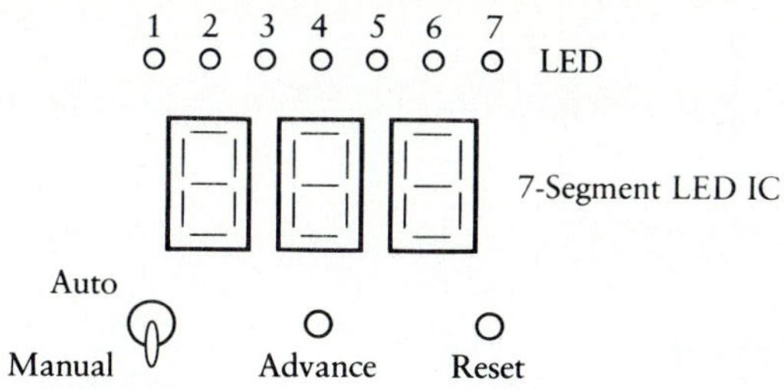

Figure P17.12

Figure P17.13

modate up to 99 people in the room at any time. The entrance sensor and the exit sensor each consists of a parallel light beam. Assume these sensors normally provide a 0-V output, which changes to 5 V when the beam is broken. Provide a reset switch to set the count to 0. Assume that the entrance is narrow enough to allow only one person to enter at a time and that the light beams break only once per person.

17.13 Design a television monitor for a hotel with 14 rooms on each floor. Design the electronic system (for one floor only) to detect when any TV receiver is ON. When the TV is ON, a 5-V signal is delivered to the office from a switch installed on each TV. When the TV is OFF, the signal delivered to the office is 0 V. Each TV is monitored for 1 min before the system passes to the next TV on the floor, and the process continues with each TV being monitored once every 14 min. Use the 60-Hz power line for timing.

The readout consists of two 7-segment LED displays that indicate the room numbers from 0 to 13 and a single LED that is ON when the TV in that room is ON and OFF when the TV is OFF.

17.14 Design an electronic voting system to be used for Parliament, which consists of 250 members. Each member has two switches, as shown in Figure P17.14(a). The first switch is used to indicate that the member is present at the session. Each member in attendance turns this switch ON, thus sending a 5-V signal to the podium. When the member is absent, the signal at the podium is 0 V. A YES vote, recorded by throwing the second switch, sends 5 V on the line to the podium, and a NO vote puts 0 V on the line at the podium.

The output is composed of three single LEDs that indicate the result of the vote, i.e., either YES, TIE, or NO. In addition, three 7-segment LED displays indicate the number of members voting. This display is shown in Figure P17.14(b).

The head of Parliament, who stays at the podium, has one switch, which is used to disable the circuit. When all members have voted,

Figure P17.14

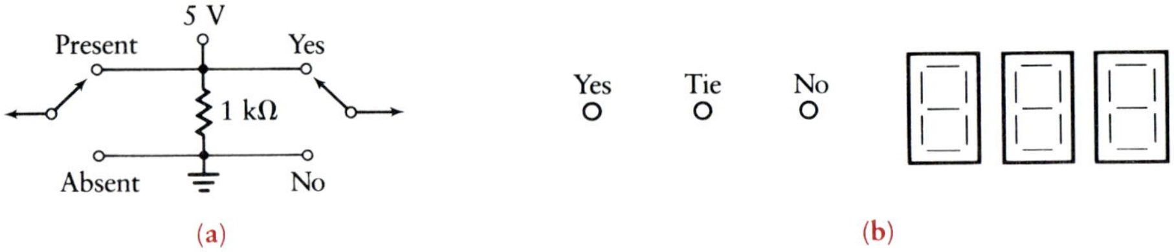

Figure P17.15

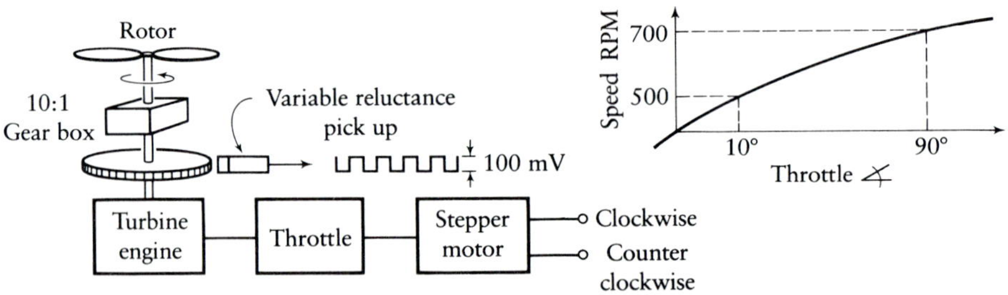

either YES or NO, the head of Parliament throws the podium switch to ON, which latches the results into the display. Once the switch is thrown ON, further vote changes do not change the results that have been latched into the display.

17.15 Design a digital control system to control the angular velocity of the main rotor of a helicopter. The system is shown in Figure P17.15. The rotor speed is displayed on two 7-segment LED displays. The desired rotor speed, which is displayed on two other 7-segment LEDs, is set on the control panel for the range of 500 to 700 rpm. The rotor, which rotates at a speed between 500 and 700 rpm, is connected through a gear train to the turbine engine, which rotates at 5000 to 7000 rpm. The difference between the desired velocity, f_D, and the actual velocity, f_a (in rpm), is used to change the setting of the throttle. Increasing the throttle will increase the speed of the engine, which increases the speed of the rotor. Each pulse into the stepper motor advances (or retards) the throttle setting and hence increases (or decreases) the angular velocity of the rotor.

The actual velocity, f_a, is generated with a 500-tooth iron gear attached to the engine shaft. This gear is used with a variable-reluctance pick-off, which produces a 100-mV peak-to-peak pulse train. Each time the rotor rotates one revolution, 5000 pulses are generated at the output.

17.16 An experimental windmill must have its blade speed monitored and displayed. The expected operating range is 60 to 240 rpm. A 100-tooth iron gear is attached to the blade shaft and is sensed by a variable-reluctance pickup, which provides a +100 mV pulse as each tooth passes the sensor.

a. The circuit should process the signal from the sensor and display the speed of the windmill in revolutions per minute using three 7-segment LED displays.

b. The circuit should energize a 400-Hz alarm if the speed exceeds 240 rpm. Assume that the IC output can directly drive a small speaker.

17.17 Design a digital control system to measure precisely the error in angular velocity of a large centrifuge. This large centrifuge rotates at a precise angular velocity and is used to apply acceleration to large objects. This is an *A-V-E* (angular-velocity-error) indicator. The output comprises five 7-segment LED displays to yield the error velocity ω_E (in rpm) to three decimal places. The *A-V-E* indicator is used to detect the error in angular velocity of the centrifuge. The large arm is to rotate at 60 rpm, and the motor rotates at 6000 rpm. An accurate reference velocity, ω_R, is obtained from a crystal oscillator (temperature controlled) that produces a 1-MHz pulse train. The actual velocity, ω_a, is generated with a 500-tooth iron gear that is attached to the motor shaft and is sensed with a variable-reluctance pickup. This produces a 100-mV pulse train with a frequency of ω_a. Use an UP-DOWN counter and two identical windows. First, up-count ω_R during a window period; then down-count ω_a during an identical window period. Assuming that $\omega_R > \omega_a$, the count left after the end of the second window is the error in revolutions per minute.

17.18 Use a comparator to design a speed control system for your automobile. The throttle is positioned with a stepper motor that operates as shown in Figure P17.16. The desired speed is set by depressing a switch on the steering wheel. This switch causes two 7-segment LED displays to advance and to indicate the set speed. When the switch is released, a latch holds this speed as the desired speed of your automobile. When the display reaches 80 mph, it resets to zero.

The actual speed is taken from the engine distributor, which outputs eight pulses for each two revolutions of the engine. In DRIVE gear, the ratio of engine speed to forward velocity is given by the following equation:

Figure P17.16

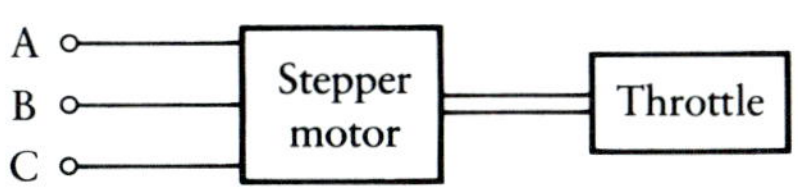

Terminal	A	B	C	
	1	0	0	Engine speed increases
	0	1	0	No change in engine speed
	0	0	1	Engine speed decreases

$$\text{Ratio} = \frac{1000 \text{ rpm}}{25 \text{ mph}} = 40 \text{ rpm/mph}$$

In the design, include a disable capability when the brake pedal is depressed and provide an ON/OFF switch.

17.19 Design an electronic system to measure the rate at which your heart beats. The system is mounted on a stationary bicycle, so you are able to use the 60-Hz power line rather than a battery. The device must read the heart rate from 150 to 250 beats per minute.

The heart rate is measured with an infrared sensor that produces a 100-mV pulse each time the heart beats. This sensor is attached to the index finger of your hand. These pulses are not clean pulses, so they must be signal-conditioned.

The readout consists of three 7-segment LED displays, which read the heart rate in beats per minute.

17.20 Design a miles-per-gallon indicator to be used on an automobile. Two inputs are available as follows:

a. Odometer signal. This signal produces a 5-V pulse for every tenth of a mile traveled.

b. Fuel tank signal. The fuel tank is instrumented so that a 5-V pulse is produced each time the fuel decreases by 0.1 gallon.

Figure P17.17

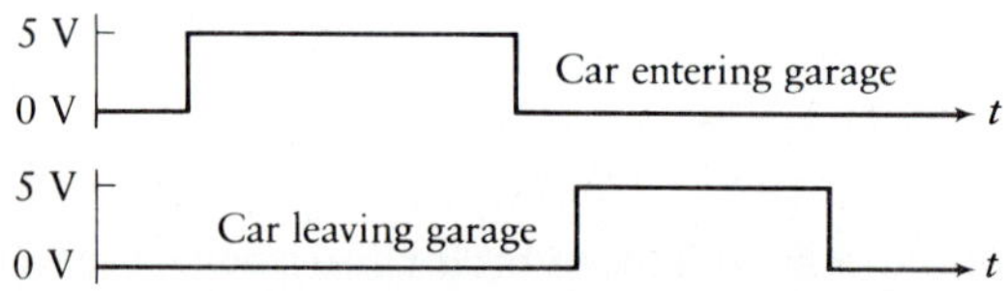

The output comprises two 7-segment LED displays that show miles per gallon to the nearest mile per gallon.

17.21 Design a wind-velocity meter. A propeller is mounted on a ball-bearing shaft so that it will turn freely, and a 100-tooth iron gear is attached to the same shaft. A variable-reluctance sensor is used to sense the teeth on the gear and produces a 100-mV signal each time a gear tooth passes the sensor. Hence 100 pulses are produced for each revolution of the propeller. When the wind velocity, V_w, is 20 mph, the propeller velocity, f_p, is 90 rpm. The relation is linear as follows:

$$f_p = 4.5V_w \text{ rpm}$$

where V_w is the wind velocity in miles per hour. The output display is composed of two 7-segment LEDs to read the wind velocity to the nearest digit.

17.22 Design a warning tachometer for a gas turbine engine that operates in the range 8000 to 9000 rpm. A 100-tooth iron gear is mounted on the engine shaft. A variable-reluctance pickup is mounted near the gear and produces 100 pulses for each revolution of the engine. The pulses are only 100 mV in magnitude, so they must be conditioned. The display consists of two 7-segment LED displays, which display the word LO when the speed is 8000 rpm or less and the word HI when the speed is 9000 rpm or more. In the range between 8000 and 9000 rpm, the LED displays are blank. When LO is displayed a 500-Hz tone is sounded, and when HI is displayed a 1-kHz tone is sounded. One-quarter watt of power is needed for each tone.

17.23 Design a system to monitor the doors on the eight cars of a subway train. Each car has a door that must be closed before the train can proceed. Eight wires, from the Hall-effect switches mounted on the eight car doors, are brought to the train operator's compartment. When a car door is open, the voltage on the wire is 0 V. When the car door is closed, the voltage on the wire is +5 V.

Energize a green LED when all doors are closed. Energize a red LED for 1 s when a door is open. Indicate, with one 7-segment LED, the number of the car with the open door. The car doors are continuously checked, for 1 s each, and the numbers 0, 1, 2, 3, 4, 5, 6, and 7 are continuously displayed on the 7-segment LED. If any door is open, the red LED will light for 1 s while the number of the car is displayed. When all doors are closed, the green LED will light, but the car numbers will still be displayed as the doors are checked.

17.24 Design an electronic system to determine the number of cars in a parking garage containing 300 spaces. Each time a car enters the garage, the entrance gate lifts and a pulse of variable width is generated. The pulse is of variable width because of the difference in time required for the gate to drop after each car of different length and different speed passes through. Each time a car leaves the garage, the exit gate, at another location, lifts and another pulse of variable width is generated. These pulses are shown in Figure 17.17. Provide a *CLEAR* for the system so that the counter can be reset to zero. The visual display is composed of the following:

1. Three 7-segment LED displays that show the number of cars in the garage at any time.
2. Four 7-segment LED displays that show the word FULL when the garage contains 300 cars.

17.25 Design a digital thermometer using a thermistor bridge as the temperature sensor and a voltage-controlled oscillator (VCO). You may wish to use a 74124 VCO. The output of the thermistor bridge is as follows:

With 110°F, the output is 50 mV.

With 90°F the output is 34.4 mV.

The readout is to be three 7-segment LED displays yielding the temperature in degrees Fahrenheit to the nearest whole number. Compute the scale factor showing that your system will yield accurate temperature readings in the 90°–110° range.

17.26 Repeat Problem 17.25 using a $3\frac{1}{2}$-digit A/D converter (ICL7107).

17.27 Design a digital bathroom scale using a strain-gauge load cell as the weight sensor. The load cell produces an output voltage that is proportional to weight. The system is shown in Figure P17.18. We select resistor values such that $R_A >> R$ and we write the equation for V_1 and V_2 as follows:

$$V_1 = \frac{(R - \Delta R)V}{R - \Delta R + R + \Delta R} = \frac{V(R - \Delta R)}{2R}$$

$$V_2 = \frac{R}{2R}V = \frac{V}{2}$$

The equation for the output of the op-amp is

$$v_o = \frac{R_F}{R_A}(V_2 - V_1) = \frac{R_F}{R_A}\left[\frac{V}{2} - \frac{V(R - \Delta R)}{2R}\right] = \frac{R_F V}{2R_A R}\Delta R$$

Select $R = 5$ kΩ, $R_A = 100$ kΩ, and $V = 9$ V. The change in resistance, ΔR, is proportional to the weight.

Figure P17.19(a) shows a schematic of the mechanical system. A strain gauge is attached on the top and bottom of the beam. The beam bends down because of the application of weight on the scale. As a result, the resistance of the upper strain gauge increases $(R + \Delta R)$ and the resistance of the lower gauge decreases $(R - \Delta R)$. This provides the input to the bridge system of Figure P17.18. Two additional strain gauges, not shown on the diagram of Figure P17.19(a), are attached to the beam in a direction that causes no change in resistance, R, as the weight is applied. These two gauges form the other two legs of the bridge and are shown as R in Figure P17.18. Since resistance is a function of temperature, the bridge will remain balanced as the temperature changes and will yield a voltage output only for a change in weight.

The scale is to respond to changes in weight from 0 to 399 lb. The load cell is linear, and when the weight is 399 lb,

$$V_o = \frac{R_F}{R_A}(34 \text{ mV})$$

Figure P17.18

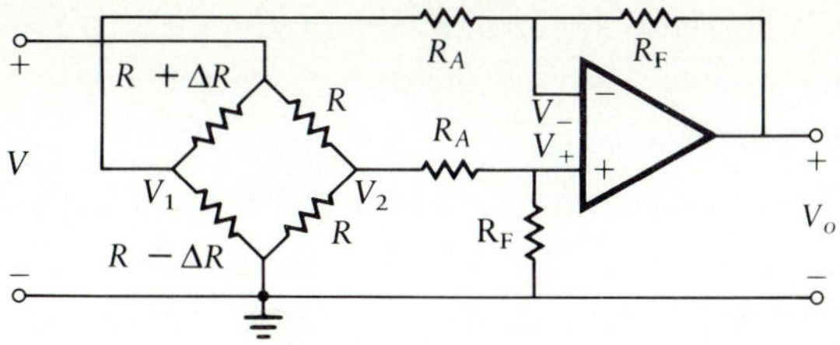

Figure P17.19

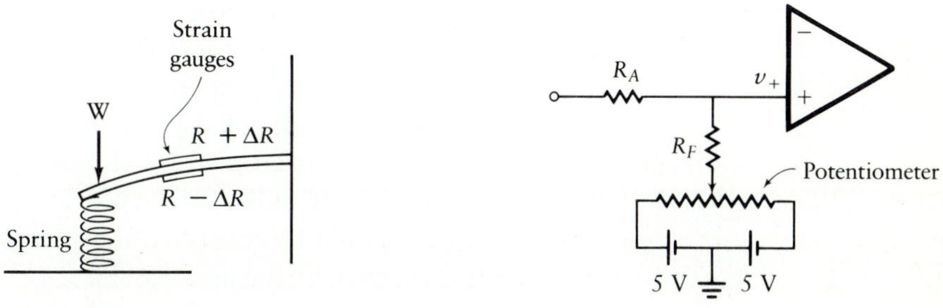

(a) (b)

Hence, over the range of 0 to 399 lb, the output voltage of the system of Figure P17.18 varies from 0 to

$$\frac{R_F}{R_A}(34 \text{ mV})$$

When someone steps on the scale, a normally open switch closes and hence applies power to the device. Three 7-segment LED displays read the weight in pounds. Calculate the scale factor, R_F/R_A, and provide a zero adjustment that can be accomplished with the circuit shown in Figure P17.19(b). The resistance of the potentiometer should be approximately 10% of the value of R_F.

17.28 Design an odometer to be used on a bicycle. A Hall-effect switch is mounted near the rear wheel. A magnet is mounted on one of the spokes of the rear wheel. As the magnet passes the generator, a 5-V pulse is generated. Hence, each time the wheel revolves once, the Hall-effect switch generates a 5-V pulse. With a nominal 26-in. wheel (including the effect of the tire), the bicycle travels 7 ft each time the wheel turns through one revolution. Thus, each pulse corresponds to 7 ft of travel by the bicycle. Since so few pulses are generated by the Hall-effect generator, use pulse multiplication as discussed in Section 17.8.

The readout is to be three 7-segment LED displays connected so that the least significant digit reads 100-ft increments and the most significant digit reads 10,000-ft increments.

Consider the changes to the design required for the following improvements:

a. Change the output to read from 00.0 to 99.9 miles rather than feet.
b. Change the odometer to operate with variable bicycle wheel diameters.

17.29 Design a run-in-place odometer to exercise by running in one place inside your home. A normally open switch closes each time either of your feet strikes the pad. The switch within the pad is a mechanical device, so it must be debounced. The controls comprise a reset button to activate the device and a potentiometer to set the length of each individual's stride in feet. The output is shown on three 7-segment LED displays with the MSD equal to 10^4 ft and the LSD equal to 10^2 ft. Use a monostable 555 to debounce the switch, and use the pulse multiplication technique as discussed in Section 17.8 to increase the number of pulses. The number of pulses generated for each pulse from the mechanical switch should be related to the stride. The stride potentiometer sets the frequency of the astable 555.

For example, if a person, when running, travels 2.8 ft every time either foot hits the ground, the stride is 2.8 ft. In this case the frequency of the astable 555, in the pulse-multiplication circuit, should be set for 28. Hence, each time either foot strikes the pad, 28 pulses are generated for the counter.

17.30 Design a timing system to measure the time for athletes to run the 100-yard dash. As the athletes leave the starting line, they break a beam of light, and when the first runner reaches the 100-yard mark at the end of the race, another beam of light is broken. The system is shown in Figure P17.20. Two wires are brought to the timing circuit: line a from the starting point and line b from the finish line. The voltage on each line is normally +5 V. When the beam is broken, the voltage drops to 0 V. Display the time on three 7-segment LEDs with the least significant digit representing 0.1 s.

17.31 Design a pulse generator to operate in the range 0 to 99 Hz. The generated pulses must have a duty cycle (time high/time low) of approximately 1. Provide two 7-segment LEDs to display the frequency of the pulse generator output.

17.32 Design a temperature control system for a home air conditioner. The control system turns ON the refrigeration unit when the actual temperature, T_a, is greater than the desired temperature, T_d. The refrigeration unit is turned OFF when $T_a < T_d$. The control system must operate from 0° to 60°C. A temperature sensor, which you are not required to design, provides the actual temperature on two 4-line BCD outputs for the MSD and the LSD digits, as shown in Figure P17.21. The desired temperature is set by use of a three-position switch, which is shown in

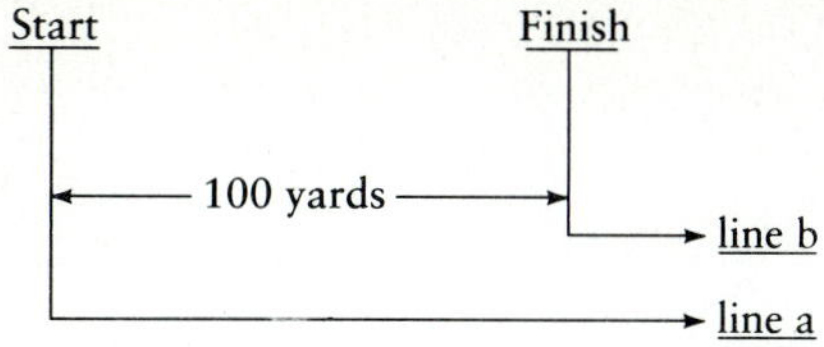

Figure P17.20

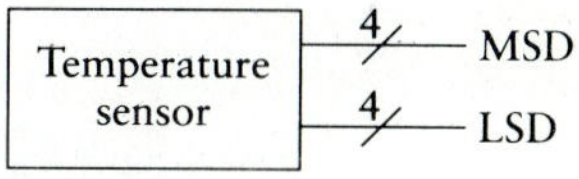

Figure P17.21

Figure P17.22

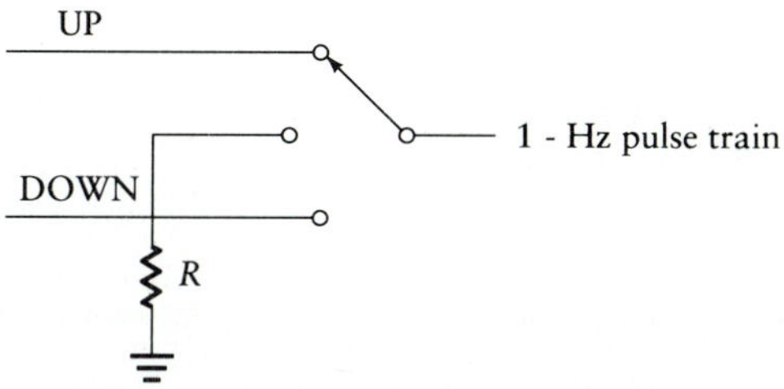

Figure P17.22. To increase the desired temperature, a 1-Hz pulse train is switched to the UP position; to decrease the desired temperature, the 1-Hz pulse is switched to the DOWN position. (The 1-Hz pulse train has a duty cycle of approximately unity.) The desired temperature is displayed on two 7-segment LEDs. The required output is a control voltage delivered to the refrigeration system. The refrigeration system is turned ON when a +5 V signal is delivered to it and OFF when 0 V is delivered to it.

SPICE

APPENDIX

A SPICE

A.0 INTRODUCTION

SPICE is the acronym for a Simulation Program with Integrated Circuit Emphasis developed by the Electronics Research Laboratory of the University of California and released to the public in July 1975. The simulation program is written in various computer languages. Through continuing research and suggestions from SPICE users, the program has been modified and updated to expand its simulation capability. The SPICE program reads input data, processes the data in batch mode on a mainframe computer, and produces the output in tabular or graphic form on an ordinary line printer. A SPICE program is also available to run on an IBM PC or equivalent.

SPICE is a computer-aided simulation program—that is, the engineer designs the circuit and then inputs it into SPICE for simulation. This, of course, is a great aid to the designer, since the results of the program illustrate how the circuit will react to certain inputs. With computer simulation, it is not necessary to build and test a circuit before completing the design. We can change circuit components and find the effects of the changes on circuit performance. We can use additional approximations to speed up the design process and then let the SPICE program determine what effect the approximations have on the desired output.

The SPICE program was designed to determine the unknown parameters by using Kirchhoff's current equations (nodal analysis). Performance of this type of analysis requires the transistors and diodes to be characterized by an equivalent circuit. All other models for the transistors and diodes are represented by using nonlinear exponential functions. For example, the BJT model used in SPICE is based on the Ebers-Moll injection model with various high-injection effects defined by Gummel and Poon. The type of equivalent circuit to be used is built into the program and is determined by the type of transistor or diode specified in the program directions. Once this is accomplished, the program can be run.

The first operation the SPICE program performs is determining the dc operating point, or Q-point, by using the Newton-Raphson method of matrix solution. After this has been completed, the program considers the ac input and performs the necessary calculations for a single frequency. If solution for more than one frequency is required, the program continues for each frequency specified in the program directions. The solution determined in the dc analysis provides the initial conditions for the transient analysis, which makes use of implicit numerical integration techniques for its solution. Since SPICE uses nodal analysis as the method of solution, it does have some limitations, as follows:

1. Only voltage-controlled current sources can be used. All other sources must be converted to voltage-controlled sources, either internally or externally.
2. Since SPICE analysis is based on a matrix operation using the admittance functions, a zero resistance cannot be included or the matrix will have a zero determinant and the solution cannot be found.
3. Multiterminal elements that have no admittance matrix (e.g., an ideal transformer) representation must be converted to allow an admittance matrix to exist.
4. Only voltage-controlled nonlinear resistors can be included in the circuit; items such as neon lamps and SCRs cannot be evaluated using nodal analysis.

Circuit behavior can be simulated with respect to time, frequency, and voltage variations. This is accomplished by specifying how the SPICE program will analyze the circuit through the use of *analysis statements.* SPICE permits the following forms of analyses: nonlinear dc, nonlinear transient, and linear ac small signal. These analyses can be conducted at various specified temperatures.

A dc analysis determines the node voltages in the circuit with the inductors shorted and capacitors open. This analysis is accomplished with only dc power applied. The dc analysis is performed prior to transient or ac analysis. The dc analysis determines the initial conditions required for the transient analysis. The dc analysis determines the linearized small-signal models of the nonlinear devices for use in ac analysis. A dc analysis can also be used to determine the small-signal value of the transfer function, the dc transfer curves, and the dc small-signal sensitivities of specified output variables. The dc analysis options are specified by the .DC, .TF, .OP, and .SENS control statements.

We compute the voltage and current output variables as a function of time over a specified time interval using a transient analysis. All sources that are not time dependent are set to their dc values. A Fourier analysis of the output waveform can be specified to obtain the frequency-domain Fourier coefficients for large-signal sinusoidal simulations. The transient time interval and Fourier analysis options are defined by the .TRAN and .FOURIER control statements.

We compute the voltage and current variables as a function of frequency using an ac small-signal analysis. The designer specifies the range of frequencies for which this analysis is to be conducted. The output of the ac analysis is a transfer function (e.g., voltage gain, transimpedance). If the circuit has only one ac input, it is convenient to set the input to unity magnitude and zero phase. In this manner, the output variables have the same value as the transfer function of the output variable with respect to the input. The ac analysis can also be used to analyze the noise generated by the resistors and semiconductor devices and the distortion characteristics of the circuit in the small-signal mode.

Once the circuit file is created by statements describing the circuit and the type of analysis of the circuit is defined, the simulation can be accomplished. The results of the simulation can be displayed in either tabular or graphic form.

MICRO-CAP II, Student Edition, is an analog circuit simulator with fully interactive graphics. Its SPICE-based circuit analysis program contains models of many popular electronic devices. From the standpoint of the user, a primary difference between MICRO-CAP II and SPICE is in the method of input of circuit information. Another difference is that SPICE is in the public domain, whereas MICRO-CAP II, Student Edition, is available from Addison-Wesley or Benjamin/Cummings Publishing Company.

MICRO-CAP II uses fully interactive graphics with a drawing board on which the circuit is constructed. For example, to draw a resistor, a set of simple commands is used to select the component and set its orientation (right, left, up, down) and its value in ohms. A similar process is used to draw other passive or active components.

The program can perform a transient analysis, ac analysis (steady state), dc analysis, or Fourier analysis. In each case, a prologue menu allows the user to set various parameters, including simulation run times, output graph scales, and various graphing options.

Most of this appendix is devoted to a description of SPICE, but we shall rerun the solution for the fourth example (the fourth-order Chebyshev lowpass filter of Section A.3.4) using MICRO-CAP immediately following the SPICE solution.

A.1 PROGRAMMING INFORMATION

Each computer has a unique set of instructions for edit programs. These instructions must be known before attempting to input circuit simulations into the program. In this appendix, we describe circuits that contain resistors, capacitors, inductors, independent voltage and current sources, four types of dependent sources, and the four most common semiconductor devices: diodes, BJTs, JFETs, and MOSFETs.

Many of the parameters of the Gummel-Poon model for transistors are not needed in the simple evaluation of circuit performance. The SPICE program sets values for the parameters not defined by the user in the model statements for the transistor. These default values should be available for review in the instructions for the particular version of the SPICE program being used. The five programming areas are introduced in the following sequence:

1. Element description
2. Source description
3. Subcircuits
4. Analysis request
5. Output request

A.1.1 Format

SPICE uses a free format for inputs. Data fields are separated by one or more delimiters (a block, a comma, an equal sign, or a left or right parenthesis). A statement may be continued by entering a plus sign in column 1 of the line immediately following the statement line to be continued. A name field must begin with a letter and must not contain any delimiters. Only the first eight characters of the name are used for identification. A number field may contain an integer (e.g., 2, −35, 77), a floating-point number (e.g., 3.1416, 5.3), either an integer or floating-point number followed by an integer exponent (1E3, 3.2E-4), or either an integer or a floating-point number followed by one of the scale factors presented next.

Scale Factors

MIL = 2.54E-6	M = 1E-3
K = 1E3	U = 1E-6
MEG = 1E6	N = 1E-9
G = 1E9	P = 1E-12
T = 1E12	F = 1E-15

Letters immediately following a number (that are not scale factors) are ignored and letters immediately following a scale factor are ignored. Hence, 10, 10V, 10 VOLTS, and 10 HZ all represent the same number. In addition, M, MA, MSEC, and MMHOS all represent the same scale factor, 10^{-3}. A comment statement may be indicated by entering an asterisk in the first column.

A.1.2 Circuit Description

Each element in the circuit to be analyzed must be defined by node numbers. All nodes in the circuit are numbered. Node numbers need not be consecutive.

The 0 node is the ground or reference node. The circuit is described to SPICE by a file of element statements, which define the circuit topology and element values, and a set of control statements, which define the model parameters and the run controls. The first statement in the file must be a title statement and the last must be the end statement (.END). Statements in between, except for continuations, may be in any order. Each element in the circuit is specified by an element statement that contains the element name, the circuit nodes to which the element is connected, and the values of the parameters that determine the electrical characteristics of the element. The first letter in the element name specifies the element type. The strings xxxxxxx and yyyyyyy denote a user-specified alphanumeric reference name for the applicable element. For example, a resistor, capacitor, and inductor may be defined as RLOAD, C3ST1, and L3ST2, respectively. Data fields enclosed in < > are optional. The element nodes must be nonnegative integers, which need not be numbered sequentially, and the ground node must always be numbered 0. Every node must have at least two connections and a dc path to ground when capacitors open and inductors short. With respect to branch voltages and currents, SPICE uses sink reference (i.e., current flows in the direction of voltage drop).

A.2 INPUT DATA

In this section, we define how the input data are formatted to be accepted by the SPICE program.

A.2.1 Element Description

Each element requires a definition of the characteristics of the device so that the SPICE program can properly analyze the circuit.

A.2.1.1 Passive Elements The element descriptions and the associated formats are shown in the following table:

Type	**Format**			
Resistor	Rxxxxxxx	n1	n2	value
Capacitor	Cxxxxxxx	n+	n−	value
Inductor	Lxxxxxxx	n+	n−	value
Diode	Dxxxxxxx	n+	n−	value

The element designations are shown in Figure A.1. The resistance dependence on temperature is given by adding TC1 and TC2 to the value (TC = 0.001, 0.013). The diode junction potential, reverse breakdown voltage, and ohmic resistance values are reset to applicable values by using the MODEL statement.

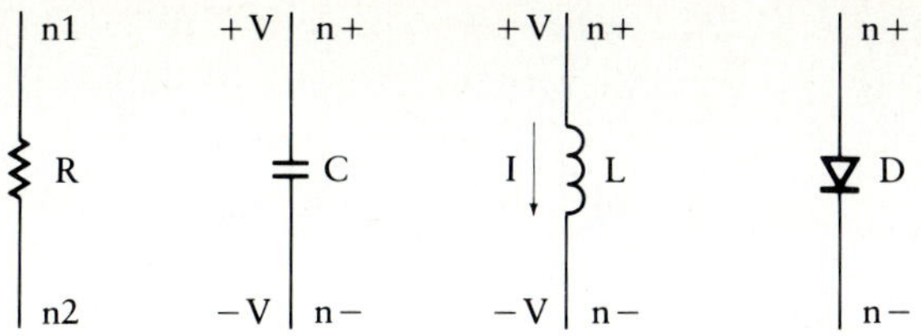

Figure A.1
Element designations.

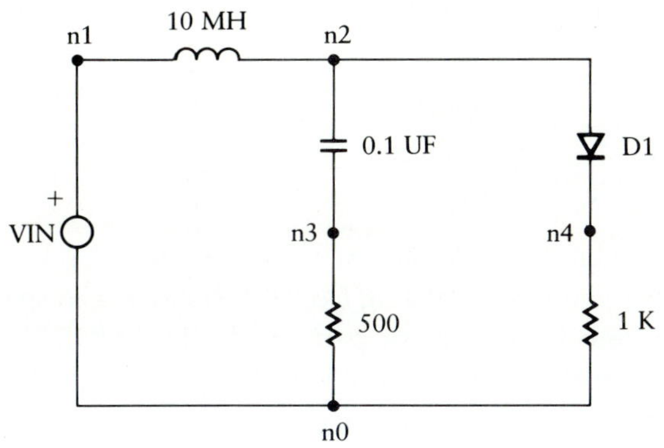

Figure A.2
Example of circuit with passive components.

A model name (MODNAME) is entered to reference the corresponding MODEL statement. (Refer to Subsection A.2.1.3 for MODEL statements and typical diode default values.)

We illustrate the procedure for an example of a passive circuit, as shown in Figure A.2. Note that the nodes have been numbered n0 to n4. The format used for entering the element descriptions for SPICE evaluation is shown as follows (we are printing element names without subscripts since that is how they appear in computers. Thus, for example, V_{in} is listed as VIN):

```
RCL SET UP
*ELEMENT DESCRIPTION
R1          3       0       500
RD          4       0       1K
LIN         1       2       10MH
CBATTERY    2       3       .1UF
D1          2       4       MOD1
.MODEL      MOD1    D       (VJ = 0.7)
*SOURCE DESCRIPTION
VIN         1       0 DC    5
.END
```

A.2.1.2 Active Components The identification of the transistor in the format for active devices is as follows: The collector (or drain) comes first, the base (or gate) is second, and the emitter (or source) is last. This is illustrated as follows:

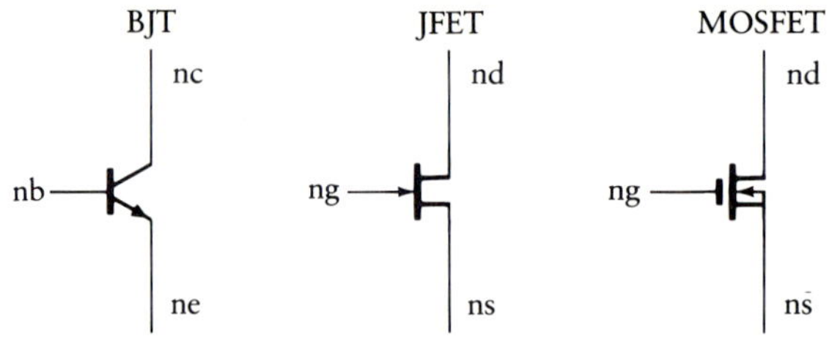

Figure A.3 Active element designations.

Type	**Format**				
BJT	Qxxxxxxx	nc	nb	ne	modname
JFET	Jxxxxxxx	nd	ng	ns	modname
MOSFET	Mxxxxxxx	nd	ng	ns	modname

Active element designations are shown in Figure A.3. A model name is entered along with the corresponding MODEL statement to specify the type of transistor (*npn*, *pnp*, *n*-channel, or *p*-channel). The MODEL statement may also be used to reset the maximum value of beta, the reverse early voltage, or the base-emitter breakdown voltage of the BJT and to reset the threshold voltage or transconductance of the FET devices. There are other parameters that can be reset, depending on what type of analysis is to be conducted. (Refer to Subsection A.2.1.3 for the MODEL statements and typical transistor default values.)

An example of a transistor amplifier is shown in Figure A.4. Note that the nodes have been numbered n0 to n6. The format used for entering the element descriptions for SPICE evaluation is shown next:

```
EMITTER FOLLOWER SET UP
*ELEMENT DESCRIPTION
R1      2    0    10K
R2      2    3    10K
RC      3    4    2K
RE      5    0    100
RL      6    0    100
CIN     2    1    22UF
C0      5    6    47UF
Q1      4    2    5 EFAM1
•MODEL EFAM1 NPN (BF=75)
*SOURCE DESCRIPTION
VCC     3    0    DC 12
VIN     1    0    AC
.END
```

A.2.1.3 Semiconductor Devices There are many parameters that can be defined in the SPICE MODEL statements—approximately 40 parameters for BJTs, 12 for JFETs, and 38 for MOSFETs. The particular version of the SPICE

Figure A.4
Example of transistor amplifier.

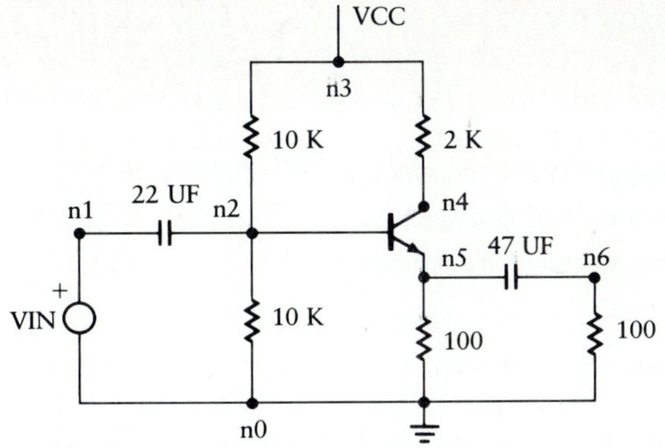

program instructions identifies the parameters that can be used and the typical default values assigned by that program. The MODEL statement abbreviations are shown next.

MODEL FORMAT

.MODEL modname type<pname1 = pvalue1 pname2 = pvalue2 pname3 = pvalue3>

Type	Applicable Device
NPN	*npn* BJT
PNP	*pnp* BJT
D	Diode
NJF	*n*-channel JFET
PJF	*p*-channel JFET
NMOS	*n*-channel MOSFET
PMOS	*p*-channel MOSFET

MODEL PARAMETER OPTIONS LIST (major parameters)

Parameter	pname	MODEL	pvalue DEFAULT
Junction potential (V_γ)	VJ	Diode	1V
Reverse breakdown voltage	BV	Diode	Infinite
Ohmic resistance (R_f)	RS	Diode	0 ohms
Forward early voltage	VAF	BJT	Infinite
Max forward beta	BF	BJT	100
Base-emitter voltage (V_{BE})	VJE	BJT	.75 V
Base-collector capacitance ($C_{B'C}$)	CJC	BJT	0
Base-emitter capacitance ($C_{B'E}$)	CJE	BJT	0
Forward transit time (τ_F)	TF	BJT	0
*Threshold voltage (V_P)	VTO	JFET	−2
Transconductance (k_n)	BETA	JFET	1E-4 A/V^2

G-S junction capacitance	CGS	JFET	0
G-D junction capacitance	CGD	JFET	0
*Threshold voltage (V_P or V_{TH})	VTO	MOSFET	0
Transconductance	KP	MOSFET	2 A/V^2
Substrate-D J. Capacitance	CBD	MOSFET	0
Substrate-S J. Capacitance	CBS	MOSFET	0

*When current in transistor is zero.

The model name (modname) included in the MODEL statement corresponds to the model name specified in the active component and diode description statements. One MODEL statement may be used to reference all devices with the same model name provided the device specifications are identical. For example, the MODEL statements for the circuit shown in Figure A.5 are presented below.

```
CSDC SET UP
*ELEMENT DESCRIPTION
R1        2        1      50
RT1       7        8      1
RT2       6        7      1
RB1       9        0      700
RB2       4        5      700
CIN       3        2      327U
C0        7        10     995U
D1        3        9      MOD1
D2        4        3      MOD1
Q1        5        4      6) MODTOP
Q2        0        9      8) MODBOT
.MODEL    MOD1            D(VJ=.7 RS=10)
.MODEL    MODTOP          NPN(BF=75 VJE=.7)
.MODEL    MODBOT          PNP(BF=75 VJE = .7)
*SOURCE DESCRIPTION
VCC       5        0      DC 12
VIN       1        0      AC    .1M 90 DEGREES
.END
```

A.2.2 Source Description

The sources that we encounter in electronic circuits are voltage and current sources. We now discuss how both linear independent and dependent voltage and current sources are entered into the SPICE program.

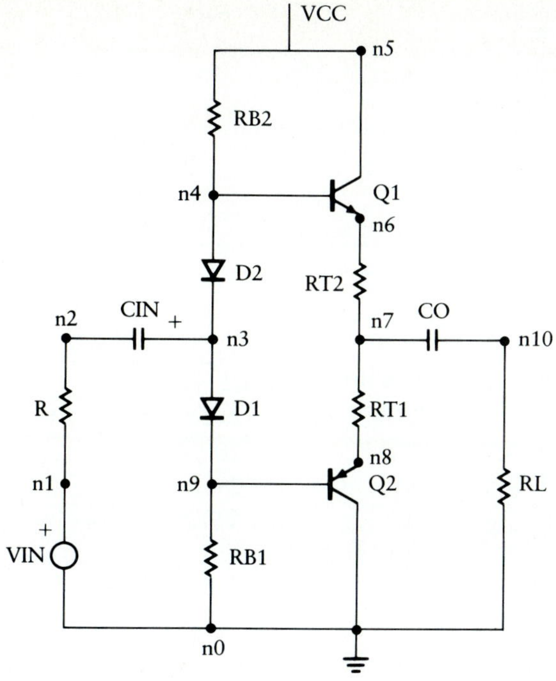

Figure A.5
Use of MODEL statements.

A.2.2.1 Dependent Linear Sources The four dependent-type voltage and current sources are listed below.

Source Type	Format
Voltage-controlled current source	Gxxxxxxx n+ n− nc+ nc− value
Voltage-controlled voltage source	Exxxxxxx n+ n− nc+ nc− value
Current-controlled current source	Fxxxxxxx n+ n− vname value
Current-controlled voltage source	Hxxxxxxx n+ n− vname value

The source control is defined by the controlling voltage nodes (nc+ and nc−) or the name of the voltage sources through which the controlling current is flowing (vname). If no voltage source is contained within the controlling current loop, a dc source with a value of zero can be used to measure the controlling current. The direction of positive controlling current flow is from the positive node, through the circuit, and to the negative node of the applicable voltage source. A current source of positive value forces current to flow from the positive node, through the source, to the negative node. The output values refer to the transconductance, voltage gain, current gain, or transresistance, respectively. We use the circuit of Figure A.6 as an example of how a dependent current source is formatted for input to SPICE. The input format is as shown next:

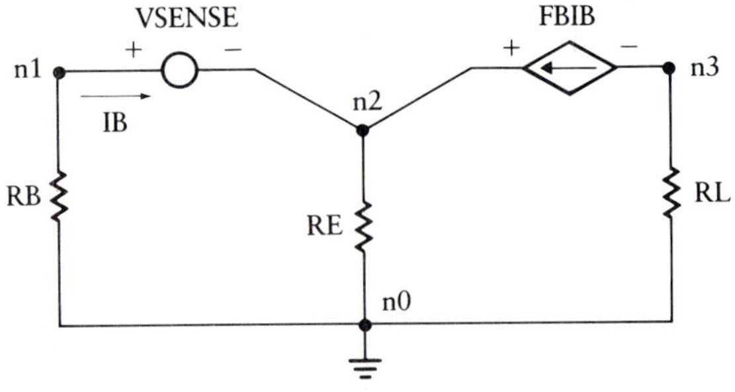

Figure A.6
Example of dependent current source.

```
DEPENDENT SOURCE SET UP
RB        1    0    20K
RE        2    0    100
RL        3    0    2K
FBIB      3    2    VSENSE 100
VSENSE    1    2
.END
```

A.2.2.2 Independent Linear Sources There are two linear sources, voltage and current, which are either ac or dc. The format for these sources is shown as follows:

Source Type	Source Input
Voltage	Vxxxxxxx n+ n− <<DC> d/t><AC mag <phase>>
Current	Ixxxxxxx n+ n− <<DC> d/t><AC mag <phase>>

An independent source is assigned an ac magnitude and phase (mag, phase), which is excited only during ac analysis, dc value (d/t) for determining the small-signal bias solution for ac and transient analysis, and/or a time-dependent function (d/t) for transient and dc (using time-zero value) analysis. If the ac magnitude is omitted following the keyword AC, a value of unity is assumed. If the ac phase is omitted, a value of zero is assumed. The time-dependent functions may be sinusoidal or pulse trains. Positive current is assumed to flow from the positive node, through the circuit, to the negative node. A current source of positive value forces current out of the positive node, through the source, and into the negative node. Voltage sources, in addition to exciting the circuit, may be used as ammeters for SPICE. That is, a zero-value voltage source may be inserted for the purpose of measuring current.

Following are examples of the format for time-dependent and pulse sources.

SINUSOIDAL FORMAT

SIN (vo va freq td theta)

Parameter	Default Values
vo (offset)	0
va (amplitude)	1
freq (frequency)	1/tstop
td (delay time)	0
theta (damping factor)	0

PULSE FORMAT

PULSE (v1 v2 td tr tf pw period)

Parameter	Default Values
v1 (initial value)	—
v2 (pulsed value)	—
td (delay time)	0
tr (rise time)	tstep
tf (fall time)	tstep
pw (pulse width)	tstop
period	tstop

The term *tstep* is defined as the printing increment and *tstop* is the final time of print. For a detailed explanation of tstep and tstop, refer to Subsection A.2.4.3.

SPICE

A.2.3 Subcircuits

Complicated circuits may be reduced by defining subcircuits and then interconnecting them with the main circuit in as many locations as needed. The subcircuit description must be placed between a SUBCKT statement and an ENDS statement. Subcircuits may be nested within other subcircuits. All subcircuit element nodes not included in the SUBCKT statement are strictly local to that subcircuit, with the exception of node 0 (ground), which is global. All device models are also local to the applicable subcircuit. The element nodes defined in the SUBCKT and CALL statement must not be labeled zero (ground). Control statements (those statements defined in the analysis request and output request sections) must not appear within the subcircuit description. The definition and CALL formats are as follows:

DEFINITION FORMAT

```
.SUBCKT     subname  n1     <n2  n3...>
                                subcircuit description
.ENDS       <subname>
```

The subcircuit name (subname) is required only for the ENDS statement where subcircuits are nested within subcircuits.

CALL FORMAT

Xyyyyyyy n1 <n2 n3...> subname

The external nodes (n1 <n2 n3...>) are the main circuit nodes, which correspond directly to the specified subcircuit nodes. Refer to the example of Section A.3.4 for demonstration of the use of subcircuits.

A.2.4 Analysis Request

Now that we have finished discussing techniques for describing the circuit, we are ready to use the program for analysis. Three types of analysis are used, as follows: dc analysis, ac analysis, and transient analysis. We discuss each of these individually.

A.2.4.1 dc Analysis The dc analysis statement defines a dc voltage or current range to an existing INDEPENDENT, ac or dc, voltage or current source (srcname). The circuit is then analyzed at every increment (incr) within the specified dc voltage or current range (start stop). Optionally, a second source (src2) may be specified with associated sweep parameters. In this case, the first source is swept over its range for each value of the second source. This analysis request is useful for obtaining semiconductor-device output characteristics. The statement .OP must be inserted if only dc analysis is being performed. This type of analysis request forces SPICE to determine the dc operating point for the specified input dc voltage with the capacitors open and the inductors shorted. Prior to transient analysis, a dc operating-point analysis is automatically performed in order to determine the initial conditions. The dc analysis is also performed prior to an ac small-signal analysis to determine the linearized, small-signal models for nonlinear devices. To obtain the dc small-signal value of the transfer function (output/input, input resistance, and output resistance), we use the .TF analysis. This is summarized as follows:

DC FORMAT

.DC srcname start stop inc <src2 start2 stop2 incr2>
.OP
.TF V (n1,n2) VIN
.TF I(VLOAD) VIN

A.2.4.2 ac Analysis The ac statement is used to perform a circuit analysis over a specified frequency range. The PLOT output statement produces a Bode plot. The frequency scale may be calibrated in decades (DEC), octaves (OCT), or linear (LIN). Before the ac analysis is performed, a small-signal bias solution is automatically performed to determine the linearized small-signal models for nonlinear devices. To make this analysis meaningful, at least one independent source must be specified with an ac value. The starting and stopping points need to be defined for this analysis. The resultant circuit is then analyzed at a number of points per decade (nd), octave (no), or linear (np) over the specified frequency range (fstart fstop). The ac analysis is insensitive to rail voltages. Therefore, to simplify work, a unity voltage or current source is used so that the output becomes the same as the transfer function. That is,

H(S) = OUTPUT(S)/INPUT(S) = OUTPUT(S)/1
H(S) = OUTPUT(S)

AC FORMAT

```
.AC   DEC   nd   fstart   fstop
.AC   OCT   no   fstart   fstop
.AC   LIN   np   fstart   fstop
```

A.2.4.3 Transient Analysis The transient analysis portion of SPICE is used to compute the transient output variables as a function of time over a specified time interval (tstop <tstart>). A small-signal bias solution is automatically performed first, using the time-zero value for time-dependent sources. Then the circuit is analyzed at each increment (tstep) within the specified time range. When tstart is omitted, it is assumed to be zero. Transient analysis is sensitive to rail voltages, and clipping therefore occurs when supply voltages are exceeded. Symbols for transient analysis are as follows:

TRANSIENT ANALYSIS

```
.TRAN   tstep   tstop   <tstart>
```

A.2.5 Output Request

After SPICE accomplishes the analysis, we record the data to be analyzed or stored for future use. This is accomplished by having the SPICE program either print or plot on a printer, using the PRINT or PLOT statements.

A.2.5.1 Print The PRINT statement produces a tabular listing of the specified output variables (0V1 <0V2 .... 0V8>), a current flowing through an INDEPENDENT voltage source, or a specified voltage to be printed. The type of analysis to be printed (prtype) (ac, dc, noise, distortion, or transient) must also be defined with the following:

PRINT FORMAT

.PRINT   prtype   ov1   <ov2...ov8>

OUTPUT VARIABLE FORMAT

Parameter	Format
Voltage	V (n+ <,n− >)
Current	I (vname)

If the negative node of the output voltage is not indicated, ground is assumed. The direction of positive current flow is from the positive node, through the circuit, to the negative node of the applicable voltage source (vname). A maximum of eight output variables may be defined in one PRINT statement, but there is no limit to the number of PRINT statements for each type of analysis.

The various ac output variable parameters that may be defined are as follows:

AC OV FORMAT

Parameter	Voltage	Current
Real part	VR (n+ <,n− >)	IR (vname)
Imaginary part	VI (n+ <,n− >)	II (vname)
Magnitude	VM (n+ <,n−>)	IM (vname)
Phase	VP (n+ <,n−>)	IP (vname)
Decibels	VDB (N+ <,n−>)	IDB (vname)

Example Formats

.PRINT AC VM(1) VP(1)
.PRINT TRAN I(VIN) V(3.4)
.PRINT DC V(1,2)

In the example formats, the first PRINT example statement prints the ac magnitude and phase values of the voltage at node 1 with respect to ground. The next statement prints the transient values of the current through the INDEPENDENT voltage source VIN and the voltage between nodes 3 and 4. The last statement prints the dc values for the voltage between nodes 1 and 2.

A.2.5.2 Plot The PLOT statement causes the value of the defined output variables (0V1, 0V2 . . . 0V8) to be plotted. The plot type (pltype) is either AC, DC, or TRANSIENT. The PLOT statement output variable syntax and sign conventions are the same as those for the PRINT statement with the exception of optional upper- and lower-bound limits (plo, phi). All output variables to the left of the limits are plotted using the specified scale. Without the optional limits, SPICE automatically determines the minimum and maximum values of all output variables and scales the plot to fit. Multiple scaled plots and a legend are produced when appropriate. When more than one output variable appears on the same plot, the first output variables are printed as well as plotted. Two or more output variables with approximately the same value are indicated by an X on the plot. The plot format is as follows:

PLOT FORMAT

.PLOT pltype ov1 <(pl01,phi1)> <ov2 <(plo2,phi2)>...ov8>

A.2.5.3 Options Statement User-defined model parameters may be suppressed (NOMOD) and/or the node (NODE) can be printed using OPTIONS statements, as follows:

OPTIONS FORMAT

.OPTIONS <NOMOD> <NODE>

SPICE

A.3 EXAMPLES OF PROGRAMS

In this section, we provide examples of circuits that show the listing of the input data and output of each example.

A.3.1 CE Amplifier

A CE amplifier, as shown in Figure A.7, is analyzed over the frequency range of 1 to 100 kHz. The voltage source, VSENSOR, is used as an ammeter to

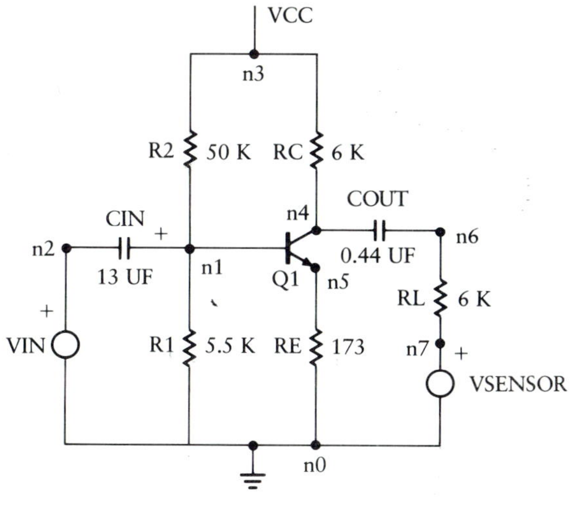

Figure A.7
CE amplifier.

measure the output current. You should study the printout and verify the following results:

$$A_v = 24 \text{ dB}$$

$$A_i = 10.5$$

$$f(\text{break}) = 30 \text{ Hz}$$

$$R_{in} = 4 \text{ k}\Omega$$

The listing is shown.


```
 FFIND,SPICE,PROJ1
1******* 86/04/30. *******  SPICE 2G.5 (10AUG81)  ******* 18.40.37.*****

0*  COMMON EMITTER

0****      INPUT LISTING                    TEMPERATURE =    27.000 DEG C

0***********************************************************************




 * ELEMENT DESCRIPTION

 RC   3   4   6K
 RL   6   7   6K
 RE   5   0   173
 R1   1   0   5.5K
 R2   3   1   50K
 CIN  2   1   13U
 COUT 6   4   .44U
 Q1  4   1   5  M2N2222A
 .MODEL M2N2222A NPN (VJE=.7,CJE=25PF,CJC=8PF,TF=2.5NS,RB=100,RC=10,BF=100)

 * SOURCE DESCRIPTION

 VCC    3   0   12
 VIN    2   0   AC
 VSENSOR   7   0

 * ANALYSIS REQUEST

 .AC  DEC    5   1   100MEGHZ

 .OPTIONS NODE

 * OUTPUT REQUESTS

 .PLOT  AC  I(VSENSOR)    I(VIN)
 .PLOT  AC  VDB(6)

 .END
1*************** 86/04/30. *********************** SPICE 2G.5 (10AUG81)  *********************** 18.40.37.**************

0*  COMMON EMITTER
0****                ELEMENT NODE TABLE                        TEMPERATURE =    27.000 DEG C
```

```
0*******************************************************************************************************************


0       0     RE        R1        VCC       VIN       VSENSOR  Q1
0       1     R1        R2        CIN       Q1
0       2     CIN       VIN
0       3     RC        R2        VCC
0       4     RC        COUT      Q1
0       5     RE        Q1
0       6     RL        COUT
0       7     RL        VSENSOR
1*************** 86/04/30. ***********************  SPICE 2G.5 (10AUG81)  *********************** 18.40.37.***************

0*  COMMON EMITTER
0****                 BJT MODEL PARAMETERS                                  TEMPERATURE =    27.000 DEG C

          M2N2222A
0TYPE       NPN
0IS       1.00E-16
0BF        100.000
0NF          1.000
0BR          1.000
0NR          1.000
0RB        100.000
0RC         10.000
0CJE      2.50E-11
0VJE          .700
0TF       2.50E-09
0CJC      8.00E-12
1*************** 86/04/30. ***********************  SPICE 2G.5 (10AUG81)  *********************** 18.40.37.***************

0*  COMMON EMITTER
0****                 SMALL SIGNAL BIAS SOLUTION                            TEMPERATURE =    27.000 DEG C

0*******************************************************************************************************************


 NODE   VOLTAGE     NODE   VOLTAGE     NODE   VOLTAGE     NODE   VOLTAGE     NODE   VOLTAGE     NODE   VOLTAGE     NODE   VOLTAGE


(  1)    1.1012    (  2)    0.0000    (  3)   12.0000    (  4)    1.3433    (  5)     .3103    (  6)    0.0000    (  7)    0.0000




    VOLTAGE SOURCE CURRENTS

    NAME        CURRENT


    VCC      -1.994E-03

    VIN        0.

    VSENSOR    0.


    TOTAL POWER DISSIPATION   2.39E-02  WATTS
1*************** 86/04/30. ***********************  SPICE 2G.5 (10AUG81)  *********************** 18.40.37.***************

0*  COMMON EMITTER
```

```
0****                  OPERATING POINT INFORMATION                      TEMPERATURE =   27.000 DEG C

0*****************************************************************************************************


0
0**** BIPOLAR JUNCTION TRANSISTORS


0            Q1
0MODEL       M2N2222A
 IB        1.78E-05
 IC        1.78E-03
 VBE           .791
 VBC          -.242
 VCE          1.033
 BETADC     100.000
 GM        6.87E-02
 RPI       1.46E+03
 RX        1.00E+02
 RO        1.00E+12
 CPI       2.16E-10
 CMU       7.33E-12
 CBX       0.
 CCS       0.
 BETAAC     100.000
 FT        4.89E+07
1*************** 86/04/30. *********************** SPICE 2G.5 (10AUG81) *********************** 18.40.37.***************

0*  COMMON EMITTER
0****                  AC ANALYSIS                                      TEMPERATURE =   27.000 DEG C

0*****************************************************************************************************


0LEGEND:

 *: I(VSENSOR)
 +: I(VIN)

     FREQ       I(VSENSOR)

  (*+)----------------   1.000E-05                1.000E-04                1.000E-03                 1.000E-02                1.000E-01
                             - - - - - - - - - - - - - - - - - - - - - - - - - - - - - - - - - - - - - - - - - - - - - - - - - - -
   1.000E+00   2.664E-05     .          *          +  .                        .                        .                        .
   1.585E+00   6.258E-05     .                   *    . +                      .                        .                        .
   2.512E+00   1.370E-04     .                        .  * +                   .                        .                        .
   3.981E+00   2.710E-04     .                        .       +  *             .                        .                        .
   6.310E+00   4.827E-04     .                        .        +       *       .                        .                        .
   1.000E+01   7.899E-04     .                        .         +           *  .                        .                        .
   1.585E+01   1.200E-03     .                        .         +              . *                      .                        .
   2.512E+01   1.669E-03     .                        .         +              .     *                  .                        .
   3.981E+01   2.088E-03     .                        .         +              .       *                .                        .
   6.310E+01   2.368E-03     .                        .         +              .        *               .                        .
   1.000E+02   2.514E-03     .                        .         +              .         *              .                        .
   1.585E+02   2.581E-03     .                        .         +              .         *              .                        .
   2.512E+02   2.609E-03     .                        .         +              .         *              .                        .
   3.981E+02   2.620E-03     .                        .         +              .         *              .                        .
   6.310E+02   2.624E-03     .                        .         +              .         *              .                        .
   1.000E+03   2.626E-03     .                        .         +              .         *              .                        .
```

```
  1.585E+03    2.627E-03    .                        .         +               .         *              .                        .
  2.512E+03    2.627E-03    .                        .         +               .         *              .                        .
  3.981E+03    2.627E-03    .                        .         +               .         *              .                        .
  6.310E+03    2.627E-03    .                        .         +               .         *              .                        .
  1.000E+04    2.627E-03    .                        .         +               .         *              .                        .
  1.585E+04    2.627E-03    .                        .         +               .         *              .                        .
  2.512E+04    2.627E-03    .                        .         +               .         *              .                        .
  3.981E+04    2.627E-03    .                        .         +               .         *              .                        .
  6.310E+04    2.627E-03    .                        .         +               .         *              .                        .
  1.000E+05    2.627E-03    .                        .          +              .         *              .                        .
  1.585E+05    2.626E-03    .                        .           +             .         *              .                        .
  2.512E+05    2.623E-03    .                        .             +           .         *              .                        .
  3.981E+05    2.617E-03    .                        .                +        .         *              .                        .
  6.310E+05    2.601E-03    .                        .                    +    .         *              .                        .
  1.000E+06    2.563E-03    .                        .                        +.         *              .                        .
  1.585E+06    2.473E-03    .                        .                         .   +     *              .                        .
  2.512E+06    2.284E-03    .                        .                         .      +*                .                        .
  3.981E+06    1.951E-03    .                        .                         .      *   +             .                        .
  6.310E+06    1.507E-03    .                        .                         .   *        +           .                        .
  1.000E+07    1.064E-03    .                        .                         .*            +          .                        .
  1.585E+07    7.110E-04    .                        .                    *    .             +          .                        .
  2.512E+07    4.653E-04    .                        .                 *       .             +          .                        .
  3.981E+07    3.066E-04    .                        .           *             .             +          .                        .
  6.310E+07    2.098E-04    .                        .       *                 .             +          .                        .
  1.000E+08    1.548E-04    .                        .    *                    .             +          .                        .
                            - - - - - - - - - - - - - - - - - - - - - - - - - - - - - - - - - - - - - - - - - - - - - - - - - -


Y
1***************  86/04/30. ***********************  SPICE 2G.5 (10AUG81)  *********************** 18.40.37.***************

0*  COMMON EMITTER
0****                AC ANALYSIS                                       TEMPERATURE =   27.000 DEG C

0****************************************************************************************************************************



     FREQ        VDB(6)

                       -2.000E+01               0.                       2.000E+01                4.000E+01                6.000E+01
                            - - - - - - - - - - - - - - - - - - - - - - - - - - - - - - - - - - - - - - - - - - - - - - - - - -
  1.000E+00   -1.593E+01    .    *                   .                        .                        .                        .
  1.585E+00   -8.509E+00    .             *          .                        .                        .                        .
  2.512E+00   -1.703E+00    .                      * .                        .                        .                        .
  3.981E+00    4.222E+00    .                        .    *                   .                        .                        .
  6.310E+00    9.237E+00    .                        .           *            .                        .                        .
  1.000E+01    1.351E+01    .                        .                *       .                        .                        .
  1.585E+01    1.715E+01    .                        .                    *   .                        .                        .
  2.512E+01    2.001E+01    .                        .                        *                        .                        .
  3.981E+01    2.196E+01    .                        .                        . *                      .                        .
  6.310E+01    2.305E+01    .                        .                        .   *                    .                        .
  1.000E+02    2.357E+01    .                        .                        .   *                    .                        .
  1.585E+02    2.380E+01    .                        .                        .    *                   .                        .
  2.512E+02    2.389E+01    .                        .                        .    *                   .                        .
  3.981E+02    2.393E+01    .                        .                        .    *                   .                        .
  6.310E+02    2.394E+01    .                        .                        .    *                   .                        .
  1.000E+03    2.395E+01    .                        .                        .    *                   .                        .
  1.585E+03    2.395E+01    .                        .                        .    *                   .                        .
  2.512E+03    2.395E+01    .                        .                        .    *                   .                        .
```

```
3.981E+03   2.395E+01  .                       .                         .    *                   .                        .
6.310E+03   2.395E+01  .                       .                         .    *                   .                        .
1.000E+04   2.395E+01  .                       .                         .    *                   .                        .
1.585E+04   2.395E+01  .                       .                         .    *                   .                        .
2.512E+04   2.395E+01  .                       .                         .    *                   .                        .
3.981E+04   2.395E+01  .                       .                         .    *                   .                        .
6.310E+04   2.395E+01  .                       .                         .    *                   .                        .
1.000E+05   2.395E+01  .                       .                         .    *                   .                        .
1.585E+05   2.395E+01  .                       .                         .    *                   .                        .
2.512E+05   2.394E+01  .                       .                         .    *                   .                        .
3.981E+05   2.392E+01  .                       .                         .    *                   .                        .
6.310E+05   2.387E+01  .                       .                         .    *                   .                        .
1.000E+06   2.374E+01  .                       .                         .    *                   .                        .
1.585E+06   2.343E+01  .                       .                         .   *                    .                        .
2.512E+06   2.274E+01  .                       .                         .  *                     .                        .
3.981E+06   2.137E+01  .                       .                         . *                      .                        .
6.310E+06   1.913E+01  .                       .                        *.                        .                        .
1.000E+07   1.610E+01  .                       .                    *    .                        .                        .
1.585E+07   1.260E+01  .                       .                *        .                        .                        .
2.512E+07   8.917E+00  .                       .           *             .                        .                        .
3.981E+07   5.295E+00  .                       .       *                 .                        .                        .
6.310E+07   2.000E+00  .                       .   *                     .                        .                        .
1.000E+08  -6.438E-01  .                      *.                         .                        .                        .
                       - - - - - - - - - - - - - - - - - - - - - - - - - - - - - - - - - - - - - - - - - - - - - - - - - -

Y
0
          JOB CONCLUDED
0          TOTAL JOB TIME                  1.63
$REVERT. SPICE COMPLETED
/
```

A.3.2 CS Amplifier

An *n*-channel JFET is used as a CS amplifier as shown in Figure A.8. You should study the printout and verify the outputs by calculations. The listings of the computer inputs and outputs are shown next.

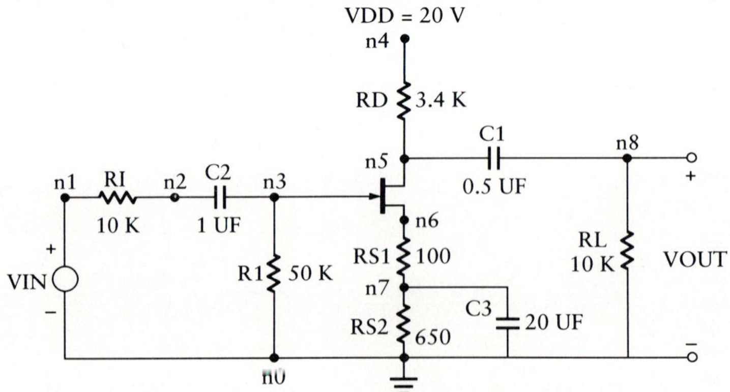

Figure A.8
JFET CS amplifier.

```
*******10/23/89 ******* IS SPICE  1.0  6/23/85 ******* 8:59: 0*****

 IS_ED CSAMP.CIR

 ****     INPUT LISTING                  TEMPERATURE =   27.000 DEG C

 ***********************************************************************



 *COMMON SOURCE AMPLIFIER

 VIN 1 0 1 AC
 VDD 4 0 20
 RI 1 2 10K
 R1 3 0 50K
 RD 4 5 3.4K
 RS1 6 7 100
 RS2 7 0 650
 RL 8 0 10K
 C1 5 8 .5U
 C2 2 3 1U
 C3 7 0 20U
 J1 5 3 6 MOD1
 .MODEL MOD1 NJF VTO=-4 BETA=.5M CGS=3P CGD=3P RD=250

 *ANALYSIS REQUESTS
 .AC DEC 5 20 5MEG

 *OUTPUT REQUESTS
 .PLOT AC VM(8)
 .END
***************10/23/89 *********************** IS SPICE  1.0  6/23/85 *********************** 8:59: 0***************

 IS_ED CSAMP.CIR
 ****               JFET MODEL PARAMETERS                              TEMPERATURE =   27.000 DEG C

 ***********************************************************************************************************



            MOD1
 TYPE       NJF
 VTO        -4.000
 BETA      5.00D-04
 RD         250.000
 CGS       3.00D-12
 CGD       3.00D-12
***************10/23/89 *********************** IS SPICE  1.0  6/23/85 *********************** 8:59: 0***************

 IS_ED CSAMP.CIR
 ****               SMALL SIGNAL BIAS SOLUTION                         TEMPERATURE =   27.000 DEG C

 ***********************************************************************************************************


 NODE   VOLTAGE     NODE   VOLTAGE     NODE   VOLTAGE     NODE   VOLTAGE     NODE   VOLTAGE     NODE   VOLTAGE     NODE   VOLTAGE
```

SPICE

```
(  1)    1.0000   (  2)    1.0000   (  3)     .0000   (  4)   20.0000   (  5)   11.8143   (  6)    1.8057   (  7)    1.5649

(  8)     .0000



    VOLTAGE SOURCE CURRENTS

    NAME       CURRENT


    VIN         .000D+00

    VDD      -2.408D-03


    TOTAL POWER DISSIPATION   4.82D-02  WATTS
***************10/23/89 *********************** IS SPICE  1.0  6/23/85 *********************** 8:59: 0***************

 IS_ED CSAMP.CIR
 ****                OPERATING POINT INFORMATION                        TEMPERATURE =   27.000 DEG C

 ********************************************************************************************************************



 **** JFETS


             J1
 MODEL      MOD1
 ID         2.41E-03
 VGS          -1.806
 VDS          10.009
 GM         2.19E-03
 GDS         .00E+00
 CGS        1.79E-12
 CGD        8.58E-13
***************10/23/89 *********************** IS SPICE  1.0  6/23/85 *********************** 8:59: 0***************

 IS_ED CSAMP.CIR
 ****                AC ANALYSIS                                        TEMPERATURE =   27.000 DEG C

 ********************************************************************************************************************



   FREQ      VM(8)

                    1.585D+00            1.995D+00            2.512D+00            3.162D+00            3.981D+00
                      - - - - - - - - - - - - - - - - - - - - - - - - - - - - - - - - - - - - - - - - - - - - - - -
  2.000D+01   1.713D+00  .     *             .                    .                    .                    .
  3.170D+01   2.493D+00  .                   .                   *.                    .                    .
  5.024D+01   3.126D+00  .                   .                    .                   *.                    .
  7.962D+01   3.497D+00  .                   .                    .                    .        *           .
```

```
 1.262D+02   3.675D+00   .                         .                         .                         .               X        .
 2.000D+02   3.752D+00   .                         .                         .                         .                  X     .
 3.170D+02   3.784D+00   .                         .                         .                         .                  X     .
 5.024D+02   3.796D+00   .                         .                         .                         .                   X    .
 7.962D+02   3.802D+00   .                         .                         .                         .                   X    .
 1.262D+03   3.804D+00   .                         .                         .                         .                   X    .
 2.000D+03   3.804D+00   .                         .                         .                         .                   X    .
 3.170D+03   3.805D+00   .                         .                         .                         .                   X    .
 5.024D+03   3.805D+00   .                         .                         .                         .                   X    .
 7.962D+03   3.805D+00   .                         .                         .                         .                   X    .
 1.262D+04   3.805D+00   .                         .                         .                         .                   X    .
 2.000D+04   3.805D+00   .                         .                         .                         .                   X    .
 3.170D+04   3.805D+00   .                         .                         .                         .                   X    .
 5.024D+04   3.804D+00   .                         .                         .                         .                   X    .
 7.962D+04   3.803D+00   .                         .                         .                         .                   X    .
 1.262D+05   3.801D+00   .                         .                         .                         .                   X    .
 2.000D+05   3.795D+00   .                         .                         .                         .                   X    .
 3.170D+05   3.780D+00   .                         .                         .                         .                 X      .
 5.024D+05   3.744D+00   .                         .                         .                         .                X       .
 7.962D+05   3.658D+00   .                         .                         .                         .              X         .
 1.262D+06   3.464D+00   .                         .                         .                         .        X               .
 2.000D+06   3.088D+00   .                         .                         .                     X   .                        .
 3.170D+06   2.508D+00   .                         .                         X                         .                        .
 5.024D+06   1.842D+00   .               X         .                         .                         .                        .
                         - - - - - - - - - - - - - - - - - - - - - - - - - - - - - - - - - - - - - - - - - - - - - - - - - - - -


         JOB CONCLUDED
          TOTAL JOB TIME            21.20
XXXXXXX10/23/89 XXXXXXX IS SPICE  1.0  6/23/85 XXXXXXX 8:59: 0XXXXX


 XXXX      INPUT LISTING                      TEMPERATURE =   27.000 DEG C

 XXXXXXXXXXXXXXXXXXXXXXXXXXXXXXXXXXXXXXXXXXXXXXXXXXXXXXXXXXXXXXXXXXXXXXX


.END
```

A.3.3 Full-Wave Rectifier

A full-wave rectifier, which is shown in Figure A.9, is designed using an op-amp subcircuit. The op-amp subcircuit, which is shown in Figure A.10, has a dependent voltage source that is a function of the differential input voltage, VIN. Two resistors of high value, RCM1 and RCM2, are used to maintain a dc path to ground. The circuit is analyzed to produce a dc transfer curve within the defined range of −5 to +5 V dc. The transient waveform is then plotted over one period. The computer inputs and outputs are shown next.

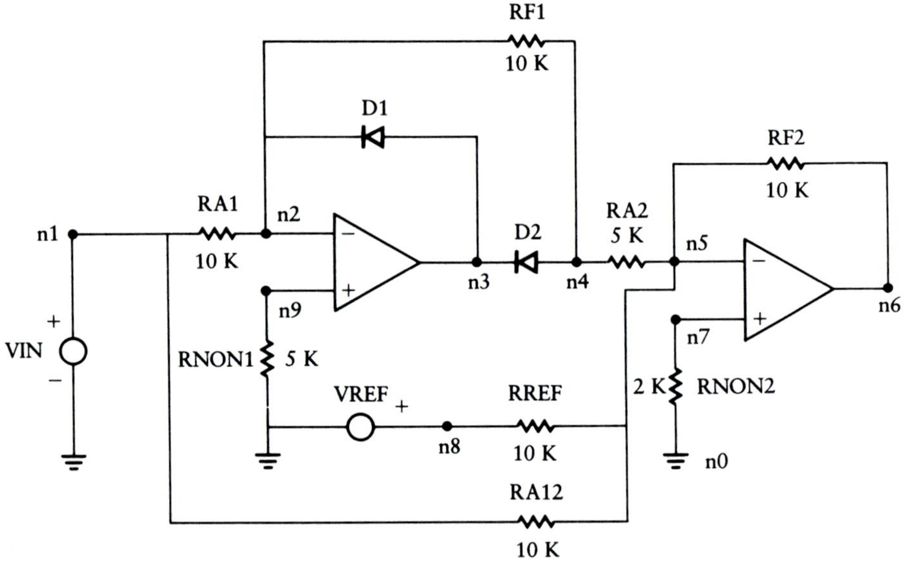

Figure A.9
Full-wave rectifier.

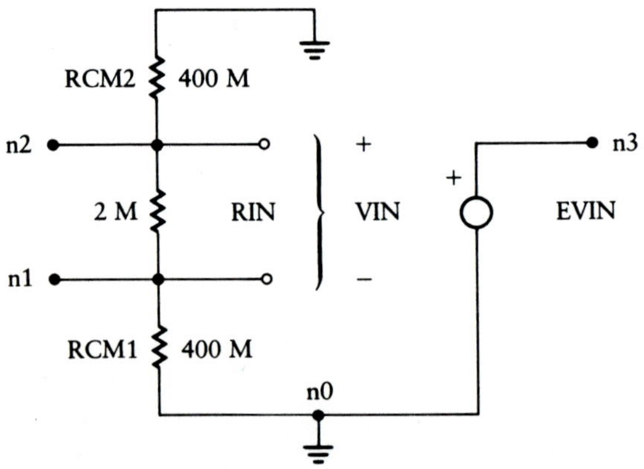

Figure A.10
Op-amp subcircuit.

```
FFIND,SPICE,PROJ4
1******* 86/04/30. *******  SPICE 2G.5 (10AUG81)  ******* 18.56.41.*****

0*  FULLWAVE RECTIFIER

0****     INPUT LISTING                    TEMPERATURE =    27.000 DEG C

0***********************************************************************


 *  SUBCIRCUIT DESCRIPTION
 .SUBCKT  OPAMP  1  2  3
 RIN    1  2  2MEG
 RCM1   1  0  400MEG
 RCM2   2  0  400MEG
```

```
EVIN  3  0  2  1  10E5
.ENDS  OPAMP
*  MAIN CIRCUIT ELEMENT DESCRIPTION
RA1    1  2  10K
RF1    4  2  10K
D1     3  2  MOD1
D2     4  3  MOD1
RA12  1  5  10K
RA2    4  5   5K
RF2    6  5  10K
RREF  8  5  10K
RNON1 9  0   5K
RNON2 7  0   2K
.MODEL  MOD1  D  (VJ=.7)
*  MAIN CIRCUIT SOURCE DESCRIPTION
VREF   8   0   2
VIN    1   0   SIN (0  10  60)
*  CALL SUBCIRCUIT OPAMP
X1  2  9  3  OPAMP
X2  5  7  6  OPAMP
*  ANALYSIS REQUESTS
.TRAN  .5M  16.5M
.DC  VIN  -5  5  .5
*  OUTPUT REQUESTS
.PLOT  TRAN  V(6)
.PLOT  DC    V(6)
.END
1*************** 86/04/30. ********************** SPICE 2G.5 (10AUG81) ********************** 18.56.41.**************

0*  FULLWAVE RECTIFIER
0****                 DIODE MODEL PARAMETERS                          TEMPERATURE =   27.000 DEG C

0*****************************************************************************************************************




            MOD1
0IS       1.00E-14
0VJ           .700

Y
1*************** 86/04/30. ********************** SPICE 2G.5 (10AUG81) ********************** 18.56.41.**************

0*  FULLWAVE RECTIFIER
0****                 INITIAL TRANSIENT SOLUTION                      TEMPERATURE =   27.000 DEG C

0*****************************************************************************************************************


 NODE   VOLTAGE     NODE   VOLTAGE     NODE   VOLTAGE     NODE   VOLTAGE     NODE   VOLTAGE     NODE   VOLTAGE     NODE   VOLTAGE


(  1)    0.0000    (  2)     .0000    (  3)    -.2636    (  4)     .0000    (  5)     .0000    (  6)   -2.0000    (  7)     .0000

(  8)    2.0000    (  9)     .0000




    VOLTAGE SOURCE CURRENTS
```

S P I C E

```
1*************** 86/04/30. *********************** SPICE 2G.5 (10AUG81) *********************** 18.56.41.***************

0* FULLWAVE RECTIFIER
0****                DC TRANSFER CURVES                                  TEMPERATURE =   27.000 DEG C

0*******************************************************************************************************************


  VIN         V(6)

                       -2.000E+00          0.                  2.000E+00           4.000E+00           6.000E+00
                      - - - - - - - - - - - - - - - - - - - - - - - - - - - - - - - - - - - - - - - - - - - - - - - - -
-5.000E+00   3.000E+00  .                   .                   .        *          .                   .
-4.500E+00   2.500E+00  .                   .                   .    *              .                   .
-4.000E+00   2.000E+00  .                   .                   *                   .                   .
-3.500E+00   1.500E+00  .                   .               *   .                   .                   .
-3.000E+00   1.000E+00  .                   .         *         .                   .                   .
-2.500E+00   5.000E-01  .                   .    *              .                   .                   .
-2.000E+00   4.058E-07  .                   *                   .                   .                   .
-1.500E+00  -5.000E-01  .              *    .                   .                   .                   .
-1.000E+00  -1.000E+00  .         *         .                   .                   .                   .
-5.000E-01  -1.500E+00  .    *              .                   .                   .                   .
-2.842E-14  -2.000E+00  *                   .                   .                   .                   .
 5.000E-01  -1.500E+00  .    *              .                   .                   .                   .
 1.000E+00  -1.000E+00  .         *         .                   .                   .                   .
 1.500E+00  -5.000E-01  .              *    .                   .                   .                   .
 2.000E+00  -1.067E-05  .                   *                   .                   .                   .
 2.500E+00   5.000E-01  .                   .    *              .                   .                   .
 3.000E+00   1.000E+00  .                   .         *         .                   .                   .
 3.500E+00   1.500E+00  .                   .               *   .                   .                   .
 4.000E+00   2.000E+00  .                   .                   *                   .                   .
 4.500E+00   2.500E+00  .                   .                   .    *              .                   .
 5.000E+00   3.000E+00  .                   .                   .        *          .                   .
                      - - - - - - - - - - - - - - - - - - - - - - - - - - - - - - - - - - - - - - - - - - - - - - - - -


     NAME        CURRENT


     VREF      -2.000E-04

     VIN        2.266E-10


     TOTAL POWER DISSIPATION   4.00E-04  WATTS
1*************** 86/04/30. *********************** SPICE 2G.5 (10AUG81) *********************** 18.56.41.***************

0* FULLWAVE RECTIFIER
0****                OPERATING POINT INFORMATION                         TEMPERATURE =   27.000 DEG C
0*******************************************************************************************************************


0
0**** VOLTAGE-CONTROLLED VOLTAGE SOURCES


0            EVIN.X1   EVIN.X2
 V-SOURCE     -.264    -2.000
 I-SOURCE  2.67E-10  2.00E-04
0
0**** DIODES
```

```
0              D1         D2
0MODEL        MOD1       MOD1
 ID       -2.74E-13  2.68E-10
 VD           -.264      .264
1*************** 86/04/30. ********************** SPICE 2G.5 (10AUG81) ********************** 18.56.41.**************

0*  FULLWAVE RECTIFIER
0****                TRANSIENT ANALYSIS                                    TEMPERATURE =   27.000 DEG C

0*****************************************************************************************************************
```

TIME	V(6)
0.	-2.000E+00
5.000E-04	-1.291E-01
1.000E-03	1.676E+00
1.500E-03	3.350E+00
2.000E-03	4.837E+00
2.500E-03	6.077E+00
3.000E-03	7.038E+00
3.500E-03	7.669E+00
4.000E-03	7.970E+00
4.500E-03	7.903E+00
5.000E-03	7.502E+00
5.500E-03	6.747E+00
6.000E-03	5.700E+00
6.500E-03	4.362E+00
7.000E-03	2.815E+00
7.500E-03	1.084E+00
8.000E-03	-7.471E-01
8.500E-03	-1.277E+00
9.000E-03	4.847E-01
9.500E-03	2.250E+00
1.000E-02	3.872E+00
1.050E-02	5.277E+00
1.100E-02	6.434E+00
1.150E-02	7.283E+00
1.200E-02	7.810E+00
1.250E-02	7.985E+00
1.300E-02	7.809E+00
1.350E-02	7.284E+00
1.400E-02	6.431E+00
1.450E-02	5.280E+00
1.500E-02	3.869E+00
1.550E-02	2.253E+00
1.600E-02	4.829E-01
1.650E-02	-1.372E+00

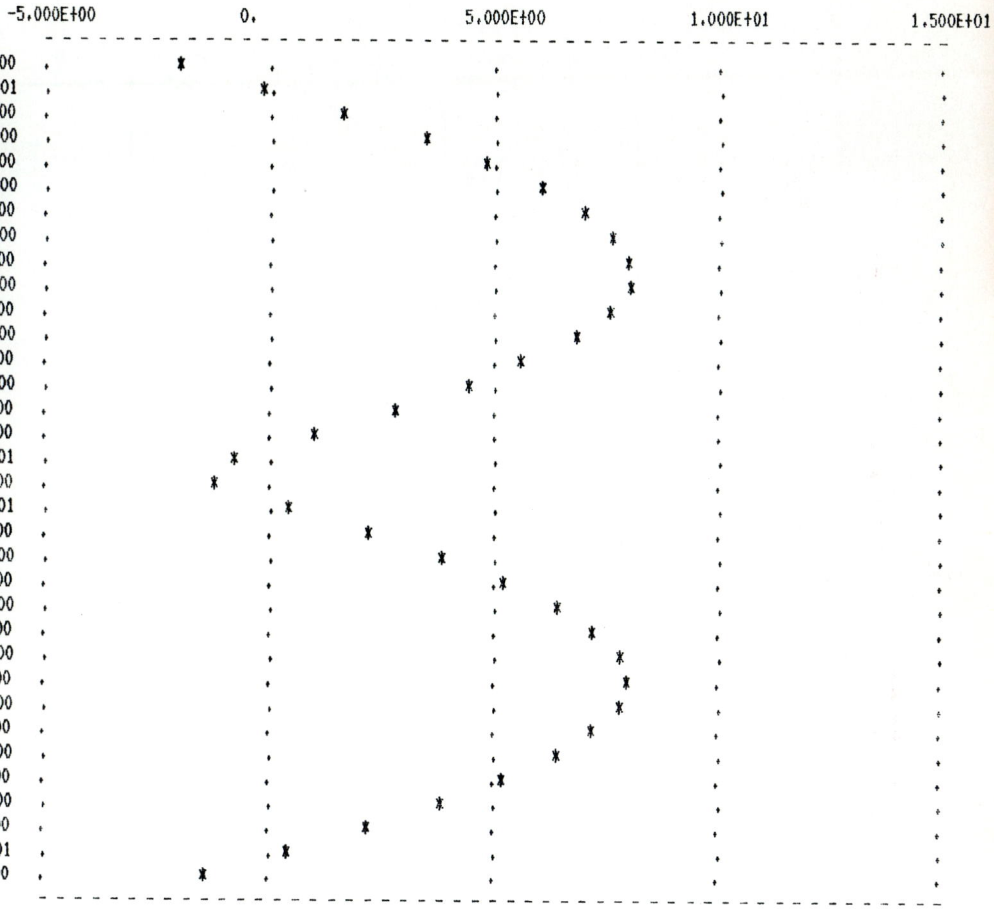


```
Y
0
         JOB CONCLUDED
0         TOTAL JOB TIME            4.52
$REVERT, SPICE COMPLETED
/
```

SPICE

A.3.4 Fourth-Order Chebyshev Low-Pass Filter

The filter shown in Figure A.11 is designed to have a 3 dB pass-band ripple, 3 dB cutoff of 1 kHz, and a maximum pass-band gain of 1. The printouts of the computer inputs and outputs are shown next.

Figure A.11

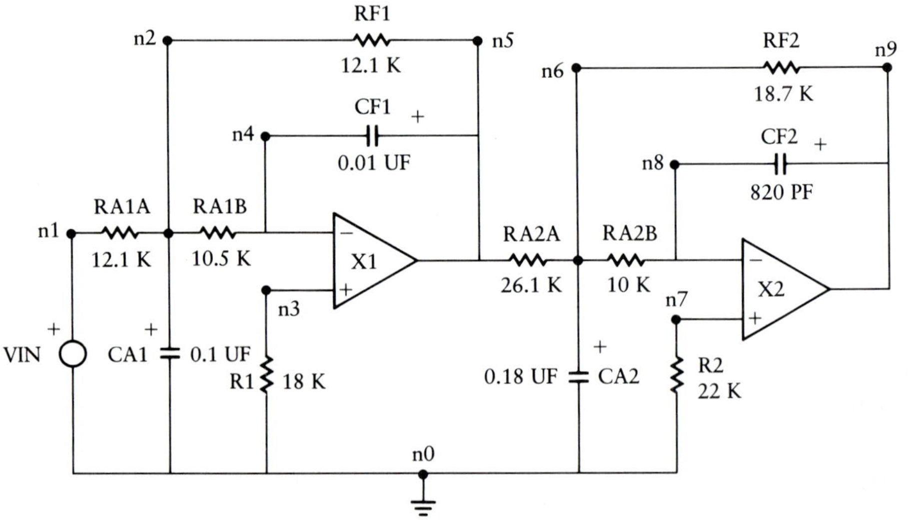


```
   PPROJ5
1******* 86/04/30. ******* SPICE 2G.5 (10AUG81) ******* 18.54.35.*****

0* FOURTH ORDER CHEBYSHEV LPF

0****      INPUT LISTING                    TEMPERATURE =   27.000 DEG C

0***********************************************************************



 * SUBCIRCUIT OPAMP DESCRIPTION
 .SUBCKT  OPAMP  1  2  3
 RIN    1   2   2MEG
 RCM1  1   0   400MEG
 RCM2  2   0   400MEG
 EVIN  3   0   2   1   10E5
 .ENDS OPAMP

 * MAIN CIRCUIT ELEMENT DESCRIPTION
 RA1A    1   2   12.1K
 RA1B    2   4   10.5K
 RF1     2   5   12.1K
 R1      3   0   18.0K
 CA1     2   0     .1U
 CF1     5   4    .01U
 RA2A    5   6   26.1K
 RA2B    6   8   10.0K
 RF2     6   9   18.7K
```

```
 R2     7   0   22.0K
 CA2    6   0     .18U
 CF2    9   8   820P

 *  CALL SUBCIRCUIT OPAMP
 X1    4    3    5    OPAMP
 X2    8    7    9    OPAMP

 *  SOURCE DESCRIPTION

 VIN  1   0   AC

 *  ANALYSIS REQUESTS

 .AC   DEC   15   100   2K

 *  OUTPUT REQUEST
 .PLOT  AC  VDB(9)  VDB(5)  VDB(9,5)
 .END
1*************** 86/04/30. ***********************  SPICE 2G.5 (10AUG81)  *********************** 18.54.35.**************

0*  FOURTH ORDER CHEBYSHEV LPF
0****                  SMALL SIGNAL BIAS SOLUTION                             TEMPERATURE =   27.000 DEG C

0*****************************************************************************************************************


  NODE   VOLTAGE     NODE   VOLTAGE     NODE   VOLTAGE     NODE   VOLTAGE     NODE   VOLTAGE     NODE   VOLTAGE     NODE   VOLTAGE


 (  1)    0.0000    (  2)    0.0000    (  3)    0.0000    (  4)    0.0000    (  5)    0.0000    (  6)    0.0000    (  7)    0.0000
 (  8)    0.0000    (  9)    0.0000


    VOLTAGE SOURCE CURRENTS

    NAME        CURRENT


    VIN       0.


    TOTAL POWER DISSIPATION   0.        WATTS
1*************** 86/04/30. ***********************  SPICE 2G.5 (10AUG81)  *********************** 18.54.35.**************

0*  FOURTH ORDER CHEBYSHEV LPF
0****                  OPERATING POINT INFORMATION                            TEMPERATURE =   27.000 DEG C

0*****************************************************************************************************************


0
0**** VOLTAGE-CONTROLLED VOLTAGE SOURCES


0            EVIN.X1   EVIN.X2
 V-SOURCE     0.000     0.000
 I-SOURCE  0.        0.
1*************** 86/04/30. ***********************  SPICE 2G.5 (10AUG81)  *********************** 18.54.35.**************

0*  FOURTH ORDER CHEBYSHEV LPF
0****                  AC ANALYSIS                                            TEMPERATURE =   27.000 DEG C
```

SPICE

```
0*****************************************************************************************************


0LEGEND:

 *: VDB(9)
 +: VDB(5)
 =: VDB(9,5)

    FREQ       VDB(9)

 (*)----------------- -6.000E+01           -4.000E+01           -2.000E+01           0.                   2.000E+01
                      - - - - - - - - - - - - - - - - - - - - - - - - - - - - - - - - - - - - - - - - - - - - - - - -

 (+=)---------------- -3.000E+01           -2.000E+01           -1.000E+01           0.                   1.000E+01
                      - - - - - - - - - - - - - - - - - - - - - - - - - - - - - - - - - - - - - - - - - - - - - - - -
  1.000E+02 -2.559E+00 .                  .                   .                *  .+       =          .
  1.166E+02 -2.440E+00 .                  .                   .                *  .+        =         .
  1.359E+02 -2.280E+00 .                  .                   .                *  .+        =         .
  1.585E+02 -2.067E+00 .                  .                   .                *  .+        =         .
  1.848E+02 -1.786E+00 .                  .                   .                 * . +        =        .
  2.154E+02 -1.426E+00 .                  .                   .                 * .  +        =       .
  2.512E+02 -9.811E-01 .                  .                   .                  *. +         =       .
  2.929E+02 -4.832E-01 .                  .                   .                  *.  +         =      .
  3.415E+02 -4.241E-02 .                  .                   .                  *   +         =      .
  3.981E+02  9.777E-02 .                  .                   .                   *  +          =     .
  4.642E+02 -3.499E-01 .                  .                   .                   *+         =        .
  5.412E+02 -1.372E+00 .                  .                   .               + * .        =          .
  6.310E+02 -2.474E+00 .                  .                   .          +     *  .    =              .
  7.356E+02 -2.837E+00 .                  .                   .    +           *  . =                 .
  8.577E+02 -1.168E+00 .                  .                   +                  *. =                 .
  1.000E+03 -2.255E+00 .                  .            +      .            =    * .                   .
  1.166E+03 -1.333E+01 .                 =.       +           .     *             .                   .
  1.359E+03 -2.210E+01 .   =              . +              *  .                   .                   .
  1.585E+03 -2.944E+01 .      =        +  .         *         .                   .                   .
  1.848E+03 -3.605E+01 .     =   +        .   *               .                   .                   .
  2.154E+03 -4.223E+01 . =  +          *  .                   .                   .                   .
                      - - - - - - - - - - - - - - - - - - - - - - - - - - - - - - - - - - - - - - - - - - - - - - - -


 Y
 0
         JOB CONCLUDED
 0        TOTAL JOB TIME            1.51
 $REVERT. SPICE COMPLETED
 /
```

A.3.4.1 MICRO-CAP II The circuit is entered as in Figure A.11. In the case of the SPICE input, we had to specify node numbers and input the various elements and op-amp subcircuits by use of these assigned node numbers. For MICRO-CAP II, we simply draw the circuit by invoking a series of instructions. For example, we begin at the center left of the screen with resistor RA1A. The actual instructions (refer to the MICRO-CAP II manual for details) are

E (to tell the computer to enter an element or device)

RES (to specify a resistor)

R (to specify the direction to the right)

N (to indicate no reflection)

12.1K (to give the parameter value)

The computer then draws the element. This process is continued until the circuit is complete. In the case of the op-amps, the parameter can be a library entry corresponding to the desired specifications or can be a specific number. We used the 741 op-amp model, which is included in MICRO-CAP.

The computer is then instructed to run the ac simulation. The program numbers the nodes and then displays a prologue menu of the type shown below.

Analysis Limits

Lowest frequency	50
Highest frequency	5000
Lowest gain (dB)	−60
Highest gain (dB)	20
Lowest phase shift	−360
Highest phase shift	90
Lowest group delay	1E-9
Highest group delay	1E-4
Input node number	1
Output node number	2
Minimum accuracy (%)	5
Auto or fixed frequency step (A, F)	A
Temperature (low/high/step)	27
Number of cases	1
Output: disk, printer, none (D, P, N)	N
Save, retrieve, normal run (S, R, N)	N
Default plotting parameters (Y, N)	N

Are these correct (Y, N, ESC = Abort)?

We chose to let the frequency range from 50 to 5000 Hz and the gain from −60 to 20 dB. The program can simultaneously plot gain, phase shift, and phase delay, but we chose only to plot the gain curve. The result is shown in Figure A.12.

Figure A.12
Gain of Chebyshev low-pass filter.

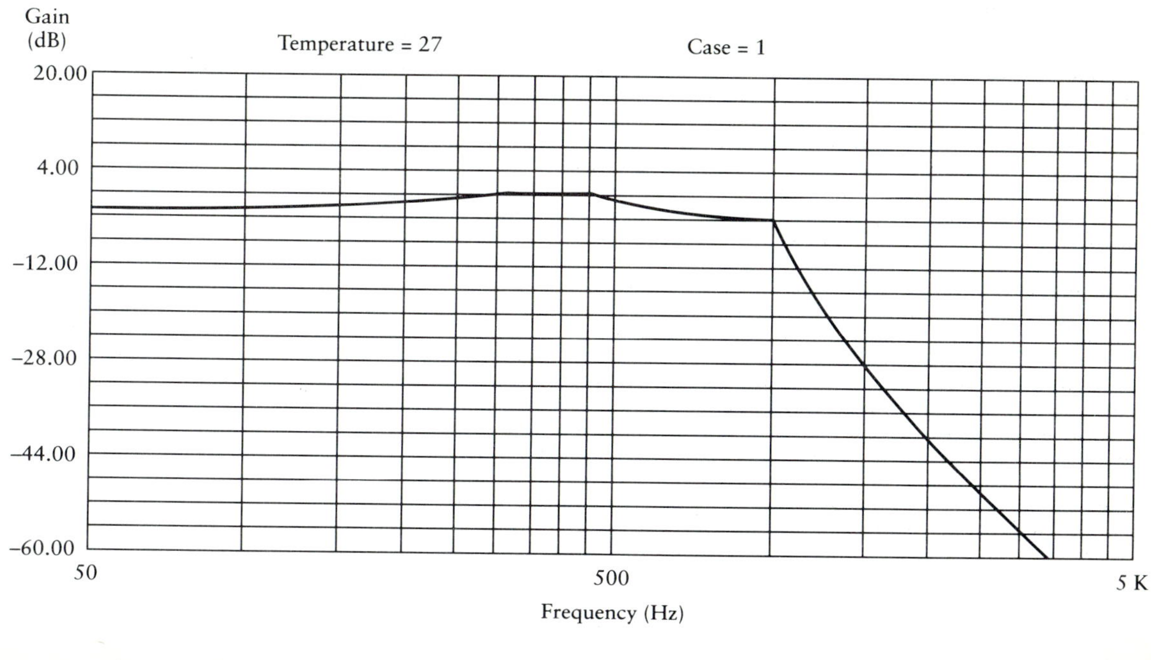

SPICE

A.3.5 Two-Input NAND Gate

Figure A.13 illustrates a two-input NAND gate. The circuit is designed to provide a high current output. This is done to provide more current to discharge and charge any parasitic capacitance that may be associated with the load.

The high-current operation decreases the transition time required to turn the device on and off. The printout of the computer inputs and outputs is shown next.

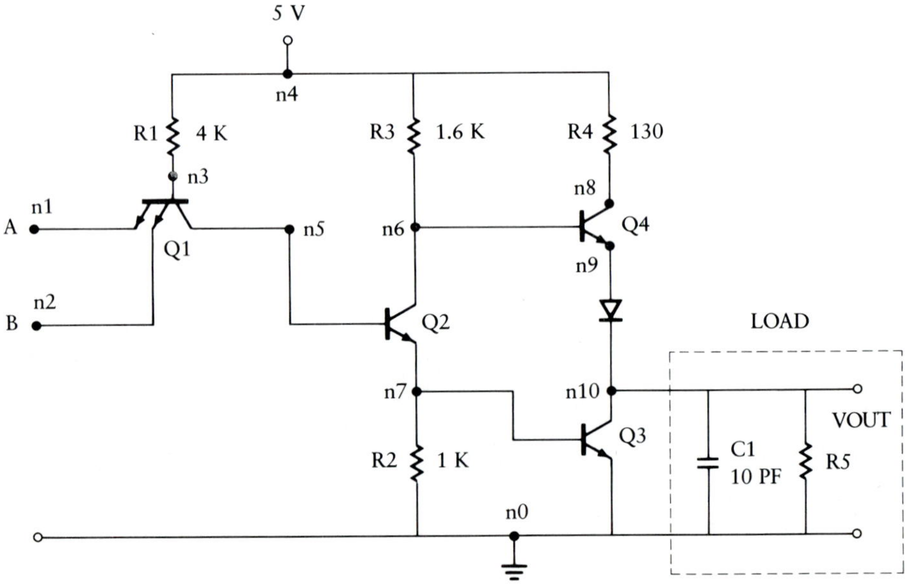

Figure A.13 Two-input NAND gate.

```
 1.700E-02    3.778E-02   .*                    REWIND,*
  9 FILES PROCESSED,
/FIND,SPICE,TEST
1******* 86/07/78, ****** SPICE 26.5 (10AUG81) ****** 16.53.57.*****

0NAND GATE

0****     INPUT LISTING                  TEMPERATURE =   27.000 DEG C

0*****************************************************************************


 C1 10  0 10P
 R5 10  0 2000
 R1 3 4 4K
 R2 0 7 1K
 R3 4 6 1.6K
 R4 4 8 130
 Q1 5 3 1 MOD1
 Q1A 5 3 2 MOD1
 Q2 6 5 7 MOD1
 Q3 10 7 0 MOD1
```

```
Q4 8 6 9 MOD1
D1 9 10 MOD2
VA 1 0 PULSE (0 5 10M 5M 5M 20M 50M)
VB 2 0 DC 5
VCC 4 0 DC 5
.MODEL MOD1 NPN (VJE=.7 BF=30 BR=.3)
.MODEL MOD2 D (VJ=.7)
*ANALYSIS REQUEST
.TRAN 1M 50M
*OUTPUT REQUEST
.PLOT TRAN V(10)
.END
1*************** 86/07/28, *********************** SPICE 26.5 (10AUG81) *********************** 16.53.57.***************

0NAND GATE
0****                DIODE MODEL PARAMETERS                          TEMPERATURE =  27.000 DEG C

0*****************************************************************************************************************



             MOD2
0IS        1.00E-14
0VJ            .700
1*************** 86/07/28, *********************** SPICE 26.5 (10AUG81) *********************** 16.53.57.***************

0NAND GATE
0****                BJT MODEL PARAMETERS                            TEMPERATURE =  27.000 DEG C

0*****************************************************************************************************************



             MOD1
0TYPE        NPN
0IS        1.00E-16
0BF          30.000
0NF           1.000
0BR            .300
0NR           1.000
0VJE           .700
1*************** 86/07/28, *********************** SPICE 26.5 (10AUG81) *********************** 16.53.57.***************

0NAND GATE
0****                INITIAL TRANSIENT SOLUTION                      TEMPERATURE =  27.000 DEG C

0 ****************************************************************************************************************



 NODE   VOLTAGE     NODE   VOLTAGE     NODE   VOLTAGE     NODE   VOLTAGE     NODE   VOLTAGE     NODE   VOLTAGE     NODE   VOLTAGE


(  1)    0.0000    (  2)    5.0000    (  3)     .7813    (  4)    5.0000    (  5)     .0559    (  6)    4.9109    (  7)     .0000

(  8)    4.7827    (  9)    4.1234    ( 10)    3.4541
```

SPICE

```
     VOLTAGE SOURCE CURRENTS

     NAME        CURRENT


     VA        1.206E-03

     VB       -1.516E-04

     VCC      -2.782E=-03


     TOTAL POWER DISSIPATION   1.47E-02  WATTS
1*************** 86/07/28. ***********************  SPICE 2G.5 (10AUG81)  ************************ 16.53.57.***************

0NAND GATE
0****                OPERATING POINT INFORMATION                        TEMPERATURE =  27.000 DEG C

0 ***************************************************************************************************************************


0
0**** DIODES



0            D1
0MODEL       MOD2
 ID        1.73E-03
 VD             .669
0
0**** BIPOLAR JUNCTION TRANSISTORS


0            Q1         Q1A        Q2         Q3         Q4
0MODEL       MOD1       MOD1       MOD1       MOD1       MOD1
 IB        5.49E-04  5.05E-04 -1.62E-11 -1,15E-11  5.57E-05
 IC        6.57E-04 -6.56E-04  2.11E-11  1.50E-11  1.67E-03
 VBE            .781     -4.219       .056       .000       .737
 VBC            .725       .725     -4.855     -3.454       .128
 VCE            .056     -4.944      4.911      3.454       .659
 BETADC        1.196     -1.300     -1.304     -1.300     30.000
1*************** 86/07/28, ***********************  SPICE 2G.5 (10AUG81)  ************************ 16.53.57.***************

0NAND GATE
0****                TRANSIENT ANALYSIS                                 TEMPERATURE =  27.000 DEG C

0 ***************************************************************************************************************************



      TIME       V(10)

                      0.                        1.000E+00                 2.000E+00                 3.000E+00                 4.000E+00
                       - - - - - - - - - - - - - - - - - - - - - - - - - - - - - - - - - - - - - - - - - - - - - - - - - - -
   0.           3.454E+00  .                          .                         .                         .          *          .
   1.000E-03    3.454E+00  .                          .                         .                         .          *          .
   2.000E-03    3.454E+00  .                          .                         .                         .          *          .
   3.000E-03    3.454E+00  .                          .                         .                         .          *          .
   4.000E-03    3.454E+00  .                          .                         .                         .          *          .
   5.000E-03    3.454E+00  .                          .                         .                         .          *          .
```

```
6.000E-03   3.454E+00   .                        .                        .                        .          *             .
7.000E-03   3.454E+00   .                        .                        .                        .          *             .
8.000E-03   3.454E+00   .                        .                        .                        .          *             .
9.000E-03   3.454E+00   .                        .                        .                        .          *             .
1.000E-02   3.454E+00   .                        .                        .                        .          *             .
1.100E-02   2.888E+00   .                        .                        .                     *  .                        .
1.200E-02   3.778E-02   .*                       .                        .                        .                        .
1.300E-02   3.778E-02   .*                       .                        .                        .                        .
1.400E-02   3.778E-09   .*                       .                        .                        .                        .
1.500E-02   3.778E-02   .*                       .                        .                        .                        .
1.600E-02   3.778E-02   .*                       .                        .                        .                        .
1.700E-02   3.778E-02   .*                       .                        .                        .                        .
1.800E-02   3.778E-02   .*                       .                        .                        .                        .
1.900E-02   3.778E-02   .*                       .                        .                        .                        .
2.000E-02   3.779E-02   .*                       .                        .                        .                        .
2.100E-02   3.778E-02   .*                       .                        .                        .                        .
2.200E-02   3.778E-02   .*                       .                        .                        .                        .
2.300E-02   3.778E-02   .*                       .                        .                        .                        .
2.400E-02   3.778E-02   .*                       .                        .                        .                        .
2.500E-02   3.778E-02   .*                       .                        .                        .                        .
2.600E-02   3.778E-02   .*                       .                        .                        .                        .
2.700E-02   3.778E-02   .*                       .                        .                        .                        .
2.800E-02   3.778E-02   .*                       .                        .                        .                        .
2.900E-02   3.778E-02   .*                       .                        .                        .                        .
3.000E-02   3.778E-02   .*                       .                        .                        .                        .
3.100E-02   3.778E-02   .*                       .                        .                        .                        .
3.200E-02   3.778E-02   .*                       .                        .                        .                        .
3.300E-02   3.778E-02   .*                       .                        .                        .                        .
3.400E-02   3.778E-02   .*                       .                        .                        .                        .
3.500E-02   3.778E-02   .*                       .                        .                        .                        .
3.600E-02   3.778E-02   .*                       .                        .                        .                        .
3.700E-02   3.778E-02   .*                       .                        .                        .                        .
3.800E-02   3.778E-02   .*                       .                        .                        .                        .
3.900E-02   2.984E+00   .                        .                        .                        *                        .
4.000E-02   3.454E+00   .                        .                        .                        .          *             .
4.100E-02   3.454E+00   .                        .                        .                        .          *             .
4.200E-02   3.454E+00   .                        .                        .                        .          *             .
4.300E-02   3.454E+00   .                        .                        .                        .          *             .
4.400E-02   3.454E+00   .                        .                        .                        .          *             .
4.500E-02   3.454E+00   .                        .                        .                        .          *             .
4.600E-02   3.454E+00   .                        .                        .                        .          *             .
4.700E-02   3.454E+00   .                        .                        .                        .          *             .
4.800E-02   3.454E+00   .                        .                        .                        .          *             .
4.900E-02   3.454E+00   .                        .                        .                        .          *             .
5.000E-02   3.454E+00   .                        .                        .                        .          *             .
                        - - - - - - - - - - - - - - - - - - - - - - - - - - - - - - - - - - - - - - - - - - - - - - - - - -


Y
0
        JOB CONCLUDED
0        TOTAL JOB TIME            7.85
$REVERT. SPICE COMPLETED
/
```

APPENDIX

B BODE PLOTS

B.1 BODE PLOTS

When a system contains capacitance and/or inductance, the response of that system is a function of the frequency of the input signal. Frequency plots of amplitude and phase are important measures of system behavior. Analysis of frequency-dependent systems can be performed using the impedance method, also known as *sinusoidal steady-state analysis.* These analyses result in formulas relating output to input. We now show how these formulas can be translated into frequency response curves that demonstrate how the amplifier behaves for varying input signal frequencies.

We begin with a simple example. Consider the series circuit of Figure B.1. The complex (phasor) output voltage of the circuit is given by

$$V_o(j\omega) = \frac{j\omega RCV_i(j\omega)}{1 + j\omega RC}$$

where $V_i(j\omega)$ is the phasor of the input voltage. It thus represents the magnitude and phase of that sinusoid. As a specific example, let us assume that the resistor and capacitor are chosen such that

$$RC = \tfrac{1}{4}$$

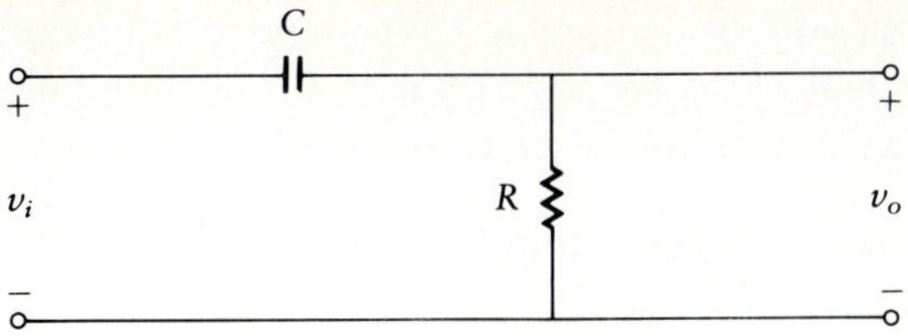

Figure B.1
A series circuit.

Then the ratio of output to input phasor, known as the *transfer function*, is given by

$$G(j\omega) = \frac{V_o(j\omega)}{V_i(j\omega)} = \frac{j\omega}{j\omega + 4}$$

This equation is a complex function of the radian frequency, ω. In dealing with complex numbers (or functions), we are actually dealing with two sets of numbers (or functions). These are commonly selected as either real and imaginary parts or magnitudes and phases. The magnitude and phase representation is commonly used in sinusoidal steady-state analysis, since it has physical significance in terms of the amplitudes and phase shifts of the sinusoidal signals.

We denote the amplitude and phase of the transfer function as $A(\omega)$ and $\phi(\omega)$, respectively. For the above example, they are given by

$$A(\omega) = |G(j\omega)| = \frac{j\omega}{|j\omega + 4|}$$

$$= \frac{\omega}{\sqrt{\omega^2 + 16}}$$

$$\phi(\omega) = \angle G(j\omega) = 90° - \tan^{-1}\left(\frac{\omega}{4}\right)$$

As an example of the application of these results, suppose that the input to the system is given by

$$v_i(t) = 3\cos(7t + 30°)$$

The steady-state output of the system is a sinusoid with the same frequency, 7 rad/s, but with altered amplitude and phase. To find the new amplitude and phase, we need only evaluate $A(\omega)$ and $\phi(\omega)$ at the given frequency. Thus,

$$A(\omega)\big|_{\omega=7} = \frac{7}{\sqrt{7^2 + 16}} = 0.87$$

$$\phi(\omega)\big|_{\omega=7} = 90° - \tan^{-1}\left(\frac{7}{4}\right) = 30°$$

The output amplitude is equal to the input amplitude multiplied by $A(\omega)$, and the output phase is the input phase shifted by $\phi(\omega)$. Therefore, the output voltage as a function of time is given by

$$v_o(t) = 2.6 \cos(7t + 60°)$$

Suppose you were asked to measure the frequency response of a system physically. You could take advantage of the amplitude and phase relationships by simply applying a sinusoidal input and observing the output amplitude and phase shift at various frequencies. Since you will be taking the ratio of output amplitude to input amplitude, the actual value of input amplitude is not critical. It need only be large enough so that the signal is not masked by noise, yet not so large that it saturates the system. In practice, the transfer function is evaluated at a particular frequency (the frequency of the input). Then the input frequency is changed to find the transfer function at other values of frequency.

The Laplace transform method is closely related to the sinusoidal steady-state transfer function, and we will use it for part of the following analysis (see Appendix C for a review of the Laplace transform method). The variable in the Laplace transform is s instead of $j\omega$. The variable s is complex $(\alpha + j\omega)$, whereas the radian frequency, ω, is a real variable. We can change an expression from the Laplace-transformed form to the sinusoidal steady state by making the substitution

$$s = j\omega$$

Although this may seem like a minor modification, it simplifies the mathematics since we can then factor polynomials in s. If $j\omega$ is used instead of s, it is necessary to deal with complex polynomials. We note that sinusoidal steady-state analysis is a special case of Laplace transform analysis where the complex variable, s, is restricted to lie on the imaginary axis. We are avoiding arguments related to existence of the transform since it is true that if the sinusoidal steady-state expression exists, the Laplace transform must converge along the imaginary axis. We do not deal with functions for which the transform does not exist along the imaginary axis, since they could not result from real circuits.

It is useful to define some new terms, as shown in Figure B.2. We define $G(s)$ as the product of all *forward-path* input-output ratios. This transfer function, $G(s)$, thus represents the transfer function of the system if all feedback paths are open-circuited (i.e., cut). We also define $H(s)$ as the product of all

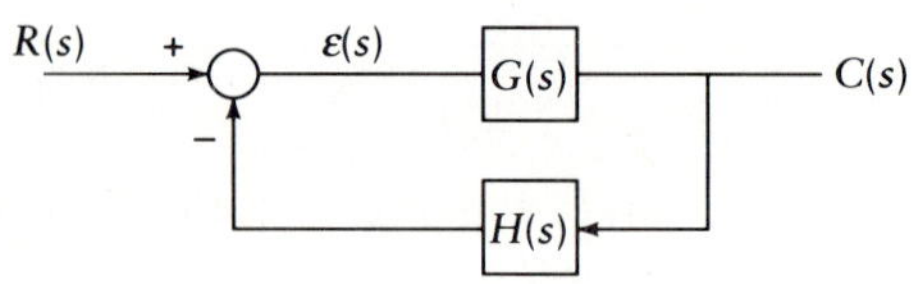

Figure B.2
Definition of feedback terms.

feedback input-output ratios. The product of these two is known as the *loop transfer function,* and it is an important characteristic of the overall system. This product function, $G(s)H(s)$, may be characterized by a ratio of polynomials for any linear system and is given in equation (B.1).

$$G(s)H(s) = \frac{b_m s^m + b_{m-1}s^{m-1} + \cdots + b_1 s + b_0}{a_n s^n + a_{n-1}s^{n-1} + \cdots + a_1 s + a_0} \tag{B.1}$$

In order to deal with the variation of amplitude and phase shift of practical systems, it is important to use the following concepts from the mathematics of complex variables. We take the time to review these concepts since they will lead to a powerful method of designing frequency-sensitive systems.

A *zero* of $G(s)H(s)$ is defined as any numerical value of s that causes the numerator polynomial in equation (B.1) to have a value of zero. Therefore, at this value of s,

$$G(s)H(s) = 0$$

A *pole* of $G(s)H(s)$ is defined as any numerical value of s that causes the denominator polynomial in equation (B.1) to have zero value. Therefore, at this value of s,

$$G(s)H(s) \to \infty$$

Our current objective is to plot the frequency response represented by $G(s)H(s)$, since $s = j\omega$. We start with the limiting values of frequency, since these will prove critical in devising the inspection method for plotting the response. As the frequency, ω, approaches infinity, all but the highest powers of s in the numerator and denominator can be ignored. That is, if ω is large, ω^n is much larger than ω^{n-1}. Thus, letting $s = j\omega$ and retaining only the highest powers, we have

$$G(j\omega)H(j\omega) = \frac{b_m(j\omega)^m}{a_n(j\omega)^n} = \frac{b_m}{a_n}(j\omega)^{m-n}$$

This shows that as the frequency approaches infinity, the amplitude curve is proportional to ω^{m-n}. The phase curve approaches $(m - n)$ times 90°, since this is the angle of the j^{m-n} term.

We now examine the other extreme, as ω approaches zero. For this case, all but the lowest power of ω are negligible. That is, if ω is small, ω^{n-1} is much larger than ω^n. Thus, as frequency approaches zero, the response function approaches

$$G(j\omega)H(j\omega) = \frac{b_0}{a_0}$$

This is true provided both a_0 and b_0 are nonzero. If either or both of these constants are equal to zero, it is necessary to include the lowest part of ω that is present.

As an example, consider the loop-transfer function,

$$G(s)H(s) = \frac{6s^3 + 2s^2 + 3s}{s^5 + 4s^4 + 2s^3 + s^2 + 10}$$

At high frequencies, this is approximately equal to

$$G(j\omega)H(j\omega) \approx \frac{6(j\omega)^3}{(j\omega)^5}$$

and the amplitude and phase are given by

$$A(\omega) = G(j\omega)H(j\omega) = 6\omega^{-2}$$

$$\phi(\omega) = \underline{/G(j\omega)H(j\omega)} \approx -180°$$

At low frequencies, we maintain the lowest powers of the numerator and denominator to get

$$G(j\omega)H(j\omega) \approx \frac{3j\omega}{10}$$

and the amplitude and phase are given by

$$A(\omega) = \frac{3\omega}{10}$$

$$\phi(\omega) \approx 90°$$

Once the behavior of the curve is found for extreme values of ω, a few more measurements should suffice to plot the entire curve. It is useful to make more measurements at frequencies where the largest changes in amplitude or phase occur and fewer measurements over ranges where the function is almost constant.

Now that we have determined the behavior of the curve for limiting values, we will examine the other critical parameters in the frequency response curves.

B.1.1 *G(s)H(s)* Function Terms

For real systems, the polynomials in s are usually in factored form, and the $G(s)H(s)$ functions we deal with can be viewed as a product of simple terms. When complex terms are multiplied together, the resulting amplitude is the product of the individual amplitudes and the resulting phase is the algebraic sum of the individual phases. We thus find it useful to learn first to deal with the simplest terms that result from factoring the $G(s)H(s)$ polynomials.

Instead of dealing with the amplitude curves directly, we use their logarithms. In this manner, multiplication of the various amplitudes is equivalent to addition of the logarithms. The *decibel,* abbreviated *dB,* is used to describe amplitude response curves as a function of frequency. The decibel is defined as in equation (B.2).

$$\text{dB} = 20 \log_{10} A(\omega) = 20 \log_{10} |G(j\omega)H(j\omega)| \tag{B.2}$$

We use the observations of the previous section to aid in the plotting of the various components of $G(s)H(s)$. We concentrate on the behavior of these curves for small and large values of frequency and note that the decibel gain is usually proportional to the log of the frequency. Therefore, if we use a logarithmic scale for the frequency axis, these curves are straight lines and are much easier to plot by inspection. For this reason, *semilog* paper is usually used to plot the components of $G(s)H(s)$. The frequency is plotted on the horizontal logarithmic scale and the dB amplitude and the phase are plotted along the linear vertical scale. The number of cycles of the semilog paper determines how many orders of magnitude of frequency variation can be plotted. Three- or four-cycle semilog paper is commonly used for general-purpose applications.

B.1.2 The Asymptotic Approximation

The numerator and denominator of $G(s)H(s)$, as given by equation (B.1), can be factored into a product of functions. The individual factors are of one of the following forms:

1. Frequency-invariant factors (constants), K.
2. Terms corresponding to simple (order 1) zeros or poles at the origin, s or $1/s$.
3. Linear terms corresponding to simple zeros (not at the origin), $s\tau_1 + 1$.
4. Linear terms corresponding to simple poles (not at the origin), $(s\tau_2 + 1)^{-1}$.
5. Quadratic terms corresponding to simple zeros and poles,

$$\left(\frac{s}{\omega_n}\right)^2 + \frac{2\zeta s}{\omega_n} + 1$$

or

$$\frac{1}{(s/\omega_n)^2 + 2\zeta s/\omega_n + 1}$$

6. Multiple zeros and poles.

We shall now see how to plot each of these factors. These could, of course, be plotted point by point by evaluating the function for various values of frequency. We avoid this time-consuming task by using straight-line asymptotes as an approximation.

B.1.2.1 Frequency-Invariant Factors The gain constant, K, which is independent of frequency, is plotted from the dB representation of equation (B.3).

$$K_{\mathrm{dB}} = 20 \log_{10} K \tag{B.3}$$

The constant, K, represents the product of all frequency-invariant terms in the $G(s)H(s)$ function. Equation (B.3) is plotted in Figure B.3 for a representative value of K. Note that the phase shift of this real constant is zero for all frequencies. On this figure, we plot the dB gain using the scale on the left vertical axis and the phase shift using the scale on the right vertical axis. Frequency is plotted using the log scale on the horizontal axis.

B.1.2.2 Zeros or Poles at the Origin The factors are of the form

$$s \quad \text{or} \quad \frac{1}{s} \tag{B.4}$$

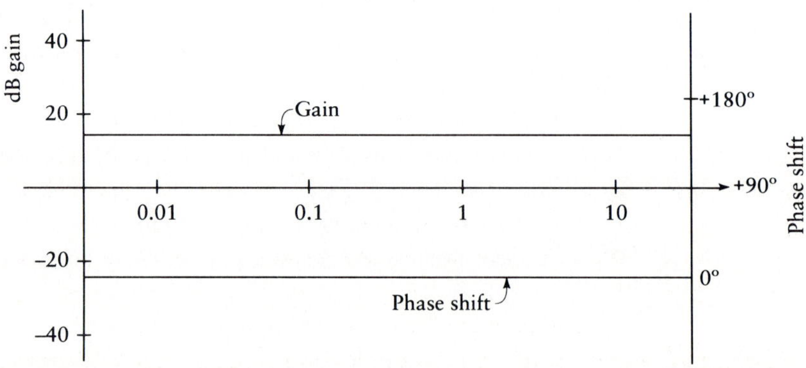

Figure B.3
Gain and phase for a constant factor.

We set $s = j\omega$ and find the amplitude and phase for the zero as follows.

$$A(\omega)_{\text{dB}} = 20 \log|j\omega| = 20 \log \omega \tag{B.5a}$$

$$\phi(\omega) = \underline{/j\omega} = 90° \tag{B.5b}$$

For the pole,

$$A(\omega)_{\text{dB}} = -20 \log \omega \tag{B.5c}$$

$$\phi(\omega) = \underline{/1/j\omega} = -90° \tag{B.5d}$$

Let us divert our attention for a moment to explore the general case of multiple zeros or poles at the origin. The equations given above are for $n = 1$. If the factor s or $1/s$ is raised to the power n, we have multiple zeros or poles. Since these factors multiply together in the expression for $G(s)H(s)$, the decibel amplitudes and the phase shifts add. Thus, the amplitude for a multiple zero is

$$20n \log \omega \text{ dB} \tag{B.6a}$$

and for a multiple pole it is

$$-20n \log \omega \text{ dB} \tag{B.6b}$$

The phase for the multiple zero is

$$n90° \tag{B.7a}$$

and for the pole it is

$$-n90° \tag{B.7b}$$

Figure B.4 illustrates the amplitude and phase plots for multiple zeros and poles. If $n = 1$, we are dealing with simple zeros and poles. Note that the amplitude curve goes through 0 dB when $\omega = 1$, since $\log 1 = 0$.

A *decade* is defined as a change in frequency by a factor of 10. When the frequency increases 10-fold, the log of the frequency increases by one. Thus, the slope of the curves in Figure B.4 is $\pm 20n$ dB/decade. An *octave* is defined as a change in frequency by a factor of 2. Since the log of 2 is approximately 0.3, a slope of $20n$ dB per decade is nearly the same as a slope of $6n$ dB per octave.

We have not yet found it necessary to make any approximations. The curves of Figure B.4 are exact representations of the amplitude and phase of the factors s^n and s^{-n}.

If a transfer function polynomial contains both a constant (frequency-invariant term) and poles and zeros at the origin, the combined plot is formed

Figure B.4
Phase and gain for multiple zeros and poles at the origin.

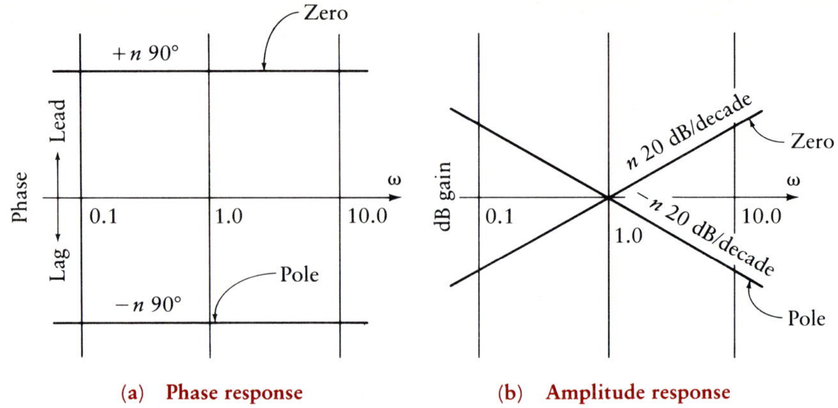

by adding together the results for forms 1 and 2. For ease in plotting, the constant portion of the transfer function, K, can be combined with the s^n terms. As an example, consider a single pole at the origin as given by equation (B.8).

$$\frac{K}{j\omega} = \frac{1}{j\omega/K} \tag{B.8}$$

We can view the frequency variable as being ω/K instead of simply ω. Multiplication of the frequency by 10 results in multiplication of ω/K by 10. Thus, the amplitude response for this term has the same slope as that corresponding to the single pole at the origin, $1/s$. The only difference is that the amplitude curve intersects the 0-dB line at a frequency of $\omega = K$ instead of $\omega = 1$. The phase remains constant at $-90°$, as is the case without the constant, K. It should be clear that these results are not any different from those that would occur if the two individual components of the amplitude curve were added together.

B.1.2.3 Zeros Let us now consider zeros that are not at the origin. If these are simple (not multiple), the factor is of the form

$$(j\omega\tau + 1) \tag{B.9}$$

The resulting amplitude plot is no longer a straight line, so we shall resort to asymptotic approximations. If the frequency is very small such that $\omega\tau << 1$, we can approximate the dB amplitude by

$$20 \log_{10}|j\omega\tau + 1| \approx 20 \log_{10} 1 = 0 \text{ dB} \tag{B.10}$$

Thus, for small values of frequency, the magnitude remains close to 0 dB. Now when the frequency becomes large, $\omega\tau >> 1$, we can neglect the constant term to obtain

$$20 \log_{10}|j\omega\tau + 1| \approx 20 \log_{10} \omega\tau \tag{B.11}$$

This is similar to the result for a zero at the origin, so the amplitude and phase plot resembles a plot for the term $j\omega\tau$. The slope (for large ω) is thus 20 dB/decade, and this straight-line asymptote intersects the 0-dB line at $\omega\tau = 1$, or $\omega = 1/\tau$. This point of intersection is known as the *corner frequency.* The two straight-line plots, one for small frequency and one for large frequency, intersect at the corner frequency and represent the asymptotic approximations to the curves. These are illustrated in Figure B.5(a).

We have shown that once the corner frequency is known, the approximate curve for the amplitude of a simple zero is easily drawn by inspection. Suppose the approximation is not good enough and additional accuracy is required. The first steps in approaching the exact curve can also be accomplished by inspec-

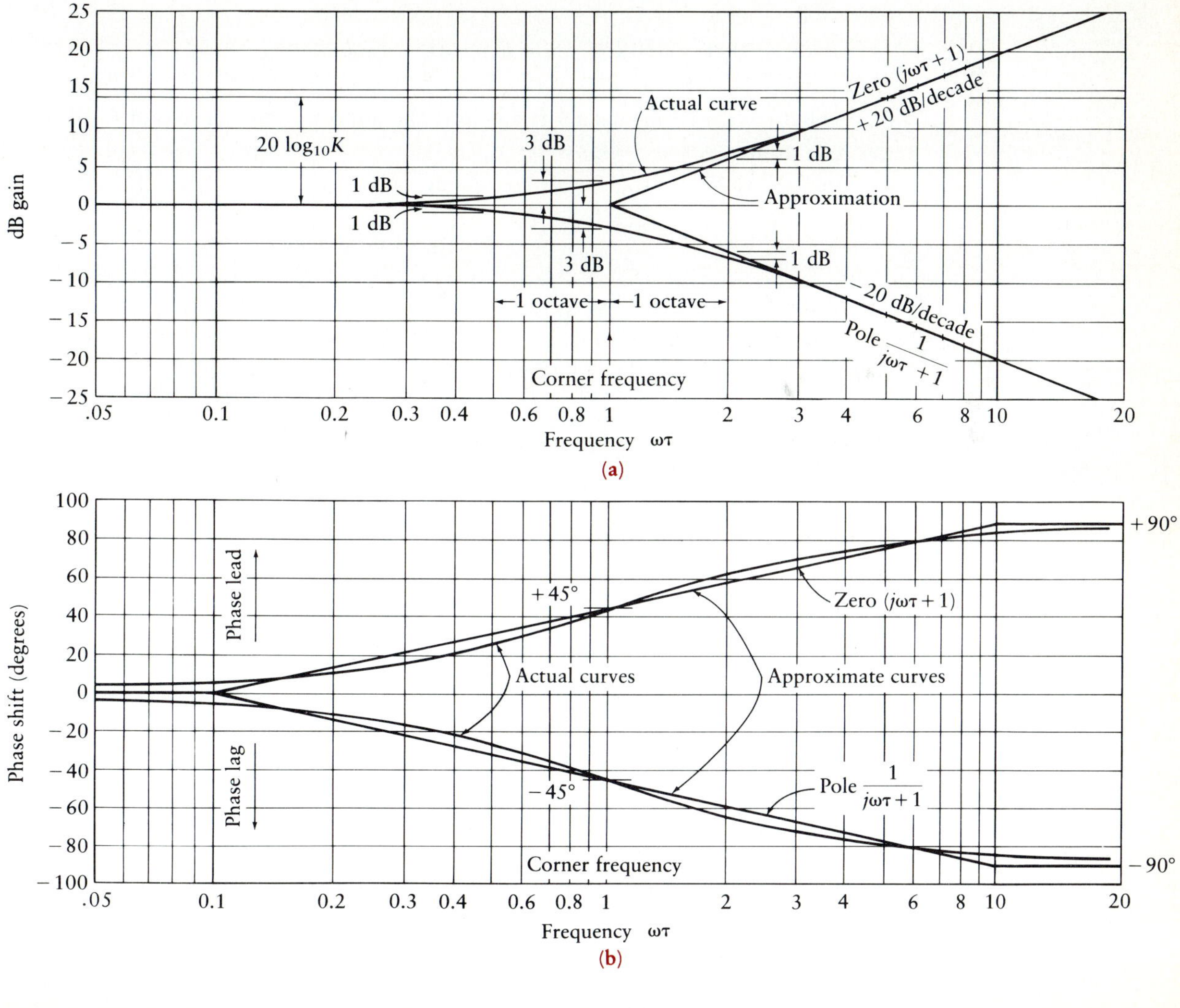

Figure B.5
Response curves for zeros and poles.

tion. In fact, the actual amplitude curve deviates only slightly from the straight-line asymptotes. Figure B.5(a) shows both the asymptotic approximation and the exact curve. At the corner frequency, the actual curve is 3 dB above the approximation. One octave away from the corner frequency, the actual curve deviates from the approximation by about 1 dB. These variations can be seen from the equation. At the corner frequency, $\omega = 1/\tau$, the value of the function is

$$20 \log|j1 + 1| = 20 \log \sqrt{2} = 3 \text{ dB}$$

Since the approximate curve is at 0 dB at the corner frequency, the deviation is 3 dB as stated above.

One octave above the corner frequency, $\omega = 2/\tau$, and the value of the function is

$$20 \log|j2 + 1| = 20 \log \sqrt{5} = 6.99 \text{ dB}$$

At one octave above the corner frequency, the approximate curve is at 6 dB, so the deviation is about 1 dB.

One octave below the corner frequency, at $\omega = 1/2\tau$, the value of the function is

$$20 \log\left|\frac{j}{2} + 1\right| = 20 \log \sqrt{\frac{5}{4}} = 0.97 \text{ dB}$$

The approximate curve is at 0 dB one octave below the corner frequency, so the deviation is once again about 1 dB.

The phase shift in equation (B.9) is simply the angle whose tangent is the ratio of the imaginary to the real part of the function. Thus,

$$\phi(\omega) = \tan^{-1}(\omega\tau) \tag{B.12}$$

In the plot of Figure B.5(b), the frequency variable, $\omega\tau$, is on the horizontal logarithmic scale. The arctangent curve has a value of 45° when $\omega\tau = 1$, which is at the corner frequency. The phase curve starts at 0°, increases to a maximum of 90°, and is symmetric about the 45° point. The complete frequency response curve for the zero comprises the amplitude curve shown in Figure B.5(a) and the phase shift curve shown in Figure B.5(b).

In sketching a first approximation to the phase curve, a straight line can be used. The curve is horizontal from low frequency to one-tenth of the corner frequency. A straight line, with a slope of 45°/decade, goes from $0.1\omega_o$ to $10\omega_o$ passing through 45° at the corner frequency (ω_o is the corner frequency). For frequencies above $10\omega_o$, the line is again horizontal through 90° at 10 times

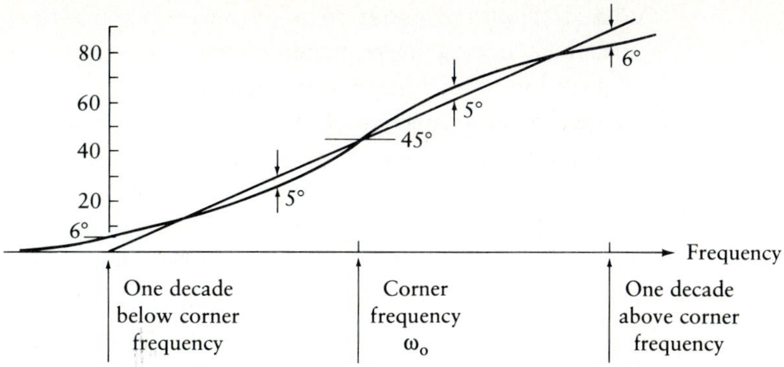

Figure B.6
Phase curve.

the corner frequency. The slope of this approximate curve is 45°/decade. The maximum amount that the straight-line approximation deviates from the actual curve is 6°. The approximation and the actual curve are shown in Figure B.6.

B.1.2.4 Poles

Simple pole factors of the form $1/(j\omega\tau + 1)$ can be treated in a fashion similar to that of simple zero factors. Since the logarithm of a reciprocal quantity is equal to the negative of the logarithm of the quantity, we simply have

$$20 \log\left|\frac{1}{j\omega\tau + 1}\right| = -20 \log|j\omega\tau + 1| \tag{B.13}$$

The curve for a simple pole factor is similar to that for a simple zero factor except that it is reflected about the 0-dB line. For small frequencies, the amplitude remains at 0 dB. For large frequencies, the asymptote is a straight line of slope −20 dB/decade. This asymptotic approximation is shown in Figure B.5(a). Note that it intersects the 0-dB axis at $\omega = 1/\tau$, which is the corner frequency. As was the case with the zero, the actual amplitude curve deviates from the straight-line approximation by −3 dB at the corner frequency and by about −1 dB at both one-half and double the corner frequency.

The phase curve of a pole is similar to that of a zero, but it is reflected about the $\phi = 0$ line. Since the pole is in the denominator of $G(s)H(s)$, the sign is changed when the angle is brought into the numerator. Thus,

$$\phi(\omega) = -\tan^{-1}(\omega\tau) \tag{B.14}$$

Equation (B.14) represents an arctangent curve that starts from zero at a frequency of zero and approaches a value of −90° for large frequencies. The −45° phase shift occurs at the corner frequency. The actual phase curve and its straight-line approximation are the negatives of those included in Figure B.6. It is important to note that the phase shift curve for the pole has an asymptote of $-90°$ and that the asymptote for zero is $+90°$.

If the transfer function has multiple zeros or poles, the various amplitude and phase curves must simply be added to themselves a number of times equal to the order. For example, for two repeated poles or zeros, the slope changes from ± 20 dB/decade to ± 40 dB/decade, and the phase angle is $\pm 90°$ at the corner frequency. The phase angle varies from 0° to $\pm 180°$ rather than from 0° to $\pm 90°$, and the slope is $\pm 90°$/decade rather than $\pm 45°$/decade. The corresponding functions are shown in Figure B.7.

B.1.2.5 Quadratic Poles Quadratic pole factors are of the form

$$G(s)H(s) = \frac{\omega_n^2}{s^2 + 2s\zeta\omega_n + \omega_n^2}$$

$$= \frac{1}{(s/\omega_n)^2 + 2s\zeta/\omega_n + 1} \tag{B.15}$$

We normalize this by making a change of variables, letting

$$u = \frac{\omega}{\omega_n}$$

This is done after substituting $s = j\omega$. The equation then becomes

$$G(u)H(u) = \frac{1}{1 - u^2 + j2\zeta u} \tag{B.16}$$

The amplitude and phase of the expression in equation (B.16) are plotted in Figure B.8.

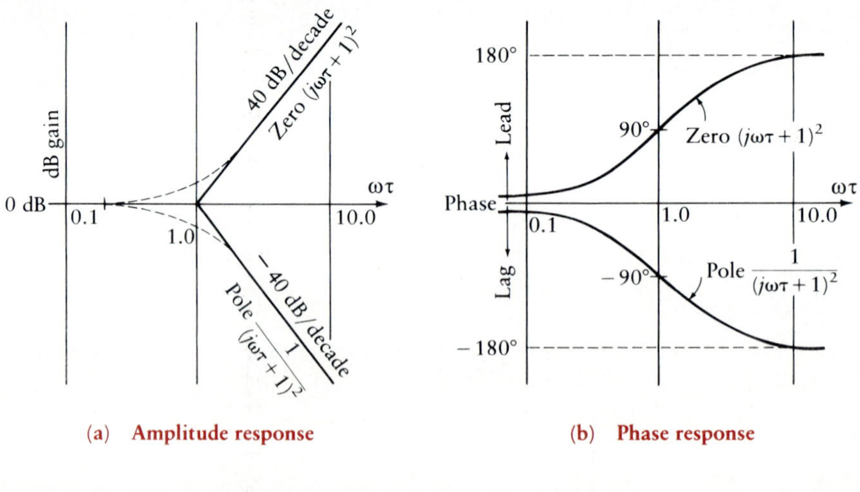

Figure B.7
Double poles and zeros.

Figure B.8
Amplitude and phase of quadratic pole.

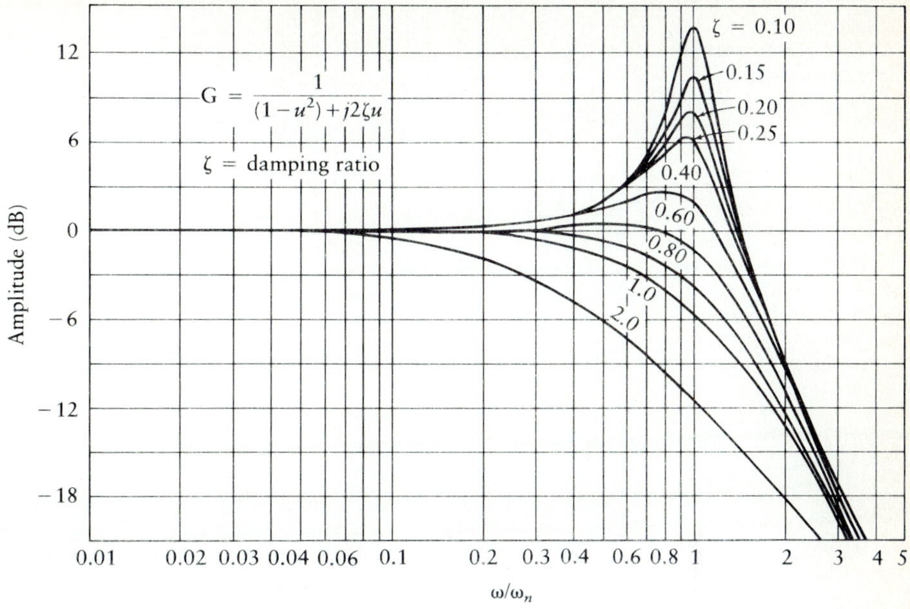

(a)

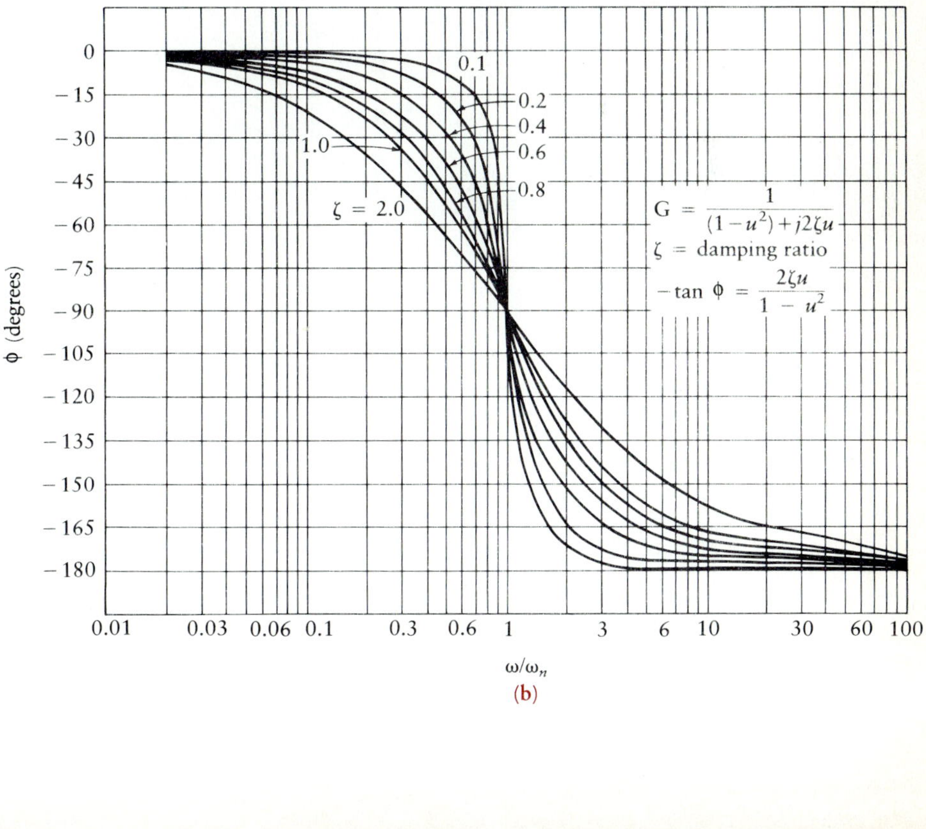

(b)

Figure B.9
Correction factors for quadratic pole.

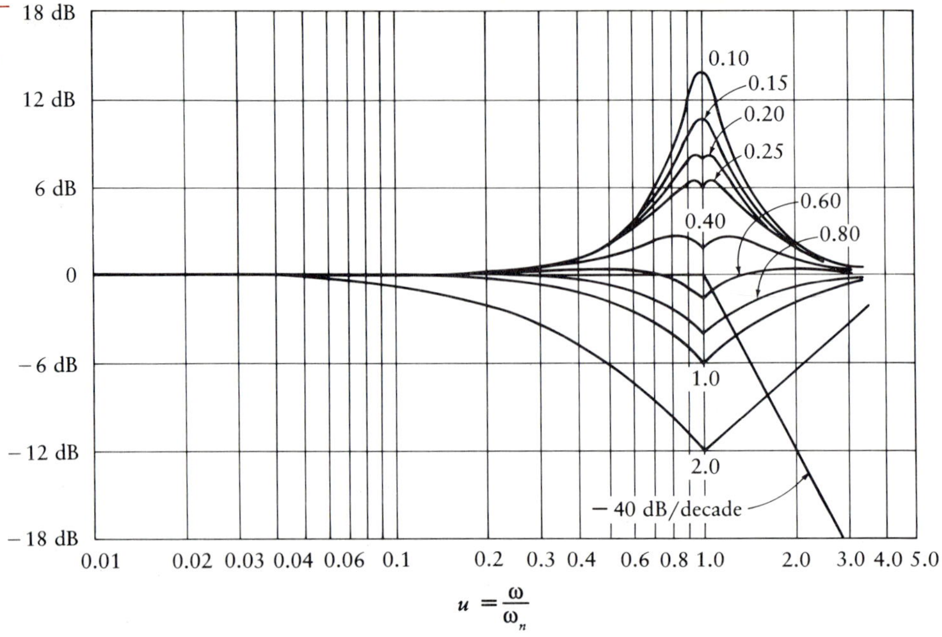

(a)

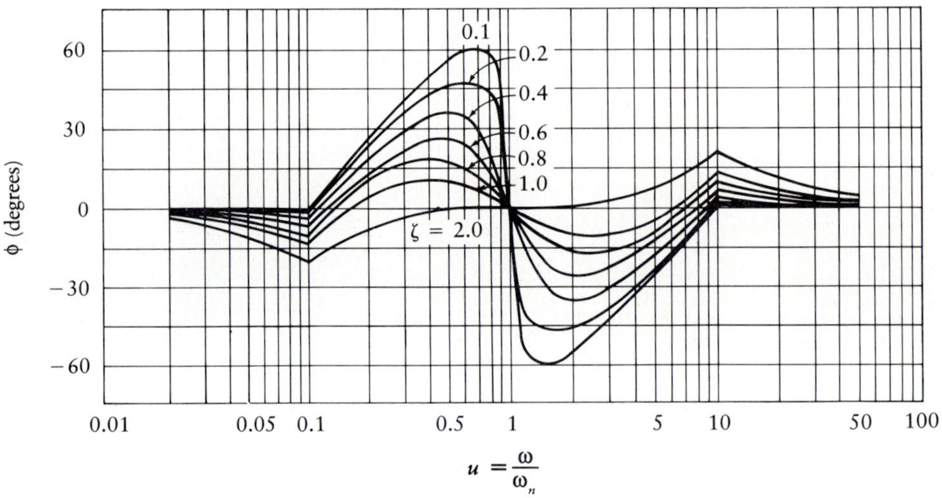

(b)

Because the amplitude and phase response for quadratic pole factors depend not only on the corner frequency but also on the *damping ratio*, ζ, a parametric chart of the form shown in Figure B.8 can be used to make the plot. The amplitude and phase response are plotted by locating the corner frequency and damping ratio for the particular quadratic factor, as found by comparison of the given expression with equation (B.15).

The results of Figure B.8 can be compared with those for a double pole with corner frequency of unity. The amplitude for the double pole is a line of slope −40 dB/decade starting at the corner frequency. The phase shift is a line of slope −90°/decade starting at $\frac{1}{10}$ and ending at 10 times the corner frequency. The difference between the double pole and the quadratic pole can be viewed as a correction factor. These differences (correction factors) are plotted as Figure B.9 for various values of the damping ratio, ζ. Thus, in solving a problem with a quadratic pole, we can first treat it as a double pole and then apply the corrections of Figure B.9.

The correction curves shown in Figure B.9(a) and (b) are used only to indicate the general characteristics of the Bode plot. In practice, we plot the straight-line asymptotes for the system and do not apply correction factors. We determine the range of frequencies of most concern in the design by examining the straight-line Bode plot. As an example, in stability analysis, the critical point is where the phase shift intersects the $\phi = -180°$ axis. Having determined the range of frequencies of greatest concern, we must perfect the plot numerically or with a frequency analysis program on a computer (e.g., SPICE, MICRO-CAP II).

B.1.3 Examples of Bode Plots

The concepts presented in the previous section are now applied to a number of examples.

Example B.1

Draw the Bode plot for the loop transfer function

$$G(s)H(s) = 100$$

SOLUTION This is a simple case of a frequency-invariant factor, and

$$K_{dB} = 20\log(100) = 40 \text{ dB} \quad \text{for all } \omega$$

The phase is zero for all frequencies since the loop transfer function is real.

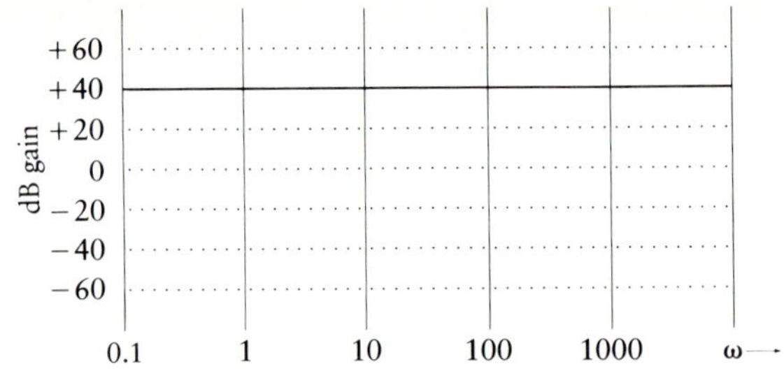

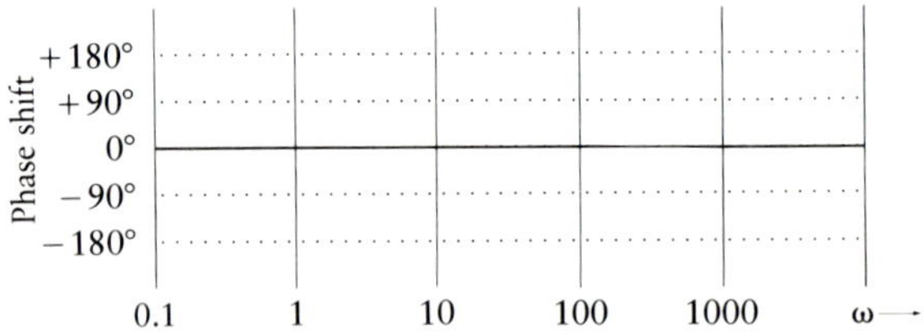

Figure B.10
Frequency response for Example B.1.

The plot is shown in Figure B.10.

Example B.2

Draw the Bode plot for the loop transfer function

$$G(s)H(s) = s$$

SOLUTION This is a case of a simple zero at the origin. The resulting amplitude plot is a straight line of slope +20 dB/decade. The line passes through

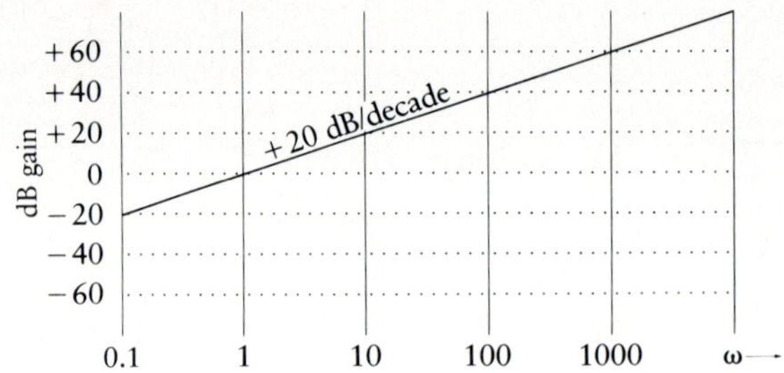

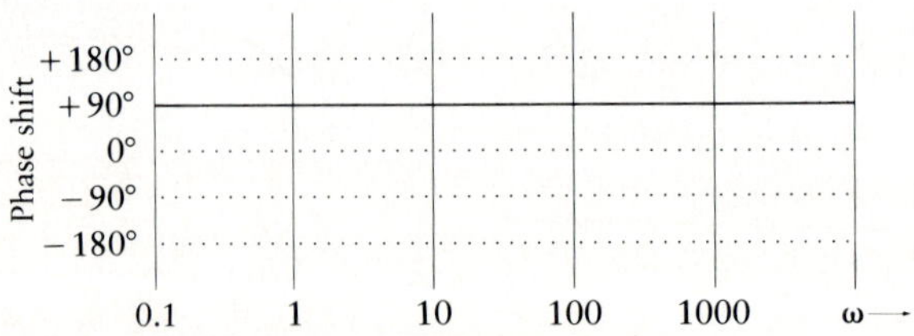

Figure B.11
Frequency response for Example B.2.

the 0-dB axis at $\omega = 1$. The phase is a constant at 90° for all frequencies since the transfer function, for $s = j\omega$, is pure imaginary. The result is shown in Figure B.11.

Example B.3

Draw the Bode plot for the loop transfer function

$$G(s)H(s) = \frac{10}{s + 10} = \frac{1}{s/10 + 1}$$

SOLUTION This expression contains a simple pole. The corner frequency is at $\omega = 10$. As the frequency approaches zero, the amplitude approaches 1, which corresponds to 0 dB. To the right of the corner frequency, the slope of the amplitude curve is -20 dB/decade. The phase curve starts at 0° and goes through $-45°$ at the corner frequency. The center portion of the phase approximation has a slope of $-45°$/decade, so the approximation starts decreasing from 0° at a frequency of 1. It reaches the maximum negative phase of $-90°$ at a frequency of 100. The resulting curves are shown in Figure B.12.

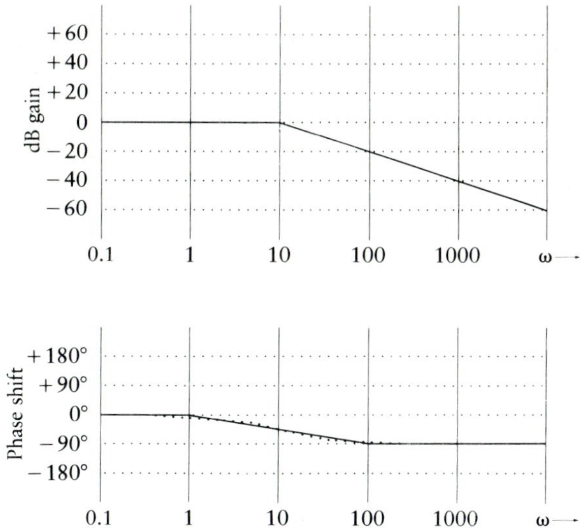

Figure B.12
Frequency response for Example B.3.

Example B.4

Draw the Bode plot for the transfer function

$$G(s)H(s) = \frac{s}{s + 10} = \frac{0.1s}{0.1s + 1}$$

SOLUTION This transfer function is the product of two factors. One of these is a simple zero at the origin as in Example B.2, and the other is a simple pole as in Example B.3. We can either plot each of these separately and add the results or plot them simultaneously. We choose the latter approach.

The zero is at the origin, so the resulting plot is a straight line. The pole has a corner frequency of 10. Therefore, if we begin the amplitude plot at small values of frequency, the zero is the dominating term. The plot will begin as a straight line with a slope of +20 dB/decade. For small values of frequency, the transfer function is approximated by $s/10$, so the plot intersects the 0-dB axis at $\omega = 10$. As the frequency increases, the next effect occurs at the corner frequency of $\omega = 10$. At this point, we add a straight line of slope −20 dB/decade. The result is to change the slope of the curve from +20 dB/decade to 0 (i.e., horizontal). The resulting curve is shown in Figure B.13. At the corner frequency, the correction is −3 dB.

The phase plot is developed in a similar manner. We again start with a small frequency, where the zero predominates. The phase is then 90° until the effects of the pole come into play. Recall that the phase plot for the pole is a

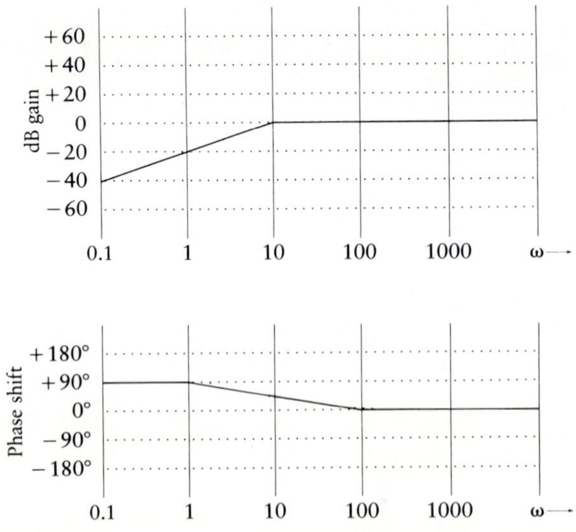

Figure B.13 Frequency response for Example B.4.

straight line with a slope of −45°/decade starting at $\frac{1}{10}$ of the corner frequency and ending at 10 times the corner frequency. The result is shown in Figure B.13.

Example B.5

Draw the Bode plot for the transfer function

$$G(s)H(s) = \frac{1}{s^2 + 3s + 10} = \frac{0.1}{0.1s^2 + 0.3s + 1}$$

SOLUTION This is an example of a quadratic pole. We begin by putting this into the form of equation (B.15) as follows:

$$G(s)H(s) = \frac{0.1}{(s/\omega_n)^2 + 2\zeta s/\omega_n + 1} \tag{B.17}$$

where

$$\omega_n = \sqrt{10} = 3.16 \text{ rad/s} \tag{B.18}$$

and

$$\zeta = 0.3\frac{\omega_n}{2} = 0.474 \tag{B.19}$$

The corner frequency is first located on the Bode plot. The first approximation is drawn in the same manner as for a double pole located at the corner frequency. That is, for the amplitude plot, we start from this frequency and draw a −40 dB/decade line extending to the right. The phase plot is an approximation to the arctangent curve extending from 0 to −180°. These asymptotic approximations are shown in Figure B.14. Because the correction from the asymptotes depends on the damping ratio, it is necessary to refer to Figure B.8 or Figure B.9. The actual shape of the amplitude plot will be as shown in Figure B.8(a) for a damping ratio of 0.474 and for a corner frequency of $\sqrt{10}$.

To perfect this plot, we use a calculator or computer to determine the dB gain and phase shift at three values of frequency, as follows.

ω	$\omega_n/2$	ω_n	$2\omega_n$
db gain	−18.9	−19.5	−31
Phase shift	−32.3°	−90°	−147.6°

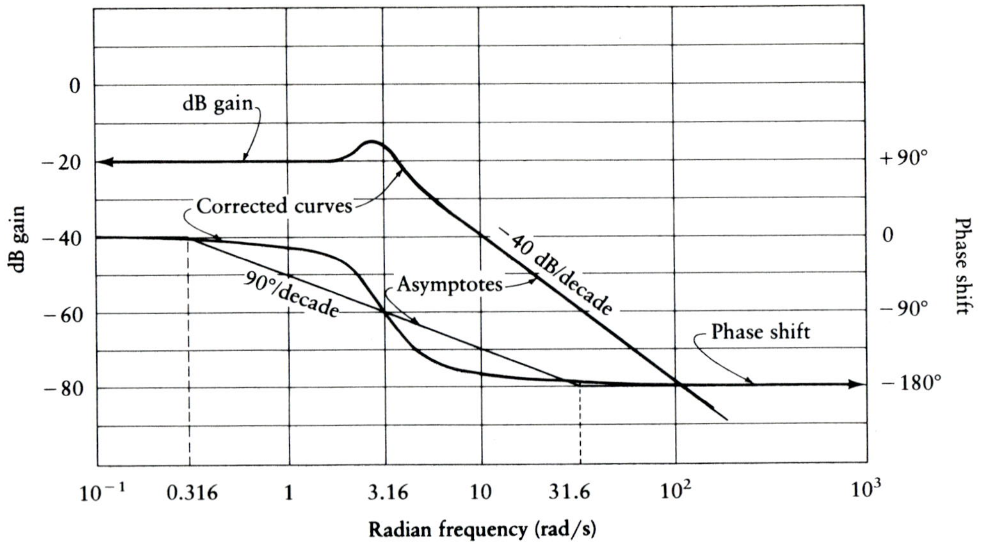

Figure B.14
Bode plots for Example B.5.

These values were calculated using the equations

$$\text{dB gain} = 20 \log \frac{0.1}{\sqrt{(1 - 0.1\omega^2)^2 + (0.3\omega)^2}}$$

$$\text{Phase shift} = -\tan^{-1}\left(\frac{0.3\omega}{1 - 0.1\omega^2}\right)$$

The resulting amplitude and phase curves are also shown in Figure B.14 with the straight-line asymptotes included on the curves. Normally, we perfect the curve only in the range of frequencies that are critical to our design.

Drill Problems

DB.1 Draw a Bode plot for the loop transfer function

$$G(s)H(s) = -0.1$$

Ans: See Figure DB.1.

DB.2 Draw a Bode plot for the loop transfer function

$$G(s)H(s) = \frac{K}{s}$$

Ans: See Figure DB.2.

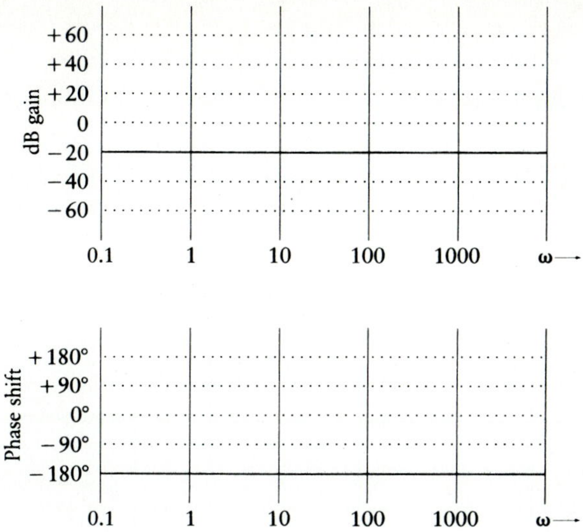

Figure DB.1

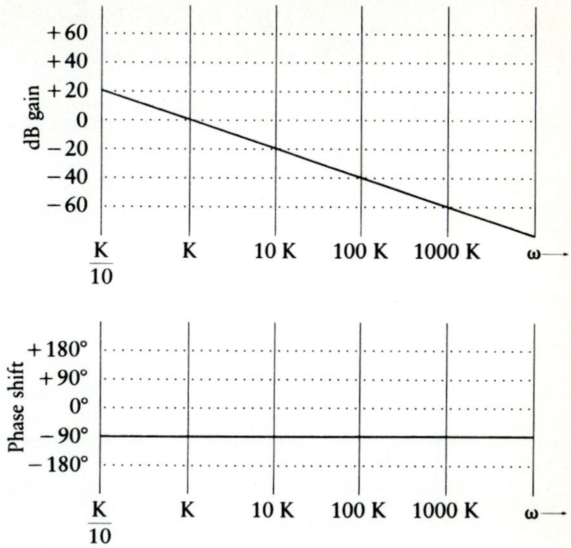

Figure DB.2

Figure DB.3

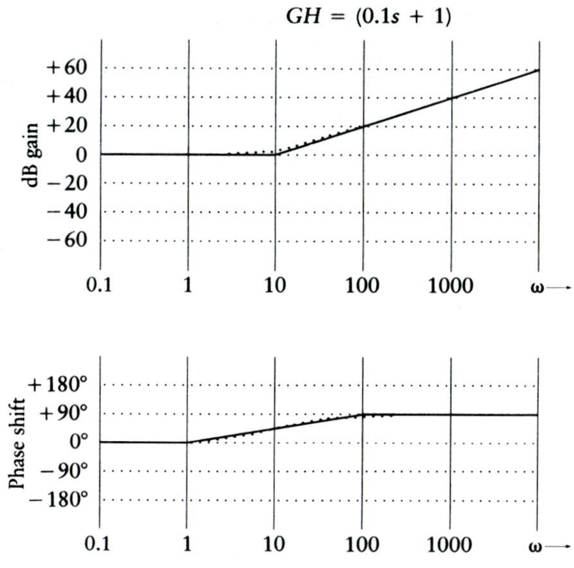

DB.3 Draw a Bode plot for the loop transfer function

$$G(s)H(s) = 0.1s + 1$$

Ans: See Figure DB.3.

DB.4 Draw the Bode plot for the system with transfer function

$$G(s)H(s) = \frac{0.03}{s(0.5s + 1)(0.06s^2 + 0.1s + 1)}$$

Ans: Note that for the quadratic term, $\omega_n = 4.08$ and $\zeta = 0.204$. The resulting plots are shown in Figure DB.4.

Figure DB.4

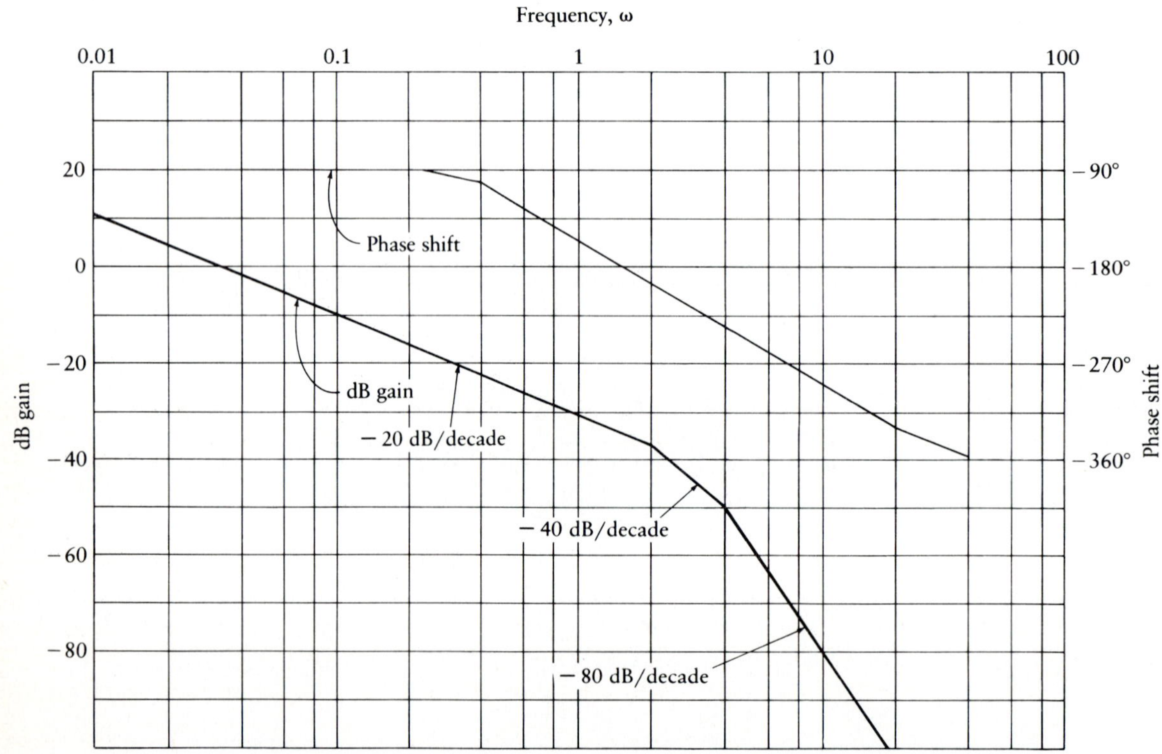

PROBLEMS

Draw the amplitude and phase curves as a function of frequency for the loop transfer functions given in Problems B.1–B.12.

B.1 $G(s)H(s) = \dfrac{100}{s^2}$

B.2 $G(s)H(s) = \dfrac{s}{s + 10}$

B.3 $G(s)H(s) = \dfrac{10(s + 10)}{s + 1}$

B.4 $G(s)H(s) = \dfrac{0.1s + 1}{s + 1}$

B.5 $G(s)H(s) = \dfrac{10s}{0.1s + 1}$

B.6 $G(s)H(s) = \dfrac{s + 1}{s(0.1s + 1)}$

B.7 $G(s)H(s) = \dfrac{2}{s}$

B.8 $G(s)H(s) = \dfrac{1}{s(s + 8)}$

B.9 $G(s)H(s) = \dfrac{1}{(s + 5)(s + 10)}$

B.10 $G(s)H(s) = \dfrac{1}{s^2 + 2s + 30}$

B.11 $G(s)H(s) = \dfrac{1}{s(s^2 + 2s + 30)}$

B.12 $G(s)H(s) = \dfrac{1}{s(s^2 + 3s + 2)}$

B.13 Construct Bode plots for V_o/V_i for the circuits of Figure PB.1 and determine the lower and upper 3-dB frequencies.

Figure PB.1

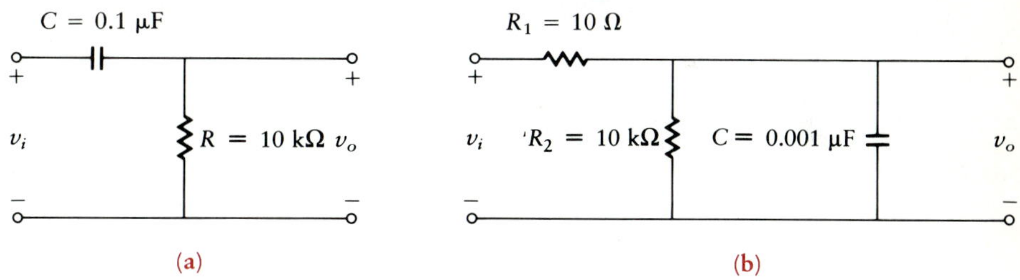

APPENDIX

C THE LAPLACE TRANSFORM

C.0 INTRODUCTION

The Laplace transformation is a mathematical operation that is useful in solving ordinary differential equations. The transformation is defined by the following integral:

$$\mathcal{L}[f(t)] = F(s) = \int_0^\infty f(t)e^{-st}\,dt \tag{C.1}$$

where $f(t)$ is a function of time that is zero for $t < 0$, s is a complex variable, $F(s)$ is the function of s that results when $f(t)$ is Laplace-transformed, and $\mathcal{L}$ is an operational symbol indicating that the function following is to be Laplace-transformed.

In system design, we often encounter a linear differential equation of the following form:

$$\frac{d^n y}{dt^n} + a_{n-1}\frac{d^{n-1}y}{dt^{n-1}} + \cdots + a_1\frac{dy}{dt} + a_0 y(t) = f(t) \tag{C.2}$$

The coefficients a_i are real constants and independent of t or y. Both the dependent variable $y(t)$ and the driving function $f(t)$ are functions of time. When equation (C.2) is Laplace-transformed, a new, or transformed, equation results. The *differential* equation is reduced to an *algebraic* equation with s as the variable. This transformed equation can be manipulated algebraically to solve for the desired quantity in transformed form.

If the time solution is required, the resulting function of s must be inverse-transformed into a function of time. This process, called 'finding the inverse Laplace transform,' is most easily accomplished with the aid of tables of Laplace transforms.

C.1 THE LAPLACE TRANSFORM OF FUNCTIONS

Application of equation (C.1) to various time functions serves as the basis for a table of transform pairs. As an example, consider the unit-step function, $u(t)$, as shown in Figure C.1. The mathematical representation is

$$u(t) = \begin{cases} 0 & t < 0 \\ 1 & t > 0 \end{cases} \tag{C.3}$$

The Laplace transform is given by

$$\mathscr{L}[u(t)] = \int_0^\infty e^{-st}\,dt = \left.\frac{e^{-st}}{s}\right|_0^\infty = \frac{1}{s} \tag{C.4}$$

Table C.1 illustrates important and common Laplace transforms.

Figure C.1 The unit-step function.

C.2 THE LAPLACE TRANSFORM OF OPERATIONS

Besides finding the Laplace transform of known functions of time, $f(t)$, it is also necessary to transform dependent variables, $y(t)$, and derivatives of $y(t)$. The Laplace transform of the derivative is given by

Table C.1 Laplace Transform Table

Ref. no.	$f(t)$	$F(s)$	Ref. no.	$f(t)$	$F(s)$
1	$u(t)$	$\frac{1}{s}$	11	$\cos \omega t$	$\frac{s}{s^2 + \omega^2}$
2	t	$\frac{1}{s^2}$	12	$\frac{K}{\omega}\sin(\omega t + \psi)$ $\psi = \tan^{-1}\frac{\omega}{a_0}$ $K = (a_0^2 + \omega^2)^{1/2}$	$\frac{s + a_0}{s^2 + \omega^2}$
3	t^n	$\frac{n!}{s^{n+1}}$			
4	$\delta(t)$ unit impulse	1	13	$\sinh \omega t$	$\frac{\omega}{s^2 - \omega^2}$
5	e^{-at}	$\frac{1}{s + a}$	14	$\cosh \omega t$	$\frac{s}{s^2 - \omega^2}$
6	$\frac{e^{-at} - e^{-bt}}{b - a}$	$\frac{1}{(s + a)(s + b)}$	15	$e^{-\alpha t}\sin \omega t$	$\frac{\omega}{(s + \alpha)^2 + \omega^2}$
7	te^{-at}	$\frac{1}{(s + a)^2}$	16	$\frac{K}{\omega}e^{-\alpha t}\sin(\omega t + \psi)$ $\psi = \tan^{-1}\frac{\omega}{\alpha_0 - \alpha}$ $K = [(\alpha_0 - \alpha)^2 + \omega^2]^{1/2}$	$\frac{s + \alpha_0}{(s + \alpha)^2 + \omega^2}$
8	$t^n e^{-at}$	$\frac{n!}{(s + a)^{n+1}}$			
9	$(Kt + 1)e^{-at}$ $K = a_0 - a$	$\frac{s + a_0}{(s + a)^2}$	17	$t\cos \omega t$	$\frac{s^2 - \omega^2}{(s^2 + \omega^2)^2}$
10	$\sin \omega t$	$\frac{\omega}{s^2 + \omega^2}$	18	$t\sin \omega t$	$\frac{2\omega^2}{(s^2 + \omega^2)^2}$

$$\mathcal{L}\left[\frac{dy}{dt}\right] = \int_0^\infty \frac{dy}{dt}e^{-st}\,dt \tag{C.5}$$

The integral of Equation (C.5) is evaluated by parts as follows:

$$\int_a^b u\,dv = uv\Big|_a^b - \int_a^b v\,du \tag{C.6}$$

where the following substitutions are made:

$$u = e^{-st} \qquad du = -se^{-st}\,dt$$

$$dv = \left(\frac{dy}{dt}\right) dt \qquad v = y(t) \tag{C.7}$$

The integral is now evaluated as follows:

$$\mathcal{L}\left[\frac{dy}{dt}\right] = y(t)e^{-st}\Big|_0^\infty + s\int_0^\infty y(t)e^{-st}\,dt \tag{C.8}$$

Since the last integral in equation (C.8) is the Laplace transform of $y(t)$, that is,

$$\mathcal{L}[y(t)] = Y(s) = \int_0^\infty y(t)e^{-st}\,dt \tag{C.9}$$

and since $y(t)e^{-st}$ is zero at the upper limit provided that the real part of s is greater than zero, we obtain the result

$$\mathcal{L}\left[\frac{dy}{dt}\right] = -y(0) + sY(s) \tag{C.10}$$

In a similar fashion, the Laplace transform of higher derivatives can be found:

$$\mathcal{L}\left[\frac{d^2y}{dt^2}\right] = s^2Y(s) - sy(0) - \frac{dy}{dt}\Big|_{t=0} \tag{C.11}$$

$$\mathcal{L}\left[\frac{d^ny}{dt^n}\right] = s^nY(s) - s^{n-1}y(0) - s^{n-2}\frac{dy}{dt}\Big|_{t=0} - \cdots - \frac{d^{n-1}y}{dt^{n-1}}\Big|_{t=0} \tag{C.12}$$

For the case of zero initial conditions, equation (C.12) reduces to

$$\mathcal{L}\left[\frac{d^ny}{dt^n}\right] = s^nY(s) \tag{C.13}$$

Hence, provided that initial conditions are equal to zero, finding the Laplace transform is equivalent to replacing the derivative operation with respect to time, d/dt, by the Laplace transform operator, s.

Another important Laplace transform pair, which is required for the solution of linear differential equations, is the transform of the sum of two time functions, as follows:

$$\mathcal{L}[a_1y_1(t) + a_2y_2(t)] = a_1Y_1(s) + a_2Y_2(s) \tag{C.14}$$

This relationship is easily proven by substituting into the defining equation, equation (C.1). The equation states that the Laplace transform is *linear.*

The Laplace transform of the definite integral of a function of time is found by integrating by parts, with the result

$$\mathcal{L}\left[\int_0^t y(\tau)\,d\tau\right] = -\frac{1}{s}e^{-st}\int_0^t y(\tau)\,d\tau\bigg|_0^\infty + \frac{1}{s}\int_0^\infty e^{-st}y(t)\,dt \qquad \text{(C.15)}$$

The first term after the equal sign is zero at the upper limit since $e^{-\infty} \rightarrow 0$. It also vanishes at the lower limit, since the integral is zero. The result then simplifies to

$$\mathcal{L}\left[\int_0^t y(\tau)\,d\tau\right] = \frac{Y(s)}{s} \qquad \text{(C.16)}$$

These transform pairs are summarized in Table C.2. The initial conditions are set equal to zero for the pairs in this table, since, in most applications, zero initial conditions are appropriate to solve the problem.

Table C.2 Operation Transform Pairs*

Function of t	Transformed Function of s
$\dfrac{dy}{dt}$	$sY(s)$
$\dfrac{d^2y}{dt^2}$	$s^2Y(s)$
$a_1y_1(t) + a_2y_2(t)$	$a_1Y_1(s) + a_2Y_2(s)$
$\int_0^t y(t)\,dt$	$\dfrac{1}{s}Y(s)$

*All initial conditions are taken equal to zero.

The Laplace transform operator, s, which has the dimension of seconds^{-1}, is often called the "complex frequency," since it has real and imaginary parts. That is,

$$s = \sigma + j\omega \qquad \text{(C.17)}$$

C.3 SOLUTION OF ORDINARY LINEAR DIFFERENTIAL EQUATIONS

As an example of the transform method of solving differential equations, consider the simple equation

$$\frac{d^2y}{dt^2} + 9y = u(t) \qquad t > 0 \tag{C.18}$$

Let us assume that all initial conditions are equal to zero; $u(t)$ is a unit-step function. The Laplace transform of both sides of the equation is taken, to yield

$$s^2Y(s) + 9Y(s) = \frac{1}{s} \tag{C.19}$$

We solve this for $Y(s)$ to obtain

$$Y(s) = \frac{1}{s(s^2 + 9)} \tag{C.20}$$

If we could find the Laplace transform of equation (C.20) in a table, our work would be finished and the time-function solution, $y(t)$, would be known. Most brief tables of transform pairs would not contain factors of this complexity. We can simplify the process by breaking equation (C.20) into a sum of two parts using a technique known as *partial-fraction expansions.* This technique is described in Section C.4. We find

$$\frac{1}{s(s^2 + 9)} = \frac{A}{s} + \frac{Bs + C}{s^2 + 9} \tag{C.21}$$

Once the constants A, B, and C are evaluated, we obtain the result

$$Y(s) = \frac{1}{9}\left[\frac{1}{s} - \frac{s}{s^2 + 9}\right] \tag{C.22}$$

Although the technique for finding A, B, and C may not yet be clear, it is easy to verify that equation (C.22) is equivalent to equation (C.21) by combining the terms of the latter equation over a common denominator:

$$\frac{1}{9}\left[\frac{1}{s} - \frac{s}{s^2 + 9}\right] = \frac{1}{9}\,\frac{s^2 + 9 - s^2}{s(s^2 + 9)} = \frac{1}{s(s^2 + 9)} \tag{C.23}$$

The inverse transforms of the terms in equation (C.22) are found in Table C.1. This process yields

$$y(t) = \frac{1}{9}[u(t) - (\cos 3t\ u(t)] = \frac{1}{9}u(t)\,(1 - \cos 3t) \tag{C.24}$$

As a second example, consider the differential equation

$$\frac{d^2y}{dt^2} + \omega^2 y = \cos \omega t \qquad \text{for } t > 0 \tag{C.25}$$

with $y(0) = 0 = dy/dt|_{t=0}$, i.e., zero initial conditions. Taking the Laplace transform of both sides and solving for $Y(s)$, we obtain

$$Y(s) = \frac{s}{(s^2 + \omega^2)^2} \tag{C.26}$$

The inverse Laplace transform is found from Table C.1.

$$y(t) = \frac{t}{2\omega} \sin \omega t \tag{C.27}$$

As a third example, consider the second-order differential equation with a step input and with zero initial conditions, as follows:

$$\frac{d^2c}{dt^2} + 2\zeta\omega_n \frac{dc}{dt} + \omega_n^2 c = \omega_n^2 u(t) \tag{C.28}$$

The Laplace transform of both sides of the equation is taken.

$$(s^2 + 2\zeta\omega_n s + \omega_n^2)C(s) = \frac{\omega_n^2}{s} \tag{C.29}$$

This is solved for the variable $C(s)$ to yield

$$C(s) = \frac{\omega_n^2}{s(s^2 + 2\zeta\omega_n s + \omega_n^2)} = \frac{\omega_n^2}{s(s + \zeta\omega_n)^2 + \omega_n^2(1 - \zeta^2)} \tag{C.30}$$

This equation is separated by partial fractions to yield

$$C(s) = \frac{1}{s} - \frac{s + 2\zeta\omega_n}{(s + \zeta\omega_n)^2 + \omega_n^2(1 - \zeta^2)} \tag{C.31}$$

The time solution is found by referring to two transform pairs (1 and 16) in Table C.1.

$$c(t) = u(t) - \frac{1}{\sqrt{1 - \zeta^2}} e^{-\zeta\omega_n t} \sin(\omega_n \sqrt{1 - \zeta^2}\, t + \phi) \tag{C.32}$$

where

$$\phi = \tan^{-1}\frac{\sqrt{1-\zeta^2}}{\zeta} \tag{C.33}$$

C.4 PARTIAL-FRACTION EXPANSION

The inverse Laplace transform of rational fractions must often be found in order to solve a differential equation. The general rational fraction is written

$$Y(s) = \frac{A(s)}{B(s)} \tag{C.34}$$

where $A(s)$ and $B(s)$ are polynomials in s. When the roots of $B(s) = 0$ are found, equation (C.34) can be written as

$$Y(s) = \frac{A(s)}{B(s)} = \frac{A(s)}{(s + s_1)(s + s_2)(s + s_3)\cdots(s + s_q)} \tag{C.35}$$

The inverse transformation is carried out by expanding equation (C.35) into partial fractions as follows:

$$\begin{aligned}\frac{A(s)}{B(s)} &= \frac{A(s)}{(s + s_1)(s + s_2)(s + s_3)\cdots(s + s_q)} \\ &= \frac{K_1}{s + s_1} + \frac{K_2}{s + s_2} + \frac{K_3}{s + s_3} + \cdots + \frac{K_q}{s + s_q}\end{aligned} \tag{C.36}$$

The inverse Laplace transform of each term in equation (C.36) is found by reference to Table C.1. The total time solution is found by summing the time solutions for the separate terms in the equation.

The partial-fraction expansion is an important step in the solution. Depending on the form of $B(s)$, this expansion is carried out as follows:

1. $B(s)$ contains only simple roots: In this case, the K_i are evaluated by multiplying each side of equation (C.36) by $s + s_i$.

$$\begin{aligned}\frac{(s + s_i)A(s)}{B(s)} &= K_1\frac{s + s_i}{s + s_1} + K_2\frac{s + s_i}{s + s_2} \\ &\quad + \cdots + K_i\frac{s + s_i}{s + s_i} + \cdots + K_q\frac{s + s_i}{s + s_q}\end{aligned} \tag{C.37}$$

Since $s + s_i$ is a factor in $B(s)$, it is divided out. When s is set equal to $-s_i$, the term on the left of the equal sign becomes a constant. All terms on the right reduce to zero except K_i. Hence, each constant can be evaluated from the equation

$$K_i = \left.\frac{A(s)\,(s + s_i)}{B(s)}\right|_{s=-s_i} \tag{C.38}$$

The procedure is unaltered if one of the roots is located at the origin. The constant K_o is evaluated similarly:

$$K_o = \frac{sA(s)}{B(s)} \tag{C.39}$$

When complex conjugate roots, $(s + \alpha)^2 + \beta^2$, exist, the procedure is similar:

$$K_{j1} = \left.\frac{(s + \alpha + j\beta)A(s)}{B(s)}\right|_{s=-\alpha-j\beta} \tag{C.40}$$

and

$$K_{j2} = \left.\frac{(s + \alpha - j\beta)A(s)}{B(s)}\right|_{s=-\alpha+j\beta} \tag{C.41}$$

Since K_{j1} and K_{j2} are complex conjugate, the sum of the imaginary parts is zero and the sum of the real parts is twice the real part of either constant. The two terms

$$\frac{K_{j1}}{s + \alpha + j\beta} \quad \text{and} \quad \frac{K_{j2}}{s + \alpha - j\beta} \tag{C.42}$$

combine into a single term, which inverse-transforms into an exponentially decaying sinusoid.

2. $B(s)$ contains multiple-order roots: If the denominator of $Y(s)$ has multiple-order zeros, the procedure of the preceding section must be altered. For example, if

$$\frac{A(s)}{B(s)} = \frac{1}{(s + s_1)\,(s + s_2)^2} \tag{C.43}$$

the partial-fraction expansion must include a second-order term.

$$\frac{1}{(s + s_1)\,(s + s_2)^2} = \frac{K_1}{s + s_1} + \frac{K_{12}}{s + s_2} + \frac{K_{22}}{(s + s_2)^2} \tag{C.44}$$

In general, an nth-order root is expanded as follows:

$$\frac{1}{(s + s_1) \cdots (s + s_i)^n} = \frac{K_1}{s + s_1} + \cdots + \frac{K_{ni}}{(s + s_i)^n}$$

$$+ \frac{K_{(n-1)i}}{(s + s_i)^{n-1}} + \cdots + \frac{K_{1i}}{s + s_i} \tag{C.45}$$

The constants K_i associated with first-order roots are evaluated as above. The constant associated with the highest power, K_{ni}, is evaluated in a manner that is the same as that used for a simple pole. That is, we multiply both sides of equation (C.45) by $(s + s_i)^n$ and let $s = -s_i$. All terms on the right side are zero except the K_{ni} term. The left side reduces to a number. K_{ni} is evaluated as follows:

$$K_{ni} = \left. \frac{(s + s_i)^n A(s)}{B(s)} \right|_{s=-s_i} \tag{C.46}$$

The procedure used for simple roots and to evaluate the constant associated with the highest power, K_{ni}, is insufficient to evaluate any of the other coefficients. These constants are evaluated using differentiation. Both sides of equation (C.45) are multiplied by $(s + s_i)^n$. The resulting equation is differentiated once with respect to s:

$$\frac{d}{ds}\frac{(s + s_i)^n A(s)}{B(s)} = \frac{d}{ds}\frac{(s + s_i)^n}{s + s_1}K_1 + \cdots + K_{(n-1)i} + 2(s + s_i)K_{(n-2)i}$$

$$+ \cdots + (n - 1)(s + s_i)^{n-2}K_{1i} \tag{C.47}$$

By letting $s = -s_i$, all terms on the right except $K_{(n-1)i}$ vanish:

$$K_{(n-1)i} = \frac{1}{(n - 1)!}\frac{d}{ds}\left. \left[\frac{(s + s_i)^n A(s)}{B(s)}\right] \right|_{s=-s_i} \tag{C.48}$$

The process of differentiating and then setting $s = -s_i$ can be repeated until all the unknown constants are determined.

As an example, consider the partial-fraction expansion of the following transfer function:

$$\frac{A(s)}{B(s)} = \frac{4s^3 + s^2 - 22s + 16}{s(s + 2)(s - 2)^2} \tag{C.49}$$

The fraction is broken up as follows:

$$\frac{4s^3 + s^2 - 22s + 16}{s(s+2)(s-2)^2} = \frac{K_1}{s} + \frac{K_2}{s+2} + \frac{K_{13}}{s-2} + \frac{K_{23}}{(s-2)^2} \tag{C.50}$$

The constants corresponding to the simple poles are found first:

$$K_1 = \left.\frac{s(4s^3 + s^2 - 22s + 16)}{s(s+2)(s-2)^2}\right|_{s=0} = 2$$

$$K_2 = \left.\frac{(s+2)(4s^3 + s^2 - 22s + 16)}{s(s+2)(s-2)^2}\right|_{s=-2} = -1 \tag{C.51}$$

K_{23} is found in a similar manner:

$$K_{23} = \left.\frac{(s-2)^2(4s^3 + s^2 - 22s + 16)}{s(s+2)(s-2)^2}\right|_{s=2} = 1 \tag{C.52}$$

K_{13} is found by multiplying equation (C.50) by $(s - 2)^2$ and then differentiating with respect to s:

$$\frac{d}{ds}\left[\frac{4s^3 + s^2 - 22s + 16}{s(s+2)}\right] = \frac{d}{ds}\left[\left(\frac{K_1}{s} + \frac{K_2}{s+2}\right)(s-2)^2\right]\Bigg|_{s=2} + \frac{d}{ds}(s-2)K_{13} + \frac{d}{ds}K_{23} \tag{C.53}$$

which reduces to

$$\begin{aligned} K_{13} &= \frac{d}{ds}\left[\frac{4s^3 + s^2 - 22s + 16}{s^2 + 2s}\right] \\ &= \left.\frac{(s^2 + 2s)(12s^2 + 2s - 22) - (4s^3 + s^2 - 22s + 16)(2s + 2)}{(s^2 + 2s)^2}\right|_{s=2} \\ &= 3 \end{aligned} \tag{C.54}$$

Hence the partial-fraction expansion of equation (C.49) is written as

$$\frac{A(s)}{B(s)} = \frac{2}{s} - \frac{1}{s+2} + \frac{1}{(s-2)^2} + \frac{3}{s-2} \tag{C.55}$$

C.5 ADDITIONAL PROPERTIES OF THE LAPLACE TRANSFORM

Some important relationships involving the Laplace transformation are included in this section.

C.5.1 Real Translation

If $F(s)$ is the Laplace transform of $f(t)$, then

$$\mathcal{L}[f(t-a)u(t-a)] = e^{-as}F(s) \tag{C.56}$$

Multiplication by e^{-as} in the complex frequency plane (s plane) results in translation in the time domain. The function $f(t-a)u(t-a)$ is shown shifted in Figure C.2. As an example of the use of this theorem, suppose it is necessary to find the Laplace transform of a truncated sinusoid representing one period of a sine wave, as shown in Figure C.3. The time function can be formed by taking the difference between a sine wave and another sinusoid that has been shifted in time by $2\pi/\omega$. This is written as

$$f(t) = A \sin \omega t\ u(t) - A \sin \omega\left(t - \frac{2\pi}{\omega}\right) u\left(t - \frac{2\pi}{\omega}\right) \tag{C.57}$$

These terms are transformed as follows:

$$\begin{aligned}\mathcal{L}[f(t)] &= \frac{A\omega}{s^2+\omega^2} - \frac{A\omega}{s^2+\omega^2}e^{-(2\pi/\omega)s} \\ &= \frac{A\omega}{s^2+\omega^2}(1 - e^{-(2\pi/\omega)s})\end{aligned} \tag{C.58}$$

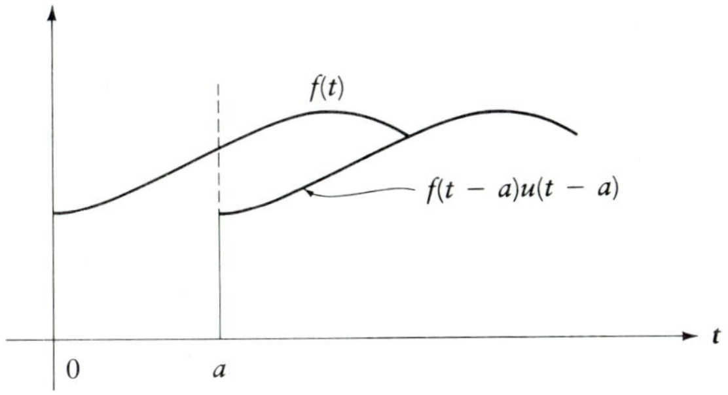

Figure C.2 Shifted function.

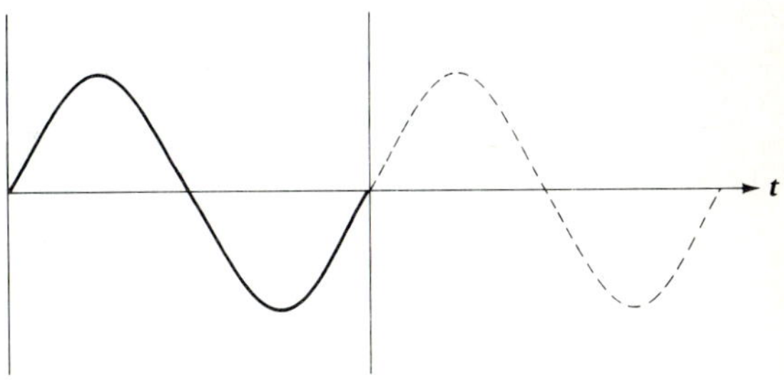

Figure C.3 Truncated sinusoid.

C.5.2 Second Independent Variable

If $F(s, a)$ is the transform of $f(t, a)$, the following relation holds:

$$\mathcal{L}_t\left[\lim_{a \to a_0} f(t, a)\right] = \lim_{a \to a_0} F(s, a) \tag{C.59}$$

where $\mathcal{L}_t$ means the Laplace transform with respect to time. As an example of the use of this equation, consider the transform pair

$$\mathcal{L}_t[e^{-at} \sin \omega t] = \frac{\omega}{(s + a)^2 + \omega^2} \tag{C.60}$$

given in Table C.1. By taking the limit as a approaches zero, a second pair results:

$$\mathcal{L}[\sin \omega t] = \frac{\omega}{s^2 + \omega^2} \tag{C.61}$$

In a similar fashion, differentiation with respect to the quantity a is permissible, yielding

$$\mathcal{L}_t\left[\frac{df(t, a)}{da}\right] = \frac{dF}{da}(s, a) \tag{C.62}$$

As an example, we differentiate the transform pair

$$\mathcal{L}_t[e^{-at}] = \frac{1}{s + a} \tag{C.63}$$

and obtain another pair

$$\mathcal{L}_t[te^{-at}] = \frac{1}{(s + a)^2} \tag{C.64}$$

C.5.3 Final Value and Initial Value Theorems

The final value and initial value theorems are valuable and unique, since they permit finding a time function at either $t = 0$ or $t \to \infty$ directly from the transform without inverting the transformed equation.

The *final value theorem* states:

$$\lim_{t \to \infty} y(t) = \lim_{s \to 0} sY(s) \tag{C.65}$$

provided $y(t)$ is stable (i.e., all poles of $sY(s)$ are in the left half-plane).

The *initial value theorem* states:

$$\lim_{t \to 0} y(t) = \lim_{s \to \infty} sY(s) \tag{C.66}$$

provided the limit exists.

C.5.4 The Convolution Theorem

The *convolution theorem* is expressed as follows: If $F_1(s)$ is the Laplace transform of $f_1(t)$ and $F_2(s)$ is the Laplace transform of $f_2(t)$, then

$$\mathcal{L}\left[\int_0^t f_1(\tau)f_2(t-\tau)\,d\tau\right] = F_1(s)F_2(s) \tag{C.67}$$

The integral on the left is called the *convolution integral.* Thus, the Laplace transform of the convolution of two time functions is the product of the individual Laplace transforms.

Convolution is extremely important in system analysis since the output of a linear system is the convolution of the input with the system's response to an impulse. Thus, if $h(t)$ is the system's response to an impulse and $f(t)$ is the input to the system, the output, $g(t)$, is given by

$$g(t) = \int_0^t h(t-\tau)f(\tau)\,d\tau \tag{C.68}$$

The Laplace transform of equation (C.68) yields the important result

$$G(s) = H(s)F(s) \tag{C.69}$$

Hence the Laplace transform of the output of a linear system is the product of the Laplace transform of the input, $F(s)$, with the Laplace transform of the impulse response, $H(s)$.

APPENDIX

D MANUFACTURERS' DATA SHEETS

This appendix contains copies of representative data sheets for diodes, transistors, voltage regulators, voltage comparators, and op-amps. The information is extracted from the manufacturers' data books. In some cases, only selected information is presented in order to give a sampling of the available data.

The appendix is not meant as a substitute for owning the appropriate data books. We once again strongly urge you to obtain copies of these books. We present only a very brief assortment of those sheets you will need to solve the problems within this text.

Contents

D.1 DIODES

D.1.1 1N4001 through 1N4007 Diode
Courtesy of Motorola Inc.

D.1.2 1N746 through 1N759 Zener Diode
Courtesy of Motorola Inc.

D.2 TRANSISTORS

D.2.1 2N3903 and 2N3904 *npn* Silicon Transistor
Courtesy of Motorola Inc.

D.2.2 2N3905 and 2N3906 *pnp* Silicon Transistor
Courtesy of Motorola Inc.

D.2.3 2N2217 through 2N2222 *npn* General Purpose Transistor
Courtesy of Texas Instruments Inc.

D.2.4 2N3821, 2N3822 and 2N3824 *n*-Channel JFET
Courtesy of Motorola Inc.

D.2.5 3N128 *n*-Channel MOSFET
Courtesy of Motorola Inc.

D.3 MISCELLANEOUS DEVICES

D.3.1 MC7800 IC Voltage Regulator
Courtesy of Motorola Inc.

D.3.2 LM139 Voltage Comparator
Courtesy of National Semiconductor Corp.

D.4 OP-AMPS

D.4.1 μA741 Op-Amp
Courtesy of Fairchild Semiconductor Corp.

D.4.2 LM101 Op-Amp
Courtesy of National Semiconductor Corp.

MOTOROLA
SEMICONDUCTORS
P.O. BOX 20912 • PHOENIX, ARIZONA 85036

1N4001 thru 1N4007

Designers▲Data Sheet

"SURMETIC"▲ RECTIFIERS

. . . subminiature size, axial lead mounted rectifiers for general-purpose low-power applications.

Designers Data for "Worst Case" Conditions

The Designers▲ Data Sheets permit the design of most circuits entirely from the information presented. Limit curves — representing boundaries on device characteristics — are given to facilitate "worst case" design.

LEAD MOUNTED SILICON RECTIFIERS

50-1000 VOLTS DIFFUSED JUNCTION

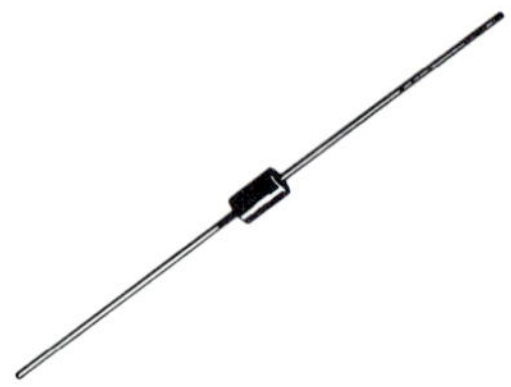

*MAXIMUM RATINGS

Rating	Symbol	1N4001	1N4002	1N4003	1N4004	1N4005	1N4006	1N4007	Unit
Peak Repetitive Reverse Voltage Working Peak Reverse Voltage DC Blocking Voltage	V_{RRM} V_{RWM} V_R	50	100	200	400	600	800	1000	Volts
Non-Repetitive Peak Reverse Voltage (halfwave, single phase, 60 Hz)	V_{RSM}	60	120	240	480	720	1000	1200	Volts
RMS Reverse Voltage	$V_{R(RMS)}$	35	70	140	280	420	560	700	Volts
Average Rectified Forward Current (single phase, resistive load, 60 Hz, see Figure 8, $T_A = 75°C$)	I_O	1.0							Amp
Non-Repetitive Peak Surge Current (surge applied at rated load conditions, see Figure 2)	I_{FSM}	30 (for 1 cycle)							Amp
Operating and Storage Junction Temperature Range	T_J, T_{stg}	-65 to +175							°C

*ELECTRICAL CHARACTERISTICS

Characteristic and Conditions	Symbol	Typ	Max	Unit
Maximum Instantaneous Forward Voltage Drop ($i_F = 1.0$ Amp, $T_J = 25°C$) Figure 1	v_F	0.93	1.1	Volts
Maximum Full-Cycle Average Forward Voltage Drop ($I_O = 1.0$ Amp, $T_L = 75°C$, 1 inch leads)	$V_{F(AV)}$	–	0.8	Volts
Maximum Reverse Current (rated dc voltage)	I_R			μA
$T_J = 25°C$		0.05	10	
$T_J = 100°C$		1.0	50	
Maximum Full-Cycle Average Reverse Current ($I_O = 1.0$ Amp, $T_L = 75°C$, 1 inch leads	$I_{R(AV)}$	–	30	μA

*Indicates JEDEC Registered Data.

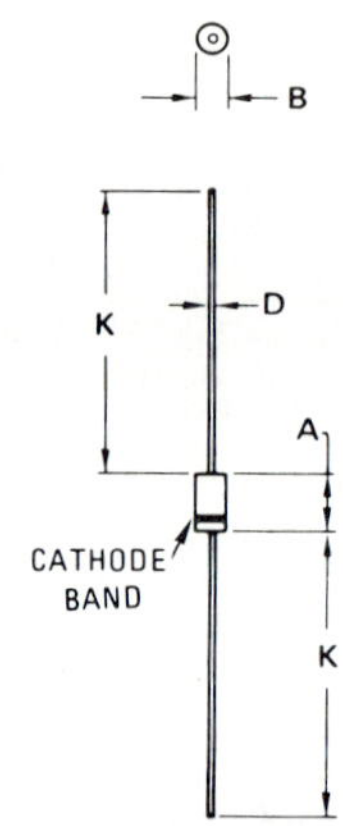

DIM	MILLIMETERS MIN	MILLIMETERS MAX	INCHES MIN	INCHES MAX
A	5.97	6.60	0.235	0.260
B	2.79	3.05	0.110	0.120
D	0.76	0.86	0.030	0.034
K	27.94	–	1.100	–

CASE 59-04
Does Not Conform to DO-41 Outline.

MECHANICAL CHARACTERISTICS

CASE: Transfer Molded Plastic
MAXIMUM LEAD TEMPERATURE FOR SOLDERING PURPOSES: 350°C, 3/8" from case for 10 seconds at 5 lbs. tension
FINISH: All external surfaces are corrosion-resistant, leads are readily solderable
POLARITY: Cathode indicated by color band
WEIGHT: 0.40 Grams (approximately)

1N4001 THRU 1N4007

FIGURE 1 — FORWARD VOLTAGE

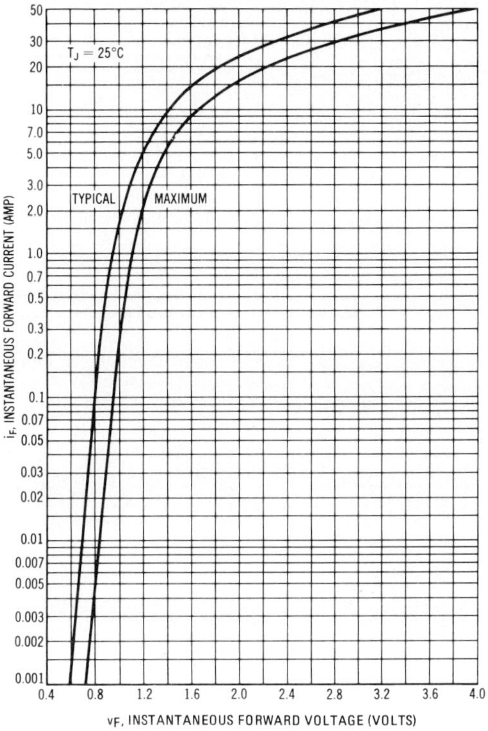

FIGURE 2 — NON-REPETITIVE SURGE CAPABILITY

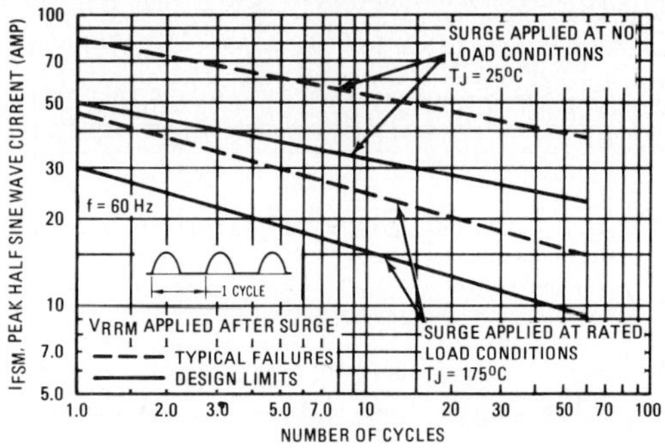

FIGURE 3 — FORWARD VOLTAGE TEMPERATURE COEFFICIENT

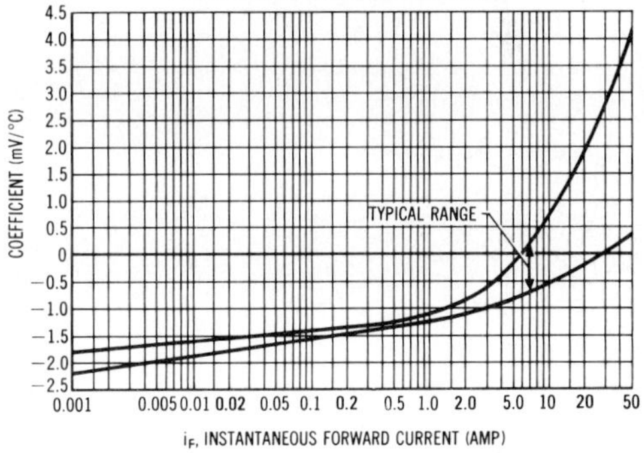

FIGURE 4 — TYPICAL TRANSIENT THERMAL RESISTANCE

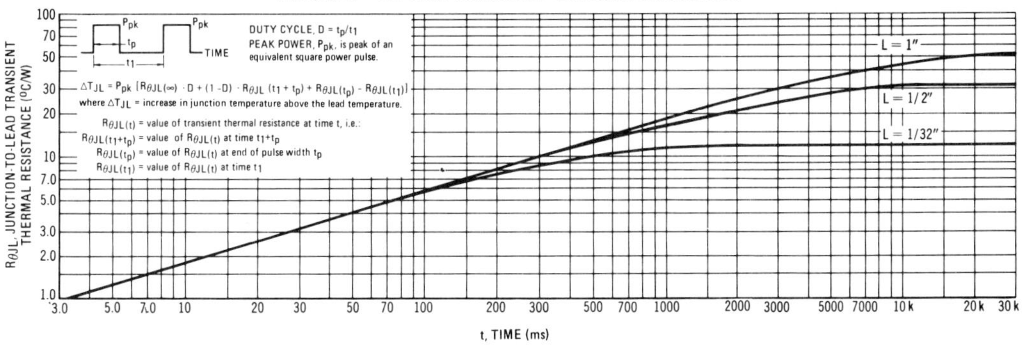

The temperature of the lead should be measured using a thermocouple placed on the lead as close as possible to the tie point. The thermal mass connected to the tie point is normally large enough so that it will not significantly respond to heat surges generated in the diode as a result of pulsed operation once steady-state conditions are achieved. Using the measured value of T_L, the junction temperature may be determined by:

$$T_J = T_L + \Delta T_{JL}.$$

1N4001 THRU 1N4007

CURRENT DERATING DATA

FIGURE 5 – FORWARD POWER DISSIPATION

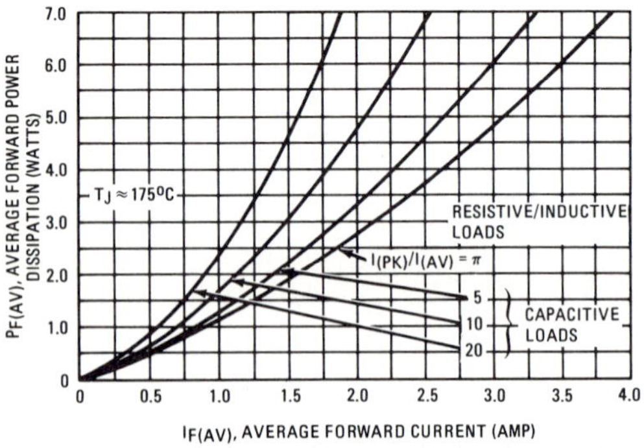

FIGURE 6 – EFFECT OF LEAD LENGTHS, RESISTIVE LOAD

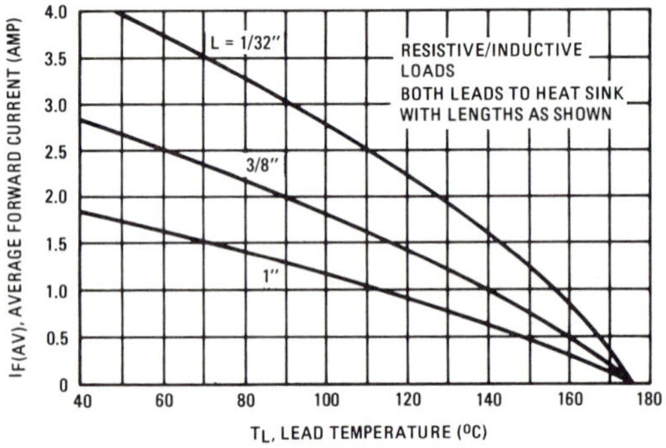

FIGURE 7 – 3/8″ LEAD LENGTH, VARIOUS LOADS

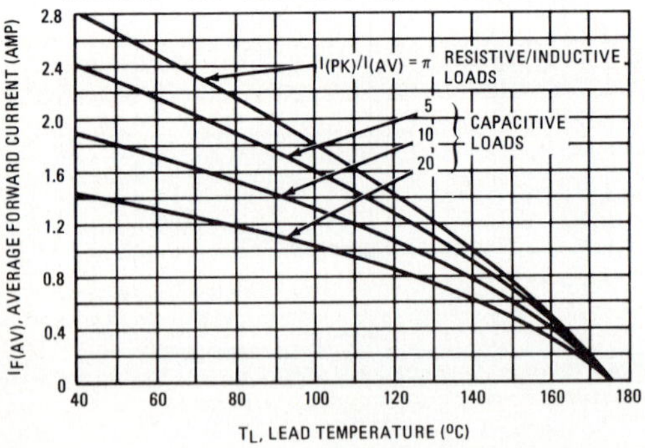

FIGURE 8 – PRINTED CIRCUIT BOARD MOUNTING – VARIOUS LOADS

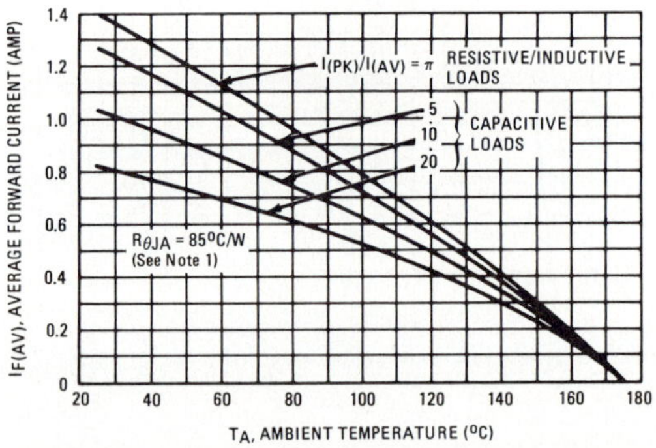

FIGURE 9 – STEADY-STATE THERMAL RESISTANCE

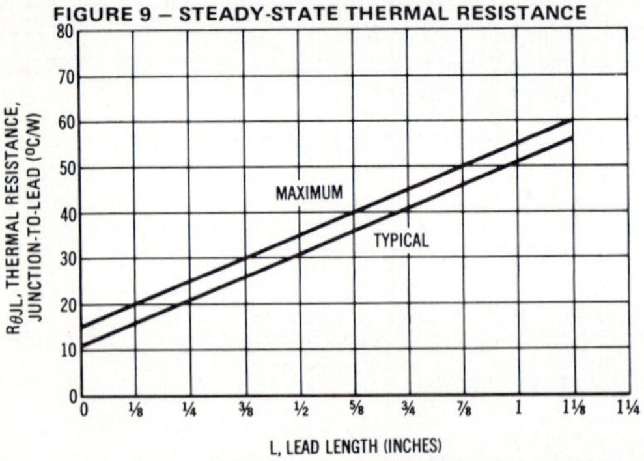

NOTE 1

Data shown for thermal resistance junction-to-ambient ($R_{\theta JA}$) for the mountings shown is to be used as typical guideline values for preliminary engineering or in case the tie point temperature cannot be measured

TYPICAL VALUES FOR $R_{\theta JA}$ IN STILL AIR

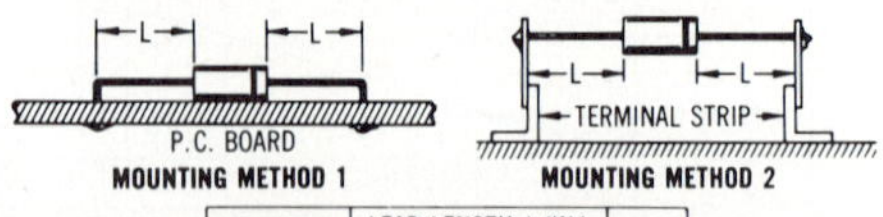

MOUNTING METHOD	LEAD LENGTH, L (IN.)			$R_{\theta JA}$
	1/32	3/8	1	
1	—	75	85	°C/W
2	55	72	85	°C/W

1N4001 THRU 1N4007

TYPICAL DYNAMIC CHARACTERISTICS

FIGURE 10 – FORWARD RECOVERY TIME

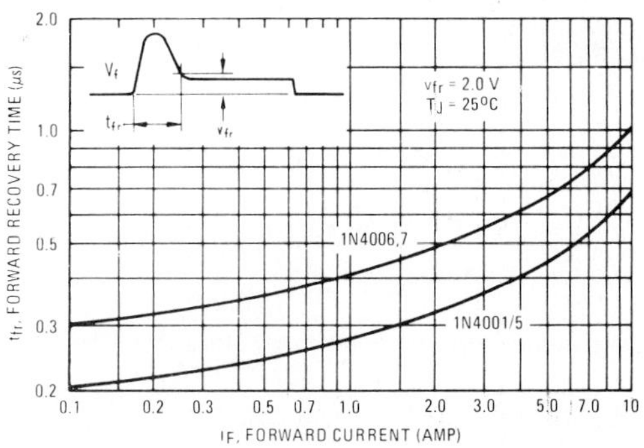

FIGURE 11 – REVERSE RECOVERY TIME

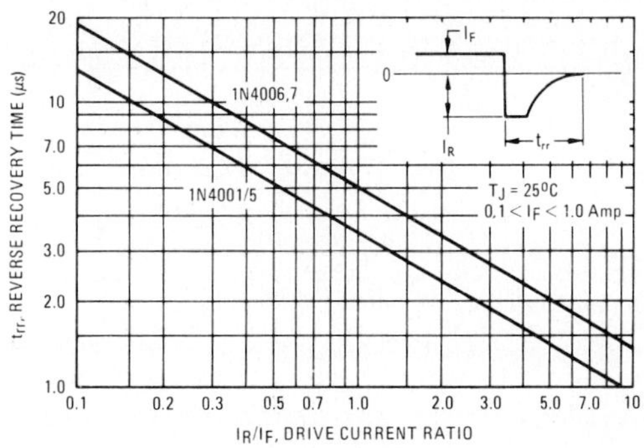

FIGURE 12 – JUNCTION CAPACITANCE

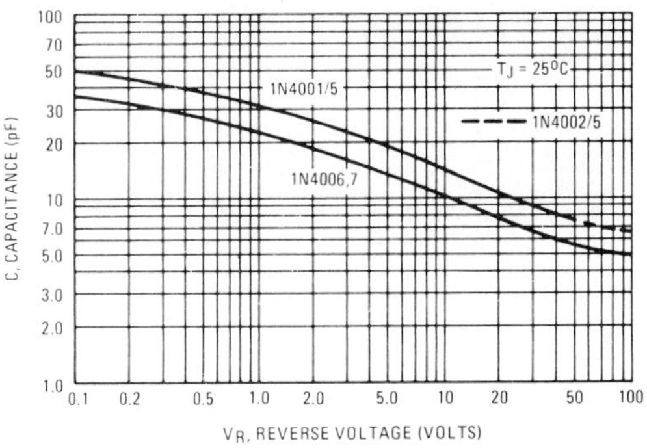

FIGURE 13 – RECTIFICATION WAVEFORM EFFICIENCY FOR SINE WAVE

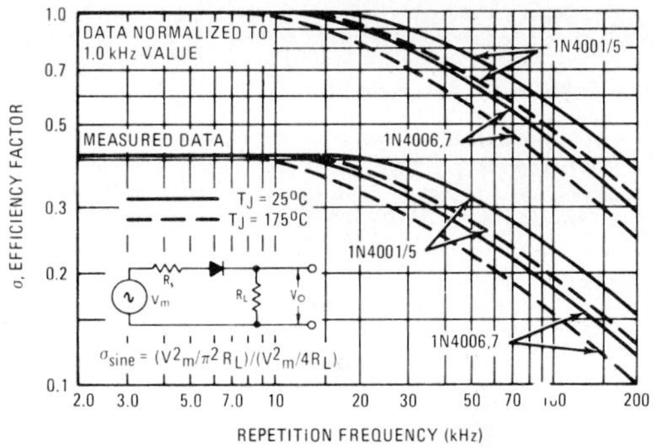

FIGURE 14 – RECTIFICATION WAVEFORM EFFICIENCY FOR SQUARE WAVE

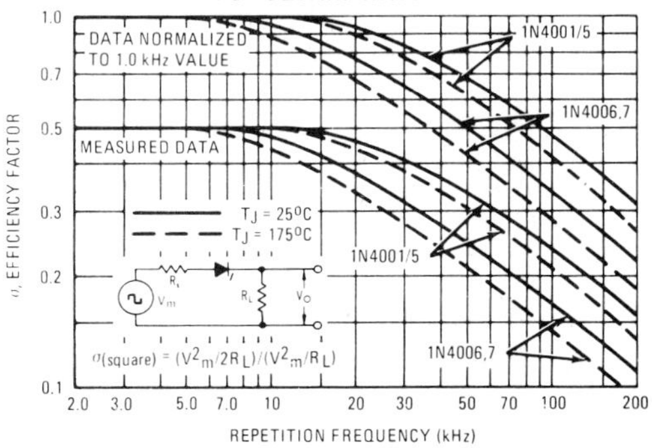

RECTIFIER EFFICIENCY NOTE

The rectification efficiency factor σ shown in Figures 13 and 14 was calculated using the formula:

$$\sigma = \frac{P_{dc}}{P_{rms}} = \frac{\dfrac{V^2_{O(dc)}}{R_L}}{\dfrac{V^2_{O(rms)}}{R_L}} \bullet 100\% = \frac{V^2_{O(dc)}}{V^2_{O(ac)} + V^2_{O(dc)}} \bullet 100\% \quad (1)$$

For a sine wave input $V_m \sin(\omega t)$ to the diode, assumed lossless, the maximum theoretical efficiency factor becomes 40%; for a square wave input of amplitude V_m, the efficiency factor becomes 50%. (A full wave circuit has twice these efficiencies).

As the frequency of the input signal is increased, the reverse recovery time of the diode (Figure 11) becomes significant, resulting in an increasing ac voltage component across R_L which is opposite in polarity to the forward current thereby reducing the value of the efficiency factor σ, as shown in Figures 13 and 14.

It should be emphasized that Figures 13 and 14 show waveform efficiency only; they do not account for diode losses. Data was obtained by measuring the ac component of V_O with a true rms voltmeter and the dc component with a dc voltmeter. The data was used in Equation 1 to obtain points for the Figures.

MOTOROLA

SEMICONDUCTORS

P.O. BOX 20912 • PHOENIX, ARIZONA 85036

Designers Data Sheet

1N746 thru 1N759

1N957A thru 1N986A

1N4370 thru 1N4372

500-MILLIWATT HERMETICALLY SEALED GLASS SILICON ZENER DIODES

- Complete Voltage Range — 2.4 to 110 Volts
- DO-35 Package — Smaller than Conventional DO-7 Package
- Double Slug Type Construction
- Metallurgically Bonded Construction
- Nitride Passivated Die

Designer's Data for "Worst Case" Conditions

The Designer's Data sheets permit the design of most circuits entirely from the information presented. Limit curves — representing boundaries on device characteristics — are given to facilitate "worst case" design.

GLASS ZENER DIODES

500 MILLIWATTS

2.4–110 VOLTS

MAXIMUM RATINGS

Rating	Symbol	Value	Unit
DC Power Dissipation @ $T_L \leq 50^oC$, Lead Length = 3/8"	P_D		
*JEDEC Registration		400	mW
*Derate above $T_L = 50^oC$		3.2	$mW/^oC$
Motorola Device Ratings		500	mW
Derate above $T_L = 50^oC$		3.33	$mW/^oC$
Operating and Storage Junction Temperature Range	T_J, T_{stg}		^oC
*JEDEC Registration		−65 to +175	
Motorola Device Ratings		−65 to +200	

*Indicates JEDEC Registered Data.

MECHANICAL CHARACTERISTICS

MAXIMUM LEAD TEMPERATURE FOR SOLDERING PURPOSES: 230°C, 1/16" from case for 10 seconds

FINISH: All external surfaces are corrosion resistant with readily solderable leads.

POLARITY: Cathode indicated by color band. When operated in zener mode, cathode will be positive with respect to anode.

MOUNTING POSITION: Any

STEADY STATE POWER DERATING

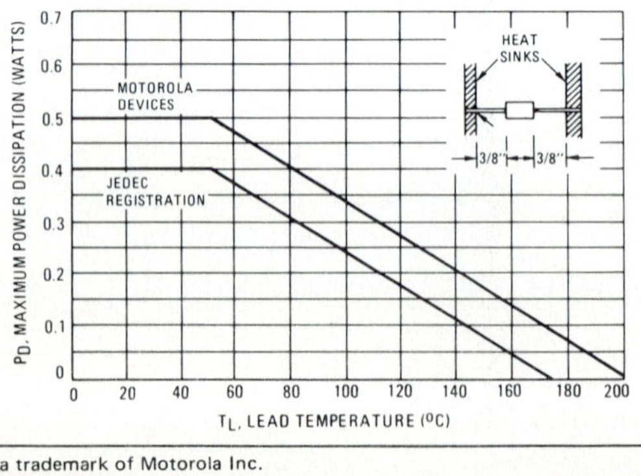

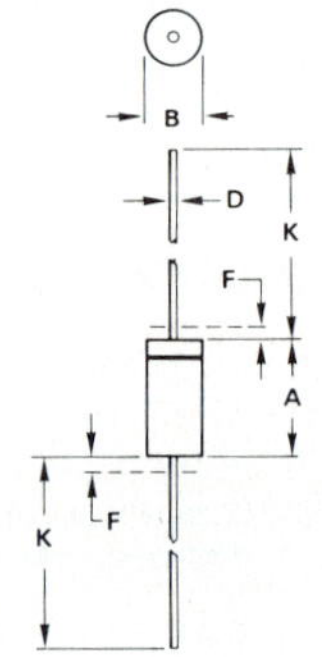

NOTES:
1. PACKAGE CONTOUR OPTIONAL WITHIN A AND B. HEAT SLUGS, IF ANY, SHALL BE INCLUDED WITHIN THIS CYLINDER, BUT NOT SUBJECT TO THE MINIMUM LIMIT OF B.
2. LEAD DIAMETER NOT CONTROLLED IN ZONE F TO ALLOW FOR FLASH, LEAD FINISH BUILDUP AND MINOR IRREGULARITIES OTHER THAN HEAT SLUGS.
3. POLARITY DENOTED BY CATHODE BAND.
4. DIMENSIONING AND TOLERANCING PER ANSI Y14.5, 1973.

DIM	MILLIMETERS MIN	MILLIMETERS MAX	INCHES MIN	INCHES MAX
A	3.05	5.08	0.120	0.200
B	1.52	2.29	0.060	0.090
D	0.46	0.56	0.018	0.022
F	–	1.27	–	0.050
K	25.40	38.10	1.000	1.500

All JEDEC dimensions and notes apply.

CASE 299-02
DO-204AH
(DO-35)

Designer's is a trademark of Motorola Inc.

DS 7021R3

ELECTRICAL CHARACTERISTICS (T_A = 25°C, V_F = 1.5 V max at 200 mA for all types)

Type Number (Note 1)	Nominal Zener Voltage V_Z @ I_{ZT} (Note 2) Volts	Test Current I_{ZT} mA	Maximum Zener Impedance Z_{ZT} @ I_{ZT} (Note 3) Ohms	*Maximum DC Zener Current I_{ZM} (Note 4) mA		Maximum Reverse Leakage Current T_A = 25°C I_R @ V_R = 1 V μA	Maximum Reverse Leakage Current T_A = 150°C I_R @ V_R = 1 V μA
1N4370	2.4	20	30	150	190	100	200
1N4371	2.7	20	30	135	165	75	150
1N4372	3.0	20	29	120	150	50	100
1N746	3.3	20	28	110	135	10	30
1N747	3.6	20	24	100	125	10	30
1N748	3.9	20	23	95	115	10	30
1N749	4.3	20	22	85	105	2	30
1N750	4.7	20	19	75	95	2	30
1N751	5.1	20	17	70	85	1	20
1N752	5.6	20	11	65	80	1	20
1N753	6.2	20	7	60	70	0.1	20
1N754	6.8	20	5	55	65	0.1	20
1N755	7.5	20	6	50	60	0.1	20
1N756	8.2	20	8	45	55	0.1	20
1N757	9.1	20	10	40	50	0.1	20
1N758	10	20	17	35	45	0.1	20
1N759	12	20	30	30	35	0.1	20

Type Number (Note 1)	Nominal Zener Voltage V_Z (Note 2) Volts	Test Current I_{ZT} mA	Maximum Zener Impedance (Note 3) Z_{ZT} @ I_{ZT} Ohms	Z_{ZK} @ I_{ZK} Ohms	I_{ZK} mA	*Maximum DC Zener Current I_{ZM} (Note 4) mA		Maximum Reverse Current I_R Maximum μA	Test Voltage Vdc V_R 5%	Test Voltage Vdc V_R 10%
1N957A	6.8	18.5	4.5	700	1.0	47	61	150	5.2	4.9
1N958A	7.5	16.5	5.5	700	0.5	42	55	75	5.7	5.4
1N959A	8.2	15	6.5	700	0.5	38	50	50	6.2	5.9
1N960A	9.1	14	7.5	700	0.5	35	45	25	6.9	6.6
1N961A	10	12.5	8.5	700	0.25	32	41	10	7.6	7.2
1N962A	11	11.5	9.5	700	0.25	28	37	5	8.4	8.0
1N963A	12	10.5	11.5	700	0.25	26	34	5	9.1	8.6
1N964A	13	9.5	13	700	0.25	24	32	5	9.9	9.4
1N965A	15	8.5	16	700	0.25	21	27	5	11.4	10.8
1N966A	16	7.8	17	700	0.25	19	37	5	12.2	11.5
1N967A	18	7.0	21	750	0.25	17	23	5	13.7	13.0
1N968A	20	6.2	25	750	0.25	15	20	5	15.2	14.4
1N969A	22	5.6	29	750	0.25	14	18	5	16.7	15.8
1N970A	24	5.2	33	750	0.25	13	17	5	18.2	17.3
1N971A	27	4.6	41	750	0.25	11	15	5	20.6	19.4
1N972A	30	4.2	49	1000	0.25	10	13	5	22.8	21.6
1N973A	33	3.8	58	1000	0.25	9.2	12	5	25.1	23.8
1N974A	36	3.4	70	1000	0.25	8.5	11	5	27.4	25.9
1N975A	39	3.2	80	1000	0.25	7.8	10	5	29.7	28.1
1N976A	43	3.0	93	1500	0.25	7.0	9.6	5	32.7	31.0
1N977A	47	2.7	105	1500	0.25	6.4	8.8	5	35.8	33.8
1N978A	51	2.5	125	1500	0.25	5.9	8.1	5	38.8	36.7
1N979A	56	2.2	150	2000	0.25	5.4	7.4	5	42.6	40.3
1N980A	62	2.0	185	2000	0.25	4.9	6.7	5	47.1	44.6
1N981A	68	1.8	230	2000	0.25	4.5	6.1	5	51.7	49.0
1N982A	75	1.7	270	2000	0.25	1.0	5.5	5	56.0	54.0
1N983A	82	1.5	330	3000	0.25	3.7	5.0	5	62.2	59.0
1N984A	91	1.4	400	3000	0.25	3.3	4.5	5	69.2	65.5
1N985A	100	1.3	500	3000	0.25	3.0	4.5	5	76	72
1N986A	110	1.1	750	4000	0.25	2.7	4.1	5	83.6	79.2

NOTE 1. TOLERANCE AND VOLTAGE DESIGNATION

Tolerance Designation

The type numbers shown have tolerance designations as follows:

1N4370 series: ±10%, suffix A for ±5% units.

1N746 series: ±10%, suffix A for ±5% units.

1N957 series: suffix A for ±10% units,
suffix B for ±5% units.

Voltage Designation

To designate units with zener voltages other than those listed, the Motorola type number should be modified as shown below. Unless otherwise specified, the electrical characteristics other than the nominal voltage (V_Z) and test voltage for leakage current will conform to the characteristics of the next higher voltage type shown in the table.

EXAMPLE: 1N746 series, 1N4370 series variations

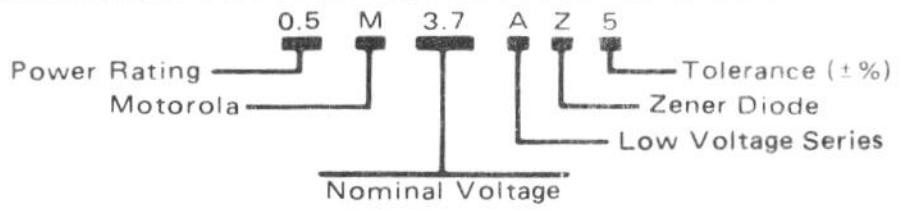

EXAMPLE: 1N957 series variations

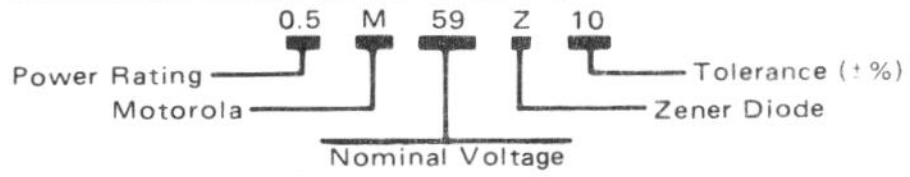

Matched Sets for Closer Tolerances or Higher Voltages

Series matched sets make zener voltages in excess of 100 volts or tolerances of less than 5% possible as well as providing lower temperature coefficients, lower dynamic impedance and greater power handling ability.

For Matched Sets or other special circuit requirements, contact your Motorola Sales Representative.

NOTE 2. ZENER VOLTAGE (V_Z) MEASUREMENT

Nominal zener voltage is measured with the device junction in thermal equilibrium at the lead temperature of 30°C ±1°C and 3/8″ lead length.

NOTE 3. ZENER IMPEDANCE (Z_Z) DERIVATION

Z_{ZT} and Z_{ZK} are measured by dividing the ac voltage drop across the device by the ac current applied. The specified limits are for $I_Z(ac) = 0.1\ I_Z(dc)$ with the ac frequency = 60 Hz.

NOTE 4. MAXIMUM ZENER CURRENT RATINGS (I_{ZM})

Maximum zener current ratings are based on the maximum voltage of a 10% 1N746 type unit or a 20% 1N957 type unit. For closer tolerance units (10% or 5%) or units where the actual zener voltage (V_Z) is known at the operating point, the maximum zener current may be increased and is limited by the derating curve.

APPLICATION NOTE

Since the actual voltage available from a given zener diode is temperature dependent, it is necessary to determine junction temperature under any set of operating conditions in order to calculate its value. The following procedure is recommended:

Lead Temperature, T_L, should be determined from:

$$T_L = \theta_{LA} P_D + T_A$$

θ_{LA} is the lead-to-ambient thermal resistance (°C/W) and P_D is the power dissipation. The value for θ_{LA} will vary and depends on the device mounting method. θ_{LA} is generally 30-40°C/W for the various clips and tie points in common use and for printed circuit board wiring.

The temperature of the lead can also be measured using a thermocouple placed on the lead as close as possible to the tie point. The thermal mass connected to the tie point is normally large enough so that it will not significantly respond to heat surges generated in the diode as a result of pulsed operation once steady-state conditions are achieved. Using the measured value of T_L, the junction temperature may be determined by:

$$T_J = T_L + \Delta T_{JL}$$

ΔT_{JL} is the increase in junction temperature above the lead temperature and may be found from Figure 1 for dc power.

$$\Delta T_{JL} = \theta_{JL} P_D$$

For worst-case design, using expected limits of I_Z, limits of P_D and the extremes of $T_J(\Delta T_J)$ may be estimated. Changes in voltage, V_Z, can then be found from:

$$\Delta V = \theta_{VZ} \Delta T_J$$

θ_{VZ}, the zener voltage temperature coefficient, is found from Figures 3 and 4.

Under high power-pulse operation, the zener voltage will vary with time and may also be affected significantly by the zener resistance. For best regulation, keep current excursions as low as possible.

Surge limitations are given in Figure 6. They are lower than would be expected by considering only junction temperature, as current crowding effects cause temperatures to be extremely high in small spots, resulting in device degradation should the limits of Figure 6 be exceeded.

FIGURE 1 – TYPICAL THERMAL RESISTANCE

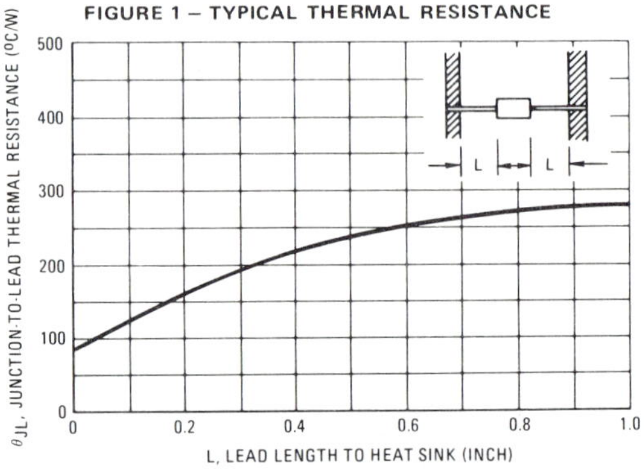

FIGURE 2 – TYPICAL LEAKAGE CURRENT

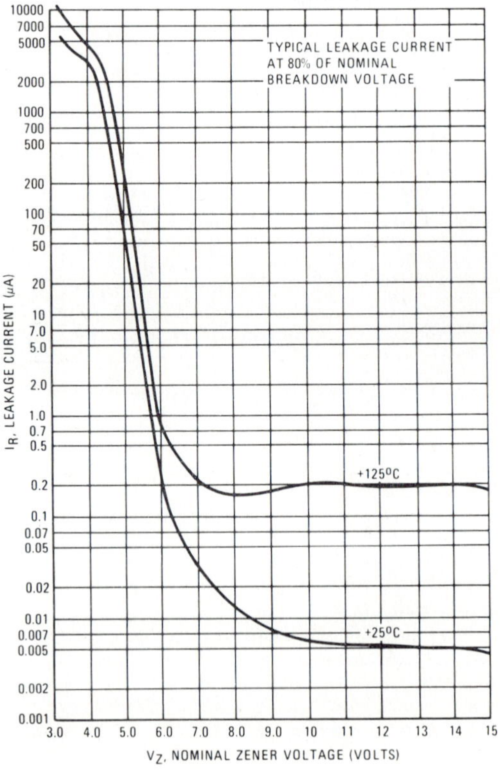

1N746–1N759 • 1N957–1N986A • 1N4370–1N4372

FIGURE 3 – TEMPERATURE COEFFICIENTS

(–55°C to +150°C temperature range; 90% of the units are in the ranges indicated.)

a – RANGE FOR UNITS TO 12 VOLTS

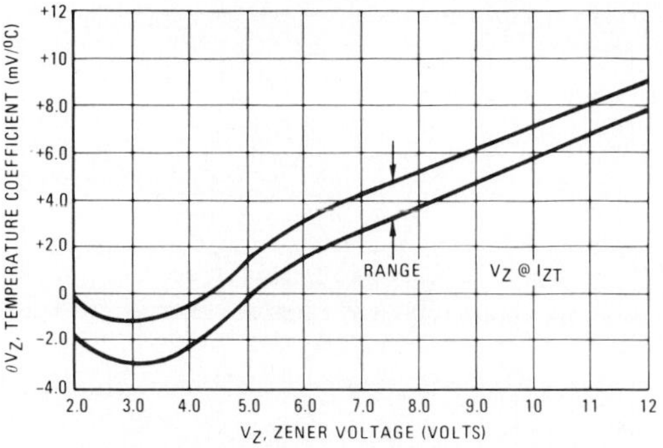

b – RANGE FOR UNITS 12 TO 100 VOLTS

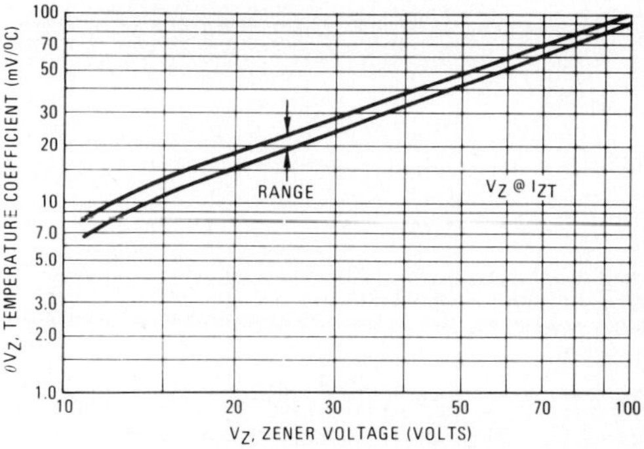

FIGURE 4 – EFFECT OF ZENER CURRENT

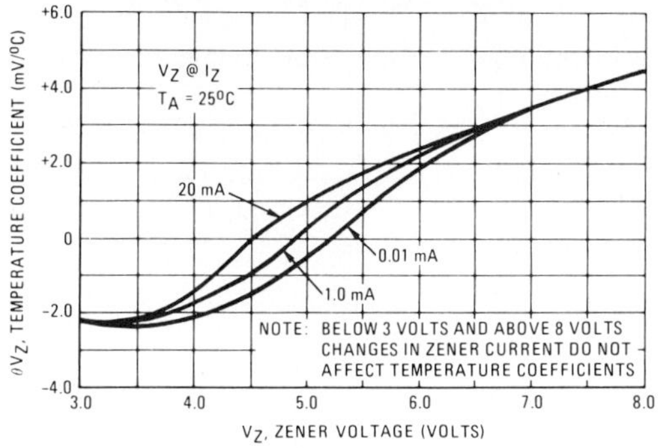

FIGURE 5 – TYPICAL CAPACITANCE

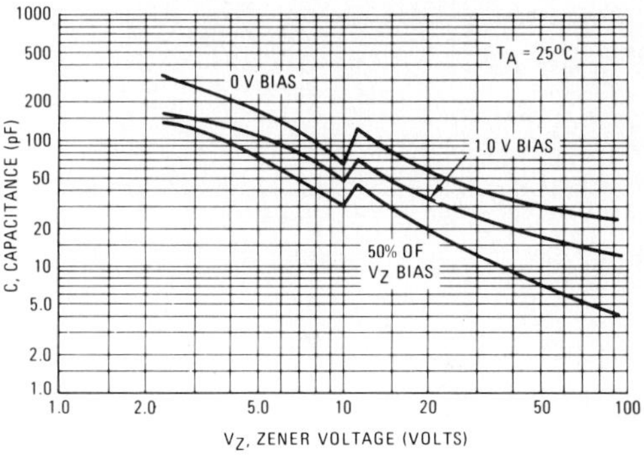

FIGURE 6 – MAXIMUM SURGE POWER

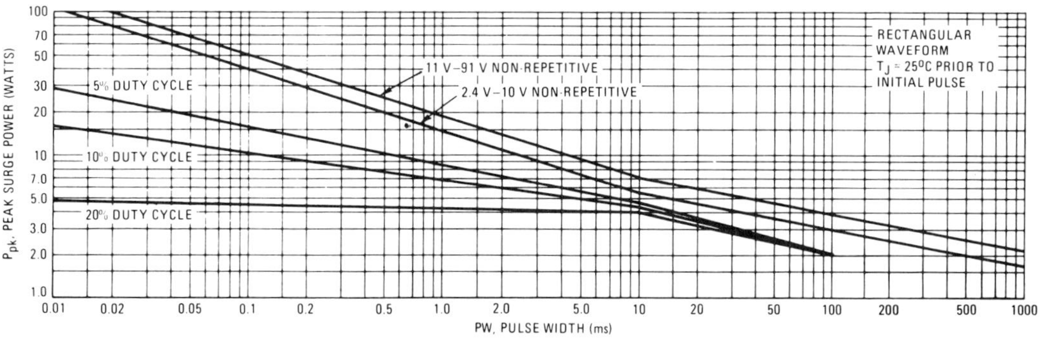

This graph represents 90 percentil data points.
For worst-case design characteristics, multiply surge power by 2/3.

FIGURE 7 – EFFECT OF ZENER CURRENT ON ZENER IMPEDANCE

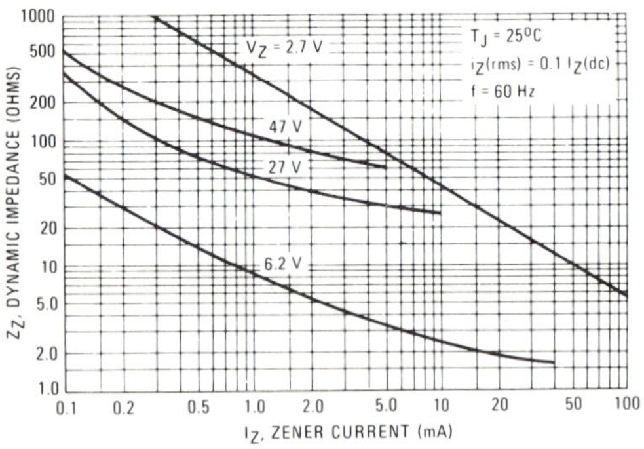

FIGURE 8 – EFFECT OF ZENER VOLTAGE ON ZENER IMPEDANCE

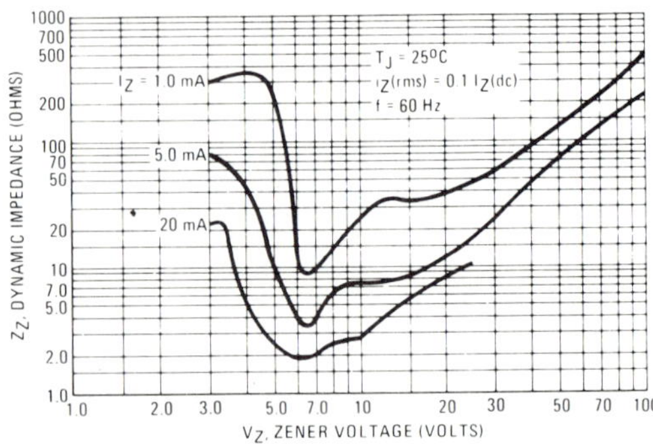

FIGURE 9 – TYPICAL NOISE DENSITY

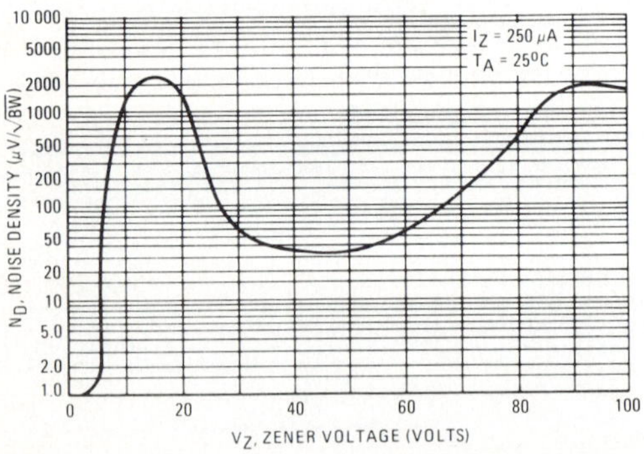

FIGURE 10 – NOISE DENSITY MEASUREMENT METHOD

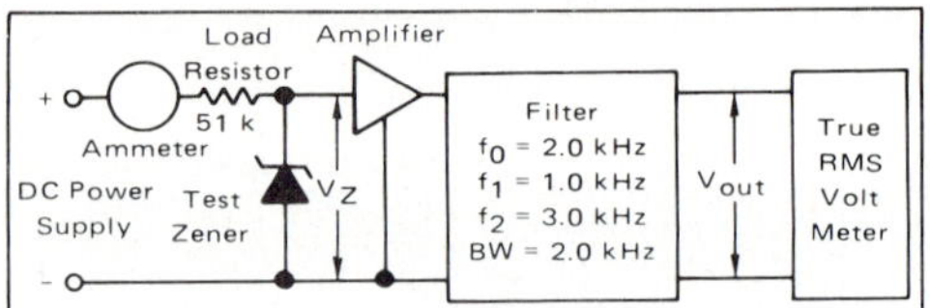

$$\text{Noise Density (Volts Per Square Root Bandwidth)} = \frac{V_{out}}{\text{Overall Gain}\sqrt{BW}}$$

Where: BW = Filter Bandwidth (Hz)
V_{out} = Output Noise (Volts RMS)

The input voltage and load resistance are high so that the zener diode is driven from a constant current source. The amplifier is low noise so that the amplifier noise is negligible compared to that of the test zener. The filter bandpass is known so that the noise density can be calculated from the formula shown.

FIGURE 11 – TYPICAL FORWARD CHARACTERISTICS

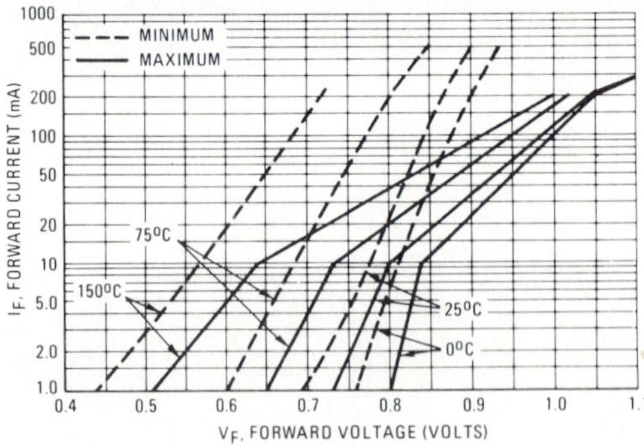

1N746–1N759 • 1N957–1N986A • 1N4370–1N4372

FIGURE 12 – ZENER VOLTAGE versus ZENER CURRENT – V_Z = 1 THRU 16 VOLTS

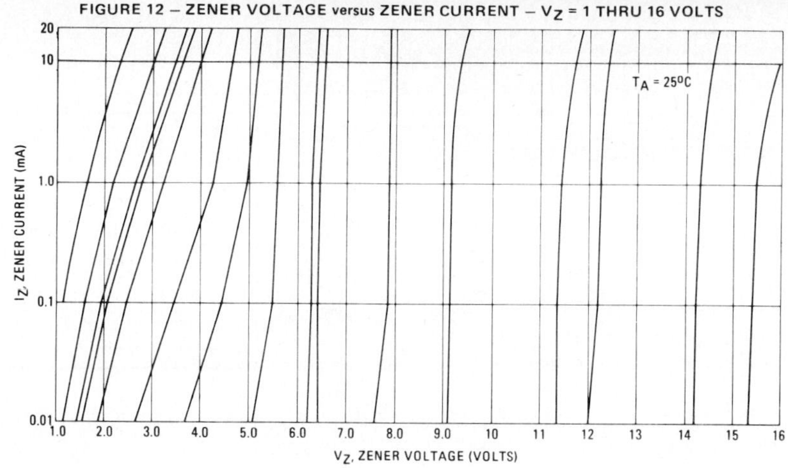

FIGURE 13 – ZENER VOLTAGE versus ZENER CURRENT – V_Z = 15 THRU 30 VOLTS

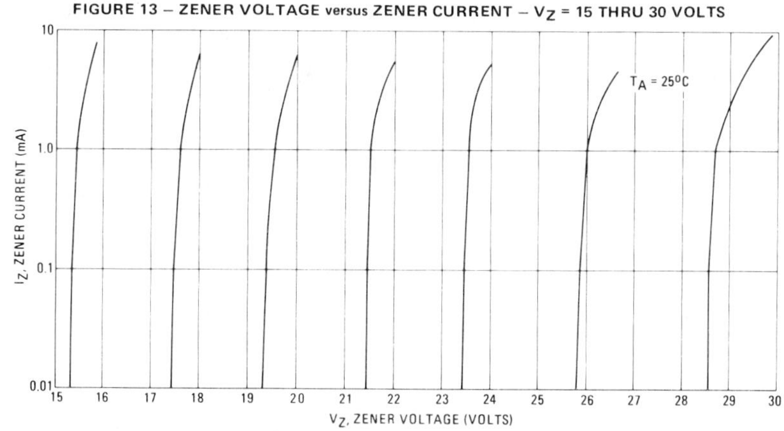

FIGURE 14 – ZENER VOLTAGE versus ZENER CURRENT – V_Z = 30 THRU 105 VOLTS

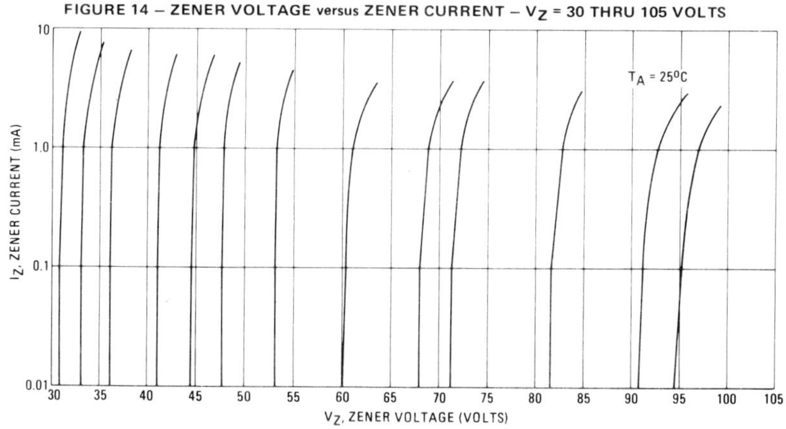

MOTOROLA

SEMICONDUCTORS

P.O. BOX 20912 • PHOENIX, ARIZONA 85036

2N3903
2N3904

NPN SILICON ANNULAR TRANSISTORS

. . . designed for general purpose switching and amplifier applications and for complementary circuitry with types 2N3905 and 2N3906.

- High Voltage Ratings — $V_{(BR)CEO}$ = 40 Volts (Min)
- Current Gain Specified from 100 μA to 100 mA
- Complete Switching and Amplifier Specifications
- Low Capacitance — C_{ob} = 4.0 pF (Max)

NPN SILICON SWITCHING & AMPLIFIER TRANSISTORS

MAXIMUM RATINGS

Rating	Symbol	Value	Unit
*Collector-Emitter Voltage	V_{CEO}	40	Vdc
*Collector-Base Voltage	V_{CBO}	60	Vdc
*Emitter-Base Voltage	V_{EBO}	6.0	Vdc
*Collector Current — Continuous	I_C	200	mAdc
**Total Device Dissipation @ T_A = 25°C Derate above 25°C	P_D	625 5.0	mW mW/°C
Total Power Dissipation @ T_A = 60°C	P_D	450	mW
**Total Device Dissipation @ T_C = 25°C Derate above 25°C	P_D	1.5 12	Watts mW/°C
**Operating and Storage Junction Temperature Range	T_J, T_{stg}	−55 to 150	°C

THERMAL CHARACTERISTICS

Characteristic	Symbol	Max	Unit
Thermal Resistance, Junction to Case	$R_{\theta JC}$	83.3	°C/W
Thermal Resistance, Junction to Ambient	$R_{\theta JA}$	200	°C/W

*Indicates JEDEC Registered Data.
**Motorola guarantees this data in addition to the JEDEC Registered Data.

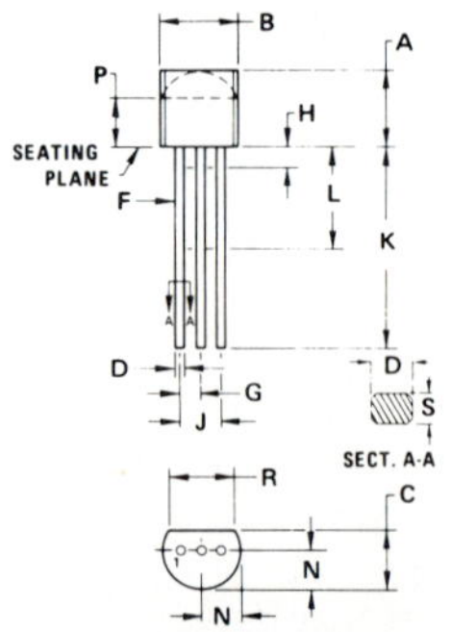

NOTES:
1. CONTOUR OF PACKAGE BEYOND ZONE "P" IS UNCONTROLLED.
2. DIM "F" APPLIES BETWEEN "H" AND "L". DIM "D" & "S" APPLIES BETWEEN "L" & 12.70 mm (0.5") FROM SEATING PLANE. LEAD DIM IS UNCONTROLLED IN "H" & BEYOND 12.70 mm (0.5") FROM SEATING PLANE.

DIM	MILLIMETERS MIN	MILLIMETERS MAX	INCHES MIN	INCHES MAX
A	4.32	5.33	0.170	0.210
B	4.44	5.21	0.175	0.205
C	3.18	4.19	0.125	0.165
D	0.41	0.56	0.016	0.022
F	0.41	0.48	0.016	0.019
G	1.14	1.40	0.045	0.055
H	–	2.54	–	0.100
J	2.41	2.67	0.095	0.105
K	12.70	–	0.500	–
L	6.35	–	0.250	–
N	2.03	2.67	0.080	0.105
P	2.92	–	0.115	–
R	3.43	–	0.135	–
S	0.36	0.41	0.014	0.016

All JEDEC dimensions and notes apply.

CASE 29-02
(TO-226AA)

EQUIVALENT SWITCHING TIME TEST CIRCUITS

FIGURE 1 — TURN-ON TIME

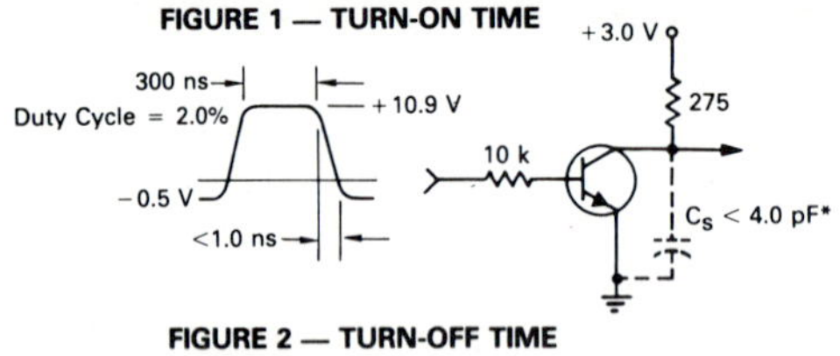

FIGURE 2 — TURN-OFF TIME

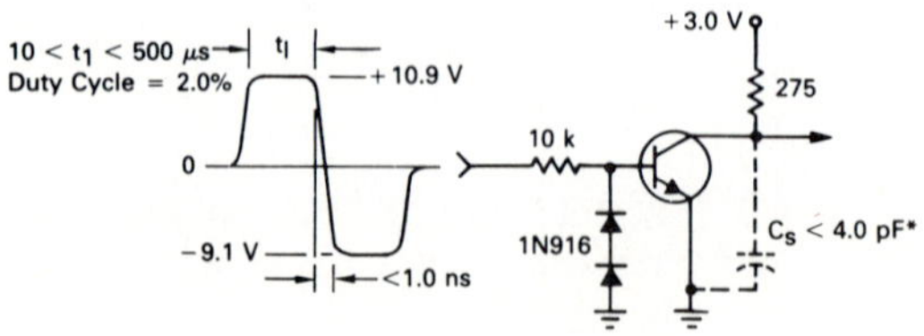

*Total shunt capacitance of test jig and connectors

*ELECTRICAL CHARACTERISTICS (T_A = 25°C unless otherwise noted.)

Characteristic		Symbol	Min	Max	Unit
OFF CHARACTERISTICS					
Collector-Emitter Breakdown Voltage(1) (I_C = 1.0 mAdc, I_B = 0)		$V_{(BR)CEO}$	40	—	Vdc
Collector-Base Breakdown Voltage (I_C = 10 μAdc, I_E = 0)		$V_{(BR)CBO}$	60	—	Vdc
Emitter-Base Breakdown Voltage (I_E = 10 μAdc, I_C = 0)		$V_{(BR)EBO}$	6.0	—	Vdc
Collector Cutoff Current (V_{CE} = 30 Vdc, $V_{EB(off)}$ = 3.0 Vdc)		I_{CEX}	—	50	nAdc
Base Cutoff Current (V_{CE} = 30 Vdc, $V_{EB(off)}$ = 3.0 Vdc)		I_{BL}	—	50	nAdc
ON CHARACTERISTICS(1)					
DC Current Gain		h_{FE}			—
(I_C = 0.1 mAdc, V_{CE} = 1.0 Vdc)	2N3903		20	—	
	2N3904		40	—	
(I_C = 1.0 mAdc, V_{CE} = 1.0 Vdc)	2N3903		35	—	
	2N3904		70	—	
(I_C = 10 mAdc, V_{CE} = 1.0 Vdc)	2N3903		50	150	
	2N3904		100	300	
(I_C = 50 mAdc, V_{CE} = 1.0 Vdc)	2N3903		30	—	
	2N3904		60	—	
(I_C = 100 mAdc, V_{CE} = 1.0 Vdc)	2N3903		15	—	
	2N3904		30	—	
Collector-Emitter Saturation Voltage		$V_{CE(sat)}$			Vdc
(I_C = 10 mAdc, I_B = 1.0 mAdc)			—	0.2	
(I_C = 50 mAdc, I_B = 5.0 mAdc)			—	0.3	
Base-Emitter Saturation Voltage		$V_{BE(sat)}$			Vdc
(I_C = 10 mAdc, I_B = 1.0 mAdc)			0.65	0.85	
(I_C = 50 mAdc, I_B = 5.0 mAdc)			—	1.0	
SMALL-SIGNAL CHARACTERISTICS					
Current-Gain — Bandwidth Product		f_T			MHz
(I_C = 10 mAdc, V_{CE} = 20 Vdc, f = 100 MHz)	2N3903		150	—	
	2N3904		200	—	
Output Capacitance (V_{CB} = 5.0 Vdc, I_E = 0, f = 100 kHz)		C_{obo}	—	4.0	pF
Input Capacitance (V_{BE} = 0.5 Vdc, I_C = 0, f = 100 kHz)		C_{ibo}	—	8.0	pF
Input Impedance		h_{ie}			kΩ
(I_C = 1.0 mAdc, V_{CE} = 10 Vdc, f = 1.0 kHz)	2N3903		0.5	8.0	
	2N3904		1.0	10	
Voltage Feedback Ratio		h_{re}			X 10^{-4}
(I_C = 1.0 mAdc, V_{CE} = 10 Vdc, f = 1.0 kHz)	2N3903		0.1	5.0	
	2N3904		0.5	8.0	
Small-Signal Current Gain		h_{fe}			—
(I_C = 1.0 mAdc, V_{CE} = 10 Vdc, f = 1.0 kHz)	2N3903		50	200	
	2N3904		100	400	
Output Admittance (I_C = 1.0 mAdc, V_{CE} = 10 Vdc, f = 1.0 kHz)		h_{oe}	1.0	40	μmhos
Noise Figure		NF			dB
(I_C = 100 μAdc, V_{CE} = 5.0 Vdc, R_S = 1.0 kΩ, f = 10 Hz to 15.7 kHz)	2N3903		—	6.0	
	2N3904		—	5.0	
SWITCHING CHARACTERISTICS					
Delay Time	(V_{CC} = 3.0 Vdc, $V_{BE(off)}$ = 0.5 Vdc, I_C = 10 mAdc, I_{B1} = 1.0 mAdc)	t_d	—	35	ns
Rise Time		t_r	—	50	ns
Storage Time	(V_{CC} = 3.0 Vdc, I_C = 10 mAdc, I_{B1} = I_{B2} = 1.0 mAdc) 2N3903	t_s	—	800	ns
	2N3904		—	900	
Fall Time		t_f	—	90	ns

(1) Pulse Test: Pulse Width ≤ 300 μs, Duty Cycle ≤ 2.0%.

TYPICAL NOISE CHARACTERISTICS
(V_{CE} = 5.0 Vdc, T_A = 25°C)

FIGURE 3 – NOISE VOLTAGE

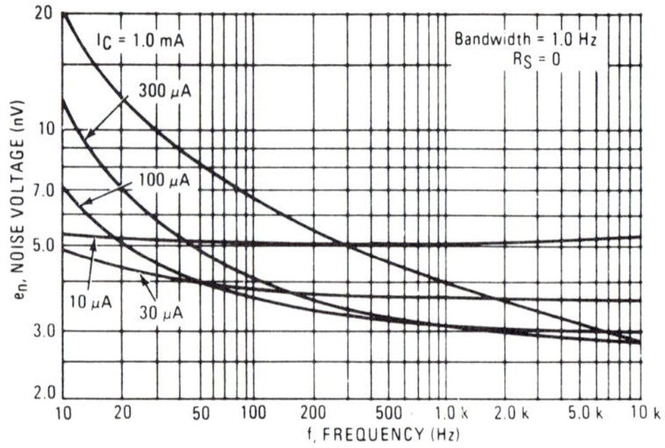

FIGURE 4 – NOISE CURRENT

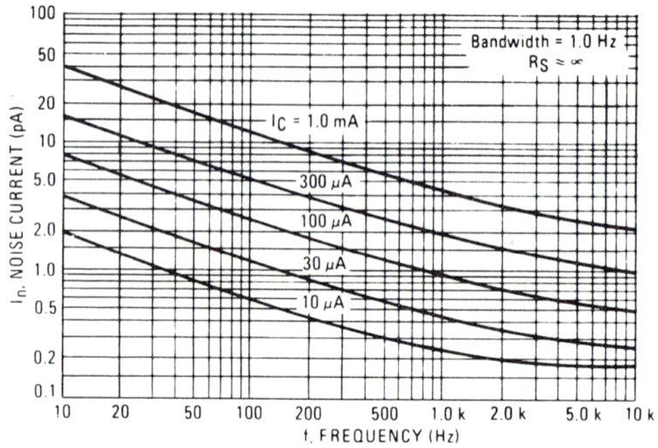

NOISE FIGURE CONTOURS
(V_{CE} = 5.0 Vdc, T_A = 25°C)

FIGURE 5 – NARROW BAND, 100 Hz

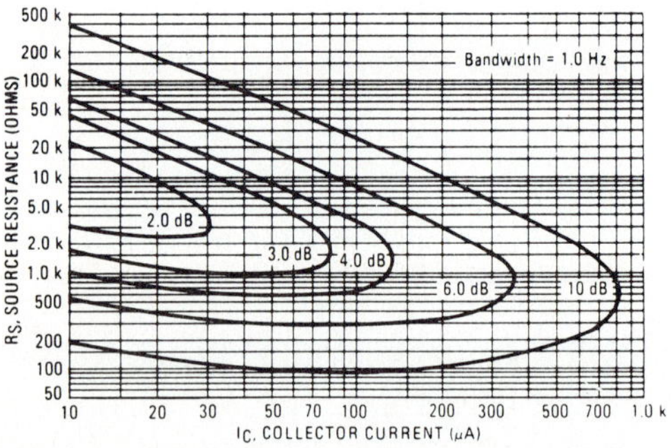

FIGURE 6 – NARROW BAND, 1.0 kHz

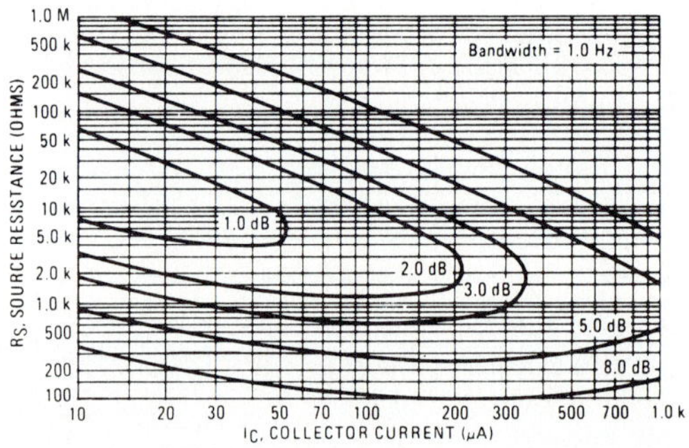

FIGURE 7 – WIDEBAND

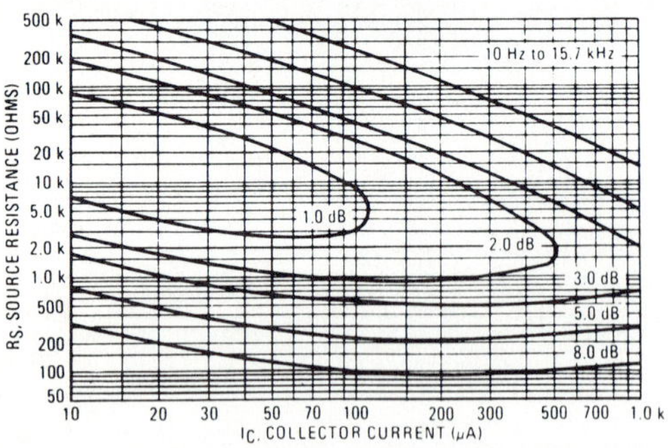

Noise Figure is Defined as:

$$NF = 20 \log_{10} \left(\frac{e_n^2 + 4KTR_S + I_n^2 R_S^2}{4KTR_S} \right)^{1/2}$$

e_n = Noise Voltage of the Transistor referred to the input (Figure 3)
I_n = Noise Current of the transistor referred to the input (Figure 4)
K = Boltzman's Constant (1.38×10^{-23} j/°K)
T = Temperature of the Source Resistance (°K)
R_S = Source Resistance (Ohms)

TYPICAL STATIC CHARACTERISTICS

FIGURE 8 – DC CURRENT GAIN

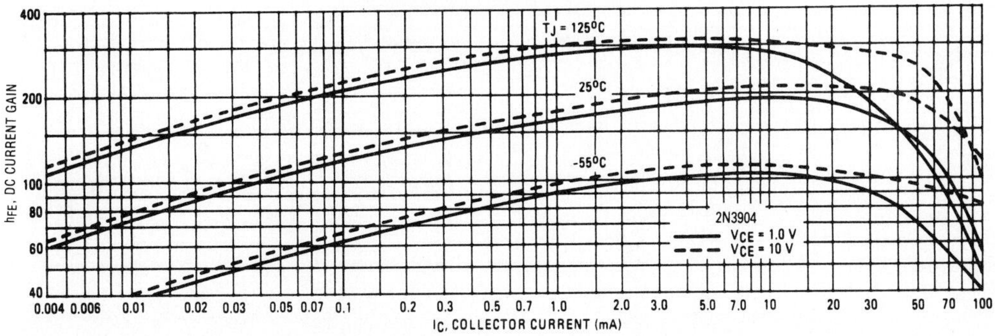

FIGURE 9 – COLLECTOR SATURATION REGION

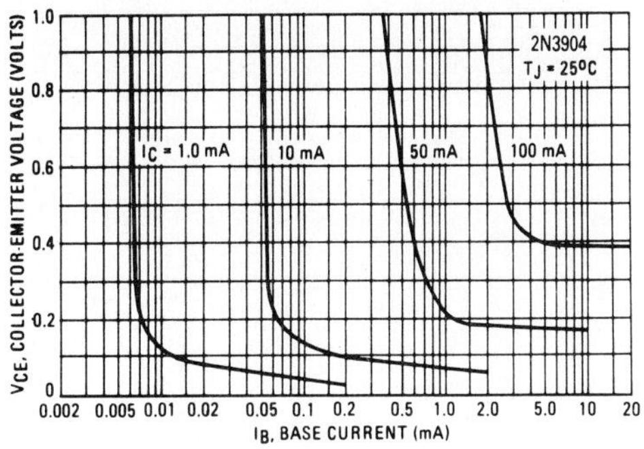

FIGURE 10 – COLLECTOR CHARACTERISTICS

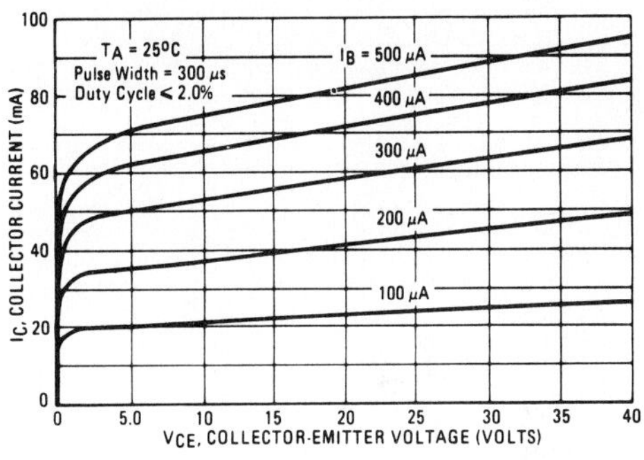

FIGURE 11 – "ON" VOLTAGES

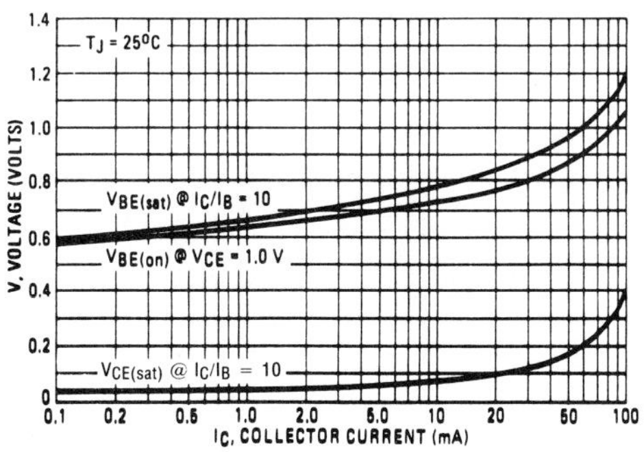

FIGURE 12 – TEMPERATURE COEFFICIENTS

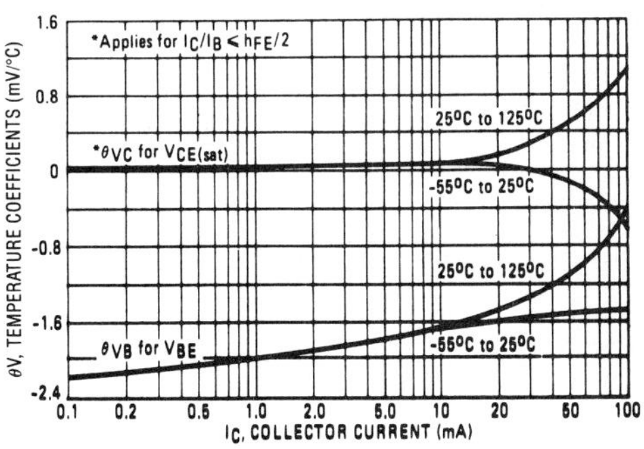

2N3903 • 2N3904

TYPICAL DYNAMIC CHARACTERISTICS

FIGURE 13 – TURN-ON TIME

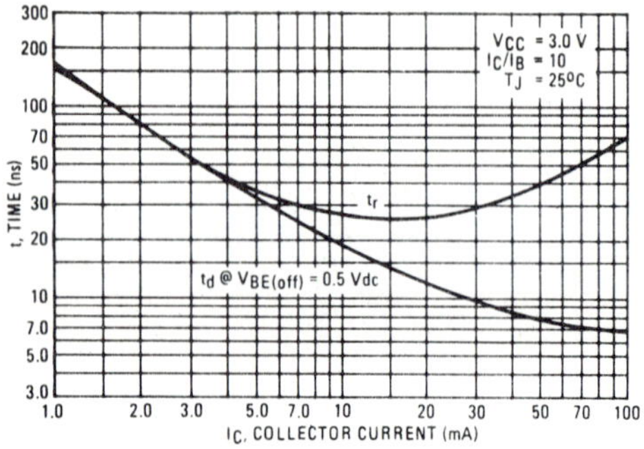

FIGURE 14 – TURN-OFF TIME

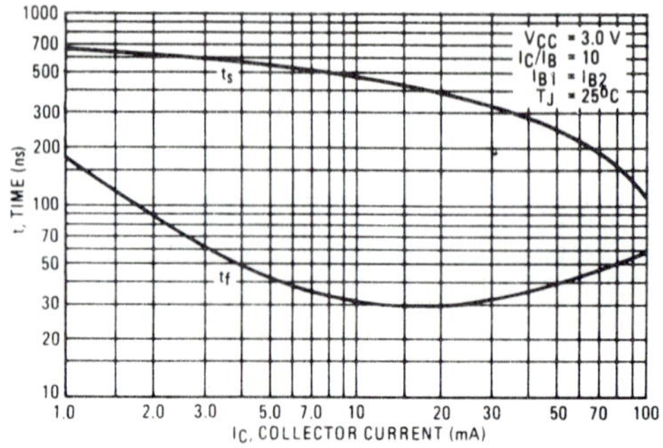

FIGURE 15 – CURRENT-GAIN – BANDWIDTH PRODUCT

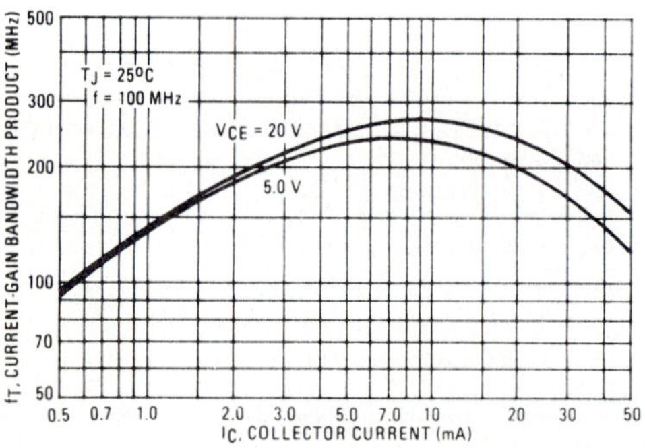

FIGURE 16 – CAPACITANCE

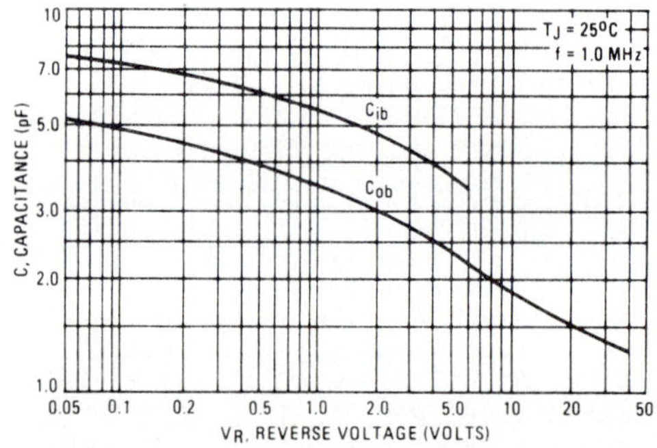

FIGURE 17 – INPUT IMPEDANCE

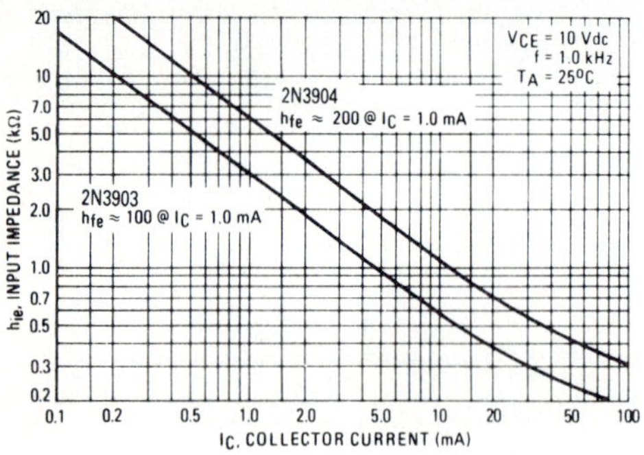

FIGURE 18 – OUTPUT ADMITTANCE

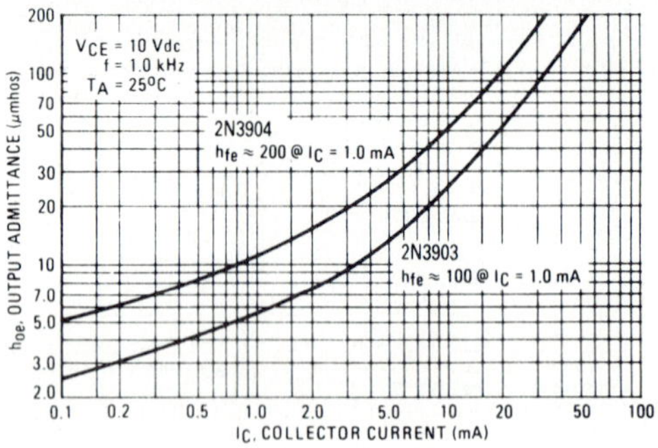

2N3903 • 2N3904

FIGURE 19 – THERMAL RESPONSE

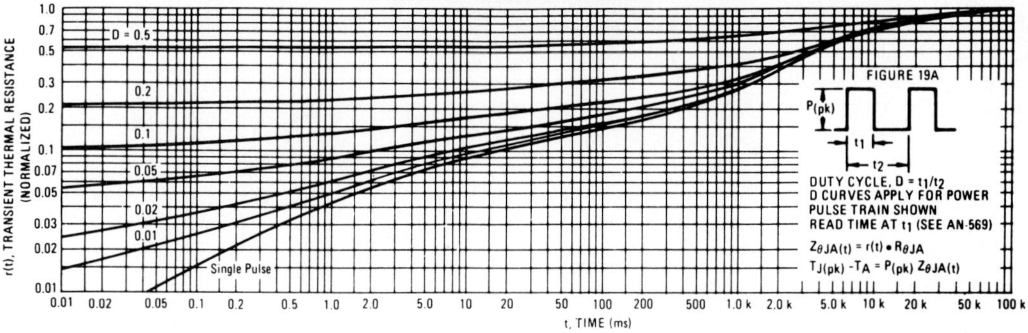

FIGURE 19A

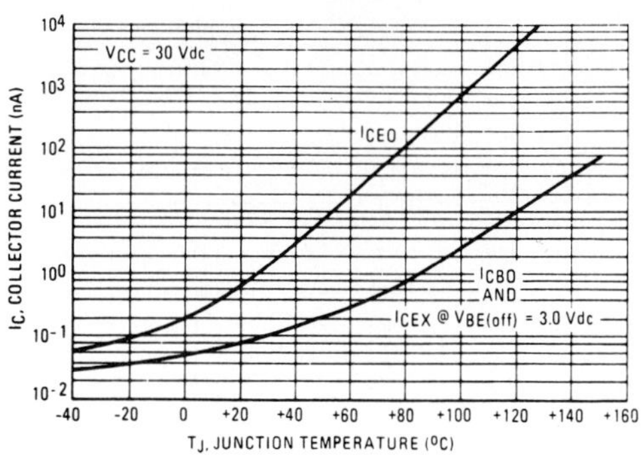

FIGURE 20

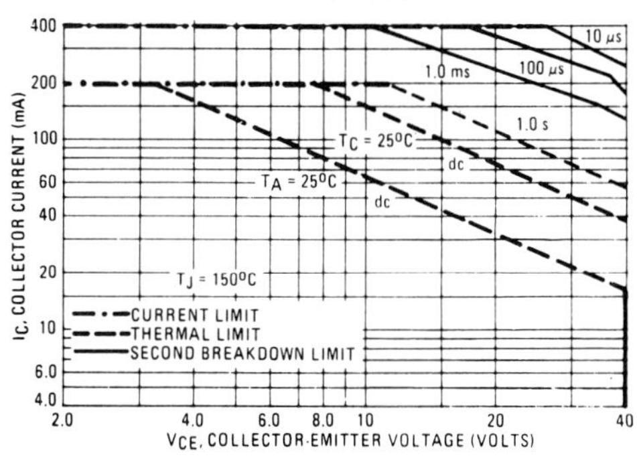

DESIGN NOTE: USE OF THERMAL RESPONSE DATA

A train of periodical power pulses can be represented by the model as shown in Figure 19A. Using the model and the device thermal response the normalized effective transient thermal resistance of Figure 19 was calculated for various duty cycles.

To find $Z_{\theta JA}(t)$, multiply the value obtained from Figure 19 by the steady state value $R_{\theta JA}$.

Example:

The 2N3903 is dissipating 2.0 watts peak under the following conditions:

$$t_1 = 1.0 \text{ ms}, t_2 = 5.0 \text{ ms}. (D = 0.2)$$

Using Figure 19 at a pulse width of 1.0 ms and D = 0.2, the reading of r(t) is 0.22.

The peak rise in junction temperature is therefore

$$\Delta T = r(t) \times P_{(pk)} \times R_{\theta JA} = 0.22 \times 2.0 \times 200 = 88°C.$$

For more information, see AN569.

The safe operating area curves indicate I_C-V_{CE} limits of the transistor that must be observed for reliable operation. Collector load lines for specific circuits must fall below the limits indicated by the applicable curve.

The data of Figure 20 is based upon $T_{J(pk)} = 150°C$; T_C or T_A is variable depending upon conditions. Pulse curves are valid for duty cycles to 10% provided $T_{J(pk)} \leq 150°C$. $T_{J(pk)}$ may be calculated from the data in Figure 19. At high case or ambient temperatures, thermal limitations will reduce the power than can be handled to values less than the limitations imposed by second breakdown. (See AN415A).

MOTOROLA Semiconductors

BOX 20912 • PHOENIX, ARIZONA 85036

2N3905
2N3906

PNP SILICON SWITCHING & AMPLIFIER TRANSISTORS

PNP SILICON ANNULAR♦ TRANSISTORS

.... designed for general purpose switching and amplifier applications and for complementary circuitry with types 2N3903 and 2N3904.

- High Voltage Ratings – BV_{CEO} = 40 Volts (Min)
- Current Gain Specified from 100 μA to 100 mA
- Complete Switching and Amplifier Specifications
- Low Capacitance – C_{ob} = 4.5 pF (Max)

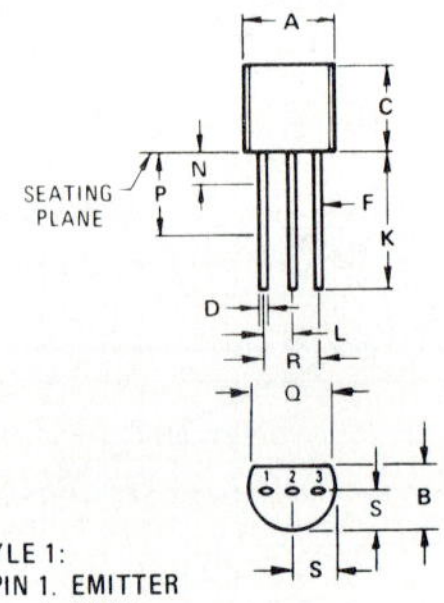

STYLE 1:
PIN 1. EMITTER
2. BASE
3. COLLECTOR

	MILLIMETERS		INCHES	
DIM	MIN	MAX	MIN	MAX
A	4.450	5.200	0.175	0.205
B	3.180	4.190	0.125	0.165
C	4.320	5.330	0.170	0.210
D	0.407	0.533	0.016	0.021
F	0.407	0.482	0.016	0.019
K	12.700	–	0.500	–
L	1.150	1.390	0.045	0.055
N	–	1.270	–	0.050
P	6.350	–	0.250	–
Q	3.430	–	0.135	–
R	2.410	2.670	0.095	0.105
S	2.030	2.670	0.080	0.105

CASE 29-02
(TO-92)

*MAXIMUM RATINGS

Rating	Symbol	Value	Unit
Collector-Base Voltage	V_{CB}	40	Vdc
Collector-Emitter Voltage	V_{CEO}	40	Vdc
Emitter-Base Voltage	V_{EB}	5.0	Vdc
Collector Current	I_C	200	mAdc
Total Power Dissipation @ $T_A = 60^oC$	P_D	250	mW
Total Power Dissipation @ $T_A = 25^oC$ Derate above 25^oC	P_D	350 2.8	mW $mW/^oC$
Total Power Dissipation @ $T_C = 25^oC$ Derate above 25^oC	P_D	1.0 8.0	Watt $mW/^oC$
Junction Operating Temperature	T_J	+150	^oC
Storage Temperature Range	T_{stg}	-55 to +150	^oC

THERMAL CHARACTERISTICS

Characteristic	Symbol	Max	Unit
Thermal Resistance, Junction to Ambient	$R_{\theta JA}$	357	$^oC/W$
Thermal Resistance, Junction to Case	$R_{\theta JC}$	125	$^oC/W$

*Indicates JEDEC Registered Data.
♦Annular semiconductors patented by Motorola Inc.

***ELECTRICAL CHARACTERISTICS** (T_A = 25°C unless otherwise noted.)

Characteristic		Fig. No.	Symbol	Min	Max	Unit
OFF CHARACTERISTICS						
Collector-Base Breakdown Voltage (I_C = 10 µAdc, I_E = 0)			BV_{CBO}	40	–	Vdc
Collector-Emitter Breakdown Voltage (1) (I_C = 1.0 mAdc, I_B = 0)			BV_{CEO}	40	–	Vdc
Emitter-Base Breakdown Voltage (I_E = 10 µAdc, I_C = 0)			BV_{EBO}	5.0	–	Vdc
Collector Cutoff Current (V_{CE} = 30 Vdc, $V_{BE(off)}$ = 3.0 Vdc)			I_{CEX}	–	50	nAdc
Base Cutoff Current (V_{CE} = 30 Vdc, $V_{BE(off)}$ = 3.0 Vdc)			I_{BL}	–	50	nAdc
ON CHARACTERISTICS (1)						
DC Current Gain			h_{FE}			
(I_C = 0.1 mAdc, V_{CE} = 1.0 Vdc)	2N3905	15		30	–	
	2N3906			60	–	
(I_C = 1.0 mAdc, V_{CE} = 1.0 Vdc)	2N3905			40	–	
	2N3906			80	–	
(I_C = 10 mAdc, V_{CE} = 1.0 Vdc)	2N3905			50	150	
	2N3906			100	300	
(I_C = 50 mAdc, V_{CE} = 1.0 Vdc)	2N3905			30	–	
	2N3906			60	–	
(I_C = 100 mAdc, V_{CE} = 1.0 Vdc)	2N3905			15	–	
	2N3906			30	–	
Collector-Emitter Saturation Voltage		16, 17	$V_{CE(sat)}$			Vdc
(I_C = 10 mAdc, I_B = 1.0 mAdc)				–	0.25	
(I_C = 50 mAdc, I_B = 5.0 mAdc)				–	0.4	
Base-Emitter Saturation Voltage		17	$V_{BE(sat)}$			Vdc
(I_C = 10 mAdc, I_B = 1.0 mAdc)				0.65	0.85	
(I_C = 50 mAdc, I_B = 5.0 mAdc)				–	0.95	
SMALL-SIGNAL CHARACTERISTICS						
Current-Gain – Bandwidth Product			f_T			MHz
(I_C = 10 mAdc, V_{CE} = 20 Vdc, f = 100 MHz)	2N3905			200	–	
	2N3906			250	–	
Output Capacitance (V_{CB} = 5.0 Vdc, I_E = 0, f = 100 kHz)		3	C_{ob}	–	4.5	pF
Input Capacitance (V_{BE} = 0.5 Vdc, I_C = 0, f = 100 kHz)		3	C_{ib}	–	1.0	pF
Input Impedance		13	h_{ie}			k ohms
(I_C = 1.0 mAdc, V_{CE} = 10 Vdc, f = 1.0 kHz)	2N3905			0.5	8.0	
	2N3906			2.0	12	
Voltage Feedback Ratio		14	h_{re}			X 10^{-4}
(I_C = 1.0 mAdc, V_{CE} = 10 Vdc, f = 1.0 kHz)	2N3905			0.1	5.0	
	2N3906			1.0	10	
Small-Signal Current Gain		11	h_{fe}			–
(I_C = 1.0 mAdc, V_{CE} = 10 Vdc, f = 1.0 kHz)	2N3905			50	200	
	2N3906			100	400	
Output Admittance		12	h_{oe}			µmhos
(I_C = 1.0 mAdc, V_{CE} = 10 Vdc, f = 1.0 kHz)	2N3905			1.0	40	
	2N3906			3.0	60	
Noise Figure		9, 10	NF			dB
(I_C = 100 µAdc, V_{CE} = 5.0 Vdc, R_S = 1.0 k ohm,	2N3905			–	5.0	
f = 10 Hz to 15.7 kHz)	2N3906			–	4.0	
SWITCHING CHARACTERISTICS						
Delay Time	(V_{CC} = 3.0 Vdc, $V_{BE(off)}$ = 0.5 Vdc	1, 5	t_d	–	35	ns
Rise Time	I_C = 10 mAdc, I_{B1} = 1.0 mAdc)	1, 5, 6	t_r	–	35	ns
Storage Time	2N3905	2, 7	t_s	–	200	ns
	(V_{CC} = 3.0 Vdc, I_C = 10 mAdc, 2N3906			–	225	
Fall Time	I_{B1} = I_{B2} = 1.0 mAdc) 2N3905	2, 8	t_f	–	60	ns
	2N3906			–	75	

*Indicates JEDEC Registered Data. (1) Pulse Width = 300 µs, Duty Cycle = 2.0 %.

FIGURE 1 – DELAY AND RISE TIME EQUIVALENT TEST CIRCUIT

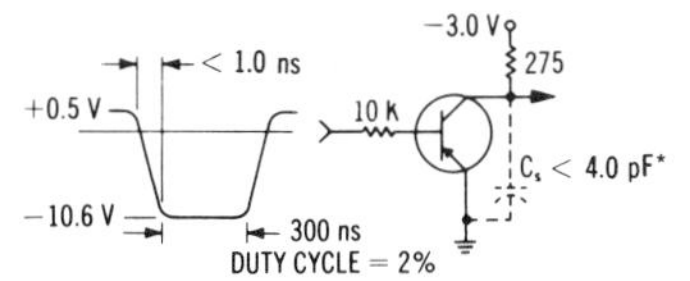

FIGURE 2 – STORAGE AND FALL TIME EQUIVALENT TEST CIRCUIT

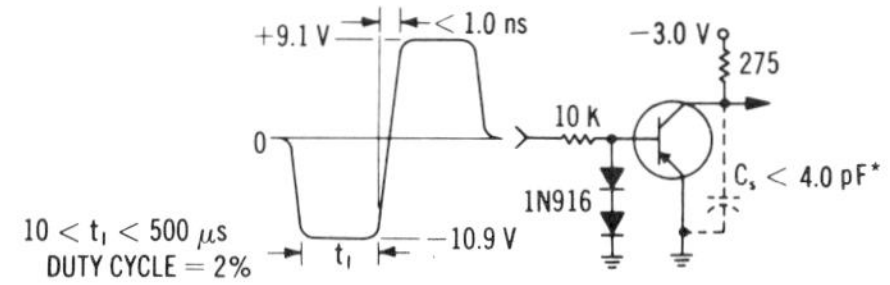

*Total shunt capacitance of test jig and connectors

TYPES 2N2217 THRU 2N2222, 2N2218A, 2N2219A, 2N2221A, 2N2222A
N-P-N SILICON TRANSISTORS

BULLETIN NO. DL-S 7311916, MARCH 1973

DESIGNED FOR HIGH-SPEED, MEDIUM-POWER SWITCHING AND GENERAL PURPOSE AMPLIFIER APPLICATIONS

- h_{FE} . . . Guaranteed from 100 μA to 500 mA
- High f_T at 20 V, 20 mA . . . 300 MHz (2N2219A, 2N2222A)
  250 MHz (all others)
- 2N2218, 2N2221 for Complementary Use with 2N2904, 2N2906
- 2N2219, 2N2222 for Complementary Use with 2N2905, 2N2906

*mechanical data

Device types 2N2217, 2N2218, 2N2218A, 2N2219, and 2N2219A are in JEDEC TO-5 packages.
Device types 2N2220, 2N2221, 2N2221A, 2N2222, and 2N2222A are in JEDEC TO-18 packages.

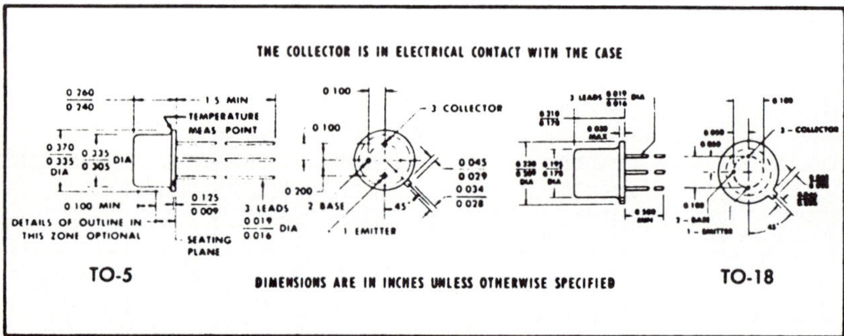

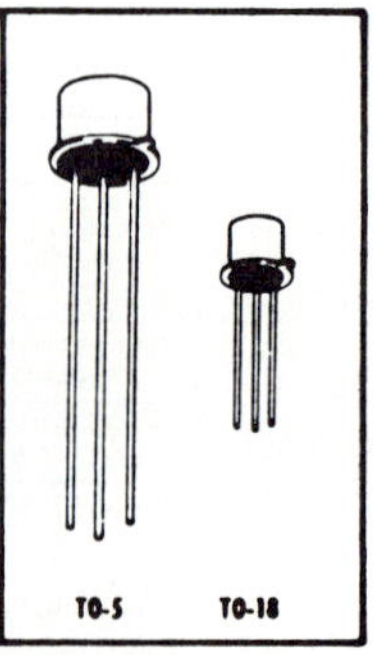

*absolute maximum ratings at 25°C free-air temperature (unless otherwise noted)

	2N2217 2N2218 2N2219	2N2218A 2N2219A	2N2220 2N2221 2N2222	2N2221A 2N2222A	UNIT
Collector-Base Voltage	60	75	60	75	V
Collector-Emitter Voltage (See Note 1)	30	40	30	40	V
Emitter-Base Voltage	5	6	5	6	V
Continuous Collector Current	0.8	0.8	0.8	0.8	A
Continuous Device Dissipation at (or below) 25°C Free-Air Temperature (See Notes 2 and 3)	0.8	0.8	0.5	0.5	W
Continuous Device Dissipation at (or below) 25°C Case Temperature (See Notes 4 and 5)	3	3	1.8	1.8	W
Operating Collector Junction Temperature Range	−65 to 175				°C
Storage Temperature Range	−65 to 200				°C
Lead Temperature 1/16 Inch from Case for 10 Seconds	230				°C

NOTES:
1. These values apply between 0 and 500 mA collector current when the base-emitter diode is open-circuited.
2. Derate 2N2217, 2N2218, 2N2218A, 2N2219, and 2N2219A linearly to 175°C free-air temperature at the rate of 5.33 mW/°C.
3. Derate 2N2220, 2N2221, 2N2221A, 2N2222, and 2N2222A linearly to 175°C free-air temperature at the rate of 3.33 mW/°C.
4. Derate 2N2217, 2N2218, 2N2218A, 2N2219, and 2N2219A linearly to 175°C case temperature at the rate of 20.0 mW/°C.
5. Derate 2N2220, 2N2221, 2N2221A, 2N2222, and 2N2222A linearly to 175°C case temperature at the rate of 12.0 mW/°C.

*JEDEC registered data. This data sheet contains all applicable registered data in effect at the time of publication.

USES CHIP N24

TYPES 2N2217 THRU 2N2222, 2N2218A, 2N2219A, 2N2221A, 2N2222A
N-P-N SILICON TRANSISTORS

2N2218A, 2N2219A, 2N2221A, 2N2222A

***electrical characteristics at 25°C free-air temperature (unless otherwise noted)**

PARAMETER		TEST CONDITIONS	TO-5 → / TO-18 →	2N2218A / 2N2221A MIN	2N2218A / 2N2221A MAX	2N2219A / 2N2222A MIN	2N2219A / 2N2222A MAX	UNIT
$V_{(BR)CBO}$	Collector-Base Breakdown Voltage	I_C = 10 μA, I_E = 0		75		75		V
$V_{(BR)CEO}$	Collector-Emitter Breakdown Voltage	I_C = 10 mA, I_B = 0,	See Note 6	40		40		V
$V_{(BR)EBO}$	Emitter-Base Breakdown Voltage	I_E = 10 μA, I_C = 0		6		6		V
I_{CBO}	Collector Cutoff Current	V_{CB} = 60 V, I_E = 0			10		10	nA
		V_{CB} = 60 V, I_E = 0,	T_A = 150°C		10		10	μA
I_{CEV}	Collector Cutoff Current	V_{CE} = 60 V, V_{BE} = −3 V			10		10	nA
I_{BEV}	Base Cutoff Current	V_{CE} = 60 V, V_{BE} = −3 V			−20		−20	nA
I_{EBO}	Emitter Cutoff Current	V_{EB} = 3 V, I_C = 0			10		10	nA
h_{FE}	Static Forward Current Transfer Ratio	V_{CE} = 10 V, I_C = 100 μA		20		35		
		V_{CE} = 10 V, I_C = 1 mA		25		50		
		V_{CE} = 10 V, I_C = 10 mA	See Note 6	35		75		
		V_{CE} = 10 V, I_C = 150 mA		40	120	100	300	
		V_{CE} = 10 V, I_C = 500 mA		25		40		
		V_{CE} = 1 V, I_C = 150 mA		20		50		
		V_{CE} = 10 V, I_C = 10 mA, T_A = −55°C		15		35		
V_{BE}	Base-Emitter Voltage	I_B = 15 mA, I_C = 150 mA	See Note 6	0.6	1.2	0.6	1.2	V
		I_B = 50 mA, I_C = 500 mA			2		2	
$V_{CE(sat)}$	Collector-Emitter Saturation Voltage	I_B = 15 mA, I_C = 150 mA	See Note 6		0.3		0.3	V
		I_B = 50 mA, I_C = 500 mA			1		1	
h_{ie}	Small-Signal Common-Emitter Input Impedance	V_{CE} = 10 V, I_C = 1 mA	f = 1 kHz	1	3.5	2	8	kΩ
		V_{CE} = 10 V, I_C = 10 mA		0.2	1	0.25	1.25	
h_{fe}	Small-Signal Forward Current Transfer Ratio	V_{CE} = 10 V, I_C = 1 mA		30	150	50	300	
		V_{CE} = 10 V, I_C = 10 mA		50	300	75	375	
h_{re}	Small-Signal Common-Emitter Reverse Voltage Transfer Ratio	V_{CE} = 10 V, I_C = 1 mA			5×10^{-4}		8×10^{-4}	
		V_{CE} = 10 V, I_C = 10 mA			2.5×10^{-4}		4×10^{-4}	
h_{oe}	Small-Signal Common-Emitter Output Admittance	V_{CE} = 10 V, I_C = 1 mA		3	15	5	35	μmho
		V_{CE} = 10 V, I_C = 10 mA		10	100	25	200	
$\|h_{fe}\|$	Small-Signal Common-Emitter Forward Current Transfer Ratio	V_{CE} = 20 V, I_C = 20 mA,	f = 100 MHz	2.5		3		
f_T	Transition Frequency	V_{CE} = 20 V, I_C = 20 mA,	See Note 7	250		300		MHz
C_{obo}	Common-Base Open-Circuit Output Capacitance	V_{CB} = 10 V, I_E = 0,	f = 100 kHz		8		8	pF
C_{ibo}	Common-Base Open-Circuit Input Capacitance	V_{EB} = 0.5 V, I_C = 0,	f = 100 kHz		25		25	pF
$h_{ie(real)}$	Real Part of Small-Signal Common-Emitter Input Impedance	V_{CE} = 20 V, I_C = 20 mA,	f = 300 MHz		60		60	Ω
$r_b'C_c$	Collector-Base Time Constant	V_{CE} = 20 V, I_C = 20 mA,	f = 31.8 MHz		150		150	ps

NOTES: 6. These parameters must be measured using pulse techniques. t_w = 300 μs, duty cycle ≤ 2%.

7. To obtain f_T, the $|h_{fe}|$ response with frequency is extrapolated at the rate of −6 dB per octave from f = 100 MHz to the frequency at which $|h_{fe}|$ = 1.

*JEDEC registered data

TYPES 2N2217 THRU 2N2222, 2N2218A, 2N2219A, 2N2221A, 2N2222A N-P-N SILICON TRANSISTORS

2N2217 THRU 2N2222

*electrical characteristics at 25°C free-air temperature (unless otherwise noted)

PARAMETER		TEST CONDITIONS		2N2217 (TO-5) / 2N2220 (TO-18) MIN	MAX	2N2218 / 2N2221 MIN	MAX	2N2219 / 2N2222 MIN	MAX	UNIT
$V_{(BR)CBO}$	Collector-Base Breakdown Voltage	$I_C = 10\ \mu A$, $I_E = 0$		60		60		60		V
$V_{(BR)CEO}$	Collector-Emitter Breakdown Voltage	$I_C = 10\ mA$, $I_B = 0$,	See Note 6	30		30		30		V
$V_{(BR)EBO}$	Emitter-Base Breakdown Voltage	$I_E = 10\ \mu A$, $I_C = 0$		5		5		5		V
I_{CBO}	Collector Cutoff Current	$V_{CB} = 50\ V$, $I_E = 0$			10		10		10	nA
		$V_{CB} = 50\ V$, $I_E = 0$,	$T_A = 150°C$		10		10		10	μA
I_{EBO}	Emitter Cutoff Current	$V_{EB} = 3\ V$, $I_C = 0$			10		10		10	nA
h_{FE}	Static Forward Current Transfer Ratio	$V_{CE} = 10\ V$, $I_C = 100\ \mu A$				20		35		
		$V_{CE} = 10\ V$, $I_C = 1\ mA$		12		25		50		
		$V_{CE} = 10\ V$, $I_C = 10\ mA$	See Note 6	17		35		75		
		$V_{CE} = 10\ V$, $I_C = 150\ mA$	See Note 6	20	60	40	120	100	300	
		$V_{CE} = 10\ V$, $I_C = 500\ mA$	See Note 6			20		30		
		$V_{CE} = 1\ V$, $I_C = 150\ mA$	See Note 6	10		20		50		
V_{BE}	Base-Emitter Voltage	$I_B = 15\ mA$, $I_C = 150\ mA$	See Note 6		1.3		1.3		1.3	V
		$I_B = 50\ mA$, $I_C = 500\ mA$	See Note 6				2.6		2.6	V
$V_{CE(sat)}$	Collector-Emitter Saturation Voltage	$I_B = 15\ mA$, $I_C = 150\ mA$	See Note 6		0.4		0.4		0.4	V
		$I_B = 50\ mA$, $I_C = 500\ mA$	See Note 6				1.6		1.6	V
$\lvert h_{fe} \rvert$	Small-Signal Common-Emitter Forward Current Transfer Ratio	$V_{CE} = 20\ V$, $I_C = 20\ mA$,	$f = 100\ MHz$	2.5		2.5		2.5		
f_T	Transition Frequency	$V_{CE} = 20\ V$, $I_C = 20\ mA$,	See Note 7	250		250		250		MHz
C_{obo}	Common-Base Open-Circuit Output Capacitance	$V_{CB} = 10\ V$, $I_E = 0$,	$f = 1\ MHz$		8		8		8	pF
$h_{ie(real)}$	Real Part of Small-Signal Common-Emitter Input Impedance	$V_{CE} = 20\ V$, $I_C = 20\ mA$,	$f = 300\ MHz$		60		60		60	Ω

NOTES: 6. These parameters must be measured using pulse techniques. $t_w = 300\ \mu s$, duty cycle $\leq 2\%$.

7. To obtain f_T, the $\lvert h_{fe} \rvert$ response with frequency is extrapolated at the rate of −6 dB per octave from f = 100 MHz to the frequency at which $\lvert h_{fe} \rvert = 1$.

switching characteristics at 25°C free-air temperature

PARAMETER		TEST CONDITIONS†	TYP	UNIT
t_d	Delay Time	$V_{CC} = 30\ V$, $I_C = 150\ mA$, $I_{B(1)} = 15\ mA$,	5	ns
t_r	Rise Time	$V_{BE(off)} = -0.5\ V$, See Figure 1	15	ns
t_s	Storage Time	$V_{CC} = 30\ V$, $I_C = 150\ mA$, $I_{B(1)} = 15\ mA$,	190	ns
t_f	Fall Time	$I_{B(2)} = -15\ mA$, See Figure 2	23	ns

†Voltage and current values shown are nominal; exact values vary slightly with transistor parameters.

*JEDEC registered data

***operating characteristics at 25°C free-air temperature**

PARAMETER	TEST CONDITIONS	TO-5 → 2N2218A / TO-18 → 2N2221A MAX	TO-5 → 2N2219A / TO-18 → 2N2222A MAX	UNIT
F Spot Noise Figure	V_{CE} = 10 V, I_C = 100 μA, R_G = 1 kΩ, f = 1 kHz		4	dB

***switching characteristics at 25°C free-air temperature**

PARAMETER	TEST CONDITIONS†	TO-5 → 2N2218A / TO-18 → 2N2221A MAX	TO-5 → 2N2219A / TO-18 → 2N2222A MAX	UNIT
t_d Delay Time	V_{CC} = 30 V, I_C = 150 mA, $I_{B(1)}$ = 15 mA, $V_{BE(off)}$ = −0.5 V, See Figure 1	10	10	ns
t_r Rise Time		25	25	ns
τ_A Active Region Time Constant‡		2.5	2.5	ns
t_s Storage Time	V_{CC} = 30 V, I_C = 150 mA, $I_{B(1)}$ = 15 mA, $I_{B(2)}$ = −15 mA, See Figure 2	225	225	ns
t_f Fall Time		60	60	ns

†Voltage and current values shown are nominal; exact values vary slightly with transistor parameters.

‡Under the given conditions τ_A is equal to $\frac{t_r}{10}$.

*PARAMETER MEASUREMENT INFORMATION

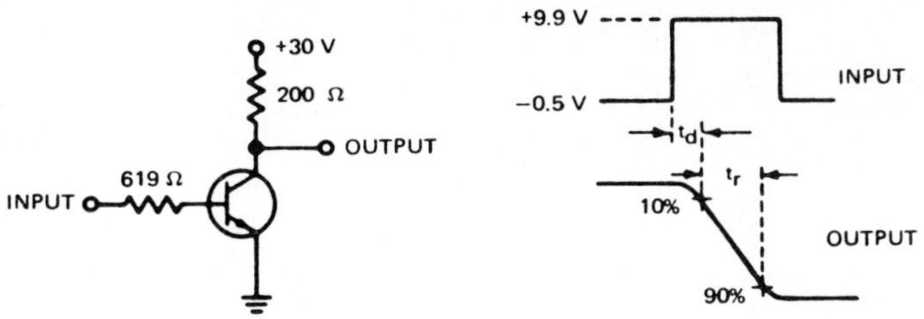

TEST CIRCUIT VOLTAGE WAVEFORMS

FIGURE 1—DELAY AND RISE TIMES

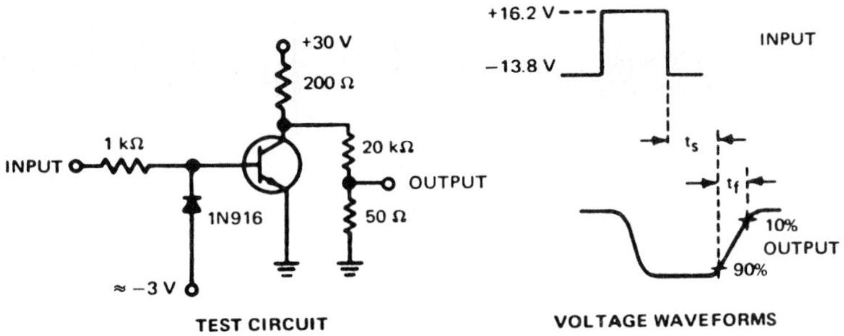

TEST CIRCUIT VOLTAGE WAVEFORMS

FIGURE 2—STORAGE AND FALL TIMES

NOTES: a. The input waveforms have the following characteristics: For Figure 1, $t_r \leq 2$ ns, $t_w \leq 200$ ns, duty cycle ≤ 2%; for Figure 2, $t_f \leq 5$ ns, $t_w \approx 100$ μs, duty cycle ≤ 17%.

b. All waveforms are monitored on an oscilloscope with the following characteristics: $t_r \leq 5$ ns, $R_{in} \geq 100$ kΩ, $C_{in} \leq 12$ pF

*JEDEC registered data

MOTOROLA SEMICONDUCTORS
P.O. BOX 20912 • PHOENIX, ARIZONA 85036

2N3821
2N3822
2N3824

JAN2N3821
JAN2N3822

N-CHANNEL JUNCTION FIELD-EFFECT TRANSISTORS

SYMMETRICAL (Type A)

FEBRUARY 1971 – DS 5148 R2

SILICON N-CHANNEL JUNCTION FIELD-EFFECT TRANSISTORS

. . . designed for audio amplifier, chopper and switching applications.

- Drain and Source Interchangeable
- Low Drain-Source Resistance – $r_{ds(on)} \leqslant 250$ Ohms (Max) – 2N3824
- Low Noise Figure – NF = 5.0 dB (Max) – 2N3821, 2N3822
- High AC Input Impedance – C_{iss} = 6.0 pF (Max)
- High DC Input Resistance – I_{GSS} = 0.1 nA (Max)
- Low Transfer Capacitance – C_{rss} = 3.0 pF (Max)
- JAN2N3821 and JAN2N3822 also Available

*MAXIMUM RATINGS

Rating	Symbol	Value	Unit
Drain-Source Voltage	V_{DS}	50	Vdc
Drain-Gate Voltage	V_{DG}	50	Vdc
Gate-Source Voltage	V_{GS}	-50	Vdc
Drain Current	I_D	10	mAdc
Total Device Dissipation @ $T_A = 25^oC$ Derate above 25^oC	P_D	300 2.0	mW mW/^oC
Operating Junction Temperature	T_J	175	^oC
Storage Temperature Range	T_{stg}	-65 to +200	^oC

*Indicates JEDEC Registered Data.

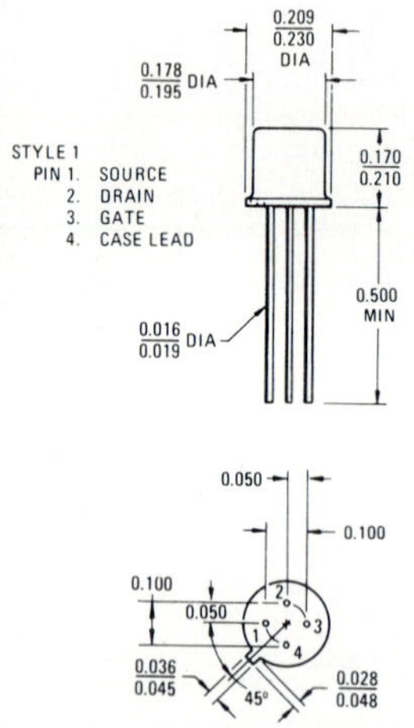

To convert inches to millimeters multiply by 25.4

All JEDEC dimensions and notes apply

CASE 20
TO-72

2N3821 • 2N3822 • 2N3824

***ELECTRICAL CHARACTERISTICS** (T_A = 25°C unless otherwise noted)

Characteristic		Symbol	Min	Max	Unit
OFF CHARACTERISTICS					
Gate-Source Breakdown Voltage (I_G = −1.0 μAdc, V_{DS} = 0)		$V_{(BR)GSS}$	-50	–	Vdc
Gate Reverse Current		I_{GSS}			nAdc
(V_{GS} = −30 Vdc, V_{DS} = 0)			–	-0.1	
(V_{GS} = −30 Vdc, V_{DS} = 0, T_A = 150°C)			–	-100	
Gate-Source Cutoff Voltage (I_D = 0.5 nAdc, V_{DS} = 15 Vdc)	2N3821	$V_{GS(off)}$	–	-4.0	Vdc
	2N3822		–	-6.0	
Gate-Source Voltage		V_{GS}			Vdc
(I_D = 50 μAdc, V_{DS} = 15 Vdc)	2N3821		-0.5	-2.0	
(I_D = 200 μAdc, V_{DS} = 15 Vdc)	2N3822		-1.0	-4.0	
Drain Cutoff Current		$I_{D(off)}$			nAdc
(V_{DS} = 15 Vdc, V_{GS} = −8.0 Vdc)	2N3824		–	0.1	
(V_{DS} = 15 Vdc, V_{GS} = −8.0 Vdc, T_A = 150°C)	2N3824		–	100	
ON CHARACTERISTICS					
Zero-Gate-Voltage Drain Current(1) (V_{DS} = 15 Vdc, V_{GS} = 0)	2N3821	I_{DSS}	0.5	2.5	mAdc
	2N3822		2.0	10	
DYNAMIC CHARACTERISTICS					
Forward Transfer Admittance		$\|y_{fs}\|$			μmhos
(V_{DS} = 15 Vdc, V_{GS} = 0, f = 1.0 kHz)(1)	2N3821		1500	4500	
	2N3822		3000	6500	
(V_{DS} = 15 Vdc, V_{GS} = 0, f = 100 MHz)	2N3821		1500	–	
	2N3822		3000	–	
Output Admittance(1) (V_{DS} = 15 Vdc, V_{GS} = 0, f = 1.0 kHz)	2N3821	$\|y_{os}\|$	–	10	μmhos
	2N3822		–	20	
Drain-Source Resistance (V_{GS} = 0, I_D = 0, f = 1.0 kHz)	2N3824	$r_{ds(on)}$	–	250	Ohms
Input Capacitance (V_{DS} = 15 Vdc, V_{GS} = 0, f = 1.0 MHz)		C_{iss}	–	6.0	pF
Reverse Transfer Capacitance		C_{rss}			pF
(V_{DS} = 15 Vdc, V_{GS} = 0, f = 1.0 MHz)	2N3821		–	3.0	
	2N3822		–	3.0	
(V_{GS} = −8.0 Vdc, V_{DS} = 0, f = 1.0 MHz)	2N3824		–	3.0	
Average Noise Figure (V_{DS} = 15 Vdc, V_{GS} = 0, R_S = 1.0 megohm, f = 10 Hz, Noise Bandwidth = 5.0 Hz)	2N3821, 2N3822	NF	–	5.0	dB
Equivalent Input Noise Voltage (V_{DS} = 15 Vdc, V_{GS} = 0, f = 10 Hz, Noise Bandwidth = 5.0 Hz)	2N3821, 2N3822	e_n	–	200	nv/Hz½

*Indicates JEDEC Registered Data.

(1)Pulse Test: Pulse Width ≤ 100 ms, Duty Cycle ≤ 10%.

MOTOROLA
SEMICONDUCTORS
P.O. BOX 20912 • PHOENIX, ARIZONA 85036

3N128

SILICON N-CHANNEL MOS FIELD-EFFECT TRANSISTOR

. . . designed for VHF amplifier and oscillator applications in communications equipment.

- High Forward Transadmittance – $|y_{fs}|$ = 5000 μmhos (Min) @ f = 1.0 kHz
- Low Input Capacitance – C_{iss} = 7.0 pF (Max) @ f = 1.0 MHz
- Low Noise Figure – NF = 5.0 dB (Max) @ f = 200 MHz
- High Power Gain – P_G = 13.5 dB (Min) @ f = 200 MHz
- Complete "y" Parameter Curves
- Third Order Intermodulation Distortion Performance Curve Provided

N-CHANNEL MOS FIELD-EFFECT TRANSISTOR

* MAXIMUM RATINGS

Rating	Symbol	Value	Unit
Drain-Source Voltage	V_{DS}	+20	Vdc
Drain-Gate Voltage	V_{DG}	+20	Vdc
Gate-Source Voltage	V_{GS}	±10	Vdc
Drain Current	I_D	50	mAdc
Power Dissipation @ T_A = 25°C Derate above 25°C	P_D	330 2.2	mW mW/°C
Operating and Storage Junction Temperature Range	T_J, T_{stg}	-65 to +175	°C

*Indicates JEDEC Registered Data.

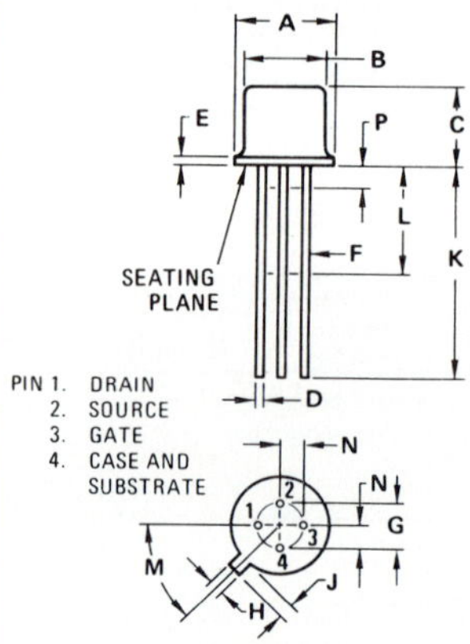

DIM	MILLIMETERS MIN	MILLIMETERS MAX	INCHES MIN	INCHES MAX
A	5.31	5.84	0.209	0.230
B	4.52	4.95	0.178	0.195
C	4.32	5.33	0.170	0.210
D	0.41	0.53	0.016	0.021
E	–	0.76	–	0.030
F	0.41	0.48	0.016	0.019
G	2.54 BSC		0.100 BSC	
H	0.91	1.17	0.036	0.046
J	0.71	1.22	0.028	0.048
K	12.70	–	0.500	–
L	6.35	–	0.250	–
M	45° BSC		45° BSC	
N	1.27 BSC		0.050 BSC	
P	–	1.27	–	0.050

CASE 20-03
TO-206AF
(TO-72)

HANDLING PRECAUTIONS

MOS field-effect transistors have extremely high input resistance. They can be damaged by the accumulation of excess static charge. Avoid possible damage to the devices while handling, testing, or in actual operation, by following the procedures outlined below:

1. To avoid the build-up of static charge, the leads of the devices should remain shorted together with a metal ring except when being tested or used.
2. Avoid unnecessary handling. Pick up devices by the case instead of the leads.
3. Do not insert or remove devices from circuits with the power on because transient voltages may cause permanent damage to the devices.

3N128

***ELECTRICAL CHARACTERISTICS** (T_A = 25°C unless otherwise noted)

Characteristic	Symbol	Min	Max	Unit
OFF CHARACTERISTICS				
Gate-Source Breakdown Voltage (1) (I_G = −10 μAdc, V_{DS} = 0)	$V_{(BR)GSS}$	-50	–	Vdc
Gate-Source Cutoff Voltage (V_{DS} = 15 Vdc, I_D = 50 μAdc)	$V_{GS(off)}$	-0.5	-8.0	Vdc
Gate Reverse Current (V_{GS} = -8.0 Vdc, V_{DS} = 0)	I_{GSS}	–	0.05	nAdc
(V_{GS} = -8.0 Vdc, V_{DS} = 0, T_A = 125°C)		–	5.0	
ON CHARACTERISTICS				
Zero-Gate-Voltage Drain Current (2) (V_{DS} = 15 Vdc, V_{GS} = 0)	I_{DSS}	5.0	25	mAdc
SMALL-SIGNAL CHARACTERISTICS				
Forward Transadmittance (V_{DS} = 15 Vdc, I_D = 5.0 mAdc, f = 1.0 kHz)	$\|y_{fs}\|$	5000	12,000	μmhos
Forward Transconductance (V_{DS} = 15 Vdc, I_D = 5.0 mAdc, f = 200 MHz)	$Re(y_{fs})$	5000	–	μmhos
Output Conductance (V_{DS} = 15 Vdc, I_D = 5.0 mAdc, f = 200 MHz)	$Re(y_{os})$	–	500	μmhos
Input Conductance (V_{DS} = 15 Vdc, I_D = 5.0 mAdc, f = 200 MHz)	$Re(y_{is})$	–	800	μmhos
Input Capacitance (V_{DS} = 15 Vdc, I_D = 5.0 mAdc, f = 1.0 MHz)	C_{iss}	–	7.0	pF
Reverse Transfer Capacitance (V_{DS} = 15 Vdc, I_D = 5.0 mAdc, f = 1.0 MHz)	C_{rss}	0.05	0.35	pF
Noise Figure (V_{DS} = 15 Vdc, I_D = 5.0 mAdc, f = 200 MHz)	NF	–	5.0	dB
Power Gain (V_{DS} = 15 Vdc, I_D = 5.0 mAdc, f = 200 MHz)	P_G	13.5	23	dB

*Indicates JEDEC Registered Data.
(1) Caution Destructive Test, can damage gate oxide beyond operation.
(2) Pulse Test: Pulse Width = 300 μs, Duty Cycle = 2.0%.

TYPICAL CHARACTERISTICS
(T_A = 25°C)

FIGURE 1 – DRAIN CHARACTERISTICS

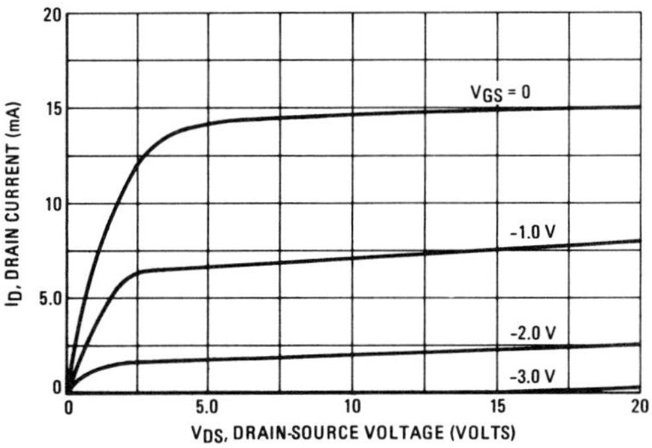

FIGURE 2 – TRANSFER CHARACTERISTICS

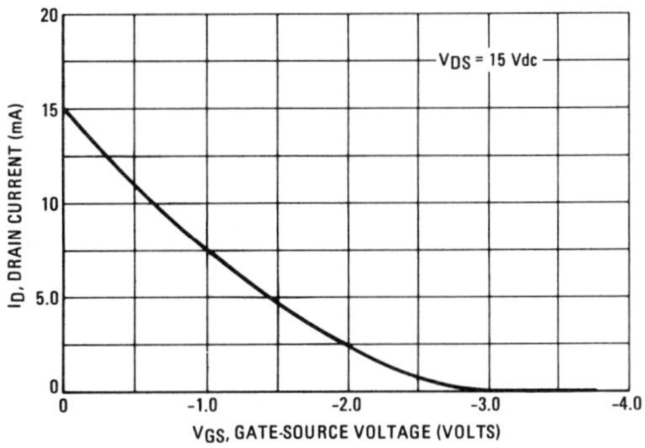

MOTOROLA
SEMICONDUCTORS
P.O. BOX 20912 • PHOENIX, ARIZONA 85036

MC7800 Series

THREE-TERMINAL POSITIVE VOLTAGE REGULATORS

These voltage regulators are monolithic integrated circuits designed as fixed-voltage regulators for a wide variety of applications including local, on-card regulation. These regulators employ internal current limiting, thermal shutdown, and safe-area compensation. With adequate heatsinking they can deliver output currents in excess of 1.0 ampere. Although designed primarily as a fixed voltage regulator, these devices can be used with external components to obtain adjustable voltages and currents.

- Output Current in Excess of 1.0 Ampere
- No External Components Required
- Internal Thermal Overload Protection
- Internal Short-Circuit Current Limiting
- Output Transistor Safe-Area Compensation
- Output Voltage Offered in 2% and 4% Tolerance

THREE-TERMINAL POSITIVE FIXED VOLTAGE REGULATORS

SILICON MONOLITHIC INTEGRATED CIRCUITS

K SUFFIX
METAL PACKAGE
CASE 1-03
TO-204AA
(TO-3)

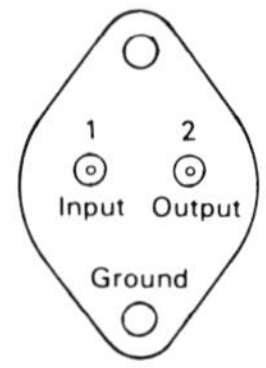

(Bottom View)

Pins 1 and 2 electrically isolated from case. Case is third electrical connection.

T SUFFIX
PLASTIC PACKAGE
CASE 221A
TO-220AB

Pin 1. Input
2. Ground
3. Output

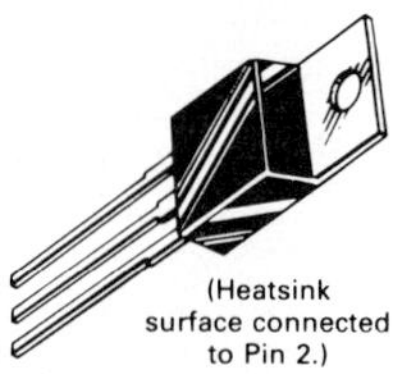

EQUIVALENT SCHEMATIC DIAGRAM

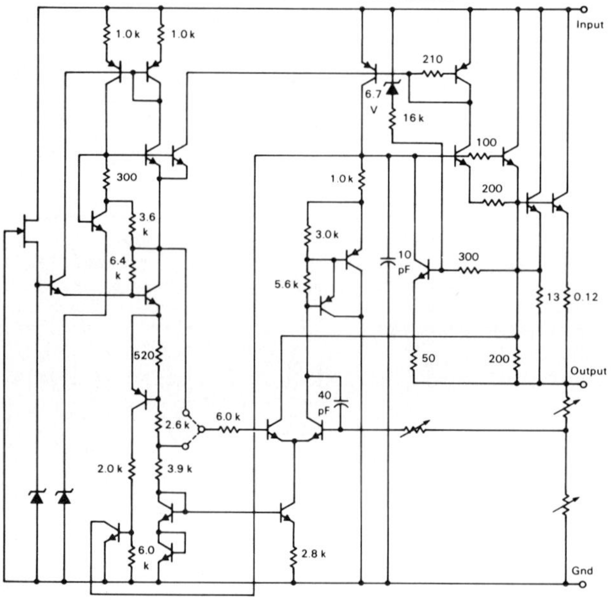

STANDARD APPLICATION

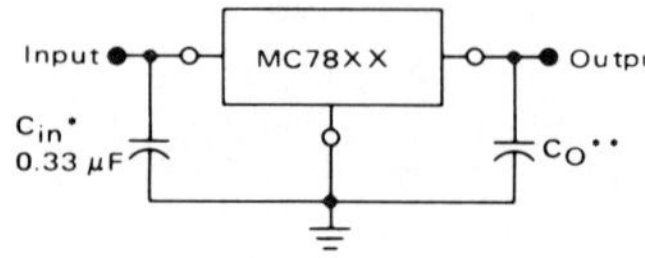

A common ground is required between the input and the output voltages. The input voltage must remain typically 2.0 V above the output voltage even during the low point on the input ripple voltage.

XX = these two digits of the type number indicate voltage.

* = C_{in} is required if regulator is located an appreciable distance from power supply filter.

** = C_O is not needed for stability; however, it does improve transient response.

XX indicates nominal voltage

ORDERING INFORMATION

Device	Output Voltage Tolerance	Temperature Range	Package
MC78XXK MC78XXAK	4% 2%	−55 to +150°C	Metal Power
MC78XXBK	4%	−40 to +125°C	
MC78XXCK MC78XXACK	4% 2%	0 to +125°C	
MC78XXCT MC78XXACT	4% 2%		Plastic Power
MC78XXBT	4%	−40 to +125°C	

TYPE NO./VOLTAGE			
MC7805	5.0 Volts	MC7815	15 Volts
MC7806	6.0 Volts	MC7818	18 Volts
MC7808	8.0 Volts	MC7824	24 Volts
MC7812	12 Volts		

DS9557R1

MC7800 Series

MC7800 Series MAXIMUM RATINGS (T_A = +25°C unless otherwise noted.)

Rating	Symbol	Value	Unit
Input Voltage (5.0 V – 18 V) (24 V)	V_{in}	35 40	Vdc
Power Dissipation and Thermal Characteristics			
Plastic Package			
T_A = +25°C	P_D	Internally Limited	Watts
Derate above T_A = +25°C	$1/\theta_{JA}$	15.4	mW/°C
Thermal Resistance, Junction to Air	θ_{JA}	65	°C/W
T_C = +25°C	P_D	Internally Limited	Watts
Derate above T_C = +75°C (See Figure 1)	$1/\theta_{JC}$	200	mW/°C
Thermal Resistance, Junction to Case	θ_{JC}	5.0	°C/W
Metal Package			
T_A = +25°C	P_D	Internally Limited	Watts
Derate above T_A = +25°C	$1/\theta_{JA}$	22.5	mW/°C
Thermal Resistance, Junction to Air	θ_{JA}	45	°C/W
T_C = +25°C	P_D	Internally Limited	Watts
Derate above T_C = +65°C (See Figure 2)	$1/\theta_{JC}$	182	mW/°C
Thermal Resistance, Junction to Case	θ_{JC}	5.5	°C/W
Storage Junction Temperature Range	T_{stg}	−65 to +150	°C
Operating Junction Temperature Range	T_J		°C
MC7800, A		−55 to +150	
MC7800C, AC		0 to +150	
MC7800, B		−40 to +150	

DEFINITIONS

Line Regulation — The change in output voltage for a change in the input voltage. The measurement is made under conditions of low dissipation or by using pulse techniques such that the average chip temperature is not significantly affected.

Load Regulation — The change in output voltage for a change in load current at constant chip temperature.

Maximum Power Dissipation — The maximum total device dissipation for which the regulator will operate within specifications.

Quiescent Current — That part of the input current that is not delivered to the load.

Output Noise Voltage — The rms ac voltage at the output, with constant load and no input ripple, measured over a specified frequency range.

Long Term Stability — Output voltage stability under accelerated life test conditions with the maximum rated voltage listed in the devices' electrical characteristics and maximum power dissipation.

OUTLINE DIMENSIONS

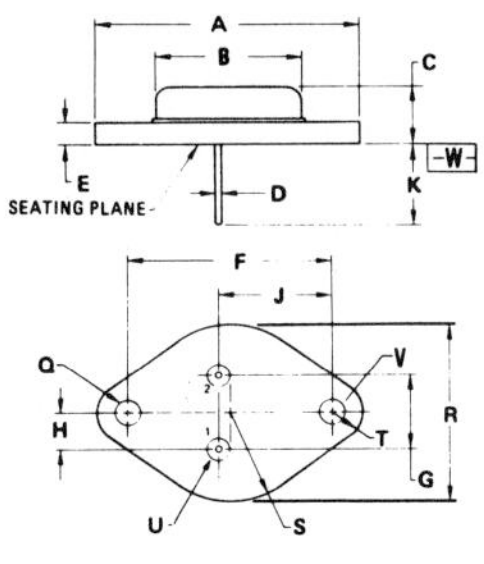

DIM	MILLIMETERS MIN	MILLIMETERS MAX	INCHES MIN	INCHES MAX
B	–	22.23	–	0.875
C	6.35	11.43	0.250	0.450
D	0.97	1.09	0.038	0.043
E	–	3.43	–	0.135
F	30.15 BSC		1.187 BSC	
G	10.92 BSC		0.430 BSC	
H	5.46 BSC		0.215 BSC	
J	16.89 BSC		0.665 BSC	
K	7.92	–	0.312	–
Q	3.84	4.09	0.151	0.161
S	–	13.34	–	0.525
T	–	4.78	–	0.188
V	3.84	4.09	0.151	0.161

CASE 1-03
(TO-204AA)
(TO-3)

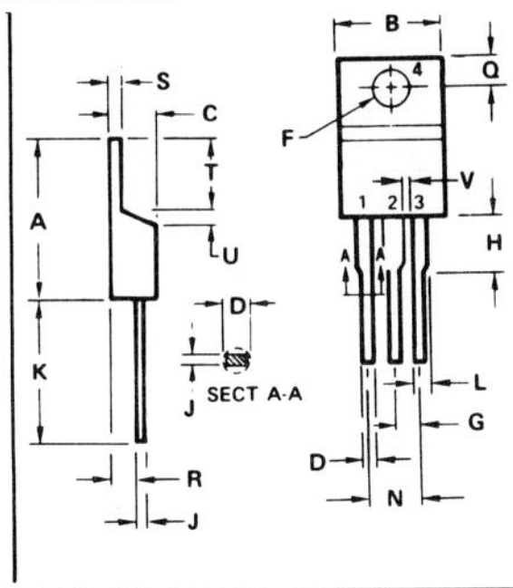

DIM	MILLIMETERS MIN	MILLIMETERS MAX	INCHES MIN	INCHES MAX
A	15.11	15.75	0.595	0.620
B	9.65	10.29	0.380	0.405
C	4.06	4.82	0.160	0.190
D	0.64	0.89	0.025	0.035
F	3.61	3.73	0.142	0.147
G	2.41	2.67	0.095	0.105
H	2.79	3.30	0.110	0.130
J	0.36	0.56	0.014	0.022
K	12.70	14.27	0.500	0.562
L	1.14	1.27	0.045	0.050
N	4.83	5.33	0.190	0.210
Q	2.54	3.04	0.100	0.120
R	2.04	2.79	0.080	0.110
S	1.14	1.39	0.045	0.055
T	5.97	6.48	0.235	0.255
U	0.76	1.27	0.030	0.050
V	1.14	·	0.045	·

CASE 221A-02
TO-220AB

THERMAL INFORMATION

The maximum power consumption an integrated circuit can tolerate at a given operating ambient temperature, can be found from the equation:

$$P_{D(T_A)} = \frac{T_{J(max)} - T_A}{R_{\theta JA}(Typ)} \geqslant V_I I_S - V_O I_O$$

Where: $P_{D(T_A)}$ = Power Dissipation allowable at a given operating ambient temperature.

$T_{J(max)}$ = Maximum Operating Junction Temperature as listed in the Maximum Ratings Section

T_A = Maximum Desired Operating Ambient Temperature

$R_{\theta JA}(Typ)$ = Typical Thermal Resistance Junction to Ambient

I_S = Total Supply Current

MC7808, B, C

ELECTRICAL CHARACTERISTICS (V_{in} = 14 V, I_O = 500 mA, T_J = T_{low} to T_{high} [Note 1] unless otherwise noted).

Characteristic	Symbol	MC7808			MC7808B			MC7808C			Unit
		Min	Typ	Max	Min	Typ	Max	Min	Typ	Max	
Output Voltage (T_J = +25°C)	V_O	7.7	8.0	8.3	7.7	8.0	8.3	7.7	8.0	8.3	Vdc
Output Voltage (5.0 mA ≤ I_O ≤ 1.0 A, P_O ≤ 15 W)	V_O										Vdc
10.5 Vdc ≤ V_{in} ≤ 23 Vdc		—	—	—	—	—	—	7.6	8.0	8.4	
11.5 Vdc ≤ V_{in} ≤ 23 Vdc		7.6	8.0	8.4	7.6	8.0	8.4	—	—	—	
Line Regulation (T_J = +25°C, Note 2)	Reg_{line}										mV
10.5 Vdc ≤ V_{in} ≤ 25 Vdc		—	3.0	80	—	12	160	—	12	160	
11 Vdc ≤ V_{in} ≤ 17 Vdc		—	2.0	40	—	5.0	80	—	5.0	80	
Load Regulation (T_J = +25°C, Note 2)	Reg_{load}										mV
5.0 mA ≤ I_O ≤ 1.5 A		—	28	100	—	45	160	—	45	160	
250 mA ≤ I_O ≤ 750 mA		—	9.0	40	—	16	80	—	16	80	
Quiescent Current (T_J = +25°C)	I_B	—	3.2	6.0	—	4.3	8.0	—	4.3	8.0	mA
Quiescent Current Change	ΔI_B										mA
10.5 Vdc ≤ V_{in} ≤ 25 Vdc		—	—	—	—	—	—	—	—	1.0	
11.5 Vdc ≤ V_{in} ≤ 25 Vdc		—	0.3	0.8	—	—	1.0	—	—	—	
5.0 mA ≤ I_O ≤ 1.0 A		—	0.04	0.5	—	—	0.5	—	—	0.5	
Ripple Rejection 11.5 Vdc ≤ V_{in} ≤ 21.5 Vdc, f = 120 Hz	RR	62	70	—	—	62	—	—	62	—	dB
Dropout Voltage (I_O = 1.0 A, T_J = +25°C)	$V_{in} - V_O$	—	2.0	2.5	—	2.0	—	—	2.0	—	Vdc
Output Noise Voltage (T_A = +25°C) 10 Hz ≤ f ≤ 100 kHz	V_n	—	10	40	—	10	—	—	10	—	µV/ V_O
Output Resistance f = 1.0 kHz	r_O	—	18	—	—	18	—	—	18	—	mΩ
Short-Circuit Current Limit (T_A = +25°C) V_{in} = 35 Vdc	I_{sc}	—	0.2	1.2	—	0.2	—	—	0.2	—	A
Peak Output Current (T_J = +25°C)	I_{max}	1.3	2.5	3.3	—	2.2	—	—	2.2	—	A
Average Temperature Coefficient of Output Voltage	TCV_O	—	±1.0	—	—	-0.8	—	—	-0.8	—	mV/ °C

MC7808A, AC

ELECTRICAL CHARACTERISTICS (V_{in} = 14 V, I_O = 1.0 A, T_J = T_{low} to T_{high} [Note 1] unless otherwise noted)

Characteristics	Symbol	MC7808A			MC7808AC			Unit
		Min	Typ	Max	Min	Typ	Max	
Output Voltage (T_J = +25°C)	V_O	7.84	8.0	8.16	7.84	8.0	8.16	Vdc
Output Voltage (5.0 mA ≤ I_O ≤ 1.0 A, P_O ≤ 15 W) 10.6 Vdc ≤ V_{in} ≤ 23 Vdc	V_O	7.7	8.0	8.3	7.7	8.0	8.3	Vdc
Line Regulation (Note 2)	Reg_{line}							mV
10.6 Vdc ≤ V_{in} ≤ 25 Vdc, I_O = 500 mA		—	4.0	13	—	12	80	
11 Vdc ≤ V_{in} ≤ 17 Vdc		—	6.0	20	—	15	80	
11 Vdc ≤ V_{in} ≤ 17 Vdc, T_J = +25°C		—	2.0	6.0	—	5.0	40	
10.4 Vdc ≤ V_{in} ≤ 23 Vdc, T_J = +25°C		—	4.0	13	—	12	80	
Load Regulation (Note 2)	Reg_{load}							mV
5.0 mA ≤ I_O ≤ 1.5 A, T_J = +25°C		—	2.0	25	—	45	100	
5.0 mA ≤ I_O ≤ 1.0 A		—	2.0	25	—	45	100	
250 mA ≤ I_O ≤ 750mA, T_J = +25°C		—	1.0	15	—	—	—	
250 mA ≤ I_O ≤ 750 mA		—	1.0	25	—	16	50	
Quiescent Current	I_B	—	—	5.0	—	—	6.0	mA
T_J = +25°C		—	3.2	4.0	—	4.3	6.0	
Quiescent Current Change	ΔI_B							mA
11 Vdc ≤ V_{in} ≤ 25 Vdc, I_O = 500 mA		—	0.3	0.5	—	—	0.8	
10.6 Vdc ≤ V_{in} ≤ 23 Vdc, T_J = +25°C		—	0.2	0.5	—	—	0.8	
5.0 mA ≤ I_O ≤ 1.0 A		—	0.04	0.2	—	—	0.5	
Ripple Rejection	RR							dB
11.5 Vdc ≤ V_{in} ≤ 21.5 Vdc, f = 120 Hz, T_J = +25°C		62	70	—	—	—	—	
11.5 Vdc ≤ V_{in} ≤ 21.5 Vdc, f = 120 Hz, I_O = 500 mA		62	70	—	—	62	—	
Dropout Voltage (I_O = 1.0 A, T_J = +25°C)	$V_{in} - V_O$	—	2.0	2.5	—	2.0	—	Vdc
Output Noise Voltage (T_A = +25°C) 10 Hz ≤ f ≤ 100 kHz	V_n	—	10	40	—	10	—	µV/V_O
Output Resistance (f = 1.0 kHz)	r_O	—	2.0	—	—	18	—	mΩ
Short-Circuit Current Limit (T_A = +25°C) V_{in} = 35 Vdc	I_{sc}	—	0.2	1.2	—	0.2	—	A
Peak Output Current (T_J = +25°C)	I_{max}	1.3	2.5	3.3	—	2.2	—	A
Average Temperature Coefficient of Output Voltage	TCV_O	—	±1.0	—	—	-0.8	—	mV/°C

NOTES: 1. T_{low} = −55°C for MC78XX, A
= 0° for MC78XXC, AC
= −40°C for MC78XXB

T_{high} = +150°C for MC78XX, A
= +125°C for MC78XXC, AC, B

2. Load and line regulation are specified at constant junction temperature. Changes in V_O due to heating effects must be taken into account separately. Pulse testing with low duty cycle is used.

MC7812, B, C

ELECTRICAL CHARACTERISTICS (V_{in} = 19 V, I_O = 500 mA, T_J = T_{low} to T_{high} [Note 1] unless otherwise noted).

Characteristic	Symbol	MC7812			MC7812B			MC7812C			Unit
		Min	Typ	Max	Min	Typ	Max	Min	Typ	Max	
Output Voltage (T_J = +25°C)	V_O	11.5	12	12.5	11.5	12	12.5	11.5	12	12.5	Vdc
Output Voltage (5.0 mA ≤ I_O ≤ 1.0 A, P_O ≤ 15 W)	V_O										Vdc
14.5 Vdc ≤ V_{in} ≤ 27 Vdc		—	—	—	—	—	—	11.4	12	12.6	
15.5 Vdc ≤ V_{in} ≤ 27 Vdc		11.4	12	12.6	11.4	12	12.6	—	—	—	
Line Regulation (T_J = +25°C, Note 2)	Reg_{line}										mV
14.5 Vdc ≤ V_{in} ≤ 30 Vdc		—	5.0	120	—	13	240	—	13	240	
16 Vdc ≤ V_{in} ≤ 22 Vdc		—	3.0	60	—	6.0	120	—	6.0	120	
Load Regulation (T_J = +25°C, Note 2)	Reg_{load}										mV
5.0 mA ≤ I_O ≤ 1.5 A		—	30	120	—	46	240	—	46	240	
250 mA ≤ I_O ≤ 750 mA		—	10	60	—	17	120	—	17	120	
Quiescent Current (T_J = +25°C)	I_B	—	3.4	6.0	—	4.4	8.0	—	4.4	8.0	mA
Quiescent Current Change	ΔI_B										mA
14.5 Vdc ≤ V_{in} ≤ 30 Vdc		—	—	—	—	—	—	—	—	1.0	
15 Vdc ≤ V_{in} ≤ 30 Vdc		—	0.3	0.8	—	—	1.0	—	—	—	
5.0 mA ≤ I_O ≤ 1.0 A		—	0.04	0.5	—	—	0.5	—	—	0.5	
Ripple Rejection 15 Vdc ≤ V_{in} ≤ 25 Vdc, f = 120 Hz	RR	61	68	—	—	60	—	—	60	—	dB
Dropout Voltage (I_O = 1.0 A, T_J = +25°C)	$V_{in} - V_O$	—	2.0	2.5	—	2.0	—	—	2.0	—	Vdc
Output Noise Voltage (T_A = +25°C) 10 Hz ≤ f ≤ 100 kHz	V_n	—	10	40	—	10	—	—	10	—	µV/ V_O
Output Resistance f = 1.0 kHz	r_O	—	18	—	—	18	—	—	18	—	mΩ
Short-Circuit Current Limit (T_A = +25°C) V_{in} = 35 Vdc	I_{sc}	—	0.2	1.2	—	0.2	—	—	0.2	—	A
Peak Output Current (T_J = +25°C)	I_{max}	1.3	2.5	3.3	—	2.2	—	—	2.2	—	A
Average Temperature Coefficient of Output Voltage	TCV_O	—	±1.5	—	—	-1.0	—	—	-1.0	—	mV/ °C

MC7812A, AC

ELECTRICAL CHARACTERISTICS (V_{in} = 19 V, I_O = 1.0 A, T_J = T_{low} to T_{high} [Note 1] unless otherwise noted)

Characteristics	Symbol	MC7812A			MC7812AC			Unit
		Min	Typ	Max	Min	Typ	Max	
Output Voltage (T_J = +25°C)	V_O	11.75	12	12.25	11.75	12	12.25	Vdc
Output Voltage (5.0 mA ≤ I_O ≤ 1.0 A, P_O ≤ 15 W) 14.8 Vdc ≤ V_{in} ≤ 27 Vdc	V_O	11.5	12	12.5	11.5	12	12.5	Vdc
Line Regulation (Note 2)	Reg_{line}							mV
14.8 Vdc ≤ V_{in} ≤ 30 Vdc, I_O = 500 mA		—	5.0	18	—	13	120	
16 Vdc ≤ V_{in} ≤ 22 Vdc		—	8.0	30	—	16	120	
16 Vdc ≤ V_{in} ≤ 22 Vdc, T_J = +25°C		—	3.0	9.0	—	6.0	60	
14.5 Vdc ≤ V_{in} ≤ 27 Vdc, T_J = +25°C		—	5.0	18	—	13	120	
Load Regulation (Note 2)	Reg_{load}							mV
5.0 mA ≤ I_O ≤ 1.5 A, T_J = +25°C		—	2.0	25	—	46	100	
5.0 mA ≤ I_O ≤ 1.0 A		—	2.0	25	—	46	100	
250 mA ≤ I_O ≤ 750mA, T_J = +25°C		—	1.0	15	—	—	—	
250 mA ≤ I_O ≤ 750 mA		—	1.0	25	—	17	50	
Quiescent Current	I_B	—	—	5.0	—	—	6.0	mA
T_J = +25°C		—	3.4	4.0	—	4.4	6.0	
Quiescent Current Change	ΔI_B							mA
15 Vdc ≤ V_{in} ≤ 30 Vdc, I_O = 500 mA		—	0.3	0.5	—	—	0.8	
14.8 Vdc ≤ V_{in} ≤ 27 Vdc, T_J = +25°C		—	0.2	0.5	—	—	0.8	
5.0 mA ≤ I_O ≤ 1.0 A		—	0.04	0.2	—	—	0.5	
Ripple Rejection	RR							dB
15 Vdc ≤ V_{in} ≤ 25 Vdc, f = 120 Hz, T_J = +25°C		61	68	—	—	—	—	
15 Vdc ≤ V_{in} ≤ 25 Vdc, f = 120 Hz, I_O = 500 mA		61	68	—	—	60	—	
Dropout Voltage (I_O = 1.0 A, T_J = +25°C)	$V_{in} - V_O$	—	2.0	2.5	—	2.0	—	Vdc
Output Noise Voltage (T_A = +25°C) 10 Hz ≤ f ≤ 100 kHz	V_n	—	10	40	—	10	—	µV/V_O
Output Resistance (f = 1.0 kHz)	r_O	—	2.0	—	—	18	—	mΩ
Short-Circuit Current Limit (T_A = +25°C) V_{in} = 35 Vdc	I_{sc}	—	0.2	1.2	—	0.2	—	A
Peak Output Current (T_J = +25°C)	I_{max}	1.3	2.5	3.3	—	2.2	—	A
Average Temperature Coefficient of Output Voltage	TCV_O	—	±1.5	—	—	-1.0	—	mV/°C

NOTES: 1. T_{low} = −55°C for MC78XX, A; = 0° for MC78XXC, AC; = −40°C for MC78XXB. T_{high} = +150°C for MC78XX, A; = +125°C for MC78XXC, AC, B

2. Load and line regulation are specified at constant junction temperature. Changes in V_O due to heating effects must be taken into account separately. Pulse testing with low duty cycle is used.

TYPICAL CHARACTERISTICS

(T_A = +25°C unless otherwise noted.)

FIGURE 1 — WORST CASE POWER DISSIPATION versus AMBIENT TEMPERATURE (Case 221A)

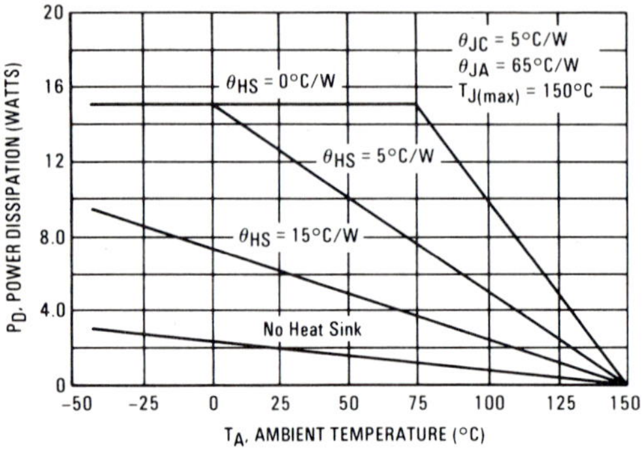

FIGURE 2 — WORST CASE POWER DISSIPATION versus AMBIENT TEMPERATURE (Case 1)

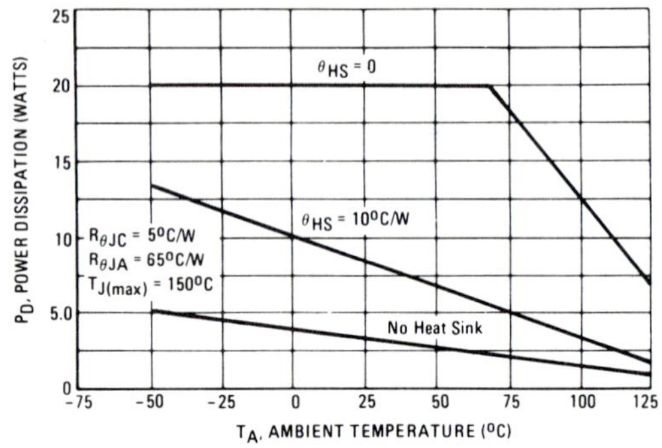

FIGURE 3 — INPUT OUTPUT DIFFERENTIAL AS A FUNCTION OF JUNCTION TEMPERATURE (MC78XXC, AC, B)

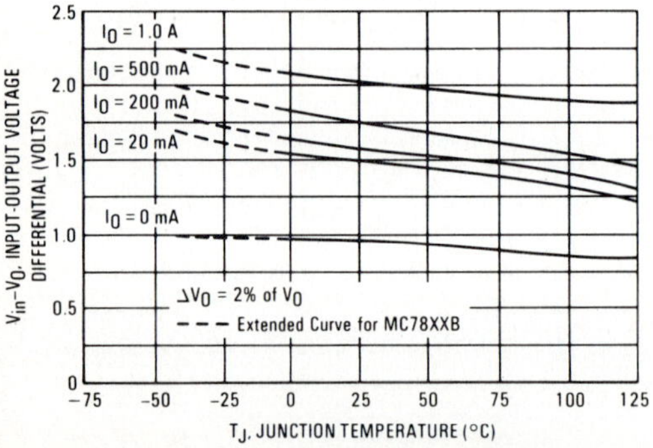

FIGURE 4 — INPUT OUTPUT DIFFERENTIAL AS A FUNCTION OF JUNCTION TEMPERATURE (MC78XX, A)

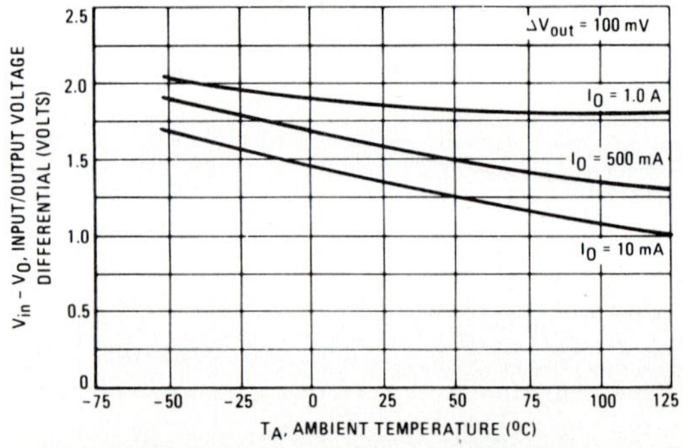

FIGURE 5 — PEAK OUTPUT CURRENT AS A FUNCTION OF INPUT-OUTPUT DIFFERENTIAL VOLTAGE (MC78XXC, AC, B)

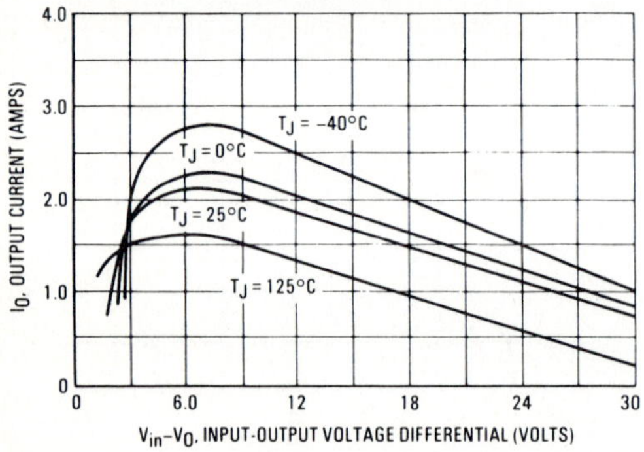

FIGURE 6 — PEAK OUTPUT CURRENT AS A FUNCTION OF INPUT-OUTPUT DIFFERENTIAL VOLTAGE (MC78XX, A)

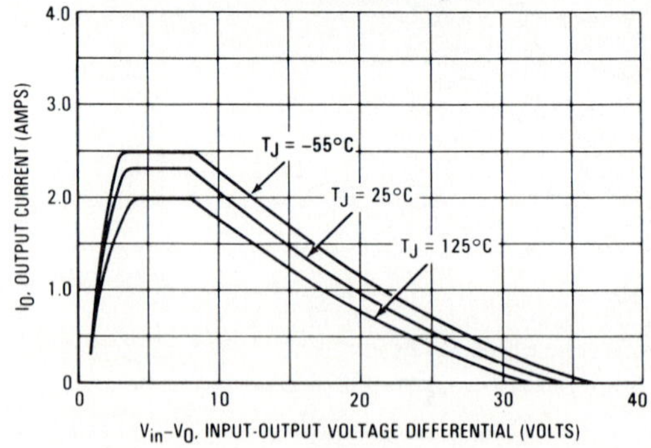

TYPICAL CHARACTERISTICS (continued)

(T_A = 25°C unless otherwise noted.)

FIGURE 7 – RIPPLE REJECTION AS A FUNCTION OF OUTPUT VOLTAGES (MC78XXC, AC)

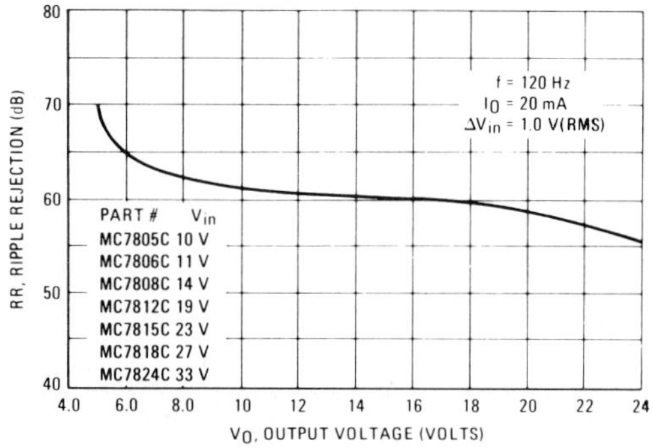

FIGURE 8 – RIPPLE REJECTION AS A FUNCTION OF FREQUENCY (MC78XXC, AC, A)

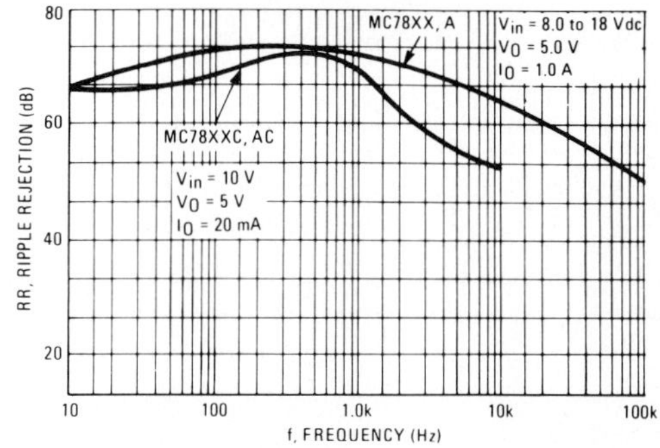

FIGURE 9 — OUTPUT VOLTAGE AS A FUNCTION OF JUNCTION TEMPERATURE (MC78XXC, AC, B)

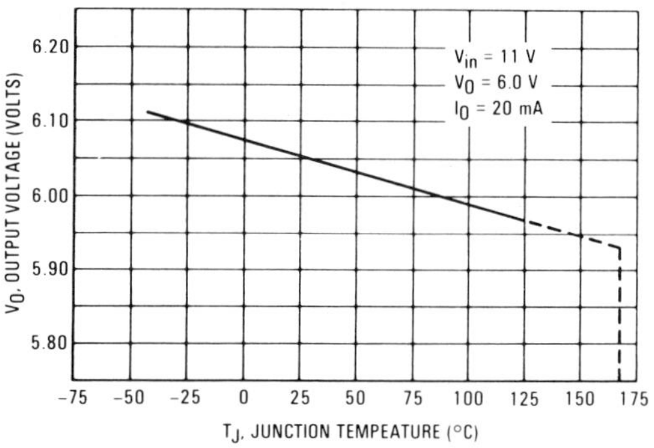

FIGURE 10 – OUTPUT IMPEDANCE AS A FUNCTION OF OUTPUT VOLTAGE (MC78XXC, AC)

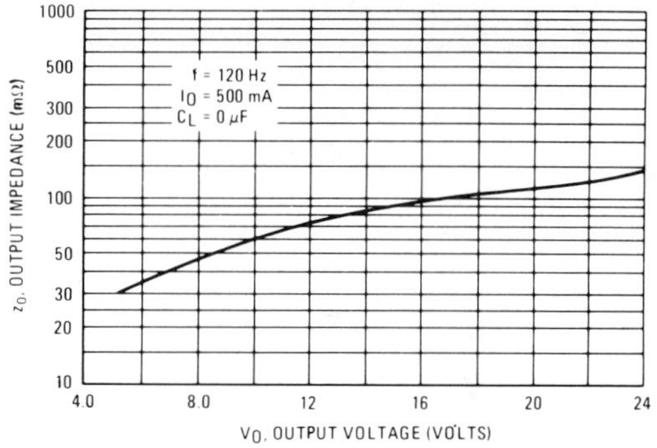

FIGURE 11 — QUIESCENT CURRENT AS A FUNCTION OF TEMPERATURE (MC78XXC, AC, B)

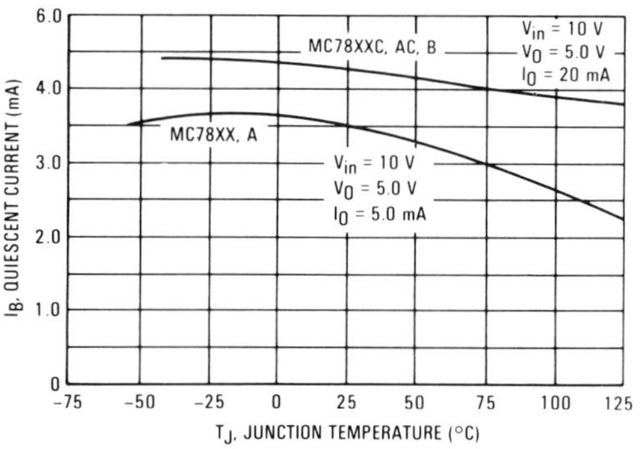

FIGURE 12 — DROPOUT CHARACTERISTICS (MC78XX, A)

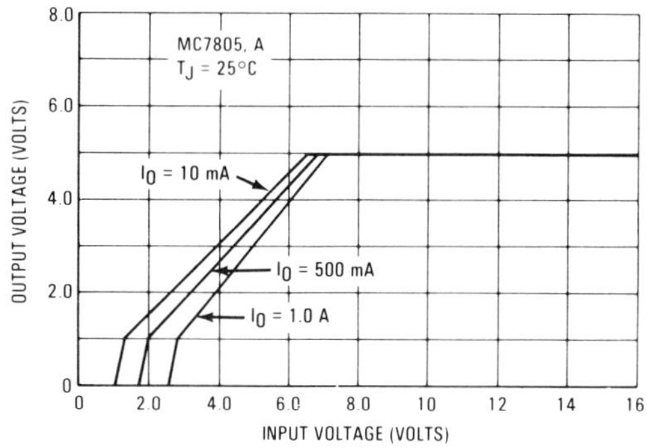

MC7800 Series

APPLICATIONS INFORMATION

Design Considerations

The MC7800 Series of fixed voltage regulators are designed with Thermal Overload Protection that shuts down the circuit when subjected to an excessive power overload condition, Internal Short-Circuit Protection that limits the maximum current the circuit will pass, and Output Transistor Safe-Area Compensation that reduces the output short-circuit current as the voltage across the pass transistor is increased.

In many low current applications, compensation capacitors are not required. However, it is recommended that the regulator input be bypassed with a capacitor if the regulator is connected to the power supply filter with long wire lengths, or if the output load capacitance is large. An input bypass capacitor should be selected to provide good high-frequency characteristics to insure stable operation under all load conditions. A 0.33 μF or larger tantalum, mylar, or other capacitor having low internal impedance at high frequencies should be chosen. The bypass capacitor should be mounted with the shortest possible leads directly across the regulators input terminals. Normally good construction techniques should be used to minimize ground loops and lead resistance drops since the regulator has no external sense lead.

FIGURE 13 — CURRENT REGULATOR

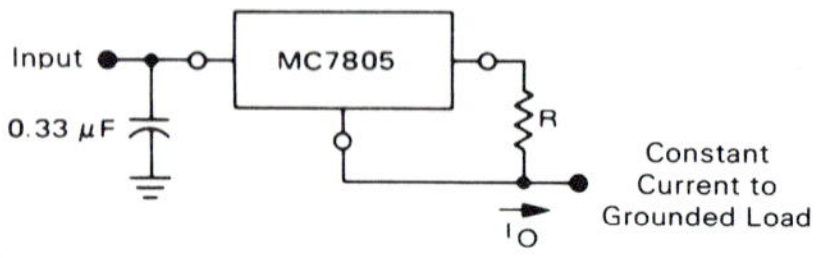

The MC7800 regulators can also be used as a current source when connected as above. In order to minimize dissipation the MC7805C is chosen in this application. Resistor R determines the current as follows:

$$I_O = \frac{5\ V}{R} + I_Q$$

$I_Q \cong 1.5$ mA over line and load changes

For example, a 1-ampere current source would require R to be a 5-ohm, 10-W resistor and the output voltage compliance would be the input voltage less 7 volts.

FIGURE 14 — ADJUSTABLE OUTPUT REGULATOR

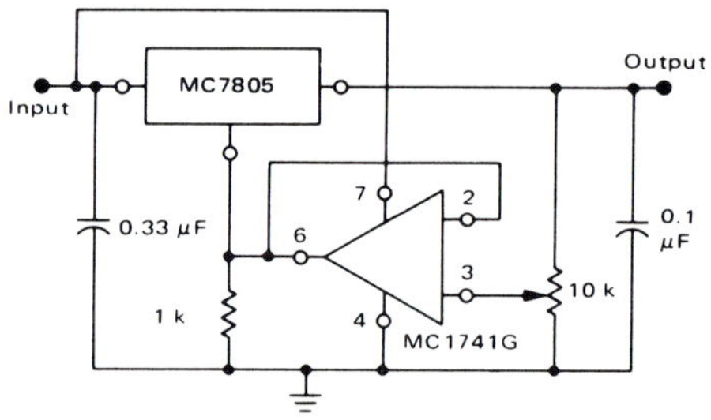

V_O, 7.0 V to 20 V
$V_{IN} - V_O \geq 2.0$ V

The addition of an operational amplifier allows adjustment to higher or intermediate values while retaining regulation characteristics. The minimum voltage obtainable with this arrangement is 2.0 volts greater than the regulator voltage.

FIGURE 15 — CURRENT BOOST REGULATOR

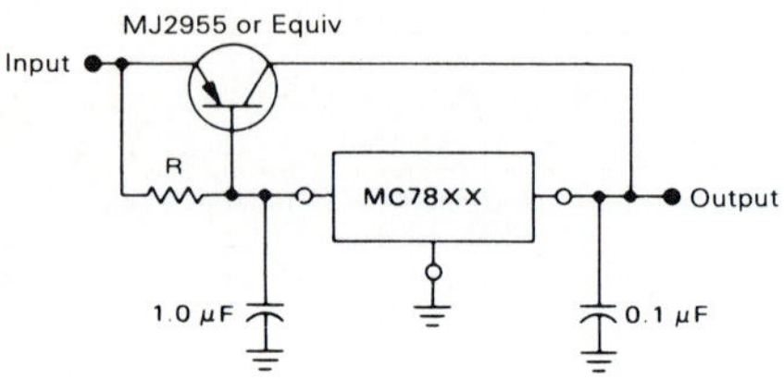

XX = 2 digits of type number indicating voltage.

The MC7800 series can be current boosted with a PNP transistor. The MJ2955 provides current to 5.0 amperes. Resistor R in conjunction with the V_{BE} of the PNP determines when the pass transistor begins conducting; this circuit is not short-circuit proof. Input-output differential voltage minimum is increased by V_{BE} of the pass transistor.

FIGURE 16 — SHORT-CIRCUIT PROTECTION

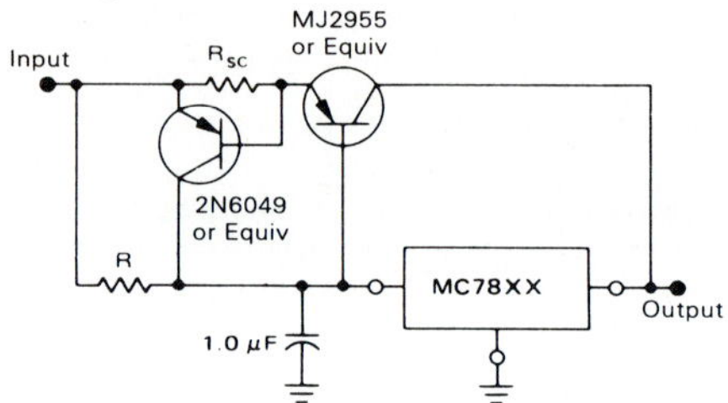

XX = 2 digits of type number indicating voltage.

The circuit of Figure 15 can be modified to provide supply protection against short circuits by adding a short-circuit sense resistor, R_{SC}, and an additional PNP transistor. The current sensing PNP must be able to handle the short-circuit current of the three-terminal regulator. Therefore, a four-ampere plastic power transistor is specified.

National Semiconductor **Voltage Comparators**

LM139/239/339, LM139A/239A/339A, LM2901,LM3302 Low Power Low Offset Voltage Quad Comparators

General Description

The LM139 series consists of four independent precision voltage comparators with an offset voltage specification as low as 2 mV max for all four comparators. These were designed specifically to operate from a single power supply over a wide range of voltages. Operation from split power supplies is also possible and the low power supply current drain is independent of the magnitude of the power supply voltage. These comparators also have a unique characteristic in that the input common-mode voltage range includes ground, even though operated from a single power supply voltage.

Application areas include limit comparators, simple analog to digital converters; pulse, squarewave and time delay generators; wide range VCO; MOS clock timers; multivibrators and high voltage digital logic gates. The LM139 series was designed to directly interface with TTL and CMOS. When operated from both plus and minus power supplies, they will directly interface with MOS logic— where the low power drain of the LM339 is a distinct advantage over standard comparators.

Advantages

- High precision comparators
- Reduced V_{OS} drift over temperature
- Eliminates need for dual supplies
- Allows sensing near gnd
- Compatible with all forms of logic
- Power drain suitable for battery operation

Features

- Wide single supply voltage range or dual supplies

LM139 series,	2 V_{DC} to 36 V_{DC} or
LM139A series, LM2901	±1 V_{DC} to ±18 V_{DC}
LM3302	2 V_{DC} to 28 V_{DC}
	or ±1 V_{DC} to ±14 V_{DC}

- Very low supply current drain (0.8 mA) — independent of supply voltage (2 mW/comparator at +5 V_{DC})
- Low input biasing current 25 nA
- Low input offset current ±5 nA and offset voltage ±3 mV
- Input common-mode voltage range includes gnd
- Differential input voltage range equal to the power supply voltage
- Low output saturation voltage 250 mV at 4 mA
- Output voltage compatible with TTL, DTL, ECL, MOS and CMOS logic systems

Schematic and Connection Diagrams

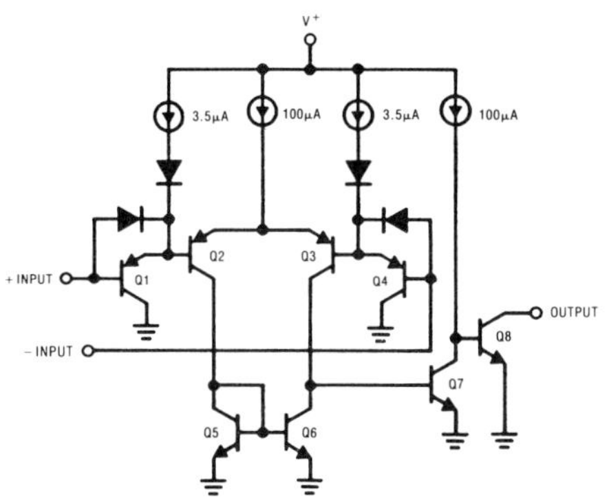

Dual-In-Line and Flat Package

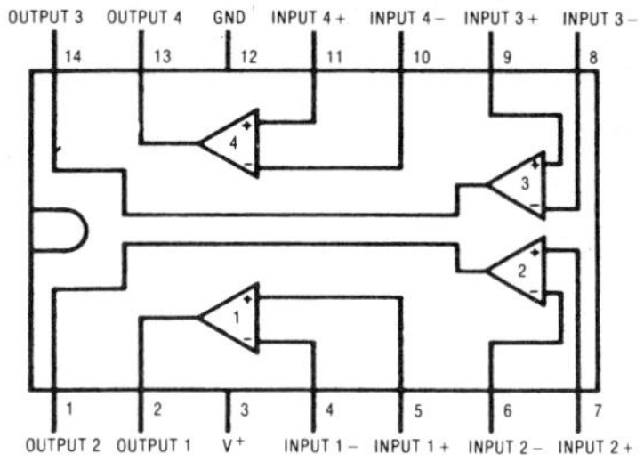

Order Number LM139J, LM139AJ, LM239J, LM239AJ, LM339J, LM339AJ, LM2901J or LM3302J
See NS Package J14A

Order Number LM339N, LM339AN, LM2901N or LM3302N
See NS Package N14A

Typical Applications ($V^+ = 5.0\ V_{DC}$)

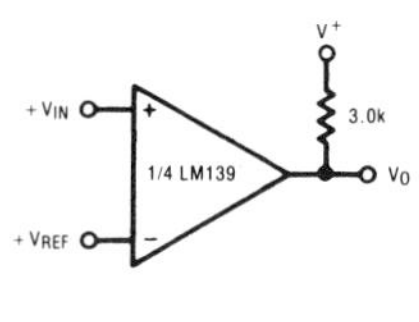

Basic Comparator

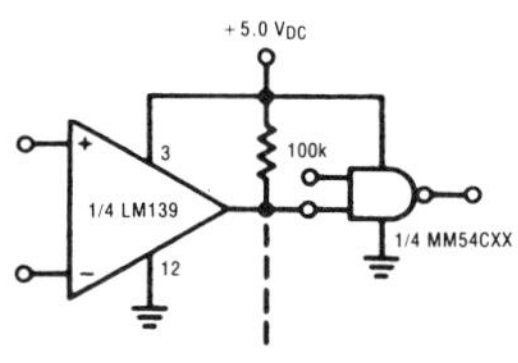

Driving CMOS

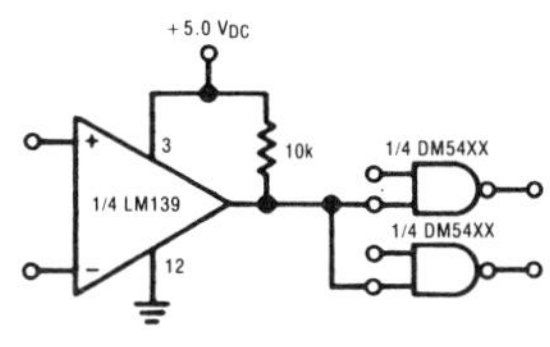

Driving TTL

Absolute Maximum Ratings

	LM139/LM239/LM339 LM139A/LM239A/LM339A LM2901	LM3302
Supply Voltage, V^+	36 V_{DC} or ±18 V_{DC}	28 V_{DC} or ±14 V_{DC}
Differential Input Voltage	36 V_{DC}	28 V_{DC}
Input Voltage	−0.3 V_{DC} to +36 V_{DC}	−0.3 V_{DC} to +28 V_{DC}
Power Dissipation (Note 1)		
Molded DIP	570 mW	570 mW
Cavity DIP	900 mW	
Flat Pack	800 mW	
Output Short-Circuit to GND, (Note 2)	Continuous	Continuous
Input Current (V_{IN} < −0.3 V_{DC}), (Note 3)	50 mA	50 mA
Operating Temperature Range		
LM339A	0°C to +70°C	−40°C to +85°C
LM239A	−25°C to +85°C	
LM2901	−40°C to +85°C	
LM139A	−55°C to +125°C	
Storage Temperature Range	−65°C to +150°C	−65°C to +150°C
Lead Temperature (Soldering, 10 seconds)	300°C	300°C

Electrical Characteristics (V^+ = 5 V_{DC}, Note 4)

PARAMETER	CONDITIONS	LM139A			LM239A, LM339A			LM139			LM239, LM339			LM2901			LM3302			UNITS
		MIN	TYP	MAX	MIN	TYP	MAX	MIN	TYP	MAX	MIN	TYP	MAX	MIN	TYP	MAX	MIN	TYP	MAX	
Input Offset Voltage	T_A = 25°C, (Note 9)		±1.0	±2.0		±1.0	±2.0		±2.0	±5.0		±2.0	±5.0		±2.0	±7.0		±3	±20	mV_{DC}
Input Bias Current	$I_{IN(+)}$ or $I_{IN(-)}$ with Output in Linear Range, T_A = 25°C, (Note 5)		25	100		25	250		25	100		25	250		25	250		25	500	nA_{DC}
Input Offset Current	$I_{IN(+)} - I_{IN(-)}$, T_A = 25°C		±3.0	±25		±5.0	±50		±3.0	±25		±5.0	±50		±5	±50		±3	±100	nA_{DC}
Input Common-Mode Voltage Range	T_A = 25°C, (Note 6)	0		V^+−1.5	0		V^+−1.5	0		V^+−1.5	0		V^+−1.5	0		V^+−1.5	0		V^+−1.5	V_{DC}
Supply Current	R_L = ∞ on all Comparators, T_A = 25°C		0.8	2.0		0.8	2.0		0.8	2.0		0.8	2.0		0.8	2.0		0.8	2	mA_{DC}
	R_L = ∞, V^+ = 30V, T_A = 25°C														1	2.5				mA_{DC}
Voltage Gain	$R_L \geq$ 15 kΩ, V^+ = 15 V_{DC} (To Support Large V_O Swing), T_A = 25°C	50	200		50	200			200			200		25	100		2	30		V/mV
Large Signal Response Time	V_{IN} = TTL Logic Swing, V_{REF} = 1.4 V_{DC}, V_{RL} = 5 V_{DC}, R_L = 5.1 kΩ, T_A = 25°C		300			300			300			300			300			300		ns
Response Time	V_{RL} = 5 V_{DC}, R_L = 5.1 kΩ, T_A = 25°C, (Note 7)		1.3			1.3			1.3			1.3			1.3			1.3		μs
Output Sink Current	$V_{IN(-)} \geq$ 1 V_{DC}, $V_{IN(+)}$ = 0, $V_O \leq$ 1.5 V_{DC}, T_A = 25°C	6.0	16		6.0	16		6.0	16		6.0	16		6.0	16		6.0	16		mA_{DC}
Saturation Voltage	$V_{IN(-)} \geq$ 1 V_{DC}, $V_{IN(+)}$ = 0, $I_{SINK} \leq$ 4 mA, T_A = 25°C		250	400		250	400		250	400		250	400			400		250	500	mV_{DC}
Output Leakage Current	$V_{IN(+)} \geq$ 1 V_{DC}, $V_{IN(-)}$ = 0, V_O = 5 V_{DC}, T_A = 25°C		0.1			0.1			0.1			0.1			0.1			0.1		nA_{DC}

Electrical Characteristics (Continued)

PARAMETER	CONDITIONS	LM139A			LM239A, LM339A			LM139			LM239, LM339			LM2901			LM3302			UNITS
		MIN	TYP	MAX	MIN	TYP	MAX	MIN	TYP	MAX	MIN	TYP	MAX	MIN	TYP	MAX	MIN	TYP	MAX	
Input Offset Voltage	(Note 9)			4.0			4.0			9.0			9.0		9	15			40	mV_{DC}
Input Offset Current	$I_{IN(+)} - I_{IN(-)}$			±100			±150			±100			±150		50	200			300	nA_{DC}
Input Bias Current	$I_{IN(+)}$ or $I_{IN(-)}$ with Output in Linear Range			300			400			300			400		200	500			1000	nA_{DC}
Input Common-Mode Voltage Range		0		$V^+-2.0$	0		$V^+-2.0$	0		$V^+-2.0$	0		$V^+-2.0$	0		$V^+-2.0$	0		$V^+-2.0$	V_{DC}
Saturation Voltage	$V_{IN(-)} \geq 1\ V_{DC}$, $V_{IN(+)} = 0$, $I_{SINK} \leq 4$ mA			700			700			700			700		400	700			700	mV_{DC}
Output Leakage Current	$V_{IN(+)} \geq 1\ V_{DC}$, $V_{IN(-)} = 0$, $V_O = 30\ V_{DC}$			1.0			1.0			1.0			1.0			1.0			1.0	μA_{DC}
Differential Input Voltage	Keep all V_{IN}'s $\geq 0\ V_{DC}$ (or V^- if used), (Note 8)			36			36			36			36	0		36			28	V_{DC}

Note 1: For operating at high temperatures, the LM339/LM339A, LM2901, LM3302 must be derated based on a 125°C maximum junction temperature and a thermal resistance of 175°C/W which applies for the device soldered in a printed circuit board, operating in a still air ambient. The LM239 and LM139 must be derated based on a 150°C maximum junction temperature. The low bias dissipation and the "ON-OFF" characteristic of the outputs keeps the chip dissipation very small ($P_D \leq 100$ mW), provided the output transistors are allowed to saturate.

Note 2: Short circuits from the output to V^+ can cause excessive heating and eventual destruction. The maximum output current is approximately 20 mA independent of the magnitude of V^+.

Note 3: This input current will only exist when the voltage at any of the input leads is driven negative. It is due to the collector-base junction of the input PNP transistors becoming forward biased and thereby acting as input diode clamps. In addition to this diode action, there is also lateral NPN parasitic transistor action on the IC chip. This transistor action can cause the output voltages of the comparators to go to the V^+ voltage level (or to ground for a large overdrive) for the time duration that an input is driven negative. This is not destructive and normal output states will re-establish when the input voltage, which was negative, again returns to a value greater than $-0.3\ V_{DC}$ (at 25°C).

Note 4: These specifications apply for $V^+ = 5\ V_{DC}$ and $-55°C \leq T_A \leq +125°C$, unless otherwise stated. With the LM239/LM239A, all temperature specifications are limited to $-25°C \leq T_A \leq +85°C$, the LM339/LM339A temperature specifications are limited to $0°C \leq T_A \leq +70°C$, and the LM2901, LM3302 temperature range is $-40°C \leq T_A \leq +85°C$.

Note 5: The direction of the input current is out of the IC due to the PNP input stage. This current is essentially constant, independent of the state of the output so no loading change exists on the reference or input lines.

Note 6: The input common-mode voltage or either input signal voltage should not be allowed to go negative by more than 0.3V. The upper end of the common-mode voltage range is $V^+ -1.5$V, but either or both inputs can go to $+30\ V_{DC}$ without damage (25V for LM3302).

Note 7: The response time specified is a 100 mV input step with 5 mV overdrive. For larger overdrive signals 300 ns can be obtained, see typical performance characteristics section.

Note 8: Positive excursions of input voltage may exceed the power supply level. As long as the other voltage remains within the common-mode range, the comparator will provide a proper output state. The low input voltage state must not be less than $-0.3\ V_{DC}$ (or $0.3\ V_{DC}$ below the magnitude of the negative power supply, if used) (at 25°C).

Note 9: At output switch point, $V_O \cong 1.4\ V_{DC}$, $R_S = 0\Omega$ with V^+ from $5\ V_{DC}$; and over the full input common-mode range ($0\ V_{DC}$ to $V^+ -1.5\ V_{DC}$).

μA741

FREQUENCY-COMPENSATED OPERATIONAL AMPLIFIER

FAIRCHILD LINEAR INTEGRATED CIRCUITS

GENERAL DESCRIPTION — The μA741 is a high performance monolithic Operational Amplifier constructed using the Fairchild Planar* epitaxial process. It is intended for a wide range of analog applications. High common mode voltage range and absence of latch-up tendencies make the μA741 ideal for use as a voltage follower. The high gain and wide range of operating voltage provides superior performance in integrator, summing amplifier, and general feedback applications.

- **NO FREQUENCY COMPENSATION REQUIRED**
- **SHORT CIRCUIT PROTECTION**
- **OFFSET VOLTAGE NULL CAPABILITY**
- **LARGE COMMON MODE AND DIFFERENTIAL VOLTAGE RANGES**
- **LOW POWER CONSUMPTION**
- **NO LATCH-UP**

ABSOLUTE MAXIMUM RATINGS

Supply Voltage	
μA741A, μA741, μA741E	±22 V
μA741C	±18 V
Internal Power Dissipation (Note 1)	
Metal Can	500 mW
Molded and Hermetic DIP	670 mW
Mini DIP	310 mW
Flatpak	570 mW
Differential Input Voltage	±30 V
Input Voltage (Note 2)	±15 V
Storage Temperature Range	
Metal Can, Hermetic DIP, and Flatpak	−65°C to +150°C
Mini DIP, Molded DIP	−55°C to +125°C
Operating Temperature Range	
Military (μA741A, μA741)	−55°C to +125°C
Commercial (μA741E, μA741C)	0°C to +70°C
Pin Temperature (Soldering)	
Metal Can, Hermetic DIPs, and Flatpak (60 s)	300°C
Molded DIPs (10 s)	260°C
Output Short Circuit Duration (Note 3)	Indefinite

CONNECTION DIAGRAMS

8-PIN METAL CAN
(TOP VIEW)
PACKAGE OUTLINE 5B
PACKAGE CODE H

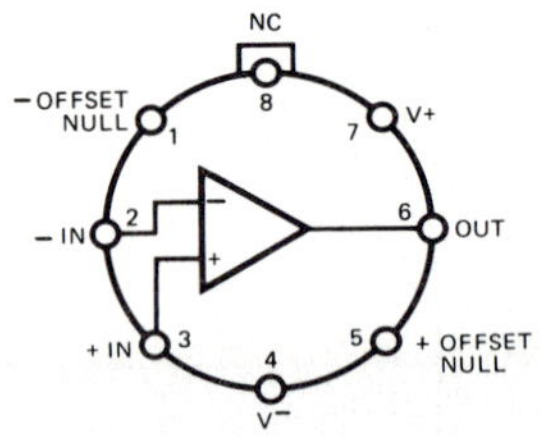

Note: Pin 4 connected to case

ORDER INFORMATION

TYPE	PART NO.
μA741A	μA741AHM
μA741	μA741HM
μA741E	μA741EHC
μA741C	μA741HC

14-PIN DIP
(TOP VIEW)
PACKAGE OUTLINES 6A, 9A
PACKAGE CODES D P

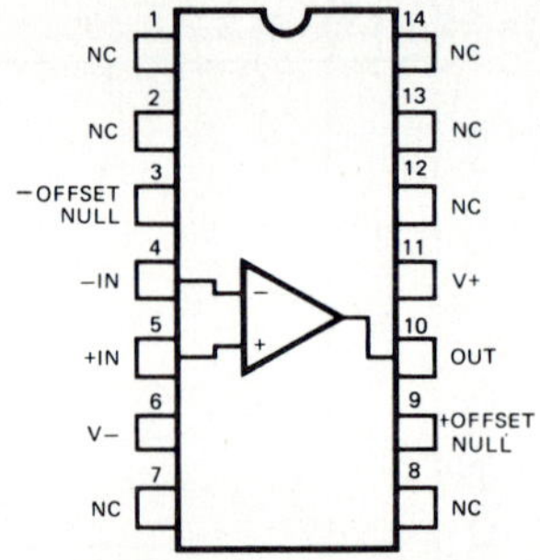

ORDER INFORMATION

TYPE	PART NO.
μA741A	μA741ADM
μA741	μA741DM
μA741E	μA741EDC
μA741C	μA741DC
μA741C	μA741PC

8-PIN MINI DIP
(TOP VIEW)
PACKAGE OUTLINES 6T 9T
PACKAGE CODES R T

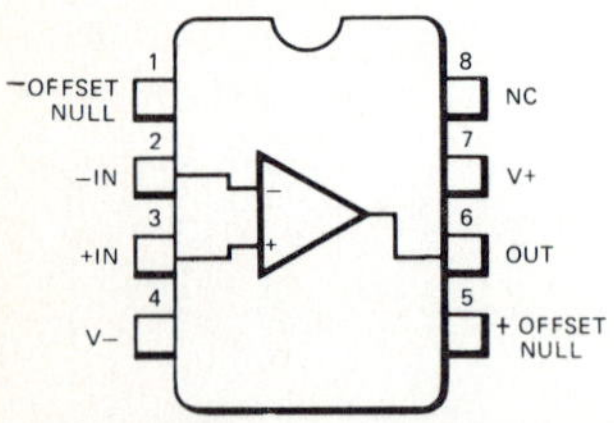

ORDER INFORMATION

TYPE	PART NO.
μA741C	μA741TC
μA741C	μA741RC

10-PIN FLATPAK
(TOP VIEW)
PACKAGE OUTLINE 3F
PACKAGE CODE F

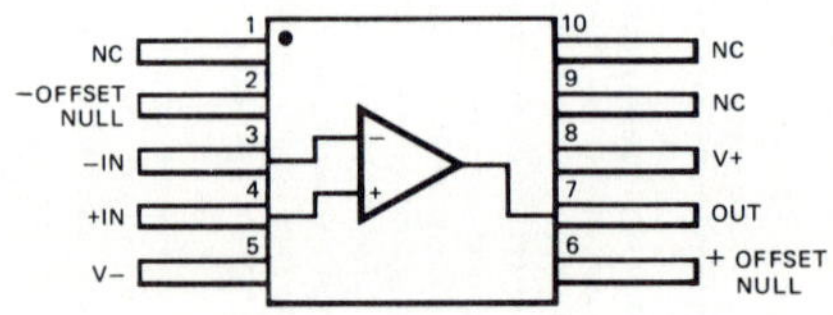

ORDER INFORMATION

TYPE	PART NO.
μA741A	μA741AFM
μA741	μA741FM

*Planar is a patented Fairchild process.

FAIRCHILD • μA741

μA741A

ELECTRICAL CHARACTERISTICS: V_S = ±15 V, T_A = 25°C unless otherwise specified.

CHARACTERISTICS (see definitions)		CONDITIONS		MIN	TYP	MAX	UNITS
Input Offset Voltage		$R_S \leq 50\Omega$			0.8	3.0	mV
Average Input Offset Voltage Drift						15	μV/°C
Input Offset Current					3.0	30	nA
Average Input Offset Current Drift						0.5	nA/°C
Input Bias Current					30	80	nA
Power Supply Rejection Ratio		V_S = +20, −20; V_S = −20, +10V, R_S = 50Ω			15	50	μV/V
Output Short Circuit Current				10	25	40	mA
Power Dissipation		V_S = ±20V			80	150	mW
Input Impedance		V_S = ±20V		1.0	6.0		MΩ
Large Signal Voltage Gain		V_S = ±20V, R_L = 2kΩ, V_{OUT} = ±15V		50			V/mV
Transient Response	Rise Time				0.25	0.8	μs
(Unity Gain)	Overshoot				6.0	20	%
Bandwidth (Note 4)				.437	1.5		MHz
Slew Rate (Unity Gain)		V_{IN} = ±10V		0.3	0.7		V/μs
The following specifications apply for −55°C ≤ T_A ≤ +125°C							
Input Offset Voltage						4.0	mV
Input Offset Current						70	nA
Input Bias Current						210	nA
Common Mode Rejection Ratio		V_S = ±20V, V_{IN} = ±15V, R_S = 50Ω		80	95		dB
Adjustment For Input Offset Voltage		V_S = ±20V		10			mV
Output Short Circuit Current				10		40	mA
Power Dissipation		V_S = ±20V	−55°C			165	mW
			+125°C			135	mW
Input Impedance		V_S = ±20V		0.5			MΩ
Output Voltage Swing		V_S = ±20V,	R_L = 10kΩ	±16			V
			R_L = 2kΩ	±15			V
Large Signal Voltage Gain		V_S = ±20V, R_L = 2kΩ, V_{OUT} = ±15V		32			V/mV
		V_S = ±5V, R_L = 2kΩ, V_{OUT} = ±2 V		10			V/mV

NOTES

1. Rating applies to ambient temperatures up to 70°C. Above 70°C ambient derate linearly at 6.3mW/°C for the metal can, 8.3mW/°C for the DIP and 7.1mW/°C for the Flatpak.
2. For supply voltages less than ±15V, the absolute maximum input voltage is equal to the supply voltage.
3. Short circuit may be to ground or either supply. Rating applies to +125°C case temperature or 75°C ambient temperature.
4. Calculated value from: $BW(MHz) = \frac{0.35}{\text{Rise Time } (\mu s)}$

FAIRCHILD • μA741

μA741

ELECTRICAL CHARACTERISTICS: $V_S = \pm 15$ V, $T_A = 25°C$ unless otherwise specified.

CHARACTERISTICS (see definitions)	CONDITIONS	MIN	TYP	MAX	UNITS
Input Offset Voltage	$R_S \leqslant 10$ kΩ		1.0	5.0	mV
Input Offset Current			20	200	nA
Input Bias Current			80	500	nA
Input Resistance		0.3	2.0		MΩ
Input Capacitance			1.4		pF
Offset Voltage Adjustment Range			±15		mV
Large Signal Voltage Gain	$R_L \geqslant 2$ kΩ, $V_{OUT} = \pm 10$ V	50,000	200,000		
Output Resistance			75		Ω
Output Short Circuit Current			25		mA
Supply Current			1.7	2.8	mA
Power Consumption			50	85	mW
Transient Response (Unity Gain) — Rise time	$V_{IN} = 20$ mV, $R_L = 2$ kΩ, $C_L \leqslant 100$ pF		0.3		μs
Transient Response (Unity Gain) — Overshoot			5.0		%
Slew Rate	$R_L \geqslant 2$ kΩ		0.5		V/μs

The following specifications apply for $-55°C \leqslant T_A \leqslant +125°C$:

CHARACTERISTICS (see definitions)	CONDITIONS	MIN	TYP	MAX	UNITS
Input Offset Voltage	$R_S \leqslant 10$ kΩ		1.0	6.0	mV
Input Offset Current	$T_A = +125°C$		7.0	200	nA
	$T_A = -55°C$		85	500	nA
Input Bias Current	$T_A = +125°C$		0.03	0.5	μA
	$T_A = -55°C$		0.3	1.5	μA
Input Voltage Range		±12	±13		V
Common Mode Rejection Ratio	$R_S \leqslant 10$ kΩ	70	90		dB
Supply Voltage Rejection Ratio	$R_S \leqslant 10$ kΩ		30	150	μV/V
Large Signal Voltage Gain	$R_L \geqslant 2$ kΩ, $V_{OUT} = \pm 10$ V	25,000			
Output Voltage Swing	$R_L \geqslant 10$ kΩ	±12	±14		V
	$R_L \geqslant 2$ kΩ	±10	±13		V
Supply Current	$T_A = +125°C$		1.5	2.5	mA
	$T_A = -55°C$		2.0	3.3	mA
Power Consumption	$T_A = +125°C$		45	75	mW
	$T_A = -55°C$		60	100	mW

TYPICAL PERFORMANCE CURVES FOR μA741A AND μA741

OPEN LOOP VOLTAGE GAIN AS A FUNCTION OF SUPPLY VOLTAGE

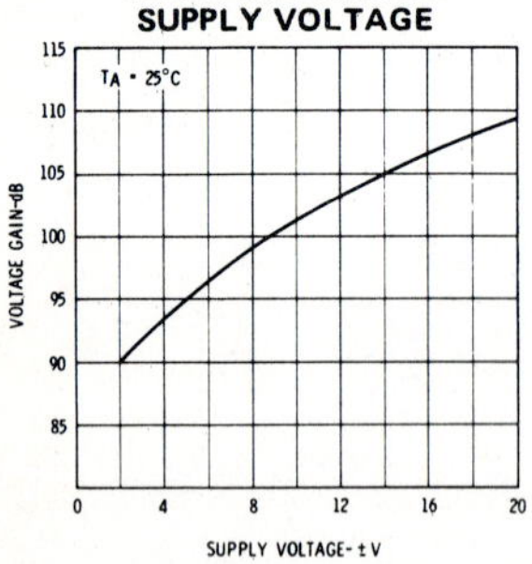

OUTPUT VOLTAGE SWING AS A FUNCTION OF SUPPLY VOLTAGE

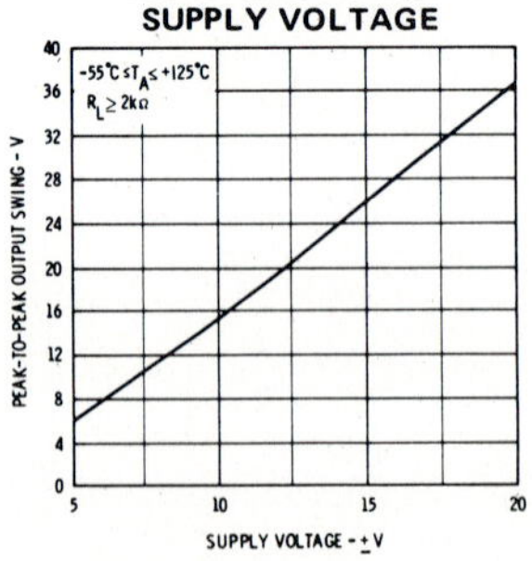

INPUT COMMON MODE VOLTAGE RANGE AS A FUNCTION OF SUPPLY VOLTAGE

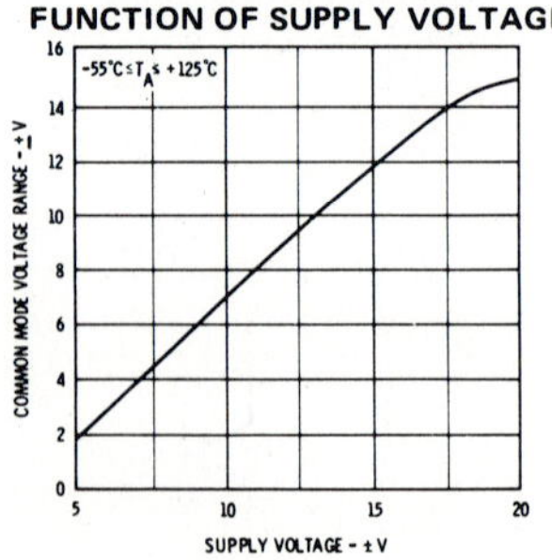

FAIRCHILD • μA741

μA741E

ELECTRICAL CHARACTERISTICS: $V_S = \pm 15$ V, $T_A = 25°C$ unless otherwise specified.

CHARACTERISTICS (see definitions)	CONDITIONS	MIN	TYP	MAX	UNITS
Input Offset Voltage	$R_S \leqslant 50\Omega$		0.8	3.0	mV
Average Input Offset Voltage Drift				15	μV/°C
Input Offset Current			3.0	30	nA
Average Input Offset Current Drift				0.5	nA/°C
Input Bias Current			30	80	nA
Power Supply Rejection Ratio	$V_S = +10, -20$; $V_S = +20, -10$V, $R_S = 50\Omega$		15	50	μV/V
Output Short Circuit Current		10	25	40	mA
Power Dissipation	$V_S = \pm 20$V		80	150	mW
Input Impedance	$V_S = \pm 20$V	1.0	6.0		MΩ
Large Signal Voltage Gain	$V_S = \pm 20$V, $R_L = 2k\Omega$, $V_{OUT} = \pm 15$V	50			V/mV
Transient Response (Unity Gain) — Rise Time			0.25	0.8	μs
Transient Response (Unity Gain) — Overshoot			6.0	20	%
Bandwidth (Note 4)		.437	1.5		MHz
Slew Rate (Unity Gain)	$V_{IN} = \pm 10$V	0.3	0.7		V/μs
The following specifications apply for $0°C \leqslant T_A \leqslant 70°C$					
Input Offset Voltage				4.0	mV
Input Offset Current				70	nA
Input Bias Current				210	nA
Common Mode Rejection Ratio	$V_S = \pm 20$V, $V_{IN} = \pm 15$V, $R_S = 50\Omega$	80	95		dB
Adjustment For Input Offset Voltage	$V_S = \pm 20$V	10			mV
Output Short Circuit Current		10		40	mA
Power Dissipation	$V_S = \pm 20$V			150	mW
Input Impedance	$V_S = \pm 20$V	0.5			MΩ
Output Voltage Swing	$V_S = \pm 20$V, $R_L = 10k\Omega$	±16			V
	$V_S = \pm 20$V, $R_L = 2k\Omega$	±15			V
Large Signal Voltage Gain	$V_S = \pm 20$V, $R_L = 2k\Omega$, $V_{OUT} = \pm 15$V	32			V/mV
	$V_S = \pm 5$V, $R_L = 2k\Omega$, $V_{OUT} = \pm 2$ V	10			V/mV

EQUIVALENT CIRCUIT

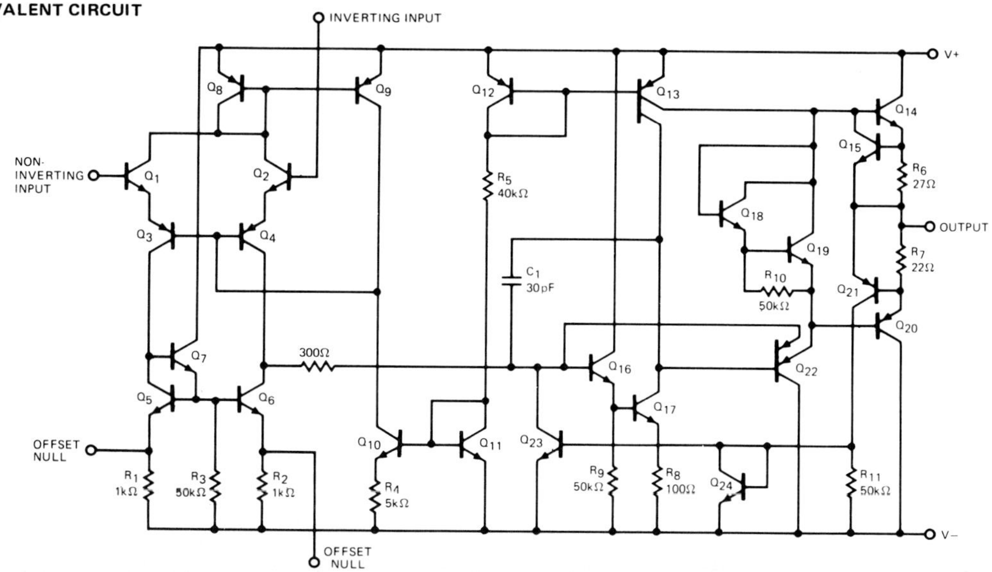

FAIRCHILD • μA741

μA741C

ELECTRICAL CHARACTERISTICS: $V_S = \pm 15$ V, $T_A = 25°C$ unless otherwise specified.

CHARACTERISTICS (see definitions)		CONDITIONS	MIN	TYP	MAX	UNITS
Input Offset Voltage		$R_S \leq 10$ kΩ		2.0	6.0	mV
Input Offset Current				20	200	nA
Input Bias Current				80	500	nA
Input Resistance			0.3	2.0		MΩ
Input Capacitance				1.4		pF
Offset Voltage Adjustment Range				±15		mV
Input Voltage Range			±12	±13		V
Common Mode Rejection Ratio		$R_S \leq 10$ kΩ	70	90		dB
Supply Voltage Rejection Ratio		$R_S \leq 10$ kΩ		30	150	μV/V
Large Signal Voltage Gain		$R_L \geq 2$ kΩ, $V_{OUT} = \pm 10$ V	20,000	200,000		
Output Voltage Swing		$R_L \geq 10$ kΩ	±12	±14		V
		$R_L \geq 2$ kΩ	±10	±13		V
Output Resistance				75		Ω
Output Short Circuit Current				25		mA
Supply Current				1.7	2.8	mA
Power Consumption				50	85	mW
Transient Response (Unity Gain)	Rise time	$V_{IN} = 20$ mV, $R_L = 2$ kΩ, $C_L \leq 100$ pF		0.3		μs
	Overshoot			5.0		%
Slew Rate		$R_L \geq 2$ kΩ		0.5		V/μs

The following specifications apply for $0°C \leq T_A \leq +70°C$:

CHARACTERISTICS	CONDITIONS	MIN	TYP	MAX	UNITS
Input Offset Voltage				7.5	mV
Input Offset Current				300	nA
Input Bias Current				800	nA
Large Signal Voltage Gain	$R_L \geq 2$ kΩ, $V_{OUT} = \pm 10$ V	15,000			
Output Voltage Swing	$R_L \geq 2$ kΩ	±10	±13		V

TYPICAL PERFORMANCE CURVES FOR μA741E AND μA741C

OPEN LOOP VOLTAGE GAIN AS A FUNCTION OF SUPPLY VOLTAGE

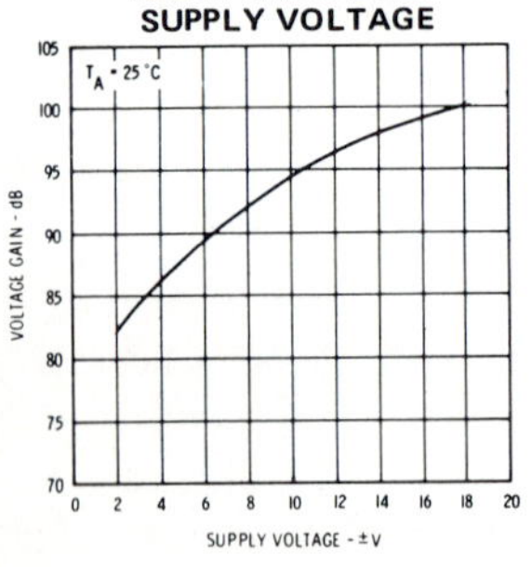

OUTPUT VOLTAGE SWING AS A FUNCTION OF SUPPLY VOLTAGE

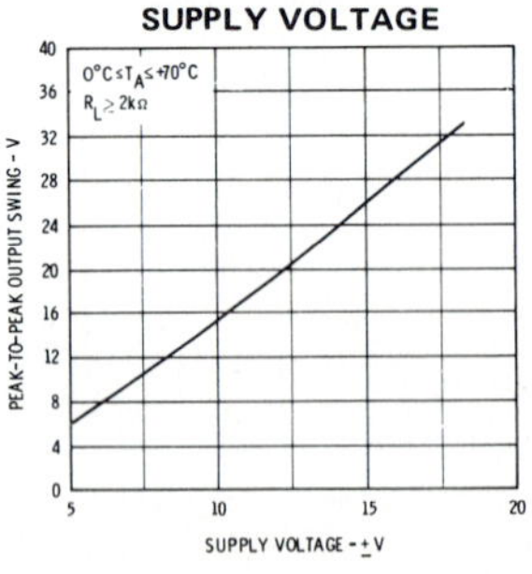

INPUT COMMON MODE VOLTAGE RANGE AS A FUNCTION OF SUPPLY VOLTAGE

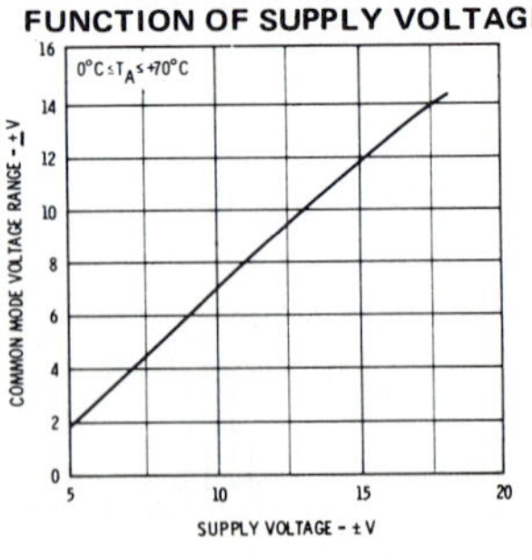

FAIRCHILD • μA741

TYPICAL PERFORMANCE CURVES FOR μA741A, μA741, μA741E AND μA741C

POWER CONSUMPTION AS A FUNCTION OF SUPPLY VOLTAGE

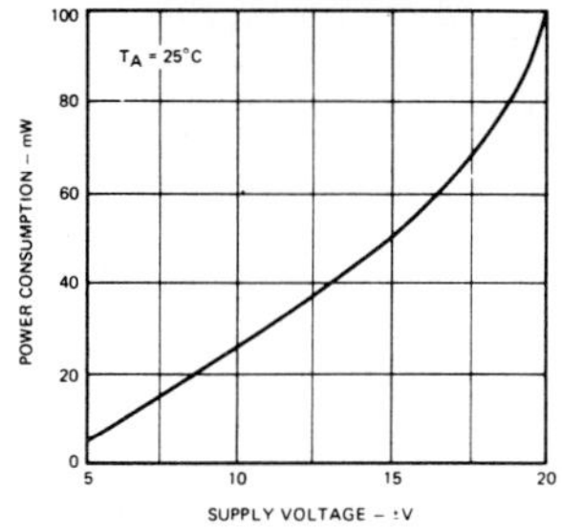

OPEN LOOP VOLTAGE GAIN AS A FUNCTION OF FREQUENCY

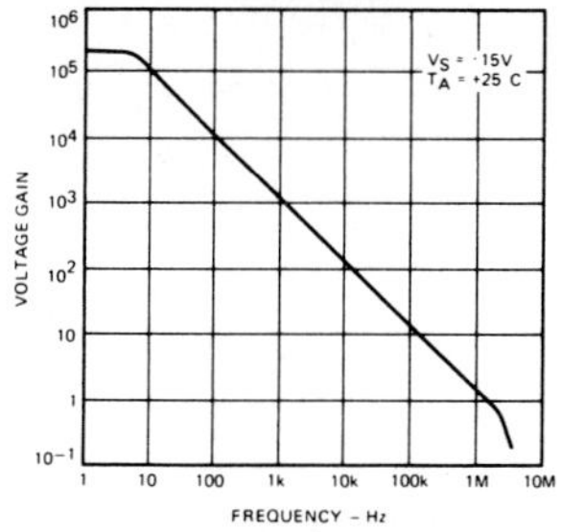

OPEN LOOP PHASE RESPONSE AS A FUNCTION OF FREQUENCY

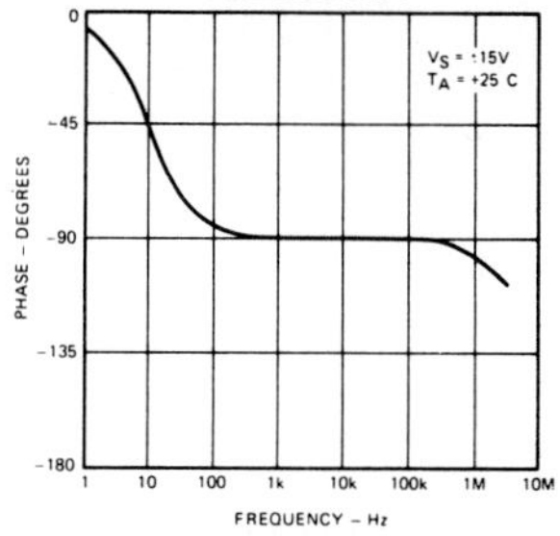

INPUT OFFSET CURRENT AS A FUNCTION OF SUPPLY VOLTAGE

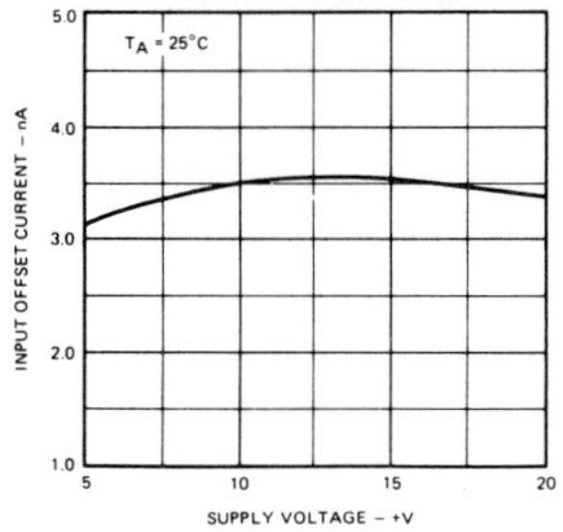

INPUT RESISTANCE AND INPUT CAPACITANCE AS A FUNCTION OF FREQUENCY

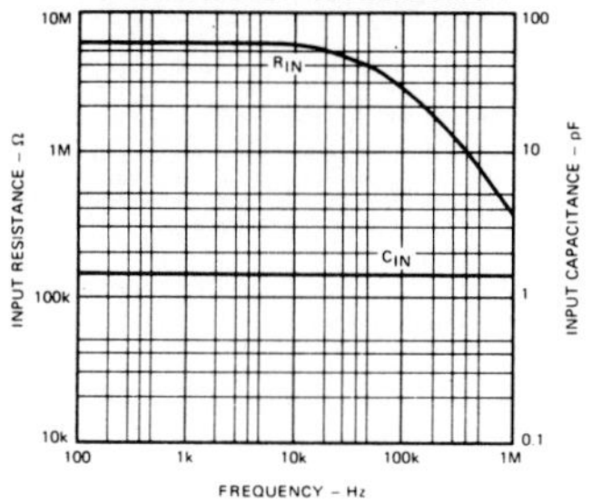

OUTPUT RESISTANCE AS A FUNCTION OF FREQUENCY

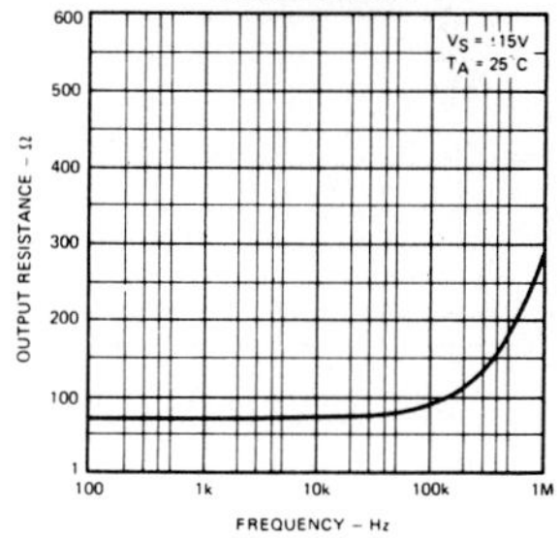

OUTPUT VOLTAGE SWING AS A FUNCTION OF LOAD RESISTANCE

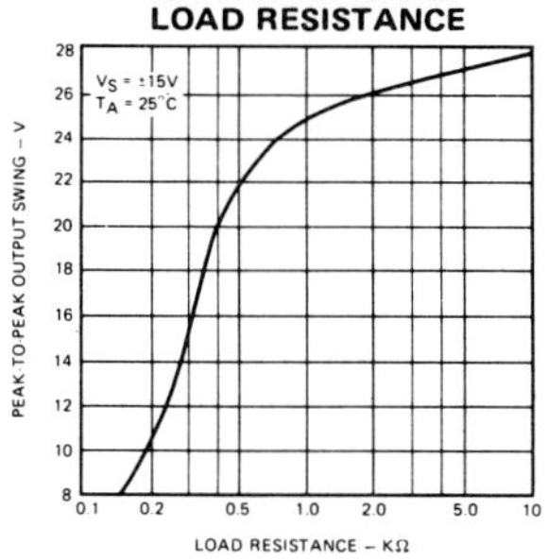

OUTPUT VOLTAGE SWING AS A FUNCTION OF FREQUENCY

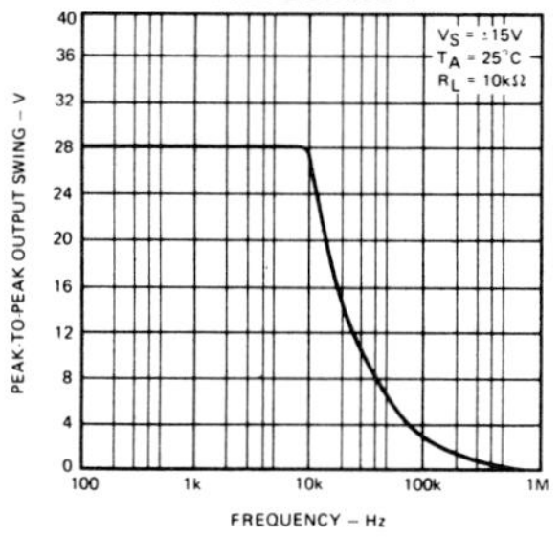

ABSOLUTE MAXIMUM POWER DISSIPATION AS A FUNCTION OF AMBIENT TEMPERATURE

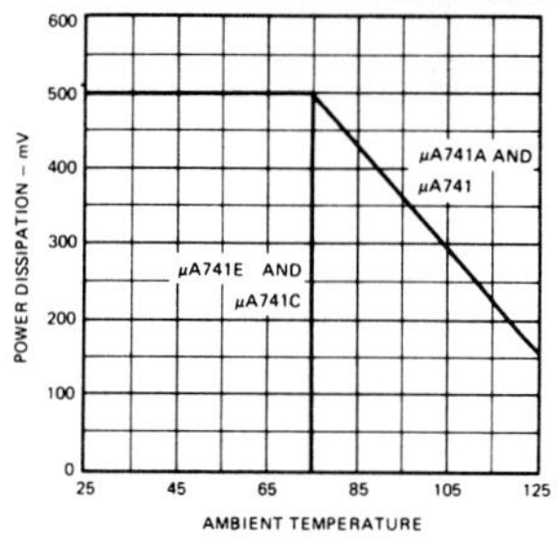

INPUT NOISE VOLTAGE AS A FUNCTION OF FREQUENCY

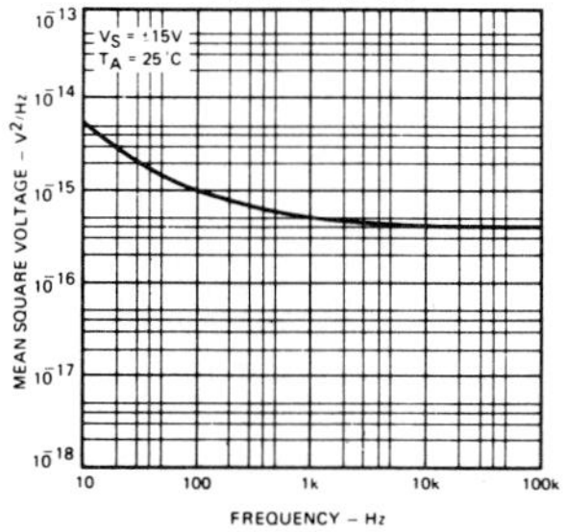

INPUT NOISE CURRENT AS A FUNCTION OF FREQUENCY

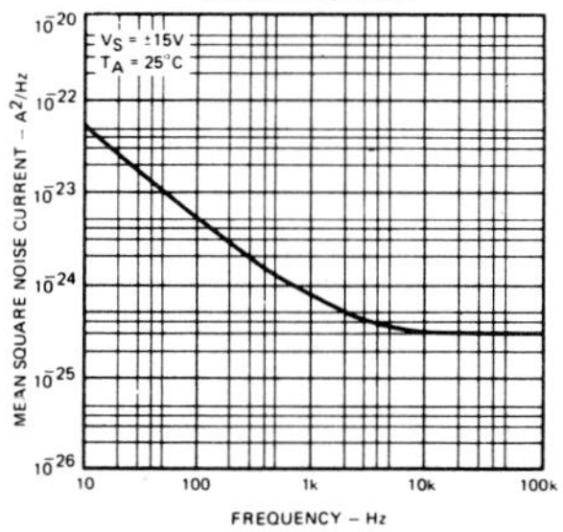

BROADBAND NOISE FOR VARIOUS BANDWIDTHS

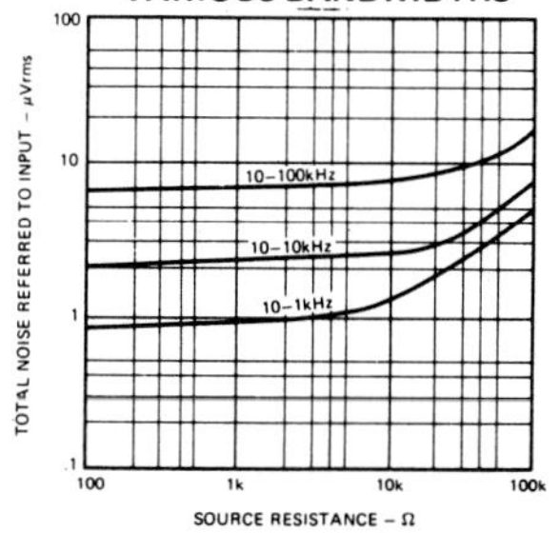

TYPICAL PERFORMANCE CURVES FOR μA741A AND μA741

INPUT BIAS CURRENT AS A FUNCTION OF AMBIENT TEMPERATURE

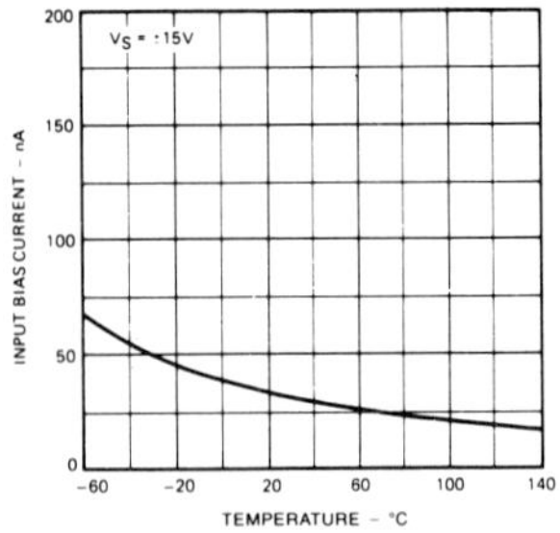

INPUT RESISTANCE AS A FUNCTION OF AMBIENT TEMPERATURE

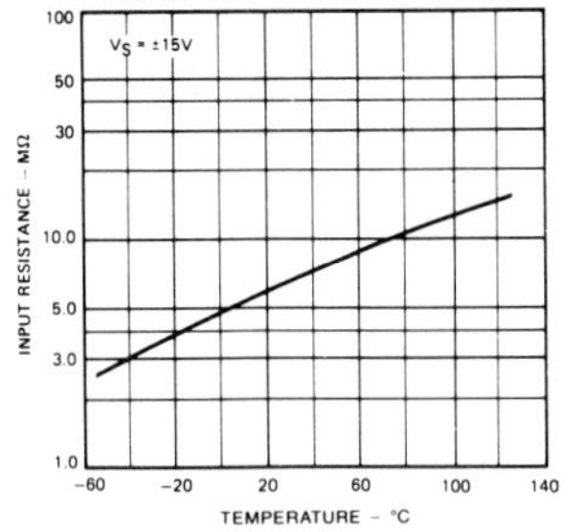

OUTPUT SHORT-CIRCUIT CURRENT AS A FUNCTION OF AMBIENT TEMPERATURE

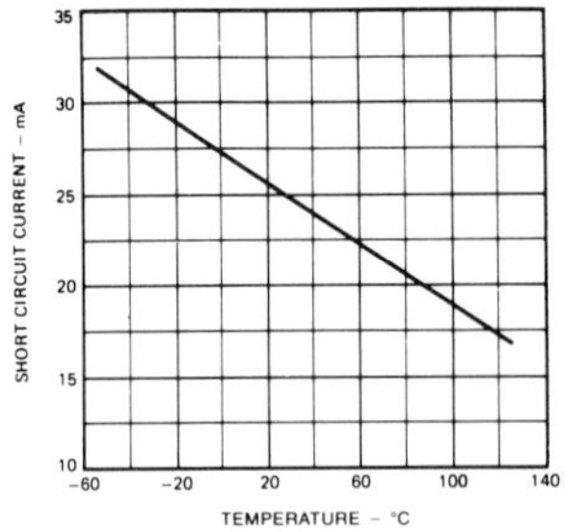

INPUT OFFSET CURRENT AS A FUNCTION OF AMBIENT TEMPERATURE

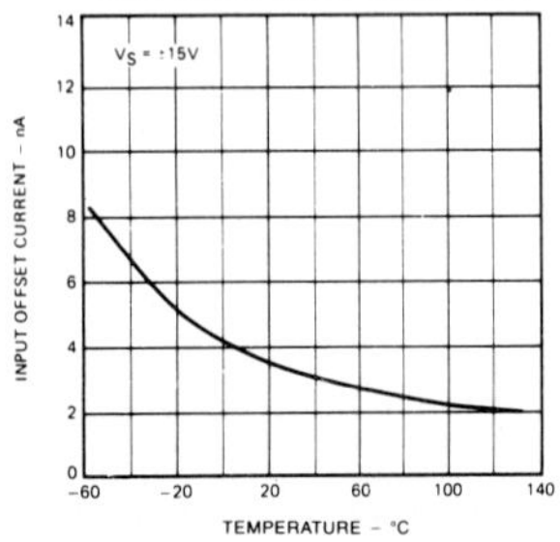

POWER CONSUMPTION AS A FUNCTION OF AMBIENT TEMPERATURE

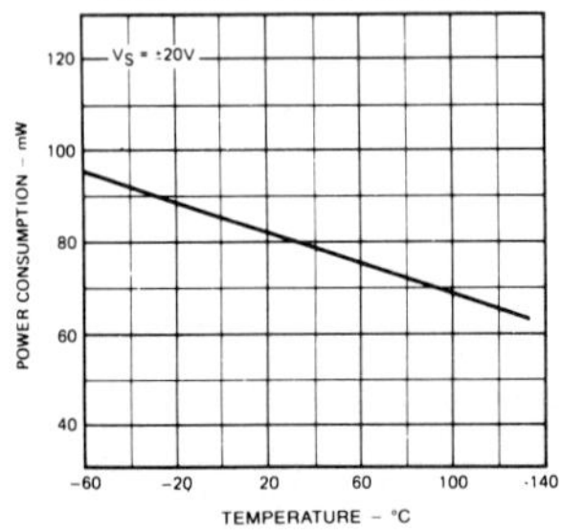

FREQUENCY CHARACTERISTICS AS A FUNCTION OF AMBIENT TEMPERATURE

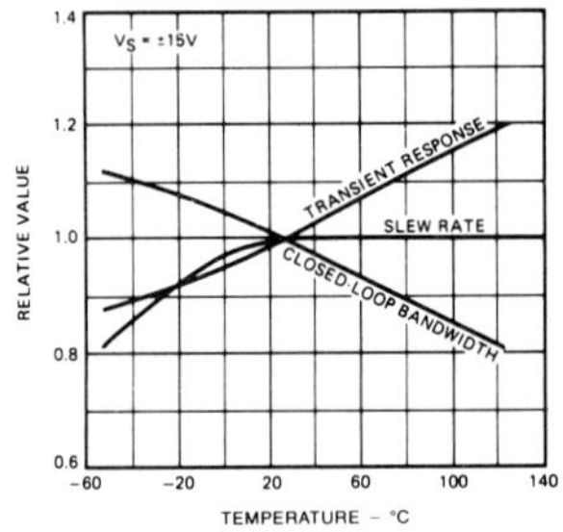

TYPICAL PERFORMANCE CURVES FOR μA741E AND μA741C

INPUT BIAS CURRENT AS A FUNCTION OF AMBIENT TEMPERATURE

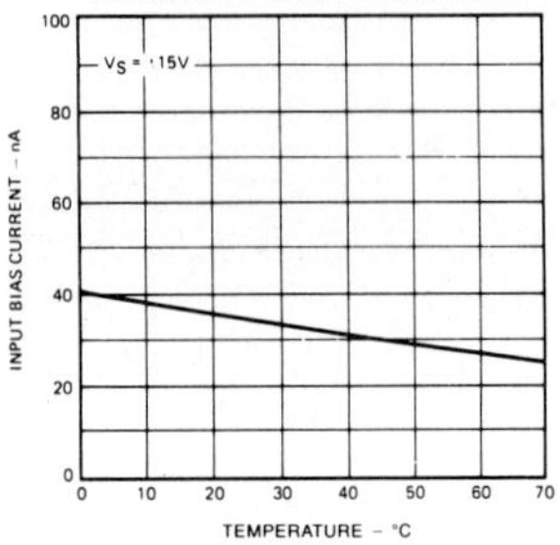

INPUT RESISTANCE AS A FUNCTION OF AMBIENT TEMPERATURE

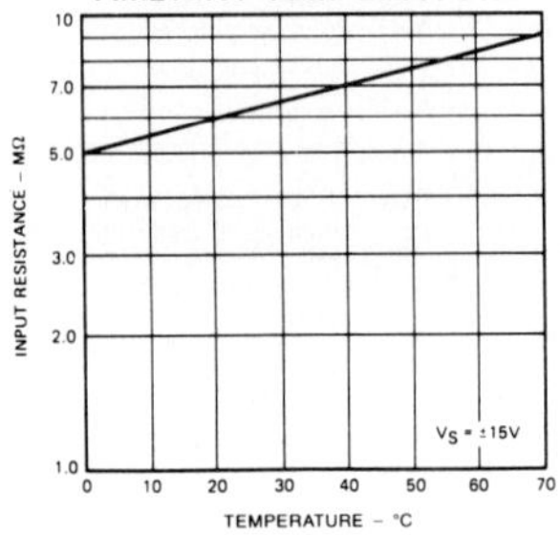

INPUT OFFSET CURRENT AS A FUNCTION OF AMBIENT TEMPERATURE

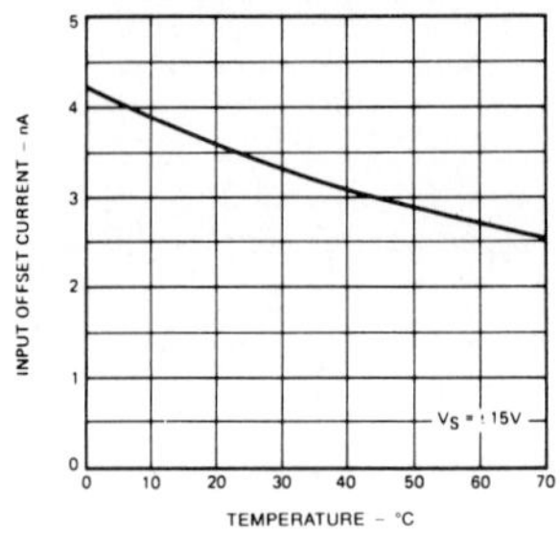

POWER CONSUMPTION AS A FUNCTION OF AMBIENT TEMPERATURE

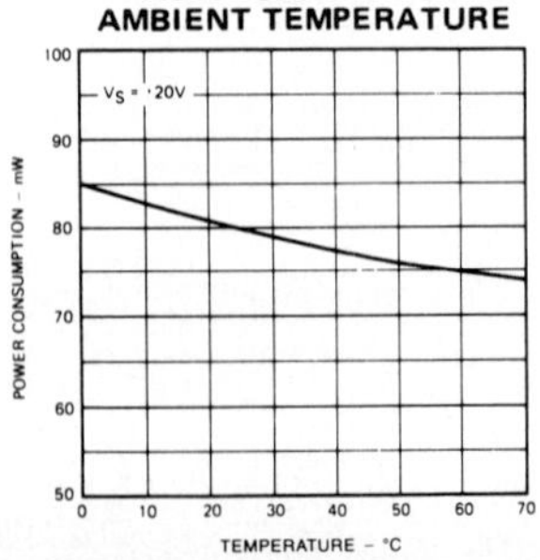

OUTPUT SHORT CIRCUIT CURRENT AS A FUNCTION OF AMBIENT TEMPERATURE

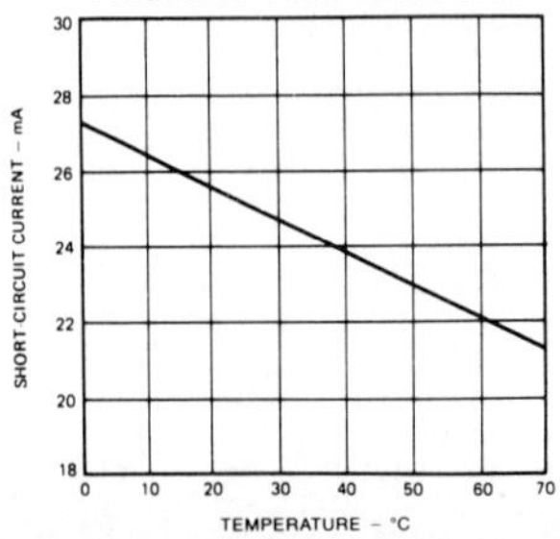

FREQUENCY CHARACTERISTICS AS A FUNCTION OF AMBIENT TEMPERATURE

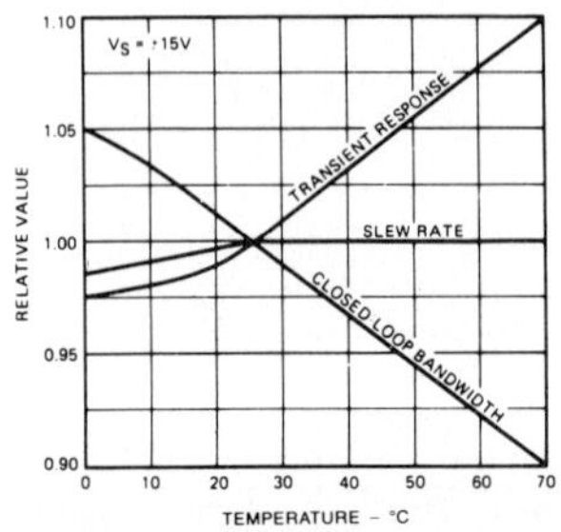

FAIRCHILD • μA741

TRANSIENT RESPONSE

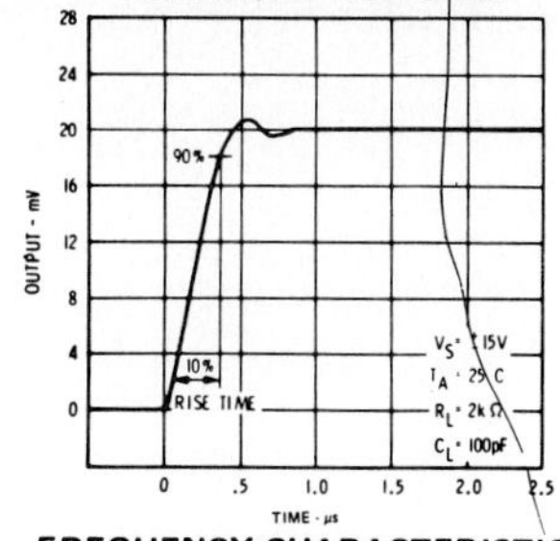

TRANSIENT RESPONSE TEST CIRCUIT

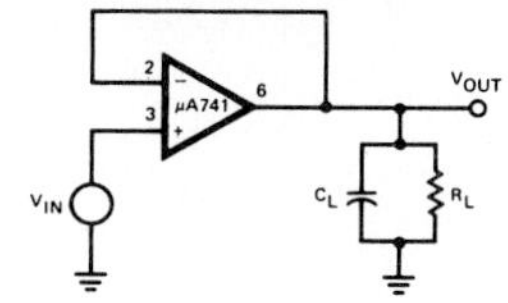

COMMON MODE REJECTION RATIO AS A FUNCTION OF FREQUENCY

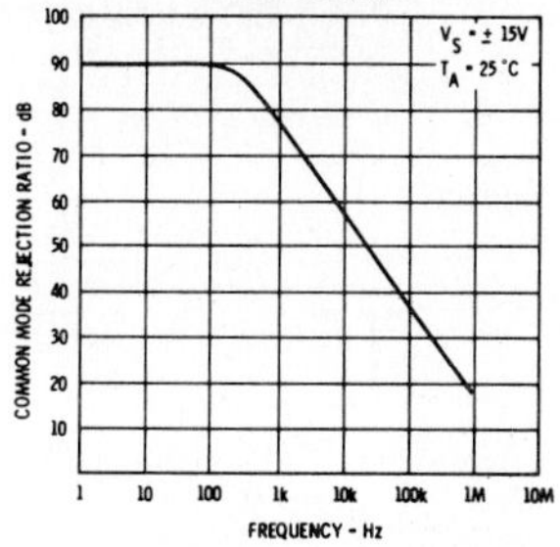

FREQUENCY CHARACTERISTICS AS A FUNCTION OF SUPPLY VOLTAGE

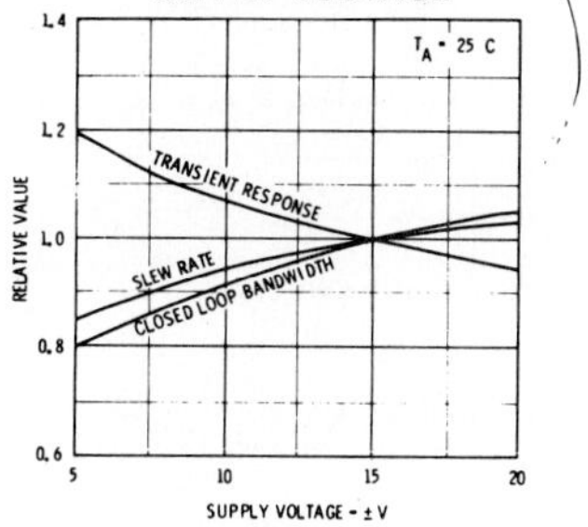

VOLTAGE OFFSET NULL CIRCUIT

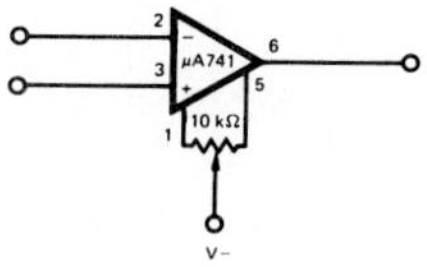

VOLTAGE FOLLOWER LARGE SIGNAL PULSE RESPONSE

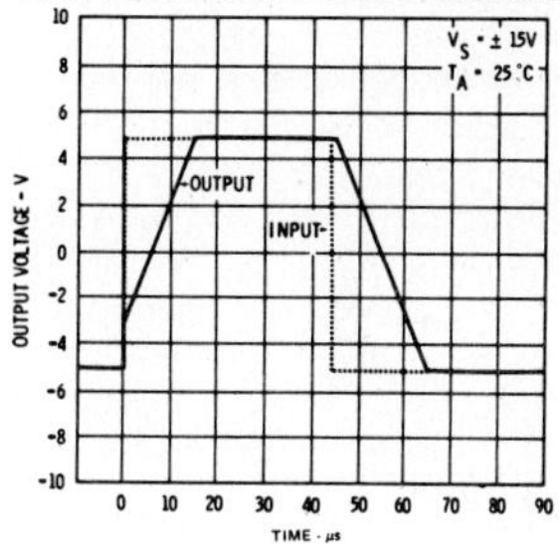

TYPICAL APPLICATIONS

UNITY-GAIN VOLTAGE FOLLOWER

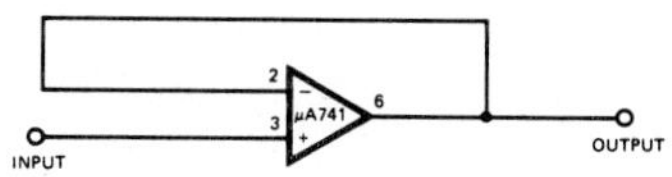

R_{IN} = 400 MΩ

C_{IN} = 1 pF

R_{OUT} << 1 Ω

B.W. = 1 MHz

NON-INVERTING AMPLIFIER

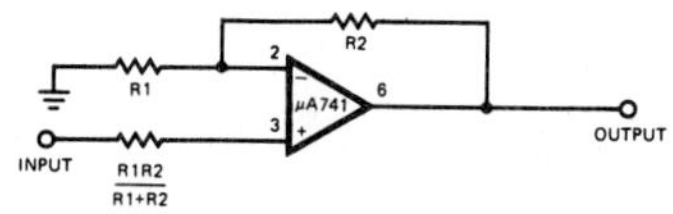

GAIN	R1	R2	B W	R_{IN}
10	1 kΩ	9 kΩ	100 kHz	400 MΩ
100	100 Ω	9.9 kΩ	10 kHz	280 MΩ
1000	100 Ω	99.9 kΩ	1 kHz	80 MΩ

INVERTING AMPLIFIER

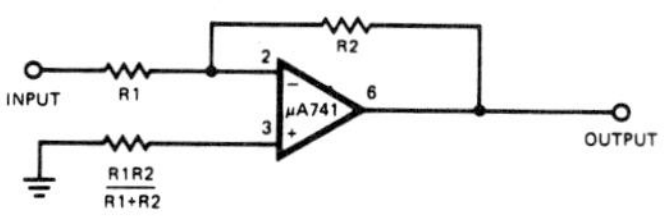

GAIN	R1	R2	B W	R_{IN}
1	10 kΩ	10 kΩ	1 MHz	10 kΩ
10	1 kΩ	10 kΩ	100 kHz	1 kΩ
100	1 kΩ	100 kΩ	10 kHz	1 kΩ
1000	100 Ω	100 kΩ	1 kHz	100 Ω

CLIPPING AMPLIFIER

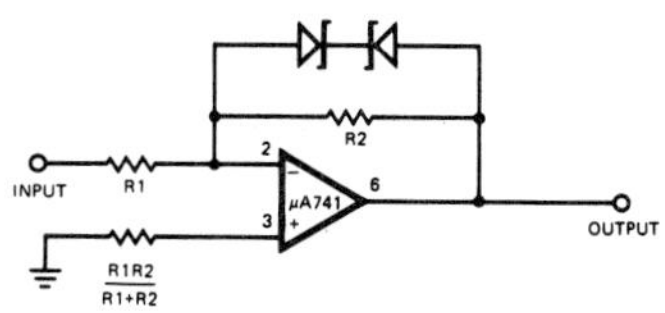

$$\frac{E_{OUT}}{E_{IN}} = \frac{R2}{R1} \text{ if } |E_{OUT}| \leqslant V_Z + 0.7 \text{ V}$$

where V_Z = Zener breakdown voltage

TYPICAL APPLICATIONS (Cont'd)

SIMPLE INTEGRATOR

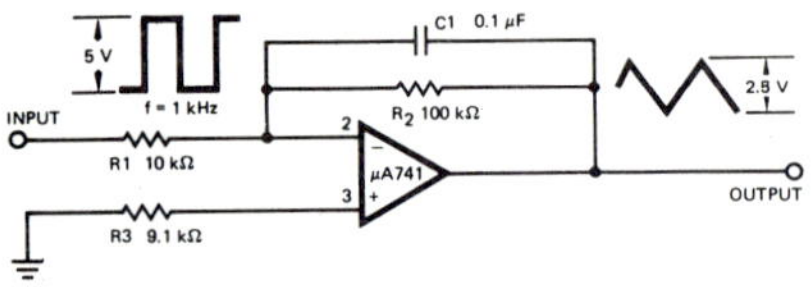

$$E_{OUT} = -\frac{1}{R_1C_1}\int E_{IN}dt$$

SIMPLE DIFFERENTIATOR

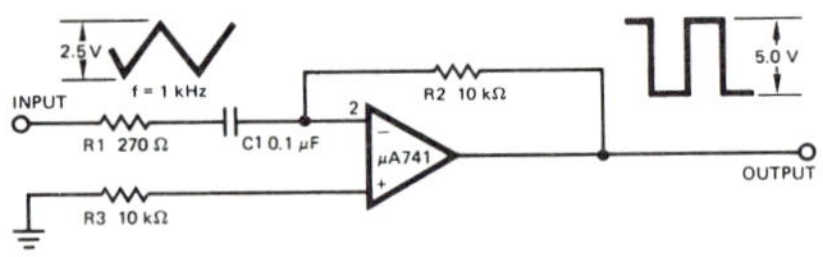

$$E_{OUT} = -R2C\frac{dE_{IN}}{dt}$$

LOW DRIFT LOW NOISE AMPLIFIER

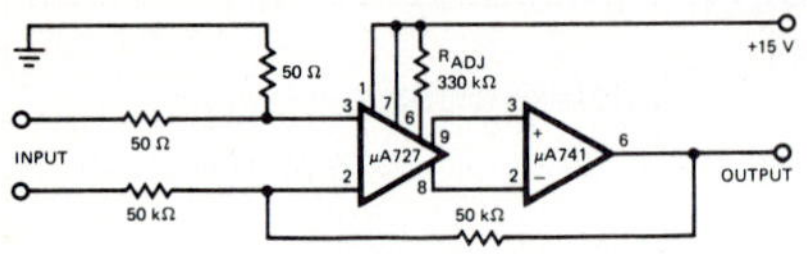

Voltage Gain = 10^3
Input Offset Voltage Drift = 0.6 μV/°C
Input Offset Current Drift = 2.0 pA/°C

HIGH SLEW RATE POWER AMPLIFIER

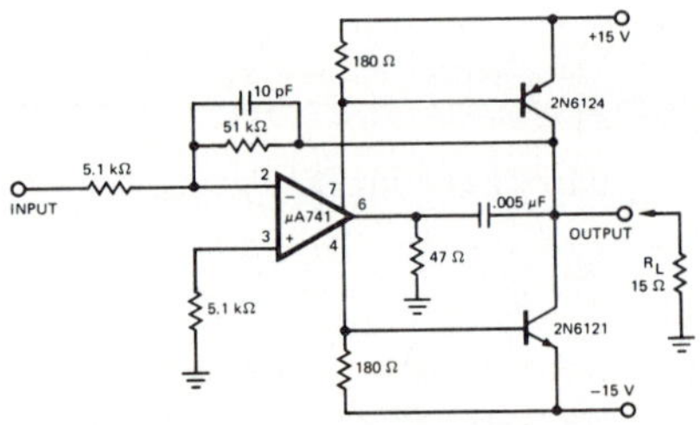

NOTCH FILTER USING THE μA741 AS A GYRATOR

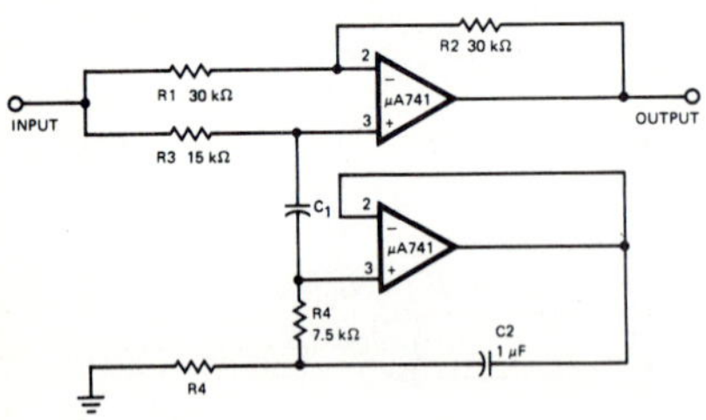

Trim R3 such that

$$\frac{R1}{R2} = \frac{R3}{2\ R4}$$

NOTCH FREQUENCY AS A FUNCTION C1

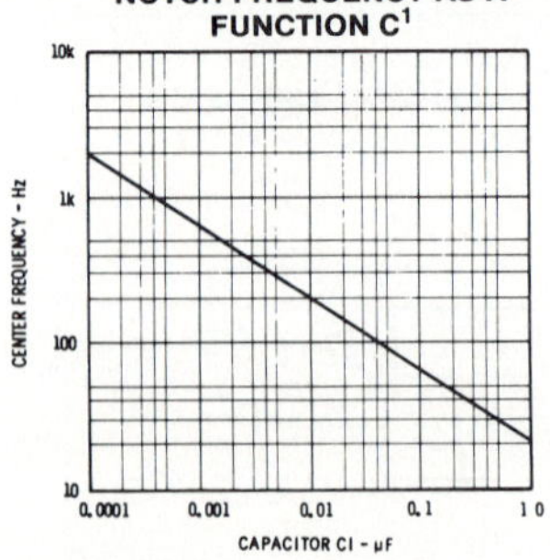

National Semiconductor

Operational Amplifiers/Buffers

LM101A/LM201A/LM301A Operational Amplifiers

General Description

The LM101A series are general purpose operational amplifiers which feature improved performance over industry standards like the LM709. Advanced processing techniques make possible an order of magnitude reduction in input currents, and a redesign of the biasing circuitry reduces the temperature drift of input current. Improved specifications include:

- Offset voltage 3 mV maximum over temperature (LM101A/LM201A)
- Input current 100 nA maximum over temperature (LM101A/LM201A)
- Offset current 20 nA maximum over temperature (LM101A/LM201A)
- Guaranteed drift characteristics
- Offsets guaranteed over entire common mode and supply voltage ranges
- Slew rate of 10V/μs as a summing amplifier

This amplifier offers many features which make its application nearly foolproof: overload protection on the input and output, no latch-up when the common mode range is exceeded, freedom from oscillations and compensation with a single 30 pF capacitor. It has advantages over internally compensated amplifiers in that the frequency compensation can be tailored to the particular application. For example, in low frequency circuits it can be overcompensated for increased stability margin. Or the compensation can be optimized to give more than a factor of ten improvement in high frequency performance for most applications.

In addition, the device provides better accuracy and lower noise in high impedance circuitry. The low input currents also make it particularly well suited for long interval integrators or timers, sample and hold circuits and low frequency waveform generators. Further, replacing circuits where matched transistor pairs buffer the inputs of conventional IC op amps, it can give lower offset voltage and drift at a lower cost.

The LM101A is guaranteed over a temperature range of −55°C to +125°C, the LM201A from −25°C to +85°C, and the LM301A from 0°C to 70°C.

Schematic** and Connection Diagrams (Top Views)

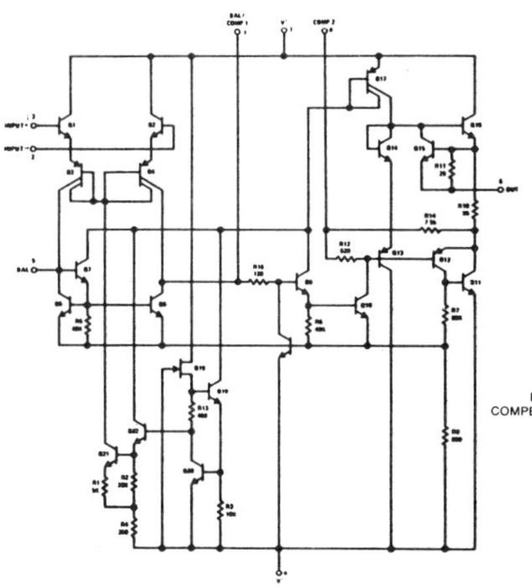

Metal Can Package

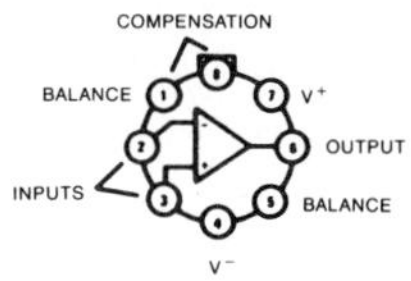

Order Number LM101AH, LM201AH or LM301AH
See NS Package H08C

Dual-In-Line Package

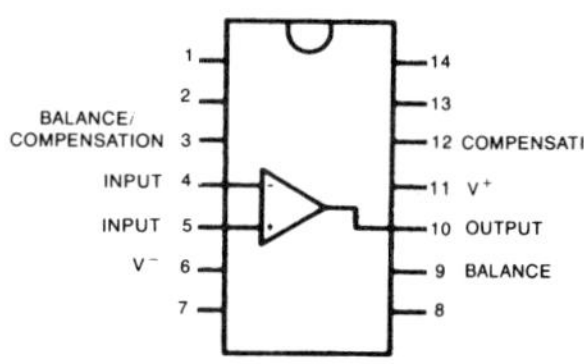

Note: Pin 6 connected to bottom of package.

Order Number LM101AJ-14
LM201AJ-14 or LM301AJ-14
See NS Package J14A

Dual-In-Line Package

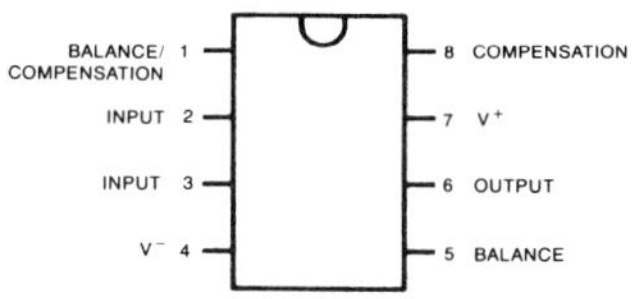

Order Number
LM101AJ, LM201AJ, LM301AJ
See NS Package J08A

Order Number LM301AN
See NS Package N08A

**Pin connections shown are for metal can.

Absolute Maximum Ratings

	LM101A/LM201A	LM301A
Supply Voltage	±22V	±18V
Power Dissipation (Note 1)	500 mW	500 mW
Differential Input Voltage	±30V	±30V
Input Voltage (Note 2)	±15V	±15V
Output Short Circuit Duration (Note 3)	Indefinite	Indefinite
Operating Temperature Range	−55°C to +125°C (LM101A) −25°C to +85°C (LM201A)	0°C to +70°C
Storage Temperature Range	−65°C to +150°C	−65°C to +150°C
Lead Temperature (Soldering, 10 seconds)	300°C	300°C

Electrical Characteristics (Note 4)

PARAMETER	CONDITIONS	LM101A/LM201A			LM301A			UNITS
		MIN	TYP	MAX	MIN	TYP	MAX	
Input Offset Voltage LM101A, LM201A, LM301A	$T_A = 25°C$ $R_S \leq 50\ k\Omega$		0.7	2.0		2.0	7.5	mV
Input Offset Current	$T_A = 25°C$		1.5	10		3.0	50	nA
Input Bias Current	$T_A = 25°C$		30	75		70	250	nA
Input Resistance	$T_A = 25°C$	1.5	4.0		0.5	2.0		MΩ
Supply Current	$T_A = 25°C$							
	$V_S = \pm 20V$		1.8	3.0				mA
	$V_S = \pm 15V$					1.8	3.0	mA
Large Signal Voltage Gain	$T_A = 25°C$, $V_S = \pm 15V$ $V_{OUT} = \pm 10V$, $R_L \geq 2\ k\Omega$	50	160		25	160		V/mV
Input Offset Voltage	$R_S \leq 50\ k\Omega$			3.0			10	mV
	$R_S \leq 10\ k\Omega$							mV
Average Temperature Coefficient of Input Offset Voltage	$R_S \leq 50\ k\Omega$		3.0	15		6.0	30	μV/°C
	$R_S \leq 10\ k\Omega$							μV/°C
Input Offset Current				20			70	nA
	$T_A = T_{MAX}$							nA
	$T_A = T_{MIN}$							nA
Average Temperature Coefficient of Input Offset Current	$25°C \leq T_A \leq T_{MAX}$		0.01	0.1		0.01	0.3	nA/°C
	$T_{MIN} \leq T_A \leq 25°C$		0.02	0.2		0.02	0.6	nA/°C
Input Bias Current				0.1			0.3	μA
Supply Current	$T_A = T_{MAX}$, $V_S = \pm 20V$		1.2	2.5				mA
Large Signal Voltage Gain	$V_S = \pm 15V$, $V_{OUT} = \pm 10V$, $R_L \geq 2k$	25			15			V/mV
Output Voltage Swing	$V_S = \pm 15V$							
	$R_L = 10\ k\Omega$	±12	±14		±12	±14		V
	$R_L = 2\ k\Omega$	±10	±13		±10	±13		V
Input Voltage Range	$V_S = \pm 20V$	±15						V
	$V_S = \pm 15V$		+15, −13		±12	+15, −13		V
Common-Mode Rejection Ratio	$R_S \leq 50\ k\Omega$	80	96		70	90		dB
	$R_S \leq 10\ k\Omega$							dB
Supply Voltage Rejection Ratio	$R_S \leq 50\ k\Omega$	80	96		70	96		dB
	$R_S \leq 10\ k\Omega$							dB

Note 1: The maximum junction temperature of the LM101A is 150°C, and that of the LM201A/LM301A is 100°C. For operating at elevated temperatures, devices in the TO-5 package must be derated based on a thermal resistance of 150°C/W, junction to ambient, or 45°C/W, junction to case. The thermal resistance of the dual-in-line package is 187°C/W, junction to ambient.

Note 2: For supply voltages less than ±15V, the absolute maximum input voltage is equal to the supply voltage.

Note 3: Continuous short circuit is allowed for case temperatures to 125°C and ambient temperatures to 75°C for LM101A/LM201A, and 70°C and 55°C respectively for LM301A.

Note 4: Unless otherwise specified, these specifications apply for C1 = 30 pF, $\pm 5V \leq V_S \leq \pm 20V$ and $-55°C \leq T_A \leq +125°C$ (LM101A), $\pm 5V \leq V_S \leq \pm 20V$ and $-25°C \leq T_A \leq +85°C$ (LM201A), $\pm 5V \leq V_S \leq \pm 15V$ and $0°C \leq T_A \leq +70°C$ (LM301A).

Guaranteed Performance Characteristics LM101A/LM201A

Input Voltage Range

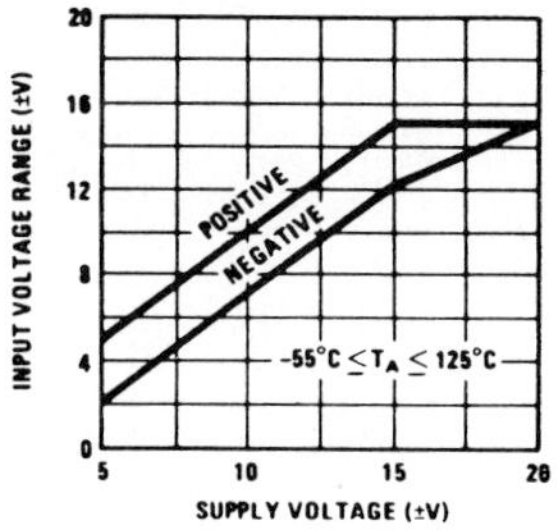

Output Swing

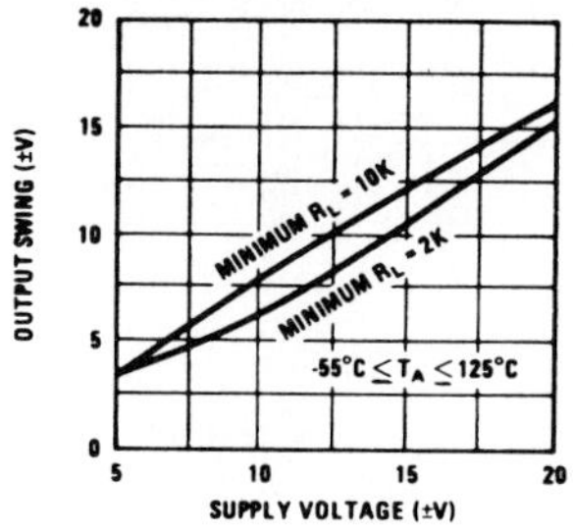

Voltage Gain

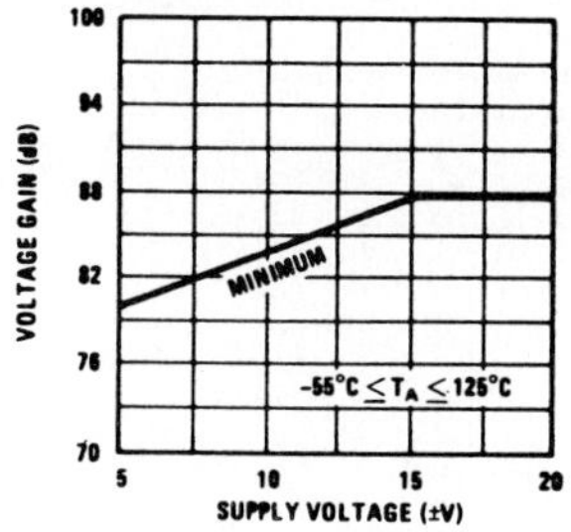

Guaranteed Performance Characteristics LM301A

Input Voltage Range

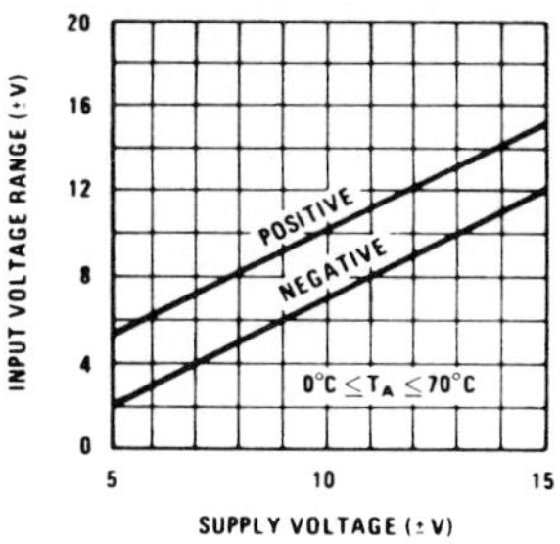

Output Swing

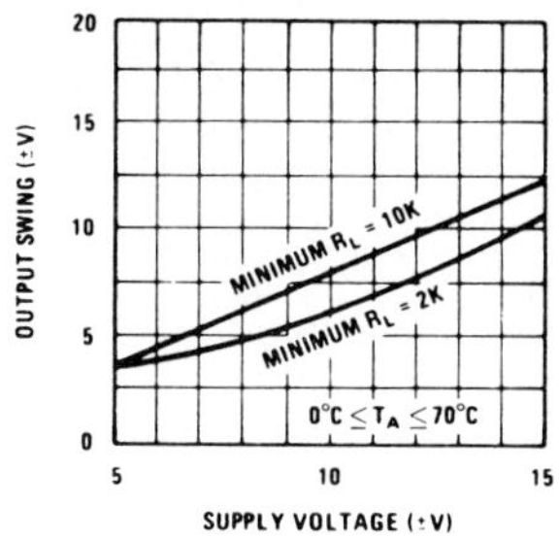

Voltage Gain

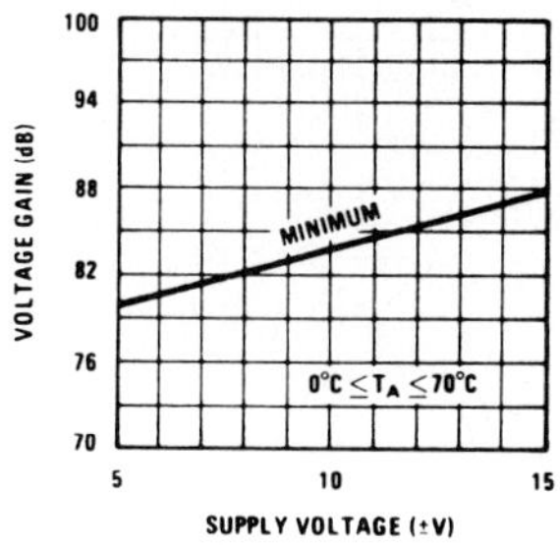

Typical Performance Characteristics

Supply Current

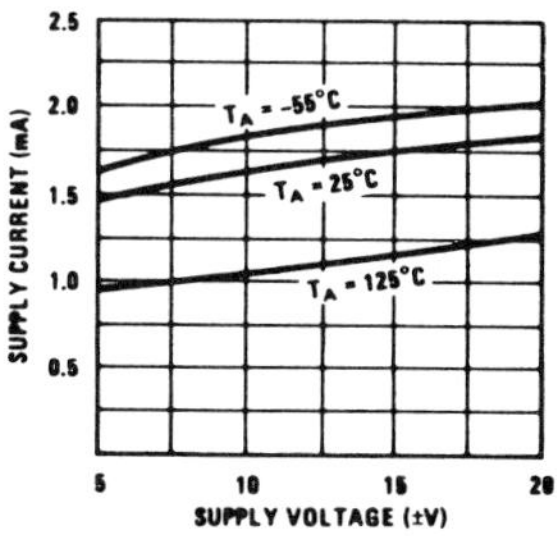

Voltage Gain

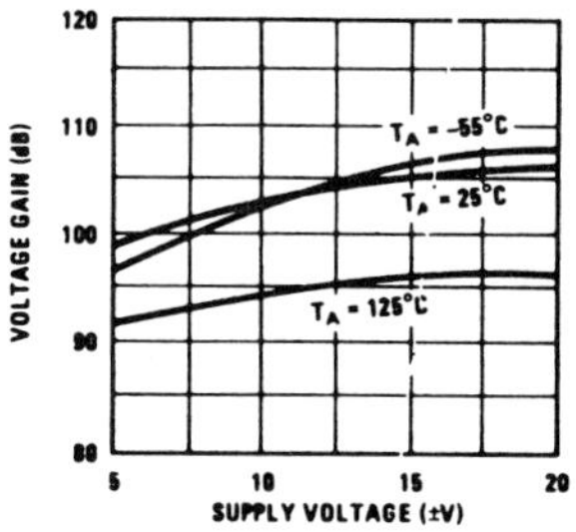

Maximum Power Dissipation

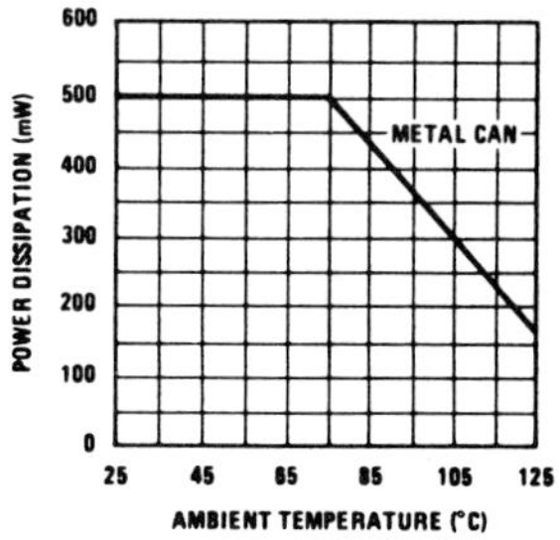

Typical Performance Characteristics (Continued)

Input Current, LM101A/LM201A/LM301A

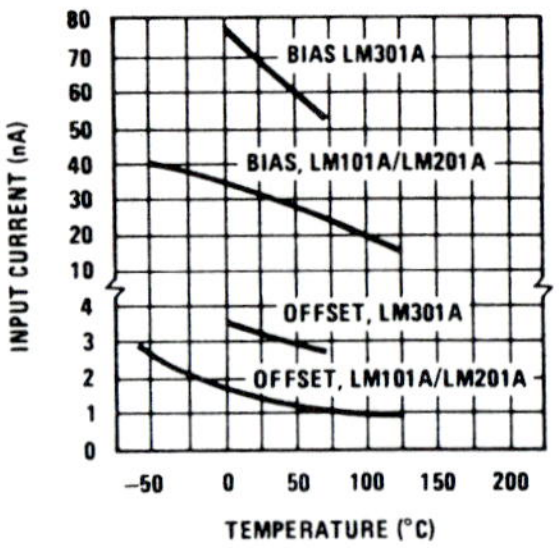

Current Limiting

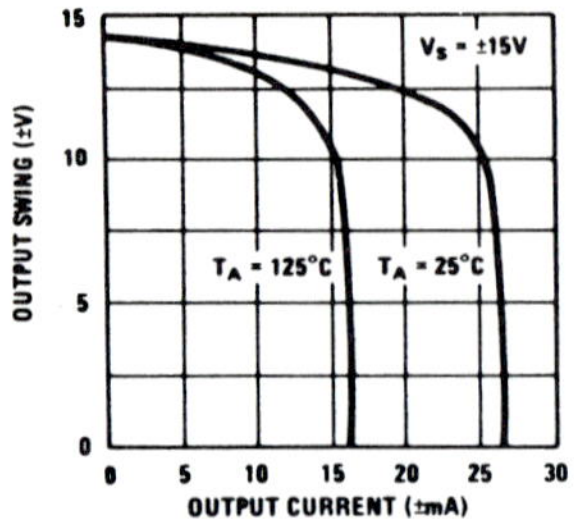

Input Noise Voltage

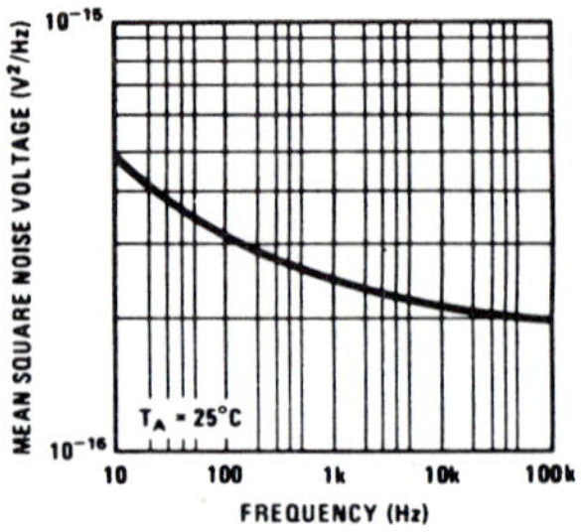

Input Noise Current

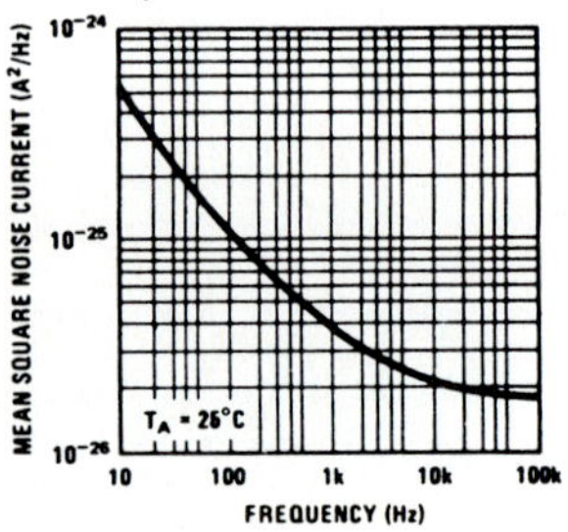

Common Mode Rejection

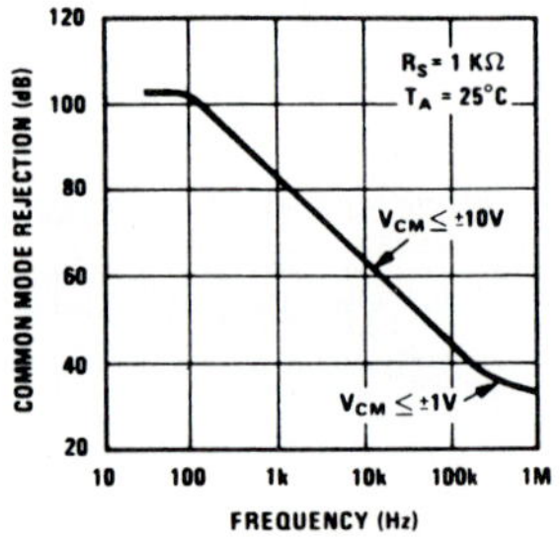

Power Supply Rejection

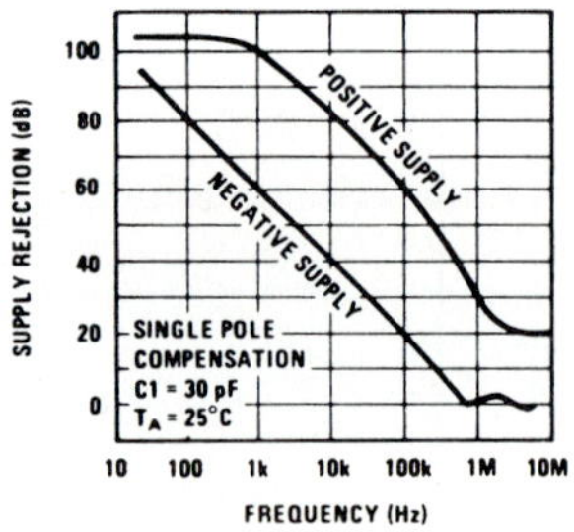

Closed Loop Output Impedance

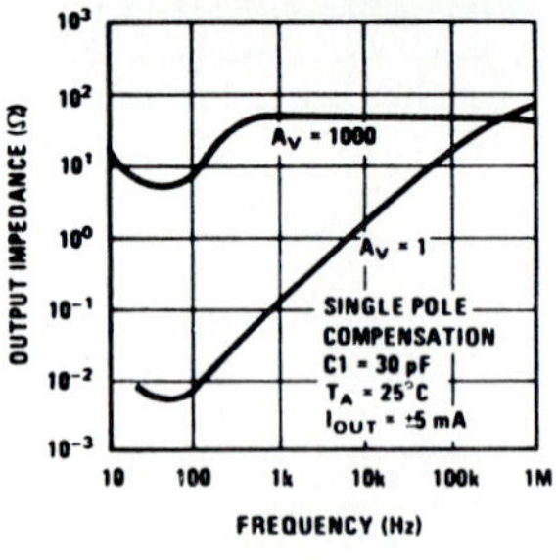

Typical Performance Characteristics for Various Compensation Circuits**

Single Pole Compensation

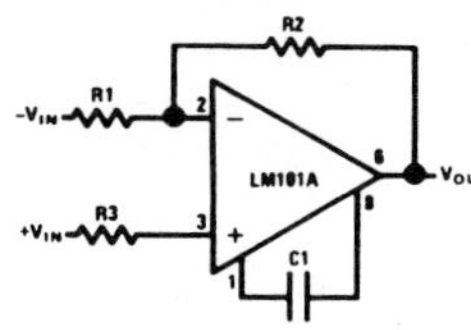

$C1 \geq \frac{R1\ C_S}{R1 + R2}$

$C_S = 30$ pF

**Pin connections shown are for metal can.

Open Loop Frequency Response

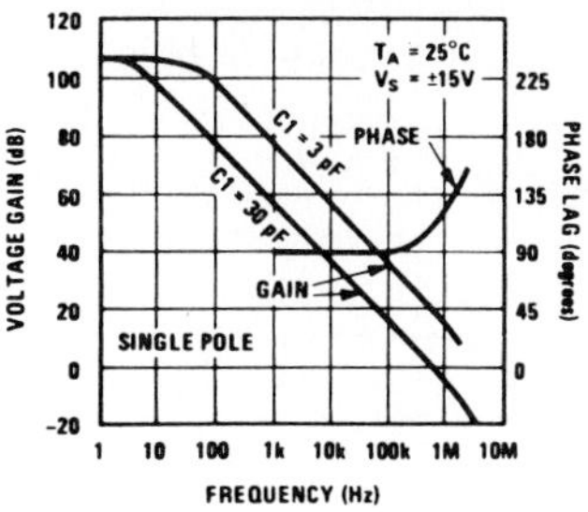

Large Signal Frequency Response

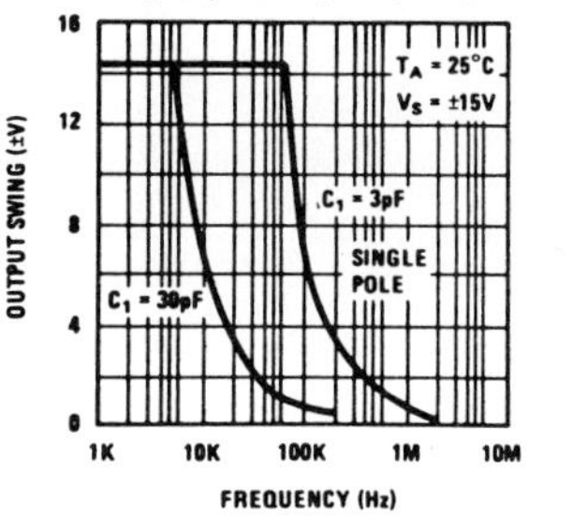

Voltage Follower Pulse Response

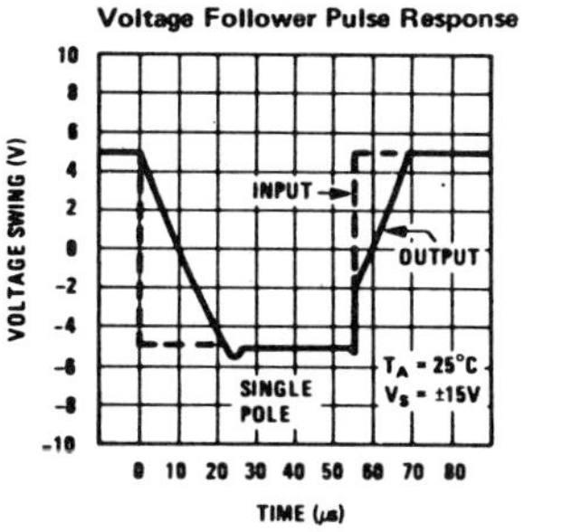

Two Pole Compensation

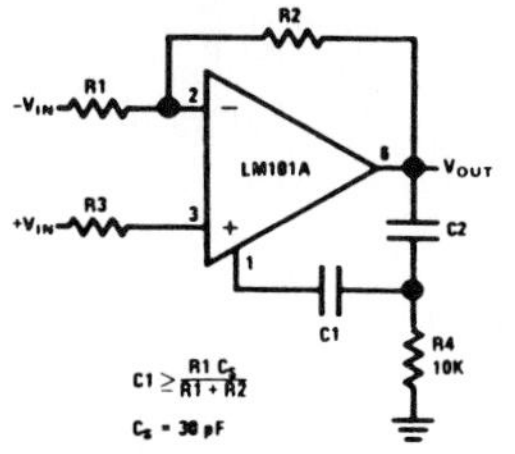

$C1 \geq \frac{R1\ C_S}{R1 + R2}$

$C_S = 30$ pF

$C2 = 10\ C1$

Open Loop Frequency Response

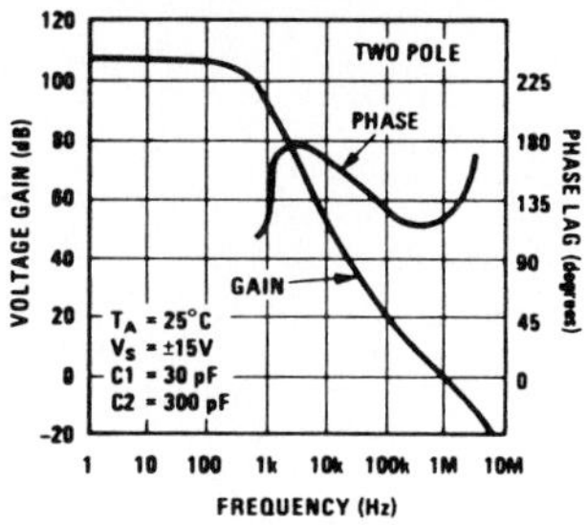

Large Signal Frequency Response

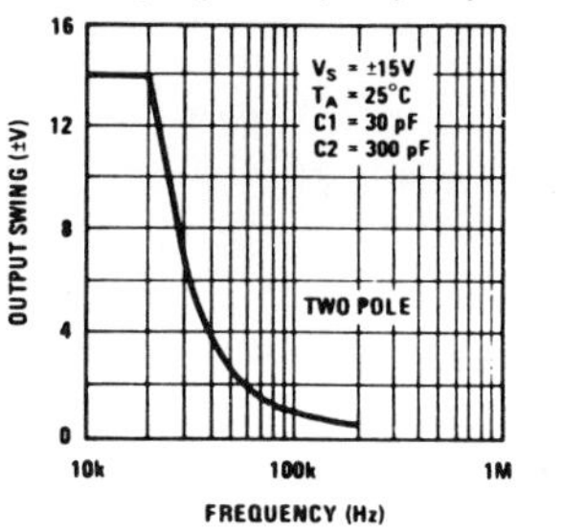

Voltage Follower Pulse Response

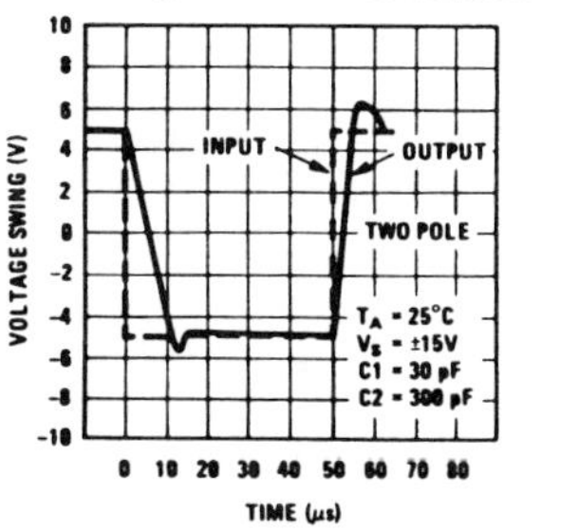

Feedforward Compensation

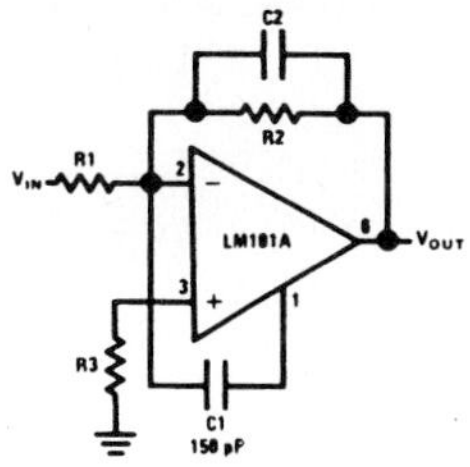

$C2 = \frac{1}{2\pi f_o R2}$

$f_o = 3$ MHz

Open Loop Frequency Response

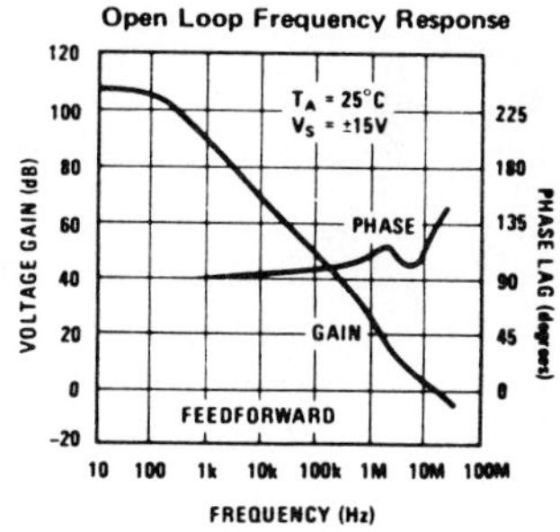

Large Signal Frequency Response

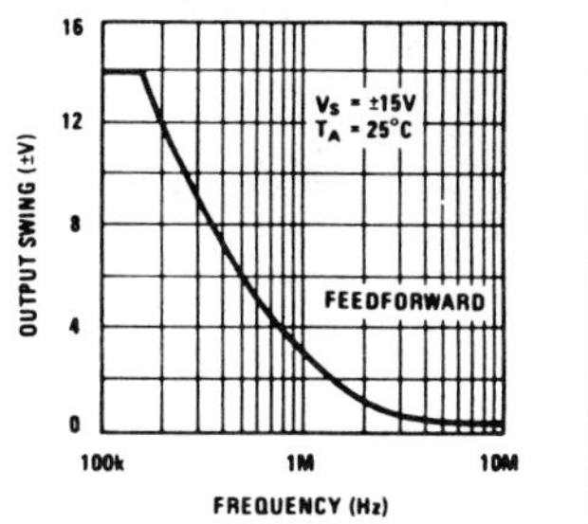

Inverter Pulse Response

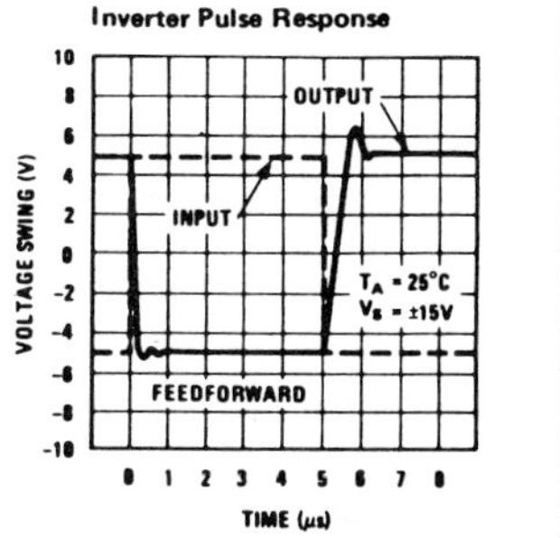

APPENDIX

E STANDARD COMPONENT VALUES

E.0 INTRODUCTION

In this appendix we list the standard component values to aid in the selection of resistor and capacitor values when designing an electronic system. There is some variation in these tabulated values from one manufacturer to another. Therefore, the tables should be considered as typical.

E.1 RESISTORS

Ten-percent tolerance carbon resistors are available in the following power ratings: $\frac{1}{4}$, $\frac{1}{2}$, 1, and 2 W. These resistors are manufactured in the sizes shown in Table E.1. Thus, for example, if your design called for a 675 Ω resistor, you would actually choose a 680 Ω resistor. If this is a 10% tolerance resistance, the manufacturer is guaranteeing the actual value to be between 612 Ω and 748 Ω.

Table E.1 Standard Resistance Values

2.7	3.3	3.9	4.7	Each of these values is multiplied
5.6	6.8	8.2	10.0	by 10^n, where $n = 0, 1, 2, 3, 4,$
12.0	15.0	18.0	22.0	$5, \ldots$

E.2 CAPACITORS

In Table E.2, we present the typical capacitor values from one manufacturer. These are 10% tolerance capacitors.

Table E.2

Ceramic-disk capacitors

3.3	30	200	560	2200
5	39	220	600	2500
6	47	240	680	2700
6.8	50	250	750	3000
7.5	51	270	800	3300
8	56	300	820	3900
10	68	330	910	4000
12	75	350	1000	4300
15	82	360	1200	4700
18	91	390	1300	5000
20	100	400	1500	5600
22	120	470	1600	6800
24	130	500	1800	7500
25	150	510	2000	8200
27	180			

Tantalum capacitors

0.0047	0.010	0.022	all $\times 10^n$, where $n = 0, 1, 2, 3, 4, 5$ (to 330 μF)
0.0056	0.012	0.027	
0.0068	0.015	0.033	
0.0082	0.018	0.039	

APPENDIX

F NOISE IN ELECTRONIC SYSTEMS

F.0 INTRODUCTION

We frequently find it necessary to measure small signals. This task is usually limited in precision and detectability due to noise. Noise limits measurement precision and signal detectability. In this appendix, we discuss some of the problems of noise and describe techniques for improving the situation.

The term *noise* includes any voltages and currents that accompany the desired signal and tend to contaminate it. Many different types of noise sources exist in electronic systems. Some of this noise is generated internally, as in resistors, diodes, and amplifiers. Other noise is generated externally, such as that generated in interconnective wiring and circuit-generated noise.

The environmental noise problem can be greatly reduced by following commonsense engineering practice. This includes proper grounding and shielding procedures. Further improvement can be realized by considering the electronic noise (thermal noise and amplifier noise), its characteristics, and the effects of this noise.

F.1 SOURCES OF NOISE

The sources of noise that must be considered include the following:

1. Thermal noise arising in the source resistance.

2. Noise produced in the instrumentation at the input to the amplifier.
3. Noise generated in the environment, such as power frequency interference, auto ignition, radio and television, lightning, and structural vibration.

These sources of noise are considered in the following sections.

F.1.1 Resistor Noise

The noise generated in a resistor is termed *thermal* or *Johnson noise* and is produced by the random motion of electrons due to thermal agitation. This noise is a function of temperature and has a mean-square value (i.e., average power) given by the equation

$$\overline{e_n^2} = 4\mathrm{k}RT(f_2 - f_1) \tag{F.1}$$

where

k = Boltzmann's constant (1.38×10^{-23} W · s)

T = resistor temperature in degrees Kelvin

R = resistance in ohms

$f_2 - f_1$ = noise bandwidth in hertz

This resistor or Johnson noise is *white* noise, since its rms value per unit bandwidth is constant from dc to high frequencies.

The equivalent circuit for a noisy resistor comprises a *noiseless* resistor in series with a noise voltage source, as shown in Figure F.1. Noise voltage is a random function with a constant root-mean-square (rms) voltage per unit bandwidth. The noise power is proportional to the bandwidth.

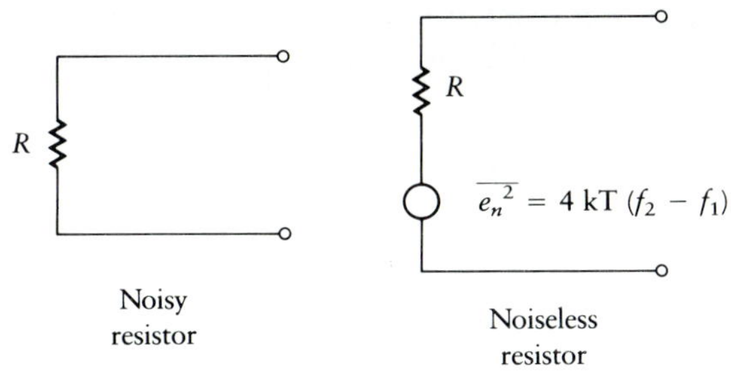

Figure F.1 Equivalent circuit for noisy resistor.

Example F.1

Determine the magnitude of the resistor noise voltage for a 10-kΩ resistor at room temperature (300°K) with $f_2 - f_1 = 50$ kHz.

SOLUTION We use equation (F.1) with the result

$$e_{n\ \text{rms}} = 2.88\ \mu\text{V}$$

With a wider bandwidth or a higher temperature, the noise voltage increases.

F.1.2 Other Types of Noise

The resistor noise of Section F.1.1 always accompanies the signal. Other types of noise *may* exist, but that from the resistance is *always* present. Other sources of noise that are present in electronic systems contribute to the total noise. Two of these types, which we consider here, are shot noise and flicker noise.

Shot noise is the result of random current fluctuations in the number of charge carriers when diffused across a semiconductor junction. The rms value of the shot-noise current is given by the following equation:

$$I_{\text{shot rms}} = 2qI_{dc}(f_2 - f_1)\ \text{A} \tag{F.2}$$

where

$$\begin{aligned} q &= \text{magnitude of the electron charge } (1.6 \times 10^{-19}\ \text{C}) \\ I_{dc} &= \text{average junction current in amps} \\ (f_2 - f_1) &= \text{noise bandwidth in hertz} \end{aligned}$$

Shot noise, like resistor noise, has a power spectrum that is approximately constant across the frequency band. Thus, this is another example of white noise.

Flicker noise, or *1/f noise,* is detectable in materials when electron conduction is present. This noise is due to surface imperfections resulting from the fabrication process. Since this type of noise has a power spectrum that decreases with increasing frequency, it is most important at low frequencies (from 0 to about 100 Hz). As a result, small-signal measurements should be made at higher frequencies, so we can avoid dc measurements. The long-term drift of transistor amplifiers is the result of flicker noise. Flicker noise power can be expressed by the equation

$$\overline{i_f^2} = KI_{dc}\frac{f_2 - f_1}{f} \tag{F.3}$$

where

$$\begin{aligned} \overline{i_f^2} &= \text{the mean-square noise current} \\ I_{dc} &= \text{the average junction current in amps} \\ f_2 - f_1 &= \text{bandwidth} \\ K &= \text{a constant of the semiconductor device} \end{aligned}$$

F.1.3 Diode Noise

The mean-square noise current, $\overline{i_D^2}$, in a diode is a combination of shot noise and flicker noise. This current is due to the series resistance of the silicon material. The mean-square value of the noise generated in a diode is given by the equation

$$\overline{i_D^2} = \left(2q + \frac{K}{f}\right) I_{dc}(f_2 - f_1) \tag{F.4}$$

Equation (F.4) represents the sum of the shot noise and the flicker noise.

F.1.4 BJT Noise

Since we now know the types of noise voltages and currents, we can develop an equivalent circuit for a single-stage BJT. We use the simplified π-model and add the noise sources as shown in Figure F.2. For the midband frequency, flicker ($1/f$) noise can be ignored. We make the assumption that the noise sources are uncorrelated, with the result that we can find the total mean-square noise voltage by adding the individual mean-square noise voltages.

The noise voltage and current generators in Figure F.2 are summarized as follows:

$$\overline{v_{bb'}^2} = 4kTr_{bb'}(f_2 - f_1) \tag{F.5a}$$

$$\overline{i_b^2} = 2qI_B(f_2 - f_1) \tag{F.5b}$$

$$\overline{i_c^2} = 2qI_C(f_2 - f_1) \tag{F.5c}$$

We find the output noise voltage for each of these noise sources separately. Then, using the fact that the sources are assumed to be uncorrelated, we can use superposition and obtain the total output noise voltage by adding each (mean-square) component of output noise voltage. Note that r_{be}' has no noise source associated with it since it is not a physical resistance.

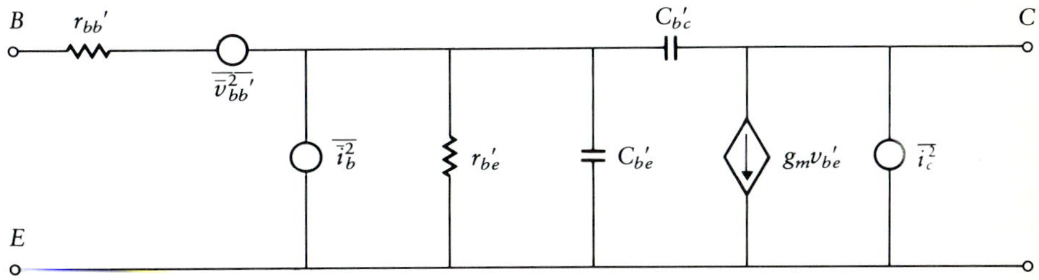

Figure F.2
Equivalent noise circuit for a BJT.

F.1.5 FET Noise

The approach for noise analysis of FETs is similar to that of BJTs. We use the equivalent circuit for the FET and add to it the noise sources, as shown in Figure F.3. These sources are summarized as follows:

$$\overline{i_g^2} = 2qI_G(f_2 - f_1) \tag{F.6a}$$

$$\overline{v_d^2} = 4kTr_{ds}(f_2 - f_1) \tag{F.6b}$$

Again, we find the total output noise voltage by superposition. That is, we add together the output mean-square voltage for each of the noise sources. The noise voltage at the output of an FET amplifier is much smaller than that for a BJT amplifier. This improved noise performance results because of the infinite input impedance of an FET.

The noise at the input to the first stage of a multistage amplifier is the most important to consider. This is true since the noise component at the input to the amplifier is multiplied by the voltage gain of the entire amplifier, whereas the noise voltage at the input to the second and later stages is amplified by a smaller gain.

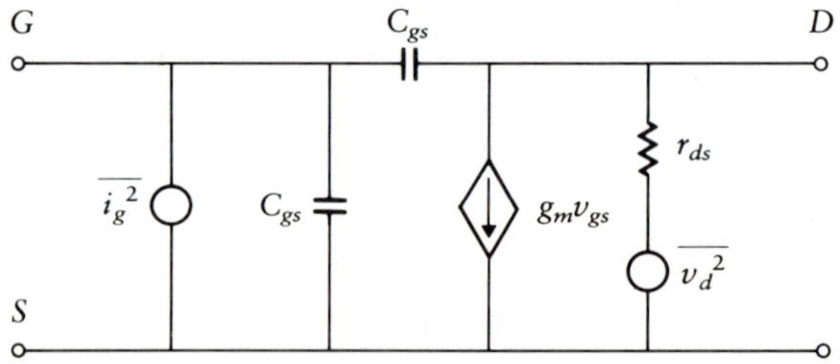

Figure F.3 Equivalent noise circuit for FET.

F.2 NOISE IN OP-AMPS

We now consider the important problem of applying the previously discussed noise sources to an op-amp system. (See [14] for further details on this topic.) We represent a noisy amplifier as a *noiseless* amplifier with a voltage-noise generator plus a current-noise generator connected to the input, as shown in Figure F.4. In this figure, we define the variables as follows:

$\overline{e_s^2}$ = mean-square signal source voltage

$\overline{e_i^2} = 4kTR_i(f_2 - f_1)$

R_i = noiseless source resistance

$\overline{e_n^2}$ = equivalent amplifier noise mean-square voltage generator

$\overline{i_n^2}$ = equivalent amplifier noise mean-square current generator

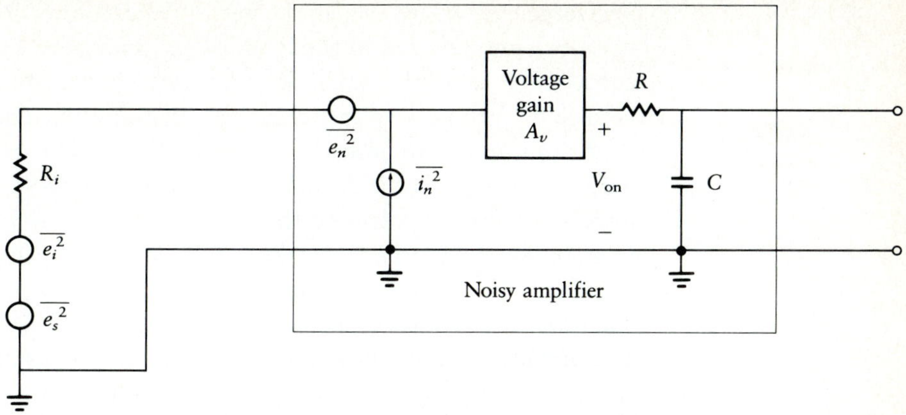

Figure F.4 Equivalent noise circuit for an amplifier.

The low-pass RC filter at the output of the amplifier is used to reduce the white noise above the corner frequency. The resistor, R, is noiseless. The dimensions of the noise generators are as follows:

e_n is in volts(rms)/hertz$^{1/2}$

i_n is in amps(rms)/hertz$^{1/2}$

We again assume that the noise sources are uncorrelated so that the individual mean-square voltages can be added together. Hence, the total rms noise voltage is the square root of the sum of the squares of each noise source output.

We wish to find the total noise voltage output for the amplifier of Figure F.4. We add together all the noise sources and multiply them by the amplifier gain, A_v, with the result

$$V_{on} = \{[4kTR_i + \overline{e_n^2} + \overline{(i_nR_i)^2}](f_2 - f_1)\}^{1/2} A_{vm} \tag{F.7}$$

The dimensions of the expression in equation (F.7) are volts (rms), and A_{vm} is the midrange voltage gain of the amplifier.

F.2.1 Signal-to-Noise Ratio

We can rate the "goodness" of an amplifier by its ratio of signal output voltage to noise output voltage. The signal-to-noise voltage ratio at the output is represented by the symbol $(\mathrm{SNR})_o$. We find $(\mathrm{SNR})_o$ for the circuit of Figure F.4 by first calculating the signal voltage at the output, which is given by the equation

$$v_{os} = e_s A_{vm} \tag{F.8}$$

Hence, the output signal-to-noise voltage ratio, $(\mathrm{SNR})_o$, is found by dividing equation (F.8) by equation (F.7), with the result

$$(\mathrm{SNR})_o = \frac{e_s}{[4kTR_i + \overline{e_n^2} + \overline{(i_nR_i)^2}]^{1/2}(f_2 - f_1)^{1/2}} \tag{F.9}$$

The larger the $(SNR)_o$ value, the better the amplifier. Note that the equivalent noise voltage, referred to the input of the amplifier, can be found by dividing equation (F.7) by the amplifier gain, A_{vm}. It is not possible to measure this voltage by placing a voltmeter across the amplifier input.

At this point, we observe that the $(SNR)_o$ can be increased if we can eliminate any term from the denominator of equation (F.9). For example, we maximize the $(SNR)_o$ by driving the amplifier with a zero source resistance, i.e., $R_i = 0$. We can further increase the $(SNR)_o$ by limiting the bandwidth of the amplifier. We select the RC time constant of the low-pass filter in Figure F.4 to cut off any unnecessary frequencies, and in this manner we increase the $(SNR)_o$. The amplifier $(SNR)_o$ can be further increased by using a higher-order filter.

F.2.2 Noise Figure

Noise figure is used to classify the noise performance of the amplifier. The noise figure is expressed in decibels and is derived from Figure F.4. We define the noise figure (NF) as follows:

$$\text{NF} = 20 \log_{10}\left[\frac{\text{input voltage SNR (amplifier disconnected)}}{\text{voltage SNR at amplifier output}}\right] \tag{F.10}$$

We use Figure F.4 and substitute into equation (F.10), with the result

$$\text{NF} = 10 \log_{10}\left[\frac{\overline{e_s^2}/[4kTR_i(f_2 - f_1)]}{\overline{e_s^2}/[4kTR_i + \overline{e_n^2} + \overline{(i_n R_i)^2}](f_2 - f_1)}\right] \tag{F.11}$$

Equation (F.11) is further simplified to the following:

$$\text{NF} = 10 \log_{10}\left[1 + \frac{\overline{e_n^2} + \overline{(i_n R_i)^2}}{4kTR_i}\right] \tag{F.12}$$

A noiseless amplifier, with $\overline{e_n^2} = \overline{i_n^2} = 0$, has a noise figure of 0 dB. For a fixed R_i, the least noisy amplifier has the smallest noise figure. As the input resistance approaches zero, the noise figure increases without limit.

F.2.3 Noise-Reduction Considerations

In this section, we consider optimizing the source resistance so that we can minimize the noise figure. The optimum source resistance, R_{iop}, is found by differentiating equation (F.12) as follows:

$$\frac{\partial \text{NF}}{\partial R_i} = 0 = \frac{4kT[2\overline{(i_n R_i)^2} - \overline{e_n^2} - \overline{(i_n R_i)^2}]}{\text{denominator in } R_i} \tag{F.13}$$

with the result that

$$R_{iop} = \frac{e_n}{i_n} \tag{F.14}$$

The minimum noise figure is found by substituting equation (F.14) into equation (F.12), with the result

$$\mathrm{NF}_{\min} = 10 \log_{10}\left[1 + \frac{e_n i_n}{2kT}\right] \tag{F.15}$$

The optimum input resistance, which is the ratio of e_n to i_n, causes a minimum noise figure, as shown in equation (F.15). We can achieve a significant improvement in signal-to-noise ratio by matching the source resistance, R_i, to the R_{iop} of equation (F.14). This can be accomplished with a matching transformer.

We summarize the steps necessary to maximize the signal-to-noise ratio as follows:

1. Match the input resistance, R_i, to R_{iop}.
2. Use a filter to limit the system bandwidth, thus reducing the noise while permitting a tolerable signal distortion.
3. Use a high-input resistance amplifier with a low e_n and i_n. FETs should be used at the input to amplifiers.
4. Reduce source shot noise.
5. Reduce the input lead lengths as much as possible.

APPENDIX

G ANSWERS TO SELECTED PROBLEMS

In this appendix, we present answers to selected end-of-chapter problems. The more complex the problem, the less likely you will find the answer here. For example, the design problems in the later text chapters are intended to be *projects* involving many trade-off decisions. In most cases, there is more than one correct answer. To give one solution, and imply that you have achieved success if your answer agrees with ours, would be counterproductive.

The solutions to problems have been obtained using a calculator set for three significant digits. As such, rounding has occurred during complex mathematical operations.

1.6 $I_D = 3$ mA, $R_f = 13.9\ \Omega$ **1.9** $C = 33.4\ \mu$F **1.12** $a = 5$, $R_L = 12\ \Omega$, 2.36 mF **1.16** (a) $R_i = 9.76\ \Omega$; (b) $P_Z = 11.2$ W **1.19** $I_{Z\max} = 125$ mA, $R_i = 93.7\ \Omega$, $V_{s\min} = 12.9$ V, $C = 141\ \mu$F, % Reg = 2.81% **1.26** (a) $v_o(v_i = -9) = -7.33$ V, $v_o(v_i > 0) = 0$, $v_o(-4 < v_i < 0) = v_i$; (b) $v_o(v_i = +9\text{ V}) = +6.5$ V, $v_o(v_i = -9) = -8$ V, $v_o(-6 < v_i < 5) = v_i$ **1.33** 630 Ω

2.1 (a) $R_1 = 19$ kΩ, $R_2 = 70.9$ kΩ; (b) $R_1 = 12.7$ kΩ, $R_2 = 47.3$ kΩ; (c) $R_1 = 6.34$ kΩ, $R_2 = 23.6$ kΩ **2.4** (a) $R_1 = 2.9$ kΩ, $R_2 = 6.45$ kΩ; (b) 15 V **2.7** (a) $P_{VCC} = 17.1$ mW; (b) $P_{R1} = 0.238$ mW, $P_{R2} = 2.38$ mW, $P_{RE} = 0.439$ mW, $P_{RC} = 7.32$ mW; (c) $P_{\text{trans}} = 6.75$ mW **2.10** (a) $R_1 = 2.28$ kΩ, $R_2 = 16.3$ kΩ; (b) 4.8 V; (d) $P_{\text{trans}} = 19.2$ mW, $P_L = 1.44$ mW **2.13** $R_C = 1863\ \Omega$, $v_o = 5.97$ V (p-p) **2.16** $I_{CQ} = 2.13$ mA, $V_{CEQ} = 3.61$ V, $V_o = 2.87$ V **2.19** (a) $I_{CQ} = -4.04$ mA, $V_{CEQ} = -3.10$ V; (b) $R_1 = 4.57$ kΩ, $R_2 = 32.2$ kΩ; (c) 5.39 V; (d) $P_{\text{trans}} = 12.5$ mW; (e) $P_o = 3.62$ mW **2.22** (a) $R_1 = 369$ kΩ, $R_2 = 21.1$ kΩ; (c) 8 V; (d) $P_{\text{trans}} = 40$ mW, $P_L = 4$ mW **2.25** (a) $I_{CQ} = -2.12$ mA, $V_{CEQ} = -18.6$ V; (c) -4.24 V; (d) V'_{CC} decreases to -20.8 V and V_o stays the same. **2.28** $P_{\text{trans}} = 40$ mW, $P_L = 16$ mW

3.2 (a) $R_{in} = 4.88\ k\Omega$, $A_v = -5$, $A_i = -4.88$; (b) $R_{in} = 4.76\ k\Omega$, $A_v = -5$, $A_i = -4.76$; (c) $R_{in} = 3.33\ k\Omega$, $A_v = -5$, $A_i = -3.33$ **3.5** -400 **3.8** $A_i = -41$, $R_{in} = 463\ \Omega$, $A_v = -33.4$ **3.11** $I_{CQ} = -3.19$ mA, $R_B = 40.1\ k\Omega$, $R_1 = 43.2\ k\Omega$, $R_2 = 563\ k\Omega$, $A_i = -75.2$ **3.14** (c) 82.3 mW; (d) -9.69 **3.17** $I_{CQ} = 5.2$ mA, $R_B = 2\ k\Omega$, $R_1 = 2.23\ k\Omega$, $R_2 = 19.3\ k\Omega$, 1.53 V (*p-p*) undistorted **3.20** $I_{CQ} = 2$ A, $R_B = 120\ \Omega$, $R_1 = 873\ \Omega$, $R_2 = 139\ \Omega$, 14.4 V (*p-p*), $R_{in} = 80\ \Omega$ **3.23** $I_{CQ} = 1.6$ A, $R_B = 111\ \Omega$, $R_1 = 410\ \Omega$, $R_2 = 152\ \Omega$ **3.26** $I_{CQ} = 2.57$ mA, $R_B = 990\ \Omega$, $R_1 = 1.11\ k\Omega$, $R_2 = 8.87\ k\Omega$, 9.25 V (*p-p*) **3.29** $V_1 = 2$ V, $V_2 = 7.4$ V, $V_3 = 8.1$ V, $V_4 = 5.7$ V **3.32** (a) $6 - 3.91 \sin 1000t$ V; (b) $-3.91 \sin 1000t$ V **3.35** $A_i = 1070$, $A_v = 20{,}980$

4.1 (a) -0.84 V; (b) 10 V; (c) 7.5 V; (d) $R_1 = 109\ k\Omega$, $R_2 = 1.2\ M\Omega$ **4.4** $R_{Sac} = 338\ \Omega$, $R_{Sdc} = 500\ \Omega$, $R_D = 9.5\ k\Omega$, $A_i = -21.1$, $R_2 \to \infty$ **4.7** $R_{Sac} = 208\ \Omega$, $R_{Sdc} = 234\ \Omega$, $R_D = 966\ \Omega$, $A_i = -15$, $R_2 \to \infty$ **4.10** $V_{GSQ} = 1.36$ V, $g_m = 2.55$ mS, $R_{Sdc} = 648\ \Omega$, $R_{Sac} = 174\ \Omega$, $P_{trans} = 16.8$ mW, $R_2 \to \infty$ **4.13** $R_D = 940\ \Omega$, $R_{Sac} = 25\ \Omega$, $R_{Sdc} = 60\ \Omega$, $R_2 \to \infty$ **4.17** $V_{GSQ} = -0.94$ V, $A_v = -3.67$, $A_i = -612$ **4.20** (a) -8.33; (b) -11.5; (c) -14.3 **4.23** (a) $V_{GSQ} = 1.55$; (b) -3.84; (c) $88.9\ k\Omega$; (d) -88.9 **4.26** $R_S = 18.2\ k\Omega$, $A_i = 0.92$, $R_1 = 37\ k\Omega$, $R_2 = 43.5\ k\Omega$ **4.29** $R_1 = 556\ k\Omega$, $R_2 = 5\ M\Omega$ **4.32** $R_G = 50.9\ k\Omega$, $A_i = 21.1$, $A_v = 0.844$, $R_{S1} = 200\ \Omega$, $R_{S2} = 1400\ \Omega$

5.5 4.77 V (*p-p*) **5.11** 9.31 V (*p-p*) for high temperature, 11.7 V (*p-p*) for low temperature (worst case is high temperature) **5.14** 9.94 V (*p-p*) **5.17** 10.3 V (*p-p*) **5.19** Must be reduced by 2 W

6.1 $R_1 = 713\ \Omega$, $R_2 = 3.79\ k\Omega$, $P_{trans} = 119$ mW, 19.6 V (*p-p*) **6.4** $R_1 = 18.1\ k\Omega$, $R_2 = 40.5\ k\Omega$, $P_L = 58.3$ mW **6.7** $R_2 = 720\ \Omega$, $R_1 = 60.8\ \Omega$, $R_{in} = 240\ \Omega$, $C_1 = 663\ \mu F$, $P_L = 3.38$ W, $P_{trans} = 0.684$ W **6.10** (b) 563 mW; (c) $R_2 = 490\ \Omega$, $R_{in} = 182\ \Omega$ **6.13** (a) $P_{trans} = 20.3$ W; (b) $R_2 = 158\ \Omega$, $R_{in} = 100\ \Omega$, $C_1 = 199\ \mu F$, $C_2 = 159\ \mu F$; (c) 9.3 **6.16** (a) $P_{trans} = 20.3$ W; (b) $R_2 = 3.79\ k\Omega$, $R_{in} = 1.85\ k\Omega$, $C_1 = 199\ \mu F$; (c) 228 **6.19** (a) $R_E = 23.6\ \Omega$, $R_1 = 8.76\ k\Omega$, $R_2 = 96.4\ k\Omega$, $R_{in} = 8\ k\Omega$; (b) 7.86 mW; (c) 8.87 V (*p-p*) **6.26** If $R_A = 5\ k\Omega$, then $R_F = 15\ k\Omega$.

7.5 $I_{CQ} = 0.75$ mA, $R_B = 17.6\ k\Omega$, $R_1 = 19.5\ k\Omega$, $R_2 = 185\ k\Omega$ (a) $C_1 = 2\ \mu F$, $C_2 = 0.2\ \mu F$; (b) $C_1 = 0.31\ \mu F$, $C_2 = 0.31\ \mu F$ **7.11** (a) $R_D = 1.25\ k\Omega$, $R_{Sdc} = 1\ k\Omega$, $R_{Sac} = 241\ \Omega$, $R_2 \to \infty$; (b) $C_2 = 57\ \mu F$, $C_1 = 10.2\ \mu F$, $C_3 = 3.18\ \mu F$ **7.14** (a) $R_D = 10\ k\Omega$, $R_S = 2\ k\Omega$, $R_2 \to \infty$; (c) $C_1 = 0.2\ \mu F$, $C_3 = 0.4\ \mu F$ **7.20** $I_{CQ} = 1.45$ mA, $R_{in} = 17.2\ \Omega$, $A_v = 139$, $C_{b'e} = 444\ \mu F$, $f_1 = 15.9$ MHz, $f_2 = 28$ MHz. Low frequency $C_1 = 7.96\ \mu F$, $C_2 = 118\ \mu F$ **7.24** $V_{GSQ} = -1.47$ V, $g_m = 2.53$ mS, $f_1 = 23.8$ Hz, $f_2 = 37.9$ Hz, $f_3 = 73.7$ Hz, $f_4 = 4$ Hz, $A_v = -3.71$, $f_H = 396$ kHz. Frequency response of amplifier is approximately 86 Hz to 396 kHz. **7.26** High frequency: $f_1 = 93.6$ kHz, $f_2 = 1.4$ MHz. Low frequency: $f_1 = 83.3$ Hz, $f_2 = 0.53$ Hz, $f_3 = 31.8$ Hz, $f_4 = 1.5$ Hz. Frequency response of amplifier is 83.3 Hz to 93.6 kHz.

8.1 $v_o = v_1/2 + v_2$ **8.4** $v_o = -v_1/5 + v_2$ **8.7** $v_o = \frac{22}{7}v_1 + \frac{11}{7}v_2$ **8.10** $v_o = -v_1 + v_2$ **8.13** Using Example 8.4: $R_F = 600\ k\Omega$, $R_1 = 600\ k\Omega$, $R_2 = 60\ k\Omega$, $R_a = 20\ k\Omega$, $R_b = 6\ k\Omega$, $R_x = 5\ k\Omega$. Using Example 8.5:

$R_F = 655$ kΩ, $R_1 = 655$ kΩ, $R_2 = 65.5$ kΩ, $R_a = 21.8$ kΩ, $R_b = 6.6$ kΩ, $R_x = 5.5$ kΩ. **8.16** Using Example 8.4: $R_F = 72$ kΩ, $R_1 = 24$ kΩ, $R_2 = 72$ kΩ, $R_3 = 12$ kΩ, $R_a = 18$ kΩ, $R_b = 14.4$ kΩ. Using Example 8.5: $R_F = 120$ kΩ, $R_1 = 40$ kΩ, $R_2 = 120$ kΩ, $R_3 = 20$ kΩ, $R_a = 30$ kΩ, $R_b = 24$ kΩ. **8.22** $f_1(t)$ output of amplifier tied to input of $f_2(t)$ amplifier through R_2. R_1 is in parallel with 1-μF capacitor as feedback network. Output of $f_1(t)$ amplifier feeds negative-input op-amp with output X. $f_2(t)$ output of amplifier tied to input of $f_1(t)$ amplifier through R_2'. R_1' is in parallel with 1-μF capacitor as feedback network. Output of $f_2(t)$ amplifier feeds negative-input op-amp with output Y. Both $f_1(t)$ and $f_2(t)$ feed to amplifier through 1-MΩ resistor. $R_1 = 10^6/a_1$, $R_2 = 10^6/a_2$, $R_1' = 10^6/b_2$, $R_2' = 10^6/b_1$ **8.25** $-j15/\omega$ **8.31** $V_o = \frac{11}{2}V_i$; square wave high of +1 and low of −5.5

9.1 $A_d = 275$, $A_c = -0.5$ **9.4** $V_{o1} = -200.2$ mV, $V_{o2} = 199.8$ mV **9.7** $-1990v_1 + 2010v_2$ **9.10** $A_c = -0.05$, $A_d = 116$, CMRR = 67.3 dB **9.13** $I_{C1} = I_{C2} = 0.376$ mA **9.16** CMRR = 63.7 dB **9.19** $R_2 = 11.2$ kΩ **9.22** 11.6 MΩ **9.25** dc offset = 0, CMRR = 76.2 dB **9.28** $R_E = 1.01$ kΩ, $R_E' = 1$ kΩ, $R_1 = 23.3$ kΩ, $R_2 = 17.5$ kΩ

10.1 0.101 Ω **10.4** $R_{in}(v_1) = 40$ kΩ, $R_{in}(v_2) = 47.6$ kΩ, $R_{in}(v_3) = 62.5$ kΩ, $R_o' = 0.015$ Ω, BW = 50 kHz **10.7** Two op-amps; total gain = 700, each with gain of 50 or less. **10.10** One inverting amplifier and a unity-gain buffer. **10.13** Three op-amps: one summing amplifier fed by noninverting amplifier at negative terminal and unity-gain buffer to positive terminal which has voltage divider. **10.16** Summing amplifier needed to meet gain requirements, then unity-gain buffers for each input to meet high input resistance requirement. **10.19** One amplifier to invert input and two op-amps in series to noninverting input for total of four op-amps. **10.22** Two op-amps to decouple inputs at noninverting input for total of three op-amps. **10.25** Use 3 pF since gain needed is greater than 10. $R_1 = 24.5$ kΩ, $R_2 = 16.3$ kΩ, $R_y = R_A = 10$ kΩ, $R_{out} = 0.05$ Ω, $R_{in}(v_1) = 40.8$ kΩ $= R_{in}(v_2)$, BW = 200 kHz **10.28** Use 3-pF op-amp at output for gain of 10 and use 30-pF unity-gain buffer. BW = 1 MHz. **10.31** Figure 10.48(c): two op-amps. Input: $R_A = 5$ kΩ, $R_F = 250$ kΩ, $R_1 = 10$ kΩ, $R_x = 9.6$ kΩ. Output: $R_x = 5$ kΩ, $R_A = 10$ kΩ, $R_F = 10$ kΩ, BW = 19.6 kHz, $R_{o1}' = 0.04$ Ω $= R_{o2}'$. **10.34** Unbalanced output, balanced input with gain of 10. Two op-amps using 30 pF, 101 op-amps. $v_i = 0.075$ V and 0.125 V, $V_{CC} = 15.6$ V

11.1 $A_v = -500$, $A_i = -9.26$ **11.4** 0.3 **11.7** $A_{vo} = -83.3$, $A_v = -8.93$ **11.11** 0.04% **11.21** Open-loop response:

$$GH(s) = \frac{s}{62.8}\frac{1}{(1 + s/62.8)^3}$$

11.26 Amplitude starts at −20 dB/dec until $\omega = 8$, where it levels off until $\omega = 10$, then drops at −40 dB/dec. Phase starts at −90° until $\omega = 0.8$, then goes up at +45°/dec until $\omega = 1$, then drops at 45°/dec until $\omega = 80$, then drops at 90°/dec until $\omega = 100$, where it levels off at a phase of −180°. System is always stable since phase never crosses −180°.
11.29

$$GH(s) = \frac{G_o}{(s/4 + 1)(s/40 + 1)^2}$$

Gain margin ($\omega = 44$) = 27.8 dB or gain of 24.4. **11.32** If $C_1 = 0.2\ \mu F$, then $C_2 = 0.205\ \mu F$.

12.1 If $C = 1\ \mu F$, then $R = 100\ k\Omega$. **12.4** Using Table 12.1(e), if $C = 1\ \mu F$ then $R_1 = 1\ M\Omega$, $R_A = 100\ k\Omega$, $R_2 = 111\ k\Omega$, $R_x = 92\ k\Omega$. **12.7** Inverting op-amp. If $C = 10\ \mu F$, then $R_1 = 100\ k\Omega$, $R_2 = 10\ k\Omega$, $R_3 = 189\ k\Omega$. **12.10** Use a standard integrator when in one position of switch. Select an amplifier with negative fixed voltage for other position of switch. Then let $C = 1\ \mu F$, $R_A = 1\ M\Omega$. When not integrating, $R_F = 1\ M\Omega$ and $V = -10$ V. **12.13** Figure 12.11(c): select $C = 1\ \mu F$, $R_2 = 400\ k\Omega$. Then $R_1 = 400\ \Omega$ and $R_1 \parallel R_2 = 400\ \Omega$. **12.17** Figure 12.15(a): let $C = 0.01\ \mu F$; then $R = 79.6\ k\Omega$, $R_A = 88.4\ k\Omega$, $R_F = 796\ k\Omega$. **12.20** High-pass filter. If $C = 1\ \mu F$, then $R_1 = 75\ k\Omega$ and $R_2 = 250\ k\Omega$. Low-pass filter. If $C = 0.1\ \mu F$, then $R_1 = 1\ M\Omega$, $R_2 = 750\ k\Omega$, $R_x = 429\ k\Omega$. Combine in summing amplifier with all resistors = 20 kΩ. **12.23** Figure 12.11(a): if $C = 0.01\ \mu F$, then $R_1 = 15.9\ k\Omega$, $R_A = 16.8\ k\Omega$, $R_F = 320\ k\Omega$. **12.26** Figure 12.18: if $C = 0.01\ \mu F$, then $R_1 = 79.6\ k\Omega$, $R_2 = 159\ k\Omega$, $R_3 = R_4 = 26.5\ k\Omega$. **12.29** Figure 12.18: if $C = 0.01\ \mu F$, then $R_2 = 159\ k\Omega$, $R_1 = 159\ k\Omega$, $R_3 = 20\ k\Omega$, $R_4 = 40\ k\Omega$, $R_F = 386\ k\Omega$ **12.35** Figure 12.29: with ratio of 3.33 and −45 dB, $n = 4$. If $C = 0.01\ \mu F$, then $R_1 = 7.17\ k\Omega$, $R_2 = 12.4\ k\Omega$, $R_3 = 2.97\ k\Omega$, $R_4 = 76.4\ k\Omega$. **12.38** $n_B = 4.98$, so use 5. If $R = 10\ k\Omega$, scale factor is 3.98×10^{-9}. $C_1 = 0.007\ \mu F$, $C_2 = 0.0054\ \mu F$, $C_3 = 0.0017\ \mu F$, $C_4 = 0.013\ \mu F$, $C_5 = 0.0012\ \mu F$. High-pass filter. Order 5. If $C = 0.05\ \mu F$, scale factor = 5305. $R_1 = 3.03\ k\Omega$, $R_2 = 3.92\ k\Omega$, $R_3 = 12.6\ k\Omega$, $R_4 = 1.64\ k\Omega$, $R_5 = 17.2\ k\Omega$ **12.41** If $R = 10\ k\Omega$, capacitor scale factor is 3.183×10^{-9}. If $C = 0.005\ \mu F$, resistor scale factor is 10.61 kΩ. Low-pass filter: $C_1 = 0.0156\ \mu F$, $C_2 = 0.00635\ \mu F$, $C_3 = 0.0377\ \mu F$, $C_4 = 745$ pF. High-pass filter: $R_1 = 3.4\ k\Omega$, $R_2 = 8.36\ k\Omega$, $R_3 = 1.41\ k\Omega$, $R_4 = 71.3\ k\Omega$.

13.1 Half-wave rectifier with $v_o/v_1 = -0.1$ in right half-plane. **13.4** Full-wave rectifier only for negative v_o with $v_o/v_i = +8$ in left half-plane and $v_o/v_i = -2$ in right half-plane. **13.7** Use circuit of basic negative output inverting in Table 13.1 and inverter. $R_F/R_A = 1$, $R_A = 10\ k\Omega$, $R_F = 10\ k\Omega$. Axis shift at input needs resistor of 20 kΩ to reduce −10 V to −5 V. **13.10** Need half-wave rectifier with axis shift, then amplifier with level shift, and finally an inverter. **13.13** Need level and axis shifting on half-wave rectifier. **13.16** First amplifier rectifies signal. Second amplifier combines rectified signal and input, and third amplifier inverts and level shifts. **13.22** If $R_A = 10\ k\Omega$, then $R_F = 15\ k\Omega$, $R_1 = 9.8\ k\Omega$, $R_2 = 4.85\ k\Omega$. **13.25** If $R_A = 100\ k\Omega$, then $R_F = 500\ k\Omega$, $R_1 = R_3 = 2.14\ k\Omega$, $R_2 = R_4 = 860\ \Omega$, $R_x = 83.3\ k\Omega$. **13.34** Limiting comparator first op-amp. Second op-amp is an inverter and level shifter. $R_1 = R_3 = 3.18\ k\Omega$, $R_2 = R_4 = 388\ \Omega$ when $R_A = 10\ k\Omega$. **13.37** Schmitt trigger with limiter and axis shift with second op-amp as level shifter. If $R_1 = 20\ k\Omega$, then $R_2 = 40\ k\Omega$. If $R_3 = 2\ k\Omega$, then $R_4 = 804\ \Omega$. Need voltage dividers for both axis shift and level shift to provide proper voltages.

14.1 $V_1 = 4.15$, $V_2 = 0.85$ V **14.6** $V_1 = 0.459$ V, $V_2 = 0$ V, $V_3 = -4.51$ V, $V_4 = -0.0855$ V **14.8** $V_1 = 0.0989$ V, $V_2 = 0.0788$ V, $V_3 = -8.221$ V, $V_4 = -8.019$ V **14.10** $R_1 = 1\ k\Omega$, $R_2 = 6.7\ k\Omega$, and the capacitor between pin 2 and ground is switchable among 0.001 μF, 0.01 μF, and 0.1 μF. **14.18** Result is a pulse train with the widths of the pulses increasing with time. The highest frequency is 102 Hz with duty cycle of 2.

15.2

A	$\overline{A}$	B	C	$\overline{C}$	D	E	Y
0	1	0	0	1	0	0	0
0	1	0	1	0	0	0	0
0	1	1	0	1	1	0	1
0	1	1	1	0	0	0	0
1	0	0	0	1	0	0	0
1	0	0	1	0	0	0	0
1	0	1	0	1	0	1	1
1	0	1	1	0	0	0	0

15.6

A	B	C	$A + B + C$	Y
0	0	0	0	1
0	0	1	1	0
0	1	0	1	0
0	1	1	1	0
1	0	0	1	0
1	0	1	1	0
1	1	0	1	0
1	1	1	1	0

15.9 $Y = A \cdot B + B \cdot C$
15.11 LUNCH SPEC = HAMBURGER · (SOUP ⊕ SALAD)
15.14 RENTED APART = COUPLE ⊕ SINGLE
15.17 MUST ATTEND = (DAY OF CONF) · (NIGHT + AFTERNOON)
15.19 GAME = 1 · 2 · 3 · 4
15.21

D	H	K	ALARM
0	0	0	0
0	0	1	0
0	1	0	0
0	1	1	0
1	0	0	0
1	0	1	1
1	1	0	1
1	1	1	1

15.23 Alarm is armed with a key-operated switch located in the fender. Once the alarm is armed and a burglar opens the door, a trigger pulse activates the 555 monostable. This timer provides a 10-min pulse. The pulse enables the astable 555 for 10 min, with the output amplified by a 2N2222 transistor driving the speaker located under the hood. Astable 555 frequency is 686 Hz.

15.30

A	B	I_{CQ1}	Q_2	Q_3	Q_4	Y
0	0	+	F	F	N	1
0	1	+	F	F	N	1
1	0	+	F	F	N	1
1	1	−	N	N	F	0

15.33

A	B	1	2	3	4	5	6	7	8	Y
0	0	N	N	F	F	F	N	N	F	1
0	1	N	F	F	N	F	N	N	F	1
1	0	F	N	N	F	F	N	N	F	1
1	1	F	F	N	N	N	F	F	N	0

15.35 See Figure 14.25 of text, with $R_1 = 1\ \text{k}\Omega$; R_2 is a series combination of a 1-kΩ fixed resistor and a 25-kΩ pot; the capacitance between pin 2 and ground is 0.01 μF; the capacitance between pin 5 and ground is 0.1 μF; and pin 3 feeds the LED display through an inverter to pin 4 of the 7447 decoder.

16.2	CMOS 74C150	TTL 74150
SUPPLY V	3–15 V	7 V
Speed	120–250 ns	8–15 ns
Power Diss	20 mW	200 mW

16.7 For a base frequency f_o of 10 kHz, $C_{ext} = 0.05\ \mu F$. Output frequency range is 5.7 to 11.3 kHz. **16.9** Output Q_1 is connected to the clock B input and the counter is reset when the count reaches 10. When the outputs are 1010, the output of the AND gate is high and the zero set inputs are high. This clears the counter.
16.12 Use the 555 astable to develop a frequency that is twice the required frequency. This signal passes through a *JK* flip-flop that toggles on the rising edge of the pulse. The *JK* divides the frequency by 2 and maintains a duty cycle of 1.
16.13 A 555 astable is set to operate at 1 kHz with a duty cycle of 20, so the output at pin 3 is suitable to trigger the 555 monostable. It is a good design practice to use a fixed resistor for R_1 and a 10-kΩ ten-turn potentiometer for R_2 so that the frequency can be adjusted to 1 kHz. The variable duty cycle is achieved with a 555 monostable.
16.16 A 1-kHz 555 astable is used to clock a $\div N$ counter. The counter is enabled by pressing a button signifying a roll of the dice. The start signal is debounced before enabling the counter. A decoder circuit is designed using logic gates. The decoder outputs are fed into an LED display that resembles the face of the die as shown in Fig. P16.4.
16.17 The rpm meter accepts the TTL-compatible pulses and ANDs them with the window from a 555 astable. The output of the AND gate is input to three decade counters (74160), three latches (74175), three decoders (7447), and the three 7-segment displays. The clear signal is delayed with inverters so that the data is latched into the display before the counters are cleared.

17.2 A 555 astable IC is designed to output pulses at a frequency of 1 kHz. These pulses are divided down until the output is one pulse per 5 s. This output clocks a 7493 binary counter. The outputs of the counter are used as select lines to a 16-to-1 multiplexer. The output of a 74150 MUX is active low. If a lamp is good, the output of the MUX is LOW, and if a lamp is faulty, the MUX output is HIGH. The MUX output is ANDed with the 1-kHz clock. The clock pulse triggers the 555 monostable IC, which is designed to output a 5-s pulse. This pulse is ANDed with the 1KH3 signal for the speaker. **17.5** We want to use a down-counter, since 9 represents full and 0 represents empty. Preset a 74193 UP/DOWN binary counter to 15 by connecting the transmitted pulse to the load of the counter. Also use the transmitted pulse to trigger a one-shot 555 IC to output a window pulse of 170 ms. This window pulse is ANDed with a 10-ms pulse train from a 555 astable IC. This gate, when enabled, begins the counter, which counts down from 15. After 6 counts the display will read 9, and after 16 counts it will read 0. The received pulse latches the display and clears the counters. If the countdown does not reach 9 or goes past 0, the tank is either over 90% full or under 10% empty, and the alarm sounds.
17.13 Use a 555 astable IC, designed to output one pulse per minute, as the circuit clock. Two cascaded 7490 decade counters are clocked by the 555 IC. To output the room number simultaneously, a 7493 binary counter is clocked and used to address a

16-to-1, 74150 multiplexer. After counting to 14, each counting system is cleared. Each room's TV set is monitored by the multiplexer for 1 minute every 14 minutes to determine whether it is on or off. The output of the multiplexer is connected through a resistor to an LED to display whether a TV is on or off. **17.24** The design requires an UP/DOWN counter to keep track of the number of cars in the garage. Initially (at the beginning of a workday), the counters must be cleared to zero. Every time a car enters the counter counts up, and every time a car leaves the counter counts down. Since the pulse output from the entrance and exit gates is held high until a car passes through, the 74193 UP/DOWN BCD counters are ideally suited. Pins 4 and 5 of these counters are used to count down and up. The count can occur only if the pin not being clocked is held high, thus fitting the design specifications. The gate pulses need to be conditioned with an RC circuit to trigger the counters. Let the RC time constant equal 0.1 s. When the most significant counter reaches 3 (300 is the full count for the garage), display the word "FULL" in the four 7-segment LED displays.

REFERENCES AND SOURCES FOR FURTHER STUDY

1. Alley, C. L., and K.W. Atwood. *Microelectronics.* Englewood Cliffs, N.J.: Prentice-Hall, Inc., 1986.

2. Ankrum, P. D. *Semiconductor Electronics.* Englewood Cliffs, N.J.: Prentice-Hall, Inc., 1971.

3. Boylestad, Robert, and Louis Nashelsky. *Electronic Devices and Circuit Theory.* Englewood Cliffs, N.J.: Prentice-Hall, Inc., 1987.

4. Casasent, David. *Electronic Circuits.* New York: Quantum, 1973.

5. Chirlian, P. M. *Analysis and Design of Integrated Electronic Circuits.* New York: Harper and Row, 1987.

6. Clayton, David. *Operational Amplifiers.* London: Newnes-Butterworths, 1979.

7. *CMOS Databook.* Santa Clara, Calif.: National Semiconductor Corp., 1988.

8. Ghausi, Mohammed S. *Electronic Devices and Circuits.* New York: Holt, Rinehart and Winston, 1985.

9. Gray, P. R., and R. G. Meyer. *Analysis and Design of Analog Integrated Circuits.* New York: John Wiley, 1984.

10. Grinich, V. H. *Introduction to Integrated Circuits.* New York: McGraw-Hill, 1975.

11. Gummel, H. K., and H. C. Poon. "An Integrated Charge Control Model of Bipolar Transistor." *Bell System Technical Journal* 49, no. 5 (1970): 827–852.

12. Hamilton, D.J., and W. G. Howard. *Basic Integrated Circuit Engineering.* New York: McGraw-Hill, 1975.

13. Hodges, D. A., and H. G. Jackson. *Analysis and Design of Digital Integrated Circuits.* New York: McGraw-Hill, 1983.

14. Irvine, R. G. *Operational Amplifier Characteristics and Applications.* Englewood Cliffs, N.J.: Prentice-Hall, Inc., 1987.

15. Jacob, J. M. *Application and Design with Integrated Circuits.* Reston, Va.: Reston Publishing Co., 1982.

16. Jung, W. *IC Op-Amp Cookbook.* Indianapolis, Ind.: Howard Sams and Co., 1986.

17. Lancaster, Don. *TTL Cookbook.* Indianapolis, Ind.: Howard Sams and Co., 1982.

18. Lancaster, Don. *CMOS Cookbook.* Indianapolis, Ind.: Howard Sams and Co., 1988.

19. Letzter, S., and N. Webster. "Noise in Amplifiers." *IEEE Spectrum* (August, 1970).

20. *Linear and Data Acquisition Products.* Melbourne, Fla.: Harris Semiconductor Products Division, 1988.

21. *Linear Databook.* Santa Clara, Calif.: National Semiconductor Corp., 1988.

22. Mano, M. M. *Digital Design.* Englewood Cliffs, N.J.: Prentice-Hall, Inc., 1984.

23. *Master Selection Guide.* Phoenix, Ariz.: Motorola Semiconductor Products, Inc.

24. *MECL Data Book.* Mesa, Ariz.: Motorola Semiconductor Products, Inc., 1987.

25. Millman, J. and A. Grabel. *Microelectronics.* New York: McGraw-Hill, 1987.

26. Milnes, A. G. *Semiconductor Devices and Integrated Circuits.* New York: Van Nostrand-Reinhold, 1980.

27. Mitchell, F. H., Jr., and F. H. Mitchell, Sr. *Introduction to Electronics Design.* Englewood Cliffs, N.J.: Prentice-Hall, Inc., 1988.

28. Peatman, J. B. *Digital Hardware Design.* New York: McGraw-Hill, 1980.

29. Pierret, R. F., and G.W. Neudeck. *Semiconductor Fundamentals.* Reading, Mass.: Addison-Wesley, 1983.

30. Ruthowski, G. B. *Solid State Electronics.* Indianapolis, Ind.: Bobbs-Merrill, 1980.

31. Sah, C.T., R. N. Noyce, and W. Shockley. "Carrier Generation and Recombination in P-N Junction Characteristics." *Proc. IRE* (1957): 1228.

32. Schilling, D. L., Charles Belove, T. Apelewicz and R. C. Saccardi. *Electronic Circuits, Discrete and Integrated.* New York: McGraw-Hill, 1989.

33. Sedra, A. S., and K. C. Smith. *Microelectronic Circuits.* New York: Holt, Rinehart and Winston, 1987.

34. Soclof, S. *Analog Integrated Circuits.* Englewood Cliffs, N.J.: Prentice-Hall, Inc., 1985.

35. Soclof, S. *Applications of Analog Integrated Circuits.* Englewood Cliffs, N.J.: Prentice-Hall, Inc., 1985.

36. Stanley, W. D. *Electronic Devices, Circuits and Applications.* Englewood Cliffs, N.J.: Prentice-Hall, Inc., 1989.

37. Streetman, B. G. *Solid State Electronic Devices.* Englewood Cliffs, N.J.: Prentice-Hall, Inc., 1980.

38. Tocci, R. J. *Fundamentals of Electronic Devices.* Columbus, Ohio: Merrill, 1982.

39. *TTL Logic Data Book*. Dallas: Texas Instruments, Inc., 1988.

40. Wang, S. *Fundamentals of Semiconductor Theory and Device Physics*. Englewood Cliffs, N.J.: Prentice-Hall, Inc., 1989.

41. Widlar, R. S. "Some Circuit Design Techniques for Linear Integrated Circuits." *IEEE Transactions on Circuit Theory*. CT-12 (1968).

42. Williams, A. B. *Electronic Filter Design Handbook*. New York: McGraw-Hill, 1981.

43. Wilson, G. R. "A Monolithic Junction FET-NPN Operational Amplifier." *IEEE Journal of Solid State Devices*. SC-2 (December 1968).

44. Wolfe, C. M., N. Holonyak, Jr., and G. E. Stillman. *Physical Properties of Semiconductors*. Englewood Cliffs, N.J.: Prentice-Hall, Inc., 1989.

45. Yang, E. S. *Microelectronic Devices*. New York: McGraw-Hill, 1988.

46. Young, Thomas. *Linear Integrated Circuits*. New York: John Wiley, 1981.

INDEX

ac load line, 32
Acceptor, 10
Active element, 70
Active filter, 527
Active pull-up, 699
Actuating error, 410
Advanced low-power Schottky TTL (ALS-TTL), 697
Advanced Schottky TTL (AS-TTL), 697
All-pass filter, 532
Always gate, 736
Amplitude selector, 51
Analog computer, 354
Analog-to-digital converter (A/D), 763
  dual-slope, 765
Analysis Examples:
  Bode plot for CE amplifier, 288
  Butterworth low-pass filter, 552
  CB amplifier, capacitor-coupled, 147
  CB amplifier, frequency response, 314
  CC Q-point determination, 108
  CE amplifier, high frequency cutoff, 310
  CE Q-point determination, 99
  CS amplifier, 189
  CS amplifier, high-frequency cutoff, 318
  Differential amplifier, 378
  Diode circuit, 655
  Enhancement MOSFET, drain current, 186
  Feedback amplifier, 474
  Feedback limiter, 605
  555 circuit, 668
  Frequency response from transfer function, 530
  Frequency response of band-pass filter, 531
  Frequency response of high-pass filter, 531
  Frequency response of low-pass filter, 530
  Frequency response of notch filter, 532
  High-pass RC filter, 641, 644
  Inverting amplifier, 332
  JFET amplifier, temperature effects, 231 heat sink, 234
  JFET CS amplifier, 189
  JFET transconductance, 181
  Low-pass RC filter, 650, 651
  MOSFET, drain current, 183
  Multistage amplifier, 158
  Noninverting amplifier, 333
  Op-amp circuit output voltage, 418
  Op-amp open-loop function, 411
  Open- and closed-loop voltage gain, 477
  Schmitt trigger, 620
  Stability given transfer function, 491
  Stability of op-amp circuit, 496
  Stability of three-pole amplifier, 503

Analysis Examples: *(continued)*
Stability of three-stage amplifier, 490
Switched-capacitor integrator, 577
Transfer function from differential equation, 529
Transformer-coupled amplifier, 154
Unity-gain buffer, output impedance, 425
Unity-gain follower, gain, 430
Unity-gain follower, input resistance, 428
AND gate, 685
Arithmetic logic unit (ALU), 774
Astable multivibrator, 660
Asynchronous counter, 747
Avalanche, 44

Balanced input and output, 456
Band-pass filter,
design, 571
Band-stop filter, 532
design, 571
Base, 71
Base-spreading resistance, 79, 304
BCD-to-binary converter, 727
Bias stability, 89, 217
design for, 229
Biasing, 80
types of, 213
Binary to single output, 727
Bipolar junction transistor (BJT), 69, 71
amplifier analysis, 97
amplifier design, 100
noise, A-31
placement of Q-point, 101
Bleeder resistor, 49
Bode plot, 284, A-36
Boltzmann's constant, 9
Boolean algebra, 686
Breakdown, 44
Breakdown voltage, 23
Bridge rectifier, 38
Bridge-power op-amp, 462
Buffer, 149, 684
Bulk resistance, 10
Butterworth filter, 550
design procedure, 558
Butterworth polynomials, 551

Capacitive loading, 506
Capacitors,
bypass, 92
junction, 26
standard values, A-127
Carrier concentration, 11
Carrier lifetime, 13
Cascode configuration, 161
CB Amplifier, 76
current gain, 144
high-frequency response, 313
input resistance, 143
output resistance, 145
parameters, table, 124
voltage gain, 145
CC amplifier, 79
current gain, 139
input resistance, 138
parameters, table, 124
output resistance, 140
voltage gain, 139
CD amplifier
analysis, 199
bootstrap amplifier, 203
design, 200
CE amplifier, 79, 129
ac load line, 93
current gain, 126
input resistance, 123
output resistance, 129
parameters, (Table), 124
voltage gain, 127
with emitter resistor, 86
Characteristic curves, 80
Charge density, 11
Chebyshev filter, 553
design procedure, 558
Chebyshev polynomials, 553
Clamper, 56, 652
Class-A operation, 239
Power amplifier circuits, 242
Class-AB operation, 241
Class-B operation, 239
power amplifier circuits, 248
Class-C operation, 242
Classes of amplifiers, 238
Classical filters, 549
Clipper, 51
Clocked logic, 682, 736
Clocks, 756
Closed-loop gain, 329, 473
Closed-loop transfer function, 470

CMOS analog switch, 711
CMOS device listing, table, 714
Collector, 71
Colpitts oscillator, 507
Common mode, 371
Common-base, See CB
Common-collector, see CC
Common-emitter, See CE
Common-mode rejection, 418
  ratio (CMRR), 573
Comparator, 612
Compensation,
  external, 479
  internal, 406
Complement, 684
Complementary MOS (CMOS), 709
Complementary symmetry, 239
  power amplifier, 248
Component selection, 194
Computer simulations, 27
Conditional stability, 484
Conduction electron, 2
Conduction in materials, 4
Conduction in semiconductor materials, 5
Conversion efficiency, 92, 244
  transformer coupling, 246
Cooling, 232
Counters, 747
Coupling, 459
  amplifier, 149
  capacitive, 150
  capacitor, 93
  direct, 150
  inductive, 243
  transformer, 152, 245
Covalent bond, 2
Crossover distortion, 252
Crystal oscillator, 511
CS amplifier,
  analysis, 188
  design, 191
Current biasing, 214
Current feedback, 214
  -current subtraction, 471
  -voltage subtraction, 472
Current gain, 70
Current mirror, 387
Current-switching R-2R ladder, 762
Current-to-voltage converter, 351
Cutoff region, 82

Damping ratio, A-50
Darlington circuit, 262
Data selector/multiplexer, 729
dc load line, 32
Debouncing, 731
Decade, 875
Decibel, definition, 250
Decoder, 725
Delay equalizer, 532
DeMorgan's theorem, 687
Demultiplexer, 728
Dependent source, 70
Dependent-current generator, 350
Depletion MOSFET, 182
Depletion region, 18, 173
Derating, 90
  curve, 61
Design Examples:
  All-pass filter, 548
  Balanced limiting comparator, 616
  Butterworth band-pass filter, 557, 572
  Butterworth high-pass filter, 555
  Butterworth low-pass filter, 566
  CB amplifier, capacitor coupled, 145
  CC amplifier, capacitor coupled, 141
  CC amplifier, for maximum output swing, 107
  CD amplifier, 201
  CD amplifier, specified low frequency response, 303
  CE amplifier, capacitor coupled, 132
  CE amplifier, for maximum output swing, 103
  CE amplifier, specified low-frequency response, 291
  CE amplifier, temperature variations, 222
  Chebyshev high-pass filter, 556
  Chebyshev low-pass filter, 567
  Class-B push-pull amplifier, 257
  Current source, 378
  Darlington Pair class-A amplifier, 266
  Differential amplifier, 388
  Digital tachometer, 794
  Direct-coupled amplifier, 151
  Display brightness controller, 735
  EF amplifier, temperature variations, 225
  Faulty-lamp indicator, 799
  FET amplifier, specified low-frequency response, 300
  555 monostable mode, 672
  555 square-wave generator, 667

Design Examples: *(continued)*
First-order active low-pass filter, 535
First-order adjustable low-pass filter, 539
Frequency counter, 754
High-pass filter, 541
IC Butterworth filter, 580
Inverting amplifier, 441
JFET CS amplifier, 195
Lead network for three-pole amplifier, 501
Level shifter, 392
Limiter, 609
Low-pass filter, 561
Multiple op-amp, 448
Multiple-input amplifier, 434
Network with given transfer function, 526
Op-amp circuit for specified transfer characteristic, 595
Op-amp for given voltage gain, 432
Op-amp summer, 346
Reactor chamber pressure detector, 801
Regulated power supply, 269
Regulator, IC, 273
Schmitt trigger, 623
Seven-segment LED driver, 744, 752, 753
SF bootstrap amplifier, 205
Switching regulator supply, 278
Tape controller, 809
Temperature sensor, 759
Theft alarm for automobile, 716
Three-input inverting amplifier, 444
Three-input summer, 437
Transformer-coupled amplifier, 246
Trigger pulse generator, 659
Unbalanced limiter comparator, 622
Warning speedometer, 805
Widlar Current Source, 384
Zener Regulator, 47
Differencing differentiator, 524
Differencing integrator, 522
Differential amplifier, 369
Differential input voltage, 326
Differential mode, 371
Differential summing, 445
Differentiator, 520
Diffused capacitor, 368
Diffusion, 7
capacitance, 26
current, 7, 19
Digital-to-analog converter (D/A), 762
Diode, 1
circuit models, 25
compensation, 227
construction, 19
current/voltage relationship, 20
load lines, 30
noise, 963
Diode-transistor logic (DTL), 694
Divide-by-n counter, 748
Donor, 9
Doping, 9, 19
Double-diffused process fabricated FET (DMOS), 208
Double-ended output, 372
Drain-source saturation current, 174
Drift, 7, 14
current, 7, 20
velocity, 7, 14
Driver IC, 733
Dual in-line package (DIP), 394, 688
Duty cycle, 667
Dynamic memory, 771
Dynamic resistance, 22, 81

Early voltage, 83
Ebers-Moll model, 76
ECL device listing, 706
Edge triggering, 736
EF amplifier, see CC amplifier
Electric field intensity, 14
Electron concentration, 12
Electron mobility, 15
Emitter, 71
Emitter follower—See CC
Emitter-coupled logic (ECL), 704
Encoder, 725
Energy gap, 4
Energy level diagram, 4
Enhancement MOSFET, 184
EPROM, 773
Excess carriers, 12
Exclusive OR gate, 685

Feedback, 468
definition, 214
Feedback amplifiers, 472
frequency response, 493
Feedback transmittance, 410
Field-effect transistor (FET), 171
advantages, 172

biasing, 186
disadvantages, 172
noise, 964
temperature effects, 229
First-order active filter, 534
555 timer
monostable mode, 671
oscillator, 664
pulse generator, 660
Flat pack, 395
Flicker noise, A-130
Flip flop, 737
clocked, 740
D, 741
JK, 739
SR, 737
T, 741
Forbidden band, 4
Forward transconductance, 177
Forward transmittance, 410
Forward-biasing, 20
Frequency counter, 753
Frequency measurement, 753
Full adder, 776
Full-wave rectification, 38
Full-wave rectifier, 590
table, 594
Full-wave zener regulator, 48

Gain impedance formula, 128
Gain-bandwidth product, 407
Graded junction, 19
Ground loop, 38

h-parameters, 119
model, 78
H-TTL gates, 700
Half-power point, 250, 284
Half-wave rectifier, 36, 357
configurations, 589
Hall-effect switch, 813
Hartley oscillator, 508
Heat sink, 233
High-frequency response
BJT amplifier, 303
CB amplifier, 313
CD amplifier, 318
CE amplifier, 304
CS amplifier, 315
EF amplifier, 312
FET amplifier, 315
High-pass network
steady-state response, 641
step response, 639
High-pass RC network, 637
ramp response, 640
High-power TTL (H-TTL), 695
High-threshold logic (HTL), 695
Hole, 6
concentration, 12
current, 7
mobility, 15
Hybrid parameters, 118
Hybrid-π model, 78, 121

IC fabrication, 367, 688
capacitors, 368
diodes, 367
resistors, 368
transistors, 367
IC packaging, 688
Injection model, 76
Input bias current, 415
temperature coefficient, 415
Input offset current
temperature coefficient, 415
Input offset voltage, 414
Insulator, 4
Integrated circuit (IC), 325
filter, 575
Schmitt trigger, 627
Integrator, 520
Intercom, 463
Interrupter/reflector module, 157
Intrinsic concentration, 9
Intrinsic semiconductor, 7, 10
Inverse dynamic resistance, 179
Inverter, 684
Inverting amplifier, 328, 438
Inverting differentiator, 354
Inverting half-wave rectifier, 587
Inverting input, 326
Inverting integrator, 354

JFET, 172
active region, 176
avalanch breakdown, 176

JFET *(continued)*
 gate-to-source voltage variation, 174
 transfer characteristics, 175

Keyboard encoder, 730

Laplace transformation, A-60
Large-scale integration (LSI), 325, 682
Large-signal amplification factor, 75
Latch memory, 742
Lateral transistor, 368
Leakage current, 23
Level shifter, 390, 590
Light-emitting diode (LED), 59
Limiter, 51, 597
 applications, 606
 configurations, 601
Limiting comparator, 614
Liquid crystal display (LCD), 767
LM317 regulator, 274
Load line, 83
Logic gate summary, 683
Look-ahead carry generator, 776
Low-frequency response
 BJT amplifier, 285
 CB amplifier, 296
 CD amplifier, 302
 CE amplifier, 285
 CS amplifier, 297
 EF amplifier, 295
 FET amplifiers, 297
Low-pass RC network, 647
Low-power Schottky TTL (LS-TTL), 696
Low-power TTL (LP-TTL), 696
LP-TTL gates, 700

Magnitude comparator, 778
Magnitude operator, 590
Majority carrier, 9
Marginally stable, 484
Maximally flat magnitude, 550
Medium-scale integration (MSI), 325, 682
Memories, 768
 serial, 768
MESFET, 206
MICRO-CAP II, 27
Miller
 capacitor, 316
 compensation, 506
 integrator, 354
 theorem, 307
Minority carrier, 9
MOS capacitor, 368
MOSFET, 181
Moving-surface memory, 769
Multiple op-amp design, 448
Multiple-input amplifier, 434, 442
Multiplexer, 728
Multistage amplifier analysis, 158
Multistage feedback amplifiers, 478
Multivibrator
 free-running, 660

n-channel MOS (NMOS), 708
n-channel enhancement MOSFET, 706
n-type semiconductor, 9
NAND gate, 685
Negative-edge trigger, 657
Negative feedback, 469
Negative-impedance circuit, 349
Negative logic, 681
Noise, A-128
Noise figure, A-134
Noninverting amplifier, 331, 426
Noninverting buffer, 340
Noninverting Miller integrator, 356
Nonlinearities, BJT transistor, 135
NOR gate, 685
NOT gate, 684
Notch filter, 532

101 op-amp, 452
 inverting amplifier, 455
 noninverting amplifier, 454
Op-amp circuit
 design, 342
 feedback, 479
 frequency response, 405
 ideal, 326
 input resistance, 338
 noise, 964
 packaging, 394
 power requirements, 395
 practical, 412
Open-loop configuration, 327
Open-circuit saturation current, 77
Optical coupling, 154
Optocoupler, 157

Optoelectronic detectors and emitters, 155
OR gate, 684
Oscillators, 506
  Phase shift, 509
Output admittance, 121

p-Channel enhancement MOSFET, 707
p-channel MOS (PMOS), 708
p-type semiconductor, 10
Parallel-biased clipper, 52
Parity checker, 732
Pass transistor, 268
Passive element, 70
Passive phase-lag network, 504
Peak inverse voltage, 23
Percent regulation, 49
Phase splitter, 157
Photocurrent, 60
Photodiode, 60
Phototransistor, 155
Pinch-off voltage, 175
Positive logic, 681
Positive-edge trigger, 657
Potential barrier, 18
Power audio op-amp, 460
Power bandwidth, 422
Power capability, 35
Power rating, 90
Power supply
  discrete components, 267
  IC regulator, 270
  rejection ratio, 419
  three-terminal adjustable regulator, 274
Programmable array logic (PAL), 779
Programmable ROM (PROM), 772
Propagation delay, 691
Prototyping, 791
Push-pull configuration, 239

Quasi-linear, 586
Quiescent operating point, 30

Random number generator, 812
Random-access memory (RAM), 771
Read-only memory (ROM), 772
Recombination, 8, 13
Rectification, 36
Rectifier, 586
Relaxation oscillator, 660
Resistor-transistor logic (RTL), 694
Resistors, standard values, A-126
Reverse saturation current, 21, 25, 218
Reverse voltage gain, 121
Reverse-bias resistance, 25
Ripple, 39
  factor, 43
  voltage, 42

Saturation comparator, 613
Saturation region, 82
  JFET, 176
Schmitt trigger, 616
Schottky diode, 58
Schottky TTL (S-TTL), 696
Schottky TTL gates, 700
Semiconductor, 4
Series regulator, 271
Series-biased clipper, 53
741 op-amp, 396
78XX regulator, 271
  table, 273
SF amplifier, see CD amplifier
Shell, 2
Shift register, 745
Short-circuit input resistance, 122
Shot noise, 962
Single-ended output, 372
Single-pole amplifier, 493
Single-pole compensation, 452
Slew rate, 420
Slicer, 51
Small-scale integration (SSI), 682
Space charge region, 36
SPICE, 27, A-1
Stability, 483
  Bode plots, 488
  frequency response, 485
Substrate, 183
Summing inverting integrator, 522
Switched-capacitor filter, 575
Switching regulator, 276
  efficiency, 278
Synchronous counter, 748
Synchronous logic, 682, 736

Temperature coefficient, 24
Temperature effects, 24
Thermal noise, A-128

Thermal runaway, 252
Thermal speed of carriers, 14
Thermometer, electronic, 358
3-dB point, 250
3- to 8-line decoder, 725
Three-pole amplifier with lead equalizer, 499
Time constant of RC network, 638
Time-dependent logic, 682, 736
Time-independent logic, 682
Timing diagram, 692
Timing windows, 814
Touch-tone generator, 512
Transconductance, 122
Transistor models, 75
Transistor-transistor logic (TTL), 695
  device listings, 703
Transport model, 76
Tri-state gate, 701
Trigger circuit, 657
Turn-on voltage, 24
Two-pole amplifier, 494

Ultra large scale integration (ULSI), 683
Unclocked logic, 682
Unstable, 484
Up/down counter, 748

Valence electron, 2
Variable-reluctance pickup, 813
Variation constants
  CE configuration, 219
  EF configuration, 224
Very large scale integration (VLSI), 325, 682
Virtual ground, 329
VMOSFET, 207
Voltage doubler, 43
Voltage feedback, 215
  -current subtraction, 475
  -voltage subtraction, 471
Voltage gain, 70
Voltage restorer, 653
Voltage-controlled oscillator (VCO), 757
Voltage-to-current converter, 352

White noise, 961
Widlar current source, 381
Wien bridge oscillator, 508
Wilson current source, 384
Wirewrap, 792

XOR gate, 685

Zener diode, 44
  breakdown, 45
  regulator, 46